STATISTICS: LEARNING FROM DATA, AP* EDITION

Roxy Peck

California Polytechnic State University, San Luis Obispo, CA

Chris Olsen

Grinnell College, Grinnell, IA, and George Washington High School, Cedar Rapids, IA

BROOKS/COLE
CENGAGE Learning

Australia • Brazil • Japan • Korea • Mexico • Singapore • Spain • United Kingdom • United States

BROOKS/COLE
CENGAGE Learning®

Statistics: Learning from Data, AP* Edition
Roxy Peck
Chris Olsen

Publisher: Richard Stratton

Senior Sponsoring Editor: Molly Taylor

Senior Development Editor: Jay Campbell

Assistant Editor: Shaylin Walsh Hogan

Editorial Assistant: Danielle Hallock

Media Editor: Andrew Coppola

Brand Manager: Gordon Lee

Associate Market Development Director:
Ryan Ahern

Marketing Communications Manager: Linda Yip

Content Project Manager: Alison Eigel Zade

Senior Art Director: Linda May

VP, Director of Advanced and Elective Product
Programs: Alison Zetterquist

Editorial Coordinator, Advanced and Elective
Product Programs: Jean Woy

Manufacturing Planner: Sandee Milewski

Rights Acquisition Specialist: Shalice Shah-
Caldwell

Production Service: MPS Limited

Text Designer: Diane Beasley

Cover Designer: Rokusek Design

Cover Image: ©Shutterstock

Compositor: MPS Limited

For product information and technology assistance, contact us at
Cengage Learning Customer & Sales Support, 1-800-354-9706

For permission to use material from this text or product,
submit all requests online at **www.cengage.com/permissions**
Further permissions questions can be emailed to
permissionrequest@cengage.com

Library of Congress Control Number: 2012947327

ISBN-13: 978-1-285-08524-1

ISBN-10: 1-285-08524-8

Brooks/Cole
20 Channel Center Street
Boston, MA 02210
USA

Cengage Learning is a leading provider of customized learning solutions with office locations around the globe, including Singapore, the United Kingdom, Australia, Mexico, Brazil and Japan. Locate your local office at
international.cengage.com/region

Cengage Learning products are represented in high schools by Holt McDougal, a division of Houghton Mifflin Harcourt.

Cengage Learning products are represented in Canada by Nelson Education, Ltd.

For your course and learning solutions, visit **www.cengage.com**

To find online supplements and other instructional support, please visit
www.cengagebrain.com

Instructors: Please visit **login.cengage.com** and log in to access instructor-specific resources.

Holt McDougal is pleased to distribute Cengage Learning college-level materials to high schools for Advanced Placement*, honors, and college-prep courses. Our Advanced & Elective Programs department is dedicated to serving teachers and students in these courses. To contact your Advanced & Elective Programs representative, please call us toll-free at 1-800-479-9799 or visit us at www.HoltMcDougal.com.

*AP and Advanced Placement Program are registered trademarks of the College Entrance Examination Board, which was not involved in the production of, and does not endorse, this product.

Printed in the United States of America
1 2 3 4 5 6 7 16 15 14 13

Twitter words: Act 2.1
Valentine's Day business: Ex 7.7
Video game performance: Ex 3.12; Ex 3.13; Ex 3.14

Manufacturing and Industry

Concrete strength: Ex 6.16
Garbage truck processing times: Ex 6.32
Glass panels for flat screen tvs: Ex 6.14; Ex 6.15

Marketing and Consumer Behavior

Back-to-college spending: Ex 2.24
Credit card use: Ex 6.23
Life insurance for cartoon characters: Ex 2.21

Medicine

Benefits of ultrasound: Ex 13.3; Ex 13.4; Ex 13.8
Blocked artery treatments: Ex 10.7
Blood platelet volume: Ex 12.1
Cancer tests: Ex 10.8; Ex 10.9
Diagnosing tuberculosis: Ex 5.13
Dogs detecting cancer: Ex 9.6
Drive-through medicine: Preview Ex 12; Ex 12.7
Effect of talking on blood pressure: Ex 14.3
Effects of vertebroplasty: Preview Ex 14
Evaluating a medical treatment: Ex 10.5
Forensic science, age identification: Preview Ex 4; Ex 4.15
Head circumference at birth: Ex 3.20
Heart rate variability in heart attack patients: Text, Sec 1.4
Hepatitis and blood transfusions: Ex 8.3; Ex 8.5
Medical errors: Ex 3.1
Medical resident errors: Ex 3.10; Ex 3.16
Newborn birth complications: Ex 1.9
Pomegranate juice and tumor growth: Ex 4.4
Premature babies: Ex 6.36
Prescription drug overdoses: Text, Sec 10.7
Quitting smoking: Ex 7.5
Stroke mortality and education: Ex 15.10
Surgery wait times: Ex 12.8
Surviving a heart attack: Ex 5.12; Ex 5.17
Value of daily vitamins: Text, Sec 1.4
Video games and pain management: Act 1.3
Wart removal: Ex 14.2

Nutrition

Fat content of hot dogs: Ex 12.5
Food tastes: Ex 15.1; Ex 15.3
Health impact of fast food: Act 1.2
Impact of food labels: Ex 10.10

Politics and Public Policy

Political action: Act 9.3
State supreme court decisions: Ex 9.1

Psychology, Sociology, and Social Issues

Connection between music and memory: Ex 16.5
Depression and chocolate: Preview Ex 13

Effects of prayer: Ex 14.4
Facial characteristics and trustworthiness: Ex 4.2
Facial expressions: Ex 15.9
Impact of Technology on Lying Behavior: Ex 1.3
Internet addiction: Ex 5.14; Ex 5.22
Motivation for revenge: Ex 1.8
Reading emotions: Ex 14.1
Recognizing a person's scent: Ex 6.20
Visual aspects of word clouds: Ex 12.11

Public Health and Safety

Bike helmet safety: Ex 4.3
Cell phone use and traffic fatalities: Text, Sec 4.6
Cell phone use while driving: Ex 7.6; Ex 7.8
Effect of crime drama television shows on jurors: Text, Sec 1.1
Hurricane evacuation: Preview Ex 9
Legal drinking age: Ex 9.2
Motorcycle helmet use: Ex 2.2; Ex 2.3

Sports

Baseball Salaries: Ex 3.2
Basketball player salaries: Ex 3.17
Hockey goals: Ex 12.2
Physiological Characteristics and Performance of Top U.S. Biathletes: Ex 16.8
Predicting sports outcomes: Preview Ex 15
Soccer goalkeepers: Ex 5.21
Sports and neurocognitive performance: Ex 15.6; Ex 15.7

Surveys and Opinion Polls

Charity and fundraising: Preview Ex 11
Family relocation: Ex 9.4
Opinion on gas prices: Ex 11.3; Ex 11.4
People's financial situation: Ex 10.2
Young adult residences: Ex 10.1

Today's World

Behavior on social networking sites: Act 14.1
Cell phone use: Ex 11.1; Ex 11.5
Children television viewing habits: Ex 2.14; Ex 2.15
Cost of fast food: Ex 3.4; Ex 3.5; Ex 3.15
Dangerous driving: Ex 9.5
Education and income: Ex 2.20
Effects of Facebook: Ex 10.13
Family planning: Ex 5.28
Gender and texting behavior: Ex 13.6
Holding hands: Ex 5.26
Online dating profiles: Preview Ex 7; Ex 7.1
Wireless phone service: Ex 2.8; Ex 2.9

Transportation

Auto insurance discounts: Ex 2.7
Distracted driving: Ex 1.6
Lost luggage: Ex 5.20
On-time airplane arrivals: Ex 10.6

Index of Applications in Examples and Activities

Act: Activity; Ex: Example; Preview Ex: Preview Example

Agriculture

Grape Production: Ex 2.23

Biology

Age of bears: Ex 4.9
Ankle motion and balance: Ex 4.5
Apgar scores: Ex 6.13
Baby birth weight: Ex 16.2
Baby gender predictions using ultrasound: Preview Ex 5; Ex 5.15
Baby gender: Preview Ex 10; Ex 10.12
Barking treefrog behavior: Ex 16.10
Bat behavior: Ex 3.6
Biometric responses: Act 7.1
Births and the lunar cycle: Ex 15.2; Ex 15.4
Bison survival and reproduction: Ex 16.4
Chimpanzee behavior: Ex 12.9; Ex 13.5
Cricket mating behavior: Ex 12.14
Effect of human activity on bears: Ex 4.16; Ex 4.17
Egg weights: Ex 6.33; Ex 6.34
Estimating elk weight: Ex 16.3; Ex 16.6
Honey bee behavior: Ex 2.10; Ex 2.11; Ex 2.12
Lead exposure and brain volume: Ex 4.12
Lobster age: Ex 4.23
Loon chick survival: Ex 4.20; 4.21
Mouse behavior: Ex 4.6; Ex 4.7; Ex 4.10; Ex 4.11
Newborn birth weights: Ex 6.29
Pediatric trachea tube use: Ex 16.9
Predicting IQ: Ex 4.14
Premature babies: Preview Ex 16
Salamander behavior: Ex 4.22
Scorpion mating behavior: Ex 12.3
Sources of flavonols: Ex 7.9
Women heights and weights: Ex 4.8

Business and Economics

Airline passenger weights: Act 3.2
Christmas price index: Ex 2.19
Customer satisfaction on car purchases: Preview Ex 2
Employee efficiency: Ex 12.13
Hybrid car purchases: Ex 15.5
Online security: Ex 6.22
Quality rating of men's shows: Ex 2.18
Salary and gender: Ex 13.2

College Life

Academic procrastination: Ex 8.1
College graduation rates: Ex 2.5; Ex 2.6
Cost of textbooks: Ex 12.10
Distance college students living from home: Ex 2.4
Facebook and academic performance: Ex 13.1
Gender of college students: Ex 8.2
Graduation rates and student-related expenditures: Ex 4.1
Graduation rates: Ex 4.13
Higher education demographics: Ex 3.11
How college seniors spend their time: Ex 2.22
Misreporting grade point average: Ex 2.16
The "Freshman 15": Ex 12.4; Ex 13.7

Education and Child Development

Child television viewing: Ex 7.3
Children's learning styles: Ex 1.5; Ex 1.7
Children's television viewing habits and test scores: Ex 1.2
IQ scores: Ex 6.30
Public university enrollments: Ex 2.13
Student freedom of the press: Ex 11.2
Test score improvement: Ex 12.15

Environmental Science

Cosmic radiation: Ex 12.6
Electromagnetic radiation: Ex 4.19
Rainfall in Albuquerque: Ex 2.17
River water velocity: Ex 4.18
Snow cover and temperature: Ex 16.7

Health and Wellness

Customer behavior at fast-food restaurants: Ex 1.4
Effects of nicotine withdrawal: Ex 12.12
Exercise and calorie consumption: Preview Ex 3; Ex 3.3
Memory and psychological wellness in older adults: Ex 1.1
Teen physical fitness: Ex 10.11
Weight loss: Ex 15.8; Ex 16.1

Leisure and Popular Culture

Belief in extraterrestrial visitors: Ex 8.4
Effect of reality television: Ex 10.4
Euro coins: Act 5.2
Facebook friending: Act 1.1
Facebook usage: Ex 7.4
Health information from television shows: Ex 5.5; Ex 5.16
Hershey's kisses: Act 5.1
iPod song shuffling: Preview Ex 6; Ex 6.7
Reality television: Ex 7.2
Should you believe everything you read? Preview Ex 1
Tattoos: Ex 5.6; Ex 5.18
Teens and gambling: Ex 9.3

Dedication

To my friends and colleagues in the Cal Poly Statistics Department.
Roxy Peck

To my wife, Sally, and my daughter, Anna.
Chris Olsen

Author Bio

 ROXY PECK is a professor emerita of statistics at California Polytechnic State University, San Luis Obispo. She was a faculty member in the Statistics Department for thirty years, serving for six years as chair of the Statistics Department and thirteen years as associate dean of the College of Science and Mathematics. Nationally known in the area of statistics education, Roxy was made a fellow of the American Statistical Association in 1998, and in 2003 she received the American Statistical Association's Founders Award in recognition of her contributions to K-12 and undergraduate statistics education. In 2009, she received the USCOTS Lifetime Achievement Award in Statistics Education. In addition to coauthoring the textbooks *Statistics: Learning from Data; Introduction to Statistics and Data Analysis*; and *Statistics: The Exploration and Analysis of Data*; she is also editor of *Statistics: A Guide to the Unknown*, a collection of expository papers that showcases applications of statistical methods. Roxy served from 1999 to 2003 as the chief faculty consultant for the Advanced Placement Statistics exam, and she is a past chair of the joint ASA/NCTM Committee on Curriculum in Statistics and Probability for Grades K-12 and of the ASA Section on Statistics Education. Outside the classroom, Roxy enjoys travel and has visited all seven continents. She collects Navajo rugs and heads to Arizona and New Mexico whenever she can find the time.

 CHRIS OLSEN taught statistics in Cedar Rapids, Iowa, for over 25 years, and has also taught at Cornell College and Grinnell College. Chris is a past member (twice) of the Advanced Placement Statistics Test Development Committee and has been a table leader and question leader at the AP Statistics reading for 11 years. As a long-time consultant to the College Board, Chris has led workshops and institutes for AP Statistics teachers in the United States and internationally. Chris was the Iowa recipient of the Presidential Award for Excellence in Science and Mathematics Teaching in 1986. He was a regional winner of the IBM Computer Teacher of the Year award in 1988 and received the Siemens Award for Advanced Placement in mathematics in 1999. Chris is a frequent contributor to the AP Statistics Electronic Discussion Group. He is a member of the editorial board of the journal Teaching Statistics and has reviewed materials for *The Mathematics Teacher*, the AP Central web site, *The American Statistician*, and the *Journal of the American Statistical Association*. He also wrote a column for *Stats* magazine and was a member of the editorial board of that journal. Chris graduated from Iowa State University with a major in mathematics and, while acquiring graduate degrees at the University of Iowa, concentrated on statistics, computer programming, psychometrics, and test development. In his spare time, he enjoys reading and hiking. He and his wife have a daughter, Anna, a Cal-Tech graduate in Civil Engineering, now working on a postdoctorate at the United States Geological Survey in Golden, Colorado.

Brief Contents

Overview

SECTION I   Collecting Data

CHAPTER 1   Collecting Data in Reasonable Ways   8

SECTION II   Describing Data Distributions

CHAPTER 2   Graphical Methods for Describing Data Distributions   56

CHAPTER 3   Numerical Methods for Describing Data Distributions   128

CHAPTER 4   Describing Bivariate Numerical Data   180

SECTION III   A Foundation for Inference: Reasoning About Probability

CHAPTER 5   Probability   266

CHAPTER 6   Random Variables and Probability Distributions   334

SECTION IV   Learning from Sample Data

CHAPTER 7   An Overview of Statistical Inference—Learning from Data   410

CHAPTER 8   Sampling Variability and Sampling Distributions   430

CHAPTER 9   Estimating a Population Proportion   454

CHAPTER 10   Asking and Answering Questions About a Population Proportion   494

CHAPTER 11   Asking and Answering Questions About the Difference Between Two Population Proportions   540

CHAPTER 12   Asking and Answering Questions About a Population Mean   566

CHAPTER 13   Asking and Answering Questions About the Difference Between Two Population Means   616

SECTION V   Additional Opportunities to Learn from Data

CHAPTER 14   Learning from Experiment Data   666

CHAPTER 15   Learning from Categorical Data   692

CHAPTER 16   Understanding Relationships—Numerical Data Part 2   734

CHAPTER 17   Asking and Answering Questions about More Than Two Means (online)

Contents

Preface   xii

AP* Statistics Curriculum Correlation Chart   xxi

Letter from the Authors   xxiv

A Brief Note about the AP* Statistics Course and Exam   xxv

Preparing for the AP* Statistics Examination   xxvi

Taking the AP* Statistics Exam   xxvii

Overview   Learning from Data   2

Statistics—It's All About Variability   3

Overview Example: Monitoring Water Quality   3

The Data Analysis Process   4

Goals for Student Learning   6

The Structure of the Chapters That Follow   7

SECTION I   COLLECTING DATA

CHAPTER 1   Collecting Data in Reasonable Ways   8

Preview   8

Chapter Learning Objectives   9

1.1   Statistical Studies: Observation and Experimentation   9

1.2   Collecting Data: Planning an Observational Study   13

1.3   Collecting Data: Planning an Experiment   24

1.4   The Importance of Random Selection and Random Assignment: What Types of Conclusions are Reasonable?   41

1.5   Avoid These Common Mistakes   45

Chapter Activities   46

Exploring the Big Ideas   47

Are You Ready to Move On? Chapter 1 Review Exercises   49

AP* Review Questions for Chapter 1   53

SECTION II   DESCRIBING DATA DISTRIBUTIONS

CHAPTER 2   Graphical Methods for Describing Data Distributions   56

Preview   56

Chapter Learning Objectives   57

2.1   Selecting an Appropriate Graphical Display   58

2.2   Displaying Categorical Data: Bar Charts and Comparative Bar Charts   62

2.3   Displaying Numerical Data: Dotplots, Stem-and-Leaf Displays, and Histograms   68

2.4   Displaying Bivariate Numerical Data: Scatterplots and Time Series Plots   96

AP* and Advanced Placement Program are registered trademarks of the College Entrance Examination Board, which is not involved in the production of, and does not endorse, this product.

2.5 Graphical Displays in the Media   101
2.6 Avoid These Common Mistakes   107
Chapter Activities   110
Exploring the Big Ideas   111
Are You Ready to Move On? Chapter 2 Review Exercises   113
Technology Notes   117
AP* Review Questions for Chapter 2   124

CHAPTER 3 **Numerical Methods for Describing Data Distributions**   **128**

Preview   128
Chapter Learning Objectives   129
3.1 Selecting Appropriate Numerical Summaries   130
3.2 Describing Center and Spread for Data Distributions That Are Approximately Symmetric   133
3.3 Describing Center and Spread for Data Distributions That Are Skewed or Have Outliers   142
3.4 Summarizing a Data Set: Boxplots   149
3.5 Measures of Relative Standing: z-scores and Percentiles   160
3.6 Avoid These Common Mistakes   164
Chapter Activities   165
Exploring the Big Ideas   166
Are You Ready to Move On? Chapter 3 Review Exercises   168
Technology Notes   170
AP* Review Questions for Chapter 3   176

CHAPTER 4 **Describing Bivariate Numerical Data**   **180**

Preview   180
Chapter Learning Objectives   181
4.1 Correlation   182
4.2 Linear Regression: Fitting a Line to Bivariate Data   198
4.3 Assessing the Fit of a Line   211
4.4 Describing Linear Relationships and Making Predictions—Putting It All Together   229
4.5 Modeling Nonlinear Relationships   231
4.6 Avoid These Common Mistakes   249
Chapter Activities   250
Exploring the Big Ideas   251
Are You Ready to Move On? Chapter 4 Review Exercises   252
Technology Notes   256
AP* Review Questions for Chapter 4   261

Want to Know More? See Chapter 4 Online Materials for coverage of Logistic Regression.

SECTION III **A FOUNDATION FOR INFERENCE: REASONING ABOUT PROBABILITY**

CHAPTER 5 **Probability**   **266**

Preview   266
Chapter Learning Objectives   267
5.1 Interpreting Probabilities   268
5.2 Computing Probabilities   273

5.3 Probabilities of More Complex Events: Unions, Intersections, and Complements 278
5.4 Conditional Probability 290
5.5 Calculating Probabilities—A More Formal Approach 302
5.6 Probability as a Basis for Making Decisions 314
5.7 Estimating Probabilities Empirically and Using Simulation 317
Chapter Activities 326
Are You Ready to Move On? Chapter 5 Review Exercises 327
AP* Review Questions for Chapter 5 331

CHAPTER **6**

Random Variables and Probability Distributions 334

Preview 334
Chapter Learning Objectives 335
6.1 Random Variables 336
6.2 Probability Distributions for Discrete Random Variables 339
6.3 Probability Distributions for Continuous Random Variables 345
6.4 Mean and Standard Deviation of a Random Variable 352
6.5 Binomial and Geometric Distributions 363
6.6 Normal Distributions 373
6.7 Checking for Normality 391
6.8 Using the Normal Distribution to Approximate a Discrete Distribution 395
Chapter Activities 399
Are You Ready to Move On? Chapter 6 Review Exercises 400
Technology Notes 403
AP* Review Questions for Chapter 6 407

Want to know more? See Chapter 6 Online Materials for coverage of Counting Rules and the Poisson Distribution.

SECTION **IV**

LEARNING FROM SAMPLE DATA

CHAPTER **7**

An Overview of Statistical Inference—Learning from Data 410

Preview 410
Chapter Learning Objectives 411
7.1 Statistical Inference—What You Can Learn from Data 412
7.2 Selecting an Appropriate Method—Four Key Questions 418
7.3 A Five-Step Process for Statistical Inference 424
Chapter Activities 425
Are You Ready to Move On? Chapter 7 Review Exercises 426
AP* Review Questions for Chapter 7 428

CHAPTER **8**

Sampling Variability and Sampling Distributions 430

Preview 430
Chapter Learning Objectives 431
8.1 Statistics and Sampling Variability 433
8.2 The Sampling Distribution of a Sample Proportion 438
8.3 How Sampling Distributions Support Learning from Data 443

Chapter Activities   448
Exploring the Big Ideas   448
Are You Ready to Move On? Chapter 8 Review Exercises   449
AP* Review Questions for Chapter 8   452

CHAPTER 9   Estimating a Population Proportion   454

Preview   454
Chapter Learning Objectives   455
9.1   Selecting an Estimator   456
9.2   Estimating a Population Proportion—Margin of Error   460
9.3   A Large Sample Confidence Interval for a Population Proportion   467
9.4   Choosing a Sample Size to Achieve a Desired Margin of Error   480
9.5   Avoid These Common Mistakes   483
Chapter Activities   485
Exploring the Big Ideas   487
Are You Ready to Move On? Chapter 9 Review Exercises   488
Technology Notes   490
AP* Review Questions for Chapter 9   491

CHAPTER 10   Asking and Answering Questions About a Population Proportion   494

Preview   494
Chapter Learning Objectives   495
10.1   Hypotheses and Possible Conclusions   496
10.2   Potential Errors in Hypothesis Testing   502
10.3   The Logic of Hypothesis Testing—An Informal Example   507
10.4   A Procedure for Carrying Out a Hypothesis Test   511
10.5   Large-Sample Hypothesis Test for a Population Proportion   514
10.6   Power and the Probability of Type II Error   526
10.7   Avoid These Common Mistakes   531
Chapter Activities   532
Exploring the Big Ideas   532
Are You Ready to Move On? Chapter 10 Review Exercises   533
Technology Notes   535
AP* Review Questions for Chapter 10   537

CHAPTER 11   Asking and Answering Questions About the Difference Between Two Population Proportions   540

Preview   540
Chapter Learning Objectives   541
11.1   Estimating the Difference between Two Population Proportions   542
11.2   Testing Hypotheses About the Difference Between Two Population Proportions   550
11.3   Avoid These Common Mistakes   559
Chapter Activities   559
Exploring the Big Ideas   560

Are You Ready to Move On? Chapter 11 Review Exercises   561
Technology Notes   561
AP* Review Questions for Chapter 11   564

CHAPTER 12 Asking and Answering Questions About a Population Mean   566

Preview   566
Chapter Learning Objectives   567
12.1   The Sampling Distribution of the Sample Mean   568
12.2   A Confidence Interval for a Population Mean   578
12.3   Testing Hypotheses About a Population Mean   591
12.4   Avoid These Common Mistakes   604
Chapter Activities   605
Exploring the Big Ideas   606
Are You Ready to Move On? Chapter 12 Review Exercises   607
Technology Notes   609
AP* Review Questions for Chapter 12   612

CHAPTER 13 Asking and Answering Questions About the Difference Between Two Population Means   616

Preview   616
Chapter Learning Objectives   617
13.1   Testing Hypotheses About the Difference Between Two Population Means Using Independent Samples   618
13.2   Testing Hypotheses About the Difference Between Two Population Means Using Paired Samples   633
13.3   Estimating the Difference between Two Population Means   644
13.4   Avoid These Common Mistakes   654
Chapter Activities   654
Exploring the Big Ideas   655
Are You Ready to Move On? Chapter 13 Review Exercises   655
Technology Notes   658
AP* Review Questions for Chapter 13   664

SECTION V ADDITIONAL OPPORTUNITIES TO LEARN FROM DATA

CHAPTER 14 Learning from Experiment Data   666

Preview   666
Chapter Learning Objectives   667
14.1   Variability and Random Assignment   668
14.2   Testing Hypotheses About Differences in Treatment Effects   670
14.3   Estimating the Difference in Treatment Effects   679
14.4   Avoid These Common Mistakes   686
Chapter Activities   687
Are You Ready to Move On? Chapter 14 Review Exercises   688
AP* Review Questions for Chapter 14   690

CHAPTER 15 **Learning from Categorical Data 692**

Preview 692

Chapter Learning Objectives 693

15.1 Chi-Square Tests for Univariate Categorical Data 694

15.2 Tests for Homogeneity and Independence in a Two-Way Table 708

15.3 Avoid These Common Mistakes 723

Chapter Activities 723

Are You Ready to Move On? Chapter 15 Review Exercises 724

Technology Notes 727

AP* Review Questions for Chapter 15 731

CHAPTER 16 **Understanding Relationships—Numerical Data Part 2 734**

Preview 734

Chapter Learning Objectives 735

16.1 The Simple Linear Regression Model 736

16.2 Inferences Concerning the Slope of the Population Regression Line 748

16.3 Checking Model Adequacy 759

Are You Ready to Move On? Chapter 16 Review Exercises 773

Technology Notes 775

AP* Review Questions for Chapter 16 777

Want to Know More? See Online Materials for the following additional chapter:

CHAPTER 17
ONLINE

Asking and Answering Questions About More Than Two Means

Preview

Chapter Learning Objectives

17.1 The Analysis of Variance—Single-Factor ANOVA and the *F* Test

17.2 Multiple Comparisons

Appendix: ANOVA Computations

Are You Ready to Move On? Chapter 17 Review Exercises

Technology Notes

Appendix A Statistical Tables 780

Table 1 Random Numbers 780

Table 2 Standard Normal Probabilities (Cumulative *z* Curve Areas) 782

Table 3 *t* Critical Values 784

Table 4 Tail Areas for *t* Curves 785

Table 5 Upper-Tail Areas for Chi-Square Distributions 788

Table 6 Binomial Probabilities 790

AP* Free-Response Questions 792

Answers 801

Index 846

Preface

Statistics is about learning from data and the role that variability plays in drawing conclusions from data. To be successful, it is not enough for students to master the computational aspects of descriptive and inferential statistics—they must also develop an understanding of the data analysis process at a conceptual level. *Statistics: Learning from Data* is informed by careful and intentional thought about how the conceptual and the mechanical should be integrated in order to promote three key types of learning objectives for students:

- conceptual understanding
- mastery of the mechanics
- the ability to demonstrate these by "putting it into practice"

A Unique Approach

A number of innovative features distinguish this text from other introductory statistics books:

- **Multiple Approaches to Probability:**
  Research has demonstrated how students develop an understanding of probability and chance. The book employs both traditional and modern methods, including the use of natural frequencies and hypothetical 1,000 tables, to make probability much easier for students to understand. The content of the probability chapters is consistent with the AP* Statistics course description.

- **Chapter on Overview of Statistical Inference (Chapter 7)**
  This short chapter focuses on the things students need to think about in order to select an appropriate method of analysis. In most texts, these considerations are "hidden" in the discussion that occurs when a new method is introduced. Discussing these considerations up front in the form of four key questions that need to be answered before choosing an inference method makes it easier for students to make correct choices.

- **An Organization That Reflects the Data Analysis Process**
  Students are introduced early to the idea that data analysis is a process that begins with careful planning, followed by data collection, data description using graphical and numerical summaries, data analysis, and finally interpretation of results. The ordering of topics in the text book mirrors this process: data collection, then data description, then statistical inference.

- **Inference for Proportions Before Inference for Means**
  Inference for proportions is covered before inference for means for the following reasons:
  - This makes it possible to develop the concept of a sampling distribution via simulation, an approach that is more accessible to students than a more formal, theoretical approach. Simulation is simpler in the context of proportions, where it is easy to construct a hypothetical population from which to sample (it is more complicated to create a hypothetical population in the context of means because this requires making assumptions about shape and spread).
  - Large-sample inferential procedures for proportions are based on the normal distribution and don't require the introduction of a new distribution (the *t* distribution). Students can focus on the new concepts of estimation and hypothesis testing without having to grapple at the same time with the introduction of a new probability distribution.

- **Separate Treatment of Inference Based on Experiment Data (Chapter 14)**
  Many statistical studies involve collecting data via experimentation. The same inference procedures used to estimate or test hypotheses about population parameters also are used to estimate or test hypotheses about treatment effects. However, the necessary assumptions are slightly different (for example, random assignment replaces the assumption of random selection), and the wording of hypotheses and conclusions is also different. Trying to treat both cases together tends to confuse students. This text makes the distinction clear.

Features That Support Student Engagement and Success

The text also includes a number of features that support conceptual understanding, mastery of mechanics, and putting ideas into practice.

- **Multiple AP* aids:**
  In this AP* edition, there are multiple choice exercises (in AP* format) at the end of each chapter. These questions allow students to check their understanding and help them prepare for the AP* exam. In addition there are 24 AP* style free response questions at the end of the text. This AP edition also contains margin *AP Tips* that highlight important AP content.
- **Simple Design**
  There is now research showing that many of the "features" in current textbooks are not really helpful to students. In fact, cartoons, sidebars, historical notes, fake Post-it notes in the margins, and the like, actually distract students and interfere with learning. *Statistics: Learning from Data* has a simple, clean design in order to minimize clutter and maximize student understanding.
- **Chapter Learning Objectives—Keeping Students Informed About Expectations**
  Chapter learning objectives explicitly state the expected student outcomes. Learning objectives fall under three headings: Conceptual Understanding, Mastery of Mechanics, and Putting It Into Practice.
- **Preview—Motivation for Learning**
  Each chapter opens with a *Preview* and *Preview Example* that provide motivation for studying the concepts and methods introduced in the chapter. They address why the material is worth learning, provide the conceptual foundation for the methods covered in the chapter, and connect to what the student already knows. A relevant and current example provides a context in which one or more questions are proposed for further investigation. This context is revisited in the chapter once students have the necessary understanding to more fully address the questions posed.
- **Real Data**
  Examples and exercises with overly simple settings do not allow students to practice interpreting results in authentic situations or give students the experience necessary to be able to use statistical methods in real settings. The exercises and examples are a particular strength of this text, and we invite you to compare the examples and exercises with those in other introductory statistics texts.

  Many students are skeptical of the relevance and importance of statistics. Contrived problem situations and artificial data often reinforce this skepticism. Examples and exercises that involve data extracted from journal articles, newspapers, and other published sources and that are of interest to today's students are used to motivate and engage students. Most examples and exercises in the book are of this nature; they cover a very wide range of disciplines and subject areas. These include, but are not limited to, health and fitness, consumer research, psychology and aging, environmental research, law and criminal justice, and entertainment.
- **Exercises Organized into a Developmental Structure—Structuring the Out-of-Class Experience**
  End-of-section exercises are organized into developmental sets. At the end of each section, there are two grouped problem sets. The exercises in each set work together to

assess all of the learning objectives for that section. In addition to the two exercise sets, each section also has additional exercises for those who want more practice.

- **Are You Ready to Move On?—Students Test Their Understanding**
  Prior to moving to the next chapter, "Are You Ready to Move On?" exercises allow students to confirm that they have achieved the chapter learning objectives. Like the developmental problem sets of the individual sections, this collection of exercises is developmental in nature. These exercises assess all of the chapter learning objectives and serve as a comprehensive end-of-chapter review.

- **Exploring the Big Ideas—Real Data Algorithmic Sampling Exercises**
  Most chapters contain extended sampling-based, real-data exercises at the end of the chapter. These exercises appear in CourseMate and Aplia, where each student gets a different random sample for the same exercise. These unique exercises
  - address the tension between the desire to use real data and the desire to have algorithmically generated exercises.
  - address the tension between the role of interpretation and communication in data analysis and the desire for exercises that can be machine scored.
  - are designed to teach about sampling variability.

- **Data Analysis Software**
  Each new textbook comes with free JMP data analysis software. See *Student Resources* for more information.

- **Technology Notes**
  Technology Notes appear at the end of most chapters and give students helpful hints and guidance on completing tasks associated with a particular chapter. The following technologies are included in the notes: JMP, Minitab, SPSS, Microsoft Excel 2007, TI-83/84, and TI-nspire. They include display screens to help students visualize and better understand the steps. More complete technology manuals are also available on the text web site.

- **Chapter Activities—Engaging Students in Hands-On Activities**
  There is a growing body of evidence that students learn best when they are actively engaged. Chapter activities guide students' thinking about important ideas and concepts.

Consistent with Recommendations for the Introductory Statistics Course Endorsed by the American Statistical Association

In 2005, the American Statistical Association endorsed the report "College Guidelines in Assessment and Instruction for Statistics Education (GAISE Guidelines)," which included the following six recommendations for the introductory statistics course:

1. Emphasize statistical literacy and develop statistical thinking.
2. Use real data.
3. Stress conceptual understanding rather than mere knowledge of procedures.
4. Foster active learning in the classroom.
5. Use technology for developing conceptual understanding and analyzing data.
6. Use assessments to improve and evaluate student learning.

Statistic: Learning from Data is consistent with these recommendations and supports the GAISE guidelines in the following ways:

1. **Emphasize Statistical Literacy and Develop Statistical Thinking.**
   Statistical literacy is promoted throughout the text in the many examples and exercises that are drawn from the popular press. In addition, a focus on the role of variability, consistent use of context, and an emphasis on interpreting and communicating results in context work together to help students develop skills in statistical thinking.

2. **Use Real Data.**

   The examples and exercises are context driven, and the reference sources include the popular press as well as journal articles.

3. **Stress Conceptual Understanding Rather Than Mere Knowledge of Procedures.**

   Nearly all exercises in the text are multipart and ask students to go beyond just computation, with a focus on interpretation and communication. The examples and explanations are designed to promote conceptual understanding. Hands-on activities in each chapter are also constructed to strengthen conceptual understanding. Which brings us to ...

4. **Foster Active Learning in the Classroom.**

   While this recommendation speaks more to pedagogy and classroom practice, *Statistics: Learning from Data* provides more than 30 hands-on activities in the text and additional activities in the accompanying instructor resources that can be used in class or assigned to be completed outside of class.

5. **Use Technology for Developing Conceptual Understanding and Analyzing Data.**

   The computer has brought incredible statistical power to the desktop of every investigator. The wide availability of statistical computer packages, such as JMP, Minitab, and SPSS, and the graphical capabilities of the modern microcomputer have transformed both the teaching and learning of statistics. To highlight the role of the computer in contemporary statistics, sample output is included throughout the book. In addition, numerous exercises contain data that can easily be analyzed using statistical software. JMP data analysis software is provided with the text, and technology manuals for JMP and for other software packages, such as Minitab and SPSS, and for the graphing calculator are available in the online materials that accompany this text.

6. **Use Assessments to Improve and Evaluate Student Learning.**

   Comprehensive chapter review exercises that are specifically linked to chapter learning objectives are included at the end of each chapter. In addition, assessment materials in the form of a test bank, quizzes, and chapter exams are available in the instructor resources that accompany this text. The items in the test bank reflect the data-in-context philosophy of the text's exercises and examples.

Teacher and Student Resources

Aplia™
Content

Aplia™ is an online interactive learning solution that improves comprehension and outcomes by increasing student effort and engagement. Founded by a professor to enhance his own courses, Aplia provides automatically graded assignments with detailed, immediate explanations for every question, along with innovative teaching materials. Our easy-to-use system has been used by more than 1,000,000 students at over 1,800 institutions. Exercises are taken directly from text.

Aplia homework engages students in critical thinking, requiring them to synthesize and apply knowledge, not simply recall it. The diverse types of questions reflect the types of exercises that help students learn. All homework is written by subject-matter experts in the field who have taught the course before.

Aplia contains a robust course management system with powerful analytics, enabling teachers to track student performance easily.

JMP Statistical Software

Ask your sales representative for details.

JMP is a statistics software for Windows and Macintosh computers from SAS, the market leader in analytics software and services for industry. JMP Student Edition is a streamlined, easy-to-use version that provides all the statistical analysis and graphics covered in this textbook. Once data is imported, students will find that most procedures require just two or three mouse clicks. JMP can import data from a variety of formats, including Excel and other statistical packages, and you can easily copy and paste graphs and output into documents.

JMP also provides an interface to explore data visually and interactively, which will help your students develop a healthy relationship with their data, work more efficiently with data, and tackle difficult statistical problems more easily. Because its output provides both statistics and graphs together, the student will better see and understand the application of concepts covered in this book as well. JMP Student Edition also contains some unique platforms for student projects, such as mapping and scripting. JMP functions in the same way on both Windows and Mac platforms and instructions contained with this book apply to both platforms.

JMP software is made available to all adopters of this text; see your sales representative for details. Find out more at www.jmp.com.

Student Resources
Digital

To access additional course materials and companion resources, please visit www.cengagebrain.com. At the CengageBrain.com home page, search for the ISBN of your title (from the back cover of your book) using the search box at the top of the page. This will take you to the product page where free companion resources can be found.

If your text includes a printed access card, you will have instant access to the following resources:

- Complete step-by-step instructions for JMP, TI-84 Graphing Calculators, Excel, Minitab, and SPSS.
- Data sets in JMP, TI-84, Excel, Minitab, SPSS, SAS, and ASCII file formats.
- Applets used in the Activities found in the text.

Print

Fast Track to a 5: Preparing for the AP* Statistics Examination. This AP* test preparation guide includes strategies for taking the exam, a diagnostic test so that students can assess their level of preparedness, chapter-review sections with self-study questions in AP* format, and two complete practice tests in AP* format. It is keyed to this text and can be purchased with the text or separately. Prepared by Brendan Murphy of John Bapst High School, Bangor, Maine, and Stephen Dartt of Shiloh High School, Snellville, Georgia.

Teacher Resources

Print

Teacher's Resource Binder (ISBN: 9781285094632): The Teacher's Resource Binder is full of wonderful resources for both AP Statistics teachers and college professors. These include:

- Recommendations for instructors on how to teach the course, including sample syllabi, pacing guides, and teaching tips.
- Recommendations for what students should read and review for a particular class period or set of class periods.
- Extensive notes on preparing students to take the AP exam.
- Additional examples from published sources (with references), classified by chapter in the text. These examples can be used to enrich your classroom discussions.
- Model responses—examples of responses that can serve as a model for work that would be likely to receive a high mark on the AP exam.
- A collection of data explorations that can be used throughout the year to help students prepare for the types of questions that they may encounter on the investigative task on the AP Statistics Exam.
- Activity worksheets that can be duplicated and used in class.
- A test bank that includes assessment items, quizzes, and chapter exams.

Digital

- Solution Builder: This online instructor database offers complete worked-out solutions to all exercises in the text, allowing you to create customized, secure solutions printouts (in PDF format) matched exactly to the problems you assign in class. Sign up for access at www.cengage.com/solutionbuilder.
- E-book: This new premium eBook has highlighting, note-taking, and search features as well as links to multimedia resources.

We would like to express our thanks and gratitude to the following people who made this book possible:

Molly Taylor, our editor at Cengage, for her support and encouragement.

Jay Campbell, our development editor, for his unfailing good humor and his ability to field just about any curve ball we threw at him.

Shaylin Walsh, for leading the development of all of the ancillary materials that support the text.

Alison Eigel Zade, the content project manager.

Teresa Christie, Lindsay Schmonsees and Liah Rose, project managers at MPS Limited.

Andrew Coppola, media editor, for managing the media content and for creating the implementation of the "Exploring the Big Ideas" activities.

Cameron Troxell, for his careful review of the manuscript and many helpful suggestions for improving the readability of the book.

Stephen Miller, for his great work on the huge task of creating the student and instructor solutions manuals.

Michael Allwood, for his detailed work in checking the accuracy of examples and solutions.

Kathy Fritz, for creating the interactive PowerPoint presentations that accompany the text.

Melissa Sovak, for creating the Technology Notes sections.

Nicole Mollica, the developmental editor for market strategies.

Anna Andersen, the associate brand development manager.

Beth Chance and Francisco Garcia for producing the applet used in the confidence interval activities.

Gary McClelland for producing the applets from *Seeing Statistics* used in the regression activities.

Chris Sabooni, the copy editor for the book.

MPS, for producing the artwork used in the book.

We would also like to give a special thanks to those who served on the Editorial Board for the book and those who class tested some of the chapters with their students:

Editorial Board

John Climent, Cecil College
Ginger Holmes Rowell, Middle Tennessee State University
Karen Kinard, Tallahassee Community College
Glenn Miller, Borough of Manhattan Community College
Ron Palcic, Johnson County Community College

Class Testers

Rob Eby, Blinn College, Bryan Campus
Stephen Miller, Winchester Thurston School
Sumona Mondal, Clarkson University
Cameron Troxell, Mount San Antonio College
Cathy Zucco-Teveloff, Rowan University

Many people provided invaluable comments and suggestion as this text was being developed. We would like to thank the following people for their contributions:

AP Reviewers

David Dimsdale, Brookwood High School, Snellville, GA
Kathie Emerson, Timberline High School, Boise, ID
Jonathan Gemmill, Billings West High School, Billings, MT
Chuck Pack, Tahlequah High School, Tahlequah, OK
Shaw Zaidins, Palm Desert High School, Palm Desert, CA

Other Reviewers

Gregory Bloxom, Pensacola State College
Denise Brown, Collin College

Jerry Chen, Suffolk County Community College
Monte Cheney, Central Oregon Community College
Ivette Chuca, El Paso Community College
Mary Ann Connors, Westfield State University
George Davis, Georgia State University
Rob Eby, Blinn College, Bryan Campus
Karen Estes, St. Petersburg College
Larry Feldman, Indiana University, Pennsylvania
Kevin Fox, Shasta College
Melinda Holt, Sam Houston State University
Kelly Jackson, Camden County College
Clarence Johnson, Cuyahoga Community College

Hoon Kim, Calforina State Polytechnic University, Pomona
Jackie MacLaughlin, Central Piedmont
Nola McDaniel, McNeese State
Glenn Miller, Borough of Manhattan Community College
Philip Miller, Indiana University SE
Kathy Mowers, Owensboro Community and Technical College
Linda Myers, Harrisburg Area Community College
Ron Palcic, Johnson County Community College
Nishant Patel, Northwest Florida State College
Maureen Petkewich, Univiversity of South Carolina
Nancy Pevey, Pellissippi State Community College
Chandler Pike, University of Georgia
Blanche Presley, Macon State College
Daniel Rowe, Heartland Community College
Fary Sami, Harford Community College
Laura Sather, St. Cloud State University
Sean Simpson, Westchester Community College
Sam Soleymani, Santa Monica College
Cameron Troxell, Mount San Antonio College
Diane Van Deusen, Napa Valley College
Richard Watkins, Tidewater Community College
Jane-Marie Wright, Suffolk Community College
Cathy Zucco-Teveloff, Rider University

Focus Group Participants (JSM and AMATYC)

Raoul Amstelveen, Johnson & Wales University
Norma Biscula, University of Maine at Augusta
Patricia Blus, National Louis University
Andy Chang, Youngstown State University
Adrienne Chu, Suffolk County Community College
Julie DePree, University of New Mexico
Debbie Garrison, Valencia Community College
Brian Gill, Seattle Pacific University
Kathy Gray, California State University, Chico
David Gurney, Southeastern Louisiana University
Patricia Humphrey, Georgia Southern University
Jack Keating, Massasoit Community College
Alex Kolesnik, Ventura College
Doug Mace, Kirtland Community College
Glenn Miller, Borough of Manhattan Community College
Philip Miller, Indiana University SE
Megan Mocko, University of Florida
Adam Molnar, Bellarmine University
Sumona Mondal, Clarkson University
Calandra Moore, CUNY – College of Staten Island
Penny Morris, Polk State College
Kathy Mowers, Owensboro Community and Technical College
Blanche Presley, Macon State College
Patricia Rhodes, Treasure Valley Community College
Ned Schillow, Lehigh-Carbon Community College
Gail St. Jacques, Johnson & Wales University
Jenting Wang, SUNY-Oneonta
Rebecca Wong, West Valley College

Webinar Participants

Bilal Abu Bakr, Alcorn State University
Cheryl Adams, York College, CUNY
Josiah Alamu, University of Illinois-Springfield
Raoul Amstelveen, Johnson & Wales University
Gokarna Aryal, Purdue University, Calumet
Tatiana Arzivian, Edison State College
Christopher Barat, Stevenson University
J.B. Boren, Wayland Baptist University-Amarillo
Monica Brown, St. Catherine University
Brenda Burger, University of South Florida-St. Petersburg
Rose Elaine Carbone, Clarion University
Julie Clark, Hollins University
Vera Dauffenbach, Bellin College
Boyan Dimitrov, Kettering University
William Dougherty, SCCC
Robert Ellis, New York University
Michelle Everson, University of Minnesota
Mark Goldstein, West Virginia Northern Community College
David Gurney, Southeastern Louisiana University
Julie Hanson, Clinton Community College
Debra Hydorn, University of Mary Washington
Marc Isaacson, Augsburg College
Grazyna Kamburowska, SUNY – Oneonta
Alex Kolesnik, Ventura College
Jeff Kollath, Oregon State University
Philip Larson, Greenville Technical College
Pam Lowry, Bellvue College
John McKenzie, Babson College
Michael Moran, Shoreline Community College
Jerry Moreno, John Carroll Univeristy
Kathy Mowers, Owensboro Community and Technical College
Rebecca Nichols, American Statistical Association
Leesa Phol, Donnelly College
Rebecca Pierce, Ball State University
Gautam Pillay, Rowan University
Wendy Pogoda, Hillsborough Community College
Gyaneswor Pokharel, Mountain State University
Krish Revuluri, Harper College
Helen Roberts, Montclair State University
Dave Saha, Florida Gulf Coast University
Jennifer Sinclair, Foothill College
Thomas Strommer, Fayetteville Technical Community College
Lori Thomas, Midland College
Elizabeth Uptegrove, Felician College
Jennifer Ward, Clark College
Joan Weinstein, Harvard Extension School
Pamela Weston, Hiwassee College
Igor Woiciechowski, Alderson-Broaddus College
Barry Woods, Unity College

Content Survey Participants

Gregg Bell, University of Alabama – Tuscaloosa
Gulhan Bourget, California State University Fullerton
Pete Bouzar, Golden West College
Ralph Burr, Mesa Community College
Joe Castille, Broward County Community College
Lyle Cook, Eastern Kentucky University
Amy Curry, College of Lake County
Charles De Vogelaere, De Anza College
Ola Disu, Tarrant County College, NE
James Dulgeroff, San Bernardino Valley College
Debbie Garrison, Valencia Community College
Donna Gorton, Butler Community College
Fred Gruhl, Washtenaw Community College
Greg Henderson, Hillsborough Community College
Leslie Hendrix, University of South Carolina
Robert Indik, University of Arizona
Jay Jahangiri, Kent State University at Geauga
Jack Keating, Massasoit Community College
JoAnn Kump, West Chester University
Rudy Maglio, Northeastern Illinois University
Richard McGowan, Boston College
Wendy Miao, El Camino College
Jackie Miller, The Ohio State University
Kathy Mowers, Owensboro Community and Technical College

Alfredo Rodriguez, El Paso Community College
Thomas Roe, South Dakota State University
Rosa Seyfried, Harrisburg Area Community College
Aileen Solomon, Trident Technical College
T. C. Sun, Wayne State University
Mahbobeh Vezvaei, Kent State University
Dennis Walsh, Middle Tennessee State University
Sheila Weaver, University of Vermont
Dave Wilson, SUNY Buffalo State
Zhiwei Zhu, University of Louisiana at Lafayette

Marketing Survey Participants

Jerry Chen, Suffolk County Community College
Ivette Chuca, El Paso Community College
Rob Eby, Blinn College, Bryan Campus
Larry Feldman, Indiana University, Pennsylvania
Patricia Humphrey, Georgia Southern University
Kelly Jackson, Camden County College
Glenn Miller, Borough of Manhattan Community College
Megan Mocko, University of Florida
Sumona Mondal, Clarkson University
Ron Palcic, Johnson County Community College
Aileen Solomon, Trident Technical College
Jenting Wang, SUNY-Oneonta
Rebecca Wong, West Valley College
Cathy Zucco-Teveloff, Rider University

And last, but certainly not least, we thank our family, friends, and colleagues for their continued support.

Roxy Peck
Chris Olsen

AP* Statistics Curriculum Correlation Chart

Below is a chart of the AP Statistics curriculum correlated to the appropriate sections of *Statistics: Learning from Data, AP Edition.*

AP Statistics Topic	Corresponding Text Sections
I. Exploring data: Describing patterns and departures from patterns	
IA. Constructing and interpreting graphical displays of distributions of univariate data (dotplot, stemplot, histogram, cumulative frequency plot)	**2.1–2.3**
IA1. Center and spread	2.3
IA2. Clusters and gaps	2.3
IA3. Outliers and other unusual features	2.3
IA4. Shape	2.3
IB. Summarizing distributions of univariate data	**3.1–3.6**
IB1. Measuring center: median, mean	3.2-3.3
IB2. Measuring spread: range, interquartile range, standard deviation	3.2-3.3
IB3. Measuring position: quartiles, percentiles, standardized scores (z-scores)	3.5
IB4. Using Boxplots	3.4
IB5. The effect of changing units on summary measures	3.2
1C. Comparing distributions of univariate data (dotplots, back-to-back stemplots, parallel boxplots)	**2.3; 3.4**
IC1. Comparing center and spread: within group, between group variation	2.3; 3.4
IC2. Comparing clusters and gaps	2.3
IC3. Comparing outliers and other unusual features	2.3
IC4. Comparing shapes	2.3
1D. Exploring bivariate data	**2.4; 4.1–4.6**
ID1. Analyzing patterns in scatterplots	2.4
ID2. Correlation and linearity	4.1
ID3. Least-squares regression line	4.2
ID4. Residual plots, outliers, and influential points	4.3
ID5. Transformations to achieve linearity: logarithmic and power transformations	4.5

(Continued)

AP Statistics Topic	Corresponding Text Sections
IE. Exploring categorical data	**2.2; 5.3–5.4**
IE1. Frequency tables and bar charts	2.2
IE2. Marginal and joint frequencies for two-way tables	5.3; 5.4
IE3. Conditional relative frequencies and associations	5.3; 5.4
IE4. Comparing distributions using bar charts	2.2
II. Sampling and experimentation: Planning and conducting a study	**1.1–1.5**
IIA. Overview of methods of data collection	**1.1–1.3**
IIA1. Census	1.2
IIA2. Sample survey	1.2
IIA3. Experiment	1.1; 1.3
IIA4. Observational study	1.1-1.2
IIB. Planning and conducting surveys	**1.2; 1.4**
IIB1. Characteristics of a well-designed and well-conducted survey	1.2
IIB2. Populations, samples, and random selection	1.2; 1.4
IIB3. Sources of bias in sampling and surveys	1.2
IIB4. Sampling methods, including simple random sampling, stratified random sampling, and cluster sampling	1.2
IIC. Planning and conducting experiments	**1.3–1.4**
IIC1. Characteristics of a well-designed and well-conducted experiment	1.3
IIC2. Treatments, control groups, experimental units, random assignments, and replication	1.3
IIC3. Sources of bias and confounding, including placebo effect and blinding	1.3
IIC4. Completely randomized design	1.3
IIC5. Randomized block design	1.3
IID. Generalizability of results and types of conclusions that can be drawn from observational studies, experiments, and surveys	**1.4**
III. Anticipating patterns: Exploring random phenomena using probability and simulation	**5.1–5.6; 6.1–6.8**
IIIA. Probability	
IIIA1. Interpreting probability, including long-run relative frequency interpretation	5.1
IIIA2. "Law of Large Numbers" concept	5.1
IIIA3. Addition rule, multiplication rule, conditional probability, and independence	5.3-5.5
IIIA4. Discrete random variables and their probability distributions, including binomial and geometric	6.2; 6.5
IIIA5. Simulation of random behavior and probability distributions	5.7
IIIA6. Mean (expected value) and standard deviation of a random variable, and linear transformation of a random variable	6.4
IIIB. Combining independent random variables	**6.4**
IIIB1. Notion of independence versus dependence	6.4
IIIB2. Mean and standard deviation for sums and differences of independent random variables	6.4

AP Statistics Topic	Corresponding Text Sections
IIIC. The normal distribution	6.6
IIIC1. Properties of the normal distribution	6.6
IIIC2. Using tables of the normal distribution	6.6
IIIC3. The normal distribution as a model for measurements	6.6
IIID. Sampling distributions	**8.1**
IIID1. Sampling distribution of a sample proportion	8.2
IIID2. Sampling distribution of a sample mean	12.1
IIID3. Central Limit Theorem	12.1
IIID4. Sampling distribution of a difference between two independent sample proportions	11.1
IIID5. Sampling distribution of a difference between two independent sample means	13.1
IIID6. Simulation of sampling distributions	8.2-8.2; 12.1
IIID7. t-distribution	12.2
IIID8. Chi-square distribution	15.1
IV. Statistical inference: Estimating population parameters and testing hypotheses	**Chapter 9–16**
IVA. Estimation (point estimators and confidence intervals)	**9.3**
IVA1. Estimating population parameters and margins of error	9.1; 9.2
IVA2. Properties of point estimators, including unbiasedness and variability	9.1
IVA3. Logic of confidence intervals, meaning of confidence level and confidence intervals, and properties of confidence intervals	9.3
IVA4. Large sample confidence interval for a proportion	9.3
IVA5. Large sample confidence interval for a difference between two proportions	11.1
IVA6. Confidence interval for a mean	12.2
IVA7. Confidence interval for a difference between two means (unpaired and paired)	13.3
IVA8. Confidence interval for the slope of a least-squares regression line	16.2
IVB. Tests of significance	**Chapter 10–16**
IVB1. Logic of significance testing, null and alternative hypotheses; p-values; one- and two-sided tests; concepts of Type I and Type II errors; concept of power	10.1-10.4; 10.6
IVB2. Large sample test for a proportion	10.5
IVB3. Large sample test for a difference between two proportions	11.2
IVB4. Test for a mean	12.3
IVB5. Test for a difference between two means (unpaired and paired)	13.1-13.2
IVB6. Chi-square test for goodness of fit, homogeneity of proportions, and independence (one- and two-way tables)	15.1-15.2
IVB7. Test for the slope of a least-squares regression line	16.2

Letter to Teachers from the Authors

We have designed this book with a particular eye toward the syllabus of the Advanced Placement Statistics course and the needs of high school teachers and students. Concerns expressed and questions asked in teacher workshops and on the AP Statistics Electronic Discussion Group have strongly influenced our coverage of certain topics, especially in the area of experimental design and probability. Because the AP Statistics course places a great deal of emphasis on interpretation and communication, all problem sets contain questions focusing on those two skills. Chapters also end with sections titled *Avoid These Common Mistakes*, which are particularly helpful in data analysis and interpretation. It reminds students of considerations that ensure statistical methods are employed in reasonable and appropriate ways.

The AP Edition also includes a page with advice to students on how they can be successful in the course and a mapping of the topics from the AP Statistics Course Description to the material in this text. You will also find AP-style exercises in the book. Five-option multiple-choice questions (the AP Statistics format) are at the end of each chapter. In addition, a collection of free-response questions with a focus on clear communication, a hallmark of AP Statistics examination free-response items, can be found at the end of the book.

Teaching AP Statistics, especially for the first time, can be a challenge. The Instructor's Resource Binder, prepared by Chris Olsen, is full of wonderful resources for the AP Statistics teacher. These include:

- Recommendations for teachers on how to teach the course, including sample syllabi, pacing guides, and teaching tips.
- Recommendations for what students should read and review for a particular class period or set of class periods.
- Extensive notes on preparing students to take the AP exam.
- Additional examples from published sources (with references), classified by chapter in the text. These examples can be used to enrich your classroom discussions.
- Model responses—examples of responses that can serve as a model for work that would be likely to receive a high mark on the AP exam.
- A collection of data explorations that can be used throughout the year to help students prepare for the types of questions that they may encounter on the investigative task on the AP Statistics Exam.
- Activity worksheets that can be duplicated and used in class.
- A test bank that includes assessment items, quizzes, and chapter exams.

We hope that you will find that these supporting materials and this text provide you with the tools you need for an enjoyable and rewarding AP Statistics teaching experience.

A Brief Note About the AP* Statistics Course and Exam

Taking an Advanced Placement course can be one of your most rewarding and challenging high school experiences. AP Statistics will provide you with the opportunity to develop both concrete data analysis skills and a fundamental understanding of statistical concepts. The process of "thinking statistically" will be developed in this course, and it will serve you well, both in everyday life and in future college pursuits.

There is one important thing that you need to know if you are going to be successful in AP Statistics. Statistics is different from mathematics. Even though this class is most likely offered through the mathematics department at your school, AP Statistics is not like other math classes you have taken, and you need to be prepared for this. One primary purpose of this course is to provide you with the tools you need to collect data in reasonable ways and then learn something from the data. Being able to interpret statistical results in real contexts is an essential part of learning from data, and this is the main reason that communication is an important aspect of statistical problems. In statistics, meaning comes from context, and it is the interpretation of the analysis *in context* that is ultimately the desired outcome of analyzing data. So, don't be surprised if your teacher asks you to interpret, explain, justify, and draw conclusions! Communication in context is an important component of the scoring of the AP Statistics exam, and you need to practice this throughout the course in order to do well on the exam.

The AP Statistics exam consists of two sections, each designed to be completed in 90 minutes. A 40-question multiple-choice section is weighted equally with a 6-question free-response section. Each multiple-choice question has five possible answers, only one of which is correct. This section tests understanding of key statistical concepts and ideas. Although you will see many test prep guides that give tips on how to take multiple-choice exams, the best advice for scoring well on the multiple-choice part of the exam is this: Know the course material!

The free-response section of the exam consists of six questions, one of which is an investigative task. The investigative task (question 6 on the exam) is a longer question that will require you to apply what you have learned in your AP Statistics course. The investigative task will ask you to stretch your knowledge and integrate your statistical and communication skills while tackling a lengthy problem. The recommended time for the investigative task is 25 minutes, which means that you should plan to spend about 10 minutes on each of the other five questions in this section.

Your answers to the free-response questions will be evaluated on how you demonstrate general statistics knowledge and on the clarity and correctness of your interpretation and communication. To be successful on these questions, be sure to provide clear, organized, and complete explanations. This isn't easy, but it is a skill that can be developed through practice. This is something that you should keep in mind and work on throughout the entire year. It is not possible to achieve top scores on the free-response section of the AP Statistics exam if communication is weak.

Most of all, the key to success in your AP Statistics course is to start with a positive attitude and resolve to invest a reasonable amount of time and effort in the course. It won't always be easy, but with the right attitude and commitment, we think that understanding, enjoyment, and a sense of accomplishment will follow.

Preparing for the AP* Statistics Examination

Advanced Placement Statistics is an exciting, practical course that is also very challenging academically. It is designed to help the student delve deeply into the methods and concepts of the appropriate use and analysis of research. Whether you are taking AP Statistics in a classroom or online, this year will prove to be both challenging and intellectually stimulating as you progress in your study of statistical analysis, inference, and experimental design.

The closer the school year gets to May, the more intimidating the looming Advanced Placement test may seem. Learning how to take all the information you have learned this year and apply it in a way that will demonstrate your expertise can be very overwhelming at times.

The best way to achieve success on any AP exam is to master it, rather than let it master you. If you think of these tests as a way to demonstrate how well your mind works, you will have an advantage—attitude does help.

Before the Exam

By February, long before the exam, you should make sure that you are registered to take it. Many schools take care of the paperwork and handle the fees for their AP students, but check with your teacher or the AP coordinator to be certain that you are on the list. This is especially important if you have a documented disability and need test accommodations. If you are studying AP independently, call AP Services at the College Board for the name of an AP coordinator at a local school who will help you through the registration process.

The evening before the exam is not a great time for partying—nor is it a great time for cramming. If you like, look over class notes or skim through your textbook, concentrating on the broad outlines rather than the small details of the course. This is a great time to get your things together for the next day. Sharpen a fistful of number 2 pencils with new erasers; be sure your calculator is in good working order with either a spare or extra batteries in case you have a malfunction of some sort; get a watch and make sure the alarm is off if it has one; pack a healthy snack for the break like a piece of fruit, a granola bar, and bottled water; make sure you have whatever identification is required as well as the admission ticket. Then relax, and get a good night's sleep.

On the day of the examination, it is wise to eat breakfast—studies show that students who eat a healthy breakfast before testing generally do better. Be careful not to drink a lot of liquids, thereby necessitating a trip to the bathroom during the test. Breakfast will give you the energy you need to power you through the test—and more. Remember, cell phones and other electronic devices, other than your approved calculator(s), are not allowed in the testing room. You will spend some time waiting while everyone is seated in the right room for the right test. That's before the test has even begun. With a short break between Section I and Section II, the statistics exam lasts well over 3 hours. Be prepared for a long test time. You don't want to be distracted by hunger pangs. Be sure to wear comfortable clothes; take along a sweater in case the heating or air-conditioning is erratic.

You have been on the fast track. Now go get a 5!

Taking the AP* Statistics Exam

The AP Statistics exam has two sections: Section I consists of 40 multiple-choice questions; Section II contains six free-response questions. You will have 90 minutes to complete the multiple-choice portion. The exam is then collected, and you will be given a short break. Then you have 90 minutes for the free-response questions. You should answer all six questions. Some AP exams allow you to choose among the free-response questions, but statistics is *not* one of them. Keep an eye on your watch and devote about 13 minutes to each of questions 1–5 and about 25 minutes to question number 6. Since question number 6 carries more weight in determining your overall score, you want to be sure to at least try the first couple of parts of this question. Please note that watch alarms are *not* allowed.

Below is a chart to help you visualize the breakdown of the exam:

Section	Multiple-Choice Questions	Six Free-Response Questions	
Weight	50% of the exam	50% of the exam (weighted as follows)	
		Questions 1–5	Question 6
Number of Questions	40	37.5% of exam	12.5% of exam
Time Allowed	90 minutes	90 minutes	
Suggested Pace	2 minutes per question	13 minutes per question	25 minutes for this question

Strategies for the Multiple-Choice Section

Here are some strategies to help you work your way through the multiple-choice questions:

Scoring of multiple-choice: There are five possible answers for each question. There is no penalty on the AP Statistics exam for wrong answers. You will simply not get credit for the wrong answer. Since there is no penalty for guessing, be sure to answer every question prior to the end of the 90 minutes allotted. This means you should bubble in a solution for each of the 40 questions.

Find questions you know first: Find questions you are confident of and work those first. (Generally the easier questions appear first on the exam.) Then return to the questions you skipped. Make a mark in the booklet on questions you are unsure of, then return to those questions later. If you skip a question to return to it later, be very careful to skip that line on the answer sheet.

Read each question carefully: Pressed for time, many students make the mistake of reading the questions too quickly or merely skimming them. By reading each question carefully, you may already have some idea about the correct answer. You can then look for that answer in the responses. Careful reading is especially important in EXCEPT questions (see the next section that describes the types of multiple-choice questions).

Eliminate any answer you know is wrong: You can write on the multiple-choice questions in the test book. As you read through the responses, draw a line through every answer you know is wrong. This will help you in choosing correct solutions on questions you aren't sure about.

Read each response, then choose the most accurate one: AP examinations are written to test your precise knowledge of a subject. Sometimes there are a few answers that seem correct, but one of them is more specific and therefore the correct response.

Avoid absolute responses: These answers often include the words "always" or "never." For example, the statement "the data **are** normal" is an absolute answer. Instead, look for the phrase, "the data appear to be approximately normal."

Mark and skip tough questions: If you are hung up on a question, mark it in the margin of the question book. You can come back to it later if you have time. Make sure you skip that question on your answer sheet as well. In the end, be sure to answer ALL questions prior to the end of the 90 minutes.

Working backward: Some students find it beneficial to glance through the exam to get an idea of how to pace their per-question time allocation. Other students find it beneficial to work backward, as they may find that some of the more difficult questions help identify questions found at the beginning. If you decide to work backward, be sure to fill in the correct ovals. If you do decide to start from the beginning of the test, be careful not to spend too much time on the early questions, since questions toward the end of the exam become more difficult. Remember, your goal is to finish and answer the entire exam.

Types of Multiple-Choice Questions

There are various kinds of multiple-choice questions. Here are some suggestions for how to approach each one.

Classic/Best-Answer Questions

This is the most common type of multiple-choice question. It simply requires you to read the question and select the correct answer. For example:

1. A firm wants to determine if the proportion of workers who are satisfied with their job is greater than the 0.58 under the previous supervisors. Which of the following hypotheses should the firm test?

    (A) $H_0: \mu = 0.58, H_a: \mu > 0.58$
    (B) $H_0: p = 0.58, H_a: p > 0.58$
    (C) $H_0: \mu = 0.58, H_a: \mu < 0.58$
    (D) $H_0: p = 0.58, H_a: p < 0.58$
    (E) $H_0: p > 0.58, H_a: p < 0.58$

Answer: B. This is the only correct answer as the question is about proportions, the null is written as an $=$ statement, and the alternative is $>$ since they want to know if the proportion satisfied is higher since the change in supervisors.

EXCEPT Questions

In EXCEPT-style questions, you will notice all of the answer choices but one are correct. The best way to approach these questions is to treat them as true/false questions. Mark a T or an F in the margin next to each possible answer. There should be only one false answer, and that is the answer you should select. For example:

1. Researchers were interested in comparing students at private universities and students at public universities with respect to how much time they spent working in a typical week. They emailed a survey to 1,000 students enrolled at a particular private university and to 2,000 students enrolled at a particular public university. Data from the 400 private university students who responded to the survey and the 900 public university students who responded to the survey was used to determine that there was a significant difference in

the mean number of hours worked in a typical week. Which of the following does NOT limit the researchers' ability to generalize to the two populations of interest?

(A) The two sample sizes are not equal.

(B) There may be bias introduced due to nonresponse.

(C) The samples were not randomly selected.

(D) All students in the private university sample attended the same university and all students in the public university sample attended the same university.

(E) All of the above limit the researchers' ability to generalize to the populations of interest.

Answer: A. If the samples are chosen in a reasonable way, having samples of different sizes is not a problem. Answer choices B, C, and D point to problems with the way in which the sample was selected.

Analysis/Application Questions

These questions will require you to apply your knowledge of statistics to a given situation or problem. For example:

1. To estimate the proportion of students who plan to purchase tickets to an upcoming school fundraiser, a high school decides to sample 100 students as they register for the spring semester. There are 2,000 students at the school. Which of the following sampling plans would result in a stratified random sample?

(A) Number the students from 1 to 2,000 and then use random numbers to select 100 students.

(B) Survey the first 100 students to register.

(C) Randomly select 100 students from a list of the 950 female students at the school.

(D) Divide the students into early registrants (the first 1,000 to register) and late registrants (the last 1,000 to register). Use random numbers to identify 50 of the early registrants and 50 of the late registrants to survey.

(E) Select one of the first 20 students to register using a random number table and then select every 20th student to register thereafter.

Answer: D. This is the only correct solution, since stratified sampling involves dividing the population into strata and then taking a random sample from each stratum.

Free-Response Questions

There are six free-response questions and all of them should be answered as completely as possible, especially Question 6, the investigative task. These six questions account for 50% of the final score you receive on this exam and are weighted as follows. The first five free-response questions combine to make up 75% of your free-response score and the sixth question, the investigative task, will account for the remaining 25% of the grade. This means the first five make up 37.5% and the investigative task makes up 12.5% of the entire test. So it is important to answer all questions as completely as possible.

These questions can cover multiple content areas and connections between several topics. Each of the four key topics of the AP exam will be found in the free-response section. The following hints may help you prepare for this half of the exam.

You should plan to do a quick read of the six questions, since you won't know which question is on your strongest topic and questions are not placed in a particular topic order. As you are reading, you may want to make a quick note as to an idea you have on the subject of each question. Once you have done this, you should go to the easiest question you find in the first five questions. Answer it as completely as you can and then jump to question number six, the investigative task.

Now spend a little time, but no more than about 15–25 minutes, working on the investigative task. Since this carries $\frac{1}{4}$ of the weight for this portion of the test, it is very important that you take the time to make a serious attempt to answer this question. While this task is intended to stretch you beyond the course, typically it will begin with 1 or 2 parts that should be familiar to an AP statistics student. Frequently, however, students don't get to this question, and that means they throw away a question that weighs heavily on their grade.

At this point, you should go back and answer the remaining four questions. If time permits, you should reread and note any key phrases or points that are made in a given question. Once you have answered the questions, be sure you read the prompt again to insure you have actually answered the question asked. Examination readers comment routinely that students don't actually answer the question that was asked. While the information you give may be correct, if it doesn't answer the question, you won't get credit.

Don't assume any question is asking for a "cookie cutter" type response. The AP readers who will review your answer will be reading for a correct answer in the context of the question, and not just a memorized definition or response. In that light, also be sure to use statistical vocabulary correctly. If you are not sure about the correct vocabulary, it is permissible to use a short phrase in place of the word you are unsure about. Correct notation is also important. It won't hurt to define your notation for clarity, but be sure you are using the notation correctly. For example, students routinely lose credit for using notation for a statistic when they should be using parameter notation and vice versa.

You will have 90 minutes for this section of the exam, so watch your time. As recommended by the College Board, you should allow about 10 minutes for each of the first 5 questions and 25 minutes for the investigative task. Then use any remaining time to read over your responses. However, earlier we recommended only about 15 minutes to begin with for the investigative task. This allows you to plan once you have begun answering the last question. In either case, it is important you take a watch and keep a close eye on your time. You want to be sure you have enough time to answer the questions that you are sure about and don't get bogged down giving more than is needed for any one question.

Key Points for Answering Free-Response Questions

In answering the free-response questions, it is important to note the word choices used in the questions. Focus your writing for the AP Statistics Exam with these points in mind.

- **Read:** Carefully read the question and circle key phrases and comments you are to address. Watch for the specific wording and don't make assumptions about what you are being asked before completely reading the question.
- **Context:** Be sure to answer in the context of the problem. For a complete answer you are required to write the conclusions and interpretations in context.
- **Define:** You are expected to use standard statistical notation correctly. Be clear on parameter vs. statistic notation when answering. For example, when discussing a mean value, use the correct population and sample notation in your written responses. A sample mean notation $\bar{x}$ will be considered incorrect if you are supposed to use a population mean notation μ.
- **Vocabulary:** Use statistical vocabulary correctly. If you are unsure of the correct term, use a defining phrase in place of the term. The AP readers don't need to hear fancy statistical lingo, especially if you use it incorrectly. This will weaken your response.
- **Check Assumptions:** Unless the question states that all assumptions are reasonable or have been checked, you should state and check assumptions whenever possible. All assumptions need to be checked for *all* samples for hypothesis tests and confidence intervals.
- **Answer:** Did you actually answer the question? Reread to be sure you have answered what was asked.
- **Interpret:** Be ready to justify your work in a variety of statistical ways. Know how to read computer output, graphical displays, and tables. Provide correct analysis and comparisons as required.

Scoring Free-Response Questions

All six free-response questions will be scored using a four-point scale. However, as stated earlier, question six will be weighted more heavily than questions one through five. The

Overview

Learning From Data

There is an old saying that "without data, you are just another person with an opinion." While anecdotes and coincidences may make for interesting stories, you wouldn't want to make important decisions on the basis of anecdotes alone. For example, just because a friend of a friend ate 16 apricots and then experienced relief from joint pain doesn't mean that this is all you need to know to help one of your parents choose a treatment for arthritis. Before recommending apricots, you would definitely want to consider relevant data on the effectiveness of apricots as a treatment for arthritis.

It is difficult to function in today's world without a basic understanding of statistics. For example, here are a few headlines from articles in a single issue of *USA Today* (June 29, 2009) that draw conclusions based on data:

- **"Poll Finds Generation Gap Biggest Since Vietnam War"** summarized a study that explored opinions regarding social values and political views. Not surprisingly, large behavioral differences between young and old were noted in the use of the Internet, cell phones, and text messaging.
- **"Few See Themselves as 'Old' No Matter What Their Age"** described results from a large survey of 2,969 adults. Those surveyed were asked at what age a person would be considered old. The resulting data revealed that there were notable differences in the answer to this question depending on the age of the responder. The average age identified as old by young adults (ages 18 to 29) was 60, while the average was 69 for those who were ages 30 to 49, 72 for those ages 50 to 64, and 74 for those ages 65 and older.
- **"If You Were Given $1,000, What Would You Do?"** reported on one aspect of a study of consumer purchasing and saving behavior. Something was definitely amiss in this report, however—the percentages for the response categories (such as save it, pay off credit card debt, and use it for a vacation) added up to 107%!
- **"Many Adults Can't Name a Scientist"** summarized the results of a survey of 1,000 adults. Of those surveyed, 23% were unable to name a single famous scientist. Of those who did come up with a name, Albert Einstein was the scientist of choice, named by 47% of those surveyed.

To be an informed consumer of reports like these, you must be able to:

- Extract information from tables, charts, and graphs.
- Follow numerical arguments.
- Understand how data should be gathered, summarized, and analyzed to draw valid conclusions.

In addition to preparing you to critically evaluate the work of others, studying statistics will enable you to plan statistical studies, collect data in a sensible way, and use data to answer questions of interest.

Throughout your personal and professional life, you will need to use data to make informed decisions. Should you go out for a sport that involves the risk of injury? Will your college club do better by trying to raise funds with a benefit concert or with a direct appeal for donations? If you choose a particular major, what are your chances of finding a job when you graduate? How should you select a graduate program based on guidebook

STATISTICS: LEARNING FROM DATA

to test if the population mean bubble content is consistent with the proper 100-unit setting. Any day that his results imply that the population mean bubble content might not be 100 bubble units, he has to contact the plant manager.

One day, a sample's mean and standard deviation are 99.06 and 5.81 respectively. Do these data warrant a call to the plant manager? Give appropriate statistical evidence to support your conclusion.

Answer: There are generally four parts in a hypothesis test that must be included in order to receive full credit.

Part I—State the hypothesis.
The null hypothesis is that the population mean bubble content is 100 bubble units. The alternative hypothesis is that the population mean bubble content is not 100 bubble units.

H_0: $\mu = 100$ bubble units,　　　　μ = average bubble content level
H_a: $\mu \neq 100$ bubble units
$\alpha = .05$　　　　　　　　　　　Since no alpha given, we will use .05.

Part II—Identify the test procedure and any conditions or assumptions that must be met.
Given this alternative hypothesis and since the population standard deviation is unknown, I will conduct a two-sided, one-sample *t*-test for a population mean. The problem states that the sample of cans is a random sample. Because the 25 measurements are not given, I can't construct a plot to check the assumption of normality, but I will assume that the distribution of bubble unit measurements is normally distributed.

Part III—Calculate the value of the test statistic and the P-value.
For a sample mean of 99.06, a sample standard deviation of 5.81, and a sample size of 25, I get a *t*-statistic of 20.809 on 24 degrees of freedom and a *P*-value of .4265 for this two-sided test.

$$t = \frac{99.06 - 100}{\frac{5.18}{\sqrt{25}}} = -0.809, \text{ with } df = 24$$

$$P\text{-value} = 0.4265$$

*Part IV—Give your conclusion **both** in context and linked to alpha.*
Since the *P*-value is quite large, much larger than 0.05 (0.4265 > 0.05), I do not reject the claim that the population mean bubble content is 100 bubble units.

Therefore, these data do not warrant a call to the plant manager.

Here are some things to keep in mind as you answer the free-response questions:

- Given an experimental design question, be sure to identify the treatments and clearly explain a method for random assignment. Think about whether you need to incorporate blocking, blinding, or a control group.
- For hypothesis testing questions, be sure to state the null and alternative hypotheses, identify the test procedure you have selected and check any assumptions required for use of the selected method, perform any necessary calculations, and give a conclusion in context.
- Finally, be careful **not** to include parallel solutions. If more than one solution is given, AP readers are instructed to read each solution, but your score will be based on the weakest solution. So try not to lose points you have already scored because you doubt yourself. Be confident in your choice and cross out any dual answer that you don't want considered.

four-point score is holistic in nature. This means the readers will consider your complete solution in addition to each of the parts. Therefore, it is important that you answer the question and not contradict yourself. You should NOT attempt more than one solution; the reader will read each solution and your score will be that of your weakest solution.

Keep in mind that both statistical knowledge as well as verbal communication is essential in the AP Statistics exam. To score a four, you must demonstrate that you are able to both correctly compute statistical values as well as properly interpret those values in the context of the question. Here are some general guidelines regarding the differences between zero and four on any free-response question. More specific guidelines can be found on the College Board website (http://www.collegeboard.com/student/testing/ap/sub_stats.html?stats) where you can download the PDF file of the course description.

Score	Statistical Knowledge (Be able to correctly use the statistical ideas and techniques.)	Communication (Demonstrate a clear and concise explanation of what work you performed and why.)
4	Demonstrates a complete understanding of the statistics and is able to perform correct statistical calculations. Even if you make minor arithmetic errors, the solution is plausible.	Your thoughts and explanations are correct, with all aspects of the explanation utilizing correct procedures, terminology, appropriate conclusions, and so on.
3	Able to demonstrate clear understanding of appropriate techniques but may have some minor gaps. Overall, this solution is substantially correct even if there are some errors.	Overall, the rationale in this solution is correct, but there may be some missing assumptions, caveats, or even a bit unfinished in the final work. While there is a conclusion, it may be incomplete.
2	Has some parts correct but others have either incorrect or even possibly unreasonable answers. There is some understanding of the pieces, yet little understanding of the relationship between them is demonstrated.	Vague or inappropriate explanations. May be missing components or diagrams needed for the explanation to be complete. Conclusions that are drawn are not complete.
1	Misuses the statistical components and develops unreasonable solutions. Very limited knowledge of statistics given.	Response is jumbled, messy, unclear, or difficult to follow. Conclusions may not match the statistical work provided. Fails to provide a conclusion or answer the question posed. Incorrect diagrams or displays.
0	Very little, if any understanding of the statistical pieces needed for the problem.	No strategy that will lead to a solution or no communication given at all.

Free-Response Sample Question

Here is an example of a hypothesis test question.

The average bubble content level for cans of a new soda must be near 100 "bubble units." If the amount is significantly less, the soda tastes flat. If it is significantly more, the soda tastes too bubbly. Each day, a quality control inspector takes a random sample of 25 cans ratings that include information on percentage of applicants accepted, time to obtain a degree, and so on? Your statistics course will provide you with a systematic approach to making decisions based on data.

Statistics—It's All About Variability

Statistical methods allow you to collect, describe, analyze, and draw conclusions from data. If you lived in a world where all measurements were identical for every individual, these tasks would be simple. For example, consider a population consisting of all of the students at a local college. Suppose that *every* student is enrolled in the same number of courses, spent exactly the same amount of money on textbooks this semester, and favors increasing student fees to support expanding library services. For this population, there is *no* variability in number of courses, amount spent on books, or student opinion on the fee increase. A person studying students from this population to draw conclusions about any of these three variables would have a particularly easy task. It would not matter how many students were studied or how the students were selected. In fact, you could collect information on number of courses, amount spent on books, and opinion on the fee increase by just stopping the next student who happened to walk by the library. Because there is no variability in the population, this one individual would provide complete and accurate information about the population, and you could draw conclusions with no risk of error.

The situation just described is obviously unrealistic. Populations with no variability are exceedingly rare, and they are of little statistical interest because they present no challenge. In fact, variability is almost universal. It is variability that makes life interesting. You need to develop an understanding of variability to be able to collect, describe, analyze, and draw conclusions from data in a sensible way.

The following example illustrates how describing and understanding variability provide the foundation for learning from data.

Overview Example: Monitoring Water Quality

As part of its regular water quality monitoring efforts, an environmental control board selects five water specimens from a particular well each day. The concentration of contaminants in parts per million (ppm) is measured for each of the five specimens, and then the average of the five measurements is calculated. The following graph is an example of a histogram. (You will learn how to construct and interpret histograms in Chapter 2.) This histogram summarizes the average contamination values for 200 days.

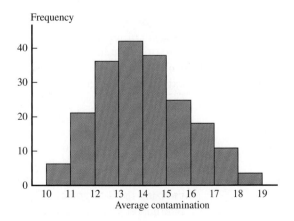

Suppose that a chemical spill has occurred at a manufacturing plant 1 mile from the well. It is not known whether a spill of this nature would contaminate groundwater in the area of the spill and, if so, whether a spill this distance from the well would affect the quality of well water.

One month after the spill, five water specimens are collected from the well, and the average contamination is 15.5 ppm. Considering the variation before the spill shown in the histogram, would you interpret this as evidence that the well water was affected by the spill? What if the calculated average was 17.4 ppm? How about 22.0 ppm?

Before the spill, the average contaminant concentration varied from day to day. An average of 15.5 ppm would not have been an unusual value, so seeing an average of 15.5 ppm after the spill isn't necessarily an indication that contamination has increased. On the other hand, an average as large as 17.4 ppm is less common, and an average as large as 22.0 ppm is not at all typical of the pre-spill values. In this case, you would probably conclude that the well contamination level has increased.

Reaching a conclusion in the water quality example required an understanding of variability. Understanding variability allows you to distinguish between usual and unusual values. The ability to recognize unusual values in the presence of variability is an important aspect of most statistical methods and is also what enables you to quantify the chance of being incorrect when a conclusion is based on available data.

The Data Analysis Process

Statistics involves collecting, summarizing, and analyzing data. All three tasks are critical. Without summarization and analysis, raw data are of little value, and even sophisticated analyses can't produce meaningful information from data that were not collected in a sensible way.

Statistical studies are undertaken to answer questions about our world. Is a new flu vaccine effective in preventing illness? Is the use of bicycle helmets on the rise? Are injuries that result from bicycle accidents less severe for riders who wear helmets than for those who do not? How many credit cards do college students have? Data collection and analysis allow you to answer questions like these.

Data analysis is a process that can be viewed as a sequence of steps leading from planning to data collection to making informed conclusions based on the resulting data. The process can be organized into the following six steps:

1. **Understand the questions of interest.** Effective data analysis requires an understanding of the research problem. You must know the goal of the research and what questions you hope to answer. It is important to have a clear direction before gathering data to ensure that you will be able to answer the questions of interest using the data collected.

2. **Decide what to measure and how to measure it.** The next step in the process is deciding what information is needed to answer the questions of interest. In some cases, the choice is obvious (for example, in a study of the relationship between the weight of a Division I football player and position played, you would need to collect data on player weight and position). In other cases, the choice of information is not as straightforward (for example, in a study of the relationship between preferred learning style and intelligence, how would you define learning style and measure it, and what measure of intelligence would you use?). It is important to carefully define the variables to be studied and to develop appropriate methods for determining their values.

3. **Collect data.** The data collection step is crucial. You must first decide whether an existing data source is adequate or whether new data must be collected. Even if a decision is made to use existing data, it is important to understand how the data were collected and for what purpose, so that any resulting limitations are also fully understood and judged to be acceptable. If new data are to be collected, a careful plan must be developed because the type of analysis that is appropriate and the conclusions that can be drawn depend on how the data are collected.

4. **Summarize the data.** After data are collected, the next step usually involves summarizing the data graphically and numerically. This summarization provides insight into important characteristics of the data and can also provide guidance in selecting appropriate methods for further analysis.

5. **Analyze the data.** This step involves selecting and applying appropriate statistical methods. Much of this textbook is devoted to methods that can be used to carry out this step.

6. **Interpret results.** The final step in the data analysis process involves communicating what you have learned from the data. This step often leads to the formulation of new research questions, which, in turn, leads back to the first step. In this way, good data analysis is often an iterative process.

The data analysis process is summarized in the following figure.

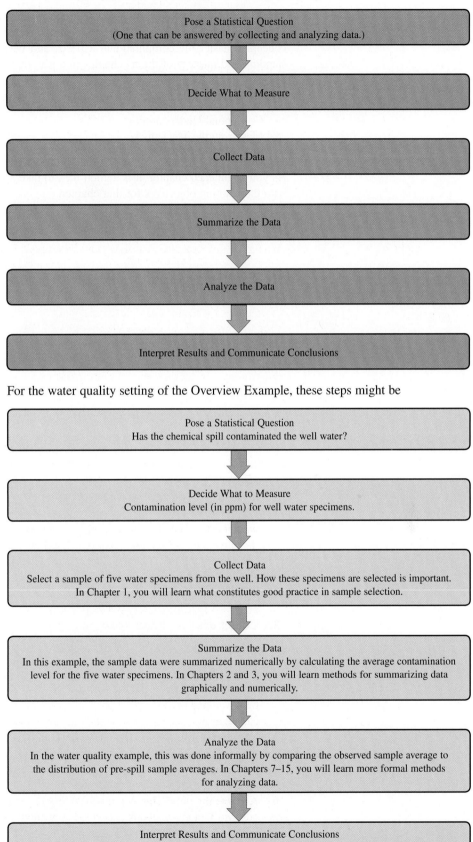

For the water quality setting of the Overview Example, these steps might be

Goals for Student Learning

Statistics is about learning from data and the role that variability plays in drawing conclusions from data. To be successful, you need to develop an understanding of the important concepts, master the computational aspects, and be able to combine conceptual understanding and mastery of the mechanics in a way that allows you to learn from data. This textbook has been written with these three broad goals in mind. At the beginning of each chapter, you will find chapter learning objectives organized under the following headings:

- Conceptual Understanding
- Mastering the Mechanics
- Putting It into Practice

As you begin each chapter, it is a good idea to take a few minutes to familiarize yourself with the learning objectives for that chapter.

The Structure of the Chapters That Follow

The examples and exercises in each chapter have been carefully selected with the stated chapter learning objectives in mind. At the end of each section, you will find two exercise sets. Each of these sets is designed to assess all of the learning objectives addressed in that section.

In addition, at the end of each chapter, you will find a comprehensive set of review exercises titled "Are You Ready to Move On?" This collection of exercises assesses all of the chapter learning objectives, and the learning objectives assessed are identified for each exercise. As you complete your work in any given chapter, it is a good idea to test yourself by working the "Are You Ready to Move On?" exercises. This will help you to solidify your knowledge and ensure that you are ready to move on to the next chapter.

Collecting Data in Reasonable Ways

Preview

Chapter Learning Objectives

1.1 Statistical Studies: Observation and Experimentation

1.2 Collecting Data: Planning an Observational Study

1.3 Collecting Data: Planning an Experiment

1.4 The Importance of Random Selection and Random Assignment: What Types of Conclusions Are Reasonable?

1.5 Avoid These Common Mistakes

Chapter Activities

Exploring the Big Ideas

Are You Ready to Move On? Chapter 1 Review Exercises

AP* Review Questions for Chapter 1

Phil Date/Shutterstock.com

PREVIEW

*Statistical methods help you to make sense of data and gain insight into the world around you. The ability to learn from data is critical for success in your personal and professional life. Data and conclusions based on data are everywhere—in newspapers, magazines, online resources, and professional publications. But should you believe what you read? For example, should you eat garlic to prevent a cold? Will doing tai chi (exercises designed for relaxation and balance) one hour per week increase the effectiveness of your flu shot? Will eating cheese before going to bed help you sleep better? These are just three recommendations out of many that appeared in one issue of **Woman's World** (September 27, 2010), a magazine with 1.6 million readers. In fact, if you followed all of the recommendations in that issue, you would also be loading up on sweet potatoes, black tea, yogurt, walnuts, grape juice, olive oil, peanuts, and strawberries! Some of these recommendations are supported by evidence (data) from research studies, but how reliable is this evidence? Are the conclusions drawn reasonable, and do they apply to you? These are important questions that will be explored in this chapter.*

CHAPTER LEARNING OBJECTIVES

Conceptual Understanding

After completing this chapter, you should be able to

C1 Understand the difference between an observational study and an experiment.

C2 Understand that the conclusions that can be drawn from a statistical study depend on the way in which the data are collected.

C3 Explain the difference between a census and a sample.

C4 Explain the difference between a statistic and a population characteristic.

C5 Understand why random selection is an important component of a sampling plan.

C6 Understand why random assignment is important when collecting data in an experiment.

C7 Understand the difference between random selection and random assignment.

C8 Explain why volunteer response samples and convenience samples are unlikely to produce reliable information about a population.

C9 Understand the limitations of using volunteers as subjects in an experiment.

C10 Explain the purpose of a control group in an experiment.

C11 Explain the purpose of blinding in an experiment.

C12 Understand the purpose of blocking in an experiment.

Mastering the Mechanics

After completing this chapter, you should be able to

M1 Create a sampling plan that would result in a simple random sample from a given population.

M2 Describe a procedure for randomly assigning experimental units to experimental conditions (for example, subjects to treatments) given a description of an experiment, the experimental conditions, and the experimental units.

M3 Identify the sampling method (simple random sampling, stratified random sampling, cluster sampling, systematic sampling or convenience sampling) used in a given study.

M4 Use diagrams to enhance a description of an experimental design.

Putting It into Practice

After completing this chapter, you should be able to

P1 Distinguish between an observational study and an experiment.

P2 Evaluate the design of an observational study.

P3 Evaluate the design of a simple comparative experiment.

P4 Evaluate whether conclusions drawn from the study are appropriate, given a description of a statistical study.

SECTION 1.1 Statistical Studies: Observation and Experimentation

If the goal is to make good decisions based on data, it should come as no surprise that the way you obtain the data is very important. It is also important to know what questions you hope to answer with the data. Depending on what you want to learn, two types of statistical studies are common—*observational studies* and *experiments*.

Sometimes you are interested in answering questions about characteristics of a single **population** or in comparing two or more well-defined populations. To accomplish this, you select a **sample** from each population and use information from the samples to learn about characteristics of the populations.

> **DEFINITION**
>
> **Population:** The population is the entire collection of individuals or objects that you want to learn about.
>
> **Sample:** A sample is a part of the population that is selected for study.

For example, many people, including the author of **"The 'CSI Effect': Does It Really Exist?"** (*National Institute of Justice* [2008]: 1–7), have speculated that watching crime scene investigation TV shows (such as *CSI*, *Cold Case*, *Bones*, or *Numb3rs*) may be associated with the kind of high-tech evidence that jurors expect to see in criminal trials. Do people who watch such shows on a regular basis have higher expectations than those who do not watch them? To answer this question, you would want to learn about two populations, one consisting of people who watch crime scene investigation shows on a regular basis and the other consisting of people who do not. You could select a sample of people from each population and interview these people to determine their levels of expectation for high-tech evidence in a criminal case. This would be an example of an **observational study**. In an observational study, it is important to obtain samples that are representative of the corresponding population.

Sometimes the questions you are trying to answer cannot be answered using data from an observational study. Such questions are often of the form, "What happens when…?" or "What is the effect of …?" For example, a teacher may wonder what happens to student test scores if the lab time for a chemistry course is increased from 3 hours to 6 hours per week. To answer this question, she could conduct an **experiment**. In such an experiment, the value of a *response* (test score) would be recorded under different experimental conditions (3-hour lab and 6-hour lab). The person carrying out the experiment creates the experimental conditions and also determines which people will be assigned to each experimental condition.

AP* EXAM TIP

These definitions are important when deciding what conclusions can be drawn from a statistical study. If asked "What conclusions can be drawn from this study?," the first consideration should be whether the study is an observational study or an experiment.

> **DEFINITION**
>
> An **observational study** is a study in which the person conducting the study observes characteristics of a sample selected from one or more existing populations. The goal of an observational study is to use data from the sample to learn about the corresponding population. In an observational study, it is important to obtain a sample that is representative of the population.
>
> An **experiment** is a study in which the person conducting the study looks at how a response variable behaves under different experimental conditions. The person carrying out the study determines who will be in each experimental group and what the experimental conditions will be. In an experiment, it is important to have comparable experimental groups.

Observational studies and experiments can each be used to compare groups. In an observational study, the person carrying out the study does not control who is in which population. However, in an experiment, the person conducting the study *does* control who is in which experimental group. For example, in the observational study to compare expectations of those who watch crime scene investigation shows and those who do not, the person conducting the study does not determine which people will watch crime scene investigation shows. However, in the chemistry experiment, the person conducting the study *does* determine which students will be in the 3-hour lab group and which students will be in the 6-hour lab group. This seemingly small difference is critical when it comes to drawing conclusions from a statistical study. This is why it is important to determine whether data are from an observational study or from an experiment. We will return to this important distinction in Section 1.4.

The following two examples illustrate how to determine whether a study is an observational study or an experiment.

> **Example 1.1** The Benefits of Acting Out

A number of studies have concluded that stimulating mental activities can lead to improved memory and psychological wellness in older adults. The article **"A Short-Term Improvement to Enhance Cognitive and Affective Functioning in Older Adults"** (*Journal of Aging and Health* [2004]: 562–585) describes a study designed to investigate whether training in acting has similar benefits. Acting requires a person to think about the character being portrayed, to remember lines of dialogue, and to move on stage as scripted, all at the same time. Does this kind of activity lead to improved functioning in daily life?

Adults ages 60 to 86 were assigned to one of three groups. One group took part in an acting class for 4 weeks; one group spent a similar amount of time in a class on visual arts; and the third group was a comparison group that did not take either class. Each participant took several tests measuring problem solving, memory span, self-esteem, and psychological well-being at the beginning and at the end of the 4-week study period. After analyzing the data from this study, the researchers concluded that the acting group showed greater gains than both the visual arts group and the comparison group in problem solving and psychological well-being.

Is this study an observational study or an experiment? To answer this question, you need to consider how the three groups in the study were formed. Because the study participants were assigned to one of the three groups by the researchers conducting the study, the study is an experiment. As you will see in Section 1.3, the way the researchers decide which people go into each group is an important aspect of the study design.

Example 1.2   Watching TV and Reading Scores

The article **"Television's Value to Kids: It's All in How They Use it"** (*Seattle Times*, July 6, 2005) described a study in which researchers analyzed standardized test results and television viewing habits of 1,700 children. They found that children younger than 3 years old who averaged more than 2 hours of television viewing per day tended to score lower on measures of reading achievement and short-term memory.

In the study described, two populations of children younger than 3 were compared—children who watched more than 2 hours of television per day and children who did not. Did the researchers determine which children were in each population? The researchers had no control over how much television the children watched, so the study is an observational study and not an experiment. It is still possible to make reasonable comparisons between the two populations, as long as the groups of children in the study were chosen to be representative of the two populations of interest—*all* children younger than 3 who watch more than 2 hours of television per day and *all* children younger than 3 who do not. The way in which the children in the two study groups are chosen is an important aspect of the design of this observational study.

In the next sections, the design of observational studies and experiments will be considered in more detail.

SECTION 1.1   EXERCISES

Each Exercise Set assesses the following chapter learning objectives: C1, P1

SECTION 1.1   Exercise Set 1

For each of the statistical studies described in Exercises 1.1 to 1.5, indicate whether the study is an observational study or an experiment. Give a brief explanation for your choice.

1.1 The following conclusion from a statistical study appeared in the article **"Smartphone Nation"** (*AARP Bulletin*, **September 2009**): "If you love your smart phone, you are not alone. Half of all boomers sleep with their cell phone within arm's length. Two of three people ages 50 to 64 use a cell phone to take photos, according to a 2010 Pew Research Center report."

1.2 The article **"Tots TV-Watching May Spur Attention Problems"** (*San Luis Obispo Tribune*, **April 4, 2004**) described a study that appeared in the journal *Pediatrics*. Researchers looked at records of 2500 children who were participating in a long-term health study. They found that 10% of these children had attention disorders at age 7 and that the number of hours of television watched at ages 1 and 3 was associated with an increased risk of having an attention disorder at age 7.

1.3 An article in *USA Today* (**October 19, 2010**) describes a study of how young children learn. Sixty-four 18-month-old toddlers participated in the study. The toddlers were allowed to play in a lab equipped with toys, which also had a robot hidden behind a screen. The article states: "After allowing the infants play time, the team removed the screen and let the children see the robot. In some tests, an adult talked to the robot and played with it. In others the adult ignored the robot. After the adult left the room, the robot beeped and then turned its head to look at a toy to the side of the infant. In cases where the adult had played with the robot, the infant was four times more likely to follow the robot's gaze to the toy."

1.4 Researchers at Stanford University surveyed 4,113 adult Americans and concluded that Internet use leads to increased social isolation (*San Luis Obispo Tribune*, **February 28, 2000**).

1.5 A paper appearing in *The Journal of Pain* (March 2010, 199–209) described a study to determine if meditation has an effect on sensitivity to pain. Study participants were assigned to one of three groups. One group meditated for 20 minutes; one group performed a distraction task (working math problems!) for 20 minutes; and one group practiced a relaxation technique for 20 minutes. Sensitivity to pain was measured both before and after the 20-minute session.

SECTION 1.1 Exercise Set 2

For each of the statistical studies described in Exercises 1.6–1.10, indicate whether the study is an observational study or an experiment. Give a brief explanation for your choice.

1.6 A study of more than 50,000 nurses found that those who drank just one soda or fruit juice a day tended to gain more weight and had an increased risk of developing diabetes than those who drank less than one soda or fruit juice per day (*The Washington Post*, August 25, 2004).

1.7 The paper **"Health Halos and Fast-Food Consumption"** (*Journal of Consumer Research* [2007]: 301–314) described a study in which 46 college students volunteered to participate. Half of the students were given a coupon for a McDonald's Big Mac sandwich and the other half were given a coupon for a Subway 12-inch Italian BMT sandwich. (For comparison, the Big Mac has 600 calories, and the Subway 12-inch Italian BMT sandwich has 900 calories.) The researchers were interested in how the perception of Subway as a healthy fast-food choice and McDonald's as an unhealthy fast-food choice would influence what additional items students would order with the sandwich. The researchers found that those who received the Subway coupon were less likely to order a diet soft drink, more likely to order a larger size drink, and more likely to order cookies than those who received the Big Mac coupon.

1.8 The article **"Workers Grow More Dissatisfied"** (*San Luis Obispo Tribune*, August 22, 2002) states that "a study of 5,000 people found that while most Americans continue to find their jobs interesting, and are even satisfied with their commutes, a bare majority like their jobs." This statement was based on the fact that only 51% of those responding to the survey indicated they were satisfied with their jobs.

1.9 In a study of whether taking a garlic supplement reduces the risk of getting a cold, 146 participants were assigned to either a garlic supplement group or to a group that did not take a garlic supplement (**"Garlic for the Common Cold,"** Cochrane Database of Systematic Reviews, 2009). Based on the study, it was concluded that the proportion of people taking a garlic supplement who get a cold is lower than the proportion of those not taking a garlic supplement who get a cold.

1.10 A study to evaluate whether vitamins can help prevent recurrence of blocked arteries in patients who have had surgery to clear a blocked artery was described in the article **"Vitamins Found to Help Prevent Blocked Arteries"** (*Associated Press*, September 1, 2002). The study involved 205 patients who were assigned to one of two groups. One group received a vitamin supplement for 6 months, while the other group did not receive the vitamin supplement.

Additional Exercises

1.11 The article **"How Dangerous Is a Day in the Hospital?"** (*Medical Care* [2011]: 1068–1075) describes a study to determine if the risk of an infection is related to the length of a hospital stay. The researchers looked at a large number of hospitalized patients and compared the proportion who got an infection for two groups of patients—those who were hospitalized overnight and those who were hospitalized for more than one night. Indicate whether the study is an observational study or an experiment. Give a brief explanation for your choice.

1.12 The authors of the paper **"Fudging the Numbers: Distributing Chocolate Influences Student Evaluations of an Undergraduate Course"** (*Teaching in Psychology* [2007]: 245–247) carried out a study to see if events unrelated to an undergraduate course could affect student evaluations. Students enrolled in statistics courses taught by the same instructor participated in the study. All students attended the same lectures and one of six discussion sections that met once a week. At the end of the course, the researchers chose three of the discussion sections to be the "chocolate group." Students in these three sections were offered chocolate prior to having them fill out course evaluations. Students in the other three sections were not offered chocolate. The researchers concluded that "Overall, students offered chocolate gave more positive evaluations than students not offered chocolate." Indicate whether the study is an observational study or an experiment. Give a brief explanation for your choice.

1.13 The article **"Why We Fall for This"** (*AARP Magazine*, May/June 2011) described a study in which a business professor divided his class into two groups. He showed students a mug and then asked students in one of the groups how much they would pay for the mug. Students in the other group were asked how much they would sell the mug for if it belonged to them. Surprisingly, the average value assigned to the mug was quite different for the two groups! Indicate whether the study is an observational study or an experiment. Give a brief explanation for your choice.

1.14 The same article referenced in Exercise 1.13 also described a study which concluded that people tend to respond differently to the following questions:

Question 1: Would you rather have $50 today or $52 in a week?

Question 2: Imagine that you could have $52 in a week. Would you rather have $50 now?

The article attributes this to the question wording: the second question is worded in a way that makes you feel that you are "losing" $2 if you take the money now. Do you think that the study which led to the conclusion that people respond differently to these two questions was an observational study or an experiment? Explain why you think this.

Collecting Data: Planning an Observational Study

In Section 1.1, two types of statistical studies were described—observational studies and experiments. In this section, you will look at some important considerations when planning an observational study or when deciding whether an observational study performed by others was well planned.

Planning an Observational Study—Collecting Data by Sampling

The purpose of an observational study is to collect data that will allow you to learn about a single population or about how two or more populations might differ. For example, you might want to answer the following questions about students at your school:

> What proportion of the students at your school support raising student fees to provide improved recreational facilities?

> What is the average number of hours per month that students at your school devote to community service?

In each case, the population of interest is all students at your school. The "ideal" study would involve carrying out a **census** of the population. In a census, data is collected from everyone in the population, so every student at the school would be included in the study. If you were to ask every student whether he or she supported the fee increase or how many hours per month he or she devotes to community service, you would be able to easily answer the questions above.

Unfortunately, very few observational studies involve a census of the population. It is usually not practical to get data from every individual in the population of interest. Instead, data are obtained from just a part of the population, called a sample. Then **statistics** calculated from the sample are used to answer questions about population characteristics.

> **DEFINITION**
>
> **Population characteristic (also sometimes called a parameter):** A population characteristic is a number that describes the entire population.
>
> **Statistic:** A statistic is a number that describes a sample.

There are many reasons for studying a sample rather than the entire population. Sometimes the process of measuring what you are interested in is destructive. For example, if you are interested in studying the breaking strength of glass bottles or the lifetime of batteries used to power watches, it would be foolish to study the entire population. But the most common reasons for selecting a sample are limitations on the time or money available to carry out the study.

If you hope to learn about a population by studying a sample selected from that population, it is important that the sample be representative of the population. To be reasonably sure of this, you must carefully consider the way the sample is selected. It is sometimes tempting to take the easy way out and gather data in the most convenient way. But if a sample is chosen on the basis of convenience alone, what you see in the sample can't be generalized to the population.

For example, it might be easy to use the students in your statistics class as a sample of students at your school. But not all majors are required to take a statistics course, and so students from some majors may not be represented in the sample. Also, more students take statistics in the senior year than in the freshman year, so using the students in your statistics class may over represent seniors and under represent freshmen. It isn't clear if or how these factors (and others that you may not have thought about) will affect your ability to generalize from this sample to the population of all students at the school. There isn't any way to tell just by looking at a sample whether it is representative of the population. The only assurance comes from the method used to select the sample.

So how do you get a representative sample? One strategy is to use a sampling method called simple random sampling. One way to select a **simple random sample** is to make sure that every individual in the population has the same chance of being selected *each* time an individual is selected into the sample. Such a selection method then ensures that each different possible sample of the desired size has an equal chance of being the sample chosen. For example, suppose that you want a simple random sample of 10 employees chosen from all those who work at a large design firm. For the sample to be a simple random sample, each of the many different subsets of 10 employees must be equally likely to be the one selected. A sample taken from only full-time employees would not be a simple random sample of *all* employees, because someone who works part-time has no chance of being selected. Although a simple random sample may, by chance, include only full-time employees, it must be *selected* in such a way that each possible sample, and therefore *every* employee, has the same chance of being selected. *It is the selection process, not the final sample, which determines whether the sample is a simple random sample.*

The letter *n* is used to denote sample size; it is the number of individuals or objects in the sample. For the design firm scenario of the previous paragraph, $n = 10$.

> **DEFINITION**
>
> **Simple random sample of size *n*:** A simple random sample of size *n* is a sample selected from a population in such a way that every different possible sample of this same size *n* has an equal chance of being selected.

AP* EXAM TIP

Pay attention to the text in italics here. Not every sampling method that gives each individual in the population an equal chance of selection results in a simple random sample.

The definition of a simple random sample implies that every individual member of the population has an equal chance of being selected. *However, the fact that every individual has an equal chance of selection, by itself, does not imply that the sample is a simple random sample.* For example, suppose that a class is made up of 100 students, 60 of whom are female. You decide to select 6 of the female students by writing all 60 names on slips of paper, mixing the slips, and then picking 6. You then select 4 male students from the class using a similar procedure. Even though every student in the class has an equal chance of being included in the sample (6 of 60 females are selected and 4 of 40 males are chosen), the resulting sample is *not* a simple random sample because not all different possible samples of 10 students from the class have an equal chance of selection. Many possible samples of 10 students—for example, a sample of 7 females and 3 males or a sample of all females—have no chance of being selected. The sample selection method described here is not necessarily a bad choice (in fact, it is an example of stratified sampling, to be discussed in more detail shortly), but it does not produce a simple random sample. This must be considered when choosing a method for analyzing the data.

Selecting a Simple Random Sample

A number of different methods can be used to select a simple random sample. One way is to put the name or number of each member of the population on different but otherwise identical slips of paper. The process of thoroughly mixing the slips and then selecting *n* of them yields a simple random sample of size *n*. This method is easy to understand, but it has obvious drawbacks. The mixing must be adequate, and producing all the necessary slips of paper can be extremely tedious, even for relatively small populations.

A commonly used method for selecting a random sample is to first create a list, called a **sampling frame**, of the objects or individuals in the population. Each item on the list can then be identified by a number, and a table of random digits or a random number generator can be used to select the sample. Most statistics computer packages include a random number generator, as do many calculators. A small table of random digits can be found in Appendix A, Table 1.

For example, suppose a car dealership wants to learn about customer satisfaction. The dealership has a list containing the names of the 738 customers who purchased a new car from the dealership during 2012. The owner wants you to interview a sample of 20 customers. Because it would be tedious to write all 738 names on slips of paper, you can use random numbers to select the sample. To do this, you use three-digit numbers, starting with 001 and ending with 738, to represent the individuals on the list.

The random digits from rows 6 and 7 of Appendix A, Table 1 are shown here:

093	876	799	562	565	842	64
410	102	204	751	194	797	51

You can use blocks of three digits (underlined in the lists above) to identify the individuals who should be included in the sample. The first block of three digits is 093, so the 93rd person on the list will be included in the sample. The next two blocks of three digits (876 and 799) do not correspond to anyone on the list, so you ignore them. The next block that corresponds to a person on the list is 562. This process would continue until 20 people have been selected for the sample. You would ignore any three-digit repeats since a person should only be selected once for the sample.

Another way to select the sample would be to use computer software or a graphing calculator to generate 20 random numbers. For example, when 20 random numbers between 1 and 738 were requested, JMP (the statistics software package that accompanies this text) produced the following:

| 71 | 193 | 708 | 582 | 64 | 62 | 624 | 336 | 187 | 455 |
| 349 | 204 | 619 | 683 | 658 | 183 | 39 | 606 | 178 | 610 |

These numbers could be used to determine which 20 customers to include in the sample.

When selecting a random sample, you can choose to do the sampling with or without replacement. **Sampling with replacement** means that after each successive item is selected for the sample, the item is "replaced" back into the population, so the same item may be selected again. In practice, sampling with replacement is rarely used. Instead, the more common method is **sampling without replacement**. This method does not allow the same item to be included in the sample more than once. After being included in the sample, an individual or object would not be considered for further selection.

The goal of random sampling is to produce a sample that is likely to be representative of the population. Although random sampling does not *guarantee* that the sample will be representative, you will see in later chapters that it does allow you to assess the risk of an unrepresentative sample. It is the ability to quantify this risk that enables you to generalize with confidence from a random sample to the corresponding population.

DEFINITIONS

Sampling with replacement: Sampling in which an individual or object, once selected, is put back into the population before the next selection. A sample selected with replacement might include any particular individual from the population more than once.

Sampling without replacement: Sampling in which an individual or object, once selected, is *not* put back into the population before the next selection. A sample selected without replacement always includes n distinct individuals from the population.

Although these two forms of sampling are different, when the sample size n is small relative to the population size, as is often the case, there is little practical difference between them. In practice, the two forms can be viewed as equivalent if the sample size is not more than 10% of the population size.

An Important Note Concerning Sample Size

Here is a common misconception: If the size of a sample is relatively small compared to the population size, the sample cannot possibly accurately reflect the population. Critics of polls often make statements such as, "There are 14.6 million registered voters in California. How can a sample of 1000 registered voters possibly reflect public opinion when only about 1 in every 14,000 people is included in the sample?"

These critics do not understand the power of random selection! Consider a population consisting of 5000 applicants to a state university, and suppose that you are interested in math SAT scores for this population. A graph of the values in this population is shown in Figure 1.1(a). Figure 1.1(b) shows graphs of the math SAT scores for individuals in five different random samples from the population, ranging in sample size from $n = 50$

to $n = 1000$. Notice that the samples all tend to reflect the distribution of scores in the population. If we were interested in using the sample to estimate the population average or the variability in SAT scores, even the smallest of the samples pictured ($n = 50$) would provide reliable information.

Although it is possible to obtain a simple random sample that does not represent the population, this is likely only when the sample size is very small, and unless the population itself is small, this risk does not depend on what fraction of the population is sampled. The random selection process allows you to be confident that the resulting sample will adequately reflect the population, even when the sample consists of only a small fraction of the population.

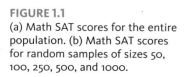

AP* EXAM TIP

This is an important concept. You should understand that even if the sample is only a small fraction of a large population, it can still be representative if it is selected in a reasonable way.

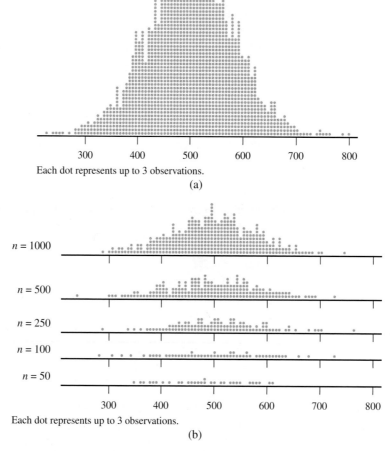

Each dot represents up to 3 observations.

(a)

Each dot represents up to 3 observations.

(b)

FIGURE 1.1
(a) Math SAT scores for the entire population. (b) Math SAT scores for random samples of sizes 50, 100, 250, 500, and 1000.

Other Reasonable Sampling Strategies

Simple random sampling tends to produce representative samples by ensuring that every different sample has an equal chance of being selected. But sometimes actually selecting a simple random sample can be difficult and impractical. For example, consider trying to take a simple random sample of leaves from a tree in order to learn about the proportion of leaves that show damage from cold weather. Or consider trying to take a simple random sample of all those attending a football game at a large stadium in order to learn about how far people traveled to get to the stadium. In situations such as these, you might want to consider other strategies for selecting the sample. Above all, you want to select a sample that can be reasonably regarded as representative of the population of interest. Most good sampling plans involve random selection in some way.

Three other common methods for selecting a sample from a population are stratified random sampling, cluster sampling, and systematic sampling.

Stratified Random Sampling   When the entire population can be divided into a set of nonoverlapping subgroups, a method known as **stratified random sampling** often proves easier to implement and more cost-effective than simple random sampling. In stratified random sampling, separate simple random samples are independently selected from each subgroup. For example, to estimate the average cost of malpractice insurance, a researcher might find it convenient to view the population of all doctors practicing in a particular metropolitan area as being made up of four subgroups: (1) surgeons, (2) internists and family practitioners, (3) obstetricians, and (4) all other doctors. Rather than taking a simple random sample from the population of all doctors, the researcher could take four separate simple random samples— one from each subgroup. These four samples could then be used to learn about the four subgroups as well as the overall population of doctors.

When the population is divided in this way, the subgroups are called **strata** and each individual subgroup is called a stratum (the singular of strata). Stratified random sampling selects a separate simple random sample from each stratum. Stratified random sampling can be used instead of simple random sampling if it is important to obtain information about characteristics of the individual strata as well as of the entire population, although a stratified sample is not required to do this—subgroup estimates can also be obtained by using an appropriate subset of data from a simple random sample.

The real advantage of stratified sampling is that it often allows you to make more accurate inferences about a population than does simple random sampling. In general, it is much easier to produce relatively accurate estimates of characteristics of a homogeneous group than of a heterogeneous group. For example, even with a small sample, it is possible to obtain an accurate estimate of the average grade point average (GPA) of students graduating with high honors from a university. The individual GPAs of these students are all quite similar (a homogeneous group), and even a sample of three or four individuals from this group should be representative. On the other hand, producing a reasonably accurate estimate of the average GPA of *all* seniors at the university, a much more diverse group of GPAs, is a more difficult task. This means that if a varied population can be divided into strata, with each stratum being much more homogeneous than the population with respect to the characteristic of interest, then a stratified random sample can produce more accurate estimates of population characteristics than a simple random sample of the same size.

Cluster Sampling   Sometimes it is easier to select groups of individuals from a population than it is to select individuals themselves. **Cluster sampling** divides the population of interest into nonoverlapping subgroups, called **clusters**. Clusters are then selected at random, and *all* individuals in the selected clusters are included in the sample. For example, suppose that a large urban high school has 600 senior students, all of whom are enrolled in a first period homeroom. There are 24 senior homerooms, each with approximately 25 students. If school administrators want to select a sample of roughly 75 seniors to participate in an evaluation of the college and career placement advising available to students, they might find it much easier to select three of the senior homerooms at random and then include all the students in the selected homerooms in the sample. In this way, an evaluation survey could be administered to all students in the selected homerooms at the same time, which is easier than randomly selecting 75 students and then administering the survey to those individual seniors.

Because whole clusters are selected, the ideal situation for cluster sampling is when each cluster mirrors the characteristics of the population. When this is the case, a small number of clusters results in a sample that is representative of the population. If it is not reasonable to think that the variability present in the population is reflected in each cluster, as is often the case when the cluster sizes are small, then it becomes important to ensure that a large number of clusters are included in the sample.

Be careful not to confuse clustering and stratification. Even though both of these sampling strategies involve dividing the population into subgroups, both the way in which the subgroups are sampled and the best strategy for creating the subgroups are different. In stratified sampling, you sample from every stratum, whereas in cluster sampling, you include only selected whole clusters in the sample. Because of this difference, to increase the chance of obtaining

> **AP* EXAM TIP**
>
> It is easy to confuse strata and clusters. Read the text in italic again to make sure you understand this difference.

a sample that is representative of the population, you want to create homogeneous groups for strata and heterogeneous (reflecting the variability in the population) groups for clusters.

Systematic Sampling   **Systematic sampling** is a procedure that can be used when the population is or can be organized sequentially. A value k is specified (for example, $k = 50$ or $k = 200$). Then one of the first k individuals is selected at random, after which every kth individual in the sequence is included in the sample. A sample selected in this way is called a **1 in k systematic sample**.

For example, a sample of faculty members at a university might be selected from the faculty phone directory. One of the first $k = 20$ faculty members listed could be selected at random, and then every 20th faculty member after that on the list would also be included in the sample. This would result in a 1 in 20 systematic sample.

The value of k for a 1 in k systematic sample is generally chosen to achieve a desired sample size. For example, in the faculty directory scenario just described, if there were 900 faculty members at the university, the 1 in 20 systematic sample described would result in a sample size of 45. If a sample size of 100 was desired, a 1 in 9 systematic sample could be used (because 900/100 = 9).

As long as there are no repeating patterns in the population list, systematic sampling works reasonably well. However, if there are such patterns, systematic sampling can result in an unrepresentative sample. For example, suppose that workers at the entry station of a state park have recorded the number of visitors to the park each day for the past 10 years. In a 1 in 70 systematic sample of days from this list, you would pick one of the first 70 days at random and then every 70th day after that. But if the first day selected happened to be a Wednesday, *every* day selected in the entire sample would also be a Wednesday (because there are 7 days a week and 70 is a multiple of 7). It is unlikely that such a sample would be representative of the entire collection of days, because no weekend days (when the number of visitors is likely to be higher) would be included in the sample.

OTHER SAMPLING METHODS

Stratified random sampling divides a population into subgroups (strata) and then takes a separate random sample from each subgroup (stratum). Stratified random sampling is most effective when the strata are homogeneous.

Cluster Sampling divides the population into subgroups (clusters) and then selects a random sample of clusters. All individuals in the selected cluster are included in the sample. Cluster sampling is most effective when the clusters are heterogeneous.

1 in k systematic sampling selects from an ordered arrangement of a population by choosing a starting point at random from the first k individuals and then selecting every kth individual after that.

When you are considering how to select a sample or evaluating an observational study conducted by someone else, if simple random sampling was not used, you should think about whether it would still result in a representative sample. Later in this text, you will see statistical methods that allow you to learn about a population using sample data. These methods are based on simple random sampling. If a study does not use a simple random sample, ask yourself the following question: Can this sample reasonably be regarded as if it were a random sample? In other words, is the sampling strategy used likely to produce a sample that can be regarded as representative of the population?

Convenience Sampling—Don't Go There!

It is often tempting to resort to "convenience" sampling, which is using an easily available group to form a sample. This is a recipe for disaster! Results from such samples are rarely informative, and it is a mistake to try to generalize from a convenience sample to any larger population.

One common form of convenience sampling is sometimes called **voluntary response sampling**. Such samples rely entirely on individuals who volunteer to be a part of the sample, often by responding to an advertisement, calling a publicized telephone number to register an opinion, or logging on to an Internet site to complete a survey. People who are motivated to volunteer responses often hold strong opinions, and it is extremely unlikely such individuals are representative of any larger population of interest.

One Other Consideration for Observational Studies—Avoiding Bias

Bias in sampling is the tendency for samples to differ from the corresponding population in some systematic way. Bias can result from the way in which the sample is selected or from the way in which information is obtained once the sample has been chosen. The most common types of bias encountered in sampling situations are selection bias, measurement or response bias, and nonresponse bias.

Selection bias is introduced when the way the sample is selected systematically excludes some part of the population of interest. For example, you may wish to learn about the population consisting of all residents of a particular city, but the method of selecting a sample from this population may exclude the homeless or those without telephones. If those who are excluded differ in some systematic way from those who are included, the sample is virtually guaranteed to be unrepresentative of the population. If the difference between the included and the excluded occurs consistently on a variable that is important to the study, conclusions based on the sample data may not be valid for the population of interest. Selection bias also occurs if only volunteers or self-selected individuals are used in a study, because those who choose to participate (for example, in a call-in telephone poll) may differ from those who do not choose to participate.

Measurement or **response bias** occurs when the method of observation tends to produce values that differ systematically from the true value in some way. This might happen if an improperly calibrated scale is used to weigh items or if questions on a survey are worded in ways that tend to influence the response. For example, a Gallup survey sponsored by the American Paper Institute **(Wall Street Journal, May 17, 1994)** included the following question: "It is estimated that disposable diapers account for less than 2 percent of the trash in today's landfills. In contrast, beverage containers, third-class mail and yard waste are estimated to account for about 21 percent of trash in landfills. Given this, in your opinion, would it be fair to tax or ban disposable diapers?" It is likely that the wording of this question prompted people to respond in a particular way.

Other things that might contribute to response bias include the appearance or behavior of the person asking the question, the group or organization conducting the study, and the tendency for people not to be completely honest when asked about illegal behavior or unpopular beliefs.

Nonresponse bias occurs when responses are not obtained from all individuals selected into the sample. As with selection bias, nonresponse bias can distort results if those who respond differ in important ways from those who do not respond. Although some level of nonresponse is unavoidable in most surveys, the biasing effect on the resulting sample is highest when the response rate is low. To minimize nonresponse bias, a serious effort should be made to follow up with individuals who do not respond to an initial request for information.

Types of Bias

Selection Bias

Tendency for samples to differ from the corresponding population as a result of systematic exclusion of some part of the population.

Measurement or **Response Bias**

Tendency for samples to differ from the corresponding population because the method of observation tends to produce values that differ from the true value.

Nonresponse Bias

Tendency for samples to differ from the corresponding population because data are not obtained from all individuals selected for inclusion in the sample.

It is important to remember that bias is introduced by the way in which a sample is selected or by the way in which the data are collected for the sample. Increasing the size of the sample, although possibly desirable for other reasons, does nothing to reduce this bias if the method of selecting the sample is still flawed or if the nonresponse rate remains high.

As you consider the following examples, ask yourself these questions about the observational studies described:

What is the population of interest?

Was the sample selected in a reasonable way?

Is the sample likely to be representative of the population of interest?

Are there any obvious sources of bias?

Example 1.3 Telling Lies

The paper **"Deception and Design: The Impact of Communication Technology on Lying Behavior"** (*Computer-Human Interaction* **[2009]: 130–136**) describes a study designed to learn whether college students lie less often in face-to-face communication than in other forms of communication such as phone conversations or e-mail. Participants in this study were 30 students in an upper-division communications course at Cornell University who received course credit for participation. Participants were asked to record all of their social interactions for a week, making note of any lies they told. Based on data from these records, the authors of the paper concluded that students lie more often in phone conversations than in face-to-face conversations and more often in face-to-face conversations than in e-mail.

Let's consider the four questions posed just prior to this example.

Question	Response
What is the population of interest?	Because the authors wanted to learn about the behavior of college students, the population of interest is all college students.
Was the sample selected in a reasonable way?	In the study described, the authors used a convenience sample. This is problematic for two reasons—students taking the communications course may not be representative of *all* students enrolled at Cornell, and students at Cornell may not be representative of the population of all college students.
Is the sample likely to be representative of the population of interest?	Because the sample was a convenience sample, it is not likely to be representative of the population of interest. Even if the population of interest had been students at Cornell rather than all college students, it is still unlikely that the sample would be representative of the population.
Are there any obvious sources of bias?	The authors should be concerned about response bias. The fact that students were asked to monitor their behavior and record any lies that they told may have influenced their behavior during the week. Also, students may not have been completely truthful in reporting lies or thorough in recording their responses.

Example 1.4 Think Before You Order That Burger!

The article **"What People Buy from Fast-Food Restaurants: Caloric Content and Menu Item Selection"** (*Obesity* [2009]: 1369–1374) reported that the average number of calories consumed at lunch in New York City fast-food restaurants was 827. The researchers selected 267 fast-food locations at random. The paper states that at each of these locations "adult customers were approached as they entered the restaurant and asked to provide their food receipt when exiting and to complete a brief survey."

Let's again consider the four questions posed previously.

Question	Response
What is the population of interest?	Because the researchers wanted to learn about the behavior of people who eat at fast-food restaurants in New York City, this is the population of interest.
Was the sample selected in a reasonable way?	Because it would not be feasible to select a simple random sample from the population of interest, the researchers made an attempt to obtain a representative sample by randomly selecting the fast-food restaurants that would be used in the study. This means that every fast-food location in the city had the same chance of being included in the study. This is a reasonable strategy given the difficulty of sampling from this population.
Is the sample likely to be representative of the population of interest?	Because the locations were selected at random from all locations in the city, the researchers believed that it was reasonable to regard the people eating at these locations as representative of the population of interest.
Are there any obvious sources of bias?	Approaching customers as they entered the restaurant and before they ordered may have influenced their purchases. This introduces the potential for response bias. In addition, some people chose not to participate when approached. If those who chose not to participate differed from those who did participate, the researchers also need to be concerned about nonresponse bias. Both of these potential sources of bias limit the researchers' ability to generalize conclusions based on data from this study.

SECTION 1.2 EXERCISES

Each Exercise Set assesses the following chapter learning objectives: C3, C4, C8, M1, M3, P2

SECTION 1.2 Exercise Set 1

1.15 The article **"Bicyclists and Other Cyclists"** (*Annals of Emergency Medicine* [2010]: 426) reported that in 2008, 716 bicyclists were killed on public roadways in the United States and that the average age of the cyclists killed was 41 years. These figures were based on an analysis of the records of all traffic-related deaths of bicyclists on U.S. public roadways (this information is kept by the National Highway Traffic Safety Administration).

a. Does the group of 716 bicycle fatalities represent a census or a sample of the bicycle fatalities in 2008?

b. If the population of interest is bicycle traffic fatalities in 2008, is the given average age of 41 years a statistic or a population characteristic?

1.16 Based on a study of 2,121 children between the ages of 1 and 4, researchers at the Medical College of Wisconsin concluded that there was an association between iron deficiency and the length of time that a child is bottle-fed (*Milwaukee Journal Sentinel*, **November 26, 2005**). Describe the sample and the population of interest for this study.

1.17 The article **"High Levels of Mercury Are Found in Californians"** (*Los Angeles Times*, **February 9, 2006**) describes a study in which hair samples were tested for mercury. The hair samples were obtained from more than 6,000 people who voluntarily sent hair samples to researchers at Greenpeace and The Sierra Club. The researchers found that nearly one-third of those tested had mercury levels that exceeded the concentration thought to be safe. Is it reasonable to generalize this result to the larger population of U.S. adults? Explain why or why not.

1.18 The article **"I'd Like to Buy a Vowel, Drivers Say"** (*USA Today*, **August 7, 2001**) speculates that young people prefer automobile names that consist of just numbers and letters and that do not form a word (such as Hyundai's XG300, Mazda's 626, and BMW's 325i). The article goes on to state that Hyundai had planned to identify the car that was eventually marketed as the XG300 with the name Concerto, until they determined that consumers hated it and thought that XG300 sounded more "technical" and deserving of a higher price. Do the students at your school feel the same way? Suppose that a list of all the students at your school is available. Describe how you would use the list to select a simple random sample of 150 students.

1.19 The article **"Teenage Physical Activity Reduces Risk of Cognitive Impairment in Later Life"** (*Journal of the American Geriatrics Society* [2010]) describes a study of more than 9,000 women over 50 years old from Maryland, Minnesota, Oregon, and Pennsylvania. The women were asked about their physical activity as teenagers and at ages 30 and 50. A press release about this study (www.wiley.com) generalized the study results to all American women. In the press release, the researcher who conducted the study is quoted as saying

> Our study shows that women who are regularly physically active at any age have lower risk of cognitive impairment than those who are inactive, but that being physically active at teenage is most important in preventing cognitive impairment.

Answer the following four questions for this observational study. (Hint: Reviewing Examples 1.3 and 1.4 might be helpful.)
a. What is the population of interest?
b. Was the sample selected in a reasonable way?
c. Is the sample likely to be representative of the population of interest?
d. Are there any obvious sources of bias?

1.20 For each of the situations described, state whether the sampling procedure is simple random sampling, stratified random sampling, cluster sampling, systematic sampling, or convenience sampling.
a. All first-year students at a university are enrolled in one of 30 sections of a seminar course. To select a sample of freshmen at this university, a researcher selects four sections of the seminar course at random from the 30 sections and all students in the four selected sections are included in the sample.
b. To obtain a sample of students, faculty, and staff at a university, a researcher randomly selects 50 faculty members from a list of faculty, 100 students from a list of students, and 30 staff members from a list of staff.
c. A university researcher obtains a sample of students at his university by using the 85 students enrolled in his Psychology 101 class.
d. To obtain a sample of the seniors at a particular high school, a researcher writes the name of each senior on a slip of paper, places the slips in a box and mixes them, and then selects 10 slips. The students whose names are on the selected slips of paper are included in the sample.
e. To obtain a sample of those attending a basketball game, a researcher selects the 24th person through the door. Then, every 50th person after that is also included in the sample.

SECTION 1.2 **Exercise Set 2**

1.21 Data from a poll conducted by Travelocity led to the following estimates: Approximately 40% of travelers check their work e-mail while on vacation; about 33% take their cell phones on vacation to stay connected with work; and about 25% bring their laptops on vacation (*San Luis Obispo Tribune*, **December 1, 2005**). Are the given percentages statistics or population characteristics?

1.22 With the increasing popularity of online shopping, many consumers use Internet access at work to browse and shop online. In fact, the Monday after Thanksgiving has been nicknamed "Cyber Monday" because of the large increase in online purchases that occurs that day. Data from a large-scale survey by a market research firm (*Detroit Free Press*, **November 26, 2005**) were used to compute estimates of the percentage of men and women who shop online while at work. The resulting estimates probably won't make most employers happy—42% of the men and 32% of the women surveyed were shopping online at work! If the population of interest is working men and women, does the group of people surveyed represent a census or a sample? Are the percentages quoted (42% for men and 32% for women) statistics or population characteristics?

1.23 The authors of the paper **"Illicit Use of Psychostimulants among College Students"** (*Psychology, Health & Medicine* [2002]: 283–287) surveyed college students about their use of legal and illegal stimulants. The sample of students surveyed consisted of students enrolled in a psychology class at a small, competitive college in the United States.

a. Was this sample a simple random sample, a stratified sample, a systematic sample, or a convenience sample? Explain.

b. Explain why an estimate of the proportion of students who reported using illegal stimulants based on data from this survey should not be generalized to all U.S. college students.

1.24 A New York psychologist recommends that if you feel the need to check your e-mail in the middle of a movie or if you sleep with your cell phone next to your bed, it might be time to "power off" (*AARP Bulletin*, **September 2010**). Suppose that you want to learn about the proportion of students at your college who feel the need to check e-mail during the middle of a movie, and that you have access to a list of all students enrolled at your college. Describe how you would use this list to select a simple random sample of 100 students.

1.25 The hand-washing behavior of adults using public restrooms at airports was the subject of a study conducted by the **American Society of Microbiology**. A press release issued by the Society (September 15, 2003) included the following description:

> Although illnesses as deadly as SARS and as troublesome as the common cold or gastric distress can be spread hand-to-hand, the survey sponsored by the American Society of Microbiology (ASM) found that many people passing through major U.S. airports don't wash their hands after using the public facilities. More than 30 percent of people using restrooms in New York airports, 19 percent of those in Miami's airport, and 27 percent of air travelers in Chicago aren't stopping to wash their hands. The survey, conducted by Wirthlin Worldwide in August 2003, observed 7,541 people in public washrooms in New York, Chicago, San Francisco, Dallas, Miami, and Toronto.

These results were then generalized to people who use public restrooms. Answer the following four questions for this observational study. (Hint: Reviewing Examples 1.3 and 1.4 might be helpful.)

a. What is the population of interest?

b. Was the sample selected in a reasonable way?

c. Is the sample likely to be representative of the population of interest?

d. Are there any obvious sources of bias?

1.26 A sample of pages from this book is to be obtained, and the number of words on each selected page will be determined. For the purposes of this exercise, equations are not counted as words and a number is counted as a word only if it is spelled out—that is, *ten* is counted as a word, but *10* is not.

a. Describe a sampling procedure that would result in a simple random sample of pages from this book.

b. Describe a sampling procedure that would result in a stratified random sample. Explain why you chose the specific strata used in your sampling plan.

c. Describe a sampling procedure that would result in a systematic sample.

d. Describe a sampling procedure that would result in a cluster sample.

Additional Exercises

1.27 The supervisors of a rural county are interested in the proportion of property owners who support the construction of a sewer system. Because it is too costly to contact all 7,000 property owners, a survey of 500 owners (selected at random) is undertaken. Describe the population and sample for this problem.

1.28 A consumer group conducts crash tests of new cars. To determine the severity of damage to 2012 Toyota Camrys resulting from a 10-mph crash into a concrete wall, the research group tests six cars of this type and assesses the amount of damage. Describe the population and sample for this problem.

1.29 A building contractor has a chance to buy an odd lot of 5,000 used bricks at an auction. She is interested in determining the proportion of bricks in the lot that are cracked and therefore unusable for her current project, but she does not have enough time to inspect all 5,000 bricks. Instead, she checks 100 bricks to determine whether each is cracked. Describe the population and sample for this problem.

1.30 In 2000, the chairman of a California ballot initiative campaign to add "none of the above" to the list of ballot options in all candidate races was quite critical of a Field poll that showed his measure trailing by 10 percentage points. The poll was based on a random sample of 1,000 registered voters in California. He is quoted by the **Associated Press (January 30, 2000)** as saying, "Field's sample in that poll equates to one out of 17,505 voters." This was so dishonest, he added, that Field should get out of the polling business! If you worked on the Field poll, how would you respond to this criticism?

1.31 Whether or not to continue a Mardi Gras Parade through downtown San Luis Obispo, California, is a hotly debated topic. The parade is popular with students and many residents, but some celebrations have led to complaints and a call to eliminate the parade. The local newspaper conducted online and telephone surveys of its readers and was surprised by the results. The online survey received more than 400 responses, with more than 60% favoring continuing the parade, while the telephone response line received more than 120 calls, with more than 90% favoring banning the parade (*San Luis Obispo Tribune*, **March 3, 2004**). What factors may have contributed to these very different results?

1.32 Suppose that you were asked to help design a survey of adult city residents in order to estimate the proportion that would support a sales tax increase. The plan is to use a stratified random sample, and three stratification schemes have been proposed.

Scheme 1: Stratify adult residents into four strata based on the first letter of their last name (A–G, H–N, O–T, U–Z).

Scheme 2: Stratify adult residents into three strata: college students, nonstudents who work full time, nonstudents who do not work full time.

Scheme 3: Stratify adult residents into five strata by randomly assigning residents into one of the five strata.

Which of the three stratification schemes would be best in this situation? Explain.

SECTION 1.3 Collecting Data: Planning an Experiment

Experiments are usually conducted to answer questions such as "What happens when …?" or "What is the effect of …?" For example, tortilla chips are made by frying tortillas, and they are best when they are crisp, not soggy. One measure of crispness is moisture content. What happens to moisture content when the frying time of a chip is 30 seconds as compared to a frying time of 45 seconds or 60 seconds? An experiment could be designed to investigate the effect of frying time on moisture content.

It would be nice if you could just take three tortillas and fry one for 30 seconds, one for 45 seconds, and one for 60 seconds and then compare the moisture content of the three chips. However, there will be variability in moisture content even if we take two tortillas and fry each one for 30 seconds. There are many reasons why this might occur: small variations in environmental conditions, small changes in the temperature of the oil used to fry the chips, slight differences in the composition of the tortillas, and so on. This creates chance-like variability in the moisture content of chips—even for chips that are fried for the same amount of time. To be able to compare different frying times, you need to distinguish the variability in moisture content that is caused by differences in the frying time from the chance-like variability. A well-designed experiment produces data that allow you to do this.

In a **simple comparative experiment**, the value of some **response variable** (such as moisture content) is measured under different **experimental conditions** (for example, frying times). Experimental conditions are also sometimes called **treatments**, and the terms are used interchangeably. **Experimental units** are the smallest unit to which a treatment is applied, such as individual people in an experiment to investigate the effect of room temperature on exam performance or the individual chips in the tortilla chip example.

> **DEFINITIONS**
>
> An **experiment** is a study in which one or more explanatory variables are manipulated in order to observe the effect on a response variable.
>
> The **explanatory variables** are those variables that have values that are controlled by the experimenter. Explanatory variables are also called **factors**.
>
> The **response variable** is the variable that the experimenter thinks may be affected by the explanatory variables. It is measured as part of the experiment but is not controlled by the experimenter.
>
> An **experimental condition** is any particular combination of values for the explanatory variables. Experimental conditions are also called **treatments**.
>
> An **experimental unit** is the smallest entity to which a treatment is applied.

The design of an experiment is the overall plan for conducting an experiment. A good design makes it possible to obtain unambiguous answers to the questions that the experiment was designed to answer. It does this by allowing you to distinguish the variability in the response that is attributable to differing experimental conditions from variability due to other sources.

Where you can get into trouble in an experiment is if the design allows some systematic (as opposed to chance-like) source of variation in the response that you can't isolate. For example, suppose that in the tortilla chip experiment you used three different batches of tortillas made from different recipes. One batch was used for all chips fried for 30 seconds, the second batch for all chips fried for 45 seconds, and the third for all chips fried for 60 seconds. If there are differences in the composition of the tortillas in the three batches, this could have an effect on the response (moisture content) that is not distinguishable from variability in the response due to the experimental conditions (the frying times).

When you cannot distinguish between the effects of two variables on the response, the two variables are **confounded**. In the situation where three different batches of tortillas were used, the tortilla batch is called a **confounding variable**, and you would say that tortilla batch and frying time are confounded. If you observe a difference in moisture content for the three frying times, you can't tell if the difference is due to differences in frying time, differences in the tortilla batches, or some combination of both. A well-designed experiment will protect against potential confounding variables.

> Two variables are **confounded** if their effects on the response variable cannot be distinguished.

Design Strategies for Simple Comparative Experiments

The goal of a simple comparative experiment is to determine if the experimental conditions (treatments) have an effect on the response variable of interest. To do this, you must consider other potential sources of variability in the response and then design your experiment to either eliminate them or ensure that they produce chance-like, as opposed to systematic, variability.

Eliminating Sources of Variability through Direct Control

An experiment can be designed to eliminate the effect of some variable on the response through **direct control**. Direct control means holding a potential source of variability constant at some fixed level. For example, in the tortilla chip experiment, you might suspect that the oil temperature and the shape of the pan used for frying are possible sources of variability in the moisture content. You could eliminate these sources of variability through direct control by using just a single pan and by having a fixed oil temperature that is maintained for all chips. In general, if there are potential sources of variability that are easy to control, this control should be incorporated into the experimental design.

Taking Variability into Account: Blocking

The effects of some variables can be filtered out by **blocking**. Blocking creates groups of experimental units (called blocks) that are similar with respect to one or more potentially confounding variables. For example, in the tortilla chip experiment, if tortillas came from three different batches, you would want to make sure that each frying time was tried with tortillas from all three batches. Then, if you were to see a difference in moisture content for the three frying times, tortilla batch could be ruled out as a possible explanation for this difference.

Ensuring That Remaining Sources of Variability Produce Chance-Like Variability: Random Assignment

What about other variables that might affect the response that can't be controlled or addressed by blocking, such as small differences in tortilla size in the tortilla chip example? This is handled by **random assignment** to experimental groups. Random assignment ensures that an experiment does not favor one experimental condition over any other and tries to create "equivalent" experimental groups (groups that are as much alike as possible). Random assignment is an *essential* part of a good experimental design.

To see how random assignment tends to create similar groups, suppose 50 first-year college students participate in an experiment to investigate whether completing an online review of course material prior to an exam improves exam performance. The 50 subjects vary quite a bit in achievement, as shown in their math and verbal SAT scores, which are displayed in Figure 1.2.

FIGURE 1.2
Math SAT and Verbal SAT Scores for 50 First-Year Students

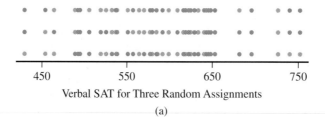

If these 50 students are to be assigned to the two experimental groups (one that will complete the online review and one that will not), you want to make sure that the assignment of students to groups does not place the higher achieving students in one group and the lower achieving students in the other.

Random assignment makes it easy to create groups of students with similar achievement levels in both verbal and math SAT scores simultaneously. Figure 1.3 (a) shows the verbal SAT scores of the students assigned to each of the two experimental groups (one shown in orange and one shown in blue) for each of three different random assignments of students to groups. Figure 1.3 (b) shows the math SAT scores for the two experimental groups for each of the same three random assignments. Notice that each of the three random assignments produced groups that are quite similar with respect to *both* verbal and math SAT scores. So, if any of these three assignments were used and the two groups differed on exam performance, you could rule out differences in math or verbal SAT scores as possible competing explanations for the difference.

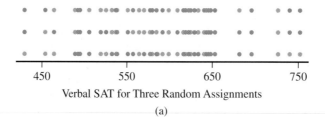

Verbal SAT for Three Random Assignments

(a)

FIGURE 1.3
Three Different Random Assignments to Two Groups, One Shown in Orange, One Shown in Blue (a) Verbal SAT Score (b) Math SAT Score

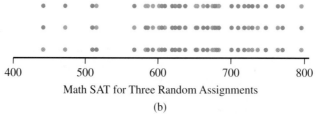

Math SAT for Three Random Assignments

(b)

Not only will random assignment create groups that are similar in verbal and math SAT scores, but you can also count on it to even out the groups with respect to other potentially confounding variables as well. As long as the number of subjects is not very small, you can rely on the random assignment to produce comparable experimental groups. This is why random assignment is a part of all well-designed experiments.

As you can see from the preceding discussion, random assignment in an experiment is different from random selection of experimental units. Random assignment is a way of determining which experimental units go in each of the different experimental groups. Random selection is one possible way of choosing the experimental units. The ideal situation would be to have both random selection of experimental units and random assignment of these units to experimental conditions. This would allow conclusions from the experiment to be generalized to a larger population. However, for many experiments the random selection of experimental units is not possible. Even so, as long as experimental units are randomly assigned to experimental conditions, it is still possible to assess treatment effects.

Not all experiments require the use of human subjects to evaluate different experimental conditions. For example, a researcher might be interested in comparing the performance of three different gasoline additives as measured by gasoline mileage. The experiment might involve using a single car with an empty gas tank. One gallon of gas with one of the additives could be put in the tank and the car driven along a standard route at a constant speed until it runs out of gas. The total distance traveled could then be recorded. This could be repeated a number of times, 10 for example, with each additive.

This gas mileage experiment can be viewed as a sequence of trials. Because there are a number of potentially confounding variables that might have an effect on gas mileage (such as variations in environmental conditions, like wind speed or humidity, and small variations in the condition of the car), it would not be a good idea to use additive 1 for the first 10 trials, additive 2 for the next 10, and so on. An approach that would not unintentionally favor any of the additives would be to randomly assign additive 1 to ten of the 30 planned trials, and then randomly assign additive 2 to ten of the remaining 20 trials. The resulting plan might look as follows:

Trial	1	2	3	4	5	6	7 ...	30
Additive	2	2	3	3	2	1	2 ...	1

When an experiment can be viewed as a sequence of trials, you randomly assign treatments to trials. **Random assignment—either of subjects to treatments or of treatments to trials—is a critical component of a good experiment.**

Replication

Random assignment can be effective only if the number of subjects or observations in each experimental condition (treatment) is large enough for each experimental group to reliably reflect variability in the population. For example, in the tortilla experiment, if there were only six tortillas, it is unlikely that you would get equivalent groups for comparison, even with random assignment to the three frying times. **Replication** is the design strategy of making multiple observations for each experimental condition. Together, replication and random assignment allow the researcher to be reasonably confident of comparable experimental groups.

Key Concepts in Experimental Design

Random Assignment

Random assignment (of subjects to treatments or of treatments to trials) ensures that the experiment does not systematically favor one experimental condition (treatment) over another.

Blocking

Using potentially confounding variables to create groups (blocks) that are similar. All experimental conditions (treatments) are then tried in each block.

Direct Control

Holding potentially confounding variables constant so that their effects are not confounded with those of the experimental conditions (treatments).

Replication

Ensuring that there is an adequate number of observations for each experimental condition.

> **DEFINITION**
>
> An experiment in which experimental units are randomly assigned to treatments is called a **completely randomized experiment** and the design is referred to as a **completely randomized design**.
>
> An experiment that incorporates blocking by dividing the experimental units into homogeneous blocks and then randomly assigns the individuals within each block to treatments is called a **randomized block experiment** and the design is referred to as a **randomized block design**.

To illustrate the design of a simple experiment, consider the dilemma of Anna, a waitress in a local restaurant. She would like to increase the amount of her tips, and her strategy is simple: She will write "Thank you" on the back of some of the checks before giving them to the patrons and on others she will write nothing. She plans to calculate the percentage of the tip as her measure of success (for example, a 15% tip is common). She will compare the average percentage of the tips calculated from checks with and without the handwritten "Thank you." If writing "Thank you" does not produce higher tips, she may try a different strategy.

Anna wonders whether writing "Thank you" on the customers' bills will have an effect on the amount of her tip. In the language of experiments, you would refer to the writing of "Thank you" and the not writing of "Thank you" as **treatments** (the two experimental conditions to be compared in the experiment). The tipping percentage is the **response variable**. The idea behind this terminology is that the tipping percentage is a *response* to the treatments *writing "Thank you"* or *not writing "Thank you."* You can think of Anna's experiment as an attempt to explain the variability in the response variable in terms of its presumed cause, the different treatments. That is, as she manipulates the explanatory variable, she expects the response by her customers to vary. Anna has a good start, but now she should consider the four fundamental design principles.

Replication. Anna cannot run a successful experiment by gathering tipping information on only one person for each treatment. There is no reason to believe that any single tipping incident is representative of what would happen in other incidents, and therefore it would be impossible to evaluate the two treatments with only two subjects. To interpret the effects of a particular treatment, she must **replicate** each treatment in the experiment.

Blocking. Suppose that Anna works on both Thursdays and Fridays. Because day of the week might affect tipping behavior, Anna could block on day of the week and make sure that observations for both treatments are made on each of the two days.

Direct Control and Random Assignment. There are a number of potentially confounding variables that might have an effect on the size of tip. Some restaurant patrons will be seated near the window with a nice view. Some will have to wait for a table, whereas others may be seated immediately. Some patrons may be on a fixed income and cannot afford a large tip. Some of these variables can be directly controlled. For example, Anna may choose to use only window tables in her experiment, eliminating table location as a potential confounding variable. Other variables, such as length of wait and customer income, cannot be easily controlled. As a result, it is important that Anna use random assignment to decide which of the window tables will be in the "Thank you" group and which will be in the "No thank you" group. She might do this by flipping a coin as she prepares the check for each window table. If the coin lands with the head side up, she could write "Thank you" on the bill, omitting the "Thank you" when a tail is observed.

The accompanying box summarizes how experimental designs deal with potentially confounding variables.

Taking Potentially Confounding Variables into Account

Potentially confounding variables are variables other than the explanatory variables in an experiment that may also have an effect on the response variable. There are several strategies for dealing with potentially confounding variables in order to avoid confounding.

Potentially confounding variables that you know about and choose to incorporate into the experimental design:

Strategies

Direct control—holds potentially confounding variables fixed so that they can't affect the response variable

Blocking—allows for valid comparisons because each treatment is tried in each block

Potentially confounding variables that you don't know about or choose not to incorporate into the experimental design through direct control or blocking:

Strategy

Random assignment

A Note on Random Assignment

There are several strategies that can be used to perform random assignment of subjects to treatments or treatments to trials. Two common strategies are:

- Write the name of each subject or a unique number that corresponds to a subject on a slip of paper. Place all of the slips in a container and mix well. Then draw out the desired number of slips to determine those who will be assigned to the first treatment group. This process of drawing slips of paper then continues until all treatment groups have been determined.

- Assign each subject a unique number from 1 to n, where n represents the total number of subjects. Use a random number generator or table of random numbers to obtain numbers that will identify which subjects will be assigned to the first treatment group. This process would be repeated, ignoring any random numbers generated that correspond to subjects that have already been assigned to a treatment group, until all treatment groups have been formed.

The two strategies above work well and can be used for experiments in which the desired number of subjects in each treatment group has been predetermined.

Another strategy that is sometimes employed is to use a random mechanism (such as tossing a coin or rolling a die) to determine which treatment will be assigned to a particular subject. For example, in an experiment with two treatments, you might toss a coin to determine if the first subject is assigned to treatment 1 or treatment 2. This could continue for each subject—if the coin lands H, the subject is assigned to treatment 1, and if the coin lands T, the subject is assigned to treatment 2. This strategy is fine, but may result in treatment groups of unequal size. For example, in an experiment with 100 subjects, 53 might be assigned to treatment 1 and 47 to treatment 2. If this is acceptable, the coin flip strategy is a reasonable way to assign subjects to treatments.

But, suppose you want to ensure that there is an equal number of subjects in each treatment group. Is it acceptable to use the coin flip strategy until one treatment group is complete and then just assign all of the remaining subjects to groups that are not yet full? The answer to this question is that it is probably not acceptable. For example, suppose a list of 20 subjects is in order by age from youngest to oldest and that you want to form two treatment groups each consisting of 10 subjects. Tossing a coin to make the assignments might result in the following (based on using the first row of random digits in Appendix A, Table 1, with an even number representing H and an odd number representing T):

Subject	Random Number	Coin Toss Equivalent	Treatment Group
1	4	H	1
2	5	T	2
3	1	T	2
4	8	H	1
5	5	T	2
6	0	H	1
7	3	T	2
8	3	T	2
9	7	T	2
10	1	T	2
11	2	H	1
12	8	H	1
13	4	H	1
14	5	T	2
15	1	T	2
16			1
17	Treatment group 2 filled. Assign all others		1
18	to treatment group 1.		1
19			1
20			1

If the list of subjects was ordered by age, treatment group 1 would end up with a disproportionate number of older people. This strategy usually results in one treatment group drawing disproportionately from the end of the list. So, the only time the strategy of assigning at random until groups fill up and then assigning the remaining subjects to the group that is not full is reasonable is if you can be sure that the list is in random order with respect to all variables that might be related to the response variable. Because of this, it is best to avoid this strategy.

On the other hand, if the number of subjects is large, it may not be important that every treatment group has exactly the same number of subjects. If this is the case, it is reasonable to use a coin flip strategy (or other strategies of this type) that does not involve stopping assignment of subjects to a group that becomes full.

Visualizing the Underlying Structure of Some Common Experimental Designs

Simple diagrams are sometimes used to highlight important features of some common experimental designs. The structure of an experiment that is based on random assignment of experimental units to one of two treatments is displayed in Figure 1.4.

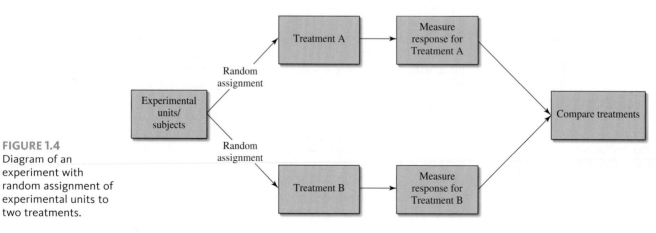

FIGURE 1.4
Diagram of an experiment with random assignment of experimental units to two treatments.

The diagram can be easily adapted for an experiment with more than two treatments. In any particular setting, you would also want to customize the diagram by indicating what the treatments are and what response will be measured. This is illustrated in Example 1.5.

Example 1.5   A Helping Hand

Can moving their hands help children learn math? This is the question investigated by the authors of the paper **"Gesturing Gives Children New Ideas about Math"** (*Psychological Science* [2009]: 267–272). An experiment was conducted to compare two different methods for teaching children how to solve math problems of the form $3 + 2 + 8 = $ ____ $+ 8$. One method involved having students point to the $3 + 2$ on the left side of the equal sign with one hand and then point to the blank on the right side of the equal sign before filling in the blank to complete the equation. The other method did not involve using these hand gestures. The paper states that the study used children ages 9 and 10 who were given a pretest containing six problems of the type described above. Only children who answered all six questions incorrectly became subjects in the experiment. There were a total of 128 subjects.

To compare the two methods, the 128 children were assigned at random to the two experimental conditions. Children assigned to one experimental condition were taught a method that used hand gestures and children assigned to the other experimental condition were taught a similar strategy that did not involve using hand gestures. Each child then took a test with six problems and the number correct was determined for each child. The researchers used the resulting data to reach the conclusion that the average number correct for children who used the method that incorporated hand gestures was significantly higher than the average number correct for children who were taught the method that did not use hand gestures.

The basic structure of this experiment can be diagramed as shown in Figure 1.5. This type of diagram provides a nice summary of the experiment, but notice that several important characteristics of the experiment are not captured in the diagram. For example, the diagram does not show that some potentially confounding variables were considered by the researchers and directly controlled. In this example, both age and prior math knowledge were directly controlled by using only children who were 9 and 10 years old and who were not able to solve any of the questions on the pretest correctly. *So, be aware that while a diagram of an experiment may be a useful tool, it usually cannot stand alone in describing an experimental design.*

AP* EXAM TIP

Diagrams are a useful tool for organizing your thinking about an experiment, but a diagram cannot stand alone as a description of an experimental design.

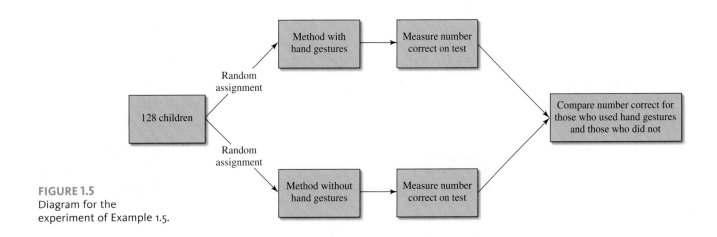

FIGURE 1.5
Diagram for the experiment of Example 1.5.

Some experiments consist of a sequence of trials, and treatments are assigned at random to the trials. The diagram in Figure 1.6 illustrates the underlying structure of such an experiment. Example 1.6 shows how this diagram can be customized to describe a particular experiment.

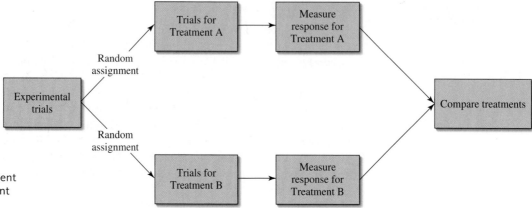

FIGURE 1.6
Diagram of an experiment with random assignment of treatments to trials.

Example 1.6 Distracted? Watch Out for Those Cars!

The paper **"Effect of Cell Phone Distraction on Pediatric Pedestrian Injury Risk" (***Pediatrics*** [2009]: e179–e185)** describes an experiment to investigate whether pedestrians who are talking on a cell phone are at greater risk of an accident when crossing the street than when not talking on a cell phone. No children were harmed in this experiment—a virtual interactive pedestrian environment was used! One possible way of conducting such an experiment would be to have a person cross 20 streets in this virtual environment. The person would talk on a cell phone for some crossings and would not use the cell phone for others. It would be important to randomly assign the two treatments (talking on the phone, not talking on the phone) to the 20 trials (the 20 simulated street crossings). This would result in a design that did not favor one treatment over the other because the pedestrian became more careful with experience or more tired and, therefore, easily distracted over time. The basic structure of this experiment is diagramed in Figure 1.7.

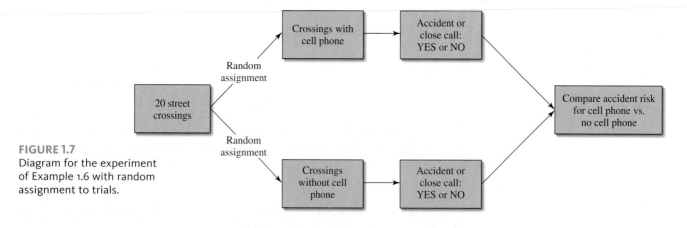

FIGURE 1.7
Diagram for the experiment of Example 1.6 with random assignment to trials.

Example 1.7 A Helping Hand Revisited

Let's return to the experiment described in Example 1.5. Take a minute to go back and read that example again. The experiment described in Example 1.5, a completely randomized design with 128 subjects, was used to compare two different methods for teaching kids how to solve a particular type of math problem. Age and prior math knowledge were potentially confounding variables that the researchers thought might be related to performance on the math test given at the end of the lesson, so the researchers chose to directly control these variables. The 128 children were assigned at random to the two experimental conditions (treatments). The researchers relied on random assignment to create treatment groups that would be roughly equivalent with respect to other extraneous variables.

But suppose that you were worried that gender might also be related to performance on the math test. One possibility would be to use direct control of gender—that is, you might use only boys or only girls as subjects in the experiment. Then if you saw a difference in test performance for the two teaching methods, it could not be due to one experimental group containing more boys and fewer girls than the other group. The downside to this strategy is that if you use only boys in the experiment, there is no basis for also generalizing any conclusions from the experiment to girls.

Another strategy for dealing with extraneous variables is to incorporate blocking into the design. In the case of gender, you could create two blocks, one consisting of girls and one consisting of boys. Then, once the blocks are formed, you would randomly assign the girls to the two treatments and randomly assign the boys to the two treatments. In the actual study, the group of 128 children included 81 girls and 47 boys. A diagram that shows the structure of an experiment that includes blocking using gender is shown in Figure 1.8.

FIGURE 1.8
Diagram for the experiment of Example 1.7 using gender to form blocks.

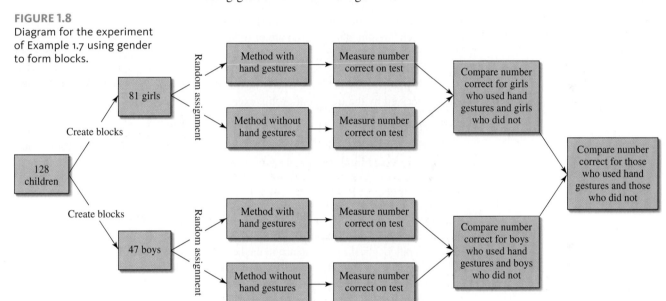

When blocking is used, the design is called a **randomized block design**. Note that one difference between the diagram that describes the experiment in which blocking is used (Figure 1.8) and the diagram of the original experiment (Figure 1.5) is where the random assignment occurs. *When blocking is incorporated in an experiment, the random assignment to treatments occurs after the blocks have been formed and is done separately for each block.*

Other Considerations When Planning an Experiment

The goal of an experimental design is to provide a method of data collection that

1. Minimizes potentially confounding sources of variability in the response so that any differences in response due to experimental conditions are easy to assess.

2. Creates experimental groups that are similar with respect to potentially confounding variables that cannot be controlled.

We now examine some additional considerations that you may need to think about when planning an experiment.

Use of a Control Group

If the purpose of an experiment is to determine whether some treatment has an effect, it is important to include an experimental group that does not receive the treatment. Such a group is called a **control group**. A control group allows the experimenter to assess how the response variable behaves when the treatment is not used. This provides a baseline against which the treatment groups can be compared to determine if the treatment has an effect.

Use of a Placebo

In experiments that use human subjects, the use of a control group may not be enough to determine if a treatment really does have an effect. People sometimes respond merely to the power of suggestion! For example, consider a study designed to determine if a particular herbal supplement is effective in promoting weight loss. Suppose the study is designed with an experimental group that takes the herbal supplement and a control group that takes nothing. It is possible that those taking the herbal supplement may believe that it will help them to lose weight. As a result, they may be more motivated and may unconsciously change their eating behavior or activity level, resulting in weight loss.

Although there is debate about the degree to which people respond, many studies have shown that people sometimes respond to treatments with no active ingredients, such as sugar pills or solutions that are nothing more than colored water, and that they often report that such "treatments" relieve pain or reduce symptoms such as nausea or dizziness. To determine if a treatment *really* has an effect, comparing a treatment group to a control group may not be enough. To address the problem, many experiments use what is called a placebo. A **placebo** is something that is identical (in appearance, taste, feel, etc.) to the treatment received by the treatment group, except that it contains no active ingredients.

> **DEFINITION**
>
> A **placebo** is something that is identical (in appearance, taste, feel, etc.) to the treatment received by the treatment group, except that it contains no active ingredients.

In the herbal supplement study, rather than using a control group that receives *no* treatment, the researchers might want to use a placebo group. Individuals in the placebo group would take a pill identical in appearance to the herbal supplement but which contains no herb or any other active ingredient. Because the subjects don't know if they are taking the herb or the placebo, the placebo group provides a better basis for comparison. This allows the researchers to determine if the herbal supplement has any real effect over and above the "placebo effect."

Single-Blind and Double-Blind Experiments

Because people often have their own beliefs about the effectiveness of various treatments, it is preferable to conduct experiments in which subjects do not know what treatment they are receiving. Such an experiment is called **single-blind**. Of course, not all experiments can be made single-blind. For example, in an experiment comparing the effect of two different types of exercise on blood pressure, it is not possible for participants to be unaware of whether they are in the swimming group or the jogging group! However, when it is possible, "blinding" the subjects in an experiment is generally a good strategy.

In some experiments, someone other than the subject is responsible for measuring the response. To ensure that they do not let personal beliefs influence their measurements, it is also preferable for this person to be unaware of which treatment an individual subject receives. For example, in a medical experiment to determine if a new vaccine reduces the risk of getting the flu, doctors must decide whether a particular individual who is not feeling well actually has the flu or has some other unrelated illness. If the doctor making this assessment knows that a participant with flu-like symptoms has been vaccinated with the new flu vaccine, she might be less likely to determine that the participant has the flu and more likely to interpret the symptoms to be the result of some other illness.

There are two ways in which "blinding" might occur in an experiment. One involves blinding the participants, while the other involves blinding the individuals who measure the response. If both participants and those measuring the response do not know which treatment was received, the experiment is described as **double-blind**. If only one of the two types of blinding is present, the experiment is single-blind.

DEFINITION

A **double-blind** experiment is one in which neither the subjects nor the individuals who measure the response know which treatment was received.

A **single-blind** experiment is one in which the subjects do not know which treatment was received but the individuals measuring the response do know which treatment was received, or one in which the subjects do know which treatment was received but the individuals measuring the response do not know which treatment was received.

Experimental Units and Replication

An **experimental unit** is the smallest unit to which a treatment is applied. In the language of experimental design, treatments are assigned at random to experimental units, and replication means that each treatment is applied to more than one experimental unit.

Replication is necessary for random assignment to be an effective way to create similar experimental groups and to get a sense of the variability in the values of the response for individuals who receive the same treatment. As you will see in the chapters that follow, this enables you to use statistical methods to decide whether differences in the responses in different treatment groups can be attributed to the treatment received or whether they can be explained by chance variation (the natural variability seen in the responses to a single treatment).

Be careful when designing an experiment to ensure that there is replication. For example, suppose that children in two third-grade classes are available to participate in an experiment to compare two different methods for teaching arithmetic. It might at first seem reasonable to select one class at random to use one method and then assign the other method to the remaining class. But what are the experimental units here? If treatments are randomly assigned to classes, classes are the experimental units. Because only one class is assigned to each treatment, this is an experiment with no replication, even though there are many children in each class. You would *not* be able to determine whether there was a difference between the two methods based on data from this experiment, because you would have only one observation per treatment.

One last note on replication: Do not confuse replication in an experimental design with replicating an experiment. Replicating an experiment means conducting a new experiment using the same experimental design as a previous experiment. It is a way of testing conclusions based on a previous experiment, but it does not eliminate the need for replication in each of the individual experiments themselves.

Using Volunteers as Subjects in an Experiment

Although using volunteers in a study that involves collecting data through sampling is never a good idea, it is common practice to use volunteers as subjects in an experiment. Even though using volunteers limits the researcher's ability to generalize to a larger population, random assignment of the volunteers to treatments should result in comparable groups, and so treatment effects can still be assessed.

As you consider the following examples, ask yourself these questions about the experiments described:

- What question is the experiment trying to answer?
- What are the experimental conditions (treatments) for the experiment?
- What is the response variable?

- What are the experimental units and how were they selected?
- Does the design incorporate random assignment of experimental units to the different experimental conditions? If not, are there potentially confounding variables that would make it difficult to draw conclusions based on data from the experiment?
- Does the experiment incorporate a control group and/or a placebo group? If not, would the experiment be improved by including them?
- Does the experiment involve blinding? If not, would the experiment be improved by making it single- or double-blind?

Example 1.8   Revenge Is Sweet

The article **"The Neural Basis of Altruistic Punishment"** (*Science*, August 27, 2004) described a study that examined motivation for revenge. Subjects in the study were healthy, right-handed men. Subjects played a game with another player in which they could both earn money by trusting each other or one player could double-cross the other player and keep all of the money. In some cases the double cross was required by the rules of the game, while in other cases the double cross was the result of a deliberate choice. The victim of a double cross was then given the opportunity to retaliate by imposing a fine, but sometimes it would cost him money to do this. Subjects were randomly assigned to one of four experimental conditions:

1. Double cross not deliberate (double cross dictated by the rules of the game) and no cost to the victim to retaliate
2. Double cross deliberate and no cost to the victim to retaliate
3. Double cross not deliberate and a cost to the victim to retaliate
4. Double cross deliberate and a cost to the victim to retaliate

All subjects chose revenge (imposed a fine on the double-crosser) when the double cross was deliberate and retaliation was free, and 86% of the subjects chose revenge when the double cross was deliberate, even when it cost them money. Only 21% imposed a fine if the double cross was dictated by the rules of the game and was not deliberate.

This study is an experiment that incorporated random assignment to treatments and direct control (controlled gender, health, and handedness by using only healthy, right-handed males as subjects). Before leaving this example, let's consider the seven questions posed just prior to this example:

Question	Response
What question is the experiment trying to answer?	The researchers wanted to find out if it was more likely that a fine would be imposed if a double cross in the game was deliberate and not dictated by the rules of the game.
What are the experimental conditions (treatments) for the experiment?	There are four experimental conditions in this experiment. They are (1) double cross not deliberate and no cost to retaliate; (2) double cross deliberate and no cost to retaliate; (3) double cross not deliberate and cost associated with retaliation; and (4) double cross deliberate and cost associated with retaliation.
What is the response variable?	The response variable is retaliation choice, with possible values *retaliate* and *do not retaliate*.
What are the experimental units and how were they selected?	The experimental units are the healthy, right-handed men who participated in the study. Although the description does not say how the subjects were selected, it is most likely that they were volunteers.

(continued)

Question	Response
Does the design incorporate random assignment of experimental units to the different experimental conditions? If not, are there potentially confounding variables that would make it difficult to draw conclusions based on data from the experiment?	Yes, the subjects were randomly assigned to one of the four experimental conditions.
Does the experiment incorporate a control group and/or a placebo group? If not, would the experiment be improved by including them?	No, but there is no need for a control group in this experiment.
Does the experiment involve blinding? If not, would the experiment be improved by making it single- or double-blind?	No. It would not be possible to include blinding of subjects in this experiment, and there is no need to blind the person recording the response.

Example 1.9   Chilling Newborns? Then You Need a Control Group...

Researchers for the National Institute of Child Health and Human Development studied 208 infants whose brains were temporarily deprived of oxygen as a result of complications at birth (*The New England Journal of Medicine*, **October 13, 2005**). These babies were subjects in an experiment to determine if reducing body temperature for three days after birth improved their chances of surviving without brain damage. The experiment was summarized in a paper that stated "infants were randomly assigned to usual care (control group) or whole-body cooling." Including a control group in the experiment provided a basis for comparison of death and disability rates for the proposed cooling treatment and those for usual care. Some variables that might also affect death and disability rates, such as duration of oxygen deprivation, could not be directly controlled. To ensure that the experiment did not unintentionally favor one experimental condition over the other, random assignment of the infants to the two groups was critical. Because this was a well-designed experiment, the researchers were able to use the resulting data and methods that you will see in Chapter 11 to conclude that cooling did reduce the risk of death and disability for infants deprived of oxygen at birth.

Let's consider the seven questions previously posed as a way of organizing our thinking about this experiment:

Question	Response
What question is the experiment trying to answer?	Does chilling newborns whose brains were temporarily deprived of oxygen at birth improve the chance of surviving without brain damage?
What are the experimental conditions (treatments) for the experiment?	The experiment compared two experimental conditions: usual care and whole-body cooling.
What is the response variable?	This experiment used two response variables—survival (yes, no) and brain damage (yes, no).
What are the experimental units and how were they selected?	The experimental units were the 208 newborns used in the study. The description of the experiment does not say how the newborns were selected, but it is unlikely that they were a random sample from some larger population.
Does the design incorporate random assignment of experimental units to the different experimental conditions? If not, are there potentially confounding variables that would make it difficult to draw conclusions based on data from the experiment?	Yes, the babies were randomly assigned to one of the two experimental conditions.

(continued)

Question	Response
Does the experiment incorporate a control group and/or a placebo group? If not, would the experiment be improved by including them?	In this experiment, the usual care group serves as a control group. This group did not receive any cooling treatment.
Does the experiment involve blinding? If not, would the experiment be improved by making it single- or double-blind?	This experiment did not involve blinding. It might have been possible to blind the person who makes the assessment of whether or not there was brain damage, but this probably isn't necessary in this experiment.

SECTION 1.3   EXERCISES

Each Exercise Set assesses the following chapter learning objectives: C6, C10, C11, C12, M2, M4, P3

SECTION 1.3   Exercise Set 1

1.33 Does playing action video games provide more than just entertainment? The authors of the paper "Action-Video-Game Experience Alters the Spatial Resolution of Vision" (*Psychological Science* [2007]: 88–94) concluded that spatial resolution, an important aspect of vision, is improved by playing action video games. They based this conclusion on data from an experiment in which 32 volunteers who had not played action video games were "equally and randomly divided between the experimental and control groups." Subjects in each group played a video game for 30 hours over a period of 6 weeks. Those in the experimental group played *Unreal Tournament 2004*, an action video game. Those in the control group played *Tetris*, a game that does not require the user to process multiple objects at once. Explain why it was important for the researchers to randomly assign the subjects to the two groups.

1.34 Is doing tai chi, a traditional Chinese martial art, good for the immune system? The authors of a paper that appeared in the *Journal of the American Medical Association* (2007, pages 511–517) describe an experiment in which 112 healthy older adults were assigned at random to one of two groups. One group was the tai chi group. People in this group did tai chi for 40 minutes, three times a week, for 16 weeks. At the end of the 16 weeks, they received a vaccine designed to protect against a certain virus. Nine weeks later, a measurement of immunity to the virus was made. People in the second group served as a control group. They did not do tai chi during the first 16 weeks but received the virus vaccine at the same time as the tai chi group, and their immunity was also measured nine weeks later.
a. Briefly describe why it was important for the researchers to assign participants to one of the two groups rather than letting the participants choose which group they wanted to be in.
b. Explain why it was important to include a control group in this experiment.

1.35 Draw a diagram that summarizes the basic structure of the experiment described in the previous exercise.

1.36 In an experiment to compare two different surgical procedures for hernia repair ("A Single-Blinded, Randomized Comparison of Laparoscopic Versus Open Hernia Repair in Children," *Pediatrics* [2009]: 332–336), 89 children were assigned at random to one of the two surgical methods. The methods studied were laparoscopic repair and open repair. In laparoscopic repair, three small incisions are made, and the surgeon works through these incisions with the aid of a small camera that is inserted through one of the incisions. In the open repair, a larger incision is used to open the abdomen. One of the response variables was the amount of medication given after the surgery to control pain and nausea. The paper states, "For postoperative pain, rescue fentanyl (1 mg/kg) and for nausea, ondansetron (0.1 mg/kg) were given as judged necessary by the attending nurse blinded to the operative approach."
a. Why do you think it was important that the nurse who administered the medications did not know which type of surgery was performed?
b. Explain why it was not possible for this experiment to be double-blind.

1.37 Suppose that the researchers who carried out the experiment described in the previous exercise thought that gender might be a potentially confounding variable. If the children participating in the experiment included 60 boys and 29 girls, describe how blocking could be incorporated into the experiment. Be specific about how you would assign the children to treatment groups.

1.38 A study of college students showed a temporary gain of up to nine IQ points after listening to a Mozart piano sonata. This conclusion, dubbed the Mozart effect, has since been criticized by a number of researchers who have been unable to confirm the result in similar studies. Suppose that you want to determine if there is really is a Mozart effect. You decide to carry out an experiment with three experimental groups. One group will listen to a Mozart piano sonata that lasts 24 minutes. The second group will listen to popular music for the same length of time, and the third group will relax for 24 minutes with no music playing. You will measure IQ before and after the 24 minute period. Suppose that you have 45 volunteers who have agreed to participate in the experiment. Describe a method of randomly assigning each of the volunteers to one of the experimental groups.

1.39 Do ethnic group and gender influence the type of care that a heart patient receives? The following passage is from the article **"Heart Care Reflects Race and Sex, Not Symptoms"** (*USA Today,* **February 25, 1999**):

> Previous research suggested blacks and women were less likely than whites and men to get cardiac catheterization or coronary bypass surgery for chest pain or a heart attack. Scientists blamed differences in illness severity, insurance coverage, patient preference, and health care access. The researchers eliminated those differences by videotaping actors—two black men, two black women, two white men, and two white women—describing chest pain from identical scripts. They wore identical gowns, used identical gestures, and were taped from the same position. Researchers asked 720 primary care doctors at meetings of the American College of Physicians or the American Academy of Family Physicians to watch a tape and recommend care. The doctors thought the study focused on clinical decision making.

Which video a particular doctor watched was determined by the roll of a four-sided die.

Answer the following seven questions for the described experiment. (Hint: Reviewing Examples 1.5 and 1.6 might be helpful.)

1. What question is the experiment trying to answer?
2. What are the experimental conditions (treatments) for this experiment?
3. What is the response variable?
4. What are the experimental units, and how were they selected?
5. Does the design incorporate random assignment of experimental units to the different experimental conditions? If not, are there potentially confounding variables that would make it difficult to draw conclusions based on data from the experiment?
6. Does the experiment incorporate a control group and/or a placebo group? If not, would the experiment be improved by including them?
7. Does the experiment involve blinding? If not, would the experiment be improved by making it single- or double-blind?

SECTION 1.3 **Exercise Set 2**

1.40 The Institute of Psychiatry at King's College London found that dealing with "infomania" has a temporary, but significant, negative effect on IQ (*Discover,*

November 2005). To reach this conclusion, researchers divided volunteers into two groups. Each subject took an IQ test. One group had to check e-mail and respond to instant messages while taking the test, and the other group took the test without any distraction. The distracted group had an average score that was 10 points lower than the average for the control group. Explain why it is important that the researchers use random assignment to create the two experimental groups.

1.41 Swedish researchers concluded that viewing and discussing art soothes the soul and helps relieve medical conditions such as high blood pressure and constipation (**AFP International News Agency, October 14, 2005**). This conclusion was based on a study in which 20 elderly women gathered once a week to discuss different works of art. The study also included a control group of 20 elderly women who met once a week to discuss their hobbies and interests. At the end of four months, the art discussion group was found to have a more positive attitude and lower blood pressure and used fewer laxatives than the control group.
a. Why is it important to determine if the researchers randomly assigned the women to one of the two groups?
b. Explain why you think that the researchers included a control group in this study.

1.42 Draw a diagram that summarizes the basic structure of the experiment described in the previous exercise.

1.43 The article **"Doctor Dogs Diagnose Cancer by Sniffing It Out"** (*Knight Ridder Newspapers,* **January 9, 2006**) refers to an experiment described in the journal *Integrative Cancer Therapies.* In this experiment, dogs were trained to distinguish between people with breast and lung cancer and people without cancer by sniffing exhaled breath. Dogs were trained to lie down if they detected cancer in a breath sample. After training, the dogs' ability to detect cancer was tested using breath samples from people whose breath had not been used in training the dogs. The paper states, "The researchers blinded both the dog handlers and the experimental observers to the identity of the breath samples." Explain why this blinding is an important aspect of the design of this experiment.

1.44 Suppose that you would like to know if keyboard design has an effect on wrist angle, as shown in the accompanying figure.

You have 40 volunteers who have agreed to participate in an experiment to compare two different keyboards. Describe a method for randomly assigning each of the volunteers to one of the experimental groups.

1.45 Suppose that the researchers who carried out the experiment described in the previous exercise thought that age might potentially be a confounding variable. If 20 of the 40 volunteers were between 20 and 30 years old and the other 20 were between 30 and 40 years old, describe how blocking could be incorporated into the experiment. Be specific about how you would assign the volunteers to treatment groups.

1.46 An advertisement for a sweatshirt that appeared in *SkyMall Magazine* (a catalog distributed by some airlines) stated the following:

> This is not your ordinary hoody! Why? Fact: Research shows that written words on containers of water can influence the water's structure for better or worse depending on the nature and intent of the word. Fact: The human body is 70% water. What if positive words were printed on the inside of your clothing?

For only $79, you could purchase a hooded sweatshirt that had over 200 positive words (such as hope, gratitude, courage, and love) in 15 different languages printed on the inside of the sweatshirt so that you could benefit from being surrounded by these positive words. The "fact" that written words on containers of water can influence the water's structure appears to be based on the work of Dr. Masaru Emoto, who typed words on paper and then pasted them on bottles of water. He noted how the water reacted to the words by observing crystals formed in the water. He describes several of his experiments in his self-published book, *The Message from Water.* If you were going to interview Dr. Emoto, what questions would you want to ask him about his experiment?

Additional Exercises

Use the following information to answer Exercises 1.47–1.50.
Many surgeons play music in the operating room. Does the type of music played have an effect on the surgeons' performance? The report **"Death Metal in the Operating Room" (NPR, Dec. 24, 2009, www.npr.org)** describes an experiment in which surgeons used a simulator to perform a surgery. Some of the surgeons listened to music with vocal elements while performing the surgery, and others listened to music that did not have vocal elements. The researchers concluded that the average time to complete the surgery was greater when music with vocal elements is played than when music without vocal elements is played.

1.47 What are the experimental conditions for the experiment described? What is the response variable?

1.48 Explain why it is important to control each of the following potentially confounding variables in the experiment.
a. the type of surgery performed
b. operating room temperature
c. volume at which the music was played

1.49 Explain why it is important that the surgeons be assigned at random to the two music conditions.

1.50 Could this experiment have been double-blind? Explain why or why not.

1.51 The report "Comparative Study of Two Computer Mouse Designs" (Cornell Human Factors Laboratory Technical Report RP7992) included the following description of the subjects used in an experiment:

> Twenty-four Cornell University students and staff (12 males and 12 females) volunteered to participate in the study. Three groups of 4 men and 4 women were selected by their stature to represent the 5th percentile (female 152.1 ± 0.3 cm, male 164.1 ± 0.4 cm), 50th percentile (female 162.4 ± 0.1 cm, male 174.1 ± 0.7 cm), and 95th percentile (female 171.9 ± 0.2 cm, male 185.7 ± 0.6 cm) ranges . . . All subjects reported using their right hand to operate a computer mouse.

> This experimental design incorporated direct control and blocking.

a. Are the possible effects of the potentially confounding variable stature (height) addressed by blocking or direct control?
b. Whether the right or left hand is used to operate the mouse was considered to be a potentially confounding variable. Are the possible effects of this variable addressed by blocking or direct control?

1.52 In an experiment comparing two different surgical procedures for hernia repair (**"A Single-Blinded, Randomized Comparison of Laparoscopic Versus Open Hernia Repair in Children,"** *Pediatrics* **[2009]: 332–336**), 89 children were assigned at random to one of the two surgical methods. Because there were potentially confounding variables that they could not control, the researchers relied on the random assignment of subjects to treatments to create comparable groups. One such variable was age. After random assignment to treatments, the researchers looked at the age distribution of the children in each of the two experimental groups (laparoscopic repair [LR] and open repair [OR]). The figure on the next page is from the paper. Based on this figure, has the random assignment of subjects to experimental groups been successful in creating groups that are similar in age? Explain.

1.53 Pismo Beach, California, has an annual clam festival that includes a clam chowder contest. Judges rate clam chowders from local restaurants. The judges are not aware of which chowder is from which restaurant when they assign the ratings. One year, much to the dismay of the seafood restaurants on the waterfront, Denny's chowder was declared the winner! (When asked what the ingredients were, the cook at Denny's said he wasn't sure—he just had to add the right amount of nondairy creamer to the soup stock that he got from Denny's distribution center!)

a. Do you think that Denny's chowder would have won the contest if the judging had not been "blind?" Explain.

b. Although this was not an experiment, your answer to Part (a) helps to explain why those measuring the response in an experiment are often blinded. Using your answer in Part (a), explain why experiments are often blinded in this way.

FIGURE FOR EXERCISE 1.52

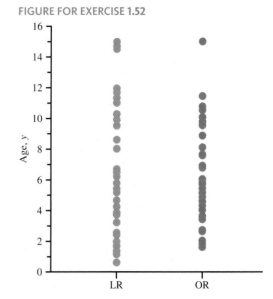

SECTION 1.4 The Importance of Random Selection and Random Assignment: What Types of Conclusions Are Reasonable?

On September 25, 2009, results from a study of the relationship between spanking and a child's IQ were reported by a number of different news media. Some of the headlines that appeared that day were:

> **"Spanking lowers a child's IQ"** (*Los Angeles Times*)
> **"Do you spank? Studies indicate it could lower your kid's IQ"** (*SciGuy, Houston Chronicle*)
> **"Spanking can lower IQ"** (*NBC4i, Columbus, Ohio*)
> **"Smacking hits kids' IQ"** (*newscientist.com*)

In this study, the investigators followed 806 kids ages 2 to 4 and 704 kids ages 5 to 9 for 4 years. IQ was measured at the beginning of the study and again 4 years later. The researchers found that at the end of the study, the average IQ of the younger kids who were not spanked was 5 points higher than that of kids who were spanked. For the older group, the average IQ of kids who were not spanked was 2.8 points higher. These headlines all imply that spanking was the cause of the observed difference in IQ. Is this conclusion reasonable?

The type of conclusion that can be drawn from a statistical study depends on the study design. In an observational study, the goal is to use sample data to draw conclusions about one or more populations. In a well-designed observational study, the sample is selected in a way that you hope will make it representative of the population of interest. When this is the case, it is reasonable to generalize from the sample to the larger population. This is why good observational studies include random selection from the population of interest whenever possible.

A well-designed experiment can result in data that provide evidence for a cause-and-effect relationship. This is an important difference between an observational study and an experiment. In an observational study, it is not possible to draw clear cause-and-effect conclusions, because you cannot rule out the possibility that the observed effect is due to some variable other than the explanatory variable being studied. Consider the following two studies, which illustrate why it is not reasonable to draw cause-and-effect conclusions from an observational study.

● The article **"Panel Can't Determine the Value of Daily Vitamins"** (*San Luis Obispo Tribune*, July 1, 2003) summarized the conclusions of a government advisory panel that investigated the benefits of vitamin use. The panel looked at a large number of studies on vitamin use and concluded that the results were "inadequate or conflicting." A major concern was that many of the studies were observational in nature. The panel worried that people who take vitamins might just be healthier because they tend to take better care of themselves in general. The observational nature of these studies prevented the panel from concluding that taking vitamins is the *cause* of observed better health.

- The article **"Heartfelt Thanks to Fido"** (*San Luis Obispo Tribune*, July 5, 2003) summarized a study that appeared in the *American Journal of Cardiology* (March 15, 2003). In this study, researchers measured heart rate variability (a measure of the heart's ability to handle stress) in patients who had recovered from a heart attack. They found that heart rate variability was higher (which is good and means the heart can handle stress better) for those who owned a dog than for those who did not. Should someone who suffers a heart attack immediately go out and get a dog? Well, maybe not yet. The American Heart Association recommends additional studies to determine if the improved heart rate variability is attributable to dog ownership or due to dog owners getting more exercise. This possibility prevents you from concluding that owning a dog is the *cause* of improved heart rate variability.

Let's return to the study on spanking and children's IQ described at the beginning of this section. Is this study an observational study or an experiment? Two groups were compared (children who were spanked and children who were not spanked), but the researchers did not randomly assign children to the spanking or no-spanking groups. The study is observational, and so cause-and-effect conclusions such as "spanking lowers IQ" are not justified. What you can say is that there is evidence that, as a group, children who are spanked tend to have a lower IQ than children who are not spanked. What you cannot say is that spanking is the *cause* of the lower IQ. It is possible that other variables—such as home or school environment, socioeconomic status, or parents' education levels—are related to both IQ and whether or not a child was spanked.

Fortunately, not everyone made the same mistake as the writers of the headlines given earlier in this section. Some examples of headlines that got it right are:

> **"Lower IQ's measured in spanked children"** (*world-science.net*)
> **"Children who get spanked have lower IQs"** (*livescience.com*)
> **"Research suggests an association between spanking and lower IQ in children"** (*CBSnews.com*)

Drawing Conclusions from Statistical Studies

In this section, two different types of conclusions have been described. One type involves generalizing from what you have seen in a sample to some larger population, and the other involves reaching a cause-and-effect conclusion about the effect of an explanatory variable on a response. When is it reasonable to draw such conclusions? The answer depends on the way the data were collected. The following table summarizes when each of these types of conclusions is reasonable.

Type of Conclusion	Reasonable When
Generalize from sample to population	Random selection is used to obtain the sample
Change in response is caused by experimental conditions (cause-and-effect conclusion)	There is random assignment of experimental units to experimental conditions

This table implies the following:

- For observational studies, it is not possible to reach cause-and-effect conclusions, but it is possible to generalize from the sample to the population of interest *if* the study design incorporated random selection.

- For experiments, it is possible to reach cause-and-effect conclusions *if* the study design incorporated random assignment to experimental conditions.
- If an experiment incorporates both random assignment to experimental conditions *and* random selection of experimental units from some population, it is possible to reach both cause-and-effect conclusions and generalize these conclusions to the population from which the experimental units were selected. If an experiment does not include random selection of experimental units (for example, if volunteers are used as subjects), then it is not reasonable to generalize results unless a strong case can be made that the experimental units are representative of some larger group.

It is important to keep these three points in mind when drawing conclusions from a statistical study or when deciding if conclusions that others have made based on a statistical study are reasonable.

SECTION 1.4   EXERCISES

Each Exercise Set assesses the following chapter learning objectives: C2, C5, C6, C7, C9

SECTION 1.4   Exercise Set 1

1.54 An article titled **"Guard Your Kids Against Allergies: Get Them a Pet"** (*San Luis Obispo Tribune*, **August 28, 2002**) described a study that led researchers to conclude that "babies raised with two or more animals are about half as likely to have allergies by the time they turned six." Explain why it is not reasonable to conclude that being raised with two or more animals is the cause of the observed lower allergy rate.

1.55 Based on a survey conducted on the DietSmart.com website, investigators concluded that women who regularly watched *Oprah* were only one-seventh as likely to crave fattening foods as those who watched other daytime talk shows (*San Luis Obispo Tribune*, October 14, 2000).
a. Is it reasonable to conclude that watching *Oprah* causes a decrease in cravings for fattening foods? Explain.
b. Is it reasonable to generalize the results of this survey to all women in the United States? To all women who watch daytime talk shows? Explain why or why not.

1.56 The following is from an article titled **"After the Workout, Got Chocolate Milk?"** that appeared in the *Chicago Tribune* (**January 18, 2005**):

> Researchers at Indiana University at Bloomington have found that chocolate milk effectively helps athletes recover from an intense workout. They had nine cyclists bike, rest four hours, then bike again, three separate times. After each workout, the cyclists downed chocolate milk or energy drinks Gatorade or Endurox (two to three glasses per hour); then, in the second workout of each set, they cycled to exhaustion. When they drank chocolate milk, the amount of time they could cycle until they were exhausted was similar to when they drank Gatorade and longer than when they drank Endurox.

a. For the experiment to have been well designed, it must have incorporated random assignment. Briefly explain where the researcher would have needed to use random assignment for the conclusion of the experiment to be valid.

b. Would it have been possible to blind the subjects in this experiment? Explain why or why not.

1.57 "Pecans Lower Cholesterol" is a headline that appeared in the magazine *Woman's World* (November 1, 2010). Consider the following five study descriptions. For each of the study descriptions, answer these five questions:

> Question 1: Is the described study an observational study or an experiment?
> Question 2: Did the study use random selection from some population?
> Question 3: Did the study use random assignment to experimental groups?
> Question 4: Would the conclusion "pecans lower cholesterol" be appropriate given the study description? Explain.
> Question 5: Would it be reasonable to generalize conclusions from this study to some larger population? If so, what population?

Study 1: Five hundred students were selected at random from those enrolled at a large college in Florida. Each student in the sample was asked whether they ate pecans more than once in a typical week, and their cholesterol levels were also measured. The average cholesterol level was significantly lower for the group who ate pecans more than once a week than for the group that did not.

Study 2: One hundred people who live in Los Angeles volunteered to participate in a statistical study. The volunteers were divided based on gender, with women in group 1 and men in group 2. Those in group 1 were asked to eat 3 ounces of pecans daily for 1 month. Those in group 2 were asked not to eat pecans for 1 month. At the end of the month, the average cholesterol level was significantly lower for group 1 than for group 2.

Study 3: Two hundred people volunteered to participate in a statistical study. Each person was asked how often he or she ate pecans, and their cholesterol levels were also measured. The average cholesterol level for those who ate pecans more than once a week was significantly lower than the average cholesterol level for those who did not eat pecans.

Study 4: Two hundred people volunteered to participate in a statistical study. For each volunteer, a coin was tossed. If the coin landed heads up, the volunteer was assigned to group 1. If the coin landed tails up, the volunteer was assigned to group 2. Those in group 1 were asked to eat 3 ounces of pecans daily for 1 month. Those in group 2 were asked not to eat pecans for 1 month. At the end of the month, the average cholesterol level was significantly lower for group 1 than for group 2.

Study 5: One hundred students were selected at random from those enrolled at a large college. Each of the selected students was asked to participate in a study, and all agreed to participate. For each student, a coin was tossed. If the coin landed heads up, the student was assigned to group 1. If the coin landed tails up, the student was assigned to group 2. Those in group 1 were asked to eat 3 ounces of pecans daily for 1 month. Those in group 2 were asked not to eat pecans for 1 month. At the end of the month, the average cholesterol level was significantly lower for group 1 than for group 2.

SECTION 1.4   **Exercise Set 2**

1.58 A survey of affluent Americans (those with incomes of $75,000 or more) indicated that 57% would rather have more time than more money (*USA Today*, January 29, 2003).
a. What condition on how the data were collected would make it reasonable to generalize this result to the population of affluent Americans?
b. Would it be reasonable to generalize this result to the population of *all* Americans? Explain why or why not.

1.59 Does living in the South cause high blood pressure? Data from a group of 6,278 people questioned in the Third National Health and Nutritional Examination Survey between 1988 and 1994 indicate that a greater percentage of Southerners have high blood pressure than do people living in any other region of the United States (**"High Blood Pressure Greater Risk in U.S. South, Study Says," January 6, 2000, cnn.com**). This difference in rate of high blood pressure was found in every ethnic group, gender, and age category studied. What are two possible reasons we cannot conclude that living in the South causes high blood pressure?

1.60 The paper **"Effect of Cell Phone Distraction on Pediatric Pedestrian Injury Risk"** (*Pediatrics* **[2009]: e179–e185**) describes an experiment examining whether people talking on a cell phone are at greater risk of an accident when crossing the street than when not talking on a cell phone. (No people were harmed in this experiment—a virtual interactive pedestrian environment was used.) One possible way of conducting such an experiment would be to have a person cross 20 streets in this virtual environment. The person would talk on a cell phone for some crossings

and would not use the cell phone for others. Explain why it would be important to randomly assign the two treatments (talking on the phone, not talking on the phone) to the 20 trials (the 20 simulated street crossings).

1.61 "Strengthen Your Marriage with Prayer" is a headline that appeared in the magazine *Woman's World* (**November 1, 2010**). The article went on to state that couples who attend religious services and pray together have happier, stronger marriages than those who do not. For each of the following study descriptions, answer these five questions:

Question 1: Is the study described an observational study or an experiment?

Question 2: Did the study use random selection from some population?

Question 3: Did the study use random assignment to experimental groups?

Question 4: Would the conclusion that you can "strengthen your marriage with prayer" be appropriate given the study description? Explain.

Question 5: Would it be reasonable to generalize conclusions from this study to some larger population? If so, what population?

Study 1: Married viewers of an afternoon television talk show were invited to participate in an online poll. A total of 1,700 viewers chose to go online and participate in the poll. Each was asked whether they attended church regularly and whether they would rate their marriage as happy. The proportion of those attending church regularly who rated their marriage as happy was significantly higher than the proportion of those not attending church regularly who rated their marriage as happy.

Study 2: Two hundred women were selected at random from the membership of the American Association of University Women (a large professional organization for women). Each woman selected was asked if she was married. The 160 married women were asked questions about church attendance and whether or not they would rate their marriage as happy. The proportion of those attending church regularly who rated their marriage as happy was significantly higher than the proportion of those not attending church regularly who rated their marriage as happy.

Study 3: A researcher asked each man of the first 200 men arriving at a baseball game if he was married. If a man said he was married, the researcher also asked if he attended church regularly and if he would rate his marriage as happy. The proportion of those attending church regularly who rated their marriage as happy was significantly higher than the proportion of those not attending church regularly who rated their marriage as happy.

Additional Exercises

1.62 A study described in *Food Network Magazine* (**January 2012**) concluded that people who push a shopping cart at a grocery store are less likely to purchase junk food than those who use a hand-held basket.

a. Do you think this study was an observational study or an experiment?

b. Is it reasonable to conclude that pushing a shopping cart causes people to be less likely to purchase junk food? Explain why or why not.

1.63 Researchers at the University of Utah carried out a study to see if the size of the fork used to eat dinner has an effect on how much food is consumed (**Food Network Magazine, January 2012**). The researchers assigned people to one of two groups. One group ate dinner using a small fork, and the other group ate using a large fork. The researchers found that those who ate with a large fork ate less of the food on the plate than those who ate with the small fork. The title of the article describing this study was **"Dieters Should Use a Big Fork."** This title implies a cause-and-effect relationship between fork size and amount eaten and also generalizes this finding to the population of dieters. What would you need to know about the study design to determine if the conclusions implied by the headline are reasonable?

Use the following information to answer Exercises 1.64–1.68.
The paper **"Turning to Learn: Screen Orientation and Reasoning from Small Devices"** (**Computers in Human Behavior** [2011]: 793–797) describes a study that investigated whether cell phones with small screens are useful for gathering information. The researchers wondered if the ability to reason using information read on a small screen was affected by the screen orientation. The researchers assigned 33 undergraduate students who were enrolled in a psychology course at a large public university to one of two groups at random. One group read material that was displayed on a small screen in portrait orientation, and the other group read material on the same size screen but turned to display the information in landscape orientation (see the following figure).

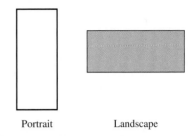

Portrait   Landscape

The researchers found that performance on a reasoning test based on the displayed material was better for the group that read material in the landscape orientation.

1.64 Is the study described an observational study or an experiment?

1.65 Did the study use random selection from some population?

1.66 Did the study use random assignment to experimental groups?

1.67 Is the conclusion that reasoning using information displayed on a small screen is improved by turning the screen to landscape orientation appropriate, given the study design? Explain.

1.68 Is it reasonable to generalize the conclusions from this study to some larger population? If so, what population?

SECTION 1.5 Avoid These Common Mistakes

It is a big mistake to begin collecting data before thinking carefully about research objectives and then developing a plan. A poorly designed plan for data collection may result in data that will not allow you to answer key questions of interest or to generalize conclusions based on the data to the desired populations of interest.

Be sure to avoid these common mistakes:

1. Drawing a cause-and-effect conclusion from an observational study. Don't do this, and don't believe it when others do it!

2. Generalizing results of an experiment that uses volunteers as subjects. Only do this if it can be convincingly argued that the group of volunteers is representative of the population of interest.

3. Generalizing conclusions based on data from a poorly designed observational study. Though it is sometimes reasonable to generalize from a sample to a population, on other occasions it is not reasonable. Generalizing from a sample to a population is justified only when the sample is likely to be representative of the population. This would be the case if the sample was a random sample from the population, and there were no major potential sources of bias. If the sample was not selected at random or if potential sources of bias were present, these issues would have to be addressed before deciding if it is reasonable to generalize the study results.

For example, the **Associated Press (January 25, 2003)** reported on the high cost of housing in California. Instead of using a random sample of counties in California, the report gave the median home price for the 10 counties with the highest home prices. Although these 10 counties are a sample of the counties in California, it would not be reasonable to generalize from this sample to all California counties.

4. Generalizing conclusions based on an observational study that used voluntary response or convenience sampling to a larger population. This is almost never reasonable!

CHAPTER ACTIVITIES

ACTIVITY 1.1 FACEBOOK FRIENDING

Background: The article "Professors Prefer Face Time to Facebook" appeared in the student newspaper at Cal Poly, San Luis Obispo (*Mustang Daily*, **August 27, 2009**). The article examined how professors and students felt about using Facebook as a means of faculty-student communication. The student who wrote this article got mixed opinions when she asked students whether they wanted to become Facebook friends with their professors. Two student comments included in the article were

"I think the younger the professor is, the more you can relate to them and the less awkward it would be if you were to become friends on Facebook. The older the professor, you just would have to wonder, 'Why are they friending me?'"

and

"I think becoming friends with professors on Facebook is really awkward. I don't want them being able to see into my personal life, and frankly, I am not really interested in what my professors do in their free time."

Even if the students interviewed had expressed a consistent opinion, it would still be unreasonable to think this represented how all students felt about this issue because only four students were interviewed, and it is not clear from the article how these students were selected.

In this activity, you will work with a partner to develop a plan to assess student opinion about being Facebook friends with teachers at your school.

1. Suppose you will select a sample of 50 students at your school to participate in a survey. Write one or more questions to ask each student in the sample.

2. Discuss with your partner whether you think it would be easy or difficult to obtain a simple random sample of 50 students at your school and to obtain the desired information from all the students selected for the sample. Write a summary of your discussion.

3. With your partner, decide how you might go about selecting a sample of 50 students from your school that reasonably could be considered representative of the population of interest, even if it may not be a simple random sample. Write a brief description of your sampling plan, and point out the aspects of your plan that you think will result in a representative sample.

4. Explain your plan to another pair of students. Ask them to critique your plan. Write a brief summary of the comments you received. Now reverse roles, and provide a critique of the plan devised by the other pair.

5. Based on the feedback you received in Step 4, would you modify your original sampling plan? If not, explain why this is not necessary. If so, describe how the plan would be modified.

ACTIVITY 1.2 McDONALD'S AND THE NEXT 100 BILLION BURGERS

Background: The article "Potential Effects of the Next 100 Billion Hamburgers Sold by McDonald's" (*American Journal of Preventative Medicine* [2005]: 379–381) estimated that 992.25 million pounds of saturated fat would be consumed as McDonald's sells its next 100 billion hamburgers. This estimate was based on the assumption that the average weight of a burger sold would be 2.4 oz. This is the average of the weight of a regular hamburger (1.6 oz.) and a Big Mac (3.2 oz.). The authors took this approach because

McDonald's does not publish sales and profits of individual items. Thus, it is not possible to estimate how many of McDonald's first 100 billion beef burgers sold were 1.6 oz hamburgers, 3.2 oz. Big

Macs (introduced in 1968), 4.0 oz. Quarter Pounders (introduced in 1973), or other sandwiches.

This activity can be completed by an individual or by a team. Your instructor will specify which approach (individual or team) you should use.

1. The authors of the article believe that the use of 2.4 oz. as the average size of a burger sold at McDonald's is "conservative," which would result in the estimate of 992.25 million pounds of saturated fat being lower than the actual amount that would be consumed. Explain why the authors' belief might be justified.

2. Do you think it would be possible to collect data that could lead to an estimate of the average burger size that would be better than 2.4 oz.? If so, explain how you would go about collecting such data. If not, explain why you think it is not possible.

ACTIVITY 1.3 VIDEO GAMES AND PAIN MANAGEMENT

Background: Video games have been used for pain management by doctors and therapists who believe that the attention required to play a video game can distract the player and thereby decrease the sensation of pain. The paper **"Video Games and Health"** (*British Medical Journal* [2005]: 122–123) makes the following statement:

However, there has been no long term follow-up and no robust randomized controlled trials of such interventions. Whether patients eventually tire of such games is also unclear. Furthermore, it is not known whether any distracting effect depends simply on concentrating on an interactive task or whether the content of games is also an important factor as there have been no controlled trials comparing video games with other distracters. Further research should examine factors within games such as novelty, users' preferences, and relative levels of challenge and should compare video games with other potentially distracting activities.

1. Working with a partner, select one of the areas of potential research suggested in the paper. Formulate a specific question that you could address by performing an experiment.

2. Propose an experiment that would provide data to address the question from Step 1. Be specific about how subjects might be selected, what the experimental conditions (treatments) would be, and what response would be measured.

EXPLORING THE BIG IDEAS

In the "Exploring the Big Ideas" sections of this text, you and your class will have the opportunity to work with several different data sets.

In the two exercises below, each student in your class will go online to select a random sample from a small population consisting of 300 adults between the ages of 18 and 64.

To learn about how the people in this population use text messaging, go online at www.cengage.com/stats/peck1e and click on the link labeled "Exploring the Big Ideas." Then locate the link for Chapter 1. When you click on this link, it will take you to a web page where you can select a random sample of 20 people from the population.

Click on the sample button. This selects a random sample and will display information for the 20 people in your sample. You should see the following information for each person selected:

1. An ID number that identifies the person selected.

2. The number of text messages sent in a typical day.

3. The response to the question "Have you ever sent or read a text message while driving?" The response is coded as 1 for yes and 2 for no.

Each student in your class will receive data from a different random sample.

Use the data for your sample to complete the following exercises.

1. This exercise uses the number of text messages sent data.
   a. A graph of the number of text messages sent data for the entire population is shown on the next page. (Note that because there were 300 people in the population, some of the dots in the graph represent as many as 6 people in the population.)
   b. Construct a similar graph with 20 dots for the number of text messages sent by the 20 people in your sample.

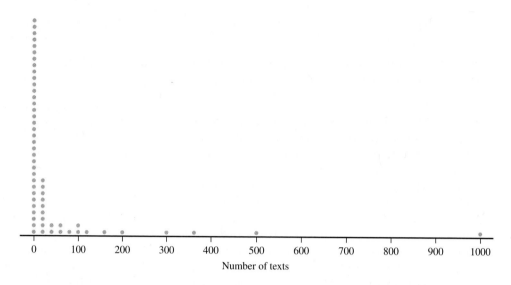

Number of texts

Each symbol represents up to 6 observations.

   c. How is the graph for your sample similar to the population graph? How is it different?

   d. If you had taken a random sample of 50 people from this population instead of 20, do you think your graph would have looked more like the population graph or less like the population graph? Why do you think this?

   e. Calculate the average number of text messages sent in a typical day by the 20 people in your sample.

   f. The average number of text messages sent for the entire population is 24.28. Did you get this same value for your sample average? Does this surprise you? Explain why or why not.

If asked to do so by your teacher, bring your graph with you to class. Your teacher will lead a class discussion of the following:

   g. Did everyone in the class get the same value of the sample average? Is this surprising?

   h. After comparing your graph to the graphs produced by several other students in your class, think about the following questions:

     Why aren't all the sample graphs the same?

     Even though not all of the sample graphs are the same, in what ways are they similar?

2. This exercise uses the text message while driving data.

   a. Calculate the proportion of people in your sample who responded yes by counting the number of yes responses and dividing that number by 20.

   b. For the entire population, the proportion who responded yes is 120/300 = 0.40. Did you get this same value for your sample proportion? Does this surprise you? Explain why or why not.

If asked to do so by your teacher, bring your answers to parts (a) and (b) to class. Your teacher will lead a class discussion of the following:

   c. Did everyone in the class get the same sample proportion? Why did this happen?

   d. Of all the sample proportions, what was the smallest? What was the largest? Does this sample-to-sample variability surprise you?

   e. Suppose everyone had selected a random sample of 50 people instead of 20 people. Do you think the sample proportions for these new samples would have varied more or varied less from sample to sample than what you saw for samples of size 20?

These two exercises explore the "big idea" of **sampling variability**. This important concept will be revisited often in the following chapters. Understanding sampling variability is the key to understanding how you can use sample data to learn about a population.

ARE YOU READY TO MOVE ON?   CHAPTER 1 REVIEW EXERCISES

All chapter learning objectives are assessed in these exercises. The learning objectives assessed
in each exercise are given in parentheses.

1.69   (C1, P1)

For each of the following, determine whether the statistical
study described is an observational study or an experiment.
Give a brief explanation of your choice.

a. Can choosing the right music make a beverage taste
better? This question was investigated by a researcher
at a university in Edinburgh (www.decanter.com/news).
Each of 250 volunteers was assigned at random to one
of five rooms where they were asked to taste and rate a
particluar beverage. No music was playing in one of the
rooms, and a different style of music was playing in each
of the other four rooms. The researcher concluded that
the beverage is rated more highly when bold music is
played than when no music is played.

b. The article **"Display of Health Risk Behaviors on MySpace
by Adolescents"** (**Archives of Pediatrics and Adolescent
Medicine** [2009]: 27–34) described a study of 500 publicly
accessible MySpace web profiles posted by 18-year-
olds. The content of each profile was analyzed and the
researchers concluded that those who indicated involvement in sports or a hobby were less likely to have references to risky behavior (such as sexual references or
references to substance abuse or violence).

c. *USA Today* (January 29, 2003) reported that in a study of
affluent Americans (defined as those with incomes of
$75,000 or more per year) 57% indicated that they would
rather have more time than more money.

d. The article **"Acupuncture for Bad Backs: Even Sham
Therapy Works"** (*Time*, May 12, 2009) summarized a study
conducted by researchers at the Group Health Center for
Health Studies in Seattle. In this study, 638 adults with
back pain were randomly assigned to one of four groups.
People in group 1 received the usual care for back pain.
People in group 2 received acupuncture at a set of points
tailored specifically for each individual. People in group
3 received acupuncture at a standard set of points typically used in the treatment of back pain. Those in group
4 received fake acupuncture—they were poked with a
toothpick at the same set of points used for the people in
group 3. Two notable conclusions from the study were:
(1) patients receiving real or fake acupuncture experienced a greater reduction in pain than those receiving
usual care; and (2) there was no significant difference
in pain reduction between those who received real acupuncture (groups 2 and 3) and those who received fake
acupuncture toothpick pokes.

1.70   (C3)

The student senate at a college with 15,000 students is
interested in the proportion of students who favor a change
in the grading system to allow for plus and minus grades
(for example, B+, B, B-, rather than just B). Two hundred
students are interviewed to determine their attitude toward
this proposed change. What is the population of interest?
What group of students constitutes the sample in this
problem?

1.71   (C4)

For each of the following statements, identify the number
that appears in boldface type as the value of either a population characteristic or a statistic:

a. A department store reports that **84%** of all customers
who use the store's credit plan pay their bills on time.

b. A sample of 100 students at a large university had a mean
age of **24.1** years.

c. The Department of Motor Vehicles reports that **22%** of
all vehicles registered in a particular state are imports.

d. A hospital reports that, based on the 10 most recent cases,
the mean length of stay for surgical patients is **6.4** days.

e. A consumer group, after testing 100 batteries of a certain
brand, reported an average life of **63** hours of use.

1.72   (C8)

According to the article **"Effect of Preparation Methods
on Total Fat Content, Moisture Content, and Sensory
Characteristics of Breaded Chicken Nuggets and Beef Steak
Fingers"** (*Family and Consumer Sciences Research Journal*
[1999]: 18–27), sensory tests were conducted using 40 college
student volunteers at Texas Women's University. Give three
reasons, other than the relatively small sample size, why it
would not be a good idea to generalize any study results to
the population of all college students.

1.73   (M1)

A petition with 500 signatures is submitted to a college's
student council. The council president would like to determine what proportion of those who signed the petition are
actually registered students at the college. There is not
enough time to check all 500 names with the registrar, so the
council president decides to select a simple random sample
of 30 signatures. Describe how she might do this.

1.74   (P2, P4)

The paper **"From Dr. Kildare to Grey's Anatomy"** (*Annals
of Emergency Medicine* [2010]: 21A–23A) describes several
studies of how the way in which doctors are portrayed on
television might influence public perception of doctors. One
study was described as follows:

> Rebecca Chory, Ph.D., now an associate professor of
> communication at West Virginia University, began study-
> ing the effect of such portrayals on patients' attitudes
> toward physicians. Using a survey of 300 undergraduate

students, she compared perceptions of physicians in 1992—the end of the era when physicians were shown as all-knowing, wise father figures—with those in 1999, when shows such as *ER* and *Chicago Hope* (1994–2000) were continuing the transformation to showing the private side and lives of physicians, including vivid demonstrations of their weaknesses and insecurities.

Dr. Chory found that, regardless of the respondents' personal experience with physicians, those who watched certain kinds of television had declining perceptions of physicians' composure and regard for others. Her results indicated that the more prime time physician shows that people watched in which physicians were the main characters, the more uncaring, cold, and unfriendly the respondents thought physicians were.

a. Answer the following four questions for the observational study described in this exercise. (Hint: Reviewing Examples 1.3 and 1.4 might be helpful.)
   1. What is the population of interest?
   2. Was the sample selected in a reasonable way?
   3. Is the sample likely to be representative of the population of interest?
   4. Are there any obvious sources of bias?
b. Based on the study design, do you think that the stated conclusions are reasonable?

1.75   (M3)

A researcher would like to select a sample of students at a large high school.

a. Describe a sampling plan that would result in a simple random sample.
b. Describe a sampling plan that would result in a systematic sample.
c. One possible sampling plan is to divide the students into four subgroups based on grade (9th, 10th, 11th, and 12th) and then to take a simple random sample from each grade. Would this result in a stratified random sample or a cluster sample? Explain your choice.

1.76   (C6)

In many digital environments, users are allowed to choose how they are represented visually online. Does the way in which people are represented online affect online behavior? This question was examined by the authors of the paper **"The Proteus Effect: The Effect of Transformed Self-Representation on Behavior"** (*Human Communication Research* [2007]: 271–290). Participants were randomly assigned either an attractive avatar (a graphical image that represents a person) to represent them or an unattractive avatar. The researchers concluded that when interacting with a person of the opposite gender in an online virtual environment, those assigned an attractive avatar moved significantly closer to the other person than those who had been assigned an unattractive avatar. This difference was attributed to the attractiveness of the avatar. Explain why the researchers would not have been able to reach this conclusion if participants had been allowed to choose one of the two avatars (attractive, unattractive) themselves.

1.77

For the experiment described in the previous exercise, draw a diagram to represent the underlying structure of the experiment.

1.78   (C10)

The article **"Yes that Miley Cyrus Biography Helps Learning"** (*The Globe and Mail*, August 5, 2010) describes an experiment investigating whether providing summer reading books to low-income children would affect school performance. Subjects in the experiment were 1,300 children randomly selected from first and second graders at low-income schools in Florida. A group of 852 of these children were selected at random from the group of 1,300 participants to be in the book group. The other 478 children were assigned to the control group. Children in the book group were invited to a book fair in the spring to choose any 12 reading books that they could then take home. Children in the control group were not given any reading books, but were given some activity and puzzle books. These children received books each year for three years until the children reached third and fourth grade. The researchers then compared reading test scores of the two groups.

a. Explain why randomly selecting 852 of the 1,300 children to be in the book group is equivalent to random assignment of the children to the two experimental groups.
b. Explain the purpose of including a control group in this experiment.

1.79

Suppose that the researchers who carried out the experiment described in the previous exercise thought that gender might be a potentially confounding variable. If 700 of the children participating in the experiment were girls and 600 were boys, describe how blocking could be incorporated into the experiment. Be specific about how you would assign the children to treatment groups.

1.80   (C6, C11)

The article **"Super Bowls: Serving Bowl Size and Food Consumption"** (*Journal of the American Medical Association* [2005]: 1727–1728) describes an experiment investigating how the size of serving bowls influences the amount a person eats. In this experiment, graduate students at a university were recruited to attend a Super Bowl party. The paper states that as the students arrived, they were "led in an alternating order to 1 of 2 identical buffet tables on opposite sides of an adjoining room. The tables had identical amounts of snacks, such as nuts, pretzels, and chips. All of the snacks contained approximately the same number of calories per gram. On one of the tables the snacks were set out in large serving bowls and on the second table the snacks were set out in smaller serving bowls. The students were given a plate and invited to serve themselves before going to another room to watch the game. When they arrived at the game room, their plates were weighed and the number of calories in the food on the plate was estimated." The researchers concluded that serving bowl size does make a difference, with those using large serving bowls tending to take more food.

a. Do you think that the alternate assignment to the experimental groups (large serving bowls, small serving bowls) based on arrival time is "close enough" to random assignment? That is, do you think it would tend to create comparable experimental groups?

b. In this study, the research assistant who weighed the plates and estimated the calorie content of the food on the plate was blinded as to which experimental group the plate belonged to and was also blinded as to the purpose of the experiment. Why do you think the researchers chose to incorporate this type of blinding?

1.81 (M2)

According to the article **"Rubbing Hands Together Under Warm Air Dryers Can Counteract Bacteria Reduction"** (*Infectious Disease News*, September 22, 2010), washing your hands isn't enough—good "hand hygiene" also includes drying hands thoroughly. The article described an experiment to compare bacteria reduction for three different hand-drying methods. In this experiment, subjects handled uncooked chicken for 45 seconds, then washed their hands with a single squirt of soap for 60 seconds, and then used one of the three hand-drying methods. The bacteria count on their hands was then measured. Suppose you want to carry out a similar experiment with 30 subjects who are willing to participate. Describe a method for randomly assigning each of the 30 subjects to one of the hand-drying methods.

1.82 (P3, P4)

Can moving their hands help children learn math? This is the question investigated by the authors of the paper **"Gesturing Gives Children New Ideas about Math"** (*Psychological Science* [2009]: 267–272). An experiment was conducted to compare two different methods for teaching children how to solve math problems of the form $3 + 2 + 8 = \underline{\hphantom{xx}} + 8$. One method involved having students point to the $3 + 2$ on the left side of the equal sign with one hand and then point to the blank on the right side of the equal sign before filling in the blank to complete the equation. The other method did not involve using these hand gestures. The paper states that the study used children ages 9 and 10 who were given a pretest containing six problems of the type described. Only children who answered all six questions incorrectly became subjects in the experiment. There were a total of 128 subjects. To compare the two methods, the 128 children were assigned at random to the two experimental conditions. Children in one group were taught a method that used hand gestures, and children in the other group were taught a similar strategy that did not involve hand gestures. Each

child then took a test with six problems and received a score based on the number correct. From the resulting data, the researchers concluded that the average score for children who used hand gestures was significantly higher than the average score for children who did not use hand gestures.

a. Answer the following seven questions for the experiment described above. (Hint: Reviewing Examples 1.5 and 1.6 might be helpful.)

1. What question is the experiment trying to answer?
2. What are the experimental conditions (treatments) for this experiment?
3. What is the response variable?
4. What are the experimental units and how were they selected?
5. Does the design incorporate random assignment of experimental units to the different experimental conditions? If not, are there potentially confounding variables that would make it difficult to draw conclusions based on data from the experiment?
6. Does the experiment incorporate a control group and/or a placebo group? If not, would the experiment be improved by including them?
7. Does the experiment involve blinding? If not, would the experiment be improved by making it single- or double-blind?

b. Based on the study design, do you think that the conclusions are reasonable?

1.83 (C2)

The article **"Display of Health Risk Behaviors on MySpace by Adolescents"** (*Archives of Pediatrics and Adolescent Medicine* [2009]: 27–34) described a study in which researchers looked at a random sample of 500 publicly accessible MySpace web profiles posted by 18-year-olds. The content of each profile was analyzed. One of the conclusions reported was that displaying sport or hobby involvement was associated with decreased references to risky behavior (sexual references or references to substance abuse or violence).

a. Is it reasonable to generalize the stated conclusion to all 18-year-olds with a publicly accessible MySpace web profile? What aspect of the study supports your answer?

b. Not all MySpace users have a publicly accessible profile. Is it reasonable to generalize the stated conclusion to all 18-year-old MySpace users? Explain.

c. Is it reasonable to generalize the stated conclusion to *all* MySpace users with a publicly accessible profile? Explain.

1.84 (C2, C5)

The authors of the paper **"Popular Video Games: Quantifying the Presentation of Violence and Its Context"** (*Journal of Broadcasting & Electronic Media* [2003]: 58–76) investigated the relationship between video game rating—suitable for everyone (E), suitable for 13 years of age and older (T), and suitable for 17 years of age and older (M)—and the number of violent interactions per minute of play. The sample consisted of 60 video games—the 20 most popular (by sales) for each of three game systems. The researchers concluded that video games rated for older children had significantly more violent interactions per minute than video games rated for more general audiences.

a. Do you think that the sample of 60 games was selected in a way that makes it representative of the population of all video games?

b. Is it reasonable to generalize the researchers' conclusion to all video games? Explain why or why not.

1.85 (C6)
To examine the effect of exercise on body composition, healthy women aged 35 to 50 were classified as either active (nine or more hours of physical activity per week) or sedentary (**"Effects of Habitual Physical Activity on the Resting Metabolic Rates and Body Composition of Women aged 35 to 50 Years," *Journal of the American Dietetic Association* [2001]: 1181–1191**). Body fat percentage was measured, and the researchers found that this percentage was significantly lower for women who were active than for sedentary women.

a. Is the study described an experiment? If so, what is the explanatory variable and what is the response variable? If not, explain why it is not an experiment.

b. From this study alone, is it reasonable to conclude that physical activity is the cause of the observed difference in body fat percentage? Justify your answer.

1.86 (C7, C9)
The article **"Rethinking Calcium Supplements" (*U.S. Airways Magazine*, October 2010)** describes a study investigating whether taking calcium supplements increases the risk of heart attack. Consider the following four study descriptions. For each study, answer the following five questions:

> Question 1: Is the study described an observational study or an experiment?
>
> Question 2: Did the study use random selection from some population?
>
> Question 3: Did the study use random assignment to experimental groups?
>
> Question 4: Based on the study description, would it be reasonable to conclude that taking calcium supplements is the cause of the increased risk of heart attack?

> Question 5: Would it be reasonable to generalize conclusions from this study to some larger population? If so, what population?

Study 1: Every heart attack patient and every patient admitted for an illness other than heart attack during the month of December, 2010, at a large urban hospital was asked if he or she took calcium supplements. The proportion of heart attack patients who took calcium supplements was significantly higher than the proportion of patients admitted for other illnesses who took calcium supplements.

Study 2: Two hundred people were randomly selected from a list of all people living in Minneapolis who receive Social Security. Each person in the sample was asked whether or not they took calcium supplements. These people were followed for five years, and whether or not they had had a heart attack during the five-year period was noted. The proportion of heart attack victims in the group taking calcium supplements was significantly higher than the proportion of heart attack victims in the group not taking calcium supplements.

Study 3: Two hundred people were randomly selected from a list of all people living in Minneapolis who receive Social Security. Each person was asked to participate in a statistical study, and all agreed to participate. Those who had no previous history of heart problems were instructed not to take calcium supplements. Those with a previous history of heart problems were instructed to take calcium supplements. The participants were followed for five years, and whether or not they had had a heart attack during the five-year period was noted. The proportion of heart attack victims in the calcium supplement group was significantly higher than the proportion of heart attack victims in the no calcium supplement group.

Study 4: Four hundred people volunteered to participate in a 10-year study. Each volunteer was assigned at random to either group 1 or group 2. Those in group 1 took a daily calcium supplement. Those in group 2 did not take a calcium supplement. The proportion who suffered a heart attack during the 10-year study period was noted for each group. The proportion of heart attack victims in group 1 was significantly higher than the proportion of heart attack victims in group 2.

AP* Review Questions for Chapter 1

1. The term used to describe the bias that occurs if some segment of a population is systematically excluded from a sample is

    (A) selection bias.
    (B) measurement bias.
    (C) response bias.
    (D) exclusion bias.
    (E) visibility bias.

2. In utilizing direct control, which of the following are held constant?

    (A) values of a potentially confounding variable
    (B) values of a blocking variable
    (C) values of a response variable
    (D) values of an explanatory variable
    (E) the experimental conditions

3. In a study to determine if the color of the label might affect whether potential customers would buy a particular brand of bottled tea, 100 volunteer subjects were recruited. They were shown two different label designs and asked which design they would be most likely to purchase. One label was blue with a photograph of a woman drinking the tea. The other label was green and did not include the photo. The 100 volunteers were divided at random into two groups. One group was shown the blue label first, followed by the green label. The other group was shown the green label first followed by the blue label. Which of the following is confounded with the treatments?

    (A) color of cover (blue or green)
    (B) cover composition (photo or no photo)
    (C) preferred label (which label the person was most likely to purchase)
    (D) group membership (blue first or green first)
    (E) There are no confounding variables.

4. In order to estimate the proportion of students at a college who spend more than 2 hours per day on Facebook, a random sample of students at the college is selected and each student is interviewed about his or her use of Facebook. The students conducting the survey are worried that people who spend a lot of time on Facebook might be embarrassed to admit it and that their responses to the survey might not be honest. What type of bias are the students conducting the survey worried about?

    (A) selection bias
    (B) nonresponse bias
    (C) measurement or response bias
    (D) bias due to confounding
    (E) They shouldn't worry—there is no obvious source of bias.

5. To estimate the proportion of students who plan to purchase tickets to an upcoming school fundraiser, a high school decides to sample 100 students as they register for the spring semester. There are 2000 students at the school. Which of the following sampling plans would result in a stratified random sample?

    (A) Number the students from 1 to 2000 and then use random numbers to select 100 students.
    (B) Survey the first 100 students to register.
    (C) Randomly select 100 students from a list of the 950 female students at the school.
    (D) Divide the students into early registrants (the first 1000 to register) and late registrants (the last 1000 to register). Use random numbers to identify 50 of the early registrants and 50 of the late registrants to survey.
    (E) Select one of the first 20 students to register using a random number table and then select every 20th student to register thereafter.

6. Which of the following describes a situation in which it is reasonable to reach a cause-and-effect conclusion based on data from a statistical study?

    (A) The study is based on a random sample from a population of interest.
    (B) The study is observational, and the sample used is not a convenience sample.
    (C) The study is an experiment that uses random assignment to assign volunteers to experimental conditions (treatments).
    (D) The study is observational, and the two samples used are not convenience samples.
    (E) It is always reasonable to reach a cause-and-effect conclusion based on data from a statistical study.

7. A reporter for a local newspaper wants to survey county residents about their opinions on a proposal to raise property taxes to benefit the county library. He decides to ask 30 county residents whether they support this tax increase. He will select his sample by asking every third person entering the library starting at noon on a Friday. He will continue until he has asked a total of 30 county residents. Which type of bias is likely to be introduced by the way the sample will be selected?

    (A) nonresponse bias
    (B) response bias
    (C) selection bias
    (D) measurement bias
    (E) No bias is introduced.

8. A random sample of 100 students at a particular college is to be selected, and each person selected will be asked how many times he or she went to a movie in a theater during the last year. The sample mean will then be used as an estimate of the mean number of times students at this college went to a movie in the last year. Which of the following is a reason to consider increasing the sample size for this study from 100 to 200?

(A) A larger sample size will reduce nonresponse bias.

(B) A larger sample size will reduce the response bias due to people not being able to accurately remember how many times they went to a movie in the last year.

(C) A larger sample size will reduce sampling variability—the differences in the sample mean that occur from sample to sample due to chance.

(D) It is less likely that one of the high values in the population (corresponding to a person who goes to a very large number of movies) will be included in a larger sample.

(E) All of the above are valid reasons to increase the sample size.

9. Researchers were interested in comparing students at private universities and students at public universities with respect to how much time they spent working in a typical week. They e-mailed a survey to 1,000 students enrolled at a particular private university and to 2,000 students enrolled at a particular public university. Data from the 400 private university students who responded to the survey and the 900 public university students who responded to the survey were used to determine that there was a significant difference in the mean number of hours worked in a typical week. Which of the following does *not* limit the researchers' ability to generalize to the two populations of interest?

(A) The two sample sizes are not equal.

(B) There may be bias introduced due to nonresponse.

(C) The samples were not randomly selected.

(D) All students in the private university sample attended the same university, and all students in the public university sample attended the same public university.

(E) All of the above limit the researchers' ability to generalize to the populations of interest.

Use the following to answer questions 10 and 11.

One hundred volunteer subjects participated in a study to determine if room temperature affects people's ability to concentrate. Female volunteers were given 10 minutes to try to memorize the words on a list of 50 nonsense words. The room temperature was controlled at 65 degrees (a cold room) while they completed the task. Male volunteers were also given 10 minutes to try to memorize the same list of words, but for the males, room temperature was controlled at 85 degrees (a hot room). At the end of the 10 minutes, each subject was asked to list as many of the words as he or she could remember, and the number correct was recorded. The resulting data were then used to determine if the mean number of words differed for the cold room and hot room conditions.

10. Which of the following is a confounding variable—that is, which of the following is confounded with the treatments?

(A) room temperature (cold or hot)

(B) gender (male or female)

(C) number of nonsense words recalled

(D) length of time given to memorize the words

(E) can't tell because volunteers were used

11. The poor design of this experiment results in a variable that is confounded with the treatments. Which of the following changes to the design would be effective in eliminating this confounding?

(A) Use only one room temperature.

(B) Use only male subjects in the study and assign the males to one of the two room-temperature conditions at random.

(C) Assign the volunteers at random to one of the two room-temperature conditions.

(D) Create two blocks by putting all of the females in one block and all of the males in the other block. Then, within each block, assign subjects at random to one of the room-temperature conditions.

(E) B, C, and D are all strategies that would be effective.

12. In a study of shopping preferences, a sample of young adults and a sample of people over age 50 were selected. Each person was asked where he or she preferred to shop for groceries. Based on the resulting data, the investigators concluded that a higher proportion of older adults prefer to shop at chain supermarkets. Which of the following is true for this study?

(A) The study is an experiment with two treatments, young and old.

(B) The study is an experiment with two treatments, prefer chain supermarkets and do not prefer chain supermarkets.

(C) The study is an experiment that incorporates blocking by age.

(D) Based on this study, it is reasonable to conclude that there is a cause-and-effect relationship between age and shopping preference.

(E) None of the above is true for this study.

Extreme values
  normal distribution, 386–387
  standard normal distribution, 380–381

F

Finite population correction factor, 484
Fit of least squares regression line, 211–229
Five-number summary, 151, 152, 154–155
Five-step process, 469–473
  for estimation problems, 424
  for hypothesis testing problems, 425
Four key questions, 418–422, 469–473, 749
Frequency, 62
Frequency distribution
  for categorical data, 62
  for continuous numerical data, 80–81
  for discrete numerical data, 76–80

G

Galton, Sir Francis, 206
Geometric probability distribution, 369–371
Geometric random variable, 370
Goodness-of-fit statistic, 697
Goodness-of-fit test, 698–705
Graphical display
  bivariate numerical data, 96–101
  categorical data, 62–68
  in the media, 101–107
  mistakes in, 107–110
  numerical data, 68–96
  purpose of, 59–60
  selection of, 58–62

H

Histogram, 76–87
  for comparing groups, 83–84
  of continuous numerical data, 80–81
  of discrete numerical data, 76–80
  mistakes in, 107–109
  shape of, 86–87
  technology and, 121–122
  unequal width intervals, 84–85
Homogeneity, testing for, 708–709, 711, 723
Hypergeometric probability distribution, 367–368
Hypotheses, 496–497, 531
Hypothesis test, 497
Hypothesis testing
  difference in population means, independent samples, 618–633
  difference in population means, paired samples, 633–644
  difference in population proportions, 514–526, 550–559
  errors in, 502–507, 531, 559
  five-step process for, 425, 511–514
  four key questions, 514–515
  independent samples, 618–633
  logic of, 507–511
  population mean, 591–604
  population proportion, 507–511, 550–559
  slope of the population regression line, 752–756
Hypothetical 1,000 table, 281–283

I

Independence, testing for, 708–709, 714–718, 723
Independent events, 285
  multiplication rule for, 285–287, 303
Independent samples, 542, 618
  hypothesis testing, 618–633
Influential observation, 215–216, 217
Intercept of a line, 198
Interpreting a confidence interval, 474, 483–484, 559, 654, 680
Interpreting confidence level, 474, 483, 682
Interpreting probabilities, 268–272
Interquartile range (iqr), 144–147
  technology and, 174
Intersection events, 279–280
  multiplication rule, 306–308

L

Law of Large Numbers, 271, 276
Law of Total Probability, 308–310
Least squares regression line, 199–205
  fit of, 211–229
  outliers, 217
  slope, 201
  standard deviation about, 220–225
Left skewed histogram, 86
Linear combination, 360–361
  mean and variance of, 358–360
Linear function, 360–361
  mean and variance of, 357–358
Linear regression, 198–211
  bivariate numerical data, 229–231
  simple linear regression model, 736–747
Linear relationship, 198–199
  correlation coefficient, 192–193
  least squares, 199–205
  and prediction, 229–231
  scatterplots, 184–187
Line
  equations of, 198
  and linear relationship, 198–199
Long-run proportion, 269
Lower quartile, 144
  and outliers, 154
Lower tail of histogram, 86
Lower-tailed test, 516, 594

M

Margin of error, 443–444
  and confidence intervals, 468–469
  estimating a population mean, 587–588
  estimating a population proportion, 460–467
  and sample size, 480–482, 587–588
  and standard error, 461
Marginal proportion, 710
Marginal total, 709
Mean
  approximately symmetric data distribution, 147
  binomial distribution, 369
  of continuous random variable, 356–357
  deviations from, 136–137
  of discrete random variable, 352–355
  of linear functions and combinations, 357–360

of normal probability distribution, 374–375
  of a random variable, 352–363
  of a sample, 134–135
  technology and, 170–171
Measurement bias, 19
Measures of center, 130
  mistakes in, 164
  skewed data distributions, 142–149
  symmetric data distributions, 133–142
Measures of relative standing, 160–164
Measures of spread, 130
  skewed data distributions, 142–149
  symmetric data distributions, 133–142
Median, 143
  technology and, 171–172
Model utility test for simple linear regression, 753–754
Mode, 86
Modified boxplot, 153–156
Multimodal histogram, 86
Multiplication rule, 285–287
  for independent events, 303
  intersection probabilities, 306–308
Multivariate data set, 58
Mutually exclusive events, 283–294
  addition rule for, 284, 303

N

Negative skewness, 87
Negatively skewed histogram, 86
Nonlinear function, 231–234
Nonlinear relationship, 185–186
  choosing among models for, 243–246
  modeling, 231–248
  quadratic regression model, 234–236
  technology and, 260
  using transformations, 236–243
Nonresponse bias, 19
Normal curve, 87, 382
  See also Normal probability distribution
Normal probability distribution, 349–350, 373–390
  binomial distribution, approximating, 396–398
  and the Central Limit Theorem, 572, 573
  checking for, 391–395
  and discrete distributions, approximating, 395–399
  extreme values, 386–387
  mean, 374–375
  other than standard normal, 381–385
  standard deviation, 374–375
  standard normal distribution, 375–380
  technology and, 403–404
Normal probability plot, 391
  technology and, 406
Normal score, 391
Null hypothesis, 497–498, 499–500
  and experiment data, 687
Numerical univariate data set, 58

O

Observational study, 9, 10
  collecting data, 13–24
Observed cell count, 708

Observed significance level, 512
1 in *k* systematic sample, 18
One-sample *t* test, 594–595
Outlier, 73, 153–156

P

Paired data, 71
Paired samples, 618
  t test, 635–636
Paired *t* tests, 635–639
Pearson's sample correlation coefficient.
  See Correlation coefficient *(r)*
Percentile, 161–162
Pie chart, 101–102, 103–104
  mistakes in, 107–108
Placebo, 34
Plotting residuals, 214–217
Pooled *t* test, 627
Population mean, 135
  confidence interval for, 578–591, 644–649
  independent samples, 618–633, 644–649
  margin of error, 587–588
  one-sample *t* test for, 594–595
  paired samples, 633–644, 649–651
  testing hypotheses, 591–604
  two-sample *t* test for difference in, 621
Population proportion, 433, 455
  confidence interval, 467–480
  estimating, 460–467
  hypothesis testing, 507–511, 526–530
  large-sample hypothesis test, 514–526
  margin of error, 463
Population proportions, difference in
  confidence interval for, 543–544, 546–548
  hypothesis testing, 550–559
  large-sample hypothesis test, 553
  test statistic, 551–552
Population regression line, 737
  estimating, 741–745
  slope of, 748–759
Population standard deviation, 138–139
Population variance, 138
Population, 9
Positive skewness, 87
Positively skewed histogram, 86
Power model, 240–241
Power of a test, 527–530, 601–602
Practical significance, 601
Predicted value, 211–213
Prediction, 229–231
Predictor variable, 198
Probabilistic model, 736
Probability
  of an event *E*, 276
  binomial, 367, 368
  calculating. *See* Calculating probabilities
  of complex event, 278–290
  conditional, 290–302
  and decision making, 314–317
  estimating, 271
  estimating empirically, 317–325
  estimating using simulation, 319–323
  histogram, 341, 345, 368
  interpreting, 268–272
  properties of, 271
  and relative frequency, 269–271, 276

Probability distribution
 for continuous random variable, 345–351
 for discrete random variable, 339–344
Probability histogram, 341
P-values, 512, 515–516, 552

Q

Quadratic regression model, 234–236
Qualitative univariate data set, 58
Quantitative univariate data set, 58
Quartile, 144
 technology and, 173–174

R

r. See Correlation coefficient *(r)*
Random assignment
 in experiments, 26–27, 28, 29–33, 35–37
Random deviation, 737–738
Random number table, 320
Random sample, 13–14
Random selection, 26–27, 41–45
Random variable, 336–339
 binomial, 364
 geometric, 370
 mean and standard deviation of, 352–363
Randomized block design, 28, 33
Randomized block experiment, 28
Range, 135–136
Regression, 205–206
 technology and, 257–258, 775–776
Regression analysis, 206, 229
Relative frequency, 62
 and bar charts, 65–66
 cumulative, 87–88
 and probabilities, 269–271, 276
Relative frequency distribution, 62
Replication, 27–29
 experimental units and, 35
Residual plot, 214–217, 762–763
 technology and, 259
Residual sum of squares, 218–219
Residuals, 211–213
 analysis of, 759–762
 plotting, 214–217, 762–767
 technology and, 258–259
Response bias, 19, 20, 21
Response variable, 24, 28, 198
Right skewed histogram, 86

S

Sample data
 and population characteristics, 483
 and statistical inference, 412–417
 from two or more populations, 413–414
 variability in, 415–416
Sample mean, 134–135
 sampling distribution of the, 568–578
Sample median, 143
Sample proportion, 433
 margin of error, 463
Sample size, 15–16
 and margin of error, 480–482, 587–588
 and the power of a test, 528
Sample space, 273
Sample standard deviation, 137

Sample variance, 137
Samples, 9
 independent, 542, 633
 paired, 633
 and statistical inference, 419
Sampling
 data collection by, 13–14
 reasonable strategies for, 16–18
 replacement, with and without, 15
 and volunteers, 35–36
Sampling distribution, 432, 435–436
 of the difference in sample means, 619
 of the difference in sample proportions, 542–543
 properties of, 441–442
 of the sample mean, 568–578
 of the sample proportion, 438–443
 of the slope of the population regression line, 748
Sampling frame, 14
Sampling variability, 430
 and statistics, 433–438
Scatterplot, 96–98
 bivariate numerical data, 182–187
 linear relationship, 184–187
 mistakes in, 108, 110
 nonlinear relationship, 185–186, 234
 patterns, common, 740–741
 technology and, 122
Segmented bar chart, 102–103
Selection bias, 19
Significance level, 504
 observed, 512
 and the power of the test, 528
Significant difference, 668
Simple comparative
 experiment, 24
 design strategies for, 25
Simple event, 273–274
Simple linear regression model, 736–747
 basic assumptions of, 738
 checking model adequacy, 759–773
Simple random sample, 13–14
 selecting, 14–15
Simulation, estimating probabilities with, 319–323
Single-blind experiment, 34–35
Skewed data distribution, 142–149
Skewed histogram, 86
Skewness, 86
Slope of a line, 198
 least squares regression line, 201
 population regression line, 748–759
Spread, measures of, 130
 skewed data distributions, 142–149
 symmetric data distributions, 133–142
Standard deviation, 137–139, 144
 about the least squares regression line, 220–225
 approximately symmetric data distributions, 147
 binomial distribution, 369
 of a random variable, 352–363
 of continuous random variables, 356–357
 of discrete random variable, 355–356
 of linear functions and combinations, 357–360

of normal probability distribution, 374–375
 technology and, 172–173
Standard error, 456
 and the margin of error, 461
Standard normal distribution, 375–380
 areas, 376–380
 extreme values, 380–381, 386–387
 probability distribution, 381–385
 and *t* distributions, 580–581
 table of areas, 376–377
Standardization, 160
 of normal curves, 382
Standardized residual plot, 762–763
Standardized residual, 760
Standardized score, 160
Statistical inference
 five-step process for, 424–425
 methods for, appropriate, 418–424
 risks of, 415
 and sample data, 412–417
Statistical studies, 9–12
 conclusions from, 42–43
 See also Experiments; Observational studies
Statistically significant, 601
Statistic, 13
 and sampling variability, 433–438
 standard error of, 456
Stem-and-leaf display, 72–76
 comparative, 75–76
 technology and, 120–121
Strata, 17
Stratified random sampling, 17
Stratified sampling, 14
Sum of squared deviations, 200
Symmetric data distribution, 133–142
Symmetric unimodal histogram, 86
Systematic sampling, 18

T

t confidence interval, 582, 644–645, 649
t critical value, 581
t distribution, 580–581
 and the standard normal distribution, 580
t test
 paired, 635–639
 pooled, 627
 population mean, 591–604
 for difference in population means, 625–626
 for difference in treatment means, 671–672
 P-values for, 593–595, 626
 for the slope of the population regression line, 752
 technology and, 610–611, 659–661, 662–663
Test statistic, 512, 515, 531
Time series plot, 98–99
 mistakes in, 108–109
 technology and, 122–123
Total sum of squares, 218–219
Transformations, 238
 models transforming *x*, 238–240
 models transforming *y*, 240–243
 nonlinear regression models, 236–238
 technology and, 260

Treatment effect
 differences in, 670–679
 differences in, estimating, 679–686
Treatment means
 conclusions about, 670
 confidence interval, 679–680
 t test for, 671–672
Treatment proportions
 confidence intervals, 682
 differences in, estimating, 682–684
 hypothesis testing, 674–676
Treatments
 in experiments, 24, 28, 416
 and statistical inference, 419
Two-sample *t* test, 621
Two-tailed test, 517, 594
Two-way frequency table, 708
Type I error, 502, 504–505
 probability of, 504
Type II error, 502, 505
 probability of, 504, 526–531

U

Unbiased estimator, 456
Unexplained variability, measure of, 219
Uniform distribution, 346
Unimodal histogram, 86
Union events, 280
 addition rule for, 303–304
Univariate data set, 58
Upper quartile, 144
 and outliers, 154
Upper tail of histogram, 86
Upper-tailed test, 516, 594

V

Variability
 blocking, 25
 in data, 415–416
 in experiment data, 668–670
Variability, measuring, 135–136
Variable, 58
Variance, 137–139
 of discrete random variable, 356
 of linear functions and combinations, 357–360
 technology and, 172
 See also Standard deviation
Vertical intercept, 198
Voluntary response sampling, 18–19
Volunteers, in experiments, 35–38

W

Weighted least squares regression, 762

Y

Y-intercept, 198

Z

z critical value, 473, 477, 543–544
z curve, 375
 and the *t* distribution, 580–581
 See also Standard normal distribution
z-score, 160–164
z-test
 large sample, 674–676
 technology and, 535–536, 563

Use the following to answer questions 13 and 14.

Fifty-two volunteers participated in an experiment to investigate the effect of noise on reaction time. The volunteers were divided into two groups based on age, with the 26 youngest in the "young" group and the 26 oldest in the "old" group. Each person in the young group was assigned at random to have his or her reaction time measured in a quiet room or in a noisy room. Each person in the old group was also assigned at random to the quiet room or the noisy room.

13. Which of the following are true for this experiment?

   I. This experiment has a control group.
   II. This experiment is double-blind.
   III. This experiment incorporates blocking.

(A) I only
(B) II only
(C) III only
(D) II and III only
(E) I, II, and III

14. Which of the following are the treatments in this experiment?

   (A) Age (young and old)
   (B) Room condition (quiet and noisy)
   (C) Reaction time
   (D) The two groups of volunteers
   (E) There are no treatments in this experiment.

2

Graphical Methods for Describing Data Distributions

Preview

Chapter Learning Objectives

2.1 Selecting an Appropriate Graphical Display

2.2 Displaying Categorical Data: Bar Charts and Comparative Bar Charts

2.3 Displaying Numerical Data: Dotplots, Stem-and-Leaf Displays, and Histograms

2.4 Displaying Bivariate Numerical Data: Scatterplots and Time Series Plots

2.5 Graphical Displays in the Media

2.6 Avoid These Common Mistakes

Chapter Activities

Exploring the Big Ideas

Are You Ready to Move On? Chapter 2 Review Exercises

Technology Notes

AP* Review Questions for Chapter 2

pollacky/Shutterstock.com

PREVIEW

When you carry out a statistical study, you hope to learn from the data you collect. But it is often difficult to "see" the information in data if it is presented as just a list of observations. An important step in the data analysis process involves summarizing the data graphically and numerically. This summarization makes it easier to see important characteristics of the data and is an effective way to communicate what you have learned.

CHAPTER LEARNING OBJECTIVES

Conceptual Understanding

After completing this chapter, you should be able to

C1 Distinguish between categorical and numerical data.

C2 Distinguish between discrete and continuous numerical data.

C3 Understand that selecting an appropriate graphical display depends on the number of variables in the data set, the data type, and the purpose of the graphical display.

Mastering the Mechanics

After completing this chapter, you should be able to

M1 Select an appropriate graphical display for a given data set.

M2 Construct and interpret bar charts and comparative bar charts.

M3 Construct and interpret dotplots and comparative dotplots.

M4 Construct and interpret stem-and-leaf and comparative stem-and-leaf displays.

M5 Construct and interpret histograms.

M6 Construct and interpret scatterplots.

M7 Construct and interpret time-series plots.

M8 Construct and interpret cumulative relative frequency plots.

Putting It into Practice

After completing this chapter, you should be able to

P1 Describe a numerical data distribution in terms of center, spread, shape, gaps, and outliers.

P2 Use graphical displays to compare groups on the basis of a categorical variable.

P3 Use graphical displays to compare groups on the basis of a numerical variable.

P4 Use a scatterplot to investigate the relationship between two numerical variables.

P5 Use a time-series plot to investigate trend over time for a numerical variable.

P6 Critically evaluate graphical displays that appear in newspapers, magazines, and advertisements.

PREVIEW EXAMPLE   Love Those New Cars?

Each year, the marketing and information firm J.D. Power and Associates surveys new car owners 90 days after they have purchased their cars. The data collected are used to rate auto brands (Toyota, Ford, and others) on initial quality and initial customer satisfaction. *USA Today* **(June 16, 2010 and July 17, 2010)** reported both the number of manufacturing defects per 100 vehicles and a satisfaction score for all 33 brands sold in the United States. The data on number of defects per 100 vehicles are shown here:

86	111	113	114	111	111	122	130	93	126	95	102	107
130	129	126	170	88	106	114	87	113	133	146	111	83
110	114	121	122	117	135	109

There are several things that you might learn from these data. What is a typical value for number of defects per 100 vehicles? Is there a lot of variability in number of defects for the different brands, or are most brands about the same? Are some brands much better or much worse than most? If you were to look at the defect data and the customer satisfaction data together, would you find a relationship between the number of manufacturing defects and customer satisfaction? Questions like these are more easily answered if the data can be organized in a sensible way.

In this chapter, you will learn how to summarize data graphically. The preview example will be revisited later in this chapter where you will see how various graphical displays can help answer the questions posed. ●

SECTION 2.1 Selecting an Appropriate Graphical Display

The first step in learning from a data set is usually to construct a graph that will enable you to see the important features of the data distribution. This involves selecting an appropriate graphical display. The choice of an appropriate graphical display will depend on three things:

- The number of variables in the data set.
- The data type.
- The purpose of the graphical display.

Let's consider each of these in more detail.

The Number of Variables in the Data Set

The individuals in any particular population typically possess many characteristics that might be of interest. Consider the group of students currently enrolled at a local college. One characteristic of the students in the population is political affiliation (Democrat, Republican, Independent, Green Party, other). Another characteristic is the number of textbooks purchased this semester, and yet another is the distance from home to the college. A **variable** is any characteristic whose value may change from one individual to another. For example, *political affiliation* is a variable, and so are *number of textbooks purchased* and *distance to the college*. **Data** result from making observations either on a single variable or simultaneously on two or more variables.

A **univariate data set** consists of observations on a single variable made on individuals in a sample or population. In some studies, however, two different characteristics are measured simultaneously. For example, both height (in inches) and weight (in pounds) might be recorded for each person. This results in a **bivariate data** set. Such a data set can be represented as ordered pairs of numbers, such as (68, 146). **Multivariate data** result when you obtain a category or value for each of two or more attributes (so bivariate data are a special case of multivariate data). For example, a multivariate data set might include height, weight, pulse rate, and systolic blood pressure for each person in a sample.

The Data Type

Variables in a statistical study can be classified as either categorical or numerical. For the population of students at your college, the variable *political affiliation* is categorical because each student's response to the question "What is your political affiliation?" is a category (Democrat, Republican, and so on). *Number of textbooks purchased* and *distance to the college* are both numerical variables. Sometimes you will see the terms qualitative and quantitative used in place of categorical and numerical.

AP* EXAM TIP

The selection of an appropriate data analysis method depends on whether the data are categorical or numerical. This is one of the first things you should think about when you encounter a new data set.

DEFINITION

A data set consisting of observations on a single variable is a **univariate data set.**

A univariate data set is **categorical** (or **qualitative**) if each observation is a categorical response.

A univariate data set is **numerical** (or **quantitative**) if each observation is a number.

Numerical variables can be further classified into two types—*discrete* and *continuous*—depending on their possible values. Suppose that the variable of interest is the number of courses a student is taking. If no student is taking more than eight courses, the possible values are 1, 2, 3, 4, 5, 6, 7, and 8. These values are identified in Figure 2.1(a) by the isolated dots on the number line. On the other hand, the line segment in Figure 2.1(b) identifies a plausible set of possible values for the time (in seconds) it takes for the first kernel in a bag of microwave popcorn to pop. Here the possible values make up an entire interval on the number line. No possible value is isolated from other possible values.

FIGURE 2.1
Possible values of a variable:
(a) number of courses
(b) time to pop

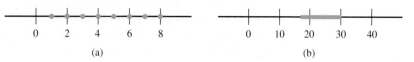

(a)   (b)

> **DEFINITION**
>
> A numerical variable is **discrete** if its possible values correspond to isolated points on the number line.
>
> A numerical variable is **continuous** if its possible values form an entire interval on the number line.

Example 2.1   Do U Txt?

The number of text messages sent on a particular day is recorded for each of 12 students. The resulting data set is

  23   0   14   13   15   0   60   82   0   40   41   22

Possible values for the variable *number of text messages sent* are 0, 1, 2, 3… . These are isolated points on the number line, so this is an example of discrete numerical data. Suppose that instead of the number of text messages sent, the *time spent texting* is recorded. This would result in continuous data. Even though time spent may be reported rounded to the nearest minute, the actual time spent could be 6 minutes, 6.2 minutes, 6.28 minutes, or any other value in an entire interval. ■

Discrete data usually arise when you count (for example, the number of roommates a student has or the number of petals on a certain type of flower). Usually, data are continuous when you measure. In practice, measuring instruments do not have infinite accuracy; so strictly speaking, the possible reported values do not form a continuum on the number line. For example, weight may be reported to the nearest pound. But even though reported weights are then whole numbers such as 140 and 141, an actual weight might be something between 140 and 141. The distinction between discrete and continuous data will be important when you select an appropriate graphical display and also later when you consider probability models.

The Purpose of the Graphical Display

One last thing to consider when selecting an appropriate graphical display is the purpose of the graphical display. For univariate data, the purpose is usually to show the data distribution. For a categorical variable, this means showing how the observations are distributed among the different possible categories that might be observed. For a numerical variable, showing the data distribution means displaying how the observations are distributed along a numerical scale.

Another common purpose for constructing a data display is so that two or more groups can be visually compared on the basis of some variable of interest. For example, you might want to compare response times for two different Internet service providers or to compare full-time and part-time students with respect to where they purchase textbooks (a categorical variable with categories campus bookstore, off-campus bookstore, or online bookseller). Some types of graphical displays are more informative than others when making comparisons.

For bivariate numerical data, the purpose of a constructing a graphical display is usually to see if there is a relationship between the two variables that define the data set. For example, if both a quality rating and a customer service rating are available for the 33 brands of cars sold in the United States (as described in the chapter preview), you might want to construct a display that would help explore the relationship between quality rating and customer satisfaction.

The later sections of this chapter introduce a number of different graphical displays. Section 2.2 covers graphical displays for univariate categorical data, Section 2.3 covers graphical displays for univariate numerical data, and Section 2.4 covers the most common graphical display for bivariate numerical data.

Figure 2.2 shows how knowing the number of variables in the data set, the data type, and the purpose of the graphical display lead to an appropriate choice for a graphical display.

Graphical Display	Number of Variables	Data Type	Purpose
Bar Chart	1	Categorical	Display Data Distribution
Comparative Bar Chart	1 for two or more groups	Categorical	Comparing 2 or More Groups
Dotplot	1	Numerical	Display Data Distribution
Comparative Dotplot	1 for two or more groups	Numerical	Comparing 2 or More Groups
Stem-and-Leaf Display	1	Numerical	Display Data Distribution
Comparative Stem-and-Leaf Display	1 for two groups	Numerical	Comparing 2 Groups
Histogram	1	Numerical	Display Data Distribution (Can also be used for comparing groups, if done carefully)
Cumulative Relative Frequency Plot	1	Numerical	Display Data Distribution (in a way that shows the approximate proportion of data at or below any given value)
Scatterplot	2	Numerical	Investigate Relationship Between 2 Numerical Variables
Time Series Plot	1, collected over time	Numerical	Investigate Trend Over Time

FIGURE 2.2
Choosing an Appropriate Graphical Display

SECTION 2.1 EXERCISES

Each Exercise Set assesses the following chapter learning objectives: C1, C2, C3, M1

SECTION 2.1 Exercise Set 1

2.1 Classify each of the following variables as either categorical or numerical. For those that are numerical, determine whether they are discrete or continuous.

a. Number of students in a class of 35 who turn in a term paper before the due date

b. Gender of the next baby born at a particular hospital

c. Amount of fluid (in ounces) dispensed by a machine used to fill bottles with soda pop

d. Thickness (in mm) of the gelatin coating of a vitamin E capsule

e. Birth order classification (only child, firstborn, middle child, lastborn) of a math major

2.2 For the following numerical variables, state whether each is discrete or continuous.

a. The number of insufficient-funds checks received by a grocery store during a given month

b. The amount by which a 1-pound package of ground beef decreases in weight (because of moisture loss) before purchase

c. The number of New York Yankees during a given year who will not play for the Yankees the next year

d. The number of students in a class of 35 who have purchased a used copy of the textbook

2.3 For each of the five data sets described, answer the following three questions and then use Figure 2.2 (on this page) to select an appropriate graphical display.

Question 1: How many variables are in the data set?

Question 2: Is the data set categorical or numerical?

Question 3: Would the purpose of a graphical display be to summarize the data distribution, to compare groups, or to investigate the relationship between two numerical variables?

Data Set 1: To learn about the reason parents believe their child is heavier than the recommended weight for children of the same age, each person in a sample of parents of overweight children was asked what they thought was the most important contributing factor. Possible responses were lack of exercise, easy access to junk food, unhealthy diet, medical condition, and other.

Data Set 2: To compare commute distances for full-time and part-time students at a large college, commute distance was determined for each student in a random sample of 50 full-time students and for each student in a random sample of 50 part-time students.

Data Set 3: To learn about how number of years of education and income are related, each person in a random sample of 500 residents of a particular city was asked how many years of education he or she had completed and what his or her annual income was.

Data Set 4: To see if there is a difference between faculty and students at a particular college with respect to how they commute to campus (drive, walk, bike, and so on), each person in a random sample of 50 faculty members and each person in a random sample of 100 students was asked how he or she usually commutes to campus.

Data Set 5: To learn about how much money students at a particular college spend on textbooks, each student in a random sample of 200 students was asked how much he or she spent on textbooks for the current semester.

SECTION 2.1 Exercise Set 2

2.4 Classify each of the following variables as either categorical or numerical. For those that are numerical, determine whether they are discrete or continuous.
a. Brand of computer purchased by a customer
b. State of birth for someone born in the United States
c. Price of a textbook
d. Concentration of a contaminant (micrograms per cubic centimeter) in a water sample
e. Zip code (Think carefully about this one.)
f. Actual weight of coffee in a 1-pound can

2.5 For the following numerical variables, state whether each is discrete or continuous.
a. The length of a 1-year-old rattlesnake
b. The altitude of a location in California selected by throwing a dart at a map of the state
c. The distance from the left edge at which a 12-inch plastic ruler snaps when bent far enough to break
d. The price per gallon paid by the next customer to buy gas at a particular station

2.6 For each of the five data sets described, answer the following three questions and then use Figure 2.2 (on page 60) to select an appropriate graphical display.

Question 1: How many variables are in the data set?
Question 2: Is the data set categorical or numerical?
Question 3: Would the purpose of a graphical display be to summarize the data distribution, to compare groups, or to investigate the relationship between two numerical variables?

Data Set 1: To learn about the heights of five-year-old children, the height of each child in a sample of 40 five-year-old children was measured.

Data Set 2: To see if there is a difference in car color preferences of men and women, each person in a sample of 100 males and each person in a sample of 100 females was shown pictures of a new model car in five different colors and asked to select which color they would choose if they were to purchase the car.

Data Set 3: To learn how GPA at the end of the freshman year in college is related to high school GPA, both high school GPA and freshman year GPA were determined for each student in a sample of 100 students who had just completed their freshman year at a particular college.

Data Set 4: To learn how the amount of money spent on a fast-food meal might differ for men and women, the amount spent on lunch at a particular fast-food restaurant was determined for each person in a sample of 50 women and each person in a sample of 50 men.

Data Set 5: To learn about political affiliation (Democrat, Republican, Independent, and so on) of students at a particular college, each student in a random sample of 200 students was asked to indicate his or her political affiliation.

Additional Exercises

2.7 Classify each of the following variables as either categorical or numerical.
a. Weight (in ounces) of a bag of potato chips
b. Number of items purchased by a grocery store customer
c. Brand of cola purchased by a convenience store customer
d. Amount of gas (in gallons) purchased by a gas station customer
e. Type of gas (regular, premium, diesel) purchased by a gas station customer

2.8 For the *numerical* variables in the previous exercise, which are discrete and which are continuous?

2.9 Classify each of the following variables as either categorical or numerical.
a. Color of an M&M candy selected at random from a bag of M&M's
b. Number of green M&M's in a bag of M&M's
c. Weight (in grams) of a bag of M&M's
d. Gender of the next person to purchase a bag of M&M's at a particular grocery store

2.10 For the *numerical* variables in the previous exercise, which are discrete and which are continuous?

2.11 Classify each of the following variables as either categorical or numerical.
a. Number of text messages sent by a college student in a typical day
b. Amount of time a high school senior spends playing computer or video games in a typical day
c. Number of people living in a house
d. A student's type of residence (dorm, apartment, house)
e. Dominant color on the cover of a book
f. Number of pages in a book
g. Rating (G, PG, R, X) of a movie

Use the following instructions for Exercises 2.12–2.15.

Answer the following three questions and then use Figure 2.2 (on page 60) to select an appropriate graphical display.

> Question 1: How many variables are in the data set?
> Question 2: Is the data set categorical or numerical?
> Question 3: Would the purpose of a graphical display be to summarize the data distribution, to compare groups, or to investigate the relationship between two numerical variables?

2.12 To learn about what super power middle school students would most like to have, each person in a sample of middle school students was asked to choose among invisibility, extreme strength, the ability to freeze time, and the ability to fly.

2.13 To compare the number of hours spent studying in a typical week for male and female students, data were collected from each person in a random sample of 50 female students and each person in a random sample of 50 male students.

2.14 To learn if there is a relationship between water consumption and headache frequency, people in a sample of young adults were asked how much water they drink in a typical day and how many days per month they experience a headache.

2.15 To learn about TV viewing habits of high school students, each person in a sample of students was asked how many hours he or she spent watching TV during the previous week.

SECTION 2.2 Displaying Categorical Data: Bar Charts and Comparative Bar Charts

In this section, you will see how bar charts and comparative bar charts can be used to summarize univariate categorical data. A bar chart is used when the purpose of the display is to show the data distribution, and a comparative bar chart (as the name implies) is used when the purpose of the display is to compare two or more groups.

Bar Charts

The first step in constructing a graphical display is often to summarize the data in a table and then use information in the table to construct the display. For categorical data, this table is called a *frequency distribution*.

> A **frequency distribution for categorical data** is a table that displays the possible categories along with the associated frequencies and/or relative frequencies.
>
> The **frequency** for a particular category is the number of times that category appears in the data set.
>
> The **relative frequency** for a particular category is calculated as
>
> $$\text{relative frequency} = \frac{\text{frequency}}{\text{number of observations in the data set}}$$
>
> The relative frequency for a particular category is the proportion of the observations that belong to that category. If a table includes relative frequencies, it is sometimes referred to as a **relative frequency distribution**.

Example 2.2 Motorcycle Helmets—Can You See Those Ears?

The U.S. Department of Transportation established standards for motorcycle helmets. To comply with these standards, helmets should reach the bottom of the motorcyclist's ears. The report **"Motorcycle Helmet Use in 2005—Overall Results" (National Highway Traffic Safety Administration, August 2005)** summarized data collected by observing 1,700 motorcyclists nationwide at selected roadway locations. Each time a motorcyclist passed by, the observer noted whether the rider was wearing no helmet, a noncompliant helmet, or a compliant helmet. Using the coding

> NH = noncompliant helmet
> CH = compliant helmet
> N = no helmet

a few of the observations were

| CH | N | CH | NH | N | CH | CH | CH | N | N |

There were 1,690 additional observations, which we didn't reproduce here! In total, there were 731 riders who wore no helmet, 153 who wore a noncompliant helmet, and 816 who wore a compliant helmet.

The corresponding relative frequency distribution is given in Table 2.1.

TABLE 2.1 Relative Frequency Distribution for Helmet Use

Helmet Use Category	Frequency	Relative Frequency
No helmet	731	0.430 ← 731/1700
Noncompliant helmet	153	0.090 ← 153/1700
Compliant helmet	816	0.480
	1,700	1.000

Total number of observations

Should total 1, but in some cases may be slightly off due to rounding

From the relative frequency distribution, you can see that a large number of the riders (43%) were not wearing a helmet, but most of those who wore a helmet were wearing one that met the Department of Transportation safety standard.

A **bar chart** is a graphical display of categorical data. Each category in the frequency distribution is represented by a bar or rectangle, and the display is constructed so that the *area* of each bar is proportional to the corresponding frequency or relative frequency.

Bar Charts

When to Use | Number of variables: 1
Data type: categorical
Purpose: displaying data distribution

How to Construct
1. Draw a horizontal axis, and write the category names or labels below the line at regularly spaced intervals.
2. Draw a vertical axis, and label the scale using either frequency or relative frequency.
3. Place a rectangular bar above each category label. The height is determined by the category's frequency or relative frequency, and all bars should have the same width. With the same width, both the height and the area of the bar are proportional to the frequency or relative frequency of the corresponding category.

What to Look For
• Which categories occur frequently, and which categories occur infrequently.

Example 2.3 Revisiting Motorcycle Helmets

Example 2.2 used data on helmet use from a sample of 1,700 motorcyclists to construct the frequency distribution in Table 2.1. A bar chart is an appropriate choice for displaying these data because

• There is one variable (helmet use category).
• The variable is categorical.
• The purpose is to display the data distribution.

You can follow the three steps in the previous box to construct the bar chart.

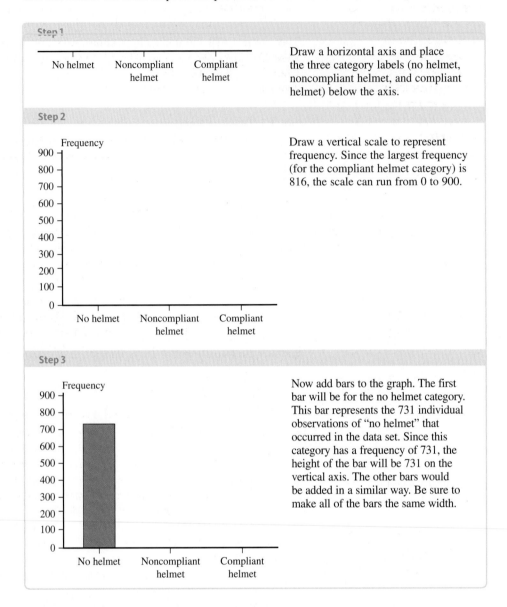

Step 1	
No helmet Noncompliant helmet Compliant helmet	Draw a horizontal axis and place the three category labels (no helmet, noncompliant helmet, and compliant helmet) below the axis.

Step 2	
	Draw a vertical scale to represent frequency. Since the largest frequency (for the compliant helmet category) is 816, the scale can run from 0 to 900.

Step 3	
	Now add bars to the graph. The first bar will be for the no helmet category. This bar represents the 731 individual observations of "no helmet" that occurred in the data set. Since this category has a frequency of 731, the height of the bar will be 731 on the vertical axis. The other bars would be added in a similar way. Be sure to make all of the bars the same width.

The completed bar chart is shown in Figure 2.3.

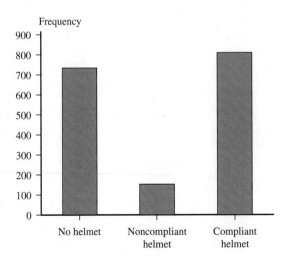

FIGURE 2.3
Bar chart of helmet use

The bar chart provides a visual representation of the distribution of the 1,700 values that make up the data set. From the bar chart, it is easy to see that the compliant helmet category occurred most often in the data set. The bar for compliant helmets is about five times as tall (and therefore has five times the area) as the bar for noncompliant helmets because approximately five times as many motorcyclists wore compliant helmets than wore noncompliant helmets.

Comparative Bar Charts

Bar charts can also be used to provide a visual comparison of two or more groups. This is accomplished by constructing two or more bar charts that use the same set of horizontal and vertical axes.

Comparative Bar Chart

When to Use   Number of variables: 1 variable with observations for two or more groups
Data type: categorical
Purpose: comparing two or more data distributions

A **comparative bar chart** is constructed by using the same horizontal and vertical axes for the bar charts of two or more groups. The bar charts are usually color-coded to indicate which bars correspond to each group. *For comparative bar charts, relative frequency should be used for the vertical axis.*

Example 2.4   How Far Is Far Enough?

Each year, *The Princeton Review* conducts surveys of high school students who are applying to college and of parents of college applicants. The report **"2009 College Hopes & Worries Survey Findings" (www.princetonreview/college-hopes-worries-2009)** included a summary of how 12,715 high school students responded to the question "Ideally how far from home would you like the college you attend to be?" Students responded by choosing one of four possible distance categories. Also included was a summary of how 3,007 parents of students applying to college responded to the question "How far from home would you like the college your child attends to be?" The accompanying relative frequency table summarizes the student and parent responses.

Ideal Distance	Frequency		Relative Frequency	
	Students	Parents	Students	Parents
Less than 250 miles	4450	1594	0.35	0.53
250 to 500 miles	3942	902	0.31	0.30
500 to 1000 miles	2416	331	0.19	0.11
More than 1000 miles	1907	180	0.15	0.06

When constructing a comparative bar graph, you use the relative frequency rather than the frequency to construct the scale on the vertical axis, so that you can make meaningful comparisons even if the sample sizes are not the same.

The same set of steps that were used to construct a bar chart are used to construct a comparative bar chart, but in a comparative bar chart each category will have a bar for each group.

Step 1

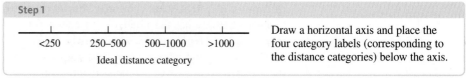

Draw a horizontal axis and place the four category labels (corresponding to the distance categories) below the axis.

(continued)

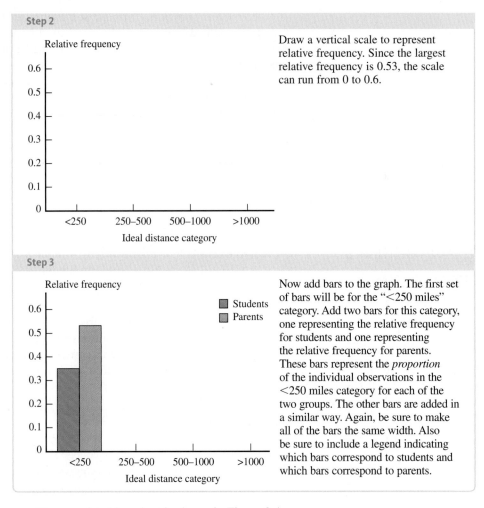

The completed bar chart is shown in Figure 2.4.

AP* EXAM TIP

When making comparative graphical displays, be sure to include a legend or labels that clearly identify the groups being compared.

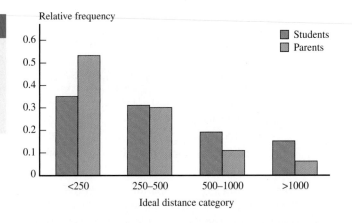

FIGURE 2.4
Comparative bar chart for ideal distance from home

It is easy to see the differences between students and parents. A higher proportion of parents than students prefer a college close to home, and a higher proportion of students than parents believe that the ideal distance from home is more than 500 miles.

To see why it is important to use relative frequencies rather than frequencies to compare groups of different sizes, consider the *incorrect* bar chart constructed using the frequencies rather than the relative frequencies (Figure 2.5). Because there were so many more students than parents who participated in the surveys (12,715 students and only 3,007 parents), the incorrect bar chart conveys a very different and misleading impression of the differences between students and parents.

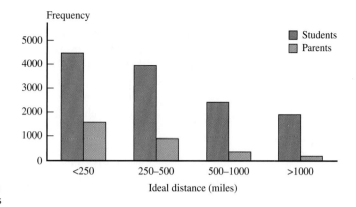

FIGURE 2.5
An INCORRECT comparative bar chart using frequency rather than relative frequency for vertical axis

---

SECTION 2.2 EXERCISES

Each Exercise Set assesses the following chapter learning objectives: M2, P2

SECTION 2.2 Exercise Set 1

2.16 The report **"Findings from the 2008 Administration of the College Senior Survey" (Higher Education Research Institute, UCLA, June 2009)** gave the following relative frequency distribution summarizing student responses to the question "If you could make your college choice over, would you still choose to enroll at your current college?"

Response	Relative Frequency
Definitely yes	0.447
Probably yes	0.373
Probably no	0.134
Definitely no	0.046

a. Use this information to construct a bar chart for the response data.
b. If you were going to use the response data and the bar chart from Part (a) as the basis for an article for your student paper, what would be a good headline for your article?

2.17 The article **"The Need to Be Plugged In" (Associated Press, December 22, 2005)** described a survey of 1,006 adults who were asked about various technologies, including personal computers, cell phones, and DVD players. The accompanying table summarizes the responses to questions about how essential these technologies were.

	Relative Frequency		
Response	Personal Computer	Cell Phone	DVD Player
Cannot imagine living without	0.46	0.41	0.19
Would miss but could do without	0.28	0.25	0.35
Could definitely live without	0.26	0.34	0.46

Construct a comparative bar chart that shows the distributions of responses for the three different technologies.

SECTION 2.2 Exercise Set 2

2.18 The report **"Trends in Education 2010: Community Colleges" (www.collegeboard.com/trends)** included the accompanying information on student debt for students graduating with an AA degree from a public community college in 2008.

Debt	Relative Frequency
None	0.62
Less than $10,000	0.23
Between $10,000 and $20,000	0.10
More than $20,000	0.05

a. Use the given information to construct a bar chart.
b. Write a few sentences commenting on student debt for public community college graduates.

2.19 The article **"Most Smokers Wish They Could Quit" (Gallup Poll Analyses, November 21, 2002)** noted that smokers and nonsmokers perceive the risks of smoking differently. The accompanying relative frequency table summarizes responses regarding the perceived harm of smoking for each of three groups: a sample of 241 smokers, a sample of 261 former smokers, and a sample of 502 nonsmokers. Construct a comparative bar chart for these data. Because the three sample sizes are different, don't forget to use relative frequencies in constructing the bar chart. Comment on how smokers, former smokers, and nonsmokers differ with respect to perceived risk of smoking.

	Frequency		
Perceived Risk of Smoking	Smokers	Former Smokers	Nonsmokers
Very harmful	145	204	432
Somewhat harmful	72	42	50
Not too harmful	17	10	15
Not at all harmful	7	5	5

Additional Exercises

2.20 Heal the Bay is an environmental organization that releases an annual beach report card based on water quality **(Heal the Bay Beach Report Card, May 2009)**. The 2009 ratings for 14 beaches in San Francisco County during wet weather were:

A+ C B A A+ A+ A A+ B D C D F F

a. Summarize the wet weather ratings by constructing a relative frequency distribution and a bar chart.

b. The dry weather ratings for these same beaches were:

A B B A+ A F A A A A A B A

Construct a bar chart for the dry weather ratings.

c. Do the bar charts from Parts (a) and (b) support the statement that beach water quality tends to be better in dry weather conditions? Explain.

2.21 The article **"Fraud, Identity Theft Afflict Consumers"** **(San Luis Obispo Tribune, February 2, 2005)** included the accompanying breakdown of identity theft complaints by type.

Type of Complaint	Percent of All Complaints
Credit card fraud	28%
Phone or utilities fraud	19%
Bank fraud	18%
Employment fraud	13%
Other	22%

Construct a bar chart for these data and write a sentence or two commenting on the most common types of identity theft complaints.

2.22 The article **"Americans Drowsy on the Job and the Road"** **(Associated Press, March 28, 2001)** summarized data from the 2001 Sleep in America poll. Each individual in a sample of 1,004 adults was asked questions about his or her sleep habits. The article states that "40 percent of those surveyed say they get sleepy on the job and their work suffers at least a few days each month, while 22 percent said the problems occur a few days each week. And 7 percent say sleepiness on the job is a daily occurrence." Assuming that everyone else reported that sleepiness on the job was not a problem, summarize the given information by constructing a relative frequency bar chart.

2.23 The article **"Ozzie and Harriet Don't Live Here Anymore"** **(San Luis Obispo Tribune, February 26, 2002)** looked at the changing makeup of America's suburbs. The article states that in the suburbs of the nation's largest cities, non-family households (for example, homes headed by a single professional or an elderly widow) now outnumber married couples with children. The article goes on to state:

> In the nation's 102 largest metropolitan areas, "non-families" comprised 29 percent of households in 2000, up from 27 percent in 1990. While the number of married-with-children homes grew too, the share did not keep pace. It declined from 28 percent to 27 percent. Married couples without children at home live in another 29 percent of suburban households. The remaining 15 percent are single-parent homes.

Use the given information on type of household in the year 2000 to construct a frequency distribution and a bar chart. (Make sure to only extract the year 2000 percentages from the given information.)

SECTION 2.3 Displaying Numerical Data: Dotplots, Stem-and-Leaf Displays, and Histograms

In this section, you will see three different types of graphical displays for univariate numerical data—dotplots, stem-and-leaf displays, and histograms.

Dotplots and Comparative Dotplots

A dotplot is a simple way to display numerical data when the data set is not too large. Each observation is represented by a dot above the location corresponding to its value on a number line. When a value occurs more than once in a data set, there is a dot for each occurrence, and these dots are stacked vertically in the plot.

> **Dotplots**
>
> **When to Use**   Number of variables: 1
> Data Type: numerical
> Purpose: display data distribution
>
> **How to Construct**
> 1. Draw a horizontal line and mark it with an appropriate measurement scale.
> 2. Locate each value in the data set along the measurement scale, and represent it by a dot. If there are two or more observations with the same value, stack the dots vertically.

(continued)

What to Look For   Dotplots convey information about
- A representative or typical value in the data set.
- The extent to which the data values spread out.
- The nature of the distribution of values along the number line.
- The presence of unusual values in the data set.

Example 2.5   Making it to Graduation

The article **"Keeping Score When It Counts: Graduation Rates and Academic Progress Rates for 2009 NCAA Division I Basketball Tournament Teams" (The Institute for Diversity and Ethics in Sport, University of Central Florida, March 2009)** included data on graduation rates of basketball players for the universities and colleges that sent teams to the 2009 Division I playoffs. The following graduation rates are the percentages of basketball players starting college in 2002 who had graduated by the end of 2008. (Note: Teams from 65 schools made it to the playoffs, but two of them—Cornell and North Dakota State—did not report graduation rates.)

Graduation Rates												
63	56	31	20	38	100	70	91	92	30	8	34	29
71	33	89	89	60	100	67	80	64	40	42	100	10
55	46	60	36	53	36	57	45	86	67	53	55	92
69	17	77	80	100	86	37	42	50	57	38	31	47
46	67	100	89	53	100	50	41	100	86	82		

One way to summarize these data graphically is to use a dotplot. A dotplot is an appropriate choice because the data set consists of data on one variable (graduation rate), the variable is numerical, and the purpose is to display the data distribution. The data set is not too large, with 63 observations. (If the data set had been much larger, a histogram, a graphical display to be introduced later in this section, might have been a better choice.)

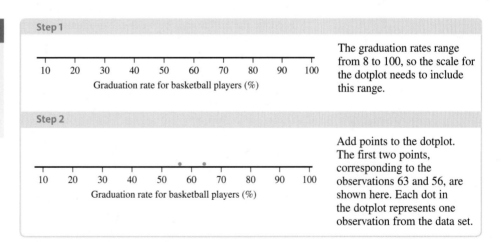

Step 1

Graduation rate for basketball players (%)

The graduation rates range from 8 to 100, so the scale for the dotplot needs to include this range.

Step 2

Graduation rate for basketball players (%)

Add points to the dotplot. The first two points, corresponding to the observations 63 and 56, are shown here. Each dot in the dotplot represents one observation from the data set.

The completed dotplot is shown in Figure 2.6. Ovals have been added to highlight one of the interesting features of the plot.

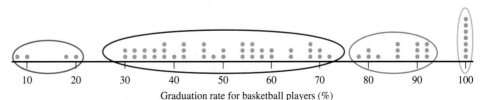

FIGURE 2.6
Dotplot of graduation rates

Graduation rate for basketball players (%)

The dotplot shows how the 63 graduation rates are distributed along the number line. You can see that basketball graduation rates vary a great deal from school to school, ranging from a low of 8% to a high of 100%. You can also see that the graduation rates seem to cluster in several groups, denoted by the colored ovals that have been added to the dotplot. There are several schools with graduation rates of 100% (excellent!) and another group of 13 schools with graduation rates that are higher than most. The majority of schools are in the large cluster, with graduation rates from about 30% to about 72%. And then there is that bottom group of four schools with embarrassingly low graduation rates for basketball players: Northridge (8%), Maryland (10%), Portland State (17%), and Arizona (20%).

Most statistical software packages and some graphing calculators can be used to construct dotplots, so you may not need to construct dotplots by hand.

Comparative Dotplots

Dotplots can also be used for comparative displays. A comparative dotplot is constructed by using the same numerical scale for two or more dotplots.

Comparative Dotplots

When to Use   Number of variables: 1 variable with observations for two or more groups
Data type: Numerical
Purpose: Comparing two or more data distributions

A comparative dotplot is constructed using the same numerical scale for two or more dotplots. Be sure to include group labels for the dotplots in the display.

Example 2.6   Graduation Rates Revisited

The article referenced in Example 2.5 also gave graduation rates for *all* student athletes at the 63 schools in the 2009 Division I basketball playoffs. The data are listed below. Also listed are the differences between the graduation rate for all student athletes and the graduation rate for basketball players.

	Graduation Rates (%)				Graduation Rates (%)		
School	Basketball	Athletes	(All–BB)	School	Basketball	Athletes	(All–BB)
1	63	75	12	17	89	97	8
2	56	57	1	18	60	67	7
3	31	86	55	19	100	80	−20
4	20	64	44	20	67	89	22
5	38	69	31	21	80	86	6
6	100	85	−15	22	64	70	6
7	70	96	26	23	40	69	29
8	91	79	−12	24	42	75	33
9	92	89	−3	25	100	94	−6
10	30	76	46	26	10	79	69
11	8	56	48	27	55	72	17
12	34	53	19	28	46	83	37
13	29	82	53	29	60	79	19
14	71	83	12	30	36	72	36
15	33	81	48	31	53	78	25
16	89	96	7	32	36	71	35

(continued)

	Graduation Rates (%)				Graduation Rates (%)		
School	Basketball	Athletes	(All–BB)	School	Basketball	Athletes	(All–BB)
33	57	70	13	49	57	71	14
34	45	57	12	50	38	78	40
35	86	85	−1	51	31	72	41
36	67	81	14	52	47	72	25
37	53	78	25	53	46	79	33
38	55	69	14	54	67	75	8
39	92	75	−17	55	100	82	−18
40	69	84	15	56	89	95	6
41	17	48	31	57	53	71	18
42	77	79	2	58	100	92	−8
43	80	91	11	59	50	83	33
44	100	95	−5	60	41	68	27
45	86	94	8	61	100	80	−20
46	37	69	32	62	86	79	−7
47	42	63	21	63	82	92	10
48	50	83	33				

How do the graduation rates of basketball players compare to those of all student athletes? Figure 2.7 shows a comparative dotplot. Notice that the comparative dotplot actually consists of two labeled dotplots that use the same numerical scale.

There are some striking differences that are easy to see when the data are displayed in this way. The graduation rates for all student athletes tend to be higher and to vary less from school to school than the graduation rates for just basketball players.

The dotplots in Figure 2.7 are informative, but we can do even better. The data given here are *paired data*. Each basketball graduation rate can be paired with the graduation rate for all student athletes from the same school. When data are paired, it is usually more informative to look at the differences. These differences (all−basketball) are also given in the data table. Figure 2.8 gives a dotplot of the 63 differences. Notice that one difference is 0. This corresponds to a school where the basketball graduation rate equals the graduation rate of all student athletes. There are 11 schools for which the difference is negative. Negative differences correspond to schools that have a higher graduation rate for basketball players than for all student athletes. The most interesting features of the difference dotplot are the very large number of positive

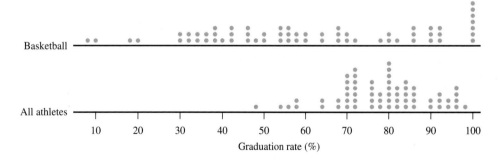

FIGURE 2.7
Comparative dotplot of graduation rates for basketball players and for all athletes

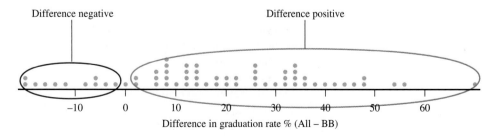

FIGURE 2.8
Dotplot of graduation rate differences (all–BB)

differences and the wide spread. Positive differences correspond to schools that have a lower graduation rate for basketball players. There is a lot of variability in the graduation rate difference from school to school, and three schools have differences that are noticeably higher than the rest. (In case you were wondering, these schools were Clemson with a difference of 53%, American University with a difference of 55%, and Maryland with a difference of 69%.)

Stem-and-Leaf Displays

A stem-and-leaf display is an effective way to summarize univariate numerical data when the data set is not too large. Each number in the data set is broken into two pieces, a stem and a leaf. The **stem** is the first part of the number and consists of the beginning digit(s). The **leaf** is the last part of the number and consists of the final digit(s). For example, the number 213 might be split into a stem of 2 and a leaf of 13 or a stem of 21 and a leaf of 3. The resulting stems and leaves are then used to construct the display.

Example 2.7   Should Doctors Get Auto Insurance Discounts?

Many auto insurance companies give job-related discounts of 5 to 15%. The article **"Auto-Rate Discounts Seem to Defy Data"** *(San Luis Obispo Tribune, June 19, 2004)* included the accompanying data on the number of automobile accidents per year for every 1,000 people in 40 occupations.

Occupation	Accidents Per 1,000	Occupation	Accidents per 1,000
Student	152	Banking-finance	89
Physician	109	Customer service	88
Lawyer	106	Manager	88
Architect	105	Medical support	87
Real estate broker	102	Computer-related	87
Enlisted military	99	Dentist	86
Social worker	98	Pharmacist	85
Manual laborer	96	Proprietor	84
Analyst	95	Teacher, professor	84
Engineer	94	Accountant	84
Consultant	94	Law Enforcement	79
Sales	93	Physical therapist	78
Military Officer	91	Veterinarian	78
Nurse	90	Clerical, secretary	77
School administrator	90	Clergy	76
Skilled labor	90	Homemaker	76
Librarian	90	Politician	76
Creative arts	90	Pilot	75
Executive	89	Firefighter	67
Insurance agent	89	Farmer	43

Figure 2.9 shows a stem-and-leaf display for the accident rate data. The numbers in the vertical column on the left of the display are the **stems**. Each number to the right of the vertical line is a **leaf** corresponding to one of the observations in the data set. The legend

    Stem: Tens
    Leaf: Ones

tells you that the observation that had a stem of 4 and a leaf of 3 corresponds to the occupation with an accident rate of 43 per 1,000. Similarly, the observation with the stem of 10 and leaf of 2 corresponds to 102 accidents per 1,000.

```
 4 | 3
 5 |
 6 | 7
 7 | 56667889
 8 | 44567788999
 9 | 000013445689
10 | 2569
11 |
12 |
13 |
14 |                Stem:  Tens
15 | 2              Leaf:  Ones
```

FIGURE 2.9

Stem-and-leaf display for accident rate per 1,000 for forty occupations

The display in Figure 2.9 suggests that a typical or representative value is in the stem 8 or 9 row, perhaps around 90. The observations are mostly concentrated in the 75 to 109 range, but there are a couple of values that stand out on the low end (43 and 67) and one observation (152) that is far removed from the rest of the data on the high end.

From the point of view of an auto insurance company, it might make sense to offer discounts to occupations with low accident rates—maybe farmers (43 accidents per 1,000) or firefighters (67 accidents per 1,000) or even some of the occupations with accident rates in the 70s. The "discounts seem to defy data" in the title of the article refers to the fact that some insurers provide discounts to doctors and engineers but not to homemakers, politicians, and other occupations with lower accident rates. Two possible explanations were offered for this apparent discrepancy. First, some occupations with higher accident rates could also have lower average costs per claim. Accident rates alone may not reflect the actual cost to the insurance company. Second, insurance companies may offer the discounted auto insurance in order to attract people who would also purchase other types of insurance, such as malpractice or liability insurance.

The leaves on each line of the display in Figure 2.9 have been arranged in order from smallest to largest. Most statistical software packages order the leaves this way, but even if this were not done, you could still see many important characteristics of the data set, such as shape and spread.

Stem-and-leaf displays can be useful for getting a sense of a typical value for the data set, as well as how spread out the values are. It is also easy to spot data values that are unusually far from the rest. Such values are called outliers. The stem-and-leaf display of the accident rate data in Figure 2.9 shows an outlier on the low end (43) and an outlier on the high end (152).

DEFINITION

An **outlier** is an unusually small or large data value.
A precise rule for deciding when an observation is an outlier is given in Chapter 3.

Stem-and-Leaf Display

When to Use   Number of variables: 1
                Data type: Numerical
                Purpose: Display data distribution

How to Construct

1. Select one or more leading digits for the stem values. The trailing digits (or sometimes just the first one of the trailing digits) become the leaves.
2. List *possible* stem values (not just the stem values that occur in the data set) in a vertical column.
3. Record the leaf for every observation beside the corresponding stem value.
4. Indicate the units for stems and leaves someplace in the display.

(continued)

What to Look For   The display conveys information about
- a representative or typical value in the data set
- the extent to which the data values spread out
- the presence of any gaps and outliers
- the extent of symmetry in the data distribution
- the number and location of peaks

Example 2.8   Going Wireless

The article **"Going Wireless"** *(AARP Bulletin, June 2009)* reported the estimated percentage of households with only wireless phone service (no landline) for the 50 U.S. states and the District of Columbia. Data for the 19 Eastern states are given here.

State	Wireless %	State	Wireless%
CN	5.6	NY	11.4
DE	5.7	NC	16.3
DC	20.0	OH	14.0
FL	16.8	PA	10.8
GA	16.5	RI	7.8
ME	13.4	SC	20.6
MD	10.8	VA	10.8
MA	9.3	VT	5.1
NH	11.6	WV	11.6
NJ	8.0		

A stem-and-leaf display is an appropriate way to summarize these data because there is one variable (wireless %), the variable is numerical, and the purpose is to display the data distribution. A dotplot would also have been a reasonable choice.

The sequence of steps given in the previous box can be used to construct a stem-and-leaf display for these data.

Step 1

Stems: 0, 1, 2

The data values range from 5.1% to 20.6%. For consistency, it is helpful to regard all of the data values as having two digits before the decimal place. For example, the observation 5.6 can be viewed as 05.6. The value 5.7 can be viewed as 05.7, and so on.

Using the first two digits would result in stems of 05, 06, …, 20. Fifteen stems are probably too many for a data set that has only 19 observations. Using the first digit as a stem results in three possible stems: 0, 1, 2; so a single digit will be used for the stem.

Step 2

```
0 |
1 |
2 |
```

Getting ready to enter leaves, list all possible stems in a column and draw a line that will separate the stems from the leaves in the display.

Step 3

```
0 | 56, 57
1 |
2 | 00
```

The first data value is 05.6, which has a stem of 0 and a leaf of 56, so 56 is entered to the right of the 0 stem. The second data value is 20.0, which has a stem of 2 and a leaf of 00. The third data value is 05.7. These values have been added in to the display. Other data values are entered in a similar way.

(continued)

AP* EXAM TIP

The legend is an important part of a stem-and-leaf display. Make sure to always include one whenever you make a stem-and-leaf display.

Step 4

0	56, 57
1	
2	00

Stem: tens
Leaves: tenths

The last step is to indicate the units for stems and leaves. The unit is determined by the place value of the rightmost digit. The stem is the tens digit, so the unit for the stem is tens. The rightmost digit of the leaves is in the tenths position, so the unit for the leaves is tenths.

The completed stem-and-leaf display is given in Figure 2.10.

0	56, 57, 93, 80, 78, 51
1	68, 65, 34, 08, 16, 14, 63, 40, 08, 08, 16
2	00, 06

Stem: tens
Leaves: tenths

FIGURE 2.10
Stem-and-leaf display of wireless percentage for eastern states

This display shows that for many eastern states, the percentage of households with only wireless phone service was in the 10% to 20% range. Two states had percentages of 20% or more.

An alternative display (Figure 2.11) results from dropping all but the first digit in each leaf. This is what most statistical computer packages do when generating a stem-and-leaf display. The computer package Minitab was used to generate the display in Figure 2.11.

```
Stem-and-leaf of Eastern States N = 19
Leaf Unit = 1.0

0   555789
1   00011134666
2   00
```

FIGURE 2.11
Minitab stem-and-leaf display of wireless percent using only the first leaf digit.

Comparative Stem-and-Leaf Displays

It is also common to see a stem-and-leaf display used to provide a visual comparison of two groups. A comparative stem-and-leaf display, in which the leaves for one group are listed to the right of the stem values and the leaves for the second group are listed to the left, can show how the two groups are similar and how they differ. This display is also sometimes called a back-to-back stem-and-leaf display.

Comparative Stem-and-Leaf Display

When to Use Number of variables: 1 variable with observations for two groups
Data type: Numerical
Purpose: Compare two data distributions

A comparative stem-and-leaf display has a single column of stems, with the leaves for one data set listed to the right of the stems and the leaves for the second data set listed to the left of the stems. Be sure to include group labels to identify which group is on the left and which is on the right.

Example 2.9 **Going Wireless Revisited**

The article referenced in Example 2.8 also gave the estimated percentage of households with only wireless phone service for 13 western states. The data are shown in the following table.

East		West	
State	Wireless %	State	Wireless %
CN	5.6	AK	11.7
DE	5.7	AZ	18.9
DC	20.0	CA	9.0
FL	16.8	CO	16.7
GA	16.5	HI	8.0
ME	13.4	ID	22.1
MD	10.8	MT	9.2
MA	9.3	NV	10.8
NH	11.6	NM	21.1
NJ	8.0	OR	17.7
NY	11.4	UT	25.5
NC	16.3	WA	16.3
OH	14.0	WY	11.4
PA	10.8		
RI	7.8		
SC	20.6		
VA	10.8		
VT	5.1		
WV	11.6		

A comparative stem-and-leaf display (using only the first digit of each leaf) is shown in Figure 2.12.

Western States   Eastern States

FIGURE 2.12
Comparative stem-and-leaf display of the percentage of households with only wireless phone service for eastern and western states

989	0	559875
1670681	1	66301164001
512	2	00

Stem: tens
Leaves: ones

From the comparative stem-and-leaf display, you can see that although there was state-to-state variability in both the western and the eastern states, the data distributions are quite similar.

Histograms

Dotplots and stem-and-leaf displays are not always effective ways to summarize numerical data. Both are awkward when the data set contains a large number of data values. Histograms are displays that don't work well for small data sets but do work well for larger numerical data sets. Histograms are constructed a bit differently, depending on whether the variable of interest is discrete or continuous.

Frequency Distributions and Histograms for Discrete Numerical Data

Discrete numerical data almost always result from counting. In such cases, each observation is a whole number. A frequency distribution for discrete numerical data lists each possible value (either individually or grouped into intervals), the associated frequency, and sometimes the corresponding relative frequency. Recall that relative frequencies are calculated by dividing each frequency by the total number of observations in the data set.

Example 2.10 Promiscuous Queen Bees

Queen honey bees mate shortly after they become adults. During a mating flight, the queen usually takes multiple partners, collecting sperm that she will store and use throughout the rest of her life. The authors of the paper **"The Curious Promiscuity of Queen Honey Bees"** **(Annals of Zoology [2001]: 255–265)** studied the behavior of 30 queen honey bees to learn about the length of mating flights and the number of partners a queen takes during a mating flight. Data on number of partners (generated to be consistent with summary values and graphs given in the paper) are shown here.

Number of Partners

12	2	4	6	6	7	8	7	8	11
8	3	5	6	7	10	1	9	7	6
9	7	5	4	7	4	6	7	8	10

The corresponding relative frequency distribution is given in Table 2.2. The smallest value in the data set is 1 and the largest is 12, so the possible values from 1 to 12 are listed in the table, along with the corresponding frequencies and relative frequencies.

TABLE **2.2** **Relative Frequency Distribution for Number of Partners**

Number of Partners	Frequency	Relative Frequency	
1	1	0.033	$\leftarrow \frac{1}{30} = .033$
2	1	0.033	
3	1	0.033	
4	3	0.100	
5	2	0.067	
6	5	0.167	
7	7	0.233	
8	4	0.133	
9	2	0.067	
10	2	0.067	
11	1	0.033	
12	1	0.033	Differs from 1
Total	**30**	**0.999**	$\leftarrow$ due to rounding

It is possible to create a more compact frequency distribution by grouping some of the possible values into intervals. For example, you might group 1, 2, and 3 partners to form an interval of 1–3, with a corresponding frequency of 3. The grouping of other values in a similar way results in the relative frequency distribution shown in Table 2.3.

TABLE **2.3** **Relative Frequency Distribution for Number of Partners Using Intervals**

Number of Partners	Frequency	Relative Frequency
1–3	3	0.100
4–6	10	0.333
7–9	13	0.433
10–12	4	0.133

A histogram for discrete numerical data is a graph that shows the data distribution. The following example illustrates the construction of a histogram for the queen bee data.

Example 2.11 Revisiting Promiscuous Queen Bees

Example 2.10 gave the following data on the number of partners during mating flights for 30 queen honey bees.

Number of Partners

12	2	4	6	6	7	8	7	8	11
8	3	5	6	7	10	1	9	7	6
9	7	5	4	7	4	6	7	8	10

Let's start by looking at the dotplot of these data in Figure 2.13.

FIGURE 2.13
Dotplot of queen bee number of partners data

A histogram for this discrete numerical data set replaces the stacks of dots that appear above each possible value with a rectangle that is centered over the value and whose height corresponds to the number of observations at that value, as shown in Figure 2.14.

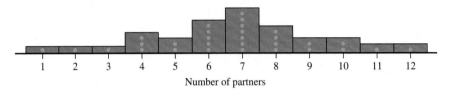

FIGURE 2.14
Histogram of queen bee data superimposed on dotplot

Notice that the area of each rectangle in the histogram is proportional to the frequency of the corresponding data value. For example, the rectangle corresponding to 4 partners is three times the area of the 1 partner rectangle because there were three times as many 4's in the data set as there were 1's.

In practice, histograms for discrete data don't show the dots corresponding to the actual data values—only the rectangles are shown, as in Figure 2.15. We started with the dotplot and used it to construct the histogram to show that the rectangles in a histogram represent a collection of individual data values. The histogram could also have been constructed directly from the frequency distribution given in Example 2.10 by following the steps in the following box.

Histogram for Discrete Numerical Data

When to Use   Number of variables: 1
                Data Type: Discrete numerical
                Purpose: Display data distribution

How to Construct
1. Draw a horizontal scale, and mark the possible values of the variable.
2. Draw a vertical scale, and add either a frequency or relative frequency scale.
3. Above each possible value, draw a rectangle centered at that value (so that the rectangle for 1 is centered at 1, the rectangle for 5 is centered at 5, and so on). The height of each rectangle is determined by the corresponding frequency or relative frequency. When possible values are consecutive whole numbers, the base width for each rectangle is 1.

What to Look For
• Center or typical value
• Extent of spread or variability
• General shape
• Location and number of peaks
• Presence of gaps and outliers

Example 2.12 Promiscuous Queen Bees One More Time

Table 2.2 summarized the queen bee data of Examples 2.10 and 2.11 in a frequency distribution. The corresponding histogram and relative frequency histogram are shown in Figure 2.15. Notice that each rectangle in the histogram is centered over the corresponding value. When relative frequency is used for the vertical scale instead of frequency, the scale on the vertical axis is different, but all essential characteristics of the graph (shape, center, spread) are unchanged.

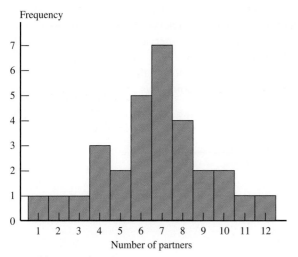

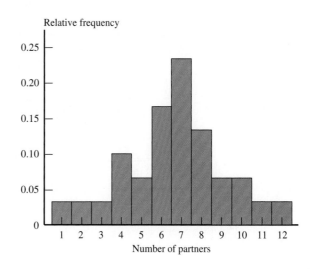

FIGURE 2.15
Histogram and relative frequency histogram of queen bee data

A histogram based on the grouped frequency distribution of Table 2.3 can be constructed in a similar fashion and is shown in Figure 2.16. A rectangle represents the frequency or relative frequency for each interval. For the interval 1–3, the rectangle extends from 0.5 to 3.5 so that there are no gaps between the rectangles of the histogram.

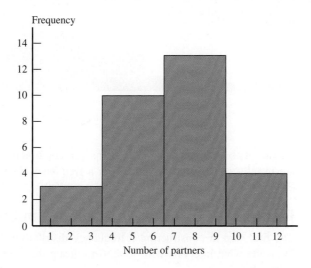

FIGURE 2.16
Histogram of queen bee data using intervals

The difficulty in constructing tabular or graphical displays with continuous data, such as observations on reaction time or weight of airline passenger carry-on luggage, is that there are no natural intervals. The way out of this dilemma is to define your own intervals. Consider the following data on carry-on luggage weight for 25 airline passengers.

Carry-on Baggage Weight

25.0 17.9 10.1 27.6 30.0 18.0 28.7 28.2 27.8 28.0 31.4

20.9 33.8 27.6 21.9 19.9 20.8 28.5 22.4 24.9 26.4 22.0

34.5 22.7 25.3

A dotplot of these data is given in Figure 2.17(a). Because the weights range from about 10 pounds to about 35 pounds, grouping the weights into 5-pound intervals is reasonable. The first interval would begin at 10 pounds and end at 15 pounds. Because you want each data value to fall in exactly one interval, you would include data values of 10 in this interval, along with any that are between 10 and 15 pounds. Data values of exactly 15 pounds would be put in the interval that starts at 15. The first interval could then be written as 10 to <15, where the less than symbol indicates that the interval will contain all data values that are greater than or equal to 10 and less than 15.

To see how the histogram is formed, take all of the dots in the interval 10 to <15 and stack them up at the center of the interval (at 12.5). Doing this for each of the intervals results in the graph shown in Figure 2.17(b).

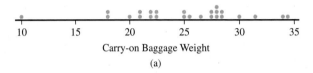

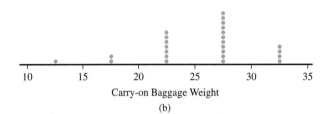

FIGURE 2.17
Graphs of carry-on baggage weight: (a) dotplot; (b) dots stacked at midpoints of intervals.

A histogram of these data is created by drawing a rectangle over each interval. The height of each rectangle is determined by the number of dots in the stack for that interval, as shown in Figure 2.18.

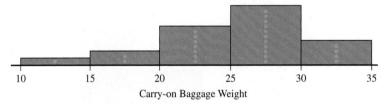

FIGURE 2.18
Histogram of carry-on baggage weight

As with discrete data, histograms for continuous data don't show the "dots" and are usually constructed directly from a frequency distribution. This is illustrated in the examples that follow.

Frequency Distributions for Continuous Numerical Data

The first step in constructing a frequency distribution for continuous numerical data is to decide what intervals will be used to group the data. These intervals are called **class intervals**.

Example 2.13 **Enrollments at Public Universities**

States differ widely in the percentage of college students who are enrolled in public institutions. The National Center for Education Statistics provided the accompanying data on this percentage for the 50 U.S. states for fall 2007.

Percentage of College Students Enrolled in Public Institutions

96	86	81	84	77	90	73	53	90	96	73
93	76	86	78	76	88	86	87	64	60	58
89	86	80	66	70	90	89	82	73	81	73
72	56	55	75	77	82	83	79	75	59	59
46	50	64	80	82	75					

The smallest observation is 46 (Massachusetts) and the largest is 96 (Alaska and Wyoming). It is reasonable to start the first class interval at 40 and let each interval have a width of 10. This gives class intervals starting with 40 to < 50 and continuing up to 90 to < 100.

Table 2.4 displays the resulting frequency distribution, along with the relative frequencies.

TABLE 2.4   Relative Frequency Distribution for Percentage of College Students Enrolled in Public Institutions

Class Interval	Frequency	Relative Frequency
40 to < 50	1	0.02
50 to < 60	7	0.14
60 to < 70	4	0.08
70 to < 80	15	0.30
80 to < 90	17	0.34
90 to < 100	6	0.12
	50	1.00

There are no set rules for selecting either the number of class intervals or the length of the intervals. Using a few relatively wide intervals will bunch the data, whereas using a great many relatively narrow intervals may spread the data over too many intervals, so that no interval contains more than a few observations. Either way, interesting features of the data set may be missed. In general, with a small amount of data, relatively few intervals, perhaps between 5 and 10, should be used. With a large amount of data, a distribution based on 15 to 20 (or even more) intervals is often recommended. The quantity

$$\sqrt{\text{number of observations}}$$

is often used as an estimate of an appropriate number of intervals: 5 intervals for 25 observations, 10 intervals when the number of observations is 100, and so on.

Histograms for Continuous Numerical Data

When the class intervals in a frequency distribution are all of equal width, you construct a histogram in a way that is very similar to what is done for discrete data.

Histogram for Continuous Numerical Data When Class Intervals Are Equal Width

When to Use   Number of variables: 1
Data Type: Continuous numerical
Purpose: Displaying data distribution

How to Construct

1. Mark the boundaries of the class intervals on a horizontal axis.
2. Use either frequency or relative frequency on the vertical axis.
3. Draw a rectangle for each class interval directly above that interval (so that the edges are at the class interval boundaries). The height of each rectangle is the frequency or relative frequency of the corresponding class interval.

(continued)

What to Look For

- Center or typical value
- Extent of spread or variability
- General shape
- Location and number of peaks
- Presence of gaps and outliers

Example 2.14   TV Viewing Habits of Children

The article **"Early Television Exposure and Subsequent Attention Problems in Children"** **(Pediatrics, April 2004)** investigated the television viewing habits of U.S. children. The data summarized in the article were obtained as part of a large-scale national survey. Table 2.5 gives approximate relative frequencies for the number of hours spent watching TV per day for a sample of 1-year-old children.

TABLE **2.5**   **Relative Frequency Distribution for Hours Spent Watching TV per Day**

TV Hours Per Day	Age 1 Year Relative Frequency
0 to < 2	0.270
2 to < 4	0.390
4 to < 6	0.190
6 to < 8	0.085
8 to < 10	0.030
10 to < 12	0.020
12 to < 14	0.010
14 to < 16	0.005

You can use the steps in the previous box to construct a histogram for the data summarized in Table 2.5.

Step 1

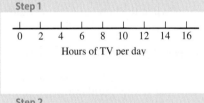

The intervals used in the frequency distribution are 0 to < 2, 2 to < 4, and so on. The boundaries of these intervals are marked on the horizontal axis.

Step 2

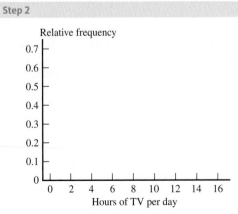

Because the frequency distribution provided gives relative frequencies, relative frequency will be used on the vertical scale. The largest relative frequency is 0.390, so the vertical scale should run from 0 to at least 0.4.

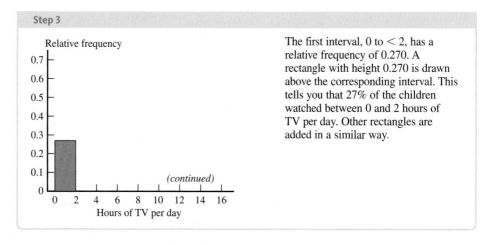

Step 3

The first interval, 0 to < 2, has a relative frequency of 0.270. A rectangle with height 0.270 is drawn above the corresponding interval. This tells you that 27% of the children watched between 0 and 2 hours of TV per day. Other rectangles are added in a similar way.

(continued)

Figure 2.19 shows the completed relative frequency histogram. Notice that the histogram has a single peak, with a majority of the children watching between 0 and 4 hours of TV per day.

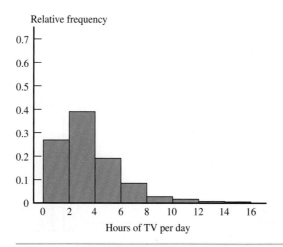

FIGURE 2.19
Histogram of TV hours per day for one-year-old children

Using Histograms to Compare Groups

Histograms can be used to compare groups, but keep two things in mind:

1. Always use relative frequency rather than frequency on the vertical axis since the number of observations in each group might be different.
2. Use the same scales on both the horizontal and vertical axes to make drawing comparisons easier.

The use of histograms to compare distributions is illustrated in Example 2.15.

Example 2.15 TV Viewing Again

The paper referenced in Example 2.14 also gave data on TV viewing for 3-year-old children. Table 2.6 gives relative frequencies for both 1-year-old and 3-year-old children.

TABLE 2.6 **Relative Frequency Distribution for Number of Hours Spent Watching TV per Day for 1-Year-Old and 3-Year-Old Children**

TV Hours Per Day	Age 1 Year Relative Frequency	Age 3 Years Relative Frequency
0 to < 2	0.270	0.630
2 to < 4	0.390	0.195
4 to < 6	0.190	0.100
6 to < 8	0.085	0.025

(continued)

TV Hours Per Day	Age 1 Year Relative Frequency	Age 3 Years Relative Frequency
8 to < 10	0.030	0.020
10 to < 12	0.020	0.015
12 to < 14	0.010	0.010
14 to < 16	0.005	0.005

Figure 2.20(a) is the relative frequency histogram for the 1-year-old children, and Figure 2.20(b) is the relative frequency histogram for 3-year-old children. Notice that both histograms have a single peak, with the majority of children in both age groups concentrated in the smaller TV hours intervals. Both histograms are quite stretched out at the upper end, indicating that some young children watch a lot of TV.

The big difference between the two histograms is at the low end, with a much higher proportion of 3-year-old children falling in the 0 to 2 TV hours interval than 1-year-old children. A typical number of TV hours per day for 1-year-old children would be somewhere between 2 and 4 hours, whereas a typical number of TV hours for 3-year-old children is in the 0- to 2-hour interval.

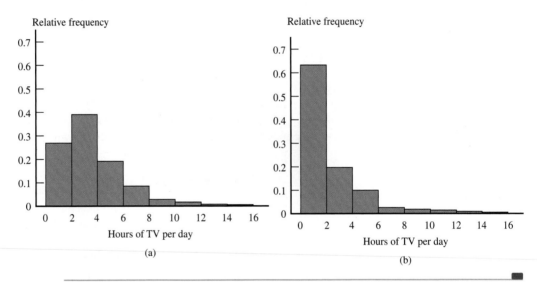

FIGURE 2.20
Histograms of TV hours per day:
(a) 1-year-old children
(b) 3-year-old children

Histograms with Unequal Width Intervals

If the intervals in the frequency distribution are not all the same width, histograms must be constructed using a different scale on the vertical axis, called the *density scale*. Using density ensures that the areas of the rectangles in the histogram will be proportional to the frequencies or relative frequencies.

CONSTRUCTING A HISTOGRAM FOR CONTINUOUS DATA WHEN CLASS INTERVAL WIDTHS ARE UNEQUAL

When class intervals are not of equal width, frequencies or relative frequencies should not be used on the vertical axis. Instead, the height of each rectangle, called the **density** for the corresponding class interval, is given by

$$\text{density} = \text{rectangle height} = \frac{\text{relative frequency of class interval}}{\text{class interval width}}$$

The vertical axis is called the **density scale**.

The formula for density can also be used when interval widths are equal. However, when the intervals are of equal width, the extra arithmetic required to obtain the densities is unnecessary.

Example 2.16   Misreporting Grade Point Average

When people are asked for the values of characteristics such as age or weight, they sometimes shade the truth in their responses. The article **"Self-Reports of Academic Performance"** *(Social Methods and Research* **[November 1981]: 165–185)** focused on SAT scores and grade point average (GPA). For each student in a sample, the difference between reported GPA and actual GPA was determined. Positive differences resulted from individuals reporting GPAs larger than the correct values. Most differences were close to 0, but there were some rather large errors. Because of this, the frequency distribution based on unequal class widths shown in Table 2.7 gives an informative yet concise summary.

TABLE **2.7** **Frequency Distribution for Errors in Reported GPA**

Class Interval	Relative Frequency	Width	Density
-2.0 to < -0.4	0.023	1.6	0.014
-0.4 to < -0.2	0.055	0.2	0.275
-0.2 to < -0.1	0.097	0.1	0.970
-0.1 to < 0	0.210	0.1	2.100
0 to < 0.1	0.189	0.1	1.890
0.1 to < 0.2	0.139	0.1	1.390
0.2 to < 0.4	0.116	0.2	0.580
0.4 to < 2.0	0.171	1.6	0.107

Figure 2.21 displays two histograms based on this frequency distribution. The histogram in Figure 2.21(a) is correctly drawn, with density used to determine the height of each bar. The histogram in Figure 2.21(b) has height equal to relative frequency and is therefore *not correct*. In particular, this second histogram exaggerates the incidence of grossly overreported and underreported values—the areas of the two most extreme rectangles are much too large. The eye is naturally drawn to large areas, so it is important that the areas correctly represent the relative frequencies.

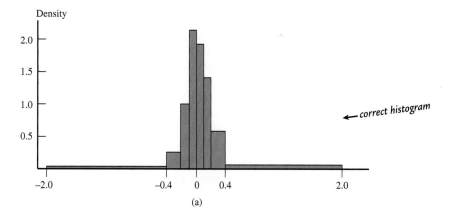

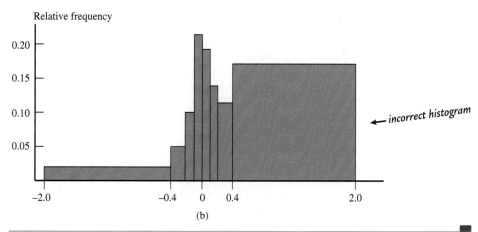

FIGURE 2.21
Histograms for errors in reporting GPA:
(a) correct histogram
(height = density)
(b) incorrect histogram
(height = relative frequency)

Histogram Shapes

General shape is an important characteristic of a histogram. In describing various shapes it is convenient to approximate the histogram itself with a smooth curve (called a *smoothed histogram*). This is illustrated in Figure 2.22.

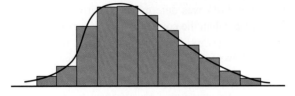

FIGURE 2.22
Approximating a histogram with a smooth curve

One description of general shape relates to the number of peaks, or **modes**.

> **DEFINITION**
>
> A histogram is said to be **unimodal** if it has a single peak, **bimodal** if it has two peaks, and **multimodal** if it has more than two peaks.

These shapes are illustrated in Figure 2.23.

FIGURE 2.23
Smoothed histograms with various numbers of modes: (a) unimodal; (b) bimodal; (c) multimodal

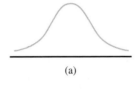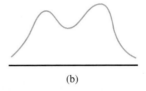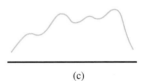

(a)   (b)   (c)

Bimodality sometimes occurs when the data set consists of observations on two quite different kinds of individuals or objects. For example, consider a large data set consisting of driving times for automobiles traveling between San Luis Obispo and Monterey, California. This histogram would show two peaks, one for those cars traveling on the inland route (roughly 2.5 hours) and another for those cars traveling up the coast highway (3.5–4 hours). However, bimodality does not automatically follow in such situations. Bimodality will occur in the histogram of the combined groups only if the centers of the two separate histograms are far apart (relative to the variability in the two data sets). For example, a large data set consisting of heights of college students would probably not produce a bimodal histogram because the typical height for males (about 69 in.) and the typical height for females (about 66 in.) are not very far apart.

A histogram is **symmetric** if there is a vertical line of symmetry such that the part of the histogram to the left of the line is a mirror image of the part to the right. Several different symmetric unimodal smoothed histograms are shown in Figure 2.24.

FIGURE 2.24
Several symmetric unimodal smoothed histograms

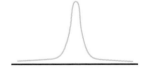

Proceeding to the right from the peak of a unimodal histogram, you move into what is called the **upper tail** of the histogram. Going in the opposite direction moves you into the **lower tail**.

> **DEFINITION**
>
> A unimodal histogram that is not symmetric is said to be **skewed**. If the upper tail of the histogram stretches out much farther than the lower tail, then the distribution of values is **positively skewed** or **right skewed**. If the lower tail is much longer than the upper tail, the histogram is **negatively skewed** or **left skewed**.

These two types of skewness are illustrated in Figure 2.25. Positive skewness is much more frequently encountered than negative skewness. An example of positive skewness would be in the distribution of single-family home prices in Los Angeles County; most homes are moderately priced (at least for California), whereas the relatively few homes in Beverly Hills and Malibu have much higher price tags.

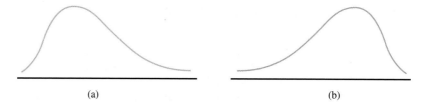

FIGURE 2.25
Two examples of skewed smoothed histograms:
(a) positive skew;
(b) negative skew

(a)   (b)

One shape, a **normal curve,** arises more frequently than any other in statistical applications. A normal curve is symmetric and bell-shaped (See Figure 2.26). Many histograms of characteristics such as arm span or weight of an apple can be well approximated by a normal curve. A more detailed discussion of normal curves is found in Chapter 6.

FIGURE 2.26
A normal curve

Cumulative Relative Frequencies and Cumulative Relative Frequency Plots

Sometimes, rather than wanting to know what proportion of the data falls in a particular class interval, you want to determine the proportion falling below a specified value. This is easily done when the value of interest is one of the class interval boundaries.

Consider the following intervals and relative frequencies for carry-on luggage weight for passengers on flights between Phoenix and New York City during October 2009:

Class Interval	0 to < 5	5 to < 10	10 to < 15	15 to < 20	. . .
Relative frequency	0.05	0.10	0.18	0.25	. . .

Then

proportion of passengers with carry-on luggage weight less than 15 lbs.
    = proportion in one of the first three classes
    = 0.05 + 0.10 + 0.18
    = 0.33

Similarly,

proportion of passengers with carry-on luggage weight less than 20 lbs.
    = 0.05 + 0.10 + 0.18 + 0.25 = 0.33 + 0.25 = 0.58

Each such sum of relative frequencies is called a **cumulative relative frequency**. Notice that the cumulative relative frequency 0.58 is the sum of the previous cumulative relative frequency 0.33 and the "current" relative frequency 0.25. The use of cumulative relative frequencies is illustrated in Example 2.17.

Example 2.17 Albuquerque Rainfall

The National Climatic Data Center has been collecting weather data for many years. Annual rainfall totals for Albuquerque, New Mexico, from 1950 to 2008 (**www .ncdc.noaa.gov/oa/climate/research/cag3/city.html**) were used to construct the relative frequency distribution shown in Table 2.8. The table also contains a column of cumulative relative frequencies.

TABLE 2.8 **Relative Frequency Distribution for Albuquerque Rainfall Data with Cumulative Relative Frequencies**

Annual Rainfall (inches)	Frequency	Relative Frequency	Cumulative Relative Frequency
4 to < 5	3	0.052	0.052
5 to < 6	6	0.103	0.155 = 0.052 + 0.103
6 to < 7	5	0.086	0.241 = 0.052 + 0.103 + 0.086 = 0.155 + 0.086
7 to < 8	6	0.103	0.344
8 to < 9	10	0.172	0.516
9 to < 10	4	0.069	0.585
10 to < 11	12	0.207	0.792
11 to < 12	6	0.103	0.895
12 to < 13	3	0.052	0.947
13 to < 14	3	0.052	0.999

The proportion of years with annual rainfall less than 10 inches is 0.585, the cumulative relative frequency for the 9 to < 10 interval. What about the proportion of years with annual rainfall less than 8.5 inches? Because 8.5 is not the endpoint of one of the intervals in the frequency distribution, you can only estimate this from the information given. The value 8.5 is halfway between the endpoints of the 8 to 9 interval, so it is reasonable to estimate that half of the relative frequency of 0.172 for this interval belongs in the 8 to 8.5 range. Then

$$\begin{pmatrix} \text{estimate of proportion of} \\ \text{years with rainfall less} \\ \text{than 8.5 inches} \end{pmatrix} = 0.052 + 0.103 + 0.086 + 0.103 + \frac{1}{2}(0.172) = 0.430$$

This proportion could also have been computed using the cumulative relative frequencies as

$$\begin{pmatrix} \text{estimate of proportion of} \\ \text{years with rainfall less} \\ \text{than 8.5 inches} \end{pmatrix} = 0.344 + \frac{1}{2}(0.172) = 0.430$$

Similarly, since 11.25 is one-fourth of the way between 11 and 12,

$$\begin{pmatrix} \text{estimate of proportion of} \\ \text{years with rainfall less} \\ \text{than 11.25 inches} \end{pmatrix} = 0.792 + \frac{1}{4}(0.103) = 0.818$$

A **cumulative relative frequency** plot is just a graph of the cumulative relative frequencies against the upper endpoint of the corresponding interval. The pairs

(upper endpoint of interval, cumulative relative frequency)

are plotted as points on a rectangular coordinate system, and successive points in the plot are connected by a line segment.

Cumulative Relative Frequency Plot

When to Use   Number of variables: 1
Data type: Numerical
Purpose: Displaying data distribution in a way that shows the approximate proportion of data at or below any given value.

(continued)

How to Construct
1. Mark the boundaries of the class intervals on a horizontal axis.
2. Add a vertical axis with a scale that goes from 0 to 1.
3. For each class interval, plot the point that is represented by (upper endpoint of interval, cumulative relative frequency).
4. Add the point represented by (lower endpoint of first interval, 0).
5. Connect consecutive points in the display with line segments.

What to Look For
• Proportion of data falling at or below any given value along the x axis

For the rainfall data of Example 2.17, the plotted points would be

(5, 0.052)	(6, 0.155)	(7, 0.241)	(8, 0.344)	(9, 0.516)
(10, 0.585)	(11, 0.792)	(12, 0.895)	(13, 0.947)	(14, 0.999)

One additional point, the pair (lower endpoint of first interval, 0), is also included in the plot (for the rainfall data, this would be the point (4, 0)), and then points are connected by line segments. Figure 2.27 shows the cumulative relative frequency plot for the rainfall data. The cumulative relative frequency plot can be used to obtain approximate answers to questions such as

What proportion of the observations is smaller than a particular value?

and

What value separates the smallest p percent from the larger values?

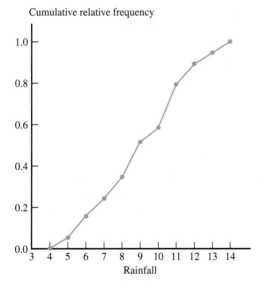

FIGURE 2.27
Cumulative relative frequency plot for the rainfall data of Example 2.17.

For example, to determine the approximate proportion of years with annual rainfall less than 9.5 inches, you would follow a vertical line up from 9.5 on the x-axis and then read across to the y-axis to obtain the corresponding relative frequency, as illustrated in Figure 2.28(a). Approximately 0.55, or 55%, of the years had annual rainfall less than 9.5 inches.

Similarly, to find the amount of rainfall that separates the 30% of years with the smallest annual rainfall from years with higher rainfall, start at 0.30 on the cumulative relative frequency axis and move across and then down to find the corresponding rainfall amount, as shown in Figure 2.28(b). Approximately 30% of the years had annual rainfall of less than about 7.6 inches.

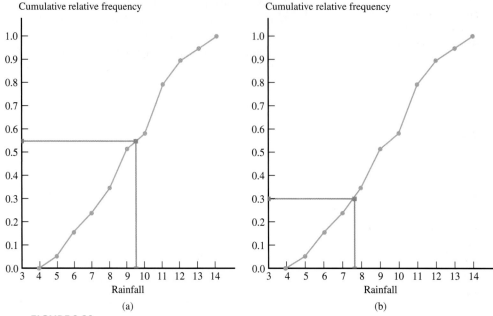

FIGURE 2.28
Using the cumulative relative frequency plot:
(a) Determining the approximate proportion of years with annual rainfall less than 9.5 inches.
(b) Finding the amount of rainfall that separates the 30% of years with the lowest rainfall from the 70% with higher rainfall.

SECTION **2.3**   **EXERCISES**

Each Exercise Set assesses the following chapter learning objectives: M3, M4, M5, M8, P1, P3

SECTION 2.3   **Exercise Set 1**

2.24   The article **"Feasting on Protein"** (*AARP Bulletin, September 2009*) gave the cost per gram of protein for 19 common food sources of protein.

Food	Cost (cents per gram of protein)	Food	Cost (cents per gram of protein)
Chicken	1.8	Yogurt	5.0
Salmon	5.8	Milk	2.5
Turkey	1.5	Peas	5.2
Soybeans	3.1	Tofu	6.9
Roast beef	2.7	Cheddar cheese	3.6
Cottage cheese	3.1	Nuts	5.2
Ground beef	2.3	Eggs	5.7
Ham	2.1	Peanut butter	1.8
Lentils	3.3	Ice cream	5.3
Beans	2.9		

a. Construct a dotplot of the cost-per-gram data.
b. Locate the cost per gram for meat and poultry items in your dotplot and highlight them in a different color. Based on the dotplot, do meat and poultry items appear to be a good value when compared to other sources of protein?

2.25   Box Office Mojo (**www.boxofficemojo.com**) tracks movie ticket sales. Ticket sales (in millions of dollars) for each of the top 20 movies in 2007 and 2008 are shown in the accompanying tables.

Movie (2007)	2007 Sales (millions of dollars)
Spider-Man 3	336.5
Shrek the Third	322.7
Transformers	319.2
Pirates of the Caribbean: At World's End	309.4
Harry Potter and the Order of the Phoenix	292.0
I Am Legend	256.4
The Bourne Ultimatum	227.5
National Treasure: Book of Secrets	220.0
Alvin and the Chipmunks	217.3
300	210.6
Ratatouille	206.4
The Simpsons Movie	183.1
Wild Hogs	168.3
Knocked Up	148.8
Juno	143.5
Rush Hour 3	140.1

(continued)

Movie (2007)	2007 Sales (millions of dollars)
Live Free or Die Hard	134.5
Fantastic Four: Rise of the Silver Surfer	131.9
American Gangster	130.2
Enchanted	127.8

Movie (2008)	2008 Sales (millions of dollars)
The Dark Knight	533.3
Iron Man	318.4
Indiana Jones and the Kingdom of the Crystal Skull	317.1
Hancock	227.9
WALL-E	223.8
Kung Fu Panda	215.4
Twilight	192.8
Madagascar: Escape 2 Africa	180.0
Quantum of Solace	168.4
Dr. Suess' Horton Hears a Who!	154.5
Sex and the City	152.6
Gran Torino	148.1
Mamma Mia!	144.1
Marley and Me	143.2
The Chronicles of Narnia: Prince Caspian	141.6
Slumdog Millionaire	141.3
The Incredible Hulk	134.8
Wanted	134.5
Get Smart	130.3
The Curious Case of Benjamin Button	127.5

Construct a comparative dotplot of the 2007 and 2008 ticket sales data. Comment on any interesting features of the dotplot. In what ways are the distributions of the 2007 and 2008 ticket sales observations similar? In what ways are they different?

2.26 *USA Today* (June 11, 2010) gave the following data on median age for each of the 50 U.S. states and the District of Columbia (DC). Construct a stem-and-leaf display using stems 28, 29, ..., 42. Comment on center, spread, and shape of the data distribution. Are there any unusual values in the data set that stand out?

State	Median Age	State	Median Age
Alabama	37.4	DC	35.1
Alaska	32.8	Florida	40.0
Arizona	35.0	Georgia	34.7
Arkansas	37.0	Hawaii	37.5
California	34.8	Idaho	34.1
Colorado	35.7	Illinois	36.2
Connecticut	39.5	Indiana	36.8
Delaware	38.4	Iowa	38.0

State	Median Age	State	Median Age
Kansas	35.9	North Dakota	36.3
Kentucky	37.7	Ohio	38.5
Louisiana	35.4	Oklahoma	35.8
Maine	42.2	Oregon	38.1
Maryland	37.7	Pennsylvania	39.9
Massachusetts	39.0	Rhode Island	39.2
Michigan	38.5	South Carolina	37.6
Minnesota	37.3	South Dakota	36.9
Mississippi	35.0	Tennessee	37.7
Missouri	37.6	Texas	33.0
Montana	39.0	Utah	28.8
Nebraska	35.8	Vermont	41.2
Nevada	35.5	Virginia	36.9
New Hampshire	40.4	Washington	37.0
New Jersey	38.8	West Virginia	40.5
New Mexico	35.6	Wisconsin	38.2
New York	38.1	Wyoming	35.9
North Carolina	36.9		

2.27 A report from **Texas Transportation Institute (Texas A&M University System, 2005)** titled "Congestion Reduction Strategies" included the following data on extra travel time during rush hour for very large and for large urban areas.

Very Large Urban Areas	Extra Hours Per Year per Traveler
Los Angeles, CA	93
San Francisco, CA	72
Washington DC, VA, MD	69
Atlanta, GA	67
Houston, TX	63
Dallas, Fort Worth, TX	60
Chicago, IL-IN	58
Detroit, MI	57
Miami, FL	51
Boston, MA, NH, RI	51
New York, NY-NJ-CT	49
Phoenix, AZ	49
Philadelphia, PA-NJ-DE-MD	38

Large Urban Areas	Extra Hours Per Year Per Traveler
Riverside, CA	55
Orlando, FL	55
San Jose, CA	53
San Diego, CA	52
Denver, CO	51
Baltimore, MD	50
Seattle, WA	46
Tampa, FL	46
Minneapolis, St Paul, MN	43
Sacramento, CA	40
Portland, OR, WA	39

(continued)

(continued)

Large Urban Areas	Extra Hours Per Year Per Traveler
Indianapolis, IN	38
St Louis, MO-IL	35
San Antonio, TX	33
Providence, RI, MA	33
Las Vegas, NV	30
Cincinnati, OH-KY-IN	30
Columbus, OH	29
Virginia Beach, VA	26
Milwaukee, WI	23
New Orleans, LA	18
Kansas City, MO-KS	17
Pittsburgh, PA	14
Buffalo, NY	13
Oklahoma City, OK	12
Cleveland, OH	10

a. Construct a back-to-back stem-and-leaf display for the two different sizes of urban areas.
b. Is the display constructed in Part (a) consistent with the following statement? Explain. Statement: The larger the urban area, the greater the extra travel time during peak period travel.

2.28 The following two relative frequency distributions were constructed from data in the report **"Undergraduate Students and Credit Cards in 2004"** (Nellie Mae, May 2005). One distribution summarizes credit bureau data for a random sample of 1,413 college students. The other distribution summarizes data from a survey completed by 132 of the 1,260 college students who received it.

Credit Card Balance (dollars)— Credit Bureau Data	Relative Frequency
0 to < 100	0.18
100 to < 500	0.19
500 to < 1000	0.14
1000 to < 2000	0.16
2000 to < 3000	0.10
3000 to < 7000	0.16
7000 or more	0.07

Credit Card Balance (dollars)— Survey Data	Relative Frequency
0 to < 100	0.18
100 to < 500	0.22
500 to < 1000	0.17
1000 to < 2000	0.22
2000 to < 3000	0.07
3000 to < 7000	0.14
7000 or more	0.00

a. Construct a histogram for the credit bureau data. Assume that no one had a balance greater than 15,000 and that

the last interval is 7,000 to < 15,000. Be sure to use the density scale.
b. Construct a histogram for the survey data. Use the same scales that you used for the histogram in Part (a) so that it will be easy to compare the two.
c. Comment on the similarities and differences in the histograms from Parts (a) and (b).
d. Do you think the high nonresponse rate for the survey may have contributed to the observed differences in the two histograms? Explain.

2.29 An exam is given to students in an introductory statistics course. Comment on the expected shape of the histogram of scores if:
a. the exam is very easy
b. the exam is very difficult
c. half the students in the class have had calculus, the other half have had no prior college math courses, and the exam emphasizes higher-level math skills
Explain your reasoning in each case.

2.30 U.S. Census data for San Luis Obispo County, California, were used to construct the following frequency distribution for commute time (in minutes) of working adults (the given frequencies were read from a graph that appeared in the **San Luis Obispo Tribune [September 1, 2002]** and so are only approximate):

CommuteTime	Frequency
0 to < 5	5,200
5 to < 10	18,200
10 to < 15	19,600
15 to < 20	15,400
20 to < 25	13,800
25 to < 30	5,700
30 to < 35	10,200
35 to < 40	2,000
40 to < 45	2,000
45 to < 60	4,000
60 to < 90	2,100
90 to < 120	2,200

a. Notice that not all intervals in the frequency distribution are equal in width. Why do you think that unequal width intervals were used?
b. Construct a table that adds a relative frequency column and a density column to the given frequency distribution (see Example 2.16).
c. Use the densities computed in Part (b) to construct a histogram for this data set. (Note: The newspaper displayed an incorrectly drawn histogram based on frequencies rather than densities!) Write a few sentences commenting on the important features of the histogram.
d. Compute the cumulative relative frequencies, and construct a cumulative relative frequency plot.
e. Use the cumulative relative frequency plot constructed in Part (d) to answer the questions on the next page.

i. Approximately what proportion of commute times were less than 50 minutes?

ii. Approximately what proportion of commute times were greater than 22 minutes?

iii. What is the approximate commute time value that separates the shortest 50% of commute times from the longest 50%?

SECTION 2.3   **Exercise Set 2**

2.31 The following data on violent crime on Florida college campuses during 2005 are from the FBI web site.

University/College	Student Enrollment	Number of Violent Crimes Reported in 2005
Florida A&M University	13,067	23
Florida Atlantic University	25,319	4
Florida Gulf Coast University	5,955	5
Florida International University	34,865	5
Florida State University	38,431	29
New College of Florida	692	1
Pensacola Junior College	10,879	2
Santa Fe Community College	13,888	3
Tallahassee Community College	12,775	0
University of Central Florida	42,465	19
University of Florida	47,993	17
University of North Florida	14,533	6
University of South Florida	42,238	19
University of West Florida	9,518	1

a. Construct a dotplot using the 14 observations on number of violent crimes reported. Which schools stand out from the rest?

b. One of the Florida schools only has 692 students, and a few of the schools are quite a bit larger than the rest. Because of this, it might make more sense to consider a crime *rate* by calculating the number of violent crimes reported *per 1,000 students*. For example, for Florida A&M University the violent crime rate would be

$$\frac{23}{13,067}(1,000) = (0.0018)(1,000) = 1.8$$

Calculate the violent crime rate for the other 13 schools. Use these values to construct a second dotplot. Do the same schools stand out as unusual in this dotplot?

c. Based on your answers from Parts (a) and (b), write a couple of sentences commenting on violent crimes reported at Florida universities and colleges in 2005.

2.32 The article **"Going Wireless"** (*AARP Bulletin, June 2009*) reported the estimated percentages of households with only wireless phone service (no landline) for the 50 states and the District of Columbia. In the accompanying data table, each state was also classified into one of three geographical regions—West (W), Middle states (M), or East (E).

Wireless %	Region	State	Wireless %	Region	State
13.9	M	AL	9.2	W	MT
11.7	W	AK	23.2	M	NE
18.9	W	AZ	10.8	W	NV
22.6	M	AR	16.9	M	ND
9.0	W	CA	11.6	E	NH
16.7	W	CO	8.0	E	NJ
5.6	E	CN	21.1	W	NM
5.7	E	DE	11.4	E	NY
20.0	E	DC	16.3	E	NC
16.8	E	FL	14.0	E	OH
16.5	E	GA	23.2	M	OK
8.0	W	HI	17.7	W	OR
22.1	W	ID	10.8	E	PA
16.5	M	IL	7.9	E	RI
13.8	M	IN	20.6	E	SC
22.2	M	IA	6.4	M	SD
16.8	M	KA	20.3	M	TN
21.4	M	KY	20.9	M	TX
15.0	M	LA	25.5	W	UT
13.4	E	ME	10.8	E	VA
10.8	E	MD	5.1	E	VT
9.3	E	MA	16.3	W	WA
16.3	M	MI	11.6	E	WV
17.4	M	MN	15.2	M	WI
19.1	M	MS	11.4	W	WY
9.9	M	MO			

a. Display the data in a comparative dotplot that makes it possible to compare wireless percent for the three geographical regions.

b. Does the graphical display in Part (a) reveal any striking differences in wireless percent for the three geographical regions or are the distributions of wireless percent similar for the three regions?

2.33 The article **"Frost Belt Feels Labor Drain"** (*USA Today, May 1, 2008*) points out that even though total population is increasing, the pool of young workers is shrinking in many states. Each entry in the accompanying table is the percent change in the population of 25- to 44-year-olds from 2000 to 2007. A negative percent change means a state had fewer 25- to 44-year-olds in 2007 than in 2000.

State	% Change	State	% Change
Alabama	−4.1	Colorado	4.1
Alaska	−2.5	Connecticut	−9.9
Arizona	17.8	Delaware	−2.2
Arkansas	0.9	DC	1.8
California	−0.4	Florida	5.8

(continued)

State	% Change	State	% Change
Georgia	7.2	New Mexico	0.6
Hawaii	−1.3	New York	−8.0
Idaho	11.1	North Carolina	2.4
Illinois	−4.6	North Dakota	−10.9
Indiana	−3.1	Ohio	−8.2
Iowa	−6.5	Oklahoma	−1.6
Kansas	−5.3	Oregon	4.4
Kentucky	−1.7	Pennsylvania	−9.1
Louisiana	−11.9	Rhode Island	−8.8
Maine	−8.7	South Carolina	0.1
Maryland	−5.7	South Dakota	−4.1
Massachusetts	−9.6	Tennessee	0.6
Michigan	−9.1	Texas	7.3
Minnesota	−4.5	Utah	19.6
Mississippi	−5.2	Vermont	−10.4
Missouri	−2.9	Virginia	−1.1
Montana	−3.7	Washington	1.6
Nebraska	−5.6	West Virginia	−5.4
Nevada	22.0	Wisconsin	−5.0
New Hampshire	−7.5	Wyoming	−2.3
New Jersey	−7.8		

a. The smallest value in the data set is −11.9, and the largest value is 22.0. One possible choice of stems for a stem-and-leaf display would be to use the tens digit, resulting in stems of −1, −0, 0, 1, and 2. Notice the two 0 stems—one is for the negative percent changes between 0 and −9.9, and one is for the positive percent changes between 0 and 9.9. Construct a stem-and-leaf plot using these five stems. (Hint: Think of each data value as having two digits before the decimal place, so 4.1 would be 04.1.)

b. The article described "the frost belt" as the cold part of the country—the Northeast and Midwest—noting that states in the frost belt generally showed a decline in the 25- to 44-year-old age group. How would you describe the group of states that saw a marked increase in the number of 25- to 44-year-olds?

2.34 The article "**A Nation Ablaze with Change**" (*USA Today*, July 3, 2001) gave the following data on percentage increase in population between 1990 and 2000 for the 50 U.S. states. Each state is classified as belonging to the eastern or western part of the United States (the states are listed in order of population size):

State	% Change	East/West	State	% Change	East/West
California	13.8	W	Ohio	4.7	E
Texas	22.8	W	Michigan	6.9	E
New York	5.5	E	New Jersey	8.9	E
Florida	23.5	E	Georgia	26.4	E
Illinois	8.6	E	North Carolina	21.4	E
Pennsylvania	3.4	E	Virginia	14.4	E

(continued)

State	% Change	East/West	State	% Change	East/West
Massachusetts	5.5	E	Kansas	8.5	W
Indiana	9.7	E	Arkansas	13.7	E
Washington	21.1	W	Utah	29.6	W
Tennessee	16.7	E	Nevada	66.3	W
Missouri	9.3	E	New Mexico	20.1	W
Wisconsin	9.6	E	West Virginia	0.8	E
Maryland	10.8	E	Nebraska	8.4	W
Arizona	40.0	W	Idaho	28.5	W
Minnesota	12.4	E	Maine	3.9	E
Louisiana	5.9	E	New Hampshire	11.4	E
Alabama	10.1	E	Hawaii	9.3	W
Colorado	30.6	W	Rhode Island	4.5	E
Kentucky	9.7	E	Montana	12.9	W
South Carolina	15.1	E	Delaware	17.6	E
Oklahoma	9.7	W	South Dakota	8.5	W
Oregon	20.4	W	North Dakota	0.5	W
Connecticut	3.6	E	Alaska	14.0	W
Iowa	5.4	E	Vermont	8.2	E
Mississippi	10.5	E	Wyoming	8.9	W

a. Construct a stem-and-leaf display for the entire data set. Regard the observations as having two digits to the left of the decimal place. For example, consider 8.5 as 08.5. Also truncate leaves to a single digit. For example, a leaf of 8.5 would be truncated to 8.

b. Comment on any interesting features of the display. Do any of the observations appear to be outliers?

c. Now construct a comparative stem-and-leaf display for the eastern and western states. Write a few sentences comparing the two distributions.

2.35 The accompanying relative frequency table is based on data from the 2007 College Bound Seniors report for California (**College Board, 2008**).

Score on SAT Reasoning Exam	Relative Frequency for Males	Relative Frequency for Females
200 to < 250	0.0404	0.0183
250 to < 300	0.0546	0.0299
300 to < 350	0.1076	0.0700
350 to < 400	0.1213	0.0896
400 to < 450	0.1465	0.1286
450 to < 500	0.1556	0.1540
500 to < 550	0.1400	0.1667
550 to < 600	0.1126	0.1550
600 to < 650	0.0689	0.1050
650 to < 700	0.0331	0.0529
700 to < 750	0.0122	0.0194
750 to < 800	0.0072	0.0105

a. Construct a relative frequency histogram for males.

b. Construct a relative frequency histogram for females.

c. Based on the histograms from Parts (a) and (b), write a few sentences commenting on the similarities and differences in the two SAT score distributions.

2.36   Using the five class intervals 100 to < 120, 120 to < 140, … , 180 to < 200, construct a frequency distribution based on 70 observations whose histogram could be described as follows:
a. symmetric
b. bimodal
c. positively skewed
d. negatively skewed

2.37   Example 2.17 used annual rainfall data for Albuquerque, New Mexico, to construct a relative frequency distribution and cumulative relative frequency plot. The National Climate Data Center also gave the accompanying annual rainfall (in inches) for Medford, Oregon, from 1950 to 2008.

28.84	20.15	18.88	25.72	16.42	20.18	28.96	20.72	23.58	10.62
20.85	19.86	23.34	19.08	29.23	18.32	21.27	18.93	15.47	20.68
23.43	19.55	20.82	19.04	18.77	19.63	12.39	22.39	15.95	20.46
16.05	22.08	19.44	30.38	18.79	10.89	17.25	14.95	13.86	15.30
13.71	14.68	15.16	16.77	12.33	21.93	31.57	18.13	28.87	16.69
18.81	15.15	18.16	19.99	19.00	23.97	21.99	17.25	14.07	

a. Construct a relative frequency distribution for the Medford rainfall data.
b. Use the relative frequency distribution of Part (a) to construct a histogram. Describe the shape of the histogram.
c. Construct a cumulative relative frequency plot for the Medford rainfall data.
d. Use the cumulative relative frequency plot of Part (c) to answer the following questions:
   i. Approximately what proportion of years had annual rainfall less than 15.5 inches?
   ii. Approximately what proportion of years had annual rainfall less than 25 inches?
   iii. Approximately what proportion of years had annual rainfall between 17.5 and 25 inches?

Additional Exercises

2.38   The **National Survey on Drug Use and Health,** conducted in 2006 and 2007 by the Office of Applied Studies, estimated the total number of people age 12 and older in each state who had used a tobacco product within the last month.

State	Number of People (in thousands)	State	Number of People (in thousands)
Alabama	1,307	Idaho	305
Alaska	161	Illinois	3,149
Arizona	1,452	Indiana	1,740
Arkansas	819	Iowa	755
California	6,751	Kansas	726
Colorado	1,171	Kentucky	1,294
Connecticut	766	Louisiana	1,138
Delaware	200	Maine	347
DC	141	Maryland	1,206
Florida	4,392	Massachusetts	1,427
Georgia	2,341	Michigan	2,561
Hawaii	239	Minnesota	1,324

State	Number of People (in thousands)	State	Number of People (in thousands)
Mississippi	763	Pennsylvania	3,170
Missouri	1,627	Rhode Island	268
Montana	246	South Carolina	1,201
Nebraska	429	South Dakota	202
Nevada	612	Tennessee	1,795
New Hampshire	301	Texas	5,533
New Jersey	1,870	Utah	402
New Mexico	452	Vermont	158
New York	4,107	Virginia	1,771
North Carolina	2,263	Washington	1,436
North Dakota	162	West Virginia	582
Ohio	3,256	Wisconsin	1,504
Oklahoma	1,057	Wyoming	157
Oregon	857		

a. Construct a stem-and-leaf display using the thousands digit as the stem and truncating the leaves to the hundreds digit. Make sure to include 0 as the thousands digit for any three-digit numbers.
b. Write a few sentences describing the shape of the distribution and any unusual observations.
c. The four largest values were for California, Texas, Florida, and New York. Does this indicate that tobacco use is more of a problem in these states than elsewhere? Explain.
d. If you wanted to compare states on the extent of tobacco use, would you use the data in the given table? If yes, explain why this would be reasonable. If no, what would you use instead?

2.39   The accompanying data on annual maximum wind speed (in meters per second) in Hong Kong for each year in a 45-year period are from an article that appeared in the journal *Renewable Energy* (March, 2007). Use the data to construct a histogram. Is the histogram approximately symmetric, positively skewed, or negatively skewed? Would you describe the histogram as unimodal, bimodal, or multimodal?

30.3	39.0	33.9	38.6	44.6	31.4	26.7	51.9	31.9
27.2	52.9	45.8	63.3	36.0	64.0	31.4	42.2	41.1
37.0	34.4	35.5	62.2	30.3	40.0	36.0	39.4	34.4
28.3	39.1	55.0	35.0	28.8	25.7	62.7	32.4	31.9
37.5	31.5	32.0	35.5	37.5	41.0	37.5	48.6	28.1

2.40   The Connecticut Agricultural Experiment Station conducted a study of the calorie content of different types of beer. The calorie contents (calories per 100 ml) for 26 brands of light beer are (from the website **brewery.org**):

29	28	33	31	30	33	30
28	27	41	39	31	29	23
32	31	32	19	40	22	34
31	42	35	29	43		

(continued)

Construct a stem-and-leaf display using stems 1, 2, 3, and 4. Write a sentence or two describing the calorie content of light beers.

2.41 The article **"Tobacco and Alcohol Use in G-Rated Children's Animated Films"** (*Journal of the American Medical Association* [1999]: 1131–1136) reported exposure to tobacco and alcohol use in all G-rated animated films released between 1937 and 1997 by five major film studios. The researchers found that tobacco use was shown in 56% of the reviewed films. Data on the total tobacco exposure time (in seconds) for G-rated films with tobacco use that were produced by Walt Disney, Inc., were as follows:

223	176	548	37	158	51	299	37
11	165	74	92	6	23	206	9

Data for 11 G-rated animated films showing tobacco use that were produced by MGM/United Artists, Warner Brothers, Universal, and Twentieth Century Fox were also given. The tobacco exposure times (in seconds) for these films were as follows:

205	162	6	1	117	5	91	155
24	55	17					

Construct a comparative stem-and-leaf display for these data. Comment on the interesting features of this display.

SECTION 2.4 Displaying Bivariate Numerical Data: Scatterplots and Time Series Plots

A bivariate data set consists of measurements or observations on two variables, x and y. For example, x might be the weight of a car and y the gasoline mileage rating of the car. When both x and y are numerical variables, each observation consists of a pair of numbers, such as (14, 5.2) or (27.63, 18.9). The first number in a pair is the value of x, and the second number is the value of y.

An unorganized list of bivariate data doesn't tell you much about the distribution of the x values or the distribution of the y values, and tells you even less about how the two variables are related to one another. Just as graphical displays are used to summarize univariate data, they can also be used to summarize bivariate data. The most important graph of bivariate numerical data is a **scatterplot**.

In a scatterplot, each observation (pair of numbers) is represented by a point on a rectangular coordinate system, like the one shown in Figure 2.29(a). Figure 2.29(b) shows the point representing the observation (4.5, 15).

FIGURE 2.29
Constructing a scatterplot:
(a) rectangular coordinate system;
(b) point corresponding to (4.5, 15)

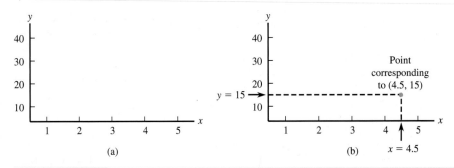

SCATTERPLOT

When to Use   Number of variables: 2
Data type: Numerical
Purpose: Investigate the relationship between variables

How to Construct
1. Draw horizontal and vertical axes. Label the horizontal axis and include an appropriate scale for the x variable. Label the vertical axis and include an appropriate scale for the y variable.
2. For each (x, y) pair in the data set, add a dot at the appropriate location in the display.

What to Look For   Relationship between x and y

In many data sets, the range of the x values, the range of the y values, or both may be far away from 0. For example, a study of how air conditioner efficiency is related to maximum daily outdoor temperature might involve observations at temperatures of 80°, 82°, ..., 98°, 100°. In such cases, the plot will be more informative if the axes intersect at some point other than (0, 0) and are marked accordingly.

Example 2.18　Worth the Price You Pay?

The accompanying table gives the cost and an overall quality rating for 10 different brands of men's athletic shoes (**www.consumerreports.org**).

Cost	Rating	Cost	Rating
65	71	110	57
45	70	30	56
45	62	80	52
80	59	110	51
110	58	70	51

Is there a relationship between x = cost and y = quality rating? A scatterplot can help answer this question.

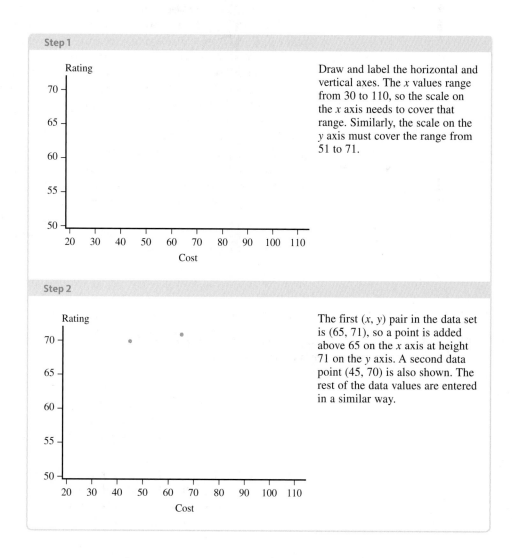

Step 1

Draw and label the horizontal and vertical axes. The x values range from 30 to 110, so the scale on the x axis needs to cover that range. Similarly, the scale on the y axis must cover the range from 51 to 71.

Step 2

The first (x, y) pair in the data set is (65, 71), so a point is added above 65 on the x axis at height 71 on the y axis. A second data point (45, 70) is also shown. The rest of the data values are entered in a similar way.

Figure 2.30 shows the completed scatterplot. There is an interesting and unexpected pattern. The larger costs tend to be paired with the lower quality ratings, suggesting that there is actually a negative association between cost and quality!

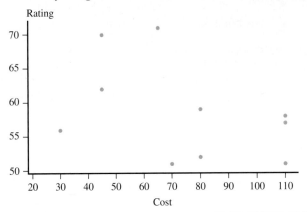

FIGURE 2.30
Scatterplot of athletic shoe data of Example 2.18

Time Series Plots

Data sets often consist of measurements collected over time at regular intervals so that you can learn about change over time. For example, stock prices, sales figures, and other socioeconomic indicators might be recorded on a weekly or monthly basis. A **time-series plot** (sometimes also called a time plot) is a simple graph of data collected over time that can help you see interesting trends or patterns.

A time series plot can be constructed by thinking of the data set as a bivariate data set, where y is the variable observed and x is the time at which the observation was made. These (x, y) pairs are plotted as in a scatterplot. Consecutive observations are then connected by a line segment.

Example 2.19 The Cost of Christmas

The Christmas Price Index is computed each year by PNC Advisors. It is a humorous look at the cost of giving all of the gifts described in the popular Christmas song "The 12 Days of Christmas." The year 2008 was the most costly year since the index began in 1984, with the "cost of Christmas" at $21,080. A plot of the Christmas Price Index over time appears on the PNC web site (**www.pncchristmaspriceindex.com**), and the data given there were used to construct the time series plot of Figure 2.31. The plot shows an upward trend in the index from 1984 until 1993. There has also been a clear upward trend in the index since 1995. You can visit the web site to see individual time series plots for each of the 12 gifts that are used to determine the Christmas Price Index (a partridge in a pear tree, two turtle doves, and so on). See if you can figure out what caused the dramatic decline from 1993 to 1995.

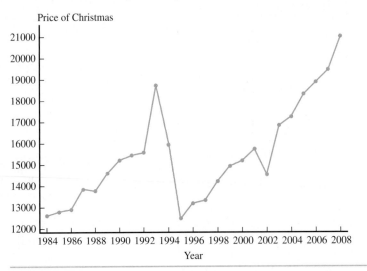

FIGURE 2.31
Time series plot for the Christmas Price Index data

Example 2.20   Education and Income—Stay in School!

The time series plot shown in Figure 2.32 appears on the U.S. Census Bureau web site. It relates the average earnings of workers by education level to the average earnings of a high school graduate and shows the change over time. For example, you can see that in 1993 the average earnings for people with bachelor's degrees was about 1.5 times the average for high school graduates. In that same year, the average earnings for those who were not high school graduates was only about 75% of the average for high school graduates. The time series plot also shows that the gap between the average earnings for high school graduates and those with a bachelor's degree or an advanced degree widened during the 1990s.

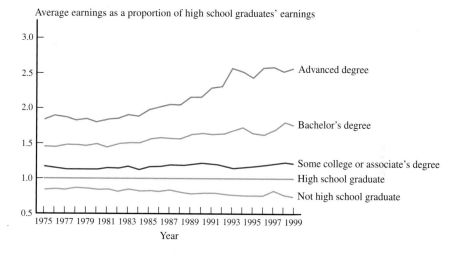

FIGURE 2.32
Time series plot for average earnings as a proportion of the average earnings of high school graduates

---

EXERCISES

Each Exercise Set assesses the following chapter learning objectives: M6, M7, P4, P5

SECTION 2.4   Exercise Set 1

2.42   *Consumer Reports Health* (www.consumerreports.org) gave the following data on saturated fat (in grams), sodium (in mg), and calories for 36 fast-food items.

a. Construct a scatterplot using y = calories and x = fat. Does it look like there is a relationship between fat and calories? Is the relationship what you expected? Explain.

b. Construct a scatterplot using y = calories and x = sodium. Write a few sentences commenting on the difference between this scatterplot and the scatterplot from Part (a).

c. Construct a scatterplot using y = sodium and x = fat. Does there appear to be a relationship between fat and sodium?

d. Add a vertical line at x = 3 and a horizontal line at y = 900 to the scatterplot in Part (c). This divides the scatterplot into four regions, with some points falling into each region. Which of the four regions corresponds to healthier fast-food choices? Explain.

Fat	Sodium	Calories	Fat	Sodium	Calories
2	1042	268	1	635	180
5	921	303	6	440	290
3	250	260	4.5	490	290
2	770	660	5	1160	360

Fat	Sodium	Calories	Fat	Sodium	Calories
3.5	970	300	1	1620	340
1	1120	315	4	660	380
2	350	160	3	840	300
3	450	200	1.5	1050	490
6	800	320	3	1440	380
3	1190	420	9	750	560
2	1090	120	1	500	230
5	570	290	1.5	1200	370
3.5	1215	285	2.5	1200	330
2.5	1160	390	3	1250	330
0	520	140	0	1040	220
2.5	1120	330	0	760	260
1	240	120	2.5	780	220
3	650	180	3	500	230

2.43   The report **"Trends in Higher Education" (www.college board.com)** gave the accompanying data on smoking rates for people age 25 and older by education level.

a. Construct a time series plot for people who did not graduate from high school, and comment on any trend over time.

(continued)

	Percent Who Smoke			
Year	Not a High School Graduate	High School Graduate But no College	Some College	Bachelor's Degree or Higher
1960	44	46	48	44
1965	44	45	47	41
1970	42	44	45	37
1975	41	42	43	33
1980	38	40	39	28
1985	36	37	36	23
1990	34	34	32	19
1995	32	31	29	16
2000	30	28	26	14
2005	29	27	21	10

b. Construct a time series plot that shows trend over time for each of the four education levels. Graph each of the four time series on the same set of axes, using different colors to distinguish the different education levels. Either label the time series in the plot or include a legend to indicate which time series corresponds to which education level.

c. Write a paragraph based on your plot from Part (b). Discuss the similarities and differences for the four different education levels.

SECTION 2.4   **Exercise Set 2**

2.44   The accompanying table gives the cost and an overall quality rating for 15 different brands of bike helmets (**www.consumerreports.org**).

Cost	Rating	Cost	Rating
35	65	40	42
20	61	28	41
30	60	20	40
40	55	25	32
50	54	30	63
23	47	30	63
30	47	40	53
18	43		

a. Construct a scatterplot using y = quality rating and x = cost.

b. Based on the scatterplot from Part (a), does there appear to be a relationship between cost and quality rating? Does the scatterplot indicate that the more expensive bike helmets tended to receive higher quality ratings?

2.45   The article "**Medicine Cabinet Is a Big Killer**" (*The Salt Lake Tribune*, **August 1, 2007**) looked at the number of prescription drug overdose deaths in Utah from 1991 to 2006. Construct a time-series plot for these data and describe the trend over time. Has the number of overdose deaths increased at a fairly steady rate?

Year	Number of Overdose Deaths	Year	Number of Overdose Deaths
1991	32	1999	88
1992	52	2000	109
1993	73	2001	153
1994	61	2002	201
1995	68	2003	237
1996	64	2004	232
1997	85	2005	308
1998	89	2006	307

2.46   The report "**Wireless Substitution: Early Release of Estimates from the National Health Interview Survey**" (**Center for Disease Control, 2009**) estimated the percentage of homes in the United States that had only wireless phone service at 6-month intervals from June 2005 to December 2008.

Date	Percent with Only Wireless Phone Service
June 2005	7.3
December 2005	8.4
June 2006	10.5
December 2006	12.8
June 2007	13.6
December 2007	15.8
June 2008	17.5
December 2008	20.2

Construct a time-series plot for these data and describe the trend over time. Has the percent increased at a fairly steady rate?

Additional Exercises

2.47   The National Telecommunications and Information Administration published a report titled "**Falling Through the Net: Toward Digital Inclusion**" (**U.S. Department of Commerce, October 2000**) that included the following information on access to computers in the home:

Year	Percentage of Households with a Computer
1985	8.2
1990	15.0
1994	22.8
1995	24.1
1998	36.6
1999	42.1
2000	51.0

a. Construct a time series plot for these data. Be careful—the observations are not equally spaced in time. The points in the plot should not be equally spaced along the x axis.

b. Comment on any trend over time.

2.48 2005 was a record year for hurricane devastation in the United States. Of the 26 tropical storms and hurricanes in the season, four hurricanes hit the mainland: Dennis, Katrina, Rita, and Wilma. The approximate insured catastrophic losses nationwide each year from 1989 to 2005 (*San Luis Obispo Tribune,* **November 30, 2005**) were as follows:

Year	Cost (in billions of dollars)	Year	Cost (in billions of dollars)
1989	7.5	1998	10.0
1990	2.5	1999	9.0
1991	4.0	2000	3.0
1992	22.5	2001	27.0
1993	5.0	2002	5.0
1994	18.0	2003	12.0
1995	9.0	2004	28.5
1996	8.0	2005	56.8
1997	2.6		

Construct a time series plot that shows the insured catastrophic loss over time. What do you think causes the peaks in the graph?

2.49 Data on x = poverty rate (%) and y = high school dropout rate (%) for the 50 U.S. states and the District of Columbia were used to construct the following scatterplot (*Chronicle of Higher Education,* **August 31, 2001**):

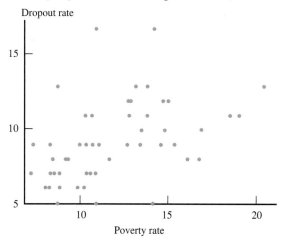

Write a few sentences commenting on this scatterplot. Would you describe the relationship between poverty rate and dropout rate as positive (y tends to increase as x increases), negative (y tends to decrease as x increases), or is there no clear relationship between x and y?

2.50 One cause of tennis elbow, a malady that strikes fear into the hearts of all serious tennis players, is the impact-induced vibration of the racket-and-arm system at ball contact. The likelihood of getting tennis elbow depends on various properties of the racket used. Consider the accompanying scatterplot of x = racket resonance frequency (in hertz) and y = sum of peak-to-peak accelerations (a characteristic of arm vibration, in meters per second per second) for 23 different rackets (**"Transfer of Tennis Racket Vibrations into the Human Forearm,"** *Medicine and Science in Sports and Exercise* **[1992]: 1134–1140).** Discuss interesting features of the scatterplot.

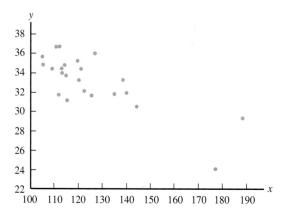

SECTION 2.5 Graphical Displays in the Media

There are several types of graphical displays that appear frequently in newspapers and magazines. In this section, we will see some alternative uses of bar charts as well as two other types of graphical displays—pie charts and segmented bar charts. Although we won't go into detail on how to construct these graphs, they pop up often enough to justify taking a quick look at them.

Pie Charts

A pie chart is another way of displaying the distribution of a categorical data set. A circle is used to represent the whole data set, with "slices" of the pie representing the categories. The size of a particular category's slice is proportional to its frequency or relative frequency. Pie charts are most effective for summarizing data sets when there are not too many categories.

Example 2.21 Life Insurance for Cartoon Characters?

The article **"Fred Flintstone, Check Your Policy"** (*The Washington Post,* October 2, 2005) summarized a survey of 1,014 adults conducted by the Life and Health Insurance Foundation for Education. Each person surveyed was asked to select which of five fictional characters had the greatest need for life insurance: Spider-Man, Batman, Fred Flintstone, Harry Potter, or Marge Simpson. The data are summarized in the pie chart of Figure 2.33.

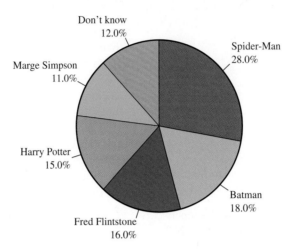

FIGURE 2.33
Pie chart of which fictional character most needs life insurance

The survey results were quite different from the assessment of an insurance expert. His opinion was that Fred Flintstone, a married father with a young child, was by far the one with the greatest need for life insurance. Spider-Man, unmarried with an elderly aunt, would need life insurance only if his aunt relied on him to supplement her income. Batman, a wealthy bachelor with no dependents, wouldn't need life insurance, in spite of his dangerous job!

A Different Type of "Pie" Chart: Segmented Bar Charts

A pie chart can be difficult to construct by hand, and the circular shape sometimes makes it difficult to compare areas for different categories, particularly when the relative frequencies are similar. The **segmented bar chart** (also called a stacked bar chart) avoids these difficulties by using a rectangular bar rather than a circle to represent the entire data set. The bar is divided into segments, with different segments representing different categories. As with pie charts, the area of the segment for a particular category is proportional to the relative frequency for that category. Example 2.22 illustrates the use of a segmented bar chart.

Example 2.22 How College Seniors Spend Their Time

Each year, the Higher Education Research Institute conducts a survey of college seniors. In 2008, approximately 23,000 seniors participated in the survey (**"Findings from the 2008 Administration of the College Senior Survey," Higher Education Research Institute, June, 2009**). The accompanying relative frequency table summarizes student responses to the question: "During the past year, how much time did you spend studying and doing homework in a typical week?"

Studying/Homework	
Amount of Time	Relative Frequency
2 hours or less	0.074
3 to 5 hours	0.227
6 to 10 hours	0.285
11 to 15 hours	0.181
16 to 20 hours	0.122
Over 20 hours	0.111

The corresponding segmented bar chart is shown in Figure 2.34.

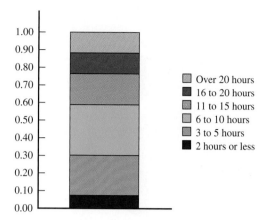

The same report also gave data on the amount of time spent on exercise or sports in a typical week. Figure 2.35 shows horizontal segmented bar charts (segmented bar charts can
be displayed either vertically or horizontally) for both time spent studying and time spent
exercising. Viewing these displays side-by-side makes it easy to see differences in time
spent on these two types of activities.

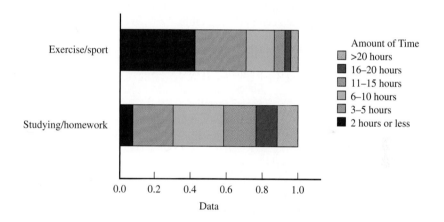

FIGURE 2.35
Segmented bar charts for time
spent studying and time spent
exercising per week

Other Uses of Bar Charts and Pie Charts

Bar charts and pie charts are used to summarize categorical data sets. However, they are
also occasionally used for other purposes, as illustrated in Examples 2.23 and 2.24.

Example 2.23 Grape Production

The 2008 Grape Crush Report for California gave the following information on grape
production for each of four different types of grapes **(California Department of Food and
Agriculture, March 10, 2009):**

Type of Grape	Tons Produced
Red Wine Grapes	1,715,000
White Wine Grapes	1,346,000
Raisin Grapes	494,000
Table Grapes	117,000
Total	**3,672,000**

Although this table is not a frequency distribution, it is common to represent information of this type graphically using a pie chart, as shown in Figure 2.36. The pie represents

the total grape production, and the slices show the proportion of the total production for each of the four types of grapes.

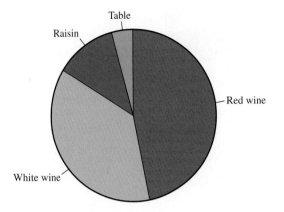

FIGURE 2.36
Pie chart for grape production data

Example 2.24 Back-to-College Spending

The National Retail Federation's 2008 Back to College Consumer Intentions and Actions Survey (**www.nrf.com**) asked college students how much they planned to spend in various categories during the upcoming academic year. The average amounts of money (in dollars) that men and women planned to spend for five different types of purchases are shown in the accompanying table.

Type of Purchase	Average for Men	Average for Women
Clothing and Accessories	$207.46	$198.15
Shoes	$107.22	$88.65
School Supplies	$86.85	$81.56
Electronics and Computers	$533.17	$344.90
Dorm or Apartment Furnishings	$266.69	$266.98

Even though this table is not a frequency distribution, this type of information is often represented graphically in the form of a bar chart, as illustrated in Figure 2.37. Here you can see that the average amounts are similar for all of the types of purchases except for electronics and computers, where the average for men is quite a bit higher.

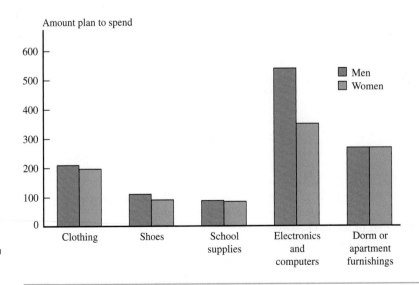

FIGURE 2.37
Comparative bar chart for the back-to-college spending for men and women

SECTION 2.5 EXERCISES

Each Exercise Set assesses the following chapter learning objective: P6

SECTION 2.5 **Exercise Set 1**

2.51 The following display appeared in *USA Today* (June 29, 2009). It is meant to be a bar graph of responses to the question shown in the display.

a. Is response to the question a categorical or numerical variable?

b. Explain why a bar chart rather than a dotplot was used to display the response data.

c. There must have been an error made in constructing this display. How can you tell that it is not a correct representation of the response data?

USA TODAY Snapshots®

If you were given $1,000, what would you do?

48%	Put it in general savings
27%	Pay off credit card debt
12%	Put it in children's education fund
10%	Use it for health care expenses
10%	Put it toward vacation

Source: Bank of America Consumer Purchasing and Savings Habits survey of 1,000 adults 18 and older. Margin of error ±4 percentage points.

USA Today. June 29, 2009, Reprinted with Permission.

2.52 The Center for Science in the Public Interest evaluated school cafeterias in 20 school districts across the United States. Each district was assigned a numerical score on the basis of rigor of food codes, frequency of food safety inspections, access to inspection information, and the results of cafeteria inspections. Based on the score assigned, each district was also assigned one of four grades. The scores and grades are summarized in the accompanying table, which appears in the report **"Making the Grade: An Analysis of Food Safety in School Cafeterias"** (cspi.us/new/pdf/makingthegrade.pdf, 2007).

a. Two variables are summarized in the table: grade and overall score. Is overall score a numerical or categorical variable? Is grade (indicated by the different colors in the table) a numerical or categorical variable?

b. Explain how the accompanying table is equivalent to a segmented bar chart of the grade data.

c. Construct a dotplot of the overall score data. Based on the dotplot, suggest an alternate assignment of grades (top of class, passing, and so on) to the 20 school districts. Explain the reasoning you used to make your assignment.

■ **Top of the Class**   ■ **Passing**   ■ **Barely Passing**   ■ **Failing**

School District	Overall Score (out of 100)
City of Fort Worth, TX	80
King County, WA	79
City of Houston, TX	78
Maricopa County, AZ	77
City and County of Denver, CO	75
Dekalb County, GA	73
Farmington Valley Health District, CT	72
State of Virginia	72
Fulton County, GA	68
City of Dallas, TX	67
City of Philadelphia, PA	67
City of Chicago, IL	65
City and County of San Francisco, CA	64
Montgomery County, MD	63
Hillsborough County, FL	60
City of Minneapolis, MN	60
Dade County, FL	59
State of Rhode Island	54
District of Columbia	46
City of Hartford, CT	37

2.53 The accompanying comparative bar chart is from the report **"More and More Teens on Cell Phones" (August 19, 2009, Pew Research Center, www.pewresearch.org).**

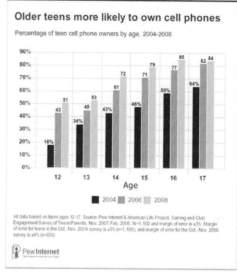

Suppose that you plan to include this graph in an article that you are writing for your school newspaper. Write a few paragraphs that could accompany the graph. Be sure to address how teen cell phone ownership is related to age and how it has changed over time.

SECTION 2.5 **Exercise Set 2**

2.54 The accompanying graphical display appeared in *USA Today* (October 22, 2009). It summarizes survey responses to a question about visiting social networking sites while at work. Which of the graph types introduced in this section is used here? (*USA Today* frequently adds artwork and text to their graphs to make them look more interesting.)

USA TODAY Snapshots®

No Facebooking for me

Does your company policy prohibit employees from visiting social-networking sites while at work?

No, but some limits apply **35%**

No **10%**

Don't know **1%**

Yes, completely **54%**

facebook   Home   Profile

Source: Robert Half Technology survey of 1,400 chief information officers. Weighted to represent actual population.

USA Today. September 03. 2002. Reprinted with Permission.

2.55 The accompanying graph appeared in *USA Today* (August 5, 2008). This graph is a modified comparative bar graph. The modifications (incorporating hands and the earth) were most likely made to construct a display that readers would find more interesting.

a. Use the information in the *USA Today* graph to construct a traditional comparative bar chart.

b. Explain why the modifications made in the *USA Today* graph may make interpretation more difficult than for the traditional comparative bar chart.

USA TODAY Snapshots®

Dreams for changing the world

When asked where they'd most like to make a difference in the world, more women than men focus closest to home:

■ **Men** ■ **Women**

42% **32%** Nation

36% **32%** World

36% **22%** Local community

Source: American Express survey of 1,000 adults conducted online by ICR and Authentic Response

USA Today. August 5, 2008. Reprinted with Permission.

2.56 The accompanying time series plot of movie box office totals (in millions of dollars) over 18 weeks of summer for both 2001 and 2002 appeared in *USA Today* (September 3, 2002).

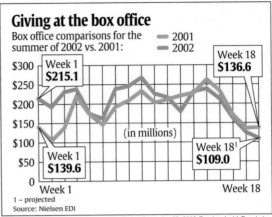

Giving at the box office

Box office comparisons for the summer of 2002 vs. 2001:

━ 2001
━ 2002

Week 1 **$215.1**

Week 18 **$136.6**

(in millions)

Week 1 **$139.6**

Week 18¹ **$109.0**

Week 1   Week 18

1 – projected
Source: Nielsen EDI

USA Today. September 03. 2002. Reprinted with Permission.

Patterns that tend to repeat on a regular basis over time are called seasonal patterns. Describe any seasonal patterns that you see in the summer box office data. (Hint: Look for patterns that seem to be consistent from year to year.)

Additional Exercises

2.57 The following display appeared in *USA Today* (May 16, 2011). It is meant to display responses to the question shown.

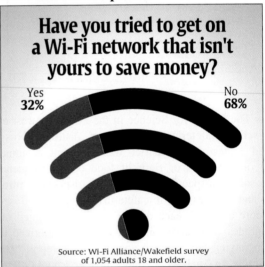

USA TODAY Snapshots®

Have you tried to get on a Wi-Fi network that isn't yours to save money?

Yes **32%**

No **68%**

Source: Wi-Fi Alliance/Wakefield survey of 1,054 adults 18 and older.

USA Today. May 16, 2011. Reprinted with Permission.

a. Explain how this display is misleading.

b. Construct a pie chart or a bar chart that accurately summarizes the responses.

2.58 The following graphical display is meant to be a comparative bar graph (*USA Today*, August 3, 2009). Do you think it is an effective summary of the data? If so, explain why. If not, explain why not and construct a display that makes it easier to compare the ice cream preferences of men and women.

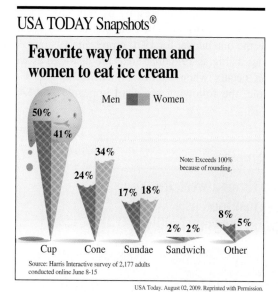

USA TODAY Snapshots®

Favorite way for men and women to eat ice cream

Men / Women

50% 41% Cup
34% 24% Cone
17% 18% Sundae
2% 2% Sandwich
8% 5% Other

Note: Exceeds 100% because of rounding.

Source: Harris Interactive survey of 2,177 adults conducted online June 8-15

USA Today. August 02, 2009. Reprinted with Permission.

2.59 An article that appeared in *USA Today* (September 3, 2003) included a graph similar to the one shown here summarizing responses from polls conducted in 1978, 1991, and 2003, in which samples of American adults were asked whether or not it was a good time or a bad time to buy a house.

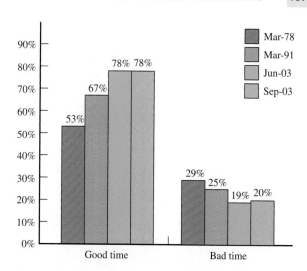

Mar-78, Mar-91, Jun-03, Sep-03

Good time: 53% 67% 78% 78%
Bad time: 29% 25% 19% 20%

a. Construct a time series plot that shows how the percentage that thought it was a good time to buy a house has changed over time.

b. Add a new line to the plot from Part (a) showing the percentage that thought it was a bad time to buy a house over time. Be sure to label the lines.

c. Which graph, the given bar chart or the time series plot, best shows the trend over time?

SECTION 2.6 Avoid These Common Mistakes

When constructing or evaluating graphical displays, you should keep the following in mind:

1. ***Areas should be proportional to frequency, relative frequency, or magnitude of the number being represented.*** The eye is naturally drawn to large areas in graphical displays, and it is natural for the observer to make informal comparisons based on area. Correctly constructed graphical displays, such as pie charts, bar charts, and histograms, are designed so that the areas of the pie slices or the bars are proportional to frequency or relative frequency. Sometimes, in an effort to make graphical displays more interesting, designers lose sight of this important principle, and the resulting graphs are misleading. For example, consider the following graph (*USA Today*, October 3, 2002):

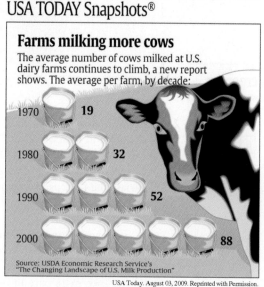

USA TODAY Snapshots®

Farms milking more cows

The average number of cows milked at U.S. dairy farms continues to climb, a new report shows. The average per farm, by decade:

1970 19
1980 32
1990 52
2000 88

Source: USDA Economic Research Service's "The Changing Landscape of U.S. Milk Production"

USA Today. August 03, 2009. Reprinted with Permission.

In trying to make the graph more visually interesting by replacing the bars of a bar chart with milk buckets, areas are distorted. For example, the two buckets for 1980 represent 32 cows, whereas the one bucket for 1970 represents 19 cows. This is misleading because 32 is not twice as big as 19. Other areas are distorted as well.

Another common distortion occurs when a third dimension is added to bar charts or pie charts. For example, the following pie chart appeared in *USA Today* (September 17, 2009).

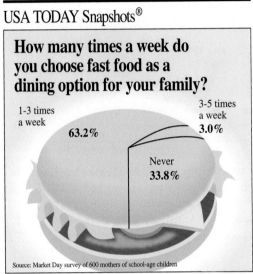

USA Today. September 17, 2009. Reprinted with Permission.

Adding the third dimension distorts the areas and makes it much more difficult to interpret correctly. A correctly drawn pie chart follows.

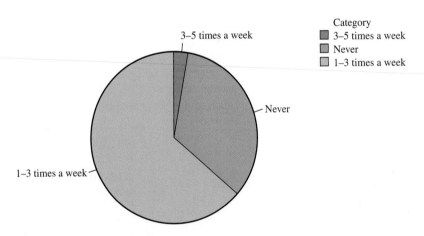

2. *Be cautious of graphs with broken axes (axes that don't start at 0).* Although it is common to see scatterplots with broken axes, be extremely cautious of time series plots, bar charts, or histograms with broken axes. The use of broken axes in a scatterplot does not result in a misleading picture of the relationship in the bivariate data set used to construct the display. On the other hand, in time series plots, broken axes can sometimes exaggerate the magnitude of change over time. Although it is not always a bad idea to break the vertical axis in a time series plot, it is something you should watch for. If you see a time series plot with a broken axis, as in the accompanying time series plot of mortgage rates (*USA Today*, October 25, 2002), you should pay particular attention to the scale on the vertical axis and take extra care in interpreting the graph.

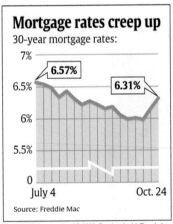

USA Today. October 25, 2002. Reprinted with Permission.

In bar charts and histograms, the vertical axis (which represents frequency, relative frequency, or density) should *never* be broken. If the vertical axis is broken in this type of graph, the resulting display will violate the "proportional area" principle, and the display will be misleading. For example, the following bar chart is similar to one in an advertisement for a software product designed to raise student test scores. By starting the vertical axis at 50, the gain from using the software is exaggerated. Areas of the bars are not proportional to the magnitude of the numbers represented—the area for the rectangle representing 68 is more than three times the area of the rectangle representing 55!

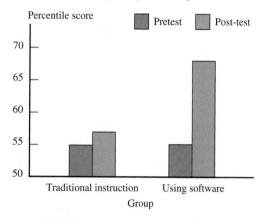

3. ***Watch out for unequal time spacing in time series plots.*** If observations over time are not made at regular time intervals, special care must be taken in constructing the time series plot. Consider the accompanying time series plot, which is similar to one in a *San Luis Obispo Tribune* (September 22, 2002) article on online banking:

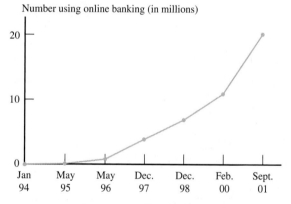

Notice that the intervals between observations are irregular, yet the points in the plot are equally spaced along the time axis. This makes it difficult to assess the rate of change over time. This could have been remedied by spacing the observations appropriately along the time axis, as shown here:

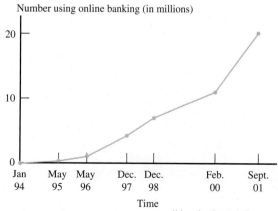

4. ***Be careful how you interpret patterns in scatterplots.*** A strong pattern in a scatterplot means that the two variables tend to vary together in a predictable way, but it does not mean that there is a cause-and-effect relationship. We will consider this point further in Chapter 4, but in the meantime, when describing patterns in scatterplots, you can't say that changes in one variable *cause* changes in the other.

5. ***Make sure that a graphical display creates the right first impression.*** For example, consider the following graph from *USA Today* (June 25, 2001). Although this graph does not violate the proportional area principle, the way the "bar" for the none category is displayed makes this graph difficult to read, and a quick glance at this graph may leave the reader with an incorrect impression.

Source: CarMax.com   USA Today. June 25, 2001. Reprinted with Permission.

CHAPTER ACTIVITIES

ACTIVITY 2.1 TWITTER WORDS

This activity requires Internet access.

TweetVolume is a web site that allows you to enter up to three words to produce a bar chart based on how often those words appear on Twitter.

1. Go to www.tweetvolume.com and spend a few minutes experimenting with different words to see how the site works. For example, in July 2010, the words statistics, sample, and population resulted in the following bar chart.

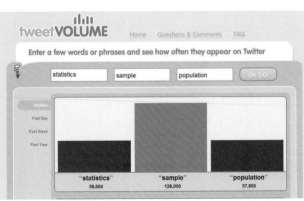

Courtesy of Tweet Volume

2. Find a set of three words that result in a bar chart in which all three bars are approximately the same height.

3. Find a set of three words that satisfy the following:
   i. One word begins with the letter *a*, one word begins with the letter *b*, and one word begins with the letter *c*.
   ii. The word that begins with the letter *a* is more common on Twitter (has a higher bar in the bar chart) than the other two words.
   iii. The word that begins with the letter *b* is more common on Twitter than the word that begins with the letter *c*.

ACTIVITY 2.2 BEAN COUNTERS!

Materials needed: A large bowl of dried beans (or marbles, plastic beads, or any other small, fairly regular objects) and a coin.

In this activity, you will investigate whether people can hold more in their right hand or in their left hand.

1. Flip a coin to determine which hand you will measure first. If the coin lands heads side up, start with the right hand. If the coin lands tails side up, start with the left hand. With the designated hand, reach into the bowl and grab as many beans as possible. Raise the hand over the bowl and count to four. If no beans drop during the count to four, drop the beans onto a piece of paper and record the number of beans grabbed. If any beans drop during the count, restart the count. Repeat the process with the other hand, and then record the following information: (1) right-hand number, (2) left-hand number, and (3) dominant hand (left or right, depending on whether you are left- or right-handed).

2. Create a class data set by recording the values of the three variables listed in Step 1 for each student in your class.

3. Using the class data set, construct a comparative stem-and-leaf display with the right-hand numbers displayed on the right and the left-hand numbers displayed on the left. Comment on the interesting features of the display and include a comparison of the right-hand number and left-hand number distributions.

4. Now construct a comparative stem-and-leaf display that allows you to compare dominant-hand count to nondominant-hand count. Does the display support the theory that dominant-hand count tends to be higher than nondominant-hand count?

5. For each person, compute the difference (dominant number − nondominant number). Construct a stem-and-leaf display of these differences. Comment on the interesting features of this display.

6. Explain why looking at the distribution of the differences (Step 5) provides more information than the comparative stem-and-leaf display (Step 4). What information is lost in the comparative stem-and-leaf display?

EXPLORING THE BIG IDEAS

In the two exercises below, each student in your class will go online to select a random sample from a small population consisting of 300 adults between the ages of 18 and 64.

To learn about the age distribution of the people in this population, go online at www.cengage.com/stats/peck1e and click on the link labeled "Exploring the Big Ideas." Then locate the link for Chapter 2. When you click on this link, it will take you to a web page where you can select a random sample of 50 people from the population.

Click on the sample button. This selects a random sample and will display information for the 50 people in your sample. You should see the following information for each person selected:

1. An ID number that identifies the person selected
2. Sex
3. Age

Each student in your class will receive data from a different random sample.

Use the data from your random sample to complete the following two exercises.

1. This exercise uses the age data.
   a. Flip a coin two times and record the outcome (H or T) for each toss. These outcomes will determine which of the following class intervals you will use to create a frequency distribution and histogram of your age data.

Coin Toss Outcomes	Use Class Intervals
HH	15 to < 25, 25 to < 35, 35 to < 45, 45 to < 55, 55 to < 65
HT	18 to < 28, 28 to < 38, 38 to < 48, 48 to < 58, 58 to < 68
TH	15 to < 20, 20 to < 25, 25 to < 30, ..., 55 to < 60, 60 to < 65
TT	18 to < 23, 23 to < 28, 28 to < 33, ..., 58 to < 63, 63 to < 68

What class intervals will you be using?
b. Using the age data for the people in your sample, construct a frequency distribution.
c. Draw a histogram of the sample age data.
d. Describe the shape of the age histogram from Part (c).
e. Here is a histogram that displays the age distribution of the entire population. How is your histogram similar to the population histogram? How is it different?

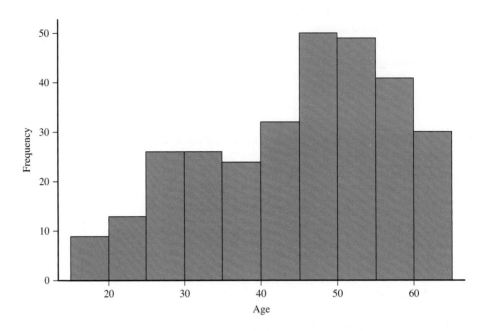

If asked to do so by your teacher, bring your histogram with you to class. Your teacher will lead a class discussion of the following:
f. Compare your histogram to a histogram produced by another student who used different class intervals. How are the two histograms similar? How are the two histograms different?
g. How do the histograms produced by the class support the following statement: Histograms based on random samples from a population tend to look like the population histogram.

2. In this exercise, you will compare the age distribution of the males and the age distribution of the females in your sample.
   a. Flip a coin to determine what type of comparative graphical display you will construct. If your flip results in a head, you will use a comparative dotplot. If your flip results in a tail, you will use a comparative stem-and-leaf display. Which type of graphical display will you be constructing?
   b. Using the data in your sample, construct either a comparative dotplot or a comparative stem-and-leaf display—depending on the outcome of Part (a)—that allows you to compare males and females. You will probably have different numbers of males and females in your sample, but this is not something to worry about.
   c. Based on your graphical display, write a few sentences comparing the two age distributions.

   If asked to do so by your teacher, bring your graphical display to class. Your teacher will lead a class discussion of the following:
   d. Compare your graphical display to that of another student who used a different type of display. Did you both make similar statements when you compared the two age distributions in Part (c), even though you each used different random samples and constructed different types of displays?

ARE YOU READY TO MOVE ON?   CHAPTER 2 REVIEW EXERCISES

All chapter learning objectives are assessed in these exercises. The learning objectives assessed in each exercise are given in parentheses.

2.60 (C1, C2)

In a survey of 100 people who had recently purchased motorcycles, data on the following variables were recorded:

Gender of purchaser
Brand of motorcycle purchased
Number of previous motorcycles owned by purchaser
Telephone area code of purchaser
Weight of motorcycle as equipped at purchase

a. Which of these variables are categorical?
b. Which of these variables are discrete numerical?
c. Which type of graphical display would be an appropriate choice for summarizing the gender data, a bar chart or a dotplot?
d. Which type of graphical display would be an appropriate choice for summarizing the weight data, a bar chart or a dotplot?

2.61 (C3, M1)

For each of the five data sets described, answer the following three questions and then use Figure 2.2 (on page 60) to choose an appropriate graphical display for summarizing the data.

Question 1: How many variables are in the data set?
Question 2: Is the data set categorical or numerical?
Question 3: Would the purpose of the graphical display be to summarize the data distribution, to compare groups, or to investigate the relationship between two numerical variables?

Data Set 1: To learn about credit card debt of students at a college, the financial aid office asks each student in a random sample of 75 students about his or her amount of credit card debt.

Data Set 2: To learn about how number of hours worked per week and number of hours spent watching television in a typical week are related, each person in a sample of size 40 was asked to keep a log for one week of hours worked and hours spent watching television. At the end of the week, each person reported the total number of hours spent on each activity.

Data Set 3: To see if satisfaction level differs for airline passengers based on where they sit on the airplane, all passengers on a particular flight were surveyed at the end of the flight. Passengers were grouped based on whether they sat in an aisle seat, a middle seat, or a window seat. Each passenger was asked to indicate his or her satisfaction with the flight by selecting one of the following choices: very satisfied, satisfied, dissatisfied, and very dissatisfied.

Data Set 4: To learn about where students purchase textbooks, each student in a random sample of 200 students at a particular college was asked to select one of the following responses: campus bookstore, off-campus bookstore, purchased all books online, or used a combination of online and bookstore purchases.

Data Set 5: To compare the amount of money men and women spent on their most recent haircut, each person in a sample of 20 women and in a sample of 20 men was asked how much was spent on his or her most recent haircut.

2.62 (M2)

The article **"Where College Students Buy Textbooks"** (*USA Today*, October 14, 2010) gave data on where students purchased books. The accompanying frequency table summarizes data from a sample of 1,200 full-time college students.

Where Books Purchased	Frequency
Campus bookstore	576
Campus bookstore web site	48
Online bookstore other than campus bookstore	240
Off-campus bookstore	168
Rented textbooks	36
Purchased mostly eBooks	12
Didn't buy any textbooks	72

Construct a bar chart to summarize the data distribution. Write a few sentences commenting on where students are buying textbooks.

2.63 (M2, P2)

An article about college loans (**"New Rules Would Protect Students,"** *USA Today*, June 16, 2010) reported the percentage of students who had defaulted on a student loan within 3 years of when they were scheduled to begin repayment. Information was given for public colleges, private nonprofit colleges, and for-profit colleges.

	Relative Frequency		
Loan Status	Public Colleges	Private Nonprofit Colleges	For-Profit Colleges
Good Standing	0.928	0.953	0.833
In Default	0.072	0.047	0.167

a. Construct a comparative bar chart that would allow you to compare loan status for the three types of colleges.
b. The article states, "Those who attended for-profit schools were more likely to default than those who attended public or private non-profit schools." What aspect of the comparative bar chart supports this statement?

2.64 (M3)

The article **"Fliers Trapped on Tarmac Push for Rules on Release"** (*USA Today*, July 28, 2009) gave the following data for 17 airlines on the number of flights that were delayed on the tarmac for at least 3 hours for the period between October 2008 and May 2009:

Airline	Number of Delays	Rate Per 10,000 Flights
ExpressJet	93	4.9
Continental	72	4.1
Delta	81	2.8
Comair	29	2.7
American Eagle	44	1.6
US Airways	46	1.6
JetBlue	18	1.4
American	48	1.3
Northwest	24	1.2
Mesa	17	1.1
United	29	1.1
Frontier	5	0.9
SkyWest	29	0.8
Pinnacle	13	0.7
Atlantic Southeast	11	0.6
AirTran	7	0.4
Southwest	11	0.1

The figure at the bottom of the page shows two dotplots: one displays the number of delays data, and one displays the rate per 10,000 flights data.

a. If you were going to rank airlines based on flights delayed on the tarmac for at least 3 hours, would you use the *total number of flights* data or the *rate per 10,000 flights* data? Explain the reason for your choice.

b. Write a short paragraph that could be part of a newspaper article on flight delays that could accompany the dotplot of the *rate per 10,000 flights* data.

2.65 (M3, P1)

The following gasoline tax per gallon data for each of the 50 U.S. states and the District of Columbia (DC) were listed in the article **"Paying at the Pump"** (*AARP Bulletin*, June 2010).

a. Construct a stem-and-leaf display of these data.

State	Gasoline Tax (cents per gallon)	State	Gasoline Tax (cents per gallon)
Alabama	20.9	Montana	27.8
Alaska	8.0	Nebraska	27.7
Arizona	19.0	Nevada	33.1
Arkansas	21.8	New Hampshire	19.6
California	48.6	New Jersey	14.5
Colorado	22.0	New Mexico	18.8
Connecticut	42.6	New York	44.9
Delaware	23.0	North Carolina	30.2
DC	23.5	North Dakota	23.0
Florida	34.4	Ohio	28.0
Georgia	20.9	Oklahoma	17.0
Hawaii	45.1	Oregon	25.0
Idaho	25.0	Pennsylvania	32.3
Illinois	40.4	Rhode Island	33.0
Indiana	34.8	South Carolina	16.8
Iowa	22.0	South Dakota	24.0
Kansas	25.0	Tennessee	21.4
Kentucky	22.5	Texas	20.0
Louisiana	20.0	Utah	24.5
Maine	31.0	Vermont	24.7
Maryland	23.5	Virginia	19.6
Massachusetts	23.5	Washington	37.5
Michigan	35.8	West Virginia	32.2
Minnesota	27.2	Wisconsin	32.9
Mississippi	18.8	Wyoming	14.0
Missouri	17.3		

b. Based on the stem-and-leaf display, what do you notice about the center and spread of the data distribution?

c. Do any values in the data set stand out as unusual? If so, which states correspond to the unusual observations and how do these values differ from the rest?

2.66 (M5, P1, P2)

The report **"Trends in College Pricing 2009"** (www.college board.com) included the information in the accompanying relative frequency distributions for public and private non-profit four-year college students.

a. Construct a relative frequency histogram for tuition and fees for students at public four-year colleges. Write a few sentences describing the distribution of tuition and fees, commenting on center, spread and shape.

FIGURE FOR EXERCISE **2.64**

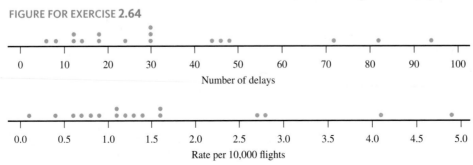

b. Construct a relative frequency histogram for tuition and fees for students at private nonprofit four-year colleges. Be sure to use the same scale for the vertical and horizontal axes as you used for the histogram in Part (a). Write a few sentences describing the distribution of tuition and fees for students at private nonprofit four-year colleges.

c. Write a few sentences describing the differences in the distributions.

Tuition and Fees	Proportion of Students (Relative Frequency)	
	Public Four-Year College Students	Private Nonprofit Four-Year College Students
0 to < 3,000	0.01	0.00
3,000 to < 6,000	0.32	0.05
6,000 to < 9,000	0.41	0.02
9,000 to < 12,000	0.13	0.02
12,000 to < 15,000	0.04	0.04
15,000 to < 18,000	0.03	0.06
18,000 to < 21,000	0.02	0.08
21,000 to < 24,000	0.02	0.11
24,000 to < 27,000	0.02	0.12
27,000 to < 30,000	0.00	0.14
30,000 to < 33,000	0.00	0.09
33,000 to < 36,000	0.00	0.07
36,000 to < 39,000	0.00	0.13
39,000 to < 42,000	0.00	0.07

2.67 (M3, M5, M6, P1, P4)

The chapter preview example introduced data from a survey of new car owners conducted by the J.D. Power and Associates marketing firm (*USA Today*, June 16 and July 17, 2010). For each brand of car sold in the United States, data on a quality rating (number of defects per 100 vehicles) and a customer satisfaction rating (called the APEAL rating) are given in the accompanying table. The APEAL rating is a score between 0 and 1000, with higher values indicating greater satisfaction. In this exercise, you will use graphical displays of these data to answer the questions posed in the preview example.

Brand	Quality Rating	APEAL Rating
Acura	86	822
Audi	111	832
BMW	113	845
Buick	114	802
Cadillac	111	818
Chevrolet	111	789
Chrysler	122	748
Dodge	130	751
Ford	93	794
GMC	126	792
Honda	95	766
Hyundai	102	760

Brand	Quality Rating	APEAL Rating
Infiniti	107	805
Jaguar	130	854
Jeep	129	727
Kia	126	761
Land Rover	170	836
Lexus	88	822
Lincoln	106	820
Mazda	114	774
Mercedes-Benz	87	842
Mercury	113	769
Mini Cooper	133	815
Mitsubishi	146	767
Nissan	111	763
Porsche	83	877
Ram	110	780
Scion	114	764
Subaru	121	755
Suzuki	122	750
Toyota	117	745
Volkswagen	135	797
Volvo	109	795

a. Construct a dotplot of the quality rating data.
b. Based on the dotplot, what is a typical value for quality rating?
c. Is there a lot of variability in the quality rating for the different brands? What aspect of the dotplot supports your answer?
d. Are some brands much better or much worse than most in terms of quality rating? Write a few sentences commenting on the brands that appear to be different from most.
e. Construct a histogram of the APEAL ratings and comment on center, spread, and shape of the data distribution.
f. Construct a scatterplot of x = quality rating and y = APEAL rating. Does customer satisfaction (as measured by the APEAL rating) appear to be related to car quality? Explain.

2.68 (M7, P5)

The article **"Cities Trying to Rejuvenate Recycling Efforts"** (*USA Today*, October 27, 2006) states that the amount of waste collected for recycling has grown slowly in recent years. This statement is supported by the data in the accompanying table. Use these data to construct a time series plot. Explain how the plot is consistent or is not consistent with the given statement.

Year	Recycled Waste (in millions of tons)
1990	29.7
1991	32.9
1992	36.0
1993	37.9
1994	43.5

(continued)

Year	Recycled Waste (in millions of tons)
1995	46.1
1996	46.4
1997	47.3
1998	48.0
1999	50.1
2000	52.7
2001	52.8
2002	53.7
2003	55.8
2004	57.2
2005	58.4
2006	56.8

2.69

The authors of the paper **"Myeloma in Patients Younger than Age 50 Years Presents with More Favorable Features and Shows Better Survival"** (*Blood* [2008]: 4039–4047) studied patients who had been diagnosed with stage 2 multiple myeloma prior to the age of 50. For each patient who received high dose chemotherapy, the number of years that the patient lived after the therapy (survival time) was recorded. The cumulative relative frequencies in the accompanying table were approximated from survival graphs that appeared in the paper.

Years Survived	Cumulative Relative Frequency
0 to < 2	0.10
2 to < 4	0.52
4 to < 6	0.54
6 to < 8	0.64
8 to < 10	0.68
10 to < 12	0.70
12 to < 14	0.72
14 to < 16	1.00

a. Use the given information to construct a cumulative relative frequency plot.

b. Use the cumulative relative frequency plot from Part (a) to answer the following questions:
   i. What is the approximate proportion of patients who lived fewer than 5 years after treatment?
   ii. What is the approximate proportion of patients who lived fewer than 7.5 years after treatment?
   iii. What is the approximate proportion of patients who lived more than 10 years after treatment?

2.70 (P6)

The accompanying graphical display is from the Fall 2008 Census Enrollment Report at Cal Poly, San Luis Obispo. It uses both a pie chart and a segmented bar graph to summarize ethnicity data for students enrolled in Fall 2008.

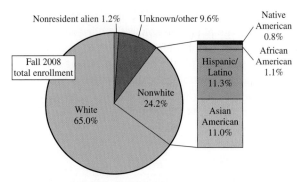

a. Use the information in the graphical display to construct a single segmented bar graph for the ethnicity data.

b. Do you think that the original graphical display or the one you created in Part (a) is more informative? Explain your choice.

c. Why do you think that the original graphical display format (combination of pie chart and segmented bar graph) was chosen over a single pie chart with seven slices?

2.71 (P6)

The following two graphical displays appeared in *USA Today* (June 1, 2009 and July 28, 2009). One is an appropriate representation and the other is not. For each of the two, explain why it is or is not drawn appropriately.

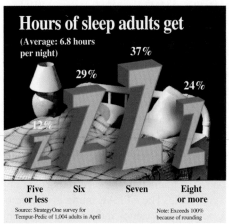

2.72 (P6)

Explain why the graphical display on the right (*USA Today, September 17, 2009*) is misleading.

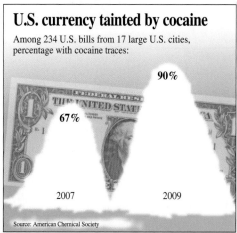

USA TODAY Snapshots®

U.S. currency tainted by cocaine

Among 234 U.S. bills from 17 large U.S. cities, percentage with cocaine traces:

67% — 2007

90% — 2009

Source: American Chemical Society

USA Today. September 17, 2009. Reprinted with Permission.

TECHNOLOGY NOTES

Bar Charts

TI-83/84

The TI-83/84 does not have the functionality to produce bar charts.

TI-Nspire

1. Enter the category names into a list (to access data lists select the spreadsheet option and press **enter**)

Note: In order to correctly enter category names, you will input them as text. Begin by pressing ?!> and then select " and press **enter**. Now type the category name and press **enter**.

Note: Be sure to title the list by selecting the top row of the column and typing a title.

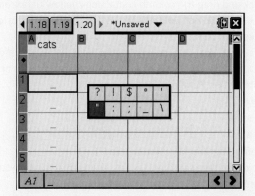

2. Enter the counts for each category into a second list (be sure to title this list as well!)
3. Press the **menu** key and navigate to **3:Data** and then **5:Frequency Plot** and select this option

4. For **Data List** select the list containing your category names from the drop-down menu
5. For **Frequency List** select the list containing the counts for each category from the drop-down menu
6. Press **OK**

JMP

1. Enter the raw data into a column (**Note:** To open a new data table, click **File** then select **New** then **Data Table**)
2. Click **Graph** and select **Chart**
3. Click and drag the column name containing your stored data from the box under **Select Columns** to the box next to **Categories, X, Levels**
4. Click **OK**

MINITAB

1. Enter the raw data into C1
2. Select **Graph** and choose **Bar Chart...**
3. Highlight Simple
4. Click **OK**
5. Double click C1 to add it to the **Categorical Variables** box
6. Click **OK**

Note: You may add or format titles, axis titles, legends, etc., by clicking on the **Labels...** button prior to performing step 6 above.

SPSS

1. Enter the raw data into a column
2. Select **Graph** and choose **Chart Builder...**
3. Under **Choose from** highlight **Bar**

4. Click and drag the first bar chart in the right box (Simple Bar) to the Chart preview area

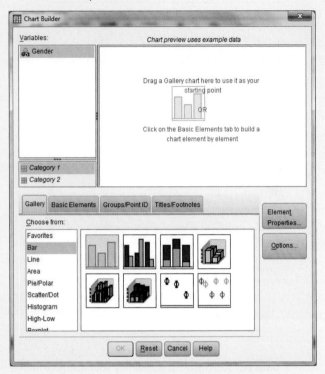

5. Click and drag the variable from the **Variables** box to the **X-Axis?** box in the chart preview area
6. Click **OK**

Note: By default, SPSS displays the count of each category in a bar chart. To display the percentage for each category, follow the above instructions to Step 5. Then, in the Element Properties dialog box, select Percentage from the drop-down box under Statistic. Click **Apply** then click **OK** on the Chart Builder dialog box.

Excel 2007
1. Enter the category names into column A (you may input the title for the variable in cell A1)
2. Enter the count or percent for each category into column B (you may put the title "Count" or "Percentage" in cell B1)
3. Select all data (including column titles if used)
4. Click on the **Insert** Ribbon

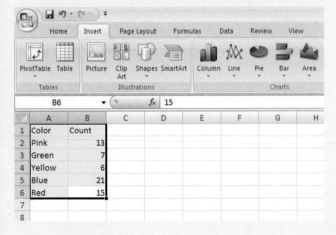

5. Choose **Column** and select the first chart under 2-D Column (Clustered Column)
6. The chart will appear on the same worksheet as your data

Note: You may add or format titles, axis titles, legends, and so on, by right-clicking on the appropriate piece of the chart.

Note: Using the Bar Option on the Insert Ribbon will produce a horizontal bar chart.

Comparative Bar Charts

TI-83/84
The TI-83/84 does not have the functionality to produce comparative bar charts.

TI-Nspire
The TI-Nspire does not have the functionality to produce comparative bar charts.

JMP
1. Enter the raw data into one column
2. In a separate column enter the group information

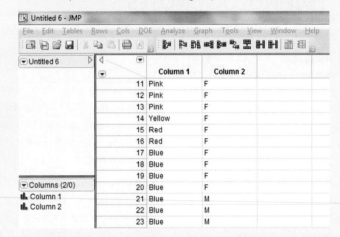

3. Click **Graph** then select **Chart**
4. Click and drag the column containing the raw data from the box under **Select Columns** to the box next to **Categories, X, Levels**
5. Click and drag the column containing the group information from the box under **Select Columns** to the box next to **Categories, X, Labels**
6. Click **OK**

MINITAB
1. Input the group information into C1
2. Input the raw data into C2 (be sure to match the response from the first column with the appropriate group)

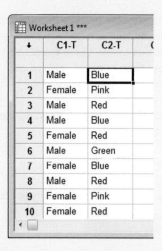

3. Select **Graph** then choose **Bar Chart...**
4. Highlight Cluster
5. Click **OK**
6. Double click C1 to add it to the **Categorical Variables** box
7. Then double click C2 to add it to the **Categorical Variables** box
8. Click **OK**

Note: Be sure to list the grouping variable first in the Categorical Variables box.

Note: You may add or format titles, axis titles, legends, and so on, by clicking on the **Labels...** button prior to performing Step 8 above.

SPSS

1. Enter the raw data into one column
2. Enter the group information into a second column (be sure to match the response from the first column with the appropriate group)

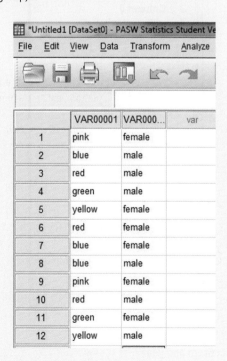

3. Select **Graph** and choose **Chart Builder...**
4. Under **Choose from** highlight Bar
5. Click and drag the second bar chart in the first column (Clustered Bar) to the Chart preview area
6. Click and drag the group variable (second column) into the **Cluster on X** box in the upper right corner of the chart preview area
7. Click and drag the data variable (first column) into the **X-Axis?** box in the chart preview area

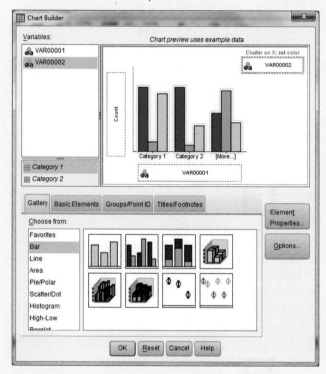

8. Click **Ok**

Excel 2007

1. Enter the category names into column A (you may input the title for the variable in cell A1)
2. Enter the count or percent for different groups into a new column, starting with column B
3. Select all data (including column titles if used)

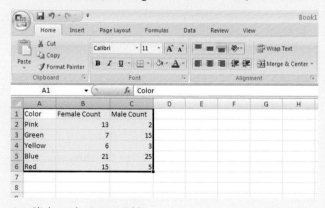

4. Click on the **Insert** Ribbon
5. Choose **Column** and select the first chart under 2-D Column (Clustered Column)

6. The chart will appear on the same worksheet as your data

Note: You may add or format titles, axis titles, legends, and so on, by right-clicking on the appropriate piece of the chart.

Note: Using the Bar Option on the **Insert** Ribbon will produce a horizontal bar chart.

Dotplots

TI-83/84
The TI-83/84 does not have the functionality to create dotplots.

TI-Nspire
1. Enter the data into a data list (to access data lists select the spreadsheet option and press **enter**)

Note: Be sure to title the list by selecting the top row of the column and typing a title.

2. Press the **menu** key then select **3:Data** then select **6:QuickGraph** and press **enter** (a dotplot will appear)

JMP
1. Input the raw data into a column
2. Click **Graph** and select **Chart**
3. Click and drag the column name containing the data from the box under **Select columns** to the box next to **Categories, X, Levels**
4. Select **Point Chart** from the second drop-down box in the **Options** section

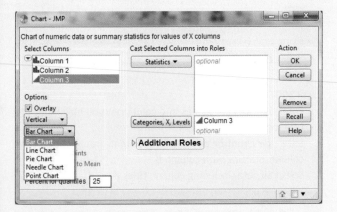

5. Click **OK**

MINITAB
1. Input the raw data into C1
2. Select **Graph** and choose **Dotplot**
3. Highlight Simple under One Y
4. Click **OK**
5. Double-click C1 to add it to the **Graph Variables** box
6. Click **OK**

Note: You may add or format titles, axis titles, legends, and so on, by clicking on the **Labels...** button prior to performing Step 6 above.

SPSS
1. Input the raw data into a column
2. Select **Graph** and choose **Chart Builder...**
3. Under **Choose from** highlight Scatter/Dot
4. Click and drag the second option in the second row (Simple Dot Plot) to the Chart preview area
5. Click and drag the variable name from the **Variables** box into the **X-Axis?** box
6. Click **OK**

Excel 2007
Excel 2007 does not have the functionality to create dotplots.

Stem-and-leaf plots

TI-83/84
The TI-83/84 does not have the functionality to create stem-and-leaf plots.

TI-Nspire
The TI-Nspire does not have the functionality to create stem-and-leaf plots.

JMP
1. Enter the raw data into a column
2. Click **Analyze** then select **Distribution**
3. Click and drag the column name containing the data from the box under **Select Columns** to the box next to **Y, Columns**
4. Click **OK**
5. Click the red arrow next to the column name

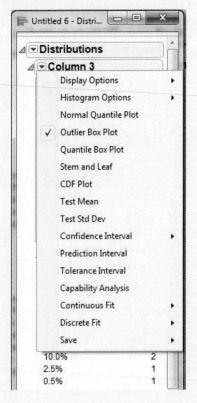

6. Select **Stem and Leaf**

MINITAB

1. Input the raw data into C1
2. Select **Graph** and choose **Stem-and-leaf...**
3. Double-click C1 to add it to the **Graph Variables** box
4. Click **OK**

Note: You may add or format titles, axis titles, legends, and so on, by clicking on the **Labels...** button prior to performing Step 4 above.

SPSS

1. Input the raw data into a column
2. Select **A̲nalyze** and choose **Descriptive Statistics** then **Explore**
3. Highlight the variable name from the box on the left
4. Click the arrow to the left of the Dependent List box to add the variable to this box

5. Click **OK**
6. **Note:** The stem-and-leaf plot is one of the plots produced along with several other descriptive statistics and plots.

Excel 2007

Excel 2007 does not have the functionality to create stem-and-leaf plots automatically.

Histograms

TI-83/84

1. Enter the data into **L1** (In order to access lists press the **STAT** key, highlight the option called **Edit...** then press **ENTER**)
2. Press the **2ⁿᵈ** key then the **Y =** key
3. Highlight the option for Plot1 and press **ENTER**
4. Highlight **On** and press **ENTER**
5. For **Type**, select the histogram shape and press **ENTER**

6. Press **GRAPH**

Note: If the graph window does not display appropriately, press the **WINDOW** button and reset the scales appropriately.

TI-Nspire

1. Enter the data into a data list (In order to access data lists select the spreadsheet option and press **enter**)

Note: Be sure to title the list by selecting the top row of the column and typing a title.

2. Press the **menu** key then select **3:Data** then select **6:QuickGraph** and press **enter** (a dotplot will appear)
3. Press the **menu** key and select **1:Plot Type**
4. Select **3:Histogram** and press **enter**

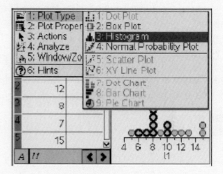

JMP

1. Enter the raw data into a column
2. Click **Analyze** then select **Distribution**
3. Click and drag the column name containing the data from the box under **Select Columns** to the box next to **Y, Columns**
4. Click **OK**
5. Click the red arrow next to the column name
6. Select **Histogram Options** and click **Vertical**

MINITAB

1. Input the raw data into C1
2. Select **Graph** and choose **H̲istogram...**
3. Highlight Simple
4. Click **OK**
5. Double-click C1 to add it to the **Graph Variables** box
6. Click **OK**

Note: You may add or format titles, axis titles, legends, and so on, by clicking on the **Labels...** button prior to performing Step 5 above.

SPSS

1. Input the raw data into a column
2. Select **Graph** and choose **C̲hart Builder...**
3. Under **Choose from** highlight Histogram
4. Click and drag the first option (Simple Histogram) to the Chart preview area
5. Click and drag the variable name from the **Variables** box into the **X-Axis?** box
6. Click **OK**

Excel 2007

Note: In order to produce histograms in Excel 2007, we will use the Data Analysis Add-On. For instructions on installing this add-on, please see the note at the end of this chapter's Technology Notes.

1. Input the raw data in column A (you may use a column heading in cell A1)
2. Click on the **Data** ribbon
3. Choose the **Data Analysis** option from the **Analysis** group
4. Select **Histogram** from the Data Analysis dialog box and click **OK**
5. Click in the **Input Range:** box and then click and drag to select your data (including the column heading)
   a. If you used a column heading, click the checkbox next to **Labels**
6. Click the checkbox next to **Chart Output**
7. Click **OK**

Note: If you have specified bins manually in another column, click in the box next to **Bin Range:** and select your bin assignments.

Scatterplots

TI-83/84

1. Input the data for the dependent variable into L2 (In order to access lists press the **STAT** key, highlight the option called **Edit…** then press **ENTER**)
2. Input the data for the dependent variable into L1
3. Press the **2nd** key then press the **Y=** key
4. Select the **Plot1** option and press **ENTER**
5. Select **On** and press **ENTER**
6. Select the scatterplot option and press **ENTER**

7. Press the **GRAPH** key

Note: If the graph window does not display appropriately, press the **WINDOW** button and reset the scales appropriately.

TI-Nspire

1. Enter the data for the independent variable into a data list (In order to access data lists, select the spreadsheet option and press **enter**)

Note: Be sure to title the list by selecting the top row of the column and typing a title.

2. Enter the data for the dependent variable into a separate data list
3. Highlight both columns of data (by scrolling to the top of the screen to highlight one list, then press shift and use the arrow key to highlight the second list)
4. Press the **menu** key and select **3:Data** then select **6:QuickGraph** and press **enter**

JMP

1. Input the data for the independent variable into the first column
2. Input the data for the dependent variable into the second column
3. Click **Analyze** then select **Fit Y by X**
4. Click and drag the column name for the independent data from the box under **Select Columns** to the box next to **Y, Response**
5. Click and drag the column name for the dependent data from the box under **Select Columns** to the box next to **X, Factor**
6. Click **OK**

MINITAB

1. Input the raw data for the independent variable into C1
2. Input the raw data for the dependent variable into C2
3. Select **Graph** and choose **Scatterplot**
4. Highlight Simple
5. Click **OK**
6. Double-click on C2 to add it to the **Y Variables** column of the spreadsheet
7. Double-click on C1 to add it to the **X Variables** column of the spreadsheet
8. Click **OK**

Note: You may add or format titles, axis titles, legends, and so on, by clicking on the **Labels…** button prior to performing Step 7 above.

SPSS

1. Input the raw data into two columns
2. Select **Graph** and choose **Chart Builder…**
3. Under **Choose from** highlight Scatter/Dot
4. Click and drag the first option (Simple Scatter) to the Chart preview area
5. Click and drag the variable name representing the independent variable to the **X-Axis?** box
6. Click and drag the variable name representing the dependent variable to the **Y-Axis?** box
7. Click **OK**

Excel 2007

1. Input the raw data into two columns (you may enter column headings)
2. Select both columns of data
3. Click on the **Insert** ribbon
4. Click **Scatter** and select the first option (Scatter with Markers Only)

Note: You may add or format titles, axis titles, legends, and so on, by right-clicking on the appropriate piece of the chart.

Time series plots

TI-83/84

1. Input the time increment data into **L1**
2. Input the data values for each time into **L2**
3. Press the **2nd** key then the **Y=** key
4. Select **1:Plot1** and press **ENTER**

5. Highlight **On** and press **ENTER**
6. Highlight the second option in the first row and press **ENTER**
7. Press **GRAPH**

Note: If the graph window does not display appropriately, press the **WINDOW** button and reset the scales appropriately.

TI-Nspire

1. Enter the data for the time increments into a data list (In order to access data lists select the spreadsheet option and press **enter**)

Note: Be sure to title the list by selecting the top row of the column and typing a title.

2. Enter the data for each time increment into a separate data list
3. Highlight both columns of data (by scrolling to the top of the screen to highlight one list, then press shift and use the arrow key to highlight the second list)
4. Press the **menu** key and select **3:Data** then select **6:QuickGraph** and press **enter**
5. Press the **menu** key and select **1:Plot Type** then select **6:XY Line Plot** and press **enter**

JMP

1. Input the raw data into one column
2. Input the time increment data into a second column

Column 4	Column 5
1	1900
2	1901
3	1902
4	1903
5	1904
6	1905
7	1906
8	1907
9	1908
1	1909
2	1910
3	1911
4	1912

3. Click **Graph** then select **Control Chart** then select **Run Chart**
4. Click and drag the name of the column containing the raw data from the box under **Select Columns** to the box next to **Process**
5. Click and drag the name of the column containing the time increment data from the box under **Select Columns** to the box next to **Sample Label**
6. Click **OK**

MINITAB

1. Input the raw data into C1
2. Select **Graph** and choose **Time Series Plot...**
3. Highlight Simple
4. Click **OK**
5. Double click C1 to add it to the **Series** box
6. Click the Time/Scale button

7. Choose the appropriate time scale under Time Scale (Choose Calendar if you want to use Days, Years, Months, Quarters, etc.)
8. Click to select one set for each variable
9. In the spreadsheet, fill in the start value for the time scale
10. Input the increment value into the box next to **Increment** (i.e., 1 to move by one year or one month, etc.)
11. Click **OK**
12. Click **OK**

Note: You may add or format titles, axis titles, legends, and so on, by clicking on the **Labels...** button prior to performing Step 12 above.

SPSS

1. Input the raw data into two columns: one representing the data and one representing the time increments
2. Select **Analyze** then choose **Forecasting** then **Sequence Charts...**
3. Highlight the column name containing the raw data and add it to the **Variables** box
4. Highlight the column name containing the time data and add it to the **Time Axis Labels** box
5. Click **OK**

Excel 2007

1. Input the data into two columns: one representing the data and one representing the time increments
2. Select an empty cell
3. Click on the **Insert** ribbon
4. Click **Line** and select the first option under 2-D Line (Line)
5. Right-click on the empty chart area that appears and select **Select Data...**
6. Under **Legend Entries (Series)** click **Add**
7. In the **Series name:** box select the column title for the data
8. In the **Series values:** box select the data values
9. Click **OK**
10. Under **Horizontal (Category) Axis Labels** click **Edit**
11. In the **Axis label range:** select the time increment data (do NOT select the column title)
12. Click **OK**
13. Click **OK**

Installing Excel 2007's Data Analysis Add-On

1. Click the **Microsoft Office Button** and then Click **Excel Options**
2. Click **Add-Ins**, then in the **Manage** box, select **Excel Add-ins**.
3. Click **Go**
4. In the **Add-Ins available** box, select the **Analysis ToolPak** checkbox, and then click **OK**

Note: If this option is not listed, click **Browse** to locate it.

Note: If you get prompted that the Analysis ToolPak is not currently installed on your computer, click **Yes** to install it.

5. After you load the Analysis ToolPak, the **Data Analysis** command is available in the **Analysis** group on the **Data** ribbon.

AP* Review Questions for Chapter 2

1. The comparative bar chart shown here is potentially misleading because

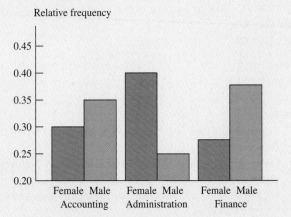

Relative frequency

(A) Frequency, not relative frequency, should have been used for the vertical axis.
(B) The vertical axis does not start at zero.
(C) The three blue bars should appear next to each other, and the three orange bars should appear next to each other.
(D) The vertical axis has increments of 0.05 instead of 0.10.
(E) Two separate bar charts should have been used.

2. *Time* magazine carried out a national online poll of 501 13-year-olds. One question asked 13-year-olds at what age they thought it was appropriate for them to begin dating, and another question asked at what age their parents thought it was appropriate for them to begin dating. The results are shown in the comparative bar chart below. Based on this graph, which of the following statements is correct?

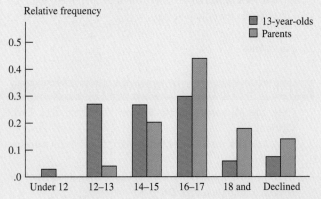

Relative frequency

(A) The proportion of parents is higher in the 16–17, 18 and older, and the declined-to-answer categories than the proportion of 13-year-olds.
(B) The proportion of parents who believe it is appropriate for their child to start dating at age 12–13 is higher than the proportion of 13-year-olds who believe this.

(C) A smaller proportion of 13-year-olds than their parents believe it is appropriate for them to start dating when they are under 12 years of age.
(D) The same proportion of 13 year-olds and their parents believe it is appropriate for them to start dating when they are 16–17 years of age.
(E) Parents and children generally agree on when it is appropriate to begin dating.

3. The stem-and-leaf display below shows the service times (in seconds) for a sample of 25 customers at Frank's Hotdogs.

Leaf Unit = 1.0

8	0113467789
9	1233
10	245
11	34
12	05
13	23
14	
15	9
16	
17	
18	
19	
20	0

The smallest value in the data set corresponds to what value?

(A) 80 seconds   (D) 0.8 seconds
(B) 8 seconds   (E) 0.08 seconds
(C) 0 seconds

4. The lengths (in feet) of 142 roller coasters are summarized in the following histogram.

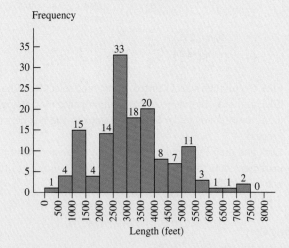

What is the proportion of roller coasters with a length less than 2500 feet?

(A) 0.500   (D) 0.232
(B) 0.380   (E) 0.099
(C) 0.268

5. Which of the following stem-and-leaf plots corresponds to a data distribution that is negatively (left) skewed?

I.

Leaf Unit = 0.10

```
1  | 467789
2  | 23557
3  | 2256
4  | 0238
5  | 08
6  | 011
7  | 7
```

II.

Leaf Unit = 0.10

```
 4 | 8
 5 | 3569
 6 | 01789
 7 | 1222778
 8 | 02567
 9 | 16
10 | 1
```

III.

Leaf Unit = 0.10

```
 3 | 3
 4 | 8
 5 | 35
 6 | 1789
 7 | 22778
 8 | 0567
 9 | 16678
10 | 1234669
```

(A) I only
(B) II only
(C) III only
(D) I and III only
(E) None of these stem-and-leaf plots is negatively skewed.

6. The resting heart rates (in beats per minute) of 228 adults are summarized in the histogram below.

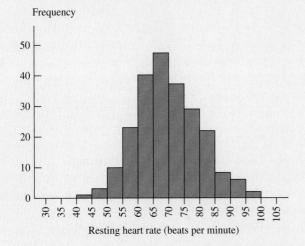

Frequency

Resting heart rate (beats per minute)

Which of the following values best describes a typical resting heart rate for this sample?

(A) 48 beats per minute
(B) 60 beats per minute
(C) 65 beats per minute
(D) 75 beats per minute
(E) 80 beats per minute

7. If a frequency distribution is based on intervals of unequal widths, the vertical axis used to construct the corresponding histogram must use which of the following?

(A) a frequency scale
(B) a relative frequency scale
(C) a density scale
(D) Either a frequency or a relative frequency scale can be used.
(E) A frequency, a relative frequency, or a density scale may be used.

8. The frequency distribution shown summarizes data on the percentage of college students enrolled in public institutions for the 50 U.S. states.

Percentage Enrolled	Frequency
40 to < 50	1
50 to < 60	5
60 to < 70	2
70 to < 80	11
80 to < 90	25
90 to < 100	6

What is the relative frequency for the 50 to < 60 interval?

(A) 0.05   (D) 10
(B) 0.10   (E) 50
(C) 5

9. The frequency distribution shown summarizes data on the percentage of college students enrolled in public institutions for the 50 U.S. states.

Percentage Enrolled	Frequency
40 to < 50	1
50 to < 60	5
60 to < 70	2
70 to < 80	11
80 to < 90	25
90 to < 100	6

What is the *cumulative* relative frequency for the 50 to < 60 interval?

(A) 0.05   (D) 5
(B) 0.06   (E) 6
(C) 0.12

10. The cumulative relative frequency plot shown was constructed using data on the percentage of college students enrolled in public institutions for the 50 U.S. states. Which of the following is closest to the approximate *proportion* of states that have fewer than 60 percent of college students enrolled in public institutions?

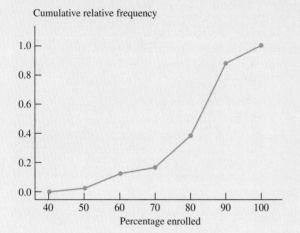

Cumulative relative frequency

(A) 0   (B) 0.20
(C) 0.40   (D) 0.60
(E) 0.80

11. Which of the following three scatterplots suggest that there is not a relationship between the two variables used to construct the plot?

(A) I only
(B) II only
(C) III only
(D) I and II only
(E) II and III only

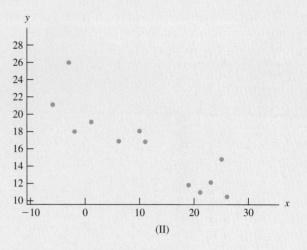

(II)

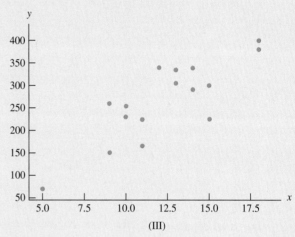

(III)

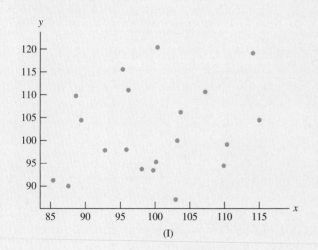

(I)

12. Which of the following best describes the time series plot below?

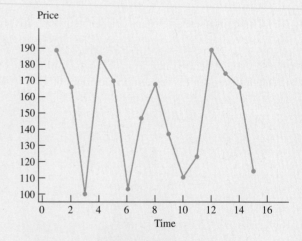

(A) There is a positive linear relationship.
(B) There is a negative linear relationship.
(C) There is an upward trend over time.
(D) There is a downward trend over time.
(E) There is no apparent long-term trend.

13. In a study of the birds in a national park, three species (labeled A, B, and C) were observed. Each bird in a sample of size 36 was classified by species. The species data were then used to construct the pie chart below. Approximately how many birds in the sample were species C?

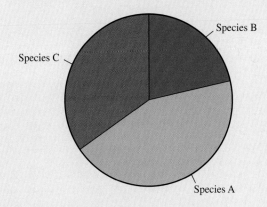

(A) 8   (D) 24
(B) 12   (E) 30
(C) 20

14. For which of the following variables would it be appropriate to construct a histogram?

(A) hair color
(B) stereo brand
(C) tree species
(D) gender
(E) phone call length

15. Which of the following variables would have a distribution that is negatively (left) skewed?

(A) number of textbooks in students' backpacks
(B) weights of strawberries
(C) scores on an easy test
(D) scores on a hard test
(E) None of the above variables would have a distribution that is negatively skewed.

3

Numerical Methods for Describing Data Distributions

Preview

Chapter Learning Objectives

3.1 Selecting Appropriate Numerical Summaries

3.2 Describing Center and Spread for Data Distributions That Are Approximately Symmetric

3.3 Describing Center and Spread for Data Distributions That Are Skewed or Have Outliers

3.4 Summarizing a Data Set: Boxplots

3.5 Measures of Relative Standing: z-scores and Percentiles

3.6 Avoid These Common Mistakes

Chapter Activities

Exploring the Big Ideas

Are You Ready to Move On? Chapter 3 Review Exercises

Technology Notes

AP* Review Questions for Chapter 3

© Blend Images/Alamy

PREVIEW

In Chapter 2, graphical displays were used to summarize data. By creating a visual display of the data distribution, it is easier to see and describe its important characteristics, such as shape, center, and spread. In this chapter, you will see how numerical measures are used to describe important characteristics of a data distribution.

CHAPTER LEARNING OBJECTIVES

Conceptual Understanding

After completing this chapter, you should be able to

C1 Understand how numerical summary measures are used to describe characteristics of a numerical data distribution.

C2 Understand how the variance and standard deviation describe variability in a data distribution.

C3 Understand the impact that outliers can have on measures of center and spread.

C4 Understand how the shape of a data distribution is considered when selecting numerical summary measures.

C5 Understand how the relative values of the mean and median are related to the shape of a data distribution.

C6 Understand the difference between sample statistics and population characteristics.

Mastering the Mechanics

After completing this chapter, you should be able to

M1 Select appropriate summary statistics for describing center and spread of a numerical data distribution.

M2 Compute and interpret the value of the sample mean.

M3 Compute and interpret the value of the sample standard deviation.

M4 Compute and interpret the value of the sample median.

M5 Compute and interpret the value of the sample interquartile range.

M6 Compute and interpret the values in the five-number summary.

M7 Given a numerical data set, construct a boxplot.

M8 Identify outliers in a numerical data set.

Putting It into Practice

After completing this chapter, you should be able to

P1 Interpret measures of center in context.

P2 Interpret measures of spread in context.

P3 Use boxplots to make comparisons between two or more groups.

P4 Use percentiles and *z*-scores to describe relative standing.

PREVIEW EXAMPLE

Just Thinking About Exercise??

Studies have shown that in some cases people overestimate the extent to which physical exercise can compensate for food consumption. When this happens, people increase food intake more than what is justified based on the exercise performed. The authors of the paper **"Just Thinking About Exercise Makes Me Serve More Food: Physical Activity and Calorie Compensation"** (*Appetite* [2011]: 332 – 335) wondered if even *thinking* about exercise would lead to increased food consumption. They carried out an experiment in which people were offered snacks as a reward for participating in the experiment. People read a short essay and then answered a few questions about the essay. Some participants read an essay that was unrelated to exercise (the control group), some read an essay that described listening to music while taking a 30 minute walk (the fun group), and some read an essay that described strenuous exercise (the exercise group). Participants were then provided with two plastic bags and invited to help themselves to two types of snacks—Chex Mix and M&Ms. After the participants served themselves, the bags were weighed so that the researchers could determine the number of calories in snacks taken.

Data on number of calories, consistent with summary values in the paper, were used to construct the comparative dotplot shown in Figure 3.1. From the dotplots, it is clear that the number of calories tends to be quite a bit higher for those who read about exercise than for those in the control group! If you want to compare the distributions with more precision, the first step is to describe them numerically.

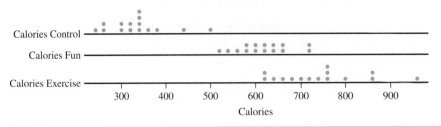

FIGURE 3.1
Comparative dotplot of calories

In Section 3.1, you will see how the shape of the data distribution affects the choice of numerical summary measures. In Section 3.2, the data distributions of the preview example are revisited to see how they can be described numerically.

SECTION 3.1 Selecting Appropriate Numerical Summaries

Suppose that you have just received your score on an exam in one of your classes. What would you want to know about the distribution of scores for this exam? Like most students, you would probably want to know two things: "What was the class average?" and "What were the high and the low scores?"

These questions consider important characteristics of the data distribution. By asking for the class average, you want to know what a "typical" exam score was. What number best describes the entire set of scores? By asking for the high and low scores on the exam, you want to know about the variability in the data set. Were the scores similar or did they differ quite a bit from student to student?

When describing a numerical data set, it is common to report both a value that describes where the data distribution is centered along the number line and a value that describes how spread out the data distribution is.

Measures of center describe where the data distribution is located along the number line. A measure of center provides information about what is "typical."

Measures of spread describe how much variability there is in a data distribution. A measure of spread provides information about how much individual values tend to differ from one another.

There is more than one way to measure center and spread in a data distribution. The two most common choices are to either use the mean and standard deviation or the median and interquartile range. These terms will be defined in Sections 3.2 and 3.3, where you will see how these summary measures are computed and interpreted. For now, we will focus on which of these two options is the best choice in a given situation.

The key factor in deciding how to measure center and spread is the shape of the data distribution, as described in Table 3.1. Because this decision is based on the shape of the distribution, the best place to start is with a graphical display of the data.

TABLE 3.1 **Choosing Appropriate Measures for Describing Center and Spread**

If the Shape of the Data Distribution Is...	Describe Center and Spread Using...
Approximately symmetric	Mean and standard deviation
Skewed or has outliers	Median and interquartile range

Example 3.1 Medical Errors

The stress of the final years of medical training can contribute to depression and burn-out. It has been estimated that the percentage of medical students completing a hospital residency who suffer from depression might be as high as 56%. The authors of the paper **"Rates of Medication Errors Among Depressed and Burnt Out Residents"** (*British Medical Journal* **[2008]: 488**) studied 24 residents in pediatrics who were classified as depressed based on their score on the Harvard National Depression Screening Scale. Medical records of patients treated by these residents during a fixed time period were examined for errors in ordering or administering medications. The accompanying data are the total number of medication errors for each of the 24 residents.

Number of Errors

0	0	0	0	0	0	0	0	0	0	0	0	0	0
0	0	0	1	1	1	2	3	5	11				

A dotplot of these data is shown in Figure 3.2.

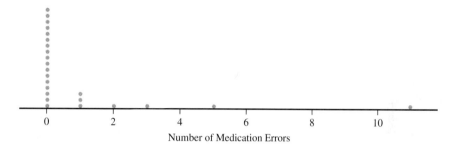

FIGURE 3.2
Dotplot of medication error data

From the dotplot, you can see that the data distribution is not symmetric and that there are outliers. For these reasons, you would choose the median and interquartile range to describe center and spread for this data set.

Before leaving this section, let's look at why the shape of the data distribution affects the choice of the summary measures that would be used to describe center and spread.

Example 3.2 Baseball Salaries

USA Today publishes professional baseball players' salaries at the beginning of each season (**content.usatoday.com/sports/baseball/salaries**). The following are the 2010 salaries for the San Francisco Giants players.

Player	Salary (in dollars)	Player	Salary (in dollars)
Jeremy Affeldt	4,000,000	Sergia Romo	416,500
John Bowker	410,000	Aaron Rowand	13,600,000
Emmanuel Burriss	410,000	Dan Runzler	400,500
Matt Cain	4,583,333	Freddy Sanchez	6,000,000
Mark De Rosa	6,000,000	Johnathan Sanchez	2,100,000
Aubrey Huff	3,000,000	Pablo Sandoval	465,000
Travis Ishikawa	417,000	Nate Schierholtz	416,500
Waldis Joaquin	400,500	Andres Torres	426,000
Fred Lewis	455,000	Juan Uribe	3,250,000
Tim Lincecum	9,000,000	Eugenio Velez	416,000
Brandom Medders	820,000	Todd Wellemeyer	1,000,000
Bengie Molina	4,500,000	Eli Whiteside	405,000
Guillermo Mota	750,000	Brian Wilson	6,500,000
Edgar Renteria	10,000,000	Barry Zito	18,500,000

A dotplot of these data is given in Figure 3.3.

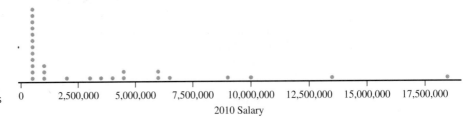

FIGURE 3.3
Dotplot of 2010 salaries for players
on the San Francisco Giants

2010 Salary

Suppose that you calculated the average salary for this team's players to describe a "typical" value. The average salary is $3,522,905. Does this seem typical for players on this team? The 12 players (43% of the team) with salaries around $400,000 probably don't think so! Because this distribution is quite skewed and has outliers, the average is pulled up by the small number of very large salaries. When this happens, the average is not the best choice to describe the center. The median (to be introduced in Section 3.3) would be a better choice.

SECTION 3.1 EXERCISES

Each Exercise Set assesses the following chapter learning objectives: C1, C2, M1

SECTION 3.1 Exercise Set 1

For each data set described in Exercises 3.1–3.3, construct a graphical display of the data distribution and then indicate what summary measures you would use to describe center and spread.

3.1 The accompanying data are consistent with summary statistics from a paper investigating the effect of the shape of drinking glasses (*British Medical Journal* [2005]: 1512–1514). They represent the actual amount (in ml) poured into a short, wide glass by individuals asked to pour 1.5 ounces (44.3 ml) into the glass.

89.2	68.6	32.7	37.4	39.6	46.8	66.1	79.2
66.3	52.1	47.3	64.4	53.7	63.2	46.4	63.0
92.4	57.8						

3.2 Data on tipping percent for 20 restaurant tables, consistent with summary statistics given in the paper **"Beauty and the Labor Market: Evidence from Restaurant Servers" (unpublished manuscript by Matt Parrett, 2007)**, are:

0.0	5.0	45.0	32.8	13.9	10.4	55.2	50.0
10.0	14.6	38.4	23.0	27.9	27.9	105.0	19.0
10.0	32.1	11.1	15.0				

3.3 Data on manufacturing defects per 100 cars for the 33 brands of cars sold in the United States (*USA Today*, June 16, 2010) are:

86	111	113	114	111	111	122	130	93	126	95
102	107	130	129	126	170	88	106	114	87	113
133	146	111	83	110	114	121	122	117	135	109

3.4 The accompanying data on number of cell phone minutes used in one month are consistent with summary statistics published in a report of a marketing study of San Diego residents **(Tele-Truth, March, 2009)**:

189	0	189	177	106	201	0	212	0	306
0	0	59	224	0	189	142	83	71	165
236	0	142	236	130					

Explain why the average may not be the best measure of a typical value for this data set.

SECTION 3.1 Exercise Set 2

For each data set described in Exercises 3.5–3.7, construct a graphical display of the data distribution and then indicate what summary measures you would use to describe center and spread.

3.5 The accompanying data are a subset of data that appeared in the paper **"Ladies First? A Field Study of Discrimination in Coffee Shops" (*Applied Economics* [April, 2008])**. The data are the times (in seconds) between ordering and receiving coffee for 19 male customers at a Boston coffee shop.

40	60	70	80	85	90	100	100	110	120
125	125	140	140	160	160	170	180	200	

3.6. For each brand of car sold in the United States, data on a customer satisfaction rating (called the APEAL rating) are given (*USA Today*, July 17, 2010). The APEAL rating is a score between 0 and 1,000, with higher values indicating greater satisfaction.

822	832	845	802	818	789	748	751	794
792	766	760	805	854	727	761	836	822

(continued)

820	774	842	769	815	767	763	877	780
764	755	750	745	797	795			

3.7 Data on weekday exercise time (in minutes) for 20 males, consistent with summary quantities given in the paper "**An Ecological Momentary Assessment of the Physical Activity and Sedentary Behaviour Patterns of University Students**" (*Health Education Journal* [2010]: 116–125), are:

Male—Weekday

43.5	91.5	7.5	0.0	0.0	28.5	199.5	57.0	142.5
8.0	9.0	36.0	0.0	78.0	34.5	0.0	57.0	151.5
8.0	0.0							

3.8 *USA Today* (**May 9, 2006**) published the accompanying average weekday circulation for the top 20 newspapers in the country for the 6-month period ending March 31, 2006:

2,272,815	2,049,786	1,142,464	851,832	724,242
708,477	673,379	579,079	513,387	438,722
427,771	398,329	398,246	397,288	365,011
362,964	350,457	345,861	343,163	323,031

Explain why the average may not be the best measure of a typical value for this data set.

Additional Exercises

For each data set described in Exercises 3.9–3.10, construct a graphical display of the data distribution and then indicate what summary measures you would use to describe center and spread.

3.9 Data on weekend exercise time (in minutes) for 20 males, consistent with summary quantities given in the paper "**An Ecological Momentary Assessment of the Physical Activity and Sedentary Behaviour Patterns of University Students**" (*Health Education Journal* [2010]: 116–125), are:

Male—Weekend

0	0	15	0	5	0	1	1	57	95
73	155	125	7	27	97	61	0	73	155

3.10 Data on weekday exercise time (in minutes) for 20 females, consistent with summary quantities given in the paper referenced in the previous exercise, are:

Female—Weekday

10.0	90.6	48.5	50.4	57.4	99.6
0.0	5.0	0.0	0.0	5.0	2.0
10.5	5.0	47.0	0.0	5.0	54.0
0.0	48.6				

3.11 Increasing joint extension is one goal of athletic trainers. In a study to investigate the effect of therapy that uses ultrasound and stretching (**Trae Tashiro, Masters Thesis, University of Virginia, 2004**), passive knee extension was measured after treatment. Passive knee extension (in degrees) is given for each of 10 study participants.

59	46	64	49	56	70	45	52	63	52

Which would you choose to describe center and spread—the mean and standard deviation or the median and interquartile range? Justify your choice.

Describing Center and Spread for Data Distributions That Are Approximately Symmetric

When the shape of a numerical data distribution is approximately symmetric, the best choice for describing center and spread is the mean and standard deviation.

The Mean

The mean of a numerical data set is just the familiar arithmetic average—the sum of all of the observations in the data set divided by the total number of observations. Before a formula for computing the mean is introduced, it is helpful to have some simple notation for the variable of interest and for the data set.

Notation

x = the variable of interest

n = the number of observations in the data set (the sample size)

x_1 = the first observation in the data set

x_2 = the second observation in the data set

$\vdots$

x_n = the nth (last) observation in the data set

For example, data consisting of the number of hours spent doing community service during the fall semester for $n = 10$ students might be

$$x_1 = 25 \qquad x_2 = 4 \qquad x_3 = 0 \qquad x_4 = 90 \qquad x_5 = 0$$

$$x_6 = 10 \qquad x_7 = 20 \qquad x_8 = 0 \qquad x_9 = 16 \qquad x_{10} = 36$$

Notice that the value of the subscript on x doesn't tell you anything about how small or large the data value is. In this example, x_1 is just the first observation in the data set and not necessarily the smallest observation. Similarly, x_n is the last observation but not necessarily the largest.

The sum of $x_1, x_2, \ldots, x_n$ can be written $x_1 + x_2 + \ldots x_n$, but this can be shortened by using the Greek letter Σ, which is used to denote summation. In particular, Σx is used to denote the sum of all of the x values in a data set.

DEFINITION

The **sample mean** is the arithmetic average of the values in a sample. It is denoted by the symbol $\bar{x}$ (pronounced as *x-bar*).

$$\bar{x} = \frac{\text{sum of all observations in the sample}}{\text{number of observations in the sample}} = \frac{\Sigma x}{n}$$

Example 3.3   **Thinking About Exercise Again**

The example of the Chapter Preview described a study that investigated how thinking about exercise might affect food intake. Data on calories in snacks taken for people in each of three groups (control, read about fun walk, and read about strenuous exercise) are given in Table 3.2. Dotplots of these three data sets were given in the Chapter Preview and are reproduced here as Figure 3.4.

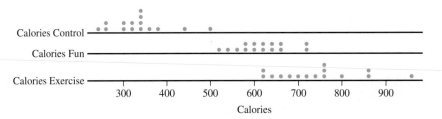

FIGURE 3.4
Dotplot of calories

TABLE 3.2 **Calorie Data**

Control Group	Fun Group	Exercise Group
340	668	626
300	585	802
329	637	768
381	588	738
331	529	751
445	597	715
320	719	670
256	612	630
252	622	862
342	553	761
332	600	648
296	658	956
357	542	671
505	711	854
242	641	703

Because each of the three data distributions is approximately symmetric, the mean is a reasonable choice for describing center. For the control group, the sum of the sample data values is

$$\Sigma x = 340 + 300 + \cdots = 242 = 5{,}030$$

and the sample mean number of calories is

$$\bar{x} = \frac{\Sigma x}{n} = \frac{5{,}030}{15} = 335.33$$

The value of the sample mean describes where number of calories for the control group is centered along the number line. It can be interpreted as a typical number of calories for people in the control group. For the fun group, the sample mean number of calories is

$$\bar{x} = \frac{\Sigma x}{n} = \frac{9{,}261}{15} = 617.40$$

and for the exercise group the sample mean number of calories is

$$\bar{x} = \frac{\Sigma x}{n} = \frac{14{,}291}{15} = 952.73$$

Notice that the sample mean number of calories for the control group was much smaller than the means for the other two groups.

It is customary to use Roman letters to denote sample statistics, as we have done with the sample mean $\bar{x}$. Population characteristics are usually denoted by Greek letters. The population mean is denoted by the Greek letter μ.

> **DEFINITION**
>
> The **population mean** is denoted by μ. It is the arithmetic average of all the x values in an entire population.

For example, the mean fuel efficiency for *all* 600,000 cars of a certain make and model might be $\mu = 27.5$ miles per gallon. A particular sample of 5 cars might have efficiencies of 27.3, 26.2, 28.4, 27.9, and 26.5, which results in $\bar{x} = 27.26$ (somewhat smaller than μ). A second sample of 5 cars might result in $\bar{x} = 28.52$, a third $\bar{x} = 26.85$, and so on. The value of $\bar{x}$ varies from sample to sample, whereas there is just one value of μ. In later chapters, you will see how the value of $\bar{x}$ from a particular sample can be used to draw conclusions about the value of μ.

Measuring Variability

Reporting a measure of center gives only partial information about a data set. It is also important to describe how much the observations in the data set differ from one another. For example, consider the three sets of six exam scores displayed in Figure 3.5. Each of these data sets has a mean of 75, but as you go from Set 1 to Set 2 to Set 3, they decrease in variability.

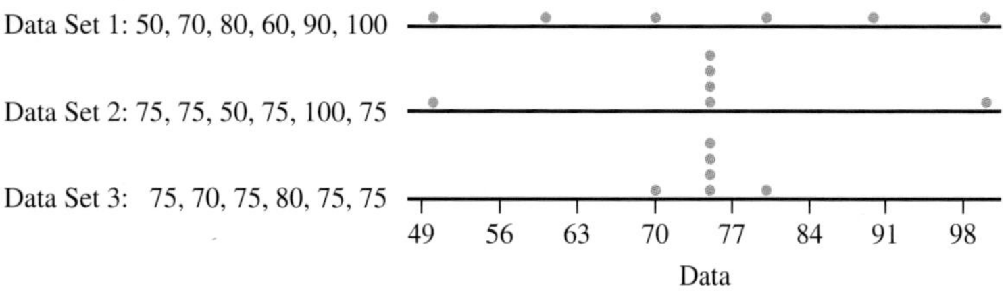

FIGURE 3.5
Three samples with the same mean but different spread

Data Set 1: 50, 70, 80, 60, 90, 100

Data Set 2: 75, 75, 50, 75, 100, 75

Data Set 3: 75, 70, 75, 80, 75, 75

49   56   63   70   77   84   91   98

Data

The simplest numerical measure of variability is the **range**, which is just the difference between the largest and smallest observations in the data set. While the range is simple to compute, it is not a very good measure of variability. For the data sets shown in Figure 3.5, both Data Set 1 and Data Set 2 have a range of $100 - 50 = 50$, but Data Set 1 has more variability than Data Set 2. For this reason, measures of variability that are based on all of the observations in the data set (not just the two most extreme values) are usually preferred over the range.

Deviations from the Mean

The most widely used measures of variability are based on how far each observation deviates from the sample mean. Subtracting $\bar{x}$ from each observation gives the set of **deviations from the mean**.

> **DEFINITION**
>
> There is a deviation from the mean for each observation x in a data set. To compute a **deviation from the mean** associated with an observation x, subtract the sample mean $\bar{x}$ from the observation to get $(x - \bar{x})$.

A deviation from the mean is positive if the corresponding observation is greater than $\bar{x}$ and negative if the observation is less than $\bar{x}$.

Example 3.4 The Big Mac Index

McDonald's fast-food restaurants are found in many countries around the world. Table 3.3 shows the cost of a Big Mac for eight countries (converted to U.S. dollars using July 2010 exchange rates) from the article **"Big Mac Index 2010" (The Economist, July 24, 2010).**

TABLE 3.3 **Big Mac Prices for Eight Countries**

Country	Big Mac Price in U.S. Dollars
Argentina	3.56
Brazil	4.91
Chile	3.34
Colombia	4.39
Costa Rica	3.83
Mexico	2.50
Peru	3.54
Uruguay	3.74

Notice that there is quite a bit of variability in the Big Mac prices. For this data set, $\Sigma x = 29.81$ and $\bar{x} = \$3.73$. Table 3.4 displays the data along with the corresponding deviations, formed by subtracting $\bar{x} = \$3.73$ from each observation. Four of the deviations are positive because four of the observations are larger than $\bar{x}$. Some of the deviations are quite large in magnitude (-1.23 and 1.18, for example), indicating observations that are far from the sample mean.

TABLE 3.4 **Deviations from the Mean for the Big Mac Data**

Price x	Deviation from the Mean $(x - \bar{x})$
3.56	−0.17
4.91	1.18
3.34	−0.39
4.39	0.66
3.83	0.10
2.50	−1.23
3.54	−0.19
3.74	0.01

In general, when there is more variability in the sample, the observations will tend to fall farther away from the mean. This will be reflected in the deviations from the mean, and this is why the deviations from the mean can be combined to obtain an overall measure of variability. You might think that a reasonable way to combine the deviations from the mean into a measure of variability is to find the average deviation from the mean. The problem with this approach is that some of the deviations are positive and some are negative, and when the deviations are added together, the positive and negative deviations offset each other. In fact, except for small differences due to rounding, the sum of the deviations from the mean is always equal to 0. The way around this is to use a measure of variability that is based on the squared deviations.

The Variance and Standard Deviation

When the data distribution is approximately symmetric, the two most widely used measures of spread are the variance and the standard deviation. Both of these measures of variability are based on the squared deviations from the mean.

DEFINITION

The **sample variance**, denoted by s^2, is the sum of the squared deviations from the mean divided by $n - 1$.

$$s^2 = \frac{\Sigma (x - \bar{x})^2}{n - 1}$$

The **sample standard deviation**, denoted by s, is the square root of the sample variance.

$$s = \sqrt{s^2} = \sqrt{\frac{\Sigma (x - \bar{x})^2}{n - 1}}$$

In computing the value of the variance or the standard deviation, notice that the deviations from the mean are first squared and then added together. When the values in a data set are spread out, some will fall far from the mean. For these values, the squared deviations from the mean will be large, resulting in a large sample variance. This is why data distributions that are quite spread out will have a large variance and standard deviation.

People find the standard deviation to be a more natural measure of variability than the variance because it is expressed in the same units as the original data values. For example, if the observations in a data set represent the price of a Big Mac in dollars, the mean and the deviations from the mean are also in dollars. But when the deviations are squared and then combined to obtain the variance, the units are dollars squared—not something that is familiar to most people! This makes it difficult to interpret the value of the variance and to decide whether the variance is large or small. Taking the square root of the variance to obtain the standard deviation results in a measure of variability that is expressed in the same units as the original data values, making it easier to interpret.

Example 3.5   Big Mac Revisited

Let's use the Big Mac data and the deviations from the mean computed in Example 3.4 to calculate the values of the sample variance and the sample standard deviation. Table 3.5 shows the original observations and the deviations from the mean, along with the squared deviations.

TABLE **3.5** **Deviations and Squared Deviations for the Big Mac Data**

Price x	Deviation from the Mean $(x - \bar{x})$	Squared Deviation from the Mean $(x - \bar{x})^2$
3.56	−0.17	0.0289
4.91	1.18	1.3924
3.34	−0.39	0.1521
4.39	0.66	0.4356
3.83	0.10	0.0100
2.50	−1.23	1.5129
3.54	−0.19	0.0361
3.74	0.01	0.0001
		$\Sigma(x-\bar{x})^2 = 3.5681$

Combining the squared deviations to compute the values of s^2 and s gives

$$s^2 = \frac{\Sigma(x - \bar{x})^2}{n - 1} = \frac{3.5681}{8 - 1} = \frac{3.5681}{7} = 0.5097$$

$$s = \sqrt{0.5097} = 0.7139$$

The calculation of s can be a bit tedious, especially if the sample size is large. Fortunately, many calculators and computer software packages can easily compute the standard deviation. One commonly used statistical computer package is JMP. The output resulting from using the JMP Analyze command with the Big Mac data is shown here.

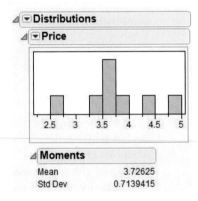

The standard deviation can be informally interpreted as the size of a "typical" or "representative" deviation from the mean. In Example 3.5, a typical deviation from $\bar{x}$ is about 0.714. Some observations are closer to $\bar{x}$ than 0.714, and others are farther away. The standard deviation is often used to compare the variability in data sets. For example, if the standard deviation of Big Mac prices for a larger group of 44 countries is $s = 1.25$, then you would conclude that the original sample has much less variability than the data set consisting of 44 countries.

The **population variance** and the **population standard deviation** are measures of variability for the entire population. They are denoted by σ^2 and σ, respectively. (Again, lowercase Greek letters are used for population characteristics.)

Notation

s^2 sample variance
σ^2 population variance
s sample standard deviation
σ population standard deviation

In many statistical procedures, you would like to use the value of σ, but you don't usually know this value. In this case, the sample standard deviation s is used as an estimate of σ. Using $n - 1$ rather than n (which might seem to be a more natural choice) as the divisor in the calculation of the sample variance and standard deviation results in estimates that tend to be closer to the corresponding population values.

Putting It Together

There are three simple steps to complete when summarizing a numerical data distribution:

1. **Select** appropriate measures of center and spread. This will involve looking at the shape of the distribution.
2. **Calculate** the values of the selected measures.
3. **Interpret** the values in context.

Let's look at one more example that illustrates this process.

Example 3.6　Thirsty Bats

The short article **"How to Confuse Thirsty Bats" (nature.com)** summarized a study that was published in the journal **Nature Communications ("Innate Recognition of Water Bodies in Echolocating Bats," November 2, 2010)**. The article states

> Echolocating bats have a legendary ability to find prey in the dark—so you'd think they would be able to tell the difference between water and a sheet of metal. Not so, report Greif and Siemers in *Nature Communications*. They have found that bats identify any extended, echo-acoustically smooth surface as water, and will try to drink from it.

This conclusion was based on a study where bats were placed in a room that had two large plates on the floor. One plate was made of wood and had an irregular surface. The other plate was made of metal and had a smooth surface. The researchers found that the bats never attempted to drink from the irregular surface, but that they made repeated attempts to drink from the smooth, metal surface. The number of attempts to drink from the smooth metal surface for 11 bats are shown here:

　　66　144　13　26　　94　163　8　125　1　64　56

These data will be used to select, compute, and interpret appropriate summary measures of center and spread.

Step	Explanation
Select The data distribution is approximately symmetric, so use sample mean and sample standard deviation to describe center and spread.	Although there is a lot of variability from bat to bat in number of drinking attempts, the data distribution does not appear to be skewed, and there are no observations that seem to be outliers. So, the sample mean and sample standard deviation are appropriate for describing center and spread of this data set.
Calculate For these data $$\Sigma x = 760$$ $$\bar{x} = 69.09$$ $$\Sigma(x-\bar{x})^2 = 31{,}754.9$$ $$s^2 = \frac{\Sigma(x-\bar{x})^2}{n-1} = \frac{31{,}754.9}{10} = 3{,}175.49$$ $$s = \sqrt{s^2} = \sqrt{3{,}175.49} = 56.35$$	Calculation of the values of $\bar{x}$ and s can be done using a calculator with statistics functions, a computer software package, or by hand. If computing by hand, first find the value of the sample mean, $\bar{x}$. Then use $\bar{x}$ to compute the deviations from the mean and the sum of the squared deviations. Finally, use this sum to compute the variance and the standard deviation.

(continued)

Step	Explanation
Interpret	
On average, the bats in the sample made 69.09 attempts to drink from the smooth, metal surface. The sample standard deviation is 56.35. This is relatively large compared to the values in the data set, indicating a lot of variability from bat to bat in number of drinking attempts.	Always finish with an interpretation of the sample mean and sample standard deviation in context. The mean represents a typical or representative value for the data set, and the sample standard deviation describes how much the values in the data set spread out around the mean. It can be informally interpreted as a representative distance from the mean.

SECTION 3.2 EXERCISES

Each Exercise Set assesses the following chapter learning objectives: C3, C4, M2, M3, P1, P2

SECTION 3.2 Exercise Set 1

3.12 The accompanying data are consistent with summary statistics in a paper investigating the effect of the shape of drinking glasses (*British Medical Journal* [2005]: 1512–1514). The data are the actual amount (in ml) poured into a tall, slender glass by individuals asked to pour 1.5 ounces (44.3 ml) into the glass. Compute and interpret the values of the mean and standard deviation.

44.0	49.6	62.3	28.4	39.1	39.8	60.5	73.0
57.5	56.5	65.0	56.2	57.7	73.5	66.4	32.7
40.4	21.4						

3.13 The paper referenced in the previous exercise also gave data on the actual amount (in ml) poured into a short, wide glass by individuals asked to pour 1.5 ounces (44.3 ml) into the glass.

89.2	68.6	32.7	37.4	39.6	46.8	66.1	79.2
66.3	52.1	47.3	64.4	53.7	63.2	46.4	63.0
92.4	57.8						

a. Compute and interpret the values of the mean and standard deviation.

b. What do the values of the means for the two types of glasses (the mean for a tall, slender glass was calculated in Exercise 3.12) suggest about the shape of glasses used?

3.14 A report from **Texas Transportation Institute (Texas A&M University System 2005)** on congestion reduction strategies included data on the extra travel time during rush hour for different urban areas. The following are the data for urban areas that had a population of over 3 million.

Urban Area	Extra Hours Per Traveler Per Year
Los Angeles	98
San Francisco	75
Washington, D.C.	66
Atlanta	64
Houston	65
Dallas, Fort Worth	61
Chicago	55
Detroit	54
Boston	53
New York	50
Phoenix	49
Miami	48
Philadelphia	40

a. Compute the mean and standard deviation for this data set.

b. Delete the observation for Los Angeles and recompute the mean and standard deviation. Compare this mean and standard deviation to the values computed in Part (a). What does this suggest about using the mean and standard deviation as measures of center and spread for a data set with outliers?

3.15 Children going back to school can be expensive for parents—second only to the Christmas holiday season in terms of spending (*San Luis Obispo Tribune*, August 18, 2005). Parents spend an average of $444 on their children at the beginning of the school year, stocking up on clothes, notebooks, and even iPods. However, not every parent spends the same amount of money. Imagine a data set consisting of the amount spent at the beginning of the school year for each student at a particular elementary school. Would it have a large or a small standard deviation? Explain.

3.16 The U.S. Department of Transportation reported the number of speed-related crash fatalities for the 20 dates that had the highest number of these fatalities between 1994 and 2003 (*Traffic Safety Facts*, July 2005).

a. Compute and interpret the mean and standard deviation for this data set.

Date	Speed-Related Fatalities	Date	Speed-Related Fatalities
Jan 1	521	Aug 17	446
Jul 4	519	Dec 24	436
Aug 12	466	Aug 25	433
Nov 23	461	Sep 2	433
Jul 3	458	Aug 6	431
Dec 26	455	Aug 10	426
Aug 4	455	Sept 21	424
Aug 31	446	Jul 27	422
May 25	446	Sep 14	422
Dec 23	446	May 27	420

b. Explain why it is not reasonable to generalize from this sample of 20 days to the other 345 days of the year.

3.17 Costs per serving (in cents) for 15 high-fiber cereals rated very good or good by *Consumer Reports* are shown below.

46 49 62 41 19 77 71 30 53 53 67 43 48 28 54

Compute and interpret the mean and standard deviation for this data set.

3.18 The accompanying data are a subset of data read from a graph in the paper **"Ladies First? A Field Study of Discrimination in Coffee Shops"** (*Applied Economics* [April, 2008]). The data are the waiting times (in seconds) between ordering and receiving coffee for 19 female customers at a Boston coffee shop.

60 80 80 100 100 100 120 120
120 140 140 150 160 180 200 200
220 240 380

a. Compute the mean and standard deviation for this data set.
b. Delete the observation of 380 and recompute the mean and standard deviation. How do these values compare to the values computed in Part (a)? What does this suggest about using the mean and standard deviation as measures of center and spread for a data set with outliers?

3.19 The article **"Rethink Diversification to Raise Returns, Cut Risk"** (*San Luis Obispo Tribune, January* 21, 2006) included the following paragraph:

In their research, Mulvey and Reilly compared the results of two hypothetical portfolios and used actual data from 1994 to 2004 to see what returns they would achieve. The first portfolio invested in Treasury bonds, domestic stocks, international stocks, and cash. Its 10-year average annual return was 9.85% and its volatility—measured as the standard deviation of annual returns—was 9.26%. When Mulvey and Reilly shifted some assets in the portfolio to include funds that invest in real estate, commodities, and options, the 10-year return rose to 10.55% while the standard deviation fell to 7.97%. In short, the more diversified portfolio had a slightly better return and much less risk.

Explain why, in this context, the standard deviation is a reasonable measure of volatility and why a smaller standard deviation means less risk.

Additional Exercises

3.20 The article **"Caffeinated Energy Drinks—A Growing Problem"** (*Drug and Alcohol Dependence* [2009]: 1–10) gave the accompanying data on caffeine concentration (mg/ounce) for eight top-selling energy drinks.

a. What is the mean caffeine concentration for this set of energy drinks?
b. Coca-Cola has 2.9 mg/ounce of caffeine and Pepsi Cola has 3.2 mg/ounce of caffeine. Write a sentence explaining how the caffeine concentration of top-selling energy drinks compares to that of these colas.

Energy Drink	Caffeine Concentration (mg/ounce)
Red Bull	9.6
Monster	10.0
Rockstar	10.0
Full Throttle	9.0
No Fear	10.9
Amp	8.9
SoBe Adrenaline Rush	9.5
Tab Energy	9.1

3.21 Research by the U.S. Food and Drug Administration (FDA) shows that acrylamide (a possible cancer-causing substance) forms in high-carbohydrate foods cooked at high temperatures, and that acrylamide levels can vary widely even within the same brand of food (**Associated Press, December 6, 2002**). FDA scientists analyzed McDonald's french fries purchased at seven different locations and found the following acrylamide levels:

497 193 328 155 326 245 270

a. Compute the mean acrylamide level. For each data value, calculate the deviation from the mean.
b. Verify that, except for the effect of rounding, the sum of the seven deviations from the mean is equal to 0 for

this data set. (If you rounded the sample mean or the deviations, your sum may not be exactly zero, but it should still be close to zero.)

c. Use the deviations from Part (a) to calculate the variance and standard deviation for this data set.

3.22   Although bats are not known for their eyesight, they are able to locate prey (mainly insects) by emitting high-pitched sounds and listening for echoes. A paper appearing in *Animal Behaviour* (**"The Echolocation of Flying Insects by Bats"** [1960]: 141–154) gave the following distances (in centimeters) at which a bat first detected a nearby insect:

    62   23   27   56   52   34   42   40   68   45   83

a. Compute and interpret the mean distance at which the bat first detected an insect.

b. Compute the sample variance and standard deviation for this data set. Interpret these values.

3.23   For the data in Exercise 3.22, subtract 10 from each sample observation. For the new set of values, compute the mean and all the deviations from the mean. How do these deviations compare to the deviations from the mean for the original sample? How will the new s^2 compare to s^2 for the old values? In general, what effect does subtracting the same number from (or adding the same number to) every value have on s^2 and s? Explain.

3.24   For the data of Exercise 3.22, multiply each data value by 10, then calculate the standard deviation. How does this value compare to s for the original data? More generally, what happens to s if each observation is multiplied by the same positive constant c?

Describing Center and Spread for Data Distributions That Are Skewed or Have Outliers

This section considers measures of center and spread that are more appropriate when the data distribution is noticeably skewed or when there are outliers (unusually large or small observations).

Describing Center

One potential drawback to the mean as a measure of center is that if the data set is skewed or has outliers, its value can be greatly affected. Consider the following example.

AP* EXAM TIP

For distributions that are skewed or that have outliers, you should consider using the median to describe center.

Example 3.7   Number of Visits to a Class Web Site

Forty students were enrolled in a statistical reasoning course at a California college. The instructor made course materials, grades, and lecture notes available to students on a class web site, and course management software kept track of how often each student accessed any of these web pages. One month after the course began, the instructor requested a report on how many times each student had accessed a class web page. The 40 observations were:

20	37	4	20	0	84	14	36	5	331	19	0
0	22	3	13	14	36	4	0	18	8	0	26
4	0	5	23	19	7	12	8	13	16	21	7
13	12	8	42								

The sample mean for this data set is $\bar{x} = 23.10$. Figure 3.6 is a dotplot of the data. Many would argue that 23.10 is not a very representative value for this sample, because only 7 of 40 observations are larger than 23.10. The two outlying values of 84 and 331 (no, that was not a typo!) have a big impact on the value of $\bar{x}$.

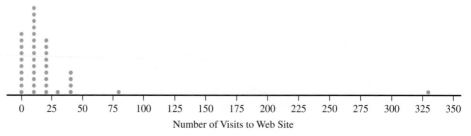

FIGURE 3.6
Dotplot of the web site visit data
of Example 3.7

Number of Visits to Web Site

Because the data distribution is skewed and there are outliers, the mean is not the best choice for describing center. A better choice here is a measure called the median.

The Median

The median strip of a highway divides the highway in half. The median of a numerical data set works the same way. Once the data values have been listed in order from smallest to largest, the **median** is the middle value in the list, dividing it into two equal parts. The process of determining the median of a sample is slightly different depending on whether the sample size n is even or odd. When n is an odd number, the sample median is the single middle value. But when n is even, there are two middle values in the ordered list, and you average these two middle values to obtain the sample median.

> **DEFINITION**
>
> The **sample median** is obtained by first ordering the n observations from smallest to largest (with any repeated values included, so that every sample observation appears in the ordered list). Then
>
> $$\text{sample median} = \begin{cases} \text{the single middle value} & \text{if } n \text{ is odd} \\ \text{the average of the two middle values} & \text{if } n \text{ is even} \end{cases}$$

Example 3.8 Web Site Data Revisited

The sample size for the web site visit data of Example 3.7 was $n = 40$, an even number. The median is the average of the 20th and 21st values in an ordered list of the data. Arranging the data from smallest to largest produces the following list (with the two middle values highlighted).

0	0	0	0	0	0	3	4	4	4	5	5
7	7	8	8	8	12	12	13	13	13	14	14
16	18	19	19	20	20	21	22	23	26	36	36
37	42	84	331								

The median can now be computed.

$$median = \frac{13 + 13}{2} = 13$$

Usually, half of the values in a data set are smaller than the median and half are greater. This isn't quite the case here because the value 13 occurs three times. Even so, it is common to interpret the median as the value that divides the data set in half. With a median of 13, you would say that half of the students in the class visited the web site fewer than 13 times, and half of the students visited the web site more than 13 times. Looking at the dotplot (Figure 3.6), you can see that 13 is more typical of the values in the data set than $\bar{x} = 23.10$.

The sample mean can be sensitive to even a single outlier. The median, on the other hand, is quite *in*sensitive to outliers. For example, the largest observations in Example 3.8 can be increased by any amount without changing the value of the median. Similarly, a decrease in several of the smallest observations would not affect the value of the median.

Measuring Variability—The Interquartile Range

As with $\bar{x}$, the value of s can also be greatly affected by the presence of even one outlier. A better choice for describing spread in this case is the *interquartile range*, a measure of variability that is resistant to the effects of outliers. It is based on quantities called *quartiles*. The *lower quartile* separates the smallest 25% of the observations from the largest 75%, and the *upper quartile* separates the largest 25% from the smallest 75%. The *middle quartile* is the median, and it separates the smallest 50% from the largest 50%. Figure 3.7 illustrates the locations of these quartiles for a smoothed histogram.

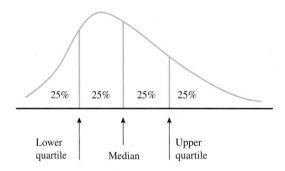

FIGURE 3.7
Quartiles for a smoothed histogram

The quartiles for sample data are obtained by dividing the n ordered observations into a lower half and an upper half. If n is odd, the median is excluded from both halves. The lower and upper quartiles are then the medians of the two halves. (The median is only temporarily excluded when computing quartiles. It is not excluded from the data set.)

DEFINITION*

> **lower quartile** = median of the lower half of the data set
>
> **upper quartile** = median of the upper half of the data set
>
> If n is odd, the median of the entire data set is excluded from both halves when computing quartiles.
>
> The **interquartile range (iqr)** is defined by
>
> **iqr = upper quartile − lower quartile**

*There are several other sensible ways to define quartiles. Some calculators and software packages use an alternate definition.

The standard deviation measures variability in terms of how far the observations in a data set tend to fall from the center of the data set (the mean). The interquartile range also measures variability in a data set, but it does so by looking at how spread out the middle half of the data set is. If the interquartile range is small, the values that make up the middle half of the data set are tightly clustered, indicating small variability. On the other hand, if the interquartile range is large, the values in the middle half of the data set are more spread out. By focusing on the middle half of the data set rather than all of the data values, the interquartile range is not influenced by extreme values.

Example 3.9 Web Site Visits One More Time

In Example 3.8, the median was used to describe the center of the web site visit data distribution. The interquartile range can be used to describe variability—how much the number of web site visits varies from student to student. You start by dividing the ordered data list from Example 3.8 into two parts at the median. The lower half of the data set consists of the 20 smallest observations and the upper half consists of the 20 largest observations, as shown below.

Lower Half:

0   0   0   0   0   0   3   4   4   4

5   5   7   7   8   8   8   12   12   13

Upper Half:

13   13   14   14   16   18   19   19   20   20

21   22   23   26   36   36   37   42   84   331

The lower quartile is just the median of the lower half. There are 20 observations (an even number) in the lower half, so the median is the average of the two middle observations (highlighted in the preceding data set):

$$\text{lower quartile} = \frac{4 + 5}{2} = 4.5$$

The upper quartile is found in the same way, using the upper half of the data set:

$$\text{upper quartile} = \frac{20 + 21}{2} = 20.5$$

The interquartile range is the difference between the upper quartile and the lower quartile:

$$\text{iqr} = 20.5 - 4.5 = 16.0$$

Because the lower quartile separates the smallest 25% of the data from the rest, you can say that 25% of the students visited the web site fewer than 4.5 times. You can also say that 50% of the students visited the web site between 4.5 and 20.5 times, and 25% visited more than 20.5 times. An iqr of 16.0 tells you there was quite a bit of variability in the data set, but not nearly as much as suggested by the value of the standard deviation $s = 52.33$. The values of the mean and standard deviation for this data set are large because of the two outliers.

Example 3.10   Some Strange Things Can Happen with Quartiles...

Example 3.1 gave data on the number of medication errors made by 24 medical residents who were classified as depressed (**"Rates of Medication Errors Among Depressed and Burnt Out Residents,"** *British Medical Journal* **[2008]: 488**). This data set is interesting because some data values appear multiple times. For example, there are 17 observations that all have the value 0.

Because the distribution is skewed and has outliers, the median and interquartile range should be used to describe center and spread. The ordered data values are

0   0   0   0   0   0   0   0   0   0   0   0   0   0   0   0   0   1   1   1   2   3   5   11

The median is the average of the two middle values. Here, both middle values are 0, and so the median is also 0.

Dividing the data set into the lower and upper halves and then finding the median of each half leads to the following quartiles.

Lower half:

0   0   0   0   0   0   0   0   0   0   0   0

lower quartile = 0

Upper half:

0   0   0   0   0   1   1   1   2   3   5   11

upper quartile = 1

You can now compute the value of the interquartile range.

$$\text{iqr} = \text{upper quartile} - \text{lower quartile} = 1 - 0 = 1$$

Notice here that the lower quartile and the median are both 0. This is because more than half of the observations in the data set have a value of 0. For this data set, the smallest

25% are all 0, and even the next 25% in the ordered list are all 0. One way to see this is to take another look at the ordered data list:

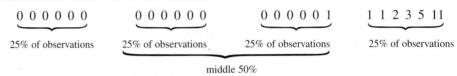

| 0 0 0 0 0 0 | 0 0 0 0 0 0 | 0 0 0 0 1 | 1 1 2 3 5 11 |
| 25% of observations | 25% of observations | 25% of observations | 25% of observations |

middle 50%

The small value of the interquartile range indicates that there is not much variability in the middle half of the data distribution. ▪

Putting It Together

When describing center and spread of a data distribution that is skewed or has outliers, the following three steps can be used.

1. **Select** appropriate measures of center and spread. This will involve taking the shape of the data distribution into account.
2. **Calculate** the values of the selected measures.
3. **Interpret** the values in context.

These are the same steps that were used when describing center and spread of a data distribution that is approximately symmetric. The key difference is that in this case, you choose the median and interquartile range as the appropriate measures of center and spread in the first step.

Let's look at one more example that illustrates this process.

Example 3.11 Higher Education

The Chronicle of Higher Education **(Almanac Issue, 2009–2010)** published the accompanying data on the percentage of the population with a bachelor's degree or graduate degree in 2007 for each of the 50 U.S. states and the District of Columbia. The 51 data values are:

```
21  27  26  19  30  35  35  26  47  26  27  30
24  29  22  24  29  20  20  27  35  38  25  31
19  24  27  27  23  34  34  25  32  26  26  24
22  28  26  30  23  25  22  25  29  33  34  30
17  25  23
```

These data will be used to select, compute, and interpret appropriate summary measures of center and spread.

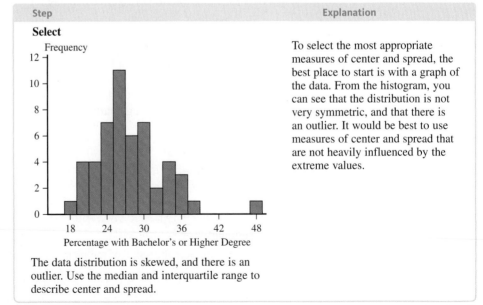

Step	Explanation
Select	To select the most appropriate measures of center and spread, the best place to start is with a graph of the data. From the histogram, you can see that the distribution is not very symmetric, and that there is an outlier. It would be best to use measures of center and spread that are not heavily influenced by the extreme values.

The data distribution is skewed, and there is an outlier. Use the median and interquartile range to describe center and spread.

(continued)

Step	Explanation
Calculate	Calculation of the median, quartiles, and interquartile range can be done using a calculator with statistics functions, using a computer package, or by hand. If computing by hand, you would arrange the data in order from smallest to largest and then find the median, the quartiles and the interquartile range, as shown to the left.

Calculate

To compute the quartiles and the interquartile range, first order the data and use the median to divide the data into a lower half and an upper half. Because there are an odd number of observations ($n = 51$), the median is excluded from both the upper and lower halves when computing the quartiles.

Ordered Data

Lower Half:

17 19 19 20 20 21 22 22 22 23

23 23 **24** 24 24 24 25 25 25 25

25 26 26 26 26

Median: 26

Upper Half:

26 27 27 27 27 27 28 29 29 29

30 30 **30** 30 31 32 33 34 34 34

35 35 35 38 47

Each half of the sample contains 25 observations. The lower quartile, 24, is the median of the lower half. The upper quartile, 30, is the median of the upper half. Then

iqr = 30 − 24 = 6

Calculation of the median, quartiles, and interquartile range can be done using a calculator with statistics functions, using a computer package, or by hand. If computing by hand, you would arrange the data in order from smallest to largest and then find the median, the quartiles and the interquartile range, as shown to the left.

An alternative to hand calculation would be to use a software package such as JMP. JMP produces the following output:

◢ ▾ **Distributions**
   ◢ ▾ **Percentage with Degree**
      ◢ **Quantiles**

100.0%	maximum	47
99.5%		47
97.5%		44.3
90.0%		34.8
75.0%	quartile	30
50.0%	median	26
25.0%	quartile	24
10.0%		20.2
2.5%		17.6
0.5%		17
0.0%	minimum	17

Notice that the median and quartiles agree with the calculations shown to the left. In the JMP output, the upper quartile is the one labeled 75% (because 75% of the data values are below the upper quartile) and the lower quartile is the one labeled 25%.

Interpret

The median for this data set is 26. For half of the states, the percentage of the population with a bachelor's or graduate degree is 26% or less. For the other half, 26% or more of the population have a bachelor's or graduate degree. (Notice that we said "26% or less" instead of just less than 26% because there were several states with 26%). The interquartile range of 6 indicates that the percentages for the states in the middle half were spread out over an interval of 6 percentage points.

Always finish with an interpretation of the sample median and sample interquartile range in context. The median represents a typical or representative value for the data set in the sense that half of the values in the data set are smaller and half are larger. The sample interquartile range describes how much the values in the middle half of the data set spread out.

The region with the largest percentage, at 47%, was the District of Columbia. The second-highest, at 38%, was Massachusetts, and the smallest, at 17%, was West Virginia.

One final note. When a data distribution is approximately symmetric, we have recommended using the mean and standard deviation to describe center and spread because these are the most commonly used summary measures. However, the median and interquartile range can also be used for approximately symmetric data distributions. In fact, when the data distribution is approximately symmetric, the values of the mean and median will be similar and the interquartile range still describes spread. The mistake to avoid is using the mean and standard deviation to describe center and spread for data distributions that are very skewed or that have notable outliers. Because the values of the mean and standard deviation can be greatly affected by extreme values, it can be misleading to use them in these situations.

SECTION 3.3 EXERCISES

Each Exercise Set assesses the following chapter learning objectives: C5, M4, M5, P1, P2

Exercise Set 1

3.25 *USA Today* (May 9, 2006) published the average weekday circulation for the top 20 newspapers in the country. Here are the data for the 6-month period ending March 31, 2006:

2,272,815	2,049,786	1,142,464	851,832	724,242
708,477	673,379	579,079	513,387	438,722
427,771	398,329	398,246	397,288	365,011
362,964	350,457	345,861	343,163	323,031

a. Compute and interpret the value of the median of this data set.
b. Explain why the median is preferable to the mean for describing center for this data set.
c. Explain why it would be unreasonable to generalize from this sample of 20 newspapers to the population of daily newspapers in the United States.

3.26 The paper **"Total Diet Study Statistics on Element Results" (Food and Drug Administration, April 25, 2000)** gave information on sodium content for various types of foods. Twenty-six brands of tomato catsup were analyzed. Data consistent with summary quantities in the paper are

Sodium content (mg/kg)

12,148	10,426	10,912	9,116	13,226	11,663
11,781	10,680	8,457	10,788	12,605	10,591
11,040	10,815	12,962	11,644	10,047	
10,478	10,108	12,353	11,778	11,092	
11,673	8,758	11,145	11,495		

Compute and interpret the values of the quartiles and the interquartile range.

3.27 The accompanying data on number of cell phone minutes used in one month are consistent with summary statistics from a marketing study of San Diego residents **(Tele-Truth, March, 2009).**

189	0	189	177	106	201	0	212	0	306
0	0	59	224	0	189	142	83	71	165
236	0	142	236	130					

Compute and interpret the values of the median and the interquartile range.

3.28 Data on tipping percent for 20 restaurant tables, consistent with summary statistics given in the paper **"Beauty and the Labor Market: Evidence from Restaurant Servers" (unpublished manuscript by Matt Parrett, 2007),** are:

0.0	5.0	45.0	32.8	13.9	10.4
55.2	50.0	10.0	14.6	38.4	23.0
27.9	27.9	105.0	19.0	10.0	32.1
11.1	15.0				

Compute and interpret the values of the median and the interquartile range.

Exercise Set 2

3.29 The **Insurance Institute for Highway Safety (www.iihs .org, June 11, 2009)** published data on repair costs for cars involved in different types of accidents. In one study, seven different 2009 models of mini- and micro-cars were driven at 6 mph straight into a fixed barrier. The following table gives the cost of repairing damage to the bumper for each of the seven models.

Model	Repair Cost
Smart Fortwo	$1,480
Chevrolet Aveo	$1,071
Mini Cooper	$2,291
Toyota Yaris	$1,688
Honda Fit	$1,124
Hyundai Accent	$3,476
Kia Rio	$3,701

a. Compute and interpret the value of the median for this data set.
b. Explain why the median is preferable to the mean for describing center in this situation.

3.30 A report from **Texas Transportation Institute (Texas A&M University System 2005)** on congestion reduction strategies included data on extra travel time during rush hour (in hours per year) for different urban areas. The following data are for urban areas that had a population of over 3 million.

Urban Area	Extra Hours Per Traveler Per Year
Los Angeles	98
San Francisco	75
Washington, D.C.	66
Atlanta	64
Houston	65
Dallas, Fort Worth	61
Chicago	55
Detroit	54
Boston	53
New York	50
Phoenix	49
Miami	48
Philadelphia	40

Compute and interpret the values of the quartiles and the interquartile range.

3.31 Data on weekday exercise time (in minutes) for 20 males, consistent with summary quantities given in the paper "An Ecological Momentary Assessment of the Physical Activity and Sedentary Behaviour Patterns of University Students" (*Health Education Journal* [2010]: 116–125), are shown below. Compute and interpret the values of the median and interquartile range.

Male—Weekday

43.5 91.5   7.5   0.0   0.0 28.5 199.5 57.0 142.5
 8.0   9.0 36.0   0.0 78.0 34.5   0.0 57.0 151.5
 8.0   0.0

3.32 Data on weekday exercise time (in minutes) for 20 females, consistent with summary quantities given in the paper "An Ecological Momentary Assessment of the Physical Activity and Sedentary Behaviour Patterns of University Students" (*Health Education Journal* [2010]: 116–125), are shown below.

Female—Weekday

10.0   90.6   48.5   50.4   57.4   99.6
 0.0   5.0   0.0   0.0   5.0   2.0
10.5   5.0   47.0   0.0   5.0   54.0
 0.0   48.6

a. Compute and interpret the values of the median and interquartile range.
b. How do the values of the median and interquartile range for women compare to those for men computed in the previous exercise?

Additional Exercises

3.33 In August 2009, Harris Interactive released the results of the Great Schools Survey, in which 1,086 parents of children attending a public or private school were asked approximately how much they had spent on school supplies over the last school year. For this sample, the mean amount spent was $235.20 and the median amount spent was $150.00. What does the large difference between the mean and median tell you about this data set?

3.34 The *San Luis Obispo Telegram-Tribune* (October 1, 1994) reported the following monthly salaries for supervisors from six different counties: $5,354 (Kern), $5,166 (Monterey), $4,443 (Santa Cruz), $4,129 (Santa Barbara), $2,500 (Placer), and $2,220 (Merced). San Luis Obispo County supervisors are supposed to be paid the average of the two counties in the middle of this salary range. Which measure of center determines this salary, and what is its value? Find the value of the other measure of center featured in this chapter. Why is it not as favorable to the San Luis Obispo County supervisors (although it might appeal to taxpayers)?

3.35 A sample of 26 offshore oil workers took part in a simulated escape exercise, resulting in the following data on time (in seconds) to complete the escape ("Oxygen Consumption and Ventilation During Escape from an Offshore Platform," *Ergonomics* [1997]: 281–292):

389 356 359 363 375 424 325 394 402
373 373 370 364 366 364 325 339 393
392 369 374 359 356 403 334 397

a. Construct a dotplot of the data. Will the mean or the median be larger for this data set?
b. Calculate the values of the mean and median.
c. By how much could the largest time be increased without affecting the value of the sample median? By how much could this value be decreased without affecting the value of the median?

SECTION 3.4 Summarizing a Data Set: Boxplots

In Sections 3.2 and 3.3, you saw ways to describe center and spread of a data distribution. It is sometimes also helpful to have a method of summarizing data that gives more information than just a measure of center and a measure of spread. A boxplot is a compact display that provides information about the center, spread, and symmetry or skewness of a data distribution.

The construction of a boxplot is illustrated in the following example.

Example 3.12 Beating That High Score

The authors of the paper "Striatal Volume Predicts Level of Video Game Skill Acquisition" (*Cerebral Cortex* [2010]: 2522–2530) studied a number of factors that affect performance in a complex video game. One factor was practice strategy. Forty college students who all reported playing video games less than 3 hours per week over the past two years and who had never played the game *Space Fortress* were assigned at random to one of two groups. Each person completed 20 two-hour practice sessions. Those in the fixed priority group were told to work on improving their total score at each practice session. Those in the variable priority group were told to focus on a different aspect of the game, such as improving speed score, in each practice session. The investigators were interested in whether practice strategy makes a difference. They measured the improvement in total score from the first practice session to the last.

Improvement scores (approximated from a graph in the paper) for the 20 people in each practice strategy group are given here.

Fixed Priority	1,100	1,900	2,000	2,400	2,600	2,600	2,550	3,500
Practice Group	3,500	3,500	3,700	3,800	3,900	4,000	4,000	4,100
	4,100	4,500	5,000	6,000				
Variable Priority	1,200	1,300	2,300	3,200	3,300	3,800	4,000	4,100
Practice Group	4,300	4,800	5,500	5,700	5,700	5,800	6,000	6,300
	6,800	6,800	6,900	7,700				

Begin by looking at the fixed priority improvement scores. From the accompanying dot-plot of these data, you can locate the median and the lower and upper quartiles.

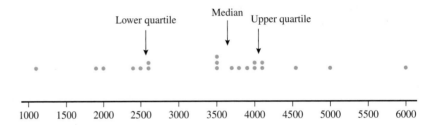

Notice that the lower quartile, the median, and the upper quartile divide this data set into four parts, each with five observations.

To construct a boxplot, first draw a box around the middle 50% of the data, as shown.

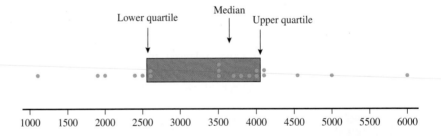

To complete the boxplot, draw a vertical line at the median. Then draw two lines—one that extends from the box to the smallest observation and one that extends from the box to the largest observation. These lines are called "whiskers."

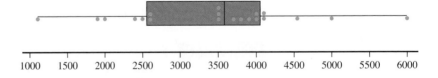

This is the boxplot for the fixed priority improvement scores. You can think of the boxplot as being made up of four parts—the lower whisker, the lower part of the box, the upper part of the box, and the upper whisker. Notice that each of these four parts represents the same number of observations (25% of the data values). If a part of the box is narrow or a whisker is short, it means the data values fall close together in this region. If a part of the box is wide or a whisker is long, then the data values are more spread out.

We started with the dotplot to illustrate the relationship between the parts of the boxplot and the data distribution. However, boxplots are usually constructed without a dotplot and the underlying data values (the dots). We will return to this example to construct a boxplot for the variable priority group improvement scores after looking at how boxplots are usually constructed.

To construct a boxplot without first drawing a dotplot, you need the following information: the smallest observation, the lower quartile, the median, the upper quartile, and the largest observation. This collection of measures is called the **five-number summary**.

> **DEFINITION**
>
> The **five-number summary** consists of the following:
>
> 1. Smallest observation in the data set
> 2. Lower quartile
> 3. Median
> 4. Upper quartile
> 5. Largest observation in the data set

A boxplot is a graph of the five-number summary. To construct a boxplot, follow the steps in the accompanying box.

> **Constructing a Boxplot**
>
> 1. Compute the values in the five-number summary.
> 2. Draw a horizontal line and add an appropriate scale.
> 3. Draw a box above the line that extends from the lower quartile to the upper quartile.
> 4. Draw a line segment inside the box at the location of the median.
> 5. Draw two line segments, called whiskers, which extend from the box to the smallest observation and from the box to the largest observation.

Example 3.13 Video Game Practice Strategies

Let's return to the video game improvement scores of Example 3.12. For the variable priority practice strategy, the ordered improvement scores are

| 1,200 | 1,300 | 2,300 | 3,200 | 3,300 | 3,800 | 4,000 | 4,100 | 4,300 | 4,800 |
| 5,500 | 5,700 | 5,700 | 5,800 | 6,000 | 6,300 | 6,800 | 6,800 | 6,900 | 7,700 |

The median is the average of the middle two observations, so

$$\text{median} = \frac{4,800 + 5,500}{2} = 5,150$$

The lower half of the data set is

| 1,200 | 1,300 | 2,300 | 3,200 | 3,300 | 3,800 | 4,000 | 4,100 | 4,300 | 4,800 |

so the lower quartile is

$$\text{lower quartile} = \frac{3,300 + 3,800}{2} = 3,550$$

The upper half of the data set is

| 5,500 | 5,700 | 5,700 | 5,800 | 6,000 | 6,300 | 6,800 | 6,800 | 6,900 | 7,700 |

so the upper quartile is

$$\text{upper quartile} = \frac{6,000 + 6,300}{2} = 6,150$$

You now have what you need for the five-number summary and are ready to construct the boxplot.

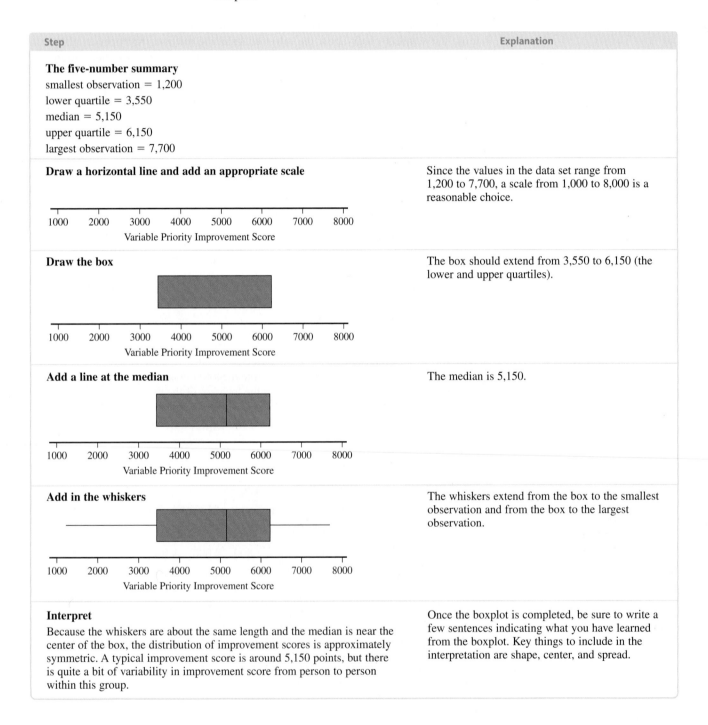

Step	Explanation
The five-number summary smallest observation = 1,200 lower quartile = 3,550 median = 5,150 upper quartile = 6,150 largest observation = 7,700	
Draw a horizontal line and add an appropriate scale	Since the values in the data set range from 1,200 to 7,700, a scale from 1,000 to 8,000 is a reasonable choice.
Draw the box	The box should extend from 3,550 to 6,150 (the lower and upper quartiles).
Add a line at the median	The median is 5,150.
Add in the whiskers	The whiskers extend from the box to the smallest observation and from the box to the largest observation.
Interpret Because the whiskers are about the same length and the median is near the center of the box, the distribution of improvement scores is approximately symmetric. A typical improvement score is around 5,150 points, but there is quite a bit of variability in improvement score from person to person within this group.	Once the boxplot is completed, be sure to write a few sentences indicating what you have learned from the boxplot. Key things to include in the interpretation are shape, center, and spread.

Boxplots can be tedious to construct by hand, especially when the data set is large. Most graphing calculators and statistical software packages will construct boxplots. Figure 3.8 shows a boxplot for the data of this example that was made using JMP. Boxplots can be displayed either vertically or horizontally, and JMP displays them vertically.

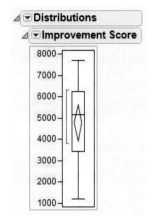

FIGURE 3.8
JMP boxplot for the variable priority group improvement scores

Using Boxplots for Comparing Groups

With two or more data sets consisting of observations on the same variable (for example, the video game improvement scores for the two different practice strategies described in Examples 3.12 and 3.13), a **comparative boxplot** is often used to compare the data sets.

A **comparative boxplot** is two or more boxplots drawn using the same numerical scale.

Example 3.14 **Comparing Practice Strategies**

The video game improvement scores from Examples 3.12 and 3.13 were used to construct the comparative boxplot shown in Figure 3.9.

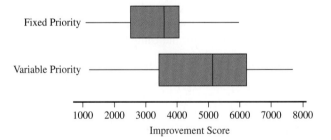

FIGURE 3.9
Comparative boxplot for improvement scores of the fixed priority practice group and the variable priority practice group.

From the comparative boxplot, you can see that both data distributions are approximately symmetric. The improvement scores tend to be higher for the variable priority group. However, there is more consistency in the improvement scores for the fixed priority group. Improvement scores in the variable priority group are more spread out, indicating more variability in improvement scores.

Outliers and Modified Boxplots

Up to this point, the term *outlier* has been used informally to describe an observation that stands out as unusual. Now you are ready to consider a formal definition of an outlier. The rules in the accompanying box are used to determine if an observation is "unusual enough" to officially qualify as an outlier.

> **DEFINITION**
>
> An observation is an **outlier** if it is
>
> $$\textbf{greater than } \text{upper quartile} + 1.5(\text{iqr})$$
>
> or
>
> $$\textbf{less than } \text{lower quartile} - 1.5(\text{iqr})$$

A **modified boxplot** is a boxplot that shows outliers. The steps for constructing a modified boxplot are described in the following box. The first steps are the same as those for constructing a regular boxplot. The differences are in how the whiskers are added and in the identification of outliers.

> **Constructing a Modified Boxplot**
>
> 1. Compute the values in the five-number summary.
> 2. Draw a horizontal line and add an appropriate scale.
> 3. Draw a box above the line that extends from the lower quartile to the upper quartile.
> 4. Draw a line segment inside the box at the location of the median.
> 5. Determine if there are any outliers in the data set.
> 6. Add a whisker to the plot that extends from the box to the smallest observation in the data set *that is not an outlier.* Add a second whisker that extends from the box to the largest observation *that is not an outlier.*
> 7. If there are outliers, add dots to the plot to indicate the positions of the outliers.

The following examples illustrate the construction of a modified boxplot and the use of modified boxplots to compare groups.

Example 3.15 **Another Look at Big Mac Prices**

Big Mac prices in U.S. dollars for 44 different countries were given in the article **"Big Mac Index 2010,"** first introduced in Example 3.4. The following 44 Big Mac prices are arranged in order from the lowest price (Ukraine at $1.84) to the highest (Norway at $7.20). The price for the United States was $3.73.

1.84	1.86	1.90	1.95	2.17	2.19	2.19	2.28	2.33	2.34	2.45
2.46	2.50	2.51	2.60	2.62	2.67	2.71	2.80	2.82	2.99	3.08
3.33	3.34	3.43	3.48	3.54	3.56	3.59	3.67	3.73	3.74	3.83
3.84	3.86	3.89	4.00	4.33	4.39	4.90	4.91	6.19	6.56	7.20

The Big Mac price data will be used to construct a modified boxplot.

Step	Explanation
The five-number summary smallest observation $= 1.84$ lower quartile $= 2.455$ median $= \dfrac{3.08 + 3.33}{2} = 3.205$ upper quartile $= 3.835$ largest observation $= 7.20$	The median and quartiles can be computed using the highlighted data values in the ordered data list.
Draw a horizontal line and add an appropriate scale 1 2 3 4 5 6 7 8 Big Mac Price	Since the values in the data set range from 1.84 to 7.20, a scale from 1 to 8 is a reasonable choice.

(continued)

Step	Explanation

Draw the box

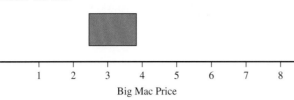

The box should extend from 2.455 to 3.835 (the lower and upper quartiles).

Add a line at the median

The median is 3.205.

Determine if there are outliers

iqr = 3.835 − 2.455 = 1.4

1.5(*iqr*) = 1.5 (1.4) = 2.1

lower quartile − 1.5(*iqr*) = 2.455 − 2.1 = 0.355

upper quartile + 1.5(*iqr*) = 3.835 + 2.1 = 5.935

There are no outliers on the low end because no country had a Big Mac price less than $0.355. On the high end, three countries had Big Mac prices greater than $5.935, so the values 6.19, 6.56, and 7.20 are outliers.

To decide if there are outliers, you need to compute the two outlier boundaries. In this example, the two boundaries are 0.355 and 5.935, so to identify outliers you are looking for values in the data set that are smaller than 0.355 or that are greater than 5.935.

Add in the whiskers

The whiskers extend from the box to the smallest observation that is not an outlier and from the box to the largest observation that is not an outlier. There are no outliers on the low side, so the whisker extends to the smallest observation, which is 1.84. The largest observation in the data set that is not an outlier is 4.91, so the whisker on the high end extends from the box to 4.91.

Add the outliers

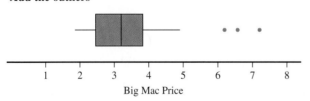

Three dots are added to the plot to indicate the positions of the three outliers at 6.19, 6.56, and 7.20.

Interpret

The most noticeable aspect of the boxplot is the presence of the three outliers, indicating that Big Mac prices in three countries were much greater than the rest. Because the whiskers are about the same length and the median is near the center of the box, the rest of the data distribution (excluding the three outliers) is approximately symmetric. A typical Big Mac price is around $3.20, but there is quite a bit of variability in Big Mac price from country to country. The price of a Big Mac in the United States was listed as $3.73, which is near the upper quartile, indicating that about 75% of the 44 countries had lower prices than the United States.

Once the boxplot is completed, be sure to write a few sentences indicating what you have learned from the boxplot. Key things to include in the interpretation are shape, center, spread, and outliers.

Example 3.16 An Odd-looking Boxplot...

Example 3.10 gave the following data on the number of medication errors made by 24 medical residents who were classified as depressed.

Number of Errors

0 0 0 0 0 0 0 0 0 0 0 0 0 0 0 0

0 0 1 1 1 2 3 5 11

Figure 3.10 shows a modified boxplot constructed using these data.

FIGURE 3.10
Boxplot of the medication error data

Three outliers are clearly shown in the boxplot, corresponding to residents that made 3, 5, and 11 medication errors. The boxplot looks a little odd. It appears that the median line and the lower whisker have been left out. But there were 17 observations in this data set that all had the value 0. As a result, the smallest value, the lower quartile, and the median are all equal to 0. The median line is actually at the left edge of the box, and there is no lower whisker shown.

Example 3.17 NBA Player Salaries

The web site HoopsHype (**hoopshype.com/salaries**) publishes salaries of NBA players. The 2009–2010 salaries of players on five teams were used to construct the comparative boxplot shown in Figure 3.11.

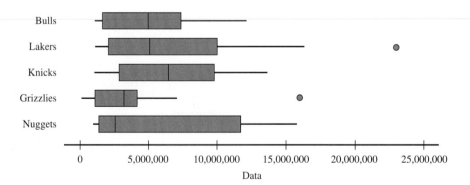

FIGURE 3.11
Comparative boxplot for salaries of five NBA teams

The comparative boxplot reveals some interesting similarities and differences in the salary distributions of the five teams. The minimum salary is lower for the Grizzlies but is about the same for the other four teams. The median salary was lowest for the Nuggets—in fact, the median for the Nuggets is about the same as the lower quartile for the Knicks and the Lakers, indicating that half of the Nuggets have salaries less than about $2.5 million, whereas only about 25% of the Knicks and the Lakers have salaries less than about $2.5 million. The player with by far the highest salary was a Laker. The Grizzlies and the Lakers were the only teams with salary outliers. Except for one highly paid player, salaries for the Grizzlies were noticeably lower than those for the other four teams.

SECTION 3.4 EXERCISES

Each Exercise Set assesses the following chapter learning objectives: M6, M7, M8, P3

SECTION 3.4 Exercise Set 1

3.36 The technical report **"Ozone Season Emissions by State" (U.S. Environmental Protection Agency, 2002)** gave the following nitrous oxide emissions (in thousands of tons) for the 48 states in the continental United States:

76	22	40	7	30	5	6	136	72	33	89
136	39	92	40	13	27	1	63	33	60	0
27	16	63	32	20	2	15	36	19	39	0
130	40	4	85	38	7	68	151	32	34	0
6	43	89	34							

Use these data to compute the values in the five-number summary.

3.37 The data below are manufacturing defects per 100 cars for the 33 brands of cars sold in the United States (*USA Today*, June 16, 2010). Many of these values are larger than 100 because one car might have many defects.

86	111	113	114	111	111	122
130	93	126	95	102	107	109
130	129	126	170	88	106	114
87	113	133	146	111	83	135
110	114	121	122	117		

Use these data to construct a boxplot. Write a few sentences describing the important characteristics of the boxplot.

3.38 Data on the gasoline tax per gallon (in cents) as of April 2010 for the 50 U.S. states and the District of Columbia (DC) are shown below (*AARP Bulletin*, June 2010).

State	Gasoline Tax Per Gallon	State	Gasoline Tax Per Gallon
Alabama	20.9	Kentucky	22.5
Alaska	8.0	Louisiana	20.0
Arizona	19.0	Maine	31.0
Arkansas	21.8	Maryland	23.5
California	48.6	Massachusetts	23.5
Colorado	22.0	Michigan	35.8
Connecticut	42.6	Minnesota	27.2
Delaware	23.0	Mississippi	18.8
DC	23.5	Missouri	17.3
Florida	34.4	Montana	27.8
Georgia	20.9	Nebraska	27.7
Hawaii	45.1	Nevada	33.1
Idaho	25.0	New Hampshire	19.6
Illinois	40.4	New Jersey	14.5
Indiana	34.8	New Mexico	18.8
Iowa	22.0	New York	44.9
Kansas	25.0	North Carolina	30.2

(continued)

State	Gasoline Tax Per Gallon	State	Gasoline Tax Per Gallon
North Dakota	23.0	Texas	20.0
Ohio	28.0	Utah	24.5
Oklahoma	17.0	Vermont	24.7
Oregon	25.0	Virginia	19.6
Pennsylvania	32.3	Washington	37.5
Rhode Island	33.0	West Virginia	32.2
South Carolina	16.8	Wisconsin	32.9
South Dakota	24.0	Wyoming	14.0
Tennessee	21.4		

a. The smallest value in the data set is 8.0 (Alaska), and the largest value is 48.6 (California). Are these values outliers? Explain.
b. Construct a boxplot for this data set and comment on the interesting features of the plot.

3.39 The National Climate Data Center gave the accompanying annual rainfall (in inches) for Medford, Oregon, from 1950 to 2008 (**www.ncdc.noaa.gov/oa/climate/research/cag3/city**):

28.84	20.15	18.88	25.72	16.42
20.18	28.96	20.72	23.58	10.62
20.85	19.86	23.34	19.08	29.23
18.32	21.27	18.93	15.47	20.68
23.43	19.55	20.82	19.04	18.77
19.63	12.39	22.39	15.95	20.46
16.05	22.08	19.44	30.38	18.79
10.89	17.25	14.95	13.86	15.30
13.71	14.68	15.16	16.77	12.33
21.93	31.57	18.13	28.87	16.69
18.81	15.15	18.16	19.99	19.00
23.97	21.99	17.25	14.07	

a. Compute the quartiles and the interquartile range.
b. Are there outliers in this data set? If so, which observations are outliers?
c. Draw a modified boxplot for this data set and comment on the interesting features of this plot.

3.40 The accompanying data are consistent with summary statistics that appeared in a paper investigating the effect of the shape of drinking glasses (*British Medical Journal* [2005]: 1512–1514). Data represent the actual amount (in ml) poured into a glass for individuals asked to pour 1.5 ounces (44.3 ml) into either a tall, slender glass or a short, wide glass.

Tall, slender glass

44.0	49.6	62.3	28.4	39.1	39.8	60.5	73.0	57.5
56.5	65.0	56.2	57.7	73.5	66.4	32.7	40.4	21.4

Short, wide glass

89.2	68.6	32.7	37.4	39.6	46.8	66.1	79.2
66.3	52.1	47.3	64.4	53.7	63.2	46.4	63.0
92.4	57.8						

Construct a comparative boxplot and comment on the differences and similarities in the two data distributions.

SECTION 3.4 Exercise Set 2

3.41 *USA Today* (June 11, 2010) gave the accompanying data on median age for residents of each of the 50 U.S. states and the District of Columbia (DC).

State	Median Age	State	Median Age
Alabama	37.4	Montana	39.0
Alaska	32.8	Nebraska	35.8
Arizona	35.0	Nevada	35.5
Arkansas	37.0	New Hampshire	40.4
California	34.8	New Jersey	38.8
Colorado	35.7	New Mexico	35.6
Connecticut	39.5	New York	38.1
Delaware	38.4	North Carolina	36.9
DC	35.1	North Dakota	36.3
Florida	40.0	Ohio	38.5
Georgia	34.7	Oklahoma	35.8
Hawaii	37.5	Oregon	38.1
Idaho	34.1	Pennsylvania	39.9
Illinois	36.2	Rhode Island	39.2
Indiana	36.8	South Carolina	37.6
Iowa	38.0	South Dakota	36.9
Kansas	35.9	Tennessee	37.7
Kentucky	37.7	Texas	33.0
Louisiana	35.4	Utah	28.8
Maine	42.2	Vermont	41.2
Maryland	37.7	Virginia	36.9
Massachusetts	39.0	Washington	37.0
Michigan	38.5	West Virginia	40.5
Minnesota	37.3	Wisconsin	38.2
Mississippi	35.0	Wyoming	35.9
Missouri	37.6		

Use these data to compute the values in the five-number summary.

3.42 The accompanying data are a subset of data read from a graph in the paper "Ladies First? A Field Study of Discrimination in Coffee Shops" (*Applied Economics* [April, 2008]). The data are the waiting time (in seconds) between ordering and receiving coffee for 19 male customers at a Boston coffee shop.

40	60	70	80	85	90	100	100	110	120
125	125	140	140	160	160	170	180	200	

Use these data to construct a boxplot. Write a few sentences describing the important characteristics of the boxplot.

3.43 The accompanying data are the percentage of babies born prematurely in 2008 for the 50 U.S. states and the District of Columbia (DC) (**USAToday.com**).

State	Premature Percent	State	Premature Percent
Alabama	15.6	Montana	11.5
Alaska	10.4	Nebraska	11.8
Arizona	12.9	Nevada	13.5
Arkansas	13.5	New Hampshire	9.6
California	10.5	New Jersey	12.5
Colorado	11.4	New Mexico	12.3
Connecticut	10.4	New York	12.0
Delaware	12.9	North Carolina	12.9
DC	15.5	North Dakota	11.1
Florida	13.8	Ohio	12.6
Georgia	13.4	Oklahoma	13.4
Hawaii	12.8	Oregon	10.1
Idaho	9.8	Pennsylvania	11.6
Illinois	12.7	Rhode Island	11.2
Indiana	12.4	South Carolina	14.3
Iowa	11.5	South Dakota	11.9
Kansas	11.2	Tennessee	13.5
Kentucky	14.0	Texas	13.3
Louisiana	15.4	Utah	11.0
Maine	10.3	Vermont	9.5
Maryland	13.0	Virginia	11.3
Massachusetts	10.8	Washington	10.7
Michigan	12.7	West Virginia	13.7
Minnesota	10.0	Wisconsin	11.1
Mississippi	18.0	Wyoming	11.2
Missouri	12.3		

a. The smallest value in the data set is 9.5 (Vermont), and the largest value is 18.0 (Mississippi). Are these values outliers? Explain.
b. Construct a boxplot for this data set and comment on the interesting features of the plot.

3.44 Shown here are the annual number of auto accidents per 1,000 people in each of 40 occupations (*Knight Ridder Tribune*, June 19, 2004):

Occupation	Accidents Per 1,000	Occupation	Accidents Per 1,000
Student	152	Banking, finance	89
Physician	109	Customer service	88
Lawyer	106	Manager	88
Architect	105	Medical support	87
Real estate broker	102	Computer-related	87
Enlisted military	99	Dentist	86
Social worker	98	Pharmacist	85

(continued)

Occupation	Accidents Per 1,000	Occupation	Accidents Per 1,000
Manual laborer	96	Proprietor	84
Analyst	95	Teacher, professor	84
Engineer	94	Accountant	84
Consultant	94	Law enforcement	79
Sales	93	Physical therapist	78
Military officer	91	Veterinarian	78
Nurse	90	Clerical, secretary	77
School administrator	90	Clergy	76
Skilled laborer	90	Homemaker	76
Librarian	90	Politician	76
Creative arts	90	Pilot	75
Executive	89	Firefighter	67
Insurance agent	89	Farmer	43

a. Are there outliers in this data set? If so, which observations are outliers?

b. Draw a modified boxplot for this data set.

c. If you were asked by an insurance company to decide which, if any, occupations should be offered a professional discount on auto insurance, which occupations would you recommend? Explain.

3.45 The following data on weekend exercise time (in minutes) for 20 males and 20 females are consistent with summary quantities in the paper **"An Ecological Momentary Assessment of the Physical Activity and Sedentary Behaviour Patterns of University Students"** (*Health Education Journal* [2010]: 116–125).

Male—Weekday

43.5	91.5	7.5	0.0	0.0	28.5	199.5
57.0	142.5	8.0	9.0	36.0	0.0	78.0
34.5	0.0	57.0	151.5	8.0	0.0	

Female—Weekday

10.0	90.6	48.5	50.4	57.4	99.6
0.0	5.0	0.0	0.0	5.0	2.0
10.5	5.0	47.0	0.0	5.0	54.0
0.0	48.6				

Construct a comparative boxplot and comment on the differences and similarities in the two data distributions.

Additional Exercises

3.46 The accompanying data on annual maximum wind speed (in meters per second) in Hong Kong for each year in a 45-year period were given in an article that appeared in the journal **Renewable Energy (March, 2007)**. Use the annual maximum wind speed data to construct a boxplot. Is the boxplot approximately symmetric?

30.3	39.0	33.9	38.6	44.6
31.4	26.7	51.9	31.9	27.2
52.9	45.8	63.3	36.0	64.0
31.4	42.2	41.1	37.0	34.4
35.5	62.2	30.3	40.0	36.0
39.4	34.4	28.3	39.1	55.0
35.0	28.8	25.7	62.7	32.4
31.9	37.5	31.5	32.0	35.5
37.5	41.0	37.5	48.6	28.1

3.47 The article **"Going Wireless"** (*AARP Bulletin*, June 2009) reported the estimated percentage of households with only wireless phone service (no landline) for the 50 states and the District of Columbia. In the accompanying data table, each state was also classified into one of three geographical regions—West (W), Middle (M), and East (E).

Wireless %	Region	State	Wireless %	Region	State
13.9	M	AL	9.2	W	MT
11.7	W	AK	23.2	M	NE
18.9	W	AZ	10.8	W	NV
22.6	M	AR	16.9	M	ND
9.0	W	CA	11.6	E	NH
16.7	W	CO	8.0	E	NJ
5.6	E	CN	21.1	W	NM
5.7	E	DE	11.4	E	NY
20.0	E	DC	16.3	E	NC
16.8	E	FL	14.0	E	OH
16.5	E	GA	23.2	M	OK
8.0	W	HI	17.7	W	OR
22.1	W	ID	10.8	E	PA
16.5	M	IL	7.9	E	RI
13.8	M	IN	20.6	E	SC
22.2	M	IA	6.4	M	SD
16.8	M	KA	20.3	M	TN
21.4	M	KY	20.9	M	TX
15.0	M	LA	25.5	W	UT
13.4	E	ME	10.8	E	VA
10.8	E	MD	5.1	E	VT
9.3	E	MA	16.3	W	WA
16.3	M	MI	11.6	E	WV
17.4	M	MN	15.2	M	WI
19.1	M	MS	11.4	W	WY
9.9	M	MO			

a. Construct a comparative boxplot that makes it possible to compare wireless percent for the three geographical regions.

b. Does the graphical display in Part (a) reveal any striking differences, or are the distributions similar for the three regions?

3.48 The accompanying data on percentage of juice lost after thawing for 19 different strawberry varieties appeared in the article **"Evaluation of Strawberry Cultivars with Different Degrees of Resistance to Red Scale"** (*Fruit Varieties Journal* [1991]: 12–17).

46 51 44 50 33 46 60 41 55 46 53
53 42 44 50 54 46 41 48

a. Are there any outliers? If so, which values are outliers?
b. Construct a boxplot, and comment on the important features of the plot.

3.49 The *Los Angeles Times* (**July 17, 1995**) reported that for a sample of 364 lawsuits in which punitive damages were awarded, the median damage award was $50,000, and the mean was $775,000. What does this suggest about the distribution of values in the sample?

SECTION 3.5 Measures of Relative Standing: *z*-scores and Percentiles

When you obtain your score after taking a test, you probably want to know how it compares to the scores of others. Is your score above or below the mean, and by how much? Does your score place you among the top 5% of the class or only among the top 25%? Answering these questions involves measuring the position, or relative standing, of a particular value in a data set. One measure of relative standing is a *z-score*.

> **DEFINITION**
>
> The *z*-**score** corresponding to a particular data value is
>
> $$z\text{-score} = \frac{\text{data value} - \text{mean}}{\text{standard deviation}}$$
>
> The *z*-score tells you how many standard deviations the data value is from the mean. The *z*-score is positive when the data value is greater than the mean and negative when the data value is less than the mean.

The process of subtracting the mean and then dividing by the standard deviation is sometimes referred to as *standardizing*. A *z*-score is one example of a *standardized score*.

Example 3.18 Relatively Speaking, Which Is the Better Offer?

Suppose that two graduating seniors, one a marketing major and one an accounting major, are comparing job offers. The accounting major has an offer for $45,000 per year, and the marketing student has an offer for $43,000 per year. Summary information about the distribution of offers is shown here:

    Accounting: mean = 46,000   standard deviation = 1,500
    Marketing: mean = 42,500   standard deviation = 1,000

Then,

$$\text{accounting } z\text{-score} = \frac{45,000 - 46,000}{1,500} = -0.67$$

so $45,000 is 0.67 standard deviations below the mean, whereas

$$\text{marketing } z\text{-score} = \frac{43,000 - 42,500}{1,000} = 0.5$$

which is above the mean. Relative to their distributions, the marketing offer is actually more attractive than the accounting offer (although this may not offer much solace to the marketing major).

The *z*-score is particularly useful when the data distribution is mound shaped and approximately symmetric. In this case, a *z*-score outside the interval from -2 to $+2$ occurs in about 5% of all cases, whereas a *z*-score outside the interval from -3 to $+3$ occurs only about 0.3% of the time. So in this type of distribution, data values with *z*-scores less than -2 or greater than $+2$ can be thought of as being among the smallest or largest values in the data set.

Sometimes, in published articles, you see a graphical display of data along with summary measures such as the mean and standard deviation, but you don't see the individual data values. A rule called the **Empirical Rule** allows you to use the mean and standard deviation to make statements about the data values. You can use this rule when the shape of the data distribution is mound shaped (has a single peak) and is approximately symmetric.

> **The Empirical Rule**
>
> If a data distribution is mound shaped and approximately symmetric, then
>
> Approximately 68% of the observations are within 1 standard deviation of the mean.
> Approximately 95% of the observations are within 2 standard deviations of the mean.
> Approximately 99.7% of the observations are within 3 standard deviations of the mean.

Figure 3.12 illustrates the percentages given by the Empirical Rule.

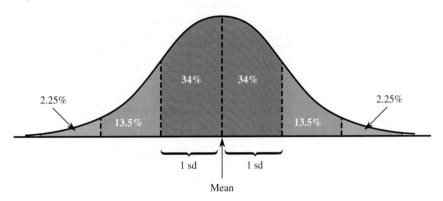

FIGURE 3.12
Approximate percentages implied by the Empirical Rule

Example 3.19 Heights of Mothers and the Empirical Rule

One of the earliest articles to promote using the normal distribution was **"On the Laws of Inheritance in Man. I. Inheritance of Physical Characters" (*Biometrika* [1903]: 375–462)**. One of the data sets discussed in this article contained 1,052 measurements of the heights of mothers. The mean and standard deviation were

$$\bar{x} = 62.484 \text{ inches} \qquad s = 2.390 \text{ inches}$$

The data distribution was described as being mound shaped and approximately symmetric. Table 3.6 contrasts actual percentages from the data set with those obtained from the Empirical Rule. Notice how closely the Empirical Rule approximates the percentages of observations falling within 1, 2, and 3 standard deviations of the mean.

TABLE 3.6 **Summarizing the Distribution of Mothers' Heights**

Number of Standard Deviations	Interval	Actual	Empirical Rule
1	60.094 to 64.874	72.1%	Approximately 68%
2	57.704 to 67.264	96.2%	Approximately 95%
3	55.314 to 69.654	99.2%	Approximately 99.7%

Percentiles

A particular observation can be located even more precisely by giving the percentage of the data that fall at or below that value. For example, if 95% of all test scores are at or below 650, which also means only 5% are above 650, then 650 is called the *95th percentile* of the data set (or of the distribution of scores). Similarly, if 10% of all scores are at or below 400, then the value 400 is the 10th percentile.

> **DEFINITION**
>
> For a number r between 0 and 100, the ***r*th percentile** is a value such that r percent of the observations in the data set fall at or below that value.

Figure 3.13 illustrates the 90th percentile. You have already seen several percentiles in disguise. The median is the 50th percentile, and the lower and upper quartiles are the 25th and 75th percentiles, respectively.

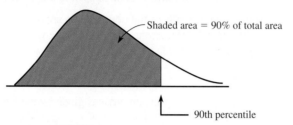

FIGURE 3.13
90th percentile for a smoothed histogram

Example 3.20 | **Head Circumference at Birth**

In addition to weight and length, head circumference is another measure of health in newborn babies. The National Center for Health Statistics reports the following summary values for head circumference (in cm) at birth for boys (approximate values from graphs on the Center for Disease Control web site):

Percentile	5	10	25	50	75	90	95
Head Circumference (cm)	32.2	33.2	34.5	35.8	37.0	38.2	38.6

Interpreting these percentiles, half of newborn boys have head circumferences of less than 35.8 cm, because 35.8 is the 50th percentile (the median). The middle 50% have head circumferences between 34.5 cm and 37.0 cm, with 25% less than 34.5 cm and 25% greater than 37.0 cm. You can tell that this distribution for newborn boys is not symmetric, because the 5th percentile is 3.6 cm below the median, whereas the 95th percentile is only 2.8 cm above the median. This means that the bottom part of the distribution stretches out more than the top part, which creates a negatively skewed distribution, like the one shown in Figure 3.14.

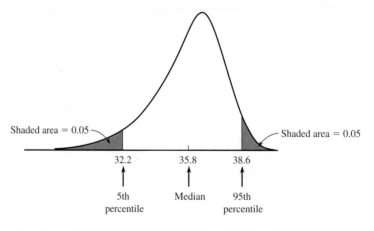

FIGURE 3.14
Negatively skewed distribution

SECTION 3.5 | EXERCISES

Each Exercise Set assesses the following chapter learning objective: P4

SECTION 3.5 | Exercise Set 1

3.50 A student took two national aptitude tests. The mean and standard deviation were 475 and 100, respectively, for the first test, and 30 and 8, respectively, for the second test. The student scored 625 on the first test and 45 on the second test. Use z-scores to determine on which exam the student performed better relative to the other test takers.

3.51 The mean playing time for a large collection of compact discs is 35 minutes, and the standard deviation is 5 minutes.
a. What value is 1 standard deviation above the mean? One standard deviation below the mean? What values are 2 standard deviations away from the mean?
b. Assuming that the distribution of times is mound shaped and approximately symmetric, approximately what percentage of times are between 25 and 45 minutes? Less than 20 minutes or greater than 50 minutes? Less than 20 minutes?

3.52 The report **"Who Borrows Most? Bachelor's Degree Recipients with High Levels of Student Debt"** (**www.collegeboard.com/trends**) included the following percentiles for the amount of student debt for students graduating with a bachelor's degree in 2010:

10th percentile = $0   25th percentile = $0
50th percentile = $11,000   75th percentile = $24,600
90th percentile = $39,300

For each of these percentiles, write a sentence interpreting the value of the percentile.

3.53 The paper **"Modeling and Measurements of Bus Service Reliability"** (*Transportation Research* [1978]: 253–256) presented data on bus travel times (in minutes) for several different routes. The following frequency distribution is for bus travel time from origin to destination on one particular route in Chicago during peak morning traffic:

Travel Time	Frequency	Relative Frequency
15 to < 16	4	0.02
16 to < 17	0	0.00
17 to < 18	26	0.13
18 to < 19	99	0.49
19 to < 20	36	0.18
20 to < 21	8	0.04
21 to < 22	12	0.06
22 to < 23	0	0.00
23 to < 24	0	0.00
24 to < 25	0	0.00
25 to < 26	16	0.08

a. Construct the histogram corresponding to this frequency distribution.
b. Find the approximate values of the following percentiles:
  i. 86th   iv. 95th
  ii. 15th   v. 10th
  iii. 90th

SECTION 3.5   Exercise Set 2

3.54 Suppose that your statistics professor returned your first midterm exam with only a z-score written on it. She also told you that a histogram of the scores was mound shaped and approximately symmetric. How would you interpret each of the following z-scores?
a. 2.2   d. 1.0
b. 0.4   e. 0
c. 1.8

3.55 In a study investigating the effect of car speed on accident severity, the vehicle speed at impact was recorded for 5,000 fatal accidents. For these accidents, the mean speed was 42 mph and the standard deviation was 15 mph. A histogram revealed that the vehicle speed distribution was mound shaped and approximately symmetric.
a. Approximately what percentage of the vehicle speeds were between 27 and 57 mph?
b. Approximately what percentage of the vehicle speeds exceeded 57 mph?

3.56 Suppose that your younger sister is applying to college and has taken the SAT exam. She scored at the 83rd percentile on the verbal section of the test and at the 94th percentile on the math section. Because you have been studying statistics, she asks you for an interpretation of these values. What would you tell her?

3.57 The following data values are the 2009 per capita expenditures on public libraries for each of the 50 U.S. states and Washington, D.C. (from **www.statemaster.com**):

16.84	16.17	11.74	11.11	8.65	7.69	7.48
7.03	6.20	6.20	5.95	5.72	5.61	5.47
5.43	5.33	4.84	4.63	4.59	4.58	3.92
3.81	3.75	3.74	3.67	3.40	3.35	3.29
3.18	3.16	2.91	2.78	2.61	2.58	2.45
2.30	2.19	2.06	1.78	1.54	1.31	1.26
1.20	1.19	1.09	0.70	0.66	0.54	0.49
0.30	0.01					

a. Summarize this data set with a frequency distribution. Construct the corresponding histogram.
b. Use the histogram from Part (a) to find the approximate values of the following percentiles:
  i. 50th   iv. 90th
  ii. 70th   v. 40th
  iii. 10th

Additional Exercises

3.58 The mean number of text messages sent per month by customers of a cell phone service provider is 1,650, and the standard deviation is 750. Find the z-score associated with each of the following numbers of text messages sent.
  a. 0
  b. 10,000
  c. 4,500
  d. 300

3.59 Suppose that the distribution of weekly water usage for single-family homes in a particular city is mound shaped and approximately symmetric. The mean is 1,400 gallons, and the standard deviation is 300 gallons.

a. What is the approximate value of the 16th percentile?

b. What is the approximate value of the median?

c. What is the approximate value of the 84th percentile?

3.60 The mean reading speed of students completing a speed-reading course is 450 words per minute (wpm). If the standard deviation is 70 wpm, find the z-score associated with each of the following reading speeds.

a. 320 wpm   **c.** 420 wpm

b. 475 wpm   **d.** 610 wpm

3.61 Suppose that the distribution of scores on an exam is mound shaped and approximately symmetric. The exam scores have a mean of 100 and the 16th percentile is 80.

a. What is the 84th percentile?

b. What is the approximate value of the standard deviation of exam scores?

c. What is the z-score for an exam score of 90?

d. What percentile corresponds to an exam score of 140?

e. Do you think there were many scores below 40? Explain.

SECTION 3.6 Avoid These Common Mistakes

When computing or interpreting summary measures, keep the following in mind.

1. Watch out for categorical data that look numerical! Often, categorical data is coded numerically. For example gender might be coded as 0 = female and 1 = male, but this does not make gender a numerical variable. Also, sometimes a categorical variable can "look" numerical, as is the case for variables like zip code or telephone area code. But zip codes and area codes are really just labels for geographic regions and don't behave like numbers. Categorical data cannot be summarized using the mean and standard deviation or the median and interquartile range.

2. Measures of center don't tell all. Although measures of center, such as the mean and the median, do give you a sense of what might be considered a typical value for a variable, this is only one characteristic of a data set. Without additional information about variability and distribution shape, you don't really know much about the behavior of the variable.

3. Data distributions with different shapes can have the same mean and standard deviation. For example, consider the following two histograms:

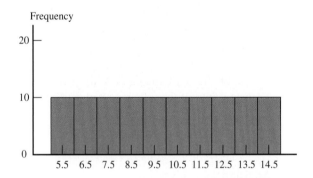
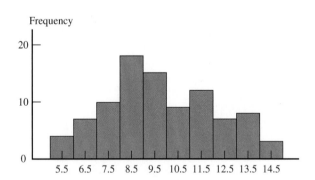

Both histograms summarize data sets that have a mean of 10 and a standard deviation of 2, yet they have different shapes.

4. Both the mean and the standard deviation are sensitive to extreme values in a data set, especially if the sample size is small. If a data distribution is markedly skewed or if the data set has outliers, the median and the interquartile range are better choices for describing center and spread.

5. Measures of center and measures of variability describe values of a variable, not frequencies in a frequency distribution or heights of the bars in a histogram. For example, consider the following two frequency distributions and histograms:

Frequency Distribution A		Frequency Distribution B	
Value	Frequency	Value	Frequency
1	10	1	5
2	10	2	10
3	10	3	20
4	10	4	10
5	10	5	5

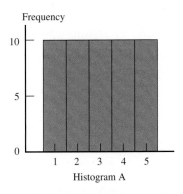

Histogram A

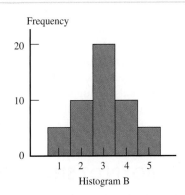

Histogram B

There is more variability in the data summarized by Frequency Distribution and Histogram A than in the data summarized by Frequency Distribution and Histogram B. This is because the values of the variable described by Histogram and Frequency Distribution B are more concentrated near the mean. Don't be misled by the fact that there is no variability in the frequencies in Frequency Distribution A or the heights of the bars in Histogram A.

6. Be careful with boxplots based on small sample sizes. Boxplots convey information about center, variability, and shape, but interpreting shape information is problematic when the sample size is small. For example, it is not really possible to decide whether a data distribution is symmetric or skewed based on a small sample.

7. Not all distributions are mound shaped. Using the Empirical Rule in situations where you are not convinced that the data distribution is mound shaped and approximately symmetric can lead to incorrect statements.

8. Watch for outliers! Unusual observations in a data set often provide important information about the variable under study, so it is important to consider outliers in addition to describing what is typical. Outliers can also be problematic because the values of some summaries are influenced by outliers and because some methods for drawing conclusions from data are not appropriate if the data set has outliers.

CHAPTER ACTIVITIES

ACTIVITY 3.1 COLLECTING AND SUMMARIZING NUMERICAL DATA

In this activity, you will work in groups to collect data on how many hours per week students at your school spend engaged in a particular activity. You will use the sampling plan designed in Activity 2.1 to collect the data.

1. With your group, pick one of the following activities to be the focus of your study (or you may choose a different activity with the approval of your teacher):
    i.   Using the Internet
    ii. Studying or doing homework
    iii. Watching TV
    iv. Exercising
    v.   Sleeping

2. Use the plan from Activity 2.1 to collect data on the variable you have chosen.

3. Summarize your data using both numerical and graphical summaries. Be sure to address both center and variability.

4. Write a short article for your school paper summarizing your findings regarding student behavior. Your article should include both numerical and graphical summaries.

ACTIVITY 3.2 AIRLINE PASSENGER WEIGHTS

The article **"Airlines Should Weigh Passengers, Bags, NTSB Says"** (*USA Today,* **February 27, 2004**) reported on a recommendation from the National Transportation Safety Board that airlines weigh passengers and their bags to prevent planes from being overloaded. This recommendation was the result of an investigation into the crash of a small commuter plane in 2003, which determined that too much weight contributed to the crash.

Rather than weighing passengers, airlines currently use estimates of mean passenger and luggage weights. After the 2003 accident, this estimate was increased by 10 pounds for passengers and 5 pounds for luggage. Although a somewhat overweight airplane can fly if all systems are working

properly, if one of its engines fails, it becomes difficult for the pilot to control.

Assuming that the new estimate of the mean passenger weight is accurate, discuss the following questions with a partner and then write a paragraph that answers these questions.

1. What role does variability in passenger weights play in creating a potentially dangerous situation for an airline?

2. Would an airline have a lower risk of a potentially dangerous situation if the variability in passenger weight is large or if it is small?

ACTIVITY 3.3 BOXPLOT SHAPES

In this activity, you will investigate the relationship between a boxplot shape and its corresponding five-number summary. The following table gives five-number summaries labeled I–IV. Also shown are four boxplots labeled A–D. Notice that scales are not included on the boxplots. Match each five-number summary to its boxplot.

Five-Number Summaries

	I	II	III	IV
Minimum	40	4	0.0	10
Lower quartile	45	8	0.1	34
Median	71	16	0.9	44
Upper quartile	88	25	2.2	82
Maximum	106	30	5.1	132

 ## EXPLORING THE BIG IDEAS

In Exercise 1, each student in your class will go online to select two random samples, one of size 10 and one of size 30, from a small population consisting of 300 adults.

1. To learn about how the values of sample statistics vary from sample to sample, go online at www.cengage.com/stats/peck1e and click on the link labeled "Exploring the Big Ideas" and locate the link for Chapter 3. Then click on the Exercise 1 link. This link will take you to a web page where you can select two random samples from the population.

Click on the samples button. This selects a random sample of size 10 and a random sample of size 30 and will display the ages of the people in your samples. Each student in your class will receive data from different random samples.

Use the data from your random samples to complete the following.

a. Calculate the mean age of the people in your sample of size 10.

b. The mean age for the entire population is 44.4 years. Did you get 44.4 years for your sample mean? How far was your sample mean from 44.4?

c. Now calculate the mean age for the people in your sample of size 30. Is the mean for your sample of size 30 closer to 44.4 years or farther away than the mean from your sample of size 10? Does this surprise you? Explain why or why not.

If asked to do so by your teacher, bring your answers to Parts (a)–(c) to class. Your teacher will lead the class through the rest of this exercise.

Have one student draw a number line on the board or on a piece of poster-size chart paper. The scale on the number line should range from 20 to 60. The class should now create a dotplot of the sample means for the samples of size 10 by having each student enter a dot that corresponds to his or her sample mean. Label this dotplot "Means of Random Samples of Size 10."

d. Based on the class dotplot for samples of size 10, how would you describe the sample-to-sample variability in the sample means? Was there a lot of variability, or were the sample means fairly similar?

e. Locate 44.4 on the scale for the class dotplot. Does it look like distribution of the sample means is centered at about 44.4?

f. Here is a dotplot of the ages for the entire population. How is the dotplot of sample means for samples of size 10 similar to the population dotplot? How is it different?

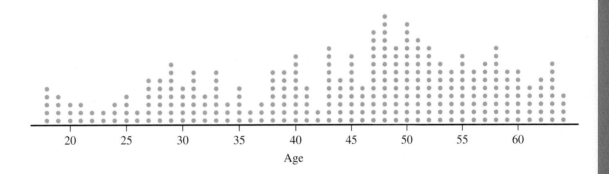

g. Now create a class dotplot of the sample means for the samples of size 30. Label this dotplot "Means of Random Samples of Size 30." Based on the class dotplots for samples of size 10 and samples of size 30, was there more sample-to-sample variability in the values of the sample mean for samples of size 10 or for samples of size 30?

h. How do the dotplots of the sample means for samples of size 10 and the sample means for samples of size 30 compare in terms of shape, center, and spread?

i. How do the dotplots support the statement that the sample mean will tend to be closer to the actual value of the population mean for a random sample of size 30 than for a random sample of size 10?

In Exercise 2, each student in your class will go online to select a random sample of 25 airplanes from a population of airplanes.

2. The cost of jet fuel is one factor that contributes to the rising cost of airfares. In a study of aircraft fuel consumption that was part of an environmental impact report, different types of planes that use the Phoenix airport were studied, and the amount of fuel consumed from the time the plane left the gate until the time the plane reached an altitude of 3,000 feet was determined. Go online at www.cengage.com/stats/peck1e and locate the link labeled "Chapter 3." Then click on the Exercise 2 link. This link will take you to a web page where you can select a random sample of 25 planes. Click on the sample button. This selects the sample and displays the amount of fuel consumed (in gallons) for each plane in your sample.

a. Compute the mean, median, and standard deviation for this sample.

b. Construct a dotplot of the fuel consumption values.

c. Write a few sentences commenting on the dotplot. What does the dotplot tell you about the distribution of fuel consumption values in the sample?

d. Which of the following best describes the distribution of fuel consumption?

 i. The distribution is approximately symmetric. A typical value for fuel consumption is approximately 130,000 gallons, and there is not much variability in fuel consumption for the different plane types.

 ii. The distribution is mound shaped. A typical value for fuel consumption is approximately 130,000 gallons, and there is quite a bit of variability in fuel consumption for the different plane types.

 iii. The distribution is markedly skewed, with a long upper tail (on the right-hand side) indicating that there are some plane types with a fuel consumption that is much higher than for most plane types.

 iv. The distribution is markedly skewed, with a long lower tail (on the left-hand side) indicating that there are some plane types with a fuel consumption that is much lower than for most plane types.

e. Which of the following do you think would have the biggest impact on overall fuel consumption at this airport?

 i. Modifications were made that would reduce the fuel consumption of every type of plane by 20 gallons.

 ii. Changes were implemented that would result in a 10% reduction in the fuel consumption of the 10 types of planes with the largest consumption.

f. Write a few sentences explaining your choice in Part (e).

ARE YOU READY TO MOVE ON? CHAPTER 3 REVIEW EXERCISES

All chapter learning objectives are assessed in these exercises. The learning objectives assessed in each exercise are given in parentheses.

3.62 (C1, C2, M1)

For each of the following data sets, construct a graphical display of the data distribution and then indicate what summary measures you would use to describe center and spread.

a. The following are data on weekend exercise time (in minutes) for 20 females consistent with summary quantities given in the paper "An Ecological Momentary Assessment of the Physical Activity and Sedentary Behaviour Patterns of University Students" (*Health Education Journal* [2010]: 116–125).

Female—Weekend

84.0	27.0	82.5	0.0	5.0	13.0	44.5
3.0	0.0	14.5	45.5	39.5	6.5	34.5
0.0	14.5	40.5	44.5	54.0	0.0	

b. The accompanying data are consistent with summary statistics that appeared in a paper investigating the effect of the shape of drinking glasses (*British Medical Journal* [2005]: 1512–1514). Data represent the actual amount (in ml) poured into a tall, slender glass by individuals asked to pour 1.5 ounces (44.3 ml) into the glass.

44.0	49.6	62.3	28.4	39.1	39.8	60.5
73.0	57.5	56.5	65.0	56.2	57.7	73.5
66.4	32.7	40.4	21.4			

c. The accompanying data are from a graph that appeared in the paper "Ladies First? A Field Study of Discrimination in Coffee Shops" (*Applied Economics* [April, 2008]). The data are the wait times (in seconds) between ordering and receiving coffee for 19 female customers at a Boston coffee shop.

60	˙80	80	100	100	100	120
120	120	140	140	150	160	180
200	200	220	240	380		

3.63 (M2, M3, P1, P2)

Data on a customer satisfaction rating (called the APEAL rating) are given for each brand of car sold in the United States (*USA Today*, July 17, 2010). The APEAL rating is a score between 0 and 1,000, with higher values indicating greater satisfaction.

822	832	845	802	818	789	748	751	794
792	766	760	805	854	727	761	836	822
820	774	842	769	815	767	763	877	780
764	755	750	745	797	795			

Compute and interpret the mean and standard deviation for this data set.

3.64 (C3, M3, P2)

The paper **"Caffeinated Energy Drinks—A Growing Problem"** (*Drug and Alcohol Dependence* [2009]: 1–10) reported caffeine per ounce for 8 top-selling energy drinks and for 11 high-caffeine energy drinks:

Top Selling Energy Drinks

| 9.6 | 10.0 | 10.0 | 9.0 | 10.9 | 8.9 | 9.5 | 9.1 |

High-Caffeine Energy Drinks

| 21.0 | 25.0 | 15.0 | 21.5 | 35.7 | 15.0 |
| 33.3 | 11.9 | 16.3 | 31.3 | 30.0 | |

The mean caffeine per ounce is clearly higher for the high-caffeine energy drinks, but which of the two groups of energy drinks is the most variable? Justify your choice.

3.65 (C4)

Data on tipping percent for 20 restaurant tables, consistent with summary statistics given in the paper **"Beauty and the Labor Market: Evidence from Restaurant Servers"** (unpublished manuscript by Matt Parrett, 2007), are:

0.0	5.0	45.0	32.8	13.9	10.4	55.2
50.0	10.0	14.6	38.4	23.0	27.9	27.9
105.0	19.0	10.0	32.1	11.1	15.0	

a. Compute the mean and standard deviation for this data set.
b. Delete the observation of 105.0 and recompute the mean and standard deviation. How do these values compare to the values from Part (a)? What does this suggest about using the mean and standard deviation as measures of center and spread for a data set with outliers?

3.66 (C5)

The **Insurance Institute for Highway Safety** (www.iihs.org, June 11, 2009) published data on repair costs for cars involved in different types of accidents. In one study, seven different 2009 models of mini- and micro-cars were driven at 6 mph straight into a fixed barrier. The following table gives the cost of repairing damage to the bumper for each of the seven models.

Model	Repair Cost
Smart Fortwo	$1,480
Chevrolet Aveo	$1,071
Mini Cooper	$2,291
Toyota Yaris	$1,688
Honda Fit	$1,124
Hyundai Accent	$3,476
Kia Rio	$3,701

Compute the values of the mean and median. Why are these values so different? Which of the two—mean or median—appears to be a better description of a typical value for this data set?

3.67 (M4, M5, P1, P2)

The accompanying data are a subset of data read from a graph in the paper **"Ladies First? A Field Study of Discrimination in Coffee Shops"** (*Applied Economics* [April, 2008]). The data are wait times (in seconds) between ordering and receiving coffee for 19 female customers at a Boston coffee shop.

60	80	80	100	100	100	120
120	120	140	140	150	160	180
200	200	220	240	380		

a. Compute and interpret the values of the median and interquartile range.
b. Explain why the median and interquartile range are appropriate choices of summary measures to describe center and spread for this data set.

3.68 (M6, M7, M8)

The report **"Who Moves? Who Stays Put? Where's Home?"** (*Pew Social and Demographic Trends*, December 17, 2008) gave the accompanying data on the percentage of the population in a state that was born in the state and is still living there for each of the 50 U.S. states.

75.8	71.4	69.6	69.0	68.6	67.5	66.7	66.3
66.1	66.0	66.0	65.1	64.4	64.3	63.8	63.7
62.8	62.6	61.9	61.9	61.5	61.1	59.2	59.0
58.7	57.3	57.1	55.6	55.6	55.5	55.3	54.9
54.7	54.5	54.0	54.0	53.9	53.5	52.8	52.5
50.2	50.2	48.9	48.7	48.6	47.1	43.4	40.4
35.7	28.2						

a. Find the values of the median, the lower quartile, and the upper quartile.
b. The two smallest values in the data set are 28.2 (Alaska) and 35.7 (Wyoming). Are these two states outliers?
c. Construct a modified boxplot for this data set and comment on the interesting features of the plot.

3.69 (M4, M5, M7, M8, P3)
Fiber content (in grams per serving) and sugar content (in grams per serving) for 18 high-fiber cereals (www.consumerreports.com) are shown.

Fiber Content

7	10	10	7	8	7	12
12	8	13	10	8	12	7
14	7	8	8			

Sugar Content

11	6	14	13	0	18	9
10	19	6	10	17	10	10
0	9	5	11			

a. Find the median, quartiles, and interquartile range for the fiber content data set.
b. Find the median, quartiles, and interquartile range for the sugar content data set.
c. Are there any outliers in the sugar content data set?
d. Explain why the minimum value and the lower quartile are equal for the fiber content data set.
e. Construct a comparative boxplot and use it to comment on the differences and similarities in the fiber and sugar distributions.

3.70 (P4)
The accompanying table gives the mean and standard deviation of reaction times (in seconds) for each of two different stimuli:

Suppose your reaction time is 4.2 seconds for the first stimulus and 1.8 seconds for the second stimulus. Compared to other people, to which stimulus are you reacting more quickly?

	Stimulus 1	Stimulus 2
Mean	6.0	3.6
Standard deviation	1.2	0.8

3.71 (P4)
The report **"Earnings by Education Level and Gender" (U.S. Census Bureau, 2009)** gave the following percentiles for annual earnings of full-time female workers age 25 and older with an associate degree:

25th percentile = $26,800   50th percentile = $36,800
75th percentile = $51,100

a. For each of these percentiles, write a sentence interpreting the value of the percentile.
b. The report also gave percentiles for men age 25 and older with an associate degree:

25th percentile = $35,700   50th percentile = $50,100
75th percentile = $68,000

Write a few sentences commenting on how these values compare to the percentiles for women.

TECHNOLOGY NOTES

Mean

TI-83/84
1. Input the raw data into **L1** (To access lists press the **STAT** key, highlight the option called **Edit...** then press **ENTER**)
2. Press the **STAT** key
3. Use the arrows to highlight **CALC**
4. Highlight **1-Var Stats** and press **ENTER**
5. Press the **2nd** key and then the **1** key
6. Press **ENTER**

Note: You may need to scroll to view all of the statistics. This procedure also produces the standard deviation, minimum, Q1, median, Q3, maximum.

TI-Nspire
1. Enter the data into a data list (To access data lists select the spreadsheet option and press **enter**)

   Note: Be sure to title the list by selecting the top row of the column and typing a title.

2. Press the **menu** key and select **4:Statistics** then select **1:Stat Calculations** then **1:One-Variable Statistics...**

3. Press **OK**
4. For **X1 List,** select the title for the column containing your data from the drop-down menu
5. Press **OK**

Note: You may need to scroll to view all of the statistics. This procedure also produces the standard deviation, minimum, Q1, median, Q3, maximum.

JMP
1. Input the raw data into a column
2. Click **Analyze** and select **Distribution**
3. Click and drag the column name containing the data from the box under **Select Columns** to the box next to **Y, Columns**
4. Click **OK**

Note: These commands also produce the following statistics: standard deviation, minimum, quartile 1, median, quartile 3, and maximum.

Minitab
1. Input the raw data into C1
2. Select **Stat** and choose **Basic Statistics** then choose **Display Descriptive Statistics...**

3. Double-click C1 to add it to the **Variables** list
4. Click **OK**

Note: These commands also produce the following statistics: standard deviation, minimum, quartile 1, median, quartile 3, and maximum.

SPSS

1. Input the raw data into the first column
2. Select **Analyze** and choose **Descriptive Statistics** then choose **Explore...**
3. Highlight the column name for the variable
4. Click the arrow to move the variable to the **Dependent List** box
5. Click **OK**

Note: These commands also produce the following statistics: median, variance, standard deviation, minimum, maximum, interquartile range, and several plots.

Excel 2007

1. Input the raw data into the first column
2. Click on the **Data** ribbon and select **Data Analysis**

   Note: If you do not see **Data Analysis** listed on the Ribbon, see the Technology Notes for Chapter 2 for instructions on installing this add-on.

3. Select **Descriptive Statistics** from the dialog box
4. Click **OK**
5. Click in the box next to **Input Range:** and select the data (if you used and selected column titles, check the box next to **Labels in First Row**)
6. Check the box next to **Summary Statistics**
7. Click **OK**

Note: These commands also produce the following statistics: standard error, median, mode, standard deviation, sample variance, range, minimum, and maximum.

Note: You can also find the mean using the Excel function average.

Median

TI-83/84

1. Input the raw data into **L1** (To access lists press the **STAT** key, highlight the option called **Edit...** then press **ENTER**)
2. Press the **STAT** key
3. Use the arrows to highlight **CALC**
4. Highlight **1-Var Stats** and press **ENTER**
5. Press the **2nd** key and then the **1** key
6. Press **ENTER**

Note: You may need to scroll to view all of the statistics. This procedure also produces the mean, standard deviation, minimum, Q1, Q3, maximum.

TI-Nspire

1. Enter the data into a data list (To access data lists select the spreadsheet option and press **enter**)

Note: Be sure to title the list by selecting the top row of the column and typing a title.

2. Press the **menu** key and select **4:Statistics** then select **1:Stat Calculations** then **1:One-Variable Statistics...**
3. Press **OK**
4. For **X1 List**, select the title for the column containing your data from the drop-down menu
5. Press **OK**

Note: You may need to scroll to view all of the statistics. This procedure also produces the mean, standard deviation, minimum, Q1, Q3, maximum.

JMP

1. Input the raw data into a column
2. Click **Analyze** and select **Distribution**
3. Click and drag the column name containing the data from the box under **Select Columns** to the box next to **Y, Columns**
4. Click **OK**

Note: These commands also produce the following statistics: mean, standard deviation, minimum, quartile 1, quartile 3, and maximum.

Minitab

1. Input the raw data into C1
2. Select **Stat** and choose **Basic Statistics** then choose **Display Descriptive Statistics...**
3. Double-click C1 to add it to the **Variables** list
4. Click **OK**

Note: These commands also produce the following statistics: mean, standard deviation, minimum, quartile 1, quartile 3, maximum.

SPSS

1. Input the raw data into the first column
2. Select **Analyze** and choose **Descriptive Statistics** then choose **Explore...**
3. Highlight the column name for the variable
4. Click the arrow to move the variable to the **Dependent List** box
5. Click **OK**

Note: These commands also produce the following statistics: mean, variance, standard deviation, minimum, maximum, interquartile range, and several plots.

Excel 2007

1. Input the raw data into the first column
2. Click on the **Data** ribbon and select **Data Analysis**
3. **Note:** If you do not see **Data Analysis** listed on the Ribbon, see the Technology Notes for Chapter 2 for instructions on installing this add-on.
4. Select **Descriptive Statistics** from the dialog box
5. Click **OK**
6. Click in the box next to **Input Range:** and select the data (if you used and selected column titles, check the box next to **Labels in First Row**)

7. Check the box next to **Summary Statistics**
8. Click **OK**

Note: These commands also produce the following statistics: mean, standard error, mode, standard deviation, sample variance, range, minimum, and maximum.

Note: You can also find the median using the Excel function **median**.

Variance

TI-83/84
The TI-83/84 does not automatically produce the variance; however, this can be determined by finding the standard deviation and squaring it.

TI-Nspire
The TI-Nspire does not automatically produce the variance; however, this can be determined by finding the standard deviation and squaring it.

JMP
1. Input the raw data into a column
2. Click **Analyze** and select **Distribution**
3. Click and drag the column name containing the data from the box under **Select Columns** to the box next to **Y, Columns**
4. Click **OK**
5. Click the red arrow next to the column name
6. Click **Display Options** then select **More Moments**

Note: These commands also produce the following statistics: standard deviation, minimum, quartile 1, median, quartile 3, and maximum.

Minitab
1. Input the raw data into C1
2. Select **Stat** and choose **Basic Statistics** then choose **Display Descriptive Statistics...**
3. Double-click C1 to add it to the **Variables** list
4. Click the **Statistics** button
5. Check the box next to **Variance**
6. Click **OK**
7. Click **OK**

SPSS
1. Input the raw data into the first column
2. Select **Analyze** and choose **Descriptive Statistics** then choose **Explore...**
3. Highlight the column name for the variable
4. Click the arrow to move the variable to the **Dependent List** box
5. Click **OK**

Note: These commands also produce the following statistics: mean, median, standard deviation, minimum, maximum, interquartile range, and several plots.

Excel 2007
1. Input the raw data into the first column
2. Click on the **Data** ribbon and select **Data Analysis**
3. **Note:** If you do not see **Data Analysis** listed on the Ribbon, see the Technology Notes for Chapter 2 for instructions on installing this add-on.
4. Select **Descriptive Statistics** from the dialog box
5. Click **OK**
6. Click in the box next to **Input Range:** and select the data (if you used and selected column titles, check the box next to **Labels in First Row**)
7. Check the box next to **Summary Statistics**
8. Click **OK**

Note: These commands also produce the following statistics: mean, standard error, median, mode, standard deviation, range, minimum, and maximum.

Note: You can also find the variance using the Excel function **var**.

Standard Deviation

TI-83/84
1. Input the raw data into **L1** (To access lists press the **STAT** key, highlight the option called **Edit...** then press **ENTER**)
2. Press the **STAT** key
3. Use the arrows to highlight **CALC**
4. Highlight **1-Var Stats** and press **ENTER**
5. Press the **2nd** key and then the **1** key
6. Press **ENTER**

Note: You may need to scroll to view all of the statistics. This procedure also produces the mean, minimum, Q1, median, Q3, maximum.

TI-Nspire
1. Enter the data into a data list (To access data lists select the spreadsheet option and press **enter**)

   Note: Be sure to title the list by selecting the top row of the column and typing a title.

2. Press the **menu** key and select **4:Statistics** then select **1:Stat Calculations** then **1:One-Variable Statistics...**
3. Press **OK**
4. For **X1 List**, select the title for the column containing your data from the drop-down menu
5. Press **OK**

Note: You may need to scroll to view all of the statistics. This procedure also produces the mean, minimum, Q1, median, Q3, maximum.

JMP
1. Input the raw data into a column
2. Click **Analyze** and select **Distribution**
3. Click and drag the column name containing the data from the box under **Select Columns** to the box next to **Y, Columns**
4. Click **OK**

Note: These commands also produce the following statistics: mean, minimum, quartile 1, median, quartile 3, and maximum.

Minitab
1. Input the raw data into C1
2. Select **Stat** and choose **Basic Statistics** then choose **Display Descriptive Statistics...**
3. Double-click C1 to add it to the **Variables** list
4. Click **OK**

Note: These commands also produce the following statistics: standard deviation, minimum, quartile 1, median, quartile 3, maximum.

SPSS
1. Input the raw data into the first column
2. Select **Analyze** and choose **Descriptive Statistics** then choose **Explore...**
3. Highlight the column name for the variable
4. Click the arrow to move the variable to the **Dependent List** box
5. Click **OK**

Note: These commands also produce the following statistics: mean, median, variance, minimum, maximum, interquartile range, and several plots.

Excel 2007
1. Input the raw data into the first column
2. Click on the **Data** ribbon and select **Data Analysis**
3. **Note:** If you do not see **Data Analysis** listed on the Ribbon, see the Technology Notes for Chapter 2 for instructions on installing this add-on.
4. Select **Descriptive Statistics** from the dialog box
5. Click **OK**
6. Click in the box next to **Input Range:** and select the data (if you used and selected column titles, check the box next to **Labels in First Row**)
7. Check the box next to **Summary Statistics**
8. Click **OK**

Note: These commands also produce the following statistics: mean, standard error, median, mode, sample variance, range, minimum, and maximum.

Note: You can also find the standard deviation using the Excel function **sd**.

Quartiles

TI-83/84
1. Input the raw data into **L1** (To access lists press the **STAT** key, highlight the option called **Edit...** then press **ENTER**)
2. Press the **STAT** key
3. Use the arrows to highlight **CALC**
4. Highlight **1-Var Stats** and press **ENTER**
5. Press the **2nd** key and then the **1** key
6. Press **ENTER**

Note: You may need to scroll to view all of the statistics. This procedure also produces the mean, standard deviation, minimum, median, maximum.

TI-Nspire
1. Enter the data into a data list (To access data lists select the spreadsheet option and press **enter**)
   Note: Be sure to title the list by selecting the top row of the column and typing a title.
2. Press the **menu** key and select **4:Statistics** then select **1:Stat Calculations** then **1:One-Variable Statistics...**
3. Press **OK**
4. For **X1 List**, select the title for the column containing your data from the drop-down menu
5. Press **OK**

Note: You may need to scroll to view all of the statistics. This procedure also produces the mean, standard deviation, minimum, median, maximum.

JMP
1. Input the raw data into a column
2. Click **Analyze** and select **Distribution**
3. Click and drag the column name containing the data from the box under **Select Columns** to the box next to **Y, Columns**
4. Click **OK**

Note: These commands also produce the following statistics: mean, standard deviation, minimum, median, and maximum.

Minitab
1. Input the raw data into C1
2. Select **Stat** and choose **Basic Statistics** then choose **Display Descriptive Statistics...**
3. Double-click C1 to add it to the **Variables** list
4. Click **OK**

Note: These commands also produce the following statistics: standard deviation, minimum, quartile 1, median, quartile 3, maximum.

SPSS
1. Input the raw data into the first column
2. Select **Analyze** and choose **Descriptive Statistics** then choose **Frequencies...**
3. Highlight the column name for the variable
4. Click the arrow to move the variable to the **Dependent List** box
5. Click the **Statistics** button
6. Check the box next to Quartiles
7. Click **Continue**
8. Click **OK**

Excel 2007
1. Input the raw data into the first column
2. Select the cell where you would like to place the first quartile results
3. Click the **Formulas** Ribbon

4. Click **Insert Function**
5. Select the category **Statistical** from the drop-down menu
6. In the **Select a function:** box click **Quartile**
7. Click **OK**
8. Click in the box next to **Array** and select the data
9. Click in the box next to **Quart** and type 1
10. Click **OK**

Note: To find the third quartile, type 3 into the box next to Quart in Step 9.

IQR

TI-83/84

The TI-83/84 does not have the functionality to produce the IQR automatically.

TI-Nspire

The TI-Nspire does not have the functionality to produce the IQR automatically.

JMP

JMP does not have the functionality to produce the IQR automatically.

Minitab

1. Input the raw data into C1
2. Select **Stat** and choose **Basic Statistics** then choose **Display Descriptive Statistics...**
3. Double-click C1 to add it to the **Variables** list
4. Click the **Statistics** button
5. Check the box next to **Interquartile range**
6. Click **OK**
7. Click **OK**

SPSS

1. Input the raw data into the first column
2. Select **Analyze** and choose **Descriptive Statistics** then choose **Explore...**
3. Highlight the column name for the variable
4. Click the arrow to move the variable to the **Dependent List** box
5. Click **OK**

Note: These commands also produce the following statistics: mean, median, variance, standard deviation, minimum, maximum, and several plots.

Excel 2007

1. Use the steps under the **Quartiles** section to find both the first and third quartiles
2. Click on an empty cell where you would like the result for IQR to appear
3. Type = into the cell
4. Click on the cell containing the third quartile
5. Type −
6. Click on the cell containing the first quartile
7. Press **Enter**

Boxplot

TI-83/83

1. Input the raw data into **L1** (To access lists press the **STAT** key, highlight the option called **Edit...** then press **ENTER**)
2. Press the **2nd** key and then press the **Y =** key
3. Select **Plot1** and press **ENTER**
4. Highlight **On** and press **ENTER**
5. Highlight the graph option in the second row, second column, and press **ENTER**
6. Press **GRAPH**

Note: If the graph window does not display appropriately, press the **WINDOW** button and reset the scales appropriately.

TI-Nspire

1. Enter the data into a data list (To access data list select the spreadsheet option and press **enter**)

   Note: Be sure to title the list by selecting the top row of the column and typing a title.

2. Press **menu** and select **3:Data** then select **6:QuickGraph**
3. Press **menu** and select **1:Plot Type** then select **2:Box Plot** and press **enter**

JMP

1. Input the raw data into a column
2. Click **Analyze** and select **Distribution**
3. Click and drag the column name containing the data from the box under **Select Columns** to the box next to **Y, Columns**
4. Click **OK**

Minitab

1. Input the raw data into C1
2. Select **Graph** then choose **Boxplot...**
3. Highlight **Simple** under **One Y**
4. Click **OK**
5. Double-click C1 to add it to the **Graph Variables** box
6. Click **OK**

Note: You may add or format titles, axis titles, legends, and so on by clicking on the **Labels**... button prior to performing Step 6 above.

SPSS

1. Enter the raw data in the first column
2. Select **Graph** and choose **Chart Builder...**
3. Under **Choose from** highlight Boxplot
4. Click and drag the first boxplot (Simple Boxplot) to the Chart preview area
5. Click and drag the data variable into the **Y-Axis?** box in the chart preview area
6. Click **OK**

Note: Boxplots can also be produced when summary statistics such as mean, median, standard deviation, and so on are produced.

Excel 2007

Excel 2007 does not have the functionality to create boxplots.

Side-by-side Boxplots

TI-83/84
The TI-83/84 does not have the functionality to create side-by-side boxplots.

TI-Nspire
The TI-Nspire does not have the functionality to create side-by-side boxplots.

JMP
1. Enter the raw data for both groups into a column
2. Enter the group information into a second column

Column 2	Column 3
F	8
F	3
F	6
F	3
F	9
F	2
F	3
F	4
F	8
M	3
M	5
M	4
M	6
M	7
M	2

3. Click **Analyze** then select **Fit Y by X**
4. Click and drag the column name containing the raw data from the box under **Select Columns** to the box next to **Y, Response**
5. Click and drag the column name containing the group information from the box under **Select Columns** to the box next to **X, Factor**
6. Click **OK**
7. Click the red arrow next to **Oneway Analysis of...**
8. Click **Quantiles**

Minitab
1. Input the raw data for the first group into C1
2. Input the raw data for the second group into C2

3. Continue to input data for each group into a separate column
4. Select **Graph** then choose **Boxplot...**
5. Highlight Simple under Multiple Y's
6. Click **OK**
7. Double-click the column names for each column to be graphed to add it to the **Graph Variables** box
8. Click **OK**

SPSS
1. Enter the raw data the first column
2. Enter the group data into the second column

3. Select **Graph** and choose **Chart Builder...**
4. Under **Choose from** highlight Boxplot
5. Click and drag the first boxplot (Simple Boxplot) to the Chart preview area
6. Click and drag the data variable into the **Y-Axis?** box in the chart preview area
7. Click and drag the group variable into the **X-Axis?** box in the chart preview area
8. Click **OK**

Excel 2007
Excel 2007 does not have the functionality to create side-by-side boxplots.

AP* Review Questions for Chapter 3

1. A study has been carried out to compare the effectiveness of four different diets on weight loss. Each of 200 people are randomly assigned to one of the four diets for a period of 6 weeks. The distributions of the amounts of weight lost for each diet are summarized in the histograms at the bottom of the page.

   One way to compare the diets would be to compare the mean weight loss for each diet. Order the diets from *largest* to *smallest* using the mean weight lost.

   (A) 2, 1, 4, 3   (D) 2, 1, 3, 4
   (B) 3, 2, 4, 1   (E) 3, 2, 1, 4
   (C) 2, 4, 1, 3

2. The commute times (in minutes) for a sample of 23 commuters in a metropolitan area are summarized in the stem-and-leaf display below.

   Leaf Unit = 1.0

   ```
   0 | 6
   1 | 258
   2 | 235
   3 | 01257
   4 | 01224455678
   ```

   Determine the mean and median commute time for this sample.

   (A) $\bar{x} = 2$, median = 2
   (B) $\bar{x} = 3.91$, median = 2
   (C) $\bar{x} = 3.91$, median = 4
   (D) $\bar{x} = 33.48$, median = 37
   (E) $\bar{x} = 37$, median = 37

Diet 1

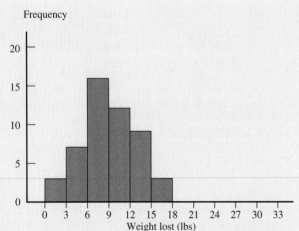

Diet 2

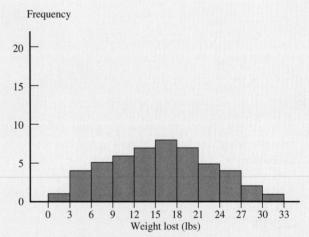

Diet 3

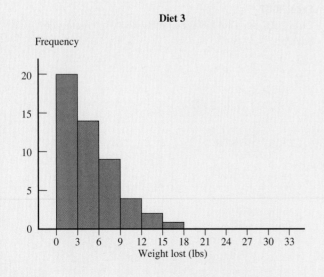

Diet 4

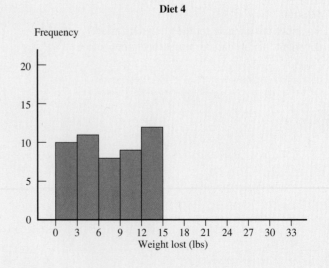

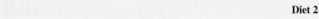

3. Fishermen fishing for spiny lobster are allowed to keep only lobsters with a carapace length of 3.25 inches or longer. (The carapace length is measured from the rear edge of the eye socket to the rear edge of the body shell.) Any lobster smaller than 3.25 inches must be returned to the sea. Suppose that lobster carapace lengths have a distribution that is mound shaped and approximately symmetric with a mean of 5.50 inches and a standard deviation of 2.25 inches. Approximately what proportion of lobsters will have to be returned to the sea?

(A) 16%
(B) 32%
(C) 50%
(D) 68%
(E) 84%

4. Which of the following is closest to the value of the upper quartile for the histogram shown?

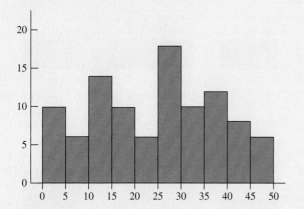

(A) 0
(B) 12
(C) 27
(D) 37
(E) 50

5. A data set consists of five numbers: -6.0, -4.5, 0, 5.0, and an unknown 5th number. For these five data points, which of the following statistics can *never* be greater than zero?

(A) the mean
(B) the sample standard deviation
(C) the interquartile range
(D) the median
(E) the upper quartile

6. The values shown are observations on the number of cars rented at a rental car facility on 20 randomly selected days last year.

0	2	6	9	10
10	18	22	22	23
24	25	25	25	28
30	33	37	40	42

Which observations would be displayed as outliers in a boxplot of these data?

(A) 0 only
(B) 42 only
(C) 0 and 42 are the only outliers
(D) 0, 2, 40, and 42 are the only outliers
(E) None of the observations is an outlier.

7. The five-number summary for a data set is

Minimum: 60
Lower Quartile: 75
Median: 78
Upper Quartile: 82
Maximum: 89

Which of the following is true for this data set?

(A) There are no outliers in the data set.
(B) This data set has at least one outlier on the high end and at least one outlier on the low end.
(C) This data set has at least one outlier on the high end and no outliers on the low end.
(D) This data set has no outliers on the high end and at least one outlier on the low end.
(E) From the information provided, it is not possible to determine if the data set contains outliers.

8. Which of the following statements are true?

I. It is impossible for a data set to have a standard deviation that is larger than the mean.
II. If two data sets have different ranges, the data set with the larger range will always have the larger standard deviation.
III. The interquartile range of a data set can never be larger than the range.

(A) I only
(B) II only
(C) III only
(D) II and III only
(E) I, II, and III

9. Which of the five boxplots shown in Figure 9 on the next page corresponds to a data set with the 5-number summary below?

Minimum: 55
Lower Quartile: 66
Median: 74
Upper Quartile: 83
Maximum: 94

(A) Boxplot 1
(B) Boxplot 2
(C) Boxplot 3
(D) Boxplot 4
(E) Boxplot 5

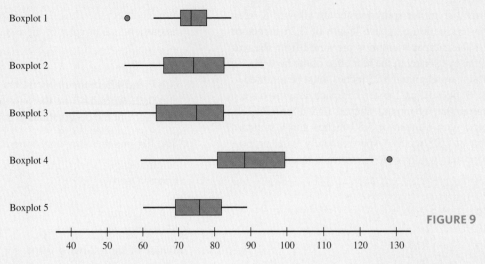

FIGURE 9

10. Each person in a sample of 75 unemployed people who were looking for work was asked how many weeks he or she had been unemployed. The data are summarized in the accompanying frequency distribution.

Weeks Unemployed	Frequency
1–3	22
4–6	31
7–9	11
10–12	8
13–15	1
>15	2

Which of the following could have been the median of the data set?

(A) 2
(B) 5
(C) 8
(D) 11
(E) 16

11. Consider the two sets of rectangles shown below:

Set 1:

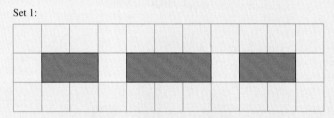

Set 2:

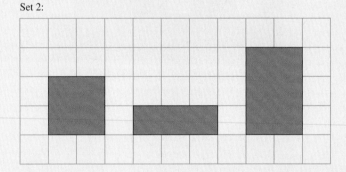

Which of the following statements are true?

I. The standard deviation of horizontal length for the rectangles in Set 1 is equal to the standard deviation of horizontal length for the rectangles in Set 2.
II. The standard deviation of vertical length for the rectangles in Set 1 is 0.
III. The standard deviation of vertical length is greater than the standard deviation of horizontal length for the rectangles of Set 2.

(A) I only
(B) II only
(C) III only
(D) I and III only
(E) I, II, and III

12. The mean and standard deviation of a data set with 50 observations are 47 and 10, respectively. A new data set is formed by adding 3 to the smallest observation in the data set and subtracting 3 from the largest observation in the data set. How will the mean and standard deviation of the new data set compare to the mean and standard deviation of the original data set?

(A) The mean and standard deviation of the new data set will be equal to the mean and standard deviation of the original data set.

(B) The mean of the new data set will be equal to the mean of the original data set, but the standard deviation of the new data set will be smaller than the standard deviation of the original data set.

(C) The mean of the new data set will be smaller than the mean of the original data set, but the standard deviation of the new data set will be equal to the standard deviation of the original data set.

(D) The mean of the new data set will be larger than the mean of the original data set, but the standard deviation of the new data set will be smaller than the standard deviation of the original data set.

(E) The mean of the new data set will be smaller than the mean of the original data set, and the standard deviation of the new data set will be smaller than the standard deviation of the original data set.

13. Fifty observations on waiting time (rounded to the nearest minute) for patients at a walk-in medical clinic were used to compute a sample mean waiting time. The two largest observations in the data set were 75 minutes and 120 minutes. It was later determined that there was an error in recording the data and the observation of 120 minutes should have been 12 minutes. How would correcting this error affect the sample mean?

(A) The corrected sample mean will be smaller.

(B) The correction will not change the value of the sample mean.

(C) The corrected sample mean will be larger.

(D) The corrected sample mean will be smaller only if the original sample mean (computed with 120 as one of the observations) was greater than 12.

(E) It is not possible to determine if the corrected sample mean will be larger or smaller than the original sample mean.

14. Each student in a sample of size 40 selected at random from those attending a particular university was asked how much they spent on housing per week. The smallest value in the data set was $70 and the largest was $620. If this data set were summarized using a stem-and-leaf display, which of the following would be the most appropriate choice of stems?

(A) 6, 7

(B) 0, 1, 2, ..., 6

(C) 70, 71, 71, ..., 619, 620

(D) 07, 08, 09, ..., 61, 62

(E) 0 to <75, 75 to <150, 150 to <225, ..., 525 to <675

15. A data set consisting of 100 observations was used to construct the following boxplot:

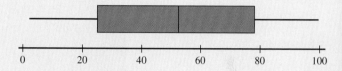

Based only on this boxplot, which of the following can be concluded about the shape of the sample distribution?

I.   The sample distribution is approximately symmetric.
II. The sample distribution is mound shaped.
III. The sample distribution is extremely skewed to the right (positively skewed).

(A) I only

(B) II only

(C) III only

(D) I and II

(E) I and III

4

Describing Bivariate Numerical Data

Preview

Chapter Learning Objectives

4.1 Correlation

4.2 Linear Regression: Fitting a Line to Bivariate Data

4.3 Assessing the Fit of a Line

4.4 Describing Linear Relationships and Making Predictions—Putting It All Together

4.5 Modeling Nonlinear Relationships

4.6 Avoid These Common Mistakes

Chapter Activities

Exploring the Big Ideas

Are You Ready to Move On? Chapter 4 Review Exercises

Technology Notes

AP* Review Questions for Chapter 4

Want to Know More? See Chapter 4 Online materials to learn about Logistic Regression.

Pablo Paul/Alamy Limited

PREVIEW

What can you learn from bivariate numerical data? A good place to start is with a scatterplot of the data. If it appears that the two variables that define the data set are related, it may be possible to describe the relationship in a way that allows you to predict the value of one variable based on the value of the other. For example, if there is a relationship between a measurement taken from a blood sample and age and you could describe that relationship mathematically, it might be possible to predict the age of a crime victim. If you can describe the relationship between fuel efficiency and the weight of a car, you could predict the fuel efficiency of a car based on its weight. In this chapter, you will see how this can be accomplished.

CHAPTER LEARNING OBJECTIVES

Conceptual Understanding

After completing this chapter, you should be able to

C1 Understand that the relationship between two numerical variables can be described in terms of form, direction, and strength.

C2 Understand that the correlation coefficient is a measure of the strength and direction of a linear relationship.

C3 Understand the difference between a statistical relationship and a causal relationship (the difference between correlation and causation).

C4 Understand how a line is used to describe the relationship between two numerical variables.

C5 Understand the meaning of *least squares* in the context of fitting a regression line.

C6 Understand how a least squares regression line is used to make predictions.

C7 Explain why it is risky to use the least squares line to make predictions for values of the predictor variable that are outside the range of the data.

C8 Explain why it is important to consider both the standard deviation about the regression line, s_e, and the value of r^2 when assessing the usefulness of the least squares regression line.

C9 Explain why it is desirable to have a small value of s_e and a large value of r^2 in a linear regression setting.

C10 Understand the role of a residual plot in assessing whether a line is the most appropriate way to describe the relationship between two numerical variables.

C11 Describe the effect of an influential observation on the equation of the least squares regression line.

C12 Understand how a curve can be used to describe a nonlinear relationship between two numerical variables.

Mastering the Mechanics

After completing this chapter, you should be able to

M1 Identify linear and nonlinear patterns in scatterplots.

M2 Distinguish between positive and negative linear relationships.

M3 Informally describe form, direction, and strength of a linear relationship, given a scatterplot.

M4 Compute and interpret the value of the correlation coefficient.

M5 Know the properties of the correlation coefficient r.

M6 Identify the response variable and the predictor variable in a linear regression setting.

M7 Find the equation of the least squares regression line.

M8 Interpret the slope of the least squares regression line in context.

M9 Interpret the intercept of the least squares regression line in context, when appropriate.

M10 Use the least squares regression line to make predictions.

M11 Compute and interpret the value of s_e, the standard deviation about the regression line.

M12 Compute and interpret the value of r^2, the coefficient of determination.

M13 Compute and interpret residuals, given a bivariate numerical data set and the equation of the least squares regression line.

M14 Construct a residual plot in a linear regression setting.

M15 Use a residual plot to comment on the appropriateness of a line for summarizing the relationship between two numerical variables.

M16 Identify outliers and potentially influential observations in a linear regression setting.

M17 Select an appropriate model to describe a nonlinear relationship.

M18 Use nonlinear regression models to make predictions.

Putting It into Practice

After completing this chapter, you should be able to

P1 Describe the relationship between two numerical variables (using a scatterplot and the correlation coefficient).

P2 Investigate the usefulness of a linear regression model for describing the relationship between two numerical variables (using s_e, r^2, and a residual plot).

P3 Use the least squares regression line to make predictions, when appropriate.

P4 Describe the anticipated accuracy of predictions based on the least squares regression line.

P5 Investigate the usefulness of a nonlinear regression model for describing the relationship between two numerical variables.

PREVIEW EXAMPLE **Help for Crime Scene Investigators**

Forensic scientists must often estimate the age of an unidentified crime victim. Prior to 2010, this was usually done by analyzing teeth and bones, and the resulting estimates were not very reliable. A groundbreaking study described in the paper **"Estimating Human Age from T-Cell DNA Rearrangements"** (*Current Biology* [2010]) examined the relationship between age and a measurement of thymus function based on a blood test. Age and the blood measurement were recorded for 195 people ranging in age from a few weeks to 80 years. A scatterplot of the data appears in Figure 4.1.

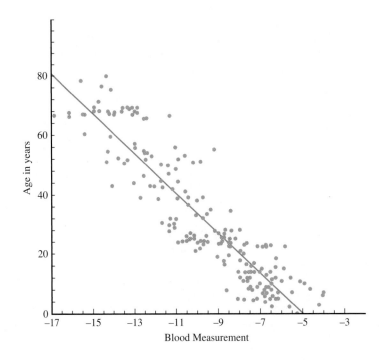

FIGURE 4.1
Scatterplot of age versus blood measurement

Based on the scatterplot, it does look like there is a relationship between the blood measurement and age. Smaller values of the blood measurement tend to be paired with larger values of age. Because the pattern in the scatterplot is linear, the line shown in the plot was proposed as a way to summarize the relationship between age and the blood measurement. This line can be used to estimate the age of a crime victim based on a measurement from a blood sample.

Later in this chapter, you will revisit the preview example to see how a line that summarizes the relationship between two numerical variables can be used to make predictions and what can be said about the accuracy of predictions based on this line.

SECTION 4.1 Correlation

A bivariate data set consists of measurements or observations on two variables, x and y. When both x and y are numerical variables, each observation is an (x, y) pair, such as (14, 5.2) or (27.63, 18.9). When investigating the relationship between two numerical variables, looking at a scatterplot of the data is the best place to start.

Look at the scatterplots of Figure 4.2 and consider these questions:

1. Does it look like there is a relationship between the two variables?
2. If there is a relationship, does it appear to be linear?

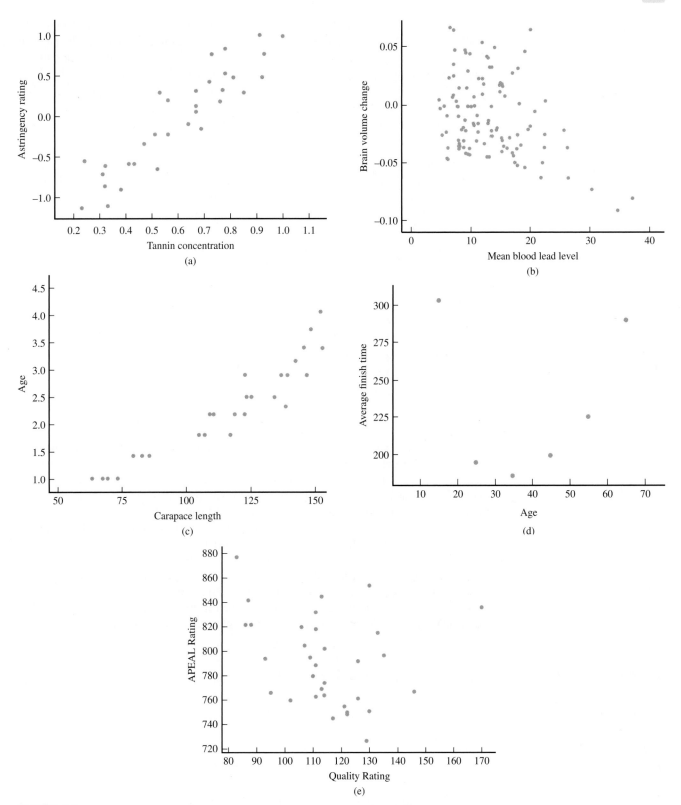

FIGURE 4.2
Scatterplots of five bivariate
data sets

To answer the first question, look for patterns in the scatterplot. Notice that in the scatterplot of Figure 4.2(e), there is no obvious pattern. The points appear to be scattered at random. In this case, you would answer that there does not appear to be a relationship between the two variables.

All of the other scatterplots in Figure 4.2 show some sort of pattern, although the pattern is more pronounced in some of the plots than in others. Notice that the patterns in the scatterplots of Figure 4.2(a) and (b) look roughly linear, whereas the patterns in the scatterplots of Figure 4.2(c) and (d) are curved. The accompanying table summarizes how the two relationship questions would be answered for each of the five scatterplots of Figure 4.2.

Scatterplot	Variables	Does It Look Like There Is a Relationship?	Does the Relationship Look Linear?
4.2 (a)	x = tannin concentration y = astringency rating	Yes	Yes
4.2 (b)	x = mean blood lead level y = brain volume change	Yes	Yes
4.2 (c)	x = carapace length y = age	Yes	No
4.2 (d)	x = age y = average finish time	Yes	No
4.2 (e)	x = quality rating y = APEAL rating	No	

Linear relationships can be either positive or negative in direction. In a positive linear relationship, larger values of x tend to be paired with larger values of y. When this is the case, the linear pattern in the scatterplot slopes upward as you move from left to right. The scatterplot in Figure 4.2(a) shows a positive linear relationship between x = tannin concentration and y = astringency rating. The scatterplot of Figure 4.2(b) shows a negative linear relationship between x = blood lead level and y = brain volume change. How would you describe a negative linear relationship? Do larger values of x tend to be paired with larger or smaller values of y? Does the linear pattern in the scatterplot slope upward or downward as you move from left to right?

Next, consider the four scatterplots of Figure 4.3. These scatterplots were constructed using data from graphs in *Archives of General Psychiatry* (June, 2010). The variables used to create these scatterplots will be described in more detail in Example 4.14. For now, just look at the general patterns in these plots. All four of these scatterplots were

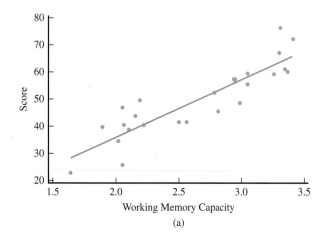

(a)

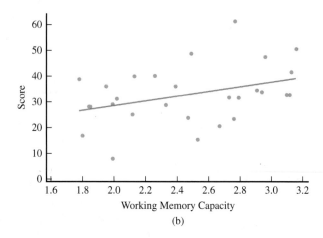

(b)

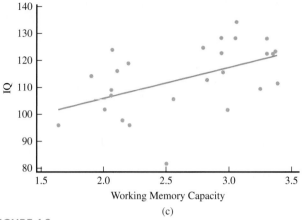
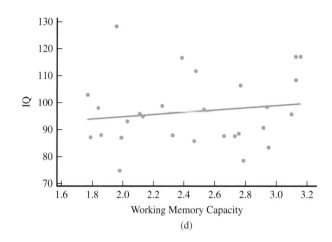

FIGURE 4.3
Four scatterplots showing positive linear relationships

described as showing a positive linear relationship. The line shown in each scatterplot was used to describe the relationship between *x* and *y*.

When the points in a scatterplot tend to cluster tightly around a line, the relationship is described as strong. Try to order the scatterplots in Figure 4.3 from the one that shows the strongest relationship to the one that shows the weakest. For which of the four scatterplots does the line provide the most information about how *x* and *y* are related? For which one is the line least informative?

Did you order the scatterplots (a), (c), (b), (d)? Scatterplots (b) and (c) are pretty close, so don't worry if you had the order of those two reversed. It can sometimes be difficult to judge the strength of a linear relationship from a scatterplot. The correlation coefficient is a statistic that can be used to assess the strength and direction of a linear relationship.

Pearson's Sample Correlation Coefficient

Pearson's sample correlation coefficient, denoted by *r*, measures the strength and direction of a linear relationship between two numerical variables. It is usually referred to as just the correlation coefficient. Let's begin by investigating some general properties of *r*.

Consider Table 4.1, which shows eight scatterplots and the associated values of the correlation coefficient. Keeping in mind that *r* measures the strength and direction of a linear relationship, look at the table and try to answer the following questions:

- When is the value of *r* positive?
- When is the value of *r* negative?
- For positive linear relationships, as the value of *r* increases, is the linear relationship getting stronger or weaker?
- What is the largest possible value of *r*? Why?
- What is the smallest possible value of *r*? Why?
- If $r = -0.7$ for one data set and $r = 0.7$ for a different data set, which data set has the stronger linear relationship?

Let's look at one more scatterplot, shown in Figure 4.4. The correlation coefficient for this data set is $r = 0.038$, which is close to zero. Because *r* is a measure of the strength of *linear* relationship, this indicates that there is not a linear relationship between age and average finish time. The scatterplot, however, shows a definite *nonlinear* pattern. This is an important point—when the value of *r* is near 0, you should not conclude that there is no relationship whatsoever between the two variables. Be sure to look at the scatterplot of the data to see if there might be a nonlinear relationship.

TABLE **4.1** **Scatterplots and Correlation Coefficients**

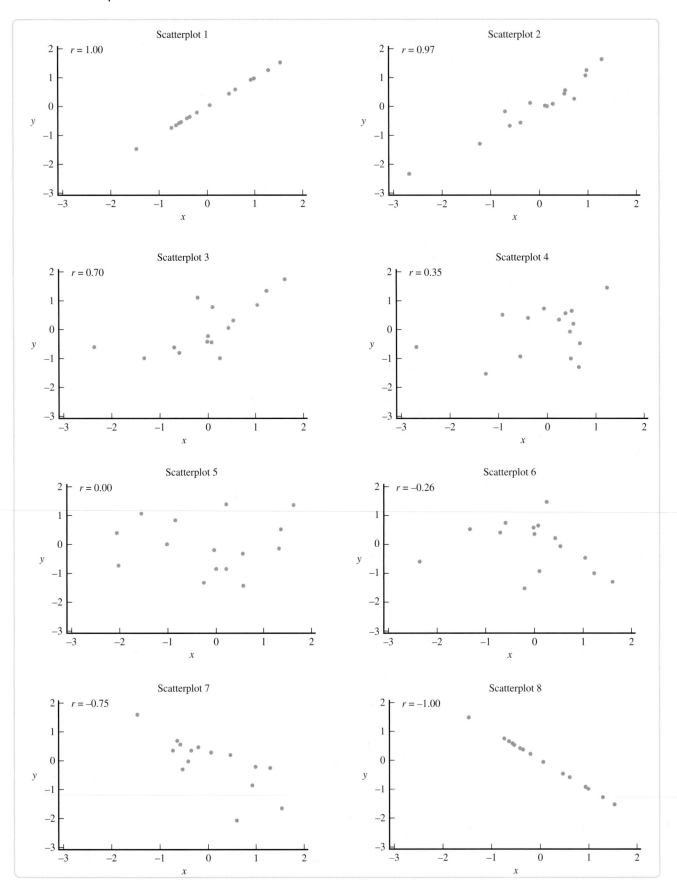

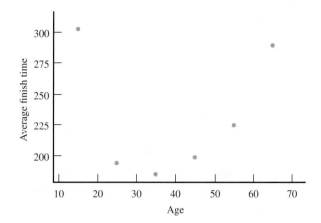

FIGURE 4.4
Scatterplot showing strong but nonlinear relationship

Properties of *r*

By looking at the scatterplots and correlation coefficients in Table 4.1, you probably discovered most of the general properties of *r* that follow.

1. The sign of *r* matches the direction of the linear relationship: *r* is positive when the linear relationship is positive and negative when the linear relationship is negative.
2. *The value of r is always greater than or equal to* -1 *and less than or equal to* $+1$. A value near the upper limit, $+1$, indicates a strong positive relationship, whereas a value close to the lower limit, -1, suggests a strong negative relationship. Figure 4.5 shows a useful way to describe the strength of relationship based on the value of *r*. It may seem surprising that a value as far away from 0 as -0.5 or $+0.5$ is in the weak category. An explanation for this is given later in the chapter.

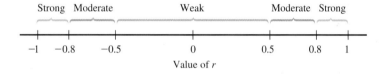

FIGURE 4.5
Describing the strength of a linear relationship

3. $r = 1$ only when all the points in the scatterplot fall on a straight line that slopes upward. Similarly, $r = -1$ only when all the points fall on a downward-sloping line.
4. *r is a measure of the extent to which x and y are linearly related*—that is, the extent to which the points in the scatterplot fall close to a straight line. Even if the value of *r* is close to 0, there could still be a strong nonlinear relationship.
5. *The value of r does not depend on the unit of measurement for either variable.* For example, if *x* is height, the value of *r* is the same whether height is expressed in inches, meters, or miles.

Computing and Interpreting the Value of the Correlation Coefficient

The first step in computing the value of the correlation coefficient is to calculate a *z*-score for each *x* value (by subtracting $\bar{x}$ and then dividing by s_x):

$$z_x = \frac{x - \bar{x}}{s_x}$$

Next, calculate a *z*-score for each *y* value (by subtracting $\bar{y}$ and then dividing by s_y):

$$z_y = \frac{y - \bar{y}}{s_y}$$

Then use the formula in the following box to calculate the value of *r*.

The correlation coefficient r is calculated using the following formula:

$$r = \frac{\sum z_x z_y}{n-1}$$

$\sum z_x z_y$ is found by multiplying z_x and z_y for each observation in the data set and then adding the $z_x z_y$ values.

Example 4.1 Graduation Rates and Student-Related Expenditures

The web site **www.collegeresults.org (The Education Trust)** publishes data on U.S. colleges and universities. The following six-year graduation rates and student-related expenditures per full-time student for 2007 were reported for the seven primarily undergraduate public universities in California with enrollments between 10,000 and 20,000.

University	Graduation Rate (percentage)	Student Related Expenditure (dollars)
1	66.1	8,810
2	52.4	7,780
3	48.9	8,112
4	48.1	8,149
5	42.0	8,477
6	38.3	7,342
7	31.3	7,984

Figure 4.6 is a scatterplot of these data.

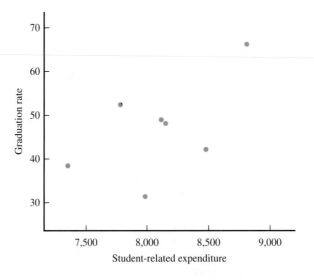

FIGURE 4.6
Scatterplot for the data of
Example 4.1

These data will be used to illustrate the calculation of the correlation coefficient. You can use x to denote the student-related expenditure and y to denote the graduation rate. For these data,

$$\bar{x} = 8{,}093.43 \qquad s_x = 472.39 \qquad \bar{y} = 46.73 \qquad s_y = 11.15$$

To calculate the correlation coefficient, begin by computing z-scores for each (x, y) pair in the data set. For example, the first observation is $(8{,}810, 66.1)$. The corresponding z scores are:

$$z_x = \frac{8{,}810 - 8{,}093.43}{472.39} = 1.52 \qquad z_y = \frac{66.1 - 46.73}{11.15} = 1.74$$

The following table shows the z scores and the product $z_x z_y$ for each observation:

x	y	z_x	z_y	$z_x z_y$
8,810	66.1	1.52	1.74	2.64
7,780	52.4	−0.66	0.51	−0.34
8,112	48.9	0.04	0.19	0.01
8,149	48.1	0.12	0.12	0.01
8,477	42.0	0.81	−0.42	−0.34
7,342	38.3	−1.59	−0.76	1.21
7,984	31.3	−0.23	−1.38	0.32

Then

$$\sum z_x z_y = 3.51$$

and

$$r = \frac{\sum z_x z_y}{n-1} = \frac{3.51}{7-1} = \frac{3.51}{6} = 0.585$$

Based on the scatterplot and the value of the correlation coefficient, you would conclude that for these seven universities, there is a moderate positive linear relationship between student-related expenditure and graduation rate.

Hand calculation of the value of the correlation coefficient is quite tedious. Fortunately, all statistical software packages and most scientific and graphing calculators can compute r. For example, SPSS, a widely used statistics package, was used to compute the correlation coefficient for the data of Example 4.1, producing the following computer output.

Correlations		
Gradrate	Pearson Correlation	.583

The difference between the correlation coefficient reported by SPSS and the value from the hand calculations in Example 4.1 is the result of rounding the z-scores when carrying out the calculations by hand.

Example 4.2   A Face You Can Trust

The article **"How to Tell if a Guy Is Trustworthy"** (*LiveScience*, **March 8, 2010**) described an interesting research study published in the journal *Psychological Science* (**"Valid Facial Cues to Cooperation and Trust: Male Facial Width and Trustworthiness,"** *Psychological Science* **[2010]: 349–354**). This study investigated whether there is a relationship between facial characteristics and people's assessment of trustworthiness. Sixty-two students were told that they would play a series of two-person games in which they could earn real money. In each game, the student could choose between two options:

1. They could end the game immediately and a total of $10 would be paid out, with each player receiving $5.

2. They could "trust" the other player. In this case, $3 would be added to the money available, but the other player could decide to either split the money fairly with $6.50 to each player or could decide to keep $10 and only give $3 to the first player.

Each student was shown a picture of the person he or she would be playing with. The student then indicated whether they would end the game immediately and take $5 or trust the other player to do the right thing in hopes of increasing the payout to $6.50. This process was repeated for a series of games, each with a photo of a different second player. For each photo, the researchers recorded the width-to-height ratio of the face in the photo and the percentage of the students who chose the trust option when shown that photo. A representative subset of the data (from a graph that appeared in the paper) is given here:

Face Width-to-Height Ratio	Percentage Choosing Trust Option	Face Width-to-Height Ratio	Percentage Choosing Trust Option
1.75	58	2.10	62
1.80	80	2.15	42
1.90	30	2.15	31
1.95	50	2.15	15
1.95	65	2.15	65
2.00	10	2.20	42
2.00	28	2.20	22
2.00	62	2.20	14
2.00	70	2.25	9
2.00	63	2.25	20
2.05	8	2.25	35
2.05	35	2.30	43
2.05	50	2.30	8
2.05	58	2.35	41
2.10	64	2.40	53
2.10	25	2.40	35
2.10	35	2.60	55
2.10	50	2.68	20
2.10	66	2.70	11

A scatterplot of these data is shown in Figure 4.7.

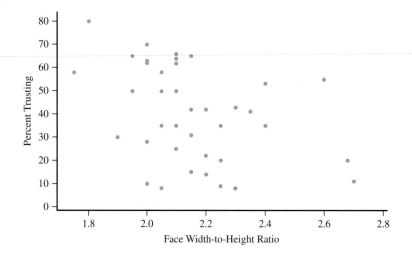

FIGURE 4.7
Scatterplot of percent choosing trust option versus facial width-to-height ratio

JMP was used to compute the value of the correlation coefficient, with the following result:

⊿ **Correlation**					
Variable	Mean	Std Dev	Correlation	Signif. Prob	Number
Percentage Choosing Trust Option	40.26316	20.58691	-0.39169	0.0150*	38
Face Width-to-Height Ratio	2.153421	0.209598			

The value of the correlation coefficient (-0.39169) is found in the JMP output under the heading "Correlation." This indicates a weak negative linear relationship between face width-to-height ratio and the percentage of people who chose to trust the face when playing the game. The relationship is negative, indicating that those with larger face width-to-height ratios (wider faces) tended to be trusted less. Based on this observation, the researchers concluded, "It is clear that male facial width ratio is a cue to male trustworthiness and that it predicts trust placed in male faces."

An interesting side note: In another study where subjects were randomly assigned to the player 1 and player 2 roles, the researchers found that those with larger face width-to-height ratios *were* less trustworthy. They found a positive correlation between player 2 face width-to-height ratio and the percentage of the time that player 2 decided to keep $10 and only give $3 to player 1.

Example 4.3   Does It Pay to Pay More for a Bike Helmet?

Are more expensive bike helmets safer than less expensive ones? The accompanying data on x = price and y = quality rating for 11 different brands of bike helmets is from the *Consumer Reports* web site **(www.consumerreports.org/health)**. Quality rating was a number from 0 (the worst possible rating) to 100 and was determined using factors that included how well the helmet absorbed the force of an impact, the strength of the helmet, ventilation, and ease of use. Figure 4.8 shows a scatterplot of the data.

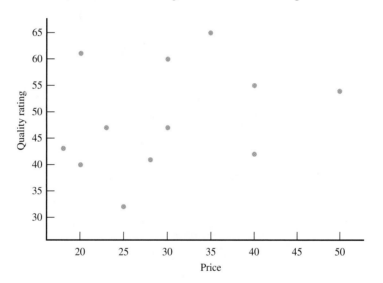

FIGURE 4.8
Scatterplot for the bike helmet data of Example 4.3

Price	Quality Rating
35	65
20	61
30	60
40	55
50	54
23	47
30	47
18	43
40	42
28	41
20	40
25	32

From the scatterplot, it appears that there is only a weak positive relationship between price and quality rating. The correlation coefficient, obtained using Minitab, is

Correlations: Price, Quality Rating

Pearson correlation of Price and Quality Rating = 0.303

A correlation coefficient of $r = 0.303$ indicates that although there is a tendency for higher quality ratings to be associated with higher priced helmets, the relationship is not very strong. In fact, the highest quality rating was for a moderately priced helmet.

How the Correlation Coefficient Measures the Strength of a Linear Relationship

The correlation coefficient measures the strength of a linear relationship by using z-scores in a clever way. Because $z_x = \frac{x - \bar{x}}{s_x}$, x values that are larger than $\bar{x}$ will have positive z-scores and those smaller than $\bar{x}$ will have negative z-scores. Also y values larger than $\bar{y}$ will have positive z-scores and those smaller will have negative z-scores. Keep this in mind as you look at the scatterplots in Figure 4.9. The scatterplot in Figure 4.9(a) indicates a strong positive relationship. A vertical line through $\bar{x}$ and a horizontal line through $\bar{y}$ divide the plot into four regions. In Region I, x is greater than $\bar{x}$ and y is greater than $\bar{y}$. So, in this region, the z-score for x and the z-score for y are both positive.

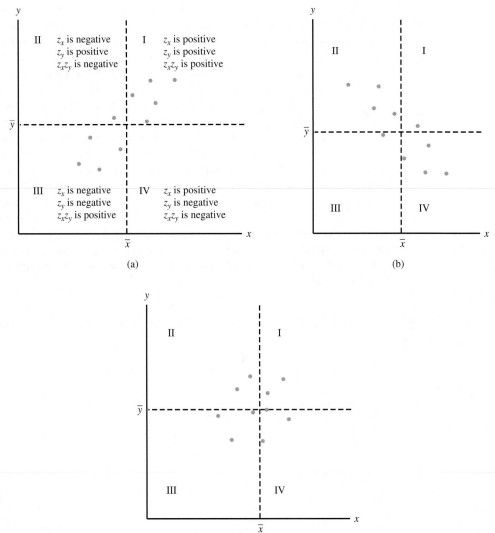

FIGURE 4.9
Viewing a scatterplot according to the signs of z_x and z_y:
(a) a positive relationship
(b) a negative relationship
(c) no strong relationship

It follows that the product $z_x z_y$ is positive. The product of the z-scores is also positive for any point in Region III, because in Region III both z-scores are negative, and multiplying two negative numbers results in a positive number. In each of the other two regions, one z-score is positive and the other is negative, so $z_x z_y$ is negative. But because the points in Figure 4.9(a) generally fall in Regions I and III, the products of z-scores tend to be positive. Thus, the *sum* of the products will be a relatively large positive number.

Similar reasoning for the data displayed in Figure 4.9(b), which exhibits a strong negative relationship, implies that $\sum z_x z_y$ will be a relatively large (in magnitude) negative number. When there is no strong relationship, as in Figure 4.9(c), positive and negative products tend to counteract one another, producing a value of $\sum z_x z_y$ that is close to zero. The sum $\sum z_x z_y$ can be a large positive number, a large negative number, or a number close to 0, depending on whether there is a strong positive, a strong negative, or no strong linear relationship. Dividing this sum by $n - 1$ results in a value between -1 and $+1$.

Correlation and Causation

A value of r close to 1 indicates that the larger values of one variable tend to be associated with the larger values of the other variable. This does not mean that a large value of one variable *causes* the value of the other variable to be large. Correlation measures the extent of association, but *association does not imply causation*. It frequently happens that two variables are highly correlated not because a change in one causes a change in the other but because they are both strongly related to a third variable. For example, among all elementary schoolchildren, the relationship between the number of cavities in a child's teeth and the size of his or her vocabulary is strong and positive. Yet no one advocates eating foods that result in more cavities to increase vocabulary size. Number of cavities and vocabulary size are both strongly related to age, so older children tend to have higher values of both variables. In the ABCNews.com series **"Who's Counting?" (February 1, 2001)**, John Paulos reminded readers that correlation does not imply causation and gave the following example: Consumption of hot chocolate is negatively correlated with crime rate (high values of hot chocolate consumption tend to be paired with lower crime rates), but both are responses to cold weather.

Scientific experiments can frequently make a strong case for causality by carefully controlling the values of all variables that might be related to the ones under study. Then, if y is observed to change in a "smooth" way as the experimenter changes the value of x, there may be a causal relationship between x and y. In the absence of such control and ability to manipulate values of one variable, there is always the possibility that an unidentified third variable is influencing both variables under investigation. A high correlation in many uncontrolled studies carried out in different settings can also marshal support for causality—as in the case of cigarette smoking and cancer—but proving causality is an elusive task.

SECTION 4.1 EXERCISES

Each Exercise Set assesses the following chapter learning objectives: C1, C2, C3, M1, M2, M3, M4, M5, P1

SECTION 4.1 Exercise Set 1

4.1 For each of the four scatterplots shown, answer the following questions:
  i. Does there appear to be a relationship between x and y?
  ii. If so, does the relationship appear to be linear?
  iii. If so, would you describe the linear relationship as positive or negative?

Scatterplot 1:

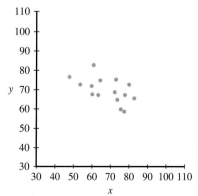

Scatterplot 2:

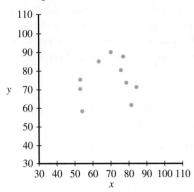

Scatterplot 3:

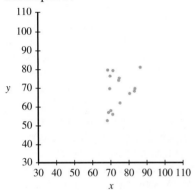

Scatterplot 4:

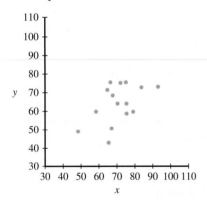

4.2 For each of the following pairs of variables, indicate whether you would expect a positive correlation, a negative correlation, or a correlation close to 0. Explain your choice.

a. Interest rate and number of loan applications
b. Height and IQ
c. Height and shoe size
d. Minimum daily temperature and cooling cost

4.3 The paper **"School Achievement Strongly Predicts Midlife IQ"** (*Intelligence* [2007]: 563–567) examined the relationship between school achievement test scores in grades 3 to 8 and IQ measured many years later. The authors reported an association between school achievement test

score and midlife IQ, with a correlation coefficient of $r = 0.64$.

a. Interpret the value of the correlation coefficient.
b. A science writer commenting on this paper (*LiveScience*, **April 16, 2007**) said that the correlation between school achievement test scores and IQ was very strong—even stronger than the correlation between height and weight in adults. Which of the following values for correlation between height and weight in adults is consistent with this statement? Justify your choice.

$$r = -0.8 \quad r = -0.3 \quad r = 0.0 \quad r = 0.6 \quad r = 0.8$$

4.4 The accompanying data on x = number of acres burned in forest fires in the western United States and y = timber sales are from graphs that appeared in the article **"Bush Timber Proposal Runs Counter to the Record"** (*San Luis Obispo Tribune*, **September 22, 2002**).

Year	Number of Acres Burned (thousands)	Timber Sales (billions of board feet)
1945	200	2.0
1950	250	3.7
1955	260	4.4
1960	380	6.8
1965	80	9.7
1970	450	11.0
1975	180	11.0
1980	240	10.2
1985	440	10.0
1990	400	11.0
1995	180	3.8

a. Compute and interpret the value of the correlation coefficient.
b. The article concludes that "heavier logging led to large forest fires." Do you think this conclusion is justified based on the given data? Explain.

4.5 Each year, marketing firm J.D. Power and Associates surveys new car owners 90 days after they purchase their cars. These data are used to rate auto brands (Toyota, Ford, etc.) on quality and customer satisfaction. *USA Today* (**June 16, 2010 and July 17, 2010**) reported a quality rating (number of defects per 100 vehicles) and a satisfaction score for all 33 brands sold in the United States.

a. Construct a scatterplot of y = satisfaction rating versus x = quality rating. How would you describe the relationship between x and y?
b. Compute and interpret the value of the correlation coefficient.

Brand	Quality Rating	Satisfaction Rating
Acura	86	822
Audi	111	832
BMW	113	845
Buick	114	802
Cadillac	111	818
Chevrolet	111	789
Chrysler	122	748
Dodge	130	751
Ford	93	794
GMC	126	792
Honda	95	766
Hyundai	102	760
Infiniti	107	805
Jaguar	130	854
Jeep	129	727
Kia	126	761
Land Rover	170	836
Lexus	88	822
Lincoln	106	820
Mazda	114	774
Mercedes-Benz	87	842
Mercury	113	769
Mini Cooper	133	815
Mitsubishi	146	767
Nissan	111	763
Porsche	83	877
Ram	110	780
Scion	114	764
Subaru	121	755
Suzuki	122	750
Toyota	117	745
Volkswagen	135	797
Volvo	109	795

4.6 Is the following statement correct? Explain why or why not.

A correlation coefficient of 0 implies that there is no relationship between two variables.

4.7 The article **"The Relationship Between Personality Traits and Political Ideologies" (*American Journal of Political Science* [2012]: 34–51)** reported that there was a negative correlation between a measure of the tendency to overestimate one's perceived social desirability and a measure of how conservative one's views on social issues are. Is it reasonable to conclude that a tendency to overestimate social desirability causes less conservative views on social issues? Explain why or why not.

4.8 The paper **"Noncognitive Predictors of Student Athletes' Academic Performance" (*Journal of College Reading and Learning* [2000]: e167)** summarizes a study of 200 Division I athletes. It was reported that the correlation coefficient for college grade point average (GPA) and a measure of academic self-worth was $r = 0.48$. Also reported were the correlation

coefficient for college GPA and high school GPA ($r = 0.46$) and the correlation coefficient for college GPA and a measure of tendency to procrastinate ($r = -0.36$). Write a few sentences summarizing what these correlation coefficients tell you about GPA for the 200 athletes in the sample.

Exercise Set 2

4.9 For each of the four scatterplots shown, answer the following questions:

   i. Does there appear to be a relationship between x and y?

   ii. If so, does the relationship appear to be linear?

   iii. If so, would you describe the linear relationship as positive or negative?

Scatterplot 1:

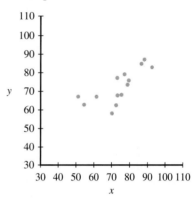

Scatterplot 2:

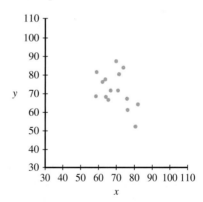

Scatterplot 3:

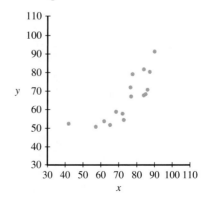

Scatterplot 4:

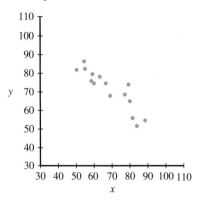

4.10 For each of the following pairs of variables, indicate whether you would expect a positive correlation, a negative correlation, or a correlation close to 0. Explain your choice.
a. Weight of a car and gas mileage
b. Size and selling price of a house
c. Height and weight
d. Height and number of siblings

4.11 The paper **"Digit Ratio as an Indicator of Numeracy Relative to Literacy in 7-Year-Old British Schoolchildren"** (*British Journal of Psychology* [2008]: 75–85) investigated a possible relationship between x = digit ratio (the ratio of the length of the second finger to the length of the fourth finger) and y = difference between numeracy score and literacy score on a national assessment. (The digit ratio is thought to be inversely related to the level of prenatal testosterone exposure.) The authors concluded that children with smaller digit ratios tended to have larger differences in test scores, meaning that they tended to have a higher numeracy score than literacy score. This conclusion was based on a correlation coefficient of $r = -0.22$. Does the value of the correlation coefficient indicate that there is a strong linear relationship? Explain why or why not.

4.12 Draw two scatterplots, one for which $r = 1$ and a second for which $r = -1$.

4.13 Data from the **U.S. Federal Reserve Board (Household Debt Service Burden, 2002)** on consumer debt (as a percentage of personal income) and household debt (also as a percentage of personal income) for selected years are shown in the following table:

Consumer Debt	Household Debt	Consumer Debt	Household Debt
7.88	6.22	6.24	5.73
7.91	6.14	6.09	5.95
7.65	5.95	6.32	6.09
7.61	5.83	6.97	6.28
7.48	5.83	7.38	6.08
7.49	5.85	7.52	5.79
7.37	5.81	7.84	5.81
6.57	5.79		

a. What is the value of the correlation coefficient for this data set?
b. Is it reasonable to conclude that there is no strong relationship between the variables (linear or otherwise)? Use a graphical display to support your answer.

4.14 The paper **"Religiosity and Teen Birth Rate in the United States"** (*Reproductive Health* [2009]: 14–20) included data on teen birth rate and on a measure of conservative religious beliefs for 49 of the U.S. states. Birth rate was measured as births per 1,000 teenage females. The authors of the paper created a "religiosity" score for each state by using responses to a large national survey conducted by the Pew Forum on Religion and Public Life. Higher religiosity scores correspond to more conservative religious beliefs. The authors reported that "teen birth rate is very highly correlated with religiosity at the state level, with more religious states having a higher rate of teen birth." The following data for 49 states are from a graph in the paper.
a. Construct a scatterplot of y = birth rate versus x = religiosity score. How would you describe the relationship between x and y? What aspects of the scatterplot support what the authors concluded about teen birth rate and religiosity?
b. Compute and interpret the value of the correlation coefficient.

Religiosity	Teen Birth Rate	Religiosity	Teen Birth Rate
26.5	10.7	42.1	24.5
28.9	8.5	42.6	37.5
29.7	16.2	42.8	58.3
30.1	36.7	43.3	32.0
30.5	11.2	43.3	31.3
32.1	13.7	44.1	26.4
32.7	18.3	44.4	32.4
34.8	16.1	44.7	35.6
35.0	36.1	45.5	38.0
35.9	21.7	46.1	34.0
36.8	15.1	47.5	30.9
36.9	27.1	49.1	37.1
37.1	31.9	50.1	57.2
38.3	18.5	51.1	47.2
38.6	24.5	51.1	53.3
39.4	49.1	51.3	47.7
39.5	56.1	53.5	42.4
39.8	24.0	54.2	56.2
39.9	31.5	54.7	25.0
40.7	17.0	55.3	46.9
40.7	25.0	55.5	47.7
40.9	21.8	55.7	45.9
41.2	34.0	58.1	46.5
41.4	24.7	62.3	62.8
41.4	31.9		

Additional Exercises

4.15 Breeders of thoroughbred horses are interested in foal weight at birth because it is an indicator of health. Is foal weight related to the weight of the mare (mother)? The accompanying data are from the paper **"Suckling Behaviour Does Not Measure Milk Intake in Horses"** (*Animal Behaviour* [1999]: 673–678).

Observation	Mare Weight (x, in kg)	Foal Weight (y, in kg)
1	556	129.0
2	638	119.0
3	588	132.0
4	550	123.5
5	580	112.0
6	642	113.5
7	568	95.0
8	642	104.0
9	556	104.0
10	616	93.5
11	549	108.5
12	504	95.0
13	515	117.5
14	551	128.0
15	594	127.5

a. Construct a scatterplot of these data.
b. Compute the value of the correlation coefficient.
c. Write a few sentences describing the relationship between mare weight and foal weight that refer both to the value of the correlation coefficient and the scatterplot.

4.16 The following time series plot is based on data from the article **"Bubble Talk Expands: Corporate Debt Is Latest Concern Turning Heads"** (*San Luis Obispo Tribune*, September 13, 2002). It shows how household debt and corporate debt have changed between 1991 (year 1 in the graph) and 2002:

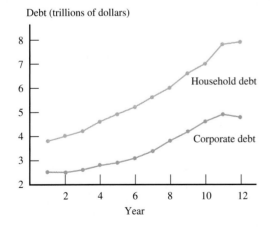

Based on the time series plot, would the correlation coefficient between household debt and corporate debt be positive or negative? Weak or strong? What aspect of the time series plot supports your answer?

4.17 Here are the debt values from the time series plot that appeared in Exercise 4.16. Compute the value of the correlation coefficient for these data. Explain how this value is consistent with your answer to Exercise 4.16.

Household Debt	Consumer Debt
3.9	2.5
4.0	2.5
4.2	2.6
4.5	2.7
4.9	2.8
5.0	3.0
5.5	3.2
6.0	3.5
6.5	4.0
6.9	4.5
7.9	4.9
8.0	4.7

4.18 An auction house released a list of 25 recently sold paintings. The artist's name and the sale price of each painting appear on the list. Would the correlation coefficient be an appropriate way to summarize the relationship between artist and sale price? Why or why not?

4.19 A sample of automobiles traveling on a particular segment of a highway is selected. Each one travels at roughly a constant rate of speed, although speed does vary from auto to auto. Let x = speed and y = time needed to travel this segment. Would the sample correlation coefficient be closest to 0.9, 0.3, −0.3, or −0.9? Explain.

4.20 It may seem odd, but biologists can tell how old a lobster is by measuring the concentration of pigment in the lobster's eye. The authors of the paper **"Neurolipofuscin Is a Measure of Age in *Panulirus argus*, the Caribbean Spiny Lobster, in Florida"** (*Biological Bulletin* [2007]: 55–66) wondered if it was sufficient to measure the pigment in just one eye, which would be the case if there is a strong relationship between the concentration in the right eye and the concentration in the left eye. Pigment concentration (as a percentage of tissue sample) was measured in both eyes for 39 lobsters, resulting in the following summary quantities (based on data from a graph in the paper):

$$n = 39 \qquad \sum x = 88.8 \qquad \sum y = 86.1$$
$$\sum xy = 281.1 \qquad \sum x^2 = 288.0 \qquad \sum y^2 = 286.6$$

An alternative formula for computing the correlation coefficient that doesn't involve calculating the z-scores is

$$r = \frac{\sum xy - \dfrac{\left(\sum x\right)\left(\sum y\right)}{n}}{\sqrt{\sum x^2 - \dfrac{\left(\sum x\right)^2}{n}}\sqrt{\sum y^2 - \dfrac{\left(\sum y\right)^2}{n}}}$$

Use this formula to compute the value of the correlation coefficient, and interpret this value.

Linear Regression: Fitting a Line to Bivariate Data

When there is a relationship between two numerical variables, you can use information about one variable to learn about the value of the second variable. For example, you might want to predict y = product sales for a month when the amount spent on advertising is x = $10,000. The letter y is used to denote the variable you would like to predict, and this variable is called the **response variable** (also sometimes called the dependent variable). The other variable, denoted by x, is the **predictor variable** (also sometimes called the independent or explanatory variable).

Scatterplots frequently exhibit a linear pattern. When this is the case, the relationship can be summarized by finding a line that is as close as possible to the points in the plot. Before seeing how this is done, let's review some elementary facts about lines and linear relationships.

Lines and Linear Relationships

The equation of a line is

$$y = a + bx$$

A particular line is specified by choosing values of a and b. For example, one line is $y = 10 + 2x$ and another is $y = 10 - 5x$. If we choose some x values and compute $y = a + bx$ for each value, the points corresponding to the resulting (x, y) pairs will fall exactly on a straight line.

DEFINITION

The equation of a line is

$$y = a + bx$$

The value of b, called the **slope** of the line, is the amount by which y increases when x increases by 1 unit.

The value of a, called the **intercept** (also called the **y-intercept** or **vertical intercept**) of the line, is the height of the line above the value $x = 0$.

The line $y = 10 + 2x$ has slope $b = 2$, so y will increase by 2 units for each 1-unit increase in x. When $x = 0$, $y = 10$, so the line crosses the vertical axis (where $x = 0$) at a height of 10. This is illustrated in Figure 4.10(a). The slope of the line $y = 100 - 5x$ is -5, so y decreases by 5 when x increases by 1. The height of this line above $x = 0$ is $a = 100$. This line is pictured in Figure 4.10(b).

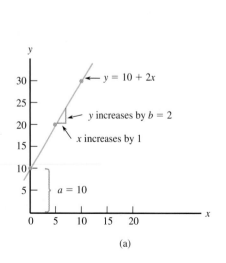

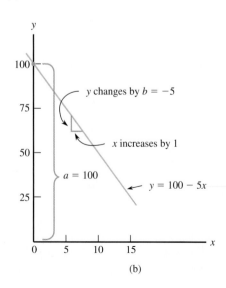

FIGURE 4.10
Graphs of two lines
(a) slope $b = 2$, intercept $a = 10$;
(b) slope $b = -5$, intercept $a = 100$

It is easy to draw the graph of any particular line. Choose any two x values and substitute them into the equation to obtain the corresponding y values. Plot the resulting two (x, y) pairs and then draw the line that passes through these points.

Choosing a Line to Summarize a Linear Relationship: The Principle of Least Squares

Consider the scatterplot shown in Figure 4.11(a). This scatterplot shows a positive linear relationship between x and y. You can think about using a line to summarize this relationship, but what line should you use? Two possible lines are shown in Figure 4.11(b). It is easy to choose between these two lines—Line 1 provides a better "fit" for these data.

But now consider the two lines shown in Figure 4.11(c). Which of these two lines is a better choice for describing the relationship? Here the choice isn't obvious. You need a way to judge how well a line fits the data in order to find the "best" line.

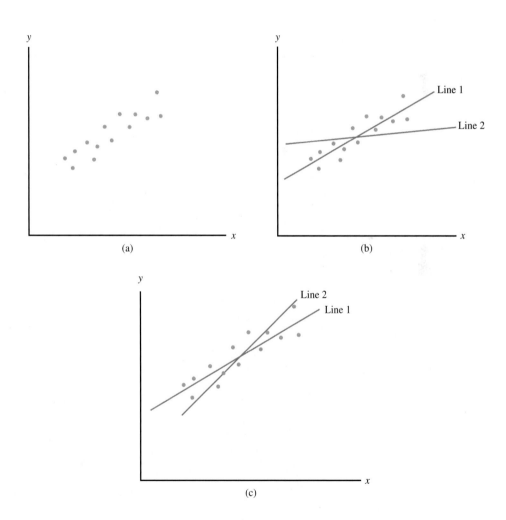

FIGURE 4.11
Comparing the fit of two lines

To measure how well a particular line fits a given data set, you use the vertical deviations from the line. There is a deviation associated with each point in the scatterplot. To calculate the deviation for a point (x, y), find the height of the line for that x value (by substituting x into the equation of the line) and then subtract this value from y.

For example, Figure 4.12 shows a scatterplot that also includes the line

$$y = 10 + 2x$$

The labeled points in the plot are the fifth and sixth points from the left. These two points are $(x_5, y_5) = (15, 44)$ and $(x_6, y_6) = (20, 45)$. For these two points, the vertical deviations from the line are

$$\text{5th deviation} = y_5 - \text{height of the line at } x_5$$
$$= 44 - [10 + 2(15)]$$
$$= 4$$

and

$$\text{6th deviation} = y_6 - \text{height of the line at } x_6$$
$$= 45 - [10 + 2(20)]$$
$$= -5$$

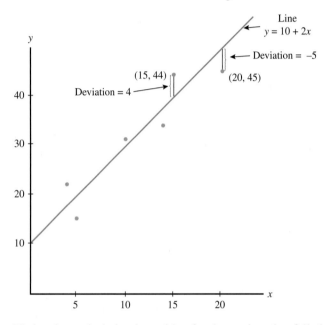

FIGURE 4.12
Scatterplot with line and two
deviations shown

Notice that a deviation is positive for data points that fall above the line and negative for data points that fall below the line.

A particular line is a good fit if the data points tend to be close to the line. This means that the deviations from the line will be small in magnitude. Looking back at Figure 4.11(b), Line 2 does not fit as well as Line 1 because the deviations from Line 2 tend to be larger than the deviations from Line 1.

To assess the overall fit of a line, you need a way to combine the n deviations into a single measure of fit. The standard approach is to square the deviations (to obtain nonnegative numbers) and then to sum these squared deviations.

DEFINITION

The most widely used measure of the fit of a line $y = a + bx$ to bivariate data $(x_1, y_1), (x_2, y_2), \ldots, (x_n, y_n)$ is the **sum of squared deviations** about the line:

$$\sum [y - (a + bx)]^2 = [y_1 - (a + bx_1)]^2 + [y_2 - (a + bx_2)]^2 + \ldots + [y_n - (a + bx_n)]^2$$

You can use the sum of squared deviations to judge the fit of a line. For a given data set, you would like to find the line that makes this sum as small as possible. This line is called the **least squares regression line**.

AP* EXAM TIP

The least squares regression line has a smaller sum of squared deviations than *any* other line.

DEFINITION

The **least squares regression line** is the line that minimizes the sum of squared deviations.

Fortunately, the equation of the least squares regression line can be obtained without having to calculate deviations from any particular line. The accompanying box gives relatively simple formulas for the slope and intercept of the least squares regression line.

The slope of the least squares regression line is

$$b = \frac{\sum(x - \bar{x})(y - \bar{y})}{\sum(x - \bar{x})^2}$$

and the y intercept is

$$a = \bar{y} - b\bar{x}$$

The equation of the least squares regression line is

$$\hat{y} = a + bx$$

where $\hat{y}$ (read as y-hat) is the predicted value of y that results from substituting a particular x value into the equation.

Statistical software packages and many graphing calculators can compute the slope and intercept of the least squares regression line. If the slope and intercept are to be computed by hand, the following computational formula for the slope can be used to reduce the amount of time required.

Calculating Formula for the Slope of the Least Squares Regression Line

$$b = \frac{\sum xy - \dfrac{(\sum x)(\sum y)}{n}}{\sum x^2 - \dfrac{(\sum x)^2}{n}}$$

Example 4.4   Pomegranate Juice and Tumor Growth

Pomegranate, a fruit native to the Middle East, has been used in the folk medicines of many cultures to treat various ailments. Researchers are now investigating if pomegranate's antioxidant properties are useful in the treatment of cancer. One such study, described in the paper "**Pomegranate Fruit Juice for Chemoprevention and Chemotherapy of Prostate Cancer**" (*Proceedings of the National Academy of Sciences* [Oct. 11, 2005]: 14813–14818), investigated whether pomegranate fruit extract (PFE) was effective in slowing the growth of prostate cancer tumors. In this study, 24 mice were injected with cancer cells. The mice were then randomly assigned to one of three treatment groups. One group of eight mice received normal drinking water, the second group of eight mice received drinking water supplemented with 0.1% PFE, and the third group received drinking water supplemented with 0.2% PFE. The average tumor volume for the mice in each group was recorded at several points in time. The following data (approximated from a graph in the paper) are for the mice that received plain drinking water. Here, y = average tumor volume (in mm^3) and x = number of days after injection of cancer cells.

x	11	15	19	23	27
y	150	270	450	580	740

A scatterplot of these data (Figure 4.13) shows that the relationship between these two variables can reasonably be summarized by a straight line.

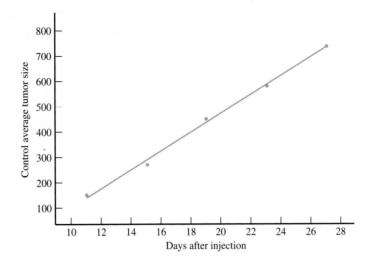

FIGURE 4.13
Scatterplot for the data of
Example 4.4

The summary quantities necessary to compute the equation of the least squares regression line are

$$\sum x = 95 \quad \sum x^2 = 1{,}965 \quad \sum xy = 47{,}570 \quad \sum y = 2{,}190 \quad \sum y^2 = 1{,}181{,}900$$

From these quantities,

$$\bar{x} = 19 \qquad \bar{y} = 438$$

$$b = \frac{\sum xy - \dfrac{\left(\sum x\right)\left(\sum y\right)}{n}}{\sum x^2 - \dfrac{\left(\sum x\right)^2}{n}} = \frac{47{,}570 - \dfrac{(95)(2{,}190)}{5}}{1{,}965 - \dfrac{(95)^2}{5}} = \frac{5{,}960}{160} = 37.25$$

and

$$a = \bar{y} - b\bar{x} = 438 - (37.25)(19) = -269.75$$

The least squares regression line is then

$$\hat{y} = -269.75 + 37.25x$$

Computer software or a graphing calculator could also be used to find the equation of the least squares regression line. For example, JMP produces the following output. The equation of the least squares line appears at the bottom of the output.

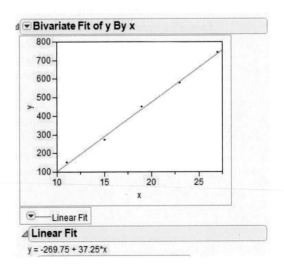

To predict average tumor volume 20 days after injection of cancer cells, use the y value of the point on the least squares regression line above $x = 20$:

$$\hat{y} = -269.75 + 37.25(20) = 475.25$$

You could predict average tumor volumes for a different number of days in the same way.

Figure 4.14 shows a scatterplot for both the group of mice that drank only water and the group that drank water supplemented by 0.2% PFE. Notice that the tumor growth seems to be much slower for the 0.2% PFE group. In fact, for this group, the relationship between average tumor volume and number of days after injection of cancer cells appears to be curved rather than linear. Because this relationship is not linear, it would be better to use one of the nonlinear models to be introduced in Section 4.5 to describe this relationship.

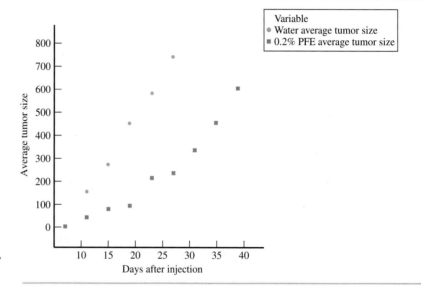

FIGURE 4.14
Scatterplot of average tumor volume versus number of days after injection of cancer cells for the water group and for the 0.2% PFE group

A word of caution: Think twice about using the least squares regression line to predict y for an x value that is outside the range of x values in the data set, because you don't know whether the observed linear pattern continues outside this range. Making predictions outside the range of the x values in the data set can produce misleading predictions if the pattern does not continue. This is sometimes referred to as the **danger of extrapolation**.

Consider the least-squares line in Example 4.4. You can see that using this line to predict average tumor volume for fewer than 10 days after injection of cancer cells can lead to nonsensical predictions. For example, if the number of days after injection is five, the predicted average tumor volume is negative:

$$\hat{y} = -269.75 + 37.25(5) = -83.5$$

Because it is impossible for average tumor volume to be negative, it is a clear that the pattern observed for x values from 11 to 27 days does not apply to values less than 11 days, and it may not apply to values greater than 27 days either. Even so, the least squares line can be a useful tool for making predictions within the 11 to 27-day range.

The Danger of Extrapolation

The least squares line should not be used to make predictions outside the range of the x values in the data set because there is no evidence that the linear pattern continues outside this range.

Example 4.5   Ankle Motion and Balance

Good balance is important in preventing falls, especially in older adults. The paper "**Correlation Between Ankle Range of Motion and Balance in Community-Dwelling Elderly Population**" (*Indian Journal of Physiotherapy and Occupational Therapy* [20012]: 127 – 129) describes a study of the relationship between ankle range of motion and balance. The accompanying data on

x = ankle range of motion (in degrees) and y = function balance test score for 30 people ages 65 to 85 are consistent with a scatterplot and summary measures given in the paper.

x = Ankle Range of Motion	y = Balance Score	x = Ankle Range of Motion	y = Balance Score
85	12.1	82	13.6
74	14.0	78	14.7
76	10.4	86	14.2
88	11.8	86	13.4
49	6.0	101	16.8
75	9.6	75	12.7
70	8.2	52	10.3
60	9.9	70	10.9
81	14.2	77	9.2
54	6.8	53	8.1
50	7.0	87	13.7
96	15.5	65	6.6
60	8.3	78	8.6
65	9.5	66	9.5
61	11.7	74	11.6

Minitab was used to construct a scatterplot of the data (Figure 4.15) and to compute the value of the correlation coefficient.

Correlations: Ankle Range of Motion, Balance Score

Pearson correlation of Ankle Range of Motion and Balance Score = 0.807
P-Value = 0.000

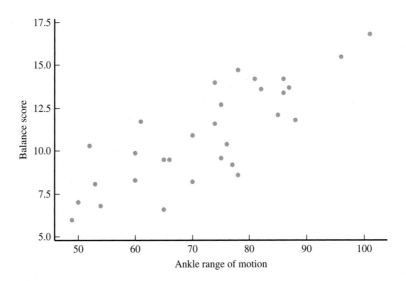

FIGURE 4.15
Minitab scatterplot for the data
of example 4.5

Notice that there is a linear pattern in the scatterplot, suggesting that a line is a reasonable way to summarize the relationship between balance score and ankle range of motion. The correlation coefficient of $r = 0.807$ is consistent with the pattern in the scatterplot and indicates a strong positive linear relationship.

Figure 4.16 shows part of the Minitab regression output. Instead of x and y, the variable labels "Balance Score" and "Ankle Range of Motion" are used. In the table just below the equation of the regression line, the first row gives information about the intercept, a, and the second row gives information about the slope, b. Notice that the coefficient column, labeled "Coef", contains more precise values for a and b than those that appear in the equation.

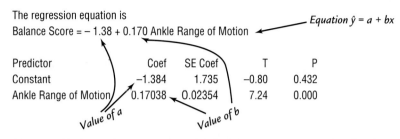

FIGURE 4.16
Partial Minitab output for example 4.5

Figure 4.17 shows the scatterplot with the least squares regression line included. The equation of the regression line

$$\hat{y} = -1.38 + 0.170x$$

Can be used to predict the balance score for a given ankle range of motion. For example, for a person with ankle range of motion of 85 degrees, you would predict a balance score of

$$\hat{y} = -1.38 + 0.170(85) = 13.07$$

The least squares regression line should not be used to predict the balance score for people with ankle ranges of motion such as $x = 30$ or $x = 120$. These x values are well outside the range of the data, and you do not know if a linear relationship is appropriate outside the observed range.

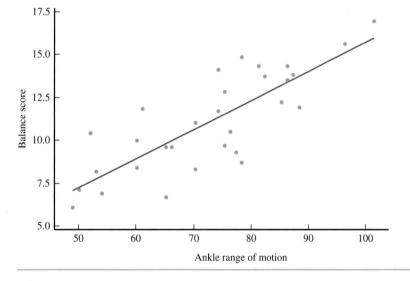

FIGURE 4.17
Scatterplot and least squares regression line for the range of motion and balance data of Example 4.5

Regression

Why is the line used to summarize a linear relationship called the least squares regression line? The "least squares" part of the name means the line has a smaller sum of squared deviations than any other line. But why is "regression" included in the name? This terminology comes from the relationship between the least squares line and Pearson's correlation coefficient. To understand this relationship, alternative expressions for the slope b and the equation of the regression line are needed. With s_x and s_y denoting the sample standard deviations of the x's and y's, respectively, a bit of algebraic manipulation results in

$$b = r\left(\frac{s_y}{s_x}\right)$$

$$\hat{y} = \bar{y} + r\left(\frac{s_y}{s_x}\right)(x - \bar{x})$$

You do not need to use these formulas in any computations, but several of their implications are important for appreciating what the least squares regression line does.

1. When $x = \bar{x}$ is substituted into the equation of the line, $\hat{y} = \bar{y}$ results. That is, the least squares regression line passes through the point of averages $(\bar{x}, \bar{y})$.

2. Suppose for the moment that $r = 1$. Then all the data points fall on the line whose equation is

$$\hat{y} = \bar{y} + \left(\frac{s_y}{s_x}\right)(x - \bar{x})$$

Now substitute $x = \bar{x} + s_x$, which is one standard deviation above $\bar{x}$:

$$\hat{y} = \bar{y} + \frac{s_y}{s_x}(\bar{x} + s_x - \bar{x}) = \bar{y} + s_y$$

That is, with $r = 1$, when x is one standard deviation above its mean, you predict that the associated y value will be one standard deviation above its mean. Similarly, if $x = \bar{x} - 2s_x$ (two standard deviations below its mean), then

$$\hat{y} = \bar{y} + \frac{s_y}{s_x}(\bar{x} - 2s_x - \bar{x}) = \bar{y} - 2s_y$$

which is also two standard deviations below the mean. If $r = -1$, then $x = \bar{x} + s_x$ results in $\hat{y} = \bar{y} - s_y$, so the predicted y is also one standard deviation from its mean but on the opposite side of $\bar{y}$ from where x is relative to $\bar{x}$. In general, if x and y are perfectly correlated, the predicted y value will be the same number of standard deviations from its mean, $\bar{y}$, as x is from its mean, $\bar{x}$.

3. Now suppose that x and y are not perfectly correlated. For example, suppose $r = 0.5$. Then the equation of the least squares regression line is

$$\hat{y} = \bar{y} + 0.5\left(\frac{s_y}{s_x}\right)(x - \bar{x})$$

Substituting $x = \bar{x} + s_x$ gives

$$\hat{y} = \bar{y} + 0.5\left(\frac{s_y}{s_x}\right)(\bar{x} + s_x - \bar{x}) = \bar{y} + 0.5s_y$$

That is, for $r = 0.5$, when x is one standard deviation above its mean, you predict that y will be only 0.5 standard deviation above its mean. Similarly, if $r = -0.5$ then the predicted y value will be only half the number of standard deviations from $\bar{y}$ than x is from $\bar{x}$, but on the opposite side.

> Consider using the least squares line to predict the value of y associated with an x value that is a certain number of standard deviations away from $\bar{x}$. The predicted y value will be only r times this number of standard deviations from $\bar{y}$. Except when $r = 1$ or $r = -1$, the predicted y will always be closer to $\bar{y}$ (in standard deviations) than x is to $\bar{x}$.

Using the least squares line for prediction results in a predicted y that is pulled back in, or *regressed*, toward the mean of y compared to where x is relative to its mean. This regression effect was first noticed by Sir Francis Galton (1822–1911), a famous biologist, when he was studying the relationship between the heights of fathers and their sons. He found that sons whose fathers were taller than average were predicted to be taller than average (because r is positive here) but not as tall as their fathers. He found a similar relationship for sons whose fathers were shorter than average. This "regression effect" led to the term **regression analysis**, which involves fitting lines, curves, and more complicated functions to bivariate and multivariate data.

The alternative form of the least squares regression line emphasizes that predicting y from knowledge of x is not the same problem as predicting x from knowledge of y. The slope of the least squares line for predicting x is $r\left(\frac{s_x}{s_y}\right)$ rather than $r\left(\frac{s_y}{s_x}\right)$ and the intercepts of the lines are almost always different. For purposes of prediction, it makes a difference whether y is regressed on x, as we have done, or x is regressed on y. *The regression line of y on x should not be used to predict x, because it is not the line that minimizes the sum of squared deviations in the x direction.*

SECTION 4.2 EXERCISES

Each Exercise Set assesses the following chapter learning objectives: C4, C5, C6, C7, M6, M7, M8, M9, M10, P3

SECTION 4.2 **Exercise Set 1**

4.21 Two scatterplots are shown below. Explain why it makes sense to use the least squares regression line to summarize the relationship between x and y for one of these data sets but not the other.

Scatterplot 1

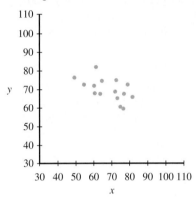

Scatterplot 2

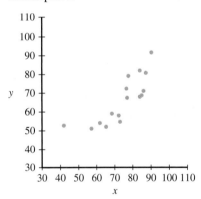

4.22 Data on y = time to complete a task (in minutes) and x = number of hours of sleep on previous night were used to find the least squares regression line. The equation of the line was $\hat{y} = 12 - 0.36x$. For this data set, would the sum of squared deviations from the line $y = 12.5 - 0.5x$ be larger or smaller than the sum of squared deviations from the least squares regression line? Explain your choice.

4.23 Data on x = size of a house (in square feet) and y = amount of natural gas used (therms) during a specified period were used to fit the least squares regression line. The slope was 0.017 and the intercept was -5.0. Houses in this data set ranged in size from 1,000 to 3,000 square feet.
a. What is the equation of the least squares regression line?

b. What would you predict for gas usage for a 2,100 sq. ft. house?
c. What is the approximate change in gas usage associated with a 1 sq. ft. increase in size?
d. Would you use the least squares regression line to predict gas usage for a 500 sq. ft. house? Why or why not?

4.24 Medical researchers have noted that adolescent females are much more likely to deliver low-birth-weight babies than are adult females. Because low-birth-weight babies have a higher mortality rate, a number of studies have examined the relationship between birth weight and mother's age. One such study is described in the article **"Body Size and Intelligence in 6-Year-Olds: Are Offspring of Teenage Mothers at Risk?"** (*Maternal and Child Health Journal* [2009]: 847–856). The following data on maternal age (in years) and birth weight of baby (in grams) are consistent with summary values given in the article and also with data published by the National Center for Health Statistics.

Mother's age	15	17	18	15	16	19
Birth weight	2289	3393	3271	2648	2897	3327

Mother's age	17	16	18	19
Birth weight	2970	2535	3138	3573

a. If the goal is to learn about how birth weight is related to mother's age, which of these two variables is the response variable and which is the predictor variable?
b. Construct a scatterplot of these data. Would it be reasonable to use a line to summarize the relationship between birth weight and mother's age?
c. Find the equation of the least squares regression line.
d. Interpret the slope of the least squares regression line in the context of this study.
e. Does it make sense to interpret the intercept of the least squares regression line? If so, give an interpretation. If not, explain why it is not appropriate for this data set.
f. What would you predict for birth weight of a baby born to an 18-year-old mother?
g. What would you predict for birth weight of a baby born to a 15-year-old mother?
h. Would you use the least squares regression equation to predict birth weight for a baby born to a 23-year-old mother? If so, what is the predicted birth weight? If not, explain why.

4.25 The article **"Air Pollution and Medical Care Use by Older Americans"** (*Health Affairs* [2002]: 207–214) gave data on a measure of pollution (in micrograms of particulate matter per cubic meter of air) and the cost of medical care per person over age 65 for six geographical regions of the United States:

Region	Pollution	Cost of Medical Care
North	30.0	915
Upper South	31.8	891
Deep South	32.1	968
West South	26.8	972
Big Sky	30.4	952
West	40.0	899

a. If the goal is to learn how cost of medical care is related to pollution, which of these two variables is the response variable and which is the predictor variable?

b. Construct a scatterplot of these data. Describe any interesting features of the scatterplot.

c. Find the equation of the least squares regression line for predicting medical cost using pollution.

d. Is the slope of the least squares line positive or negative? Is this consistent with the scatterplot in Part (b)?

e. Do the scatterplot and the equation of the least squares regression line support the researchers' conclusion that elderly people who live in more polluted areas have higher medical costs? Explain.

f. What is the predicted cost of medical care for a region that has a pollution measure of 35?

g. Would you use the least squares regression equation to predict cost of medical care for a region that has a pollution measure of 60? If so, what is the predicted cost of medical care? If not, explain why.

4.26 The authors of the paper "Statistical Methods for Assessing Agreement Between Two Methods of Clinical Measurement" (*International Journal of Nursing Studies* [2010]: 931–936) compared two different instruments for measuring a person's ability to breathe out air. (This measurement is helpful in diagnosing various lung disorders.) The two instruments considered were a Wright peak flow meter and a mini-Wright peak flow meter. Seventeen people participated in the study, and for each person air flow was measured once using the Wright meter and once using the mini-Wright meter.

Subject	Mini-Wright Meter	Wright Meter	Subject	Mini-Wright Meter	Wright Meter
1	512	494	10	428	434
2	430	395	11	500	476
3	520	516	12	600	557
4	364	413	13	260	267
5	380	442	14	477	478
6	658	650	15	259	178
7	445	433	16	350	423
8	432	417	17	451	427
9	626	656			

a. Suppose that the Wright meter is thought to provide a better measure of air flow, but the mini-Wright meter is easier to transport and to use. If the two types of meters produce different readings but there is a strong relationship between the readings, it would be possible to use a reading from the mini-Wright meter to predict the reading that the larger Wright meter would have given. Use the given data to find an equation to predict Wright meter reading using a reading from the mini-Wright meter.

b. What would you predict for the Wright meter reading for a person whose mini-Wright meter reading was 500?

c. What would you predict for the Wright meter reading for a person whose mini-Wright meter reading was 300?

SECTION 4.2   Exercise Set 2

4.27 Two scatterplots are shown below. Explain why it makes sense to use the least squares regression line to summarize the relationship between x and y for one of these data sets but not the other.

Scatterplot 1

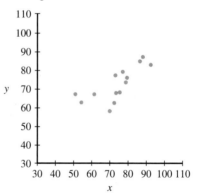

Scatterplot 2

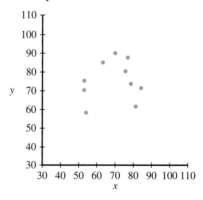

4.28 What does it mean when we say that the regression line is the *least squares* line?

4.29 The accompanying data on x = head circumference z-score (a comparison score with peers of the same age) at 6 to 14 months of age and y = volume of cerebral grey matter (in ml) at 2 to 5 years of age are from a graph that appeared in the *Journal of the American Medical Association* (2003).

Cerebral Grey Matter (ml) at 2–5 yr	Head Circumference z-Scores at 6–14 Months
680	−0.75
690	1.20
700	−0.30
720	0.25
740	0.30
740	1.50
750	1.10
750	2.00
760	1.10
780	1.10
790	2.00
810	2.10
815	2.80
820	2.20
825	0.90
835	2.35
840	2.30
845	2.20

a. Construct a scatterplot for these data.
b. What is the value of the correlation coefficient?
c. Find the equation of the least squares regression line.
d. Predict the volume of cerebral grey matter for a child whose head circumference z-score was 1.8.
e. Explain why it would not be a good idea to use the least squares regression line to predict the volume of grey matter for a child whose head circumference z-score was 3.0.

4.30 The authors of the paper **"Evaluating Existing Movement Hypotheses in Linear Systems Using Larval Stream Salamanders"** (*Canadian Journal of Zoology* [2009]: 292–298) investigated whether water temperature was related to how far a salamander would swim and to whether it would swim upstream or downstream. Data for 14 streams with different mean water temperatures appear below (from a graph in the paper). The two variables of interest are mean water temperature (°C) and net directionality, which was defined as the difference in the relative frequency of released larvae moving upstream and the relative frequency of released larvae moving downstream. A positive value of net directionality means a higher proportion were moving upstream than downstream. A negative value means a higher proportion were moving downstream.

Mean Temperature (x)	Net Directionality (y)	Mean Temperature (x)	Net Directionality (y)
6.17	−0.08	17.98	0.29
8.06	0.25	18.29	0.23
8.62	−0.14	19.89	0.24
10.56	0.00	20.25	0.19
12.45	0.08	19.07	0.14
11.99	0.03	17.73	0.05
12.50	−0.07	19.62	0.07

a. If the goal is to learn about how net directionality is related to mean temperature, which of these two variables is the response variable and which is the predictor variable?
b. Construct a scatterplot of the data. How would you describe the relationship between these two variables?
c. Find the equation of the least squares line describing the relationship between y = net directionality and x = mean water temperature.
d. Interpret the slope of the least squares regression line in the context of this study.
e. Does it make sense to interpret the intercept of the least squares regression line? If so, give an interpretation. If not, explain why.
f. What value of net directionality would you predict for a stream that had a mean water temperature of 15°C?
g. The authors state that "when temperatures were warmer, more larvae were captured moving upstream, but when temperatures were cooler, more larvae were captured moving downstream." Do the scatterplot and least squares line support this statement?

4.31 The article **"California State Parks Closure List Due Soon"** (*The Sacramento Bee,* August 30, 2009) gave the following data on number of visitors in fiscal year 2007–2008 and percentage of operating costs covered by park revenues for the 20 state park districts in California.

a. Use a statistical software package or a graphing calculator to construct a scatterplot of these data. Describe any interesting features of the scatterplot.
b. Find the equation of the least squares regression line (use software or a graphing calculator).
c. Is the slope of the least squares line positive or negative? Is this consistent with your description in Part (a)?
d. Based on the scatterplot, do you think that the correlation coefficient for this data set would be less than 0.5 or greater than 0.5? Explain.

Number of Visitors	Percentage of Operating Costs Covered by Park Revenues
2,755,849	37
1,124,102	19
1,802,972	32
1,757,386	80
1,424,375	17
1,524,503	34
1,943,208	36
819,819	32
1,292,942	38
3,170,290	40
3,984,129	53
1,575,668	31
1,383,898	35
14,519,240	108

(continued)

Number of Visitors	Percentage of Operating Costs Covered by Park Revenues
3,983,963	34
14,598,446	97
4,551,144	62
10,842,868	36
1,351,210	36
603,938	34

4.32 The article **"No-Can-Do Attitudes Irk Fliers"** (*USA Today*, **September 15, 2010**) included the accompanying data on airline quality score and a ranking based on the number of passenger complaints per 100,000 passengers boarded. In the complaint ranking, a rank of 1 is best, corresponding to the fewest complaints. Similarly, for the J.D. Power quality score, lower scores correspond to higher quality.

Airline	Passenger Complaint Rank	J.D. Power Quality Score
Airtran	6	3
Alaska	2	4
American	8	7
Continental	7	6
Delta	12	8
Frontier	5	5
Hawaiian	3	Not rated
JetBlue	4	1
Northwest	9	Not rated
Southwest	1	2
United	11	9
US Airways	10	10

Two of the airlines were not scored by J.D. Power. Use the ranks for the other 10 airlines to fit a least squares regression line and predict the J.D. Power Quality Score for Hawaiian Airlines and Northwest Airlines.

Additional Exercises

4.33 The following table gives data on age, number of cell phone calls made in a typical day, and number of text messages sent in a typical day for a random sample of 10 people selected from those enrolled in adult education classes offered by a school district.

Age	Number of Cell Phone Calls	Number of Text Messages Sent	Age	Number of Cell Phone Calls	Number of Text Messages Sent
55	6	20	19	15	14
33	5	35	30	10	30
60	1	3	33	3	2
38	2	1	37	3	1
55	4	2	52	5	20

Use the given data to find the equation of the least squares regression line for predicting y = number of cell phone calls using x = age as a predictor.

4.34 Use the data given in the previous exercise to find the equation of the least squares regression line for predicting y = number of text messages sent using x = age as a predictor.

4.35 Use the data given in Exercise 4.33 to construct two scatterplots—one of number of cell phone calls versus age and the other of number of text messages sent versus age. Based on the scatterplots, do you think age is a better predictor of number of cell phone calls or number of text messages sent? Explain why you think this.

4.36 In a study of the relationship between TV viewing and eating habits, a sample of 548 ethnically diverse students from Massachusetts was followed over a 19-month period (*Pediatrics* [2003]: 1321–1326). For each additional hour of television viewed per day, the number of fruit and vegetable servings per day was found to decrease on average by 0.14 serving.
a. For this study, what is the response variable? What is the predictor variable?
b. Would the least squares regression line for predicting number of servings of fruits and vegetables using number of hours spent watching TV have a positive or negative slope? Justify your choice.

4.37 An article on the cost of housing in California (*San Luis Obispo Tribune*, **March 30, 2001**) included the following statement: "In Northern California, people from the San Francisco Bay area pushed into the Central Valley, benefiting from home prices that dropped on average $4000 for every mile traveled east of the Bay." If this statement is correct, what is the slope of the least squares regression line, $\hat{y} = a + bx$, where y = house price (in dollars) and x = distance east of the Bay (in miles)? Justify your answer.

4.38 The data below on runoff sediment concentration for plots with varying amounts of grazing damage are representative values from a graph in the paper **"Effect of Cattle Treading on Erosion from Hill Pasture: Modeling Concepts and Analysis of Rainfall Simulator Data"** (*Australian Journal of Soil Research* [2002]: 963–977). Damage was measured by the percentage of bare ground in the plot. Data are given for gradually sloped and for steeply sloped plots.

Gradually Sloped Plots

Bare ground (%)	5	10	15	25	30	40	
Concentration		50	200	250	500	600	500

Steeply Sloped Plots

Bare ground (%)	5	5	10	15	20	25	20
Concentration		100	250	300	600	500	500	900

(continued)

Bare ground (%)	30	35	40	35
Concentration	800	1100	1200	1000

a. *Using the data for steeply sloped plots*, find the equation of the least squares regression line for predicting $y =$ runoff sediment concentration using $x =$ percentage of bare ground.

b. What would you predict runoff sediment concentration to be for a steeply sloped plot with 18% bare ground?

c. Would you recommend using the least squares regression line from Part (a) to predict runoff sediment concentration for gradually sloped plots? Explain.

SECTION 4.3   Assessing the Fit of a Line

Once the least squares regression line has been obtained, the next step is to examine how effectively the line summarizes the relationship between x and y. Important questions to consider are:

1. Is a line an appropriate way to summarize the relationship between the two variables?
2. Are there any unusual aspects of the data set that you need to consider before using the least squares regression line to make predictions?
3. If you decide that it is reasonable to use the regression line as a basis for prediction, how accurate can you expect the predictions to be?

This section looks at graphical and numerical methods that will help you to answer these questions. Most of these methods are based on the vertical deviations of the data points from the regression line, which represent the differences between actual y values and the corresponding predicted $\hat{y}$ values from the regression line.

Predicted Values and Residuals

The predicted value corresponding to the first observation in a data set, (x_1, y_1), is obtained by substituting x_1 into the regression equation to obtain $\hat{y}_1$, so

$$\hat{y}_1 = a + bx_1$$

The difference between the actual y value, y_1, and the corresponding predicted value, $\hat{y}_1$ is

$$y_1 - \hat{y}_1$$

This difference, called a *residual*, is the vertical deviation of a point in the scatterplot from the least squares regression line. A point that is above the line results in a positive residual, whereas a point that is below the line results in a negative residual. This is shown in Figure 4.18.

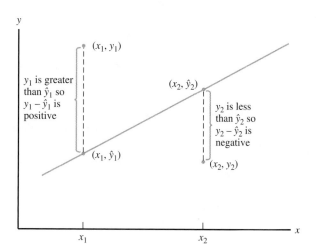

FIGURE 4.18
Positive and negative residuals

> **DEFINITION**
>
> The **predicted values** result from substituting each x value into the equation for the regression line. This gives
>
> $$\hat{y}_1 = \text{first predicted value} = a + bx_1$$
> $$\hat{y}_2 = \text{second predicted value} = a + bx_2$$
> $$\vdots$$
> $$\hat{y}_n = n^{th} \text{ predicted value} = a + bx_n$$
>
> The **residuals** are the n quantities
>
> $$y_1 - \hat{y}_1 = \text{first residual}$$
> $$y_2 - \hat{y}_2 = \text{second residual}$$
> $$\vdots$$
> $$y_n - \hat{y}_n = n^{th} \text{ residual}$$
>
> Each residual is the difference between an observed y value and the corresponding predicted y value.

Example 4.6 It May Be a Pile of Debris to You, but It Is Home to a Mouse

The accompanying data is a subset of data from a scatterplot that appeared in the paper **"Small Mammal Responses to Fine Woody Debris and Forest Fuel Reduction in Southwest Oregon"** (*Journal of Wildlife Management* [2005]: 625–632). The authors of the paper were interested in how the distance a deer mouse will travel for food is related to the distance from the food to the nearest pile of fine woody debris. Distances were measured in meters. The data are given in Table 4.2.

TABLE 4.2 Data, Predicted Values, and Residuals for Example 4.6

Distance From Debris (x)	Distance Traveled (y)	Predicted Distance Traveled ($\hat{y}$)	Residual ($y - \hat{y}$)
6.94	0.00	14.76	-14.76
5.23	6.13	9.23	-3.10
5.21	11.29	9.16	2.13
7.10	14.35	15.28	-0.93
8.16	12.03	18.70	-6.67
5.50	22.72	10.10	12.62
9.19	20.11	22.04	-1.93
9.05	26.16	21.58	4.58
9.36	30.65	22.59	8.06

JMP was used to fit the least squares regression line. The resulting output is shown in Figure 4.19.

Rounding the slope and intercept, the equation of the regression line is

$$\hat{y} = -7.69 + 3.23x.$$

A scatterplot of the data showing the least squares regression line is included in the JMP output of Figure 4.19. The residuals for this data set are the signed vertical distances from the points to the line.

For the mouse with the smallest x value ($x_3 = 5.21$ and $y_3 = 11.29$), the corresponding predicted value and residual are

predicted value $= \hat{y}_3 = -7.69 + 3.234(x_3) = -7.69 + 3.234(5.21) = 9.16$

residual $= y_3 - \hat{y}_3 = 11.29 - 9.16 = 2.13$

The other predicted values and residuals are computed in a similar manner and are included in Table 4.2.

This is the least squares regression line.

This is r^2 (discussed below).

In AP Statistics the "Adjusted r^2" is not used.

The "RMSE" is JMP's term for the standard deviation about the least squares regression line (also discussed below).

The "Mean of response" is the value of $\bar{y}$, the mean of the y values in the data.

The "Observations" is the number of data points.

The "Analysis of Variance" is not used in simple regression. (It is valuable to know what parts of the computer output you can safely ignore, and this is one of those ignorable parts.)

The Parameter Estimates are the intercept and slope. In Chapter 16, you will see how the rest of the information in this section is used for.

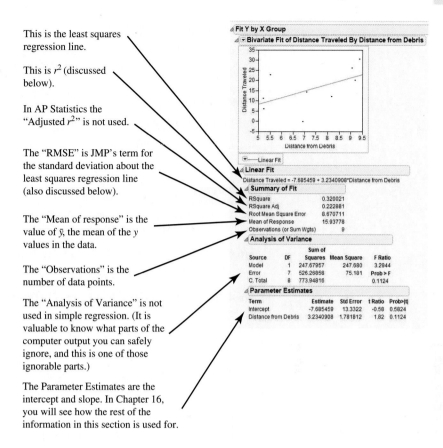

FIGURE 4.19
JMP output for the data of Example 4.6

Computing the predicted values and residuals by hand can be tedious, but JMP and other statistical software packages, as well as many graphing calculators, can compute and save the residuals. When requested, JMP adds the predicted values and residuals to the data table, as shown here:

	Distance from Debris	Distance Traveled	Predicted Distance Traveled	Residuals Distance Traveled
1	6.94	0	14.759131352	-14.75913135
2	5.23	6.13	9.2288360798	-3.09883608
3	5.21	11.29	9.1641542638	2.1258457362
4	7.1	14.35	15.27658588	-0.92658588
5	8.16	12.03	18.704722131	-6.674722131
6	5.5	22.72	10.102040596	12.617959404
7	9.19	20.11	22.035835658	-1.925835658
8	9.05	26.16	21.583062945	4.5769370548
9	9.36	30.65	22.585631094	8.0643689061

Plotting the Residuals

A careful look at the residuals can reveal many potential problems. To assess the appropriateness of the regression line, a *residual plot* is a good place to start.

> **DEFINITION**
>
> A **residual plot** is a scatterplot of the (*x*, residual) pairs.
>
> Isolated points or a pattern of points in the residual plot indicate potential problems.

Example 4.7   Revisiting the Deer Mice

In the previous example, the nine residuals for the deer mice data were computed. From Table 4.2 (or the JMP output), the nine (*x*, residual) pairs are

$$(6.94, -14.76) \quad (5.23, -3.10) \quad (5.21, 2.13) \quad (7.10, -0.93)$$
$$(8.16, -6.67) \quad (5.50, 12.62) \quad (9.19, -1.93) \quad (9.05, 4.58) \quad (9.36, 8.06)$$

Figure 4.20 shows the corresponding residual plot.

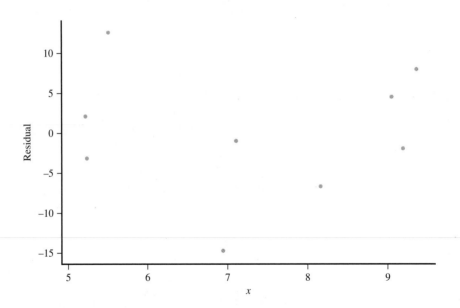

FIGURE 4.20
Residual plot for the deer mice data

There is no strong pattern in the residual plot. The points in the residual plot appear scattered at random.

A desirable residual plot is one that exhibits no particular pattern, such as curvature. Curvature in the residual plot indicates that the relationship between *x* and *y* is not linear. In this case, a curve would be a better choice for describing the relationship. Looking at a residual plot is the best way to begin to assess the fit of the least squares regression line. It is often easier to detect curvature in a residual plot than in a scatterplot, as illustrated in Example 4.8.

Example 4.8   Heights and Weights of American Women

Consider the accompanying data on *x* = height (in inches) and *y* = average weight (in pounds) for American females, ages 30–39 (from *The World Almanac and Book of Facts*).

x	58	59	60	61	62	63	64	65
y	113	115	118	121	124	128	131	134
x	66	67	68	69	70	71	72	
y	137	141	145	150	153	159	164	

The scatterplot, displayed in Figure 4.21(a), appears rather straight. However, even though $r = 0.99$, when the residuals from the least squares regression line are plotted in Figure 4.21(b), there is a definite curved pattern. Because of this, it is not accurate to say that weight increases linearly with height.

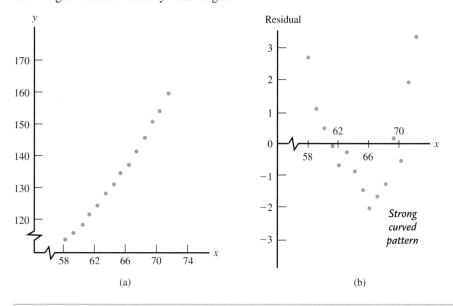

FIGURE 4.21
Plots for the data of Example 4.8.
(a) scatterplot; (b) residual plot

It is also important to look for unusual values in the scatterplot or in the residual plot. In the residual plot, a point falling far above or below the horizontal line at height 0 corresponds to a large residual. This may indicate some type of unusual behavior, such as a recording error. A point whose x value differs greatly from others in the data set may have excessive influence in determining the equation of the regression line. To assess the impact of such an isolated point, delete it from the data set and then recalculate the equation of the regression line. You can then evaluate the extent to which the equation has changed.

Example 4.9 **Older Than Your Average Bear**

The accompanying data on x = age (in years) and y = weight (in kg) for 12 black bears appeared in the paper **"Habitat Selection by Black Bears in an Intensively Logged Boreal Forest"** (*Canadian Journal of Zoology* [2008]: 1307–1316). A scatterplot and residual plot are shown in Figures 4.22(a) and 4.22(b), respectively. One bear in the sample was much older than the others (bear 3 with an age of $x = 28.5$ years and a weight of $y = 62$ kg). This results in a point in the scatterplot that is far to the right of the other points. Because the least squares regression line minimizes the sum of squared residuals, the line is pulled toward this observation. As a result, this single observation plays a big role in determining the slope of the least squares regression line, and it is therefore called an *influential observation*. Notice that this influential observation does not have a large residual, because the least squares regression line actually passes near this point. Figure 4.23 shows what happens when the influential observation is removed from the data set. Both the slope and intercept of the new least squares regression line are quite different from the slope and intercept of the original line.

Bear	Age	Weight	Bear	Age	Weight
1	10.5	54	7	6.5	62
2	6.5	40	8	5.5	42
3	28.5	62	9	7.5	40
4	10.5	51	10	11.5	59
5	6.5	55	11	9.5	51
6	7.5	56	12	5.5	50

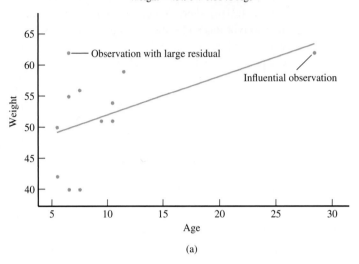

(a)

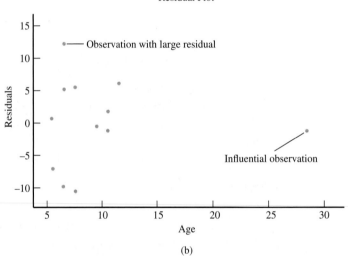

FIGURE 4.22
Minitab plots for the bear data of
Example 4.9. (a) scatterplot;
(b) residual plot

(b)

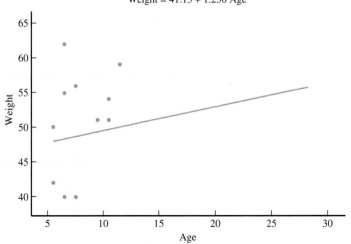

FIGURE 4.23
Scatterplot and least squares
regression line with bear 3
removed from the data set

Some points in the scatterplot may fall far from the least squares regression line in the y direction, resulting in a large residual. These points are sometimes referred to as *outliers*. In this example, the observation with the largest residual is bear 7 with an age of $x = 6.5$ years and a weight of $y = 62$ kg (labeled in Figure 4.22). Even though this observation has a large residual, it is not influential. The equation of the least squares line for the complete data set is $\hat{y} = 45.90 + 0.6141x$, which is not much different from $\hat{y} = 43.81 + 0.7131x$, the equation that results when bear 7 is omitted.

Unusual points in a bivariate data set are those that are far away from most of the other points in the scatterplot in either the x direction or the y direction.

An observation is potentially an **influential observation** if it has an x value that is far away from the rest of the data (separated from the rest of the data in the x direction). To decide if an observation *is* influential, determine if removal of the observation has a large impact on the value of the slope or intercept of the least squares regression line. (The decision about whether the impact is large is subjective and depends on the context of the given problem.)

An observation is an **outlier** if it has a large residual. Outliers fall far away from the least squares line in the y direction.

Careful examination of a scatterplot and a residual plot can help you evaluate whether a line is an appropriate way to summarize a relationship. If you decide that a line is appropriate, the next step is to assess the accuracy of predictions based on the least squares line and whether these predictions tend to be better than those made without knowledge of the value of x. Two numerical measures that help with this assessment are the coefficient of determination, r^2, and the standard deviation about the regression line, s_e.

Coefficient of Determination, r^2

Suppose that you would like to predict the price of houses in a particular city. A random sample of 20 houses that are for sale is selected, and $y =$ price and $x =$ size (in square feet) are recorded for each house. There will be variability in house price, and it is this variability that makes accurate price prediction a challenge. But if you know that differences in house size account for a large proportion of the variability in house price, then knowing the size of a house will help you predict its price. The proportion of variability in house price that can be "explained" by a linear relationship between house size and house price is called the **coefficient of determination**.

DEFINITION

The **coefficient of determination**, denoted by r^2, is the proportion of variability in y that can be attributed to an approximate linear relationship between x and y.

The value of r^2 is often converted to a percentage (by multiplying by 100).

To understand how r^2 is computed, consider variability in the y values. Variability in y can be effectively explained by an approximate straight-line relationship when the points in the scatterplot tend to fall close to the least squares regression line—that is, when the residuals are small. A natural measure of variability about the least squares regression line is the sum of the squared residuals. (Squaring before combining prevents negative and positive residuals from offsetting one another.) A second sum of squares assesses the total amount of variability in the observed y values by considering how spread out the y values are around the mean y value.

The **residual sum of squares** (sometimes also referred to as the error sum of squares), denoted by **SSResid**, is defined as

$$\text{SSResid} = (y_1 - \hat{y}_1)^2 + (y_2 - \hat{y}_2)^2 + \cdots + (y_n - \hat{y}_n)^2 = \sum(y - \hat{y})^2$$

The **total sum of squares**, denoted by **SSTo**, is defined as

$$\text{SSTo} = (y_1 - \bar{y})^2 + (y_2 - \bar{y})^2 + \cdots + (y_n - \bar{y})^2 = \sum(y - \bar{y})^2$$

These sums of squares can be found as part of the regression output from most statistical packages or can be obtained using the following computational formulas:

$$\text{SSTo} = \sum y^2 - \frac{\left(\sum y\right)^2}{n}$$

$$\text{SSResid} = \sum y^2 - a\sum y - b\sum xy$$

Example 4.10 Deer Mice Again

Figure 4.24 displays part of the Minitab output that results from fitting the least squares line to the data on y = distance traveled for food and x = distance to nearest woody debris pile from Example 4.6. From the output,

$$\text{SSTo} = 773.95 \text{ and } \text{SSResid} = 526.27$$

Regression Analysis: Distance Traveled versus Distance to Debris

The regression equation is
Distance Traveled = − 7.7 + 3.23 Distance to Debris

Predictor	Coef	SE Coef	T	P
Constant	−7.69	13.33	−0.58	0.582
Distance to Debris	3.234	1.782	1.82	0.112

S = 8.67071   R-Sq = 32.0%   R-Sq(adj) = 22.3%

Analysis of Variance

Source	DF	SS	MS	F	P
Regression	1	247.68	247.68	3.29	0.112
Residual Error	7	526.27	75.18		
Total	8	773.95			

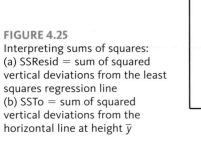

FIGURE 4.24
Minitab output for the deer mice data

The residual sum of squares is the sum of the squared vertical deviations from the least squares line. As Figure 4.25 illustrates, SSTo is also a sum of squared vertical deviations from a line—the horizontal line at height $\bar{y}$. The least squares line is, by definition, the line that has the smallest sum of squared deviations. It follows that SSResid $\leq$ SSTo. These two sums of squares are equal only when the least squares line *is* the horizontal line $y = \bar{y}$.

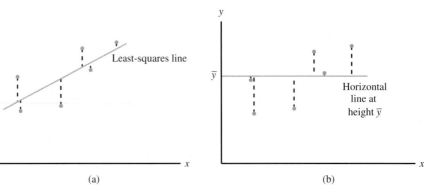

FIGURE 4.25
Interpreting sums of squares:
(a) SSResid = sum of squared vertical deviations from the least squares regression line
(b) SSTo = sum of squared vertical deviations from the horizontal line at height $\bar{y}$

SSResid is often referred to as a measure of unexplained variability—the variability in y that cannot be attributed to the linear relationship between x and y. The more the points in the scatterplot deviate from the least squares line, the larger the value of SSResid. Similarly, SSTo is interpreted as a measure of total variability. The greater the amount of variability in the y values, the larger the value of SSTo.

The ratio SSResid/SSTo is the fraction or proportion of total variability that can't be explained by a linear relationship. Subtracting this ratio from 1 gives the proportion of total variability that is explained, which is the coefficient of determination.

> The coefficient of determination is computed as
>
> $$r^2 = 1 - \frac{\text{SSResid}}{\text{SSTo}}$$

Multiplying r^2 by 100 gives the percentage of y variability that can be explained by the approximate linear relationship. The closer this percentage is to 100%, the more successful the relationship is in explaining variability in y.

Example 4.11 r^2 for the Deer Mice Data

For the data on distance traveled for food and distance to nearest debris pile from Example 4.10, SSTo = 773.95 and SSResid = 526.27. From these values,

$$r^2 = 1 - \frac{\text{SSResid}}{\text{SSTo}} = 1 - \frac{526.27}{773.95} = 0.32$$

This means that only 32% of the observed variability in distance traveled for food can be explained by an approximate linear relationship between distance traveled for food and distance to nearest debris pile. Notice that the r^2 value can be found in the Minitab output of Figure 4.24, labeled "R-Sq" and in the JMP output of Figure 4.19 labeled "RSquare."

The symbol r was used in Section 4.1 to denote the sample correlation coefficient. This notation suggests how the correlation coefficient and the coefficient of determination are related:

$$(\text{correlation coefficient})^2 = \text{coefficient of determination}$$

If $r = 0.8$ or $r = -0.8$, then $r^2 = 0.64$, so 64% of the observed variability in y can be explained by the linear relationship. Because the value of r does not depend on which variable is labeled x, the same is true of r^2. The coefficient of determination is one of the few quantities computed in a regression analysis whose value remains the same when the roles of response and predictor variables are interchanged. When $r = 0.5$, $r^2 = 0.25$, so only 25% of the observed variability in y is explained by the linear relationship. This is why a value of r between -0.5 and 0.5 is interpreted as indicating only a weak linear relationship.

Example 4.12 Lead Exposure and Brain Volume

The authors of the paper **"Decreased Brain Volume in Adults with Childhood Lead Exposure"** (*Public Library of Science Medicine* **[May 27, 2008]: e112**) studied the relationship between childhood environmental lead exposure and volume change in a particular region of the brain. Data on x = mean childhood blood lead level (mg/dL) and y = brain volume change (percent) from a graph in the paper were used to produce the scatterplot in Figure 4.26. The least squares line is also shown on the scatterplot.

Notice that although there is a slight tendency for smaller y values (corresponding to a brain volume decrease) to be paired with higher values of mean blood lead levels, the relationship is weak. The points in the plot are widely scattered around the least squares line.

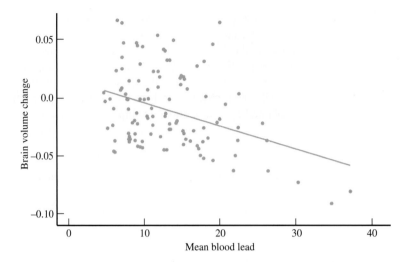

FIGURE 4.26
Scatterplot and least squares regression line for the data of Example 4.12

Figure 4.27 displays part of the Minitab output that results from fitting the least squares regression line to the data.

Regression Analysis: Brain Volume Change versus Mean Blood Lead

The regression equation is
Brain Volume Change = 0.01559 – 0.001993 Mean Blood Lead

S = 0.0310931   R-Sq = 13.6%   R-Sq(adj) = 12.9%

Analysis of Variance

Source	DF	SS	MS	F	P
Regression	1	0.016941	0.0169410	17.52	0.000
Error	111	0.107313	0.0009668		
Total	112	0.124254			

FIGURE 4.27
Minitab output for the data of Example 4.12

From the computer output, $r^2 = 0.136$. This means that differences in childhood mean blood lead level explain only 13.6% of the variability in adult brain volume change. You can compute the value of the correlation coefficient by taking the square root of r^2. In this case, you know that the correlation coefficient is negative (because there is a negative relationship between x and y), so you choose the negative square root:

$$r = -\sqrt{0.136} = -0.369$$

Based on the values of r and r^2, you would conclude that there is a weak negative linear relationship and that childhood mean blood lead level explains only about 13.6% of adult change in brain volume.

Standard Deviation About the Least Squares Regression Line

The coefficient of determination (r^2) measures the extent of variability about the least squares regression line *relative* to overall variability in y. A large value of r^2 does not necessarily imply that the deviations from the line are small in an absolute sense. A typical observation could deviate from the line by quite a bit.

Recall that the sample standard deviation

$$s = \sqrt{\frac{\sum(x - \bar{x})^2}{n - 1}}$$

is a measure of variability in a single sample. The value of s can be interpreted as a typical amount by which a sample observation deviates from the sample mean. There is an analogous measure of variability around the least squares regression line.

DEFINITION

The **standard deviation about the least squares regression line** is

$$s_e = \sqrt{\frac{\text{SSResid}}{n - 2}}$$

The value of s_e can be interpreted as the typical amount by which an observation deviates from the least squares regression line.

Example 4.13 Predicting Graduation Rates

Consider the accompanying data from 2007 on six-year graduation rate (%), student-related expenditure per full-time student, and median SAT score for the 38 primarily undergraduate public universities and colleges in the United States with enrollments between 10,000 and 20,000 **(Source: College Results Online, The Education Trust).**

Graduation Rate	Expenditure	Median SAT
81.2	7,462	1,160
66.8	7,310	1,115
66.4	6,959	1,070
66.1	8,810	1,205
64.9	7,657	1,135
63.7	8,063	1,060
62.6	8,352	1,130
62.5	7,789	1,200
61.2	8,106	1,015
59.8	7,776	1,100
56.6	8,515	990
54.8	7,037	1,085
52.7	8,715	1,040
52.4	7,780	1,040
52.4	7,198	1,105
50.5	7,429	975
49.9	7,551	1,030
48.9	8,112	1,030
48.1	8,149	950
46.5	6,744	1,010
45.3	8,842	1,223
45.2	7,743	990
43.7	5,587	1,010
43.5	7,166	1,010
42.9	5,749	950
42.1	6,268	955
42.0	8,477	985
38.9	7,076	990
38.8	8,153	990
38.3	7,342	910
35.9	8,444	1,075
32.8	7,245	885
32.6	6,408	1,060
32.3	4,981	990
31.8	7,333	970
31.3	7,984	905
31.0	5,811	1,010
26.0	7,410	1,005

Figure 4.28 displays scatterplots of graduation rate versus student-related expenditure and graduation rate versus median SAT score. The least squares lines and the values of r^2 and s_e are also shown.

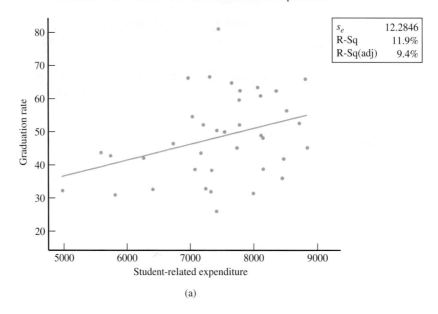

(a)

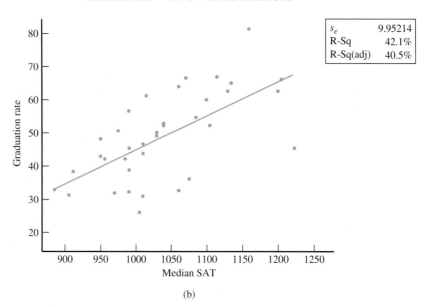

(b)

FIGURE 4.28
Scatterplots for the data of Example 4.13: (a) graduation rate versus student-related expenditure (b) graduation rate versus median SAT

Notice that while there is a positive linear relationship between student-related expenditure and graduation rate, the relationship is weak. The value of r^2 is only 0.119, indicating that only about 11.9% of the variability in graduation rate from university to university can be explained by differences in student-related expenditures. The standard deviation about the regression line is $s_e = 12.2846$, which is larger than s_e for median SAT. This means that the points in the first scatterplot tend to fall farther from the least squares regression line than the points in the second scatterplot. The value of r^2 for the second scatterplot is 0.421 and $s_e = 9.95214$, indicating that median SAT does a better job of explaining variability in graduation rates. You would expect the regression line that uses median SAT as a predictor to produce more accurate predictions of graduation rates than the line that uses student-related expenditure as a predictor. Based on the values of r^2 and s_e, median SAT would be a better choice for predicting graduation rates than student-related expenditures.

When evaluating the usefulness of the least squares regression line for making predictions, it is important to consider both the value of r^2 *and* the value of s_e. These two measures assess different aspects of the fit of the line. In general, you would like to have a small value for s_e (which indicates that deviations from the line tend to be small) and a large value for r^2 (which indicates that the linear relationship explains a large proportion of the variability in the y values).

Interpreting the Values of s_e and r^2

A small value of s_e indicates that residuals tend to be small. Because residuals represent the difference between a predicted y value and an observed y value, the value of s_e tells you about the accuracy you can expect when using the least squares regression line to make predictions.

A large value of r^2 indicates that a large proportion of the variability in y can be explained by the approximate linear relationship between x and y. This tells you that knowing the value of x is helpful for predicting y.

A useful regression line will have a reasonably small value of s_e and a reasonably large value of r^2.

Example 4.14   Predicting IQ

On May 25, 2010, *LiveScience* published an article titled **"Simple Memory Test Predicts Intelligence."** The article summarized a study that found the score on a test of working memory capacity was correlated with a number of different measures of intelligence. The actual study (***Archives of General Psychiatry*** [2010]: 570–577) looked at how working memory capacity was related to scores on a test of cognitive functioning and to scores on an IQ test. Two groups were studied—one group consisted of patients diagnosed with schizophrenia and the other group consisted of healthy control subjects.

There are several interesting things to note about the linear relationships in Figure 4.29. Looking first at the cognitive functioning score scatterplots, you can see that the relationship between cognitive functioning score and working memory capacity is much stronger for the healthy control group (Figure 4.29(b)) than for the schizophrenic patient group (Figure 4.29(a)). For the patient group, $s_e = 10.7362$ and $r^2 = 0.140$. If you were to use the least squares regression line $\hat{y} = 10.24 + 9.229x$ to predict $y =$ cognitive functioning score based on $x =$ working memory capacity, a typical prediction error would be around 10.7362. Also, for this group, the approximate linear relationship between the variables explains only about 14% of the variability in cognitive functioning score. The value of the correlation coefficient is

$$r = \sqrt{r^2} = \sqrt{0.140} = 0.374$$

There is a weak positive linear relationship between working memory capacity and cognitive functioning score for the patient group.

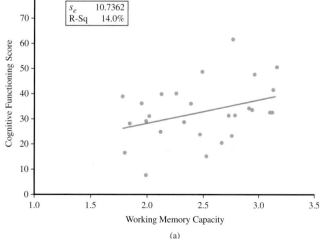

FIGURE 4.29
Scatterplots for Example 4.14. (a) cognitive functioning score versus working memory capacity for patients

(continued)

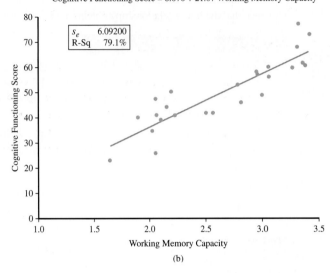

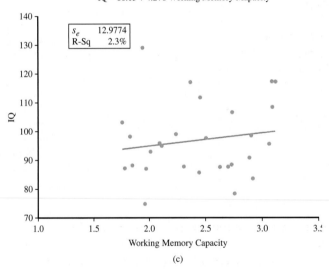

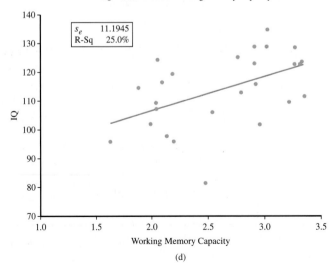

FIGURE 4.29 Scatterplots for Example 4.14. (b) cognitive functioning score versus working memory capacity for controls, (c) IQ versus working memory capacity for patients, (d) IQ versus working memory capacity for controls

For the healthy control group, $s_e = 6.092$ and $r^2 = 0.791$. This means that using the least squares regression line $\hat{y} = 6.070 + 21.07x$ to predict $y =$ cognitive functioning score for healthy controls would result in smaller prediction errors, typically around 6.092. For the healthy control group, the approximate linear relationship between the two variables explains about 79.1% of the variability in cognitive functioning score, a much larger percentage than for the patient group.

Working memory capacity is not nearly as good a predictor of IQ as it is for cognitive functioning score. For both groups, the linear relationship between working memory capacity and IQ is much weaker than the relationship between working memory capacity and cognitive functioning score. Also, notice that the relationship between working memory capacity and IQ is stronger for the healthy control group than for the patient group.

SECTION 4.3 EXERCISES

Each Exercise Set assesses the following chapter learning objectives: C8, C9,, C10, C11, M11, M12, M13, M14, M15, M16, P2, P4

SECTION 4.3 Exercise Set 1

4.39 The paper "Accelerated Telomere Shortening in Response to Life Stress" (*Proceedings of the National Academy of Sciences* [2004]: 17312–17315) describes a study that examined whether stress accelerates aging at a cellular level. The accompanying data on a measure of perceived stress (x) and telomere length (y) for 38 people are from a scatterplot in the paper. Telomere length is a measure of cell longevity.

Perceived Stress	Telomere Length	Perceived Stress	Telomere Length
5	1.25	20	1.22
6	1.32	20	1.30
6	1.50	20	1.32
7	1.35	21	1.24
10	1.30	21	1.26
11	1.00	21	1.30
12	1.18	22	1.18
13	1.10	22	1.22
14	1.08	22	1.24
14	1.30	23	1.18
15	0.92	24	1.12
15	1.22	24	1.50
15	1.24	25	0.94
17	1.12	26	0.84
17	1.32	27	1.02
17	1.40	27	1.12
18	1.12	28	1.22
18	1.46	29	1.30
19	0.84	33	0.94

a. Compute the equation of the least squares regression line.
b. What is the value of r^2 for this data set? Write a sentence interpreting this value in context.
c. What is the value of s_e for this data set? Write a sentence interpreting this value in context.
d. Is the linear relationship between perceived stress and telomere length positive or negative? Is it weak, moderate, or strong? Justify your answer.

4.40 Briefly explain why a small value of s_e is desirable in a regression setting.

4.41 Briefly explain why it is important to consider the value of r^2 in addition to the value of s_e when evaluating the usefulness of the least squares regression line.

4.42 The data in the accompanying table are from the paper "Six-Minute Walk Test in Children and Adolescents" (*The Journal of Pediatrics* [2007]: 395–399). Boys completed a test that measures the distance that the subject can walk on a flat, hard surface in 6 minutes. For each age group shown in the table, the median distance walked by the boys in that age group is given.

Age Group	Representative Age (midpoint of age group)	Median Six-minute Walk Distance (meters)
3–5	4.0	544.3
6–8	7.0	584.0
9–11	10.0	667.3
12–15	13.5	701.1
16–18	17.0	727.6

a. With $x =$ representative age and $y =$ median distance walked in 6 minutes, construct a scatterplot. Does the pattern in the scatterplot look linear?
b. Find the equation of the least squares regression line.
c. Compute the five residuals and construct a residual plot. Are there any unusual features in the residual plot?

4.43 The paper referenced in the previous exercise also gave the 6-minute walk distances for girls ages 3 to 18 years. The median distances for the five age groups were

492.4	578.3	655.8	657.6	660.9

a. With $x =$ representative age and $y =$ median distance walked in 6 minutes, construct a scatterplot. How does the pattern for girls differ from the pattern for boys from Exercise 4.42?

b. Find the equation of the least squares regression line that describes the relationship between median distance walked in 6 minutes and representative age for girls.

c. Compute the five residuals and construct a residual plot. The authors of the paper decided to use a curve rather than a straight line to describe the relationship between median distance walked in 6 minutes and age for girls. What aspect of the residual plot supports this decision?

4.44 Northern flying squirrels eat lichen and fungi, which makes for a relatively low quality diet. The authors of the paper "**Nutritional Value and Diet Preference of Arboreal Lichens and Hypogeous Fungi for Small Mammals in the Rocky Mountains**" (*Canadian Journal of Zoology* [2008]: 851–862) measured nitrogen intake and nitrogen retention in six flying squirrels that were fed the fungus *Rhizopogon*. Data from a graph in the paper are given in the following table. (The negative value for nitrogen retention for the first squirrel represents a net loss in nitrogen.)

Nitrogen Intake x (grams)	Nitrogen Retention y (grams)
0.03	−0.04
0.10	0.00
0.07	0.01
0.06	0.01
0.07	0.04
0.25	0.11

a. Construct a scatterplot of these data.

b. Find the equation of the least squares regression line. Based on this line, what would you predict nitrogen retention to be for a flying squirrel whose nitrogen intake is 0.06 grams? What is the residual associated with the observation (0.06, 0.01)?

c. Look again at the scatterplot from Part (a). Which observation is potentially influential? Explain the reason for your choice.

d. When the potentially influential observation is deleted from the data set, the equation of the least squares regression line fit to the remaining five observations is $\hat{y} = -0.037 + 0.627x$. Use this equation to predict nitrogen retention for a flying squirrel whose nitrogen intake is 0.06. Compare this prediction to the prediction made in Part (b).

4.45 Researchers have observed that bears hunting salmon in a creek often carry the salmon away from the creek before eating it. The relationship between x = total number of salmon in a creek and y = percentage of salmon killed by bears and transported away from the stream prior to being eaten by a bear was examined in the paper "**Transportation of Pacific Salmon Carcasses from Streams to Riparian Forests by Bears**" (*Canadian Journal of Zoology* [2009]: 195–203). Data

for the 10 years from 1999 to 2008 are given in the accompanying table.

Total Number	Percentage Transported
19,504	77.8
3,460	28.7
1,976	28.9
8,439	27.9
11,142	55.3
3,467	20.4
3,928	46.8
20,440	76.3
7,850	40.3
4,134	24.1

a. Construct a scatterplot of these data. Does there appear to be a relationship between the total number of salmon and the percentage of transported salmon?

b. Find the equation of the least squares regression line.

c. The residuals from the least squares line are shown in the accompanying table. The observation (3,928, 46.8) has a large residual. Is this data point also an influential observation?

Total Number	Percentage Transported	Residual
19,504	77.8	3.43
3,460	28.7	0.30
1,976	28.9	4.76
8,439	27.9	−14.76
11,142	55.3	4.89
3,467	20.4	−8.02
3,928	46.8	17.06
20,440	76.3	−0.75
7,850	40.3	−0.68
4,134	24.1	−6.23

d. The two points with unusually large x values (19,504 and 20,440) were not thought to be influential observations even though they are far removed from the rest of the points. Explain why these two points are not influential.

e. Partial Minitab output resulting from fitting the least squares regression line is shown here. What is the value of s_e? Write a sentence interpreting this value.

Regression Analysis: Percent Transported versus Total Number

The regression equation is
Percent Transported = 18.5 + 0.00287 Total Number

Predictor	Coef
Constant	18.483
Total Number	0.0028655

S = 9.16217   R-Sq = 83.2%   R-Sq(adj) = 81.1%

f. What is the value of r^2? Write a sentence interpreting this value.

4.46 Some types of algae have the potential to cause damage to river ecosystems. The accompanying data on algae colony density (y) and rock surface area (x) for nine rivers are a subset of data that appeared in a scatterplot in a paper in the journal *Aquatic Ecology* (2010: 33–40).

x	50	55	50	79	44	37	70	45	49
y	152	48	22	35	38	171	13	185	25

a. Compute the equation of the least squares regression line.
b. What is the value of r^2 for this data set? Write a sentence interpreting this value in context.
c. What is the value of s_e for this data set? Write a sentence interpreting this value in context.
d. Is the linear relationship between rock surface area and algae colony density positive or negative? Is it weak, moderate, or strong? Justify your answer.

4.47 Briefly explain why a large value of r^2 is desirable in a regression setting.

4.48 Briefly explain why it is important to consider the value of s_e in addition to the value of r^2 when evaluating the usefulness of the least squares regression line.

4.49 The accompanying data resulted from an experiment in which weld diameter x and shear strength y (in pounds) were determined for five different spot welds on steel.

x	200.1	210.1	220.1	230.1	240.0
y	813.7	785.3	960.4	1118.0	1076.2

a. With x = weld diameter and y = shear strength, construct a scatterplot. Does the pattern in the scatterplot look linear?
b. Find the equation of the least squares regression line.
c. Compute the five residuals and construct a residual plot. Are there any unusual features in the residual plot?

4.50 The article "Master's Performance in the New York City Marathon" (*British Journal of Sports Medicine* [2004]: 408–412) gave the following data on the average finishing time by age group for female participants in the New York City Marathon.

Age Group	Representative Age	Average Finish Time
10–19	15	302.38
20–29	25	193.63
30–39	35	185.46
40–49	45	198.49
50–59	55	224.30
60–69	65	288.71

a. Find the equation of the least squares regression line that describes the relationship between average finish time (y) and representative age (x).

b. Compute the six residuals and construct a residual plot. The authors of the paper decided to use a curve rather than a straight line to describe the relationship between average finish time and representative age. What aspect of the residual plot supports this decision?

4.51 The article **"Air Pollution and Medical Care Use by Older Americans"** (*Health Affairs* [2002]: 207–214) gave data on a measure of pollution (in micrograms of particulate matter per cubic meter of air) and the cost of medical care per person over age 65 for six geographical regions of the United States:

Region	Pollution	Cost of Medical Care
North	30.0	915
Upper South	31.8	891
Deep South	32.1	968
West South	26.8	972
Big Sky	30.4	952
West	40.0	899

The equation of the least-squares regression line for this data set is $\hat{y} = 1082.2 - 4.691x$, where y = medical cost and x = pollution.
a. Compute the six residuals.
b. What is the value of the correlation coefficient for this data set? Does this value indicate that the linear relationship between pollution and medical cost is strong, moderate, or weak? Explain.
c. Construct a residual plot. Are there any unusual features of the residual plot?
d. The observation for the West, (40.0, 899), is far removed from the other values in the sample. Is this observation influential in determining the values of the slope and/or intercept of the least squares line? Justify your answer.

Additional Exercises

4.52 The article **"Examined Life: What Stanley H. Kaplan Taught Us About the SAT"** (*The New Yorker* [December 17, 2001]: 86–92) included a summary of findings regarding the use of SAT I scores, SAT II scores, and high school grade point average (GPA) to predict first-year college GPA. The article states that "among these, SAT II scores are the best predictor, explaining 16 percent of the variance in first-year college grades. GPA was second at 15.4 percent, and SAT I was last at 13.3 percent."
a. If the data from this study were used to fit a least squares regression line with y = first-year college GPA and x = high school GPA, what would be the value of r^2?
b. The article states that SAT II was the best predictor of first-year college grades. Do you think that predictions based on a least-squares line with y = first-year college GPA and x = SAT II score would be very accurate? Explain why or why not.

4.53 The paper **"Effects of Age and Gender on Physical Performance"** (*Age* [2007]: 77–85) describes a study investigating the relationship between age and swimming performance. Data on age and 1-hour swim distance for over 10,000 men participating in a national long-distance 1-hour swimming competition are summarized in the accompanying table.

Age Group	Representative Age (midpoint of age group)	Average Swim Distance (meters)
20–29	25	3,913.5
30–39	35	3,728.8
40–49	45	3,579.4
50–59	55	3,361.9
60–59	65	3,000.1
70–79	75	2,649.0
80–89	85	2,118.4

a. Find the equation of the least squares regression line using x = representative age and y = average swim distance.
b. Compute the seven residuals and use them to construct a residual plot. What does the residual plot suggest about the appropriateness of using a line to describe the relationship between representative age and average swim distance?
c. Would it be reasonable to use the least squares regression line from Part (a) to predict the average swim distance for women ages 40 to 49 by substituting the representative age of 45 into the equation of the least squares regression line? Explain why or why not.

4.54 The article **"California State Parks Closure List Due Soon"** (*The Sacramento Bee*, **August 30, 2009**) gave the following data on y = number of employees in fiscal year 2007–2008 and x = park size (in acres) for the 20 state park districts in California:

Number of Employees, y	Total Park Size, x	Number of Employees, y	Total Park Size, x
95	39,334	96	103,289
95	324	71	130,023
102	17,315	76	16,068
69	8,244	112	3,286
67	620,231	43	24,089
77	43,501	87	6,309
81	8,625	131	14,502
116	31,572	138	62,595
51	14,276	80	23,666
36	21,094	52	35,833

a. Construct a scatterplot of these data.
b. Find the equation of the least squares regression line. Do you think this line would give accurate predictions? Explain.

c. Delete the observation with the largest x value from the data set and recalculate the equation of the least squares regression line. Does this observation have a big effect on the equation of the line?

4.55 The article referenced in the previous exercise also gave data on the percentage of operating costs covered by park revenues for the 2007–2008 fiscal year.

Number of Employees, x	Percent of Operating Cost Covered by Park Revenues, y	Number of Employees, x	Percent of Operating Cost Covered by Park Revenues, y
95	37	96	53
95	19	71	31
102	32	76	35
69	80	112	108
67	17	43	34
77	34	87	97
81	36	131	62
116	32	138	36
51	38	80	36
36	40	52	34

a. Find the equation of the least squares regression line relating y = percent of operating costs covered by park revenues and x = number of employees.
b. Calculate the values of r^2 and s_e. Based on these values, do you think that the least squares regression line does a good job of describing the relationship between percentage of operating costs covered by park revenues and number of employees? Explain.
c. The following graph is a scatterplot of these data. The least squares line is also shown. Which observations are outliers? Do the observations with the largest residuals correspond to the park districts with the largest number of employees?

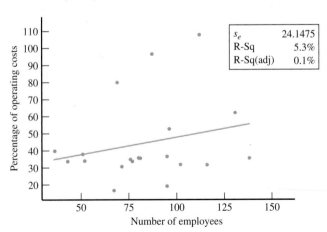

Fitted Line Plot
Percentage of operating costs = 27.71 + 0.2011 number of employees

s_e	24.1475
R-Sq	5.3%
R-Sq(adj)	0.1%

4.56 No tortilla chip lover likes soggy chips, so it is important to identify characteristics of the production process that produce chips with an appealing texture. The accompanying data on x = frying time (in seconds) and y = moisture content (%) appeared in the paper, **"Thermal and Physical Properties of Tortilla Chips as a Function of Frying Time"** (*Journal of Food Processing and Preservation* **[1995]: 175–189**):

Frying time (x):	5	10	15	20	25	30	45	60
Moisture content (y):	16.3	9.7	8.1	4.2	3.4	2.9	1.9	1.3

Construct a scatterplot of these data. Would you recommend using the least squares regression line to summarize the relationship between moisture content and frying time? Explain why or why not.

SECTION 4.4   Describing Linear Relationships and Making Predictions—Putting It All Together

Now that you have considered all of the parts of a linear regression analysis, you can put all the parts together. The steps in a linear regression analysis are summarized in the accompanying box.

AP* EXAM TIP

In Section 4.5, you will see methods for modeling relationships that are not approximately linear.

Steps in a Linear Regression Analysis

Given a bivariate numerical data set consisting of observations on a response variable y and a predictor variable x:

Step 1 Summarize the data graphically by constructing a scatterplot.
Step 2 Based on the scatterplot, decide if it looks like there is an approximate linear relationship between x and y. If so, proceed to the next step.
Step 3 Find the equation of the least squares regression line.
Step 4 Construct a residual plot and look for any patterns or unusual features that may indicate that a line is not the best way to summarize the relationship between x and y. If none are found, proceed to the next step.
Step 5 Compute the values of s_e and r^2 and interpret them in context.
Step 6 Based on what you have learned from the residual plot and the values of s_e and r^2, decide whether the least squares regression line is useful for making predictions. If so, proceed to the last step.
Step 7 Use the least squares regression line to make predictions.

Let's return to the Chapter Preview example to see how following these steps can help you learn about the age of an unidentified crime victim.

Example 4.15   Revisiting Help for Crime Scene Investigators

One of the tasks that forensic scientists face is estimating the age of an unidentified crime victim. Prior to 2010, this was usually done by analyzing teeth and bones, and the resulting estimates were not very reliable. In a groundbreaking study described in the paper **"Estimating Human Age from T-Cell DNA Rearrangements"** (*Current Biology* **[2010]**), scientists examined the relationship between age and a blood measurement. They recorded age and the blood measurement for 195 people ranging in age from a few weeks to 80 years. Because the scientists were interested in predicting age using the blood measurement, the response variable is y = age, and the predictor variable is x = blood measurement.

Step 1: The scientists first constructed a scatterplot of the data, which is shown in Figure 4.30.

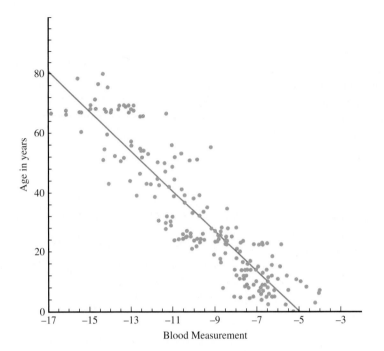

FIGURE 4.30
Scatterplot of age versus blood measurement

Step 2: Based on the scatterplot, it does appear that there is a reasonably strong negative linear relationship between age and the blood measurement. The scientists also reported that the correlation coefficient for this data set was $r = -0.92$, which is consistent with the strong negative linear pattern in the scatterplot.

Step 3: The scientists calculated the equation of the least squares regression line to be

$$\hat{y} = -33.65 - 6.74x$$

Step 4: A residual plot constructed from these data showed a few observations with large residuals, but these observations were not far removed from the rest of the data in the x direction. These observations were not judged to be influential. Also, there were no unusual patterns in the residual plot that would suggest a nonlinear relationship between age and the blood measurement.

Step 5: The values of s_e and r^2 were

$$s_e = 8.9 \qquad \text{and} \qquad r^2 = 0.835$$

This means that approximately 83.5% of the variability in age can be explained by the linear relationship between age and blood measurement. The value of s_e tells you that if the least squares regression line is used to predict age from the blood measurement, a typical difference between the predicted age and the actual age would be about 9 years.

Step 6: Based on the residual plot, the large value of r^2, and the relatively small value of s_e, the scientists proposed using the blood measurement and the least squares regression line as a way to estimate ages of crime victims.

Step 7: To illustrate predicting age, suppose that a blood sample is taken from an unidentified crime victim and that the value of the blood measurement is determined to be -10. The predicted age of the victim would be

$$\hat{y} = -33.65 - 6.74\,(-10)$$
$$= -33.65 - (-67.4)$$
$$= 33.75 \text{ years}$$

Because a typical prediction error is about 9 years, remember that this is just an estimate of the actual age.

SECTION 4.5 Modeling Nonlinear Relationships

When the points in a scatterplot exhibit a linear pattern and the residual plot does not reveal any problems with the linear fit, the least squares regression line is a sensible way to summarize the relationship between two numerical variables. But what should you do if a scatterplot or residual plot exhibits a curved pattern, indicating a more complicated relationship between x and y? When this happens, you should consider using a nonlinear function to summarize the relationship between x and y. The process you will use is similar to what you have done for linear models, but now you will use a curve instead of a line. After choosing an appropriate curve, you will evaluate the fit of the model and use a residual plot to assess the appropriateness of the nonlinear model. If you decide the model is a useful summary of the relationship between x and y, you can then use the model equation to make predictions.

Choosing a Nonlinear Function to Describe a Relationship

Once you have decided to use a nonlinear function to describe the relationship between two variables x and y, you need to decide what type of function should be considered. Some common functions that are used to model relationships are shown in Figure 4.31.

Example 4.16 Fishing Bears

The article "Quantifying Spatiotemporal Overlap of Alaskan Brown Bears and People" (*Journal of Wildlife Management* [2005]: 810–817) describes a study of the effect of human activity on bears. The researchers wondered if sport fishing and boating might be limiting bears' access to salmon. They collected data on

x = number of days from the beginning of June 2003 (June 1st = 1)

and

y = total fishing time (in bear-hours) for the day at a particular location in Alaska

For example, the data pair (33, 45.7) corresponds to the day that is 32 days after June 1 (which is July 2). On this day, the total fishing time for bears was 45.7 bear-hours. This is the sum of the number of hours spent fishing for all bears fishing at this location on July 2, 2003.

Date (x) (June 1 = 1)	Fishing Time (bear-hours)	Date (x) (June 1 = 1)	Fishing Time (bear-hours)	Date (x) (June 1 = 1)	Fishing Time (bear-hours)
11	11.3	24	11.3	40	22.0
12	15.1	25	18.4	41	26.0
13	6.6	27	16.2	44	10.5
14	12.9	28	19.5	45	18.6
15	12.1	31	35.8	46	21.1
17	18.1	32	37.1	49	11.9
18	20.9	33	45.7	50	13.7
19	17.6	36	34.8	51	13.7
20	11.0	37	25.6	54	6.3
21	24.6	38	26.7	55	1.8

Function	Equation	Looks Like

(a) Quadratic $\hat{y} = a + b_1 x + x^2$

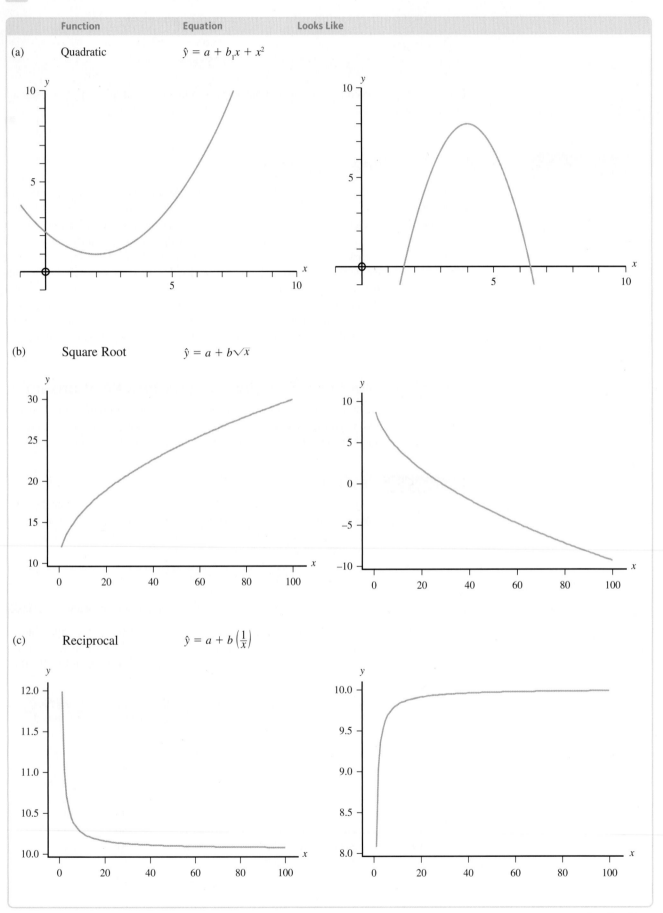

(b) Square Root $\hat{y} = a + b\sqrt{x}$

(c) Reciprocal $\hat{y} = a + b\left(\frac{1}{x}\right)$

(continued)

(d)   Log $\hat{y} = a + b\,\ln(x)$

(e)   Exponential $\hat{y} = e^{a+bx}$

(f)   Power $\hat{y} = ax^{b}$

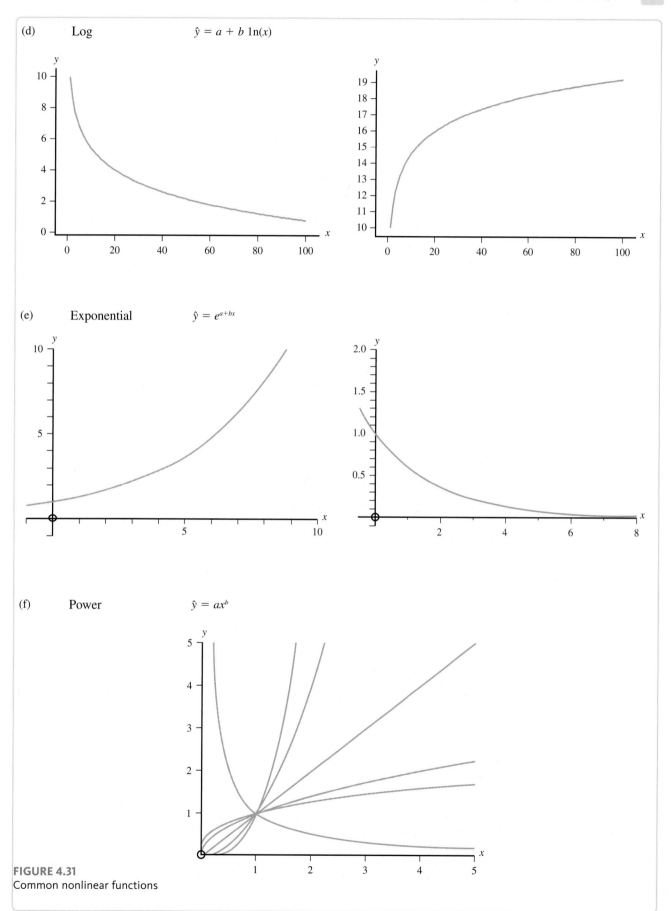

FIGURE 4.31
Common nonlinear functions

Figure 4.32 is a scatterplot of these data.

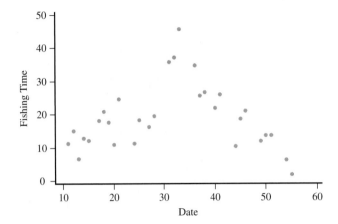

FIGURE 4.32
Scatterplot of fishing time
versus date

The pattern in the scatterplot is clearly nonlinear. Because the shape of the pattern looks like a quadratic curve (see Figure 4.31(a)), a quadratic function could be used to describe this relationship. The model to consider is

$$\hat{y} = a + b_1 x + b_2 x^2$$

Once you have decided on a potential nonlinear model, such as quadratic or exponential, the next step is to estimate the coefficients in the model. For example, for the quadratic model $\hat{y} = a + b_1 x + b_2 x^2$, you would need to use the sample data to determine appropriate values for a, b_1, and b_2. The process used to estimate the coefficients in a nonlinear model is a bit different for quadratic models than for the other nonlinear models listed in Figure 4.31. Let's begin by considering quadratic models.

Quadratic Regression Models

The general form of the quadratic regression model is $\hat{y} = a + b_1 x + b_2 x^2$. What are the best choices for the values of a, b_1, and b_2? In fitting a line to data, the principle of least squares was used to determine the values of the slope and intercept for the "best fit" line. Least squares can also be used to fit a quadratic function. The deviations, $y - \hat{y}$, are still represented by vertical distances in the scatterplot, but now they are vertical distances from the points to a parabola (the graph of a quadratic function) rather than to a line, as shown in Figure 4.33. Values for the coefficients in the quadratic function are then chosen to make the sum of squared deviations as small as possible.

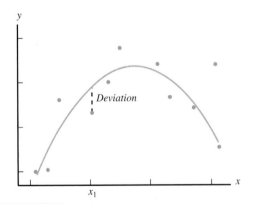

FIGURE 4.33
Deviation for a quadratic function

For a quadratic regression, the least squares estimates of a, b_1 and b_2 are those values that minimize the sum of squared deviations, $\sum(y - \hat{y})^2$, where $\hat{y} = a + b_1 x + b_2 x^2$.

For quadratic regression, a measure that is useful for assessing fit is

$$R^2 = 1 - \frac{SSResid}{SSTo}$$

where $SSResid = \sum(y - \hat{y})^2$. The measure R^2 is defined in a way similar to r^2 for linear regression model and is interpreted in a similar fashion. The lowercase notation r^2 is used only with linear regression to emphasize the relationship between r^2 and the correlation coefficient, r, in the linear case. For nonlinear models, an uppercase R^2 is used.

The general expressions for computing the least squares quadratic regression estimates are somewhat complicated, so a statistical software package or a graphing calculator will be used to do the computations.

Example 4.17   Fishing Bears Revisited—Fitting a Quadratic Model

For the bear fishing data of Example 4.16, the scatterplot showed a curved pattern. If a least squares line is fit to these data, it is not surprising that the line does not do a good job of describing the relationship ($r^2 = 0.000$, and $s_e = 10.1$). Both the scatterplot and the residual plot show a distinct curved pattern (see Figure 4.34).

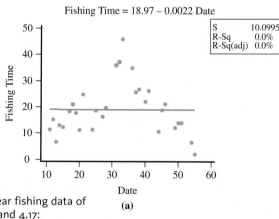

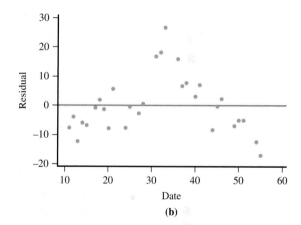

FIGURE 4.34
Plots for the bear fishing data of Examples 4.16 and 4.17:
(a) least squares regression line;
(b) residual plot.

JMP output resulting from fitting the least squares quadratic regression to these data is shown in Figure 4.35.

Polynomial Fit Degree=2

Usage = −20.96713 + 2.9957567* Date − 0.0463151* Date^2

Summary of Fit

RSquare	0.555541
RSquare Adj	0.522618
Root Mean Square Error	6.856681
Mean of Response	18.89667
Observations (or Sum Wgts)	30

Analysis of Variance

Parameter Estimates

Term	Estimate	Std Error	t Ratio	Prob>[t]
Intercept	−20.96713	7.567052	−2.77	0.0100*
Date	2.9957567	0.524223	5.71	<.0001*
Date^2	−0.046315	0.007973	−5.81	<.0001*

FIGURE 4.35
JMP output for the quadratic regression of Example 4.17

From the JMP output the least squares quadratic is

$$\hat{y} = -20.967 + 2.996x - 0.046x^2$$

The curve and the corresponding residual plot for the quadratic regression are shown in Figure 4.36. Notice that there is no strong pattern in this residual plot, unlike the linear case. For the quadratic regression, $R^2 = 0.556$ (as opposed to essentially zero for the linear model). This means that 55.6% of the variability in the bear fishing time can be explained by an approximate quadratic relationship between fishing time and date.

FIGURE 4.36
Quadratic regression of Example 4.17:
(a) scatterplot with curve;
(b) residual plot.

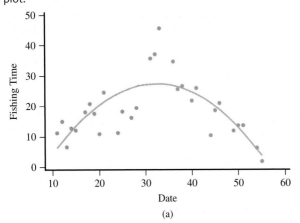

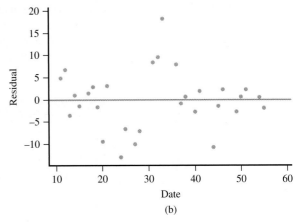

(a)

(b)

To use the quadratic regression model to make a prediction, you would substitute a value of x into the regression equation. For example, the predicted fishing time for June 18 ($x = 18$) is

$$\hat{y} = -20.967 + 2.996x - 0.046x^2$$
$$= -20.967 + 2.996(18) - 0.046(18)^2$$
$$= -20.967 + 53.928 - 14.904$$
$$= 18.057 \text{ bear-hours}$$

Other Nonlinear Regression Models: Using Transformations

Now that you know how to use a quadratic regression model to describe relationships that look like the curves in Figure 4.31(a), let's consider some other nonlinear models. The square root function of Figure 4.31(b) is

$$\hat{y} = a + b\sqrt{x}$$

Notice that if you define a new variable x' where

$$x' = \sqrt{x}$$

the square root model $\hat{y} = a + b\sqrt{x}$ can be written as $\hat{y} = a + bx'$, so y is a linear function of x'. This suggests that if a scatterplot of (x, y) pairs looks like the curve in Figure 4.31(b), a scatterplot of (x', y) pairs should look linear. You can then use a line to describe the relationship between y and x'—something you already know how to do. This is illustrated in the following example.

Example 4.18   River Water Velocity and Distance from Shore

As fans of white-water rafting know, a river flows more slowly close to its banks (because of friction between the riverbank and the water). To study the nature of the relationship between water velocity and the distance from the shore, data were gathered on velocity (in centimeters per second) of a river at different distances (in meters) from the bank. Suppose that the resulting data were as follows:

Distance	0.5	1.5	2.5	3.5	4.5	5.5	6.5	7.5	8.5	9.5
Velocity	22.00	23.18	25.48	25.25	27.15	27.83	28.49	28.18	28.50	28.63

A scatterplot of the data exhibits a curved pattern, as seen in both the scatterplot (Figure 4.37(a)) and the residual plot from a linear fit (Figure 4.37(b)).

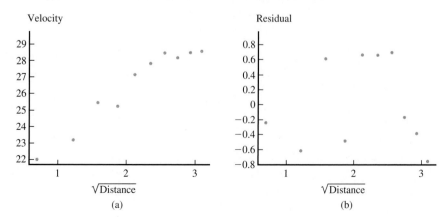

FIGURE 4.37
Plots for the data of
Example 4.18:
(a) scatterplot of the river data;
(b) residual plot from linear fit.

Because the scatterplot of the (x, y) pairs looks like the curve for the square root function in Figure 4.31(b), begin by replacing each x value by its square root:

$$x' = \sqrt{x}$$

This is called transforming the x values. The original and transformed data are shown in Table 4.3.

TABLE **4.3** **Original and transformed
data of Example 4.18**

Original Data		Transformed Data	
x	y	x'	y
0.5	22.00	0.7071	22.00
1.5	23.18	1.2247	23.18
2.5	25.48	1.5811	25.48
3.5	25.25	1.8708	25.25
4.5	27.15	2.1213	27.15
5.5	27.83	2.3452	27.83
6.5	28.49	2.5495	28.49
7.5	28.18	2.7386	28.18
8.5	28.50	2.9155	28.50
9.5	28.63	3.0822	28.63

Figure 4.38(a) shows a scatterplot of y versus x' (or equivalently y versus $\sqrt{x}$). The pattern of points in this plot looks linear, and so you can fit the least squares regression line using the transformed data. The Minitab output from this regression is as follows:

Regression Analysis
The regression equation is
velocity = 20.1 + 3.01 sqrt distance

Predictor	Coef	SE Coef	T	P
Constant	20.1103	0.6097	32.99	0.000
sqrt distance	3.0085	0.2726	11.03	0.000

S = 0.629242   R-Sq = 93.8%   R-Sq(adj) = 93.1%

FIGURE 4.38
Plots for the transformed data of
Example 4.18:
(a) scatterplot of y versus x';
(b) residual plot resulting from a
linear fit to the transformed data.

The resulting regression equation is

$$\hat{y} = 20.1 + 3.01x'$$

or, equivalently,

$$\hat{y} = 20.1 + 3.01\sqrt{x}$$

This model is evaluated using R^2, s_e, and a residual plot. The residual plot (shown in Figure 4.38(b)) does not have any patterns or unusual features. The values of R^2 and s_e (see the Minitab output) indicate that a line is a reasonable way to describe the relationship between y and x'.

To predict velocity of the river at a distance of 9 meters from shore, you first compute $x' = \sqrt{x} = \sqrt{9} = 3$ and then use the least squares regression line to obtain a predicted y value:

$$\hat{y} = 20.1 + 3.01x' = 20.1 + (3.01)(3) = 29.13 \text{ cm per second}$$

A general strategy for fitting a nonlinear model is to find a way to transform the x and/or y values so that a scatterplot of the transformed data has a linear appearance. A **transformation** (sometimes called a re-expression) involves using a simple function of a variable in place of the variable itself (like $x' = \sqrt{x}$ in Example 4.18). Common transformations involve taking square roots, logarithms, or reciprocals.

The nonlinear models (other than the quadratic model) given in Figure 4.31 (square root, reciprocal, log, exponential, and power) can all be fit by transforming variables and then fitting a line to the transformed data. It is convenient to separate these models into two types:

1. Models that involve transforming *only* the x variable (square root, reciprocal, and log).

2. Models that involve transforming the y variable (exponential and power).

Models That Involve Transforming Only *x*

The square root, reciprocal, and log models all have the form

$$\hat{y} = a + b(\text{some function of } x)$$

where the function of x is square root, reciprocal, or log. Notice that these models all describe a linear relationship between y and a function of x. This suggests that if the pattern in the scatterplot of (x, y) pairs looks like one of the curves in Figure 4.31(b) – (d), an appropriate transformation of the x values should result in transformed data that shows a linear pattern.

Model	Transformation
Square root	$x' = \sqrt{x}$
Reciprocal	$x' = \dfrac{1}{x}$
Log	$x' = \ln(x)$

Example 4.19 Electromagnetic Radiation

Is electromagnetic radiation from phone antennae associated with declining bird populations? This is one of the questions investigated by the authors of the paper "The Urban Decline of the House Sparrow (*Passer domesticus*): A Possible Link with Electromagnetic Radiation" (*Electromagnetic Biology and Medicine* **[2007]: 141–151**). The accompanying data on x = electromagnetic field strength (Volts per meter) and y = sparrow density (sparrows per hectare) were read from a graph in the paper.

Field Strength	Sparrow Density	Field Strength	Sparrow Density
0.11	41.71	0.90	16.29
0.20	33.60	1.20	16.97
0.29	24.74	1.30	12.83
0.40	19.50	1.41	13.17
0.50	19.42	1.50	4.64
0.61	18.74	1.80	2.11
1.01	24.23	1.90	0.00
1.10	22.04	3.01	0.00
0.70	16.29	3.10	14.69
0.80	14.69	3.41	0.00

A scatterplot of these data is shown in Figure 4.39.

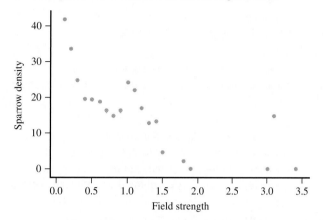

FIGURE 4.39
Scatterplot of the sparrow data

Notice that there is a curved pattern in the scatterplot that has a shape that is similar to the first graph in Figure 4.31(d). This suggests that a reasonable choice for describing the relationship between sparrow density and field strength is a log model

$$\hat{y} = a + b\ln(x)$$

To fit this model, begin by transforming the x values using

$$x' = \ln(x)$$

In this text, natural logs (denoted by ln or $\log_e$) are used, but in practice, either the natural log or log base 10 (denoted by log or $\log_{10}$) can be used. The transformed data is given here.

ln(Field Strength)	Sparrow Density	ln(Field Strength)	Sparrow Density
−2.207	41.71	−0.105	16.29
−1.609	33.60	0.182	16.97
−1.238	24.74	0.262	12.83
−0.916	19.50	0.344	13.17
−0.693	19.42	0.405	4.64
−0.494	18.74	0.588	2.11
0.001	24.23	0.642	0.00
0.095	22.04	1.102	0.00
−0.357	16.29	1.131	14.69
−0.223	14.69	1.227	0.00

A scatterplot of the (x', y) pairs is shown in Figure 4.40. The pattern in this plot looks linear, so you can find the least squares regression line using the transformed data. From the resulting Minitab output (Figure 4.41), the equation of the least squares regression line is

$$\hat{y} = 14.8 - 10.5x'$$

or, in terms of x

$$\hat{y} = 14.8 - 10.5\ln(x)$$

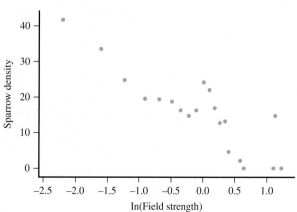

FIGURE 4.40
Scatterplot of the transformed
data of Example 4.19.

Regression Analysis: Sparrow Density Versus In (Field Strength)

The regression equation is
Sparrow Density = 14.8 − 10.5 ln (Field Strength)

Predictor	Coef	SE Coef	T	P
Constant	14.805	1.238	11.96	0.000
ln (Field Strength)	−10.546	1.389	−7.59	0.000

S = 5.50641   R−Sq = 76.2%   R−Sq(adj) = 74.9%

FIGURE 4.41
Minitab output for the log model of Example 4.19.

The value of R^2 for this model is 0.762 and $s_e = 5.5$. A residual plot from the least squares regression line fit to the transformed data is shown in Figure 4.42. There are no apparent patterns or unusual features in the residual plot, so it appears that the log model is a reasonable choice for describing the relationship between sparrow density and field strength.

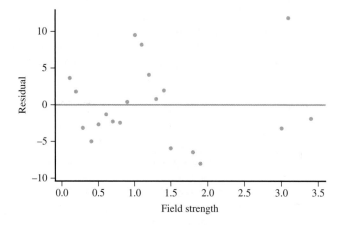

FIGURE 4.42
Residual plot for the log model fit to the sparrow data of Example 4.19.

This model can now be used to predict sparrow density from field strength. For example, if the field strength is 1.6 Volts per meter, you would predict that sparrow density would be

$$\hat{y} = 14.8 - 10.5x'$$
$$= 14.8 - 10.5 \ln(x)$$
$$= 14.8 - 10.5 \ln(1.6)$$
$$= 14.8 - 10.5 (0.470)$$
$$= 9.685 \text{ sparrows per hectare}$$

Models That Involve Transforming y

There are two nonlinear models in Figure 4.31 that have not yet been considered—the exponential and power models. Let's begin by considering the exponential model

$$y = e^{a + bx}$$

Using properties of logarithms, it follows that

$$\ln(y) = \ln e^{a + bx}$$
$$= a + bx$$

For the exponential model, $\ln(y)$ is a linear function of x. This means that if the pattern in a scatterplot of the (x, y) pairs looks like the exponential curves of Figure 4.31(e), a scatterplot of the (x, y') pairs, where

$$y' = \ln y$$

should look linear.

Now consider the power model of Figure 4.31(f),

$$y = ax^b$$

Again using properties of logarithms,

$$\ln y = \ln(ax^b)$$
$$= \ln a + \ln x^b (ax^b)$$
$$= \ln a + b\ln x$$

If a power model is an appropriate way to describe the relationship between x and y, a linear model would describe the relationship between $x' = \ln x$ and $y' = \ln y$.

Model	Transformation
Exponential	$y' = \ln y$
Power	$x' = \ln x$
	$y' = \ln y$

Example 4.20   Loons in Acidic Lakes

A study of factors that affect the survival of loon chicks is described in the paper "Does Prey Biomass or Mercury Exposure Affect Loon Chick Survival in Wisconsin?" (*The Journal of Wildlife Management* [2005]: 57–67). In this study, a relationship between the pH of lake water and blood mercury level in loon chicks was observed. The researchers thought that it is possible that the pH of the lake water could be related to the type of fish that the loons ate. The accompanying data (Table 4.4) on x = lake pH and y = blood mercury level (mg/g) for 37 loon chicks from different lakes in Wisconsin were read from a graph in the paper. A scatterplot is shown in Figure 4.43(a).

TABLE 4.4   Data and Transformed Loon Data from Example 4.20

Lake pH (x)	Blood Mercury Level (y)	$y' = \ln(y)$
5.28	1.10	0.0953
5.69	0.76	−0.2744
5.56	0.74	−0.3011
5.51	0.60	−0.5108
4.90	0.48	−0.7340
5.02	0.43	−0.8440
5.02	0.29	−1.2379
5.04	0.09	−2.4079
5.30	0.10	−2.3026
5.33	0.20	−1.6094
5.64	0.28	−1.2730
5.83	0.17	−1.7720
5.83	0.18	−1.7148
6.17	0.55	−0.5978
6.22	0.43	−0.8440
6.15	0.40	−0.9163
6.05	0.33	−1.1087
6.04	0.26	−1.3471
6.24	0.18	−1.7148
6.30	0.16	−1.8326
6.80	0.45	−0.7985
6.58	0.30	−1.2040
6.65	0.28	−1.2730
7.06	0.22	−1.5141
6.99	0.21	−1.5606
6.97	0.13	−2.0402
7.03	0.12	−2.1203
7.20	0.15	−1.8971
7.89	0.11	−2.2073
7.93	0.11	−2.2073
7.99	0.09	−2.4079
7.99	0.06	−2.8134
8.30	0.09	−2.4079
8.42	0.09	−2.4079
8.42	0.04	−3.2189
8.95	0.12	−2.1203
9.49	0.14	−1.9661

The pattern in this scatterplot is typical of exponential decay and looks like the second curve in Figure 4.31(e). The change in y as x increases is much smaller for large x values than for small x values. You can see that a change of 1 in pH is associated with a much larger change in blood mercury level in the part of the plot where the x values are small than in the part of the plot where the x values are large.

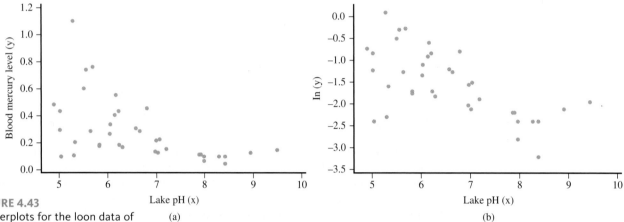

FIGURE 4.43
Scatterplots for the loon data of Example 4.20: (a) scatterplot of original data; (b) scatterplot of transformed data with $y' = \ln y$.

To fit an exponential model, you begin by transforming the y values using $y' = \ln y$. In addition to the original data, Table 4.4 also displays the transformed data. A scatterplot of the (x, y') pairs is shown in Figure 4.43(b). The pattern in the scatterplot of the transformed data looks linear, so it is reasonable to use the least squares regression line to describe the relationship between y' and x.

The following Minitab output shows the result of fitting the least-squares line to the transformed data:

The regression equation is
$\ln(y) = 1.06 - 0.396$ Lake pH (x)

Predictor	Coef	SE Coef	T	P
Constant	1.0550	0.5535	1.91	0.065
Lake pH (x)	−0.39564	0.08264	−4.79	0.000

$S = 0.605645$ R-Sq = 39.6%   R-Sq(adj) = 37.8%

The resulting least squares regression line is

$$\hat{y}' = 1.06 - 0.396x$$

or equivalently

$$\widehat{\ln y} = 1.06 - 0.396x$$

Notice that the linear model in Example 4.20 could be used to predict values of $\ln(y)$. To translate this linear model back to the form of the exponential model, you can take antilogarithms of each side of the equation. This is illustrated in Example 4.21.

Example 4.21   Revisiting the Loon Data

For the loon data of Example 4.20, $y' = \ln y$ and the least squares regression line describing the relationship between y' and x was $\hat{y}' = 1.06 - 0.396x$ or $\widehat{\ln y} = 1.06 - 0.396x$. To "undo" the transformation, you can take the antilog of both sides of this equation:

$$e^{\ln y} = e^{1.06 - 0.396x}$$

Using properties of logs and exponents, $e^{\ln y} = e^{1.06 - 0.396x} = (e^{1.06})(e^{-0.396x})$. Then

$$\hat{y} = e^{1.06} e^{-0.396x} = 2.886 e^{-0.396x}$$

The equation $\hat{y} = 2.886e^{-0.396x}$ can be used to predict the y value (blood mercury level) for a given x (lake pH). For example, the predicted blood mercury level when lake pH is 6 is

$$\hat{y} = 2.886e^{-0.396x} = 2.886e^{\,0.396(6)} = 2.886e^{\,-2.376} = 2.886(0.093) = 0.268$$

The process of transforming data, fitting a line to the transformed data, and then undoing the transformation to get an equation for a curved relationship between x and y usually results in a curve that provides a reasonable fit to the sample data. But when y is transformed, this process does not result in the least squares curve for the data. For example, in Example 4.21, a transformation was used to fit the curve $\hat{y} = 2.886e^{-0.396x}$. However, there may be another equation of the form $\hat{y} = ae^{bx}$ that has a smaller sum of squared residuals for the *original* data than the one obtained using transformations. Finding the least squares estimates for a and b for an exponential or power model is difficult. Fortunately, the curves found using transformations usually provide reasonable predictions of y.

Choosing Among Different Possible Nonlinear Models

Often there is more than one reasonable model that could be used to describe a nonlinear relationship between two variables. When this is the case, how do you choose a model? The choice often involves considering how well the model fits the data and what scientific theory suggests the form of the relationship should be. In the absence of any scientific theory, you will want to choose a model that has small residuals (small s_e) and accounts for a large proportion of the variability in y (large R^2). On the other hand, if scientific theory suggests that the relationship should have a particular form, you would choose a model that is consistent with the suggested form. The following two examples illustrate these two different situations.

Example 4.22   Look for Salamanders Under Those Rocks

The paper "The Relationship Between Rock Density and Salamander Density in a Mountain Stream" (*Herpetologica* [1987]: 357–361) describes a study of factors that influence population density of salamanders. In this study, researchers created a range of salamander habitats at different locations in a small stream in the Appalachian Mountains by using different sized rocks and pebbles. Three months later, they returned to measure salamander density. This resulted in data on

$$x = \text{rock density (rocks per 1.4 square meters)}$$

and

$$y = \text{salamander density (salamanders per 1.4 square meters)}$$

A scatterplot of these data is shown in Figure 4.44.

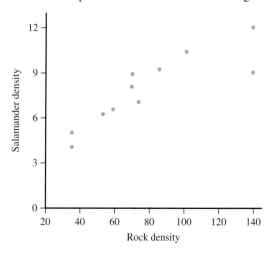

FIGURE 4.44
Scatterplot of salamander density versus rock density.

Looking at the curves in Figure 4.31, you might think about using a square root model or a reciprocal model. The researchers chose the reciprocal regression model, $y = a + b\left(\frac{1}{x}\right)$, for two reasons: (1) it provided a good fit to the data, and (2) the reciprocal model made sense scientifically. The investigators felt that there would be an upper limit to the population density, since the streambed is a nonrenewable resource and the stream therefore had a limit in the number of salamanders that could be sustained. This limit is known as the "carrying capacity" of an environment and is estimated by the value of the intercept, a, in the reciprocal model.

Using the transformation $x' = \frac{1}{x}$, a least squares regression line was fit using the transformed data. The resulting model was

$$y = 12.37 - 292.6x' = 12.37 - 292.6\frac{1}{x}$$

The value of R^2 for this model is $R^2 = 0.82$. Figure 4.45(a) shows a scatterplot of the transformed data and the least squares regression line. Figure 4.45(b) is a scatterplot of the original data that also shows the curve corresponding to the reciprocal model.

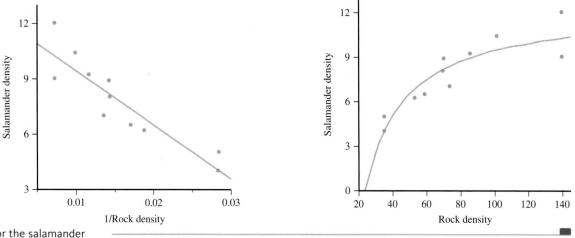

FIGURE 4.45
Scatterplots for the salamander data of Example 4.22:
(a) transformed data;
(b) original data.

Example 4.23   How Old Is That Lobster?

Can you tell how old a lobster is by its size? This question was investigated by the authors of a paper that appeared in the *Biological Bulletin* (August, 2007). Researchers measured carapace (the exterior shell) length (in mm) of 27 laboratory-raised lobsters of known age. The data on $x =$ carapace length and $y =$ age (in years) in Table 4.5 were read from a graph in the paper.

TABLE 4.5   Original and Transformed Data for Example 4.23

y	x	ln(x)	ln(y)
1.00	63.32	4.148	0.000
1.00	67.50	4.212	0.000
1.00	69.58	4.242	0.000
1.00	73.41	4.296	0.000
1.42	79.32	4.373	0.351
1.42	82.80	4.416	0.351
1.42	85.59	4.450	0.351
1.82	105.07	4.655	0.599
1.82	107.16	4.674	0.599
1.82	117.25	4.764	0.599
2.18	109.24	4.694	0.779
2.18	110.64	4.706	0.779
2.17	118.99	4.779	0.775
2.17	122.81	4.811	0.775
2.33	138.47	4.931	0.846
2.50	133.95	4.897	0.916
2.51	125.25	4.830	0.920
2.50	123.51	4.816	0.916
2.93	146.82	4.989	1.075
2.92	139.17	4.936	1.072
2.92	136.73	4.918	1.072
2.92	122.81	4.811	1.072
3.17	142.30	4.958	1.154
3.41	152.73	5.029	1.227
3.42	145.78	4.982	1.230
3.75	148.21	4.999	1.322
4.08	152.04	5.024	1.406

The scatterplot of the data in Figure 4.46 shows a clear curved pattern. Considering the curves in Figure 4.31, there are several models that could be considered, including the exponential model and the power model.

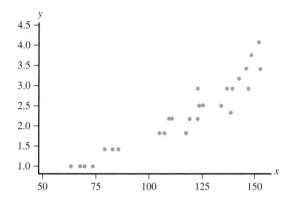

FIGURE 4.46
Scatterplot of age versus carapace length

In this situation, there is no scientific theory to suggest what the form of the relationship might be. A reasonable way to proceed is to fit both models and then choose the one that provides the best fit. Let's start with the exponential model. Using the transformation

$$y' = \ln y$$

results in the transformed y values given in Table 4.5. A scatterplot of the (x, y') pairs is shown in Figure 4.47(a). Notice that the relationship between y' and x looks linear. Minitab was used to fit the least squares regression line to the (x, y') data, resulting in the following output:

Regression Analysis: ln(y) versus x
The regression equation is
$\ln(y) = -0.927 + 0.0145 x$

Predictor	Coef	SE Coef	T	P
Constant	−0.92704	0.08692	−10.67	0.000
x	0.0144903	0.0007311	19.82	0.000

$S = 0.105917$ $R\text{-}Sq = 94.0\%$ $R\text{-}Sq(adj) = 93.8\%$

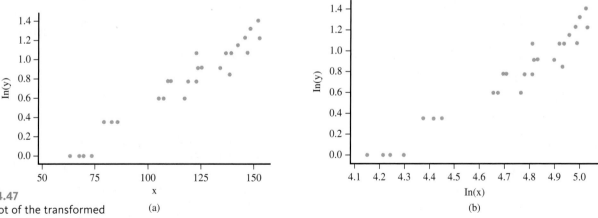

FIGURE 4.47
Scatterplot of the transformed lobster data: (a) x = carapace length, y' = ln(age); (b) x' = ln(carapace length), y' = ln(age).

(a)

(b)

Figure 4.47(b) shows a scatterplot of the transformed data that would be used to fit a power model ($x' = \ln$, $y' = \ln y$). The pattern in this scatterplot also looks linear. Minitab output from fitting a least squares regression line using the (x', y') data is shown here:

Regression Analysis: ln(y) versus ln(x)

The regression equation is

$\ln(y) = -6.32 + 1.50 \ln(x)$

Predictor	Coef	SE Coef	T	P
Constant	-6.3205	0.3687	-17.14	0.000
ln(x)	1.49866	0.07805	19.20	0.000

$S = 0.109120$ R-Sq = 93.6%   R-Sq(adj) = 93.4%

Notice that both the exponential model and the power model have large R^2 values. Comparing the fit of the two models results in

Model	R^2	S_e
Exponential	0.940	0.106
Power	0.936	0.109

Both models do a good job in describing the relationship between age and carapace length. In the absence of any scientific theory suggesting the form of the model, you would choose the exponential model because it provides a (slightly) better fit.

The exponential model is

$$\hat{y}' = -0.927 + 0.0145x$$

or equivalently

$$\ln \hat{y} = -0.927 + 0.0145x$$

or

$$\hat{y} = e^{-0.927}e^{0.0145x}$$

SECTION 4.5 EXERCISES

Each Exercise Set assesses the following chapter learning objectives: C12, M17, M18, P5

SECTION 4.5 Exercise Set 1

4.57 The following data on x = frying time (in seconds) and y = moisture content (%) appeared in the paper "Thermal and Physical Properties of Tortilla Chips as a Function of Frying Time" (*Journal of Food Processing and Preservation* [1995]: 175–189):

x	5	10	15	20	25	30	45	60
y	16.3	9.7	8.1	4.2	3.4	2.9	1.9	1.3

a. Construct a scatterplot of the data. Would a straight line provide an effective summary of the relationship?

b. Here are the values of $x' = \log(x)$ and $y' = \log(y)$:

x'	0.70	1.00	1.18	1.30	1.40	1.48	1.65	1.78
y'	1.21	0.99	0.91	0.62	0.53	0.46	0.28	0.11

Construct a scatterplot of these transformed data, and comment on the pattern.

c. Based on the accompanying MINITAB output, does the least squares line effectively summarize the relationship between y' and x'?

The regression equation is

$\log(\text{moisture}) = 2.02 - 1.05 \log(\text{time})$

Predictor	Coef	SE Coef	T	P
Constant	2.01780	0.09584	21.05	0.000
log(time)	-1.05171	0.07091	-14.83	0.000

$S = 0.0657067$ R-Sq = 97.3%   R-Sq(adj) = 96.9%

Analysis of Variance

Source	DF	SS	MS	F	P
Regression	1	0.94978	0.94978	219.99	0.000
Residual Error	6	0.02590	0.00432		
Total	7	0.97569			

d. Use the MINITAB output to predict moisture content when frying time is 35 sec.

4.58 The paper "Aspects of Food Finding by Wintering Bald Eagles" (*The Auk* [1983]: 477–484) examined the relationship between the time that eagles spend aloft searching for food (indicated by the percentage of eagles soaring) and relative food availability. The accompanying data were taken from a scatterplot that appeared in this paper. Salmon availability is denoted by x and the percentage of eagles in the air is denoted by y.

x	0	0	0.2	0.5	0.5	1.0
y	28.2	69.0	27.0	38.5	48.4	31.1

x	1.2	1.9	2.6	3.3	4.7	6.5
y	26.9	8.2	4.6	7.4	7.0	6.8

a. Draw a scatterplot for this data set. Would you describe the plot as linear or curved?

b. One possible transformation that might lead to a linear pattern involves taking the square root of both the x and y values. Construct a scatterplot using the variables $\sqrt{x}$ and $\sqrt{y}$. Is this scatterplot straighter than the scatterplot in Part (a)?

c. After considering your scatterplot in Part (a), suggest another transformation that might be used to straighten the original plot. Use your suggestion to construct a scatterplot for your x' and/or y'. Which transformation (the one from this part or the square root transformation from Part (b)) do you prefer? Explain.

4.59 Food intake of grazing animals is limited by the rate grass can be chewed and swallowed, as well as the rate at which food can be digested. The authors of the paper "What Constrains Daily Intake in Thomson's Gazelles?" (*Ecology* [1999]: 2338–2347) observed the grazing activity of captive Thomson's gazelles. They recorded grazing rate (amount of grass eaten, in grams per minute) and biomass of the grazing area (food density, in grams per square meter). Scatterplots of these data with four possible functions that might be used to describe the relationship between grazing rate and biomass are shown in the figure at the bottom of the page. Which of these functions would you recommend? Explain your reasoning in a few sentences.

SECTION 4.5 **Exercise Set 2**

4.60 A study, described in the paper "Prediction of Defibrillation Success from a Single Defibrillation Threshold Measurement" (*Circulation* [1988]: 1144–1149) investigated the relationship between defibrillation success and the energy of the defibrillation shock (expressed as a multiple of the defibrillation threshold). The accompanying data are from this study.

Energy of Shock	Success (%)
0.5	33.3
1.0	58.3
1.5	81.8
2.0	96.7
2.5	100.0

Figure for 4.59

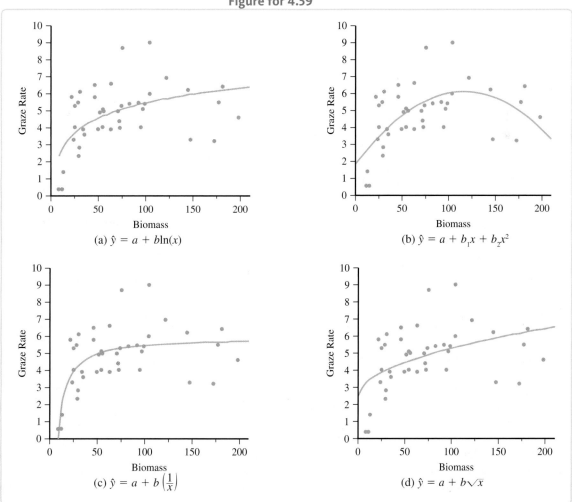

(a) $\hat{y} = a + b\ln(x)$

(b) $\hat{y} = a + b_1 x + b_2 x^2$

(c) $\hat{y} = a + b\left(\frac{1}{x}\right)$

(d) $\hat{y} = a + b\sqrt{x}$

a. Construct a scatterplot of y = success and x = energy of shock. Does the relationship appear to be linear or nonlinear?

b. Fit a least squares line to the given data, and construct a residual plot. Does the residual plot support your conclusion in Part (a)? Explain.

c. Consider transforming the data by leaving y unchanged and using either $x' = \sqrt{x}$ or $x'' = \ln(x)$. Which of these transformations would you recommend? Justify your choice by appealing to appropriate graphical displays.

d. Using the transformation you recommended in Part (c), find the equation of the least squares line that describes the relationship between y and the transformed x.

e. What would you predict success to be when the energy of shock is 1.75 times the threshold level? When it is 0.8 times the threshold level?

4.61 The article "Organ Transplant Demand Rises Five Times as Fast as Existing Supply" (*San Luis Obispo Tribune*, February 23, 2001) included a graph that showed the number of people waiting for organ transplants each year from 1990 to 1999. The following data are approximate values and were read from the graph in the article:

Year	Number Waiting for Transplant (in thousands)
1 (1990)	22
2	25
3	29
4	33
5	38
6	44
7	50
8	57
9	64
10 (1999)	72

a. Construct a scatterplot of the data with y = number waiting for transplant and x = year. Describe how the number of people waiting for transplants has changed over time from 1990 to 1999.

b. Find a transformation of x and/or y that straightens the plot. Construct a scatterplot for your transformed variables.

c. Using the transformed variables from Part (b), fit a least squares line and use it to predict the number waiting for an organ transplant in 2000 (Year 11).

d. The prediction made in Part (c) involves prediction for an x value that is outside the range of the x values in the sample. What assumption must you be willing to make for this to be reasonable? Do you think this assumption is reasonable in this case? Would your answer be the same if the prediction had been for the year 2010 rather than 2000? Explain.

Additional Exercises

4.62 The paper "Population Pressure and Agricultural Intensity" (*Annals of the Association of American Geographers* [1977]: 384–396) reported a positive relationship between population density and agricultural intensity. The following data consist of measures of population density (x) and agricultural intensity (y) for 18 different subtropical locations:

x	1.0	26.0	1.1	101.0	14.9	134.7
y	9	7	6	50	5	100

x	3.0	5.7	7.6	25.0	143.0	27.5
y	7	14	14	10	50	14

x	103.0	180.0	49.6	140.6	140.0	233.0
y	50	150	10	67	100	100

a. Construct a scatterplot of y versus x. Is the scatterplot compatible with the statement of positive relationship made in the paper?

b. Try a scatterplot that uses y and x^2. Does this transformation straighten the plot?

c. Draw a scatterplot that uses $\log(y)$ and x. The $\log(y)$ values, given in order corresponding to the y values, are 0.95, 0.85, 0.78, 1.70, 0.70, 2.00, 0.85, 1.15, 1.15, 1.00, 1.70, 1.15, 1.70, 2.18, 1.00, 1.83, 2.00, and 2.00. How does this scatterplot compare with that of Part (b)?

d. Now consider a scatterplot that uses transformations on both x and y: $\log(y)$ and x^2. Is this effective in straightening the plot? Explain.

4.63 Determining the age of an animal can sometimes be a difficult task. One method of estimating the age of harp seals is based on the width of the pulp canal in the seal's canine teeth. To investigate the relationship between age and the width of the pulp canal, researchers recorded age and canal width in seals of known age. The following data on x = age (in years) and y = canal length (in millimeters) are a portion of a larger data set that appeared in the paper "Validation of Age Estimation in the Harp Seal Using Dentinal Annuli" (*Canadian Journal of Fisheries and Aquatic Science* [1983]: 1430–1441):

x	0.25	0.25	0.50	0.50	0.50	0.75	0.75	1.00
y	700	675	525	500	400	350	300	300

x	1.00	1.00	1.00	1.00	1.25	1.25	1.50	1.50
y	250	230	150	100	200	100	100	125

x	2.00	2.00	2.50	2.75	3.00	4.00	4.00	5.00
y	60	140	60	50	10	10	10	10

x	5.00	5.00	5.00	6.00	6.00
y	15	10	10	15	10

Construct a scatterplot for this data set. Would you describe the relationship between age and canal length as linear? If not, suggest a transformation that might straighten the plot.

SECTION 4.6 Avoid These Common Mistakes

There are a number of ways you can get into trouble when analyzing bivariate numerical data! Here are some things to keep in mind when conducting your own analyses or when reading reports of such analyses:

1.   Correlation does not imply causation. A common media blunder is to infer a cause-and-effect relationship between two variables simply because there is a strong correlation between them. Don't fall into this trap! A strong correlation implies only that the two variables tend to vary together in a predictable way, but there are many possible explanations for why this is occurring other than one variable causing change in the other.

For example, the article **"Ban Cell Phones? You May as Well Ban Talking Instead" (*USA Today*. April 27, 2000)** gave data that showed a strong negative correlation between the number of cell phone subscribers and traffic fatality rates. During the years from 1985 to 1998, the number of cell phone subscribers increased from 200,000 to 60,800,000, and the number of traffic deaths per 100 million miles traveled decreased from 2.5 to 1.6 over the same period. However, based on this correlation alone, the conclusion that cell phone use is the cause of improved road safety is not reasonable!

Similarly, the ***Calgary Herald* (April 16, 2002)** reported that heavy and moderate drinkers earn more than light drinkers or those who do not drink. Based on the correlation between number of drinks consumed and income, the author of the study concluded that moderate drinking "causes" higher earnings. This is obviously a misleading statement, but at least the article goes on to state that there are many possible reasons for the correlation. It could be because better-off men have more discretionary income. Or it might have something to do with stress: Maybe those with stressful jobs tend to earn more after controlling for age, occupation, etc., and maybe they also drink more in response to the stress."

2.   A correlation coefficient near 0 does not necessarily imply that there is no relationship between two variables. Before drawing such a conclusion, carefully examine a scatterplot of the data. Although the variables may be unrelated, it is also possible that there is a strong but nonlinear relationship.

3.   The least squares regression line for predicting y from x is not the same line as the least squares regression line for predicting x from y. The least squares regression line is, by definition, the line that has the smallest possible sum of squared deviations of points from the line *in the y direction* — it minimizes $\sum(y - \hat{y})^2$. This is not generally the same line as the line that minimizes the sum of the squared deviations in the x direction. It is not appropriate, for example, to fit a line to data using $y =$ house price and $x =$ house size and then use the resulting least squares line Price $= a + b$(Size) to predict the size of a house by substituting in a price and then solving for size. Make sure that the response and predictor variables are clearly identified and that the appropriate line is fit.

4.   Beware of extrapolation. You can't assume that a regression line is valid over a wider range of x values. Using the least squares regression line to make predictions outside the range of x values in the data set often leads to poor predictions.

5.   Be careful in interpreting the value of the intercept of the least squares regression line. In many instances interpreting the intercept as the value of y that would be predicted when $x = 0$ is equivalent to extrapolating way beyond the range of the x values in the data set, and this should be avoided unless $x = 0$ is within the range of the data.

6.   Remember that the least squares regression line may be the "best" line (in that it has a smaller sum of squared deviations than any other line), but that doesn't necessarily mean that the line will produce good predictions. Be cautious of predictions based on a least squares regression line without any information about the appropriateness of the regression line, such as s_e and r^2.

7. It is not enough to look at just r^2 or just s_e when evaluating the regression line. Remember to consider both values. These two measures address different aspects of the fit of the line. In general, you would like to have both a small value for s_e (which indicates that deviations from the line tend to be small) and a large value for r^2 (which indicates that the linear relationship explains a large proportion of the variability in the y values).

8. The value of the correlation coefficient, as well as the values for the intercept and slope of the least squares regression line, can be sensitive to influential observations in the data set, particularly if the sample size is small. Because potentially influential observations are those whose x values are far away from most of the x values in the data set, it is important to *always* start with a plot of the data.

CHAPTER ACTIVITIES

ACTIVITY 4.1 EXPLORING CORRELATION AND REGRESSION TECHNOLOGY ACTIVITY (APPLETS)

Open the applet (available at www.cengage.com/states/peck1e) called CorrelationPoints. You should see a screen like the one shown.

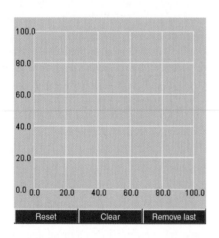

1. Using the mouse, you can click to add points to form a scatterplot. The value of the correlation coefficient for the data in the plot at any given time (once you have two or more points in the plot) will be displayed at the top of the screen. Try to add points to obtain a correlation coefficient that is close to −0.8. Briefly describe the factors that influenced where you placed the points.

2. Reset the plot (by clicking on the Reset bar in the lower-left corner of the screen) and add points trying to produce a data set with a correlation that is close to +0.4. Briefly describe the factors that influenced where you placed the points.

Now open the applet called RegDecomp. You should see a screen that looks like the one shown here:

The black points in the plot represent the data set, and the blue line represents a possible line to describe the relationship between the two variables. The pink lines in the graph represent the deviations of points from the line, and the pink bar on the left-hand side of the display depicts the sum of squared deviations for the line shown.

3. Using your mouse, you can move the line and see how the deviations from the line and the sum of squared deviations change as the line changes. Try to move the line into a position that you think will minimize the sum of squared deviations. When you think you are close, click on the bar that says "Find Best Model" to see how close you were to the actual least squares line.

ACTIVITY 4.2 AGE AND FLEXIBILITY

Materials needed: Yardsticks.

In this activity, you will investigate the relationship between age and a measure of flexibility. Flexibility will be measured by asking a person to bend at the waist as far as possible, extending his or her arms toward the floor. Using a yardstick, measure the distance from the floor to the fingertip closest to the floor.

1. Age and the measure of flexibility just described will be recorded for a group of individuals. Your goal is to determine if there is a relationship between age and this measure of flexibility. What are two reasons why it would not be a good idea to use just the students in your class as the subjects for your study?

2. Working as a class, decide on a reasonable way to collect data.

3. After your class has collected appropriate data, use these data to construct a scatterplot. Comment on the interesting features of the plot. Does it look like there is a relationship between age and flexibility?

4. If there appears to be a linear relationship between age and flexibility, calculate the equation of the least squares regression line.

5. *In the context of this activity*, write a brief description of the danger of extrapolation.

EXPLORING THE BIG IDEAS

In the following exercise, each student in your class will go online to select a random sample of size 20 from a small population consisting of 300 adults.

To learn about the relationship between age and the number of text messages sent in a typical day for this population, go online at www.cengage.com/stats/peck1e and click on the link labeled "Exploring the Big Ideas." Then locate the link for Chapter 4. This link will take you to a web page where you can select a random sample from the population.

Click on the sample button. This selects a random sample of size 20 and will display the age and number of text messages sent for each of the people in your sample. Each student in your class will receive data from a different random sample.

Use the data from your random sample to complete the following.
a. Calculate the value of the correlation coefficient.
b. Based on the value of the correlation coefficient, which of the following best describes the linear relationship between age and the number of text messages sent?
  i. Strong and negative
  ii. Moderate and negative
  iii. Weak and negative
  iv. Weak and positive
  v. Moderate and positive
  vi. Strong and positive
c. What is the value of the slope of the least squares regression line? Round your answer to two decimal places.
d. What is the value of the intercept of the least squares regression line? Round your answer to two decimal places.
e. What is the value of r^2 for this data set? Express your answer as a decimal and round it to two decimal places.
f. Write a sentence interpreting the value of r^2.
g. What is the value of s_e? Round you answer to two decimal places.
h. Based on the values of r^2 and s_e, would you recommend using this regression equation to predict number of text messages sent? Explain why or why not.

If asked to do so by your teacher, bring the equation of the regression line for your sample to class. Your teacher will lead the class through the rest of this exercise.

After students have had a chance to compare regression equations, consider the following questions.
i. Why didn't every student get the same regression equation?
j. If two different students were to use their regression equations to predict the number of text messages sent for a person who is 28 years old, would the predictions be the same? Do you think they would be similar?

k. If every student had obtained a random sample of 50 people instead of 20, do you think the regression equations would have differed more or less from student to student than they did for samples of size 20?

ARE YOU READY TO MOVE ON? CHAPTER 4 REVIEW EXERCISES

All chapter learning objectives are assessed in these exercises. The learning objectives assessed in each exercise are given in parentheses for each exercise.

4.64 (M1, M2)

For each of the scatterplots shown, answer the following questions:

i. Does there appear to be a relationship between x and y?
ii. If so, does the relationship appear to be linear?
iii. If so, would you describe the linear relationship as positive or negative?

Scatterplot 1:

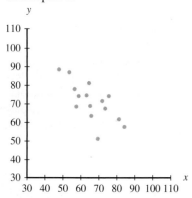

Scatterplot 2:

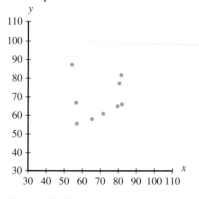

Scatterplot 3:

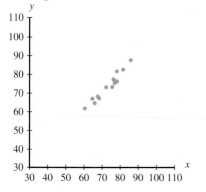

Scatterplot 4:

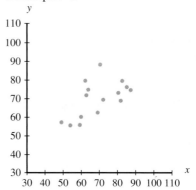

4.65 (M1, M2)

For each of the following pairs of variables, indicate whether you would expect a positive correlation, a negative correlation, or a correlation close to 0. Explain your choice.

a. Price and weight of an apple
b. A person's height and the number of pets he or she has
c. Time spent studying for an exam and score on the exam
d. A person's weight and the time it takes him or her to run one mile

4.66 (C2)

The authors of the paper **"Flat-Footedness Is Not a Disadvantage for Athletic Performance in Children Aged 11 to 15 Years"** (*Pediatrics* [2009]: e386–e392) studied the relationship between y = arch height and scores on five different motor ability tests. They reported the following correlation coefficients:

Motor Ability Test	Correlation between Test Score and Arch Height
Height of counter movement jump	−0.02
Hopping: average height	−0.10
Hopping: average power	−0.09
Balance, closed eyes, one leg	0.04
Toe flexion	0.05

a. Interpret the value of the correlation coefficient between average hopping height and arch height. What does the fact that the correlation coefficient is negative say about the relationship? Do larger arch heights tend to be paired with larger or smaller average hopping heights?

b. The title of the paper suggests that having a small value for arch height (flat-footedness) is not a disadvantage when it comes to motor skills. Do the given correlation coefficients support this conclusion? Explain.

4.67 (C1, C2, M3, M4)
The accompanying data are x = cost (cents per serving) and y = fiber content (grams per serving) for 18 high-fiber cereals rated by *Consumer Reports* (www.consumerreports.org/health).

Cost Per Serving	Fiber Per Serving	Cost Per Serving	Fiber Per Serving
33	7	53	13
46	10	53	10
49	10	67	8
62	7	43	12
41	8	48	7
19	7	28	14
77	12	54	7
71	12	27	8
30	8	58	8

a. Construct a scatterplot of fiber content versus cost. Based on the scatterplot, how would you describe the relationship between fiber content and cost?
b. Compute and interpret the value of the correlation coefficient.
c. The serving size differed for the different cereals, with serving sizes varying from a half cup to one and a quarter cups. Converting price and fiber content to "per cup" rather than "per serving" results in the following data. Is the correlation coefficient for the per cup data greater than or less than the correlation coefficient for the per-serving data?

Cost per Cup	Fiber per Cup	Cost per Cup	Fiber per Cup
44.0	9.3	53.0	13.0
46.0	10.0	53.0	10.0
49.0	10.0	67.0	8.0
62.0	7.0	43.0	12.0
32.8	6.4	48.0	7.0
19.0	7.0	56.0	28.0
77.0	12.0	54.0	7.0
56.8	9.6	54.0	16.0
30.0	8.0	77.3	10.7

4.68 (C3)
Based on data from six countries, the paper "A Cross-National Relationship Between Sugar Consumption and Major Depression?" (*Depression and Anxiety* [2002]: 118–120) concluded that there was a correlation between refined sugar consumption (calories per person per day) and annual rate of major depression (cases per 100 people). The following data are from a graph in the paper:

Country	Sugar Consumption	Depression Rate
Korea	150	2.3
United States	300	3.0
France	350	4.4
Germany	375	5.0
Canada	390	5.2
New Zealand	480	5.7

a. Compute and interpret the correlation coefficient for this data set.
b. Is it reasonable to conclude that increasing sugar consumption leads to higher rates of depression? Explain.
c. What concerns do you have about this study that would make you hesitant to generalize these conclusions to other countries?

4.69 (P1)
The article "Air Pollution and Medical Care Use by Older Americans" (*Health Affairs* [2002]: 207–214) gave data on a measure of pollution (in micrograms of particulate matter per cubic meter of air) and the cost of medical care per person over age 65 for six geographical regions of the United States:
a. Construct a scatterplot for these data.
b. Compute and interpret the value of the correlation coefficient.
c. Write a few sentences summarizing what the scatterplot and value of the correlation coefficient tell you about the relationship between pollution and medical cost per person over age 65.

Region	Pollution	Cost of Medical Care
North	30.0	915
Upper South	31.8	891
Deep South	32.1	968
West South	26.8	972
Big Sky	30.4	952
West	40.0	899

4.70 (C4)
The relationship between hospital patient-to-nurse ratio and various characteristics of job satisfaction and patient care has been the focus of a number of research studies. Suppose x = patient-to-nurse ratio is the predictor variable. For each of the following response variables, indicate whether you expect the slope of the least squares line to be positive or negative and give a brief explanation for your choice.
a. y = a measure of nurse's job satisfaction (higher values indicate higher satisfaction)
b. y = a measure of patient satisfaction with hospital care (higher values indicate higher satisfaction)
c. y = a measure of quality of patient care (higher values indicate higher quality)

4.71 (C5)

For a given data set, the sum of squared deviations from the line $y = 40 + 6x$ is 529.5. For this same data set, which of the following might be the sum of squared deviations from the least squares regression line? Explain your choice.

    i. 308.6
    ii. 529.6
    iii. 617.4

4.72 (C7)

Explain why it can be dangerous to use the least squares regression line to obtain predictions for x values that are substantially larger or smaller than the x values in the data set.

4.73 (M7, M8, M9, M10)

Studies have shown that people who suffer sudden cardiac arrest have a better chance of survival if a defibrillator shock is administered very soon after cardiac arrest. How is survival rate related to the length of time between cardiac arrest and the defibrillator shock being delivered? This question is addressed in the paper "Improving Survival from Sudden Cardiac Arrest: The Role of Home Defibrillators" (www.heartstarthome.com). The accompanying data give y = survival rate (percent) and x = mean call-to-shock time (minutes) for a cardiac rehabilitation center (in which cardiac arrests occurred while victims were hospitalized and so the call-to-shock time tended to be short) and for four other communities of different sizes.

Mean call-to-shock time, x	2	6	7	9	12
Survival rate, y	90	45	30	5	2

a. Find the equation of the least squares line.
b. Interpret the slope of the least squares line in the context of this study.
c. Does it make sense to interpret the intercept of the least squares regression line? If so, give an interpretation. If not, explain why it is not appropriate for this data set.
d. Use the least squares line to predict survival rate for a community with a mean call-to-shock time of 10 minutes.

4.74 (M6, M7)

The following data on sale price, size, and land-to-building ratio for 10 large industrial properties appeared in the paper "Using Multiple Regression Analysis in Real Estate Appraisal" (*Appraisal Journal* [2002]: 424–430):

Property	Sale Price (millions of dollars)	Size (thousands of sq. ft.)	Land-to-Building Ratio
1	10.6	2,166	2.0
2	2.6	751	3.5
3	30.5	2,422	3.6
4	1.8	224	4.7

(continued)

Property	Sale Price (millions of dollars)	Size (thousands of sq. ft.)	Land-to-Building Ratio
5	20.0	3,917	1.7
6	8.0	2,866	2.3
7	10.0	1,698	3.1
8	6.7	1,046	4.8
9	5.8	1,108	7.6
10	4.5	405	17.2

a. If you wanted to predict sale price and you could use either size or land-to-building ratio as the basis for making predictions, which would you use? Explain.
b. Based on your choice in Part (a), find the equation of the least squares regression line for predicting sale price.

4.75 (P3)

Does it pay to stay in school? The report *Trends in Higher Education* (The College Board, 2010) looked at the median hourly wage gain per additional year of schooling in 2007. The report states that workers with a high school diploma had a median hourly wage that was 10% higher than those who had only completed 11 years of school. Workers who had completed 1 year of college (13 years of education) had a median hourly wage that was 11% higher than that of the workers who had completed only 12 years of school. The gain in median hourly wage for each additional year of school is shown in the accompanying table. The entry for 15 years of schooling has been intentionally omitted from the table.

Years of Schooling	2007 Median Hourly Wage Gain for the Additional Year (percent)
12	10
13	11
14	13
16	16
17	18
18	19

a. Use the given data to predict the median hourly wage gain for the fifteenth year of schooling.
b. The actual wage gain for fifteenth year of schooling was 14%. How close was the predicted wage gain percent from Part (a) to the actual value?

4.76 (C10, M13, M14, M15)

The table on the following page gives the number of organ transplants performed in the United States each year from 1990 to 1999 (The Organ Procurement and Transplantation Network, 2003).

a. Construct a scatterplot of these data, and then find the equation of the least squares regression line that describes the relationship between y = number of transplants performed and x = year. Describe how the number of transplants performed has changed over time from 1990 to 1999.

b. Compute the 10 residuals, and construct a residual plot. Are there any features of the residual plot that indicate that the relationship between year and number of transplants performed would be better described by a curve rather than a line? Explain.

Year	Number of Transplants (in thousands)
1 (1990)	15.0
2	15.7
3	16.1
4	17.6
5	18.3
6	19.4
7	20.0
8	20.3
9	21.4
10 (1999)	21.8

4.77   (C10, M14, M15)
Can you tell how old a lobster is by its size? This question was investigated by the authors of a paper that appeared in the *Biological Bulletin* (**August 2007**). Researchers measured carapace (the exterior shell) length of 27 laboratory-raised lobsters of known age. The data on x = carapace length (in mm) and y = age (in years) in the following table were read from a graph that appeared in the paper.

y	x	y	x
1.00	63.32	2.33	138.47
1.00	67.50	2.50	133.95
1.00	69.58	2.51	125.25
1.00	73.41	2.50	123.51
1.42	79.32	2.93	146.82
1.42	82.80	2.92	139.17
1.42	85.59	2.92	136.73
1.82	105.07	2.92	122.81
1.82	107.16	3.17	142.30
1.82	117.25	3.41	152.73
2.18	109.24	3.42	145.78
2.18	110.64	3.75	148.21
2.17	118.99	4.08	152.04
2.17	122.81		

a. Construct a scatterplot of these data, and then find the equation of the least squares regression line that describes the relationship between y = age and x = carapace length. Describe how age varies with carapace length.

b. Using computer software or a graphing calculator, construct a residual plot. Are there any features of the

residual plot that indicate that the relationship between age and carapace length would be better described by a curve rather than a line? Explain.

4.78   (C11, M16)
Cost-to-charge ratio (the percentage of the amount billed that represents the actual cost) for inpatient and outpatient services at 11 Oregon hospitals is shown in the following table (**Oregon Department of Health Services, 2002**). A scatterplot of the data is also shown.

Hospital	Cost-to-Charge Ratio	
	Outpatient Care	Inpatient Care
1	62	80
2	66	76
3	63	75
4	51	62
5	75	100
6	65	88
7	56	64
8	45	50
9	48	54
10	71	83
11	54	100

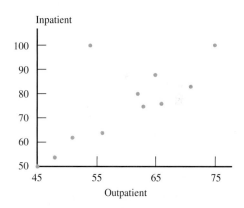

The least-squares regression line with y = inpatient cost-to-charge ratio and x = outpatient cost-to-charge ratio is $\hat{y} = -1.1 + 1.29x$.

a. Is the observation for Hospital 11 an influential observation? Justify your answer.

b. Is the observation for Hospital 11 an outlier? Explain.

c. Is the observation for Hospital 5 an influential observation? Justify your answer.

d. Is the observation for Hospital 5 an outlier? Explain.

4.79   (C8, C9, M11, M12, P2, P4)
The accompanying data on x = average energy density (calories per 100 grams) and y = average cost (in dollars per 100 grams) for eight different food groups are from the paper **"The Cost of U.S. Foods as Related to Their**

Nutritional Value" (*American Journal of Clinical Nutrition* [2010]: 1181–1188). The authors were interested in the relationship between average cost and the average energy density of foods.

Food Group	Average Cost	Average Energy Density
Milk and milk products	0.40	182
Eggs	0.32	171
Dry beans, legumes, nuts, and seeds	0.50	330
Grain products	0.47	337
Fruit	0.28	67
Vegetables	0.33	83
Fats, oils, and salad dressings	0.37	390
Sugars, sweets and beverages	0.40	242

a. Construct a scatterplot of these data. Does the relationship between cost and energy density look approximately linear?
b. Find the equation of the least squares regression line.
c. Construct a residual plot and comment on any unusual features of the residual plot.
d. Compute and interpret the value of r^2 in the context of this study.
e. Explain why it would also be informative to look at the value of s_e in addition to the value of r^2 in evaluating the usefulness of the least squares regression line.
f. Compute and interpret the value of s_e.
g. Based on you answers to Parts (c) – (f), do you think it is appropriate to use the least squares regression line to predict average cost for a food group based on its average energy density? Explain.
h. The meat, poultry, and fish food group has an average energy density of 196. What average cost would you predict for this food group?

4.80   (C12, M17, M18, C5)
The report "Older Driver Involvement in Injury Crashes in Texas" (Texas Transportation Institute, 2004) included a scatterplot of y = fatality rate (percent of drivers killed in injury crashes) versus x = driver age. The accompanying data are approximate values read from the scatterplot.

Age	Fatality Rate
20	1.00
25	0.99
30	0.80
35	0.80
40	0.75
45	0.75
50	0.95
55	1.05
60	1.15
65	1.20
70	1.30
75	1.65
80	2.20
85	3.00
90	3.20

a. Construct a scatterplot of these data. Does this relationship look linear?
b. What nonlinear model would you suggest to describe the relationship between fatality rate and age?
c. Using computer software or a graphing calculator, fit the model you suggested in Part (b).
d. Do you think this model will be useful for predicting y = fatality rate? Explain why or why not.
e. Use your nonlinear model to predict the fatality rate for 78-year-old drivers.

TECHNOLOGY NOTES

Correlation

TI-83/84

Note: Before beginning this chapter, press the **2nd** key then the **0** key and scroll down to the entry DiagnosticOn. Press **ENTER** twice. After doing this, the regression correlation coefficient r will appear as output with the linear regression line.

1. Enter the data for the independent variable into **L1** (to access lists press the **STAT** key, highlight the option called **Edit...** then press **ENTER**)
2. Input the data for the dependent variable into **L2**
3. Press the **STAT** key
4. Highlight **CALC** then select **LinReg(a+bx)** and press **ENTER**
5. Press the **2nd** key then the **1** key

6. Press ,
7. Press the **2nd** key then the **2** key
8. Press **ENTER**

TI-Nspire

1. Enter the data for the independent variable into a data list (to access data lists select the spreadsheet option and press **enter**)

   Note: Be sure to title the list by selecting the top row of the column and typing a title.

2. Enter the data for the dependent variable into a separate data list

3. Press the **menu** key and select **4:Statistics** then **1:Stat Calculations** then **3:Linear Regression(mx+b)...** and press **enter**
4. For **X List**: select the column with the independent variable data from the drop-down menu
5. For **Y List:** select the column with the dependent variable data from the drop-down menu
6. Press **OK**

Note: You may need to scroll to view the correlation coefficient in the list of output.

JMP
1. Input the data for the dependent variable into one column
2. Input the data for the independent variable into another column
3. Click **Analyze** then select **Multivariate Methods** then select **Multivariate**
4. Click and drag the column name containing the dependent variable from the box under **Select Columns** to the box next to **Y, Columns**
5. Click and drag the column name containing the independent variable from the box under **Select Columns** to the box next to **Y, Columns**
6. Click **OK**

Note: This produces a table of correlations. The correlation between the two variables can be found in the first row, second column.

MINITAB
1. Input the data for the dependent variable into the first column
2. Input the data for the independent variable into the second column
3. Select **Stat** then **Basic Statistics** then **Correlation...**
4. Double-click each column name in order to move it to the box under **Variables:**
5. Click **OK**

SPSS
1. Input the data for the dependent variable into the first column
2. Input the data for the independent variable into the second column
3. Select **Analyze** then choose **Correlate** then choose **Bivariate...**
4. Highlight the name of both columns by holding the **ctrl** key and clicking on each name
5. Click the arrow button to move both variables to the **Variables** box
6. Click **OK**

Note: This produces a table of correlations. The correlation between the two variables can be found in the first row, second column.

Note: The correlation can also be produced by following the steps to produce the regression equation.

Excel 2007
1. Input the data into two separate columns
2. Click the **Data** ribbon and select **Data Analysis**

3. **Note:** If you do not see **Data Analysis** listed on the Ribbon, see the Technology Notes for Chapter 2 for instructions on installing this add-on.
4. Select **Correlation** from the dialog box and click **OK**
5. Click in the box next to **Input Range:** and select BOTH columns of data (if you input and selected titles for the columns, click the box next to **Labels in First Row**)
6. Click **OK**

Note: The correlation between the variables can be found in the first column, second row of the table that is output.

Regression

TI-83/84
1. Enter the data for the independent variable into **L1** (to access lists press the **STAT** key, highlight the option called **Edit...** then press **ENTER**)
2. Input the data for the dependent variable into **L2**
3. Press the **STAT** key
4. Highlight **CALC** then select **LinReg(a+bx)** and press **ENTER**
5. Press the **2nd** key then the **1** key
6. Press **,**
7. Press the **2nd** key then the **2** key
8. Press **ENTER**

TI-Nspire
1. Enter the data for the independent variable into a data list (to access data lists select the spreadsheet option and press **enter**)

   Note: Be sure to title the list by selecting the top row of the column and typing a title.

2. Enter the data for the dependent variable into a separate data list
3. Press the **menu** key and select **4:Statistics** then **1:Stat Calculations** then **3:Linear Regression(mx+b)...** and press **enter**
4. For **X List:** select the column with the independent variable data from the drop-down menu
5. For **Y List:** select the column with the dependent variable data from the drop-down menu
6. Press **OK**

JMP
1. Enter the data for the dependent variable into the first column
2. Input the data for the independent variable into the second column
3. Click **Analyze** then select **Fit Y by X**
4. Click and drag the column name containing the dependent data from the box under **Select Columns** to the box next to **Y, Response**
5. Click and drag the column name containing the independent data from the box under **Select Columns** to the box next to **X, Factor**
6. Click **OK**
7. Click the red arrow next to **Bivariate Fit...**
8. Click **Fit Line**

MINITAB

1. Input the data for the dependent variable into the first column
2. Input the data for the independent variable into the second column
3. Select **Stat** then **Regression** then **Regression...**
4. Highlight the name of the column containing the dependent variable and click **Select**
5. Highlight the name of the column containing the independent variable and click **Select**
6. Click **OK**

Note: You may need to scroll up in the Session window to view the regression equation.

SPSS

1. Input the data for the dependent variable into the first column
2. Input the data for the independent variable into the second column
3. Select **Analyze** then choose **Regression** then choose **Linear...**
4. Highlight the name of the column containing the dependent variable
5. Click the arrow button next to the **Dependent** box to move the variable to this box
6. Highlight the name of the column containing the independent variable
7. Click the arrow button next to the **Independent** box to move the variable to this box
8. Click **OK**

Note: The regression coefficients can be found in the **Coefficients** table. The intercept value can be found in the first column of the (Constant) row. The value of the slope can be found in the first column of the row labeled with the independent variable name.

Excel 2007

1. Input the data into two separate columns
2. Click the **Data** ribbon and select **Data Analysis**
3. **Note:** If you do not see Data Analysis listed on the Ribbon, see the Technology Notes for Chapter 2 for instructions on installing this add-on.
4. Select **Regression** from the dialog box and click **OK**
5. Click in the box next to **Y:** and select the dependent variable data
6. Click in the box next to **X:** and select the independent variable data (if you input and selected titles for BOTH columns, check the box next to **Labels**)
7. Click **OK**

Note: The regression coefficients can be found in the third table under the **Coefficients** column.

Residuals

TI 83/84

Note: When the TI 83/84 performs a regression analysis, the residuals are automatically stored in a list called "RESID." To prevent loss of the residuals, copy them into a named list.

1. Press the **2nd** key then the List key. **Select RESID**. (You may have to scroll down to find it.) Your screen should now show "LRESID."

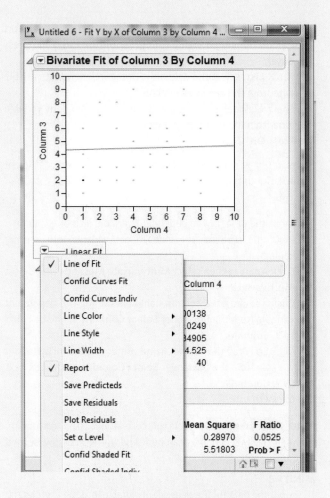

2. Click on **STO**, then the **2nd** key, and then **L3**.
3. Click on Enter.

TI-Nspire

1. Enter the data for the independent variable into a data list (to access data lists select the spreadsheet option and press **enter**)
   Note: Be sure to title the list by selecting the top row of the column and typing a title.
2. Enter the data for the dependent variable into a separate data list
3. Press the menu key and select **4:Statistics** then **1:Stat Calculations** then **3:Linear Regression(mx+b)...** and press **enter**
4. For **X List:** select the column with the independent variable data from the drop-down menu
5. For **Y List:** select the column with the dependent variable data from the drop-down menu
6. Press **OK**

JMP

1. Enter the data for the dependent variable into the first column
2. Input the data for the independent variable into the second column
3. Click **Analyze** then select **Fit Y by X**
4. Click and drag the column name containing the dependent data from the box under **Select Columns** to the box next to **Y, Response**

5. Click and drag the column name containing the independent data from the box under **Select Columns** to the box next to **X, Factor**
6. Click **OK**
7. Click the red arrow next to **Bivariate Fit...**
8. Click **Fit Line**
9. Click the red arrow next to **Linear Fit**
10. Click **Save Residuals**

MINITAB

1. Input the data for the dependent variable into the first column
2. Input the data for the independent variable into the second column
3. Select **Stat** then **Regression** then **Regression...**
4. Highlight the name of the column containing the dependent variable and click **Select**
5. Highlight the name of the column containing the independent variable and click **Select**
6. Click **Storage...**
7. Check the box next to **Residuals**
8. Click **OK**
9. Click **OK**

SPSS

1. Input the data for the dependent variable into the first column
2. Input the data for the independent variable into the second column
3. Select **Analyze** then choose **Regression** then choose **Linear...**
4. Highlight the name of the column containing the dependent variable
5. Click the arrow button next to the **Dependent** box to move the variable to this box
6. Highlight the name of the column containing the independent variable
7. Click the arrow button next to the **Independent** box to move the variable to this box
8. Click on the **Save...** button
9. Click the check box next to **Unstandardized** under the **Residuals** section
10. Click **Continue**
11. Click **OK**

Note: The residuals will be saved in the SPSS worksheet in a new column.

Excel 2007

1. Input the data into two separate columns
2. Click the **Data** ribbon and select **Data Analysis**
    Note: If you do not see Data Analysis listed on the Ribbon, see the Technology Notes for Chapter 2 for instructions on installing this add-on.
3. Select **Regression** from the dialog box and click **OK**
4. Click in the box next to **Y:** and select the dependent variable data
5. Click in the box next to **X:** and select the independent variable data (if you input and selected titles for BOTH columns, check the box next to **Labels**)
6. Check the box next to **Residuals** under the **Residuals** section of the dialog box
7. Click **OK**

Residual Plot

TI-83/84

The TI-83/84 does not have the functionality to produce a residual plot automatically. After using Linreg(a+bx), select 2nd, Statplot. Set the first plot as a scatter with XList: L1 and YList: RESID. Select Zoom, Stats, and a residual plot is displayed.

TI-Nspire

The TI-Nspire does not have the functionality to produce a residual plot automatically.

JMP

1. Begin by saving the residuals as described in the previous section
2. Form a scatterplot of the independent variable versus the residuals using the procedures described in Chapter 2 for scatterplots

MINITAB

1. Input the data for the dependent variable into the first column
2. Input the data for the independent variable into the second column
3. Select **Stat** then **Regression** then **Regression...**
4. Highlight the name of the column containing the dependent variable and click **Select**
5. Highlight the name of the column containing the independent variable and click **Select**
6. Click **Graphs...**
7. Click the box under **Residuals versus the variables**:
8. Double-click the name of the independent variable
9. Click **OK**
10. Click **OK**

SPSS

1. Begin by saving the residuals as described in the previous section
2. Form a scatterplot of the independent variable versus the residuals using the procedures described in Chapter 2 for scatterplots

Excel 2007

1. Input the data into two separate columns
2. Click the **Data** ribbon and select **Data Analysis**
    Note: If you do not see **Data Analysis** listed on the Ribbon, see the Technology Notes for Chapter 2 for instructions on installing this add-on.
3. Select **Regression** from the dialog box and click **OK**
4. Click in the box next to **Y:** and select the dependent variable data
5. Click in the box next to **X:** and select the independent variable data (if you input and selected titles for BOTH columns, check the box next to **Labels**)
6. Check the box next to **Residuals Plots** under the **Residuals** section of the dialog box
7. Click **OK**

Transforming Variables

TI 83/84

1. Enter the data for the variable you wish to transform into **L1** (to access lists press the **STAT** key, highlight the option called Edit...then press **ENTER**).
2. Click on **Stat** and select **Edit**.
3. Using the arrow keys (directly above the **VARS** and **CLEAR** Keys) position the cursor at the top of **List2**. That is, the cursor should be covering the "**L2**." In addition, "**L2=**" should appear in the bottom line.

Now choose the transformation you wish to make. For example, to set the values of **L2** equal to the natural log of **L1**, do the following:

4. Click the **LN** button; the formula at the bottom of the screen will change to "**L2=ln(**".
5. Click on the **2nd** key, then the **STO** key, and finally **L1**; the formula at the bottom of the screen will change to "**L2=ln(L1**".
6. Click on the right parenthesis key; the formula at the bottom of the screen will change to "**L2=ln(L1)**".
7. Click on **Enter**.

JMP

1. Enter the pre-transformed data into **Column1**.
2. Double-click at the top of **Column2**. A window with properties for **Column2** will pop up.
3. In that window, click on **Column Properties**, select **Formula**, and click on **Edit Formula**.

Now choose the transformation you wish to make. For example, to set the values of Column2 equal to the natural logarithm of Column1, do the following:

5. Click on Transcendental, then click on **Log**; the formula box will be updated to "**Log(   )**.

6. In the TableColumns panel, click on **Column1**; the formula box will be updated to "**Log(Column1)**".
7. Click **Ok** and close the column properties window.
8. Respond with "**Yes**" to the "**Apply changes**" question.

MINITAB

1. Enter the pre-transformed data into column "**C1**."
2. Click on Calc in the main menu, and then Calculator
3. Type **C2** in the "Store result in variable" box

Now choose the transformation you wish to make. For example, to set the values of C2 equal to the natural log of C1:

4. Click on Natural log (log bas e) in the Functions drop-down menu and click on **Select**. The "Expression:" window will be updated to "**LN(number)**".
5. Double-click on **C1** in the variables list panel; the "Expression:" window will be updated to "**LN(C1)**".
6. Click **Ok**.

Automatic Nonlinear Regression

TI 83/84

1. Enter the data for the independent variable into **L1** (to access lists press the **STAT** key, highlight the option called **Edit...** then press **ENTER**)
2. Input the data for the dependent variable into **L2**
3. Press the **STAT** key
4. Select **CALC**.

Now choose the nonlinear option you wish; for example, to perform exponential regression, do the following:

5. Select **ExpReg** and press **ENTER**.
6. Press the **2nd** key then the **1** key
7. Press,
8. Press the **2nd** key then the **2** key
9. Press **ENTER**

AP* Review Questions for Chapter 4

1. Which of the following five scatterplots displays a data set that would have a correlation coefficient farthest from 0?

(A)

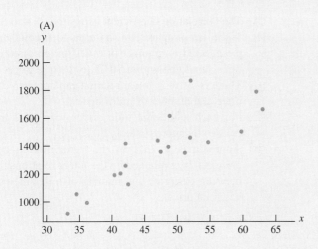

(B)

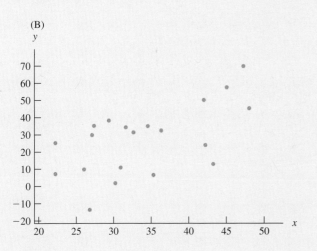

(C)

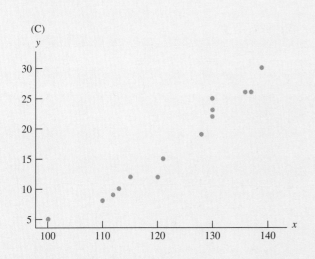

(D)

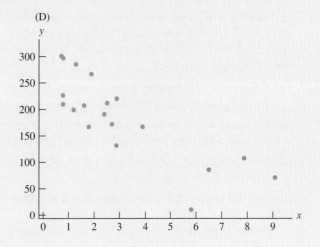

(E)

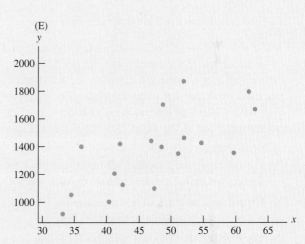

2. Data on x = weight (in ounces) and y = overnight shipping cost (dollars) for 20 items were used to fit the least squares regression line, $\hat{y} = 22 + 1.505x$. Which of the following statements about different lines, ($y = 10 + 0.75x$ and $y = 10 + 1.5x$), is true?

(A) The line $y = 10 + 0.75x$ will have a smaller sum of squared deviations than the regression line.
(B) The line $y = 10 + 0.75x$ will have the same sum of squared deviations as the regression line.
(C) The line $y = 10 + 0.75x$ will have a larger sum of squared deviations than the regression line.
(D) The line $y = 10 + 1.5x$ will have the same sum of squared deviations as the regression line.
(E) It is not possible to tell how the sum of squared deviations for the line $y = 10 + 0.75x$ will compare to the sum of squared deviations for the regression line.

3. Data on x = the weight of a pickup truck (pounds) and y = distance (in feet) required for a truck traveling 30 miles per hour to come to a complete stop for 16 trucks was used to fit the least squares regression line $\hat{y} = 23 + 0.02x$. Which of the following statements is a correct interpretation of the value 0.02 in the equation of the regression line?

   (A) On average, the stopping distance goes up 0.02 foot for each 1-pound increase in truck weight.

   (B) On average, the truck weight goes up 0.02 pound for each additional foot required to stop the truck.

   (C) On average, the stopping distance is 0.02 foot when the truck weight is 0.

   (D) Approximately 2% of the variation in the stopping distances can be explained by the linear relationship between stopping distance and truck weight.

   (E) The correlation coefficient for this data set is 0.02.

4. Data on x = the weight of a pickup truck (pounds) and y = distance (in feet) required for the truck traveling 30 miles per hour to come to a complete stop for 16 trucks was used to fit the least squares regression line, $\hat{y} = 23 + 0.02x$. Which of the following statements is a correct interpretation of the value 23 in the equation of the regression line?

   (A) On average, the stopping distance increases by 23 feet for each 1-pound increase in truck weight.

   (B) On average, the truck weight increases by 23 pounds for each additional foot in stopping distance.

   (C) On average, the stopping distance is 23 feet when the car weight is 0.

   (D) Approximately 2% of the variation in the stopping distances can be explained by the linear relationship between stopping distance and truck weight.

   (E) It is not reasonable to interpret the intercept in this setting because a weight of 0 is outside the range of the data used to fit the regression line.

5. A realtor is interested in being able to predict the selling price of houses in a particular city. She plans to choose one predictor variable and then find the least-squares regression line and use it to make her predictions. Five possible predictor variables have been identified. These variables and the correlation coefficient for each of these variables and house prices are given. Which of these five predictors has the strongest linear relationship with selling price?

   (A) house age, $r = -0.46$
   (B) number of other homes for sale on same block, $r = -0.34$
   (C) size of kitchen, $r = 0.12$
   (D) number of bedrooms, $r = 0.42$
   (E) house size, $r = 0.45$

6. Data were collected on y = price of car (in dollars) and x = age of car (in years) for each car in a sample of 50 used Toyota Camrys. A scatterplot showed a negative linear relationship between x and y. The least squares regression line was fit and the r^2 value was computed. If $r^2 = 0.25$, which of the following is a correct statement?

   (A) The correlation coefficient is positive, $r = 0.5$.
   (B) If the least-squares line is used to predict car price based on number of miles driven, predictions should be within $0.25 of the true price.
   (C) There is a strong linear relationship between car price and number of miles driven.
   (D) For each additional mile driven, car price increases by approximately $0.25.
   (E) Approximately 25% of the variability in car price can be explained by the linear relationship between car price and number of miles the car has been driven.

7. Consider the following three scatterplots. All three scatterplots are drawn to the same scale on both the x and y axes. Which of the following correctly orders these from smallest to largest based on the value of the correlation coefficient?

(A) I, II, III
(B) II, III, I
(C) III, II, I
(D) II, I, III
(E) III, I, II

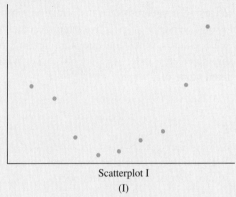

Scatterplot I
(I)

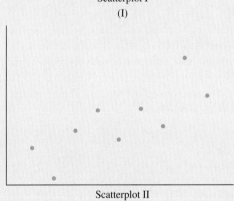

Scatterplot II
(II)

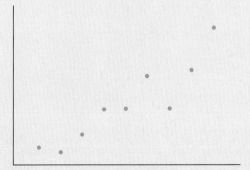

Scatterplot III
(III)

8. Consider the following five scatterplots (scatterplot (E) is on the next page). All are drawn to the same scale on both the x and y axes. For which scatterplot is the standard deviation about the least-squares regression line the smallest?

(A)

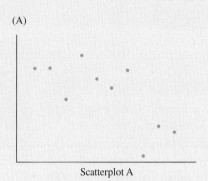

Scatterplot A

(B)

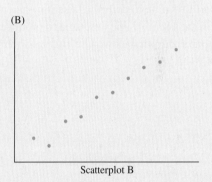

Scatterplot B

(C)

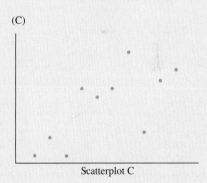

Scatterplot C

(D)

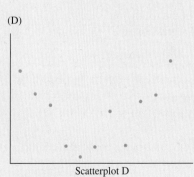

Scatterplot D

(E)

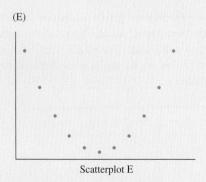

Scatterplot E

The following applies to questions 9–11:
Twenty-five assembly-line workers participated in a study to investigate the relationship between experience and the amount of time required to complete an assembly task. Assembly time (in minutes) and number of months the worker had been employed on the assembly line were measured for each worker. The resulting data on y = time to complete assembly and x = number of months on the assembly line were used to produce the scatterplot and computer output below. Use this information to answer questions 9–11.

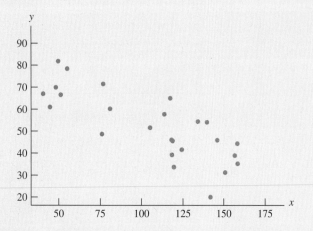

Predictor	Coef	SE Coef	T	P
Constant	84.683	5.602	15.12	0.000
x	−0.30411	0.04963	−6.13	0.000

S = 9.79097   R-Sq = 62.0%   R-Sq(adj) = 60.4%

9. Which of the following is the value of the slope of the least-squares regression line?

(A) −0.30411
(B) 0.04963
(C) 0.620
(D) 9.79097
(E) 84.683

10. What is the value of the correlation coefficient for this data set?

(A) −0.787
(B) −0.620
(C) −0.30411
(D) 0.620
(E) 0.787

11. What proportion of variability in assembly time can be explained by the linear relationship between assembly time and number of months on the assembly line?

(A) 0.4963
(B) 0.620
(C) 6.602
(D) 9.97097
(E) 84.683

12. A scatterplot showed a nonlinear relationship between y and an independent variable x. The x values were transformed using a square-root transformation, and a scatterplot of y versus $x' = \sqrt{x}$ was approximately linear. The least squares regression line summarizing the relationship between y and x' was $\hat{y} = 14 - 2x'$. What is the predicted value of y when x = 9?

(A) −4
(B) $\sqrt{-4}$
(C) 8
(D) $\sqrt{8}$
(E) We cannot say, because it is not possible to take the square root of a negative number.

13. A study was carried out to investigate the relationship between x = time spent working each week (hours) and y = time spent on math homework on a typical weekday (in minutes) for a random sample of 20 students in a first-quarter calculus course. The scatterplot of the data is shown below.

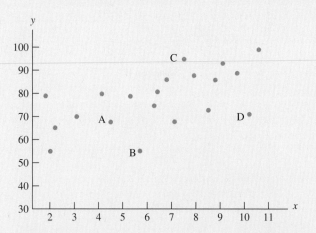

When the simple linear regression model is fit to the sample data, which point will have the smallest (in magnitude) residual?

(A) A
(B) B
(C) C
(D) D
(E) All four points will have the same positive value for the residual.

14. Consider the following three scatterplots. For which of these scatterplots would it be appropriate to use the least squares regression line to describe the relationship pictured?
    (A) I only
    (B) II only
    (C) III only
    (D) II and III only
    (E) I, II, and III

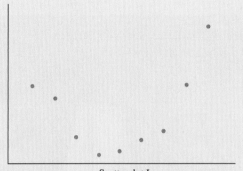

Scatterplot I
(I)

Scatterplot II
(II)

Scatterplot III
(III)

15. Of the following, which is *not* true of *r*?
    (A) The value of r does not depend on the units of y and x.
    (B) r is always between 0 and 1.
    (C) r measures the strength of the linear relation between x and y.
    (D) The value of r does not depend on which of two variables is labeled x.
    (E) $|r|$ is less than 1.0 unless the points on a scatterplot line up exactly.

5

Probability

Preview

Chapter Learning Objectives

5.1 Interpreting Probabilities

5.2 Computing Probabilities

5.3 Probabilities of More Complex Events: Unions, Intersections, and Complements

5.4 Conditional Probability

5.5 Calculating Probabilities—A More Formal Approach

5.6 Probability as a Basis for Making Decisions

5.7 Estimating Probabilities Empirically and Using Simulation

Chapter Activities

Are You Ready to Move On?

Chapter 5 Review Exercises

AP* Review Questions for Chapter 5

Baris Simsek/istockphoto.com

PREVIEW

In many situations, you need to assess risks in order to make an informed decision. Should you purchase an extended warranty for a new laptop? An extended warranty can be expensive, but it would result in significant savings if the laptop failed during the period covered by the extended warranty. How likely is it that your laptop will fail during this period? Suppose that the deadline to apply for a summer internship is five days away, and you are just mailing your application. Should you pay for priority mail service, which promises two-day delivery, or is it likely that your application will arrive on time by regular mail? Suppose that a prescription drug is available that will reduce pain caused by arthritis, but there are possible side effects. How serious are the side effects, and how likely are they to occur? Each of the situations just described involves making a decision in the face of uncertainty. Making good decisions in situations that involve uncertainty is easier when you can be more precise about the meaning of terms such as "likely" or "unlikely."

CHAPTER LEARNING OBJECTIVES

Conceptual Understanding

After completing this chapter, you should be able to

C1 Interpret a probability as a long-run relative frequency of occurrence.
C2 Understand what it means for two events to be mutually exclusive.
C3 Understand what it means for two events to be independent.
C4 Understand the difference between an unconditional probability and a conditional probability.
C5 Understand how probabilities can be estimated using simulation.

Mastering the Mechanics

After completing this chapter, you should be able to

M1 Interpret probabilities in context.
M2 Compute the probability of an event when outcomes in the sample space are equally likely.
M3 Use information in a two-way table to compute probabilities of events, unions of two events, and intersections of two events.
M4 Given the probabilities of two events E and F and the probability of the intersection $E \cap F$, construct a "hypothetical 1000" table, and use the table to compute other probabilities of interest.
M5 Given the probabilities of two events E and F and the probability of the union $E \cup F$, construct a "hypothetical 1000" table and use the table to compute other probabilities of interest.
M6 Compute probabilities of unions for mutually exclusive events.
M7 Compute probabilities of intersections for independent events.
M8 Given the probabilities of two independent events E and F, construct a "hypothetical 1000" table and use the table to compute other probabilities of interest.
M9 Use information in a two-way table to compute conditional probabilities.
M10 Given probability and conditional probability information, construct a "hypothetical 1000" table and use the table to compute other probabilities of interest.
M11 Use probability formulas to calculate probabilities of unions and intersections and to calculate conditional probabilities.
M12 Carry out a simulation to estimate a probability.

Putting It into Practice

After completing this chapter, you should be able to

P1 Use basic properties of probabilities to reason in a given context.
P2 Distinguish between questions that can be answered by computing an unconditional probability and questions that can be answered by computing a conditional probability.
P3 Given a question that can be answered by computing a probability, compute and interpret an appropriate probability to answer the question.
P4 Use probability to make decisions and justify conclusions.

PREVIEW EXAMPLE   Should You Paint the Nursery Pink?

Ultrasound is a medical imaging technique routinely used to assess the health of a baby prior to birth. It is sometimes possible to determine the baby's gender during an ultrasound examination. How accurate are gender identifications made during the first trimester (3 months) of pregnancy?

The paper **"The Use of Three-Dimensional Ultrasound for Fetal Gender Determination in the First Trimester" (*The British Journal of Radiology* [2003]: 448–451)** describes a study of ultrasound gender predictions. An experienced radiologist looked at 159 first trimester ultrasound images and made a gender prediction for each one. When each baby was born, the ultrasound gender prediction was compared to the baby's actual gender. The following table summarizes the resulting data.

	Radiologist 1	
	Predicted Male	**Predicted Female**
Baby is Male	74	12
Baby is Female	14	59

Notice that the gender prediction based on the ultrasound image is not always correct. Several questions come to mind:

1. How likely is it that a predicted gender is correct?
2. Is a predicted gender more likely to be correct when the baby is male than when the baby is female?
3. If the predicted gender is female, should you paint the nursery pink? If you do, how likely is it that you will need to repaint?

The paper also included gender predictions made by a second radiologist, who looked at 154 first trimester ultrasound images. The data are summarized in the following table.

	Radiologist 2	
	Predicted Male	**Predicted Female**
Baby is Male	81	8
Baby is Female	7	58

In addition to the questions posed previously, you can also compare the accuracy of gender predictions for the two radiologists. Does the skill of the radiologist make a difference?

All of these questions can be answered using the methods introduced in this chapter. This example will be revisited in Section 5.4.

From its roots in the analysis of games of chance, probability has evolved into a science that enables you to make informed decisions with confidence. In this chapter, you will be introduced to the basic ideas of probability, explore strategies for computing probabilities, and consider ways to estimate probabilities when it is difficult to compute them directly.

SECTION 5.1 Interpreting Probabilities

People often find themselves in situations where the outcome is uncertain. For example, when a ticketed passenger shows up at the airport, she faces two possible outcomes: (1) she is able to take the flight, or (2) she is denied a seat as a result of overbooking by the airline, and must take a later flight. Based on her past experience with this particular flight, the passenger may know that one outcome is more likely than the other. She may believe that the chance of being denied a seat is quite small. Although this outcome is possible, she views it as unlikely.

To quantify the likelihood of its occurrence, a number between 0 and 1 can be assigned to an outcome. This number is called a *probability*. Assigning a probability to an outcome is an attempt to quantify what is meant by "*likely*" or "*unlikely*."

> A **probability** is a number between 0 and 1 that reflects the likelihood of occurrence of some outcome.

There are several ways to interpret a probability. One is a subjective interpretation, in which a probability is interpreted as a personal measure of the strength of belief that an outcome will occur. A probability of 1 represents a belief that the outcome will certainly occur. A probability of 0 represents a belief that the outcome will certainly *not* occur—that it is impossible. All other probabilities fall between these two extremes. This interpretation is common in ordinary speech. For example, you might say, "There's about a 50–50 chance," or "My chances are nil."

The subjective interpretation, however, presents some difficulties. Because different people may have different subjective beliefs, they may assign different probabilities to the same outcome. Whenever possible, we will use an objective *relative frequency* approach to probability. With this approach, a probability specifies the *long-run proportion* of the time that an outcome will occur. A probability of 1 corresponds to an outcome that occurs 100% of the time—a certain outcome. A probability of 0 corresponds to an outcome that occurs 0% of the time—an impossible outcome.

Relative Frequency Interpretation of Probability

The probability of an outcome, denoted by P(outcome), is interpreted as the proportion of the time that the outcome occurs *in the long run*.

Consider the following situation. A package delivery service promises 2-day delivery between two cities in California but is often able to deliver packages in just 1 day. The company reports that the probability of next-day delivery is 0.3. This implies that in the long run, 30% of all packages arrive in 1 day. An equivalent way to interpret this probability would be to say that in the long run, about 30 out of 100 packages shipped would arrive in 1 day.

Suppose that you track the delivery of packages shipped with this company. With each new package shipped, you could compute the relative frequency of packages shipped so far that have arrived in 1 day:

$$\frac{\text{number of packages that arrived in 1 day}}{\text{total number of packages shipped}}$$

The results for the first 15 packages might be as follows:

Package Number	Did the Package Arrive in 1 Day?	Relative Frequency of Packages Shipped so Far That Arrived in 1 Day
1	Yes	$\frac{1}{1} = 1.00$
2	Yes	$\frac{2}{2} = 1.00$
3	No	$\frac{2}{3} = 0.67$
4	Yes	$\frac{3}{4} = 0.75$
5	No	$\frac{3}{5} = 0.60$
6	No	$\frac{3}{6} = 0.50$
7	Yes	$\frac{4}{7} = 0.57$
8	No	$\frac{4}{8} = 0.50$
9	No	$\frac{4}{9} = 0.44$
10	No	$\frac{4}{10} = 0.40$
11	No	$\frac{4}{11} = 0.36$
12	Yes	$\frac{5}{12} = 0.42$
13	No	$\frac{5}{13} = 0.38$
14	No	$\frac{5}{14} = 0.36$
15	No	$\frac{5}{15} = 0.33$

Figure 5.1 shows how this relative frequency of packages arriving in 1 day changes over the first 15 packages.

Figure 5.2 illustrates how this relative frequency fluctuates during a sequence of 50 shipments. As the number of packages in the sequence increases, the relative frequency does not continue to fluctuate wildly but instead settles down and approaches a specific value, which is the probability of interest. Figure 5.3 illustrates how the relative frequency settles down over

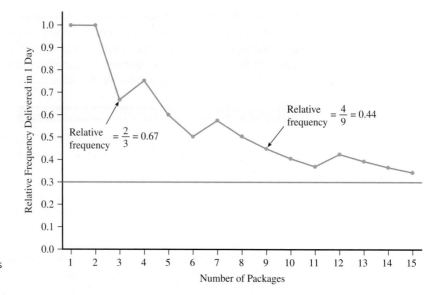

FIGURE 5.1
Relative frequency of packages
delivered in 1 day for 15 packages

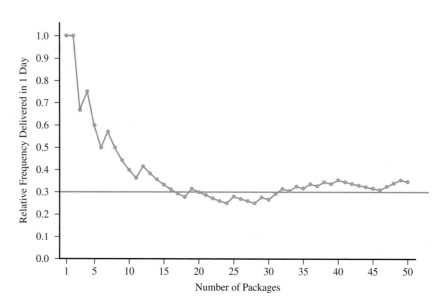

FIGURE 5.2
Relative frequency of packages
delivered in 1 day for 50 packages

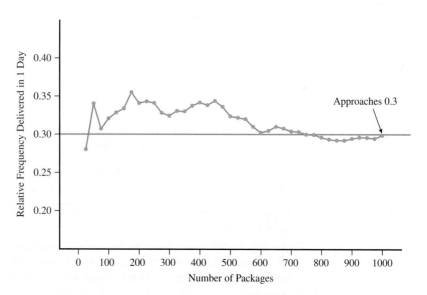

FIGURE 5.3
Relative frequency of
packages delivered in
1 day for 1000 packages

a sequence of 1000 shipments. What you see happening in this figure is not unique to this particular example. This "settling down" is a consequence of the Law of Large Numbers.

> **Law of Large Numbers**
>
> **As the number of observations increases, the proportion of the time that an outcome occurs gets close to the probability of that outcome.**

The Law of Large Numbers is the basis for the relative frequency interpretation of probabilities.

Some Basic Properties of Probabilities

Keep the relative frequency interpretation of probability in mind as you consider the following basic properties of probabilities:

1. *The probability of any outcome is a number between 0 and 1.* A relative frequency, which is the number of times the outcome occurs divided by the total number of observations, cannot be less than 0 (there can't be a negative number of occurrences) or greater than 1 (the outcome can't occur more often than the total number of observations).

2. *If outcomes can't occur at the same time, then the probability that any one of them will occur is the sum of their individual probabilities.* For example, *The Chronicle of Higher Education* **(August 26, 2011)** reports that of students enrolled full time at 4-year colleges in 2011, 62% were enrolled in public colleges, 26% were enrolled in private nonprofit colleges, and 12% were enrolled in private for-profit colleges. Imagine selecting a student at random from this group of students. If you observe the type of college the selected student attends, there are three possible outcomes: (1) public, (2) private nonprofit, and (3) private for-profit. Interpreting the given percentages as probabilities, you can say that

    $$P(\text{selected student attends a private nonprofit}) = 0.26$$

    and

    $$P(\text{selected student attends a private for-profit}) = 0.12$$

    Assuming that no person is enrolled *full-time* at more than one college, the probability that the selected student attends a private nonprofit *or* a private for-profit college is

    $$P(\text{private nonprofit or private for-profit}) = 0.26 + 0.12 = 0.38$$

    About 38 in 100 full-time 4-year college students attend a private college (nonprofit or for-profit).

3. *The probability that an outcome will not occur is equal to 1 minus the probability that the outcome will occur.* Continuing with the 4-year college student example, you know that about 38 in 100 full-time students attend a private college. This means that the others do not attend a private college, so about 62 in 100 full-time students are not at private colleges. Here

    $$P(\text{not private}) = 1 - P(\text{private}) = 1 - 0.38 = 0.62$$

    Notice that this is equal to $P(\text{public})$, because there are only three possible outcomes. Also notice that because there are only three possible outcomes that might be observed,

    $$P(\text{public } or \text{ private nonprofit } or \text{ private for-profit}) = 0.62 + 0.26 + 0.12 = 1$$

Because a probability represents a long-run relative frequency, in situations where exact probabilities are not known, it is common to estimate probabilities based on observation. For example, a shipping company's reported 1-day delivery probability of 0.3 is most likely based on historical data by computing the proportion of packages delivered in 1 day over a large number of shipments.

SECTION 5.1 EXERCISES

Each Exercise Set assesses the following chapter learning objectives: C1, M1, P1

SECTION 5.1 Exercise Set 1

5.1 An article in the *New York Times* reported that people who suffer cardiac arrest in New York City have only a 1 in 100 chance of survival. Using probability notation, an equivalent statement would be $P(\text{survival}) = 0.01$ for people who suffer cardiac arrest in New York City. (The article attributed this poor survival rate to factors common in large cities: traffic congestion and difficulty finding victims in large buildings. Similar studies in smaller cities showed higher survival rates.)
a. Give a relative frequency interpretation of the given probability.
b. The basis for the *New York Times* article was a research study of 2,329 consecutive cardiac arrests in New York City. To justify the "1 in 100 chance of survival" statement, how many of the 2,329 cardiac arrest sufferers do you think survived? Explain.

5.2 An airline reports that for a particular flight operating daily between Phoenix and Atlanta, the probability of an on-time arrival is 0.86. Give a relative frequency interpretation of this probability.

5.3 For a monthly subscription fee, a video download site allows people to download and watch up to five movies per month. Based on past download histories, the following table gives the estimated probabilities that a randomly selected subscriber will download 0, 1, 2, 3, 4 or 5 movies in a particular month.

Number of downloads	0	1	2	3	4	5
Estimated probability	0.03	0.45	0.25	0.10	0.10	0.07

If a subscriber is selected at random, what is the estimated probability that this subscriber downloads
a. three or fewer movies?
b. at most three movies?
c. four or more movies?
d. zero or one movie?
e. more than one movie?

SECTION 5.1 Exercise Set 2

5.4 What does it mean to say that the probability that a coin toss will land head side up is 0.5?

5.5 In a particular state, automobiles that are more than 10 years old must pass a vehicle inspection to be registered.

This state reports the probability that a car more than 10 years old will fail the vehicle inspection is 0.09. Give a relative frequency interpretation of this probability.

5.6 A bookstore sells books in several formats—hardcover, paperback, digital, and audio.

Based on past sales, the table below gives the estimated probability that a randomly selected purchase will be of a particular type.

Hardcover	Paperback	Digital	Audio
0.16	0.36	0.40	0.08

If a purchase is selected at random, what is the probability that this purchase is for a book that is
a. digital or audio?
b. not digital?
c. a printed book?

Additional Exercises

5.7 If you were to roll a fair die 1,000 times, about how many sixes do you think you would observe? What is the probability of observing a six when a fair die is rolled?

5.8 The probability of getting a king when a card is selected at random from a standard deck of 52 playing cards is $\frac{1}{13}$.
a. Give a relative frequency interpretation of this probability.
b. Express the probability as a decimal rounded to three decimal places. Then complete the following statement: If a card is selected at random, I would expect to see a king about _____ times in 1,000.

5.9 At a particular college, students have several options for purchasing textbooks. The options and the proportion of students choosing each option are shown in the table at the bottom of this page.

If a student at this college is selected at random from this college, what is the probability that the student
a. purchased all print books online?
b. purchased all print books?
c. purchased some e-books or all e-books?
d. did not purchase all print books from the campus bookstore?

5.10 Give a relative frequency interpretation for each probability calculated in the previous exercise.

TABLE FOR EXERCISE **5.9**

Option	All Print Books From Campus Bookstore	All Print Books From Off-Campus Bookstore	All Print Books From Online Booksellers	All Print Books, but Purchased Both From Campus Bookstore and Online	All e-Books Downloaded From Publishers	Mix of Print and e-Books
Proportion Choosing Option	0.25	0.12	0.17	0.31	0.04	0.11

SECTION 5.2 Computing Probabilities

In the previous section, you saw how probabilities are interpreted and learned some basic properties. In this section, you will learn two ways that probabilities can be calculated. Before you begin, a few key terms need to be introduced.

Chance Experiments and Sample Spaces

In most probability settings, there is uncertainty about what the outcome will be. Such situations are called *chance experiments*.

> **DEFINITION**
>
> A **chance experiment** is the process of making an observation when there is uncertainty about which of two or more possible outcomes will result.

For example, in an opinion poll, there is uncertainty about whether an individual selected at random from some population supports a school bond, and when a die is rolled there is uncertainty about which side will land face up. Observing an outcome in either of these situations would be a chance experiment. Notice that a *chance* experiment is different from the kind of experiment considered in Chapter 1, where the term experiment was used to describe a type of statistical study.

Consider a chance experiment to investigate whether men or women are more likely to choose a hybrid engine over a traditional internal combustion engine when purchasing a Honda Civic at a particular car dealership. A customer will be selected at random from those who purchased a Honda Civic. The type of vehicle purchased (hybrid or traditional) will be determined, and the customer's gender will be recorded. Before the customer is selected, there is uncertainty about the outcome, so this is a chance experiment. You do know, however, what the possible outcomes are. This set of possible outcomes is called the *sample space*.

> **DEFINITION**
>
> The collection of all possible outcomes of a chance experiment is the **sample space** for the experiment.

The sample space of a chance experiment can be represented in many ways. One representation is a simple list of all the possible outcomes. For the chance experiment where gender and engine type were observed, the possible outcomes are the following:

1. male, hybrid
2. female, hybrid
3. male, traditional
4. female, traditional

To simplify, you could also use M and F to indicate gender and H and T to indicate engine type. Using set notation, the sample space is then

$$\text{sample space} = \{\text{MH, FH, MT, FT}\}$$

Events

In the car-purchase chance experiment, you might be interested in a particular outcome. Or you might focus on a group of outcomes that involve the purchase of a hybrid—the group consisting of MH and FH. When you combine one or more individual outcomes into a collection, you are creating what is known as an *event*.

> **DEFINITION**
>
> An **event** is any collection of outcomes from the sample space of a chance experiment.
>
> A **simple event** is an event consisting of exactly one outcome.

An event can be represented by a name, such as *hybrid*, or by an uppercase letter, such as *A*, *B*, or *C*. Sometimes different events are denoted using the same letter with different numerical subscripts, such as E_1, E_2, and E_3.

Example 5.1   Car Preferences

Reconsider the situation in which a person who purchased a Honda Civic was categorized by gender (M or F) and type of car purchased (H or T). The sample space is

$$\text{sample space} = \{MH, FH, MT, FT\}$$

Because there are four outcomes, there are four simple events. One event of interest consists of all outcomes in which a hybrid is purchased. The event *hybrid* is

$$hybrid = \{MH, FH\}$$

Another event is the event that the purchaser is male,

$$male = \{MH, MT\}$$

Because it consists of only one outcome, a simple event occurs whenever a chance experiment is performed. You also say that a given event occurs whenever one of the outcomes making up the event occurs. For example, if the outcome in Example 5.1 is MH, then the simple event *male purchasing a hybrid* has occurred, and so has the non-simple event *hybrid*.

Example 5.2   Simple Coin Toss Game

Consider a game in which two players each toss a coin. If both coins land heads up, Player 1 is declared the winner. If both coins land tails up, Player 2 is declared the winner. If the coins land with one showing heads and the other showing tails, the game is considered a tie. This game can be viewed as a chance experiment. Before the coins are tossed, there is uncertainty about what outcome will occur.

For this chance experiment, one possible outcome is that Player 1 tosses a head and Player 2 tosses a tail. This outcome can be abbreviated as HT. Using this notation, the sample space is

$$\text{sample space} = \{HH, HT, TH, TT\}$$

Using W_1 to denote the event *player 1 wins*,

$$W_1 = \{HH\}$$

For the event *game results in a tie*, denoted by *T*,

$$T = \{HT, TH\}$$

Computing Probabilities

Some chance experiments have outcomes that are equally likely to occur. This would be the case when, for example, a fair coin is tossed (two equally likely outcomes, H and T) or when a fair die is rolled (six equally likely outcomes). When outcomes are equally likely, it is easy to calculate the probability of an event using what is known as the classical approach.

The Classical Approach to Computing Probabilities

When a chance experiment has equally likely outcomes, in the long run you would expect each of these outcomes to occur the same proportion of the time. For example, if there are 10 equally likely outcomes, each one would occur about $\frac{1}{10}$ or 10% of the time in the long run. If an event consisted of 3 of these 10 outcomes, this event would occur about $\frac{3}{10}$ or 30% of the time.

In general, if there are N equally likely outcomes, the probability of each outcome is $\frac{1}{N}$, and the probability of an event can be determined if you know how many of the possible outcomes are included in the event.

> **Classical Approach to Calculating Probabilities for Equally Likely Outcomes**
>
> When the outcomes in the sample space of a chance experiment are equally likely, the **probability of an event** E, denoted by $P(E)$, is the ratio of the number of outcomes favorable to E to the total number of outcomes in the sample space:
>
> $$P(E) = \frac{\text{number of outcomes favorable to E}}{\text{number of outcomes in the sample space}}$$

Example 5.3   Calling the Toss

On some football teams, the honor of calling the toss at the beginning of a football game is determined by random selection. Suppose that, this week, a member of the offense will call the toss. There are 5 linemen on the 11-player offense. Because a player will be selected at random, each of these 11 players is equally likely to be chosen. If you define the event L as the event that a lineman is selected to call the toss, 5 of the 11 possible outcomes are included in L. The probability that a lineman will be selected is then

$$P(L) = \frac{5}{11} = 0.455$$

In Section 5.1, probabilities were interpreted as long-run relative frequencies. This interpretation is appropriate even when probabilities are computed using the classical approach. The probability computed indicates that in a long sequence of selections, a lineman will be chosen about 45.5% of the time.

Example 5.4   Math Contest

Four students (Adam, Bettina, Carlos, and Debra) submitted correct solutions to a math contest that had two prizes. The contest rules specify that if more than two correct responses are submitted, the winners will be selected at random from those submitting correct responses. You can use AB to denote the outcome that Adam and Bettina are the two selected. The other outcomes can be denoted in a similar way. Then the sample space for selecting the two winners from the four correct responses is

$$\text{sample space} = \{\text{AB, AC, AD, BC, BD, CD}\}$$

Because the winners are selected at random, the six possible outcomes are equally likely and the probability of each individual outcome is $\frac{1}{6}$.

If E is the event that both selected winners are the same gender,

$$E = \{\text{AC, BD}\}$$

Because E contains two outcomes, $P(E) = \frac{2}{6} = 0.333$. In the long run, if two winners are selected at random from this group, both winners will be the same gender about 33.3% of the time. If F denotes the event that at least one of the selected winners is female, then F consists of all outcomes except AC and $P(F) = \frac{5}{6} = 0.833$. In the long-run, about 83.3% of the selections would include at least one female winner.

The classical approach to probability works well for chance experiments that have a finite set of outcomes that are equally likely. However, many chance experiments do not have equally likely outcomes. For example, consider the chance experiment of selecting a student from those enrolled at a particular school and observing whether the student is a freshman, sophomore, junior, or senior. If there are more seniors than sophomores at the school, the four possible outcomes for this chance experiment are not equally likely. In this situation, it would be a mistake to think that the probability of each outcome is $\frac{1}{4}$. To compute or estimate probabilities in situations where outcomes are not equally likely, you need an alternate approach.

Relative Frequency Approach to Computing Probabilities

When a chance experiment is performed, some events may be likely to occur, whereas others may not be as likely. For a specified event E, its probability indicates how frequently E occurs when the chance experiment is performed many times. For example, recall the package delivery example from Section 5.1. Figures 5.1–5.3 showed that the relative frequency (proportion) of packages delivered in 1 day fluctuated quite a bit over the short run, but in the long run this relative frequency settled down and stayed close to 0.3, the reported probability of a next-day delivery.

The Law of Large Numbers tells you that, as the number of repetitions of a chance experiment increases, the proportion of the time an event occurs gets close to the actual probability of the event, *even if the value of this probability is not known.* This means that you can observe outcomes from a chance experiment and then use the observed outcomes to estimate probabilities.

Relative Frequency Approach to Computing Probabilities

The **probability of an event** E, denoted by $P(E)$, is defined to be the value approached by the relative frequency of occurrence of E in a very long series of observations from a chance experiment. If the number of observations is large,

$$P(E) \approx \frac{\text{number of times } E \text{ occurs}}{\text{number of repetitions}}$$

The relative frequency definition of probability depends on being able to repeat a chance experiment many times. For example, suppose that you perform a chance experiment that consists of flipping a cap from a 20-ounce bottle of soda and noting whether the cap lands with the top up or down. Unlike tossing a coin, there is no particular reason to believe the cap is equally likely to land top up or top down. You can flip the cap many times and compute the relative frequency of the event $T = top\ up$:

$$\frac{\text{number of times the event } top\ up \text{ occurs}}{\text{total number of flips}}$$

This relative frequency is an estimate of the probability of the event T. If the bottle cap was flipped 1,000 times and it landed top up 694 times, the estimate of $P(T)$ is

$$P(T) \approx \frac{694}{1,000} = 0.694$$

In some situations, probabilities are estimated based on past history. For example, an insurance company may use past claims to estimate the probability that a 20-year-old male driver will submit a claim for a car accident in a given year. If 45,000 20-year-old males were insured in the previous year and 3,200 of them submitted a car accident claim, an estimate of this probability is $\frac{3,200}{45,000} = 0.071$. This probability is interpreted as meaning that for 20-year-old males insured by this company, about 71 out of 1,000 will submit a car accident claim in a given year.

The relative frequency approach to probability is based on observation. By observing many outcomes, you can obtain stable long-run relative frequencies that provide reasonable estimates of the probabilities of different events. The relative frequency approach to probability is intuitive, and it can be used in situations where the classical approach is not appropriate. In situations where outcomes *are* equally likely, however, either the classical or the relative frequency approach can be used.

SECTION 5.2 EXERCISES

Each Exercise Set assesses the following chapter learning objectives: M1, M2, P3

SECTION 5.2 Exercise Set 1

5.11 Phoenix is a hub for a large airline. Suppose that on a particular day, 8,000 passengers arrived in Phoenix on this airline. Phoenix was the final destination for 1,800 of these passengers. The others were all connecting to flights to other cities. On this particular day, several inbound flights were late, and 480 passengers missed their connecting flight. Of these 480 passengers, 75 were delayed overnight and had to spend the night in Phoenix. Consider the chance experiment of choosing a passenger at random from these 8,000 passengers. Calculate the following probabilities:

a. the probability that the selected passenger had Phoenix as a final destination.

b. the probability that the selected passenger did not have Phoenix as a final destination.

c. the probability that the selected passenger was connecting and missed the connecting flight.

d. the probability that the selected passenger was a connecting passenger and did not miss the connecting flight.

e. the probability that the selected passenger either had Phoenix as a final destination or was delayed overnight in Phoenix.

f. An independent customer satisfaction survey is planned. Fifty passengers selected at random from the 8,000 passengers who arrived in Phoenix on the day described above will be contacted for the survey. The airline knows that the survey results will not be favorable if too many people who were delayed overnight are included. Write a few sentences explaining whether or not you think the airline should be worried, using relevant probabilities to support your answer.

5.12 A professor assigns five problems to be completed as homework. At the next class meeting, two of the five problems will be selected at random and collected for grading. You have only completed the first three problems.

a. What is the sample space for the chance experiment of selecting two problems at random? (Hint: You can think of the problems as being labeled A, B, C, D, and E. One possible selection of two problems is A and B. If these two problems are selected and you did problems A, B and C, you will be able to turn in both problems. There are nine other possible selections to consider.)

b. Are the outcomes in the sample space equally likely?

c. What is the probability that you will be able to turn in both of the problems selected?

d. Does the probability that you will be able to turn in both problems change if you had completed the last three problems instead of the first three problems? Explain.

e. What happens to the probability that you will be able to turn in both problems selected if you had completed four of the problems rather than just three?

5.13 Suppose you want to estimate the probability that a patient will develop an infection while hospitalized at a particular hospital. In the past year, this hospital had 6,450 patients, and 712 of them developed an infection. What is the estimated probability that a patient at this hospital will develop an infection?

SECTION 5.2 Exercise Set 2

5.14 A college job placement center has requests from five students for employment interviews. Three of these students are math majors, and the other two students are statistics majors. Unfortunately, the interviewer has time to talk to only two of the students. These two will be randomly selected from among the five.

a. What is the sample space for the chance experiment of selecting two students at random? (Hint: You can think of the students as being labeled A, B, C, D, and E. One possible selection of two students is A and B. There are nine other possible selections to consider.)

b. Are the outcomes in the sample space equally likely?

c. What is the probability that both selected students are statistics majors?

d. What is the probability that both students are math majors?

e. What is the probability that at least one of the students selected is a statistics major?

f. What is the probability that the selected students have different majors?

5.15 Roulette is a game of chance that involves spinning a wheel that is divided into 38 equal segments, as shown in the accompanying picture.

Anna Baburkina/Shutterstock.com

A metal ball is tossed into the wheel as it is spinning, and the ball eventually lands in one of the 38 segments. Each segment has an associated color. Two segments are green. Half of the other 36 segments are red, and the others are black. When a balanced roulette wheel is spun, the ball is equally likely to land in any one of the 38 segments.

a. When a balanced roulette wheel is spun, what is the probability that the ball lands in a red segment?

b. In the roulette wheel shown, black and red segments alternate. Suppose instead that all red segments were grouped together and that all black segments were together. Does this increase the probability that the ball will land in a red segment? Explain.

c. Suppose that you watch 1000 spins of a roulette wheel and note the color that results from each spin. What would be an indication that the wheel was not balanced?

5.16 Suppose you want to estimate the probability that a randomly selected customer at a particular grocery store will pay by credit card. Over the past 3 months, 80,500 purchases were made, and 37,100 of them were paid for by credit card. What is the estimated probability that a randomly selected customer will pay by credit card?

Additional Exercises

5.17 According to *The Chronicle for Higher Education* **(Aug. 26, 2011)**, there were 787,325 Associate degrees awarded by U.S. community colleges in the 2008-2009 academic year. A total of 488,142 of these degrees were awarded to women.

a. If a person who received an Associate degree in 2008-2009 is selected at random, what is the probability that the selected person will be female?

b. What is the probability that the selected person will be male?

5.18 The same issue of *The Chronicle for Higher Education* referenced in Exercise 5.17 also reported the following information for degrees awarded to Hispanic students by U.S. colleges in the 2008-2009 academic year:

- A total of 274,515 degrees were awarded to Hispanic students.
- 97,921 of these degrees were Associate degrees.
- 129,526 of these degrees were Bachelor's degrees.
- The remaining degrees were either graduate or professional degrees.

What is the probability that a randomly selected Hispanic student who received a degree in 2008-2009

a. received an Associate degree?

b. received a graduate or professional degree?

c. did not receive a Bachelor's degree?

5.19 Refer to the following information on births in the United States over a given period of time:

Type of Birth	Number of Births
Single birth	41,500,000
Twins	500,000
Triplets	5000
Quadruplets	100

Use this information to estimate the probability that a randomly selected pregnant woman who gives birth

a. delivers twins

b. delivers quadruplets

c. gives birth to more than a single child

5.20 A deck of 52 cards is mixed well, and 5 cards are dealt.

a. It can be shown that (disregarding the order in which the cards are dealt) there are 2,598,960 possible hands, of which only 1,287 are hands consisting entirely of spades. What is the probability that a hand will consist entirely of spades? What is the probability that a hand will consist entirely of a single suit?

b. It can be shown that exactly 63,206 of the possible hands contain only spades and clubs, with both suits represented. What is the probability that a hand consists entirely of spades and clubs with both suits represented?

5.21 Six people hope to be selected as a contestant on a TV game show. Two of these people are younger than 25 years old. Two of these six will be chosen at random to be on the show.

a. What is the sample space for the chance experiment of selecting two of these people at random? (Hint: You can think of the people as being labeled A, B, C, D, E, and F. One possible selection of two people is A and B. There are 14 other possible selections to consider.)

b. Are the outcomes in the sample space equally likely?

c. What is the probability that both the chosen contestants are younger than 25?

d. What is the probability that both the chosen contestants are not younger than 25?

e. What is the probability that one is younger than 25 and the other is not?

SECTION 5.3 Probabilities of More Complex Events: Unions, Intersections, and Complements

In many situations, two or more different events are of interest. For example, consider a chance experiment that consists of selecting a student at random from those enrolled at a particular college. If there are 9000 students enrolled at the college, the sample space would consist of 9000 different possible outcomes, each corresponding to a student who might be selected.

In this situation, here are some possible events:

F = event that the selected student is female
O = event that the selected student is older than 30
A = event that the selected student favors expanding the athletics program
S = event that the selected student is majoring in one of the lab sciences

Because a student is to be selected at random, the outcomes in the sample space are equally likely. This means that you can use the classical approach to compute probabilities. For example, if 6,000 of the 9,000 students at the college are female,

$$P(F) = \frac{6,000}{9,000} = 0.67$$

Similarly, if 4,300 of the 9,000 students favor expanding the athletics program,

$$P(A) = \frac{4,300}{9,000} = 0.48$$

Once a number of events have been specified, it is possible to use these events to create new events. For example, you might be interested in the event that the selected student is *not* majoring in a lab science or the event that the selected student is female *and* older than 30.

Complements

The complement of an event E is a new event denoted by E^c. E^c is the event that E does *not* occur. For the event A described previously, the complement of A is

A^C = *not A* = event that the selected student does *not* favor expanding
the athletics program

In terms of outcomes, the event A^C includes all the possible outcomes in the sample space that are not in the event A. This implies

$$P(A^C) = 1 - P(A)$$

Since $P(A)$ was computed to be $P(A) = 0.48$, you know $P(A^C) = 1.0 - 0.48 = 0.52$

Complement Probabilities

If E is an event, the **complement of E**, denoted by E^c, is the event that E does *not* occur.

The probability of E^c can be computed from the probability of E as follows:

$$P(E^C) = 1 - P(E)$$

Intersections

The intersection of two events E and F is denoted by $E \cap F$. The intersection $E \cap F$ is the event that E *and* F both occur. For example, consider the two events previously described

O = event that the selected student is over 30

and

S = event that the selected student is majoring in one of the lab sciences

The event $O \cap S$ is the event that the selected student is over 30 *and* is majoring in one of the lab sciences. In the following table, the 9,000 students have been classified into one of four cells. The rows of the table correspond to whether or not the event O occurs and the columns correspond to whether or not the event S occurs.

	S (Majoring in Lab Science)	Not S (Not Majoring in Lab Science)	Total
O (Over 30)	400	1,700	**2,100**
Not O (Not Over 30)	1,100	5,800	**6,900**
Total	**1,500**	**7,500**	**9,000**

From this table, you can see that

$$P(O) = \frac{2,100}{9,000} = 0.23$$

$$P(S) = \frac{1,500}{9,000} = 0.17$$

$$P(O \cap S) = \frac{400}{9,000} = 0.04$$

The 400 in the numerator of the fraction used to compute $P(O \cap S)$ comes from the upper-left cell of the table—these 400 students are both over 30 and majoring in a lab science.

> **Intersection Events**
>
> If E and F are events, the **intersection of E and F** is denoted by $E \cap F$ and is the new event that *both E and F* occur.

Unions

The union of two events E and F is denoted by $E \cup F$. The event $E \cup F$ is the event that either E *or* F (or both) occur. In the chance experiment of selecting a student at random, you might be interested in the event that the selected student favors expanding the athletics program or is over 30 years old. This event is the union of the events A and O described earlier, and would be denoted by $A \cup O$.

Suppose that the following table describes the students at the college:

	A (Favors Expanding Athletics)	Not A (Does Not Favor Expanding Athletics)	Total
O (Over 30)	1,600	500	**2,100**
Not O (Not Over 30)	2,700	4,200	**6,900**
Total	**4,300**	**4,700**	**9,000**

From this table you can see that

$$P(O) = \frac{2,100}{9,000} = 0.23$$

$$P(A) = \frac{4,300}{9,000} = 0.48$$

$$P(O \cup A) = \frac{1,600 + 500 + 2,700}{9,000} = \frac{4,800}{9,000} = 0.53$$

Notice that the 4,800 in the numerator of the fraction used to compute $P(O \cup A)$ comes from adding the numbers in three cells of the table. These cells represent outcomes for which at least one of the events O or A occurs. Also notice that just adding the total for event O (2,100) and the total for event A (4,300) does not result in the correct numerator because the 1,600 students who are both over 30 and favor expanding athletics have been counted twice.

> **Union Events**
>
> If E and F are events, the **union** of these events is denoted by $E \cup F$. The event $E \cup F$ is the new event that E *or* F occurs.

Working with "Hypothetical 1000" Tables to Compute Probabilities

In the previous discussion, you were able to use tables to compute the probability of an intersection of two events and the probability of a union of two events. This was possible because a student was to be selected at random (making each of the 9000 possible outcomes equally likely) and because the numbers of students falling into each of the cells of the appropriate table were given. In many situations, you may only know the probabilities of some events. In this case, it is often possible to create a "hypothetical 1000" table and then use the table to compute probabilities. The following examples show how this is done.

Example 5.5 Health Information From TV Shows

The report **"TV Drama/Comedy Viewers and Health Information"** (**www.cdc.gov/ healthmarketing**) describes a large survey that was conducted for the Centers for Disease Control (CDC). The CDC believed that the sample used was representative of adult Americans. One question on the survey asked respondents if they had learned something new about a health issue or disease from a TV show in the previous 6 months. Consider the following events:

L = event that a randomly selected adult American reports learning something new about a health issue or disease from a TV show in the previous 6 months

and

F = event that a randomly selected adult American is female

Data from the survey were used to estimate the following probabilities:

$$P(L) = 0.58 \quad P(F) = 0.50 \quad P(L \cap F) = 0.31$$

From the given information, you can create a table for a "hypothetical 1000" people. You would start by labeling rows and columns of the table as follows:

	F (Female)	Not F (Not Female)	Total
L (Learned from TV)			
Not L (Did Not Learn from TV)			
Total			

Because you are creating a table of 1,000 individuals, $P(L) = 0.58$ tells you that 58% of the 1,000 people should be in the L row: $(0.58)(1,000) = 580$. This means that $1,000 - 580 = 420$ should be in the *not L* row. You can add these row totals to the table:

	F (Female)	Not F (Not Female)	Total
L (Learned from TV)			580
Not L (Did Not Learn from TV)			420
Total			1,000

Similarly, you can figure out what the column totals should be. $P(F) = 0.50$, so the total for the F column should be $(0.50)(1,000) = 500$. This means that the column total for the *not F* column should also be 500 because $1,000 - 500 = 500$. Adding these column totals to the table gives the following:

	F (Female)	Not F (Not Female)	Total
L (Learned from TV)			580
Not L (Did Not Learn from TV)			420
Total	500	500	1,000

One other probability was given:

$$P(L \cap F) = 0.31$$

This means that you would expect there to be $(0.31)(1,000) = 310$ people who were female and who learned something about a health issue or disease from TV in the previous 6 months. This number would go in the upper-left cell of the table. The remaining cells of the table can then be filled in to obtain a complete "hypothetical 1,000" table.

	F (Female)	Not F (Not Female)	Total
L (Learned from TV)	310	270	580
Not L (Did Not Learn from TV)	190	230	420
Total	500	500	1,000

This table can now be used to compute other probabilities from those that were given. For example,

$$P(L \cup F) = \frac{310 + 270 + 190}{1,000} = \frac{770}{1,000} = 0.770$$

Example 5.6 Tattoos

The article **"Chances Are You Know Someone with a Tattoo, and He's Not a Sailor"** (*Associated Press, June 11, 2006*) reported the following approximate probabilities based on a survey of adults ages 18 to 50:

$$P(\text{age 18 to 29}) = 0.50$$

$$P(\text{tattoo}) = 0.24$$

$$P(\text{tattoo } or \text{ age 18 to 29}) = 0.56$$

Consider a chance experiment that consists of selecting a person at random from a population of adults age 18 to 50. Here is some notation for the events of interest:

$$T = \text{the event that the selected person has a tattoo}$$

and

$$A = \text{the event that the selected person is age 18 to 29}$$

Assuming that the survey participants are representative of the population of adults age 18 to 50, you know

$$P(T) = 0.24 \quad P(A) = 0.50 \quad P(T \cup A) = 0.56$$

Using the given probability information, you can now create a "hypothetical 1000" table. Setting up the table and labeling the rows and columns gives

	T (Tattoo)	Not T (No Tattoo)	Total
A (Age 18 to 29)			
Not A (Age 30 to 50)			
Total			1,000

Notice that "not age 18 to 29" is equivalent to age 30 to 50 because everyone in the population of interest is age 18 to 50. $P(18 \text{ to } 29) = 0.50$ tells you that the hypothetical 1,000 people are divided evenly between the two age groups. Because $P(tattoo) = 0.24$, you expect $(0.24)(1,000) = 240$ of the hypothetical 1,000 people to be in the tattoo column and the rest $(1,000 - 240 = 760)$ to be in the no tattoo column, as shown in the following table.

	T (Tattoo)	Not T (No Tattoo)	Total
A (Age 18 to 29)	Cell 1	Cell 2	**500**
Not A (Age 30 to 50)	Cell 3		**500**
Total	**240**	**760**	**1,000**

The next step is to try to fill in the cells of the table using the given information. Because you know that $P(T \cup A) = 0.56$, you know that the sum of the counts in the cells labeled Cell 1, Cell 2, and Cell 3 in the preceding table must total $(0.56)(1,000) = 560$ because these cells correspond to the union of the events *tattoo* and *age 18 to 29*. Adding the total for the *tattoo* column and the *age 18 to 29* row gives

$$240 + 500 = 740$$

Because this counts the people in Cell 1 twice, the difference between 740 and the union count of 560 must be the number of people in Cell 1. This means that the Cell 1 count is $740 - 560 = 180$. Once you know the marginal totals and this cell count, you can fill in the rest of the cells to get the following table:

	T (Tattoo)	Not T (No Tattoo)	Total
A (Age 18 to 29)	180	320	**500**
Not A (Age 30 to 50)	60	440	**500**
Total	**240**	**760**	**1,000**

This table can now be used to compute other probabilities. For example,

$$P(T \cap A) = \frac{180}{1,000} = 0.180$$

Completing a "hypothetical 1000" table for two events E and F usually requires that you know $P(E)$, $P(F)$, and either $P(E \cup F)$ or $P(E \cap F)$. However, there are two special cases where it is possible to complete a "hypothetical 1000" table from just $P(E)$ and $P(F)$. These special cases are when the events E and F are *mutually exclusive* or when the events E and F are *independent*.

Mutually Exclusive Events

Two events E and F are *mutually exclusive* if they can't occur at the same time. For example, consider the chance experiment that involves selecting a student at random from those enrolled at a particular college. The two events

$$F = \text{event that the selected student is a freshman}$$

and

$$S = \text{event that the selected student is a sophomore}$$

are mutually exclusive because no outcome could result in both of these events occurring. This implies that in the following "hypothetical 1000" table, the number in the upper-left cell must be 0.

	S (Sophomore)	Not S (Not Sophomore)	Total
F (Freshman)	0		
Not F (Not Freshman)			
Total			**1,000**

In this case, knowing $P(F)$ and $P(S)$ would provide enough information to complete the table.

Notice that because the upper-left cell of the table has a value of 0, $P(E \cap F) = 0$. Also, if E and F are mutually exclusive, $P(E \cup F)$ will be the sum of the two individual event probabilities $P(E)$ and $P(F)$.

Example 5.7 Calls to 9-1-1

Sometimes people call the emergency 9-1-1 number to report situations that are not considered emergencies (such as to report a lost dog). Suppose that 30% of the calls to the emergency 9-1-1 number in a particular county are for medical emergencies and that 20% are calls that are not considered emergencies. Consider the two events

M = event that the next call to the emergency line is for a medical emergency

and

N = event that the next call to the emergency line is not considered an emergency

Based on the given information, you know $P(M) = 0.30$ and $P(N) = 0.20$. In this situation, M and N are mutually exclusive because the next call to the emergency line can't be both a medical emergency *and* a call that is not considered an emergency.

Setting up the "hypothetical 1000" table for these two events gives

	N (Non-Emergency Call)	Not N (Emergency Call)	Total
M (Medical Emergency)	0		300
Not M (Not Medical Emergency)			700
Total	200	800	1,000

Completing the table by filling in the remaining cells gives

	N (Non-Emergency Call)	Not N (Emergency Call)	Total
M (Medical Emergency)	0	300	300
Not M (Not Medical Emergency)	200	500	700
Total	200	800	1,000

From the table, you can see that

$$P(M \cap N) = 0$$

and

$$P(M \cup N) = \frac{300 + 200}{1000} = 0.500$$

which is equal to $P(M) + P(N) = 0.3 + 0.2 = 0.5$.

As you probably noticed in working through Example 5.7, in the case of mutually exclusive events, you don't really need to complete the "hypothetical 1000" table to compute the probability of a union or an intersection.

Addition Rule for Mutually Exclusive Events

If E and F are mutually exclusive events, then

$$P(E \cap F) = 0$$

and

$$P(E \cup F) = P(E) + P(F)$$

Independent Events

Suppose that the incidence rate for a particular disease in a certain population is known to be 1 in 1,000. Then the probability that a randomly selected individual from this population has the disease is 0.001. Further suppose that a diagnostic test for the disease is given to the selected individual, and the test result is positive. Even if the diagnostic test sometimes returns a positive result for someone who does not have the disease, a positive result will make you want to revise the probability that this person has the disease upward from 0.001. Knowing that one event has occurred (the selected person has a positive test result) can change your assessment of the probability of another event (the selected person has the disease).

As another example, consider a university's course registration process, which divides students into 12 priority groups. Overall, only 10% of all students receive all requested classes, but 75% of those in the first priority group receive all requested classes. Interpreting these figures as probabilities, you would say that the probability that a randomly selected student at this university receives all requested classes is 0.10. However, if you know that the selected student is in the first priority group, you revise the probability that the selected student receives all requested classes to 0.75. Knowing that the event *selected person is in the first priority group* has occurred changes your assessment of the probability of the event *selected person received all requested classes*. These two events are said to be **dependent**.

It is often the case, however, that the likelihood of one event is not affected by the occurrence of another event. For example, suppose that you purchase a desktop computer system with a separate monitor and keyboard. Two possible events of interest are

Event 1: The monitor needs service while under warranty
Event 2: The keyboard needs service while under warranty

Because the two components operate independently of one another, learning that the monitor has needed warranty service would not affect your assessment of the likelihood that the keyboard will need repair. If you know that 1% of all keyboards need repair while under warranty, you say that the probability that a keyboard needs warranty service is 0.01. Knowing that the monitor needed warranty service does not affect this probability. You say that *monitor failure* (Event 1) and *keyboard failure* (Event 2) are **independent** events.

DEFINITION

Independent events: Two events are **independent** if the probability that one event occurs is not affected by knowledge of whether the other event has occurred.

Dependent events: Two events are **dependent** if knowing that one event has occurred changes your assessment of the probability that the other event occurs.

In the previous examples,

- the events *has disease* and *tests positive* are dependent.
- the events *receives all requested classes* and *is in the first priority group* are dependent.
- the events *monitor needs warranty service* and *keyboard needs warranty service* are independent.

Multiplication Rule for Independent Events

An individual who purchases a computer system might wonder how likely it is that *both* the monitor and the keyboard will need service while under warranty. A student who must take either a chemistry course or a physics course to fulfill a science requirement might be concerned about the chance that both courses are full before the student has a chance to register. In each of these cases, the focus is on the probability that two different events occur together. For independent events, a simple multiplication rule relates the individual event probabilities to the probability that the events occur together.

> ### Multiplication Rule for Independent Events
>
> If two events are independent, the probability that *both* events occur is the product of the individual event probabilities. Denoting the events as E and F, we write
>
> $$P(E \cap F) = P(E)P(F)$$
>
> More generally, if there are k independent events, the probability that all the events occur is the product of all the individual event probabilities.

Example 5.8 Nuclear Power Plant Warning Sirens

The Diablo Canyon nuclear power plant in Avila Beach, California, has a warning system that includes a network of sirens. Sirens are located approximately 0.5 miles from each other. When the system is tested, individual sirens sometimes fail. The sirens operate independently of one another, so knowing that a particular siren has failed does not change the probability that any other siren fails.

Imagine that you live near Diablo Canyon and that there are two sirens (Siren 1 and Siren 2) that can be heard from your home. You might be concerned about the probability that *both* Siren 1 *and* Siren 2 fail. Suppose that when the siren system is activated, about 5% of the individual sirens fail. Then

$$P(\text{Siren 1 fails}) = 0.05$$

$$P(\text{Siren 2 fails}) = 0.05$$

and the two events *Siren 1 fails* and *Siren 2 fails* are independent. Using the multiplication rule for independent events,

$$P(\text{Siren 1 fails} \cap \text{Siren 2 fails}) = P(\text{Siren 1 fails})P(\text{Siren 2 fails})$$

$$= (0.05)(0.05)$$

$$= 0.0025$$

Even though the probability that any individual siren will fail is 0.05, the probability that *both* of the two sirens will fail is much smaller. This would happen in the long run only about 2.5 times for every 1000 times the system is activated.

If two events E and F are known to be independent, it is possible to complete a "hypothetical 1000" table using just $P(E)$ and $P(F)$. This is illustrated in the following example.

Example 5.9 Satellite Backup Systems

A satellite has both a main and a backup solar power system. (It is common to design such redundancy into products to increase their reliability.) Suppose that the probability of failure during the 10-year lifetime of the satellite is 0.05 for the main power system and 0.08 for the backup system. What is the probability that both systems will fail? Because one system failing has no effect on whether the other one does, it is reasonable to assume that the following two events are independent:

  $M =$ event that main system fails
  $B =$ event that backup system fails

From the given information, you know

  $P(M) = 0.05$
  $P(B) = 0.08$

Because M and B are independent,

$$P(M \cap B) = P(M)P(B) = (0.05)(0.08) = 0.004$$

This gives the information needed to complete a "hypothetical 1000" table:

	B (Backup Fails)	Not B (Backup Does Not Fail)	Total
M (Main Fails)	4	46	50
Not M (Main Does Not Fail)	76	874	950
Total	80	920	1,000

Using this table allows you to compute $P(M \cup B)$ as

$$P(M \cup B) = \frac{4 + 46 + 76}{1,000} = \frac{126}{1,000} = 0.126$$

You can also compute the probability of the satellite having at least one functioning system as

$P(\text{at least one system works}) = P(\text{main system does not fail} \cup \text{backup system does not fail})$

$$= P(M^C \cup B^C)$$
$$= \frac{76 + 874 + 46}{1,000}$$
$$= \frac{996}{1,000}$$
$$= 0.996$$

The probability that the main system fails but the backup system does not fail is

$$P(M \cap B^C) = \frac{46}{1,000} = 0.046$$

SECTION 5.3 EXERCISES

Each Exercise Set assesses the following chapter learning objectives: C2, C3, M3, M4, M5, M6, M7, M8

SECTION 5.3 Exercise Set 1

5.22 A small college has 2,700 students enrolled. Consider the chance experiment of selecting a student at random. For each of the following pairs of events, indicate whether or not you think they are mutually exclusive and explain your reasoning.
a. the event that the selected student is a senior and the event that the selected student is majoring in computer science.
b. the event that the selected student is female and the event that the selected student is majoring in computer science.
c. the event that the selected student's college residence is more than 10 miles from campus and the event that the selected student lives in a college dormitory.
d. the event that the selected student is female and the event that the selected student is on the college football team.

5.23 A Gallup survey found that 46% of women and 37% of men experience pain on a daily basis (*San Luis Obispo Tribune*, **April 6, 2000**). Suppose that this information is

representative of U.S. adults. If a U.S. adult is selected at random, are the events *selected adult is male* and *selected adult experiences pain on a daily basis* independent or dependent? Explain.

5.24 A bank offers both adjustable-rate and fixed-rate mortgage loans on residential properties, which are classified into three categories: single-family houses, condominiums, and multifamily dwellings. Each loan made in 2010 was classified according to type of mortgage and type of property, resulting in the following table. Consider the chance experiment of selecting one of these 3,750 loans at random.

	Single-Family	Condo	Multifamily	Total
Adjustable	1,500	788	337	2,625
Fixed-Rate	375	377	373	1,125
Total	1,875	1,165	710	3,750

a. What is the probability that the selected loan will be for an adjustable rate mortgage?
b. What is the probability that the selected loan will be for a multifamily property?

c. What is the probability that the selected loan will not be for a single-family property?

d. What is the probability that the selected loan will be for a single-family property or a condo?

e. What is the probability that the selected loan will be for a multifamily property or for an adjustable rate loan?

f. What is the probability that the selected loan will be a fixed-rate loan for a condo?

5.25 There are two traffic lights on Shelly's route from home to work. Let E denote the event that Shelly must stop at the first light, and define the event F in a similar manner for the second light. Suppose that $P(E) = 0.4$, $P(F) = 0.3$, and $P(E \cap F) = 0.15$.

a. Use the given probability information to set up a "hypothetical 1000" table with columns corresponding to E and *not E* and rows corresponding to F and *not F*.

b. Use the table from Part (a) to find the following probabilities:
   i. the probability that Shelly must stop for at least one light (the probability of $E \cup F$).
   ii. the probability that Shelly does not have to stop at either light.
   iii. the probability that Shelly must stop at exactly one of the two lights.
   iv. the probability that Shelly must stop only at the first light.

5.26 A large cable company reports that 80% of its customers subscribe to its cable TV service, 42% subscribe to its Internet service, and 97% subscribe to at least one of these two services.

a. Use the given probability information to set up a "hypothetical 1000" table.

b. Use the table from Part (a) to find the following probabilities:
   i. the probability that a randomly selected customer subscribes to both cable TV and Internet service.
   ii. the probability that a randomly selected customer subscribes to exactly one of these services.

5.27 **a.** Suppose events E and F are mutually exclusive with $P(E) = 0.41$ and $P(F) = 0.23$.
   i. What is the value of $P(E \cap F)$?
   ii. What is the value of $P(E \cup F)$?

b. Suppose that for events A and B, $P(A) = 0.26$, $P(B) = 0.34$, and $P(A \cup B) = 0.47$. Are A and B mutually exclusive? How can you tell?

5.28 Each time a class meets, the professor selects one student at random to explain the solution to a homework problem. There are 40 students in the class, and no one ever misses class. Luke is one of these students. What is the probability that Luke is selected both of the next two times that the class meets?

5.29 A large retail store sells MP3 players. A customer who purchases an MP3 player can pay either by cash or credit card. An extended warranty is also available for purchase. Suppose that the events

   M = event that the customer pays by cash
   E = event that the customer purchases an extended warranty

are independent with $P(M) = 0.47$ and $P(E) = 0.16$.

a. Construct a "hypothetical 1000" table with columns corresponding to cash or credit card and rows corresponding to whether or not an extended warranty is purchased.

b. Use the table to find $P(M \cup E)$. Give a long-run relative frequency interpretation of this probability.

SECTION 5.3 **Exercise Set 2**

5.30 Consider a chance experiment that consists of selecting a student at random from a high school with 3,000 students.

a. In the context of this chance experiment, give an example of two events that would be mutually exclusive.

b. In the context of this chance experiment, give an example of two events that would not be mutually exclusive.

5.31 The Associated Press (*San Luis Obispo Telegram-Tribune*, **August 23, 1995**) reported the results of a study in which schoolchildren were screened for tuberculosis (TB). It was reported that for Santa Clara County, California, the proportion of all tested kindergartners who were found to have TB was 0.0006. The corresponding proportion for recent immigrants (thought to be a high-risk group) was 0.0075. Suppose that a Santa Clara County kindergartner is to be selected at random. Are the events *selected student is a recent immigrant* and *selected student has TB* independent or dependent events? Justify your answer using the given information.

5.32 The accompanying data are from the article **"Characteristics of Buyers of Hybrid Honda Civic IMA: Preferences, Decision Process, Vehicle Ownership, and Willingness-to-Pay" (Institute for Environmental Decisions, November 2006).** Each of 311 people who purchased a Honda Civic was classified according to gender and whether or not the car purchased had a hybrid engine.

	Hybrid	Not Hybrid
Male	77	117
Female	34	83

Suppose one of these 311 individuals is to be selected at random.

a. What is the probability that the selected individual purchased a hybrid?

b. What is the probability that the selected individual is male?

c. What is the probability that the selected individual is male and purchased a hybrid?

d. What is the probability that the selected individual is female or purchased a hybrid?

e. What is the probability that the selected individual is female and purchased a hybrid?

f. Explain why the probabilities computed in Parts (d) and (e) are not equal.

5.33 The paper **"Predictors of Complementary Therapy Use Among Asthma Patients: Results of a Primary Care Survey"** (*Health and Social Care in the Community* [2008]: 155–164) described a study in which each person in a large sample of asthma patients responded to two questions:

> Question 1: Do conventional asthma medications usually help your symptoms?
>
> Question 2: Do you use complementary therapies (such as herbs, acupuncture, aroma therapy) in the treatment of your asthma?

Suppose that this sample is representative of asthma patients. Consider the following events:

E = event that the patient uses complementary therapies

F = event that the patient reports conventional medications usually help

The data from the sample were used to estimate the following probabilities:

$$P(E) = 0.146 \quad P(F) = 0.879 \quad P(E \cap F) = 0.122$$

a. Use the given probability information to set up a "hypothetical 1000" table with columns corresponding to E and *not E* and rows corresponding to F and *not F*.

b. Use the table from Part (a) to find the following probabilities:

i. The probability that an asthma patient responds that conventional medications do not help and that the patient uses complementary therapies.

ii. The probability that an asthma patient responds that conventional medications do not help and that the patient does not use complementary therapies.

iii. The probability that an asthma patient responds that conventional medications usually help or the patient uses complementary therapies.

c. Are the events E and F independent? Explain.

5.34 An appliance manufacturer offers extended warranties on its washers and dryers. Based on past sales, the manufacturer reports that of customers buying both a washer and a dryer, 52% purchase the extended warranty for the washer, 47% purchase the extended warranty for the dryer, and 59% purchase at least one of the two extended warranties.

a. Use the given probability information to set up a "hypothetical 1000" table.

b. Use the table from Part (a) to find the following probabilities:

i. the probability that a randomly selected customer who buys a washer and a dryer purchases an extended warranty for both the washer and the dryer.

ii. the probability that a randomly selected customer purchases an extended warranty for neither the washer nor the dryer.

5.35 **a.** Suppose events E and F are mutually exclusive with $P(E) = 0.14$ and $P(F) = 0.76$.

i. What is the value of $P(E \cap F)$?

ii. What is the value of $P(E \cup F)$?

b. Suppose that for events A and B, $P(A) = 0.24$, $P(B) = 0.24$, and $P(A \cup B) = 0.48$. Are A and B mutually exclusive? How can you tell?

5.36 "N.Y. Lottery Numbers Come Up 9-1-1 on 9/11" was the headline of an article that appeared in the *San Francisco Chronicle* **(September 13, 2002)**. More than 5,600 people had selected the sequence 9-1-1 on that date, many more than is typical for that sequence. A professor at the University of Buffalo was quoted as saying, "I'm a bit surprised, but I wouldn't characterize it as bizarre. It's randomness. Every number has the same chance of coming up. People tend to read into these things. I'm sure that whatever numbers come up tonight, they will have some special meaning to someone, somewhere." The New York state lottery uses balls numbered 0–9 circulating in three separate bins. One ball is chosen at random from each bin. What is the probability that the sequence 9-1-1 would be selected on any particular day?

5.37 A rental car company offers two options when a car is rented. A renter can choose to pre-purchase gas or not and can also choose to rent a GPS device or not. Suppose that the events

$$A = \text{event that gas is pre-purchased}$$

$$B = \text{event that a GPS is rented}$$

are independent with $P(A) = 0.20$ and $P(B) = 0.15$.

a. Construct a "hypothetical 1000" table with columns corresponding to whether or not gas is pre-purchased and rows corresponding to whether or not a GPS is rented.

b. Use the table to find $P(A \cup B)$. Give a long-run relative frequency interpretation of this probability.

Additional Exercises

5.38 The table at the top of the next page describes (approximately) the distribution of students by gender and college at a midsize public university in the West. Suppose you were to randomly select one student from this university.

a. What is the probability that the selected student is a male?

b. What is the probability that the selected student is in the College of Agriculture?

c. What is the probability that the selected student is a male in the College of Agriculture?

d. What is the probability that the selected student is a male who is not from the College of Agriculture?

TABLE FOR EXERCISE **5.38**

			College				
Gender	Education	Engineering	Liberal Arts	Science and Math	Agriculture	Business	Architecture
Male	200	3,200	2,500	1,500	2,100	1,500	200
Female	300	800	1,500	1,500	900	1,500	300

5.39 A Nielsen survey of teens between the ages of 13 and 17 found that 83% use text messaging and 56% use picture messaging (**"How Teens Use Media," Nielsen, June 2009**). Use these percentages to explain why the two events

T = event that a randomly selected teen uses text messaging

and

P = event that a randomly selected teen uses picture messaging

cannot be mutually exclusive.

5.40 The report **"Improving Undergraduate Learning"** (**Social Science Research Council, 2011**) summarizes data from a survey of several thousand college students. These students were thought to be representative of the population of all college students in the United States. When asked about an upcoming semester, 68% said they would be taking a class that is reading-intensive (requires more than 40 pages of reading per week). Only 50% said they would be taking a class that is writing-intensive (requires more than 20 pages of writing over the course of the semester). The percentage who said that they would be taking both a reading-intensive course and a writing-intensive course was 42%.
a. Use the given information to set up a "hypothetical 1000" table.
b. Use the table to find the following probabilities:
   i. the probability that a randomly selected student would be taking at least one of these intensive courses.
   ii. the probability that a randomly selected student would be taking one of these intensive courses, but not both.
   iii. the probability that a randomly selected student would be taking neither a reading-intensive nor a writing-intensive course.

5.41 Airline tickets can be purchased online, by telephone, or by using a travel agent. Passengers who have a ticket sometimes don't show up for their flights. Suppose a person who purchased a ticket is selected at random. Consider the following events:

O = event selected person purchased ticket online
N = event selected person did not show up for flight

Suppose $P(O) = 0.70$, $P(N) = 0.07$, and $P(O \cap N) = 0.04$.
a. Are the events N and O independent? How can you tell?
b. Construct a "hypothetical 1000" table with columns corresponding to N and *not* N and rows corresponding to O and *not* O.
c. Use the table to find $P(O \cup N)$. Give a relative frequency interpretation of this probability.

5.42 The following statement is from a letter to the editor that appeared in **USA Today (September 3, 2008)**: "Among Notre Dame's current undergraduates, our ethnic minority students (21%) and international students (3%) alone equal the percentage of students who are children of alumni (24%). Add the 43% of our students who receive need-based financial aid (one way to define working-class kids), and more than 60% of our student body is composed of minorities and students from less affluent families."

Do you think that the statement that more than 60% of the student body is composed of minorities and students from less affluent families is likely to be correct? Explain.

SECTION 5.4 Conditional Probability

In the previous section, you saw that sometimes the knowledge that one event has occurred changes the assessment of the chance that another event occurs. In this section, we begin by revisiting one of the examples from Section 5.3 in more detail.

Consider a population in which 0.1% of all individuals have a certain disease. The presence of the disease cannot be discerned from outward appearances, but there is a diagnostic test available. Unfortunately, the test is not always correct. Of those with positive test results, 80% actually have the disease and the other 20% who show positive test results are false-positives. To put this in probability terms, consider the chance experiment in which an individual is randomly selected from the population. Define the following events:

E = event that the individual has the disease

F = event that the individual's diagnostic test is positive

We will use $P(E|F)$ to denote the probability of the event E given that the event F is known to have occurred. This new notation is used to indicate that a probability calculation has been made conditional on the occurrence of another event. The standard symbol for this is a vertical line, and it is read "given." You would say, "the probability that an individual has the disease *given* that the diagnostic test is positive," and represent this symbolically as P(has disease|positive test) or $P(E|F)$. This probability is called a **conditional probability**.

The information provided above implies that

$$P(E) = 0.001$$

$$P(E|F) = 0.8$$

Before you have diagnostic test information, you would think that the occurrence of E is unlikely, but once it is known that the test result is positive, your assessment of the likelihood of the disease increases dramatically. (Otherwise the diagnostic test would not be very useful!)

Example 5.10 After-School Activities

After-school activities at a large high school can be classified into three types: athletics, fine arts, and other. The following table gives the number of students participating in each of these types of activities by grade:

	9th	10th	11th	12th	Total
Athletics	150	160	140	150	**600**
Fine Arts	100	90	120	125	**435**
Other	125	140	150	150	**656**
Total	**375**	**300**	**410**	**425**	**1,600**

For purposes of this example, you can assume that *every* student at the school is in *exactly one* of these after-school activities. Look to see how each of the following statements follows from the information in the table:

1. There are 160 10th-grade students participating in athletics.
2. The number of seniors participating in fine arts activities is 125.
3. There are 435 students participating in fine arts activities.
4. There are 410 juniors.
5. The total number of students at the school is 1,600.

Each week during the school year, the principal plans to select a student at random and invite this student to have lunch with her to discuss various student concerns. She feels that random selection will give her the greatest chance of hearing from a representative cross-section of the student body. Because the student will be selected at random, each student is equally likely to be selected. What is the probability that a randomly selected student is a senior athlete? You can calculate this probability as follows:

$$P(\text{senior athlete}) = \frac{\text{number of senior athletes}}{\text{total number of students}} = \frac{150}{1,600} = 0.09375$$

Now suppose that the principal's secretary records not only the student's name but also the student's grade level. The secretary has indicated that the selected student is a senior. Does this information change the assessment of the likelihood that the selected student is an athlete? Because 150 of the 425 seniors participate in athletics, this suggests that

$$P(\text{athlete}|\text{senior}) = \frac{\text{number of senior athletes}}{\text{total number of seniors}} = \frac{150}{425} = 0.3529$$

The probability is calculated in this way because you know that the selected student is one of 425 seniors, each of whom is equally likely to have been selected. If you were to repeat the chance experiment of selecting a student at random, about 35.29% of the times *that resulted in a senior being selected* would also result in the selection of someone who participates in athletics.

Example 5.11 GFI Switches

A GFI (ground fault interrupt) switch turns off power to a system in the event of an electrical malfunction. A spa manufacturer currently has 25 spas in stock, each equipped with a single GFI switch. Two different companies supply the switches, and some of the switches are defective, as summarized in the following table:

	Not Defective	Defective	Total
Company 1	10	5	**15**
Company 2	8	2	**10**
Total	**18**	**7**	**25**

A spa is randomly selected for testing. Consider the following events:

$$E = \text{event that GFI switch in the selected spa is from Company 1}$$

$$F = \text{event that GFI switch in the selected spa is defective}$$

Using the table, you can compute the following probabilities:

$$P(E) = \frac{15}{25} = 0.60$$

$$P(F) = \frac{7}{25} = 0.28$$

$$P(E \cap F) = \frac{5}{25} = 0.20$$

Now suppose that testing reveals a defective switch. This means that the chosen spa is one of the seven in the "defective" column. How likely is it that the switch came from the first company? Because five of the seven defective switches are from Company 1,

$$P(E|F) = P(\text{company 1}|\text{defective}) = \frac{5}{7} = 0.714$$

Notice that this is larger than the unconditional probability $P(E)$. This is because Company 1 has a much higher defective rate than Company 2.

Example 5.12 Surviving a Heart Attack

Medical guidelines recommend that a hospitalized patient who suffers cardiac arrest should receive defibrillation (an electric shock to the heart) within 2 minutes. The paper **"Delayed Time to Defibrillation After In-Hospital Cardiac Arrest"** (*The New England Journal of Medicine* [2008]: 9–17) describes a study of the time to defibrillation for hospitalized patients in hospitals of various sizes. The authors examined medical records of 6,716 patients who suffered cardiac arrest while hospitalized, recording the size of the hospital and whether or not defibrillation occurred within 2 minutes. Data from this study are summarized here:

Hospital Size	Time to Defibrillation		Total
	Within 2 Minutes	More Than 2 Minutes	
Small (Fewer Than 250 Beds)	1,124	576	**1,700**
Medium (250–499 Beds)	2,178	886	**3,064**
Large (500 or More Beds)	1,387	565	**1,952**
Total	**4,689**	**2,027**	**6,716**

In this example, assume that these data are representative of the larger group of all hospitalized patients who suffer a cardiac arrest. Suppose that a hospitalized patient who suffered a cardiac arrest is selected at random. The following events are of interest:

S = event that the selected patient is at a small hospital
M = event that the selected patient is at a medium-sized hospital
L = event that the selected patient is at a large hospital
D = event that the selected patient receives defibrillation within 2 minutes

Using the table, you can compute

$$P(D) = \frac{4,689}{6,716} = 0.698$$

This probability is interpreted as the proportion of hospitalized patients suffering cardiac arrest that receive defibrillation within 2 minutes. That is, 69.8% of these patients receive timely defibrillation.

Now suppose it is known that the selected patient was at a small hospital. How likely is it that this patient received defibrillation within 2 minutes? To answer this question, you need to compute $P(D|S)$, the probability of defibrillation within 2 minutes, given that the patient is at a small hospital. From the information in the table, you know that of the 1,700 patients in small hospitals, 1,124 received defibrillation within 2 minutes, so

$$P(D|S) = \frac{1,124}{1,700} = 0.661$$

Notice that this is smaller than the unconditional probability, $P(D) = 0.698$. This tells you that there is a smaller probability of timely defibrillation at a small hospital.

Two other conditional probabilities of interest are

$$P(D|M) = \frac{2,178}{3,064} = 0.711$$

and

$$P(D|L) = \frac{1,387}{1,952} = 0.711$$

From this, you see that the probability of timely defibrillation is the same for patients at medium-sized and large hospitals, and that this probability is higher than that for patients at small hospitals.

It is also possible to compute $P(L|D)$, the probability that a patient is at a large hospital given that the patient received timely defibrillation:

$$P(L|D) = \frac{1,387}{4,689} = 0.296$$

Let's look carefully at the interpretation of some of these probabilities:

1. $P(D) = 0.698$ is interpreted as the proportion of *all* hospitalized patients suffering cardiac arrest who receive timely defibrillation. Approximately 69.8% of these patients receive timely defibrillation.
2. $P(D \cap L) = \frac{1,387}{6,716} = 0.207$ is the proportion of *all* hospitalized patients who suffer cardiac arrest who are at a large hospital *and* who receive timely defibrillation.
3. $P(D|L) = 0.711$ is the proportion *of patients at large hospitals* suffering cardiac arrest who receive timely defibrillation.
4. $P(L|D) = 0.296$ is the proportion *of patients who receive timely defibrillation* who were at large hospitals.

Notice the difference between the unconditional probabilities in Interpretations 1 and 2 and the conditional probabilities in Interpretations 3 and 4. The unconditional probabilities involve the entire group of interest (all hospitalized patients suffering cardiac arrest), whereas the conditional probabilities only involve a subset of the entire group defined by the "given" event.

AP* EXAM TIP

If two events E and F are independent, then $P(E|F) = P(E)$. One way to check for independence is to compare these two probabilities. In the example here, $P(D) = 0.698$ and $P(D|S) = 0.661$. The fact that these two probabilities are not equal tells you that the events D and S are not independent.

As mentioned in the opening paragraphs of this section, one of the most important applications of conditional probability is in making diagnoses. Your mechanic diagnoses your car by hooking it up to a machine and reading the pressures and speeds of the various components. A meteorologist diagnoses the weather by looking at temperatures, isobars, and wind speeds. Medical doctors diagnose the state of a person's health by performing various tests and gathering information, such as weight and blood pressure. Doctors also observe characteristics of their patients in an attempt to determine whether or not their patients have a certain disease. Many diseases are not actually observable—or at least not easily so—and often the doctor must make a probabilistic judgment. As you will see, conditional probability plays a large role in evaluating diagnostic techniques.

Normally, a criterion exists to determine with certainty whether or not a person has a disease. This criterion is colloquially known as the "gold standard." The gold standard might be an invasive surgical procedure, which can be expensive or dangerous. Or, the gold standard might be a test that takes a long time to perform in the lab or that is costly. For a diagnostic test to be preferred over the gold standard, it would need to be faster, less expensive, or less invasive and yet still produce results that agree with the gold standard. In this context, *agreement* means (1) that the test generally comes out positive when the patient has the disease and (2) that the test generally comes out negative when the patient does not have the disease. These statements involve conditional probabilities, as illustrated in the following example.

Example 5.13 Diagnosing Tuberculosis

To illustrate the calculations involved in evaluating a diagnostic test, consider the case of tuberculosis (TB), an infectious disease that typically attacks lung tissue. Before 1998, culturing was the existing gold standard for diagnosing TB. This method took 10 to 15 days to get a result. In 1998, investigators evaluated a DNA technique that turned out to be much faster (**"LCx: A Diagnostic Alternative for the Early Detection of *Mycobacterium tuberculosis* Complex," *Diagnostic Microbiology and Infectious Diseases* [1998]: 259–264**). The DNA technique for detecting tuberculosis was evaluated by comparing results from the test to the existing gold standard, with the following results for 207 patients exhibiting symptoms:

	Has Tuberculosis (Gold Standard)	Does Not Have Tuberculosis (Gold Standard)	Total
DNA Test Positive	14	0	**14**
DNA Test Negative	12	181	**193**
Total	**26**	**181**	**207**

Looking at the table, the DNA technique seems to be effective as a diagnostic test. Patients who tested positive with the DNA test were also positive with the gold standard in every case. Patients who tested negative with the DNA test generally tested negative with the gold standard, but the table also indicates some false-negative results for the DNA test.

Consider a randomly selected individual who is tested for tuberculosis. The following events are of interest:

$$T = \text{event that the individual has tuberculosis}$$

$$N = \text{event that the DNA test is negative}$$

What is the probability that a person has tuberculosis, given that the DNA test was negative? To answer this question, you need to compute $P(T|N)$. From the table, you see that of the 193 negative DNA tests, 12 actually did have tuberculosis. So,

$$P(T|N) = P(\text{tuberculosis}|\text{negative DNA test}) = \frac{12}{193} = 0.0622$$

Notice that 12.56% $\left(\frac{26}{207} = 0.1256\right)$ of those tested had tuberculosis. The added information provided by the diagnostic test has altered the probability—and provided some measure of relief for the patients who test negative. Once it is known that the test result is negative, the estimated probability of having TB is cut in half.

A probability computed from information in a table is a fraction. The key to computing these probabilities is having the right numerator and the right denominator when you put your fraction together. The denominator for a conditional probability will be a row or column total, whereas the denominator for an unconditional probability is the total for the entire table.

In the examples considered so far, you have started with information about a population that was already in table form. When this is not the case, it is sometimes possible to use given probability information to construct a "hypothetical 1000" table. This approach usually makes computing conditional probabilities relatively easy, as illustrated in the following two examples.

Example 5.14   Internet Addiction

Internet addiction has been defined by researchers as a disorder characterized by excessive time spent on the Internet, impaired judgment and decision-making ability, social withdrawal, and depression. The paper **"The Association between Aggressive Behaviors and Internet Addiction and Online Activities in Adolescents"** (*Journal of Adolescent Health* [2009]: 598–605) describes a study of a large number of adolescents. Each participant in the study was assessed using the Chen Internet Addiction Scale to determine if he or she suffered from Internet addiction. The following statements are based on the survey results:

1. 51.8% of the study participants were female and 48.2% were male.
2. 13.1% of the females suffered from Internet addiction.
3. 24.8% of the males suffered from Internet addiction.

Consider the chance experiment that consists of selecting a study participant at random, and define the following events:

F = the event that the selected participant is female

M = the event that the selected participant is male

I = the event that the selected participant suffers from Internet addiction

The three statements from the paper define the following probabilities:

$$P(F) = 0.518$$

$$P(M) = 0.482$$

$$P(I|F) = 0.131$$

$$P(I|M) = 0.248$$

Let's create a "hypothetical 1000" table based on the given probabilities. You can set up a table with rows that correspond to whether or not the selected person is female and columns that correspond to whether or not the selected person is addicted to the Internet.

	I (Internet Addiction)	Not *I* (No Internet Addiction)	Total
F (Female)			
Not F (Male)			
Total			1,000

Because $P(F) = 0.518$, you would expect $(0.518)(1000) = 518$ females and $(0.482)(1000) = 482$ males in the hypothetical 1000. This gives the row totals for the table:

	I (Internet Addiction)	Not *I* (No Internet Addiction)	Total
F (Female)			**518**
Not *F* (Male)			**482**
Total			**1,000**

How do you use the information provided by the given conditional probabilities? The probability $P(I|F) = 0.131$ tells you that 13.1% *of the females* are addicted to the Internet. So, you would expect the 518 females to be divided into

$$(0.131)(518) \approx 68$$

who are addicted and $518 - 68 = 450$ who are not. For males, you would expect

$$(0.248)(482) \approx 120$$

who are addicted and the remaining $482 - 120 = 362$ who are not. This allows you to complete the "hypothetical 1000" table as shown here:

	I (Internet Addiction)	Not *I* (No Internet Addiction)	Total
F (Female)	68	450	**518**
Not *F* (Male)	120	362	**482**
Total	**188**	**812**	1,000

You can now use this table to compute some probabilities of interest that were not given. For example,

$$P(I) = \frac{188}{1,000} = 0.188$$

means that 18.8% of the participants were addicted to the Internet. Also, the probability that a selected participant is female given that the selected person is an Internet addict is

$$P(F|I) = \frac{68}{188} = 0.362$$

Notice that this probability is *not* the same as $P(I|F)$. $P(I|F)$ is the proportion *of females* who are addicted, whereas $P(F|I)$ is the proportion *of Internet addicts* who are female.

Example 5.15   Should You Paint the Nursery Pink? Revisited

The example in the Chapter Preview section dealt with gender predictions based on ultra-sounds performed during the first trimester of pregnancy (**"The Use of Three-Dimensional Ultrasound for Fetal Gender Determination in the First Trimester,"** *The British Journal of Radiology* [2003]: 448–451). Radiologist 1 looked at 159 first trimester ultrasound images and made a gender prediction for each one. The predictions were then compared with the actual gender when the baby was born. The table below summarizes the data for Radiologist 1.

	Radiologist 1		
	Predicted Male	Predicted Female	Total
Baby is Male	74	12	**86**
Baby is Female	14	59	**73**
Total	**88**	**71**	**159**

Several questions were posed in the Preview Example:

1. How likely is it that a predicted gender is correct?
2. Is a predicted gender more likely to be correct when the baby is male than when the baby is female?
3. If the predicted gender is female, should you paint the nursery pink? If you do, how likely is it that you will need to repaint?

The first question posed can be answered by looking for cells in the table where the gender prediction is correct—the upper-left and the lower-right cells. Then

$$P(\text{predicted gender is correct}) = \frac{74 + 59}{159} = \frac{133}{159} = 0.836$$

Assuming that these 159 ultrasound images are representative of all first trimester ultrasound images, you can interpret this probability as a long-run relative frequency. For Radiologist 1, the gender prediction is correct about 83.6% of the time.

The second question posed asks if Radiologist 1 is more likely to be correct when the baby is male than when the baby is female. You can answer this question by comparing two conditional probabilities,

$$P(\text{predicted gender is male}|\text{baby is male})$$

and

$$P(\text{predicted gender is female}|\text{baby is female})$$

From the table, you can see that when the baby was male, 74 of the 86 predictions were correct, so

$$P(\text{predicted gender is male}|\text{baby is male}) = \frac{74}{86} = 0.86$$

Also

$$P(\text{predicted gender is female}|\text{baby is female}) = \frac{59}{73} = 0.81$$

Radiologist 1 is correct slightly more often when the baby is male than when the baby is female (about 86% of the time for males compared with 81% of the time for females).

The third question posed is an interesting one: If Radiologist 1 predicts that the gender is female and you paint the nursery pink, how likely is it that you will need to repaint? To answer this question, you need to look at a conditional probability that is different from the one computed above. There you looked at the probability that the predicted gender is female given that the baby was female. To answer question 3, the "given" part is that the prediction is female. That is, you want to know the probability that the baby is female given that the prediction is female:

$$P(\text{baby is female}|\text{prediction is female})$$

From the table, you can see that for the 71 babies that were predicted to be female, 59 were actually female. This gives

$$P(\text{baby is female}|\text{prediction is female}) = \frac{59}{71} = 0.831$$

For Radiologist 1, when the predicted gender is female, about 83.1% of the time the baby is actually female. This means that if the nursery was painted pink based on a predicted first trimester ultrasound image, the probability that it would need to be repainted is $1 - 0.831 = 0.169$, or about 17% of the time.

The preview example also gave the following data for a second radiologist who made gender predictions based on 154 first trimester ultrasound images.

| | Radiologist 2 | | |
	Predicted Male	Predicted Female	Total
Baby Is Male	81	8	89
Baby Is Female	7	58	65
Total	88	66	154

An additional question was posed: Does the chance of a correct gender prediction differ for the two radiologists? For Radiologist 2

$$P(\text{correct prediction}) = \frac{81 + 58}{154} = \frac{139}{154} = 0.903$$

Notice that this is quite a bit greater than the corresponding probability of 0.836 for Radiologist 1.

Let's take this example a little further. Suppose that these two radiologists both work in the same clinic and that the probability of a correct gender prediction from a first trimester ultrasound image for each of the two Radiologists is equal to the probability previously computed—0.836 for Radiologist 1 and 0.903 for Radiologist 2. Further, suppose that Radiologist 1 works part-time and handles 30% of the ultrasounds for the clinic and that Radiologist 2 handles the remaining 70%. You can then ask the following questions:

1. What is the probability that a gender prediction based on a first trimester ultrasound at this clinic is correct?
2. If a first trimester ultrasound gender prediction is incorrect, what is the probability that the prediction was made by Radiologist 2?

To answer these questions, you can translate the given probability information into a "hypothetical 1000" table. You know the following for first trimester gender predictions made at this clinic:

$$P(\text{prediction is made by Radiologist 1}) = 0.30$$

$$P(\text{prediction is made by Radiologist 2}) = 0.70$$

$$P(\text{prediction is correct}|\text{prediction made by Radiologist 1}) = 0.836$$

$$P(\text{prediction is correct}|\text{prediction made by Radiologist 2}) = 0.903$$

Setting up the table with row totals included gives

	Prediction Correct	Prediction Incorrect	Total
Radiologist 1			300
Radiologist 2			700
Total			1,000

The row totals are computed from the probability that a gender prediction was made by a particular radiologist—you would expect 30% or 300 of the hypothetical 1000 predictions to have been made by Radiologist 1 and 70% or 700 of the hypothetical 1000 to have been made by Radiologist 2.

Next consider the 300 predictions made by Radiologist 1. Because Radiologist 1 is correct about 83.6% of the time (from $P(\text{prediction is correct}|\text{prediction made by Radiologist 1}) = 0.836$), you would expect

$$(0.836)(300) \approx 251$$

to be correct and the remaining $300 - 251 = 49$ to be incorrect. For Radiologist 2, you would expect

$$(0.903)(700) \approx 632$$

to be correct and the remaining $700 - 632 = 68$ to be incorrect.

Adding these numbers to the "hypothetical 1000" table gives

	Prediction Correct	Prediction Incorrect	Total
Radiologist 1	251	49	300
Radiologist 2	632	68	700
Total	883	117	1,000

You can now compute the two desired probabilities:

$$P(\text{Prediction is correct}) = \frac{883}{1000} = 0.883$$

and

$$P(\text{Radiologist 2}|\text{prediction incorrect}) = \frac{68}{117} = 0.581$$

This last probability tells you that about 58.1% of the incorrect gender predictions at this clinic are made by Radiologist 2. This may seem strange at first, given that Radiologist 2 is correct more often than Radiologist 1. But remember that Radiologist 2 does more than twice as many predictions as Radiologist 1. So, even though Radiologist 2 has a lower error rate, more of the errors are made by Radiologist 2 because many more gender predictions are made by Radiologist 2.

SECTION 5.4 EXERCISES

Each Exercise Set assesses the following chapter learning objectives: C4, M1, M9, M10, P2

SECTION 5.4 Exercise Set 1

5.43 Suppose that 20% of all teenage drivers in a certain county received a citation for a moving violation within the past year. Assume in addition that 80% of those receiving such a citation attended traffic school so that the citation would not appear on their permanent driving record. Consider the chance experiment that consists of randomly selecting a teenage driver from this county.

a. One of the percentages given in the problem specifies an unconditional probability, and the other percentage specifies a conditional probability. Which one is the conditional probability, and how can you tell?

b. Suppose that two events E and F are defined as follows:

E = selected driver attended traffic school

F = selected driver received such a citation

Use probability notation to translate the given information into two probability statements of the form $P(\underline{\hspace{1cm}}) =$ probability value.

5.44 The accompanying data are from the article **"Characteristics of Buyers of Hybrid Honda Civic IMA: Preferences, Decision Process, Vehicle Ownership, and Willingness-to-Pay" (Institute for Environmental Decisions, November 2006).** Each of 311 people who purchased a Honda Civic was classified according to gender and whether the car purchased had a hybrid engine or not.

	Hybrid	Not Hybrid
Male	77	117
Female	34	83

a. Find the following probabilities:
  i. $P(male)$
  ii. $P(hybrid)$
  iii. $P(hybrid|male)$
  iv. $P(hybrid|female)$
  v. $P(female|hybrid)$

b. For each of the probabilities calculated in Part (a), write a sentence interpreting the probability.

c. Are the probabilities $P(hybrid|male)$ and $P(male|hybrid)$ equal? If not, write a sentence or two explaining the difference.

5.45 The paper **"Action Bias among Elite Soccer Goalkeepers: The Case of Penalty Kicks" (** *Journal of Economic Psychology* **[2007]: 606–621)** presents an interesting analysis of 286 penalty kicks in televised championship soccer games from around the world. In a penalty kick, the only players involved are the kicker and the goalkeeper from the opposing team. The kicker tries to kick a ball into the goal from a point located 11 meters away. The goalkeeper tries to block the ball from entering the goal. For each penalty kick analyzed, the researchers recorded the direction that the goalkeeper moved (jumped to the left, stayed in the center, or jumped to the right) and whether or not the penalty kick was successfully blocked. Consider the following events:

L = the event that the goalkeeper jumps to the left

C = the event that the goalkeeper stays in the center

R = the event that the goalkeeper jumps to the right

B = the event that the penalty kick is blocked

Based on their analysis of the penalty kicks, the authors of the paper gave the following probability estimates:

$$P(L) = 0.493 \quad P(C) = 0.063 \quad P(R) = 0.444$$

$$P(B|L) = 0.142 \quad P(B|C) = 0.333 \quad P(B|R) = 0.126$$

a. For each of the given probabilities, write a sentence giving an interpretation of the probability in the context of this problem.

b. Use the given probabilities to construct a "hypothetical 1000" table with columns corresponding to whether or not a penalty kick was blocked and rows corresponding to whether the goalkeeper jumped left, stayed in the center, or jumped right.

c. Use the table to compute the probability that a penalty kick is blocked.

d. Based on the given probabilities and the probability computed in Part (c), what would you recommend to a goalkeeper as the best strategy when trying to defend against a penalty kick? How does this compare to what goalkeepers actually do when defending against a penalty kick?

5.46 The accompanying table summarizes data from a medical expenditures survey carried out by the National Center for Health Statistics (**"Assessing the Effects of Race and Ethnicity on Use of Complementary and Alternative Therapies in the USA,"** *Ethnicity and Health* [2005]: 19–32).

Use of Alternative Therapies by Education Level	
Education Level	Percent Using Alternative Therapies
High school or less	4.3%
College—1 to 4 years	8.2%
College—5 years or more	11.0%

These percentages were based on data from 7,320 people whose education level was high school or less, 4,793 people with 1 to 4 years of college, and 1,095 people with 5 or more years of college.

a. Use the information given to determine the *number* of respondents falling into each of the six cells of the table below.

	Uses Alternative Therapies	Does Not Use Alternative Therapies	Total
HS/Less			7,320
College: 1–4 Yrs			4,793
College: ≥ 5 Yrs			1,095

b. The authors of the study indicated that the sample was representative of the adult population in the United States. Estimate the following probabilities for adults in the United States.

i. The probability that a randomly selected adult has five or more years of college.

ii. The probability that a randomly selected adult uses alternative therapies.

iii. The probability that a randomly selected adult uses alternative therapies given that he or she has five or more years of college.

iv. The probability that a randomly selected adult uses alternative therapies given that he or she has an education level of high school or less.

v. The probability that a randomly selected adult who uses alternative therapies has an education level of high school or less.

vi. The probability that a randomly selected adult with some college uses alternative therapies.

c. Consider the events H = event that a randomly selected adult has an education level of high school or less and A = event that a randomly selected adult uses alternative therapies. Are these independent events? Explain.

5.47 An electronics store sells two different brands of DVD players. The store reports that 30% of customers purchasing a DVD choose Brand 1. Of those that choose Brand 1, 20% purchase an extended warranty. Consider the chance experiment of randomly selecting a customer who purchased a DVD player at this store.

a. One of the percentages given in the problem specifies an unconditional probability, and the other percentage specifies a conditional probability. Which one is the conditional probability, and how can you tell?

b. Suppose that two events B and E are defined as follows:

B = selected customer purchased Brand 1

E = selected customer purchased an extended warranty

Use probability notation to translate the given information into two probability statements of the form $P(\_\_\_\_) =$ probability value.

5.48 Delayed diagnosis of cancer is a problem because it can delay the start of treatment. The paper **"Causes of Physician Delay in the Diagnosis of Breast Cancer" (***Archives of Internal Medicine* **[2002]: 1343–1348)** examined possible causes of delayed diagnosis for women with breast cancer. The accompanying table summarizes data on the initial written mammogram report (benign or suspicious) and whether or not diagnosis was delayed for 433 women with breast cancer.

	Diagnosis Delayed	Diagnosis Not Delayed
Mammogram Report Benign	32	89
Mammogram Report Suspicious	8	304

Consider the following events:

B = event that the mammogram report says benign

S = event that the mammogram report says suspicious

D = event that diagnosis is delayed

a. Assume that these data are representative of the larger group of all women with breast cancer. Use the data in the table to estimate and interpret the following probabilities:

i. $P(B)$

ii. $P(S)$

iii. $P(D|B)$

iv. $P(D|S)$

b. Remember that all of the 433 women in this study actually had breast cancer, so benign mammogram reports were, by definition, in error. Write a few sentences explaining whether this type of error in the reading of mammograms is related to delayed diagnosis of breast cancer.

5.49 Lyme disease is the leading tick-borne disease in the United States and Europe. Diagnosis of the disease is difficult and is aided by a test that detects particular antibodies in the blood. The article **"Laboratory Considerations in the Diagnosis and Management of Lyme Borreliosis" (American Journal of Clinical Pathology [1993]: 168–174)** used the following notation:

+ represents a positive result on the blood test
− represents a negative result on the blood test
L represents the event that the patient actually has Lyme disease
L^C represents the event that the patient actually does not have Lyme disease

The following probabilities were reported in the article:

$$P(L) = 0.00207$$

$$P(L^C) = 0.99793$$

$$P(+|L) = 0.937$$

$$P(-|L) = 0.063$$

$$P(+|L^C) = 0.03$$

$$P(-|L^C) = 0.97$$

a. For each of the given probabilities, write a sentence giving an interpretation of the probability in the context of this problem.
b. Use the given probabilities to construct a "hypothetical 1000" table with columns corresponding to whether or not a person has Lyme disease and rows corresponding to whether the blood test is positive or negative.
c. Notice the form of the known conditional probabilities; for example, $P(+|L)$ is the probability of a positive test given that a person selected at random from the population actually has Lyme disease. Of more interest is the probability that a person has Lyme disease, given that the test result is positive. Use the table constructed in Part (b) to compute this probability.

5.50 The article **"Checks Halt over 200,000 Gun Sales" (San Luis Obispo Tribune, June 5, 2000)** reported that required background checks blocked 204,000 gun sales in 1999. The article also indicated that state and local police reject a higher percentage of would-be gun buyers than does the FBI, stating, "The FBI performed 4.5 million of the 8.6 million checks, compared with 4.1 million by state and local agencies. The rejection rate among state and local agencies was 3%, compared with 1.8% for the FBI."
a. Use the given information to estimate $P(F)$, $P(S)$, $P(R|F)$, and $P(R|S)$, where F = event that a randomly selected gun purchase background check is performed by the FBI, S = event that a randomly selected gun

purchase background check is performed by a state or local agency, and R = event that a randomly selected gun purchase background check results in a blocked sale.
b. Use the probabilities from Part (a) to create a "hypothetical 1000" table. Use the table to compute $P(S|R)$, and write a sentence interpreting this value in the context of this problem.

Additional Exercises

5.51 Suppose that an individual is randomly selected from the population of all adult males living in the United States. Let A be the event that the selected individual is over 6 feet in height, and let B be the event that the selected individual is a professional basketball player. Which do you think is larger, $P(A|B)$ or $P(B|A)$? Why?

5.52 The paper **"Good for Women, Good for Men, Bad for People: Simpson's Paradox and the Importance of Sex-Specific Analysis in Observational Studies" (Journal of Women's Health and Gender-Based Medicine [2001]: 867–872)** described the results of a medical study in which one treatment was shown to be better for men and better for women than a competing treatment. However, if the data for men and women are combined, it appears as though the competing treatment is better. To see how this can happen, consider the accompanying data tables constructed from information in the paper. Subjects in the study were given either Treatment A or Treatment B, and their survival was noted. Let S be the event that a patient selected at random survives, A be the event that a patient selected at random received Treatment A, and B be the event that a patient selected at random received Treatment B.
a. The following table summarizes data for men and women combined:

	Survived	Died	Total
Treatment A	215	85	**300**
Treatment B	241	59	**300**
Total	**456**	**144**	**600**

  i. Find $P(S)$.
  ii. Find $P(S|A)$.
  iii. Find $P(S|B)$.
  iv. Which treatment appears to be better?
b. Now consider the summary data for the men who participated in the study:

	Survived	Died	Total
Treatment A	120	80	**200**
Treatment B	20	20	**40**
Total	**140**	**100**	**240**

  i. Find $P(S)$.
  ii. Find $P(S|A)$.
  iii. Find $P(S|B)$.
  iv. Which treatment appears to be better?

c. Now consider the summary data for the women who participated in the study:

	Survived	Died	Total
Treatment A	95	5	**100**
Treatment B	221	39	**260**
Total	**316**	**44**	**360**

i. Find $P(S)$.
ii. Find $P(S|A)$.
iii. Find $P(S|B)$.
iv. Which treatment appears to be better?
d. You should have noticed from Parts (b) and (c) that for both men and women, Treatment A appears to be better. But in Part (a), when the data for men and women are combined, it looks like Treatment B is better. This is an example of what is called Simpson's paradox. Write a brief explanation of why this apparent inconsistency occurs for this data set. (Hint: Do men and women respond similarly to the two treatments?)

5.53 A large cable TV company reports the following:
- 80% of its customers subscribe to its cable TV service
- 42% of its customers subscribe to its Internet service
- 32% of its customers subscribe to its telephone service
- 25% of its customers subscribe to both its cable TV and Internet service
- 21% of its customers subscribe to both its cable TV and phone service
- 23% of its customers subscribe to both its Internet and phone service
- 15% of its customers subscribe to all three services

Consider the chance experiment that consists of selecting one of the cable company customers at random. Find and interpret the following probabilities:
a. P(cable TV only)
b. P(Internet|cable TV)
c. P(exactly two services)
d. P(Internet and cable TV only)

5.54 A friend who works in a big city owns two cars, one small and one large. Three-quarters of the time he drives the small car to work, and one-quarter of the time he takes the large car. If he takes the small car, he usually has little trouble parking and so is at work on time with probability 0.9. If he takes the large car, he is on time to work with

probability 0.6. Given that he was at work on time on a particular morning, what is the probability that he drove the small car?

5.55 Students at a particular university use an online registration system to select their courses for the next term. There are four different priority groups, with students in Group 1 registering first, followed by those in Group 2, and so on. Suppose that the university provided the accompanying information on registration for the fall semester. The entries in the table represent the proportion of students falling into each of the 20 priority-unit combinations.

Priority Group	Number of Units Secured During First Attempt to Register				
	0–3	4–6	7–9	10–12	More Than 12
1	0.01	0.01	0.06	0.10	0.07
2	0.02	0.03	0.06	0.09	0.05
3	0.04	0.06	0.06	0.06	0.03
4	0.04	0.08	0.07	0.05	0.01

a. What proportion of students at this university got 10 or more units during the first attempt to register?
b. Suppose that a student reports receiving 11 units during the first attempt to register. Is it more likely that he or she is in the first or the fourth priority group? Explain.
c. If you are in the third priority group next term, is it likely that you will get more than 9 units during the first attempt to register? Explain.

5.56 Consider the following events:

C = event that a randomly selected driver is observed to be using a cell phone

A = event that a randomly selected driver is observed driving a car

V = event that a randomly selected driver is observed driving a van or SUV

T = event that a randomly selected driver is observed driving a pickup truck

Based on the article **"Three Percent of Drivers on Hand-Held Cell Phones at Any Given Time"** (*San Luis Obispo Tribune,* July 24, 2001), the following probability estimates are reasonable: $P(C) = 0.03$, $P(C|A) = 0.026$, $P(C|V) = 0.048$, and $P(C|T) = 0.019$. Explain why $P(C)$ is not just the average of the three given conditional probabilities.

SECTION 5.5 Calculating Probabilities—A More Formal Approach

In the previous two sections, you saw how tables could be used to calculate event probabilities and conditional probabilities. Sometimes you started with a data table and used the data in the table to estimate probabilities. In other situations, you started with probability information and used that information to construct a hypothetical 1000 table that could be used to compute probabilities of interest. In this section, you will see that it is possible to bypass the construction of a hypothetical 1000 table and use a few key probability formulas to calculate probabilities directly.

Probability Formulas

The Complement Rule (for calculating complement probabilities)
For any event E,

$$P(E^C) = 1 - P(E)$$

The Addition Rule (for calculating union probabilities)
For any two events E and F,

$$P(E \cup F) = P(E) + P(F) - P(E \cap F)$$

For **mutually exclusive events**, this simplifies to

$$P(E \cup F) = P(E) + P(F)$$

The Multiplication Rule (for calculating intersection probabilities)
For any two events E and F,

$$P(E \cap F) = P(E \mid F)P(F)$$

For **independent events,** this simplifies to

$$P(E \cap F) = P(E)P(F)$$

Conditional Probabilities
For any two events E and F, with $P(F) \neq 0$,

$$P(E \mid F) = \frac{P(E \cap F)}{P(F)}$$

Three of the probability formulas in the preceding box were introduced in Section 5.3 (the Complement Rule, the Addition Rule for mutually exclusive events, and the Multiplication Rule for independent events). In the following examples, you will see how the other formulas follow from the work you have already done with hypothetical 1000 tables.

Union Probabilities: The Addition Rule

To see how the Addition Rule can be used to compute the probability of a union event, let's revisit the health information study first introduced in Example 5.5.

Example 5.16   Health Information from TV Shows Revisited

The report **"TV Drama/Comedy Viewers and Health Information"** (www.cdc.gov /healthmarketing) describes a large survey that was conducted for the Centers for Disease Control (CDC). The CDC believed that the sample used was representative of adult Americans. One question on the survey asked respondents if they had learned something new about a health issue or disease from a TV show in the previous 6 months. Consider the following events:

L = event that a randomly selected adult American reports learning something new about a health issue or disease from a TV show in the previous 6 months

and

F = event that a randomly selected adult American is female

Data from the survey were used to estimate the following probabilities:

$$P(L) = 0.58 \quad P(F) = 0.50 \quad P(L \cap F) = 0.31$$

In Example 5.5, the given probability information was used to construct the following hypothetical 1000 table:

	F (female)	not F (not female)	Total
L (learned from TV)	310	270	580
not L (did not learn from TV)	190	230	420
Total	500	500	1000

The table was used to calculate

$$P(L \cup F) = \frac{310 + 270 + 190}{1000} = \frac{770}{1000} = 0.770$$

This probability could have been calculated directly using the Addition Rule as follows:
$$P(L \cup F) = P(L) + P(F) - P(L \cap F)$$
$$= 0.58 + 0.50 - 0.31$$
$$= 0.77$$

If you are comfortable working with formulas, this is much quicker than setting up the hypothetical 1000 table.

To see why the Addition Rule works, look back at the calculation based on the hypothetical 1000 table:

$$P(L \cup F) = \frac{310 + 270 + 190}{1000} \qquad 0$$

This can be rewritten as

$$P(L \cup F) = \frac{310 + 270 + 190 + (310 - 310)}{1000}$$
$$= \frac{(310 + 270)}{1000} + \frac{(190 + 310)}{1000} - \frac{310}{1000}$$
$$= 0.580 + 0.500 - 0.310$$
$$= P(L) + P(F) - P(L \cap F)$$

Subtracting $P(L \cap F)$ from $P(L) + P(F)$ corrects for the fact that outcomes in $L \cap F$ are included in both L and in F.

Conditional Probabilities

A probability that takes into account whether or not another event has occurred is called a "conditional" probability. The probability of the event E given that the event F is known to have occurred is denoted by $P(E \mid F)$. To see how the formula for calculating a conditional probability is used, let's revisit the heart attack study of Example 5.12.

Example 5.17 Surviving a Heart Attack Revisited

Medical guidelines recommend that a hospitalized patient who suffers cardiac arrest should receive defibrillation (an electric shock to the heart) within 2 minutes. The paper **"Delayed Time to Defibrillation After In-Hospital Cardiac Arrest" (*The New England Journal of Medicine* [2008]: 9–17)** describes a study of the time to defibrillation for hospitalized patients in hospitals of various sizes. The authors examined medical records of 6,716 patients who suffered cardiac arrest while hospitalized, recording the size of the hospital and whether or not defibrillation occurred within 2 minutes. Data from this study are reproduced here:

Hospital Size	Time to Defibrillation		Total
	Within 2 Minutes	More Than 2 Minutes	
Small (Fewer Than 250 Beds)	1,124	576	1,700
Medium (250–499 Beds)	2,178	886	3,064
Large (500 or More Beds)	1,387	565	1,952
Total	4,689	2,027	6,716

In Example 5.12, you assumed that these data were representative of the larger group of all hospitalized patients who suffer a cardiac arrest. The following events were defined:

S = event that a randomly selected patient is at a small hospital
M = event that a randomly selected patient is at a medium-sized hospital
L = event that a randomly selected patient is at a large hospital
D = event that a randomly selected patient receives defibrillation within 2 minutes

Information in the table was used to calculate $P(D)$, the probability that a randomly selected patient receives defibrillation within 2 minutes. Because 4,689 out of the 6,716 patients received defibrillation within 2 minutes: $P(D) = \dfrac{4,689}{6,716} = 0.698$. Conditional probabilities involving event D were also calculated:

$$P(D \mid S) = \frac{1,124}{1,700} = 0.661 \quad P(D \mid M) = \frac{2,178}{3,064} = 0.711 \quad P(D \mid L) = \frac{1,387}{1,952} = 0.711$$

These probabilities are interpreted as the proportions of patients at small, medium, and large hospitals, respectively, who received timely defibrillations.

The reasoning used to calculate these conditional probabilities based on information in the given table is the same as using the formula for calculating a conditional probability. To see this, let's use the formula to calculate $P(D \mid S)$:

$$P(D \mid S) = \frac{P(D \cap S)}{P(S)}$$

$$= \frac{1,124/6,716}{1,700/6,716}$$

$$= \frac{1,124}{1,700} \qquad \longleftarrow \quad \textit{Same as calculation}$$

$$\textit{based on the table.}$$

$$= 0.661$$

Notice that when you reasoned directly from the table, you calculated this probability by looking only at patients in small hospitals, and then calculating the proportion of those patients who received timely defibrillation. The formula for calculating a conditional probability is based on this same reasoning.

In Example 5.17, using the formula to calculate a conditional probability isn't any quicker than using the information in the table because the data table was given. However, if you are starting with probability information rather than a data table, it is possible to use the formula to calculate conditional probabilities without having to set up a hypothetical 1000 table. This is illustrated in the following example.

Example 5.18   Tattoos

The article **"Chances Are You Know Someone with a Tattoo, and He's Not a Sailor"** (*Associated Press, June 11, 2006*) summarized data from a representative sample of adults ages 18 to 50. Consider the following events:

T = the event that a randomly selected person in the targeted age group has a tattoo
A = the event that a randomly selected person in the targeted age group is between 18 and 29 years old

The following probabilities were estimated based on data from the sample:

$$P(T) = 0.24$$

$$P(A) = 0.50$$

$$P(T \cap A) = 0.18$$

Suppose you are interested in the probability that a person who is between 18 and 29 years old has a tattoo. This is the conditional probability $P(T \mid A)$. One way to estimate this probability is to construct a hypothetical 1000 table using the given information. This involves completing the following table:

	T	Not T	Total
A	180		500
Not A			
Total	240		1000

You could then use the hypothetical 1000 table to calculate $P(T \mid A)$. Using the hypothetical 1000 table would give you a correct answer, but you could calculate $P(T \mid A)$ directly from the conditional probability formula, as shown here:

$$P(T \mid A) = \frac{P(T \cap A)}{P(A)} = \frac{0.18}{0.50} = 0.36$$

You could also calculate $P(A \mid T)$:

$$P(A \mid T) = \frac{P(T \cap A)}{P(T)} = \frac{0.18}{0.24} = 0.75$$

Notice the difference between the three probabilities

$$P(T) = 0.24$$
$$P(T \mid A) = 0.36$$
$$P(A \mid T) = 0.75$$

These three probabilities have different interpretations.

Probability	Interpretation
$P(T) = 0.24$	About 24% of *adults ages 18 to 50* have a tattoo.
$P(T \mid A) = 0.36$	About 36% of *adults ages 18 to 29* have a tattoo.
$P(A \mid T) = 0.75$	About 75% of *adults ages 18 to 50 who have a tattoo* are between 18 and 29 years old.

Intersection Probabilities: The Multiplication Rule

To see how the Multiplication Rule can be used to compute the probability of an intersection event, consider the following example.

Example 5.19 DVD Player Warranties

A large electronics store sells two different portable DVD players, Brand 1 and Brand 2. Based on past sales records, the store sales manager reports that 70% of the DVD players sold are Brand 1 and 30% are Brand 2. The manager also reports that 20% of the people who buy Brand 1 also purchase an extended warranty, and 40% of the people who buy Brand 2 purchase an extended warranty. Consider selecting a person at random from those who purchased a DVD player from this store, and define the following events:

B_1 = event that the selected customer bought Brand 1
B_2 = event that the selected customer bought Brand 2
E = event that the selected customer purchased an extended warranty

Suppose you would like to know the probability that the selected customer bought Brand 1 and purchased an extended warranty, $P(B_1 \cap E)$.

There are three different ways you might calculate this probability.

Method 1: Hypothetical 1000 Table

One way to compute $P(B_1 \cap E)$ is to use the given information to complete a hypothetical 1000 table. This would result in the accompanying table.

	E	Not E	Total
Brand 1	140	560	**700**
Brand 2	120	180	**300**
Total	**260**	**740**	**1,000**

The table could then be used to calculate $P(B_1 \cap E)$, as shown.

$$P(B_1 \cap E) = \frac{140}{1,000} = 0.14$$

Method 2: The Multiplication Rule

Translating the given information into probability statements results in

$P(B_1) = 0.70$ (70% buy Brand 1)
$P(B_2) = 0.30$ (30% buy Brand 2)
$P(E \mid B_1) = 0.20$ (20% *of those who buy Brand 1* purchase an extended warranty)
$P(E \mid B_2) = 0.40$ (40% *of those who buy Brand 2* purchase an extended warranty)

Using the Multiplication Rule to calculate $P(B_1 \cap E)$, you get

$$P(B_1 \cap E) = P(E \cap B_1) = P(E \mid B_1)P(B_1) = (0.20)\,(0.70) = 0.14$$

Notice that this is equal to the probability calculated from the hypothetical 1000 table and can be computed without having to construct the table.

Method 3: Using a Tree Diagram

Another way to calculate the probability of an intersection event when some conditional probabilities are known is to use a tree diagram. Many people prefer this more visual approach. The tree diagram of Figure 5.4 gives a nice display that is based on the Multiplication Rule.

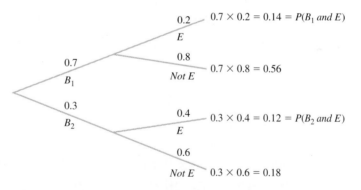

FIGURE 5.4
A tree diagram for the probability calculations of Example 5.19.

The two first-generation branches are labeled with events B_1 and B_2 along with their probabilities. Two second-generation branches extend from each first-generation branch; these correspond to the two events E and *not E*. The *conditional* probabilities $P(E \mid B_1)$, $P(not\ E \mid B_1)$, $P(E \mid B_2)$, and $P(not\ E \mid B_2)$, appear on these branches. Application of the Multiplication Rule then consists of multiplying probabilities across the branches of the tree diagram. For example,

$$P(B_2 \cap E) = P(E \mid B_2)P(B_2)$$
$$= (0.4)(0.3)$$
$$= 0.12$$

and this probability is displayed to the right of the E branch that comes from the B_2 branch. You can now also easily calculate $P(E)$, the probability that an extended warranty is purchased. The event E can occur in two different ways: Buy Brand 1 *and* warranty, or buy Brand 2 *and* warranty. Symbolically, these events are $B_1 \cap E$ and $B_2 \cap E$. Furthermore, if each customer purchased a single DVD player, he or she could not have simultaneously purchased both Brand 1 and Brand 2, so the two events $B_1 \cap E$ and $B_2 \cap E$ are mutually exclusive. Then

$$
\begin{aligned}
P(E) &= P(B_1 \cap E) + P(B_2 \cap E) \\
&= P(E \mid B_1)P(B_1) + P(E \mid B_2)P(B_2) \\
&= (0.2)(0.7) + (0.4)(0.3) \\
&= 0.14 + 0.12 \\
&= 0.26
\end{aligned}
$$

This probability is the sum of two of the probabilities shown on the right-hand side of the tree diagram. Thus 26% of all DVD player purchasers selected an extended warranty.

The general multiplication rule can be extended to give an expression for the probability that several events occur together. In the case of three events E, F, and G, we have

$$P(E \cap F \cap G) = P(E \mid F \cap G)P(F \mid G)P(G)$$

When the events are all independent, $P(E \mid F \cap G) = P(E)$ and $P(F \mid G) = P(F)$ so the right-hand side of the equation for $P(E \cap F \cap G)$ is simply the product of the three unconditional probabilities.

Example 5.20 **Lost Luggage**

Twenty percent of all passengers who fly from Los Angeles (LA) to New York (NY) do so on Airline G. This airline misplaces luggage for 10% of its passengers, and 90% of this lost luggage is subsequently recovered. If a passenger who has flown from LA to NY is randomly selected, what is the probability that the selected individual flew on Airline G (event G), had luggage misplaced (event F), and subsequently recovered the misplaced luggage (event E)? The given information implies that

$$P(G) = 0.20 \quad P(F \mid G) = 0.10 \quad P(E \mid F \cap G) = 0.90$$

Then

$$P(E \cap F \cap G) = P(E \mid F \cap G)P(F \mid G)P(G) = (.90)(.10)(.20) = .018$$

That is, about 1.8% of passengers flying from LA to NY fly on Airline G, have their luggage misplaced, and subsequently recover the lost luggage.

More Complex Probability Calculations: The Law of Total Probability and Bayes' Rule

Law of Total Probability

Let's reconsider the information on DVD player sales from Example 5.19. In this example, the following events were defined:

B_1 = event that Brand 1 is purchased
B_2 = event that Brand 2 is purchased
E = event that an extended warranty is purchased

Based on the information given in Example 5.19, the following probabilities are known:

$$P(B_1) = 0.7 \quad P(B_2) = 0.3 \quad P(E \mid B_1) = 0.2 \quad P(E \mid B_2) = 0.4$$

Notice that the conditional probabilities $P(E \mid B_1)$ and $P(E \mid B_2)$ are known but that the unconditional probability $P(E)$ is not known.

To find $P(E)$ notice that the event E can occur in two ways: (1) A customer purchases an extended warranty *and* buys Brand 1 (which can be written as $E \cap B_1$); or (2) a customer purchases an extended warranty *and* buys Brand 2 (which can be written as $E \cap B_2$). Because these are the only ways in which E can occur, you can write the event E as

$$E = (E \cap B_1) \cup (E \cap B_2)$$

Again, the two events $(E \cap B_1)$ and $(E \cap B_2)$ are mutually exclusive (since B_1 and B_2 are mutually exclusive), so using the addition rule for mutually exclusive events gives

$$P(E) = P((E \cap B_1) \cup (E \cap B_2))$$
$$= P(E \cap B_1) + P(E \cap B_2)$$

Finally, using the general multiplication rule to evaluate $P(E \cap B_1)$ and $P(E \cap B_2)$ results in

$$P(E) = P(E \cap B_1) + P(E \cap B_2)$$
$$= P(E \mid B_1)P(B_1) + P(E \mid B_2)P(B_2)$$

Substituting in the known probabilities gives

$$P(E) = P(E \mid B_1)P(B_1) + P(E \mid B_2)P(B_2)$$
$$= (0.2)(0.7) + (0.4)(0.3)$$
$$= 0.26$$

You can conclude that 26% of the DVD player customers purchased an extended warranty.

As this example illustrates, when conditional probabilities are known, they can sometimes be used to compute unconditional probabilities. The **law of total probability** formalizes this use of conditional probabilities.

The Law of Total Probability

If B_1 and B_2 are disjoint events with $P(B_1) + P(B_2) = 1$, then for any event E

$$P(E) = P(E \cap B_1) + P(E \cap B_2)$$
$$= P(E \mid B_1)P(B_1) + P(E \mid B_2)P(B_2)$$

More generally, if $B_1, B_2, \cdots, B_k$ are disjoint events with $P(B_1) + P(B_2) + \cdots + P(B_k) = 1$, then for any event E

$$P(E) = P(E \cap B_1) + P(E \cap B_2) + \cdots + P(E \cap B_k)$$
$$= P(E \mid B_1)P(B_1) + P(E \mid B_2)P(B_2) + \cdots + P(E \mid B_k)P(B_k)$$

Example 5.21 Which Way to Jump?

The paper **"Action Bias among Elite Soccer Goalkeepers: The Case of Penalty Kicks"** (*Journal of Economic Psychology* [2007]: 606–621) presents an interesting analysis of 286 penalty kicks in televised championship soccer games from around the world. In a penalty kick, the only players involved are the kicker and the goalkeeper from the opposing team. The kicker tries to kick a ball into the goal from a point located 11 meters away. The goalkeeper tries to block the ball from reaching the goal. For each penalty kick analyzed, the researchers recorded the direction that the goalkeeper moved (jumped to the left, stayed in the center, or jumped to the right) and whether or not the penalty kick was successfully blocked.

Consider the following events:

L = the event that the goalkeeper jumps to the left
C = the event that the goalkeeper stays in the center
R = the event that the goalkeeper jumps to the right
B = the event that the penalty kick is blocked

Based on their analysis of the penalty kicks, the authors of the paper gave the following probability estimates:

$$P(B \mid L) = 0.142 \qquad P(B \mid C) = 0.333 \qquad P(B \mid R) = 0.126$$
$$P(L) = 0.493 \qquad P(C) = 0.063 \qquad P(R) = 0.444$$

What proportion of penalty kicks were blocked? You can use the law of total probability to answer this question. Here, the three events L, C, and R play the role of B_1, B_2, and B_3 and B plays the role of E in the formula for the law of total probability.

Substituting into the formula, you get

$$
\begin{aligned}
P(B) &= P(B \cap L) + P(B \cap C) + P(B \cap R) \\
&= P(B \mid L)P(L) + P(B \mid C)P(C) + P(B \mid R)P(R) \\
&= (0.142)(0.493) + (0.333)(0.063) + (0.126)(0.444) \\
&= 0.070 + 0.021 + 0.056 \\
&= 0.147
\end{aligned}
$$

A tree diagram could also have been used here. Figure 5.5 is a visual display of the given probability information. Following the branches in the tree that lead to a blocked kick and adding these three probabilities gives

$$P(B) = 0.070 + 0.021 + 0.056 = 0.147$$

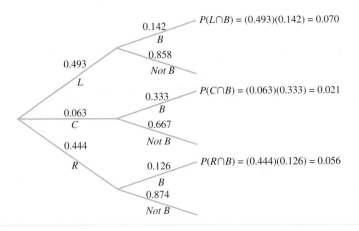

FIGURE 5.5
Tree diagram for the soccer example

This means that only 14.7% of penalty kicks were successfully blocked. Two other interesting findings of this study were

1. The direction that the goalkeeper moves appears to be independent of whether the kicker kicked the ball to the left, center, or right of the goal. This was attributed to the fact that goalkeepers have to choose their action before they can clearly observe the direction of the kick.
2. Based on the three conditional probabilities— $P(B \mid L) = 0.142$, $P(B \mid C) = 0.333$, and $P(B \mid R) = 0.126$—the optimal strategy for a goalkeeper appears to be to stay in the center of the goal. However, staying in the center was only chosen 6.3% of the time— much less often that jumping left or right. The authors believe that this is because a goalkeeper does not feel as bad about not successfully blocking a kick if some action (jumping left or right) is taken compared to if no action (staying in the center) is taken. This is the "action bias" referred to in the title of the paper.

Bayes' Rule

In this subsection, you will see a formula discovered by the Reverend Thomas Bayes (1702–1761), an English Presbyterian minister. He discovered what is now known as Bayes' Rule. This rule provides a solution to what Bayes called the converse problem. A converse problem is one where some conditional probabilities are known, such as the probability of a positive test result given cancer, but you would like to know the probability that has the conditioning "reversed." That is, you would like to know the probability of cancer given a positive test result.

You have already seen how to find such probabilities using hypothetical 1000 tables. Now you will see how Bayes' Rule might be used to calculate these probabilities without constructing a hypothetical 1000 table. Although many people find the hypothetical 1000 table to be the easier approach, either method can be used.

Bayes' Rule

If B_1 and B_2 are disjoint events with $P(B_1) + P(B_2) = 1$, then for any event E

$$P(B_1 \mid E) = \frac{P(E \mid B_1)P(B_1)}{P(E \mid B_1)P(B_1) + P(E \mid B_2)P(B_2)}$$

More generally, if $B_1, B_2, \cdots, B_k$ are disjoint events with $P(B_1) + P(B_2) + \cdots + P(B_k) = 1$ then for any event E,

$$P(B_i \mid E) = \frac{P(E \mid B_i)P(B_i)}{P(E \mid B_1)P(B_1) + P(E \mid B_2)P(B_2) + \cdots + P(E \mid B_k)P(B_k)}$$

To see how Bayes' Rule can be used to compute a conditional probability, let's revisit the Internet addiction study first described in Example 5.14.

Example 5.22   Internet Addiction Revisited

Internet addiction has been defined by researchers as a disorder characterized by excessive time spent on the Internet, impaired judgment and decision-making ability, social withdrawal, and depression. The paper **"The Association between Aggressive Behaviors and Internet Addiction and Online Activities in Adolescents" (*Journal of Adolescent Health* [2009]: 598–605)** describes a study of a large number of adolescents. Each participant in the study was assessed using the Chen Internet Addiction Scale to determine if he or she suffered from Internet addiction. The following statements are based on the survey results:

1. 51.8% of the study participants were female and 48.2% were male.
2. 13.1% of the females suffered from Internet addiction.
3. 24.8% of the males suffered from Internet addiction.

Consider the chance experiment that consists of selecting a study participant at random and define the following events:

F = the event that the selected participant is female
M = the event that the selected participant is male
I = the event that the selected participant suffers from Internet addiction

The three statements from the paper define the following probabilities:

$$P(F) = 0.518$$
$$P(M) = 0.482$$
$$P(I \mid F) = 0.131$$
$$P(I \mid M) = 0.248$$

In Example 5.14, the given information was used to construct the following hypothetical 1000 table:

	I (Internet addiction)	*not I* (no Internet addiction)	Total
F (female)	68	450	**518**
not F (male)	120	362	**482**
Total	**188**	**812**	**1000**

This table was then used to calculate

$$P(F \mid I) = \frac{68}{188} = 0.362$$

Notice that $P(I \mid F)$ was given, and then the table was used to calculate $P(F \mid I)$.

Bayes' Rule can also be used to calculate this probability. With F and M playing the role of B_1 and B_2 and I playing the role of E in the Bayes' Rule formula, you get

$$P(F \mid I) = \frac{P(I \mid F)P(F)}{P(I \mid F)P(F) + P(I \mid M)P(M)}$$

$$= \frac{(0.131)(0.518)}{(0.131)(0.518) + (0.482)(0.248)}$$

$$= \frac{0.068}{0.068 + 0.120}$$

$$= \frac{0.068}{0.188}$$

$$= 0.362$$

Notice that this probability is equal to the one computed using the hypothetical 1000 table, but was calculated without having to construct the table.

The rules for calculating probabilities introduced in this section provide an alternative to using the hypothetical 1000 table. Either the table or the rules can be used, so you can choose the method you prefer.

SECTION 5.5 EXERCISES

Each Exercise Set assesses the following chapter learning objectives: M11, P1

SECTION 5.5 Exercise Set 1

5.57 There are two traffic lights on Shelly's route from home to work. Let E denote the event that Shelly must stop at the first light, and define the event F in a similar manner for the second light. Suppose that $P(E) = 0.4$, $P(F) = 0.3$ and $P(E \cap F) = 0.15$. In Exercise 5.25, you constructed a hypothetical 1000 table to compute the following probabilities. Now use the probability formulas of this section to find these probabilities.

a. The probability that Shelly must stop for at least one light (the probability of the event $E \cup F$).

b. The probability that Shelly does not have to stop at either light.

c. The probability that Shelly must stop at exactly one of the two lights.

d. The probability that Shelly must stop only at the first light.

5.58 A large cable company reports that 80% of its customers subscribe to cable TV, 42% subscribe to Internet service, and 97% subscribe to at least one of cable TV and Internet service. In Exercise 5.26, you constructed a hypothetical 1000 table to compute the following probabilities. Now use the probability formulas of this section to find these probabilities.

a. The probability that a randomly selected customer subscribes to both cable TV and Internet service.

b. The probability that a randomly selected customer subscribes to exactly one of cable TV or Internet service.

5.59 Two different airlines have a flight from Los Angeles to New York that departs each weekday morning at a certain time. Let E denote the event that the first airline's flight is fully booked on a particular day, and let F denote the event that the second airline's flight is fully booked on that same day. Suppose that $P(E) = 0.7$, $P(F) = 0.6$ and $P(E \cap F) = 0.54$.

a. Calculate $P(E \mid F)$, the probability that the first airline's flight is fully booked given that the second airline's flight is fully booked.

b. Calculate $P(F \mid E)$.

5.60 A certain university has 10 vehicles available for use by faculty and staff. Six of these are vans and four are cars. On a particular day, only two requests for vehicles have been made. Suppose that the two vehicles to be assigned are chosen at random from among the 10 vehicles.

a. Let E denote the event that the first vehicle assigned is a van. What is the value of $P(E)$?

b. Let F denote the event that the second vehicle assigned is a van. What is the value of $P(F \mid E)$?

c. Use the results of Parts (a) and (b) to calculate $P(E \ and \ F)$ Hint: Use the definition of $P(F \mid E)$.

5.61 The paper **"Predictors of Complementary Therapy Use Among Asthma Patients: Results of a Primary Care Survey"** (*Health and Social Care in the Community* [2008]: 155–164) described a study in which each person in a large sample of asthma patients responded to two questions:

> Question 1: Do conventional asthma medications usually help your asthma symptoms?
> Question 2: Do you use complementary therapies (such as herbs, acupuncture, aroma therapy) in the treatment of your asthma?

Suppose that this sample is representative of asthma patients. Consider the following events:

> E = event that an asthma patient uses complementary therapies
> F = event that an asthma patient reports that conventional medications usually help.

The data from the sample were used to estimate the following probabilities:

$$P(E) = 0.146 \quad P(F) = 0.879 \quad P(E \cap F) = 0.122$$

In Exercise 5.33, you constructed a hypothetical 1000 table to compute the following probabilities. Now use the probability formulas of this section to find these probabilities.

a. The probability that an asthma patient responds that conventional medications do not help and uses complementary therapies.
b. The probability that an asthma patient responds that conventional medications do not help and does not use complementary therapies.
c. The probability that an asthma patient responds that conventional medications usually help or the patient uses complementary therapies.

5.62 An appliance manufacturer offers extended warranties on its washers and dryers. Based on past sales, the manufacturer reports that of customers buying both a washer and a dryer, 52% purchase the extended warranty for the washer, 47% purchase the extended warranty for the dryer, and 59% purchase at least one of the two extended warranties. In Exercise 5.34, you constructed a hypothetical 1000 table to compute the following probabilities. Now use the probability formulas of this section to find these probabilities.

a. The probability that a randomly selected customer who buys a washer and a dryer purchases an extended warranty for both the washer and the dryer.
b. The probability that a randomly selected customer does not purchase an extended warranty for either the washer or dryer.

5.63 The article **"SUVs Score Low in New Federal Roll Over Ratings"** (*San Luis Obispo Tribune, January 6, 2001*) gave information on death rates for various kinds of accidents by vehicle type for accidents reported to the police. Suppose that we randomly select an accident reported to the police and consider the following events:

> R = event that the selected accident is a single-vehicle rollover
> F = event that the selected accident is a frontal collision
> D = event that the selected accident results in a death

Information in the article indicates that the following probability estimates are reasonable:

$$P(R) = 0.06 \quad P(F) = 0.60 \quad P(R \mid D) = 0.30$$
$$P(F \mid D) = 0.54$$

a. Interpret the value of $P(R \mid D)$.
b. Interpret the value of $P(F \mid D)$.
c. Are the events R and D independent? Justify your answer.

5.64 A construction firm bids on two different contracts. Let E_1 be the event that the bid on the first contract is successful, and define E_2 analogously for the second contract. Suppose that $P(E_1) = 0.4$ and $P(E_2) = 0.3$ and that E_1 and E_2 are independent events.
a. Calculate the probability that both bids are successful (the probability of the event E_1 *and* E_2).
b. Calculate the probability that neither bid is successful (the probability of the event *not* E_1 *and not* E_2).
c. What is the probability that the firm is successful in at least one of the two bids?

5.65 The authors of the paper **"Do Physicians Know When Their Diagnoses Are Correct?"** (*Journal of General Internal Medicine* [2005]: 334–339) presented detailed case studies to medical students and to faculty at medical schools. Each participant was asked to provide a diagnosis in the case and also to indicate whether his or her confidence in the correctness of the diagnosis was high or low. Define the events C, I, and H as follows:

> C = event that diagnosis is correct
> I = event that diagnosis is incorrect
> H = event that confidence in the correctness of the diagnosis is high

a. Data appearing in the paper were used to estimate the following probabilities for medical students:

$$P(C) = 0.261$$
$$P(I) = 0.739$$
$$P(H \mid C) = 0.375$$
$$P(H \mid I) = 0.073$$

Use Bayes' Rule to compute the probability of a correct diagnosis given that the student's confidence level in the correctness of the diagnosis is high.
b. Data from the paper were also used to estimate the following probabilities for medical school faculty:

$$P(C) = 0.495$$
$$P(I) = 0.505$$
$$P(H \mid C) = 0.537$$
$$P(H \mid I) = 0.252$$

Compute $P(C \mid H)$ for medical school faculty. How does the value of this probability compare to the value of $P(C \mid H)$ for students computed in Part (a)?

Additional Exercises

5.66 The report "**Improving Undergraduate Learning**" (**Social Science Research Council, 2011**) summarizes data from a survey of several thousand college students. These students were thought to be representative of the population of all U.S. college students. When asked about a typical semester, 68% said they would be taking a class that is reading intensive (requires more than 40 pages of reading per week). Only 50% said they would be taking a class that is writing intensive (requires more than 20 pages of writing over the course of the semester). The percentage who said that they would be taking both a reading intensive course and a writing intensive course in a typical semester was 42%. In Exercise 5.40, you constructed a hypothetical 1000 table to compute the following probabilities. Now use the probability formulas of this section to find these probabilities.

a. The probability that a randomly selected student would be taking at least one reading intensive or writing intensive course.

b. The probability that a randomly selected student would be taking a reading intensive course or a writing intensive course, but not both.

c. The probability that a randomly selected student is taking neither a reading intensive nor a writing intensive course.

5.67 A large cable company reports the following:

- 80% of its customers subscribe to cable TV service
- 42% of its customers subscribe to Internet service
- 32% of its customers subscribe to telephone service
- 25% of its customers subscribe to both cable TV and Internet service
- 21% of its customers subscribe to both cable TV and phone service
- 23% of its customers subscribe to both Internet and phone service
- 15% of its customers subscribe to all three services

Consider the chance experiment that consists of selecting one of the cable company customers at random. In Exercise 5.53, you constructed a hypothetical 1000 table to compute the following probabilities. Now use the probability formulas of this section to find these probabilities.

a. P(cable TV only)

b. P(Internet | cable TV)

c. P(exactly two services)

d. P(Internet and cable TV only)

5.68 Information from a poll of registered voters in Cedar Rapids, Iowa, to assess voter support for a new school tax was the basis for the following statements (*Cedar Rapids Gazette*, **August 28, 1999**):

> The poll showed 51% of the respondents in the Cedar Rapids school district are in favor of the tax. The approval rating rises to 56% for those with children in public schools. It falls to 45% for those with no children in public schools. The older the respondent, the less favorable the view of the proposed tax: 36% of those over age 56 said they would vote for the tax compared with 72% of 18- to 25-year-olds.

Suppose that a registered voter from Cedar Rapids is selected at random, and define the following events:

F = event that the selected individual favors the school tax
C = event that the selected individual has children in the public schools
O = event that the selected individual is over 56 years old
Y = event that the selected individual is 18–25 years old

a. Use the given information to estimate the values of the following probabilities:
  i. $P(F)$
  ii. $P(F \mid C)$
  iii. $P(F \mid C^c)$
  iv. $P(F \mid O)$
  v. $P(F \mid Y)$

b. Are F and C independent? Justify your answer.

c. Are F and O independent? Justify your answer.

5.69 Consider the following events: C = event that a randomly selected driver is observed to be using a cell phone, A = event that a randomly selected driver is observed driving a passenger automobile, V = event that a randomly selected driver is observed driving a van or SUV, and T = event that a randomly selected driver is observed driving a pickup truck. Based on the article "**Three Percent of Drivers on Hand-Held Cell Phones at Any Given Time**" (*San Luis Obispo Tribune*, **July 24, 2001**), the following probability estimates are reasonable: $P(C) = 0.03$, $P(C \mid A) = 0.026$, $P(C \mid V) = 0.048$, and $P(C \mid T) = 0.019$. Explain why $P(C)$ is not just the average of the three given conditional probabilities.

SECTION 5.6 Probability as a Basis for Making Decisions

Probability plays an important role in drawing conclusions from data. This is illustrated in Examples 5.23 and 5.24.

Example 5.23 Age and Gender of College Students

Table 5.1 shows the proportions of the student body in various age–gender combinations at a university. These proportions can be interpreted as probabilities if a student is selected at random from the student body. For example, the probability (long-run proportion of the time) that a male, age 21–24, will be selected is 0.16.

TABLE 5.1   **Age and Gender Distribution**

	Age					
Gender	Under 17	17–20	21–24	25–30	31–39	40 and Older
Male	0.006	0.15	0.16	0.08	0.04	0.01
Female	0.004	0.18	0.14	0.10	0.04	0.09

Suppose that you are told that a 43-year-old student from this university is waiting to meet you. You are asked if you think the student is male or female. How would you respond? What if the student was 27 years old? What about 33 years old? Are you equally confident in all three of your choices?

Reasonable responses to these questions can be based on the probabilities in Table 5.1. You would think that the 43-year-old student was female. You cannot be certain that this is correct, but you can see that someone in the 40-and-over age group is much less likely to be male than female. You would also think that the 27-year-old was female, but you would be less confident in this conclusion. For the age group 31–39, the proportion of males and the proportion of females are equal, so you would think it equally likely that a 33-year-old student would be male or female. You could decide in favor of male (or female), but with little confidence in your conclusion; in other words, there is a good chance that you would be incorrect.

Example 5.24   **Can You Pass by Guessing?**

A professor planning to give a quiz that consists of 20 true–false questions is interested in knowing how someone who answers by guessing would do on such a test. To investigate, he asks the 500 students in his introductory psychology course to write the numbers from 1 to 20 on a piece of paper and then to arbitrarily write T or F next to each number. The students are forced to guess at the answer to each question, because they are not even told what the questions are! These answer sheets are then collected and graded using the key for the quiz. The results are summarized in Table 5.2.

TABLE 5.2   **Quiz "Guessing" Distribution**

Number of Correct Responses	Number of Students	Proportion of Students	Number of Correct Responses	Number of Students	Proportion of Students
0	0	0.000	11	79	0.158
1	0	0.000	12	61	0.122
2	1	0.002	13	39	0.078
3	1	0.002	14	18	0.036
4	2	0.004	15	7	0.014
5	8	0.016	16	1	0.002
6	18	0.036	17	1	0.002
7	37	0.074	18	0	0.000
8	58	0.116	19	0	0.000
9	81	0.162	20	0	0.000
10	88	0.176			

Because probabilities are long-run proportions, an entry in the "Proportion of Students" column of Table 5.2 can be considered an estimate of the probability of correctly guessing the answers to a specific number of questions. For example, a proportion of 0.122 (or 12.2%) of the 500 students got 12 of the 20 correct. You could estimate the probability of a student correctly guessing 12 answers to be (approximately) 0.122.

Let's use the information in Table 5.2 to answer the following questions.

1. Would you be surprised if someone guessing on a 20-question true–false quiz got only 3 correct? The approximate probability of someone getting 3 correct by guessing is

0.002. This means that, in the long run, only about 2 in 1000 guessers would get exactly 3 correct. Because it is so unlikely, this outcome would be considered surprising.

2. If a score of 15 or more correct is a passing grade on the quiz, is it likely that someone who is guessing will pass? The long-run proportion of people who are guessing who would pass is the sum of the proportions for all the passing scores (15, 16, 17, 18, 19, and 20). It follows that

$$\text{probability of passing} \approx 0.014 + 0.002 + 0.002 + 0.000 + 0.000 + 0.000$$
$$= 0.018$$

It would be unlikely that a student who is guessing would be able to pass.

3. The professor actually gives the quiz, and a student scores 16 correct. Do you think that the student was just guessing? You could begin by assuming that the student was guessing and determine whether a score at least as high as 16 is a likely or an unlikely occurrence. Table 5.2 tells you that the approximate probability of getting a score at least as high as this student's score is

$$\text{probability of scoring 16 or higher} \approx 0.002 + 0.002 + 0.000 + 0.000 + 0.000$$
$$= 0.004$$

That is, in the long run, only about 4 times in 1,000 would someone score 16 or higher by guessing. There are two possible explanations for a score of 16: (1) The student was guessing and was *really* lucky, or (2) the student was not just guessing. Given that the first possibility is highly unlikely, you could conclude that a student who scored 16 was not just guessing at the answers. Although you cannot be certain that you are correct, the evidence is compelling.

4. What score on the quiz would it take to convince you that a student was not just guessing? You should be convinced by a score that is high enough to be unlikely for a guesser. Consider the following approximate probabilities (computed from the entries in Table 5.2):

Score	Approximate Probability
20	0.000
19 or better	0.000 + 0.000 = 0.000
18 or better	0.000 + 0.000 + 0.000 = 0.000
17 or better	0.002 + 0.000 + 0.000 + 0.000 = 0.002
16 or better	0.002 + 0.002 + 0.000 + 0.000 + 0.000 = 0.004
15 or better	0.014 + 0.002 + 0.002 + 0.000 + 0.000 + 0.000 = 0.018
14 or better	0.036 + 0.014 + 0.002 + 0.002 + 0.000 + 0.000 + 0.000 = 0.054
13 or better	0.078 + 0.036 + 0.014 + 0.002 + 0.002 + 0.000 + 0.000 + 0.000 = 0.132

You might say that a score of 14 or higher is reasonable evidence that someone is not just guessing, because the approximate probability that a guesser would score this high is only 0.054. Of course, if you conclude that a student is not guessing based on a quiz score of 14 or higher, then there is a risk that you are incorrect (about 1 in 20 guessers would score this high by chance). A guesser will score 13 or more correct about 13.2% of the time, which is often enough that most people would not rule out guessing.

Examples 5.23 and 5.24 illustrate how probabilities can be used to make decisions. Later chapters look more formally at drawing conclusions based on available but often incomplete information and at assessing the reliability of such conclusions.

SECTION 5.6 EXERCISES

Each Exercise Set assesses the following chapter learning objectives: P3, P4

SECTION 5.6 Exercise Set 1

5.70 A Gallup Poll conducted in November 2002 examined how people perceived the risks associated with smoking. The following table summarizes data consistent with summary quantities published by Gallup on smoking status and perceived risk of smoking:

Smoking Status	Perceived Risk			
	Very Harmful	Somewhat Harmful	Not Too Harmful	Not at All Harmful
Current Smoker	60	30	5	1
Former Smoker	78	16	3	2
Never Smoked	86	10	2	1

Assume that it is reasonable to consider these data representative of the U.S. adult population. Consider the following conclusion: Current smokers are less likely to view smoking as very harmful than either former smokers or those who have never smoked. Provide a justification for this conclusion. Use the information in the table to calculate any relevant probabilities.

SECTION 5.6   Exercise Set 2

5.71   Researchers at UCLA were interested in whether working mothers were more likely to suffer workplace injuries than women without children. They studied 1,400 working women, and a summary of their findings was reported in the *San Luis Obispo Telegram-Tribune* (February 28, 1995). The information in the following table is consistent with summary values reported in the article:

	No Children	Children Under 6	Children, but None Under 6	Total
Injured on the Job	32	68	56	**156**
Not Injured on the Job	368	232	644	**1,244**
Total	**400**	**300**	**700**	**1,400**

The researchers drew the following conclusion: Women with children younger than age 6 are much more likely to be injured on the job than childless women or mothers with older children. Provide a justification for the researchers' conclusion. Use the information in the table to calculate any relevant probabilities.

SECTION 5.7   Estimating Probabilities Empirically and Using Simulation

In the examples presented so far, reaching conclusions required knowledge of the probabilities of various events. This is reasonable when you know the actual long-run proportion of the time that each event will occur. In other situations, these probabilities are not known and must be determined by using mathematical rules and probability properties, including the basic ones introduced in this chapter.

In this section, we change gears a bit and focus on an empirical approach to probability. When an analytical approach is impossible, impractical, or requires more than the limited probability tools of an introductory course, you can *estimate* probabilities empirically through observation or by using simulation.

Estimating Probabilities Empirically

It is fairly common practice to use observed long-run proportions to estimate probabilities. The process used to estimate probabilities is simple:

1. Observe a large number of chance outcomes under controlled circumstances.
2. Interpreting probability as a long-run relative frequency, estimate the probability of an event by using the observed proportion of occurrence.

This process is illustrated in Examples 5.25 and 5.26.

Example 5.25   Fair Hiring Practices

To recruit a new faculty member, a university biology department intends to advertise for someone with a Ph.D. in biology and at least 10 years of college-level teaching experience. A member of the department expresses the belief that requiring at least 10 years of teaching experience will exclude most potential applicants and will exclude far more female applicants than male applicants. The biology department would like to determine the probability an applicant would be excluded because of the experience requirement.

A similar university just completed a search in which there was no requirement for prior teaching experience but the information about prior teaching experience was recorded. Data from the 410 applications received is summarized in the following table:

| | Number of Applicants | | |
	Less Than 10 Years of Experience	10 Years of Experience or More	Total
Male	178	112	**290**
Female	99	21	**120**
Total	**277**	**133**	**410**

Let's assume that the populations of applicants for the two positions is the same. The available information can be used to approximate the probability that an applicant will fall into each of the four gender–experience combinations. The estimated probabilities (obtained by dividing the number of applicants in each gender–experience combination by 410) are given in Table 5.3. From Table 5.3,

P(candidate excluded because of experience requirement) $\approx 0.4341 + 0.2415 = 0.6756$

TABLE 5.3 **Estimated Probabilities for Example 5.25**

	Less Than 10 Years of Experience	10 Years of Experience or More
Male	0.4341	0.2732
Female	0.2415	0.0512

You can also assess the impact of the experience requirement separately for male and female applicants. From the given information, the proportion of male applicants who have less than 10 years of experience is $\frac{178}{290} = 0.6138$, whereas the corresponding proportion for females is $\frac{99}{120} = 0.8250$. Therefore, approximately 61% of the male and about 83% of the female applicants would be excluded.

These subgroup proportions—0.6138 for males and 0.8250 for females—are examples of conditional probabilities. A conditional probability shows how the original probability changes in light of new information. In this example, the probability that a potential candidate has less than 10 years of experience is 0.6756, but this probability changes to 0.8250 if you know that a candidate is female. These probabilities can be expressed as an unconditional probability

$$P(\text{less than 10 years of experience}) = 0.6756$$

and a conditional probability

$$P(\text{less than 10 years of experience}|\text{female}) = 0.8250$$

Example 5.26 Who Has the Upper Hand?

Men and women frequently express intimacy through touch. A common instance of this is holding hands. Some researchers have suggested that hand-holding might not be just an expression of intimacy but also might communicate status differences. For two people to hold hands, one must assume an "overhand" grip and one an "underhand" grip. Research in this area has shown that it is the male who usually assumes the overhand grip. In the view of some investigators, the overhand grip implies status or superiority. The authors of the paper **"Men and Women Holding Hands: Whose Hand Is Uppermost?"** (*Perceptual and Motor Skills* [1999]: 537–549) investigated an alternative explanation: Perhaps the positioning of hands is a function of the heights of the individuals. Because men tend to be taller than women, maybe comfort, not status, dictates the positioning. Investigators at two universities observed hand-holding male–female pairs, resulting in the following data:

| | Gender of Person with Uppermost Hand | | |
	Men	Women	Total
Man Taller	2,149	299	**2,448**
Equal Height	780	246	**1,026**
Woman Taller	241	205	**446**
Total	**3,170**	**750**	**3,920**

Assuming that these hand-holding couples are representative of hand-holding couples in general, you can use the available information to estimate various probabilities. For example, if a hand-holding couple is selected at random, then

$$\text{estimate of } P(\text{man's hand uppermost}) = \frac{3{,}170}{3{,}920} = 0.809$$

For a randomly selected hand-holding couple, if the man is taller, the probability that the male has the uppermost hand is $\frac{2{,}149}{2{,}448} = 0.878$—so to speak—if the woman is taller, the probability that the female has the uppermost hand is $\frac{205}{446} = 0.460$. Notice that these probabilities, $P(\text{male uppermost } given \text{ male taller})$ and $P(\text{female uppermost } given \text{ female taller})$, are conditional. Also, because $P(\text{male uppermost } given \text{ male taller}) = 0.878$ is not equal to $P(\text{male uppermost}) = 0.809$, the events *male uppermost* and *male taller* are not independent events. And even when the female is taller, the male is still more likely to have the upper hand!

Estimating Probabilities by Using Simulation

Simulation provides a way to estimate probabilities when you are unable (or do not have the time or resources) to determine probabilities analytically and when it is impractical to estimate them empirically by observation. Simulation involves generating "observations" in a situation that is similar to the real situation of interest.

To illustrate the idea of simulation, consider the situation in which a professor wishes to estimate the probabilities of scores resulting from students guessing on a 20-question true–false quiz. Observations could be collected by having 500 students actually guess at the answers to 20 questions and then scoring the resulting papers. This approach requires considerable time and effort. Simulation provides an alternative approach.

Because each question on the quiz is a true–false question, a person who is guessing should be equally likely to answer correctly or incorrectly. Rather than asking a student to select *true* or *false* and then comparing the choice to the correct answer, an equivalent process would be to pick a ball at random from a box that contains equal numbers of red and blue balls, with a blue ball representing a correct answer. Making 20 selections from the box (replacing each ball selected before picking the next one) and then counting the number of correct choices (the number of times a blue ball is selected) is a physical substitute for an observation from a student who has guessed at the answers to 20 true–false questions. The number of blue balls in 20 selections should have the same probability as the corresponding number of correct responses to the quiz when a student is guessing.

For example, 20 selections of balls might yield the following results (R, red ball; B, blue ball):

Selection	1	2	3	4	5	6	7	8	9	10
Result	R	R	B	R	B	B	R	R	R	B

Selection	11	12	13	14	15	16	17	18	19	20
Result	R	R	B	R	R	B	B	R	R	B

This corresponds to a quiz with eight correct responses, and it provides you with one observation for estimating the probabilities of interest. This process can then be repeated a large number of times to generate additional observations. For example, you might find the following:

Repetition	Number of Correct Responses
1	8
2	11
3	10
4	12
⋮	⋮
1,000	11

The 1,000 simulated quiz scores could then be used to construct a table of estimated probabilities.

Taking this many balls out of a box and writing down the results would be cumbersome and tedious. The process can be simplified by using random digits to substitute for drawing balls from the box. For example, a single digit could be selected at random from the 10 digits 0, 1, 2, 3, 4, 5, 6, 7, 8, 9. When using random digits, each of the 10 possibilities is equally likely to occur, so you can use the even digits (including 0) to indicate a correct response and the odd digits to indicate an incorrect response. This would maintain the important property that a correct response and an incorrect response are equally likely, because correct and incorrect are each represented by 5 of the 10 digits.

To aid in carrying out such a simulation, tables of random digits (such as Appendix Table 1) or computer-generated random digits can be used. The numbers in Appendix Table 1 were produced using a computer's random number generator. (Imagine repeatedly drawing a chip from a box containing 10 chips numbered 0, 1, . . ., 9. After each selection, the result is recorded, the chip returned to the box, and the chips mixed. Thus, any of the digits is equally likely to occur on any of the selections.)

To see how a table of random numbers can be used to carry out a simulation, let's reconsider the quiz example. A random digit is used to represent the guess on a single question, with an even digit representing a correct response. A series of 20 digits represents the answers to the 20 quiz questions. To use the random digit table, start by picking an arbitrary starting point in Appendix Table 1. Suppose that you start at Row 10 and take the 20 digits in a row to represent one quiz. The first five "quizzes" and the corresponding number correct are

Quiz	Random Digits	Number Correct
	* * * *　 * * *　 *　 * *　 * * *	
1	9 4 6 0 6 9 7 8 8 2 5 2 9 6 0 1 4 6 0 5	13
2	6 6 9 5 7 4 4 6 3 2 0 6 0 8 9 1 3 6 1 8	12
3	0 7 1 7 7 7 2 9 7 8 7 5 8 8 6 9 8 4 1 0	9
4	6 1 3 0 9 7 3 3 6 6 0 4 1 8 3 2 6 7 6 8	11
5	2 2 3 6 2 1 3 0 2 2 6 6 9 7 0 2 1 2 5 8	13

denotes correct response ⟶ (annotation pointing to the asterisks above Quiz 1)

This process can be repeated to generate a large number of observations, which could be used to construct a table of estimated probabilities.

The simulation method for generating observations must preserve the important characteristics of the actual process being considered. For example, consider a multiple-choice quiz in which each of the 20 questions on the quiz has 5 possible responses, only 1 of which is correct. For any particular question, you would expect a student to be able to guess the correct answer only one-fifth of the time in the long run. To simulate this situation, you could select at random from a box that contained four times as many red balls as blue balls. If you are using random digits for the simulation, you could use 0 and 1 to represent a correct response and 2, 3, …, 9 to represent an incorrect response.

AP* EXAM TIP

Although technology is usually used to carry out simulations in practice, on the AP* exam, random numbers are often used so that student responses can be evaluated consistently. So, it is important that you know how to use a random number table to carry out a simulation.

Using Simulation to Approximate a Probability

1. Design a method that uses a random mechanism (such as a random number generator or table, the selection of a ball from a box, or the toss of a coin) to represent an observation. Be sure that the important characteristics of the actual process are preserved.
2. Generate an observation using the method from Step 1, and determine whether the event of interest has occurred.
3. Repeat Step 2 a large number of times.
4. Calculate the estimated probability by dividing the number of observations for which the event of interest occurred by the total number of observations generated.

The simulation process is illustrated in Examples 5.27–5.29.

Example 5.27 Building Permits

Many California cities limit the number of building permits that are issued each year. Because of limited water resources, one such city plans to issue permits for only 10 dwelling units in the upcoming year. The city will decide who is to receive permits by holding a lottery. Suppose that you are 1 of 39 individuals applying for a permit. Thirty people are requesting a permit for a single-family home, 8 are requesting a permit for a duplex (which counts as two dwelling units), and 1 person is requesting a permit for a small apartment building with eight units (which counts as eight dwelling units). Each request will be entered into the lottery. Requests will be selected at random one at a time, and if there are enough permits remaining, the request will be granted. This process will continue until all 10 permits have been issued. If your request is for a single-family home, what are your chances of receiving a permit? You can use simulation to estimate this probability. (It is not easy to determine analytically.)

To carry out the simulation, you can view the requests as being numbered from 1 to 39 as follows:

01–30: Requests for single-family homes
31–38: Requests for duplexes
39: Request for eight-unit apartment

For ease of discussion, let's assume that your request is Request 01.

One method for simulating the permit lottery consists of these three steps:

1. Choose a random number between 01 and 39 to indicate which permit request is selected first, and grant this request.
2. Select a random number between 01 and 39 to indicate which permit request is considered next. (If a number that has already been selected is chosen, ignore it and select again.) Determine the number of dwelling units for the selected request. Grant the request only if there are enough permits remaining to satisfy the request.
3. Repeat Step 2 until permits for 10 dwelling units have been granted.

Minitab was used to generate random numbers between 01 and 39 to imitate the lottery drawing. (The random number table in Appendix Table 1 could also be used, by selecting 2 digits and ignoring 00 and any value over 39.) The first sequence generated by Minitab is:

Random Number	Type of Request	Total Number of Units So Far
25	Single-family home	1
07	Single-family home	2
38	Duplex	4
31	Duplex	6
26	Single-family home	7
12	Single-family home	8
33	Duplex	10

You would stop at this point, because permits for 10 units would have been issued. In this simulated lottery, Request 01 was not selected, so you would not have received a permit.

The next simulated lottery (using Minitab to generate the selections) is as follows:

Random Number	Type of Request	Total Number of Units So Far
38	Duplex	2
16	Single-family home	3
30	Single-family home	4
39	Apartment, not granted because there are not 8 permits remaining	4
14	Single-family home	5
26	Single-family home	6
36	Duplex	8
13	Single-family home	9
15	Single-family home	10

Again, Request 01 was not selected, so you would not have received a permit in this simulated lottery either.

Now that a strategy for simulating a lottery has been devised, the tedious part of the simulation begins. You now have to simulate a large number of lottery drawings, determining for each whether Request 01 is granted. Five hundred such drawings were simulated, and Request 01 was selected in 85 of the lotteries. Based on this simulation,

$$\text{estimated probability of receiving a building permit} = \frac{85}{500} = 0.17$$

Example 5.28 One-Boy Family Planning

Suppose that couples who wanted children were to continue having children until a boy is born. Assuming that each newborn child is equally likely to be a boy or a girl, would this behavior change the proportion of boys in the population? This question was posed in an article that appeared in *The American Statistician* (**1994: 290–293**), and many people answered the question incorrectly. Simulation will be used to estimate the long-run proportion of boys in the population. This proportion is an estimate of the probability that a randomly selected child from this population is a boy. Note that every sibling group would have exactly one boy.

You can use a single-digit random number to represent a child. The odd digits (1, 3, 5, 7, 9) could represent a male birth, and the even digits could represent a female birth. An observation is constructed by selecting a sequence of random digits. If the first random number obtained is odd (a boy), the observation is complete. If the first selected number is even (a girl), another digit is chosen. You would continue in this way until an odd digit is obtained. For example, reading across Row 15 of the random number table Appendix Table 1, the first 10 digits are

$$0 \quad 7 \quad 1 \quad 7 \quad 4 \quad 2 \quad 0 \quad 0 \quad 0 \quad 1$$

Using these numbers to simulate sibling groups, results in the following:

Sibling group 1	0 7	girl, boy
Sibling group 2	1	boy
Sibling group 3	7	boy
Sibling group 4	4 2 0 0 0 1	girl, girl, girl, girl, girl, boy

Continuing along Row 15 of the random number table,

Sibling group 5	3	boy
Sibling group 6	1	boy
Sibling group 7	2 0 4 7	girl, girl, girl, boy
Sibling group 8	8 4 1	girl, girl, boy

After simulating eight sibling groups, there were 8 boys among 19 children. The proportion of boys is $\frac{8}{19}$, which is close to 0.5. Continuing the simulation to obtain a large number of observations leads to the conclusion that the long-run proportion of boys in the population would still be 0.5.

Example 5.29 ESP?

Can a close friend read your mind? Try the following experiment. Write the word *blue* on one piece of paper and the word *red* on another, and place the two slips of paper in a box. Select one slip of paper from the box, look at the word written on it, then try to convey the word by sending a mental message to a friend who is seated in the same room. Ask your friend to select either red or blue, and record whether the response is correct. Repeat this 10 times and count the number of correct responses. How did your friend do? Is your friend receiving your mental messages or just guessing?

Let's investigate by using simulation to get the approximate probabilities for someone who *is* guessing. You can use a random digit to represent a response, with an even digit representing a correct response (C) and an odd digit representing an incorrect response (X). A sequence of 10 digits can be used to generate one observation. For example, using the last 10 digits in Row 25 of the random number table (Appendix Table 1) gives

5	2	8	3	4	3	0	7	3	5
X	C	C	X	C	X	C	X	X	X

which results in four correct responses. Minitab was used to generate 150 sequences of 10 random digits and the following results were obtained:

Sequence Number	Digits	Number Correct
1	3996285890	5
2	1690555784	4
3	9133190550	2
⋮	⋮	⋮
149	3083994450	5
150	9202078546	7

Table 5.4 summarizes the results of the simulation. The estimated probabilities in Table 5.4 are based on the assumption that a correct and an incorrect response are equally likely. Evaluate your friend's performance in light of this information. Is it likely that someone guessing would have been able to get as many correct as your friend did? Do you think your friend was receiving your mental messages? How are the estimated probabilities in Table 5.4 used to support your answer?

TABLE 5.4 **Estimated Probabilities for Example 5.29**

Number Correct	Number of Sequences	Estimated Probability
0	0	0.0000
1	1	0.0067
2	8	0.0533
3	16	0.1067
4	30	0.2000
5	36	0.2400
6	35	0.2333
7	17	0.1133
8	7	0.0467
9	0	0.0000
10	0	0.0000
Total	**150**	**1.0000**

SECTION 5.7 EXERCISES

Each Exercise Set assesses the following chapter learning objectives: C5, 12

SECTION 5.7 Exercise Set 1

5.72 The *Los Angeles Times* (June 14, 1995) reported that the U.S. Postal Service was getting speedier, with higher overnight on-time delivery rates than in the past. Postal Service standards call for overnight delivery within a zone of about 60 miles for any first-class letter deposited by the last posted collection time. Two-day delivery is promised within a 600-mile zone, and three-day delivery is promised for distances over 600 miles. The Price Waterhouse accounting firm conducted an independent audit by "seeding" the mail with letters and recording on-time delivery rates for these letters. Suppose that the results of the Price Waterhouse study were as follows (these numbers are fictitious, but are compatible with summary values given in the article):

Letter Mailed To	Number of Letters Mailed	Number of Letters Arriving on Time
Los Angeles	500	425
New York	500	415
Washington, D.C.	500	405
Nationwide	6,000	5,220

Use the given information to estimate the following probabilities:
a. The probability of an on-time delivery in Los Angeles
b. The probability of late delivery in Washington, D.C.
c. The probability that two letters both mailed to New York are delivered on time
d. The probability of on-time delivery nationwide

5.73 Many cities regulate the number of taxi licenses, and there is a great deal of competition for both new and existing licenses. Suppose that a city has decided to sell 10 new licenses for $25,000 each. A lottery will be held to determine who gets the licenses, and no one may request more than 3 licenses. Twenty individuals and taxi companies have entered the lottery. Six of the 20 entries are requests for 3 licenses, nine are requests for 2 licenses, and the rest are requests for a single license. The city will select requests at random, filling as much of the request as possible. For example, if there were only one license left, any request selected would only receive this single license.
a. An individual who wishes to be an independent driver has put in a request for a single license. Use simulation to approximate the probability that the request will be granted. Perform at least 20 simulated lotteries (more is better!).
b. Do you think that this is a fair way of distributing licenses? Can you propose an alternative procedure for distribution?

SECTION 5.7 Exercise Set 2

5.74 Five hundred first-year students at a state university were classified according to both high school grade point average (GPA) and whether they were on academic probation at the end of their first semester. The data are summarized in the accompanying table.

Probation	High School GPA			Total
	2.5 to <3.0	3.0 to <3.5	3.5 and Above	
Yes	50	55	30	**135**
No	45	135	185	**365**
Total	**95**	**190**	**215**	**500**

a. Use the given table to approximate the probability that a randomly selected first-year student at this university will be on academic probation at the end of the first semester.
b. What is the estimated probability that a randomly selected first-year student at this university had a high school GPA of 3.5 or above?
c. Are the two events *selected student has a high school GPA of 3.5 or above* and *selected student is on academic probation at the end of the first semester* independent events? How can you tell?
d. Estimate the proportion of first-year students with high school GPAs between 2.5 and 3.0 who are on academic probation at the end of the first semester.
e. Estimate the proportion of those first-year students with high school GPAs of 3.5 and above who are on academic probation at the end of the first semester.

5.75 Four students must work together on a group project. They decide that each will take responsibility for a particular part of the project, as follows:

Person	Maria	Alex	Juan	Jacob
Task	Survey design	Data collection	Analysis	Report writing

Because of the way the tasks have been divided, one student must finish before the next student can begin work. To ensure that the project is completed on time, a time line is established, with a deadline for each team member. If any one of the team members is late, the timely completion of the project is jeopardized. Assume the following probabilities:
1. The probability that Maria completes her part on time is 0.8.
2. If Maria completes her part on time, the probability that Alex completes on time is 0.9, but if Maria is late, the probability that Alex completes on time is only 0.6.
3. If Alex completes his part on time, the probability that Juan completes on time is 0.8, but if Alex is late, the probability that Juan completes on time is only 0.5.
4. If Juan completes his part on time, the probability that Jacob completes on time is 0.9, but if Juan is late, the probability that Jacob completes on time is only 0.7.

Use simulation (with at least 20 trials) to estimate the probability that the project is completed on time. Think carefully about this one. For example, you might use a random digit to represent each part of the project (four in all). For the first digit (Maria's part), 1–8 could represent *on time*, and 9 and 0 could represent *late*. Depending on what happened with Maria (late or on time), you would then look at the digit representing Alex's part. If Maria was on time, 1–9 would represent *on time* for Alex, but if Maria was late, only 1–6 would represent *on time*. The parts for Juan and Jacob could be handled similarly.

Additional Exercises

5.76 In Exercise 5.75, the probability that Maria completes her part on time was 0.8. Suppose that this probability is really only 0.6. Use simulation (with at least 20 trials) to estimate the probability that the project is completed on time.

5.77 Suppose that the probabilities of timely completion are as in Exercise 5.75 for Maria, Alex, and Juan, but that Jacob has a probability of completing on time of 0.7 if Juan is on time and 0.5 if Juan is late.
a. Use simulation (with at least 20 trials) to estimate the probability that the project is completed on time.
b. Compare the probability from Part (a) to the ones computed in Exercises 5.75 and 5.76. Which decrease in the probability of on-time completion (Maria's or Jacob's) resulted in the bigger change in the probability that the project is completed on time?

5.78 A medical research team wishes to evaluate two different treatments for a disease. Subjects are selected two at a time, and one is assigned to Treatment 1 and the other to Treatment 2. The treatments are applied, and each is either a success (S) or a failure (F). The researchers keep track of the total number of successes for each treatment. They plan to continue the experiment until the number of successes for one treatment exceeds the number of successes for the other by 2. For example, based on the results in the accompanying table, the experiment would stop after the sixth pair, because Treatment 1 has two more successes than Treatment 2. The researchers would conclude that Treatment 1 is preferable to Treatment 2.

Pair	Treatment 1	Treatment 2	Cumulative Number of Successes for Treatment 1	Cumulative Number of Successes for Treatment 2
1	S	F	1	0
2	S	S	2	1
3	F	F	2	1
4	S	S	3	2
5	F	F	3	2
6	S	F	4	2

Suppose that Treatment 1 has a success rate of 0.7 and Treatment 2 has a success rate of 0.4. Use simulation to estimate the probabilities requested in Parts (a) and (b). (Hint: Use a pair of random digits to simulate one pair of subjects. Let the first digit represent Treatment 1 and use 1–7 as an indication of a success and 8, 9, and 0 to indicate a failure. Let the second digit represent Treatment 2, with 1–4 representing a success. For example, if the two digits selected to represent a pair were 8 and 3, you would record failure for Treatment 1 and success for Treatment 2. Continue to select pairs, keeping track of the cumulative number of successes for each treatment. Stop the trial as soon as the number of successes for one treatment exceeds that for the other by 2. This would complete one trial. Now repeat this whole process until you have results for at least 20 trials [more is better]. Finally, use the simulation results to estimate the desired probabilities.)
a. Estimate the probability that more than five pairs must be treated before a conclusion can be reached? (Hint: $P(\text{more than } 5) = 1 - P(5 \text{ or fewer})$.)
b. Estimate the probability that the researchers will incorrectly conclude that Treatment 2 is the better treatment?

5.79 A single-elimination tournament with four players is to be held. A total of three games will be played. In Game 1, the players seeded (rated) first and fourth play. In Game 2, the players seeded second and third play. In Game 3, the winners of Games 1 and 2 play, with the winner of Game 3 declared the tournament winner. Suppose that the following probabilities are known:

$$P(\text{Seed 1 defeats Seed 4}) = 0.8$$
$$P(\text{Seed 1 defeats Seed 2}) = 0.6$$
$$P(\text{Seed 1 defeats Seed 3}) = 0.7$$
$$P(\text{Seed 2 defeats Seed 3}) = 0.6$$
$$P(\text{Seed 2 defeats Seed 4}) = 0.7$$
$$P(\text{Seed 3 defeats Seed 4}) = 0.6$$

a. How would you use random digits to simulate Game 1 of this tournament?
b. How would you use random digits to simulate Game 2 of this tournament?
c. How would you use random digits to simulate the third game in the tournament? (This will depend on the outcomes of Games 1 and 2.)
d. Simulate one complete tournament, giving an explanation for each step in the process.
e. Simulate 10 tournaments, and use the resulting information to estimate the probability that the first seed wins the tournament.
f. Ask four classmates for their simulation results. Along with your own results, this should give you information on 50 simulated tournaments. Use this information to estimate the probability that the first seed wins the tournament.
g. Why do the estimated probabilities from Parts (e) and (f) differ? Which do you think is a better estimate of the actual probability? Explain.

CHAPTER ACTIVITIES

ACTIVITY 5.1 KISSES

Background: The paper **"What Is the Probability of a Kiss? (It's Not What You Think)"** (found in the online *Journal of Statistics Education* [2002]) posed the following question: What is the probability that a Hershey's Kiss candy will land on its base (as opposed to its side) if it is flipped onto a table? Unlike flipping a coin, there is no reason to believe that this probability is 0.5.

Working as a class, develop a plan that would enable you to estimate this probability. Once you have an acceptable plan, use it to produce an estimate of the desired probability.

ACTIVITY 5.2 A CRISIS FOR EUROPEAN SPORTS FANS?

Background: The *New Scientist* (January 4, 2002) reported on a controversy surrounding the Euro coins that have been introduced as a common currency across most of Europe. Each country mints its own coins, but these coins are accepted in any of the countries that have adopted the Euro as its currency.

A group in Poland claims that the Belgium-minted Euro does not have an equal chance of landing heads or tails. This claim was based on 250 tosses of the Belgium Euro, of which 140 (56%) came up heads. Should this be cause for alarm for European sports fans, who know that "important" decisions are made by the flip of a coin?

In this activity, you will investigate whether this difference should be cause for alarm by examining whether observing 140 heads out of 250 tosses is an unusual outcome if the coin is fair.

1. For this first step, you can either (a) flip a U.S. penny 250 times, keeping a tally of the number of heads and tails observed (this won't take as long as you think) or (b) simulate 250 coin tosses by using your calculator or a statistics software package to generate random numbers (if you choose this option, give a brief description of how you carried out the simulation).

2. For your sequence of 250 tosses, calculate the proportion of heads observed.

3. Form a data set that consists of the values for the proportion of heads observed in 250 tosses of a fair coin for the entire class. Summarize this data set by constructing a graphical display.

4. Working with a partner, write a paragraph explaining why European sports fans should or should not be worried by the results of the Polish experiment. Your explanation should be based on the observed proportion of heads from the Polish experiment and the outcome of the simulation.

ACTIVITY 5.3 THE "HOT HAND" IN BASKETBALL

Background: Consider a mediocre basketball player who has consistently made only 50% of his free throws over several seasons. If you were to examine his free throw record over the last 50 free throw attempts, is it likely that you would see a "streak" of 5 in a row where he is successful in making the free throw? In this activity, you will investigate this question. Assume that the outcomes of successive free throw attempts are independent and that the probability that the player is successful on any particular attempt is 0.5.

1. Begin by simulating a sequence of 50 free throws for this player. Because this player has a probability of success of 0.5 for each attempt and the attempts are independent, you can model a free throw by tossing a coin. Using heads to represent a successful free throw and tails to represent a missed free throw, simulate 50 free throws by tossing a coin 50 times, recording the outcome of each toss.

2. For your sequence of 50 tosses, identify the longest streak by looking for the longest string of heads in your sequence. Determine the length of this streak.

3. Combine your longest streak value with those from the rest of the class, and construct a histogram or dotplot of all the longest streak values.

4. Based on the graph from Step 3, does it appear likely that a player of this skill level would have a streak of 5 or more successes at some point during a sequence of 50 free throw attempts? Justify your answer based on the graph from Step 3.

5. Use the combined class data to estimate the probability that a player of this skill level has a streak of at least 5 somewhere in a sequence of 50 free throw attempts.

6. Using basic probability rules, the probability of a player of this skill level being successful on the next 5 free throw attempts is

$$P(SSSSS) = \left(\frac{1}{2}\right)\left(\frac{1}{2}\right)\left(\frac{1}{2}\right)\left(\frac{1}{2}\right)\left(\frac{1}{2}\right) = \left(\frac{1}{2}\right)^5 = 0.031$$

which is relatively small. At first, this value might seem inconsistent with your answer in Step 5, but the estimated probability from Step 5 and the computed probability of 0.031 are really considering different situations. Explain how these situations are different.

7. Do you think that the assumption that the outcomes of successive free throws are independent is reasonable? Explain. (This is a hotly debated topic among both sports fans and statisticians!)

ARE YOU READY TO MOVE ON? **CHAPTER 5 REVIEW EXERCISES**

All chapter learning objectives are assessed in these exercises. The learning objectives assessed in each exercise are given in parentheses.

5.80 (C1)
The article **"Anxiety Increases for Airline Passengers After Plane Crash"** (*San Luis Obispo Tribune*, **November 13, 2001**) reported that air passengers have a 1 in 11 million chance of dying in an airplane crash. This probability was then interpreted as "You could fly every day for 26,000 years before your number was up." Comment on why this probability interpretation is misleading.

5.81 (M1)
A company that offers roadside assistance to drivers reports that the probability that a call for assistance will be to help someone who is locked out of his or her car is 0.18. Give a relative frequency interpretation of this probability.

5.82 (P1)
A mutual fund company offers its customers several different funds: a money market fund, three different bond funds, two stock funds, and a balanced fund. Among customers who own shares in just one fund, the percentages of customers in the different funds are as follows:

Money market	20%
Short-term bond	15%
Intermediate-term bond	10%
Long-term bond	5%
High-risk stock	18%
Moderate-risk stock	25%
Balanced fund	7%

A customer who owns shares in just one fund is to be selected at random.
a. What is the probability that the selected individual owns shares in the balanced fund?
b. What is the probability that the individual owns shares in a bond fund?
c. What is the probability that the selected individual does not own shares in a stock fund?

5.83 (M2)
The student council for a school of science and math has one representative from each of five academic departments: Biology (B), Chemistry (C), Mathematics (M), Physics (P), and Statistics (S). Two of these students are to be randomly selected for inclusion on a university-wide student committee.

a. What are the 10 possible outcomes?
b. From the description of the selection process, all outcomes are equally likely. What is the probability of each outcome?
c. What is the probability that one of the committee members is the statistics department representative?
d. What is the probability that both committee members come from laboratory science departments?

5.84 (M1, P3)
Consider the following two lottery-type games:
　Game 1: You pick one number between 1 and 50. After you have made your choice, a number between 1 and 50 is selected at random. If the selected number matches the number you picked, you win.
　Game 2: You pick two numbers between 1 and 10. After you have made your choices, two different numbers between 1 and 10 are selected at random. If the selected numbers match the two you picked, you win.
a. The cost to play either game is $1, and if you win you will be paid $20. If you can only play one of these games, which game would you pick and why? Use relevant probabilities to justify your choice.
b. For either of these games, if you plan to play the game 100 times, would you expect to win money or lose money overall? Explain.

5.85 (C2)
Consider a chance experiment that consists of selecting a customer at random from all people who purchased a car at a large car dealership during 2012.
a. In the context of this chance experiment, give an example of two events that would be mutually exclusive.
b. In the context of this chance experiment, give an example of two events that would not be mutually exclusive.

5.86 (C3)
The Australian newspaper *The Mercury* (**May 30, 1995**) reported that based on a survey of 600 reformed and current smokers, 11.3% of those who had attempted to quit smoking in the previous 2 years had used a nicotine aid (such as a

nicotine patch). It also reported that 62% of those who had quit smoking without a nicotine aid began smoking again within 2 weeks and 60% of those who had used a nicotine aid began smoking again within 2 weeks. If a smoker who is trying to quit smoking is selected at random, are the events *selected smoker who is trying to quit uses a nicotine aid* and *selected smoker who has attempted to quit begins smoking again within 2 weeks* independent or dependent events? Justify your answer using the given information.

5.87 (M3)
The following table summarizes data on smoking status and perceived risk of smoking and is consistent with summary quantities obtained in a Gallup Poll conducted in November 2002:

	Perceived Risk			
Smoking Status	Very Harmful	Somewhat Harmful	Not Too Harmful	Not at All Harmful
Current Smoker	60	30	5	1
Former Smoker	78	16	3	2
Never Smoked	86	10	2	1

Assume that it is reasonable to consider these data as representative of the American adult population. Consider the chance experiment of selecting an American adult at random.
a. What is the probability that the selected adult is a former smoker?
b. What is the probability that the selected adult perceives the risk of smoking to be very harmful?
c. What is the probability that the selected adult perceives the risk of smoking to be very harmful or somewhat harmful?
d. What is the probability that the selected adult is a current smoker who perceives the risk of smoking as very harmful?
e. What is the probability that the selected adult never smoked or perceives the risk of smoking to be very harmful?

5.88 (M4)
A study of how people are using online services for medical consulting is described in the paper **"Internet Based Consultation to Transfer Knowledge for Patients Requiring Specialized Care"** (*British Medical Journal* [2003]: 696–699). Patients using a particular online site could request one or both (or neither) of two services: specialist opinion and assessment of pathology results. The paper reported that 98.7% of those using the service requested a specialist opinion, 35.4% requested the assessment of pathology results, and 34.7% requested both a specialist opinion and assessment of pathology results. Consider the following two events:

S = event that a specialist opinion is requested

A = event that an assessment of pathology results is requested

a. What are the values of $P(S)$, $P(A)$, and $P(S \cap A)$?
b. Use the given probability information to set up a "hypothetical 1000" table with columns corresponding to S and *not S* and rows corresponding to A and *not A*.
c. Use the table to find the following probabilities:
   i. the probability that a request is for neither a specialist opinion nor assessment of pathology results.
   ii. the probability that a request is for a specialist opinion or an assessment of pathology results.

5.89 (M5)
A large cable company reports that 42% of its customers subscribe to its Internet service, 32% subscribe to its phone service, and 51% subscribe to its Internet service or its phone service (or both).
a. Use the given probability information to set up a "hypothetical 1000" table.
b. Use the table to find the following:
   i. the probability that a randomly selected customer subscribes to both the Internet service and the phone service.
   ii. the probability that a randomly selected customer subscribes to exactly one of the two services.

5.90 (M6)
a. Suppose events E and F are mutually exclusive with $P(E) = 0.64$ and $P(F) = 0.17$.
   i. What is the value of $P(E \cap F)$?
   ii. What is the value of $P(E \cup F)$?
b. Suppose that A and B are events with $P(A) = 0.3, P(B) = 0.5$, and $P(A \cap B) = 0.15$. Are A and B mutually exclusive? How can you tell?
c. Suppose that A and B are events with $P(A) = 0.65$ and $P(B) = 0.57$. Are A and B mutually exclusive? How can you tell?

5.91 (M7)
In a small city, approximately 15% of those eligible are called for jury duty in any one calendar year. People are selected for jury duty at random from those eligible, and the same individual cannot be called more than once in the same year. What is the probability that an eligible person in this city is selected in both of the next 2 years? All of the next 3 years?

5.92 (M8)
An online store offers two methods of shipping—regular ground service and an expedited 2-day shipping. Customers may also choose whether or not to have the purchase gift wrapped. Suppose that the events

E = event that the customer chooses expedited shipping

G = event that the customer chooses gift wrap

are independent with $P(E) = 0.26$ and $P(G) = 0.12$.
a. Construct a "hypothetical 1000" table with columns corresponding to whether or not expedited shipping was chosen and rows corresponding to whether or not gift wrap was selected.

b. Use the table to compute $P(E \cup G)$. Give a long-run relative frequency interpretation of this probability.

5.93 **(M9)**

USA Today **(June 6, 2000)** gave information on seat belt usage by gender. The proportions in the following table are based on a survey of a large number of adult men and women in the United States:

	Male	Female
Uses Seat Belts Regularly	0.10	0.175
Does Not Use Seat Belts Regularly	0.40	0.325

Assume that these proportions are representative of adults in the United States and that a U.S. adult is selected at random.

a. What is the probability that the selected adult regularly uses a seat belt?

b. What is the probability that the selected adult regularly uses a seat belt, given that the individual selected is male?

c. What is the probability that the selected adult does not use a seat belt regularly, given that the selected individual is female?

d. What is the probability that the selected individual is female, given that the selected individual does not use a seat belt regularly?

e. Are the probabilities from Parts (c) and (d) equal? Write a couple of sentences explaining why or why not.

5.94 **(C4, M1, M10)**

The report **"Twitter in Higher Education: Usage Habits and Trends of Today's College Faculty" (Magna Publications, September 2009)** describes a survey of nearly 2,000 college faculty. The report indicates the following:

- 30.7% reported that they use Twitter, and 69.3% said that they do not use Twitter.
- Of those who use Twitter, 39.9% said they sometimes use Twitter to communicate with students.
- Of those who use Twitter, 27.5% said that they sometimes use Twitter as a learning tool in the classroom.

Consider the chance experiment that selects one of the study participants at random.

a. Two of the percentages given in the problem specify unconditional probabilities, and the other two percentages specify conditional probabilities. Which are conditional probabilities, and how can you tell?

b. Suppose the following events are defined:

T = event that selected faculty member uses Twitter

C = event that selected faculty member sometimes uses Twitter to communicate with students

L = event that selected faculty member sometimes uses Twitter as a learning tool in the classroom

Use the given information to determine the following probabilities:

i. $P(T)$ **iii.** $P(C|T)$

ii. $P(T^C)$ **iv.** $P(L|T)$

c. Construct a "hypothetical 1000" table using the given probabilities and use it to compute $P(C)$, the probability that the selected study participant sometimes uses Twitter to communicate with students.

d. Construct a "hypothetical 1000" table using the given probabilities and use it to compute the probability that the selected study participant sometimes uses Twitter as a learning tool in the classroom.

5.95 **(M10, P2)**

In an article that appears on the website of the **American Statistical Association (www.amstat.org)**, Carlton Gunn, a public defender in Seattle, Washington, wrote about how he uses statistics in his work as an attorney. He states:

> I personally have used statistics in trying to challenge the reliability of drug testing results. Suppose the chance of a mistake in the taking and processing of a urine sample for a drug test is just 1 in 100. And your client has a "dirty" (i.e., positive) test result. Only a 1 in 100 chance that it could be wrong? Not necessarily. If the vast majority of all tests given—say 99 in 100—are truly clean, then you get one false dirty and one true dirty in every 100 tests, so that half of the dirty tests are false.

Define the following events as

TD = event that the test result is dirty

TC = event that the test result is clean

D = event that the person tested is actually dirty

C = event that the person tested is actually clean

a. Using the information in the quote, what are the values of

i. $P(TD \mid D)$ **iii.** $P(C)$

ii. $P(TD \mid C)$ **iv.** $P(D)$

b. Use the probabilities from Part (a) to construct a "hypothetical 1000" table.

c. What is the value of $P(TD)$?

d. Use the table to compute the probability that a person is clean given that the test result is dirty, $P(C|TD)$. Is this value consistent with the argument given in the quote? Explain.

5.96 **(M10, P2)**

The authors of the paper **"Do Physicians Know when Their Diagnoses Are Correct?" (***Journal of General Internal Medicine* **[2005]: 334–339)** presented detailed case studies to students and faculty at medical schools. Each participant was asked to provide a diagnosis in the case and also to indicate whether his or her confidence in the correctness of the diagnosis was high or low. Define the events C, I, and H as follows:

C = event that diagnosis is correct

I = event that diagnosis is incorrect

H = event that confidence in the correctness of the diagnosis is high

a. Data appearing in the paper were used to estimate the following probabilities for medical students:

$$P(C) = 0.261 \qquad P(I) = 0.739$$

$$P(H|C) = 0.375 \qquad P(H|I) = 0.073$$

Use the given probabilities to construct a "hypothetical 1000" table with rows corresponding to whether the diagnosis was correct or incorrect and columns corresponding to whether confidence was high or low.

b. Use the table to compute the probability of a correct diagnosis, given that the student's confidence level in the correctness of the diagnosis is high.

c. Data from the paper were also used to estimate the following probabilities for medical school faculty:

$$P(C) = 0.495 \qquad P(I) = 0.505$$

$$P(H|C) = 0.537 \qquad P(H|I) = 0.252$$

Construct a "hypothetical 1000" table for medical school faculty and use it to compute the probability of a correct diagnosis given that the faculty member's confidence level in the correctness of the diagnosis is high. How does the value of this probability compare to the value for students computed in Part (b)?

5.97 (P3, P4)

Is ultrasound a reliable method for determining the gender of an unborn baby? Consider the following data on 1,000 births, which are consistent with summary values that appeared in the online *Journal of Statistics Education* ("New Approaches to Learning Probability in the First Statistics Course" [2001]):

	Ultrasound Predicted Female	Ultrasound Predicted Male
Actual Gender Is Female	432	48
Actual Gender Is Male	130	390

Do you think that a prediction that a baby is male and a prediction that a baby is female are equally reliable? Explain, using the information in the table to calculate any relevant probabilities.

5.98 (M11)

To help ensure the safety of school classrooms, the local fire marshal does an inspection at Thomas Jefferson High School each month to check for faulty wiring, overloaded circuits, and other fire code violations. Each month, one room is selected for inspection. Suppose that the probability that the selected room is a science classroom (biology, chemistry, or physics) is 0.6 and the probability that the selected room is a chemistry room is 0.4. Use probability formulas to find the following probabilities.

a. The probability that the selected room is not a science room.

b. The probability that the selected room is a chemistry room and a science room.

c. The probability that the selected room is a chemistry room given that the room selected was a science room.

d. The probability that the selected room was a chemistry room or a science room.

5.99 (M11)

Three of the most common types of pets are cats, dogs and fish. Many families have more than one type of pet. Suppose that a family is selected at random and consider the following events and probabilities:

F = event that the selected family has at least one fish
D = event that the selected family has at least one dog
C = event that the selected family has at least one cat

$$P(F) = 0.20 \quad P(D) = 0.32 \quad P(C) = 0.35$$
$$P(F \cap D) = 0.18 \quad P(F \cap C) = 0.07 \quad P(D \mid C) = 0.30$$

Calculate the following probabilities.

a. $P(F \mid D)$

b. $P(F \cup D)$

c. $P(C \cap D)$

AP* Review Questions for Chapter 5

1. A die is weighted so that when it is rolled, an odd number will come up twice as often as an even number in the long run. Which of the following is a legitimate assignment of probabilities to the outcomes in the sample space that is consistent with the description of the die?

   (A)

Outcome	1	2	3	4	5	6
Probability	1/6	1/6	1/6	1/6	1/6	1/6

   (B)

Outcome	1	2	3	4	5	6
Probability	1/6	2/6	1/6	2/6	1/6	2/6

   (C)

Outcome	1	2	3	4	5	6
Probability	1/9	2/9	1/9	2/9	1/9	2/9

   (D)

Outcome	1	2	3	4	5	6
Probability	2/9	1/9	2/9	1/9	2/9	1/9

   (E) None of the above

Use the following to answer questions 2 and 3.

In a survey of 100 students selected at random from those enrolled at a small liberal arts college, the following data was obtained. Each student was asked how much time he or she spent on the Internet during a typical day, and gender was also noted.

	None	Less than 1 hour	1 hour or more	Total
Male	15	20	30	65
Female	10	15	10	35
Total	25	35	40	100

Assume that this data is representative of students at this college.

2. What is the approximate probability that a randomly selected student at this college is a female that spends no time on the Internet in a typical day?

   (A) 10/10
   (B) 10/15
   (C) 10/25
   (D) 10/35
   (E) 10/100

3. If a randomly selected student is male, what is the approximate probability that he spends 1 hour or more on the Internet in a typical day?

   (A) 30/30
   (B) 30/40
   (C) 30/65
   (D) 35/65
   (E) 30/100

4. Under what conditions is it true that P(A *or* B) = P(A) + P(B)

   (A) A and B must be mutually exclusive
   (B) A and B cannot be mutually exclusive
   (C) A and B must be independent
   (D) A and B must be both mutually exclusive and independent
   (E) No conditions are required. It is always true that P(A *or* B) = P(A) + P(B)

5. Under what conditions is it true that P(A *and* B) = P(B|A)P(A)

   (A) A and B must be mutually exclusive
   (B) A and B cannot be mutually exclusive
   (C) A and B must be independent
   (D) A and B must be both mutually exclusive and independent
   (E) No conditions are required. It is always true that P(A *and* B) = P(B|A)P(A)

6. A statistics class consists of 30 males and 20 females. Ten of the males and 10 of the females are also currently enrolled in a science course. If a student is selected at random from those enrolled in this statistics class, what is the probability that a female is selected?

   (A) 0.1
   (B) 0.2
   (C) 0.4
   (D) 0.5
   (E) 0.6

7. A statistics class consists of 30 males and 20 females. Ten of the males and ten of the females are also currently enrolled in a science course. If a student is selected at random from those enrolled in this statistics class, what is the probability that the selected student is a male who is also enrolled in a science course?

   (A) 0.1
   (B) 0.2
   (C) 0.4
   (D) 0.5
   (E) 0.6

8. A statistics class consists of 30 males and 20 females. Ten of the males and ten of the females are also currently enrolled in a science course. A student is selected at random from those enrolled in this statistics class. Given that the selected student is female, what is the probability that the selected student is enrolled in a science course?

(A) 0.1
(B) 0.2
(C) 0.4
(D) 0.5
(E) 0.6

9. Sixty percent of the students at an urban university carry more than one credit card. Selecting a student at random from the students at this university can be viewed as an experiment with two possible outcomes: *has more than one credit card* and *does not have more than one credit card*. Which of the following would be a correct assignment of random digits to the outcomes in order to simulate selecting a student at random from the students enrolled at the university?

(A) 0 represents more than one credit card, 1 represents not more than one credit card, ignore all other digits.
(B) Odd digits represent more than one credit card, and even digits represent not more than one credit card.
(C) 0, 1, 2, and 3 represent more than one credit card, and 4, 5, 6, 7, 8, and 9 represent not more than one credit card.
(D) 0, 1, 2, 3, 4, and 5 represent more than one credit card, and 6, 7, 8, and 9 represent not more than one credit card.
(E) C and D both represent correct assignments.

10. Which of the following would be a correct assignment of random digits to the outcomes of heads and tails in order to simulate the flip of a fair coin?

(A) 0 represents heads, 1 represents tails; ignore all other digits.
(B) Odd digits represent heads, even digits represent tails.
(C) 0, 1, 2, 3, and 4 represent heads, and 5, 6, 7, 8, and 9 represent tails.
(D) Only B and C above are correct.
(E) A, B, and C above are all correct.

11. Which of the following statements is true for two events, each with probability greater than 0?

(A) If the events are mutually exclusive, they must be independent.
(B) If the events are independent, they must be mutually exclusive.
(C) If the events are not mutually exclusive, they must be independent.
(D) If the events are not independent, they must be mutually exclusive.
(E) If the events are mutually exclusive, they cannot be independent.

12. The Air Force receives 20% of its parachutes from Company C1, 55% from Company C2, and the rest from Company C3. The probability that a randomly selected parachute will fail to open is 0.0025, 0.002, and 0.0022, depending on whether it is from Company C1, C2, or C3, respectively. If a randomly chosen parachute fails to open, what is the probability it is from company C3?

(A) 0.00055
(B) 0.00215
(C) 0.0022
(D) 0.250
(E) 0.256

13. A and B are events with $P(A) = 0.6$, $P(B) = 0.7$, and $P(A \cap B) = 0.4$. Which of the following is true?

(A) A and B are mutually exclusive.
(B) A and B are independent.
(C) $P(A \cup B) = 0.6 + 0.7$
(D) $P(A \mid B) = P(A)$
(E) None of the above is true.

14. Two coins—a nickel and a dime—are to be flipped. Consider the following two events:

A = the event that the nickel lands with the head side up
B = the event that the dime lands with the head side up

Which of the following is true?

(A) The events A and B are independent and mutually exclusive.
(B) The events A and B are neither independent nor mutually exclusive.
(C) The events A and B are independent but not mutually exclusive.
(D) The events A and B are mutually exclusive but not independent.
(E) The events A and B are independent only if the coins are fair.

15. A chance experiment involves selecting a person at random from those who purchased a new car at a particular dealership during 2010 and noting both the gender and whether or not the selected person purchased a hybrid vehicle. The sample space for this experiment is {male who purchased a hybrid, male who did not purchase a hybrid, female who purchased a hybrid, female who did not purchase a hybrid}. Under what conditions would the four outcomes in the sample space be equally likely?

(A) No conditions are necessary—they are always equally likely.
(B) There must be an equal number of males and females who purchased cars from this dealership during 2010.
(C) 50% of the males must have purchased a hybrid.
(D) 50% of the males must have purchased a hybrid and 50% of the females must have purchased a hybrid.
(E) Both conditions B and D above are necessary.

6

Random Variables and Probability Distributions

Preview

Chapter Learning Objectives

6.1 Random Variables

6.2 Probability Distributions for Discrete Random Variables

6.3 Probability Distributions for Continuous Random Variables

6.4 Mean and Standard Deviation of a Random Variable

6.5 Binomial and Geometric Distributions

6.6 Normal Distributions

6.7 Checking for Normality

6.8 Using the Normal Distribution to Approximate a Discrete Distribution

Chapter Activities

Are You Ready to Move On? Chapter 6 Review Exercises

Technology Notes

AP* Review Questions for Chapter 6

Want to Know More? See Chapter 6 Online Materials to learn about Counting Rules and the Poisson Distribution

Kathy deWitt/Alamy Limited

PREVIEW

One way to learn from data is to use information from a sample to learn about a population distribution. In this situation, you are usually interested in the distribution of one or more variables. For example, an environmental scientist who obtains an air sample from a specified location might be interested in the concentration of ozone (a major constituent of smog). Before selection of the air sample, the value of the ozone concentration is uncertain. Because the value of a variable quantity such as ozone concentration is subject to uncertainty, such variables are called random variables. *In this Chapter, you will learn how probability models are used to describe the behavior of random variables.*

CHAPTER LEARNING OBJECTIVES

Conceptual Understanding

After completing this chapter, you should be able to

C1 Understand that a probability distribution describes the long-run behavior of a random variable.

C2 Understand how the long-run behavior of a random variable is described by its mean and standard deviation.

C3 Understand that areas under a density curve for a continuous random variable are interpreted as probabilities.

Mastering the Mechanics

After completing this chapter, you should be able to

M1 Distinguish between discrete and continuous random variables.

M2 Given a probability distribution for a discrete random variable, compute and interpret probabilities.

M3 Construct the probability distribution of a discrete random variable.

M4 Compute and interpret the mean and standard deviation of a discrete random variable.

M5 Given the probability density curve for a continuous random variable, identify areas corresponding to various probabilities.

M6 Compute probabilities involving continuous random variables whose density curves have a simple form.

M7 Interpret an area under a normal curve as a probability.

M8 For a normal random variable x, use technology or tables to compute probabilities of the form $P(a < x < b)$, $P(x < b)$, and $P(x > a)$ and to find percentiles.

M9 Use technology or given normal scores to construct a normal probability plot.

M10 Given a normal probability plot, assess whether it is reasonable to think that a population distribution is approximately normal.

M11 Distinguish between binomial and geometric random variables.

M12 Compute binomial probabilities using technology or tables.

M13 Compute probabilities using the geometric distribution.

M14 Given a binomial distribution, determine if it is appropriate to use a normal approximation to estimate binomial probabilities. If appropriate, approximate probabilities using a normal distribution.

M15 Calculate the mean and standard deviation of a linear function of a random variable and of a linear combination of independent random variables.

Putting It into Practice

After completing this chapter, you should be able to

P1 Use information provided by a probability distribution to draw conclusions in context.

PREVIEW EXAMPLE   iPod Shuffles

The paper **"Does Your iPod *Really* Play Favorites?" (*American Statistician* [2009]: 263–268)** investigated the shuffle feature of the iPod. The shuffle feature takes a group of songs, called a playlist, and plays them in a random order. Some users have questioned the "randomness" of the shuffle, citing situations where several songs from the same artist were played in close proximity to each other. One such example appeared in a *Newsweek* article in 2005, where the author states that Steely Dan songs always seem to pop up in the first hour of play. (For readers unfamiliar with Steely Dan, see http://www. steelydan.com.) Is this consistent with a "random" shuffle? You will return to this example in Section 6.2 to answer this question.

SECTION 6.1 Random Variables

In most chance experiments, an investigator focuses attention on one or more variable quantities. For example, consider the chance experiment of randomly selecting a customer who is leaving a store. One numerical variable of interest to the store manager might be the number of items purchased by the customer. Let's use the letter x to denote this variable. Possible values of x are 0 (a person who left without making a purchase), 1, 2, 3, and so on. In this example, the values of x are isolated points on the number line. Until a customer is selected and the number of items counted, the value of x is uncertain. Another variable of interest might be y = number of minutes spent in a checkout line. One possible value of y is 3.0 minutes and another is 4.0 minutes, but *any* other number between 3.0 and 4.0 is also a possibility. The possible y values form an entire interval on the number line.

> **AP* EXAM TIP**
>
> Random variables are either discrete or continuous. To determine whether a random variable is discrete or continuous, think about what the possible values are.

> **DEFINITION**
>
> **Random variable:** a numerical variable whose value depends on the outcome of a chance experiment. A random variable associates a numerical value with each outcome of a chance experiment.
>
> A random variable is **discrete** if its possible values are isolated points along the number line.
>
> A random variable is **continuous** if its possible values are *all* points in some interval.

Lowercase letters, such as x and y, are used to represent random variables.* Figure 6.1 shows a set of possible values for each type of random variable. In practice, a discrete random variable almost always involves counting (for example, number of items purchased, number of gas pumps in use, or number of broken eggs in a carton). A continuous random variable is one whose value is typically obtained by measuring (for example, temperature in a freezer, weight of a pineapple, or amount of time spent in a store). Because there is a limit to the accuracy of any measuring instrument, such as a watch or a scale, it may seem that any variable should be regarded as discrete. For example, when weight is measured to the nearest pound, the observed values appear to be isolated points along the number line, such as 2 pounds, 3 pounds, and so on. But this is just a function of the accuracy with which weight is recorded and not because a weight between 2 and 3 pounds is impossible. In this case, the variable's behavior is continuous.

FIGURE 6.1
Two different types of random variables

Possible values of a
discrete random variable

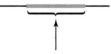

Possible values of a
continuous random variable

Example 6.1 Book Sales

Consider a chance experiment in which the type of book, print (P) or digital (D), chosen by each of three successive customers making a purchase from on an online bookstore is noted. Define a random variable x by

$$x = \text{number of customers purchasing a digital book}$$

The outcome in which the first and third customers purchase a digital book and the second customer purchases a print book can be abbreviated DPD. The associated x value is 2, because two of the three customers selected a digital book. Similarly, the x value for the

---

*In some books, uppercase letters are used to name random variables, with lowercase letters representing a particular value that the variable might assume. We have chosen to use a simpler and less formal notation.

outcome DDD (all three purchased a digital book) is 3. The following table displays each of the eight possible outcomes and the corresponding value of x.

Outcome	PPP	DPP	PDP	PPD	DDP	DPD	PDD	DDD
x value	0	1	1	1	2	2	2	3

There are only four possible x values—0, 1, 2, and 3. Because these possible values are isolated points on the number line, x is a discrete random variable.

In some situations, the random variable of interest is discrete, but the number of possible values is not finite. This is illustrated in Example 6.2.

Example 6.2 This Could Be a Long Game

Two friends agree to play a game that consists of a sequence of trials. The game continues until one player wins two trials in a row. One random variable of interest might be

$$x = \text{number of trials required to complete the game}$$

Let A denote a win for Player 1 and B denote a win for Player 2. The simplest possible outcomes are AA (Player 1 wins the first two trials and the game ends) and BB (Player 2 wins the first two trials). With either of these two outcomes, $x = 2$. There are also two outcomes for which $x = 3$: ABB and BAA. Some other possible outcomes and associated x values are

Outcomes	x Value
AA, BB	2
BAA, ABB	3
ABAA, BABB	4
ABABB, BABAA	5
⋮	⋮
ABABABABAA, BABABABABB	10

Any positive integer that is 2 or greater is a possible value. Because the values 2, 3, 4, ... are isolated points on the number line, x is a discrete random variable even though there is no upper limit to the number of possible values.

Example 6.3 Stress

In an engineering stress test, pressure is applied to a thin 1-foot-long bar until the bar snaps. The precise location where the bar will snap is uncertain. You can use x to denote the distance (in feet) from the left end of the bar to the break. Then $x = 0.25$ is one possibility, $x = 0.9$ is another, and in fact any number between 0 and 1 is a possible value of x. Figure 6.2 shows the case of the outcome $x = 0.6$. Because the set of possible values is an entire interval on the number line, x is a continuous random variable.

Depending on the accuracy of the measuring instrument, you may be able to measure the distance only to the nearest hundredth of a foot or thousandth of a foot. The *actual* distance, though, could be any number between 0 and 1, which indicates that the variable is continuous.

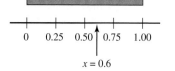

FIGURE 6.2
The bar for Example 6.3 and the outcome $x = 0.6$

In data analysis, random variables often arise in the context of summarizing data from a sample that has been selected from some population. This is illustrated in Example 6.4.

Example 6.4 College Plans

Suppose that a high school counselor plans to select a random sample of 50 seniors. She will ask each student in the sample if he or she plans to attend college after graduation. The process of selecting a sample is a chance experiment. The sample space for this experiment consists of all the different possible random samples of size 50 that might result (there are a very large number of them), and for random sampling, each of these outcomes is equally likely. Suppose

$$x = \text{number of successes in the sample}$$

where a success in this instance is defined as a student who plans to attend college. Then x is a random variable, because it associates a numerical value with each of the possible outcomes (random samples) that might occur. The possible values of x are 0, 1, 2, . . . , 50, so x is a discrete random variable. ∎

SECTION 6.1 EXERCISES

Each Exercise Set assesses the following chapter learning objective: M1

SECTION 6.1 Exercise Set 1

6.1 State whether each of the following random variables is discrete or continuous:
a. The number of defective tires on a car
b. The body temperature of a hospital patient
c. The number of pages in a book
d. The number of draws (with replacement) from a deck of cards until a heart is selected
e. The lifetime of a light bulb

6.2 Starting at a particular time, each car entering an intersection is observed to see whether it turns left (L) or right (R) or goes straight ahead (S). The experiment terminates as soon as a car is observed to go straight. Let x denote the number of cars observed. What are possible x values? List five different outcomes and their associated x values.

6.3 A box contains four slips of paper marked 1, 2, 3, and 4. Two slips are selected *without* replacement. List the possible values for each of the following random variables:
a. x = sum of the two numbers
b. y = difference between the first and second numbers (first − second)
c. z = number of slips selected that show an even number
d. w = number of slips selected that show a 4

SECTION 6.1 Exercise Set 2

6.4 Classify each of the following random variables as either discrete or continuous:
a. The fuel efficiency (mpg) of an automobile
b. The amount of rainfall at a particular location during the next year

c. The distance that a person throws a baseball
d. The number of questions asked during a 1-hour lecture
e. The tension (in pounds per square inch) at which a tennis racket is strung
f. The amount of water used by a household during a given month
g. The number of traffic citations issued by the highway patrol in a particular county on a given day

6.5 A point is randomly selected from the interior of a square, as pictured:

Let x denote the distance from the lower left-hand corner A to the selected point. What are possible values of x? Is x a discrete or a continuous variable?

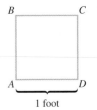

1 foot

6.6 A person stands at the corner marked A of the square pictured in the previous exercise and tosses a coin. If it lands heads up, the person moves one corner clockwise, to B. If the coin lands tails up, the person moves one corner counterclockwise, to D. This process is then repeated until the person arrives back at A. Let y denote the number of coin tosses. What are possible values of y? Is y discrete or continuous?

Additional Exercises

6.7 State whether each of the following random variables is discrete or continuous.
a. The number of courses a student is enrolled in
b. The time spent completing a homework assignment
c. The length of a person's forearm
d. The number of times out of 10 throws that a dog catches a Frisbee

6.8 A person is asked to draw a line segment that they think is 3 inches long. The length of the line segment drawn will be measured and the value of $x = $ (actual length $- 3$) will be calculated.
a. What is the value of x for a person who draws a line segment that is 3.1 inches long?
b. Is x a discrete or continuous random variable?

6.9 Two six-sided dice, one red and one white, will be rolled. List the possible values for each of the following random variables.
a. $x = $ sum of the two numbers showing
b. $y = $ difference between the number on the red die and the number on the white die (red $-$ white)
c. $w = $ largest number showing

SECTION 6.2 Probability Distributions for Discrete Random Variables

The probability distribution of a random variable is a model that describes the long-run behavior of that variable. For example, suppose that the Department of Animal Regulation in a particular county is interested in studying the variable

$x = $ number of licensed dogs or cats at a randomly selected house

County regulations prohibit more than five dogs or cats per house. Consider the chance experiment of randomly selecting a house in this county. In this case, x is a discrete random variable because it associates a numerical value (0, 1, 2, 3, 4, or 5) with each of the possible outcomes (houses) in the sample space. Although you know what the possible values for x are, it would also be useful to know how this variable would behave if it were observed for many houses. What would be the most common value? What proportion of the time would $x = 5$ be observed? $x = 3$? A probability distribution provides answers to such questions.

> **AP* EXAM TIP**
>
> The probability distribution of a random variable describes its long-run behavior.

> **DEFINITION**
>
> The **probability distribution of a discrete random variable** x gives the probability associated with each possible x value. Each probability is the long-run proportion of the time that the corresponding x value will occur.
>
> Common ways to display a probability distribution for a discrete random variable are a table, a probability histogram, or a formula.
>
> If one possible value of x is 2, it is common to write $p(2)$ in place of $P(x = 2)$. Similarly, $p(5)$ denotes the probability that $x = 5$, and so on.

Example 6.5 Energy-Efficient Refrigerators

Suppose that each of four randomly selected customers purchasing a refrigerator at a large appliance store chooses either an energy-efficient model (E) or one from a less expensive group of models (G) that do not have an energy-efficient rating. Assume that these customers make their choices independently of one another and that 40% of all customers select an energy-efficient model. This implies that for any particular one of the four customers, $P(E) = 0.4$ and $P(G) = 0.6$. One possible outcome is EGGE, where the first and fourth customers select energy-efficient models and the other two choose less expensive models. Because the customers make their choices independently, the multiplication rule for independent events implies that

$P(\text{EGGE}) = P(\text{1st chooses E } and \text{ 2nd chooses G } and \text{ 3rd chooses G } and \text{ 4th chooses E})$
$= P(E)P(G)P(G)P(E)$
$= (0.4)(0.6)(0.6)(0.4)$
$= 0.0576$

Similarly,

$$P(EGEG) = P(E)P(G)P(E)P(G)$$
$$= (0.4)(0.6)(0.4)(0.6)$$
$$= 0.0576 \text{ (which is equal to } P(EGGE))$$

and

$$P(GGGE) = (0.6)(0.6)(0.6)(0.4) = 0.0864$$

The number among the four customers who purchase energy-efficient models is a random variable. Using x to denote this random variable,

x = the number of energy efficient refrigerators purchased by the four customers.

Table 6.1 displays the 16 possible outcomes that might be observed, the probability of each outcome, and the value of the random variable x that is associated with each outcome.

TABLE 6.1 Outcomes and Probabilities for Example 6.5

Outcome	Probability	x Value	Outcome	Probability	x Value
GGGG	0.1296	0	GEEG	0.0576	2
EGGG	0.0864	1	GEGE	0.0576	2
GEGG	0.0864	1	GGEE	0.0576	2
GGEG	0.0864	1	GEEE	0.0384	3
GGGE	0.0864	1	EGEE	0.0384	3
EEGG	0.0576	2	EEGE	0.0384	3
EGEG	0.0576	2	EEEG	0.0384	3
EGGE	0.0576	2	EEEE	0.0256	4

The probability distribution of x is easily obtained from this information. Consider the smallest possible x value, 0. The only outcome for which $x = 0$ is GGGG, so

$$p(0) = P(x = 0) = P(GGGG) = 0.1296$$

There are four different outcomes for which $x = 1$, so $p(1)$ results from summing the four corresponding probabilities:

$$p(1) = P(x = 1) = P(EGGG \text{ or } GEGG \text{ or } GGEG \text{ or } GGGE)$$
$$= P(EGGG) + P(GEGG) + P(GGEG) + P(GGGE)$$
$$= 0.0864 + 0.0864 + 0.0864 + 0.0864$$
$$= 4(0.0864)$$
$$= 0.3456$$

Similarly,

$$p(2) = P(EEGG) + \cdots + P(GGEE) = 6(0.0576) = 0.3456$$
$$p(3) = 4(0.0384) = 0.1536$$
$$p(4) = 0.0256$$

The probability distribution of x is summarized in the following table:

x Value	0	1	2	3	4
p(x)	0.1296	0.3456	0.3456	0.1536	0.0256

To interpret $p(3) = 0.1536$, think of performing the chance experiment repeatedly, each time with a new group of four customers. In the long run, 15.36% of these groups will have exactly three customers purchasing an energy-efficient refrigerator.

The probability distribution can be used to determine probabilities of various events involving x. For example, the probability that at least two of the four customers choose energy-efficient models is

$$P(x \geq 2) = P(x = 2 \ or \ x = 3 \ or \ x = 4)$$
$$= p(2) + p(3) + p(4)$$
$$= 0.5248$$

This means that in the long run, a group of four refrigerator purchasers will include at least two who select energy-efficient models about 52.48% of the time.

A probability distribution table for a discrete variable shows the possible x values and the value of $p(x)$ for each one. Because $p(x)$ is a probability, it must be a number between 0 and 1, and because the probability distribution includes all possible x values, the sum of all the $p(x)$ values must equal 1. These properties of discrete probability distributions are summarized in the following box.

> **Properties of Discrete Probability Distributions**
>
> 1. For every possible x value, $0 \leq p(x) \leq 1$.
> 2. $\displaystyle\sum_{\text{all } x \text{ values}} p(x) = 1$

A *probability histogram* is a graphical representation of a discrete probability distribution. The graph has a rectangle centered above each possible value of x, and the area of each rectangle is proportional to the probability of the corresponding value. Figure 6.3 displays the probability histogram for the probability distribution of Example 6.5.

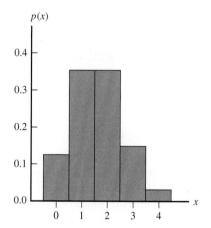

FIGURE 6.3
Probability histogram for the distribution of number of energy-efficient refrigerators purchased

In Example 6.5, the probability distribution was derived by starting with a simple chance experiment and applying basic probability rules. Sometimes this approach is not feasible because of the complexity of the chance experiment. In this case, the long-run behavior of a random variable is often described by an approximate probability distribution based on empirical evidence and prior knowledge. This distribution must still follow rules of probability. Specifically,

1. $p(x) \geq 0$ for every x value
2. $\displaystyle\sum_{\text{all } x \text{ values}} p(x) = 1$

Example 6.6 Paint Flaws

In automobile manufacturing, painting is one of the last steps in the process of assembling a new car. Some minor blemishes in the paint surface are considered acceptable, but if there are too many, it becomes noticeable to a potential customer, and the car must be repainted. Cars coming off the assembly line are carefully inspected, and the number of minor blemishes in the paint surface is determined. Let x denote the number of minor blemishes on a randomly selected car from a particular manufacturing plant. Suppose that a large number of automobiles are evaluated, leading to the following probability distribution:

x	0	1	2	3	4	5	6	7	8	9	10
$p(x)$	0.041	0.130	0.209	0.223	0.178	0.114	0.061	0.028	0.011	0.004	0.001

The corresponding probability histogram appears in Figure 6.4. The probabilities in this distribution reflect the car manufacturer's experience. For example, $p(3) = 0.223$ indicates that 22.3% of new automobiles had exactly 3 minor paint blemishes. The probability that the number of minor paint blemishes is between 2 and 5 inclusive is

$$P(2 \leq x \leq 5) = p(2) + p(3) + p(4) + p(5) = 0.724$$

If many cars of this type were examined, about 72.4% would have 2, 3, 4, or 5 minor paint blemishes.

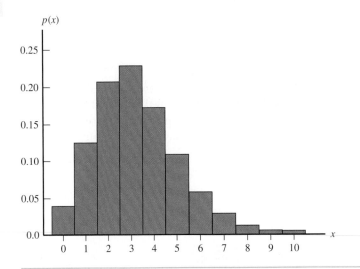

FIGURE 6.4
Probability histogram for the distribution of the number of minor paint blemishes on a randomly selected car

Example 6.7 iPod Shuffles Revisited

Recall from the Preview Example that the paper **"Does Your iPod *Really* Play Favorites?"** (*American Statistician* **[2009]: 263–268)** investigated the shuffle feature of the iPod. The shuffle feature is supposed to play the songs in a playlist in a random order. Some users have questioned the "randomness" of the shuffle, citing situations where several songs from the same artist were played in close proximity to each other. One such example appeared in a *Newsweek* article in 2005, where the author states that Steely Dan songs always seem to pop up in the first hour of play. Is this consistent with a "random" shuffle?

To investigate the alleged non-randomness, suppose you create a playlist of 3,000 songs that includes 50 songs by Steely Dan (this is the situation considered by the authors of the *American Statistician* paper). You could then carry out a simulation by creating a shuffle of 20 songs (about 1 hour of playing time) from the playlist and noting the number of songs by Steely Dan that were among the 20. This could be done by thinking of the songs in the playlist as being numbered from 1 to 3,000, with numbers 1 to 50 representing the Steely Dan songs. A random number generator could then be used to select 20 random numbers between 1 and 3,000. Counting how many times a number between 1 and 50 was included in this list would give you the number of Steely Dan songs in this particular shuffle of 20 songs. Repeating this process a large number of times would enable you to estimate the probabilities needed for the probability distribution of

x = number of Steely Dan songs in a random shuffle consisting of 20 songs

The accompanying probability distribution is from the *American Statistician* paper. Even though the possible values of x are 0, 1, 2, ... , 20, the probabilities of x values of 4 or greater are very small. For this reason, 4, 5, ... , 20 have been grouped into a single entry in the table.

Probability Distribution of x = Number of Steely Dan Songs in a 20 Song Shuffle	
x	$p(x)$
0	0.7138
1	0.2435
2	0.0387
3	0.0038
4 or more	0.0002

Notice that $P(x \geq 1) = 1 - 0.7138 = 0.2862$. This means that about 28.6% of the time, at least one Steely Dan song would occur in a random shuffle of 20 songs. Given that there are only 50 Steely Dan songs in the playlist of 3,000 songs, this result surprises many people!

You have seen examples in which the probability distribution of a discrete random variable has been given as a table and as a probability (relative frequency) histogram. It is also possible to use a formula that specifies the probability for each possible value of the random variable. Examples of this approach can be found in Section 6.6.

SECTION 6.2 EXERCISES

Each Exercise Set assesses the following chapter learning objectives: C1, M2, M3

SECTION 6.2 Exercise Set 1

6.10 Suppose y = the number of broken eggs in a randomly selected carton of one dozen eggs. The probability distribution of y is as follows:

y	0	1	2	3	4
$p(y)$	0.65	0.20	0.10	0.04	?

a. Only y values of 0, 1, 2, 3, and 4 have probabilities greater than 0. What is $p(4)$?
b. How would you interpret $p(1) = 0.20$?
c. Calculate $P(y \leq 2)$, the probability that the carton contains at most two broken eggs, and interpret this probability.
d. Calculate $P(y < 2)$, the probability that the carton contains *fewer than* two broken eggs. Why is this smaller than the probability in Part (c)?
e. What is the probability that the carton contains exactly 10 unbroken eggs?
f. What is the probability that at least 10 eggs are unbroken?

6.11 Suppose that fund-raisers at a university call recent graduates to request donations for campus outreach programs. They report the following information for last year's graduates:

Size of donation	$0	$10	$25	$50
Proportion of calls	0.45	0.30	0.20	0.05

Three attempts were made to contact each graduate. A donation of $0 was recorded both for those who were contacted but declined to make a donation and for those who were not reached in three attempts. Consider the variable x = amount of donation for a person selected at random from the population of last year's graduates of this university.

a. Write a few sentences describing what donation amounts you would expect to see if the value of x was observed for each of 1,000 graduates.
b. What is the most common value of x in this population?
c. What is $P(x \geq 25)$?
d. What is $P(x > 0)$?

6.12 A restaurant has four cartons of non-fat milk in stock. Two of these cartons (Cartons 1 and 2) have spoiled. Suppose that two cartons are ordered, and the server selects two of the four cartons at random. Consider the random variable x = the number of good cartons among these two.

a. One possible experimental outcome is (1, 2) (Cartons 1 and 2 are selected) and another is (2, 4). List all possible outcomes.

b. What is the probability of each outcome in Part (a)?
c. The value of x for the $(1, 2)$ outcome is 0 (neither selected carton is good), and $x = 1$ for the outcome $(2, 4)$. Determine the x value for each possible outcome. Then use the probabilities in Part (b) to determine the probability distribution of x.

6.13 Suppose that 20% of all homeowners in an earthquake-prone area of California are insured against earthquake damage. Four homeowners are selected at random. Define the random variable x as the number among the four who have earthquake insurance.
a. Find the probability distribution of x. (Hint: Let S denote a homeowner who has insurance and F one who does not. Then one possible outcome is SFSS, with probability $(0.2)(0.8)(0.2)(0.2)$ and associated x value of 3. There are 15 other outcomes.)
b. What is the most likely value of x?
c. What is the probability that at least two of the four selected homeowners have earthquake insurance?

SECTION 6.2 **Exercise Set 2**

6.14 Suppose $x =$ the number of courses a randomly selected student at a certain university is taking. The probability distribution of x appears in the following table:

x	1	2	3	4	5	6	7
$p(x)$	0.02	0.03	0.09	0.25	0.40	0.16	0.05

a. What is $P(x = 4)$?
b. What is $P(x \le 4)$?
c. What is the probability that the selected student is taking at most five courses?
d. What is the probability that the selected student is taking at least five courses? More than five courses?
e. Calculate $P(3 \le x \le 6)$ and $P(3 < x < 6)$. Explain in words why these two probabilities are different.

6.15 A pizza shop sells pizzas in four different sizes. The 1,000 most recent orders for a single pizza resulted in the following proportions for the various sizes:

Size	12 in.	14 in.	16 in.	18 in.
Proportion	0.20	0.25	0.50	0.05

With $x =$ the size of a pizza in a single-pizza order, the given table is an approximation to the population distribution of x.
a. Write a few sentences describing what you would expect to see for pizza sizes over a long sequence of single-pizza orders.
b. What is the approximate value of $P(x < 16)$?
c. What is the approximate value of $P(x \le 16)$?

6.16 Of all airline flight requests received by a certain ticket broker, 70% are for domestic travel (D) and 30% are for international flights (I). Define x to be the number that are for domestic flights among the next three requests received. Assuming independence of successive

requests, determine the probability distribution of x. (Hint: One possible outcome is DID, with the probability $(0.7)(0.3)(0.7) = 0.147$.)

6.17 Suppose that a computer manufacturer receives computer boards in lots of five. Two boards are selected from each lot for inspection. You can represent possible outcomes of the selection process by pairs. For example, the pair $(1, 2)$ represents the selection of Boards 1 and 2 for inspection.
a. List the 10 different possible outcomes.
b. Suppose that Boards 1 and 2 are the only defective boards in a lot of five. Two boards are to be chosen at random. Let $x =$ the number of defective boards observed among those inspected. Find the probability distribution of x.

Additional Exercises

6.18 A business has six customer service telephone lines. Let x denote the number of lines in use at a specified time. Suppose that the probability distribution of x is as follows:

x	0	1	2	3	4	5	6
$p(x)$	0.10	0.15	0.20	0.25	0.20	0.06	0.04

Write each of the following events in terms of x, and then calculate the probability of each one:
a. At most three lines are in use
b. Fewer than three lines are in use
c. At least three lines are in use
d. Between two and five lines (inclusive) are in use
e. Between two and four lines (inclusive) are not in use
f. At least four lines are not in use

6.19 A contractor is required by a county planning department to submit anywhere from one to five forms (depending on the nature of the project) in applying for a building permit. Let y be the number of forms required of the next applicant. The probability that y forms are required is known to be proportional to y; that is, $p(y) = ky$ for $y = 1, \ldots, 5$.
a. What is the value of k? (Hint: $\sum p(y) = 1$)
b. What is the probability that at most three forms are required?
c. What is the probability that between two and four forms (inclusive) are required?

6.20 A new battery's voltage may be acceptable (A) or unacceptable (U). A certain flashlight requires two batteries, so batteries will be independently selected and tested until two acceptable ones have been found. Suppose that 80% of all batteries have acceptable voltages, and let y denote the number of batteries that must be tested.
a. What are the possible values of y?
b. What is $p(2)$, that is, $P(y = 2)$?
c. What is $p(3)$? (Hint: There are two different outcomes that result in $y = 3$.)

SECTION 6.3 Probability Distributions for Continuous Random Variables

The possible values of a continuous random variable form an entire interval on the number line. One example of a continuous random variable is the weight x (in pounds) of a full-term newborn baby. Suppose for the moment that weight is recorded only to the nearest pound. Then a reported weight of 7 pounds would be used for any weight greater than or equal to 6.5 pounds and less than 7.5 pounds. If you observe the weights of a large number of newborns, you could construct a density histogram with rectangles centered at 4, 5, and so on. In this histogram, the area of each rectangle is the approximate probability of the corresponding weight value. The total area of all the rectangles is 1, and the probability that a weight (to the nearest pound) is between two values, such as 6 and 8, is the sum of the corresponding rectangular areas. Figure 6.5(a) illustrates this.

FIGURE 6.5
Probability distribution for birth weight: (a) weight measured to the nearest pound; (b) weight measured to the nearest tenth of a pound; (c) limiting curve as measurement accuracy increases: shaded area = $P(6 \leq \text{weight} \leq 8)$

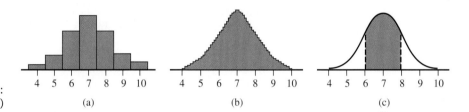

(a) (b) (c)

Now suppose that weight is measured to the nearest tenth of a pound. There are many more possible reported weight values than before, such as 5.0, 5.1, 5.7, 7.3, and 8.9. As shown in Figure 6.5(b), the rectangles in the density histogram are much narrower, giving this histogram a much smoother appearance. As before, the area of each rectangle is an approximate probability, and the total area of all the rectangles is 1.

Figure 6.5(c) shows what happens as weight is measured to a greater degree of accuracy. The resulting histogram approaches a smooth curve. The curve cannot go below the horizontal measurement scale, and the total area under the curve is 1 (because this is true of every probability histogram). The probability that x falls in an interval such as $6 \leq x \leq 8$ is represented by the area under the curve and above that interval.

AP* EXAM TIP

The probability distribution of a continuous random variable is specified by a density curve. Probabilities are calculated by finding areas under the density curve.

DEFINITION

A **probability distribution for a continuous random variable x** is specified by a curve called a **density curve**. The function that defines this curve is denoted by $f(x)$ and it is called the **density function**.

The following are properties of all continuous probability distributions:

1. $f(x) \geq 0$ (so that the curve cannot dip below the horizontal axis).
2. The total area under the density curve is equal to 1.

The probability that x falls in any particular interval is the area under the density curve and above the interval.

Many probability calculations for continuous random variables involve the following three types of events:

1. $a < x < b$, the event that the random variable x takes a value between two given numbers, a and b
2. $x < a$, the event that the random variable x takes a value less than a given number a
3. $x > b$, the event that the random variable x takes a value greater than a given number b (this can also be written as $b < x$)

Figure 6.6 shows the probabilities of these events as areas under a density curve.

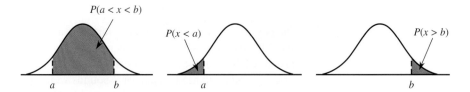

FIGURE 6.6
Probabilities as areas under a probability density curve

Example 6.8 **Application Processing Times**

Let x represent a continuous random variable defined as the amount of time (in minutes) required to process a credit application. Suppose that x has a probability distribution with density function

$$f(x) = \begin{cases} 0.5 & 4 < x < 6 \\ 0 & \text{otherwise} \end{cases}$$

The graph of $f(x)$, the density curve, is shown in Figure 6.7(a). It is especially easy to use this density curve to calculate probabilities, because it just requires finding the areas of rectangles using the formula

$$\text{area} = (\text{base})(\text{height})$$

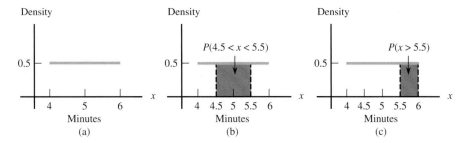

FIGURE 6.7
The uniform distribution for Example 6.8

The curve has positive height, 0.5, only between $x = 4$ and $x = 6$. The total area under the curve is just the area of the rectangle with base extending from 4 to 6 and with height 0.5. This gives

$$\text{area} = (6 - 4)(0.5) = 1$$

as required.

When the density is constant over an interval (resulting in a horizontal density curve), the probability distribution is called a *uniform distribution*.

As illustrated in Figure 6.7(b), the probability that x is between 4.5 and 5.5 is

$$\begin{aligned} P(4.5 < x < 5.5) &= \text{area of shaded rectangle} \\ &= (\text{base width})(\text{height}) \\ &= (5.5 - 4.5)(0.5) \\ &= 0.5 \end{aligned}$$

Figure 6.7(c) shows the probability that x is greater than 5.5. Because, in this context, $x > 5.5$ is equivalent to $5.5 < x < 6$,

$$P(x > 5.5) = (6 - 5.5)(0.5) = 0.25$$

According to this model, in the long run, about 25% of all credit applications will have processing times that exceed 5.5 min.

The probability that a *discrete* random variable x lies in the interval between two limits a and b depends on whether either limit is included in the interval. Suppose, for example, that x is the number of major defects on a new automobile. Then

$$P(3 \le x \le 7) = p(3) + p(4) + p(5) + p(6) + p(7)$$

whereas

$$P(3 < x < 7) = p(4) + p(5) + p(6)$$

However, if x is a *continuous* random variable, such as task completion time, then

$$P(3 \le x \le 7) = P(3 < x < 7)$$

because the area under a density curve and above a single value such as 3 or 7 is 0. Geometrically, you can think of finding the area above a single point as finding the area of a rectangle with width = 0. The area above an interval of values, therefore, does not depend on whether either endpoint is included.

> If x is a continuous random variable, then for any two numbers a and b with $a < b$,
>
> $$P(a \le x \le b) = P(a < x \le b) = P(a \le x < b) = P(a < x < b)$$

Example 6.9 Priority Mail Package Weights

Two hundred packages shipped using the Priority Mail rate for packages less than 2 pounds were weighed, resulting in a sample of 200 observations of the variable

$$x = \text{package weight (in pounds)}$$

from the population of all Priority Mail packages under 2 pounds. A histogram (using the density scale, where height = (relative frequency)/(interval width)) of the 200 weights is shown in Figure 6.8(a). Because the histogram is based on a sample of 200 packages, it provides only an approximation to the actual distribution of x. The shape of this histogram suggests that a reasonable model for the distribution of x might be the triangular distribution shown in Figure 6.8(b).

FIGURE 6.8
Graphs for Example 6.9:
(a) histogram of package weights;
(b) continuous probability
distribution for x = package
weight

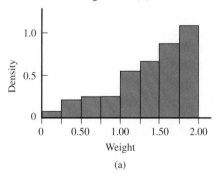

(a)

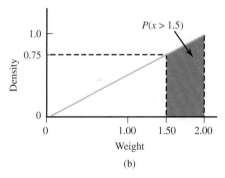

(b)

Notice that the total area under the triangular probability density curve is equal to

$$\text{total area of triangle} = \frac{1}{2}(\text{base})(\text{height}) = \frac{1}{2}(2)(1) = 1$$

This probability distribution can be used to compute the proportion of packages over 1.5 lb, $P(x > 1.5)$. This corresponds to the area of the shaded region in Figure 6.8(b). In this case, it is easier to compute the area of the unshaded triangular region (which corresponds to $P(x \le 1.5)$):

$$P(x \le 1.5) = \frac{1}{2}(1.5)(0.75) = 0.5625$$

Because the total area under a probability density curve is 1,

$$P(x > 1.5) = 1 - 0.5625 = 0.4375$$

It is also the case that

$$P(x \ge 1.5) = 0.4375$$

and that

$$P(x = 1.5) = 0$$

(because the area under the density curve above a single x value is 0).

Example 6.10 Service Times

An airline's toll-free reservation number recorded the length of time (in minutes) required to provide service to each of 500 callers. This resulted in 500 observations of the continuous numerical variable

$$x = \text{service time}$$

A density histogram is shown in Figure 6.9(a).

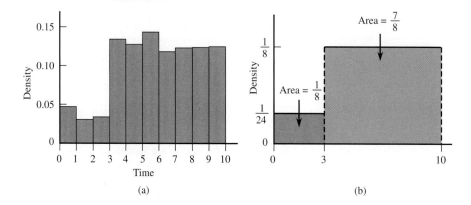

FIGURE 6.9
Graphs for Example 6.10:
(a) histogram of service times;
(b) continuous distribution of service times

The population of interest is all callers to the reservation line. Based on the density histogram, it appears that a reasonable model for the population distribution would be flat over the interval from 0 to 3 and higher but also flat over the interval from 3 to 10. Because service requests were usually one of two types: (1) requests to make a flight reservation and (2) requests to cancel a reservation, this type of model seems reasonable. Canceling a reservation, which accounted for about one-eighth of the calls to the reservation line, could usually be accomplished fairly quickly, whereas making a reservation (seven-eighths of the calls) required more time.

Figure 6.9(b) shows the probability distribution curve proposed as a model for the variable $x = $ service time. The height of the curve for each of the two segments was chosen so that the total area under the curve would be 1 and so that $P(x \le 3) = \frac{1}{8}$ (these were thought to be cancellation calls) and $P(x > 3) = \frac{7}{8}$.

Once the model has been developed, it can be used to calculate probabilities. For example,

$$P(x > 8) = \text{area under curve and above interval from 8 to 10}$$

$$= 2\left(\frac{1}{8}\right) = \frac{2}{8} = \frac{1}{4}$$

In the long run, about one-fourth of all service requests will require more than 8 minutes.
Similarly,

$$P(2 < x < 4) = \text{area under curve and above interval from 2 to 4}$$
$$= (\text{area under curve and above interval from 2 to 3})$$
$$+ (\text{area under curve and above interval from 3 to 4})$$
$$= 1\left(\frac{1}{24}\right) + 1\left(\frac{1}{8}\right)$$
$$= \frac{1}{24} + \frac{3}{24}$$
$$= \frac{4}{24}$$
$$= \frac{1}{6}$$

In each of the previous examples, the model for the population distribution was simple enough that it was possible to calculate probabilities (evaluate areas under the density curve) using simple geometry. Example 6.11 shows that this is not always the case.

Example 6.11 Online Registration Times

Students at a university use an online registration system to register for courses. The variable

$$x = \text{length of time (in minutes) required for a student to register}$$

was recorded for a large number of students using the system, and the resulting values were used to construct a probability histogram (see Figure 6.10). The general form of the histogram can be described as bell shaped and symmetric, and a smooth curve has been superimposed. This smooth curve is a reasonable model for the probability distribution of x. Although this is a common population model for many variables, it is not obvious how you could use it to calculate probabilities, because it is not clear how to calculate areas under such a curve.

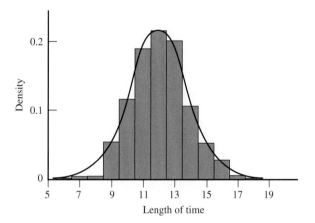

FIGURE 6.10

Histogram and continuous probability distribution for time to register for Example 6.11

The symmetric, bell-shaped probability model of Example 6.11 is known as a *normal probability distribution*. Normal distributions are widely used, and they are investigated in more detail in Section 6.5.

Probabilities for continuous random variables are often calculated using cumulative areas. A cumulative area is all of the area under the density curve to the left of a particular value. Figure 6.11 illustrates the cumulative area to the left of 0.5, which is $P(x < 0.5)$.

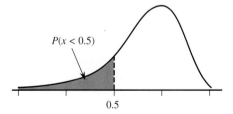

FIGURE 6.11

A cumulative area under a probability density curve

The probability that x is in any particular interval, $P(a < x < b)$, can be calculated as the difference between two cumulative areas.

> The probability that a continuous random variable x lies between a lower limit a and an upper limit b is
>
> $$P(a < x < b) = \text{(cumulative area to the left of } b) - \text{(cumulative area to the left of } a)$$
> $$= P(x < b) - P(x < a)$$

This property is illustrated in Figure 6.12 for the case of $a = 0.25$ and $b = 0.75$. You will use this result often in Section 6.5 when you calculate probabilities using the normal distribution.

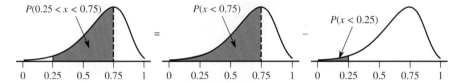

FIGURE 6.12
Calculation of $P(a < x < b)$ using cumulative areas

For some continuous distributions, cumulative areas can be calculated using methods from the branch of mathematics called integral calculus. However, because we are not assuming knowledge of calculus, you will rely on technology or tables that have been constructed for the commonly encountered continuous probability distributions. Many graphing calculators and statistical software packages will compute areas for the most widely used continuous probability distributions.

SECTION 6.3 EXERCISES

Each Exercise Set assesses the following chapter learning objectives: C1, C3, M5, M6

SECTION 6.3 Exercise Set 1

6.21 Let x denote the lifetime (in thousands of hours) of a certain type of fan used in diesel engines. The density curve of x is as pictured.

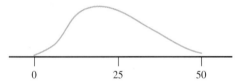

Shade the area under the curve corresponding to each of the following probabilities (draw a new curve for each part):
a. $P(10 < x < 25)$
b. $P(10 \le x \le 25)$
c. $P(x < 30)$
d. The probability that the lifetime is at least 25,000 hours

6.22 Suppose that the random variable

x = waiting time for service at a bank (in minutes)

has the probability distribution described by the density curve pictured below.

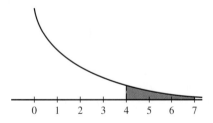

a. What probability is represented by the shaded area?
b. Suppose the shaded area = 0.26. Interpret this probability in the context of this problem.

6.23 Let x denote the time (in seconds) necessary for an individual to react to a certain stimulus. The probability distribution of x is specified by the accompanying density curve.

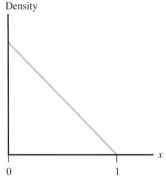

a. What is the height of the density curve above $x = 0$? (Hint: Total area = 1.)
b. What is the probability that reaction time exceeds 0.5 sec?
c. What is the probability that reaction time is at most 0.25 sec?

6.24 Consider the population that consists of all soft contact lenses made by a particular manufacturer, and define the random variable x = thickness (in millimeters). Suppose that a reasonable model for the population distribution is the one shown in the following figure.

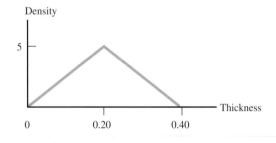

a. Verify that the total area under the density curve is equal to 1. [Hint: The area of a triangle is equal to 0.5(base)(height).]

b. What is the probability that x is less than 0.20? Less than 0.10? More than 0.30?

c. What is the probability that x is between 0.10 and 0.20? (Hint: First find the probability that x is *not* between 0.10 and 0.20.)

SECTION 6.3 **Exercise Set 2**

6.25 An online store charges for shipping based on the weight of the items in an order. Define the random variable

x = weight of a randomly selected order (in ounces)

The density curve of x is shown here:

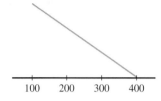

Shade the area under the curve corresponding to each of the following probabilities (draw a new curve for each part):

a. $P(x > 200)$

b. $P(100 < x < 300)$

c. $P(x \le 250)$

d. The probability that the selected order has a weight greater than 350 ounces.

6.26 Suppose that the random variable

x = weekly water usage (in gallons) for a randomly selected studio apartment in Los Angeles

has the probability distribution described by the following density curve.

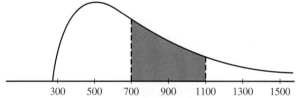

a. What probability is represented by the shaded area?

b. Suppose the shaded area = 0.35. Interpret this probability in the context of this problem.

6.27 The article **"Modeling Sediment and Water Column Interactions for Hydrophobic Pollutants"** (*Water Research* [1984]: 1169–1174) suggests the uniform distribution on the interval from 7.5 to 20 as a model for x = depth (in centimeters) of the bioturbation layer in sediment for a certain region.

a. Draw the density curve for x.

b. What is the height of the density curve?

c. What is the probability that x is at most 12?

d. What is the probability that x is between 10 and 15? Between 12 and 17? Why are these two probabilities equal?

Additional Exercises

6.28 The continuous random variable x has the probability distribution shown here:

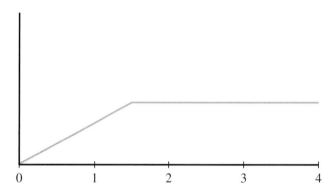

Shade the area under the curve corresponding to each of the following probabilities. (Draw a new curve for each part.)

a. $P(x < 1)$

b. $P(x > 3)$

c. $P(1 < x < 2)$

d. $P(2 < x < 3)$

6.29 Refer to the probabilities given in Parts (a)-(d) of the previous exercise. Which of these probabilities is smallest? Which is largest?

6.30 A particular professor never dismisses class early. Let x denote the amount of additional time (in minutes) that elapses before the professor dismisses class. Suppose that x has a uniform distribution on the interval from 0 to 10 minutes. The density curve is shown in the following figure:

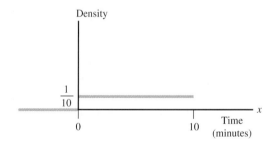

a. What is the probability that at most 5 minutes elapse before dismissal?

b. What is the probability that between 3 and 5 minutes elapse before dismissal?

6.31 Refer to the probability distribution given in the previous exercise. Put the following probabilities in order, from smallest to largest:

$P(2 < x < 3)$ $P(2 \le x \le 3)$ $P(x < 2)$ $P(x > 7)$

Explain your reasoning.

SECTION 6.4 Mean and Standard Deviation of a Random Variable

We study a random variable x, such as the number of insurance claims made by a home-owner (a discrete variable) or the birth weight of a baby (a continuous variable), to learn something about how its values are distributed along the measurement scale. For sample data, the sample mean $\bar{x}$ and sample standard deviation s summarize center and spread. Similarly, the mean value and standard deviation of a random variable describe where its probability distribution is centered and the extent to which it spreads out about the center.

> The **mean value of a random variable x**, denoted by μ_x, describes where the probability distribution of x is centered.
>
> The **standard deviation of a random variable x**, denoted by σ_x, describes variability in the probability distribution. When the value of σ_x is small, observed values of x will tend to be close to the mean value. The larger the value of σ_x, the more variability there will be in observed x values.

Figure 6.13(a) shows two discrete probability distributions with the same standard deviation (spread) but different means (center). One distribution has a mean of $\mu_x = 6$ and the other has $\mu_x = 10$. Which is which? Figure 6.13(b) shows two continuous probability distributions that have the same mean but different standard deviations. Which distribution—(i) or (ii)—has the larger standard deviation? Finally, Figure 6.13(c) shows three continuous distributions with different means and standard deviations. Which of the three distributions has the largest mean? Which has a mean of about 5? Which distribution has the smallest standard deviation? (The correct answers to these questions are the following: Figure 6.13(a)(ii) has a mean of 6, and Figure 6.13(a)(i) has a mean of 10; Figure 6.13(b)(ii) has the larger standard deviation; Figure 6.13(c)(iii) has the largest mean, Figure 6.13(c)(ii) has a mean of about 5, and Figure 6.13(c)(iii) has the smallest standard deviation.)

It is common to use the terms *mean of the random variable x* and *mean of the probability distribution of x* interchangeably. Similarly, the standard deviation of the random variable x and the standard deviation of the probability distribution of x refer to the same thing. Although the mean and standard deviation are computed differently for discrete and continuous random variables, the interpretation is the same in both cases.

Mean Value of a Discrete Random Variable

Consider an experiment consisting of randomly selecting a car licensed in a particular state. Define the discrete random variable x to be the number of low-beam headlights on the selected car that need adjustment. Possible x values are 0, 1, and 2, and the probability distribution of x might be as follows:

x value	0	1	2
probability	0.5	0.3	0.2

The corresponding probability histogram appears in Figure 6.14.

In a sample of 100 cars, the sample relative frequencies might differ somewhat from the given probabilities (which are the limiting relative frequencies). For example, we might see:

x value	0	1	2
frequency	46	33	21

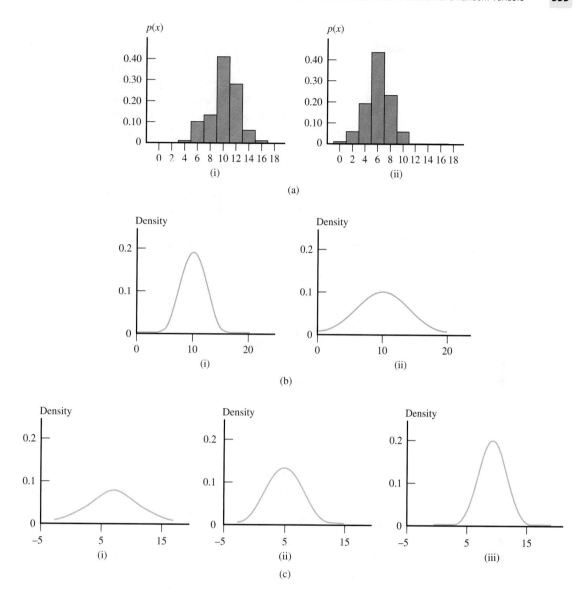

FIGURE 6.13
Some probability distributions: (a) different values of μ_x with the same value of σ_x; (b) different values of σ_x with the same value of μ_x; (c) different values of μ_x and σ_x

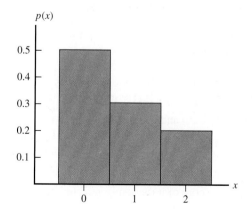

FIGURE 6.14
Probability histogram for the distribution of the number of headlights needing adjustment

The sample average value of x for these 100 observations is the sum of 46 zeros, 33 ones, and 21 twos, all divided by 100:

$$\bar{x} = \frac{(46)(0) + (33)(1) + (21)(2)}{100}$$

$$= \left(\frac{46}{100}\right)(0) + \left(\frac{33}{100}\right)(1) + \left(\frac{21}{100}\right)(2)$$

$$= (\text{rel. freq. of } 0)(0) + (\text{rel. freq. of } 1)(1) + (\text{rel. freq. of } 2)(2)$$

$$= 0.75$$

As the sample size increases, each relative frequency approaches the corresponding probability, and the value of $\bar{x}$ approaches

$$P(x = 0)(0) + P(x = 1)(1) + P(x = 2)(2) = (0.5)(0) + (0.3)(1) + (0.2)(2) = 0.70$$

Notice that the expression for $\bar{x}$ is a weighted average of possible x values (the weight of each value is the observed relative frequency). Similarly, the mean value of the random variable x is a weighted average, but now the weights are the probabilities from the probability distribution.

DEFINITION

The **mean value of a discrete random variable x,** denoted by μ_x, is computed by first multiplying each possible x value by the probability of observing that value and then adding the resulting quantities. Symbolically,

$$\mu_x = \sum_{\text{all possible } x \text{ values}} x \cdot p(x)$$

The term **expected value** is sometimes used in place of mean value, and $E(x)$ is another way to denote μ_x.

Example 6.12 **Exam Attempts**

Individuals applying for a certain license are allowed up to four attempts to pass the licensing exam. Consider the random variable

$$x = \text{the number of attempts made by a randomly selected applicant}$$

The probability distribution of x is as follows:

x	1	2	3	4
$p(x)$	0.10	0.20	0.30	0.40

Then x has mean value

$$\mu_x = \sum x \cdot p(x)$$

$$= (1)p(1) + (2)p(2) + (3)p(3) + (4)p(4)$$

$$= (1)(0.10) + (2)(0.20) + (3)(0.30) + (4)(0.40)$$

$$= 0.10 + 0.40 + 0.90 + 1.60$$

$$= 3.00$$

Example 6.13 **Apgar Scores**

At 1 minute after birth and again at 5 minutes, each newborn child is given a numerical rating called an Apgar score. Possible values of this score are 0, 1, 2, ... , 9, 10. A child's score is determined by five factors: muscle tone, skin color, respiratory effort, strength of heartbeat, and reflex, with a high score indicating a healthy infant. Consider the random variable

$$x = \text{Apgar score of a randomly selected newborn infant at a particular hospital}$$

and suppose that x has the following probability distribution:

x	0	1	2	3	4	5	6	7	8	9	10
$p(x)$	0.002	0.001	0.002	0.005	0.02	0.04	0.17	0.38	0.25	0.12	0.01

The mean value of x is

$$\mu_x = \sum x \cdot p(x)$$
$$= (0)p(0) + (1)p(1) + \cdots + (9)p(9) + (10)p(10)$$
$$= (0)(0.002) + (1)(0.001) + \cdots + (9)(0.12) + (10)(0.01)$$
$$= 7.16$$

The mean Apgar score for a *sample* of newborn children born at this hospital may be $\bar{x} = 7.05$, $\bar{x} = 8.30$, or any one of a number of other possible values between 0 and 10. However, as child after child is born and rated, the mean score will approach the value 7.16. This value can be interpreted as the mean Apgar score for the population of all babies born at this hospital.

Standard Deviation of a Discrete Random Variable

The mean value μ_x provides only a partial summary of a probability distribution. Two different distributions can have the same value of μ_x, but observations from one distribution might vary more than observations from the other distribution.

Example 6.14   **Glass Panels**

Flat screen TVs require high-quality glass with very few flaws. A television manufacturer receives glass panels from two different suppliers. Consider

x = number of flaws in a randomly selected glass panel from Supplier 1
y = number of flaws in a randomly selected glass panel from Supplier 2

Suppose that the probability distributions for x and y are as follows:

x	0	1	2	3	y	0	1	2	3
$p(x)$	0.4	0.3	0.2	0.1	$p(y)$	0.2	0.6	0.2	0

Probability histograms for x and y are shown in Figure 6.15.

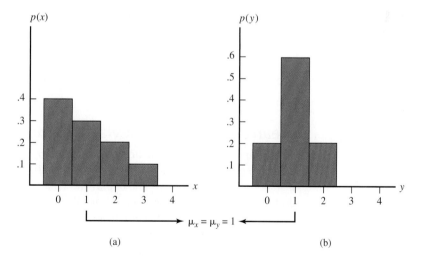

FIGURE 6.15
Probability histograms for the number of flaws in a glass panel for Example 6.14: (a) Supplier 1; (b) Supplier 2

It is easy to verify that the mean values of both x and y are 1, so for either supplier the long-run average number of flaws per panel is 1. However, the two probability histograms show that the probability distribution for the first supplier is more spread out than that of the second supplier.

The greater spread of the first distribution means there will be more variability in observed x values than in observed y values. For example, even though both distributions have a mean of 1, none of the y observations will be 3, but about 10% of x observations will be 3.

The variance and standard deviation are used to summarize variability in a probability distribution.

> **DEFINITION**
>
> The **variance of a discrete random variable x,** denoted by σ_x^2, is computed by first subtracting the mean from each possible x value to obtain the deviations, then squaring each deviation and multiplying the result by the probability of the corresponding x value, and finally adding these quantities. Symbolically,
>
> $$\sigma_x^2 = \sum_{\text{all possible } x \text{ values}} (x - \mu_x)^2 p(x)$$
>
> The **standard deviation of x,** denoted by σ_x, is the square root of the variance.

As with s^2 and s (the variance and standard deviation of a sample), the variance and standard deviation of a random variable x are based on squared deviations from the mean. A value far from the mean results in a large squared deviation. However, such a value contributes substantially to variability in x only if the probability associated with that value is not too small. For example, if $\mu_x = 1$ and $x = 25$ is a possible value, then the squared deviation is $(25 - 1)^2 = 576$. But if $P(x = 25) = .000001$, the value 25 will hardly ever be observed, so it won't contribute much to variability in observed x values. This is why each squared deviation is multiplied by its probability to obtain a measure of variability.

Example 6.15   Glass Panels Revisited

For $x =$ number of flaws in a glass panel from the first supplier in Example 6.14

$$\sigma_x^2 = (0 - 1)^2\, p(0) + (1 - 1)^2\, p(1) + (2 - 1)^2\, p(2) + (3 - 1)^2\, p(3)$$
$$= (1)(0.4) + (0)(0.3) + (1)(0.2) + (4)(0.1)$$
$$= 1.0$$

Then $\sigma_x = \sqrt{1.0} = 1.0$. For $y =$ the number of flaws in a glass panel from the second supplier,

$$\sigma_y^2 = (0 - 1)^2\, p(0) + (1 - 1)^2\, p(1) + (2 - 1)^2\, p(2)$$
$$= (0 - 1)^2(0.2) + (1 - 1)^2(0.6) + (2 - 1)^2(0.2)$$
$$= 0.4$$

Then $\sigma_y = \sqrt{0.4} = 0.632$. The fact that $\sigma_x > \sigma_y$ confirms what you saw in Figure 6.15.

Mean and Standard Deviation When x Is Continuous

For continuous probability distributions, μ_x and σ_x are defined and computed using methods from calculus, so you don't need to worry about how to compute them. However, it is important to know that μ_x and σ_x play the same role for continuous distributions as they do for discrete distributions. The mean value μ_x locates the center of the continuous distribution and gives the approximate long-run average of observed x values. The standard deviation σ_x measures the extent that the continuous distribution (density curve) spreads out about μ_x and indicates the amount of variability that can be expected in observed x values.

Example 6.16   A "Concrete" Example

A company receives batches of concrete from two different suppliers. Define random variables x and y as follows:

     $x =$ strength of a randomly selected batch from Supplier 1
     $y =$ strength of a randomly selected batch from Supplier 2

Suppose that

$$\mu_x = 4{,}650 \text{ pounds/inch}^2 \qquad\qquad \sigma_x = 200 \text{ pounds/inch}^2$$
$$\mu_y = 4{,}500 \text{ pounds/inch}^2 \qquad\qquad \sigma_y = 275 \text{ pounds/inch}^2$$

The long-run average strength per batch for many, many batches from Supplier 1 will be roughly 4,650 pounds/inch^2. This is 150 pounds/inch^2 greater than the long-run average for Supplier 2. In addition, batches from Supplier 1 will exhibit substantially less variability in strength than will batches from Supplier 2. The first supplier is preferred to the second because its concrete is stronger on average and because there is less variability from batch to batch.

Mean and Variance of Linear Functions and Linear Combinations

The mean and standard deviation of a random variable provide useful information about its long-run behavior. Sometimes you may also be interested in the behavior of some function of the variables.

For example, consider the experiment in which a customer of a propane gas company is randomly selected. Suppose that the mean and standard deviation of the random variable

x = number of gallons required to fill a customer's propane tank

are known to be 318 gallons and 42 gallons, respectively. The company is considering two different pricing models:

Model 1: $3 per gallon
Model 2: service charge of $50 + $2.80 per gallon

The company is interested in the variable

y = amount billed

For each of the two models, y can be expressed as a function of the random variable x:

Model 1: $y_{model\ 1} = 3x$
Model 2: $y_{model\ 2} = 50 + 2.8x$

Both of these equations are examples of a linear function of x. The mean and standard deviation of a linear function of x can be computed from the mean and standard deviation of x, as described in the following box.

The Mean, Variance, and Standard Deviation of a Linear Function

If x is a random variable with mean μ_x and variance σ_x and a and b are numerical constants, the random variable y defined by

$$y = a + bx$$

is called a **linear function of the random variable x.**

The mean of $y = a + bx$ is

$$\mu_y = \mu_{a+bx} = a + b\mu_x$$

The variance of y is

$$\sigma_y^2 = \sigma_{a+bx}^2 = b^2\sigma_x^2$$

from which it follows that the standard deviation of y is

$$\sigma_y = \sigma_{a+bx} = |b|\,\sigma_x$$

You can use the results in the preceding box to compute the mean and standard deviation of the billing amount variable for the propane gas example, as follows:

For Model 1:

$$\mu_{model\ 1} = \mu_{3x} = 3\mu_x = 3(318) = 954$$
$$\sigma^2_{model\ 1} = \sigma^2_{3x} = 3^2\sigma^2_x = 9(42)^2 = 15{,}876$$
$$\sigma_{model\ 1} = \sqrt{15{,}876} = 126, \text{ which is equal to } 3(42)$$

For Model 2:

$$\mu_{model\ 2} = \mu_{50+2.8x} = 50 + 2.8\mu_x = 50 + 2.8(318) = 940.40$$
$$\sigma^2_{model\ 2} = \sigma^2_{50+2.8x} = 2.8^2\sigma^2_x = (2.8)^2 (42)^2 = 13{,}829.76$$
$$\sigma_{model\ 2} = \sqrt{13{,}829.76} = 117.60, \text{ which is equal to } 2.8(42)$$

The mean billing amount for Model 1 is a bit higher than for Model 2, as is the variability in billing amounts. Model 2 results in slightly more consistency from bill to bill in the amount charged.

Linear Combinations Now consider a different type of problem. Suppose that you have three tasks that you plan to complete on the way home from school: stop at the public library to return an overdue book for which you must pay a fine, deposit your most recent paycheck at the bank, and stop by the office supply store to purchase paper for your computer printer. Define the following variables:

x_1 = time required to return book and pay fine
x_2 = time required to deposit paycheck
x_3 = time required to buy printer paper

You can then define a new variable, y, to represent the total amount of time to complete these tasks:

$$y = x_1 + x_2 + x_3$$

Defined in this way, y is an example of a linear combination of random variables.

If $x_1, x_2, \ldots, x_n$ are random variables and $a_1, a_2, \ldots, a_n$ are numerical constants, the random variable y defined as

$$y = a_1x_1 + a_2x_2 + \cdots + a_nx_n$$

is a **linear combination of the x_i's.**

For example, $y = 10x_1 - 5x_2 + 8x_3$ is a linear combination of x_1, x_2, and x_3 with $a_1 = 10$, $a_2 = -5$ and $a_3 = 8$. It is easy to compute the mean of a linear combination of x_i's if the individual means $\mu_1, \mu_2, \ldots, \mu_n$ are known. The variance and standard deviation of a linear combination of the x_i's are also easily computed *if the x_i's are independent*. Two random variables x_i and x_j are independent if any event defined solely by x_i is independent of any event defined solely by x_j. When the x_i's are not independent, computation of the variance and standard deviation of a linear combination of the x_i is more complicated and this case is not considered here.

Mean, Variance, and Standard Deviation for Linear Combinations

If $x_1, x_2, \ldots, x_n$ are random variables with means $\mu_1, \mu_2, \ldots, \mu_n$ and variances $\sigma^2_1, \sigma^2_2, \ldots, \sigma^2_n$, respectively, and

$$y = a_1 x_1 + a_2 x_2 + \cdots + a_n x_n$$

then

1. $\mu_y = \mu_{a_1 x_1 + a_2 x_2 + \cdots + a_n x_n} = a_1\mu_1 + a_2\mu_2 + \cdots + a_n\mu_n$

(continued)

This result is true regardless of whether the x_i's are independent.

2. When $x_1, x_2, \ldots, x_n$ are *independent* random variables,

$$\sigma_y^2 = \sigma_{a_1 x_1 + a_2 x_2 + \cdots + a_n x_n}^2 = a_1^2 \sigma_1^2 + a_2^2 \sigma_2^2 + \cdots + a_n^2 \sigma_n^2$$

$$\sigma_y = \sigma_{a_1 x_1 + a_2 x_2 + \cdots + a_n x_n} = \sqrt{a_1^2 \sigma_1^2 + a_2^2 \sigma_2^2 + \cdots + a_n^2 \sigma_n^2}$$

This result is true only when the x_i's are independent.

Examples 6.17–6.19 illustrate the use of these rules.

Example 6.17 Freeway Traffic

Three different roads feed into a particular freeway entrance. Suppose that during a fixed time period, the number of cars coming from each road onto the freeway is a random variable with mean values as follows:

Road	1	2	3
Mean	800	1000	600

With x_i representing the number of cars entering from road i, you can define y, the total number of cars entering the freeway, as $y = x_1 + x_2 + x_3$. The mean value of y is

$$\mu_y = \mu_{x_1 + x_2 + x_3}$$
$$= \mu_{x_1} + \mu_{x_2} + \mu_{x_3}$$
$$= 800 + 1000 + 600$$
$$= 2400$$

Example 6.18 Combining Exam Subscores

A nationwide standardized exam consists of a multiple-choice section and a free response section. For each section, the mean and standard deviation are reported to be

	Mean	Standard Deviation
Multiple Choice	38	6
Free Response	30	7

Suppose x_1 and x_2 are the multiple-choice score and the free-response score, respectively, of a student selected at random from those taking this exam. Now consider the variable $y =$ overall exam score. Suppose that the free-response score is given twice the weight of the multiple-choice score in determining the overall exam score. Then the overall score is computed as $y = x_1 + 2x_2$. What are the mean and standard deviation of y?

Because $y = x_1 + 2x_2$ is a linear combination of x_1 and x_2, the mean of y is

$$\mu_y = \mu_{x_1 + 2x_2}$$
$$= \mu_{x_1} + 2\mu_{x_2}$$
$$= 38 + 2(30)$$
$$= 98$$

What about the variance and standard deviation of y? To use Rule 2 in the preceding box, x_1 and x_2 must be independent. It is unlikely that the value of x_1 (a student's multiple-choice score) would be unrelated to the value of x_2 (the same student's free-response score), because it seems probable that students who score well on one section of the exam will also tend to score well on the other section. Therefore, it would not be appropriate to use the formulas given in Rule 2 to calculate the variance and standard deviation.

Example 6.19 Baggage Weights

A commuter airline flies small planes between San Luis Obispo and San Francisco. For small planes, the baggage weight is a concern, especially on foggy mornings, because the weight of the plane has an effect on how quickly the plane can ascend. Suppose that it is known that the variable x = weight (in pounds) of baggage checked by a randomly selected passenger has a mean and standard deviation of 42 pounds and 16 pounds, respectively. Consider a flight on which 10 passengers, all traveling alone, are flying. If you use x_i to denote the baggage weight for passenger i (for i ranging from 1 to 10), the total weight of checked baggage, y, is then

$$y = x_1 + x_2 + \cdots + x_{10}$$

Notice that y is a linear combination of the x_i's. The mean value of y is

$$\mu_y = \mu_{x_1} + \mu_{x_2} + \cdots + \mu_{x_{10}}$$
$$= 42 + 42 + \cdots + 42$$
$$= 420$$

Since the 10 passengers are all traveling alone, it is reasonable to think that the 10 baggage weights are unrelated and that the x_i's are independent. (This might not be a reasonable assumption if the 10 passengers were not traveling alone.) Then the variance of y is

$$\sigma_y^2 = \sigma_{x_1}^2 + \sigma_{x_2}^2 + \cdots + \sigma_{x_{10}}^2$$
$$= 16^2 + 16^2 + \cdots + 16^2$$
$$= 2{,}560$$

and the standard deviation of y is

$$\sigma_y = \sqrt{2{,}650} = 50.596$$

One Last Note on Linear Functions and Linear Combinations

In Example 6.19, the random variable of interest was x = weight of baggage checked by a randomly selected airline passenger. However, when we considered a flight with multiple passengers, subscripts were added to create variables like

$$x_1 = \text{baggage weight for passenger 1}$$

and

$$x_2 = \text{baggage weight for passenger 2}$$

It is important to note that the sum of the baggage weights for two different passengers does not result in the same value as doubling the baggage weight of a single passenger. That is, the linear combination $x_1 + x_2$ is different from the linear function $2x$. Although the mean of $x_1 + x_2$ and the mean of $2x$ are the same, the variances and standard deviations are different. For example, for the baggage weight distribution described in Example 6.19 (mean of 42 and standard deviation of 16), the mean of $2x$ is

$$\mu_{2x} = 2\mu_x = 2(42) = 84$$

and the mean of $x_1 + x_2$ is

$$\mu_{x_1+x_2} = \mu_{x_1} + \mu_{x_2} = 42 + 42 = 84$$

However, the variance of $2x$ is

$$\sigma_{2x}^2 = (2)^2 \sigma_x^2 = 4(16^2) = 1024$$

whereas the variance of $x_1 + x_2$ is

$$\sigma_{x_1+x_2}^2 = \sigma_{x_1}^2 + \sigma_{x_2}^2 = 16^2 + 16^2 = 256 + 256 = 512$$

This means that observations of $2x$ for randomly selected passengers will show more variability than observations of $x_1 + x_2$ for pairs of two randomly selected passengers.

AP* EXAM TIP

Be careful. $x_1 + x_2$ and $2x$ are not the same! Be sure you understand the explanation on this page.

This may seem counterintuitive, but it is true! For example, each row of the first two columns of the accompanying table show simulated baggage weights for two passengers (an x_1 and an x_2 value) that were selected at random from a probability distribution with mean 42 and standard deviation 16. The third column of the table contains the value of $2x_1$ (two times the passenger 1 baggage weight) and the last column contains the values of $x_1 + x_2$ (the sum of the passenger 1 baggage weight and the passenger 2 baggage weight).

x_1	x_2	$2x_1$	$x_1 + x_2$
56	41	112	97
30	5	60	35
32	15	64	47
55	45	110	100
40	54	80	94
41	68	82	109
42	35	84	77
33	47	66	80
46	52	92	98
73	40	146	113
44	23	88	67
48	54	96	102
52	69	104	121
56	63	112	119
79	35	158	114
18	34	36	52
8	71	16	79
51	58	102	109
18	49	36	67
41	46	82	87

Figure 7.16 shows dotplots of the 20 values of $2x_1$ and the 20 values of $x_1 + x_2$. You can see that there is less variability in the $x_1 + x_2$ values than in the values of $2x_1$.

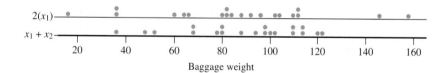

FIGURE 6.16
Dotplots of $2x_1$ and $x_1 + x_2$.

SECTION 6.4   EXERCISES

Each Exercise Set assesses the following chapter learning objectives: C2, M4, M15

SECTION 6.4   Exercise Set 1

6.32   Suppose y = the number of broken eggs in a randomly selected carton of one dozen eggs. The probability distribution of y is as follows:

y	0	1	2	3	4
$p(y)$	0.65	0.20	0.10	0.04	0.01

a. Calculate and interpret μ_y.
b. In the long run, for what percentage of cartons is the number of broken eggs less than μ_y? Does this surprise you?
c. Explain why μ_y is not equal to $\dfrac{0 + 1 + 2 + 3 + 4}{5} = 2.0$.

6.33   An appliance dealer sells three different models of freezers having 13.5, 15.9, and 19.1 cubic feet of storage space. Let x = the amount of storage space purchased by

the next customer to buy a freezer. Suppose that x has the following probability distribution:

x	13.5	15.9	19.1
$p(x)$	0.2	0.5	0.3

a. Calculate the mean and standard deviation of x.
b. Give an interpretation of the mean and standard deviation of x in the context of observing the outcomes of many purchases.

6.34   A grocery store has an express line for customers purchasing at most five items. Let x be the number of items purchased by a randomly selected customer using this line. Make two tables that represent two different possible probability distributions for x that have the same mean but different standard deviations.

6.35 A multiple-choice exam consists of 50 questions. Each question has five choices, of which only one is correct. Suppose that the total score on the exam is computed as

$$y = x_1 - \frac{1}{4}x_2$$

where x_1 = number of correct responses and x_2 = number of incorrect responses. (Calculating a total score by subtracting a term based on the number of incorrect responses is known as a correction for guessing and is designed to discourage test takers from choosing answers at random.)

a. It can be shown that if a totally unprepared student answers all 50 questions by just selecting one of the five answers at random, then $\mu_{x_1} = 10$ and $\mu_{x_2} = 40$. What is the mean value of the total score, y? Does this surprise you? Explain.

b. Explain why it is unreasonable to use the formulas given in this section to compute the variance or standard deviation of y.

6.36 Consider a large ferry that can accommodate cars and buses. The toll for cars is $3, and the toll for buses is $10. Let x and y denote the number of cars and buses, respectively, carried on a single trip. Cars and buses are accommodated on different levels of the ferry, so the number of buses accommodated on any trip is independent of the number of cars on the trip. Suppose that x and y have the following probability distributions:

x	0	1	2	3	4	5
$p(x)$	0.05	0.10	0.25	0.30	0.20	0.10

y	0	1	2
$p(y)$	0.50	0.30	0.20

a. Compute the mean and standard deviation of x.
b. Compute the mean and standard deviation of y.
c. Compute the mean and variance of the total amount of money collected in tolls from cars.
d. Compute the mean and variance of the total amount of money collected in tolls from buses.
e. Compute the mean and variance of z = total number of vehicles (cars and buses) on the ferry.
f. Compute the mean and variance of w = total amount of money collected in tolls.

SECTION 6.4 Exercise Set 2

6.37 A local television station sells 15-second, 30-second, and 60-second advertising spots. Let x denote the length of a randomly selected commercial appearing on this station, and suppose that the probability distribution of x is given by the following table:

x	15	30	60
$p(x)$	0.1	0.3	0.6

What is the mean length for commercials appearing on this station?

6.38 Suppose that for a given computer salesperson, the probability distribution of

$$x = \text{the number of systems sold in 1 month}$$

is given by the following table:

x	1	2	3	4	5	6	7	8
$p(x)$	0.05	0.10	0.12	0.30	0.30	0.11	0.01	0.01

a. Find the mean value of x (the mean number of systems sold).
b. Find the variance and standard deviation of x. How would you interpret these values?
c. What is the probability that the number of systems sold is within 1 standard deviation of its mean value?
d. What is the probability that the number of systems sold is more than 2 standard deviations from the mean?

6.39 A grocery store has an express line for customers purchasing at most five items. Let x be the number of items purchased by a randomly selected customer using this line. Make two tables that represent two different possible probability distributions for x that have the same standard deviation but different means.

6.40 To assemble a piece of furniture, a wood peg must be inserted into a predrilled hole. Suppose that the diameter of a randomly selected peg is a random variable with mean 0.25 inch and standard deviation 0.006 inch and that the diameter of a randomly selected hole is a random variable with mean 0.253 inch and standard deviation 0.002 inch. Let x_1 = peg diameter, and let x_2 = denote hole diameter.

a. Why would the random variable y, defined as $y = x_2 - x_1$, be of interest to the furniture manufacturer?
b. What is the mean value of the random variable y?
c. Assuming that x_1 and x_2 are independent, what is the standard deviation of y?
d. Is it reasonable to think that x_1 and x_2 are independent? Explain.
e. Based on your answers to Parts (b) and (c), do you think that finding a peg that is too big to fit in the predrilled hole would be a relatively common or a relatively rare occurrence? Explain.

6.41 Consider a game in which a red die and a blue die are rolled. Let x_R denote the value showing on the uppermost face of the red die, and define x_B similarly for the blue die.

a. The probability distribution of x_R is

x_R	1	2	3	4	5	6
$p(x_R)$	1/6	1/6	1/6	1/6	1/6	1/6

Find the mean, variance, and standard deviation of x_R.
b. What are the values of the mean, variance, and standard deviation of x_B? (You should be able to answer this question without doing any additional calculations.)

c. Suppose that you are offered a choice of the following two games:

Game 1: Costs \$7 to play, and you win y_1 dollars, where $y_1 = x_R + x_B$.

Game 2: Doesn't cost anything to play initially, but you "win" $3y_2$ dollars, where $y_2 = x_R - x_B$. If y_2 is negative, you must pay that amount; if it is positive, you receive that amount.

For Game 1, the net amount won in a game is $w_1 = y_1 - 7 = x_R + x_B - 7$. What are the mean and standard deviation of w_1?

d. For Game 2, the net amount won in a game is $w_2 = 3y_2 = 3(x_R - x_B)$. What are the mean and standard deviation of w_2?

e. Based on your answers to Parts (c) and (d), if you had to play, which game would you choose and why?

Additional Exercises

6.42 The probability distribution of x, the number of defective tires on a randomly selected automobile checked at a certain inspection station, is given in the following table:

x	0	1	2	3	4
$p(x)$	0.54	0.16	0.06	0.04	0.20

The mean value of x is $\mu_x = 1.2$. Calculate the values of σ_x^2 and σ_x.

6.43 A chemical supply company currently has in stock 100 pounds of a certain chemical, which it sells to customers in 5-pound lots. Let x = the number of lots ordered by a randomly chosen customer. The probability distribution of x is as follows:

x	1	2	3	4
$p(x)$	0.2	0.4	0.3	0.1

a. Calculate and interpret the mean value of x.
b. Calculate and interpret the variance and standard deviation of x.

6.44 A business has six customer service telephone lines. Let x denote the number of lines in use at any given time. Suppose that the probability distribution of x is as follows:

x	0	1	2	3	4	5	6
$p(x)$	0.10	0.15	0.20	0.25	0.20	0.06	0.04

a. Calculate the mean value and standard deviation of x.
b. What is the probability that the number of lines in use is farther than 3 standard deviations from the mean value?

SECTION 6.5 **Binomial and Geometric Distributions**

In this section, two discrete probability distributions are introduced: the binomial distribution and the geometric distribution. These distributions arise when a chance experiment consists of a sequence of trials, each with only two possible outcomes. For example, one characteristic of blood type is the Rh factor, which can be either positive or negative. You can think of a chance experiment that consists of recording the Rh factor for each of 25 blood donors as a sequence of 25 trials, where each trial consists of observing the Rh factor (positive or negative) of a single donor.

You could also conduct a different chance experiment that consists of observing the Rh factor of blood donors until a donor who is Rh-negative is encountered. This second experiment can also be viewed as a sequence of trials, but here the total number of trials in this experiment is not predetermined, as it was in the previous experiment. Chance experiments of the two types just described are characteristic of those leading to the binomial and the geometric probability distributions, respectively.

Binomial Distributions

Suppose that you decide to record the gender of each of the next 25 newborn children at a particular hospital. What is the chance that at least 15 are female? What is the chance that between 10 and 15 are female? How many among the 25 can you expect to be female? These and other similar questions can be answered by studying the *binomial probability distribution*. This distribution arises when the experiment of interest is a *binomial experiment*. Binomial experiments have the properties listed in the following box.

> **Properties of a Binomial Experiment**
>
> A binomial experiment consists of a sequence of trials with the following conditions:
>
> 1. There is a fixed number of trials.
> 2. Each trial can result in one of only two possible outcomes, labeled success (S) and failure (F).

(continued)

3. Outcomes of different trials are independent.
4. The probability of success is the same for each trial.

The **binomial random variable x** is defined as

x = number of successes observed when a binomial experiment is performed

The probability distribution of x is called the *binomial probability distribution*.

The term *success* here does not necessarily mean something positive. It is simply the "counted" outcome of a trial. For example, if the random variable counts the number of female births among the next 25 births at a particular hospital, then a female birth would be labeled a success (because this is what the variable counts). If male births were counted instead, a male birth would be labeled a success and a female birth a failure.

One example of a binomial random variable was given in Example 6.5. In that example, you considered x = number among four customers who selected an energy efficient refrigerator (rather than a less expensive model). This is a binomial experiment with four trials, where the purchase of an energy-efficient refrigerator (denoted by E) is considered a success and $P(\text{success}) = P(E) = 0.4$. The 16 possible outcomes, along with the associated probabilities, were displayed in Table 6.1.

Consider now the case of five customers, a binomial experiment with five trials. The possible values of

x = number who purchase an energy-efficient refrigerator

are 0, 1, 2, 3, 4, and 5. There are 32 possible outcomes of the binomial experiment, each one a sequence of five successes and failures. Five of these outcomes result in $x = 1$:

SFFFF, FSFFF, FFSFF, FFFSF, and *FFFFS*

Because the trials are independent, the first of these outcomes has probability

$$
\begin{aligned}
P(SFFFF) &= P(S)P(F)P(F)P(F)P(F) \\
&= (0.4)(0.6)(0.6)(0.6)(0.6) \\
&= (0.4)(0.6)^4 \\
&= 0.05184
\end{aligned}
$$

The probability calculation will be the same for any outcome with only one success ($x = 1$). It does not matter where in the sequence the single success occurs. Then

$$
\begin{aligned}
p(1) &= P(x = 1) \\
&= P(SFFFF \text{ or } FSFFF \text{ or } FFSFF \text{ or } FFFSF \text{ or } FFFFS) \\
&= 0.05184 + 0.05184 + 0.05184 + 0.05184 + 0.05184 \\
&= 5(0.05184) \\
&= 0.25920
\end{aligned}
$$

Similarly, there are 10 outcomes for which $x = 2$, because there are 10 ways to select two from among the five trials to be the S's: *SSFFF, SFSFF,* ... , and *FFFSS*. The probability of each results from multiplying together (0.4) two times and (0.6) three times. For example,

$$
\begin{aligned}
P(SSFFF) &= (0.4)(0.4)(0.6)(0.6)(0.6) \\
&= (0.4)^2(0.6)^3 \\
&= 0.03456
\end{aligned}
$$

and so

$$
\begin{aligned}
p(2) &= P(x = 2) \\
&= P(SSFFF) + \cdots + P(FFFSS) \\
&= (10)(0.4)^2(0.6)^3 \\
&= 0.34560
\end{aligned}
$$

Other probabilities can be calculated in a similar way, A general formula for calculating the probabilities associated with the different possible values of x is

$$p(x) = P(x\ S\text{'}s \text{ among the five trials})$$

$$= (\text{number of outcomes with } x\ S\text{'s})\cdot(\text{probability of any given outcome with } x\ S\text{'s})$$

To use this general formula, you need to determine the number of different outcomes that have x successes. Let n denote the number of trials in the experiment. Then the number of outcomes with x S's is the number of ways of selecting x success trials from among the n trials. A simple expression for this quantity is

$$\text{number of outcomes with } x \text{ successes} = \frac{n!}{x!(n-x)!}$$

where, for any positive whole number m, the symbol $m!$ (read "m factorial") is defined by

$$m! = m(m-1)(m-2)\cdots(2)(1)$$

and $0! = 1$.

The Binomial Distribution

Let

n = number of independent trials in a binomial experiment
p = constant probability that any particular trial results in a success

Then

$$p(x) = P(x \text{ successes among the } n \text{ trials})$$

$$= \frac{n!}{x!(n-x)!}\, p^x\,(1-p)^{n-x} \qquad x = 0,1,2,\ldots, n$$

The expressions $\binom{n}{x}$ or $_nC_x$ are sometimes used in place of $\frac{n!}{x!(n-x)!}$. Both are read as "n choose x" and represent the number of ways of choosing x items from a set of n.

Notice that here the probability distribution is specified by a formula rather than a table or a probability histogram.

Example 6.20   Recognizing Your Roommate's Scent

An interesting experiment was described in the paper **"Sociochemosensory and Emotional Functions"** (*Psychological Science* [2009]: 1118–1123). The authors of this paper wondered if a female college student could recognize her roommate by scent. They carried out an experiment in which the participants used fragrance-free soap, deodorant, shampoo, and laundry detergent for a period of time. Their bedding was also laundered using a fragrance-free detergent. Each person was then given a new t-shirt that she slept in for one night. The shirt was then collected and sealed in an airtight bag. Later, the roommate was presented with three identical t-shirts (one worn by her roommate and two worn by other women) and asked to pick the one that smelled most like her roommate. (Yes, hard to believe, but people really do research like this!) This process was repeated, with the shirts refolded and rearranged before the second trial. The researchers recorded how many times (0, 1, or 2) that the shirt worn by the roommate was correctly identified.

This can be viewed as a binomial experiment consisting of $n = 2$ trials. Each trial results in either a correct identification or an incorrect identification. Because the researchers counted the number of correct identifications, a correct identification is considered a success. You can then define

$$x = \text{number of correct identifications}$$

Suppose that a participant is not able to identify her roommate by smell. In this case, she is essentially just picking one of the three shirts at random and so the probability of success (picking the correct shirt) is $\frac{1}{3}$. Here, it would also be reasonable to regard the two trials as independent. In this case, the experiment satisfies the conditions of a binomial experiment, and x is a binomial random variable with $n = 2$ and $p = \frac{1}{3}$.

You can use the binomial distribution formula to compute the probability associated with each of the possible x values:

$$p(0) = \frac{2!}{0!2!} \left(\frac{1}{3}\right)^0 \left(\frac{2}{3}\right)^2 = (1)(1)\left(\frac{2}{3}\right)^2 = 0.4444$$

$$p(1) = \frac{2!}{1!1!} \left(\frac{1}{3}\right)^1 \left(\frac{2}{3}\right)^1 = (2)\left(\frac{1}{3}\right)^1 \left(\frac{2}{3}\right)^1 = 0.4444$$

$$p(2) = \frac{2!}{2!0!} \left(\frac{1}{3}\right)^2 \left(\frac{2}{3}\right)^0 = (1)\left(\frac{1}{3}\right)^2 (1) = 0.1111$$

Summarizing in table form gives

x	p(x)
0	0.4444
1	0.4444
2	0.1111

This means that about 44.4% of the time, a person who is just guessing would pick the correct shirt on neither trial, about 44.4% of the time the correct shirt would be identified on one of the two trials, and about 11.1% of the time the correct shirt would be identified on both trials.

The authors of the paper actually performed this experiment with 44 subjects. They reported that 47.7% of the subjects identified the correct shirt on neither trial, 22.7% identified the correct shirt on one trial, and 31.7% identified the correct shirt on both trials. These results differed quite a bit from the expected "just guessing" results given in the preceding table). This difference was interpreted as evidence that some women could identify their roommates by smell.

Example 6.21 Computer Sales

Sixty percent of all computers sold by a large computer retailer are laptops and 40% are desktop models. The type of computer purchased by each of the next 12 customers will be recorded. Define a random variable x as

$$x = \text{number of laptops among these 12}$$

Because x counts the number of laptops, you use S to denote the sale of a laptop. Then x is a binomial random variable with $n = 12$ and $p = P(S) = 0.60$. The probability distribution of x is given by

$$p(x) = \frac{12!}{x!(12 - x)!} (0.6)^x (0.4)^{12-x} \qquad x = 0, 1, 2, \ldots, 12$$

The probability that exactly four computers are laptops is

$$\begin{aligned} p(4) &= P(x = 4) \\ &= \frac{12!}{4!8!} (0.6)^4 (0.4)^8 \\ &= (495)(0.6)^4 (0.4)^8 \\ &= 0.042 \end{aligned}$$

If many groups of 12 purchases are examined, about 4.2% of them include exactly four laptops.

The probability that between four and seven (inclusive) are laptops is

$$P(4 \le x \le 7) = P(x = 4 \text{ or } x = 5 \text{ or } x = 6 \text{ or } x = 7)$$

Since these outcomes are mutually exclusive, this is equal to

$$P(4 \leq x \leq 7) = p(4) + p(5) + p(6) + p(7)$$
$$= \frac{12!}{4!8!}(0.6)^4(0.4)^8 + \cdots + \frac{12!}{7!5!}(0.6)^7(0.4)^5$$
$$= 0.042 + 0.101 + 0.177 + 0.227$$
$$= 0.547$$

Notice that

$$P(4 < x < 7) = P(x = 5 \text{ or } x = 6)$$
$$= p(5) + p(6)$$
$$= 0.278$$

so the probability depends on whether $<$ or $\leq$ appears. (This is typical of *discrete* random variables.)

___

The binomial distribution formula can be tedious to use unless n is small. Appendix Table 9 gives binomial probabilities for selected values of n in combination with selected values of p. Statistical software packages and most graphing calculators can also be used to compute binomial probabilities. The following box explains how Appendix Table 9 can be used.

Using Appendix Table 9 to Compute Binomial Probabilities

To find $p(x)$ for any particular value of x,

1. Locate the part of the table corresponding to the value of n (5, 10, 15, 20, or 25).
2. Move down to the row labeled with the value of x.
3. Go across to the column headed by the specified value of p.

The desired probability is at the intersection of the designated x row and p column. For example, when $n = 20$ and $p = 0.8$,

$p(15) = P(x = 15) = $ (entry at intersection of $n = 15$ row and $p = 0.8$ column) $= 0.175$

Although $p(x)$ is positive for every possible x value, many probabilities are zero when rounded to three decimal places, so they appear as 0.000 in the table.

Sampling Without Replacement

If sampling is done by selecting an element at random from the population, observing whether it is a success or a failure, and then returning it to the population before the next selection is made, the variable $x =$ number of successes observed in the sample would fit all the requirements of a binomial random variable. However, sampling is usually carried out without replacement. That is, once an element has been selected for the sample, it is not a candidate for future selection. When sampling is done without replacement, the trials (individual selections) are not independent. In this case, the number of successes observed in the sample does not have a binomial distribution but rather a different type of distribution called a *hypergeometric distribution*. The probability calculations for this distribution are even more tedious than for the binomial distribution. Fortunately, when the sample size n is small relative to N, the population size, probabilities calculated using the binomial distribution and the hypergeometric distribution are nearly equal. They are so close, in fact, that you can ignore the difference and use the binomial probabilities in place of the hypergeometric probabilities. The following guideline can be used to determine if the binomial probability distribution is appropriate when sampling without replacement.

> Let x denote the number of S's in a sample of size n selected at random and without replacement from a population consisting of N individuals or objects. If $(n/N) \leq 0.05$, meaning no more than 5% of the population is sampled, then the binomial distribution gives a good approximation to the probability distribution of x.

Example 6.22 Online Security

The **2009 National Cyber Security Alliance (NCSA) Symantec Online Safety Study** found that only about 25% of adult Americans change passwords on online accounts at least quarterly as recommended by the NCSA (**"New Study Shows Need for Americans to Focus on Securing Online Accounts and Backing Up Critical Data," PRNewswire, October 29, 2009**). Suppose that exactly 25% of adult Americans change passwords quarterly. Consider a random sample of $n = 20$ adult Americans (much less than 5% of the population). Then

$$x = \text{the number in the sample who change passwords quarterly}$$

has (approximately) a binomial distribution with $n = 20$ and $p = 0.25$. The probability that five of those sampled change passwords quarterly is (from Appendix Table 9)

$$p(5) = P(x = 5)$$
$$= 0.202$$

The probability that at least half of those in the sample (10 or more) change passwords quarterly is

$$P(x \geq 10) = p(10) + p(11) + \cdots + p(20)$$
$$= 0.010 + 0.003 + 0.001 + \cdots + 0.000$$
$$= 0.014$$

If $p = 0.25$, only about 1.4% of all samples of size 20 would contain at least 10 people who change passwords quarterly. Because $P(x \geq 10)$ is so small when $p = 0.25$, if $x \geq 10$ were actually observed, you would have to wonder whether the reported value of $p = 0.25$ is correct. In Chapter 10, you will see how hypothesis-testing methods can be used to decide between two contradictory claims about a population (such as $p = 0.25$ and $p > 0.25$).

Technology, Appendix Table 9, or hand calculations using the formula for binomial probabilities can be used to compute binomial probabilities. Figure 6.17 shows the probability histogram for the binomial distribution with $n = 20$ and $p = 0.25$. Notice that the distribution is slightly skewed to the right. (The binomial distribution is symmetric only when $p = 0.5$.)

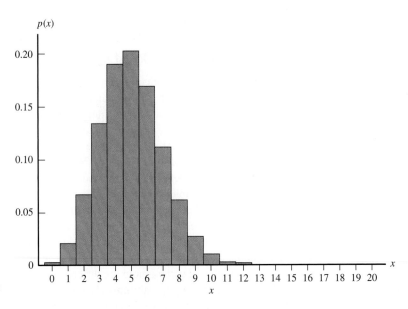

FIGURE 6.17
The binomial probability histogram when $n = 20$ and $p = 0.25$

Mean and Standard Deviation of a Binomial Random Variable

A binomial random variable x based on n trials has possible values 0, 1, 2, ... , n, so the mean value is

$$\mu_x = \sum xp(x) = (0)p(0) + (1)p(1) + \ldots + (n)p(n)$$

and the variance of x is

$$\sigma_x^2 = \sum (x - \mu_x)^2 \cdot p(x)$$
$$= (0 - \mu_x)^2 p(0) + (1 - \mu_x)^2 p(1) + \cdots (n - \mu_x)^2 p(n)$$

These expressions would be very tedious to evaluate for any particular values of n and p. Fortunately, they can be simplified to the expressions given in the following box.

> The mean value and the standard deviation of a binomial random variable are
>
> $$\mu_x = np \qquad \text{and} \qquad \sigma_x = \sqrt{np(1 - p)}$$

Example 6.23 Credit Cards Paid in Full

The paper **"Debt Literacy, Financial Experiences and Overindebtedness"** (*Social Science Research Network*, **Working Paper w14808, 2008**) reported that 37% of credit card users pay their bills in full each month. This figure represents an average over different types of credit cards issued by many different banks. Suppose that 40% of all individuals holding Visa cards issued by a certain bank pay in full each month. A random sample of $n = 25$ cardholders is to be selected. The bank is interested in the variable

$$x = \text{number in the sample who pay in full each month}$$

Even though sampling here is done without replacement, the sample size $n = 25$ is very small compared to the total number of credit card holders, so you can approximate the probability distribution of x using a binomial distribution with $n = 25$ and $p = 0.4$. "Paid in full" is defined as a success because this is the outcome counted by the random variable x. The mean value of x is then

$$\mu_x = np = 25(0.40) = 10.0$$

and the standard deviation is

$$\sigma_x = \sqrt{np(1 - p)} = \sqrt{25(0.4)(0.6)} = \sqrt{6} = 2.45$$

There are two cases where there is no variability in x. When $p = 0$, x will always equal 0. When $p = 1$, x will always equal n. In both cases, the value of σ_x is 0. It is also easily verified that $p(1 - p)$ is largest when $p = 0.5$. This means that the binomial distribution spreads out the most when sampling from a population that consists of half successes and half failures. The farther p is from 0.5, the less spread out and the more skewed the distribution.

Geometric Distributions

A binomial random variable is defined as the number of successes in n independent trials, where each trial can result in either a success or a failure and the probability of success is the same for each trial. Suppose, however, that you are not interested in the number of successes in a *fixed* number of trials but rather in the number of trials that must be carried out before a success occurs. Counting the number of boxes of cereal that must be purchased before finding one with a rare toy and counting the number of games that a professional bowler must play before achieving a score over 250 are two examples of this.

The variable

$$x = \text{number of trials to first success}$$

is called a *geometric random variable*, and the probability distribution that describes its behavior is called a *geometric probability distribution*.

> Suppose an experiment consists of a sequence of trials with the following conditions:
>
> 1. The trials are independent.
> 2. Each trial can result in one of two possible outcomes, success or failure.
> 3. The probability of success is the same for all trials.
>
> A **geometric random variable** is defined as
>
> $$x = \text{number of trials until the first success is observed (including the success trial)}$$
>
> The probability distribution of x is called the **geometric probability distribution**.

For example, suppose that 40% of the students who drive to campus at your school carry jumper cables. Your car has a dead battery and you don't have jumper cables, so you decide to ask students who are headed to the parking lot whether they have a pair of jumper cables. You might be interested in the number of students you would have to ask in order to find one who has jumper cables. Defining a success as a student with jumper cables, a trial would consist of asking an individual student for help. The random variable

$$x = \text{number of students asked in order to find one with jumper cables}$$

is an example of a geometric random variable, because it can be viewed as the number of trials to the first success in a sequence of independent trials.

The probability distribution of a geometric random variable is easy to construct. As before, p is used to denote the probability of success on any given trial. Possible outcomes can be denoted as follows:

Outcome	x = Number of Trials to First Success
S	1
FS	2
FFS	3
$\vdots$	$\vdots$
$FFFFFFS$	7
$\vdots$	$\vdots$

Each possible outcome consists of 0 or more failures followed by a single success. When this is the case,

$$p(x) = P(x \text{ trials to first success})$$
$$= P(FF \ldots FS)$$

x − 1 failures followed by a success on trial x

The probability of success is p for each trial, so the probability of failure for each trial is $1 - p$. Because the trials are independent,

$$p(x) = P(x \text{ trials to first success}) = P(FF \ldots FS)$$
$$= P(F)P(F) \cdots P(F)P(S)$$
$$= (1 - p)(1 - p) \cdots (1 - p)p$$
$$= (1 - p)^{x-1} p$$

This leads to the formula for the geometric probability distribution.

> ### Geometric Probability Distribution
>
> If x is a geometric random variable with probability of success $= p$ for each trial, then
>
> $$p(x) = (1 - p)^{x-1}p \qquad x = 1, 2, 3, \ldots$$

Example 6.24 **Jumper Cables**

Consider the jumper cable problem described previously. Because 40% of the students who drive to campus carry jumper cables, $p = 0.4$. The probability distribution of

$x = $ number of students asked in order to find one with jumper cables is

$$p(x) = (0.6)^{x-1}(0.4) \qquad x = 1, 2, 3, \ldots$$

The probability distribution can now be used to compute various probabilities. For example, the probability that the first student asked has jumper cables (that is, $x = 1$) is

$$p(1) = (0.6)^{1-1}(0.4) = (0.6)^0(0.4) = 0.4$$

The probability that three or fewer students must be asked is

$$\begin{aligned}
P(x \le 3) &= p(1) + p(2) + p(3) \\
&= (0.6)^0(0.4) + (0.6)^1(0.4) + (0.6)^2(0.4) \\
&= 0.4 + 0.24 + 0.144 \\
&= 0.784
\end{aligned}$$

SECTION 6.5 EXERCISES

Each Exercise Set assesses the following chapter learning objectives: M11, M12, M13

SECTION 6.5 Exercise Set 1

6.45 Example 6.27 described a study in which a person was asked to determine which of three t-shirts had been worn by her roommate by smelling the shirts (**"Sociochemosensory and Emotional Functions,"** *Psychological Science* [2009]: 1118–1123). Suppose that instead of three shirts, each participant was asked to choose among four shirts and that the process was performed five times. If a person can't identify her roommate by smell and is just picking a shirt at random, then $x = $ number of correct identifications is a binomial random variable with $n = 5$ and $p = \frac{1}{4}$.
a. What are the possible values of x?
b. For each possible value of x, find the associated probability $p(x)$ and display the possible x values and $p(x)$ values in a table.
c. Construct a histogram displaying the probability distribution of x.

6.46 Suppose that in a certain metropolitan area, 90% of all households have cable TV. Let x denote the number among four randomly selected households that have cable TV. Then x is a binomial random variable with $n = 4$ and $p = 0.9$.

a. Calculate $p(2) = P(x = 2)$, and interpret this probability.
b. Calculate $p(4)$, the probability that all four selected households have cable TV.
c. Determine $P(x \le 3)$.

6.47 Twenty-five percent of the customers of a grocery store use an express checkout. Consider five randomly selected customers, and let x denote the number among the five who use the express checkout.
a. Calculate $p(2)$, that is, $P(x = 2)$.
b. Calculate $P(x \le 1)$.
c. Calculate $P(2 \le x)$. (Hint: Make use of your answer to Part (b).)
d. Calculate $P(x \ne 2)$.

6.48 Industrial quality control programs often include inspection of incoming materials from suppliers. If parts are purchased in large lots, a typical plan might be to select 20 parts at random from a lot and inspect them. Suppose that a lot is judged acceptable if one or fewer of these 20 parts are defective. If more than one part is defective, the lot is rejected and returned to the supplier. Find the probability of accepting lots that have each of the following (Hint: Identify success with a defective part):

a. 5% defective parts
b. 10% defective parts
c. 20% defective parts

6.49 Suppose a playlist on an MP3 music player consisting of 100 songs includes 8 by a particular artist. Suppose that songs are played by selecting a song at random (with replacement) from the playlist. The random variable x represents the number of songs until a song by this artist is played.

a. Explain why the probability distribution of x is not binomial.
b. Find the following probabilities:
   i. $p(4)$
   ii. $P(x \leq 4)$
   iii. $P(x > 4)$
   iv. $P(x \geq 4)$
c. Interpret each of the probabilities in Part (b) and explain the difference between them.

6.50 The article **"FBI Says Fewer than 25 Failed Polygraph Test"** (*San Luis Obispo Tribune*, July 29, 2001) states that false-positives in polygraph tests (tests in which an individual fails even though he or she is telling the truth) are relatively common and occur about 15% of the time. Let x be the number of trustworthy FBI agents tested until someone fails the test.

a. Is the probability distribution of x binomial or geometric?
b. What is the probability that the first false-positive will occur when the third agent is tested?
c. What is the probability that fewer than four are tested before the first false-positive occurs?
d. What is the probability that more than three agents are tested before the first false-positive occurs?

SECTION 6.5 **Exercise Set 2**

6.51 In a press release dated October 2, 2008, **The National Cyber Security Alliance** reported that approximately 80% of adult Americans who own a computer claim to have a firewall installed on their computer to prevent hackers from stealing personal information. This estimate was based on a survey of 3,000 people. It was also reported that in a study of 400 computers, only about 40% actually had a firewall installed.

a. Suppose that the true proportion of computer owners who have a firewall installed is 0.80. If 20 computer owners are selected at random, what is the probability that more than 15 have a firewall installed?
b. Suppose that the true proportion of computer owners who have a firewall installed is 0.40. If 20 computer owners are selected at random, what is the probability that more than 15 have a firewall installed?
c. Suppose that a random sample of 20 computer owners is selected and that 14 have a firewall installed. Is it more likely that the true proportion of computer owners who

have a firewall installed is 0.40 or 0.80? Justify your answer based on probability calculations.

6.52 The *Los Angeles Times* **(December 13, 1992)** reported that what 80% of airline passengers like to do most on long flights is rest or sleep. Suppose that the actual percentage is exactly 80%, and consider randomly selecting six passengers. Then $x =$ the number among the selected six who prefer to rest or sleep is a binomial random variable with $n = 6$ and $p = 0.8$.

a. Calculate $p(4)$, and interpret this probability.
b. Calculate $p(6)$, the probability that all six selected passengers prefer to rest or sleep.
c. Calculate $P(x \geq 4)$.

6.53 Refer to the previous exercise, and suppose that 10 rather than six passengers are selected ($n = 10$, $p = 0.8$). Use Appendix Table 9 to find the following:

a. $p(8)$
b. $P(x \leq 7)$
c. The probability that more than half of the selected passengers prefer to rest or sleep.

6.54 Thirty percent of all automobiles undergoing an emissions inspection at a certain inspection station fail the inspection.

a. Among 15 randomly selected cars, what is the probability that at most 5 fail the inspection?
b. Among 15 randomly selected cars, what is the probability that between 5 and 10 (inclusive) fail the inspection?
c. Among 25 randomly selected cars, what is the mean value of the number that pass inspection, and what is the standard deviation?
d. What is the probability that among 25 randomly selected cars, the number that pass is within 1 standard deviation of the mean value?

6.55 Sophie is a dog who loves to play catch. Unfortunately, she isn't very good at this, and the probability that she catches a ball is only 0.1. Let x be the number of tosses required until Sophie catches a ball.

a. Does x have a binomial or a geometric distribution?
b. What is the probability that it will take exactly two tosses for Sophie to catch a ball?
c. What is the probability that more than three tosses will be required?

6.56 Suppose that 5% of cereal boxes contain a prize and the other 95% contain the message, "Sorry, try again." Consider the random variable x, where $x =$ number of boxes purchased until a prize is found.

a. What is the probability that at most two boxes must be purchased?
b. What is the probability that exactly four boxes must be purchased?
c. What is the probability that more than four boxes must be purchased?

Additional Exercises

6.57 An experiment was conducted to investigate whether a graphologist (a handwriting analyst) could distinguish a normal person's handwriting from that of a psychotic. A well-known expert was given 10 files, each containing handwriting samples from a normal person and from a person diagnosed as psychotic, and asked to identify the psychotic's handwriting. The graphologist made correct identifications in 6 of the 10 trials (data taken from *Statistics in the Real World*, by R. J. Larsen and D. F. Stroup [New York: Macmillan, 1976]). Does this indicate that the graphologist has an ability to distinguish the handwriting of psychotics? (Hint: What is the probability of correctly guessing 6 or more times out of 10? Your answer should depend on whether this probability is relatively small or relatively large.)

6.58 A breeder of show dogs is interested in the number of female puppies in a litter. If a birth is equally likely to result in a male or a female puppy, give the probability distribution of the variable x = number of female puppies in a litter of size 5.

6.59 You are to take a multiple-choice exam consisting of 100 questions with five possible responses to each question. Suppose that you have not studied and so must guess (randomly select one of the five answers) on each question. Let x represent the number of correct responses on the test.
a. What kind of probability distribution does x have?
b. What is your expected score on the exam? (Hint: Your expected score is the mean value of the x distribution.)
c. Compute the variance and standard deviation of x.
d. Based on your answers to Parts (b) and (c), is it likely that you would score over 50 on this exam? Explain the reasoning behind your answer.

6.60 Suppose that 20% of the 10,000 signatures on a certain recall petition are invalid. Would the number of invalid signatures in a sample of size 2,000 have (approximately) a binomial distribution? Explain.

6.61 A soft-drink machine dispenses only regular Coke and Diet Coke. Sixty percent of all purchases from this machine are diet drinks. The machine currently has 10 cans of each type. If 15 customers want to purchase drinks before the machine is restocked, what is the probability that each of the 15 is able to purchase the type of drink desired? (Hint: Let x denote the number among the 15 who want a diet drink. For which possible values of x is everyone satisfied?)

6.62 A coin is flipped 25 times. Let x be the number of flips that result in heads (H). Consider the following rule for deciding whether or not the coin is fair:

Judge the coin fair if $8 \leq x \leq 17$.
Judge the coin biased if either $x \leq 7$ or $x \geq 18$.

a. What is the probability of judging the coin biased when it is actually fair?
b. Suppose that a coin is not fair and that $P(H) = 0.9$. What is the probability that this coin would be judged fair? What is the probability of judging a coin fair if $P(H) = 0.1$?
c. What is the probability of judging a coin fair if $P(H) = 0.6$? if $P(H) = 0.4$? Why are these probabilities large compared to the probabilities in Part (b)?
d. What happens to the "error probabilities" of Parts (a) and (b) if the decision rule is changed so that the coin is judged fair if $7 \leq x \leq 18$ and unfair otherwise? Is this a better rule than the one first proposed? Explain.

6.63 The longest "run" of S's in the 10-trial sequence *SSFSSSSFFS* has length 4, corresponding to the S's on the fourth, fifth, sixth, and seventh trials. Consider a binomial experiment with $n = 4$, and let y be the length (number of trials) in the longest run of S's.
a. When $p = 0.5$, the 16 possible outcomes are equally likely. Determine the probability distribution of y in this case (first list all outcomes and the y value for each one). Then calculate μ_y.
b. Repeat Part (a) for the case $p = 0.6$.

SECTION 6.6 **Normal Distributions**

Normal distributions are continuous probability distributions that formalize the mound-shaped distributions introduced in Section 6.3. Normal distributions are widely used for two reasons. First, they approximate the distributions of many different variables. Second, they also play a central role in many of the inferential procedures that will be discussed in later chapters.

Normal distributions are bell shaped and symmetric, as shown in Figure 6.18. They are sometimes referred to as *normal curves*.

FIGURE 6.18
A normal distribution

There are many different normal distributions. They are distinguished from one another by their mean μ and standard deviation σ. The mean μ of a normal distribution describes where the corresponding curve is centered, and the standard deviation σ describes how much the curve spreads out around that center. As with all continuous probability distributions, the total area under any normal curve is equal to 1.

Three normal distributions are shown in Figure 6.19. Notice that the smaller the standard deviation, the taller and narrower the corresponding curve. There will also be a larger area concentrated around μ at the center of the curve. Because areas under a continuous probability distribution curve represent probabilities, the chance of observing a value near the mean is much greater when the standard deviation is small.

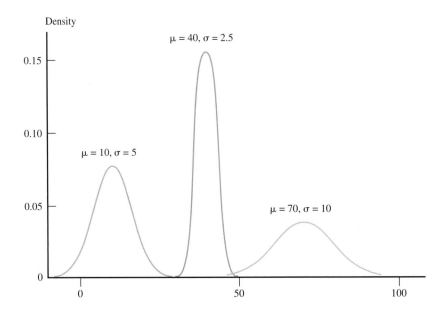

FIGURE 6.19
Three normal distributions

The value of μ is the number on the measurement axis lying directly below the top of the bell. The value of σ can also be approximated from a picture of the curve. It is the distance to either side of μ at which a normal curve changes from turning downward to turning upward. Consider the normal curve in Figure 6.20. Starting at the top of the bell (above $\mu = 100$) and moving to the right, the curve turns downward until it is above the value 110. After that point, it continues to decrease in height but begins to turn upward rather than downward. Similarly, to the left of $\mu = 100$, the curve turns downward until it reaches 90 and then begins to turn upward. Because the curve changes from turning downward to turning upward at a distance of 10 on either side of μ, $\sigma = 10$ for this normal curve.

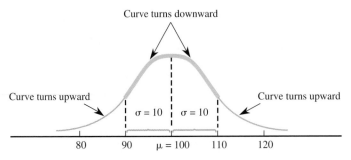

FIGURE 6.20
Mean μ and standard deviation σ
for a normal curve

To use a particular normal distribution to describe the behavior of a random variable, a mean and a standard deviation must be specified. For example, you might use a normal distribution with mean 7 pounds and standard deviation 1 pound as a model for the distribution of x = birth weight. If this model is a reasonable description of the probability distribution, you could then use areas under the normal curve with $\mu = 7$ and $\sigma = 1$ to approximate various probabilities related to birth weight. For example, the probability that a birth weight is over 8 pounds (expressed symbolically as $P(x > 8)$) corresponds to the shaded area in Figure 6.21(a). The shaded area in Figure 6.21(b) is the probability of a birth weight falling between 6.5 and 8 pounds, $P(6.5 < x < 8)$.

FIGURE 6.21
Normal distribution for birth
weight: (a) shaded area $= P(x > 8)$;
(b) shaded area $= P(6.5 < x < 8)$

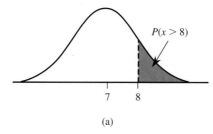

(a)

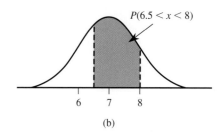

(b)

Unfortunately, calculating such probabilities (areas under a normal curve) is not simple. To overcome this difficulty, you can rely on technology or a table of areas for a reference normal distribution, called the *standard normal distribution*.

DEFINITION

The **standard normal distribution** is the normal distribution with

$$\mu = 0 \text{ and } \sigma = 1$$

The corresponding density curve is called the *standard normal curve*. It is customary to use the letter z to represent a variable whose distribution is described by the standard normal curve. The standard normal curve is also sometimes called the z **curve**.

Few naturally occurring variables have distributions that are well described by the standard normal distribution, but this distribution is important because it is used in probability calculations for other normal distributions. When you are interested in finding a probability based on some other normal curve, you either rely on technology or translate the problem into an equivalent problem that involves finding an area under the standard normal curve. A table for the standard normal distribution is then used to find the desired area. To be able to do this, you must first learn to work with the standard normal distribution.

The Standard Normal Distribution

In working with normal distributions, you need two general skills:

1. You must be able to use the normal distribution to compute probabilities, which are areas under a normal curve and above given intervals.
2. You must be able to characterize extreme values in the distribution, such as the largest 5%, the smallest 1%, and the most extreme 5% (which would include the largest 2.5% and the smallest 2.5%).

Let's begin by looking at how to accomplish these tasks for the standard normal distribution.

The standard normal or z curve is shown in Figure 6.22(a). It is centered at $\mu = 0$, and the standard deviation, $\sigma = 1$, is a measure of how much it spreads out about its mean. Notice that this picture is consistent with the Empirical Rule of Chapter 3: About 95% of the area is associated with values that are within 2 standard deviations of the mean (between -2 and 2), and almost all of the area is associated with values that are within 3 standard deviations of the mean (between -3 and 3).

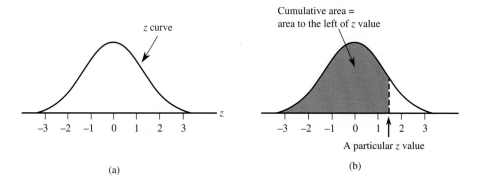

FIGURE 6.22
(a) a standard normal curve;
(b) a cumulative area

Appendix Table 2 gives cumulative z curve areas of the sort shown in Figure 6.22(b) for many different values of z. The smallest value for which the cumulative area is given is $z = -3.89$, a value far out in the lower tail of the z curve. The next smallest value for which an area is given is -3.88, then -3.87, and so on in increments of 0.01 all the way up to 3.89.

Using the Table of Standard Normal Curve Areas

For any number $z*$ between -3.89 and 3.89 and rounded to two decimal places, Appendix Table 2 gives

$$(\text{area under } z \text{ curve to the left of } z*) = P(z < z*) = P(z \le z*)$$

where the letter z is used to represent a random variable whose distribution is the standard normal distribution.

To find this probability using the table, locate the following:

1. The row labeled with the sign of $z*$ and the digit to either side of the decimal point (for example, -1.7 or 0.5)
2. The column identified with the second digit to the right of the decimal point in $z*$ (for example, .06 if $z* = -1.76$)

The number at the intersection of this row and column is the desired probability, $P(z < z*)$.

A portion of the table of standard normal curve areas appears in Figure 6.23. To find the area under the z curve to the left of 1.42, look in the row labeled 1.4 and the column labeled .02 (the highlighted row and column in Figure 6.23). From the table, the corresponding cumulative area is 0.9222. So

$$z \text{ curve area to the left of } 1.42 = 0.9222$$

z^*	0.00	0.01	0.02	0.03	0.04	0.05
0.0	0.5000	0.5040	0.5080	0.5120	0.5160	0.5199
0.1	0.5398	0.5438	0.5478	0.5517	0.5557	0.5596
0.2	0.5793	0.5832	0.5871	0.5910	0.5948	0.5987
0.3	0.6179	0.6217	0.6255	0.6293	0.6331	0.6368
0.4	0.6554	0.6591	0.6628	0.6664	0.6700	0.6736
0.5	0.6915	0.6950	0.6985	0.7019	0.7054	0.7088
0.6	0.7257	0.7291	0.7324	0.7357	0.7389	0.7422
0.7	0.7580	0.7611	0.7642	0.7673	0.7704	0.7734
0.8	0.7881	0.7910	0.7939	0.7967	0.7995	0.8023
0.9	0.8159	0.8186	0.8212	0.8238	0.8264	0.8289
1.0	0.8413	0.8438	0.8461	0.8485	0.8508	0.8531
1.1	0.8643	0.8665	0.8686	0.8708	0.8729	0.8749
1.2	0.8849	0.8869	0.8888	0.8907	0.8925	0.8944
1.3	0.9032	0.9049	0.9066	0.9082	0.9099	0.9115
1.4	0.9192	0.9207	0.9222	0.9236	0.9251	0.9265
1.5	0.9332	0.9345	0.9357	0.9370	0.9382	0.9394
1.6	0.9452	0.9463	0.9474	0.9484	0.9495	0.9505
1.7	0.9554	0.9564	0.9573	0.9582	0.9591	0.9599
1.8	0.9641	0.9649	0.9656	0.9664	0.9671	0.9678

$P(z < 1.42)$

FIGURE 6.23
Portion of the table of standard normal curve areas

You can also use the table to find the area to the right of 1.42. Because the total area under the z curve is 1, it follows that

$$(z \text{ curve area to the right of } 1.42) = 1 - (z \text{ curve area to the left of } 1.42)$$
$$= 1 - 0.9222$$
$$= 0.0778$$

These probabilities can be interpreted to mean that about 92.22% of observed z values will be smaller than 1.42, and about 7.78% will be larger than 1.42.

Example 6.25   Finding Standard Normal Curve Areas

The probability $P(z < -1.76)$ appears in the z table at the intersection of the -1.7 row and the .06 column. The result is

$$P(z < -1.76) = 0.0392$$

as shown in the following figure:

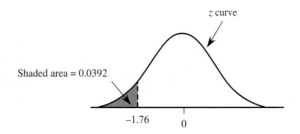

In other words, about 3.9% of observed z values will be smaller than -1.76. Similarly,

$$P(z \leq 0.58) = \text{entry in 0.5 row and .08 column of Table 2} = 0.7190$$

as shown in the following figure:

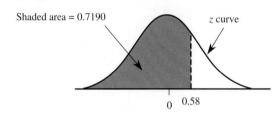

Now consider $P(z < -4.12)$. This probability does not appear in Appendix Table 2; there is no -4.1 row. However, it must be less than $P(z < -3.89)$, the smallest z value in the table, because -4.12 is farther out in the lower tail of the z curve. Since $P(z < -3.89) \approx .0000$ (that is, zero to four decimal places), it follows that

$$P(z < -4.12) \approx 0$$

Similarly,

$$P(z < 4.18) > P(z < 3.89) \approx 1.0000$$

so

$$P(z < 4.18) \approx 1$$

As illustrated in Example 6.17, you can use Appendix Table 2 to calculate other probabilities involving z. The probability that z is larger than a value c is

$$P(z > c) = \text{area under the } z \text{ curve to the right of } c = 1 - P(z \leq c)$$

In other words, the area to the right of a value (a right-tail area) is 1 minus the corresponding cumulative area. This is illustrated in Figure 6.24.

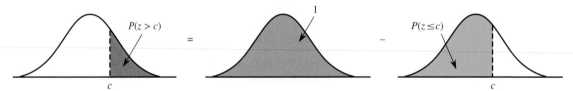

FIGURE 6.24
The relationship between an upper-tail area and a cumulative area

Similarly, the probability that z falls in the interval between a lower limit a and an upper limit b is

$$P(a < z < b) = \text{area under the } z \text{ curve and above the interval from } a \text{ to } b$$
$$= P(z < b) - P(z < a)$$

That is, $P(a < z < b)$ is the difference between two cumulative areas, as illustrated in Figure 6.25.

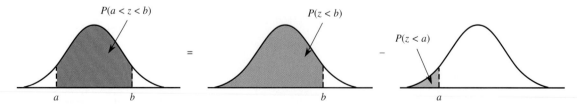

FIGURE 6.25
$P(a < z < b)$ as the difference between two cumulative areas

Example 6.26   **More About Standard Normal Curve Areas**

The probability that z is between -1.76 and 0.58 is

$$P(-1.76 < z < 0.58) = P(z < 0.58) - P(z < -1.76)$$
$$= 0.7190 - 0.0392$$
$$= 0.6798$$

as shown in the following figure:

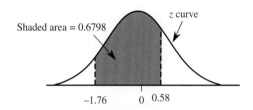

The probability that z is between -2 and $+2$ (within 2 standard deviations of its mean, since $\mu = 0$ and $\sigma = 1$) is

$$P(-2.00 < z < 2.00) = P(z < 2.00) - P(z < -2.00)$$
$$= 0.9772 - 0.0228$$
$$= 0.9544$$
$$\approx 0.95$$

as shown in the following figure:

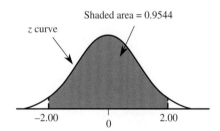

This last probability is the basis for one part of the Empirical Rule, which states that when a data distribution is well approximated by a normal curve, approximately 95% of the values are within 2 standard deviations of the mean.

The probability that the value of z exceeds 1.96 is

$$P(z > 1.96) = 1 - P(z \leq 1.96)$$
$$= 1 - 0.9750$$
$$= 0.0250$$

as shown in the following figure:

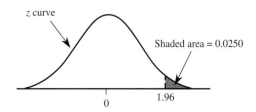

That is, 2.5% of the area under the z curve lies to the right of 1.96 in the upper tail. Similarly,

$$P(z > -1.28) = \text{area to the right of } -1.28$$
$$= 1 - P(z < -1.28)$$
$$= 1 - 0.1003$$
$$= 0.8997$$
$$\approx 0.90$$

Identifying Extreme Values

Suppose that you want to describe the values included in the smallest 2% of a distribution or the values making up the most extreme 5% (which includes the largest 2.5% and the smallest 2.5%). Examples 6.27 and 6.28 show how to identify extreme values in the z distribution.

Example 6.27   Identifying Extreme Values

Suppose that you want to describe the values that make up the smallest 2% of the standard normal distribution. Symbolically, you are trying to find a value (call it z^*), such that

$$P(z < z^*) = 0.02$$

This is illustrated in Figure 6.26, which shows that the cumulative area for z^* is 0.02. Therefore, look for a cumulative area of 0.0200 in the body of Appendix Table 2. The closest cumulative area in the table is 0.0202, in the -2.0 row and .05 column, so $z^* = -2.05$, the best approximation from the table. Variable values less than -2.05 make up the smallest 2% of the standard normal distribution. Notice that, because of the symmetry of the normal curve, variable values greater than 2.05 make up the largest 2% of the standard normal distribution.

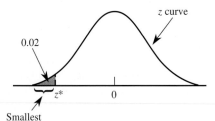

FIGURE 6.26
The smallest 2% of the standard normal distribution

Now suppose that you are interested in the largest 5% of all z values. You would then be trying to find a value of z^* for which

$$P(z > z^*) = 0.05$$

as illustrated in Figure 6.27. Because Appendix Table 2 always works with cumulative area (area to the left), the first step is to determine

$$\text{area to the left of } z^* = 1 - 0.05 = 0.95$$

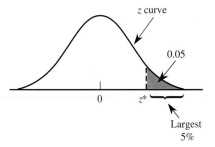

FIGURE 6.27
The largest 5% of the standard normal distribution

Looking for the cumulative area closest to 0.95 in Appendix Table 2, you find that 0.95 falls exactly halfway between 0.9495 (corresponding to a z value of 1.64) and 0.9505 (corresponding to a z value of 1.65). Because 0.9500 is exactly halfway between the two areas, use a z value that is halfway between 1.64 and 1.65. (If one value had been closer to 0.9500 than the other, you could just use the z value corresponding to the closest area). This gives

$$z^* = \frac{1.64 + 1.65}{2} = 1.645$$

Values greater than 1.645 make up the largest 5% of the standard normal distribution. By symmetry, -1.645 separates the smallest 5% of all z values from the others.

Example 6.28 More Extremes

Sometimes you are interested in identifying the most extreme (unusually large *or* small) values in a distribution. How can you describe the values that make up the most extreme 5% of the standard normal distribution? Here you want to separate the middle 95% from the most extreme 5%. This is illustrated in Figure 6.28.

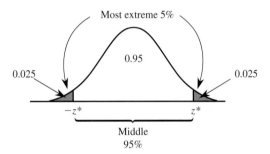

FIGURE 6.28
The most extreme 5% of the standard normal distribution

Because the standard normal distribution is symmetric, the most extreme 5% is equally divided between the high side and the low side of the distribution, resulting in an area of 0.025 for each of the tails of the z curve. To find z^*, first determine the cumulative area for z^*, which is

area to the left of $z^* = 0.95 + 0.025 = 0.975$

The cumulative area 0.9750 appears in the 1.9 row and .06 column of Appendix Table 2, so $z^* = 1.96$. Symmetry about 0 implies that if z^* denotes the value that separates the largest 2.5%, the value that separates the smallest 2.5% is simply $-z^*$, or -1.96. For the standard normal distribution, those values that are either greater than 1.96 or less than -1.96 are the most extreme 5%.

Other Normal Distributions

Areas under the z curve can be used to calculate probabilities and to describe values for *any* normal distribution. Remember that the letter z is reserved for those variables that have a standard normal distribution. Other letters, such as x, are used to denote a variable whose distribution is described by a normal curve with mean μ and standard deviation σ.

Suppose that you want to compute $P(a < x < b)$, the probability that the variable x lies in a particular range. This probability corresponds to an area under a normal curve and above the interval from a to b, as shown in Figure 6.29(a).

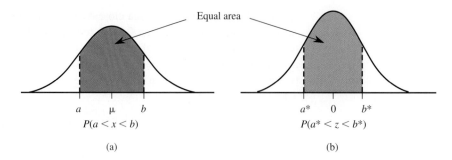

FIGURE 6.29
Equality of nonstandard and
standard normal curve areas

The strategy for obtaining this probability is to find an equivalent problem involving the standard normal distribution. Finding an equivalent problem means determining an interval (a^*, b^*) that has the same probability for z (same area under the z curve) as the interval (a, b) in the original normal distribution (see Figure 6.29(b)). The asterisk is used to distinguish a and b, the values from the original normal distribution from a^* and b^*, the values from the z distribution. To find a^* and b^*, simply calculate z-scores for a and b. This process is called **standardizing** the endpoints. For example, suppose that the variable x has a normal distribution with mean $\mu = 100$ and standard deviation $\sigma = 5$. To find

$$P(98 < x < 107)$$

first translate this problem into an equivalent one for the standard normal distribution. Recall from Chapter 3 that a z-score is calculated by first subtracting the mean and then dividing by the standard deviation. Converting the lower endpoint $a = 98$ to a z-score gives

$$a^* = \frac{98 - 100}{5} = \frac{-2}{5} = -0.40$$

and converting the upper endpoint yields

$$b^* = \frac{107 - 100}{5} = \frac{7}{5} = 1.40$$

Then

$$P(98 < x < 107) = P(-0.40 < z < 1.40)$$

The probability $P(-0.40 < z < 1.40)$ can now be evaluated using Appendix Table 2.

Finding Probabilities

To calculate probabilities for any normal distribution, standardize the relevant values and then use the table of z curve areas. More specifically, if x is a variable whose behavior is described by a normal distribution with mean μ and standard deviation σ, then

$$P(x < b) = P(z < b^*)$$
$$P(x > a) = P(z > a^*)$$
$$P(a < x < b) = P(a^* < z < b^*)$$

where z is a variable whose distribution is standard normal and

$$a^* = \frac{a - \mu}{\sigma} \qquad b^* = \frac{b - \mu}{\sigma}$$

Example 6.29   Newborn Birth Weights

Data from the paper **"Fetal Growth Parameters and Birth Weight: Their Relationship to Neonatal Body Composition"** (*Ultrasound in Obstetrics and Gynecology* **[2009]: 441–446**) suggest that a normal distribution with mean $\mu = 3,500$ grams and standard deviation

$\sigma = 600$ grams is a reasonable model for the probability distribution of $x =$ birth weight of a randomly selected full-term baby. What proportion of birth weights are between 2,900 and 4,700 grams?

To answer this question, you must find

$$P(2{,}900 < x < 4{,}700)$$

First, translate the interval endpoints to equivalent endpoints for the standard normal distribution:

$$a^* = \frac{a - \mu}{\sigma} = \frac{2{,}900 - 3{,}500}{600} = -1.00$$

$$b^* = \frac{b - \mu}{\sigma} = \frac{4{,}700 - 3{,}500}{600} = 2.00$$

Then

$$
\begin{aligned}
P(2{,}900 < x < 4{,}700) &= P(-1.00 < z < 2.00) \\
&= (z \text{ curve area to the left of } 2.00) \\
&\quad - (z \text{ curve area to the left of } -1.00) \\
&= 0.9772 - 0.1587 \\
&= 0.8185
\end{aligned}
$$

The probabilities for x and z are shown in Figure 6.30. If birth weights were observed for many babies from this population, about 82% of them would be between 2,900 and 4,700 grams.

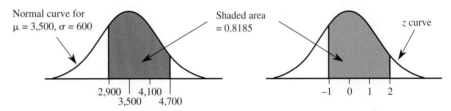

FIGURE 6.30
$P(2{,}900 < x < 4{,}700)$ and corresponding z curve area for the birth weight problem of Example 6.29

What is the probability that a randomly chosen baby will have a birth weight greater than 4,500? To evaluate $P(x > 4{,}500)$, first compute

$$a^* = \frac{a - \mu}{\sigma} = \frac{4{,}500 - 3{,}500}{600} = 1.67$$

Then (see Figure 6.31)

$$
\begin{aligned}
P(x > 4{,}500) &= P(z > 1.67) \\
&= z \text{ curve area to the right of } 1.67 \\
&= 1 - (z \text{ curve area to the left of } 1.67) \\
&= 1 - 0.9525 \\
&= 0.0475
\end{aligned}
$$

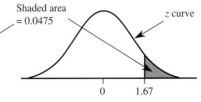

FIGURE 6.31
$P(x > 4{,}500)$ and corresponding z curve area for the birth weight problem of Example 6.29

Example 6.30 | **IQ Scores**

Although there is some controversy regarding the appropriateness of IQ scores as a measure of intelligence, they are still used for a variety of purposes. One commonly used IQ scale has a mean of 100 and a standard deviation of 15, and these IQ scores are approximately normally distributed. (Because this IQ score is based on the number of correct responses on a test, it is actually a discrete variable. However, the population distribution of IQ scores closely resembles a normal curve.) If we define the random variable

x = IQ score of a randomly selected individual

then x has approximately a normal distribution with $\mu = 100$ and $\sigma = 15$.

One way to become eligible for membership in Mensa, an organization for those of high intelligence, is to have an IQ score above 130. What proportion of the population would qualify for Mensa membership? Answering this question requires evaluating $P(x > 130)$. This probability is shown in Figure 6.32.

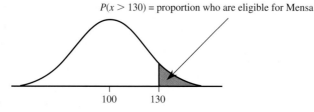

FIGURE 6.32
Normal distribution and desired proportion for Example 6.30

With $a = 130$,

$$a* = \frac{a - \mu}{\sigma} = \frac{130 - 100}{15} = 2.00$$

So (see Figure 6.31)

$$P(x > 130) = P(z > 2.00)$$
$$= z \text{ curve area to the right of } 2.00$$
$$= 1 - (z \text{ curve area to the left of } 2.00)$$
$$= 1 - 0.9772$$
$$= 0.0228$$

Only about 2.28% of the population would qualify for Mensa membership.

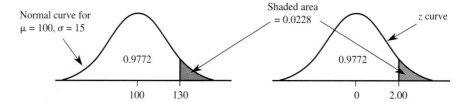

FIGURE 6.33
$P(x > 130)$ and corresponding z curve area

Suppose that you are interested in the proportion of the population with IQ scores below 80—that is, $P(x < 80)$. With $b = 80$,

$$b* = \frac{b - \mu}{\sigma} = \frac{80 - 100}{15} = -1.33$$

So

$$P(x < 80) = P(z < -1.33)$$
$$= z \text{ curve area to the left of } -1.33$$
$$= 0.0918$$

as shown in Figure 6.34. This probability tells you that just a little over 9% of the population has an IQ score below 80.

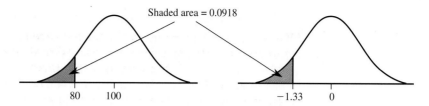

FIGURE 6.34

P(x < 80) and corresponding z curve area for the IQ problem of Example 6.30

Now consider the proportion of the population with IQs between 75 and 125. Using $a = 75$ and $b = 125$, you obtain

$$a^* = \frac{75 - 100}{15} = -1.67 \qquad\qquad b^* = \frac{125 - 100}{15} = 1.67$$

so

$$
\begin{aligned}
P(75 < x < 125) &= P(-1.67 < z < 1.67) \\
&= z \text{ curve area between } -1.67 \text{ and } 1.67 \\
&= (z \text{ curve area to the left of } 1.67) \\
&\quad - (z \text{ curve area to the left of } -1.67) \\
&= 0.9525 - 0.0475 \\
&= 0.9050
\end{aligned}
$$

This is illustrated in Figure 6.35. This probability tells you that about 90.5% of the population has an IQ score between 75 and 125. Of the 9.5% with an IQ score that is not between 75 and 125, about half of them (4.75%) have scores over 125, and about half have scores below 75.

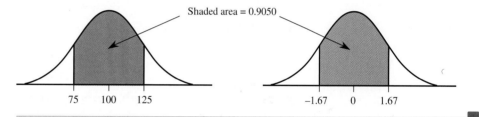

FIGURE 6.35

P(75 < x < 125) and corresponding z curve area

When translating a problem involving a normal distribution with mean μ and standard deviation σ to a problem involving the standard normal distribution, you convert to z-scores:

$$z = \frac{x - \mu}{\sigma}$$

Recall from Chapter 3 that a z-score can be interpreted as the distance of an x value from the mean in units of the standard deviation. A z-score of 1.4 corresponds to an x value that is 1.4 standard deviations above the mean, and a z-score of -2.1 corresponds to an x value that is 2.1 standard deviations below the mean.

Suppose that you are trying to evaluate $P(x < 60)$ for a variable whose distribution is normal with $\mu = 50$ and $\sigma = 5$. Converting the endpoint 60 to a z-score gives

$$z = \frac{60 - 50}{5} = 2$$

which tells you that the value 60 is two standard deviations above the mean. You then have

$$P(x < 60) = P(z < 2)$$

where z is a standard normal variable. Notice that for the standard normal distribution, which has $\mu = 0$ and $\sigma = 1$, the value 2 is two standard deviations above the mean. The value $z = 2$ is located the same distance (measured in standard deviations) from the mean of the standard normal distribution as is the value $x = 60$ from the mean in the normal distribution with $\mu = 50$ and $\sigma = 5$. This is why the translation using z-scores results in an equivalent problem involving the standard normal distribution.

Describing Extreme Values in a Normal Distribution

To describe the extreme values for a normal distribution with mean μ and standard deviation σ, first solve the corresponding problem for the standard normal distribution and then translate your answer back to the normal distribution of interest. This process is illustrated in Example 6.31.

Example 6.31 **Registration Times**

Data on the length of time (in minutes) required to register for classes using an online system suggest that the distribution of the variable

$$x = \text{time to register}$$

for students at a particular university can be well approximated by a normal distribution with mean $\mu = 12$ minutes and standard deviation $\sigma = 2$ minutes. Because some students do not log off properly, the university would like to log off students automatically after some amount of time has elapsed. The university has decided to choose this time so that only 1% of the students are logged off while they are still attempting to register.

To determine the amount of time that should be allowed before disconnecting a student, you need to describe the largest 1% of the distribution of time to register. These are the individuals who will be mistakenly disconnected. This is illustrated in Figure 6.36(a). To determine the value of x^*, first solve the analogous problem for the standard normal distribution, as shown in Figure 6.36(b).

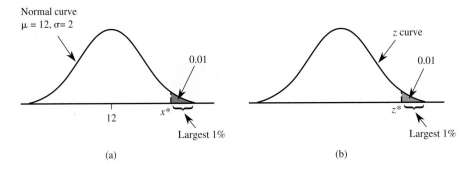

FIGURE 6.36
Capturing the largest 1% in a normal distribution

(a)   (b)

By looking for a cumulative area of 0.99 in Appendix Table 2, you find the closest entry (0.9901) in the 2.3 row and the .03 column, from which $z^* = 2.33$. For the standard normal distribution, the largest 1% of the distribution is made up of those values greater than 2.33. This implies that in the distribution of time to register x (or any other normal distribution), the largest 1% are those values with z-scores greater than 2.33 or, equivalently, those x values more than 2.33 standard deviations above the mean. Here, the standard deviation is 2, so 2.33 standard deviations is 2.33(2), and it follows that

$$x^* = 12 + 2.33(2) = 12 + 4.66 = 16.66$$

The largest 1% of the distribution for time to register is made up of values that are greater than 16.66 minutes. If the system was set to log off students after 16.66 minutes, only about 1% of the students registering would be logged off before completing their registration.

A general formula for converting a z-score back to an x value results from solving

$$z^* = \frac{x^* - \mu}{\sigma}$$

for x^*, as shown in the accompanying box.

To convert a z score z^* back to an x value, use

$$x^* = \mu + z^*\sigma$$

Example 6.32 Garbage Truck Processing Times

Garbage trucks entering a particular waste management facility are weighed before offloading garbage into a landfill. Data from the paper **"Estimating Waste Transfer Station Delays Using GPS"** (*Waste Management* **[2008]: 1742–1750**) suggest that a normal distribution with mean $\mu = 13$ minutes and $\sigma = 3.9$ minutes is a reasonable model for the probability distribution of the random variable $x =$ total processing time for a garbage truck at this waste management facility (total processing time includes waiting time as well). Suppose that you want to describe the total processing times of the trucks making up the 10% with the longest processing times. These trucks would be the 10% with times corresponding to the shaded region in the accompanying illustration.

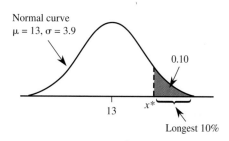

For the standard normal distribution, the largest 10% are those with z-scores greater than $z^* = 1.28$ (from Appendix Table 2, based on a cumulative area of 0.90).
Then

$$x^* = \mu + z^*\sigma$$
$$= 13 + 1.28(3.9)$$
$$= 13 + 4.992$$
$$= 17.992$$

About 10% of the garbage trucks using this facility would have a total processing time of more than 17.992 minutes.
The 5% with the fastest processing times would be those with z-scores less than $z^* = -1.645$ (from Appendix Table 2, based on a cumulative area of 0.05). Then

$$x^* = \mu + z^*\sigma$$
$$= 13 + (-1.645)(3.9)$$
$$= 13 - 6.416$$
$$= 6.584$$

About 5% of the garbage trucks processed at this facility will have total processing times of less than 6.584 minutes.

SECTION 6.6 EXERCISES

Each Exercise Set assesses the following chapter learning objectives: M7, M8, P1

SECTION 6.6 Exercise Set 1

6.64 Determine the following standard normal (z) curve areas:
a. The area under the z curve to the left of 1.75
b. The area under the z curve to the left of -0.68
c. The area under the z curve to the right of 1.20
d. The area under the z curve to the right of -2.82
e. The area under the z curve between -2.22 and 0.53
f. The area under the z curve between -1 and 1
g. The area under the z curve between -4 and 4

6.65 Let z denote a random variable that has a standard normal distribution. Determine each of the following probabilities:
a. $P(z < 2.36)$
b. $P(z \le 2.36)$
c. $P(z < -1.23)$
d. $P(1.14 < z < 3.35)$
e. $P(-0.77 \le z \le -0.55)$
f. $P(z > 2)$
g. $P(z \ge -3.38)$
h. $P(z < 4.98)$

6.66 Consider the population of all one-gallon cans of dusty rose paint manufactured by a particular paint company. Suppose that a normal distribution with mean $\mu = 5$ ml and standard deviation $\sigma = 0.2$ ml is a reasonable model for the distribution of the variable

$$x = \text{amount of red dye in the paint mixture}$$

Use the normal distribution to calculate the following probabilities:
a. $P(x < 5.0)$
b. $P(x < 5.4)$
c. $P(x \le 5.4)$
d. $P(4.6 < x < 5.2)$
e. $P(x > 4.5)$
f. $P(x > 4.0)$

6.67 Suppose that the distribution of typing speed in words per minute (wpm) for experienced typists using a new type of split keyboard can be approximated by a normal curve with mean 60 wpm and standard deviation 15 wpm (**"The Effects of Split Keyboard Geometry on Upper Body Postures, Ergonomics [2009]: 104–111).**
a. What is the probability that a randomly selected typist's speed is at most 60 wpm? Less than 60 wpm?
b. What is the probability that a randomly selected typist's speed is between 45 and 90 wpm?
c. Would you be surprised to find a typist in this population whose speed exceeded 105 wpm?
d. Suppose that two typists are independently selected. What is the probability that both their speeds exceed 75 wpm?

e. Suppose that special training is to be made available to the slowest 20% of the typists. What typing speeds would qualify individuals for this training?

6.68 A machine that cuts cork stoppers for bottles operates in such a way that the distribution of the diameter of the corks produced is well approximated by a normal distribution with mean 3 cm and standard deviation 0.1 cm. The specifications call for corks with diameters between 2.9 and 3.1 cm. A cork not meeting the specifications is considered defective. (A cork that is too small leaks and a cork that is too large doesn't fit in the bottle.) What proportion of corks produced by this machine are defective?

6.69 Refer to the previous exercise. Suppose that there are two machines available for cutting corks. The machine described in the preceding problem produces corks with diameters that are approximately normally distributed with mean 3 cm and standard deviation 0.1 cm. The second machine produces corks with diameters that are approximately normally distributed with mean 3.05 cm and standard deviation 0.01 cm. Which machine would you recommend? (Hint: Which machine would produce fewer defective corks?)

SECTION 6.6 Exercise Set 2

6.70 Determine each of the following areas under the standard normal (z) curve:
a. To the left of -1.28
b. To the right of 1.28
c. Between -1 and 2
d. To the right of 0
e. To the right of -5
f. Between -1.6 and 2.5
g. To the left of 0.23

6.71 Let z denote a random variable having a normal distribution with $\mu = 0$ and $\sigma = 1$. Determine each of the following probabilities:
a. $P(z < 0.10)$
b. $P(z < -0.10)$
c. $P(0.40 < z < 0.85)$
d. $P(-0.85 < z < -0.40)$
e. $P(-0.40 < z < 0.85)$
f. $P(z > -1.25)$
g. $P(z < -1.50 \text{ or } z > 2.50)$

6.72 According to the paper **"Commuters' Exposure to Particulate Matter and Carbon Monoxide in Hanoi, Vietnam"** (*Transportation Research* **[2008]: 206–211**), the carbon monoxide exposure of someone riding a motorbike for 5 km on a highway in Hanoi is approximately normally distributed with a mean of 18.6 ppm. Suppose that the standard deviation of carbon monoxide exposure is 5.7 ppm. Approximately what proportion of those who ride a motorbike for 5 km on a Hanoi highway will experience a carbon monoxide exposure of more than 20 ppm? More than 25 ppm?

6.73 A gasoline tank for a certain car is designed to hold 15 gallons of gas. Suppose that the random variable x = actual capacity of a randomly selected tank has a distribution that is well approximated by a normal curve with mean 15.0 gallons and standard deviation 0.1 gallon.
a. What is the probability that a randomly selected tank will hold at most 14.8 gallons?
b. What is the probability that a randomly selected tank will hold between 14.7 and 15.1 gallons?
c. If two such tanks are independently selected, what is the probability that both hold at most 15 gallons?

6.74 Consider babies born in the "normal" range of 37–43 weeks gestational age. The paper **"Fetal Growth Parameters and Birth Weight: Their Relationship to Neonatal Body Composition"** (*Ultrasound in Obstetrics and Gynecology* [2009]: 441–446) suggests that a normal distribution with mean μ = 3,500 grams and standard deviation σ = 600 grams is a reasonable model for the probability distribution of x = birth weight of a randomly selected full-term baby.
a. What is the probability that the birth weight of a randomly selected full-term baby exceeds 4,000 g? is between 3,000 and 4,000 g?
b. What is the probability that the birth weight of a randomly selected full-term baby is either less than 2,000 g or greater than 5,000 g?
c. What is the probability that the birth weight of a randomly selected full-term baby exceeds 7 pounds? (Hint: 1 lb = 453.59 g.)
d. How would you characterize the most extreme 0.1% of all full-term baby birth weights?

6.75 The time that it takes a randomly selected job applicant to perform a certain task has a distribution that can be approximated by a normal distribution with a mean of 120 seconds and a standard deviation of 20 seconds. The fastest 10% are to be given advanced training. What task times qualify individuals for such training?

Additional Exercises

6.76 The paper **"The Effect of Temperature and Humidity on Size of Segregated Traffic Exhaust Particle Emissions"** (*Atmospheric Environment* [2008]: 2369–2382) gave the following summary quantities for a measure of traffic flow (vehicles/second) during peak traffic hours. Traffic flow was recorded daily at a particular location over a long sequence of days.

Mean = 0.41 Standard Deviation = 0.26 Median = 0.45
5th percentile = 0.03   Lower quartile = 0.18
Upper quartile = 0.57   95th Percentile = 0.86

Based on these summary quantities, do you think that the distribution of the measure of traffic flow is approximately normal? Explain your reasoning.

6.77 A machine that produces ball bearings has initially been set so that the true average diameter of the bearings it produces is 0.500 inches. A bearing is acceptable if its diameter is within 0.004 inches of this target value. Suppose, however, that the setting has changed during the course of production, so that the distribution of the diameters produced is now approximately normal with mean 0.499 inch and standard deviation 0.002 inch. What percentage of the bearings produced will not be acceptable?

6.78 The paper referenced in Example 6.32 (**"Estimating Waste Transfer Station Delays Using GPS,"** *Waste Management* [2008]: 1742–1750) describing processing times for garbage trucks also provided information on processing times at a second facility. At this second facility, the mean total processing time was 9.9 minutes and the standard deviation of the processing times was 6.2 minutes. Explain why a normal distribution with mean 9.9 and standard deviation 6.2 would not be an appropriate model for the probability distribution of the variable x = total processing time of a randomly selected truck entering this second facility.

6.79 The number of vehicles leaving a highway at a certain exit during a particular time period has a distribution that is approximately normal with mean value 500 and standard deviation 75. What is the probability that the number of cars exiting during this period is
a. At least 650?
b. Strictly between 400 and 550? (*Strictly* means that the values 400 and 550 are not included.)
c. Between 400 and 550 (inclusive)?

6.80 A pizza company advertises that it puts 0.5 pound of real mozzarella cheese on its medium-sized pizzas. In fact, the amount of cheese on a randomly selected medium pizza is normally distributed with a mean value of 0.5 pound and a standard deviation of 0.025 pound.
a. What is the probability that the amount of cheese on a medium pizza is between 0.525 and 0.550 pound?
b. What is the probability that the amount of cheese on a medium pizza exceeds the mean value by more than 2 standard deviations?
c. What is the probability that three randomly selected medium pizzas each have at least 0.475 pound of cheese?

6.81 Suppose that fuel efficiency (miles per gallon, mpg) for a particular car model under specified conditions is normally distributed with a mean value of 30.0 mpg and a standard deviation of 1.2 mpg.

a. What is the probability that the fuel efficiency for a randomly selected car of this model is between 29 and 31 mpg?

b. Would it surprise you to find that the efficiency of a randomly selected car of this model is less than 25 mpg?

c. If three cars of this model are randomly selected, what is the probability that each of the three have efficiencies exceeding 32 mpg?

d. Find a number x^* such that 95% of all cars of this model have efficiencies exceeding x^* (i.e., $P(x > x^*) = 0.95$).

6.82 The amount of time spent by a statistical consultant with a client at their first meeting is a random variable that has a normal distribution with a mean value of 60 minutes and a standard deviation of 10 minutes.

a. What is the probability that more than 45 minutes is spent at the first meeting?

b. What amount of time is exceeded by only 10% of all clients at a first meeting?

6.83 When used in a particular DVD player, the lifetime of a certain brand of battery is normally distributed with a mean value of 6 hours and a standard deviation of 0.8 hour. Suppose that two new batteries are independently selected and put into the player. The player ceases to function as soon as one of the batteries fails.

a. What is the probability that the DVD player functions for at least 4 hours?

b. What is the probability that the DVD player functions for at most 7 hours?

c. Find a number x^* such that only 5% of all DVD players will function without battery replacement for more than x^* hours.

6.84 A machine producing vitamin E capsules operates so that the actual amount of vitamin E in each capsule is normally distributed with a mean of 5 mg and a standard deviation of 0.05 mg. What is the probability that a randomly selected capsule contains less than 4.9 mg of vitamin E? At least 5.2 mg of vitamin E?

6.85 The *Wall Street Journal* **(February 15, 1972)** reported that General Electric was sued in Texas for sex discrimination over a minimum height requirement of 5 feet, 7 inches. The suit claimed that this restriction eliminated more than 94% of adult females from consideration. Let x represent the height of a randomly selected adult woman. Suppose that x is approximately normally distributed with mean 66 inches (5 ft. 6 in.) and standard deviation 2 inches.

a. Is the claim that 94% of all women are shorter than 5 feet, 7 inches correct?

b. What proportion of adult women would be excluded from employment as a result of the height restriction?

6.86 Suppose that your statistics teacher tells you that the scores on a midterm exam were approximately normally distributed with a mean of 78 and a standard deviation of 7. The top 15% of all scores have been designated A's. Your score is 89. Did you receive an A? Explain.

6.87 Suppose that the pH of soil samples taken from a certain geographic region is normally distributed with a mean of 6.00 and a standard deviation of 0.10. The pH of a randomly selected soil sample from this region will be determined.

a. What is the probability that the resulting pH is between 5.90 and 6.15?

b. What is the probability that the resulting pH exceeds 6.10?

c. What is the probability that the resulting pH is at most 5.95?

d. What value will be exceeded by only 5% of all such pH values?

6.88 The light bulbs used to provide exterior lighting for a large office building have an average lifetime of 700 hours. If lifetime is approximately normally distributed with a standard deviation of 50 hours, how often should all the bulbs be replaced so that no more than 20% of the bulbs will have already burned out?

6.89 Let x denote the duration of a randomly selected pregnancy (the time elapsed between conception and birth). Accepted values for the mean value and standard deviation of x are 266 days and 16 days, respectively. Suppose that the probability distribution of x is approximately normal.

a. What is the probability that the duration of a randomly selected pregnancy is between 250 and 300 days?

b. What is the probability that the duration is at most 240 days?

c. What is the probability that the duration is within 16 days of the mean duration?

d. A "Dear Abby" column dated January 20, 1973, contained a letter from a woman who stated that the duration of her pregnancy was exactly 310 days. (She wrote that the last visit with her husband, who was in the navy, occurred 310 days before the birth of her child.) What is the probability that the duration of pregnancy is at least 310 days? Does this probability make you a bit skeptical of the claim?

e. Some insurance companies will pay the medical expenses associated with childbirth only if the insurance has been in effect for more than 9 months (275 days). This restriction is designed to ensure that benefits are only paid if conception occurred during coverage. Suppose that conception occurred 2 weeks after coverage began. What is the probability that the insurance company will refuse to pay benefits because of the 275-day requirement?

SECTION **6.7** Checking for Normality

Some of the most frequently used statistical methods are valid only when a sample is from a population distribution that is at least approximately normal. One way to see if the population distribution is approximately normal is to construct a **normal probability plot** of the data. This plot uses quantities called **normal scores**. The values of the normal scores depend on the sample size n. For example, the normal scores when $n = 10$ are as follows:

$$-1.539 \quad -1.001 \quad -0.656 \quad -0.376 \quad -0.123 \quad 0.123 \quad 0.376 \quad 0.656 \quad 1.001 \quad 1.539$$

To interpret these numbers, think of selecting sample after sample from a normal distribution, each one consisting of $n = 10$ observations. Then -1.539 is the long-run average z-score of the smallest observation from each sample, -1.001 is the long-run average z-score of the second smallest observation from each sample, and so on.

Tables of normal scores for many different sample sizes are available. Alternatively, many software packages (such as JMP and Minitab) and some graphing calculators can compute these scores and then use them to construct a normal probability plot. Not all calculators and software packages use the same algorithm to compute normal scores. However, this does not change the overall character of a normal probability plot, so either tabulated values or those given by the computer or calculator can be used.

To construct a normal probability plot, first arrange the sample observations from smallest to largest. The smallest normal score is then paired with the smallest observation, the second smallest normal score with the second smallest observation, and so on. The first number in a pair is the normal score, and the second number in the pair is the corresponding data value. A normal probability plot is just a scatterplot of the (normal score, observed value) pairs.

If the sample has been selected from a *standard* normal distribution, the second number in each pair should be reasonably close to the first number (ordered observation ≈ corresponding mean value). If this is the case, the n plotted points will fall near a line with slope equal to 1 (a 45° line) passing through (0, 0). In general, when the sample is from a normal population distribution, the plotted points should be close to *some* straight line (but not necessarily one with slope 1 and intercept 0).

> **DEFINITION**
>
> A **normal probability** plot is a scatterplot of the (normal score, observed value) pairs. A strong linear pattern in a normal probability plot suggests that the population distribution is approximately normal. On the other hand, systematic departure from a straight-line pattern (such as curvature in the plot) suggests that the population distribution is not normal.
>
> Some software packages and graphing calculators plot the (observed value, normal score) pairs. This is also an acceptable way to construct a normal probability plot.

AP* EXAM TIP

In Chapters 12 and 13, you will see some methods for learning from data that can be used when you think that the population distribution is approximately normal. Constructing a normal probability plot of the sample data is one way to determine if this is reasonable.

Example 6.33 Egg Weights

The following data represent egg weights (in grams) for a sample of 10 eggs. These data are consistent with summary quantities in the paper **"Evaluation of Egg Quality Traits of Chickens Reared under Backyard System in Western Uttar Pradesh"** (*Indian Journal of Poultry Science*, 2009).

53.04 53.50 52.53 53.00 53.07 52.86 52.66 53.23 53.26 53.16

Arranging the sample observations in order from smallest to largest results in

52.53 52.66 52.86 53.00 53.04 53.07 53.16 53.23 53.26 53.50

Pairing these ordered observations with the normal scores for a sample of size 10 (given at the beginning of this section) results in the following 10 ordered pairs that can be used to construct the normal probability plot:

(−1.539, 52.53)	(−1.001, 52.66)	(−0.656, 52.86)	(−0.376, 53.00)
(−0.123, 53.04)	(0.123, 53.07)	(0.376, 53.16)	(0.656, 53.23)
(1.001, 53.26)	(1.539, 53.50)		

The normal probability plot is shown in Figure 6.37. The linear pattern in the plot suggests that the population egg-weight distribution is approximately normal.

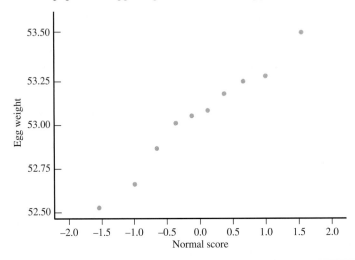

FIGURE 6.37
A normal probability plot for the egg weight data of Example 6.33

How do you know if a normal probability plot shows a strong linear pattern? Sometimes it is a matter of opinion. Particularly when n is small, normality should not be ruled out unless the pattern is clearly nonlinear. Figure 6.36 displays several plots for population distributions that are not normal. Figure 6.38(a) shows a pattern that results when the population distribution is skewed. The pattern in Figure 6.38(b) is one that occurs when the population distribution is roughly symmetric but flatter and not as bell shaped as the normal distribution. Figure 6.38(c) shows what the normal probability plot might look like if there is an outlier in a small data set. Any of these patterns suggest that the population distribution may not be normal.

FIGURE 6.38
Plots suggesting a population distribution is not normal:
(a) indication that the population distribution is skewed;
(b) indication that the population distribution has heavier tails than a normal distribution;
(c) indication of an outlier.

Using the Correlation Coefficient to Check Normality

The correlation coefficient r was introduced in Chapter 4 as a way to measure how closely the points in a scatterplot fall to a straight line. Consider calculating the value of r using the n (normal score, observed value) pairs:

(smallest normal score, smallest observation)

⋮

(largest normal score, largest observation)

The normal probability plot always slopes upward (because it is based on values ordered from smallest to largest), so r will be a positive number. A value of r quite close to 1 indicates a very strong linear relationship in the normal probability plot. If r is much smaller than 1, the population distribution is not likely to be normal.

How much smaller than 1 does r have to be before you should begin to seriously doubt normality? The answer depends on the sample size n. If n is small, an r value somewhat below 1 is not surprising, even when the population distribution is normal. However, if n is large, only an r value very close to 1 supports normality. For selected values of n, Table 6.2 gives critical values that can be used to check for normality. If the sample size is in between two tabulated values of n, use the critical value for the larger sample size. (For example, if $n = 46$, use the value 0.966 given for sample size 50.)

TABLE **6.2**   **Critical Values for r***

n	5	10	15	20	25	30	40	50	60	75
Critical r	0.832	0.880	0.911	0.929	0.941	0.949	0.960	0.966	0.971	0.976

*Source: *MINITAB User's Manual.*

If

$r <$ critical r for the corresponding n

it is not reasonable to think that the population distribution is normal.

How were the critical values in Table 6.2, such as the critical value 0.941 for $n = 25$, obtained? Consider selecting a large number of different random samples, each one consisting of 25 observations from a normally distributed population, then using the 25 (normal score, observed value) pairs to compute the value of r for each sample. Only about 1% of these samples would result in an r value less than the critical value 0.941. That is, 0.941 was chosen to guarantee a 1% error rate: In only 1% of all cases will you think normality is not reasonable when the distribution really is normal. The critical values for other sample sizes were also chosen to yield a 1% error rate.

Another type of error is also possible: obtaining a large value of r and concluding that normality is reasonable when the distribution is actually not normal. This type of error is more difficult to control than the type mentioned previously, but the procedure described here generally does a good job in controlling for both types of error.

Example 6.34   **Egg Weights Revisited**

The sample size for the egg-weight data of Example 6.33 is $n = 10$. The critical r for $n = 10$ from Table 6.2 is 0.880. From Minitab, the correlation coefficient calculated using the (normal score, observed value) pairs is $r = 0.986$. Because r is larger than the critical r for a sample of size 10, it is reasonable to think that the population distribution of egg weights is approximately normal.

Correlations: Egg Weight, Normal Score

Pearson correlation of Egg Weight and Normal Score = 0.986

Each Exercise Set assesses the following chapter learning objectives: M9, M10

SECTION 6.7 **Exercise Set 1**

6.90 The authors of the paper "Development of Nutritionally At-Risk Young Children Is Predicted by Malaria, Anemia, and Stunting in Pemba, Zanzibar" (*The Journal of Nutrition* [2009]: 763–772) studied factors that might be related to dietary deficiencies in children. Children were observed for a fixed period of time, and the amount of time spent in various activities was recorded. One variable of interest was the amount of time (in minutes) a child spent fussing. Data for 15 children, consistent with summary quantities in the paper, are given in the accompanying table. Normal scores for a sample size of 15 are also given.

Fussing Time (x)	Normal Score
0.05	−1.739
0.10	−1.245
0.15	−0.946
0.40	−0.714
0.70	−0.515
1.05	−0.335
1.95	−0.165
2.15	0.000
3.70	0.165
3.90	0.335
4.50	0.515
6.00	0.714
8.00	0.946
11.00	1.245
14.00	1.739

a. Construct a normal probability plot for the fussing time data. Does the plot look linear? Do you agree with the authors of the paper that the fussing time distribution is not normal?
b. Compute the correlation coefficient for the (normal score, x) pairs. Compare this value to the critical r value from Table 6.2 to determine if it is reasonable to think that the fussing time distribution is approximately normal.

6.91 The paper "Risk Behavior, Decision Making, and Music Genre in Adolescent Males" (Marshall University, May 2009) examined the effect of type of music playing and performance on a risky, decision-making task.
a. Participants in the study responded to a questionnaire used to assign a risk behavior score. Risk behavior scores (from a graph in the paper) for 15 participants follow. Use these data to construct a normal probability plot (the normal scores for a sample of size 15 appear in the previous exercise).

102 105 113 120 125 127 134 135
139 141 144 145 149 150 160

b. Participants also completed a positive and negative affect scale (PANAS) designed to measure emotional response to music. PANAS values (from a graph in the paper) for 15 participants follow. Use these data to construct a normal probability plot (the normal scores for a sample of size 15 appear in the previous exercise).

36 40 45 47 48 49 50 52
53 54 56 59 61 62 70

c. The author of the paper stated that it was reasonable to consider both risk behavior scores and PANAS scores to be approximately normally distributed. Do the normal probability plots from Parts (a) and (b) support this conclusion? Explain.

SECTION 6.7 **Exercise Set 2**

6.92 Consider the following 10 observations on the lifetime (in hours) for a certain type of power supply:

152.7 172.0 172.5 173.3 193.0
204.7 216.5 234.9 262.6 422.6

Construct a normal probability plot, and comment on whether it is reasonable to think that the distribution of power supply lifetime is approximately normal. (The normal scores for a sample of size 10 are −1.539, −1.001, −0.656, −0.376, −0.123, 0.123, 0.376, 0.656, 1.001, and 1.539.)

6.93 Fat contents (x, in grams) for seven randomly selected hot dog brands that were rated as very good by **Consumer Reports (www.consumerreports.org)** are shown below.

14 15 11 10 6 15 16

The normal scores for a sample of size 7 are

−1.364 −0.758 −0.353 0 0.353 0.758 1.364

a. Construct a normal probability plot for the fat content data. Does the plot look linear?
b. Compute the correlation coefficient for the (normal score, x) pairs. Compare this value to the critical r value from Table 6.2 to determine if it is reasonable to think that the fat content distribution is approximately normal.

Additional Exercises

6.94 An automobile manufacturer is interested in the fuel efficiency of a proposed new car design. Six nonprofessional drivers were selected, and each one drove a prototype of the new car from Phoenix to Los Angeles. The resulting fuel efficiencies (x, in miles per gallon) are:

27.2 29.3 31.2 28.4 30.3 29.6

The normal scores for a sample of size 6 are

−1.282 −0.643 −0.202 0.202 0.643 1.282

a. Construct a normal probability plot for the fuel efficiency data. Does the plot look linear?

b. Compute the correlation coefficient for the (normal score, x) pairs. Compare this value to the critical r value from Table 6.2 to determine if it is reasonable to think that the fuel efficiency distribution is approximately normal.

6.95 The paper "The Load-Life Relationship for M50 Bearings with Silicon Nitride Ceramic Balls" (*Lubrication Engineering* [1984]: 153–159) reported the accompanying data on $x =$ bearing load life (in millions of revolutions). The corresponding normal scores are also given.

x	Normal Score	x	Normal Score
47.1	−1.867	240.0	0.062
68.1	−1.408	240.0	0.187
68.1	−1.131	278.0	0.315
90.8	−0.921	278.0	0.448
103.6	−0.745	289.0	0.590
106.0	−0.590	289.0	0.745
115.0	−0.448	367.0	0.921
126.0	−0.315	385.9	1.131
146.6	−0.187	392.0	1.408
229.0	−0.062	395.0	1.867

a. Construct a normal probability plot.

b. Compute the correlation coefficient for the (normal score, x) pairs. Compare this value to the critical r value from Table 6.2 to determine if it is reasonable to think that the distribution of bearing load life is approximately normal.

6.96 Consider the following sample of 25 observations on $x =$ diameter (in centimeters) of CD disks produced by a particular manufacturer:

16.01	16.08	16.13	15.94	16.05	16.27	15.89
15.84	15.95	16.10	15.92	16.04	15.82	16.15
16.06	15.66	15.78	15.99	16.29	16.15	16.19
16.22	16.07	16.13	16.11			

The 13 largest normal scores for a sample of size 25 are 1.965, 1.524, 1.263, 1.067, 0.905, 0.764, 0.637, 0.519, 0.409, 0.303, 0.200, 0.100, and 0. The 12 smallest scores result from placing a negative sign in front of each of the given nonzero scores. Construct a normal probability plot. Is it reasonable to think that disk diameter is normally distributed? Explain.

SECTION 6.8 Using the Normal Distribution to Approximate a Discrete Distribution

In this section, you will see how probabilities for some discrete random variables can be approximated using a normal curve. The most important case of this is the approximation of binomial probabilities.

The Normal Curve and Discrete Variables

The probability distribution of a discrete random variable x is represented graphically by a probability histogram. The probability of a particular value is the area of the rectangle centered at that value. Possible values of x are isolated points on the number line, usually whole numbers. For example, if $x =$ IQ of a randomly selected 8-year-old child, then x is a discrete random variable, because an IQ score must be a whole number.

Often, a probability histogram can be well approximated by a normal curve, as illustrated in Figure 6.39. In such cases, the distribution of x is said to be approximately normal. A normal distribution can then be used to calculate approximate probabilities of events involving x.

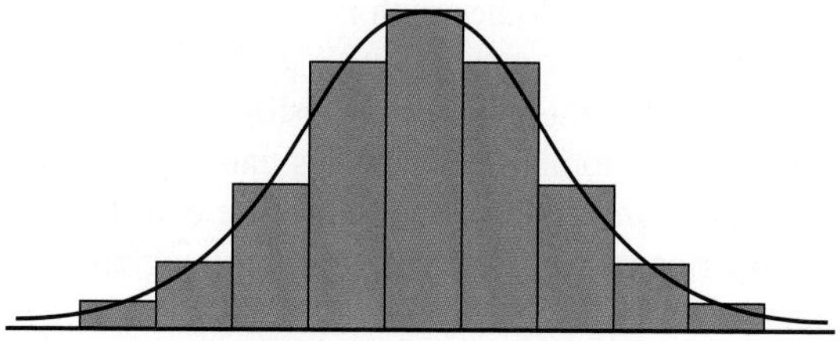

FIGURE 6.39
A normal curve approximation to a probability histogram

Example 6.35 **Express Mail Packages**

The number of express mail packages mailed at a certain post office on a randomly selected day is approximately normally distributed with mean 18 and standard deviation 6. Let's first calculate the approximate probability that $x = 20$. Figure 6.40(a) shows a portion of the probability histogram for x with the approximating normal curve superimposed. The area of the shaded rectangle is $P(x = 20)$. The left edge of this rectangle is at 19.5 on the horizontal scale, and the right edge is at 20.5. Therefore, the desired probability is approximately the area under the normal curve between 19.5 and 20.5. Standardizing these limits gives

$$\frac{20.5 - 18}{6} = 0.42 \qquad \frac{19.5 - 18}{6} = 0.25$$

so

$$P(x = 20) \approx P(0.25 < z < 0.42) = 0.6628 - 0.5987 = 0.0641$$

Figure 6.40(b) shows that $P(x \le 10)$ is approximately the area under the normal curve to the left of 10.5. It follows that

$$P(x \le 10) \approx P\left(z \le \frac{10.5 - 18}{6}\right) = P(z \le -1.25) = 0.1056$$

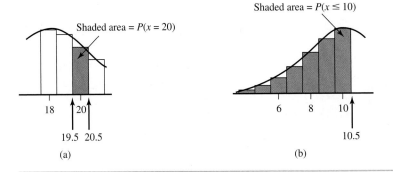

FIGURE 6.40
The normal approximation for
Example 6.35

The calculation of probabilities in Example 6.35 illustrates the use of what is known as a **continuity correction**. Because the rectangle for $x = 10$ extends to 10.5 on the right, the normal curve area to the left of 10.5 rather than 10 is used. In general, if possible x values are consecutive whole numbers, then $P(a \le x \le b)$ will be approximately the normal curve area between limits $a - \frac{1}{2}$ and $b + \frac{1}{2}$. Similarly, $P(a < x < b)$ will be approximately the area between $a + \frac{1}{2}$ and $b - \frac{1}{2}$.

Normal Approximation to a Binomial Distribution

Figure 6.41 shows the probability histograms for two binomial distributions, one with $n = 25$, $p = 0.4$, and the other with $n = 25$, $p = 0.1$. For each distribution, $\mu = np$ and $\sigma = \sqrt{np(1 - p)}$ were computed and a normal curve with this μ and σ was superimposed. A normal curve fits the probability histogram well in the first case (Figure 6.40(a)). When this happens, binomial probabilities can be accurately approximated by areas under the normal curve. In the second case (Figure 6.40(b)), the normal curve does not give a good approximation because the probability histogram is skewed, whereas the normal curve is symmetric.

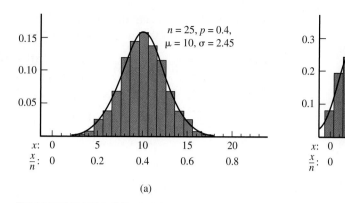

FIGURE 6.41
Normal approximations to binomial distributions

Suppose x is a binomial random variable based on n trials and success probability p, so that

$$\mu = np \text{ and } \sigma = \sqrt{np(1 - p)}$$

If n and p are such that

$$np \geq 10 \text{ and } n(1 - p) \geq 10$$

Then the distribution of x is approximately normal.

Combining this result with the continuity correction implies that

$$P(a \leq x \leq b) \approx P\left(\frac{a - \frac{1}{2} - \mu}{\sigma} \leq z \leq \frac{b + \frac{1}{2} - \mu}{\sigma}\right)$$

That is, the probability that x is between a and b inclusive is approximately the area under the approximating normal curve between $a - \frac{1}{2}$ and $b + \frac{1}{2}$.

Similarly,

$$P(x \leq b) \approx P\left(z \leq \frac{b + \frac{1}{2} - \mu}{\sigma}\right)$$

$$P(a \leq x) \approx P\left(\frac{a - \frac{1}{2} - \mu}{\sigma} \leq z\right)$$

When either $np < 10$ or $n(1 - p) < 10$, the binomial distribution is too skewed for the normal approximation to give accurate probability estimates.

Example 6.36 Premature Babies

Premature babies are those born before 37 weeks, and those born before 34 weeks are most at risk. The paper **"Some Thoughts on the True Value of Ultrasound"** (*Ultrasound in Obstetrics and Gynecology* **[2007]: 671–674**) reported that 2% of births in the United States occur before 34 weeks. Suppose that 1,000 births will be randomly selected and that the value of

$$x = \text{the number of these births that occurred prior to 34 weeks}$$

is to be determined. Because

$$np = 1{,}000\,(0.02) = 20 \geq 10$$
$$n(1 - p) = 1{,}000\,(0.98) = 980 \geq 10$$

the distribution of x is approximately normal with

$$\mu = np = 1{,}000(0.02) = 20$$
$$\sigma = \sqrt{np(1 - p)} = \sqrt{1{,}000(0.02)(0.98)} = \sqrt{19.60} = 4.427$$

You can now approximate the probability that the number of babies born prior to 34 weeks in a sample of 1,000 will be between 10 and 25 (inclusive):

$$P(10 \leq x \leq 25) \approx P\left(\frac{9.5 - 20}{4.427} \leq z \leq \frac{25.5 - 20}{4.427}\right)$$

$$= P(-2.37 \leq z \leq 1.24)$$

$$= 0.8925 - 0.0089$$

$$= 0.8836$$

as shown in the following figure:

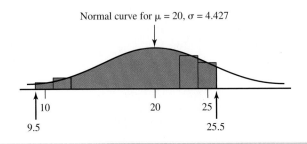

Normal curve for $\mu = 20$, $\sigma = 4.427$

SECTION 6.8 EXERCISES

Each Exercise Set assesses the following chapter learning objectives: M14

SECTION 6.8 Exercise Set 1

6.97 Let x denote the IQ of an individual selected at random from a certain population. The value of x must be a whole number. Suppose that the distribution of x can be approximated by a normal distribution with mean value 100 and standard deviation 15. Approximate the following probabilities:

a. $P(x = 100)$
b. $P(x \leq 110)$
c. $P(x < 110)$ (Hint: $x < 110$ is the same as $x \leq 109$.)
d. $P(75 \leq x \leq 125)$

6.98 Seventy percent of the bicycles sold by a certain store are mountain bikes. Among 100 randomly selected bike purchases, what is the approximate probability that

a. At most 75 are mountain bikes?
b. Between 60 and 75 (inclusive) are mountain bikes?
c. More than 80 are mountain bikes?
d. At most 30 are not mountain bikes?

6.99 Suppose that 65% of all registered voters in a certain area favor a seven-day waiting period before purchase of a handgun. Among 225 randomly selected registered voters, what is the probability that

a. At least 150 favor such a waiting period?
b. More than 150 favor such a waiting period?
c. Fewer than 125 favor such a waiting period?

SECTION 6.8 Exercise Set 2

6.100 The distribution of the number of items produced by an assembly line during an 8-hour shift can be approximated by a normal distribution with mean value 150 and standard deviation 10.

a. What is the probability that the number of items produced is at most 120?
b. What is the probability that at least 125 items are produced?
c. What is the probability that between 135 and 160 (inclusive) items are produced?

6.101 Suppose that 25% of the fire alarms in a large city are false alarms. Let x denote the number of false alarms in a random sample of 100 alarms. Approximate the following probabilities:

a. $P(20 \leq x \leq 30)$
b. $P(20 < x < 30)$
c. $P(x \geq 35)$
d. The probability that x is farther than 2 standard deviations from its mean value.

6.102 A company that manufactures mufflers for cars offers a lifetime warranty on its products, provided that ownership of the car does not change. Only 20% of its mufflers are replaced under this warranty.

a. In a random sample of 400 purchases, what is the approximate probability that between 75 and 100 (inclusive) mufflers are replaced under warranty?
b. Among 400 randomly selected purchases, what is the probability that at most 70 mufflers are replaced under warranty?
c. If you were told that fewer than 50 among 400 randomly selected purchases were replaced under warranty, would you question the 20% figure? Explain.

Additional Exercises

6.103 A symptom validity test (SVT) is sometimes used to confirm diagnosis of psychiatric disorders. The paper **"Developing a Symptom Validity Test for Posttraumatic Stress Disorder: Application of the Binomial Distribution"** (*Journal of Anxiety Disorders* [2008]: 1297–1302) investigated the use of SVTs in the diagnosis of post-traumatic stress disorder. One SVT proposed is a 60-item test (called the MENT test), where each item has only a correct or incorrect response. The MENT test is designed so that responses to the individual questions can be considered independent of one another. For this reason, the authors of the paper believe that the score on the MENT test can be viewed as a binomial random variable with $n = 60$. The MENT test is designed to help in distinguishing fictitious claims of post-traumatic stress disorder. The items on the test are written so that the correct response to an item should be relatively obvious, even to people suffering from stress disorders. Researchers have found that a patient with a fictitious claim of stress disorder will try to "fake" the test, and that the probability of a correct response to an item for these patients is 0.7 (compared to 0.96 for other patients). The authors used a normal approximation to the binomial distribution with $n = 60$ and $p = 0.70$ to compute various probabilities of interest, where x = number of correct responses on the MENT test for a patient who is trying to fake the test.

a. Verify that it is appropriate to use a normal approximation to the binomial distribution in this situation.

b. Approximate the following probabilities:
   i. $P(x = 42)$
   ii. $P(x < 42)$
   iii. $P(x \leq 42)$

c. Explain why the probabilities computed in Part (b) are not all equal.

d. The authors computed the exact binomial probability of a score of 42 or less for someone who is *not* faking the test. Using $p = 0.96$, they found

$$P(x \leq 42) = .000000000013$$

Explain why the authors computed this probability using the binomial formula rather than using a normal approximation.

e. The authors propose that someone who scores 42 or less on the MENT exam is faking the test. Explain why this is reasonable, using some of the probabilities from Parts (b) and (d) as justification.

6.104 Studies have found that women diagnosed with cancer in one breast also sometimes have cancer in the other breast that was not initially detected by mammogram or physical examination (**"MRI Evaluation of the Contralateral Breast in Women with Recently Diagnosed Breast Cancer,"** *The New England Journal of Medicine* [2007]: 1295–1303). To determine if magnetic resonance imaging (MRI) could detect missed tumors in the other breast, 969 women diagnosed with cancer in one breast had an MRI exam. The MRI detected tumors in the other breast in 30 of these women.

a. Use $p = \dfrac{30}{969} = 0.031$ as an estimate of the probability that a woman diagnosed with cancer in one breast has an undetected tumor in the other breast. Consider a random sample of 500 women diagnosed with cancer in one breast. Explain why it is reasonable to think that the random variable x = number in the sample who have an undetected tumor in the other breast has a binomial distribution with $n = 500$ and $p = 0.031$.

b. Is it reasonable to use the normal distribution to approximate probabilities for the random variable x defined in Part (a)? Explain why or why not.

c. Approximate the following probabilities:
   i. $P(x < 10)$
   ii. $P(10 \leq x \leq 25)$
   iii. $P(x > 20)$

d. For each of the probabilities computed in Part (c), write a sentence interpreting the probability.

CHAPTER ACTIVITIES

ACTIVITY 6.1   IS IT REAL?

Background: Three students were asked to complete the following:

a. Flip a coin 30 times, and note the number of heads observed in the 30 flips.

b. Repeat Step (a) 99 more times, to obtain 100 observations of the random variable x = number of heads in 30 flips.

c. Construct a dotplot of the 100 values of x.

Because this was a tedious assignment, one or more of the three did not really carry out the coin flipping and just made up 100 values of x that they thought would look "real." The three dotplots produced by these students are shown here.

1. Which students made up x values? What about the dotplot makes you think the student did not actually do the coin flipping?

2. Working as a group, each student in your class should flip a coin 30 times and note the number of heads in the 30 tosses. If there are fewer than 50 students in the class, each student should repeat this process until there are a total of at least 50 observations of x = number of heads in 30 flips. Using the data from the entire class, construct a dotplot of the x values.

3. After looking at the dotplot in Step 2 that resulted from actually flipping a coin 30 times and observing number of heads, review your answers in Question 1. For each of the three students, explain why you now think that he or she did or did not actually do the coin flipping.

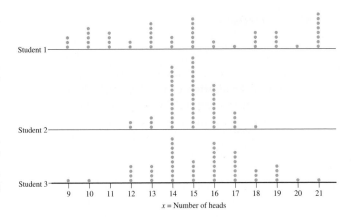

ARE YOU READY TO MOVE ON?   CHAPTER 6   REVIEW EXERCISES

All chapter learning objectives are assessed in these exercises. The learning objectives assessed in each exercise are given in parentheses.

6.105   (M1)

A point is randomly selected on the surface of a lake that has a maximum depth of 100 feet. Let x be the depth of the lake at the randomly chosen point. What are possible values of x? Is x discrete or continuous?

6.106   (M2)

A box contains five slips of paper, marked $1, $1, $1, $10, and $25. The winner of a contest selects two slips of paper at random and then gets the larger of the dollar amounts on the two slips. Define a random variable w by w = amount awarded. Determine the probability distribution of w. (Hint: Think of the slips as numbered 1, 2, 3, 4, and 5. An outcome of the experiment will consist of two of these numbers.)

6.107   (M2, P1)

Airlines sometimes overbook flights. Suppose that for a plane with 100 seats, an airline takes 110 reservations. Define the random variable x as

x = the number of people who actually show up for a sold-out flight on this plane

From past experience, the probability distribution of x is given in the following table:

x	$p(x)$	x	$p(x)$
95	0.05	103	0.03
96	0.10	104	0.02
97	0.12	105	0.01
98	0.14	106	0.005
99	0.24	107	0.005
100	0.17	108	0.005
101	0.06	109	0.0037
102	0.04	110	0.0013

a. What is the probability that the airline can accommodate everyone who shows up for the flight?

b. What is the probability that not all passengers can be accommodated?

c. If you are trying to get a seat on such a flight and you are number 1 on the standby list, what is the probability that you will be able to take the flight? What if you are number 3?

6.108   (C1, C2)

Suppose that the random variable

x = actual weight (in ounces) of a randomly selected package of cereal

has the probability distribution described by the density curve pictured here.

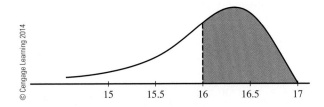

a. What probability is represented by the shaded area?

b. Suppose the shaded area = 0.50. Interpret this probability in the context of this problem.

6.109   (M6)

Let x be the amount of time (in minutes) that a particular San Francisco commuter will have to wait for a BART train. Suppose that the density curve is as pictured (a uniform distribution):

a. What is the probability that x is less than 10 minutes? More than 15 minutes?
b. What is the probability that x is between 7 and 12 minutes?
c. Find the value c for which $P(x < c) = 0.9$.

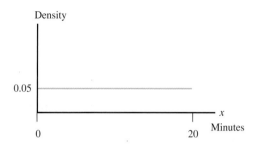

6.110 (M5)

Referring to the previous exercise, let x and y be waiting times on two independently selected days. Define a new random variable w by $w = x + y$, the sum of the two waiting times. The set of possible values for w is the interval from 0 to 40 (because both x and y can range from 0 to 20). It can be shown that the density curve of w is as pictured (this curve is called a triangular distribution, for obvious reasons!):

a. Shade the area under the density curve that corresponds to $P(w > 35)$.

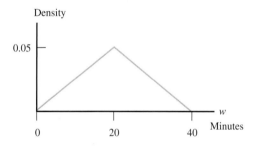

b. Put the following probabilities in order from smallest to largest:

$P(w < 10)$ $P(20 < w < 25)$ $P(25 < w < 30)$ $P(w > 30)$

Explain your reasoning.

6.111 (C2)

A company receives light bulbs from two different suppliers. Define the variables x and y as

x = lifetime of a bulb from Supplier 1
y = lifetime of a bulb from Supplier 2

Five hundred bulbs from each supplier are tested, and the lifetime of each bulb (in hours) is recorded. The density histograms at the bottom of the page are constructed from these two sets of observations. Although these histograms are constructed using data from only 500 bulbs, they can be considered approximations to the corresponding probability distributions.

a. Which probability distribution has the larger mean?
b. Which probability distribution has the larger standard deviation?
c. Assuming that the cost of the light bulbs is the same for both suppliers, which supplier would you recommend? Explain.
d. One of the two distributions pictured has a mean of approximately 1,000, and the other has a mean of about 900. Which of these is the mean of the distribution for the variable x (lifetime for a bulb from Supplier 1)?
e. One of the two distributions pictured has a standard deviation of approximately 100, and the other has a standard deviation of about 175. Which of these is the standard deviation of the distribution for the variable x (lifetime for a bulb from Supplier 1)?

6.112 (M4)

The probability distribution of x, the number of defective tires on a randomly selected automobile checked at a certain inspection station, is given in the following table:

x	0	1	2	3	4
$p(x)$	0.54	0.16	0.06	0.04	0.20

a. Calculate the mean value of x.
b. Interpret the mean value of x in the context of a long sequence of observations of number of defective tires.
c. What is the probability that x exceeds its mean value?
d. Calculate the standard deviation of x.

HISTOGRAMS FOR EXERCISE 6.111

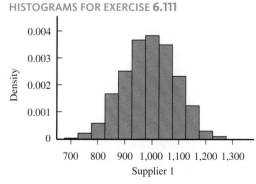

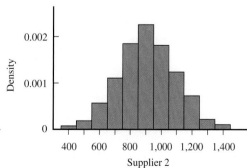

6.113   (M7, M8)

The size of the left upper chamber of the heart is one measure of cardiovascular health. When the upper left chamber is enlarged, the risk of heart problems is increased. The paper **"Left Atrial Size Increases with Body Mass Index in Children"** (*International Journal of Cardiology* [2009]: 1–7) described a study in which the left atrial size was measured for a large number of children ages 5 to 15 years. Based on these data, the authors concluded that for healthy children, left atrial diameter was approximately normally distributed with a mean of 26.4 mm and a standard deviation of 4.2 mm.

a. Approximately what proportion of healthy children have left atrial diameters less than 24 mm?

b. Approximately what proportion of healthy children have left atrial diameters greater than 32 mm?

c. Approximately what proportion of healthy children have left atrial diameters between 25 and 30 mm?

d. What is the value for which only about 20% of healthy children have a larger left atrial diameter?

6.114   (M7, M8)

The paper referenced in the previous exercise also included data on left atrial diameter for children who were considered overweight. For these children, left atrial diameter was approximately normally distributed with a mean of 28 mm and a standard deviation of 4.7 mm.

a. Approximately what proportion of overweight children have left atrial diameters less than 25 mm?

b. Approximately what proportion of overweight children have left atrial diameters greater than 32 mm?

c. Approximately what proportion of overweight children have left atrial diameters between 25 and 30 mm?

d. What proportion of overweight children have left atrial diameters greater than the mean for healthy children?

6.115   (M7, M8)

Consider the variable x = time required for a student to complete a standardized exam. Suppose that for the population of students at a particular school, the distribution of x is well approximated by a normal curve with mean 45 minutes and standard deviation 5 minutes.

a. If 50 minutes is allowed for the exam, what proportion of students at this school would be unable to finish in the allotted time?

b. How much time should be allowed for the exam if you wanted 90% of the students taking the test to be able to finish in the allotted time?

c. How much time is required for the fastest 25% of all students to complete the exam?

6.116   (P1)

Example 6.14 gave the probability distributions shown below for

   x = number of flaws in a randomly selected glass panel from Supplier 1

   y = number of flaws in a randomly selected glass panel from Supplier 2

for two suppliers of glass used in the manufacture of flat screen TVs. If the manufacturer wanted to select a single supplier for glass panels, which of these two suppliers would you recommend? Justify your choice based on consideration of both center and variability.

x	0	1	2	3	y	0	1	2	3
$p(x)$	0.4	0.3	0.2	0.1	$p(y)$	0.2	0.6	0.2	0

6.117   (M9)

The following normal probability plot was constructed using some of the data appearing in the paper **"Trace Metals in Sea Scallops"** (*Environmental Concentration and Toxicology* 19: 1326–1334).

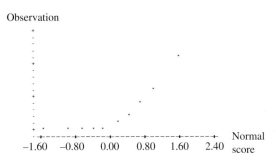

The variable under study was the cadmium concentration in North Atlantic scallops. Do the sample data suggest that the cadmium concentration distribution is not normal? Explain.

6.118   (M10)

Macular degeneration is the most common cause of blindness in people older than 60 years. One variable thought to be related to a type of inflammation associated with this disease is level of a substance called soluble Fas Ligand (sFasL) in the blood. The accompanying data are representative values of x = sFasL level for 10 patients with age-related macular degeneration. These data are consistent with summary quantities and descriptions of the data given in the paper **"Associations of Plasma-Soluble Fas Ligand with Aging and Age-Related Macular Degeneration"** (*Investigative Ophthalmology & Visual Science* [2008]: 1345–1349).

x	Normal Score
0.069	−1.539
0.074	−1.001
0.176	−0.656
0.185	−0.376
0.216	−0.123
0.287	0.123
0.343	0.376
0.343	0.656
0.512	1.001
0.729	1.539

a. Construct a normal probability plot. Does the normal probability plot appear linear or curved?

b. Compute the correlation coefficient for the (normal score, x) pairs. Compare this value to the critical r value from Table 6.2 to determine if it is reasonable to consider the distribution of sFasL levels to be approximately normal.

6.119 (M15) An appliance dealer sells three different models of upright freezers having 13.5, 15.9, and 19.1 cubic feet of storage space. Let x = the amount of storage space

purchased by the next customer to buy a freezer. Suppose that x has the following probability distribution:

x	13.5	15.9	19.1
$p(x)$	0.2	0.5	0.3

a. Calculate the mean and standard deviation of x.

b. If the price of the freezer depends on the size of the storage space, x, such that $Price = 25x - 8.5$, what is the mean value of the variable $Price$ paid by the next customer?

c. What is the standard deviation of the price paid?

TECHNOLOGY NOTES

Finding Normal Probabilities

TI-83/84
$P(x_1 < X < x_2)$
1. Press the **2nd** key then press the **VARS** key
2. Highlight **normalcdf(** and press **ENTER**
3. Type in the lower bound x_1
4. Type,
5. Type in the upper bound x_2
6. Type,
7. Type in the value of the mean
8. Type,
9. Type in the value of the standard deviation
10. Type **)**
11. Press **ENTER**

Note: If you are looking for $P(X < x)$, use a lower bound of $-10,000,000$. If you are looking for $P(X > x)$, use an upper bound of $10,000,000$.

TI-Nspire
$P(x_1 < X < x_2)$
1. Enter the Scratchpad
2. Press the **menu** key and select **5:Probablity** then **5:Distributions** then **2:Normal Cdf...** then press **enter**
3. In the box next to **Lower Bound** type the value for x_1
4. In the box next to **Upper Bound** type the value for x_2
5. In the box next to **μ** type the value of the mean
6. In the box next to **σ** type the value of the standard deviation
7. Press **OK**
Note: If you are looking for $P(X < x)$, use a lower bound of $-10,000,000$. If you are looking for $P(X > x)$, use an upper bound of $10,000,000$

JMP
1. After opening a new data table, click **Rows** then select **Add rows**
2. Type **1** in the box next to **How many rows to add:**
3. Click **OK**
4. Double-click on the **Column 1** heading
5. Click **Column Properties** and select **Formula**
6. Click **Edit Formula**

7. In the box under **Functions (grouped)** click **Probability** then select **Normal Distribution**
8. In the white box at the bottom half of the screen, double-click in the red box around x

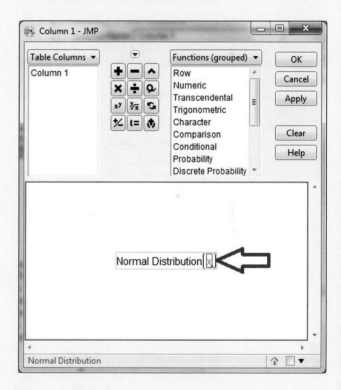

9. Type the value that you would like to find a probability for and click **OK**
10. Click **OK**

Note: This procedure outputs the value for $P(X \leq x)$. If you want to find the value for $P(X \geq x)$ you will need to subtract the output from one.

Minitab
1. Click **Calc** then click **Probability Distributions** then click **Normal...**
2. In the box next to **Mean:** type the mean of the normal distribution that you are working with

3. In the box next to **Standard deviation:** type the standard deviation of the normal distribution that you are working with
4. Click the radio button next to **Input constant:**
5. Click in the box next to **Input constant:** and type the value that you are finding the probability for
6. Click **OK**

Note: This procedure outputs the value for $P(X \leq x)$. If you want to find the value for $P(X \geq x)$ you will need to subtract the output from one.

Note: You may also type a column of values for which you would like to find probabilities (these must ALL be from the SAME distribution) and use the **Input Column** option to find probabilities for each value in the column selected.

SPSS

1. Type in the values that you would like to find the probabilities for in one column (this column will be automatically titled VAR00001)
2. Click **Transform** and select **Compute Variable...**
3. In the box under **Target Variable:** type NormProb (this will be the title of a new column that lists the cumulative probabilities for each value in VAR00001)
4. Under **Function group:** select **CDF & Noncentral CDF**
5. Under **Functions and Special Variables:** double-click **Cdf. Normal**
6. In the box under **Numeric Expression:** you should see CDF. NORMAL(?,?,?)
7. Highlight the first ? and double-click VAR00001
8. Highlight the second ? and type the mean of the normal distribution that you are working with
9. Highlight the third ? and type the standard deviation of the normal distribution that you are working with

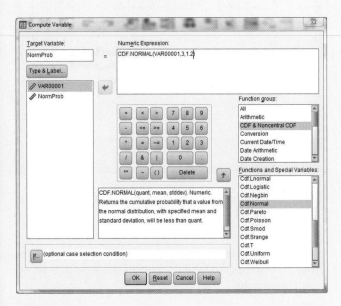

10. Click **OK**

Note: This procedure outputs the value for $P(X \leq x)$. If you want to find the value for $P(X \geq x)$ you will need to subtract the output from one.

Excel

1. Click in an empty cell
2. Click the **Formulas** ribbon
3. Select **Insert Function**
4. Select **Statistical** from the drop-down menu for category
5. Select **NORMDIST** from the **Select a function:** box
6. Click **OK**
7. Click in the box next to **X** and type the data value that you would like to find the probability for
8. Click in the box next to **Mean** and type the mean of the Normal distribution that you are working with
9. Click in the box next to **Standard_dev** and type the standard deviation of the Normal distribution that you are working with
10. Click in the box next to **Cumulative** and type **TRUE**
11. Click **OK**

Note: This procedure outputs the value for $P(X \leq x)$. If you want to find the value for $P(X \geq x)$ you will need to subtract the output from one.

Finding Binomial Probabilities

TI-83/84

$P(X \leq x)$

1. Press the **2nd** key then press the **VARS** key
2. Highlight the **binomcdf(** option and press the **ENTER** key
3. Type in the number of trials, n
4. Type,
5. Type in the success probability, p
6. Type,
7. Type the value for x
8. Type **)**
9. Press **ENTER**

$P(X = x)$

1. Press the **2nd** key then press the **VARS** key
2. Highlight the **binompdf(** option and press the **ENTER** key
3. Type in the number of trials, n
4. Type,
5. Type in the success probability, p
6. Type,
7. Type the value for x
8. Type **)**
9. Press **ENTER**

TI-Nspire

$P(x_1 \leq X \leq x_2)$

1. Enter the Scratchpad
2. Press the **menu** key and select **5:Probability** then select **5:Distributions** then select **E:Binomial Cdf...** and press the **enter** key
3. In the box next to **Num Trials, n** type in the number of trials, n
4. In the box next to **Prob Success, p** type in the success probability, p
5. In the box next to **Lower Bound** type in the value for x_1
6. In the box next to **Upper Bound** type in the value for x_2
7. Press **OK**

Note: In order to find $P(X \leq x)$ input a 0 for the lower bound.

P(X = x)
1. Enter the Calculate Scratchpad
2. Press the **menu** key and select **5:Probability** then select **5:Distributions** then select **D:Binomial Pdf...** and press the **enter** key
3. In the box next to **Num Trials, n** type in the number of trials, *n*
4. In the box next to **Prob Success, p** type in the success probability, *p*
5. In the box next to **X Value** type the value for *x*
6. Press **OK**

JMP

P(X ≤ x)
1. After opening a new data table, click **Rows** then select **Add rows**
2. Type **1** in the box next to **How many rows to add:**
3. Click **OK**
4. Double-click on the **Column 1** heading
5. Click **Column Properties** and select **Formula**
6. Click **Edit Formula**
7. In the box under **Functions (grouped)** click **Discrete Probability** then select **Binomial Distribution**
8. In the white box at the bottom half of the screen, double-click in the red box around p and type the value for the success probability, *p* and press **enter** on your keyboard
9. Double-click in the red box around n and type the value for the number of trials, *n* and press **enter** on your keyboard
10. Double-click in the red box around **k** and type the value for the number of successes for which you would like to find the probability
11. Click **OK**
12. Click **OK**

P(X = x)
1. After opening a new data table, click **Rows** then select **Add rows**
2. Type **1** in the box next to **How many rows to add:**
3. Click **OK**
4. Double-click on the **Column 1** heading
5. Click **Column Properties** and select **Formula**
6. Click **Edit Formula**
7. In the box under **Functions (grouped)** click **Discrete Probability** then select **Binomial Probability**
8. In the white box at the bottom half of the screen, double-click in the red box around **p** and type the value for the success probability, *p*, and press **enter** on your keyboard
9. Double-click in the red box around **n** and type the value for the number of trials, *n*, and press **enter** on your keyboard
10. Double-click in the red box around **k** and type the value for the number of successes for which you would like to find the probability
11. Click **OK**
12. Click **OK**

Minitab

P(X ≤ x)
1. Click **Calc** then click **Probability Distributions** then click **Binomial...**

2. In the box next to **Number of Trials:** input the value for *n*, the total number of trials
3. In the box next to **Probability of Success:** input the value for *p*, the success probability
4. Click the radio button next to **Input Constant**
5. In the box next to **Input Constant:** type the value that you want to find the probability for
6. Click **OK**

Note: You may also type a column of values for which you would like to find probabilities (these must ALL be from the SAME distribution) and use the **Input Column** option to find probabilities for each value in the column selected.

P(X = x)
1. Click **Calc** then click **Probability Distributions** then click **Binomial...**
2. Click the radio button next to **Probability**
3. In the box next to **Number of Trials:** input the value for *n*, the total number of trials
4. In the box next to **Probability of Success:** input the value for *p*, the success probability
5. Click the radio button next to **Input Constant**
6. In the box next to **Input Constant:** type the value that you want to find the probability for
7. Click **OK**

Note: You may also type a column of values for which you would like to find probabilities (these must ALL be from the SAME distribution) and use the **Input Column** option to find probabilities for each value in the column selected.

SPSS

P(X = x)
1. Type in the values that you would like to find the probabilities for in one column (this column will be automatically titled VAR00001)
2. Click **Transform** and select **Compute Variable...**
3. In the box under **Target Variable:** type **BinomProb** (this will be the title of a new column that lists the cumulative probabilities for each value in VAR00001)
4. Under **Function group:** select **CDF & Noncentral CDF**
5. Under **Functions and Special Variables:** double-click **Cdf. Binom**
6. In the box under **Numeric Expression:** you should see CDF.BINOM(?,?,?)
7. Highlight the first ? and double-click VAR00001
8. Highlight the second ? and type the value for *n*, the total number of trials
9. Highlight the third ? and type success probability, *p*
10. Click **OK**

P(X = x)
1. Type in the values that you would like to find the probabilities for in one column (this column will be automatically titled VAR00001)
2. Click **Transform** and select **Compute Variable...**
3. In the box under **Target Variable:** type BinProb (this will be the title of a new column that lists the cumulative probabilities for each value in VAR00001)

4. Under **Function group:** select **PDF & Noncentral PDF**
5. Under **Functions and Special Variables:** double-click Pdf. Binom
6. In the box under **Numeric Expression:** you should see PDF. BINOM(?,?,?)
7. Highlight the first ? and double-click VAR00001
8. Highlight the second ? and type the value for *n*, the total number of trials
9. Highlight the third ? and type success probability, *p*
10. Click **OK**

Excel

1. Click in an empty cell
2. Click the **Formulas** ribbon
3. Select **Insert Function**
4. Select **Statistical** from the drop-down menu for category
5. Select **BINOMDIST** from the **Select a function:** box
6. Click **OK**
7. Click in the box next to **Number_s** and type the number of successes that you are finding a probability for
8. Click in the box next to **Trials** and type the value for *n*, the total number of trials
9. Click in the box next to **Probability_s** and type success probability, *p*
10. Click in the box next to **Cumulative** and type **TRUE** if you are finding $P(X \leq x)$ or type **FALSE** if you are finding $P(X = x)$
11. Click **OK**

Note: This procedure outputs the value for $P(X \leq x)$ when **TRUE** is used as input for **Cumulative.** If you want to find the value for $P(X \geq x)$ you will need to subtract this output from one.

Normal Probability Plots

TI-83/84

1. Input the raw data into **L1** (In order to access lists press the **STAT** key, highlight the option called **Edit...** then press **ENTER**)
2. Press the **2ⁿᵈ** key then press the **Y** = key
3. Highlight **Plot1** and press **ENTER**
4. Highlight **On** and press **ENTER**
5. Highlight the plot type on the second row, third column and press **ENTER**
6. Press **GRAPH**

TI-Nspire

1. Enter the data into a data list (In order to access data lists select the spreadsheet option and press **enter)**

Note: Be sure to title the list by selecting the top row of the column and typing a title.

2. Press **menu** and select **3:Data** then **6:QuickGraph** then press **enter**
3. Press **menu** and select **1:Plot Type** then select **4:Normal Probability Plot** and press **enter**

JMP

1. Enter the raw data into a column
2. Click **Analyze** then select **Distribution**
3. Click and drag the column name containing the data from the box under **Select Columns** to the box next to **Y, Columns**
4. Click **OK**
5. Click the red arrow next to the column name
6. Select **Normal Quantile Plot**

Minitab

1. Input the data for which you would like to check Normality into a column
2. Click **Graph** then click **Probability Plot...**
3. Highlight the **Single** plot
4. Click **OK**
5. Double click the column name for the column that contains your data to move it into the **Graph Variables:** box
6. Click **OK**

SPSS

1. Input the data for which you would like to check Normality into a column
2. Click **Analyze** then select **Descriptive Statistics** then select **Q-Q Plots...**
3. Highlight the column name for the variable
4. Click the arrow to move the variable to the **Variables:** box
5. Click **OK**

Note: The normal probability plot is output with several other plots and statistics. This plot can be found under the title **Normal Q-Q Plot.**

Excel

Excel does not have the functionality to automatically produce Normal Probability Plots.

However, Excel can produce Normal Probability Plots in the course of running a regression using the Analysis ToolPak.

AP* Review Questions for Chapter 6

1. The discrete random variable x = the number of 911 calls in a small town in one day has the following probability distribution:

x	1	2	3	4	5	6
$p(x)$	0.20	0.35	0.15	0.10	0.10	0.10

Which of the following could be evaluated to compute the mean of this probability distribution?

(A) $\dfrac{1 + 2 + 3 + 4 + 5 + 6}{6}$

(B) $\dfrac{0.20 + 0.35 + 0.15 + 0.10 + 0.10 + 0.10}{6}$

(C) $1(0.20) + 2(0.35) + 3(0.15) + 4(0.10) + 5(0.10) + 6(0.10)$

(D) $\dfrac{1(0.20) + 2(0.35) + 3(0.15) + 4(0.10) + 5(0.10) + 6(0.10)}{6}$

(E) $1^2(0.20) + 2^2(0.35) + 3^2(0.15) + 4^2(0.10) + 5^2(0.10) + 6^2(0.10)$

Use the following for questions 2–3

A college basketball game is made up of two 20-minute halves. Each half is 20 minutes of playing time, but the "real" time elapsed from the start of play to the end of the half is much longer. This is due to time-outs and other periods of time when the game clock is stopped. Suppose that the mean and standard deviation for the "real" time (in minutes) of each half are

	Mean	Standard Deviation
1st half	48	6
2nd half	52	8

2. What is the mean "real" time for the entire game (1st half + 2nd half)?

(A) 48
(B) 50
(C) 52
(D) 100
(E) 114

3. Under what conditions is the standard deviation of the total "real" game time (1st half + 2nd half) equal to $\sqrt{6^2 + 8^2} = 10$?

(A) No conditions are necessary—the standard deviation will be 10 minutes.
(B) The standard deviation will not be 10 minutes under any circumstances.
(C) The standard deviation will be 10 minutes only if the 1st half "real" time is shorter than the 2nd half "real" time.
(D) The standard deviation will be 10 minutes only if the "real" times in the two halves are independent.
(E) The standard deviation will be 10 minutes only if the "real" times in the two halves are mutually exclusive.

4. Ten percent of the students at a particular school own a pet. Consider the chance experiment that consists of selecting a random sample of 20 students from this university. Let x = the number of students in the sample who own a pet. The random variable x has which of the following probability distributions?

(A) a binomial distribution
(B) a geometric distribution
(C) a normal distribution
(D) a uniform distribution
(E) none of the above

---

5. Which of the following four binomial histograms, all with $n = 10$, corresponds to the distribution with the largest value of p?

(A)

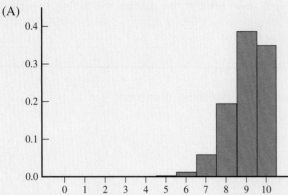

(B)

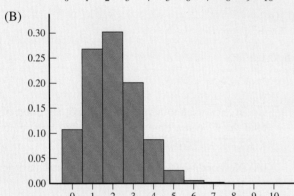

(C)

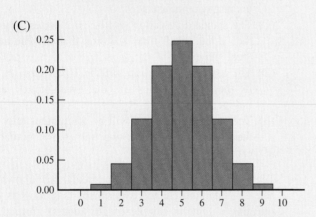

(D)

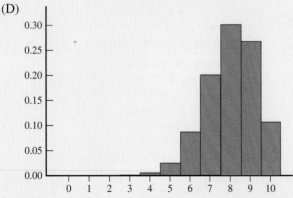

(E) Cannot be determined from the given information.

6. Let x denote the number of accidents in a given month at a certain intersection. Suppose that the probability distribution of x is:

x	0	1	2	3	4	5
$p(x)$	0.110	0.215	0.260	0.214	0.134	0.067

If the numbers of accidents in any 2 months are independent, what is the probability that no car accidents occur at this intersection for 2 consecutive months?

(A) 0.0121
(B) 0.1100
(C) 0.2200
(D) 0.8900
(E) None of the above

7. A normal probability plot suggests that a normal probability model is plausible when:

(A) No obvious patterns are present in the plot.
(B) A bell-shaped pattern is present in the plot.
(C) A substantial quadratic pattern is present in the plot.
(D) A substantial linear pattern is present in the plot.
(E) Any of the above is present in the plot.

8. For a normally distributed population with mean 0 and standard deviation 1.0, the population interquartile range is closest to which of the following values?

(A) 0.50
(B) 1.28
(C) 1.349
(D) 1.645
(E) 1.96

9. Which of the following is *not* a property of a binomial experiment?

(A) It consists of a fixed number of trials, n.
(B) Outcomes of different trials are independent.
(C) Each trial can result in one of several different outcomes.
(D) Observations consist of the number of successes for each trial of the experiment.
(E) The probability of success is constant for each trial.

10. Suppose that x has a probability distribution with density function $f(x) = \begin{cases} c, \text{ if } 6 < x < 8 \\ 0 \text{ otherwise} \end{cases}$

The value of c is:

(A) 0.5
(B) 0.6
(C) 0.7
(D) 0.8
(E) 0.9

11. The number of passengers who choose to check luggage on a flight from Los Angeles to San Francisco is a binomial random variable with n = number of passengers on the flight. If the probability that any individual passenger checks luggage is 0.3, what is the probability that exactly 40 of the 100 passengers on this flight on a particular day check luggage?

    (A) 0
    (B) 0.3
    (C) $(0.3)^{40}$
    (D) $(0.3)^{40} (0.7)^{60}$
    (E) $\binom{100}{40}(0.3)^{40}(0.7)^{60}$

12. Suppose that 65% of the students at a particular university have a Twitter account. If 100 students are selected at random from this university, what are the mean and standard deviation of the random variable x = number of selected students who have a Twitter account?

    (A) mean = 0.65, standard deviation = 0.35
    (B) mean = 0.65, standard deviation = 0.002
    (C) mean = 65, standard deviation = 22.75
    (D) mean = 65, standard deviation = 0.228
    (E) mean = 65, standard deviation = 4.770

13. Suppose that x is a random variable that has a uniform distribution over the interval from 0 to 10. Which of the following probabilities is largest?

    (A) $p(x > 8)$
    (B) $p(x \geq 8)$
    (C) $p(x = 10)$
    (D) $p(3 < x < 6)$
    (E) $p(x < 4)$

14. Suppose that 20% of cereal boxes contain a prize and the other 80% contain the message "Sorry, try again." Boxes of this cereal will be purchased and opened until a prize is found. Define the random variable x to be the number of boxes purchased. What is the probability that exactly three boxes must be purchased?

    (A) $(0.20)^3$
    (B) $0.8 + 0.8 + 0.2$
    (C) $(0.8)^2 + (0.2)$
    (D) $(0.8)^2 (0.2)$
    (E) $\dfrac{3!}{2!1!}(0.8)^2(0.2)^1$

15. For a normal distribution with mean 100 and standard deviation 10, the 95th percentile is

    (A) -1.645
    (B) 1.645
    (C) 16.45
    (D) 83.55
    (E) 116.45

7

An Overview of Statistical Inference— Learning from Data

Preview

Chapter Learning Objectives

7.1 Statistical Inference—What You Can Learn from Data

7.2 Selecting an Appropriate Method—Four Key Questions

7.3 A Five-Step Process for Statistical Inference

Chapter Activities

Are You Ready to Move On? Chapter 7 Review Exercises

AP* Review Questions for Chapter 7

Ian Dagnall/Alamy Limited

PREVIEW

Whether data are collected by sampling from populations or result from an experiment to compare treatments, the ultimate goal is to learn from the data. This chapter introduces the inferential process shared by all of the methods for learning from data that are covered in the chapters that follow. You will also see how the answers to four key questions guide the selection of an appropriate inference method.

CHAPTER LEARNING OBJECTIVES

Conceptual Understanding

After completing this chapter, you should be able to

C1 Understand the difference between questions that can be answered by using sample data to estimate population characteristics and those that can be answered by testing hypotheses about population characteristics.

C2 Understand that there is risk involved in drawing conclusions from sample data—that when generalizing from sample data, the sample may not always provide an accurate picture of the population.

C3 Understand that there is risk involved with drawing conclusions from experiment data—that when generalizing from experiment data, the observed difference in treatment effects may sometimes be due to variability in the response variable and the random assignment to treatments.

C4 Understand that data type distinguishes between questions that involve proportions and those that involve means.

C5 Know that different methods are used to draw conclusions based on categorical data and to draw conclusions based on numerical data.

C6 Know the four key questions that help identify an appropriate inferential method.

C7 Know the five-step process for estimation problems.

C8 Know the five-step process for hypothesis testing problems.

Mastering the Mechanics

After completing this chapter, you should be able to

M1 Distinguish between estimation problems and hypothesis testing problems.

M2 Distinguish between problems that involve proportions and those that involve means.

Putting It into Practice

After completing this chapter, you should be able to

P1 Given a scenario, answer the four key questions that help identify an appropriate inference method.

PREVIEW EXAMPLE   ## Deception in Online Dating Profiles

With the increasing popularity of online dating services, the truthfulness of information in the personal profiles provided by users is a topic of interest. The authors of the paper **"Self-Presentation in Online Personals: The Role of Anticipated Future Interaction, Self-Disclosure, and Perceived Success in Internet Dating"** (*Communication Research* **[2006]: 152–177**) designed a statistical study to investigate misrepresentation of personal characteristics. The researchers hoped to answer three questions:

1. What proportion of online daters believe they have misrepresented themselves in an online profile?

2. What proportion of online daters believe that others frequently misrepresent themselves?

3. Are people who place a greater importance on developing a long-term, face-to-face relationship more honest in their online profiles?

What did the researchers learn? Based on the data, they estimated that only about 6% of online daters believe that they have intentionally misrepresented themselves in online profiles. In spite of the fact that most users believe themselves to be honest, about 86% believed that others frequently misrepresented characteristics such as physical appearance in the online profile. The researchers also found that the data supported the claim that those who placed greater importance on developing a long-term, face-to-face relationship were more honest in their online profiles.

How were these researchers able to reach these conclusions? The estimates (the 6% and the 86% in the previous statements) are based on sample data. Do these estimates provide an accurate picture of the entire population of online daters? The researchers concluded that the data supported the claim that those who placed greater importance on developing a long-term, face-to-face relationship were more honest in the way they represented themselves online, but how did they reach this conclusion, and should you be convinced? These are important questions. In this chapter and those that follow, you will see how questions like these can be answered.

SECTION 7.1 Statistical Inference—What You Can Learn from Data

Statistical inference is all about learning from data. Let's begin by considering data that arise when a sample is selected from a population of interest.

Learning from Sample Data

When you obtain information from a sample selected from some population, it is usually because

1. You want to learn something about characteristics of the population. This results in an **estimation problem**. It involves using sample data to estimate population characteristics.
OR

2. You want to use the sample data to decide whether there is support for some claim or statement about the population. This results in a **hypothesis testing problem**. It involves testing a claim (hypothesis) about the population.

Example 7.1 Deception in Online Dating Profiles Revisited

Let's revisit the online dating example of the chapter preview. In that example, the population of interest was all online daters. Three questions about this population were identified:

1. What proportion of online daters believe they have misrepresented themselves in an online profile?
2. What proportion of online daters believe that others frequently misrepresent themselves?
3. Are people who place a greater importance on developing a long-term, face-to-face relationship more honest in their online profiles?

The first two of these questions are estimation problems because they involve using sample data to learn something about a population characteristic. The population characteristic of interest in the first question is the proportion of all online daters who believe they have misrepresented themselves online. In the second question, the population characteristic of interest is the proportion of all online daters who believe that others frequently misrepresent themselves. The third question is a hypothesis testing problem because it involves determining if sample data support a claim about the population of online daters.

AP* EXAM TIP

There are two types of inference problems: estimation and hypothesis testing. It is important that you be able to distinguish between them because you will use different methods for these two types of problems.

An **estimation problem** involves using sample data to estimate the value of a population characteristic.

A **hypothesis testing problem** involves using sample data to test a claim about a population.

Methods for estimation and hypothesis testing are called **statistical inference** methods because they involve generalizing (making an inference) from a sample to the population from which the sample was selected.

Statistical inference involves generalizing from a sample to a population.

Example 7.2 Whose Reality?

The article **"Who's Afraid of Reality Shows?"** (*Communication Research* **[2008]:382–397**) considers social concern over reality television shows. Researchers conducted telephone interviews with 606 individuals in a sample designed to represent the adult population of Israel. One of the things that the researchers hoped to learn was whether the data supported the theory that a majority of Israeli adults believed they were much less affected by reality shows than other people. They concluded that the sample data did provide support for this theory. This study involves generalizing from the sample to the population of Israeli adults and it is a hypothesis testing problem because it uses sample data to test a claim (that a majority consider themselves less affected than others).

Learning from Data When There Are Two or More Populations

Sometimes sample data are obtained from two or more populations of interest, and the goal is to learn about differences between the populations. Consider the following two examples.

Example 7.3 Tuned-In Babies

The director of the Kaiser Family Foundation's Program for the Study of Entertainment Media and Health said, "It's not just teenagers who are wired up and tuned in, its babies in diapers as well." A study by Kaiser Foundation provided one of the first looks at media use among the very youngest children—those from 6 months to 6 years of age **(Kaiser Family Foundation, 2003, www.kff.org)**. Because previous research indicated that children who have a TV in their bedroom spend less time reading than other children, the authors of the Foundation study were interested in learning about the proportion of kids who have a TV in their bedroom. They collected data from two samples of parents. One sample consisted of parents of children 6 months to 3 years of age. The second sample consisted of parents of children 3 to 6 years of age. Based on the resulting data, they were able to estimate the proportion of children who had a TV in their bedroom for each of the two age groups (0.30 or 30% for children in the younger group and 0.43 or 43% for children in the older group). From this information, they also estimated the difference in the proportions for the two populations (the two age groups). The proportion with TVs in the bedroom is 0.13 or 13 percentage points higher for children in the older group. This study illustrates statistical inference because it involves generalizing from samples to corresponding populations. It is an estimation problem because sample data were used to estimate the values of population characteristics. Because there were two samples, one from each of two different populations, it was also possible to learn something about how the two populations differ with respect to a population characteristic.

Example 7.4 Do Facebook Members Spend More Time Online?

College students spend a lot of time online, but do members of Facebook spend more time online than non-members? The authors of the paper **"Spatially Bounded Online Social Networks and Social Capital: The Role of Facebook" (Annual Conference of the International Communication Association, 2006)** collected data from two samples of college students. One sample consisted of Facebook members and the other consisted of non-members. One of the variables studied was the amount of time spent on the Internet in a typical day. Based on the resulting data, the authors concluded that there was no support for the claim that the mean time spent online for Facebook members was greater than the mean time for non-members. This study involves generalizing from samples, and it is a hypothesis testing problem because it involves testing a claim about the difference between the two groups (the claim that the mean time spent online is greater for Facebook members than for non-members).

Learning from Experiment Data

Statistical inference methods are also used to learn from experiment data. When data are obtained from an experiment, it is usually because

1. You want to learn about the effect of the different experimental conditions (treatments) on the measured response. This is an **estimation problem** because it involves using sample data to estimate a characteristic of the treatments, such as the mean response for a treatment or the difference in mean response for two treatments.

OR

2. You want to determine if experiment data provide support for a claim about how the effects of two or more treatments differ. This is a **hypothesis testing problem** because it involves testing a claim (hypothesis) about treatment effects.

The following two examples illustrate an estimation problem and a hypothesis testing problem in the context of learning from experiment data.

Example 7.5   Do U Smoke After Txt?

Researchers in New Zealand investigated whether mobile phone text messaging could be used to help people stop smoking. The article **"Do U Smoke After Txt? Results of a Randomized Trial of Smoking Cessation Using Mobile Phone Text Messaging"** (*Tobacco Control* [2005]: 255–261) describes an experiment designed to compare two experimental conditions (treatments). Subjects for the experiment were 1,705 smokers who were older than 15 years and owned a mobile phone and who wanted to quit smoking. The subjects were assigned at random to one of two groups. People in the first group received personalized text messages providing support and advice on stopping smoking. The second group was a control group, and people in this group did not receive any of these text messages. After 6 weeks, each person participating in the study was contacted and asked if he or she had smoked during the previous week. Data from the experiment were used to estimate the difference in the proportion who had quit for those who received the text messages and those who did not. Using statistical inference methods that you will learn in Chapters 11 and 14, the researchers estimated that the proportion of those who successfully quit smoking was higher by 0.15 (15 percentage points) for those who received the text messages. This is an example of an estimation problem. It involves generalizing from experiment data to treatment characteristics—in this case, the difference in the proportion of favorable responses for the two treatments.

Example 7.6   Cell Phones Can Slow You Down

The previous example illustrated a positive use of cell phone technology, but many believe that using a cell phone while driving isn't a good idea. In many states it is now illegal to use cell phones while driving. Is this justified? Researchers at the University of Utah designed an experiment to test the claim that people talking on a cell phone respond more slowly to a traffic signal change (**"Driven to Distraction,"** *Psychological Science* [2001]:

462–466). Subjects for the experiment were students at the university. Each subject was assigned at random to one of two treatments. One treatment involved driving in a simulator while talking on a cell phone. The other treatment involved driving in a simulator while listening to the radio. As subjects drove in the simulator, they were told to brake when a signal light changed from green to red. Subjects encountered several light changes while driving in the simulator, and the average response time (the time between when the light changed and when the subject applied the brake) was computed for each subject. Using methods that you will learn in Chapters 13 and 14, the researchers concluded that the experiment data provided support for the claim that mean reaction time is slower for those talking on a cell phone than it is for those listening to the radio. This is an example of a hypothesis testing problem. It involves using experiment data to test a claim about treatment effects—in this case a claim about the difference in mean response time for the two treatments.

Statistical Inference Involves Risk

Examples 7.1–7.6 illustrate learning from sample data and experiment data in a variety of settings. However, one important aspect has not yet been addressed. *Learning from data involves risk.* To be fully informed, you need to recognize what these risks are and consider how you can assess and describe the risks involved. Even when a reasonable sampling plan is used or a well-designed experiment is carried out, the resulting data provide only incomplete information about the population or about treatment effects.

The risks associated with statistical inference arise because you are attempting to draw conclusions on the basis of data that provide partial rather than complete information. But you can still make good decisions, even with incomplete information. The statistical inference methods that you will learn in the chapters that follow provide ways to make effective use of available data, but it is important to recognize that there are still risks involved.

What are the risks? In estimation problems, where sample data are used to estimate population or treatment characteristics, you run the risk that these estimates will be inaccurate. Even when you accompany an estimate with a description of the anticipated accuracy, sometimes estimates aren't as accurate as you say. When you produce estimates and evaluate estimates produced by others, you need to understand that there is a risk that the estimates are not as accurate as claimed. It is important to understand how likely it is that the method used to produce the estimates and accompanying measures of accuracy might mislead in this way.

In hypothesis testing situations, where you use partial rather than complete information to determine whether there is support for a claim about population or treatment characteristics, you run the risk of an incorrect conclusion. Based on the available data you might conclude that there is support for a claim that is actually not true. Or, you may decide that the data do not support a claim that really is true. When you carry out a hypothesis test, you will need to understand the consequences of reaching an incorrect conclusion. You also must understand how likely it is that the method used to decide whether or not a claim is supported might lead to an incorrect conclusion.

Variability in Data

Sampling

The most important factor in evaluating risk is understanding variability in the data. When a sample is selected from a population, the sample provides only a partial picture of the population. When there is variability in the population, you need to consider whether this partial picture is representative of the population. Even when a sample is selected in a reasonable way, if there is a lot of variability in the population, the partial pictures provided by different samples might be quite different. This sample-to-sample variability should be considered when you assess the risk associated with drawing conclusions about the population from sample data.

Experimentation

Variability also plays an important role in determining the risks involved when generalizing from experiment data. Experiments are usually conducted to investigate the effect of experimental conditions (treatments) on a response. For example, an experiment might be designed to determine if noise level has an effect on the time required to perform a task requiring concentration. Suppose there are 20 individuals available to serve as subjects in this experiment with two treatments conditions (quiet environment and noisy environment). The response variable is the time required to complete the task.

If noise level has no effect on completion time, the time observed for each of the 20 subjects would be the same whether they are in the quiet group or the noisy group. Any observed differences in the completion times for the two treatments would not be due to noise level. These differences would only be due to person-to-person variability in completion times and the random assignment of individuals to experimental groups. That is, the distribution of completion times for the two treatment groups might differ, but the difference can be explained by person-to-person variability in completion times and by the random assignment (different assignments will result in different people being assigned to the quiet and noisy conditions). You must understand how differences might result from variability in the response and the random assignment to treatment groups in order to distinguish them from differences created by a treatment effect. This important idea will be explored more fully in Chapter 14.

SECTION 7.1 EXERCISES

Each Exercise Set assesses the following chapter learning objectives: C1, C2, C3, M1

SECTION 7.1 Exercise Set 1

7.1 The report **"Teens and Distracted Driving: Texting, Talking and Other Uses of the Cell Phone Behind the Wheel"** (Pew Research Center, Nov. 16, 2009) summarizes data from a survey of a representative sample of 800 teens between the ages of 12 and 17. The following statements were made on the basis of the resulting data:

- 75% of all American teens own a cell phone.
- 66% of all American teens use a cell phone to send and receive text messages.
- 26% of American teens ages 16–17 have used a cell phone to text while driving.

Are the inferences made ones that involve estimation or ones that involve hypothesis testing?

7.2 For the study described in the previous exercise, answer the following questions.
a. What is the population of interest?
b. What population characteristics are being estimated?
c. Do you think that the actual percentage of all American teens who own a cell phone is exactly 75%? Explain why or why not.
d. Two of the estimates of population characteristics from this study were that 75% of teens own a cell phone and that 26% of teens ages 16–17 have used their phones to text while driving. Which of these two estimates do you think is more accurate and why? (Hint: What do you know about the number of teens surveyed?)

7.3 The article **"More Communities Banning 'Television on a Stick'"** (*USA Today*, March 23, 2010) describes an ongoing controversy over the distraction caused by digital billboards along highways. One study mentioned in the newspaper article is described in **"Effects of Advertising Billboards During Simulated Driving"** (*Applied Ergonomics* [2010]: 1–8). In this study, 48 people made a 9 km drive in a driving simulator. Drivers were instructed to change lanes according to roadside lane change signs. Some of the lane changes occurred near digital billboards. What was displayed on the digital billboard changed once during the time that the billboard was visible by the driver to simulate the changing digital billboards that appear along highways. Data from this study supported the theory that the time required to respond to road signs was greater when digital billboards were present. Is the inference made one that involves estimation or one that involves hypothesis testing?

7.4 For the study described in the previous exercise, answer the following questions.
a. What is the population of interest?
b. What claim was tested?
c. What additional information would you want before deciding if it is reasonable to generalize the conclusions of this study to the population of interest?
d. Assuming that the people who participated in the study are representative of the population of interest, do you

think that the risk of an incorrect conclusion would have been lower, about the same, or higher if 100 people had participated instead of 48?

7.5 Consider the population that consists of all students enrolled at your school.
a. Give an example of a question about this population that could be answered by collecting data and using it to estimate a population characteristic.
b. Give an example of a question about this population that could be answered by collecting data and using it to test a claim about this population.

SECTION 7.1 **Exercise Set 2**

7.6 Do people better remember what they learned if they are in the same physical space where they first learned it? The authors of the paper **"The Dynamics of Memory: Context-Dependent Updating"** (*Learning & Memory* (2008): 574–579) asked people to learn a set of 20 unrelated objects. Two days later, these people were asked to recall the objects learned on the first day. Some of the people were asked to recall the objects in the same room where they originally learned the objects. The others were asked to recall the objects in a different room. People were assigned at random to one of these two recall conditions. The authors found that the data on the number of objects recalled supported the claim that recall is better when people return to the original learning context. Is the inference made one that involves estimation or one that involves hypothesis testing?

7.7 The article **"The Largest Last Supper: Depictions of Food Portions and Plate Size Increase Over the Millennium"** (*International Journal of Obesity* [2010]: 1–2) describes a study in which each painting in a sample of 52 paintings of *The Last Supper* was analyzed by comparing the size of the food plates in the painting to the head sizes of the people in the painting. For paintings that were painted prior to the year 1500, the estimated average plate-to-head size ratio was smaller than this ratio for the paintings that were painted after 1500. Is the inference made one that involves estimation or one that involves hypothesis testing?

7.8 For the study described in the previous exercise, answer the following questions.
a. The original sample consisted of 52 paintings. These paintings were then divided into two samples consisting of 30 painted before 1500 and 22 painted after 1500. What are the two populations of interest?
b. What population characteristics are being estimated?
c. Suppose that the paintings selected for analysis were selected at random from all paintings that portray *The Last Supper*. Do you think that the estimate produced for average plate-to-head size ratio for paintings made before 1500 is likely to be less accurate than the corresponding estimate for paintings made after 1500? Explain.

7.9 Consider the population that consists of all employees of a large computer manufacturer.
a. Give an example of a question about this population that could be answered by collecting data and using it to estimate a population characteristic.
b. Give an example of a question about this population that could be answered by collecting data and using it to test a claim about this population.

Additional Exercises

7.10 Fans of professional soccer are probably aware that players sometimes fake injuries (called dives or flops). But how common is this practice? The articles **"A Field Guide to Fakers and Floppers"** (*Wall Street Journal*, June 28, 2010) and **"Red Card for Faking Footballers"** (*Science Daily*, Oct. 10, 2009) describe a study of deceptive behavior in soccer. Based on this study, it was possible to categorize injuries as real or fake based on movements that were characteristic of fake injuries (such as an arched back with hands raised, which is meant to attract the attention of a referee but which is not characteristic of the way people fall naturally). Data from an analysis of a sample of soccer games were then used to make the following statements:
- On average, referees stop a soccer game to deal with apparent injuries 11 times per game.
- On average, there is less than one "real" injury per soccer game.

Are the inferences made ones that involve estimation or ones that involve hypothesis testing?

7.11 **"Want to Lose More Fat? Skip Breakfast Before Workout"** (*The Tribune*, June 4, 2010) is the headline of a newspaper article describing a study comparing men who did endurance training without eating before training and men who ate before training. Twenty men were assigned at random to one of two 6-week diet and exercise programs. Both groups followed a similar diet and performed the same daily morning exercise routine. Men in one group did the exercise routine prior to eating, and those in the other group ate first and then exercised. The resulting data supported the claim that those who do not eat prior to exercising burn a higher proportion of fat than those who eat before exercising. Is the inference made one that involves estimation or one that involves hypothesis testing?

7.12 For the study described in the previous exercise, answer the following questions.
a. What is the population of interest for this study?
b. What claim was tested?
c. What additional information would you want before deciding if it is reasonable to generalize the conclusions of this study to the population of interest?
d. Assuming that the people who participated in the study are representative of the people in the population of interest, do you think that the risk of an incorrect conclusion would have been lower, about the same, or higher if 10 men had participated instead of 20?

Selecting an Appropriate Method—
Four Key Questions

This chapter and Chapter 8 lay the conceptual groundwork for the methods that are covered in the rest of this book. There are quite a few methods for learning from data, and choosing a method that is appropriate for a particular problem is one of the biggest challenges of learning statistics. Identifying characteristics of a problem that determine an appropriate method will help you to develop a systematic approach.

For each new situation that you encounter, begin by answering the four key questions that follow. The answers to these questions will guide your decision about which methods to consider.

Question Type (Q): Is the question you are trying to answer an estimation problem or a hypothesis testing problem?

In Section 7.1, you learned how to distinguish between estimation problems and hypothesis testing problems. It is important to make this distinction because you will choose different methods depending on the answer to this question.

Study Type (S): Does the situation involve generalizing from a sample to learn about a population (an observational study or survey), or does it involve generalizing from an experiment to learn about treatment effects?

The answer to this question affects the choice of method as well as the type of conclusion that can be drawn. For example, the study investigating concern over reality television shows described in Example 7.2 was an observational study and used data from a sample to learn about the population of Israeli adults. The study of Example 7.5, which examined whether the use of text messages had an effect on the proportion of smokers who successfully quit smoking, used data from an experiment to learn about treatment effects.

Type of Data (T): What type of data will be used to answer the question? Is the data set univariate (one variable) or bivariate (two variables)? Are the data categorical or numerical?

Univariate data are used to answer questions about a single variable. For example, the goal of the study in Example 7.3 was to learn how the proportion with a TV in the bedroom differed for children in two age groups. There were two samples (one consisting of parents of children ages 6 months to 3 years and one consisting of parents of children ages 3 to 6 years). The two groups were compared on the basis of one categorical variable (*TV in bedroom*, with possible values of yes or no). Bivariate data are used to learn about the relationship between two variables. For example, the study of deception in online dating profiles (see Example 7.1) investigated whether people who place a greater importance on developing a long-term, face-to-face relationship are more honest in their online profiles. Answering this question involves looking at two variables (*importance placed on developing a long-term face-to-face relationship* and a measure of *honesty in the online profile*) for each person in the sample.

Whether you are working with categorical data or numerical data is also an important consideration in selecting an appropriate method. For example, if you have a single variable and the data are categorical, the question of interest is probably about a population proportion. But if the data are numerical, the question of interest is probably about a population mean.

The methods used to learn about proportions are different from those used to learn about means. The easiest way to distinguish between a situation involving proportions and one involving means is to determine whether the data are categorical or numerical. For example, suppose the variable of interest is whether there is a TV in a child's bedroom. Possible values for this variable are yes and no, so the variable is categorical. These data can be used to estimate a proportion or to test a hypothesis about a proportion. On the other hand, if the variable of interest is the number of hours per day a child spends watching TV, possible values are numbers, so the variable is numerical. Data on this variable can be used to estimate a mean or test a hypothesis about a mean. In Chapter 2, you learned that the graphical displays used for categorical data (such as bar charts and pie charts) are different from those used for numerical data (such as dotplots, histograms, and boxplots). The same is true for statistical inference methods—different methods are used for different types of data.

Number of Samples or Treatments (N): How many samples are there, or, if the data are from an experiment, how many treatments are being compared?

For situations that involve sample data, different methods are used depending on whether there are one, two, or more than two samples. Also, you may choose a different method to analyze data from an experiment with only two treatments than you would for an experiment with more than two treatments.

One way to remember these four questions is to use the acronym **QSTN**—think of this as the word "question" without the vowels. A brief version of the four key questions is given here:

Q **Question Type**	Estimation or hypothesis testing?
S **Study Type**	Sample data or experiment data?
T **Type of Data**	One variable or two? Categorical or numerical?
N **Number of Samples or Treatments**	How many samples or treatments?

Table 7.1 shows how answering these questions can help identify a data analysis method for consideration. You will learn the methods identified in the table's "Method to Consider" column in the remaining chapters of this text.

The following examples illustrate how the four key questions (QSTN) are answered for three different studies.

AP* EXAM TIP

You may want to bookmark the page for Table 7.1. You will refer to it often to determine an appropriate method in a given situation.

Example 7.7 My Funny Valentine

It probably wouldn't surprise you to know that Valentine's Day means big business for florists, jewelry stores, and restaurants. But would it surprise you to know that it is also a big day for pet stores? In January 2008, the National Retail Federation conducted a survey of 8,447 consumers that were selected in a way that the federation believed would produce a representative sample of U.S. adults (**"Consumers Opt for Quality Time with Loved Ones Over Traditional Gifts This Valentine's Day," www.nrf.com**). One question in the survey asked, "Do you plan to spend money on a Valentine's Day gift for your pet this year?" The Federation hoped to learn about the proportion of U.S. adults that planned to buy a gift for their pet. Let's answer the four key questions that would help identify an appropriate analysis method.

Q **Question Type**	Estimation or hypothesis testing?	Estimation. The researchers wanted to estimate a population characteristic—the proportion of U.S. adults who plan to buy a gift for their pet. There is no claim made about this proportion that would lead to a hypothesis test.
S **Study Type**	Sample data or experiment data?	Sample data. The study involved a survey that was given to people in a sample selected from the population of U.S. adults.
T **Type of Data**	One variable or two? Categorical or numerical?	One categorical variable. Data were collected on one variable—people were asked whether or not they planned to purchase a gift for their pet. This variable is categorical—possible values are yes and no.
N **Number of Samples or Treatments**	How many samples or treatments?	One. There is only one sample, consisting of 8,447 adults.

TABLE 7.1 Answering Four Key Questions to Identify An Appropriate Method

Table Row	Q Question Type Estimation or Hypothesis Test?	S Study Type Sample Data or Experimental Data?	T Type of Data 1 Variable or 2? Categorical or Numerical?	N Number How Many Samples or Treatments?	Method to Consider	Chapter
1	Estimation	Sample	1 categorical variable	1	One-Sample z Confidence Interval for a Proportion	9
2	Hypothesis Test	Sample	1 categorical variable	1	One-Sample z Test for a Proportion	10
3	Estimation	Sample	1 categorical variable	2	Two-Sample z Confidence Interval for a Difference in Proportions	11
4	Hypothesis Test	Sample	1 categorical variable	2	Two-Sample z Test for a Difference in Proportions	11
5	Estimation	Sample	1 numerical variable	1	One-Sample t Confidence Interval for a Mean	12
6	Hypothesis Test	Sample	1 numerical variable	1	One-Sample t Test for a Mean	12
7	Estimation	Sample	1 numerical variable	2	Two-Sample t or Paired t Confidence Interval for a Difference in Means	13
8	Hypothesis Test	Sample	1 numerical variable	2	Two-Sample t or Paired t Test for a Difference in Means	13
9	Hypothesis Test	Sample	1 numerical variable	More than 2	ANOVA F Test	17 (online)
10	Estimation	Sample	1 numerical variable	More than 2	Multiple Comparisons	17 (online)

In **Chapter 14**, you will see that, with some minor modifications, the methods used for learning from sample data can also be applied to situations that call for learning from **experiment data**. **Chapter 15** includes methods for comparing **more than two groups** on the basis of a **proportion** and for learning about **relationships between categorical variables**. **Chapter 16** expands on what you learned in Chapter 4, providing additional methods for learning about **relationships between numerical variables**.

The answers to these four questions lead to a suggested method (see row 1 of Table 7.1). In case you are curious, the researchers estimated the proportion of U.S. adults that planned to purchase a gift for a pet to be 0.172, or 17.2%. Using methods that you will learn in Chapter 9, they were also able to say that they believed that this estimate was accurate to within one percentage point of the actual population percentage.

Example 7.8 Those Darn Cell Phones!

Example 7.6 described a study that investigated whether talking on a cell phone while driving had an effect on response time (**"Driven to Distraction,"** *Psychological Science* **[2001]: 462–466).** In the study, subjects were assigned at random to one of two experimental conditions (driving while talking on a cell phone and driving while listening to the radio). While driving in a simulator, subjects were told to brake when a traffic light changed from green to red, and response time was measured. The researchers were interested in answering the following question: Does using a cell phone while driving result in a slower reaction time?

Before you look at the answers to the four key questions that follow, think about how you would answer these questions based on the study description.

Q Question Type	Estimation or hypothesis testing?	Hypothesis test. The researchers planned to use the data to test a claim about treatment effects. Specifically, they wanted to test the claim that response time is slower for the cell phone condition than for the radio condition.
S Study Type	Sample data or experiment data?	Experiment data. This was a study to investigate the effect of cell phone use on reaction time. Subjects were randomly assigned to the two conditions considered.
T Type of Data	One variable or two? Categorical or numerical?	One numerical variable. The claim that is to be tested is about a single variable—*response time*. Measuring response time results in numerical data.
N Number of Samples or Treatments	How many samples or treatments?	Two. This experiment has two treatments (experimental conditions). In this example, the two treatments are cell phone and radio.

The answers to these questions direct you to a method listed in Table 7.1. This method can be used to confirm the researchers' conclusion that mean reaction time is slower for those talking on a cell phone than for those listening to the radio.

Example 7.9 Red Wine or Onions?

Flavonols are compounds found in some foods that are thought to have beneficial health-related properties when absorbed into the blood. Just because a food contains flavanols, however, doesn't mean that it will be absorbed into the blood. A study described in the paper **"Red Wine Is a Poor Source of Bioavailable Flavonols in Men"** (*Journal of Nutrition* [2001]: 745–748) investigated the absorption of one type of flavonol called quercetin. In this study, one group of healthy men consumed 750 mL of red wine daily for four days and another group consumed 50 grams of fried onions daily for four days. This amount of wine contains the same amount of quercetin as the onions. Blood quercetin concentration was measured at the end of the four-day period. The resulting data were then used to learn about the difference in blood quercetin concentration between the two sources of quercetin.

How would you answer the four key questions for this study? The answer to Question 1 (Q) isn't obvious, but try to answer the other questions. Then take a look at the discussion that follows.

Q Question Type	Estimation or hypothesis testing?	From the given description, it is hard to tell. If the objective was to estimate the difference in mean concentration for the two treatments, then this would be an estimation problem. On the other hand, if the goal was to test a claim about the difference—for example, to test the claim that the mean blood quercetin concentration was not the same for wine and onions—this would be a hypothesis testing problem. The answer here will depend on the research objective. This would need to be clarified before analyzing the data.
S Study Type	Sample data or experiment data?	Experiment data. Two different experimental conditions (wine and onions) were investigated, with the goal of learning about their effect on blood quercetin concentration.
T Type of Data	One variable or two? Categorical or numerical?	One numerical variable. The variable of interest is blood quercetin concentration, and this is a numerical variable.
N Number of Samples or Treatments	How many samples or treatments?	Two. This experiment involved two treatments (red wine and onions).

A football playbook is a book containing all of the plays that a team may use during a game. Players must learn how to run each of these plays, and the person calling the plays must learn which plays are appropriate in a given situation. You can think of your statistics course as helping you to develop your statistics "playbook." Each new method that you learn is like a new play that you get to add to your playbook as you master it. The four key questions of this section will help you to determine which plays in your playbook might be appropriate in a particular situation and give you a systematic way of thinking about each new method as you add it to your statistics playbook.

SECTION 7.2 EXERCISES

Each Exercise Set assesses the following chapter learning objectives: C4, C5, C6, M2, P1

SECTION 7.2 Exercise Set 1

7.13 Suppose that a study was carried out in which each person in a random sample of students at a particular college was asked how much money he or she spent on textbooks for the current semester. Would you use these data to estimate a population mean or to estimate a population proportion? How did you decide?

7.14 When you collect data to learn about a population, why do you worry about whether the data collected are categorical or numerical?

7.15 Consider the four key questions that guide the choice of an inference method. Two of these questions are

T: Type of data. One variable or two? Categorical or numerical?
N: Number of samples or treatments. How many samples or treatments?

What are the other two questions that make up the four key questions?

For each of the studies described in Exercises 7.16 to 7.18, answer the following key questions:

Q Question Type	Estimation or hypothesis testing?
S Study Type	Sample data or experiment data?
T Type of Data	One variable or two? Categorical or numerical?
N Number of Samples or Treatments	How many samples or treatments?

7.16 The article **"Smartphone Nation"** (*AARP Bulletin*, **September 2009**) described a study of how people ages 50 to 64 years use cell phones. In this study, each person in a sample of adults thought to be representative of this age group was asked about whether he or she kept a cell phone by the bed at night. The researchers conducting this study hoped to use the resulting data to learn about the proportion of people in this age group who sleep with their cell phone nearby.

7.17 Do children diagnosed with attention deficit/hyperactivity disorder (ADHD) have smaller brains than children without this condition? This question was the topic of a research study described in the paper **"Developmental Trajectories of Brain Volume Abnormalities in Children and Adolescents with Attention Deficit/Hyperactivity Disorder"** (*Journal of the American Medical Association* [2002]: 1740–1747). Brain scans were completed for 152 children with ADHD and 139 children of similar age without ADHD. The researchers wanted to see if the resulting data supported the claim that the mean brain volume of children with ADHD is smaller than the mean for children without ADHD.

7.18 The article **"Tots' TV-Watching May Spur Attention Problems"** (*San Luis Obispo Tribune*, April 4, 2004) describes a study that appeared in the journal *Pediatrics*. In this study, researchers looked at records of 2,500 children who were participating in a long-term health study. For each child, they determined if the child had attention disorders at age 7 and the number of hours of television the child watched at age 3. They hoped to use the resulting data to learn about how these variables might be related.

SECTION 7.2 **Exercise Set 2**

7.19 Suppose that a study is carried out in which each student in a random sample selected from students at a particular school is asked whether or not he or she would purchase a recycled paper product even if it cost more than the same product that was not made with recycled paper. Would you use the resulting data to estimate a population mean or to estimate a population proportion? How did you decide?

7.20 Comment on the following statement: The same statistical inference methods are used for learning from categorical data and for learning from numerical data.

7.21 Consider the four key questions that guide the choice of an inference method. Two of these questions are

Q: Question type. Estimation or hypothesis testing?
S: Study type. Sample data or experiment data?

What are the other two questions that make up the four key questions?

For each of the studies described in Exercises 7.22 to 7.24, answer the following key questions:

Q Question Type	Estimation or hypothesis testing?
S Study Type	Sample data or experiment data?
T Type of Data	One variable or two? Categorical or numerical?
N Number of Samples or Treatments	How many samples or treatments?

7.22 A study of fast-food intake is described in the paper **"What People Buy From Fast-Food Restaurants"** (*Obesity* [2009]: 1369–1374). Adult customers at three hamburger chains (McDonald's, Burger King, and Wendy's) at lunchtime in New York City were approached as they entered the restaurant and were asked to provide their receipt when exiting. The receipts were then used to determine what was purchased and the number of calories consumed. The sample mean number of calories consumed was 857, and the sample standard deviation was 677. This information was used to learn about the mean number of calories consumed in a New York fast-food lunch.

7.23 Common Sense Media surveyed 1,000 teens and 1,000 parents of teens to learn about how teens are using social networking sites such as Facebook and MySpace (**"Teens Show, Tell Too Much Online,"** *San Francisco Chronicle*, August 10, 2009). The two samples were independently selected and were chosen in a way that makes it reasonable to regard them as representative of American teens and parents of American teens. When asked if they check their online social networking sites more than 10 times a day, 220 of the teens surveyed said yes. When parents of teens were asked if their teen checks his or her site more than 10 times a day, 40 said yes. The researchers used these data to conclude that there was evidence that the proportion of all parents who think their teen checks a social networking site more than 10 times a day is less than the proportion of all teens who report that they check the sites more than 10 times a day.

7.24 Researchers at the Medical College of Wisconsin studied 2,121 children between the ages of 1 and 4 (*Milwaukee Journal Sentinel*, November 26, 2005). For each child in the study, a measure of iron deficiency and the length of time the child was bottle-fed were recorded. The resulting data were used to learn about whether there was a relationship between iron deficiency and the length of time a child is bottle fed.

Additional Exercises

For each of the studies described in Exercises 7.25 to 7.29, answer the following key questions:

Q Question Type	Estimation or hypothesis testing?
S Study Type	Sample data or experiment data?
T Type of Data	One variable or two? Categorical or numerical?
N Number of Samples or Treatments	How many samples or treatments?

7.25 The concept of a "phantom smoker" was introduced in the paper **"I Smoke but I Am Not a Smoker: Phantom Smokers and the Discrepancy Between Self-Identity and Behavior"** (*Journal of American College Health* [2010]: 117– 125). Previous studies of college students found that how students respond when asked to identify themselves as either a smoker or a nonsmoker was not always consistent with how they respond to a question about how often they smoked cigarettes. A phantom smoker is defined to be someone who self-identifies as a nonsmoker but who admits to smoking cigarettes when asked about frequency of smoking. This prompted researchers to wonder if asking college students to self-identify as being a smoker or nonsmoker might be resulting in an underestimate of the actual percentage of smokers. The researchers planned to use data from a sample of 899 students to estimate the percentage of college students who are phantom smokers.

7.26 An article in *USA Today* (October 19, 2010) described a study to investigate how young children learn. Sixty-four toddlers age 18 months participated in the study. The toddlers were allowed to play in a lab equipped with toys and which had a robot that was hidden behind a screen. The article states: "After allowing the infants playtime, the team removed the screen and let the children see the robot. In some tests, an adult talked to the robot and played with it. In others the adult ignored the robot. After the adult left the room, the robot beeped and then turned its head to look at a toy to the side of the infant." The researchers planned to see if the resulting data supported the claim that children are more likely to follow the robot's gaze to the toy when they see an adult play with the robot than when they see an adult ignore the robot.

7.27 In a study of whether taking a garlic supplement reduces the risk of getting a cold, 146 participants were assigned to either a garlic supplement group or to a group that did not take a garlic supplement **("Garlic for the Common Cold," Cochrane Database of Systematic Reviews, 2009).** Researchers planned to see if there is evidence that the proportion of people taking a garlic supplement who get a cold is less than the proportion of those not taking a garlic supplement who get a cold.

7.28 To examine the effect of exercise on body composition, data were collected from a sample of active women who engaged in 9 hours or more of physical activity per week **("Effects of Habitual Physical Activity on the Resting Metabolic Rates and Body Composition of Women Aged 35 to 50 Years," Journal of the American Dietetic Association [2001]: 1181–1191).** Percent of body fat was measured for each woman. The researchers planned to use the resulting data to learn about mean percent body fat for active women.

7.29 Can moving their hands help children learn math? This question was investigated by the authors of the paper **"Gesturing Gives Children New Ideas about Math" (Psychological Science [2009]: 267–272).** A study was conducted to compare two different methods for teaching children how to solve math problems of the form $3 + 2 + 8 = \underline{\quad} + 8$. One method involved having students point to the $3 + 2$ on the left side of the equal sign with one hand and then point to the blank on the right side of the equal sign before filling in the blank to complete the equation. The other method did not involve using these hand gestures. To compare the two methods, 128 children were assigned at random to one of the methods. Each child then took a test with six problems, and the number correct was determined for each child. The researchers planned to see if the resulting data supported the theory that the mean number correct for children who use hand gestures is higher than the mean number correct for children who do not use hand gestures.

SECTION 7.3 A Five-Step Process for Statistical Inference

In the chapters that follow, you will see quite a few different methods for learning from data. Even though you will choose different methods in different situations, all of the methods used for estimation problems are carried out using the same overall process, and all of the hypothesis testing methods follow a similar overall process.

A Five-Step Process for Estimation Problems (EMC³)

Once a problem has been identified as an estimation problem (the E in EMC³), the following steps will be used:

Step	What Is This Step?
E	**Estimate**: Explain what population characteristic you plan to estimate.
M	**Method**: Select a potential method. This step involves using the answers to the four key questions (QSTN) to identify a potential method.
C	**Check**: Check to make sure that the method selected is appropriate. Many methods for learning from data only provide reliable information under certain conditions. It will be important to verify that any such conditions are met before proceeding.
C	**Calculate**: Sample data are used to perform any necessary calculations. This step is often accomplished using technology, such as a graphing calculator or computer software.
C	**Communicate Results**: This is a critical step in the process. In this step, you will answer the question of interest, explain what you have learned from the data, and acknowledge potential risks.

A Five-Step Process for Hypothesis Testing Problems (HMC³)

Once a problem has been identified as a hypothesis testing problem (the H in HMC³), the first step in the five-step process is

| H | **Hypotheses**: Define the hypotheses that will be tested. |

This is really the only step in the process that is different from the steps described for estimation—the MC³ steps are the same, resulting in

Step	What Is This Step?
H	**Hypotheses**: Define the hypotheses that will be tested.
M	**Method**: Select a potential method. This step involves using the answers to the four key questions (QSTN) to identify a potential method.
C	**Check**: Check to make sure that the method selected is appropriate. Many methods for learning from data only provide reliable information under certain conditions. It will be important to verify that any such conditions are met before proceeding.
C	**Calculate**: Sample data are used to perform any necessary calculations. This step is often accomplished using technology, such as a graphing calculator or computer software.
C	**Communicate Results**: This is a critical step in the process. In this step, you will answer the question of interest, explain what you have learned from the data, and acknowledge potential risks.

These steps provide a common systematic approach to learning from data in a variety of situations. Returning to the playbook analogy, you can think of the four key questions of the last section as providing a strategy for selecting a play (statistical method). The five-step processes of this section describe the structure of each play in the playbook. So, as you move forward in Chapters 9 through 17, just remember—you will answer four questions to select a "play," and then each "play" has five steps.

CHAPTER ACTIVITIES

ACTIVITY 7.1 TONY'S APOLOGY

In 2010, one of the worst oil spills in history occurred along the Gulf Coast of the United States when a British Petroleum (BP) offshore oil well failed. The online article **"Are Men More Sympathetic Than Women to BP CEO Tony Hayward?"** (http://www.fastcompany.com/1671707/are-men-more-sympathetic-to-bp-ceo-tony-hayward) describes a study conducted by Innerscope. The article states

> BP's marketing response to the Gulf oil disaster has been an unquestionable failure, thanks in large part to CEO Tony Hayward and his many callous remarks ("I'd like my life back," anyone?). But according to neuromarketing research company Innerscope, men and women don't hate Hayward's halfhearted apologies equally. Innerscope measured the biometric response (skin

response, heart rate, body movement, and breathing) of 54 volunteers while they watched BP's now-infamous apology advertisements. The 27 men and 27 women were all college-educated, but otherwise had nothing in common.

The biometrics, visible in the video below, reveal that men and women had wildly different biometric responses at two different points during the ad—when Tony Hayward talks about how BP is engaging in the largest environmental response in the country's history, and when the camera zooms in on an apologetic Hayward.

Take a look at the video referred to in the quote. The video can be found at http://www.youtube.com/watch?v=eirWV0Z63YQ&feature=player_embedded

After discussing the following questions with a partner, give a brief answer to each question.

a. The video was intended to show the estimated responses of men and women at various points in time during Tony Hayward's apology. Do you think that the display does a good job of conveying this information visually? Explain.

b. Are the two points where the responses of men and women were "wildly different" obvious in the graphical display?

c. Was this study an observational study or an experiment? Did the study incorporate random selection or random assignment?

d. If the goal of the study was to generalize the estimates of the mean biometric responses for men and women to the population of American adult males and American adult females, do you have any concerns about how the samples were selected?

e. Can you suggest a better way to carry out a study so that the generalization described in Part (d) would be more reasonable?

ARE YOU READY TO MOVE ON?   CHAPTER 7 REVIEW EXERCISES

All chapter learning objectives are assessed in these exercises. The learning objectives assessed in each exercise are given in parentheses.

7.30 **(C1, M1)**
Should advertisers worry about people with digital video recorders (DVRs) fast-forwarding through their TV commercials? Recent studies by MillwardBrown and Innerscope Research indicate that when people are fast-forwarding through commercials they are actually still quite engaged and paying attention to the screen to see when the commercials end and the show they were watching starts again. If a commercial goes by that the viewer has seen before, the impact of the commercial may be equivalent to viewing the commercial at normal speed. One study of DVR viewing behavior is described in the article **"Engaging at Any Speed? Commercials Put to the Test"** (*New York Times*, July 3, 2007). For each person in a sample of adults, physical responses (such as respiratory rate and heart rate) were recorded while watching commercials at normal speed and while watching commercials at fast-forward speed. These responses were used to compute an engagement score. Engagement scores ranged from 0 to 100 (higher values indicate greater engagement). The researchers estimated that the mean engagement score for people watching at regular speed was 66, and for people watching at fast-forward speed it was 68. Is the described inference one that resulted from estimation or one that resulted from hypothesis testing?

7.31 **(C1, M1)**
Can choosing the right music make a beverage taste better? This question was investigated by a researcher at a university in Edinburgh (www.decanter.com/news). Each of 250 volunteers was assigned at random to one of five rooms where they tasted and rated a beverage. No music was playing in one of the rooms, and a different style of music was playing in each of the other four rooms. The mean rating given to the beverage under each of the five music conditions was reported. Is the described inference one that resulted from estimation or one that resulted from hypothesis testing?

7.32 **(C1, M1)**
The article **"Display of Health Risk Behaviors on MySpace by Adolescents"** (*Archives of Pediatrics and Adolescent Medicine* [2009]: 27–34) described a study of 500 publically accessible MySpace Web profiles posted by 18-year-olds. The content of each profile was analyzed, and the researchers used the resulting data to conclude that there was support for the claim that those involved in sports or a hobby were less likely to have references to risky behavior (such as sexual references or references to substance abuse or violence). Is the described inference one that resulted from estimation or one that resulted from hypothesis testing?

7.33 **(C1)**
Consider the population that consists of all people who purchased season tickets for home games of the New York Yankees.
a. Give an example of a question about this population that could be answered by collecting data and using the data to estimate a population characteristic.
b. Give an example of a question about this population that could be answered by collecting data and using the data to test a claim about this population.

7.34 **(C2)**
Data from a poll conducted by Travelocity led to the following estimates: Approximately 40% of travelers check their work e-mail while on vacation, about 33% take cell phones on vacation in order to stay connected with work, and about 25% bring laptop computers on vacation (*San Luis Obispo Tribune*, December 1, 2005).
a. What additional information about the survey would you need in order to decide if it is reasonable to generalize these estimates to the population of all American adult travelers?
b. Assuming that the given estimates were based on a representative sample, do you think that the estimates would more likely be closer to the actual population values if the sample size had been 100 or if the sample size had been 500? Explain.

7.35 **(C3)**
In a study of whether taking a garlic supplement reduces the risk of getting a cold, 146 participants were randomly assigned to either a garlic supplement group or to a group that did

not take a garlic supplement **("Garlic for the Common Cold," Cochrane Database of Systematic Reviews, 2009).** Based on the study, it was concluded that the proportion of people taking a garlic supplement who get a cold is lower than the proportion of those not taking a garlic supplement who get a cold.

a. What claim about the effect of taking garlic is supported by the data from this study?

b. Is it possible that the conclusion that the proportion of people taking garlic who get a cold is lower than the proportion for those not taking garlic is incorrect? Explain.

c. If the number of people participating in the study had been 50, do you think that the chance of an incorrect conclusion would be greater than, about the same as, or lower than for the study described?

7.36 (C4, M2)

Suppose that a study was carried out in which each student in a random sample of students at a particular college was asked if he or she was registered to vote. Would these data be used to estimate a population mean or to estimate a population proportion? How did you decide?

7.37 (C5)

Explain why the question

> **T: Type of data**—one variable or two? Categorical or numerical?

is one of the four key questions used to guide decisions about what inference method should be considered.

For each of the studies described in Exercises 7.38 to 7.40, answer the following key questions:

Q Question Type	Estimation or hypothesis testing?
S Study Type	Sample data or experiment data?
T Type of Data	One variable or two? Categorical or numerical?
N Number of Samples or Treatments	How many samples or treatments?

7.38 (C6, M2, P1)

A study of teens in Canada conducted by the polling organization Ipsos **("Untangling the Web: The Facts About Kids and the Internet," January 25, 2006)** asked each person in a sample how many hours per week he or she spent online. The responses to this question were used to learn about the mean amount of time spent online by Canadian teens.

7.39 (C6, M2, P1)

"Doctors Praise Device That Aids Ailing Hearts" (Associated Press, November 9, 2004) is the headline of an article describing a study of the effectiveness of a fabric device that acts like a support stocking for a weak or damaged heart. People who consented to treatment were assigned at random to either a standard treatment consisting of drugs or the experimental treatment that consisted of drugs plus surgery to install the stocking. After two years, 38% of the 57 patients receiving the stocking had improved, and 27% of the 50 patients receiving the standard treatment had improved. The researchers used these data to determine if there was evidence to support the claim that the proportion of patients who improve is higher for the experimental treatment than for the standard treatment.

7.40 (C6, M2, P1)

The authors of the paper **"Flat-Footedness Is Not a Disadvantage for Athletic Performance in Children Aged 11 to 15 Years" (Pediatrics [2009]: e386–e392)** collected data from 218 children on foot arch height and motor ability. The resulting data were used to investigate the relationship between arch height and motor ability.

7.41 (C7)

The process for statistical inference described in Section 7.3 consists of five steps:

a. What are the five steps in the process for estimation problems?

b. Explain how the first step differs for estimation problems and hypothesis testing problems.

AP* Review Questions for Chapter 7

1. Consider the following two statistical inference situations:

   I.   An investigator collected data from each person in a random sample of registered voters in order to learn about the value of the proportion of all registered voters who favor a particular candidate in an upcoming election.

   II. A city council member collected data from each person in a random sample of registered voters in order to determine if there is support for the claim that a majority favor an increase in the city sales tax to fund additional recreational facilities.

   Which of these inference situations are estimation problems?

   (A) I only
   (B) II only
   (C) both I and II
   (D) neither I nor II
   (E) No inference in possible in either situation because of the way the samples were selected.

2. Consider the following two statistical inference situations:

   I.   A researcher plans to carry out an experiment to compare a new medical treatment to the currently used standard treatment for a particular disease. She would like to determine if there is evidence that the proportion of patients who improve is higher for the new treatment than for the current treatment.

   II. Fifty volunteers were assigned at random to one of two groups. Those in one group watched a sequence of 10 commercials played at normal speed. Those in the second group watched the same sequence of commercials, but the commercials were played at a speed that was twice as fast as normal speed. The researcher was interested in determining if there is support for the claim that the mean number of products recalled one day later is greater for commercials viewed at normal speed than for commercials viewed at twice the normal speed.

   Which of these inference situations are hypothesis testing problems?

   (A) I only
   (B) II only
   (C) both I and II
   (D) neither I nor II
   (E) No inference in possible in either situation because both studies are experiments.

Use the following for Questions 3–5.

Consider the following two statistical studies:

Study I: Researchers studied approximately 75,000 Swedish adults and found that the proportion of those who reported eating four or more servings of low-fat dairy products per day who suffered a stroke during a 10-year period was less than the proportion for those who reported eating no servings of nonfat dairy products per day (Stroke, 2012).

Study II: A paper in the *Archives of Internal Medicine* described a study in which 77 older women with low scores on an assessment of cognitive function were randomly assigned to one of three exercise groups. Those in group 1 lifted weights twice a week, those in group 2 walked twice a week, and those in group 3 did balance and stretching exercises twice a week. After six months, the women took a memory test. The researchers concluded that those in the weight-lifting group had significantly higher scores on a memory test than those in the other two groups.

3. Which of the following describes the study type for these two studies?

   (A) The data from both Study I and Study II are sample data.
   (B) The data from both Study I and Study II are experiment data.
   (C) The data from Study I are sample data and the data from Study II are experiment data.
   (D) The data from Study I are experiment data and the data from Study II are sample data.
   (E) Based on the given study descriptions, the study types can't be determined.

4. Which of the following describes the type of data for the response variable in these two studies?

   (A) The response data from both Study I and Study II are categorical data.
   (B) The response data from both Study I and Study II are numerical data.
   (C) The response data from Study I are categorical data and the response data from Study II are numerical data.
   (D) The response data from Study I are numerical data and the response data from Study II are categorical data.
   (E) The response data from both studies are continuous numerical data.

AP* and the Advanced Placement Program are registered trademarks of the College Entrance Examination Board, which was not involved in the production of, and does not endorse, this product.

5. How many populations or treatments were compared in these two studies?

(A) Study I compared 2 populations and Study II compared 2 treatments.

(B) Study I compared 2 treatments and Study II compared 2 populations.

(C) Study I compared 2 populations and Study II compared 3 treatments.

(D) Study I compared 2 treatments and Study II compared 3 populations.

(E) Study I compared 3 populations and Study II compared 3 treatments.

8

Sampling Variability and Sampling Distributions

Preview

Chapter Learning Objectives

8.1 Statistics and Sampling Variability

8.2 The Sampling Distribution of a Sample Proportion

8.3 How Sampling Distributions Support Learning from Data

Chapter Activities

Exploring the Big Ideas

Are You Ready to Move On? Chapter 8 Review Exercises

AP* Review Questions for Chapter 8

Chabruken/The Image Bank/Getty Images

PREVIEW

When you collect sample data, it is usually to learn something about the population from which the sample was selected. To do this, you need to understand sampling variability—*the chance differences that occur from one random sample to another as a result of random selection. The following example illustrates the key ideas of this chapter in a simple setting. Even though this example is a bit unrealistic because it considers a very small population, it will help you understand what is meant by sampling variability.*

CHAPTER LEARNING OBJECTIVES

Conceptual Understanding

After completing this chapter, you should be able to

C1 Understand that the value of a sample statistic varies from sample to sample.

C2 Understand that a sampling distribution describes the sample-to-sample variability of a statistic.

C3 Understand how the standard deviation of the sampling distribution of a sample proportion is related to sample size.

C4 Understand how knowing the sampling distribution of a sample proportion enables you to use sample data to learn about a population proportion.

Mastering the Mechanics

After completing this chapter, you should be able to

M1 Define the terms statistic and sampling variability.

M2 Distinguish between a sample proportion and a population proportion and use correct notation to denote a sample proportion and a population proportion.

M3 Know that the sampling distribution of a sample proportion is centered at the actual value of the population proportion.

M4 Know the formula for the standard deviation of the sampling distribution of a sample proportion.

M5 Know what is required in order for the sampling distribution of a sample proportion to be approximately normal.

Putting It into Practice

After completing this chapter, you should be able to

P1 Determine the mean and standard deviation of the sampling distribution of a sample proportion and interpret them in context.

P2 Determine if the sampling distribution of a sample proportion is approximately normal.

P3 Use sample data and properties of the sampling distribution of a sample proportion to reason informally about a population proportion.

PREVIEW EXAMPLE   Suppose that you are interested in learning about the proportion of women in the group of students pictured in Figure 8.1 and that this group is the entire population of interest. Because 19 of the 34 students in the group are female, the proportion of women in the population is $\frac{19}{34} = 0.56$. You can compute this proportion because the picture provides complete information on gender for the entire population (a census).

FIGURE 8.1
A population of students

Digital Vision/Getty Images

But, suppose that the population information is not available. To learn about the proportion of women in the population, you decide to select a sample from the population by choosing 5 of the students at random. One possible sample is shown in Figure 8.2(a). The proportion of women in this sample (a sample statistic) is $\frac{3}{5} = 0.6$. A different sample, also resulting from random selection, is shown in Figure 8.2(b). For this second sample, the proportion of women is $\frac{2}{5} = 0.4$.

(a)   (b)

FIGURE 8.2
Two random samples of size 5

Notice the following:

1. The sample proportion is different for the two different random samples. Figure 8.2 shows only two samples, but there are many other possible samples that might result when five students are selected from this population. *The value of the sample proportion varies from sample to sample.*

2. For both of the samples considered, the value of the sample proportion was not equal to the value of the corresponding population proportion.

These observations raise the following important questions:

1. How much will the value of a sample proportion tend to differ from sample to sample?

2. How much will a sample proportion tend to differ from the actual value of the population proportion?

3. Based on what you see in a sample, can you provide an accurate estimate of a population proportion? In the context of this example, based on a sample, what can you say about the proportion students in the population who are female?

4. What conclusions can you draw based on the value of a sample proportion? For instance, in the context of this example, you might want to know if it is reasonable to conclude that a majority (more than half) of the students in the population are female.

In this chapter, you will see that these questions can be answered by looking at the *sampling distribution* of the sample proportion.

Just as the distribution of a numerical variable describes the long-run behavior of that variable, the sampling distribution of the sample proportion provides information about the long-run behavior of the sample proportion when many different random samples are selected. In this chapter, you will see how sampling distributions provide a foundation for using sample data to learn about a population proportion.

SECTION 8.1   Statistics and Sampling Variability

A number computed from the values in a sample is called a **statistic**. Statistics, such as the sample mean $\bar{x}$, the sample median, the sample standard deviation s, or the proportion of individuals in a sample that possess a particular property $\hat{p}$, provide information about population characteristics.

The most common way to learn about the value of a population characteristic is to study a sample from the population. For example, suppose you want to learn about the proportion of students at a particular school who participated in community service projects during the previous semester. A random sample of 100 students from the school could be selected. Each student in the sample could be asked whether he or she had participated in one or more community service projects in the previous semester. This would result in an observation for the variable $x =$ *community service* (with possible values yes and no) for each student in the sample. A sample proportion (denoted by $\hat{p}$) could be computed which provides an estimate of the value of the corresponding population proportion (denoted by p).

NOTATION

A **sample proportion** is denoted by $\hat{p}$.

A **population proportion** is denoted by p.

It would be nice if the value of $\hat{p}$ was equal to the value of the population proportion p, but this is an unusual occurrence. For example, $\hat{p}$ might be 0.26 for your random sample of 100 students, but this doesn't mean that exactly 26% of the students at the school participated in a community service project. The value of the sample proportion depends on which 100 students are randomly selected to be in the sample. Not only will the value of $\hat{p}$ for a particular sample differ from the population proportion p, but $\hat{p}$ values from different samples can also differ from one another. This sample-to-sample variability makes it challenging to generalize from a sample to the population from which it was selected. To meet this challenge, you need to understand sample-to-sample variability.

DEFINITION

Any quantity computed from values in a sample is called a **statistic**.

The observed value of a statistic varies from sample to sample depending on the particular sample selected. This variability is called **sampling variability**.

Example 8.1   Academic Procrastination—Exploring Sampling Variability

Consider a small population consisting of the 100 students enrolled in an introductory psychology course. Students in the class completed a survey on academic procrastination. Suppose that on the basis of their responses, 40 of the students were identified as severe procrastinators. Although these are not real data, this proportion is consistent with what was reported in a large study of university students **("Patterns of Academic Procrastination,"** *Journal of College Reading and Learning* **[2000]: 120)**. So that you have an easy way to identify the students in the population, you can assign each student a number from 1 to 100, starting with those who were classified as severe procrastinators, as shown here.

Severe Procrastinators	Students 1, 2, 3, ..., 39, 40
Not Severe Procrastinators	Students 41, 42, 43, ..., 99, 100

For this population,

$$p = \frac{40}{100} = 0.40$$

indicating that 40% are severe procrastinators.

Suppose that you don't know the value of this population proportion. You decide to estimate p by taking a random sample of 20 students. Each student in the sample completes the academic procrastination survey, and you determine how many in the sample are severe procrastinators. You then compute $\hat{p}$, the sample proportion of severe procrastinators. Is the value of $\hat{p}$ likely to be close to 0.40, the corresponding population proportion?

To answer this question, consider a simple investigation that examines the behavior of the statistic $\hat{p}$ when random samples of size 20 are selected. (This scenario is not realistic. Most populations of interest consist of many more than 100 individuals. However, this small population size is easier to work with as the idea of sampling variability is developed.)

Begin by selecting a random sample of size 20 from this population. One way to do this is by writing the numbers from 1 to 100 on slips of paper, mixing them well, and then selecting 20 slips without replacement. The numbers on the 20 slips identify which of the 100 students in the population are included in the sample. Another approach is to use either a table of random digits or a random number generator to determine which 20 students should be selected. Using JMP to obtain 20 random numbers between 1 and 100 resulted in the following:

78	16	31	86	5	67	39	28	97	70
34	65	47	89	26	79	52	4	82	6

The nine highlighted numbers correspond to students who were identified as severe procrastinators (remember, these were the students numbered 1 to 40). So, for this sample

$$\hat{p} = \frac{9}{20} = 0.45$$

This value is 0.05 larger than the population proportion of 0.40. Is this difference typical, or is this particular sample proportion unusually far away from p? Looking at some additional samples will provide some insight.

Four more random samples (Samples 2–5) from this same population are shown in the following table, along with the value of $\hat{p}$ for each sample.

Sample	Students Selected	$\hat{p}$
2	89, 63, 19, 53, 71, 62, 42, 60, 69, 43, 40, 36, 45, 90, 73, 32, 48, 56, 44, 5	$\hat{p} = \dfrac{5}{20} = 0.25$
3	73, 6, 49, 38, 21, 92, 45, 10, 93, 66, 64, 5, 59, 3, 85, 75, 81, 18, 57, 60	$\hat{p} = \dfrac{7}{20} = 0.35$
4	31, 53, 6, 64, 38, 22, 62, 92, 84, 100, 60, 17, 61, 27, 34, 55, 49, 80, 66, 8	$\hat{p} = \dfrac{8}{20} = 0.40$
5	93, 37, 25, 68, 21, 79, 44, 2, 81, 58, 17, 1, 99, 49, 3, 52, 89, 18, 11, 20	$\hat{p} = \dfrac{10}{20} = 0.50$

Remember that $p = 0.40$. You can now see the following:

1. The value of the sample proportion $\hat{p}$ varies from one random sample to another (sampling variability).

2. Some samples produced $\hat{p}$ values that were greater than p (Samples 1 and 5), some produced values less than p (Samples 2 and 3), and there was one sample with $\hat{p}$ equal to p.

3. Samples 1, 3, and 4 resulted in sample proportions that were equal to or relatively close to the value of the population proportion, but Sample 2 resulted in a value that was much farther away from p.

Continuing the investigation, 45 more random samples (each of size $n = 20$) were selected. The resulting sample proportions are shown here:

Sample	$\hat{p}$	Sample	$\hat{p}$	Sample	$\hat{p}$
6	0.40	21	0.25	36	0.65
7	0.55	22	0.25	37	0.35
8	0.55	23	0.40	38	0.50
9	0.55	24	0.35	39	0.45
10	0.30	25	0.45	40	0.30
11	0.45	26	0.35	41	0.25
12	0.50	27	0.50	42	0.30
13	0.20	28	0.40	43	0.50
14	0.35	29	0.40	44	0.35
15	0.25	30	0.30	45	0.35
16	0.40	31	0.30	46	0.55
17	0.50	32	0.55	47	0.35
18	0.35	33	0.40	48	0.50
19	0.50	34	0.45	49	0.45
20	0.25	35	0.20	50	0.35

Figure 8.3 is a relative frequency histogram of the 50 sample proportions. This histogram describes the behavior of $\hat{p}$. Many of the samples resulted in $\hat{p}$ values that are equal to or relatively close to $p = 0.40$, but occasionally the $\hat{p}$ values were as small as 0.20 or as large as 0.65. This tells you that if you were to take a sample of size 20 from this student population and use $\hat{p}$ as an estimate of the population proportion p, you should not necessarily expect $\hat{p}$ to be close to p.

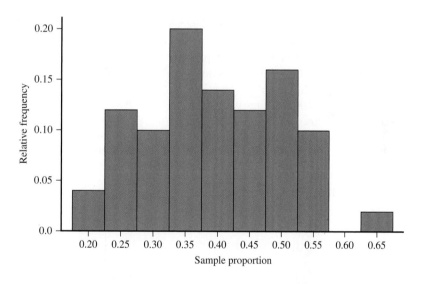

FIGURE 8.3
Histogram of $\hat{p}$ values from 50 random samples for Example 8.1

The histogram of Figure 8.3 shows the sampling variability in the statistic $\hat{p}$. It also provides an approximation to the distribution of $\hat{p}$ values that would have been observed if you had considered every different possible sample of size 20 from this population.

In the example just considered, the approximate sampling distribution of the statistic $\hat{p}$ was based on just 50 different samples. The actual sampling distribution results from considering *all* possible samples.

DEFINITION

The distribution formed by the values of a sample statistic for every possible different sample of a given size is called its **sampling distribution**.

The sampling distribution of a statistic, such as $\hat{p}$, provides important information about variability in the values of the statistic. The histogram of Figure 8.3 is an *approximation* of the sampling distribution of the statistic $\hat{p}$ for samples of size 20 from the population described in Example 8.1. To determine the actual sampling distribution of $\hat{p}$, you would need to consider every possible sample of size 20 from the population of 100 students, compute the proportion for each sample, and then construct a relative frequency histogram of these $\hat{p}$ values. But you wouldn't really want to do this—there are more than a quintillion (that's a 1 followed by 18 zeros!) different possible samples of size 20 that can be selected from a population of size 100. And, for more realistic situations with larger population and sample sizes, the situation becomes even worse! Fortunately, when more examples are considered in Section 8.2, patterns emerge that allow you to describe the sampling distribution of $\hat{p}$ without actually having to look at all possible samples.

SECTION 8.1 EXERCISES

Each Exercise Set assesses the following chapter learning objectives: C1, C2, M1, M2

SECTION 8.1 Exercise Set 1

8.1 A random sample of 1,000 students at a large college included 428 who had one or more credit cards. For this sample, $\hat{p} = \dfrac{428}{1,000} = 0.428$. If another random sample of 1,000 students from this university were selected, would you expect that $\hat{p}$ for that sample would also be 0.428? Explain why or why not.

8.2 Consider the two relative frequency histograms at the bottom of this page. The histogram on the left was constructed by selecting 100 different random samples of size 50 from a population consisting of 55% females and 45% males. For each sample, the sample proportion of females, $\hat{p}$, was computed. The 100 $\hat{p}$ values were used to construct the histogram. The histogram on the right was constructed in a similar way, except that the samples were of size 75 instead of 50.

a. Which of the two histograms shows more sample-to-sample variability? How can you tell?
b. For which of the two sample sizes, $n = 50$ or $n = 75$, do you think the value of $\hat{p}$ is more likely to be close to 0.55? What about the given histograms supports your choice?

8.3 Consider the following statement:
The Department of Motor Vehicles reports that the proportion of all vehicles registered in California that are imports is **0.22**.
a. Is the number that appears in boldface in this statement a sample proportion or a population proportion?
b. Which of the following use of notation is correct, $p = 0.22$ or $\hat{p} = 0.22$?

8.4 Consider the following statement:
A sample of size 100 was selected from those admitted to a particular college in fall 2012. The proportion of these 100 who were transfer students is **0.38**.

HISTOGRAMS FOR EXERCISE **8.2**

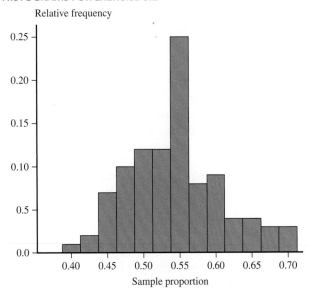

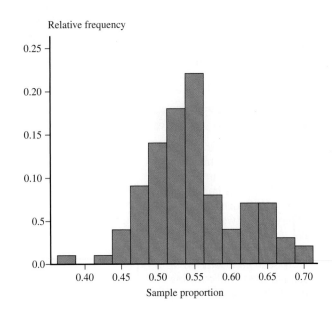

a. Is the number that appears in boldface in this statement a sample proportion or a population proportion?

b. Which of the following use of notation is correct, $p = 0.38$ or $\hat{p} = 0.38$?

8.5 Explain the difference between p and $\hat{p}$.

SECTION 8.1 **Exercise Set 2**

8.6 A random sample of 100 employees of a large company included 37 who had worked for the company for more than one year. For this sample, $\hat{p} = \dfrac{37}{100} = 0.37$. If a different random sample of 100 employees were selected, would you expect that $\hat{p}$ for that sample would also be 0.37? Explain why or why not.

8.7 Consider the two relative frequency histograms at the bottom of this page. The histogram on the left was constructed by selecting 100 different random samples of size 40 from a population consisting of 20% part-time students and 80% full-time students. For each sample, the sample proportion of part-time students, $\hat{p}$, was computed. The 100 $\hat{p}$ values were used to construct the histogram. The histogram on the right was constructed in a similar way, but using samples of size 70.

a. Which of the two histograms indicates that the value of $\hat{p}$ has smaller sample-to-sample variability? How can you tell?

b. For which of the two sample sizes, $n = 40$ or $n = 70$, do you think the value of $\hat{p}$ would be less likely to be close to 0.20? What about the given histograms supports your choice?

8.8 Explain what the term sampling variability means in the context of using a sample proportion to estimate a population proportion.

8.9 Consider the following statement:
Fifty people were selected at random from those attending a football game. The proportion of these 50 who made a food or beverage purchase while at the game was **0.83**.

a. Is the number that appears in boldface in this statement a sample proportion or a population proportion?

b. Which of the following use of notation is correct, $p = 0.83$ or $\hat{p} = 0.83$?

8.10 Consider the following statement:
The proportion of all calls made to a county 9-1-1 emergency number during the year 2011 that were nonemergency calls was **0.14**.

a. Is the number that appears in boldface in this statement a sample proportion or a population proportion?

b. Which of the following use of notation is correct, $p = 0.14$ or $\hat{p} = 0.14$?

Additional Exercises

8.11 Explain what it means when we say the value of a sample statistic varies from sample to sample.

8.12 Consider the two relative frequency histograms at the top of the next page. The histogram on the left was constructed by selecting 100 different random samples of size 35 from a population in which 17% donated to a nonprofit organization. For each sample, the sample proportion $\hat{p}$ was computed and then the 100 $\hat{p}$ values were used to construct the histogram. The histogram on the right was constructed in a similar way but used samples of size 110.

HISTOGRAMS FOR EXERCISE 8.7

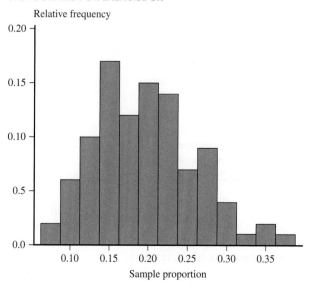

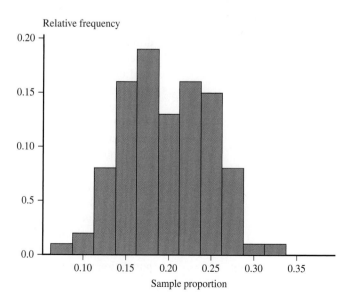

HISTOGRAMS FOR EXERCISE 8.12

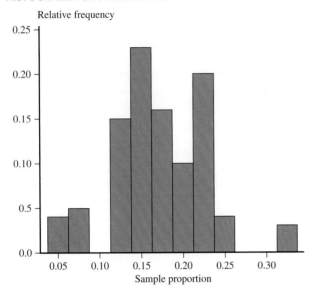

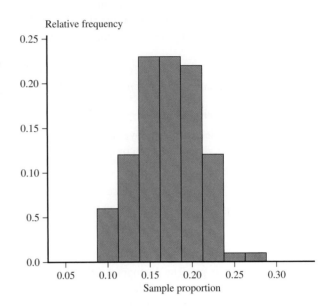

a. Which of the two histograms indicates that the value of $\hat{p}$ has more sample-to-sample variability? How can you tell?

b. For which of the two sample sizes, $n = 35$ or $n = 110$, do you think the value of $\hat{p}$ would be more likely to be close to the actual population proportion of $p = 0.17$? What about the given histograms supports your choice?

8.13 Consider the following statement:
In a sample of 20 passengers selected from those who flew from Dallas to New York City in April 2012, the proportion who checked luggage was **0.45**.

a. Is the number that appears in boldface in this statement a sample proportion or a population proportion?

b. Which of the following use of notation is correct, $p = 0.45$ or $\hat{p} = 0.45$?

8.14 Consider the following statement:
A county tax assessor reported that the proportion of property owners who paid 2012 property taxes on time was **0.93**.

a. Is the number that appears in boldface in this statement a sample proportion or a population proportion?

b. Which of the following use of notation is correct, $p = 0.93$ or $\hat{p} = 0.93$?

SECTION 8.2 The Sampling Distribution of a Sample Proportion

Sometimes you want to learn about the proportion of individuals or objects in a population that possess a particular characteristic. For example, you might be interested in the proportion of MP3 players that possess a particular manufacturing flaw, the proportion of registered voters who support a particular political candidate, or the proportion of coffee drinkers who prefer decaffeinated coffee. Because proportions are of interest in so many different settings, it helps to introduce some general terminology and notation that can be used in any context. An individual or object that possesses the characteristic of interest is called a *success*. One that does not possess the characteristic is called a *failure*. The terms success and failure mean different things in different settings. For example, in a setting where you are interested in learning about the proportion of MP3 players that have a particular manufacturing flaw, a flawed MP3 player is a "success." The letter p denotes the proportion of successes in the population. The value of p is a number between 0 and 1, and $100p$ is the percentage of successes in the population. For example, if $p = 0.75$, the population consists of 75% successes and 25% failures. If $p = 0.01$, the population consists of 1% successes and 99% failures.

The value of the population proportion p is usually unknown. When a random sample of size n is selected from the population, some of the individuals in the sample are successes, and the others are failures. The statistic that is used to draw conclusions about the population proportion is $\hat{p}$, the **sample proportion of successes**:

$$\hat{p} = \frac{\text{number of successes in the sample}}{n}$$

To learn about the sampling distribution of the statistic $\hat{p}$, we begin by considering the results of some sampling investigations. In the examples that follow, we start with a specified population, fix a sample size n, and select 500 different random samples of this size. The value of $\hat{p}$ is computed for each sample and a histogram of these 500 $\hat{p}$ values is constructed. Because 500 is a reasonably large number of samples, the histogram of the $\hat{p}$ values should resemble the actual sampling distribution of $\hat{p}$ (which would be obtained by considering *all* possible samples). This process is repeated for several different values of n to see how the choice of sample size affects the sampling distribution. Careful examination of the resulting histograms will help you to understand the general properties to be stated shortly.

| **Example 8.2** | **Gender of College Students** |

In fall 2008, there were 18,516 undergraduate students enrolled at California Polytechnic State University, San Luis Obispo. Of these students, 8,091 (43.7%) were female. What would you expect to see for the sample proportion of females if you were to take a random sample of size 10 from this population? To investigate, you can simulate sampling from this student population. With success denoting a female student, the proportion of successes in the population is $p = 0.437$. A statistical software package was used to select 500 random samples of size $n = 10$ from the population. The sample proportion of females $\hat{p}$ was computed for each sample, and these 500 values of $\hat{p}$ were used to construct the relative frequency histogram shown in Figure 8.4.

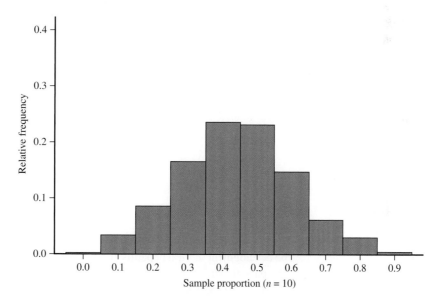

FIGURE 8.4

Histogram of 500 $\hat{p}$ values based on random samples of size $n = 10$ from a population with $p = 0.437$

The histogram in Figure 8.4 describes the behavior of the sample proportion $\hat{p}$ for samples of size $n = 10$ from this population. Notice that there is a lot of sample-to-sample variability in the value of $\hat{p}$. For some samples, $\hat{p}$ was as small as 0 or 0.1, and for other samples, $\hat{p}$ was as large as 0.9! This tells you that a sample of size 10 from this population of students may not provide very accurate information about the proportion of women in the population. What if a larger sample were selected? To investigate the effect of sample size on the behavior of $\hat{p}$, 500 samples of size 25, 500 samples of size 50, and 500 samples of size 100 were selected. Histograms of the resulting $\hat{p}$ values, along with the histogram for the samples of size 10, are displayed in Figure 8.5.

The most noticeable feature of the histogram shapes in Figure 8.5 is that they are all relatively symmetric. All four histograms appear to be centered at roughly 0.437, the value of p, the population proportion. Had the histograms been based on all possible samples rather than just 500, each histogram would have been centered at exactly 0.437. Finally, notice that the histograms spread out more for small sample sizes than for large sample sizes. Not surprisingly, the value of $\hat{p}$ based on a large sample tends to be closer to p, the population proportion of successes, than does the value of $\hat{p}$ from a small sample.

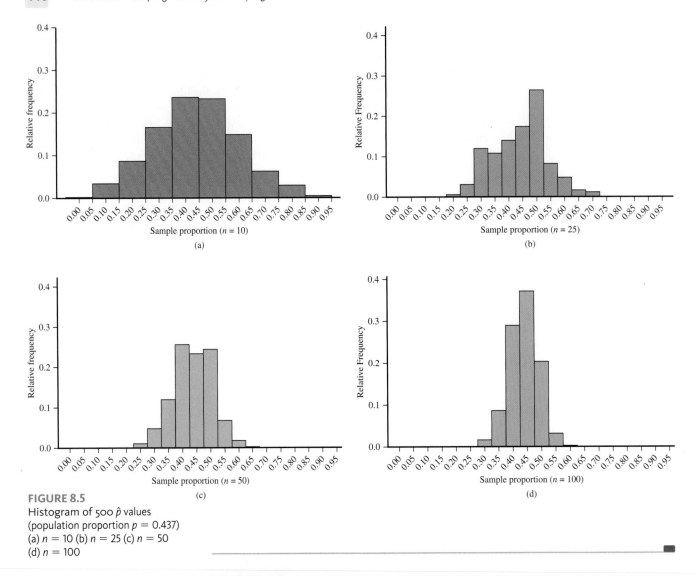

FIGURE 8.5
Histogram of 500 $\hat{p}$ values
(population proportion $p = 0.437$)
(a) $n = 10$ (b) $n = 25$ (c) $n = 50$
(d) $n = 100$

Example 8.3 Contracting Hepatitis from Blood Transfusion

The development of viral hepatitis after a blood transfusion can cause serious complications for a patient. The article **"Lack of Awareness Results in Poor Autologous Blood Transfusion"** (***Health Care Management*, May 15, 2003)** reported that hepatitis occurs in 7% of patients who receive blood transfusions during heart surgery. Let's simulate sampling from the population of patients who receive blood transfusions. For this example, a patient who contracts hepatitis will be considered a success, so $p = 0.07$. Figure 8.6 displays histograms of 500 values of $\hat{p}$ for each of the four sample sizes $n = 10, 25, 50,$ and 100.

As was the case in Example 8.2, all four histograms are centered at approximately the value of p, the population proportion. If the histograms had been based on all possible samples, they would all have been centered at exactly $p = 0.07$. The sample-to-sample variability decreases as the sample size increases. For example, the distribution is much less spread out in the histogram for $n = 100$ than for $n = 25$. The larger the value of n, the closer the sample proportion $\hat{p}$ tends to be to the value of the population proportion p.

Another thing to notice about the histograms in Figure 8.6 is the progression toward the shape of a normal curve as n increases. The histograms for $n = 10$ and $n = 25$ are quite skewed, and the skew of the histogram for $n = 50$ is still moderate (compare Figure 8.6(c) with Figure 8.5(c)). Only the histogram for $n = 100$ looks approximately normal in shape. It appears that whether a normal curve provides a good approximation to the sampling distribution of $\hat{p}$ depends on the values of both n and p. Knowing only that $n = 50$ is not enough to guarantee that the shape of the $\hat{p}$ histogram will be approximately normal.

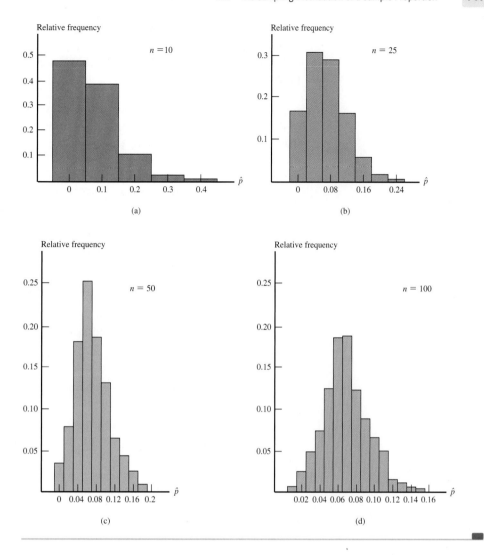

FIGURE 8.6
Histogram of 500 $\hat{p}$ values
(population proportion $p = 0.07$)
(a) $n = 10$ (b) $n = 25$ (c) $n = 50$
(d) $n = 100$

General Properties of the Sampling Distribution of $\hat{p}$

Examples 8.2 and 8.3 suggest that the center of the $\hat{p}$ distribution (the mean value of $\hat{p}$) is equal to the value of the population proportion for any sample size. Also, the spread of the $\hat{p}$ distribution decreases as n increases. The sample histograms of Figure 8.6 also suggest that in some cases, the $\hat{p}$ distribution is approximately normal in shape. These observations are stated more formally in the following general rules.

> **General Properties of the Sampling Distribution of $\hat{p}$**
>
> $\hat{p}$ is the proportion of successes in a random sample of size n from a population where the proportion of successes is p. The mean value of $\hat{p}$ is denoted by $\mu_{\hat{p}}$, and the standard deviation of $\hat{p}$ is denoted by $\sigma_{\hat{p}}$.
>
> The following rules hold.
>
> **Rule 1.** $\mu_{\hat{p}} = p$
>
> **Rule 2.** $\sigma_{\hat{p}} = \sqrt{\dfrac{p(1-p)}{n}}$. This rule is exact if the population is infinite and is approximately correct if the population is finite and no more than 10% of the population is included in the sample.
>
> **Rule 3.** When n is large and p is not too near 0 or 1, the sampling distribution of $\hat{p}$ is approximately normal.

AP* EXAM TIP

These general properties tell you about shape, center, and variability of the sampling distribution of a sample proportion, $\hat{p}$.

Rule 1, $\mu_{\hat{p}} = p$, states that the sampling distribution of $\hat{p}$ is always centered at the value of the population success proportion p. This means that the $\hat{p}$ values from many different random samples will tend to cluster around the actual value of the population proportion.

Rule 2, $\sigma_{\hat{p}} = \sqrt{\dfrac{p(1-p)}{n}}$, implies that, because the sample size n is in the denominator of the expression, sample-to-sample variability in $\hat{p}$ decreases as n increases. This means that for larger samples, $\hat{p}$ values tend to cluster more tightly around the actual value of the population proportion. This rule also gives a precise relationship between the standard deviation of $\hat{p}$ and both the population proportion p and the sample size n.

Rule 3 states that in some cases the distribution is approximately normal. Examples 8.2 and 8.3 suggest that both p and n must be considered when deciding if the sampling distribution of $\hat{p}$ is approximately normal.

When is the sampling distribution of $\hat{p}$ approximately normal?

The farther the value of p is from 0.5, the larger n has to be in order for $\hat{p}$ to have a sampling distribution that is approximately normal.

A conservative rule is that if both $np \geq 10$ and $n(1-p) \geq 10$, then the sampling distribution of $\hat{p}$ is approximately normal.

An equivalent way of stating this rule is to say that the sampling distribution of $\hat{p}$ is approximately normal if the sample size is large enough to expect at least 10 successes and at least 10 failures in the sample.

A sample size of $n = 100$ is not by itself enough to say that the sampling distribution of $\hat{p}$ is approximately normal. If $p = 0.01$, the sampling distribution is still quite skewed, so a normal curve is not a good approximation. Similarly, if $n = 100$ and $p = 0.99$ (so that $n(1-p) = 1 < 10$), the distribution of $\hat{p}$ is negatively skewed. The conditions $np \geq 10$ and $n(1-p) \geq 10$ ensure that the sampling distribution of $\hat{p}$ is not too skewed. If $p = 0.5$, the sampling distribution is approximately normal for n as small as 20, whereas for $p = 0.05$ or 0.95, n must be at least 200.

In the next section, you will see how properties of the sampling distribution of $\hat{p}$ allow you to use sample data to draw conclusions about a population proportion.

SECTION 8.2 EXERCISES

Each Exercise Set assesses the following chapter learning objectives: M2, M3, M4, P1, P2

SECTION 8.2 Exercise Set 1

8.15 A random sample is to be selected from a population that has a proportion of successes $p = 0.65$. Determine the mean and standard deviation of the sampling distribution of $\hat{p}$ for each of the following sample sizes:

a. $n = 10$ **d.** $n = 50$
b. $n = 20$ **e.** $n = 100$
c. $n = 30$ **f.** $n = 200$

8.16 For which of the sample sizes given in the previous exercise would the sampling distribution of $\hat{p}$ be approximately normal if $p = 0.65$? If $p = 0.2$?

8.17 The article **"Unmarried Couples More Likely to Be Interracial"** (*San Luis Obispo Tribune, March 13, 2002*) reported that 7% of married couples in the United States are mixed racially or ethnically. Consider the population consisting of all married couples in the United States.

a. A random sample of $n = 100$ couples will be selected from this population and $\hat{p}$, the proportion of couples that are mixed racially or ethnically, will be computed. What are the mean and standard deviation of the sampling distribution of $\hat{p}$?

b. Is the sampling distribution of $\hat{p}$ approximately normal for random samples of size $n = 100$? Explain.

c. Suppose that the sample size is $n = 200$ rather than $n = 100$. Does the change in sample size affect the mean and standard deviation of the sampling distribution of $\hat{p}$? If so, what are the new values for the mean and standard deviation? If not, explain why not.

d. Is the sampling distribution of $\hat{p}$ approximately normal for random samples of size $n = 200$? Explain.

8.18 The article referenced in the previous exercise also reported that for unmarried couples living together, the proportion that are racially or ethnically mixed is 0.15. Answer the questions posed in Parts (a)–(d) of the previous exercise for the population of unmarried couples living together.

SECTION 8.2 **Exercise Set 2**

8.19 A random sample is to be selected from a population that has a proportion of successes $p = 0.25$. Determine the mean and standard deviation of the sampling distribution of $\hat{p}$ for each of the following sample sizes:

a. $n = 10$ d. $n = 50$
b. $n = 20$ e. $n = 100$
c. $n = 30$ f. $n = 200$

8.20 For which of the sample sizes given in the previous exercise would the sampling distribution of $\hat{p}$ be approximately normal if $p = 0.25$? If $p = 0.6$?

8.21 A certain chromosome defect occurs in only 1 in 200 adult Caucasian males. A random sample of 100 adult Caucasian males will be selected. The proportion of men in this sample who have the defect, $\hat{p}$, will be computed.
a. What are the mean and standard deviation of the sampling distribution of $\hat{p}$?
b. Is the sampling distribution of $\hat{p}$ approximately normal? Explain.
c. What is the smallest value of n for which the sampling distribution of $\hat{p}$ is approximately normal?

8.22 The article **"Should Pregnant Women Move? Linking Risks for Birth Defects with Proximity to Toxic Waste Sites"** (*Chance* [1992]: 40–45) reported that in a large study carried out in the state of New York, approximately 30% of the study subjects lived within 1 mile of a hazardous waste site. Let p denote the proportion of all New York residents who live within 1 mile of such a site, and suppose that $p = 0.3$.
a. Would $\hat{p}$ based on a random sample of only 10 residents have a sampling distribution that is approximately normal? Explain why or why not.
b. What are the mean and standard deviation of the sampling distribution of $\hat{p}$ if the sample size is 400?
c. Suppose that the sample size is $n = 200$ rather than $n = 400$. Does the change in sample size affect the mean and

standard deviation of the sampling distribution of $\hat{p}$? If so, what are the new values for the mean and standard deviation? If not, explain why not.

Additional Exercises

8.23 For which of the following combinations of sample size and population proportion would the standard deviation of $\hat{p}$ be smallest?

$n = 40$ $p = 0.3$
$n = 60$ $p = 0.4$
$n = 100$ $p = 0.5$

8.24 Explain why the standard deviation of $\hat{p}$ is equal to 0 when the population proportion is equal to 1.

8.25 For which of the following sample sizes would the sampling distribution of $\hat{p}$ be approximately normal when $p = 0.2$? When $p = 0.8$? When $p = 0.6$?

$n = 10$ $n = 25$
$n = 50$ $n = 100$

8.26 A random sample of size 300 is to be selected from a population. Determine the mean and standard deviation of the sampling distribution of $\hat{p}$ for each of the following population proportions.
a. $p = 0.20$
b. $p = 0.45$
c. $p = 0.70$
d. $p = 0.90$

8.27 For which of the population proportions given in the previous exercise would the sampling distribution of $\hat{p}$ be approximately normal if $n = 40$? If $n = 75$?

8.28 The article **"Facebook Etiquette at Work"** (*USA Today*, **March 24, 2010**) reported that 56% of people participating in a survey of social network users said it was not OK for someone to "friend" his or her boss. Let p denote the proportion of all social network users who feel this way and suppose that $p = 0.56$.
a. Would $\hat{p}$ based on a random sample of 50 social network users have a sampling distribution that is approximately normal?
b. What are the mean and standard deviation of the sampling distribution of $\hat{p}$ if the sample size is 100?

SECTION 8.3 How Sampling Distributions Support Learning from Data

Before moving on, let's take a quick and informal look at the role that sampling distributions play in learning about a population characteristic. In an estimation situation, you need to understand sampling variability to assess how close an estimate is likely to be to the actual value of the corresponding population characteristic. Published reports often include statements about a margin of error. For example, a report based on a survey of randomly selected registered voters might state that the proportion of registered voters in California who support increasing state funding for community colleges is 0.55 (or 55%) with a margin of error of 0.03 (or 3%). The reported margin of error acknowledges that the actual population proportion is not likely to be exactly 0.55. It indicates that the estimate of

55% support is likely to be within 3 percentage points of the actual population percentage. This margin of error is based on an assessment of sampling variability as described by a sampling distribution. This is illustrated in Example 8.4.

Example 8.4 Visitors from Outer Space?

In this example, a situation in which investigators were interested in estimating a population proportion is considered. The authors of the article **"Believe It or Not: Religious and Other Paranormal Beliefs in the United States"** (*Journal for the Scientific Study of Religion* **[2003]: 95–106**) asked each person in a random sample of U.S. adults the following question: Do you believe that extraterrestrial beings have visited Earth at some time in the past? Of the 1,255 people in the sample, 442 answered yes to this question, resulting in a sample proportion of $\hat{p} = \dfrac{442}{1,255} = 0.35$. The population proportion who believe that extraterrestrial beings have visited Earth probably isn't exactly 0.35. How accurate is this estimate likely to be?

To answer this question, you can use what you know about the sampling distribution of $\hat{p}$ for random samples of size $n = 1,255$. Using the general results given in Section 8.2, you know three things about the sampling distribution of $\hat{p}$:

What You Know	How You Know It
The sampling distribution of $\hat{p}$ is centered at p. This means that the $\hat{p}$ values from random samples cluster around the actual value of the population proportion.	Rule 1 states that $\mu_{\hat{p}} = p$. This is true for random samples, and the description of the study says that the sample was a random sample.
Values of $\hat{p}$ will cluster fairly tightly around the actual value of the population proportion. The standard deviation of $\hat{p}$, which describes how much the $\hat{p}$ values spread out around the population proportion p, is $\sqrt{\dfrac{p(1-p)}{n}}$. An estimate of this standard deviation is $$\sqrt{\dfrac{\hat{p}(1-\hat{p})}{1,255}} = \sqrt{\dfrac{0.35(1-0.35)}{1,255}} = 0.013$$	Rule 2 states that $\sigma_{\hat{p}} = \sqrt{\dfrac{p(1-p)}{n}}$. In this example, $n = 1,255$. The actual value of the population proportion p is not known. However, $\hat{p}$ provides an estimate of p that can be used to estimate the standard deviation of the sampling distribution. The estimated standard deviation provides information about how tightly the $\hat{p}$ values from different random samples will cluster around p.
The sampling distribution of $\hat{p}$ is approximately normal.	Rule 3 states that the sampling distribution of $\hat{p}$ is approximately normal if n is large and p is not too close to 0 or 1. Here the sample size is 1,255. The sample includes 442 successes and 813 failures, which are both much greater than 10. So, it is reasonable to think that the sampling distribution of $\hat{p}$ is approximately normal.

Summarizing, you know that the $\hat{p}$ distribution is centered at the actual population proportion, has a standard deviation of about 0.013, and is approximately normal. By using this information and what you know about normal distributions, you can now get a sense of the accuracy of the estimate $\hat{p} = 0.35$. For any variable described by a normal distribution, about 95% of the values are within two standard deviations of the center. Since the sampling distribution of $\hat{p}$ is approximately normal and is centered at the actual population proportion p, you now know that about 95% of all possible random samples will produce a sample proportion that is within $2(0.013) = 0.026$ of the actual value of the population proportion. So, a margin of error of 0.026 can be reported. This tells you that the sample estimate $\hat{p} = 0.35$ is likely to be within 0.026 of the actual proportion of U.S. adults who believe that extraterrestrial beings have visited Earth at some time in the past. You could also say that plausible values for the actual proportion of U.S. adults who believe this are those between 0.324 and 0.376. This example shows that the sampling distribution of the sample proportion is the key to assessing the accuracy of the estimate.

Sample data can also be used to evaluate whether or not a claim about a population is believable. For example, the study **"Digital Footprints" (Pew Internet & American Life Project, www.pewinternet.org, 2007)** reported that 47% of Internet users have searched for information about themselves online. The value of 47% was based on a random sample of Internet users. Is it reasonable to conclude that fewer than half of *all* Internet users have searched online for information about themselves? This requires careful thought. There are two reasons why the sample proportion might be less than 0.50. One reason is sampling variability—the value of the sample proportion varies from sample to sample and won't usually be exactly equal to the value of the population proportion. Maybe the population proportion is 0.50 (or even greater) and a sample proportion of 0.47 is "explainable" just due to sampling variability. If this is the case, you can't interpret the sample proportion of 0.47 as convincing evidence that fewer than half of *all* Internet users have searched online for information about themselves. Another reason that the sample proportion might be less than 0.50 is that the actual population proportion is less than 0.50. Is 0.47 enough less than 0.50 that the difference can't be explained by sampling variability alone? If so, the sample data provide convincing evidence that fewer than half of Internet users have performed such a search. This determination is made based on an assessment of sampling variability, which is illustrated in Example 8.5.

Example 8.5 | Blood Transfusions Continued

In the article referenced in Example 8.3, the proportion of all cardiac patients receiving blood transfusions who contract hepatitis was given as 0.07. Suppose that a new blood screening procedure is believed to reduce the incidence of hepatitis. Blood screened using this procedure is given to $n = 200$ randomly selected blood recipients. Only 6 of the 200 patients contract hepatitis. This appears to be a favorable result, because $\hat{p} = \dfrac{6}{200} = 0.03$. The question of interest to medical researchers is whether this result indicates that the actual proportion of patients who contract hepatitis when the new screening procedure is used is less than 0.07, or if the smaller value of the sample proportion could be attributed to just sampling variability.

If the screening procedure is not effective, here is what you know about sample-to-sample variability in $\hat{p}$:

What You Know	How You Know It
The sampling distribution of $\hat{p}$ is centered at p. This means that if the screening procedure is not effective, the $\hat{p}$ values from random samples cluster around 0.07.	Rule 1 states that $\mu_{\hat{p}} = p$. This is true for random samples, and the description of the study says that the sample was selected at random. If the screening procedure is not effective, the proportion who get hepatitis will be the same as it was without screening, which was 0.07.
Values of $\hat{p}$ will cluster fairly tightly around the actual value of the population proportion. If the screening procedure is not effective, the standard deviation of $\hat{p}$, which describes how much the $\hat{p}$ values spread out around the population proportion p, is $$\sqrt{\frac{p(1 - p)}{n}} = \sqrt{\frac{(0.07)(0.93)}{200}} = 0.018.$$	Rule 2 states that $\sigma_{\hat{p}} = \sqrt{\dfrac{p(1 - p)}{n}}$. In this example, $n = 200$. If the screening is not effective, the value of p is 0.07. These values of n and p can be used to compute the standard deviation of the $\hat{p}$ distribution. This standard deviation provides information about how tightly the $\hat{p}$ values from different random samples will cluster around p.
The sampling distribution of $\hat{p}$ is approximately normal.	Rule 3 states that the sampling distribution of $\hat{p}$ is approximately normal if n is large and p is not too close to 0 or 1. Here the sample size is 200 and if the screening procedure is not effective, $p = 0.07$. Because $np = 200(0.07) = 14 \geq 10$ and $n(1 - p) = 200(0.93) = 186 \geq 10$, the sampling distribution of $\hat{p}$ is approximately normal.

Summarizing, you know that if the screening is not effective, the $\hat{p}$ distribution is centered at 0.07, has a standard deviation of 0.18, and is approximately normal. By using this information and what you know about normal distributions, you can now determine how likely it is that a sample proportion as small as 0.03 would be observed.

Step	Explanation
Determine what probability will answer the question posed: $P(\hat{p} \le 0.03)$	You are interested in whether a sample proportion as small as 0.03 would be unusual if the screening is not effective. This can be evaluated by computing the probability of observing a sample proportion of 0.03 or smaller just by chance (due to sampling variability alone).
$P(\hat{p} \le 0.03) = P\left(z \le \dfrac{0.03 - 0.07}{0.018}\right)$ $= P(z \le -2.22)$	You know that if the screening is not effective, the $\hat{p}$ distribution for random samples of size $n = 200$ is centered at 0.07, has standard deviation 0.018, and is approximately normal. Since the $\hat{p}$ distribution is approximately normal, but not *standard* normal, to compute the desired probability, translate the problem into an equivalent problem for the standard normal by using a z-score. (See Chapter 6 if you don't remember how to do this.)
$P(z \le -2.22) = 0.0132$	The standard normal table, a calculator with statistical functions, or computer software can then be used to obtain the desired probability.

Based on the probability calculation, you now know that if the screening is *not* effective, it would be very unlikely (probability 0.0132) that a sample proportion as small as 0.03 would be observed for a sample of size $n = 200$. The new screening procedure does appear to lead to a smaller incidence rate for hepatitis. The sampling distribution of $\hat{p}$ played a key role in reaching this conclusion.

Examples 8.4 and 8.5 used informal reasoning based on the sampling distribution of $\hat{p}$ to learn about a population proportion. This type of reasoning is considered more formally in Chapters 9 and 10.

SECTION 8.3 EXERCISES

Each Exercise Set assesses the following chapter learning objectives: C4, P3

SECTION 8.3 Exercise Set 1

8.29 Suppose that a particular candidate for public office is favored by 48% of all registered voters in the district. A polling organization will take a random sample of 500 of these voters and will use $\hat{p}$, the sample proportion, to estimate p.

a. Show that $\sigma_{\hat{p}}$, the standard deviation of $\hat{p}$, is equal to 0.0223.

b. If for a different sample size, $\sigma_{\hat{p}} = 0.0500$, would you expect more or less sample-to-sample variability in the sample proportions than when $n = 500$?

c. Is the sample size that resulted in $\sigma_{\hat{p}} = 0.0500$ larger than 500 or smaller than 500? Explain your reasoning.

8.30 The article **"Students Increasingly Turn to Credit Cards"** (*San Luis Obispo Tribune*, July 21, 2006) reported that 37% of college freshmen carry a credit card balance from month to month. Suppose that the reported percentage was based on a random sample of 1,000 college freshmen. Suppose you are interested in learning about the value of p, the proportion of all college freshmen who carry a credit card balance from month to month.

The following table is similar to the table that appears in Examples 8.4 and 8.5, and is meant to summarize what you know about the sampling distribution of $\hat{p}$ in the situation just described. The "What You Know" information has been provided. Complete the table by filling in the "How You Know It" column.

What You Know	How You Know It
The sampling distribution of $\hat{p}$ is centered at the actual (but unknown) value of the population proportion.	
An estimate of the standard deviation of $\hat{p}$, which describes how much the $\hat{p}$ values spread out around the population proportion p, is 0.0153.	
The sampling distribution of $\hat{p}$ is approximately normal.	

8.31 **"Tongue Piercing May Speed Tooth Loss, Researchers Say"** is the headline of an article that appeared in the *San Luis Obispo Tribune* (June 5, 2002). The article describes a study of 52 young adults with pierced tongues. The researchers believed that it was reasonable to regard this sample as a random sample from the population of all young adults with pierced tongues. The researchers found receding gums, which can lead to tooth loss, in 18 of the study participants.

a. Suppose you are interested in learning about the value of p, the proportion of all young adults with pierced tongues who have receding gums. This proportion can be estimated using the sample proportion, $\hat{p}$. What is the value of $\hat{p}$ for this sample?

b. Based on what you know about the sampling distribution of $\hat{p}$, is it reasonable to think that this estimate is within 0.05 of the actual value of the population proportion? Explain why or why not.

8.32 The report **"California's Education Skills Gap: Modest Improvements Could Yield Big Gains" (Public Policy Institute of California, April 16, 2008, www.ppic.org)** states that nationwide, 61% of high school graduates go on to attend a two-year or four-year college the year after graduation. The proportion of high school graduates in California who go on to college was estimated to be 0.55. Suppose that this estimate was based on a random sample of 1,500 California high school graduates. Is it reasonable to conclude that the proportion of California high school graduates who attend college the year after graduation is different from the national figure? (Hint: Use what you know about the sampling distribution of $\hat{p}$. You might also refer to Example 8.5.)

SECTION 8.3 Exercise Set 2

8.33 Suppose that 20% of the customers of a cable television company watch the Shopping Channel at least once a week. The cable company does not know the actual proportion of all customers who watch the Shopping Channel at least once a week and is trying to decide whether to replace this channel with a new local station. The company plans to take a random sample of 100 customers and to use $\hat{p}$ as an estimate of the population proportion.

a. Show that $\sigma_{\hat{p}}$, the standard deviation of $\hat{p}$, is equal to 0.0400.

b. If for a different sample size, $\sigma_{\hat{p}} = 0.0231$, would you expect more or less sample-to-sample variability in the sample proportions than when $n = 100$?

c. Is the sample size that resulted in $\sigma_{\hat{p}} = 0.0231$ larger than 100 or smaller than 100? Explain your reasoning.

8.34 The article **"CSI Effect Has Juries Wanting More Evidence"** (*USA Today*, **August 5, 2004**) examines how the popularity of crime-scene investigation television shows is influencing jurors' expectations of what evidence should be produced at a trial. In a random sample of 500 potential jurors, one study found that 350 were regular watchers of at least one crime-scene forensics television series. Suppose you are interested in learning about the value of p, the proportion of all potential jurors who are regular watchers of crime-scene shows. The following table is similar to the table that appears in Examples 8.4 and 8.5 and is meant to summarize what you know about the sampling distribution of $\hat{p}$ in the situation just described. The "What You Know" information has been provided. Complete the table by filling in the "How You Know It" column.

What You Know	How We Know It
The sampling distribution of $\hat{p}$ is centered at the actual (but unknown) value of the population proportion.	
An estimate of the standard deviation of $\hat{p}$, which describes how much the $\hat{p}$ values spread out around the population proportion p, is 0.0205.	
The sampling distribution of $\hat{p}$ is approximately normal.	

8.35 The article **"Career Expert Provides DOs and DON'Ts for Job Seekers on Social Networking" (CareerBuilder.com, August 19, 2009)** included data from a survey of 2,667 hiring managers and human resource professionals. The article noted that many employers are using social networks to screen job applicants and that this practice is becoming more common. Of the 2,667 people who participated in the survey, 1,200 indicated that they use social networking sites (such as Facebook, MySpace, and LinkedIn) to research job applicants. For the purposes of this exercise, assume that the sample can be regarded as a random sample of hiring managers and human resource professionals.

a. Suppose you are interested in learning about the value of p, the proportion of all hiring managers and human resource managers who use social networking sites to research job applicants. This proportion can be estimated using the sample proportion, $\hat{p}$. What is the value of $\hat{p}$ for this sample?

b. Based on what you know about the sampling distribution of $\hat{p}$, is it reasonable to think that this estimate is within 0.02 of the actual value of the population proportion? Explain why or why not.

8.36 In a random sample of 1,000 adult Americans, only 430 could name at least one justice who is currently serving on the U.S. Supreme Court **(Ipsos, January 10, 2006).** Is it reasonable to conclude that the proportion of adult Americans who could name at least one justice is less than 0.5? (Hint: Use what you know about the sampling distribution of $\hat{p}$. You might also refer to Example 8.5.)

Additional Exercises

8.37 Some colleges now allow students to rent textbooks for a semester. Suppose that 38% of all students enrolled at a particular college would rent textbooks if that option were available to them. If the campus bookstore uses a random sample of size 100 to estimate the proportion of students at the college who would rent textbooks, is it likely that this estimate would be within 0.05 of the actual population proportion? Use what you know about the sampling distribution of $\hat{p}$ to support your answer.

8.38 In a study of pet owners, it was reported that 24% celebrate their pet's birthday **(Pet Statistics, Bissell Homecare, Inc., 2010).** Suppose that this estimate was from a random sample of 200 pet owners. Is it reasonable to conclude that the proportion of all pet owners who celebrate their pet's birthday is less than 0.25? Use what you know about the sampling distribution of $\hat{p}$ to support your answer.

8.39 The article **"Facebook Etiquette at Work" (USA Today, March 24, 2010)** reported that 56% of 1,200 social network users surveyed indicated that they thought it was not OK for someone to "friend" his or her boss. Suppose that this sample can be regarded as a random sample of social network users. Is it reasonable to conclude that more than half of social network users feel this way? Use what you know about the sampling distribution of $\hat{p}$ to support your answer.

8.40 The article referenced in the previous exercise also reported that 38% of the 1,200 social network users surveyed said it was OK to ignore a coworker's friend request. If $\hat{p} = 0.38$ is used as an estimate of the proportion of *all* social network users who believe this, is it likely that this estimate is within 0.05 of the actual population proportion? Use what you know about the sampling distribution of $\hat{p}$ to support your answer.

CHAPTER ACTIVITIES

ACTIVITY 8.1   DEFECTIVE M&Ms

Materials needed: 100 plain M&Ms for each group of four students

In this activity, you will work in a group with three other students as M&M inspectors. You will investigate the distribution of the proportion of "defective" M&Ms in a sample of size 100.

1. Some M&Ms are defective. Examples of possible defects include a broken M&M, a misshapen M&M, and an M&M that does not have the "m" on the candy coating.

    Suppose that a claim has been made that 10% of plain M&Ms are defective. Assuming this claim is correct, describe the sampling distribution of $\hat{p}$, the proportion of defective M&Ms in the sample, for a sample of size 100. (That is, if $p = 0.10$ and $n = 100$, what is the sampling distribution of $\hat{p}$?)

2. Data collection: With your teammates, carefully inspect 100 M&Ms. Sort them into two piles, with nondefective M&Ms in one pile and defective M&Ms in a second pile. Compute the sample proportion of defective M&Ms.

3. If the claim of 10% defective is true, would a sample proportion as extreme as what you observed in Step 2 be unusual, or is your sample proportion from Step 2 consistent with a 10% population defective rate? Justify your answer based on what you said about the sampling distribution of $\hat{p}$ in Step 1.

EXPLORING THE BIG IDEAS

In the exercise below, you will go online to select random samples from a population of adults between the ages of 18 and 64.

Each person in this population was asked if they ever slept with their cell phone. The proportion of people who responded yes to this question was $p = 0.64$. This is a population proportion. Suppose that you didn't know the value of this proportion and that you planned to take a random sample of $n = 20$ people in order to estimate p.

Go online to **www.cengage.com/stats/peck1e** and click on the link labeled "Exploring the Big Ideas." Then locate the link for Chapter 8. This link will take you to a web page where you can select random samples of 20 people from the population.

Click on the Select Sample button. This selects a random sample and will display the following information:

1. The ID number that identifies the person selected.
2. The response to the question "Do you sometimes sleep with your cell phone?" These responses are coded numerically—a 1 indicates a yes response and a 2 indicates a no response.

The value of $\hat{p}$ for this sample will also be computed and displayed.

Record the value of $\hat{p}$ for your first sample. Then click the Select Sample button again to get another random sample. Record the value of $\hat{p}$ for your second sample. Continue to select samples until you have recorded $\hat{p}$ values for 25 different random samples.

Use the 25 $\hat{p}$ values to complete the following:
a. Construct a dotplot of the 25 $\hat{p}$ values. Is the dotplot centered at about 0.64, the value of the population proportion?
b. One way to estimate $\sigma_{\hat{p}}$, the standard deviation of the sampling distribution of $\hat{p}$, is to calculate the standard deviation of the 25 $\hat{p}$ values. Compute this estimate.
c. Because $p = 0.64$, the standard deviation of the sampling distribution of $\hat{p}$ is
$$\sigma_{\hat{p}} = \sqrt{\frac{p(1-p)}{n}} = \sqrt{\frac{(0.64)(0.36)}{20}} = 0.11.$$ Was your estimate of $\sigma_{\hat{p}}$ from Part (b) close to this value?
d. Explain how the dotplot you constructed in Part (a) is or is not consistent with the following statement: Because $p = 0.64$ and $n = 20$, it is not reasonable to think that the sampling distribution of $\hat{p}$ is approximately normal.

ARE YOU READY TO MOVE ON? CHAPTER 8 REVIEW EXERCISES

All chapter learning objectives are assessed in these exercises. The learning objectives assessed in each exercise are given in parentheses.

8.41 (C1)
A random sample of 50 registered voters in a particular city included 32 who favored using city funds for the construction of a new recreational facility. For this sample, $\hat{p} = \dfrac{32}{50} = 0.64$. If a second random sample of 50 registered voters was selected, would it surprise you if $\hat{p}$ for that sample was not equal to 0.64? Why or why not?

8.42 (C1, C2)
Consider the following three relative frequency histograms. Each histogram was constructed by selecting 100 random samples from a population composed of 40% women and 60% men. For each sample, the sample proportion of women, $\hat{p}$, was computed and the 100 $\hat{p}$ values were used to construct the histogram. For each histogram, a different sample size was used. One histogram was constructed using 100 samples of size 20, one was constructed using 100 samples of size 40, and one was constructed using 100 samples of size 100.

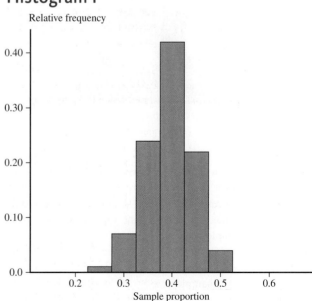

Histogram I

Histogram II

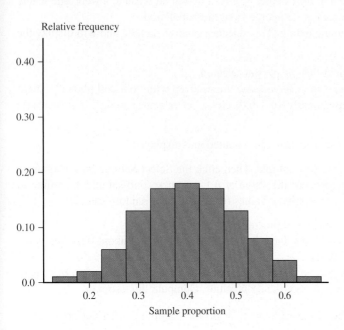

Histogram III

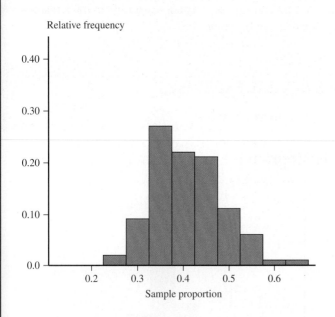

a. These three histograms are approximate sampling distributions of $\hat{p}$ for the three different sample sizes. Which histogram corresponds to the sample size that would produce sample proportions that varied the most from sample to sample?

b. Match the three sample sizes to the three histograms. That is, which histogram is the approximate sampling distribution for sample size = 20, which is for sample size = 40, and which is for sample size = 100? Explain how you made your choices.

8.43 (M1)
Explain why there is sample-to-sample variability in $\hat{p}$ but not in p.

8.44 (M1, M2)
Consider the following statement:
An inspector selected 20 eggs at random from the eggs processed at a large egg production facility. These 20 eggs were tested for salmonella, and the proportion of the eggs that tested positive for salmonella was **0.05**.
a. Is the number that appears in boldface in this statement a sample proportion or a population proportion?
b. Which of the following use of notation is correct, $p = 0.05$ or $\hat{p} = 0.05$?

8.45 (M1, M2)
Consider the following statement:
The proportion of all students enrolled at a particular university during 2012 who lived on campus was **0.21**.
a. Is the number that appears in boldface in this statement a sample proportion or a population proportion?
b. Which of the following use of notation is correct, $p = 0.21$ or $\hat{p} = 0.21$?

8.46 (M2, M3)
A random sample will be selected from a population that has a proportion of successes $p = 0.70$. Determine the mean and standard deviation of the sampling distribution of $\hat{p}$ for each of the following sample sizes:
a. $n = 10$ d. $n = 50$
b. $n = 20$ e. $n = 100$
c. $n = 30$ f. $n = 200$

8.47 (M4)
For which of the sample sizes given in the previous exercise would the sampling distribution of $\hat{p}$ be approximately normal if $p = 0.70$? If $p = 0.30$?

8.48 (P1, P2)
The article **"Thrillers"** (**Newsweek, April 22, 1985**) stated, "Surveys tell us that more than half of America's college graduates are avid readers of mystery novels." Let p denote the actual proportion of college graduates who are avid readers of mystery novels. Consider a sample proportion $\hat{p}$ that is based on a random sample of 225 college graduates. If $p = 0.5$, what are the mean value and standard deviation of the sampling distribution of $\hat{p}$? Answer this question for $p = 0.6$. Is the sampling distribution of $\hat{p}$ approximately normal in both cases? Explain.

8.49 (P1, P2)
The report **"New Study Shows Need for Americans to Focus on Securing Online Accounts and Backing Up Critical Data"** (**PRNewswire, October 29, 2009**) reported that only 25% of Americans change computer passwords quarterly, in spite of a recommendation from the National Cyber Security

Alliance that passwords be changed at least once every 90 days. For purposes of this exercise, assume that the 25% figure is correct for the population of adult Americans.

a. A random sample of size $n = 200$ will be selected from this population and $\hat{p}$, the proportion who change passwords quarterly, will be computed. What are the mean and standard deviation of the sampling distribution of $\hat{p}$?

b. Is the sampling distribution of $\hat{p}$ approximately normal for random samples of size $n = 200$? Explain.

c. Suppose that the sample size is $n = 50$ rather than $n = 200$. Does the change in sample size affect the mean and standard deviation of the sampling distribution of $\hat{p}$? If so, what are the new values of the mean and standard deviation? If not, explain why not.

d. Is the sampling distribution of $\hat{p}$ approximately normal for random samples of size $n = 50$? Explain.

8.50 (C4)

Suppose that the actual proportion of students at a particular college who use public transportation to travel to campus is 0.15. In a study of parking needs at the campus, college administrators would like to estimate this proportion. They plan to take a random sample of 75 students and use the sample proportion who use public transportation, $\hat{p}$, as an estimate of the population proportion.

a. Show that the standard deviation of $\hat{p}$ is equal to $\sigma_{\hat{p}} = 0.0412$.

b. If for a different sample size, $\sigma_{\hat{p}} = 0.0319$, would you expect more or less sample-to-sample variability in the sample proportions than for when $n = 75$?

c. Is the sample size that resulted in $\sigma_{\hat{p}} = 0.0319$ larger than 75 or smaller than 75? Explain your reasoning.

8.51 (C4)

In a survey of 1,000 randomly selected adults in the United States, participants were asked what their most favorite and least favorite subjects were when they were in school **(Associated Press, August 17, 2005)**. Math was chosen by 230 of the 1,000 as their favorite subject. Suppose you are interested in learning about the value of p, the proportion of all adults for whom math was their favorite subject in school. The following table is similar to the table that appears in Examples 8.4 and 8.5 and is meant to summarize what you know about the sampling distribution of $\hat{p}$ in the situation just described. The "What You Know" information has been provided. Complete the table by filling in the "How You Know It" column.

What You Know	How You Know It
The sampling distribution of $\hat{p}$ is centered at the actual (but unknown) value of the population proportion.	
An estimate of the standard deviation of $\hat{p}$, which describes how much the $\hat{p}$ values spread out around the population proportion p, is 0.0133.	
The sampling distribution of $\hat{p}$ is approximately normal.	

8.52 (P3)

If a hurricane were headed your way, would you evacuate? The headline of a press release issued January 21, 2009, by the survey research company International Communications Research (**icrsurvey.com**) states, "Thirty-one Percent of People on High-Risk Coast Will Refuse Evacuation Order, Survey of Hurricane Preparedness Finds." This headline was based on a survey of 5,046 adults who live within 20 miles of the coast in high hurricane risk counties of eight southern states. The sample was selected to be representative of the population of coastal residents in these states, so assume that it is reasonable to regard the sample as if it were a random sample.

a. Suppose you are interested in learning about the value of p, the proportion of adults who would refuse to evacuate. This proportion can be estimated using the sample proportion, $\hat{p}$. What is the value of $\hat{p}$ for this sample?

b. Based on what you know about the sampling distribution of $\hat{p}$, is it reasonable to think that the estimate is within 0.03 of the actual value of the population proportion? Explain why or why not.

8.53 (P3)

In a national survey of 2,013 American adults, 1,283 indicated that they believe that rudeness is a more serious problem than in past years **(Associated Press, April 3, 2002)**. Assume that it is reasonable to regard this sample as a random sample of adult Americans. Is it reasonable to conclude that the proportion of adults who believe that rudeness is a worsening problem is greater than 0.5? (Hint: Use what you know about the sampling distribution of $\hat{p}$. You might also refer to Example 8.5.)

AP* Review Questions for Chapter 8

1. Which of the following best describes what is meant by the term *sampling variability*?

   (A) There are many different methods for selecting a sample.

   (B) Two different samples from the same population may have different values for a sample statistic.

   (C) The sample size can sometimes vary from sample to sample.

   (D) Convenience samples may not be representative of the population.

   (E) Population characteristics vary from population to population.

2. A random sample is to be selected from a population. For which combination of n and p is it reasonable to assume that the sampling distribution of the sample proportion $\hat{p}$ will be approximately normal?

   (A) $n = 10$ and $p = 0.4$
   (B) $n = 25$ and $p = 0.5$
   (C) $n = 30$ and $p = 0.2$
   (D) $n = 40$ and $p = 0.1$
   (E) $n = 100$ and $p = 0.05$

3. Suppose that 50% of those taking a national licensing exam pass the exam on the first attempt. A random sample of $n = 100$ people taking the exam for the first time will be selected. If $\hat{p}$ is the proportion in the sample that pass the exam on the first attempt, what are the mean and standard deviation of the sampling distribution of $\hat{p}$?

   (A) The mean is 0.5 and the standard deviation is 0.25.

   (B) The mean is 0.5 and the standard deviation is 0.025.

   (C) The mean is 0.5 and the standard deviation is $\sqrt{0.0025}$.

   (D) The mean is 0.25 and the standard deviation is 0.25.

   (E) The mean is 0.25 and the standard deviation is $\sqrt{0.0025}$.

4. For which of the following will the sample proportion tend to differ least from sample to sample?

   (A) Random samples of size 30 from a population with $p = 0.1$

   (B) Random samples of size 40 from a population with $p = 0.1$

   (C) Random samples of size 40 from a population with $p = 0.4$

   (D) Random samples of size 50 from a population with $p = 0.5$

   (E) Random samples of size 60 from a population with $p = 0.5$

5. Which of the following is true?

   I.   The sampling distribution of $\hat{p}$ has a mean equal to p.
   II. The sampling distribution of $\hat{p}$ has a standard deviation equal to $\sqrt{np(1 - p)}$.
   III. The sampling distribution of $\hat{p}$ is always symmetric

   (A) I only
   (B) II only
   (C) III only
   (D) I and II only
   (E) I and III only

9

Estimating a Population Proportion

Preview

Chapter Learning Objectives

9.1 Selecting an Estimator

9.2 Estimating a Population Proportion—Margin of Error

9.3 A Large Sample Confidence Interval for a Population Proportion

9.4 Choosing a Sample Size to Achieve a Desired Margin of Error

9.5 Avoid These Common Mistakes

Chapter Activities

Exploring the Big Ideas

Are You Ready to Move On? Chapter 9 Review Exercises

Technology Notes

AP* Review Questions for Chapter 9

Bill Manning/istockphoto.com

PREVIEW

When a sample is selected from a population, it is usually because you hope it will provide information about the population. For example, you might want to use sample data to learn about the value of a population characteristic such as the proportion of students enrolled at a college who purchase textbooks online or the mean number of hours that students at the college spend studying each week. This chapter considers how sample data can be used to estimate the value of a population proportion.

CHAPTER LEARNING OBJECTIVES

Conceptual Understanding

After completing this chapter, you should be able to

C1 Understand what makes a statistic a "good" estimator of a population characteristic.

C2 Know what it means to say that a statistic is an unbiased estimator of a population characteristic.

C3 Understand how the standard deviation of the sample proportion, $\hat{p}$, is related to sample size and to the value of the population proportion.

C4 Understand the relationship between sample size and margin of error.

C5 Understand how an interval is used to estimate a population characteristic.

C6 Know what factors affect the width of a confidence interval estimate of a population proportion.

C7 Understand the meaning of the confidence level associated with a confidence interval estimate.

Mastering the Mechanics

After completing this chapter, you should be able to

M1 Use a sample proportion $\hat{p}$ to estimate a population proportion p and compute the associated margin of error.

M2 Know the conditions for appropriate use of the margin of error and confidence interval formulas when estimating a population proportion.

M3 Know the key characteristics that lead to selection of the z confidence interval for a population proportion as an appropriate method.

M4 Use the five-step process for estimation problems (EMC³) to compute and interpret a confidence interval for a population proportion.

M5 Compute the sample size necessary to achieve a desired margin of error when estimating a population proportion.

Putting It in to Practice

After completing this chapter, you should be able to

P1 Interpret a margin of error in context.

P2 Interpret a confidence interval for a population proportion in context and interpret the associated confidence level.

P3 Determine the required sample size, given a description of a proposed study and a desired margin of error.

PREVIEW EXAMPLE   Hurricane Evacuation

If a hurricane were headed your way, would you evacuate? The headline of a press release issued January 21, 2009, by the survey research company **International Communications Research (icrsurvey.com)** states "31 Percent of People on High Risk Coast Will Refuse Evacuation Order, Survey of Hurricane Preparedness Finds." This headline was based on a survey of a representative sample of 5,046 coastal residents in eight southern states.

The survey was conducted so that officials involved in emergency planning could learn what proportion of residents might refuse to evacuate. For an estimate to be useful, these officials would need to know something about its accuracy. Because the given estimate of 0.31 (corresponding to 31%) is a sample proportion and the sample proportion will vary from sample to sample, you don't expect the actual *population* proportion to be exactly 0.31. The sample size was large, and the sample was selected to be representative of the population of coastal residents, so you might think that the sample proportion would be "close" to the actual value of the population proportion. But how close? Could the actual value of the population proportion be as large as 0.40? As small as 0.25? Being able to evaluate the accuracy of this estimate is important to the people responsible for developing emergency response plans.

This example will be revisited in Section 9.3 after you have learned how sampling distributions enable you to evaluate the accuracy of estimates.

Selecting an Estimator

The first step in estimating a population characteristic, such as a population proportion or a population mean, is to select an appropriate sample statistic. When the goal is to estimate a population proportion, p, the usual choice is the sample proportion $\hat{p}$. But what makes $\hat{p}$ a reasonable choice? More generally, what makes any particular statistic a good choice for estimating a population characteristic?

Because of sampling variability, the value of any statistic you might choose as a potential estimator will vary from one random sample to another. Taking this variability into account, two questions are of interest:

1. Will the statistic consistently tend to overestimate or to underestimate the value of the population characteristic?
2. Will the statistic tend to produce values that are close to the actual value of the population characteristic?

A statistic that does not consistently tend to underestimate or to overestimate the value of a population characteristic is said to be an **unbiased** estimator of that characteristic. Ideally, you would choose a statistic that is unbiased or for which the bias is small (meaning, for example, that if there is a tendency for the statistic to underestimate, the underestimation tends to be small). You can tell if this is the case by looking at the statistic's sampling distribution. If the sampling distribution is centered at the actual value of the population characteristic, then the statistic is unbiased.

The sampling distribution of a statistic also has a standard deviation, which describes how much the values of the statistic vary from sample to sample. If a sampling distribution is centered very close to the actual value of the population characteristic, a small standard deviation ensures that values of the statistic will cluster tightly around the actual value of the population characteristic. This means that the value of the statistic will tend to be close to the actual value. Because the standard deviation of a sampling distribution is so important in evaluating the accuracy of an estimate, it is given a special name: the **standard error** of the statistic. For example, the standard error of the sample proportion $\hat{p}$ is

$$\text{standard error of } \hat{p} = \sqrt{\frac{p(1-p)}{n}}$$

where p is the value of the population proportion and n is the sample size (see Section 8.2).

This new terminology can be used to describe what makes a statistic a good estimator of a population characteristic.

A statistic with a sampling distribution that is centered at the actual value of a population characteristic is an **unbiased estimator** of that population characteristic. In other words, a statistic is unbiased if the mean of its sampling distribution is equal to the actual value of the population characteristic.

The standard deviation of the sampling distribution of a statistic is called the **standard error** of the statistic.

What makes a statistic a good estimator of a population characteristic?

1. Unbiased (or nearly unbiased)
2. Small standard error

A statistic that is unbiased and has a small standard error is likely to result in an estimate that is close to the actual value of the population characteristic.

As an example of a statistic that is biased, consider the sample range as an estimator of the population range. Recall that the range of a population is defined as the difference between its smallest and largest values. The sample range equals the population range *only* if the sample includes both the largest and the smallest values in the whole population. In any other case, the sample range is smaller than the population range, so the sample range tends to underestimate the population range. The sampling distribution of the sample range is not centered at the actual value of the population range.

The sampling distribution of a statistic provides information about the accuracy of estimation. Figure 9.1 displays the sampling distributions of three different statistics. The value of the population characteristic, which is labeled *population value* in the figure, is marked on the measurement axis.

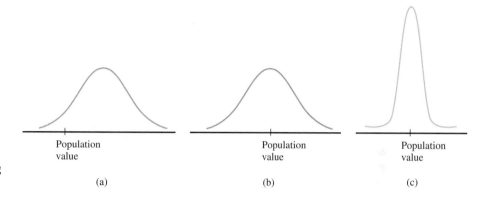

FIGURE 9.1
Sampling distributions of three different statistics for estimating a population characteristic.

The sampling distribution in Figure 9.1(a) is from a statistic that is not likely to result in an estimate close to the population value. The distribution is centered to the right of the population value, making it very likely that an estimate (the value of the statistic for a particular sample) will be larger than the population value. This statistic is not unbiased and will consistently tend to overestimate the population value.

The sampling distribution in Figure 9.1(b) is centered at the population value. Although one estimate may be smaller than the population value and another may be larger, with this statistic there will be no long-run tendency to over- or underestimate the population value. However, even though the statistic is unbiased, the sampling distribution spreads out quite a bit around the population value. Because of this, some estimates will be far above or far below the population value.

The mean value of the statistic with the distribution shown in Figure 9.1(c) is also equal to the population value, implying no systematic tendency to over- or underestimate. The narrow shape of the distribution curve implies that the statistic's standard error (standard deviation) is relatively small. Estimates based on this statistic will almost always be quite close to the population value—certainly more often than estimates from the sampling distribution shown in Figure 9.1(b). This statistic has the characteristics of a good estimator—it is unbiased and has a small standard error.

Let's return to the problem of using sample data to estimate a population proportion. In Chapter 8, you learned about properties of the sampling distribution of the statistic $\hat{p}$, the sample proportion. Two of the properties considered there were:

1. $\mu_{\hat{p}} = p$. This tells you that the mean of the sampling distribution of $\hat{p}$ is equal to the value of the population proportion p. The $\hat{p}$ values from all the different possible random samples of size n will center around the actual value of the population proportion. In other words, $\hat{p}$ is an unbiased estimator of the population proportion, p.

2. $\sigma_{\hat{p}} = \sqrt{\dfrac{p(1-p)}{n}}$. This specifies the standard error (standard deviation) of the statistic $\hat{p}$. The standard error describes how much $\hat{p}$ values will spread out around the actual value of the population proportion. If the sample size is large, the standard error will tend to be small.

These two observations suggest that the statistic $\hat{p}$ is a reasonable estimator of the corresponding population proportion as long as the sample size is large enough to ensure a relatively small standard error.

Example 9.1 State Supreme Court Decisions

The authors of the paper **"A Comparison of the Criminal Appellate Decisions of Appointed State Supreme Courts: Insights, Questions, and Implications for Judicial Independence"** (*The Fordham Urban Law Journal* [April 4, 2007]: 343–362) examined *all* criminal cases heard by the Supreme Courts of 11 states from 2000 to 2004. Of the 1,488 cases considered, 391 were decided in favor of the defendant. Using this information, you can compute

$$p = \frac{391}{1{,}488} = 0.263$$

where p is a population proportion because *all* cases were considered.

Suppose that the population proportion $p = 0.263$ was not known. To estimate this proportion, you plan to select 25 of the 1,488 cases at random and compute $\hat{p}$, the sample proportion that were decided in favor of the defendant. Because $\hat{p}$ is an unbiased estimator of p, there is no consistent tendency for $\hat{p}$ to over- or underestimate the population proportion. Also, because $p = 0.263$ and $n = 25$,

$$\text{standard error of } \hat{p} = \sqrt{\frac{p(1-p)}{n}} = \sqrt{\frac{(0.263)(1-0.263)}{25}} = \sqrt{\frac{0.1938}{25}}$$

$$= \sqrt{0.0078} = 0.088$$

In practice, the value of p is not known, and the standard error can only be estimated. You will see how to do this in Section 9.2. Even though this example is a bit unrealistic, it can help you to understand what the standard error tells you about an estimator.

If the sample size is 100 rather than 25, then

$$\text{standard error of } \hat{p} = \sqrt{\frac{p(1-p)}{n}} = \sqrt{\frac{(0.263)(1-0.263)}{100}} = \sqrt{\frac{0.1938}{100}}$$

$$= \sqrt{0.0019} = 0.044$$

Notice that the standard error of $\hat{p}$ is much smaller for a sample size of 100 than it is for a sample size of 25. This tells you that $\hat{p}$ values from random samples of size 100 will cluster more tightly around the actual value of the population proportion than the $\hat{p}$ values from random samples of size 25.

In Example 9.1, you saw that the standard error of $\hat{p}$ is smaller when $n = 100$ than when $n = 25$. The fact that the standard error of $\hat{p}$ becomes smaller as the sample size increases makes sense because a large sample provides more information about the population than a small sample. What may not be as evident is how the standard error is related to the actual value of the population proportion.

In Example 9.1, p was 0.263 and the standard error of $\hat{p}$ for random samples of size 25 was 0.088. What if the population proportion had been 0.4 instead of 0.263? Then

$$\text{standard error of } \hat{p} = \sqrt{\frac{p(1-p)}{n}} = \sqrt{\frac{(0.4)(1-0.4)}{25}} = 0.098$$

For the same sample size ($n = 25$), the standard error of $\hat{p}$ is larger when the population proportion is 0.4 than when it is 0.263. A larger standard error corresponds to less accurate estimates of the population proportion. For a fixed sample size, the standard error of $\hat{p}$ is greatest when $p = 0.5$.

It may initially surprise you that $\hat{p}$ tends to produce more accurate estimates the farther the population proportion is from 0.5. To see why this actually makes sense, consider the extreme case where $p = 0$. If $p = 0$, this means that *no* individual in the population possesses the characteristic of interest. *Every* possible sample from this population will result in $\hat{p} = 0$, since every sample will include zero individuals with the characteristic of interest. There is no sample-to-sample variability in the values of $\hat{p}$, so it makes sense that the standard error of $\hat{p}$ is 0. The same would be true in the other extreme case where $p = 1$. Every sample would result in $\hat{p} = 1$, and again there would be no sample-to-sample variability in the value of $\hat{p}$.

For populations where p is close to 0 or to 1, it is fairly easy to get a good estimate of the population proportion, because most samples will tend to produce similar results. The case where there will be the most sample-to-sample variability is where the population is most diverse. For a categorical variable with just two possible values, this is where $p = 0.5$ (half of the population possesses the characteristic of interest and half does not). Here, the values of $\hat{p}$ will vary more from sample to sample.

These ideas will be developed more formally in Sections 9.2 and 9.3.

SECTION 9.1   EXERCISES

Each Exercise Set assesses the following chapter learning objectives: C1, C2, C3

SECTION 9.1   Exercise Set 1

9.1   For estimating a population characteristic, why is an unbiased statistic with a small standard error preferred over an unbiased statistic with a larger standard error?

9.2   Three different statistics are being considered for estimating a population characteristic. The sampling distributions of the three statistics are shown in the following illustration:

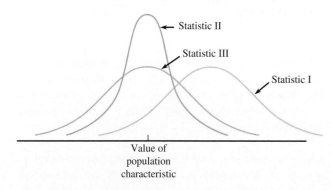

Which of these statistics are unbiased estimators of the population characteristic?

9.3   Three different statistics are being considered for estimating a population characteristic. The sampling distributions of the three statistics are shown in the following illustration:

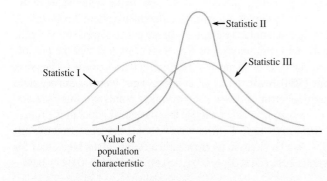

Which statistic would you recommend? Explain your choice.

9.4   A researcher wants to estimate the proportion of students enrolled at a university who eat fast food more than three times in a typical week. Would the standard error of the sample proportion $\hat{p}$ be smaller for random samples of size $n = 50$ or random samples of size $n = 200$?

9.5   Use the formula for the standard error of $\hat{p}$ to explain why
a. The standard error is greater when the value of the population proportion p is near 0.5 than when it is near 1.
b. The standard error of $\hat{p}$ is the same when the value of the population proportion is $p = 0.2$ as it is when $p = 0.8$.

9.6   A random sample will be selected from the population of all adult residents of a particular city. The sample proportion $\hat{p}$ will be used to estimate p, the proportion of all adult residents who are employed full time. For which of the following situations will the estimate tend to be closest to the actual value of p?
  i. $n = 500, p = 0.6$
  ii. $n = 450, p = 0.7$
  iii. $n = 400, p = 0.8$

SECTION 9.1   Exercise Set 2

9.7   For estimating a population characteristic, why is an unbiased statistic generally preferred over a biased statistic? Does unbiasedness alone guarantee that the estimate will be close to the true value? Explain.

9.8   Three different statistics are being considered for estimating a population characteristic. The sampling distributions of the three statistics are shown in the following illustration:

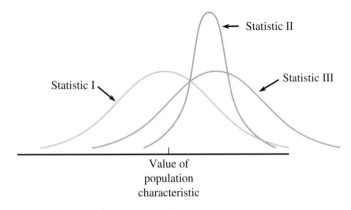

Which of these statistics are unbiased estimators of the population characteristic?

9.9 Three different statistics are being considered for estimating a population characteristic. The sampling distributions of the three statistics are shown in the following illustration:

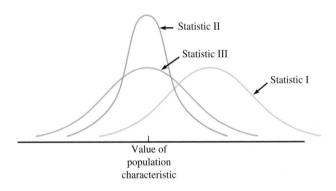

Which statistic would you recommend? Explain your choice.

9.10 A researcher wants to estimate the proportion of students enrolled at a university who are registered to vote. Would the standard error of the sample proportion $\hat{p}$ be larger if the actual population proportion was $p = 0.4$ or $p = 0.8$?

9.11 Use the formula for the standard error of $\hat{p}$ to explain why increasing the sample size decreases the standard error.

9.12 A random sample will be selected from the population of all adult residents of a particular city. The sample proportion $\hat{p}$ will be used to estimate p, the proportion of all adult residents who do not own a car. For which of the following situations will the estimate tend to be closest to the actual value of p?

    i. $n = 500, p = 0.1$
    ii. $n = 1,000, p = 0.2$
    iii. $n = 1,200, p = 0.3$

Additional Exercises

9.13 If two statistics are available for estimating a population characteristic, under what circumstances might you choose a biased statistic over an unbiased statistic?

9.14 A random sample will be selected from the population of all students enrolled at a large college. The sample proportion $\hat{p}$ will be used to estimate p, the proportion of all students who use public transportation to travel to campus. For which of the following situations will the estimate tend to be closest to the actual value of p?

    i. $n = 300, p = 0.3$
    ii. $n = 700, p = 0.2$
    iii. $n = 1,000, p = 0.1$

9.15 A researcher wants to estimate the proportion of city residents who favor spending city funds to promote tourism. Would the standard error of the sample proportion $\hat{p}$ be smaller for random samples of size $n = 100$ or random samples of size $n = 200$?

9.16 A researcher wants to estimate the proportion of property owners who would pay their property taxes one month early if given a $50 reduction in their tax bill. Would the standard error of the sample proportion $\hat{p}$ be larger if the actual population proportion were $p = 0.2$ or if it were $p = 0.4$?

SECTION 9.2 **Estimating a Population Proportion—Margin of Error**

When sample data are used to estimate a population proportion, understanding sampling variability will help you assess how close the estimate is likely to be to the actual value of the population proportion. Newspapers and journals often include statements about the *margin of error* associated with an estimate. For example, based on a representative sample of 511 U.S. teenagers ages 12 to 17, International Communications Research estimated that the proportion of teens who support keeping the legal drinking age at 21 is $\hat{p} = 0.64$ (or 64%). The press release titled **"Majority of Teens (Still) Favor the Legal Drinking Age" (www.icrsurvey.com, Jan 21, 2009)** also reported a margin of error of 0.04 (or 4 percentage points) for this estimate. By reporting a margin of error, the researchers acknowledge that the actual population proportion (the proportion of *all* U.S. teens ages 12 to 17 who favor keeping the legal drinking age at 21) is not likely to be *exactly* 0.64. The margin of error indicates that plausible values for the population proportion are values between $0.64 - 0.04 = 0.60$ and $0.64 + 0.04 = 0.68$.

The value of the sample proportion $\hat{p}$ provides an estimate of the population proportion p. If $\hat{p} = 0.426$ and $p = 0.484$, the estimate is "off" by 0.058, which is the difference between the actual value and the estimate. This difference represents the error in the estimate. A different sample might produce an estimate of $\hat{p} = 0.498$, resulting in an estimation error of 0.014. For a given the sampling method, statistic, and sample size, the maximum likely estimation error is called the **margin of error**.

> The **margin of error** of a statistic is the maximum likely estimation error. It is unusual for an estimate to differ from the actual value of the population characteristic by more than the margin of error.

In Section 9.1, the standard error (the standard deviation of the sampling distribution) of a statistic was introduced. The standard error and information about the shape of the sampling distribution of the statistic are used to find the margin of error.

Let's take a look at the reasoning that leads to calculation of a margin of error. With p denoting the proportion in the population possessing some characteristic of interest, recall that three things are known about the sampling distribution of $\hat{p}$ when a random sample of size n is selected:

1. The mean of the $\hat{p}$ sampling distribution is p. ($\mu_{\hat{p}} = p$)

2. The standard error (standard deviation) of $\hat{p}$ is $\sqrt{\dfrac{p(1-p)}{n}}$.

3. If n is large, the $\hat{p}$ sampling distribution is approximately normal.

The sampling distribution describes the behavior of the statistic $\hat{p}$. If the sample size is large, the $\hat{p}$ sampling distribution is approximately normal and you can make use of what you know about normal distributions. Consider the following line of reasoning:

Step	What You Know	How You Know It
1	If a variable has a standard normal distribution, about 95% of the time the value of the variable will be between -1.96 and 1.96.	The table of z curve areas (Appendix Table 2) can be used to show that for the standard normal distribution, a central area of 0.95 is captured between 1.96 and -1.96 (see Section 6.5). This tells you that for a variable that has a standard normal distribution, observed values will be between -1.96 and 1.96 about 95% of the time. 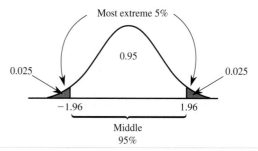
2	For any normal distribution, about 95% of the observed values will be within 1.96 standard deviations of the mean.	Because the mean of the standard normal distribution is 0 and the standard deviation is 1, the statement in Step 1 is equivalent to saying that 95% of the values are within 1.96 standard deviations of the mean. This is true for *all* normal distributions, not just the standard normal distribution.
3	If n is large, the $\hat{p}$ sampling distribution is approximately normal with mean p and standard error $\sqrt{\dfrac{p(1-p)}{n}}$.	These are the three properties of the sampling distribution of $\hat{p}$.

(continued)

Step	What You Know	How You Know It
4	If n is large, 95% of all possible random samples will produce a value of $\hat{p}$ that is within $1.96\sqrt{\frac{p(1-p)}{n}}$ of the value of the population proportion p.	This is just a rewording of the statement in Step 2 in terms of the $\hat{p}$ sampling distribution. This is reasonable because you know that if the sample size is large, the $\hat{p}$ distribution is approximately normal, and so the statement in Step 2 also applies to the $\hat{p}$ sampling distribution.
5	If n is large, it is unlikely that $\hat{p}$ will be farther than $$1.96\sqrt{\frac{p(1-p)}{n}}$$ from the actual value of the population proportion p. In other words, if n is large, it is unlikely that $\hat{p}$ will differ from p by more than 1.96(standard error of $\hat{p}$)	This is rewording the statement in Step 4 in terms of the standard error, which is the standard deviation of $\hat{p}$.

The margin of error of a statistic is the maximum likely estimation error expected when the statistic is used as an estimator. Based on the reasoning described in Steps 1–5, you know that if the sample size is large, about 95% of the time a random sample will result in a sample proportion that is closer than $1.96\sqrt{\frac{p(1-p)}{n}}$ to the actual value of the population proportion. This leads to the following definition of margin of error when $\hat{p}$ from a large sample is used to estimate a population proportion:

$$\text{margin of error} = 1.96\sqrt{\frac{p(1-p)}{n}} = 1.96(\text{standard error of } \hat{p})$$

There are just two more questions that need to be answered before looking at an example.

Question	Answer
How large does the sample need to be?	The formula for margin of error is appropriate when the sampling distribution of $\hat{p}$ is approximately normal. Recall from Chapter 8 that whether this is the case depends on the population proportion. If both $np \geq 10$ and $n(1-p) \geq 10$, it is reasonable to think that the sampling distribution of $\hat{p}$ is approximately normal.
	Since the value of the population proportion p will be unknown, you check the sample size conditions using the estimated proportion, $\hat{p}$.
	The sample size is large enough if either
	1. Both $n\hat{p} \geq 10$ and $n(1-\hat{p}) \geq 10$
	OR (equivalently)
	2. The sample includes at least 10 successes (individuals that have the characteristic of interest) and at least 10 failures.
The formula for margin of error is $$\text{margin of error} = 1.96\sqrt{\frac{p(1-p)}{n}}$$ **But p is the value of the population proportion, which is unknown. How can the margin of error be computed if you don't know the value of p?**	Since the value of p is unknown, you can estimate the margin of error by using $\hat{p}$, the sample proportion, in place of p. The estimated standard error is then $\sqrt{\frac{\hat{p}(1-\hat{p})}{n}}$ and the estimated margin of error is $$1.96\sqrt{\frac{\hat{p}(1-\hat{p})}{n}}.$$

The following box summarizes what is needed to compute and interpret a margin of error when estimating a population proportion.

MARGIN OF ERROR WHEN THE SAMPLE PROPORTION $\hat{p}$ IS USED TO ESTIMATE THE POPULATION PROPORTION p.

Appropriate when the following conditions are met

1. The sample is a random sample from the population of interest
   OR
   the sample is selected in a way that makes it reasonable to think the sample is representative of the population.
2. The sample size is large enough. This condition is met when either both
   $n\hat{p} \geq 10$ and $n(1 - \hat{p}) \geq 10$
   OR (equivalently)
   the sample includes at least 10 successes and at least 10 failures.

When these conditions are met

$$\text{margin of error} = 1.96 \sqrt{\frac{\hat{p}(1 - \hat{p})}{n}}$$

Interpretation of margin of error

It would be unusual for the sample proportion to differ from the actual value of the population proportion by more than the margin of error. For 95% of all random samples, the estimation error will be less than the margin of error.

Notes

1. The formula given for margin of error is actually for the *estimated* margin of error, but it is common to refer to it without including "estimated." Any time a margin of error is reported, it will be an estimated margin of error.
2. The value 2 is often used in place of 1.96 in the margin of error calculation. This simplifies the calculations and doesn't usually change the result by much. When calculating a margin of error, it is acceptable to use either 2 or 1.96.

The following examples illustrate the calculation and interpretation of margin of error when estimating a population proportion.

Example 9.2 | Legal Drinking Age

Earlier in this section, a study to assess opinion on the legal drinking age was described. Based on a representative sample of 511 U.S. teenagers ages 12 to 17, **International Communications Research** estimated that the proportion of teens who support keeping the legal drinking age at 21 is $\hat{p} = 0.64$ (64%). A margin of error of 0.04 (4 percentage points) was also reported. Let's see how this margin of error was computed.

Step	Explanation
Check Conditions It is appropriate to use the margin of error formula because	The sample was described in the article as representative of the population of U.S. teens. Although no information about how the sample was selected was given, assume that the researchers took reasonable steps to ensure a representative sample.
1. The sample was representative of the population.	
2. The sample size is large enough.	Since $\hat{p} = 0.64$ and $n = 511$, you can verify that the sample size is large enough: $n\hat{p} = 511(0.64) = 327$ and $n(1 - \hat{p}) = 511(0.36) = 184$. Both $n\hat{p}$ and $n(1 - \hat{p})$ are greater than 10.

(continued)

Step	Explanation
Compute margin of error margin of error $= 1.96\sqrt{\dfrac{\hat{p}(1-\hat{p})}{n}}$ $= 1.96\sqrt{\dfrac{(0.64)(0.36)}{511}}$ $= 1.96(0.021)$ $= 0.04$	Substitute $n = 511$ and $\hat{p} = 0.64$ into the formula for margin of error.

Interpretation
An estimate of the proportion of U.S. teens who favor keeping the legal drinking age at 21 is 0.64. Although this is only an estimate, it is unlikely that this estimate differs from the actual population proportion by more than 0.04.

Example 9.3   Teens and Gambling

How common is gambling among high school students? This is one of the questions that the authors of the paper **"Gambling and Alcohol Use in Adolescence"** (*Journal of Social Behavior and Personality* **[2000]: 51–66**) hoped to answer using data from a random sample of 318 students at a large New York high school. Gambling was defined as "a behavior where a valuable item (such as money, baseball cards or jewelry) is risked for the chance of winning an item of similar or greater value." Using this definition of gambling, 170 of the students surveyed reported that they had gambled at least once during the previous year. The sample proportion

$$\hat{p} = \frac{170}{318} = 0.53$$

provides an estimate of the proportion of *all* students at the high school who had gambled at least once in the previous year. How accurate is this estimate? You can describe the accuracy by computing the margin of error.

Check Conditions
It is appropriate to use the margin of error formula because

1. The sample was a random sample.
2. The sample size is large.

You know that the sample was a random sample of students at the high school. In this situation, a success is a student who reports that he or she has gambled at least once in the previous year. There were 170 successes in the sample and $318 - 170 = 148$ failures in the sample. Since both the number of successes and the number of failures are greater than 10, the sample size is large enough. Because both conditions are met, it is appropriate to use $1.96\sqrt{\dfrac{\hat{p}(1-\hat{p})}{n}}$ to compute the margin of error.

Compute margin of error

$$\text{margin of error} = 1.96\sqrt{\frac{\hat{p}(1-\hat{p})}{n}}$$

$$= 1.96\sqrt{\frac{(0.53)(0.47)}{318}}$$

$$= 1.96(0.028)$$

$$= 0.055$$

Interpretation
An estimate of the proportion of students at this high school who have gambled at least once in the last year is 0.53 (or 53%). Although this is only an estimate, it is unlikely that this estimate differs from the actual population proportion by more than 0.055 (or 5.5 percentage points).

If the margin of error is large, it indicates that you don't have very accurate information about a population characteristic. On the other hand, if the margin of error is small, you know that the sample resulted in an estimate that is relatively accurate.

In the chapters that follow, you will see that the same reasoning used to develop the margin of error for $\hat{p}$ can also be used in other situations.

In general, for an unbiased statistic that has a sampling distribution that is approximately normal,

$$\text{margin of error} = 1.96(\text{standard error})$$

The estimation error, which is the difference between the value of the statistic and the value of the population characteristic being estimated, will be greater than the margin of error for only about 5% of all possible random samples.

In the next section, you will see another approach that also conveys information about the accuracy of an estimate. This approach uses an interval of plausible values rather than just a single number estimate.

SECTION 9.2 EXERCISES

Each Exercise Set assesses the following chapter learning objectives: C4, M1, M2, P1

SECTION 9.2 Exercise Set 1

9.17 A large online retailer is interested in learning about the proportion of customers making a purchase during a particular month who were satisfied with the online ordering process. A random sample of 600 of these customers included 492 who indicated they were satisfied. For each of the three following statements, indicate if the statement is correct or incorrect. If the statement is incorrect, explain what makes it incorrect.

Statement 1: It is unlikely that the estimate $\hat{p} = 0.82$ differs from the value of the actual population proportion by more than 0.0157.

Statement 2: It is unlikely that the estimate $\hat{p} = 0.82$ differs from the value of the actual population proportion by more than 0.0307.

Statement 3: The estimate $\hat{p} = 0.82$ will never differ from the value of the actual population proportion by more than 0.0307.

9.18 Consider taking a random sample from a population with $p = 0.40$.
a. What is the standard error of $\hat{p}$ for random samples of size 100?
b. Would the standard error of $\hat{p}$ be larger for samples of size 100 or samples of size 200?
c. If the sample size were doubled from 100 to 200, by what factor would the standard error of $\hat{p}$ decrease?

9.19 The report "2005 Electronic Monitoring & Surveillance Survey: Many Companies Monitoring, Recording, Videotaping—and Firing—Employees" (American Management Association, 2005) summarized a survey of 526 U.S. businesses. The report stated that 137 of the 526 businesses had fired workers for misuse of the Internet. Assume that this sample is representative of businesses in the United States.
a. Estimate the proportion of all businesses in the U.S. that have fired workers for misuse of the Internet. What statistic did you use?
b. Use the sample data to estimate the standard error of $\hat{p}$.
c. Compute and interpret the margin of error associated with the estimate in Part (a).

9.20 The use of the formula for margin of error requires a large sample. For each of the following combinations of n and $\hat{p}$, indicate whether the sample size is large enough for use of this formula to be appropriate.
a. $n = 50$ and $\hat{p} = 0.30$
b. $n = 50$ and $\hat{p} = 0.05$
c. $n = 15$ and $\hat{p} = 0.45$
d. $n = 100$ and $\hat{p} = 0.01$

9.21 Most American college students make use of the Internet for both academic and social purposes. What proportion of students use it for more than 3 hours a day? The authors of the paper "U.S. College Students' Internet Use: Race, Gender and Digital Divides" (*Journal of Computer-Mediated Communication* [2009]: 244–264) describe a survey of 7,421 students at 40 colleges and universities. The sample was selected to reflect general demographics of U.S. college students. Of the students surveyed, 2,998 reported Internet use of more than 3 hours per day.

a. Use the given information to estimate the proportion of college students who use the Internet more than 3 hours per day.
b. Verify that the conditions needed in order for the margin of error formula to be appropriate are met.
c. Compute the margin of error.
d. Interpret the margin of error in the context of this problem.

9.22 Consumption of fast food is a topic of interest to researchers in the field of nutrition. The article **"Effects of Fast-Food Consumption on Energy Intake and Diet Quality Among Children"** (*Pediatrics* [2004]: 112–118) reported that 1,720 of those in a random sample of 6,212 American children indicated that they ate fast food on a typical day.
a. Use the given information to estimate the proportion of American children who eat fast food on a typical day.
b. Verify that the conditions needed in order for the margin of error formula to be appropriate are met.
c. Compute the margin of error.
d. Interpret the margin of error in the context of this problem.

<u>SECTION 9.2</u>   **Exercise Set 2**

9.23 A car manufacturer is interested in learning about the proportion of people purchasing one of its cars who plan to purchase another car of this brand in the future. A random sample of 400 of these people included 267 who said they would purchase this brand again. For each of the three statements below, indicate if the statement is correct or incorrect. If the statement is incorrect, explain what makes it incorrect.

> **Statement 1:** The estimate $\hat{p} = 0.668$ will never differ from the value of the actual population proportion by more than 0.0462.
>
> **Statement 2:** It is unlikely that the estimate $\hat{p} = 0.668$ differs from the value of the actual population proportion by more than 0.0235.
>
> **Statement 3:** It is unlikely that the estimate $\hat{p} = 0.668$ differs from the value of the actual population proportion by more than 0.0462.

9.24 Consider taking a random sample from a population with $p = 0.70$.
a. What is the standard error of $\hat{p}$ for random samples of size 100?
b. Would the standard error of $\hat{p}$ be smaller for samples of size 100 or samples of size 400?
c. Does decreasing the sample size by a factor of 4, from 400 to 100, result in a standard error of $\hat{p}$ that is four times as large?

9.25 The study **"Digital Footprints"** (Pew Internet & American Life Project, www.pewinternet.org, 2007) reported that 47% of Internet users have searched for information about themselves online. The 47% figure was based on a representative sample of Internet users. Suppose that the sample size was $n = 300$ (the actual sample size was much larger).
a. Use the given information to estimate the proportion of Internet users who have searched for information about themselves online. What statistic did you use?
b. Use the sample data to estimate the standard error of $\hat{p}$.
c. Compute and interpret the margin of error associated with the estimate in Part (a).

9.26 The use of the formula for margin of error requires a large sample. For each of the following combinations of n and $\hat{p}$, indicate whether the sample size is large enough for use of this formula to be appropriate.
a. $n = 100$ and $\hat{p} = 0.70$
b. $n = 40$ and $\hat{p} = 0.25$
c. $n = 60$ and $\hat{p} = 0.25$
d. $n = 80$ and $\hat{p} = 0.10$

9.27 The article **"Nine out of Ten Drivers Admit in Survey to Having Done Something Dangerous"** (*Knight Ridder Newspapers,* July 8, 2005) reported on a survey of 1,100 drivers. Of those surveyed, 990 admitted to careless or aggressive driving during the previous 6 months. Assume that this sample of 1,100 is representative of the population of drivers.
a. Use the given information to estimate the proportion of drivers who had engaged in careless or aggressive driving in the previous 6 months.
b. Verify that the conditions needed in order for the margin of error formula to be appropriate are met.
c. Compute the margin of error.
d. Interpret the margin of error in the context of this problem.

9.28 Suppose that 935 smokers each received a nicotine patch, which delivers nicotine to the bloodstream at a much slower rate than cigarettes do. Dosage was decreased to 0 over a 12-week period. Of these 935 people, 245 were still not smoking 6 months after treatment. Assume this sample is representative of all smokers.
a. Use the given information to estimate the proportion of all smokers who, when given this treatment, would refrain from smoking for at least 6 months.
b. Verify that the conditions needed in order for the margin of error formula to be appropriate are met.
c. Compute the margin of error.
d. Interpret the margin of error in the context of this problem.

Additional Exercises

9.29 The article **"Viewers Speak Out Against Reality TV"** (Associated Press, September 12, 2005) included the following statement: "Few people believe there's much reality in reality TV: a total of 82% said the shows are either 'totally made up' or 'mostly distorted.'" This statement was based on a survey of 1,002 randomly selected adults. Compute and interpret the margin of error for the reported percentage.

9.30 *USA Today* (October 14, 2002) reported that 36% of adult drivers admit that they often or sometimes talk on a cell phone when driving. This estimate was based on data from a representative sample of 1,004 adult drivers. A margin of error of 3.1% was also reported. Is this margin of error correct? Explain.

9.31 The Gallup Organization conducts an annual survey on crime. It was reported that 25% of all households experienced some sort of crime during the past year. This estimate was based on a sample of 1,002 randomly selected adults. The report states, "One can say with 95% confidence that the margin of sampling error is ± 3 percentage points." Explain how this statement is justified.

9.32 Based on a representative sample of 511 U.S. teens ages 12 to 17, International Communications Research estimates that the proportion of teens who support keeping the legal drinking age at 21 is $\hat{p} = 0.64$. The press release

titled **"Majority of Teens (Still) Favor the Legal Drinking Age"** (www.icrsurvey.com, January 21, 2009) also reported a margin of error of 0.04 for this estimate. Show how the reported value for the margin of error was computed.

9.33 An article in the *Chicago Tribune* **(August 29, 1999)** reported that in a poll of residents of the Chicago suburbs, 43% felt that their financial situation had improved during the past year. The following statement is from the article: "The findings of this Tribune poll are based on interviews with 930 randomly selected suburban residents. The sample included suburban Cook County plus DuPage, Kane, Lake, McHenry, and Will Counties. In a sample of this size, one can say with 95% certainty that results will differ by no more than 3% from results obtained if all residents had been included in the poll." Give a statistical argument to justify the claim that the estimate of 43% is within 3% of the actual percentage of all residents who feel that their financial situation has improved.

SECTION 9.3 A Large Sample Confidence Interval for a Population Proportion

The preview example of this chapter described how data from a survey were used to estimate the proportion of coastal residents who would refuse to evacuate if a hurricane was approaching. Based on a representative sample of 5,046 adults living within 20 miles of the coast in high hurricane risk counties of eight southern states, it was estimated that the proportion of residents who would refuse to evacuate was 0.31 (or 31%).

When you estimate that 31% of residents would refuse to evacuate, you don't really believe that *exactly* 31% would refuse. The estimate $\hat{p} = 0.31$ is a sample proportion, and the value of $\hat{p}$ varies from sample to sample. What the estimate tells you is that the actual population proportion is "around" 0.31. In Section 9.2 you saw one way to be more specific about what is meant by "around" by reporting both the estimate and a margin of error. For example, you could say that the estimated proportion who would refuse to evacuate is 0.31, with a margin of error of 0.01. This implies that "around 0.31" means "within 0.01 of 0.31."

Another common way to be more specific about what is meant by "around" is to report an interval of reasonable values for the population proportion rather than a single number estimate. For example, you could report that, based on data from the sample of coastal residents, you believe the actual proportion who would refuse to evacuate is a value in the interval from 0.30 to 0.32. By reporting an interval, you acknowledge that you don't know the exact value of the population proportion. The width of the interval conveys information about accuracy. The fact that the interval here is narrow implies that you have fairly precise information about the value of the population proportion. On the other hand, if you say that you think the population proportion is between 0.21 and 0.41, it would be clear that you don't know as much about the actual value. In this section, you will learn how to use sample data to construct an interval estimate, called a confidence interval.

What Is a Confidence Interval?

A **confidence interval** estimate specifies a range of plausible values for a population characteristic. For example, using sample data and what you know about the behavior of the sample proportion, it is possible to construct an interval that you think should include the actual value of the population proportion. Because a sample provides only incomplete information about the population, there is some risk involved with a confidence interval estimate. Occasionally, but hopefully not very often, a sample will lead you to an interval that does not include the value of the population characteristic. If this were to happen and you make a statement such as "the proportion of coastal residents who would refuse to evacuate is between 0.30 and 0.32," you would be wrong. It is important to know how likely it is that the method used to compute a confidence interval estimate will lead to a correct statement. So, associated with every confidence interval is a **confidence level** which specifies the "success rate" of the method used to produce the confidence interval.

> **DEFINITION**
>
> **Confidence Interval**
> A confidence interval is an interval that you think includes the value of the population characteristic.
>
> **Confidence Level**
> The confidence level associated with a confidence interval is the success rate of the *method* used to construct the interval.

What Does the Confidence Level Tell You About a Confidence Interval?

The confidence level associated with a confidence interval tells you how much "confidence" you can have in the *method* used to construct the interval (but *not* your confidence in any *one* particular interval). Usual choices for confidence levels are 90%, 95%, and 99%, although other levels are also possible. If you construct a 95% confidence interval using the method to be described shortly, you would be using a method that is "successful" in capturing the value of the population characteristic 95% of the time. Different random samples will lead to different intervals, but 95% of these intervals will include the value of the population characteristic. Similarly, a 99% confidence interval is one constructed using a method that captures the actual value of the population characteristic for 99% of all possible random samples.

Margin of Error and a 95% Confidence Interval

The margin of error when the sample proportion $\hat{p}$ is used to estimate a population proportion was introduced in Section 9.2. When the following conditions are met,

1. The sample size is large ($n\hat{p} \geq 10$ and $n(1 - \hat{p}) \geq 10$)
2. The sample is a random sample or the sample is selected in a way that should result in a representative sample from the population

the margin of error was defined as

$$\text{margin of error} = 1.96(\text{standard error of } \hat{p}) = 1.96\sqrt{\frac{\hat{p}(1 - \hat{p})}{n}}$$

The margin of error formula was based on the following reasoning:

1. When the conditions are met, the sampling distribution of $\hat{p}$ is approximately normal, is centered at the actual value of the population proportion, and has a standard error that is estimated by $\sqrt{\dfrac{\hat{p}(1 - \hat{p})}{n}}$.

2. For any variable that has a standard normal distribution, about 95% of the values will be between -1.96 and 1.96. This also means that for any normal distribution, about 95% of the values are within 1.96 standard deviations of the mean.

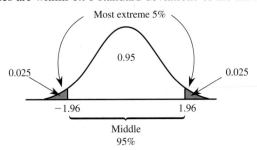

3.   Because $\hat{p}$ has a normal distribution (approximately) with mean p, 95% of all random samples have a sample proportion that is within $1.96\sqrt{\dfrac{p(1-p)}{n}}$ of the population proportion p.

Based on this reasoning, we are able to say that for 95% of all random samples, the estimation error (the difference between the value of the sample proportion $\hat{p}$ and the value of the population proportion p) will be less than the margin of error. This means that the interval

$$(\hat{p} - \text{margin of error}, \hat{p} + \text{margin of error})$$

which is often abbreviated as $\hat{p} \pm$ margin of error, should include the value of the population proportion for 95% of all possible random samples. The interval

$$\hat{p} \pm \text{margin of error}$$

is the 95% confidence interval (meaning the confidence level is 95%) for a population proportion.

> ### A Large-Sample 95% Confidence Interval for a Population Proportion
>
> Appropriate when the following conditions are met:
>
> 1. The sample is a random sample from the population of interest or the sample is selected in a way that should result in a representative sample from the population.
> 2. The sample size is large. This condition is met when both $n\hat{p} \geq 10$ and $n(1 - \hat{p}) \geq 10$ or (equivalently) the sample includes at least 10 successes and at least 10 failures.
>
> **When these conditions are met, a 95% confidence interval for the population proportion is**
>
> $$\hat{p} \pm 1.96\sqrt{\frac{\hat{p}(1-\hat{p})}{n}}$$
>
> **Note:** This formula corresponds to a confidence level of 95%. Other confidence levels will be considered later in this section.

> ### Interpretation of Confidence Interval
>
> You can be 95% confident that the actual value of the population proportion is included in the computed interval.

> ### Interpretation of 95% Confidence Level
>
> A method has been used to produce the confidence interval that is successful in capturing the actual population proportion approximately 95% of the time.

Before looking at an example of computing and interpreting a 95% confidence interval, let's do a quick review of some important ideas from Chapter 7.

Chapter 7 Review: Four Key Questions and the Five-Step Process for Estimation Problems

In Section 7.2, you learned four key questions that guide the decision about what statistical inference method to consider in any particular situation. In Section 7.3, a five-step process for estimation problems was introduced. In the playbook analogy, the four key questions provide a way to select an appropriate play from the playbook (that is, to select an appropriate statistical inference method). The five-step process provides a structure for carrying out each play in the playbook.

The four key questions of Section 7.2 were

Q Question Type	Estimation or hypothesis testing?
S Study Type	Sample data or experiment data?
T Type of Data	One variable or two? Categorical or numerical?
N Number of Samples or Treatments	How many samples or treatments?

When the answers to these questions are

Q: Estimation
S: Sample data
T: One categorical variable
N: One sample

the method you will want to consider is the large-sample confidence interval for a population proportion.

Because this is an estimation problem, you would proceed by following the five-step process for estimation problems (EMC^3). The steps in this process are (see Section 7.3):

Step	What Is This step?
E	**Estimate:** Explain what population characteristic you plan to estimate.
M	**Method:** Select a potential method. This step involves using the answers to the four key questions (QSTN) to identify a potential method.
C	**Check:** Check to make sure that the method selected is appropriate. Many methods for learning from data only provide reliable information under certain conditions. It will be important to verify that any such conditions are met before proceeding.
C	**Calculate:** Use sample data to perform any necessary calculations. This step is often accomplished using technology, such as a graphing calculator or computer software for statistical analysis.
C	**Communicate Results:** This is a critical step in the process. In this step, you answer the research question of interest, explain what you have learned from the data, and acknowledge potential risks.

You will see how this process is used in the examples that follow. Be sure to include all five steps when you use a confidence interval to estimate a population characteristic.

Example 9.4 Moving Home

When unemployment rises and the economy struggles, some people make the decision to move in with friends or family. Researchers studying relocation patterns were interested in learning about the proportion of U.S. adults 21 years of age or older who moved back home or in with friends during the previous year. In a survey of 743 U.S. adults age 21 or older, 52 reported that in the previous year they had made the change from living independently to living with friends or relatives (*USA Today*, **July 29, 2008**). Based on these data, what can you learn about the proportion of all adults who moved home or in with friends during the previous year?

Begin by answering the four key questions for this problem.

Q	Estimation or hypothesis testing?	Estimation
Question Type		
S	Sample data or experiment data?	Sample data
Study Type		
T	One variable or two? Categorical or numerical?	One categorical variable
Type of Data		
N	How many samples or treatments?	One sample
Number of Samples or Treatments		

The answers are *estimation, sample data, one categorical variable, one sample*. It is this combination of answers that indicates you should consider a confidence interval for a population proportion. This is the only statistical inference method you have considered so far, but as the number of available methods grows, the answers to the four key questions will help you decide which method to use.

Now you can use the five-step process to learn from the data.

Step	
Estimate: Explain what population characteristic you plan to estimate.	In this example, you will estimate the value of *p*, **the proportion of U.S. adults age 21 and older who have moved home or in with friends during the past year.**
Method: Select a potential method.	Because the answers to the four key questions are estimation, sample data, one categorical variable, one sample, consider constructing a **confidence interval for a population proportion** (see Table 7.1).
	Associated with every confidence interval is a confidence level. When the method selected is a confidence interval, you will also need to specify a confidence level.
	For now, you only know how to construct an interval corresponding to a **95% confidence level**. You will see how to construct intervals for other confidence levels after this example.
Check: Check to make sure that the method selected is appropriate.	There are **two conditions** that need to be met in order to use the confidence interval of this section.
	The sample size is large enough because there are more than 10 successes (52 people who have moved home or in with friends) and more than 10 failures (the other $743 - 52 = 691$ people in the sample).
	The requirement of a random or representative sample is more difficult. No information was provided regarding how the sample was selected. To proceed, you need to assume that the sample was selected in a reasonable way. You will need to keep this in mind when you get to the step that involves interpretation.
Calculate: Use the sample data to perform any necessary calculations.	$n = 743$ $\hat{p} = \dfrac{52}{743} = 0.07$ 95% confidence interval $$\hat{p} \pm 1.96\sqrt{\dfrac{\hat{p}(1 - \hat{p})}{n}}$$ $$0.07 \pm 1.96\sqrt{\dfrac{(0.07)(0.93)}{743}}$$ $0.07 \pm 1.96\sqrt{0.0001}$ $0.07 \pm 1.96(0.01)$ 0.07 ± 0.02 $(0.05, 0.09)$

(continued)

Step	
Communicate Results: Answer the research question of interest, explain what you have learned from the data, and acknowledge potential risks.	**Confidence interval:** Assuming that the sample was selected in a reasonable way, you can be 95% confident that the actual proportion of U.S. adults age 21 or older who have moved home or in with friends during the previous year is somewhere between 0.05 and 0.09.
	Confidence level: The method used to construct this interval estimate is successful in capturing the actual value of the population proportion about 95% of the time.
	In this example, no information was given about how the sample was selected. Because of this, it is important to also state that the interpretation is only valid if the sample was selected in a reasonable way.

Be Careful...

The 95% confidence interval for p calculated in Example 9.4 is (0.05, 0.09). It is tempting to say that there is a "probability" of 0.95 that p is between 0.05 and 0.09, but you can't make a chance (probability) statement concerning the value of p. The 95% refers to the percentage of *all* possible random samples that would result in an interval that includes p. In other words, if you take sample after sample from the population and use each one separately to compute a 95% confidence interval, in the long run about 95% of these intervals will capture p. Figure 9.2 illustrates this concept for intervals generated

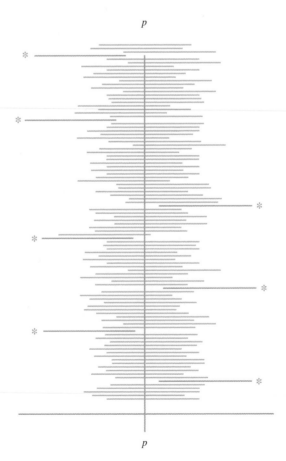

FIGURE 9.2
One hundred 95% confidence intervals for p computed from 100 different random samples (asterisks identify intervals that do not include p).

from 100 different random samples; 93 of the intervals include p, whereas 7 do not. Any specific interval, and the interval (0.05, 0.09) in particular, either includes p or it does not (remember, the value of p is fixed but not known to us). *The confidence level of 95% refers to the method used to construct the interval rather than to any particular interval, such as the one you obtained.*

Other Confidence Levels

The formula given for a 95% confidence interval can easily be adapted for other confidence levels. The choice of a 95% confidence level led to the use of the z value 1.96 (chosen to capture a central area of 0.95 under the standard normal curve) in the formula. Any other confidence level can be obtained by using an appropriate z critical value in place of 1.96. For example, suppose that you wanted to achieve a confidence level of 99%. To obtain a central area of 0.99, the appropriate z critical value would have a cumulative area (area to the left) of 0.995, as illustrated in Figure 9.3. From Appendix Table 2, the corresponding z critical value is 2.58.

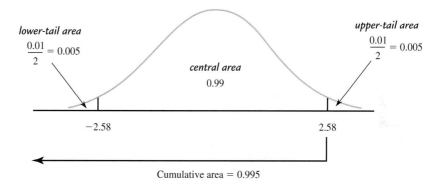

FIGURE 9.3
Finding the z critical value for a 99% confidence level

For other confidence levels, the appropriate z critical values are found in a similar way. The desired confidence level specifies a central area under the standard normal curve and then Appendix Table 2 or technology is used to determine the corresponding z critical value.

AP* EXAM TIP

A more general definition of margin of error involves associating a confidence level with the margin of error. The margin of error is then the " $\pm$ " part of the confidence interval with the associated confidence level. For example, a 90% margin of error for $\hat{p}$ is $1.645\sqrt{\dfrac{\hat{p}(1-\hat{p})}{n}}$.

| A Large-Sample Confidence Interval for a Population Proportion p |

Appropriate when the following conditions are met:

1. The sample is a random sample from the population of interest or the sample is selected in a way that should result in a representative sample from the population.
2. The sample size is large. This condition is met when both $n\hat{p} \geq 10$ and $n(1 - \hat{p}) \geq 10$ or (equivalently) the sample includes at least 10 successes and at least 10 failures.

When these conditions are met, a confidence interval for the population proportion is

$$\hat{p} \pm (z \text{ critical value})\sqrt{\frac{\hat{p}(1 - \hat{p})}{n}}$$

The desired confidence level determines which z critical value is used. The three most common confidence levels use the following z critical values:

Confidence Level	z Critical Value
90%	1.645
95%	1.96
99%	2.58

(continued)

Interpretation of Confidence Interval

You can be confident that the actual value of the population proportion is included in the computed interval. In any given problem, this statement should be worded in context.

Interpretation of Confidence Level

The confidence level specifies the approximate percentage of the time that this method is expected to be successful in capturing the actual population proportion.

A Note on Confidence Level

Why settle for 95% confidence when 99% confidence is possible? Because the higher confidence level comes at a price. The 99% confidence interval is wider than the 95% confidence interval. The width of the 95% interval is $2\left(1.96\sqrt{\dfrac{p(1-p)}{n}}\right)$ whereas the 99% interval has width $2\left(2.58\sqrt{\dfrac{p(1-p)}{n}}\right)$. The wider interval provides less precise information about the value of the population proportion. On the other hand, a 90% confidence interval would be narrower than the 95% confidence interval, but the associated risk of being incorrect is higher. In the opinion of many investigators, a 95% confidence level is a reasonable compromise between confidence and precision, and this is the most widely used confidence level.

Example 9.5 Dangerous Driving

The article **"Nine out of Ten Drivers Admit in Survey to Having Done Something Dangerous"** (**Knight Ridder Newspapers**, **July 8, 2005**) reported on a survey of 1,100 drivers. Of those surveyed, 990 admitted to careless or aggressive driving during the previous 6 months. Assuming that it is reasonable to regard this sample of 1,100 as representative of the population of drivers, you can use this information to construct an estimate of p, the proportion of all drivers who have engaged in careless or aggressive driving in the last 6 months.

   Answers to the four key questions for this problem are:

Q Question Type	Estimation or hypothesis testing?	Estimation
S Study Type	Sample data or experiment data?	Sample data. The data are from a sample of 1,100 drivers.
T Type of Data	One variable or two? Categorical or numerical?	One categorical variable. The data are responses to the question "Have you engaged in careless or aggressive driving during the last 6 months?" so there is one variable. The variable is categorical with two possible values—yes and no.
N Number of Samples or Treatments	How many samples or treatments?	One sample.

So, this is *estimation, sample data, one categorical variable, one sample*. This combination of answers indicates that you should consider a confidence interval for a population proportion.

You can now use the five-step process for estimation problems to construct a 90% confidence interval.

E Estimate:
The proportion of drivers who have engaged in careless or aggressive driving during the last six months, p, will be estimated.

M Method:
Because the answers to the four key questions are estimation, sample data, one categorical variable, one sample, a confidence interval for a population proportion will be considered. A confidence level of 90% was specified for this example.

C Check:
There are two conditions that need to be met for the confidence interval of this section to be appropriate. The large sample condition is easily verified:

$$n\hat{p} = 1{,}100 \left(\frac{990}{1{,}100} \right) = 1{,}100(0.90) = 990 \geq 10$$

and $\quad n(1 - \hat{p}) = 1{,}100(0.10) = 110 \geq 10$

The requirement of a random sample or representative sample is more difficult. You do not know how the sample was selected. In order to proceed, you need to assume that the sample was selected in a reasonable way. You will need to keep this in mind when you get to the step that involves interpretation.

C Calculate:
The appropriate z critical value is found in the box on page 473 or by using Appendix table 2 to find the z critical value that captures a central area of 0.90.

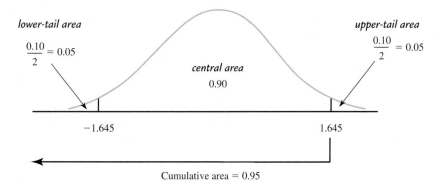

lower-tail area

$\frac{0.10}{2} = 0.05$

central area
0.90

upper-tail area

$\frac{0.10}{2} = 0.05$

-1.645 1.645

Cumulative area $= 0.95$

$n = 1{,}100$

$\hat{p} = \dfrac{990}{1{,}100} = 0.90$

z critical value $= 1.645$

90% confidence interval

$$\hat{p} \pm 1.645 \sqrt{\frac{\hat{p}(1 - \hat{p})}{n}}$$

$$0.90 \pm 1.645 \sqrt{\frac{(0.90)(0.10)}{1{,}100}}$$

$0.90 \pm 1.645(0.009)$

0.90 ± 0.015

$(0.885, 0.915)$

C Communicate Results:

Interpret Confidence interval:

Assuming that the sample was selected in a reasonable way, you can be 90% confident that the actual proportion of drivers who engaged in careless or aggressive driving in the past 6 months is somewhere between 0.885 and 0.915.

Interpret Confidence level:

The method used to construct this interval estimate is successful in capturing the actual value of the population proportion about 90% of the time.

In this example, you do not know how the sample was selected. Because of this, it is also important to state that the given interpretation is only valid if the sample was selected in a reasonable way.

___

Notice that the confidence interval in Example 9.5 is relatively narrow—it ranges from 0.885 to 0.915. In general, a narrow interval is desirable because it conveys more information about the actual value of the population proportion than a wider interval. Saying that the value of the population proportion is between 0.885 and 0.915 is more informative than saying that the value of the population proportion is between 0.82 and 0.98.

There are three things that affect the width of the interval. These are the confidence level, the sample size, and the value of the sample proportion.

What You Know	How You Know It
The higher the confidence level, the wider the interval.	The formula for the confidence interval is $\hat{p} \pm (z \text{ critical value}) \sqrt{\dfrac{\hat{p}(1 - \hat{p})}{n}}$. The higher the confidence level, the larger the z critical value will be, making the interval wider.
The larger the sample size, the narrower the interval.	The sample size n appears in the denominator of the standard error of $\hat{p}$. The larger the sample size, the smaller the standard error $\sqrt{\dfrac{\hat{p}(1 - \hat{p})}{n}}$ will be. When the standard error is smaller, the interval will be narrower.
The closer $\hat{p}$ is to 0.5, the wider the interval.	The numerator of the standard error of $\hat{p}$ is $\hat{p}(1 - \hat{p})$. This value is greatest when $\hat{p}$ is 0.5. This means that the interval will be wide for values of $\hat{p}$ close to 0.5. This is the case where the population is very diverse, making it most difficult to accurately estimate the population proportion.

A Few Final Things to Consider

1. What About Small Samples?

The confidence level associated with a z confidence interval for a population proportion is only approximate. When you report a 95% confidence interval for a population proportion, the 95% confidence level implies that you have used a method that produces an interval that includes the actual value of the population proportion about 95% of the time in the long run. However, the normal distribution is only an approximation to the sampling distribution of $\hat{p}$ and the true confidence level may differ somewhat from the reported value. If $n\hat{p} \geq 10$ and $n(1 - \hat{p}) \geq 10$, the approximation is reasonable and the actual confidence level is usually quite close to the reported level. This is why it is important to check the sample size conditions before computing and reporting a z confidence interval for a population proportion.

What should you do if these conditions are not met? If the sample size is too small to satisfy the sample size conditions, an alternative procedure can be used. Consult a statistician or a more advanced textbook in this case.

2. **General Form of a Confidence Interval**

Many confidence intervals have the same general form as the large-sample z interval for a population proportion. The confidence interval

$$\hat{p} \pm (z \text{ critical value}) \sqrt{\frac{\hat{p}(1 - \hat{p})}{n}}$$

is one example of a more general form for confidence intervals, which is

$$(\text{statistic}) \pm (\text{critical value}) \left(\begin{array}{c} \text{standard error} \\ \text{of the statistic} \end{array} \right)$$

For a population characteristic other than p, a statistic for estimating the characteristic is selected. Then the standard error of the statistic and an appropriate critical value are used to construct a confidence interval with this same general structure.

An Alternative to the Large-Sample *z* Interval

Investigators have shown that in some instances, even when the sample size conditions of the large sample z confidence interval for a population proportion are met, the actual confidence level associated with the method may be noticeably different from the reported confidence level. One way to achieve an actual confidence level that is closer to the reported confidence level is to use a modified sample proportion, $\hat{p}_{\text{mod}}$, the proportion of successes after adding two successes and two failures to the sample. Then $\hat{p}_{\text{mod}}$ is

$$\hat{p}_{\text{mod}} = \frac{\text{number of successes} + 2}{n + 4}$$

$\hat{p}_{\text{mod}}$ is used in place of $\hat{p}$ in the usual confidence interval formula. Properties of this modified confidence interval are investigated in Activity 9.2 at the end of this chapter.

SECTION 9.3 EXERCISES

Each Exercise Set assesses the following chapter learning objectives: C6, C7, M2, M3, M4, P2

SECTION 9.3 **Exercise Set 1**

9.34 Suppose that a city planning commission wants to know the proportion of city residents who support installing streetlights in the downtown area. Two different people independently selected random samples of city residents and used their sample data to construct the following confidence intervals for the population proportion:

    Interval 1: (0.28, 0.34)
    Interval 2: (0.31, 0.33)

a. Explain how it is possible that the two confidence intervals are not centered in the same place.
b. Which of the two intervals conveys more precise information about the value of the population proportion?
c. If both confidence intervals have a 95% confidence level, which confidence interval was based on the smaller sample size? How can you tell?
d. If both confidence intervals were based on the same sample size, which interval has the higher confidence level? How can you tell?

9.35 For each of the following choices, explain which one would result in a wider large-sample confidence interval for p:
a. 90% confidence level or 95% confidence level
b. $n = 100$ or $n = 400$

9.36 Based on data from a survey of $1,200$ randomly selected Facebook users (*USA Today*, **March 24, 2010**), a 95% confidence interval for the proportion of all Facebook users who say it is OK for someone to "friend" his or her boss is (0.41, 0.47). What is the meaning of the confidence level of 95% that is associated with this interval?

9.37 Appropriate use of the interval

$$\hat{p} \pm (z \text{ crital value}) \sqrt{\frac{\hat{p}(1 - \hat{p})}{n}}$$

requires a large sample. For each of the following combinations of n and $\hat{p}$, indicate whether the sample size is large enough for this interval to be appropriate.

a. $n = 50$ and $\hat{p} = 0.30$
b. $n = 50$ and $\hat{p} = 0.05$
c. $n = 15$ and $\hat{p} = 0.45$
d. $n = 100$ and $\hat{p} = 0.01$

9.38 The formula used to compute a large-sample confidence interval for p is

$$\hat{p} \pm (z \text{ crital value})\sqrt{\frac{\hat{p}(1 - \hat{p})}{n}}$$

What is the appropriate z critical value for each of the following confidence levels?
a. 90%
b. 99%
c. 80%

9.39 The article **"Career Expert Provides DOs and DON'Ts for Job Seekers on Social Networking" (CareerBuilder.com, August 19, 2009)** included data from a survey of 2,667 hiring managers and human resource professionals. The article noted that more employers are now using social networks to screen job applicants. Of the 2,667 people who participated in the survey, 1,200 indicated that they use social networking sites such as Facebook, MySpace, and LinkedIn to research job applicants. Assume that the sample is representative of hiring managers and human resource professionals. Answer the four key questions (QSTN) to confirm that the suggested method in this situation is a confidence interval for a population proportion.

9.40 For the study described in the previous exercise, use the five-step process for estimation problems (EMC³) to construct and interpret a 95% confidence interval for the proportion of hiring managers and human resource professionals who use social networking sites to research job applicants. Identify each of the five steps in your solution.

9.41 The article **"Students Increasingly Turn to Credit Cards"** (*San Luis Obispo Tribune*, **July 21, 2006**) reported that 37% of college freshmen and 48% of college seniors carry a credit card balance from month to month. Suppose that the reported percentages were based on random samples of 1,000 college freshmen and 1,000 college seniors.
a. Construct and interpret a 90% confidence interval for the proportion of college freshmen who carry a credit card balance from month to month.
b. Construct and interpret a 90% confidence interval for the proportion of college seniors who carry a credit card balance from month to month.
c. Explain why the two 90% confidence intervals from Parts (a) and (b) are not the same width.

9.42 In a survey of 1,000 randomly selected adults in the United States, participants were asked what their most favorite and least favorite subjects were when they were in school **(Associated Press, August 17, 2005)**. In what might seem like a contradiction, math was chosen more often than any other

subject in *both* categories. Math was chosen by 230 of the 1,000 as their most favorite subject and chosen by 370 of the 1,000 as their least favorite subject.
a. Construct and interpret a 95% confidence interval for the proportion of U.S. adults for whom math was their most favorite subject.
b. Construct and interpret a 95% confidence interval for the proportion of U.S. adults for whom math was their least favorite subject.

9.43 It probably wouldn't surprise you to know that Valentine's Day means big business for florists, jewelry stores, and restaurants. But did you know that it is also a big day for pet stores? In January 2010, the National Retail Federation conducted a survey of consumers in a representative sample of adult Americans **("This Valentine's Day, Couples Cut Back on Gifts to Each Other, According to NRF Survey," www.nrf.com)**. One of the questions in the survey asked if the respondent planned to spend money on a Valentine's Day gift for his or her pet.
a. The proportion who responded that they did plan to purchase a gift for their pet was 0.173. Suppose that the sample size for this survey was $n = 200$. Construct and interpret a 95% confidence interval for the proportion of all adult Americans who planned to purchase a Valentine's Day gift for their pet.
b. The actual sample size for the survey was much larger than 200. Would a 95% confidence interval computed using the actual sample size have been narrower or wider than the confidence interval computed in Part (a)?

SECTION 9.3 **Exercise Set 2**

9.44 Suppose that a college bookstore manager wants to know the proportion of students at the college who purchase some or all of their textbooks online. Two different people independently selected random samples of students at the college and used their sample data to construct the following confidence intervals for the population proportion:

> Interval 1: (0.54, 0.57)
> Interval 2: (0.46, 0.62)

a. Explain how it is possible that the two confidence intervals are not centered in the same place.
b. Which of the two intervals conveys more precise information about the value of the population proportion?
c. If both confidence intervals have a 95% confidence level, which confidence interval was based on the smaller sample size? How can you tell?
d. If both confidence intervals were based on the same sample size, which interval has the higher confidence level? How can you tell?

9.45 For each of the following choices, explain which would result in a narrower large-sample confidence interval for p:

a. 95% confidence level or 99% confidence level

b. $n = 200$ or $n = 500$

9.46 Based on data from a survey of 1,200 randomly selected Facebook users (*USA Today*, March 24, 2010), a 98% confidence interval for the proportion of all Facebook users who say it is OK to ignore a coworker's "friend" request is (0.35, 0.41). What is the meaning of the confidence level of 98% that is associated with this interval?

9.47 Appropriate use of the interval

$$\hat{p} \pm (z \text{ critial value})\sqrt{\frac{\hat{p}(1 - \hat{p})}{n}}$$

requires a large sample. For each of the following combinations of n and $\hat{p}$, indicate whether the sample size is large enough for this interval to be appropriate.

a. $n = 100$ and $\hat{p} = 0.70$

b. $n = 40$ and $\hat{p} = 0.25$

c. $n = 60$ and $\hat{p} = 0.25$

d. $n = 80$ and $\hat{p} = 0.10$

9.48 The formula used to compute a large-sample confidence interval for p is

$$\hat{p} \pm (z \text{ critial value})\sqrt{\frac{\hat{p}(1 - \hat{p})}{n}}$$

What is the appropriate z critical value for each of the following confidence levels?

a. 95%

b. 98%

c. 85%

9.49 The article **"Nine Out of Ten Drivers Admit in Survey to Having Done Something Dangerous"** (*Knight Ridder Newspapers*, July 8, 2005) reported on a survey of 1,100 drivers. Of those surveyed, 990 admitted to careless or aggressive driving during the previous 6 months. Assume that the sample is representative of the population of drivers. Answer the four key questions (QSTN) to confirm that the suggested method in this situation is a confidence interval for a population proportion.

9.50 For the study described in the previous exercise, use the five-step process for estimation problems (EMC³) to construct and interpret a 99% confidence interval for the proportion of drivers who have engaged in careless or aggressive driving in the previous 6 months. Identify each of the five steps in your solution.

9.51 The article **"CSI Effect Has Juries Wanting More Evidence"** (*USA Today*, August 5, 2004) examines how the popularity of crime-scene investigation television shows is influencing jurors' expectations of what evidence should be produced at a trial. In a survey of 500 potential jurors, one study found that 350 were regular watchers of at least one crime-scene investigation television series.

a. Assuming that it is reasonable to regard this sample as representative of potential jurors in the United States, use the given information to construct and interpret a 95% confidence interval for the proportion of *all* potential jurors who regularly watch at least one crime-scene investigation series.

b. Would a 99% confidence interval be wider or narrower than the 95% confidence interval from Part (a)?

9.52 The report **"2005 Electronic Monitoring & Surveillance Survey: Many Companies Monitoring, Recording, Videotaping—and Firing—Employees"** (American Management Association, 2005) summarized a survey of 526 U.S. businesses. The report stated that 137 of the 526 businesses had fired workers for misuse of the Internet, and 131 had fired workers for e-mail misuse. Assume that the sample is representative of businesses in the United States.

a. Construct and interpret a 95% confidence interval for the proportion of U.S. businesses that have fired workers for misuse of the Internet.

b. What are two reasons why a 90% confidence interval for the proportion of U.S. businesses that have fired workers for misuse of e-mail would be narrower than the 95% confidence interval computed in Part (a)?

9.53 The article **"Doctors Cite Burnout in Mistakes"** (*San Luis Obispo Tribune*, March 5, 2002) reported that many doctors who are completing their residency have financial struggles that could interfere with training. In a sample of 115 residents, 38 reported that they worked moonlighting jobs and 22 reported a credit card debt of more than $3,000. Suppose that it is reasonable to consider this sample as representative of all medical residents in the United States.

a. Construct and interpret a 95% confidence interval for the proportion of U.S. medical residents who work moonlighting jobs.

b. Construct and interpret a 90% confidence interval for the proportion of U.S. medical residents who have a credit card debt of more than $3,000.

c. Give two reasons why the confidence interval in Part (a) is wider than the confidence interval in Part (b).

Additional Exercises

9.54 In an AP-AOL sports poll (Associated Press, December 18, 2005), 394 of 1,000 randomly selected U.S. adults indicated that they considered themselves to be baseball fans. Of the 394 baseball fans, 272 stated that they thought the designated hitter rule should either be expanded to both baseball leagues or eliminated.

a. Construct and interpret a 95% confidence interval for the proportion of U.S. adults who consider themselves to be baseball fans.

b. Construct and interpret a 95% confidence interval for the proportion *of baseball fans* who think the designated hitter rule should be expanded to both leagues or eliminated.

c. Explain why the confidence intervals of Parts (a) and (b) are not the same width even though they both have a confidence level of 95%.

9.55 One thousand randomly selected adult Americans participated in a survey conducted by the **Associated Press (June, 2006)**. When asked "Do you think it is sometimes justified to lie, or do you think lying is never justified?" 52% responded that lying was never justified. When asked about lying to avoid hurting someone's feelings, 650 responded that this was often or sometimes OK.
a. Construct and interpret a 90% confidence interval for the proportion of adult Americans who would say that lying is never justified.
b. Construct and interpret a 90% confidence interval for the proportion of adult Americans who think that it is often or sometimes OK to lie to avoid hurting someone's feelings.
c. Using the confidence intervals from Parts (a) and (b), comment on the apparent inconsistency in the responses.

9.56 **"Tongue Piercing May Speed Tooth Loss, Researchers Say"** is the headline of an article that appeared in the *San Luis Obispo Tribune* **(June 5, 2002)**. The article describes a study of a representative sample of 52 young adults with pierced tongues. The researchers found receding gums, which can lead to tooth loss, in 18 of the participants. Construct and interpret a 95% confidence interval for the proportion of young adults with pierced tongues who have receding gums.

9.57 In a study of 1,710 schoolchildren in Australia (*Herald Sun*, **October 27, 1994**), 1,060 children indicated that they normally watch TV before school in the morning. (Interestingly, only 35% of the parents said their children watched TV before school.) Construct and interpret a 95% confidence interval for the proportion of all Australian children who say they watch TV before school. In order for the method used to construct the interval to be valid, what assumption about the sample must be reasonable?

9.58 In the article **"Fluoridation Brushed Off by Utah" (Associated Press, August 24, 1998)**, it was reported that a small but vocal minority in Utah has been successful in keeping fluoride out of Utah water supplies despite evidence that fluoridation reduces tooth decay and despite the fact that a clear majority of Utah residents favor fluoridation. To support this statement, the article included the result of a survey of Utah residents that found 65% to be in favor of fluoridation. Suppose that this result was based on a random sample of 150 Utah residents. Construct and interpret a 90% confidence interval for p, the proportion of all Utah residents who favor fluoridation. Is this interval consistent with the statement that fluoridation is favored by a clear majority of residents?

9.59 An Associated Press article on potential violent behavior reported the results of a survey of 750 workers who were employed full time (*San Luis Obispo Tribune*, **September 7, 1999**). Of those surveyed, 125 indicated that they were so angered by a coworker during the past year that they felt like hitting the coworker (but didn't). Assuming that it is reasonable to regard this sample as representative of the population of full-time workers, use this information to construct and interpret a 90% confidence interval estimate of p, the proportion of all full-time workers so angered in the last year that they wanted to hit a coworker.

9.60 High-profile legal cases have many people reevaluating the jury system. Many believe that juries in criminal trials should be able to convict on less than a unanimous vote. To assess support for this idea, investigators asked each individual in a random sample of Californians whether they favored allowing conviction by a 10–2 verdict in criminal cases not involving the death penalty. The Associated Press (*San Luis ObispoTelegram-Tribune*, **September 13, 1995**) reported that 71% favored conviction with a 10–2 verdict. Suppose that the sample size for this survey was $n = 900$. Compute and interpret a 99% confidence interval for the proportion of Californians who favor conviction with a 10–2 verdict.

SECTION 9.4 Choosing a Sample Size to Achieve a Desired Margin of Error

In the previous two sections, you saw that sample size plays a role in determining both the margin of error and the width of the confidence interval. Before collecting any data, you might want to determine a sample size that ensures a particular value for the margin of error. For example, with p representing the actual proportion of students at a college who purchase textbooks online, you may want to estimate p to within 0.03. The required sample size n is found by setting the expression for margin of error equal to 0.03 to obtain

$$1.96\sqrt{\frac{p(1-p)}{n}} = 0.03$$

and then solving for n.

In general, if M is the desired margin of error, finding the necessary sample size requires solving the equation

$$M = 1.96\sqrt{\frac{p(1-p)}{n}}$$

Solving this equation for n results in

$$n = p(1-p)\left(\frac{1.96}{M}\right)^2$$

To use this formula to determine a sample size, you need the value of p, which is unknown. One solution is to carry out a preliminary study and use the resulting data to get a rough estimate of p. In other cases, prior knowledge may suggest a reasonable estimate of p. If there is no prior information about the value of p and a preliminary study is not feasible, a more conservative solution makes use of the fact that the maximum value of $p(1-p)$ is 0.25 (its value when $p = 0.5$). Replacing $p(1-p)$ with 0.25 yields

$$n = 0.25\left(\frac{1.96}{M}\right)^2$$

Using this formula to determine n results in a sample size for which $\hat{p}$ will be within M of the population proportion at least 95% of the time for any value of p.

Using a 95% confidence level, the sample size required to estimate a population proportion p with a margin of error M is

$$n = p(1-p)\left(\frac{1.96}{M}\right)^2$$

The value of p may be estimated using prior information. If prior information is not available, using $p = 0.5$ in this formula results in a conservatively large value for the required sample size.

Example 9.6 Sniffing Out Cancer

Researchers have found biochemical markers of cancer in the exhaled breath of cancer patients, but chemical analysis of breath specimens has not yet proven effective in diagnosing cancer. The authors of the paper **"Diagnostic Accuracy of Canine Scent Detection in Early- and Late-Stage Lung and Breast Cancers"** (*Integrative Cancer Therapies* (2006): 1–10) describe a study to investigate whether dogs can be trained to identify the presence or absence of cancer by sniffing breath specimens. Suppose you want to collect data that would allow you to estimate the long-run proportion of accurate identifications for a particular dog that has completed training. The dog has been trained to lie down when presented with a breath specimen from a cancer patient and to remain standing when presented with a specimen from a person who does not have cancer. How many different breath specimens should be used if you want to estimate the long-run proportion of correct identifications for this dog with a margin of error of 0.10?

Using the conservative value of $p = 0.5$ in the formula for required sample size gives

$$n = p(1-p)\left(\frac{1.96}{M}\right)^2 = 0.25\left(\frac{1.96}{0.10}\right)^2 = 96.04$$

A sample of at least 97 breath specimens should be used. Notice that, in sample size calculations, you always round up.

SECTION 9.4 EXERCISES

Each Exercise Set assesses the following chapter learning objectives: M5, P3

9.61 A discussion of digital ethics appears in the article "**Academic Cheating, Aided by Cell Phones or Web, Shown to be Common**" (*Los Angeles Times*, June 17, 2009). One question posed in the article is: What proportion of college students have used cell phones to cheat on an exam? Suppose you have been asked to estimate this proportion for students enrolled at a large college. How many students should you include in your sample if you want to estimate this proportion with a margin of error of 0.02?

9.62 The article "**Consumers Show Increased Liking for Diesel Autos**" (*USA Today*, January 29, 2003) reported that 27% of U.S. consumers would opt for a diesel car if it ran as cleanly and performed as well as a car with a gas engine. Suppose that you suspect that the proportion might be different in your area. You decide to conduct a survey to estimate this proportion for the adult residents of your city. What is the required sample size if you want to estimate this proportion with a margin of error of 0.05? Compute the required sample size first using 0.27 as a preliminary estimate of p and then using the conservative value of 0.5. How do the two sample sizes compare? What sample size would you recommend for this study?

9.63 A manufacturer of small appliances purchases plastic handles for coffeepots from an outside vendor. If a handle is cracked, it is considered defective and can't be used. A large shipment of plastic handles is received. How many handles from the shipment should be inspected in order to estimate p, the proportion of defective handles in the shipment, with a margin of error of 0.1?

9.64 The article "**Should Canada Allow Direct-to-Consumer Advertising of Prescription Drugs?**" (*Canadian Family Physician* [2009]: 130–131) calls for the legalization of advertising of prescription drugs in Canada. Suppose you wanted to conduct a survey to estimate the proportion of Canadians who would allow this type of advertising. How large a random sample would be required to estimate this proportion with a margin of error of 0.02?

9.65 The 1991 publication of the book *Final Exit*, which includes chapters on doctor-assisted suicide, caused a great deal of controversy in the medical community. The Society for the Right to Die and the American Medical Association quoted very different figures regarding the proportion of primary-care physicians who have participated in some form of doctor-assisted suicide for terminally ill patients (*USA Today*, July 1991). Suppose that a survey of physicians will be carried out to estimate this proportion with a margin of error of 0.05. How many primary-care physicians should be included in a random sample?

Additional Exercises

9.66 In spite of the potential safety hazards, some people would like to have an Internet connection in their car. A preliminary survey of adult Americans has estimated the proportion of adult Americans who would like Internet access in their car to be somewhere around 0.30 (*USA Today*, May 1, 2009). Use the given preliminary estimate to determine the sample size required to estimate this proportion with a margin of error of 0.02.

9.67 Data from a representative sample were used to estimate that 32% of all computer users in 2011 had tried to get on a Wi-Fi network that was not their own in order to save money (*USA Today*, May 16, 2011). You decide to conduct a survey to estimate this proportion for the current year. What is the required sample size if you want to estimate this proportion with a margin of error of 0.05? Compute the required sample size first using 0.32 as a preliminary estimate of p and then using the conservative value of 0.5. How do the two sample sizes compare? What sample size would you recommend for this study?

9.68 *USA Today* (January 24, 2012) reported that ownership of tablet computers and e-readers is soaring. Suppose you want to estimate the proportion of students at your school who own at least one tablet or e-reader. What sample size would you use in order to estimate this proportion with a margin of error of 0.03?

9.69 The article "**Hospitals Dispute Medtronic Data on Wires**" (*The Wall Street Journal*, February 4, 2010) describes several studies of the failure rate of defibrillators used in the treatment of heart problems. In one study conducted by the Mayo Clinic, it was reported that failures within the first 2 years were experienced by 18 of 89 patients under 50 years old and 13 of 362 patients age 50 and older. Assume that these two samples are representative of patients who receive this type of defibrillator in the two age groups.

a. Construct and interpret a 95% confidence interval for the proportion of patients under 50 years old who experience a failure within the first 2 years.

b. Construct and interpret a 99% confidence interval for the proportion of patients age 50 and older who experience a failure within the first 2 years.

c. Suppose that the researchers wanted to estimate the proportion of patients under 50 years old who experience this type of failure with a margin of error of 0.03. How large a sample should be used? Use the given study results to obtain a preliminary estimate of the population proportion.

SECTION 9.5 Avoid These Common Mistakes

When using sample data to estimate a population characteristic, either a single number estimate or a confidence interval estimate might be used. Confidence intervals are generally preferred because a single number estimate, by itself, does not convey any information about its accuracy. For this reason, whenever you report the value of a single number estimate, it is a good idea to also include a margin of error.

Reporting and interpreting a confidence interval estimate requires a bit of care. First, always report both the confidence interval and the associated confidence level. Also remember that both the confidence interval and the confidence level should be interpreted. A good strategy is to begin with an interpretation of the confidence interval in the context of the problem and to follow that with an interpretation of the confidence level. For example, if a 90% confidence interval for p, the proportion of students at a particular college who own a computer, is (0.56, 0.78), you might say

Interpretation of interval → You can be 90% confident that between 56% and 78% of the students at this college own a computer.

explanation of "90% confidence"
interpretation of confidence level → You have used a method to produce this estimate that is successful in capturing the actual population proportion about 90% of the time.

Unfortunately, there is no customary way of reporting interval estimates of population characteristics in published sources. Possibilities include

confidence interval
estimate ± margin of error
estimate ± standard error

If the reported interval is described as a confidence interval, a confidence level should accompany it. These intervals can be interpreted just as you have interpreted the confidence intervals in this chapter, and the confidence level specifies the long-run success rate associated with the method used to construct the interval.

A form particularly common in news articles is estimate ± margin of error. The margin of error reported is usually two times the standard error of the estimate. This method of reporting is a little more informal than a confidence interval, but if the sample size is reasonably large, it is roughly equivalent to reporting a 95% confidence interval. You can interpret these intervals as you would a confidence interval with an approximate confidence level of 95%.

Be careful when interpreting intervals reported in the form of an estimate ± standard error. Recall from Section 9.3 that the general form of a confidence interval is

statistic ± (critical value)(standard error of the statistic)

The critical value in the confidence interval formula is determined by the sampling distribution of the statistic and by the confidence level. Note that the form estimate ± standard error is equivalent to a confidence interval with the critical value set equal to 1. For a statistic whose sampling distribution is approximately normal (such as a large-sample proportion), a critical value of 1 corresponds to a confidence level of about 68%. Because a confidence level of 68% is rather low, you may want to use the given information and the confidence interval formula to convert to an interval with a higher confidence level.

When working with single number estimates and confidence interval estimates, here are a few things you need to keep in mind:

1. In order for an estimate to be useful, you must know something about its accuracy. You should beware of single number estimates that are not accompanied by a margin of error or some other measure of accuracy.

2. A confidence interval estimate that is wide indicates that you don't have very precise information about the population characteristic being estimated. Don't be fooled by a high confidence level if the resulting interval is wide. High confidence, while desirable, is not the same thing as saying you have precise information about the value of a population characteristic.

   The width of a confidence interval is affected by the confidence level, the sample size, and the standard deviation of the statistic used to construct the interval. The best strategy for decreasing the width of a confidence interval is to take a larger sample. It is far better to think about this before collecting data. Use the sample size formula to determine a sample size that will result in a small margin of error or that will result in a confidence interval estimate that is narrow enough to provide useful information.

3. The accuracy of an estimate depends on the sample size, not the population size. This may be counter to intuition, but as long as the sample size is small relative to the population size (n is less than 10% of the population size), the margin of error for estimating a population proportion with 95% confidence is approximately $2\sqrt{\dfrac{p(1-p)}{n}}$. Notice that the margin of error involves the sample size n, and decreases as n increases. However, the approximate margin of error does not depend on the population size, N.

   The size of the population does need to be considered if sampling is without replacement and the sample size is more than 10% of the population size. In this case, the margin of error is adjusted by multiplying it by a **finite population correction factor** $\sqrt{\dfrac{N-n}{N-1}}$. Since this correction factor is always less than 1, the adjusted margin of error will be smaller.

4. Conditions are important. Appropriate use of the margin of error formula and the confidence interval of this chapter requires certain conditions be met:

   i. The sample is a random sample from the population of interest or is selected in a way that should result in a representative sample.
   ii. The sample size is large enough for the sampling distribution of $\hat{p}$ to be approximately normal.

   If these conditions are met, the large-sample confidence interval provides a method for using sample data to estimate the population proportion with confidence, and the confidence level is a good approximation of the success rate for the method.

   Whether the random or representative sample condition is plausible will depend on how the sample was selected and the intended population. Conditions for the sample size are the following:

   $$n\hat{p} \geq 10 \text{ and } n(1 - \hat{p}) \geq 10$$

   (so that the sampling distribution of $\hat{p}$ will be approximately normal).

5. When reading published reports, don't fall into the trap of thinking confidence interval every time you see a $\pm$ in an expression. As was discussed earlier in this section, published reports are not consistent. In addition to confidence intervals, it is common to see both estimate $\pm$ margin of error and estimate $\pm$ standard error reported.

CHAPTER ACTIVITIES

GETTING A FEEL FOR CONFIDENCE LEVEL

Technology Activity (Applet): Open the applet (www .cengage.com/stats/peck1e) called ConfidenceIntervals. You should see a screen like the one shown.

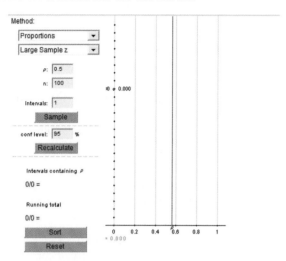

Getting Started: If the "Method" box does not say "Proportions," use the drop-down menu to select Proportions. In the box just below, select "Large Sample z" from the drop-down menu. This applet will select a random sample from a population with a specified proportion of successes and then use the sample to construct a confidence interval for the population proportion. The interval is then plotted on the display at the right, and you can see if the interval contains the actual value of the population proportion. For purposes of this activity, you will sample from a population with $p = 0.4$ and begin with a sample size of $n = 50$. In the applet window, set $p = 0.40$ and $n = 50$. Leave the conf-level box set at 95%. Click the "Recalculate" button to rescale the picture on the right. Now click on the sample button. You should see a confidence interval appear on the display on the right-hand side. If the interval contains the actual population proportion of 0.40, the interval is drawn in green. If 0.40 is not in the confidence interval, the interval is shown in red. Your screen should look something like the one on the left at the bottom of this page.

Click on the "Sample" button several more times, and notice how the confidence interval estimate changes from sample to sample. Also notice that at the bottom of the left-hand side of the display, the applet is keeping track of the proportion of all the intervals calculated so far that include the actual value of p. If you were to construct a large number of intervals, this proportion should closely approximate the success rate for the confidence interval method.

To look at more than one interval at a time, change the "Intervals" box from 1 to 100, and then click the sample button. You should see a screen similar to the one below on the right, with 100 intervals in the display on the right-hand side. Again, intervals containing 0.40 (the value of p in this case) will be green and those that do not contain 0.40 will be red. Also notice that the success proportion on the left-hand side has been updated to reflect what happened with the 100 newly generated intervals.

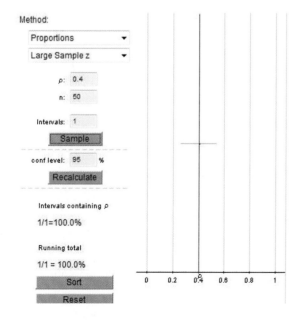

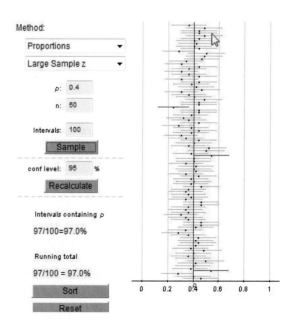

Continue generating intervals until you have seen at least 1,000 intervals, and then answer the following question:

a. How does the proportion of intervals constructed that contain $p = 0.40$ compare to the stated confidence level of 95%?

Experiment with three other confidence levels of your choice, and then answer the following question:

b. In general, is the proportion of computed confidence intervals that contain $p = 0.40$ close to the stated confidence level?

ACTIVITY 9.2 AN ALTERNATIVE CONFIDENCE INTERVAL FOR A POPULATION PROPORTION

Technology Activity (Applet): This activity presumes that you have completed Activity 9.1.

Background: As an alternative to the usual large-sample z confidence interval, consider the confidence interval

$$\hat{p}_{\text{mod}} \pm (z \text{ critical value})\sqrt{\frac{\hat{p}_{\text{mod}}(1 - \hat{p}_{\text{mod}})}{n}}$$

where $\hat{p}_{\text{mod}} = \dfrac{\text{successes} + 2}{n + 4}$. This alternative interval is preferred by many statisticians because, in repeated sampling, the proportion of intervals constructed that include the actual value of the population proportion, p, tends to be closer to the stated confidence level. In this activity, you will compare the success rates for the two different confidence interval methods.

Open the applet (www.cengage.com/stats/peck1e) called ConfidenceIntervals. You should see a screen like the one shown.

Simulating Confidence Intervals

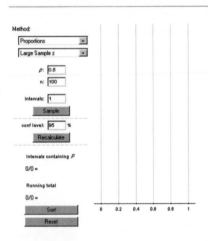

Select "Proportions" from the Method box drop-down menu, and then select "Large Sample z" from the drop-down menu of the second box. Consider sampling from a population with $p = 0.3$ using a sample size of 40. In the applet window, enter $p = 0.3$ and $n = 40$. Note that $n = 40$ is large enough to satisfy $np \geq 10$ and $n(1 - p) \geq 10$.

Set the "Intervals" box to 100, and then use the applet to construct a large number (at least 1000) of 95% confidence intervals.

1. How does the proportion of intervals constructed that include $p = 0.3$, the population proportion, compare to 95%? Does this surprise you? Explain.

Now use the drop-down menu to change "Large Sample z" to "Modified." Now the applet will construct the alternative confidence interval that is based on $\hat{p}_{\text{mod}}$. Use the applet to construct a large number (at least 1000) of 95% confidence intervals.

2. How does the proportion of intervals constructed that include $p = 0.3$, the population proportion, compare to 95%? Is this proportion closer to 95% than was the case for the large-sample z interval?

3. Experiment with different combinations of values of sample size and population proportion p. Can you find a combination for which the large-sample z interval has a success rate that is close to 95%? Can you find a combination for which it has a success rate that is even farther from 95% than it was for $n = 40$ and $p = 0.3$? How does the modified interval perform in each of these cases?

ACTIVITY 9.3 VERIFYING SIGNATURES ON A RECALL PETITION

Background: In 2003, petitions were submitted to the California Secretary of State calling for the recall of Governor Gray Davis. Each of California's 58 counties reported the number of valid signatures on the petitions from that county so that the state could determine whether there were enough to certify the recall and set a date for the recall election. The following paragraph appeared in the *San Luis Obispo Tribune* (July 23, 2003):

"In the campaign to recall Gov. Gray Davis, the secretary of state is reporting 16,000 verified signatures from San Luis Obispo County. In all, the County Clerk's Office received 18,866 signatures on recall petitions and was instructed by the state to check a random sample of 567. Out of those, 84.48% were good. The verification process includes checking whether the signer is a registered voter and whether the address and signature on the recall petition match the voter registration."

1. Use the data from the random sample of 567 San Luis Obispo County signatures to construct a 95% confidence interval for the proportion of petition signatures that are valid.

2. How do you think that the reported figure of 16,000 verified signatures for San Luis Obispo County was obtained?

3. Based on your confidence interval from Part 1, explain why you think that the reported figure of 16,000 verified signatures is or is not reasonable.

🖥 EXPLORING THE BIG IDEAS

In the exercise below, each student in your class will go online to select a random sample of 30 animated movies produced between 1980 and 2011. You will use this sample to estimate the proportion of animated movies made between 1980 and 2011 that were produced by Walt Disney Studios.

Go online at www.cengage.com/stats/peck1e and click on the link labeled "Exploring the Big Ideas." Then locate the link for Chapter 9. This link will take you to a web page where you can select a random sample from the animated movie population.

Click on the "sample" button. This selects a random sample of 30 movies and will display the movie name and the production studio for each movie in your sample. Each student in your class will receive a different random sample.

Use the data from your random sample to complete the following:

a. Calculate the proportion of movies in your sample that were produced by Disney.

b. Is the proportion you calculated in Part (a) a sample proportion or a population proportion?

c. Construct a 90% confidence interval for the proportion of animated movies made between 1980 and 2011 that were produced by Disney. Write a few sentences interpreting the confidence interval and the associated confidence level.

d. The actual population proportion is 0.41. Did your confidence interval from Part (c) include this value?

e. Which of the following is a correct interpretation of the 90% confidence level?

1. The probability that the actual population proportion is contained in the computed interval is 0.90.

2. If the process of selecting a random sample of movies and then computing a 90% confidence interval for the proportion of all animated movies made between 1980 and 2011 that were produced by Disney is repeated 100 times, exactly 90 of the 100 intervals will include the actual population proportion.

3. If the process of selecting a random sample of movies and then computing a 90% confidence interval for the proportion of all animated movies made between 1980 and 2011 that were produced by Disney is repeated a very large number of times, approximately 90% of the intervals will include the actual population proportion.

If asked to do so by your teacher, bring your confidence interval estimate of the proportion of animated movies made between 1980 and 2011 that were produced by Disney to class. Your teacher will lead the class through a discussion of the questions that follow.

Compare your confidence interval to the confidence interval obtained by another student in your class.

f. Are the two confidence intervals the same?

g. Did both intervals contain the actual population proportion of 0.41?

h. How many people in your class have a confidence interval that does not include the actual value of the population proportion? Is this surprising, given the 90% confidence level associated with the confidence intervals?

All chapter learning objectives are assessed in these exercises. The learning objectives assessed in each exercise are given in parentheses.

9.70 (C1)
Two statistics are being considered for estimating the value of a population characteristic. The sampling distributions of the two statistics are shown here. Explain why Statistic II would be preferred over Statistic I.

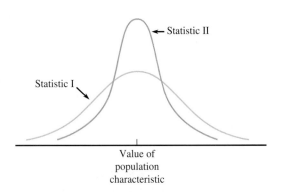

9.71 (C2)
Two statistics are being considered for estimating the value of a population characteristic. The sampling distributions of the two statistics are shown here.

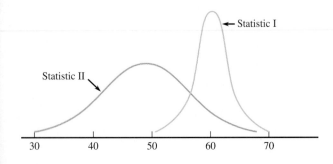

a. Suppose the actual value of the population characteristic is 50. Is Statistic I or Statistic II unbiased? If so, which one?
b. Suppose the actual value of the population characteristic is 54. In this case, neither Statistic I nor Statistic II is unbiased. Which of these two statistics would you recommend as the better estimator of the population characteristic? Explain the reasoning for your choice.

9.72 (C3)
Will $\hat{p}$ from a random sample from a population with 60% successes tend to be closer to 0.6 for a sample size of $n = 400$ or a sample size of $n = 800$? Provide an explanation for your choice.

9.73 (C3)
Will $\hat{p}$ from a random sample of size 400 tend to be closer to the actual value of the population proportion when $p = 0.4$ or when $p = 0.7$? Provide an explanation for your choice.

9.74 (C3)
A random sample will be selected from the population of all adult residents of a particular city. The sample proportion $\hat{p}$ will be used to estimate p, the proportion of all adult residents who are registered to vote. For which of the following situations will the estimate tend to be closest to the actual value of p?
I. $n = 1,000, p = 0.5$
II. $n = 200, p = 0.6$
III. $n = 100, p = 0.7$

9.75 (P1)
In response to budget cuts, county officials are interested in learning about the proportion of county residents who favor closure of a county park rather than closure of a county library. In a random sample of 500 county residents, 198 favored closure of a county park. For each of the three statements below, indicate if the statement is correct or incorrect. If the statement is incorrect, explain what makes it incorrect.

Statement 1: It is unlikely that the estimate $\hat{p} = 0.396$ differs from the value of the actual population proportion by more than 0.0429.
Statement 2: The estimate $\hat{p} = 0.396$ will never differ from the value of the actual population proportion by more than 0.0429.
Statement 3: It is unlikely that the estimate $\hat{p} = 0.396$ differs from the value of the actual population proportion by more than 0.0219.

9.76 (C4)
Consider taking a random sample from a population with $p = 0.25$.
a. What is the standard error of $\hat{p}$ for random samples of size 400?
b. Would the standard error of $\hat{p}$ be smaller for random samples of size 200 or samples of size 400?

c. Does cutting the sample size in half from 400 to 200 double the standard error of $\hat{p}$?

9.77 (C6, M1, M2, P1, P2)

In a survey on supernatural experiences, 722 of 4,013 adult Americans reported that they had seen a ghost (**"What Supernatural Experiences We've Had," USA Today, February 8, 2010**). Assume that this sample is representative of the population of adult Americans.

a. Use the given information to estimate the proportion of adult Americans who would say they have seen a ghost.

b. Verify that the conditions for use of the margin of error formula to be appropriate are met.

c. Compute the margin of error.

d. Interpret the margin of error in context.

e. Construct and interpret a 90% confidence interval for the proportion of all adult Americans who have seen a ghost.

f. Would a 99% confidence interval be narrower or wider than the interval computed in Part (e)? Justify your answer.

9.78 (C6)

Suppose that county planners are interested in learning about the proportion of county residents who would pay a fee for a curbside recycling service if the county were to offer this service. Two different people independently selected random samples of county residents and used their sample data to construct the following confidence intervals for the proportion who would pay for curbside recycling:

Interval 1: (0.68, 0.74)
Interval 2: (0.68, 0.72)

a. Explain how it is possible that the two confidence intervals are not centered in the same place.

b. Which of the two intervals conveys more precise information about the value of the population proportion?

c. If both confidence intervals are associated with a 95% confidence level, which confidence interval was based on the smaller sample size? How can you tell?

d. If both confidence intervals were based on the same sample size, which interval has the higher confidence level? How can you tell?

9.79 (C6)

Describe how each of the following factors affects the width of the large-sample confidence interval for p:

a. The confidence level

b. The sample size

c. The value of $\hat{p}$

9.80 (C7)

Based on data from a survey of 1,200 randomly selected Facebook users (**USA Today, March 24, 2010**), a 90% confidence interval for the proportion of all Facebook users who say it is not OK to "friend" someone who reports to you at work is (0.60, 0.64). What is the meaning of the 90% confidence level associated with this interval?

9.81 (M3)

The study **"Digital Footprints" (Pew Internet & American Life Project, www.pewinternet.org, 2007)** reported that 47% of Internet users have searched for information about themselves online. The 47% figure was based on a random sample of Internet users. Suppose that the sample size was $n = 300$ (the actual sample size was much larger). Answer the four key questions (QSTN) to confirm that the suggested method in this situation is a large-sample confidence interval for a population proportion.

9.82 (M4)

For the study described in the previous exercise, use the five-step process for estimation problems (EMC³) to construct and interpret a 90% confidence interval for the proportion of Internet users who have searched online for information about themselves. Identify each of the five steps in your solution.

9.83 (C6, P2)

The article **"Kids Digital Day: Almost 8 Hours" (USA Today, January 20, 2010)** summarized a national survey of 2,002 Americans ages 8 to 18. The sample was selected to be representative of Americans in this age group.

a. Of those surveyed, 1,321 reported owning a cell phone. Use this information to construct and interpret a 90% confidence interval for the proportion of all Americans ages 8 to 18 who own a cell phone.

b. Of those surveyed, 1,522 reported owning an MP3 music player. Use this information to construct and interpret a 90% confidence interval for the proportion of all Americans ages 8 to 18 who own an MP3 music player.

c. Explain why the confidence interval from Part (b) is narrower than the confidence interval from Part (a) even though the confidence levels and the sample sizes used to compute the two intervals were the same.

9.84 (M5, P3)

A consumer group is interested in estimating the proportion of packages of ground beef sold at a particular store that have an actual fat content exceeding the fat content stated on the label. How many packages of ground beef should be tested in order to have a margin of error of 0.05?

TECHNOLOGY NOTES

Confidence Intervals for Proportions

TI-83/84
1. Press the **STAT** key
2. Highlight **TESTS**
3. Highlight **1-PropZInterval** and press **ENTER**
4. Next to *x* type the number of successes
5. Next to *n* type the number of trials, *n*
6. Next to **C-Level** type the value for the confidence level
7. Highlight **Calculate** and press **ENTER**

TI-Nspire
1. Enter the Calculate Scratchpad
2. Press the **menu** key then select **6:Statistics** then select **6:Confidence Intervals** then **5:1-Prop z Interval...** then press **enter**
3. In the box next to **Successes,** *x* type the number of successes
4. In the box next to *n* type the number of trials, *n*
5. In the box next to **C Level** type the confidence level
6. Press **OK**

JMP
Summarized data
1. Enter the data table into the JMP data table with categories in the first column and counts in the second column

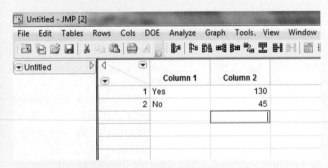

2. Click **Analyze** and select **Distribution**
3. Click and drag the first column name from the box under **Select Columns** to the box next to **Y, Columns**
4. Click and drag the second column name from the box under **Select Columns** to the box next to **Freq**
5. Click **OK**
6. Click the red arrow next to the column name and click **Confidence Interval** then select the appropriate level or select **Other** to input a level that is not listed

Raw Data
1. Enter the raw data into a column
2. Click **Analyze** and select **Distribution**
3. Click and drag the first column name from the box under **Select Columns** to the box next to **Y, Columns**
4. Click **OK**
5. Click the red arrow next to the column name and click **Confidence Interval** then select the appropriate level or select **Other** to input a level that is not listed

Minitab
Summarized data
1. Click **Stat** then click **Basic Statistics** then click **1 Proportion...**
2. Click the radio button next to **Summarized data**
3. In the box next to **Number of Trials:** type the value for *n*, the total number of trials
4. In the box next to **Number of events:** type the value for the number of successes
5. Click **Options...**
6. Input the appropriate confidence level in the box next to **Confidence Level**
7. Check the box next to **Use test and interval based on normal distribution**
8. Click **OK**
9. Click **OK**

Raw data
1. Input the raw data into a column
2. Click **Stat** then click **Basic Statistics** then click **1 Proportion...**
3. Click in the box under **Samples in columns:**
4. Double click the column name where the raw data is stored
5. Click **Options...**
6. Input the appropriate confidence level in the box next to **Confidence Level**
7. Check the box next to **Use test and interval based on normal distribution**
8. Click **OK**
9. Click **OK**

SPSS
SPSS does not have the functionality to automatically produce confidence intervals for a population proportion.

Excel
Excel does not have the functionality to automatically produce confidence intervals for a population proportion. However, you can manually type in the formula for the lower and upper limits separately into two different cells to have Excel calculate the result for you.

AP* Review Questions for Chapter 9

1. Each figure below displays the sampling distribution of a statistic used to estimate a population characteristic (parameter). The actual value of the population characteristic is marked on each sampling distribution. Which is the sampling distribution of a statistic that is an unbiased estimator of the population characteristic?

Statistic I

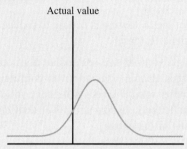

Statistic II

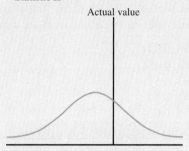

Statistic III

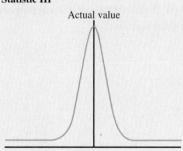

(A) Statistic I only
(B) Statistic II only
(C) Statistic III only
(D) Both Statistic I and Statistic II are unbiased.
(E) None of the three statistics is unbiased.

2. Four different statistics are being considered for estimating a population characteristic. The sampling distributions of the four statistics are shown here. Which statistic is most likely to result in an estimate that is close to the actual value of the population characteristic?

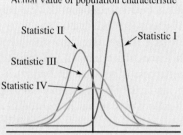

(A) Statistic I
(B) Statistic II
(C) Statistic III
(D) Statistic IV
(E) Statistics III and IV are equally likely to result in an estimate that is close to the true value.

3. Given a choice of unbiased statistics, the best statistic would be the one with a sampling distribution that

(A) is most closely approximated by a normal curve.
(B) has the smallest variance.
(C) has the largest standard error.
(D) is most nearly symmetric.
(E) is positively skewed.

4. To estimate the proportion of faculty at a state university who own a home, a random sample of faculty is selected. For which of the following combinations of n and $\hat{p}$ would it be appropriate to use the confidence interval $\hat{p} \pm (z \text{ critical value}) \sqrt{\dfrac{\hat{p}(1-\hat{p})}{n}}$ to estimate the population proportion?

(A) $n = 20$ and $\hat{p} = 0.40$
(B) $n = 40$ and $\hat{p} = 0.20$
(C) $n = 100$ and $\hat{p} = 0.05$
(D) $n = 150$ and $\hat{p} = 0.45$
(E) $n = 200$ and $\hat{p} = 0.02$

5. Suppose that a random sample of 100 high school class-rooms in the state of California is selected and a 95% confidence interval for the proportion that has Internet access is (0.62, 0.78). Which of the following is a correct interpretation of the 95% confidence <u>level</u>?

    (A) The method used to construct the interval will produce an interval that includes the value of the population proportion about 95% of the time in repeated sampling.
    (B) You can be 95% confident that the sample proportion is between 0.62 and 0.78.
    (C) There is a 95% chance that the proportion of all high school classrooms in California that have Internet access is between 0.62 and 0.78.
    (D) You can be 95% confident that the proportion of all high school classrooms in California that have Internet access is between 0.62 and 0.78.
    (E) None of the above is a correct interpretation of the confidence level.

6. Which of the following must be true of a sample in order for it to be appropriate to use a large-sample z confidence interval to estimate the population proportion?

    (A) The sample is a random sample from the population of interest.
    (B) $n\hat{p} \geq 10$ and $n(1 - \hat{p}) \geq 10$
    (C) The population distribution is normal.
    (D) All of (A), (B), and (C) are required for appropriate use of the large-sample z confidence interval to estimate the population proportion.
    (E) Only (A) and (B) are required for appropriate use of the large-sample z confidence interval to estimate the population proportion.

7. Which of the following would tend to decrease the width of a confidence interval?

    I.   Increasing the sample size
    II. Using a higher confidence level
    III. Using a lower confidence level

    (A) I only
    (B) II only
    (C) III only
    (D) I and II only
    (E) I and III only

8. Which of the following is certain to reduce the width of a confidence interval?

    (A) Larger sample size and higher confidence level
    (B) Larger sample size and lower confidence level
    (C) Smaller sample size and higher confidence level
    (D) Smaller sample size and lower confidence level
    (E) None of the above

9. A 90% confidence interval for a population proportion calculated using data from a random sample of size $n = 200$ is (0.38, 0.48). Which of the following is the 95% confidence interval calculated from the same data?

    (A) (0.43, 0.53)
    (B) (0.40, 0.46)
    (C) (0.33, 0.53)
    (D) (0.42, 0.54)
    (E) (0.35, 0.61)

10. A researcher wants to estimate a population proportion with a margin of error of 0.05. What is the smallest sample size for which the sample proportion would be within 0.05 of the actual population proportion for 95% of all random samples?

    (A) 30
    (B) 100
    (C) 193
    (D) 385
    (E) 1,000

10

Asking and Answering Questions About a Population Proportion

Preview

Chapter Learning Objectives

10.1 Hypotheses and Possible Conclusions

10.2 Potential Errors in Hypothesis Testing

10.3 The Logic of Hypothesis Testing—An Informal Example

10.4 A Procedure for Carrying Out a Hypothesis Test

10.5 Large-Sample Hypothesis Test for a Population Proportion

10.6 Power and the Probability of Type II Error

10.7 Avoid These Common Mistakes

Chapter Activities

Exploring the Big Ideas

Are You Ready to Move On? Chapter 10 Review Exercises

Technology Notes

AP* Review Questions for Chapter 10

Petro Feketa/Shutterstock.com

PREVIEW

Two types of inference problems are considered in this text. In estimation problems, sample data are used to learn about the value of a population characteristic. In hypothesis testing problems, sample data are used to decide if some claim about the value of a population characteristic is plausible. In Chapter 9, you saw how to use sample data to estimate a population proportion. In this chapter, you will see how sample data can also be used to decide whether a claim, called a hypothesis, *about a population proportion is believable.*

CHAPTER LEARNING OBJECTIVES

Conceptual Understanding

After completing this chapter, you should be able to

C1 Understand how a research question is translated into hypotheses.

C2 Understand that rejecting the null hypothesis implies strong support for the alternative hypothesis.

C3 Understand why failing to reject the null hypothesis does not imply strong support for the null hypothesis.

C4 Know the two types of errors that are possible in a hypothesis test.

C5 Understand how the significance level of a test relates to the probability of a Type I error.

C6 Understand the reasoning used to reach a decision in a hypothesis test.

C7 Understand how the power of a test is affected by sample size and choice of significance level.

Mastering the Mechanics

After completing this chapter, you should be able to

M1 Translate a research question or claim about a population proportion into null and alternative hypotheses.

M2 Describe Type I and Type II errors in context.

M3 Know the key characteristics that lead to selection of a large-sample test for a population proportion as an appropriate method.

M4 Know the conditions for appropriate use of the large-sample test for a population proportion.

M5 Select an appropriate significance level based on consideration of the consequences of Type I and Type II errors.

M6 Use the five-step process for hypothesis testing (HMC³) to carry out a large-sample test for a population proportion.

Putting It into Practice

After completing this chapter, you should be able to

P1 Recognize when a situation calls for testing hypotheses about a population proportion.

P2 Carry out a large-sample test for a population proportion and interpret the conclusion in context.

PREVIEW EXAMPLE **Gender Neutral?**

The article **"Boy or Girl: Which Gender Baby Would You Pick?"** (*LiveScience*, **March 23, 2005, www.livescience.com**) summarized a study that was published in *Fertility and Sterility*. The *LiveScience* article makes the following statements:

> "When given the opportunity to choose the sex of their baby, women are just as likely to choose pink socks as blue, a new study shows."

> "Of the 561 women who participated in the study, 229 said they would like to choose the sex of a future child. Among these 229 there was no greater demand for boys or girls."

These statements are equivalent to the claim that the proportion of women who would choose a girl is 0.50 or 50%,

Is this claim supported by the study data? The paper referenced in the *LiveScience* article (**"Preimplantation Sex Selection Demand and Preferences in an Infertility Population,"** *Fertility and Sterility* **[2005]: 649–658**) provides the following information about the study:

- A survey with 19 questions was mailed to 1,385 women who had visited the Center for Reproductive Medicine at Brigham and Women's Hospital.
- A total of 561 women returned the survey.

- Of the 561 women who responded, 229 (or 40.8%) wanted to select the sex of their next child.
- Of the 229 who wanted to select the baby's sex, 89 (or 38.9%) wanted a boy and 140 (or 61.1%) wanted a girl.

You should have a few questions at this point. What is the population of interest, and are the 561 women who responded to the survey representative of this population? There was a high nonresponse rate—only 561 of 1,385 surveys were returned. Is it possible that women who returned the survey differ in some important way from women who did not return the survey? It will be important to address these questions before deciding if it is reasonable to generalize from this sample to a larger population.

Moreover, the proportion of women wanting to choose the baby's sex who would select a girl was $\frac{140}{229} = 0.611$. However, the *LiveScience* summary indicates that there is no preference for girls over boys, implying that the proportion who would choose a girl is 0.50. Is a sample proportion of 0.611 consistent with a population proportion of 0.50? There are two possibilities to consider:

1. Even when the population proportion who would choose a girl is 0.50, the sample proportion will tend to differ from 0.50. So, a sample proportion as large as 0.611 might just be due to sampling variability. If this is the case, the sample data would be consistent with the claim that the population proportion is 0.50.

or

2. When the population proportion who would choose a girl is 0.50, observing a sample proportion as large as 0.611 might be very unlikely. In this case, the sample data would provide evidence against the claim that the population proportion is 0.50.

How should you decide between these two possibilities? This example will be revisited later in this chapter where you will apply what you learned about sampling distributions in Chapter 8 to decide which of these two possibilities is the more reasonable choice.

This chapter is the first of several that cover tests of hypotheses. In this chapter, you will see how to use sample data to choose between two competing claims about a population proportion.

SECTION 10.1 Hypotheses and Possible Conclusions

In Chapter 7, you learned to distinguish estimation problems from hypothesis testing problems. Recall that a hypothesis testing problem involves using sample data to test a claim (called a hypothesis) about a population. Later in this chapter, you will learn how to test hypotheses about a population proportion. But first let's introduce some general terminology and consider the risks involved in this type of inference.

In its simplest form, a hypothesis is a claim or statement about the value of a single population characteristic. (You will see more complicated hypotheses in later chapters.) For example, consider a population consisting of all e-mail messages sent using a college e-mail system during a particular month. The following are examples of hypotheses about population proportions:

Hypothesis	Population Proportion of Interest	The Hypothesis Says...
$p < 0.01$	where p is the proportion of e-mail messages that were returned as undeliverable	Less than 1% of the e-mail messages sent were returned as undeliverable.
$p < 0.25$	where p is the proportion of e-mail messages that included an attachment	Less than 25% of the e-mail messages sent included an attachment.

(continued)

Hypothesis	Population Proportion of Interest	The Hypothesis Says...
$p > 0.8$	where p is the proportion of e-mail messages that were longer than 500 characters	More than 80% of the e-mail messages sent were longer than 500 characters.
$p = 0.3$	where p is the proportion of e-mail messages that were sent to multiple recipients	30% of the e-mail messages sent were sent to multiple recipients.

AP* EXAM TIP

Hypotheses are about population characteristics. Never state a null or alternative hypothesis using sample statistics!

Notice that hypotheses are *always* statements about population characteristics and *never* about sample statistics. Statements like $\hat{p} < 0.50$ or $\hat{p} = 0.72$ are not hypotheses because $\hat{p}$ is a sample proportion, not a population proportion.

In a hypothesis testing situation, you want to know if the sample data provide support for a claim about a population characteristic. For example, suppose that a particular community college claims that the majority of students completing an associate's degree transfer to a 4-year college. To determine if you should believe this claim, you might begin by selecting a random sample of 100 graduates of this community college. Each graduate could be asked if he or she transferred to a 4-year college. You would then want to determine if the sample data provide convincing evidence in support of the hypothesis $p > 0.5$, where p is the proportion of graduates who transfer to a 4-year college. This hypothesis states that the population proportion is greater than 0.5, or equivalently that more than 50% (a majority) of the graduates transfer to a 4-year college.

Although it will seem a bit odd at first, the way to decide if the sample data provide convincing evidence in support of a particular hypothesis is to set up a pair of competing hypotheses that includes the hypothesis of interest. In the community college transfer example, the two competing hypotheses might be

$$p \leq 0.5 \text{ and } p > 0.5$$

The hypothesis $p > 0.5$ says that a majority of graduates transfer to a 4-year college, and the hypothesis $p \leq 0.5$ says that the proportion of graduates who transfer is less than or equal to 0.5. If it were possible to survey *every* graduate (a census), you would know the actual value of p and you would know which of these two hypotheses was correct. Unfortunately, you must usually decide between the two competing hypotheses using data from a sample, which means that you will not have complete information.

A **hypothesis test** uses sample data to choose between two competing hypotheses about a population characteristic.

A criminal trial is a familiar situation in which a choice between two contradictory claims must be made. A person accused of a crime must be judged either guilty or not guilty. Under the U.S. system of justice, the individual on trial is initially presumed not guilty. Only strong evidence to the contrary causes the not guilty claim to be rejected in favor of a guilty verdict. The burden is put on the prosecution to *convince* the jurors that a defendant is guilty.

As in a judicial proceeding, in a hypothesis test you initially assume that a particular hypothesis, called the *null hypothesis*, is the correct one. You then consider the evidence (sample data) and reject the null hypothesis in favor of a competing hypothesis, called the *alternative hypothesis*, only if there is convincing evidence against the null hypothesis.

The **null hypothesis,** denoted by H_0, is a claim about a population characteristic that is initially assumed to be true.

The **alternative hypothesis,** denoted by H_a, is a competing claim.

(continued)

In carrying out a test of H_0 versus H_a, the null hypothesis H_0 is rejected in favor of the alternative hypothesis H_a only if the sample data provide convincing evidence that H_0 is false. If the sample data do not provide such evidence, H_0 is not rejected.

The two possible conclusions in a hypothesis test are *reject H_0* and *fail to reject H_0*.

In choosing between the null and alternative hypotheses, notice that you initially assume that the null hypothesis is true. When you observe sample data that would be very unlikely to occur if the null hypothesis were true, you have convincing evidence against the null hypothesis. In this case, you would reject the null hypothesis in favor of the alternative hypothesis. This is a strong conclusion because you have *convincing* evidence against the null hypothesis.

If you are not convinced that the null hypothesis is false, you fail to reject the null hypothesis. It is important to note that this does *not* mean that you have strong or convincing evidence that the null hypothesis is true. You can only say that you were not convinced that the null hypothesis is false.

AP* EXAM TIP

If the null hypothesis is *not* rejected, the conclusion is *fail to reject* the null hypothesis. Because this does not mean strong support for the null hypothesis, you don't want to imply that you have evidence that the null hypothesis is true.

Hypothesis Test Conclusion	This Means That...
Reject H_0	There is convincing evidence against the null hypothesis. If H_0 were true, the sample data would be very surprising.
Fail to reject H_0	There is not convincing evidence against the null hypothesis. If H_0 were true, the sample data would not be considered surprising.

A statistical hypothesis test can only demonstrate strong support for the alternative hypothesis. If the goal of a study is to decide whether sample data support a particular claim about a population, this claim will determine the alternative hypothesis. In a hypothesis testing situation, you usually begin by translating the claim of interest into the alternative hypothesis.

The **alternative hypothesis** will have one of the following three forms:

H_a: population characteristic > hypothesized value

H_a: population characteristic < hypothesized value

H_a: population characteristic ≠ hypothesized value

The hypothesized value and whether >, <, or ≠ appears in the alternative hypothesis is determined by the study context and the question of interest. Some examples illustrating selection of the alternative hypothesis follow.

Example 10.1   Living at Home

The report **"Charles Schwab 2009 Young Adults & Money Survey Findings" (www.schwab .com)** summarizes a survey of 1,252 young adults. Researchers were interested in determining if sample data support the claim that more than one in four young adults live with their parents. With $p =$ the proportion of young adults who live with their parents, the question of interest becomes "Is $p > 0.25$?" The hypothesized value of 0.25 is used because "one in four" is equivalent to $\frac{1}{4} = 0.25$. The appropriate alternative hypothesis is $H_a: p > 0.25$.

| Example 10.2 | Unhappy Customers |

A study sponsored by the **American Savings Education Council ("Preparing for Their Future: A Look at the Financial State of Gen X and Gen Y," March 2008)** included data from a survey of 1,752 people ages 19 to 39. One of the survey questions asked participants how satisfied they were with their current financial situation. Suppose you want to determine if the survey data provide convincing evidence that fewer than 10% of adults ages 19 to 39 are very satisfied with their current financial situation. With p = the proportion of adults ages 19 to 39 who are very satisfied with their financial situation, the claim of interest (fewer than 10%) is "p is less than 0.10." The appropriate alternative hypothesis is $H_a: p < 0.10$.

| Example 10.3 | M&M Color Mix |

The manufacturer of M&Ms claims that 40% of plain M&M's are brown. A sample of M&M's will be used to determine if the proportion of brown M&M's is different from what the manufacturer claims. With p = the proportion of plain M&M's that are brown, the manufacturer's claim is $p = 0.40$. The question is whether there is evidence that the manufacturer's claim is false, so the alternative hypothesis would be $H_a: p \neq 0.40$.

| Example 10.4 | Whose Reality Revisited |

Example 7.2 described a study in which researchers wanted to learn about how people view reality TV shows **("Who's Afraid of Reality Shows?" Communications Research [2008]: 382–397).** The researchers hoped to learn if the study data supported the theory that a majority of adults in Israel believe they are much less affected by reality TV shows than other people. With p = the proportion of Israeli adults who believe they are much less affected than other people, the researchers claim is $p > 0.5$. To see if there is evidence to support this claim, the researchers could carry out a hypothesis test with an alternative hypothesis of

$$H_a: p > 0.5$$

Once an alternative hypothesis has been selected, the next step is to formulate a competing null hypothesis. The null hypothesis can be formed by replacing the inequality that appears in the alternative hypothesis ($<$, $>$ or $\neq$) with an equal sign. For the alternative hypothesis $H_a: p > 0.5$ from Example 10.4, the corresponding null hypothesis would be $H_0: p = 0.5$.

Because the alternative hypothesis in Example 10.4 was $H_a: p > 0.5$, it might seem sensible to state H_0 as the inequality $p \leq 0.5$ rather than $p = 0.5$. While either is acceptable, we will state H_0 as a claim of equality. There are several reasons for this. First, the development of a test statistic is most easily understood if there is only a single hypothesized value in the null hypothesis. Second, suppose that the sample data provide convincing evidence that $H_0: p = 0.5$ should be rejected in favor of $H_a: p > 0.5$. This means that you are convinced by the sample data that the population proportion is greater than 0.5. It follows that you must also be convinced that the population proportion is not 0.4 or 0.3 or any other value that is less than 0.5. As a consequence, the conclusion when testing $H_0: p = 0.5$ versus $H_0: p > 0.5$ is always the same as the conclusion for a test where the null hypothesis is $H_0: p \leq 0.5$.

Because it is customary to state the null hypothesis as a claim of equality, this convention will be followed in the examples of this text. However, it is also acceptable to write the null hypothesis as $\leq$ when the inequality in the alternative hypothesis is $>$ or as $\geq$ when the inequality in the alternative hypothesis is $<$. *Just remember that the null hypothesis must include the equal case.*

> The form of the null hypothesis is
>
> $$H_0: \text{population characteristic} = \text{hypothesized value}$$
>
> where the hypothesized value is a specific number determined by the problem context.

Example 10.5 illustrates how the selection of H_0 (the claim initially believed true) and H_a depend on the objectives of a study.

Example 10.5 Evaluating a New Medical Treatment

A medical research team has been given the task of evaluating a new laser treatment for certain types of tumors. Consider the following two scenarios:

Scenario 1: The current standard treatment is considered reasonable and safe by the medical community, is not costly, has no major side effects, and has a known success rate of 0.85 (or 85%).

Scenario 2: The current standard treatment sometimes has serious side effects, is costly, and has a known success rate of 0.30 (or 30%).

In the first scenario, the research question of interest would probably be "Does the new treatment have a higher success rate than the standard treatment?" Unless there is convincing evidence that the new treatment has a higher success rate, it is unlikely that current medical practice would change. With p representing the true proportion of success for the laser treatment, the following hypotheses would be tested:

$$H_0: p = 0.85 \text{ versus } H_a: p > 0.85$$

In this case, rejecting the null hypothesis would require convincing evidence that the success rate is higher for the new treatment.

In the second scenario, the current standard treatment does not have much to recommend it. The new laser treatment may be considered preferable because of cost or because it has fewer or less serious side effects, as long as the success rate for the new procedure is no worse than that of the standard treatment. Here, researchers might decide to test

$$H_0: p = 0.30 \text{ versus } H_a: p < 0.30$$

If the null hypothesis is rejected, the new treatment would not be put forward as an alternative to the standard treatment, because there is strong evidence that the laser method has a lower success rate. On the other hand, if the null hypothesis is not rejected, there is not convincing evidence that the success rate for the laser treatment is lower than that for the standard. This is *not* the same as saying that there is evidence that the laser treatment is as good as the standard treatment. If medical practice were to embrace the new procedure, it would not be because it has a higher success rate but rather because it costs less or because it has fewer side effects.

One last reminder—you should be careful in setting up the hypotheses for a test. *Remember that a statistical hypothesis test is only capable of demonstrating strong support for the alternative hypothesis (by rejecting the null hypothesis). When the null hypothesis is not rejected, it does not mean strong support for H_0—only lack of strong evidence against it.* When deciding what alternative hypothesis to use, be sure to keep the research objectives in mind.

SECTION 10.1 EXERCISES

Each Exercise Set assesses the following chapter learning objectives: C1, C2, C3, M1

SECTION 10.1 Exercise Set 1

10.1 Explain why the statement $\hat{p} = 0.40$ is not a legitimate hypothesis.

10.2 **CareerBuilder.com** conducted a survey to learn about the proportion of employers who had ever sent an employee home because they were dressed inappropriately (**June 17, 2008, www.careerbuilder.com**). Suppose you are interested in determining if the resulting data provide strong evidence in support of the claim that more than one-third of employers have sent an employee home to change clothes. To answer this question, what null and alternative hypotheses should you test?

10.3 The report **"How Teens Use Media" (Nielsen, June 2009)** says that 67% of U.S. teens have a Facebook page that they update at least once a week. Suppose you plan to select a random sample of 400 students at the local high school. You will ask each student in the sample if he or she has a Facebook page and if it is updated at least once per week. You plan to use the resulting data to decide if there is evidence that the proportion of students at the high school who have a Facebook page that they update at least once a week differs from the national figure given in the Nielsen report. What hypotheses should you test?

10.4 The article **"Poll Finds Most Oppose Return to Draft, Wouldn't Encourage Children to Enlist" (Associated Press, December 18, 2005)** reports that in a random sample of 1,000 American adults, 430 answered yes to the following question: "If the military draft were reinstated, would you favor drafting women as well as men?" The data were used to test $H_0: p = 0.5$ versus $H_a: p < 0.5$, and the null hypothesis was rejected.
a. Based on the result of the hypothesis test, what can you conclude about the proportion of American adults who favor drafting women if a military draft were reinstated?
b. Is it reasonable to say that the data provide strong support for the alternative hypothesis?
c. Is it reasonable to say that the data provide strong evidence against the null hypothesis?

10.5 Do state laws that allow private citizens to carry concealed weapons result in a reduced crime rate? The author of a study carried out by the Brookings Institution is reported as saying, "The strongest thing I could say is that I don't see any strong evidence that they are reducing crime" (*San Luis Obispo Tribune,* **January 23, 2003**).
a. Is this conclusion consistent with testing
H_0: concealed weapons laws reduce crime
versus
H_a: concealed weapons laws do not reduce crime
or with testing

H_0: concealed weapons laws do not reduce crime
versus
H_a: concealed weapons laws reduce crime
Explain.
b. Does the stated conclusion indicate that the null hypothesis was rejected or not rejected? Explain.

10.6 In a hypothesis test, what does it mean to say that the null hypothesis was rejected?

SECTION 10.1 Exercise Set 2

10.7 Explain why the statement $\hat{p} > 0.50$ is not a legitimate hypothesis.

10.8 *USA Today* **(March 4, 2010)** described a survey of 1,000 women ages 22 to 35 who work full time. Each woman who participated in the survey was asked if she would be willing to give up some personal time in order to make more money. To determine if the resulting data provided convincing evidence that the majority of women ages 22 to 35 who work full time would be willing to give up some personal time for more money, what hypotheses should you test?

10.9 The report **"How Teens Use Media" (Nielsen, June 2009)** says that 83% of U.S. teens use text messaging. Suppose you plan to select a random sample of 400 students at the local high school and ask each one if he or she uses text messaging. You plan to use the resulting data to decide if there is evidence that the proportion of students at the high school who use text messaging differs from the national figure given in the Nielsen report. What hypotheses should you test?

10.10 According to a *Washington Post-ABC News* poll, 331 of 502 randomly selected American adults said they would not be bothered if the National Security Agency collected records of personal telephone calls. The data were used to test $H_0: p = 0.5$ versus $H_a: p > 0.5$, and the null hypothesis was rejected.
a. Based on the hypothesis test, what can you conclude about the proportion of American adults who would not be bothered if the National Security Agency collected records of personal telephone calls?
b. Is it reasonable to say that the data provide strong support for the alternative hypothesis?
c. Is it reasonable to say that the data provide strong evidence against the null hypothesis?

10.11 Consider the following quote from the article **"Review Finds No Link Between Vaccine and Autism" (San Luis Obispo Tribune, October 19, 2005):** "'We found no evidence that giving MMR causes Crohn's disease and/or autism in the children that get the MMR,' said Tom Jefferson, one of the authors of *The Cochrane Review.* 'That does not mean it doesn't cause it. It means we could find no evidence of it.'" (MMR is a measles-mumps-rubella vaccine.) In the context of a hypothesis test with the null hypothesis being that MMR does not cause autism, explain why the author could not conclude that the MMR vaccine does not cause autism.

10.12 In a hypothesis test, what does it mean to say that the null hypothesis was not rejected?

Additional Exercises

10.13 Which of the following are legitimate hypotheses?
a. $p = 0.65$
b. $\hat{p} = 0.90$
c. $\hat{p} = 0.10$
d. $p = 0.45$
e. $p > 4.30$

10.14 Which of the following specify legitimate pairs of null and alternative hypotheses?
a. $H_0: p = 0.25$ $H_a: p > 0.25$
b. $H_0: p < 0.40$ $H_a: p > 0.40$
c. $H_0: p = 0.40$ $H_a: p < 0.65$

d. $H_0: p \neq 0.50$ $H_a: p = 0.50$
e. $H_0: p = 0.50$ $H_a: p > 0.50$
f. $H_0: \hat{p} = 0.25$ $H_a: \hat{p} > 0.25$

10.15 A college has decided to introduce the use of plus and minus with letter grades, as long as there is convincing evidence that more than 60% of the faculty favor the change. A random sample of faculty will be selected, and the resulting data will be used to test the relevant hypotheses. If p represents the proportion of *all* faculty who favor a change to plus–minus grading, which of the following pairs of hypotheses should be tested?

$$H_0: p = 0.6 \text{ versus } H_a: p < 0.6$$

or

$$H_0: p = 0.6 \text{ versus } H_a: p > 0.6$$

Explain your choice.

10.16 A television station has been providing live coverage of a sensational criminal trial. The station's program director wants to know if more than half of potential viewers prefer a return to regular daytime programming. A survey of randomly selected viewers is conducted. With p representing the proportion of all viewers who prefer regular daytime programming, what hypotheses should the program director test?

SECTION 10.2 Potential Errors in Hypothesis Testing

Once hypotheses have been formulated, a hypothesis test using sample data is carried out to decide whether H_0 should be rejected. Just as a jury may reach the wrong verdict in a trial, there is some chance that sample data may lead you to the wrong conclusion about a population characteristic in a hypothesis test. In this section, the kinds of errors that can occur are considered.

One incorrect conclusion in a criminal trial is for a jury to convict an innocent person, and another is for a guilty person to be found not guilty. Similarly, there are two different types of errors that might be made when reaching a decision in a hypothesis testing problem. One type of error involves rejecting H_0 even though the null hypothesis is true. The second type of error results from failing to reject H_0 when it is false. These errors are known as Type I and Type II errors, respectively.

> **DEFINITION**
>
> A **Type I error** is the error of rejecting H_0 when H_0 is true
>
> A **Type II error** is the error of failing to reject H_0 when H_0 is false

The only way to guarantee that neither type of error occurs is to base the decision on a census of the entire population. When you base a decision on sample data (which provide incomplete information abut the population), you run the risk of making the wrong decision—there is some chance that you will make a Type I error and some chance that you will make a Type II error.

Example 10.6 On-Time Arrivals

The U.S. Department of Transportation reports that for 2008, 65.3% of all domestic passenger flights arrived within 15 minutes of the scheduled arrival time (*Air Travel Consumer*

Report, Feb. 2009). Suppose that an airline with a poor on-time record decides to offer its employees a bonus if the airline's proportion of on-time flights exceeds the overall industry rate of 0.653 in an upcoming month. In this context, p will represent the actual proportion of the airline's flights that are on time during the month of interest. A random sample of flights could be selected and used as a basis for choosing between

$$H_0: p = 0.653 \text{ and } H_a: p > 0.653$$

In this context, a Type I error (rejecting a true H_0) is concluding that the airline on-time rate exceeds the overall industry rate, when in fact the airline does not have a better on-time record. This Type I error would result in the airline rewarding its employees when the proportion of on-time flights was not actually greater than 0.653. A Type II error (not rejecting a false H_0) is not concluding that the airline's on-time proportion is better than the industry proportion when the airline really did have a better on-time record. A Type II error would result in the airline employees *not* receiving a reward that they deserved. Notice that the consequences associated with Type I and Type II errors are different.

Example 10.7   Treating Blocked Arteries

The article **"Boston Scientific Stent Study Flawed"** (*Wall Street Journal*, **August 14, 2008**) described a research study conducted by Boston Scientific Corporation. Boston Scientific developed a new heart stent used to treat arteries blocked by heart disease. The new stent, called the Liberte, is made of thinner metal than heart stents currently in use, including a stent called the Express that is also made by Boston Scientific. Because the new stent was made of thinner metal, the manufacturer thought it would be easier for doctors to direct through a patient's arteries to get to a blockage.

One type of problem that can sometimes occur after a stent is used is that the artery can become blocked again. In order to obtain approval to manufacture and sell the new Liberte stent in the United States, the Food and Drug Administration (FDA) required Boston Scientific to provide evidence that the proportion of patients receiving the Liberte stent who experienced a re-blocked artery was less than 0.1. The proportion of 0.1 (or 10%) was determined by the FDA based on the re-block proportions for currently approved stents, including the Express stent, which had a historical re-block proportion of 0.07.

Because Boston Scientific was asked to provide evidence that the re-block proportion for the new Liberte stent was less than 0.1, an alternative hypothesis of $H_a: p < 0.1$ was selected. This resulted in the following null and alternative hypotheses:

$$H_0: p = 0.1 \quad H_a: p < 0.1$$

The null hypothesis states that the re-block proportion for the new stent is 0.1 (or equivalently that 10% of patients receiving the new stent experience a re-blocked artery). The alternative hypothesis states that the re-block proportion is less than 0.1.

Consider Type I and Type II errors in this context:

Error	Definition of Error	Description of Error in Context	Consequence of Error
Type I error	Reject a true H_0	Conclude that the re-block proportion for the new stent is less than 0.1 when it really is 0.1 (or greater).	The new stent is approved for sale. More patients will experience re-blocked arteries.
Type II error	Fail to reject a false H_0	You are not convinced that the re-block proportion is less than 0.1 when it really is less than 0.1.	The new stent is not approved for sale. Patients and doctors will not benefit from the new design that makes the new stent easier to use.

Notice that the consequences associated with Type I and Type II errors are very different. This is something to consider when evaluating the acceptable error risks.

Examples 10.6 and 10.7 illustrate the two different types of errors that might occur when testing hypotheses. Type I and Type II errors—and the associated consequences of making such errors—are different. The accompanying box introduces the terminology and notation used to describe error probabilities.

The **probability of a Type I error** is denoted by α.

The **probability of a Type II error** is denoted by β.

In a hypothesis test, the probability of a Type I error, α, is also called the **significance level**. For example, a hypothesis test with $\alpha = 0.01$ is said to have a significance level of 0.01.

Example 10.8 Testing for Ovarian Cancer

Women with ovarian cancer are usually not diagnosed until the disease is in an advanced stage, when it is most difficult to treat. A new blood biomarker test has been developed that appears to be able to identify ovarian cancer in its earliest stages. The paper **"Diagnostic Markers for Early Detection of Ovarian Cancer"** (*Clinical Cancer Research* **[2008]: 1065–1072**) included the following information from a preliminary evaluation of the biomarker test:

- The test was given to 156 women known to have ovarian cancer, and it correctly identified 151 of them as having ovarian cancer.
- The test was given to 362 women known not to have ovarian cancer, and it correctly identified 361 of these 362 as being cancer free.

You can think of using this blood test to choose between two hypotheses:

H_0: woman has ovarian cancer
H_a: woman does not have ovarian cancer

Although these are not "statistical hypotheses" (statements about a population characteristic), the possible decision errors are analogous to Type I and Type II errors.

In this situation, believing that a woman with ovarian cancer is cancer-free would be a Type I error—rejecting the hypothesis of ovarian cancer when it is, in fact, true. Believing that a cancer-free woman has ovarian cancer is a Type II error—not rejecting the null hypothesis when it is, in fact, false. Based on the preliminary study results, you can estimate the error probabilities. The probability of a Type I error, α, is approximately $5/156 = 0.032$. The probability of a Type II error, β, is approximately $1/362 = 0.003$.

The ideal hypothesis test procedure would have both $\alpha = 0$ and $\beta = 0$. However, if you must base your decision on incomplete information—a sample rather than a census—this is impossible to achieve. Standard test procedures allow you to specify a desired value for α, but they provide no direct control over β. Because α represents the probability of rejecting a true null hypothesis, selecting a significance level $\alpha = 0.05$ results in a test procedure that, used over and over with different samples, rejects a *true* null hypothesis about 5 times in 100. Selecting $\alpha = 0.01$ results in a test procedure with a Type I error rate of 1% in long-term repeated use. Choosing a small value for α implies that the user wants a procedure with a small Type I error rate.

One question arises naturally at this point: If you can select α (the probability of making a Type I error), why would you ever select $\alpha = 0.05$ rather than $\alpha = 0.01$? Why not always select a very small value for α? To achieve a small probability of making a Type I error, the corresponding test procedure must require very strong evidence against H_0 before the null hypothesis can be rejected. Although this makes a Type I error unlikely, it increases the risk of a Type II error (*not* rejecting H_0 when it should have been rejected). You must balance the consequences of Type I and Type II errors. If a Type II error has serious consequences, it may be a good idea to select a somewhat larger value for α.

In general, there is a compromise between small α and small β, leading to the following widely accepted principle for selecting the significance level to be used in a hypothesis test.

> After assessing the consequences of Type I and Type II errors, identify the largest α that is acceptable. Then carry out the hypothesis test using this maximum acceptable value as the level of significance (because using any smaller value for α will result in a larger value for β).

Example 10.9   **Heart Stents Revisited**

In Example 10.7, the following hypotheses were selected:

$$H_0: p = 0.1 \qquad H_a: p < 0.1$$

where p represents the proportion of all patients receiving a new Liberte heart stent who experience a re-blocked artery. For these hypotheses, the consequences of a Type I error were that the new stent would be approved for sale even though the re-block proportion was actually 0.10 or greater, and that more patients will experience re-blocked arteries. Consequences of a Type II error were that the new stent would not be approved for sale even though the actual re-block proportion is less than 0.10, so patients and doctors would not benefit from the design that makes the new stent easier to use. A Type I error would result in the approval of a heart stent for which 10% or more of the patients receiving the stent would experience a re-blocked artery. Because this represents an unnecessary risk to patients (given that other stents with lower re-block proportions are available), a small value for α (Type I error probability), such as $\alpha = 0.01$, would be selected. Of course, selecting a small value for α increases the risk of a Type II error (not approving the new stent when the re-block proportion is actually acceptable).

SECTION 10.2   **EXERCISES**

Each Exercise Set assesses the following chapter learning objectives: C4, C5, M2, M5

SECTION 10.2   **Exercise Set 1**

10.17   One type of error in a hypothesis test is rejecting the null hypothesis when it is true. What is the other type of error that might occur when a hypothesis test is carried out?

10.18   Suppose that for a particular hypothesis test, the consequences of a Type I error are very serious. Would you want to carry out the test using a small significance level α (such as 0.01) or a larger significance level (such as 0.10)? Explain the reason for your choice.

10.19   Occasionally, warning flares of the type contained in most automobile emergency kits fail to ignite. A consumer group wants to investigate a claim that the proportion of defective flares made by a particular manufacturer is much higher than the advertised value of 0.1. A large number of flares will be tested, and the results will be used to decide between $H_0: p = 0.1$ and $H_a: p > 0.1$, where p represents the actual proportion of defective flares made by this manufacturer. If H_0 is rejected, charges of false advertising will be filed against the manufacturer.

a. Explain why the alternative hypothesis was chosen to be $H_a: p > 0.1$.

b. Complete the last two columns of the following table. (Hint: See Example 10.7 for an example of how this is done.)

Error	Definition of Error	Description of Error in Context	Consequence of Error
Type I error	Reject a true H_0		
Type II error	Fail to reject a false H_0		

10.20 A television manufacturer states that at least 90% of its TV sets will not need service during the first 3 years of operation. A consumer group wants to investigate this statement. A random sample of $n = 100$ purchasers is selected and each person is asked if the set purchased needed repair during the first 3 years. Let p denote the proportion of all sets made by this manufacturer that will not need service in the first 3 years. The consumer group does not want to claim false advertising unless there is strong evidence that $p < 0.9$. The appropriate hypotheses are then $H_0: p = 0.9$ versus $H_a: p < 0.9$.
a. In the context of this problem, describe Type I and Type II errors, and discuss the possible consequences of each.
b. Would you recommend a test procedure that uses $\alpha = 0.01$ or one that uses $\alpha = 0.10$? Explain.

10.21 The paper **"MRI Evaluation of the Contralateral Breast in Women with Recently Diagnosed Breast Cancer"** (*New England Journal of Medicine* [2007]: 1295–1303) describes a study of MRI (magnetic resonance imaging) exams used to diagnose breast cancer. The purpose of the study was to determine if MRI exams do a better job than mammograms in determining if women who have recently been diagnosed with cancer in one breast also have cancer in the other breast. The study participants were 969 women who had been diagnosed with cancer in one breast and for whom a mammogram did not detect cancer in the other breast. These women had MRI exams of the other breast, and 121 of those exams indicated possible cancer. After undergoing biopsies, it was determined that 30 of the 121 did in fact have cancer in the other breast and that 91 did not. The 969 women were all followed for 1 year, and three of the women for whom the MRI exam did not indicate cancer in the other breast were subsequently diagnosed with cancer that the MRI did not detect. The accompanying table summarizes this information.

	Cancer Present	Cancer Not Present	Total
MRI Positive for Cancer	30	91	**121**
MRI Negative for Cancer	3	845	**848**
Total	**33**	**936**	**969**

Suppose that for women recently diagnosed with cancer in only one breast, an MRI is used to decide between the two "hypotheses"

H_0: woman has cancer in the other breast
H_a: woman does not have cancer in the other breast

(Although these are not hypotheses about a population characteristic, this exercise illustrates the definitions of Type I and Type II errors.)
a. One possible error would be deciding that a woman who does have cancer in the other breast is cancer-free.

Is this a Type I or a Type II error? Use the information in the table to approximate the probability of this type of error.
b. There is a second type of error that is possible in this setting. Describe this error and use the information in the given table to approximate the probability of this type of error.

SECTION 10.2 **Exercise Set 2**

10.22 One type of error in a hypothesis test is failing to reject a false null hypothesis. What is the other type of error that might occur when a hypothesis test is carried out?

10.23 Suppose that for a particular hypothesis test, the consequences of a Type I error are not very serious, but there are serious consequences associated with making a Type II error. Would you want to carry out the test using a small significance level α (such as 0.01) or a larger significance level (such as 0.10)? Explain the reason for your choice.

10.24 A manufacturer of handheld calculators receives large shipments of printed circuits from a supplier. It is too costly and time-consuming to inspect all incoming circuits, so when each shipment arrives, a sample is selected for inspection. A shipment is defined to be of inferior quality if it contains more than 1% defective circuits. Information from the sample is used to test $H_0: p = 0.01$ versus $H_a: p > 0.01$, where p is the actual proportion of defective circuits in the shipment. If the null hypothesis is not rejected, the shipment is accepted, and the circuits are used in the production of calculators. If the null hypothesis is rejected, the entire shipment is returned to the supplier because of inferior quality.
a. Complete the last two columns of the following table. (Hint: See Example 10.7 for an example of how this is done.)

Error	Definition of Error	Description of Error in Context	Consequence of Error
Type I error	Reject a true H_0		
Type II error	Fail to reject a false H_0		

b. From the calculator manufacturer's point of view, which type of error would be considered more serious?
c. From the printed circuit supplier's point of view, which type of error would be considered more serious?

10.25 An automobile manufacturer is considering using robots for part of its assembly process. Converting to robots is expensive, so it will be done only if there is strong evidence that the proportion of defective installations is less for the robots than for human assemblers. Let p denote the actual proportion of defective installations for the robots. It is known that the proportion of defective installations for human assemblers is 0.02.

a. Which of the following pairs of hypotheses should the manufacturer test?

$$H_0: p = 0.02 \text{ versus } H_a: p < 0.02$$

or

$$H_0: p = 0.02 \text{ versus } H_a: p > 0.02$$

Explain your choice.

b. In the context of this exercise, describe Type I and Type II errors.

c. Would you prefer a test with $\alpha = 0.01$ or $\alpha = 0.10$? Explain your reasoning.

10.26 Medical personnel are required to report suspected cases of child abuse. Because some diseases have symptoms that are similar to those of child abuse, doctors who see a child with these symptoms must decide between two competing hypotheses:

H_0: symptoms are due to child abuse
H_a: symptoms are not due to child abuse

(Although these are not hypotheses about a population characteristic, this exercise illustrates the definitions of Type I and Type II errors.) The article **"Blurred Line Between Illness, Abuse Creates Problem for Authorities" (Macon Telegraph, February 28, 2000)** included the following quote from a doctor in Atlanta regarding the consequences of making an incorrect decision: "If it's disease, the worst you have is an angry family. If it is abuse, the other kids (in the family) are in deadly danger."

a. For the given hypotheses, describe Type I and Type II errors.

b. Based on the quote regarding consequences of the two kinds of error, which type of error is considered more serious by the doctor quoted? Explain.

Additional Exercises

10.27 Describe the two types of errors that might be made when a hypothesis test is carried out.

10.28 Give an example of a situation where you would want to select a small significance level.

10.29 Give an example of a situation where you would not want to select a very small significance level.

10.30 Ann Landers, in her advice column of October 24, 1994 **(San Luis Obispo Telegram-Tribune)**, described the reliability of DNA paternity testing as follows: "To get a completely accurate result, you would have to be tested, and so would (the man) and your mother. The test is 100% accurate if the man is *not* the father and 99.9% accurate if he is."

a. Consider using the results of DNA paternity testing to decide between the following two hypotheses:

H_0: a particular man is the father
H_a: a particular man is not the father

In the context of this problem, describe Type I and Type II errors. (Although these are not hypotheses about a population characteristic, this exercise illustrates the definitions of Type I and Type II errors.)

b. Based on the information given, what are the values of α, the probability of a Type I error, and β, the probability of a Type II error?

10.31 The article **"Fewer Parolees Land Back Behind Bars"** **(Associated Press, April 11, 2006)** includes the following statement: "Just over 38% of all felons who were released from prison in 2003 landed back behind bars by the end of the following year, the lowest rate since 1979." Explain why it would not be necessary to carry out a hypothesis test to determine if the proportion of felons released in 2003 who landed back behind bars by the end of the following year was less than 0.40.

SECTION 10.3 The Logic of Hypothesis Testing—An Informal Example

Now that the necessary foundation is in place, you are ready to see how knowing about the sampling distribution of a sample proportion makes it possible to reach a conclusion in a hypothesis test. A simple example will be used to illustrate the reasoning that leads to a choice between a null and an alternative hypothesis.

With p denoting the proportion in the population that possess some characteristic of interest, recall that three things are known about the sampling distribution of the sample proportion $\hat{p}$ when a random sample of size n is selected:

1. The mean of the $\hat{p}$ sampling distribution is p. ($\mu_{\hat{p}} = p$)

2. The standard error (standard deviation) of $\hat{p}$ is $\sqrt{\dfrac{p(1-p)}{n}}$

3. If n is large, the $\hat{p}$ sampling distribution is approximately normal.

The sampling distribution describes the behavior of the statistic $\hat{p}$. If the $\hat{p}$ sampling distribution is approximately normal, you can use what you know about normal distributions.

An Informal Example

When sample data are used to choose between a null and an alternative hypothesis, there are two possible conclusions. You either decide to reject the null hypothesis in favor of the alternative hypothesis or you decide that the null hypothesis cannot be rejected. The fundamental idea behind hypothesis testing procedures is this: *You reject the null hypothesis H_0 if the observed sample is very unlikely to have occurred if H_0 is true.* Example 10.10 illustrates this in the context of testing hypotheses about a population proportion when the sample size is large.

Example 10.10   Impact of Food Labels

In June of 2006, an **Associated Press** survey was conducted to investigate how people use the nutritional information provided on food package labels. Interviews were conducted with 1,003 randomly selected adult Americans, and each participant was asked a series of questions, including the following two:

> Question 1: When purchasing packaged food, how often do you check the nutrition labeling on the package?
>
> Question 2: How often do you purchase foods that are bad for you, even after you've checked the nutrition labels?

It was reported that 582 responded "frequently" to the question about checking labels and 441 responded "very often or somewhat often" to the question about purchasing "bad" foods even after checking the label.

Let's start by considering the responses to the first question. Based on these data, is it reasonable to conclude that a majority of adult Americans frequently check nutrition labels? With p denoting the proportion of American adults who frequently check nutrition labels (a population proportion), this question can be answered by testing the following hypotheses:

$$H_0: p = 0.5$$
$$H_a: p > 0.5$$

This alternative hypothesis states that the proportion of adult Americans who frequently check nutrition labels is greater than 0.5. This would mean that more than half (a majority) frequently check nutrition labels. The null hypothesis will be rejected only if there is convincing evidence that $p > 0.5$.

For this sample,

$$\hat{p} = \frac{582}{1,003} = 0.58$$

The observed sample proportion is certainly greater than 0.5, but could this just be due to sampling variability? That is, when $p = 0.5$ (meaning H_0 is true), the sample proportion $\hat{p}$ usually differs somewhat from 0.5 simply because of chance variability from one sample to another. Is it plausible that a sample proportion of $\hat{p} = 0.58$ occurred only as a result of this sampling variability, or is it unusual to observe a sample proportion this large when $p = 0.5$?

If the null hypothesis, $H_0: p = 0.5$, is true, you know the following:

What You Know	How You Know It...
The $\hat{p}$ values will center around 0.5—that is, the mean of the $\hat{p}$ distribution is 0.5.	The sample is described as 1,003 randomly selected adult Americans. When random samples are selected from a population, $\mu_{\hat{p}} = p$. If the null hypothesis is true, $p = 0.5$, so $\mu_{\hat{p}} = 0.5$.

(continued)

What You Know	How You Know It...
The standard deviation of $\hat{p}$, which describes how much the $\hat{p}$ values tend to vary from sample to sample, is $\sqrt{\dfrac{(0.5)(1-0.5)}{1,003}}.$	When random samples of size n are selected from a population, $\sigma_{\hat{p}} = \sqrt{\dfrac{p(1-p)}{n}}$. The sample size is $n = 1,003$ and, if the null hypothesis is true, $p = 0.5$, so $$\sigma_{\hat{p}} = \sqrt{\dfrac{(0.5)(1-0.5)}{1,003}}.$$
The sampling distribution of $\hat{p}$ is approximately normal.	When random samples of size n are selected from a population and the sample size is large, the sampling distribution of $\hat{p}$ is approximately normal. When $p = 0.5$, the sample size of $n = 1,003$ is large enough for the $\hat{p}$ distribution to be approximately normal.
The statistic $$z = \dfrac{\hat{p} - 0.5}{\sqrt{\dfrac{(0.5)(1-0.5)}{1,003}}}$$ has (approximately) a standard normal distribution.	Transforming a statistic that has a normal distribution to a z statistic (by subtracting its mean and then dividing by its standard deviation) results in a statistic that has a standard normal distribution. Since $\hat{p}$ has (approximately) a normal distribution with mean 0.5 and standard deviation $\sqrt{\dfrac{(0.5)(1-0.5)}{1,003}}$ when the null hypothesis is true, the statistic $z = \dfrac{\hat{p} - 0.5}{\sqrt{\dfrac{(0.5)(1-0.5)}{1,003}}}$ will have (approximately) a standard normal distribution.

You can now use what you know about normal distributions to judge whether the observed sample proportion would be surprising if the null hypothesis were true. For the sample in the study, $\hat{p} = \dfrac{582}{1,003} = 0.58$. Using this $\hat{p}$ value in the z statistic formula

$$z = \dfrac{\hat{p} - 0.5}{\sqrt{\dfrac{(0.5)(1-0.5)}{1,003}}}$$

results in

$$z = \dfrac{p - 0.5}{\sqrt{\dfrac{(0.5)(0.5)}{n}}} = \dfrac{0.58 - 0.5}{\sqrt{\dfrac{(0.5)(0.5)}{1,003}}} = \dfrac{0.08}{0.016} = 5.00$$

This means that $\hat{p} = 0.58$ is 5 standard deviations greater than what you would expect it to be if the null hypothesis H_0: $p = 0.5$ were true. This would be *very* unusual for a normal distribution and can be regarded as inconsistent with the null hypothesis. You can evaluate just how unusual this would be by computing the following probability:

P(observing a value of z as unusual as 5.00 when H_0 is true)

$= P(z \geq 5.00$ when H_0 is true)

$=$ area under the z curve to the right of 5.00

≈ 0

When H_0 is true, there is virtually no chance of seeing a sample proportion and corresponding z value this extreme as a result of sampling variability alone. In this case, the evidence for rejecting H_0 in favor of H_a is very compelling.

Interestingly, in spite of the fact that there is strong evidence that a majority of American adults frequently check nutrition labels, the responses to the second question suggest that the percentage of people who then ignore the information on the label and purchase "bad" foods anyway is not small—the sample proportion of adults who gave a response of very often or somewhat often was 0.44.

The preceding example illustrates the logic of hypothesis testing. It is worth reading a second time, because understanding the reasoning illustrated is critical to mastering the hypothesis testing method you will see in the next section. You begin by assuming that the null hypothesis is correct. The sample is examined in light of this assumption. If the observed sample proportion would not be unusual when H_0 is true, then chance variability from one sample to another is a plausible explanation for what has been observed, and H_0 would not be rejected. On the other hand, if the observed sample proportion would have been very unlikely when H_0 is true, you would interpret this as convincing evidence against the null hypothesis and you would reject H_0. The decision to reject or to fail to reject the null hypothesis is based on an assessment of how unlikely the observed sample data would be if the null hypothesis were true.

SECTION 10.3 EXERCISES

Each Exercise Set assesses the following chapter learning objectives: C6

SECTION 10.3 Exercise Set 1

10.32 The article **"Breast-Feeding Rates Up Early"** (*USA Today*, Sept. 14, 2010) summarizes a survey of mothers whose babies were born in 2009. The Center for Disease Control sets goals for the proportion of mothers who will still be breast-feeding their babies at various ages. The goal for 12 months after birth is 0.25 or more. Suppose that the survey used a random sample of 1,200 mothers and that you want to use the survey data to decide if there is evidence that the goal is not being met. Let p denote the proportion of all mothers of babies born in 2009 who were still breast-feeding at 12 months.

a. Describe the shape, center, and spread of the sampling distribution of $\hat{p}$ for random samples of size 1,200 if the null hypothesis H_0: $p = 0.25$ is true.

b. Would you be surprised to observe a sample proportion of $\hat{p} = 0.24$ for a sample of size 1,200 if the null hypothesis H_0: $p = 0.25$ were true? Explain why or why not.

c. Would you be surprised to observe a sample proportion as small as $\hat{p} = 0.20$ for a sample of size 1,200 if the null hypothesis H_0: $p = 0.25$ were true? Explain why or why not.

d. The actual sample proportion observed in the study was $\hat{p} = 0.22$. Based on this sample proportion, is there convincing evidence that the goal is not being met, or is the observed sample proportion consistent with what you would expect to see when the null hypothesis is true? Support your answer with a probability calculation.

10.33 The article **"The Benefits of Facebook Friends: Social Capital and College Students' Use of Online Social Network**

Sites" (*Journal of Computer-Mediated Communication* [2007]: 1143–1168) describes a study of $n = 286$ undergraduate students at Michigan State University. Suppose that it is reasonable to regard this sample as a random sample of undergraduates at Michigan State. You want to use the survey data to decide if there is evidence that more than 75% of the students at this university have a Facebook page that includes a photo of themselves. Let p denote the proportion of all Michigan State undergraduates who have such a page.

a. Describe the shape, center, and spread of the sampling distribution of $\hat{p}$ for random samples of size 286 if the null hypothesis H_0: $p = 0.75$ is true.

b. Would you be surprised to observe a sample proportion of $\hat{p} = 0.83$ for a sample of size 286 if the null hypothesis H_0: $p = 0.75$ were true? Explain why or why not.

c. Would you be surprised to observe a sample proportion of $\hat{p} = 0.79$ for a sample of size 286 if the null hypothesis H_0: $p = 0.75$ were true? Explain why or why not.

d. The actual sample proportion observed in the study was $\hat{p} = 0.80$. Based on this sample proportion, is there convincing evidence that the null hypothesis H_0: $p = 0.75$ is not true, or is $\hat{p}$ consistent with what you would expect to see when the null hypothesis is true? Support your answer with a probability calculation.

SECTION 10.3 Exercise Set 2

10.34 The article **"Most Customers OK with New Bulbs"** (*USA Today*, Feb. 18, 2011) describes a survey of 1,016 randomly selected adult Americans. Each person in the sample was asked if they have replaced standard light bulbs in their home with the more energy efficient compact fluorescent

(CFL) bulbs. Suppose you want to use the survey data to determine if there is evidence that more than 70% of adult Americans have replaced standard bulbs with CFL bulbs. Let p denote the proportion of all adult Americans who have replaced standard bulbs with CFL bulbs.

a. Describe the shape, center, and spread of the sampling distribution of $\hat{p}$ for random samples of size 1,016 if the null hypothesis $H_0: p = 0.70$ is true.

b. Would you be surprised to observe a sample proportion of $\hat{p} = 0.72$ for a sample of size 1,016 if the null hypothesis $H_0: p = 0.70$ were true? Explain why or why not.

c. Would you be surprised to observe a sample proportion as large as $\hat{p} = 0.75$ for a sample of size 1,016 if the null hypothesis $H_0: p = 0.70$ were true? Explain why or why not.

d. The actual sample proportion observed in the study was $\hat{p} = 0.71$. Based on this sample proportion, is there convincing evidence that more than 70% have replaced standard bulbs with CFL bulbs, or is this sample proportion consistent with what you would expect to see when the null hypothesis is true? Support your answer with a probability calculation.

10.35 *USA Today* (Feb. 17, 2011) described a survey of 1,008 American adults. One question on the survey asked people if they had ever sent a love letter using e-mail. Suppose that this survey used a random sample of adults and that you want to decide if there is evidence that more than 20% of American adults have written a love letter using e-mail.

a. Describe the shape, center, and spread of the sampling distribution of $\hat{p}$ for random samples of size 1,008 if the null hypothesis $H_0: p = 0.20$ is true.

b. Based on your answer to Part (a), what sample proportion values would convince you that more than 20% of adults have sent a love letter via e-mail?

Additional Exercises

10.36 A *Newsweek* article titled **"America the Ignorant"** (www.newsweek.com) described a Gallup poll that asked adult Americans if they believe that there are real witches

and warlocks. Suppose that the poll used a random sample of 800 adult Americans and that you want to use the poll data to decide if there is evidence that more than 10% of adult Americans believe in witches and warlocks. Let p be the proportion of all adult Americans who believe in witches and warlocks.

a. Describe the shape, center, and spread of the sampling distribution of $\hat{p}$ for random samples of size 800 if the null hypothesis $H_0: p = 0.10$ is true.

b. Would you be surprised to observe a sample proportion of $\hat{p} = 0.16$ for a sample of size 800 if the null hypothesis $H_0: p = 0.10$ were true? Explain why or why not.

c. Would you be surprised to observe a sample proportion of $\hat{p} = 0.12$ for a sample of size 800 if the null hypothesis $H_0: p = 0.10$ were true? Explain why or why not.

d. The actual sample proportion observed in the study was $\hat{p} = 0.21$. Based on this sample proportion, is there convincing evidence that more than 10% of adult Americans believe in witches and warlocks, or is the sample proportion consistent with what you would expect to see when the null hypothesis is true? Support your answer with a probability calculation.

10.37 Every year on Groundhog Day (February 2), the famous groundhog Punxsutawney Phil tries to predict whether there will be 6 more weeks of winter. The article **"Groundhog Has Been Off Target"** (*USA Today*, Feb. 1, 2011) states that "based on weather data, there is no predictive skill for the groundhog." Suppose that you plan to take a random sample of 20 years and use weather data to determine the proportion of these years the groundhog's prediction was correct.

a. Describe the shape, center, and spread of the sampling distribution of $\hat{p}$ for samples of size 20 if the groundhog has only a 50–50 chance of making a correct prediction.

b. Based on your answer to Part (a), what sample proportion values would convince you that the groundhog's predictions have a better than 50–50 chance of being correct?

A Procedure for Carrying Out a Hypothesis Test

The first three sections of this chapter provide what is needed to develop a systematic strategy for testing hypotheses. In Section 10.1, you saw how research questions or claims about a population are translated into two competing hypotheses—the null hypothesis and the alternative hypothesis. You also considered the two possible conclusions in a hypothesis test—rejecting the null hypothesis and failing to reject the null hypothesis. In Section 10.2, you considered the two types of errors (Type I and Type II) that are possible. You also saw that evaluating the consequences of Type I and Type II errors helps in the selection of an appropriate significance level α. Finally, in Section 10.3, you saw that the reasoning process used to reach a decision in a hypothesis test is straightforward and intuitive. You first assume that the null hypothesis is true. You then determine if the sample data are consistent with the null hypothesis (these data would not be surprising if the null hypothesis were true) or if the sample data convinces you that the null hypothesis should be rejected (these data would be unlikely if the null hypothesis were true).

In Example 10.10 (Section 10.3), sample data from a survey of 1,003 adult Americans were used to decide between

$$H_0: p = 0.5$$

and

$$H_a: p > 0.5$$

where p represented the proportion of adult Americans who frequently check nutritional labels when purchasing packaged foods. The decision in Example 10.10 was based on computing the probability of seeing sample data as unusual as what was observed if the null hypothesis is true. This computed probability was approximately 0, so a decision was made to reject the null hypothesis.

The probability computed in Example 10.10 is called a **P-value**. A P-value specifies how likely it is that a sample would be as or more extreme than the one observed if H_0 were true. To find the P-value, a z statistic was computed:

$$z = \frac{\hat{p} - hypothesized\ value}{standard\ deviation\ of\ \hat{p}}$$

This z is called a **test statistic**. Knowing the value of the test statistic (for example $z = 5.0$) allows calculation of the corresponding P-value (for example, P-value ≈ 0). The P-value is used to make a decision in a hypothesis test. If the P-value is small (meaning that the sample data would be very unlikely if the null hypothesis were true), you reject the null hypothesis. If the P-value is not small, you fail to reject the null hypothesis.

A **test statistic** is computed using sample data. The value of the test statistic is used to determine the P-value associated with the test.

The **P-value** (also sometimes called the **observed significance level**) is a measure of inconsistency between the null hypothesis and the observed sample. It is the probability, assuming that H_0 is true, of obtaining a test statistic value at least as inconsistent with H_0 as what actually resulted.

You reject the null hypothesis when the P-value is small.

The test statistic appropriate for a particular hypothesis testing situation will depend on the answers to the four key questions that are used to select a method. Recall that in the playbook analogy, these are the questions you need to answer in order to select the right statistics "play."

The null hypothesis will be rejected if the P-value is small, but just how small does the P-value have to be in order to reject H_0? The answer depends on the significance level α (the probability of a Type I error) selected for the test. For example, suppose you select $\alpha = 0.05$. This implies that the probability of rejecting a true null hypothesis should be 0.05. To obtain a test procedure with this probability of a Type I error, you would reject the null hypothesis if the sample result is among the most unusual 5% of all samples when H_0 is true. This means that H_0 is rejected if the computed P-value is less than or equal to 0.05. If you select $\alpha = 0.01$, H_0 will be rejected only if you observe a sample result so extreme that it would be among the most unusual 1% if H_0 is true. This would be the case if P-value ≤ 0.01.

A decision in a hypothesis test is based on comparing the P-value to the chosen significance level α.

H_0 is rejected if P-value $\leq \alpha$.

H_0 is not rejected if P-value $> \alpha$.

For example, suppose that P-value $= 0.0352$ and that a significance level of 0.05 is chosen. Then, because

$$P\text{-value} = 0.0352 \le 0.05 = \alpha$$

H_0 would be rejected. This would not be the case, though, for $\alpha = 0.01$, because then P-value $> \alpha$.

The five-step process for hypothesis testing (HMC³) first described in Chapter 7 is used to carry out a hypothesis test. The five steps are:

Step	What Is This Step?
H	**Hypotheses:** Define the hypotheses that will be tested.
M	**Method:** Use the answers to the four key questions (QSTN) to identify a potential method.
C	**Check:** Check to make sure that the method selected is appropriate. Many methods for learning from data only provide reliable information under certain conditions. Verify that these conditions are met before proceeding.
C	**Calculate:** Use the sample data to perform any necessary calculations.
C	**Communicate Results:** This is a critical step in the process. In this step, you will answer the research question of interest, explain what you have learned from the data, and acknowledge potential risks.

The accompanying table details what each of the steps entails when carrying out a hypothesis test.

AP* EXAM TIP

When carrying out a hypothesis test on the AP exam, be sure to include all of these steps. Omitting any of these steps will reduce your score on a hypothesis test exam question.

Step	This Step Includes...
H Hypotheses	1. Describe the population and the population characteristic of interest. 2. Translate the research question or claim made about the population into null and alternative hypotheses.
M Method	1. Identify the appropriate test and test statistic. 2. Select a significance level for the test.
C Check	1. Verify that any conditions for the selected test are met.
C Calculate	1. Find the values of any sample statistics needed to calculate the value of the test statistic. 2. Calculate the value of the test statistic. 3. Determine the P-value for the test.
C Communicate Results	1. Compare the P-value to the selected significance level and make a decision to either reject H_0 or fail to reject H_0. 2. Provide a conclusion in words that is in context and that addresses the question of interest.

A statistical software program or a graphing calculator is often used to complete the calculate step in this process.

You will see how this process is used to test hypotheses about a population proportion in Section 10.5.

Each Exercise Set assesses the following chapter learning objectives: C5, C6

SECTION 10.4   **Exercise Set 1**

10.38   Use the definition of the P-value to explain the following:
a. Why H_0 would be rejected if P-value = 0.0003
b. Why H_0 would not be rejected if P-value = 0.350

10.39   The article **"Poll Finds Most Oppose Return to Draft, Wouldn't Encourage Children to Enlist" (Associated Press, December 18, 2005)** reports that in a random sample of 1,000 American adults, 700 indicated that they oppose the reinstatement of a military draft. Suppose you want to use this information to decide if there is convincing evidence that the proportion of American adults who oppose reinstatement of the draft is greater than two-thirds.
a. What hypotheses should be tested in order to answer this question?
b. The P-value for this test is 0.013. What conclusion would you reach if $\alpha = 0.05$?

10.40   Step 2 of the five-step process for hypothesis testing is selecting an appropriate method. What is involved in completing this step?

SECTION 10.4   **Exercise Set 2**

10.41   For which of the following P-values will the null hypothesis be rejected when performing a test with a significance level of 0.05?
a. 0.001   d. 0.047
b. 0.021   e. 0.148
c. 0.078

10.42   According to a survey of 1,000 adult Americans conducted by Opinion Research Corporation, 210 of those surveyed said playing the lottery would be the most practical way for them to accumulate $200,000 in net wealth in their lifetime (**"One in Five Believe Path to Riches Is the Lottery,"** *San Luis Obispo Tribune*, **January 11, 2006**). Although the article does not describe how the sample was selected, for purposes of this exercise, assume that the sample is a random sample of adult Americans. Suppose that you want to use the data from this survey to decide if there is convincing evidence that more than 20% of adult Americans believe that playing the lottery is the best strategy for accumulating $200,000 in net wealth.
a. What hypotheses should be tested in order to answer this question?
b. The P-value for this test is 0.215. What conclusion would you reach if $\alpha = 0.05$?

10.43   Step 5 of the five-step process for hypothesis testing is communication of results. What is involved in completing this step?

Additional Exercises

10.44   For which of the following combinations of P-value and significance level would the null hypothesis be rejected?
a. P-value = 0.426 $\alpha = 0.05$
b. P-value = 0.033 $\alpha = 0.01$
c. P-value = 0.046 $\alpha = 0.10$
d. P-value = 0.026 $\alpha = 0.05$
e. P-value = 0.004 $\alpha = 0.01$

10.45   Explain why a P-value of 0.0002 would be interpreted as strong evidence against the null hypothesis.

10.46   Explain why you would not reject the null hypothesis if the P-value were 0.37.

Large-Sample Hypothesis Test for a Population Proportion

In this section, you will see how the five-step process for hypothesis testing (HMC^3) is used to test hypotheses about a population proportion. We will start with a quick review of the four key questions (QSTN) introduced in Section 7.2:

Q Question Type	Estimation or hypothesis testing?
S Study Type	Sample data or experiment data?
T Type of Data	One variable or two? Categorical or numerical?
N Number of Samples or Treatments	How many samples or treatments?

When the answers to these questions are *hypothesis testing, sample data, one categorical variable, and one sample,* the method to consider is the large-sample hypothesis test for a population proportion.

> ### Test Statistic
>
> The test statistic for a large-sample hypothesis test for a population proportion is
>
> $$z = \frac{\hat{p} - p_0}{\sqrt{\dfrac{p_0(1 - p_0)}{n}}}$$
>
> where p_0 is the hypothesized value from the null hypothesis.
>
> When the null hypothesis is true and the sample size is large, this test statistic has a distribution that is approximately standard normal.
>
> The sample size is large enough for the distribution of the test statistic to be approximately standard normal if both np_0 and $n(1 - p_0)$ are greater than or equal to 10.

Computing the *P*-value

When the null hypothesis is true and the sample size is large, the test statistic

$$z = \frac{p - p_0}{\sqrt{\dfrac{p_0(1 - p_0)}{n}}}$$

has a distribution that is approximately standard normal. Because the *P*-value is the probability of observing a test statistic value that is more extreme than what was observed in the sample when the null hypothesis is true, computing the *P*-value involves finding an area under the standard normal curve.

The way you find the *P*-value depends on the form of the inequality ($<$, $>$, or $\neq$) in the alternative hypothesis, H_a. Suppose, for example, that you want to test

$$H_0: p = 0.6 \qquad \text{versus} \qquad H_a: p > 0.6$$

using a large sample. The appropriate test statistic is

$$z = \frac{\hat{p} - 0.6}{\sqrt{\dfrac{(0.6)(1 - 0.6)}{n}}}$$

Values of $\hat{p}$ contradictory to H_0 and more consistent with H_a are those that are much greater than 0.6. In fact, to reject H_0 you would need to see a value of $\hat{p}$ that is enough greater than 0.6 that it can't be explained just by sample-to-sample variability. These values of $\hat{p}$ will correspond to z values that are much greater than 0. For example, if $n = 400$ and $\hat{p} = 0.679$, then

$$z = \frac{0.679 - 0.6}{\sqrt{\dfrac{(0.6)(1 - 0.6)}{400}}} = \frac{0.079}{0.024} = 3.29$$

If H_0 is true, the value $\hat{p} = 0.679$ is more than 3 standard deviations larger than what you would have expected. This means that

$$\begin{aligned}
\textit{P}\text{-value} &= P(z \text{ at least as contradictory to } H_0 \text{ as } 3.29 \text{ when } H_0 \text{ is true}) \\
&= P(z \geq 3.29 \text{ when } H_0 \text{ is true}) \\
&= \text{area under the } z \text{ curve to the right of } 3.29 \\
&= 1 - 0.9995 \\
&= 0.0005
\end{aligned}$$

This *P*-value is illustrated in Figure 10.1. If H_0 is true, in the long run only about 5 out of 10,000 samples would result in a *z* value as or more extreme than what actually resulted. This would be quite unusual. Using a significance level of $\alpha = 0.01$, you would reject the null hypothesis because the *P*-value is less than α ($0.0005 \leq 0.01$).

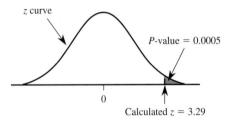

FIGURE 10.1
Calculating a *P*-value.

Now consider testing $H_0: p = 0.3$ versus $H_a: p \neq 0.3$. A value of $\hat{p}$ *either* much greater than 0.3 *or* much less than 0.3 is inconsistent with H_0 and provides support for H_a. Such a value of $\hat{p}$ corresponds to a *z* value far out in *either* tail of the *z* curve. For example, if $z = 1.75$, then (as shown in Figure 10.2)

P-value $= P(z$ value at least as inconsistent with H_0 as 1.75 when H_0 is true)
$\quad = P(z \geq 1.75$ or $z \leq -1.75$ when H_0 is true)
$\quad = (z$ curve area to the right of 1.75) $+$ (z curve area to the left of -1.75)
$\quad = (1 - 0\ 0.9599) + 0.0401$
$\quad = 0.0802$

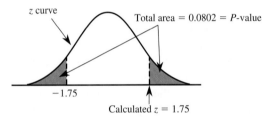

FIGURE 10.2
P-value as the sum of two tail areas.

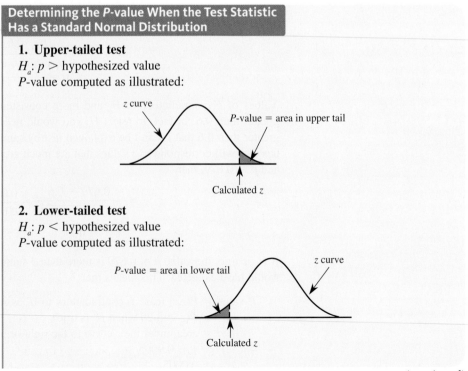

Determining the *P*-value When the Test Statistic Has a Standard Normal Distribution

1. Upper-tailed test
$H_a: p >$ hypothesized value
P-value computed as illustrated:

2. Lower-tailed test
$H_a: p <$ hypothesized value
P-value computed as illustrated:

(continued)

3. Two-tailed test
H_a: $p \neq$ hypothesized value
P-value computed as illustrated:

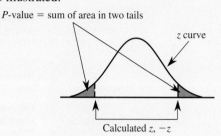

P-value = sum of area in two tails

z curve

Calculated *z*, −*z*

The symmetry of the *z* curve implies that when the test is two-tailed (the "not equal" alternative), it is not necessary to add two curve areas. Instead,

If *z* is positive, *P*-value = 2·(area to the right of *z*).
If *z* is negative, *P*-value = 2·(area to the left of *z*).

The following box summarizes the large-sample test for a population proportion.

A Large-Sample Test for a Population Proportion

Appropriate when the following conditions are met:
1. The sample is a random sample from the population of interest or the sample is selected in a way that would result in a representative sample.
2. The sample size is large. This condition is met when both $np_0 \geq 10$ and $n(1 - p_0) \geq 10$.

When these conditions are met, the following test statistic can be used:

$$z = \frac{\hat{p} - p_0}{\sqrt{\dfrac{p_0(1 - p_0)}{n}}}$$

where p_0 is the hypothesized value from the null hypothesis.

Form of the null hypothesis: H_0: $p = p_0$

Associated *P*-value:

When the Alternative Hypothesis Is…	The *P*-value Is…
H_a: $p > p_0$	Area under the *z* curve to the right of the calculated value of the test statistic
H_a: $p < p_0$	Area under the *z* curve to the left of the calculated value of the test statistic
H_a: $p \neq p_0$	2·(area to the right of *z*) if *z* is positive or 2·(area to the left of *z*) if *z* is negative

To carry out the test, follow the five-step process (HMC³) described in Section 10.4.

Example 10.11 Unfit Teens

The article **"7 Million U.S. Teens Would Flunk Treadmill Tests" (Associated Press, December 11, 2005)** summarized a study in which 2,205 adolescents ages 12 to 19 took a cardiovascular treadmill test. The researchers conducting the study believed that the sample was representative of adolescents nationwide. Of the 2,205 adolescents tested, 750 had a poor level of cardiovascular fitness. Does this sample provide support for the claim that more than thirty percent of adolescents have a poor level of cardiovascular fitness?

Begin by answering the four key questions (QSTN) for this problem.

Q **Question Type**	Estimation or hypothesis testing?	Hypothesis testing. You are asked to test a claim about a population (more than 30% of adolescents have a poor level of cardiovascular fitness).
S **Study Type**	Sample data or experiment data?	Sample data. The data are from a sample of American adolescents.
T **Type of Data**	One variable or two? Categorical or numerical?	One categorical variable. The variable here (cardiovascular fitness) is categorical, with two possible values—poor and not poor.
N **Number of Samples or Treatments**	How many samples or treatments?	One.

Because the answers are *hypothesis testing, sample data, one categorical variable, and one sample*, you should consider a large-sample test for a population proportion.

Now you can use the five-step process for hypothesis testing problems (HMC³) to carry out the test.

Process Step	
H Hypotheses	The claim is about the proportion of adolescents who have poor cardiovascular fitness, so the population characteristic of interest is a population proportion: p = proportion of American adolescents who have a poor level of cardiovascular fitness The claim (more than 30%) translates into an alternative hypothesis of $p > 0.3$. You want to know if there is evidence to support this claim, so this will be the alternative hypothesis. The null hypothesis is found by replacing the inequality ($>$) in the alternative hypothesis by an equal sign. **Hypotheses:** *Null hypothesis*: H_0: $p = 0.3$ *Alternative hypothesis*: H_a: $p > 0.3$
M Method	Because the answers to the four key questions are hypothesis testing, sample data, one categorical variable, and one sample, consider a large-sample hypothesis test for a population proportion. **Potential method:** Large-sample test for a population proportion. The test statistic for this test is $$z = \frac{\hat{p} - 0.3}{\sqrt{\dfrac{0.3(1 - 0.3)}{n}}}$$ The value of 0.3 in the test statistic is the hypothesized value from the null hypothesis. You also need to select a significance level for the test. In some cases, a significance level will be specified. If not, you should choose a significance level based on a consideration of the consequences of Type I and Type II errors. In this example, a Type I error would be deciding that more than 30% of adolescents have poor fitness levels, when in fact the actual percentage is 30% or less. A Type II error is thinking that 30% or fewer have poor fitness levels when the actual percentage is greater than 30%. In this situation, because neither type of error is much more serious than the other, you might choose a value for α of 0.05 (as opposed to something much smaller or much larger). **Significance level:** $\alpha = 0.05$
C Check	There are two conditions that need to be met in order for the large sample test for a population proportion to be appropriate. The large-sample condition is easily verified. The sample size is large enough because $np_0 = 2{,}205(0.3) = 661.5$ and $n(1 - p_0) = 2{,}205(0.7) = 1{,}543.5$ are both greater than or equal to 10. The requirement of a random or representative sample is more difficult. The researchers conducting the study indicated that they believed that the sample was representative of adolescents nationwide. If this is the case, it is reasonable to proceed.

(continued)

Process Step	
C Calculate	To calculate the value of the test statistic, you need to find the value of the sample proportion.

$$n = 2{,}205$$

$$\hat{p} = \frac{750}{2{,}205} = 0.34$$

Substituting the values for n and $\hat{p}$ into the test statistic formula results in

Test statistic:

$$z = \frac{p - 0.3}{\sqrt{\dfrac{0.3(1 - 0.3)}{n}}} = \frac{0.34 - 0.3}{\sqrt{\dfrac{(0.3)(1 - 0.3)}{2{,}205}}} = \frac{0.04}{0.01} = 4.00$$

Associated P-value:

This is an upper-tailed test (the inequality in H_a is $>$), so the P-value is the area under the z curve and to the right of the computed z value.

$$P\text{-value} = \text{area under } z \text{ curve to the right of } 4.00$$

$$= P(z > 4.00)$$

$$\approx 0$$

Since $z = 4.00$ is so far out in the upper tail of the standard normal distribution, the area to its right is approximately 0.

| C Communicate Results | Because the P-value is less than the selected significance level, you reject the null hypothesis. |

Decision: $0 < 0.05$, Reject H_0.

The final conclusion for the test should be stated in context and answer the question posed.

Conclusion: The sample provides convincing evidence that more than 30% of adolescents have a poor fitness level.

Example 10.12   Gender Neutral Revisited

The preview example at the beginning of the chapter described a study published in the journal *Fertility and Sterility* and summarized in the article **"Boy or Girl: Which Gender Baby Would You Pick?" (LiveScience, March 23, 2005)**. The LiveScience summary states that, of 229 women surveyed who said that they would like to choose the sex of a future child, "there was no greater demand for boys or girls." However, the journal article states that, of the 229 who wanted to select the baby's sex, 89 wanted a boy and 140 wanted a girl. Does this provide evidence against the statement of no preference in the LiveScience summary?

Considering the four key questions (QSTN), this situation can be described as *hypothesis testing, sample data, one categorical variable (gender preference), and one sample*. This combination suggests a large-sample test for a population proportion.

H Hypotheses   In this example, you want to use sample data to determine if there is evidence against a claim that has been made. Before you can proceed, you need to think about the appropriate population. In this study, all of the women who participated in the survey were patients at a particular fertility clinic—the Center for Reproductive Medicine at Brigham and Women's Hospital in Boston. Because of this, it is wise to limit any conclusions to the population of all female patients of this fertility clinic who would like to choose a future baby's sex.

The claim is about the proportion of women who would choose a particular sex, so the population characteristic of interest is a population proportion. Let's focus on the proportion who would choose a boy.

Population characteristic of interest:

p = proportion of female patients at the Center for Reproductive Medicine who would choose a boy, out of all female patients whose who want to choose a future baby's sex.

The claim of no preference translates into $p = 0.5$ because if there is no preference, you would expect half to choose boy and half to choose girl. Since you are asked if there is evidence against this claim, the alternative hypothesis is $p \neq 0.5$.

Hypotheses:
Null hypothesis: H_0: $p = 0.5$
Alternative hypothesis: H_a: $p \neq 0.5$

M Method   Because the answers to the four key questions are hypothesis testing, sample data, one categorical variable, and one sample, a large-sample hypothesis test for a population proportion will be considered. The test statistic for this test is

$$z = \frac{\hat{p} - 0.5}{\sqrt{\dfrac{0.5(1 - 0.5)}{n}}}$$

The value of 0.5 in the test statistic is the hypothesized value from the null hypothesis.

You also need to select a significance level for the test. Because no significance level was specified, you should choose a significance level after considering the consequences of Type I and Type II errors. In this situation, neither type of error is much more serious than the other, so you might choose a value of 0.05 for α.

Significance level: $\alpha = 0.05$

C Check   There are two conditions that must be met before using the large-sample test for a population proportion. The large-sample condition is easily verified. The sample size is large enough because

$$np_0 = 229(0.5) = 114.5 \text{ and } n(1 - p_0) = 229(1 - 0.5) = 114.5$$

are both greater than or equal to 10.

The requirement of a random or representative sample is more difficult. Although a random sample of patients from the clinic was selected, there was a very large non-response rate. Only about 40% of those who received the survey returned it (see the study description in the Chapter Preview). In order for the sample to be considered representative of the population, you need to be willing to assume that those who returned the survey did not differ in some important way from those that did not return the survey.

C Calculate   To calculate the value of the test statistic, you need to find the value of the sample proportion:

$$n = 229$$

$$\hat{p} = \frac{89}{229} = 0.389$$

Next, you calculate the value of the test statistic:

$$z = \frac{\hat{p} - 0.5}{\sqrt{\dfrac{0.5(1 - 0.5)}{n}}} = \frac{0.389 - 0.5}{\sqrt{\dfrac{(0.5)(1 - 0.5)}{229}}} = \frac{-0.111}{0.033} = -3.36$$

This is a two-tailed test (the inequality in H_a is $\neq$), so the P-value is twice the area under the z curve to the left of -3.36. This area can be found using a graphing calculator, statistical software, or the table of normal curve areas.
Associated P-value:

$$P\text{-value} = 2 \cdot (\text{area under } z \text{ curve to the left of } -3.36)$$
$$= 2 \cdot P(z < -3.36)$$
$$= 2 \cdot (0.0004)$$
$$= 0.0008$$

C Communicate Results   Because the P-value is less than the selected significance level, the null hypothesis is rejected.

Decision: $0.0008 < 0.05$, Reject H_0.

Interpretation:
There is convincing evidence that the proportion of women who would choose a boy is not equal to 0.5. This means that there is evidence that the LiveScience summary of the study findings is not correct when it states that there is no gender preference among female patients at the clinic who would like to choose the sex of a future baby. This conclusion is only valid if the sample is representative of the population—nonresponse is a concern here.

Most statistical computer packages and graphing calculators can calculate and report P-values for a variety of hypothesis tests, including the large-sample test for a population proportion. Minitab was used to carry out the test of Example 10.12, and the resulting computer output follows:

Test and CI for One Proportion
Test of p = 0.5 vs. p not = 0.5

Sample	X	N	Sample p	95% CI	Z-Value	P-Value
1	89	229	0.388646	(0.325514, 0.451779)	−3.37	0.001

From the Minitab output, $z = -3.37$, and the associated P-value is 0.001. The small differences in the values of the test statistic and P-value from Example 10.12 are the result of rounding.

Example 10.13 Be Careful What You Put on Your Facebook Page!

The article **"Career Expert Provides DOs and DON'Ts for Job Seekers on Social Networking" (CareerBuilder.com, August 19, 2009)** included data from a survey of 2,667 hiring managers and human resource professionals. The article noted that many employers are using social networks to screen job applicants and that this practice is becoming more common. Of the 2,667 people who participated in the survey, 1,200 indicated that they use social networking sites (such as Facebook, MySpace, and LinkedIn) to research job applicants. Based on the survey data, is there convincing evidence that fewer than half of hiring managers and human resource professionals use social networking sites in this way?

In terms of the four key questions, this study can be classified as *hypothesis testing, sample data, one categorical variable (use of social networking sites to screen job applicants, with possible values use or do not use), and one sample.* This combination suggests a large-sample test for a population proportion.

H Hypotheses   In this example, the claim is about the proportion of hiring managers and human resource professionals who use social networking sites to screen job applicants, so the population characteristic of interest is

$p =$ proportion of hiring managers and human resource professionals who use social networking sites to screen job applicants

The question of interest is whether fewer than half use social networking sites to screen job applicants, so the alternative hypothesis is $p < 0.5$. The null hypothesis is then $p = 0.5$.

Hypotheses:
Null hypothesis: H_0: $p = 0.5$
Alternative hypothesis: H_a: $p < 0.5$

M Method   Because the answers to the four key questions are hypothesis testing, sample data, one categorical variable, and one sample, a large-sample hypothesis test for a population proportion is considered.

Test statistic:

$$z = \frac{\hat{p} - 0.5}{\sqrt{\dfrac{0.5(1 - 0.5)}{n}}}$$

The value of 0.5 in the test statistic is the hypothesized value from the null hypothesis.

Next, you need to select a significance level for the test. As a future job applicant, you might consider a Type I error (incorrectly believing that fewer than half screen applicants using social networking sites) to be a more serious error than a Type II error. In this situation, it would make sense to choose a relatively small value for α, such as 0.01.

Significance level: $\alpha = 0.01$

C Check There are two conditions that must be met before using the large-sample test for a population proportion. The large sample condition is easily verified. The sample size is large enough because $np_0 = 2{,}667(0.5) = 1{,}333.5$ and $n(1 - p_0) = 2{,}667(1 - 0.5) = 1{,}333.5$ are both greater than or equal to 10.

The study authors indicate that they believe that the sample of hiring managers and human resource professionals was selected in a way that would result in a representative sample. Because both conditions required for the test are met, it is reasonable to proceed.

C Calculate To calculate the value of the test statistic, you need to find the value of the sample proportion.

$$n = 2{,}667$$

$$\hat{p} = \frac{1{,}200}{2{,}667} = 0.45$$

Substituting the values for n and $\hat{p}$ into the test statistic formula results in the following:

Test statistic:

$$z = \frac{\hat{p} - 0.5}{\sqrt{\dfrac{0.5(1 - 0.5)}{n}}} = \frac{0.45 - 0.5}{\sqrt{\dfrac{(0.5)(1 - 0.5)}{2{,}667}}} = \frac{-0.05}{0.01} = -5.00$$

This is a lower-tailed test (the inequality in H_a is <), so the P-value is the area under the z curve and to the left of -5.00.

Associated P-value:

$$P\text{-value} = \text{area under } z \text{ curve to the left of } -5.00$$
$$= P(z < -5.00)$$
$$\approx 0$$

C Communicate Results Because the P-value is less than the selected significance level, the null hypothesis is rejected. There is convincing evidence that the proportion of hiring managers and human resource professionals who use social networking sites to screen job applicants is less than 0.5.

Even though the survey data provide convincing evidence that less than half currently use social networking sites to screen job applicants, it was noted that the practice is becoming more common (only 22% of those surveyed in 2008, compared to 45% in 2009). Just in case you are applying for a job in the near future, here are a few of the things that hiring managers said that they found on social networking sites that caused them not to hire a candidate: provocative or inappropriate photos, references to drinking or drugs, negative comments about a previous employer or coworker, and poor communication.

A Few Final Things to Consider

1. What About Small Samples?

If $np_0 \geq 10$ and $n(1 - p_0) \geq 10$, the standard normal distribution is a reasonable approximation to the distribution of the z test statistic when the null hypothesis is true. In this case,

P-values based on the standard normal distribution can be used to reach a conclusion in the hypothesis test. If the sample size is not large enough to satisfy the large sample conditions, the distribution of the test statistic may be quite different from the standard normal distribution. In this case, you can't use the standard normal distribution to judge whether what was observed in the sample is consistent or inconsistent with the null hypothesis. There are small-sample tests for population proportions, but they are beyond the scope of this text. If you find yourself in this situation, you should consult a more advanced text or talk with a statistician who can provide advice.

2. Choosing a Potential Method

Take a look back at Table 7.1 (on page 420 and also on the inside back cover of the text). You have now seen how the answers to the four key questions lead you to the potential methods indicated in the first two rows of the table. You now have these two "plays" in your statistics playbook. As you get into the habit of answering the four key questions for each new situation that you encounter, it will become easier to use this table to select an appropriate method in a given situation.

SECTION 10.5 EXERCISES

Each Exercise Set assesses the following chapter learning objectives: C6, M3, M4, M6, P1, P2

SECTION 10.5 Exercise Set 1

For questions **10.47–10.49**, answer the following four key questions and indicate whether the method that you would consider would be a large-sample hypothesis test for a population proportion.

Q Question Type	Estimation or hypothesis testing?
S Study Type	Sample data or experiment data?
T Type of Data	One variable or two? Categorical or numerical?
N Number of Samples or Treatments	How many samples or treatments?

10.47 The paper **"College Students' Social Networking Experiences on Facebook"** (*Journal of Applied Developmental Psychology* **[2009]: 227–238**) summarized a study in which 92 students at a private university were asked how much time they spent on Facebook on a typical weekday. The researchers were interested in estimating the average time spent on Facebook by students at this university.

10.48 The article **"iPhone Can Be Addicting, Says New Survey"** (**msnbc.com, March 8, 2010**) described a survey administered to 200 college students who owned an iPhone. One of the questions on the survey asked students if they slept with their iPhone in bed with them. You would like to use the data from this survey to determine if there is convincing evidence that a majority of college students with iPhones sleep with their phones.

10.49 *USA Today* (**Feb. 17, 2011**) reported that 10% of 1,008 American adults surveyed about their use of e-mail said that they had ended a relationship by e-mail. You would like to use this information to estimate the proportion of all adult Americans who have used e-mail to end a relationship.

10.50 Assuming a random sample from a large population, for which of the following null hypotheses and sample sizes is the large-sample z test appropriate?
a. $H_0: p = 0.2$, $n = 25$
b. $H_0: p = 0.6$, $n = 200$
c. $H_0: p = 0.9$, $n = 100$
d. $H_0: p = 0.05$, $n = 75$

10.51 Let p denote the proportion of students at a large university who plan to purchase a campus meal plan in the next academic year. For a large-sample z test of $H_0: p = 0.20$ versus $H_a: p < 0.20$, find the P-value associated with each of the following values of the z test statistic.
a. -0.55
b. -0.92
c. -1.99
d. -2.24
e. 1.40

10.52 The paper **"Debt Literacy, Financial Experiences and Over-Indebtedness"** (*Social Science Research Network, Working paper W14808, 2008*) included data from a survey of 1,000 Americans. One question on the survey was: "You owe $3,000 on your credit card. You pay a minimum payment of $30 each month. At an Annual Percentage Rate of 12% (or 1% per month), how many years would it take to eliminate your credit card debt if you made no additional charges?" Answer options for this question were: (a) less than 5 years; (b) between 5 and 10 years; (c) between 10 and 15 years; (d) never—you will continue to be in debt; (e) don't know; and (f) prefer not to answer.
a. Only 354 of the 1,000 respondents chose the correct answer of never. Assume that the sample is representative

of adult Americans. Is there convincing evidence that the proportion of adult Americans who can answer this question correctly is less than 0.40 (40%)? Use the five-step process for hypothesis testing (HMC^3) described in this section and $\alpha = 0.05$ to test the appropriate hypotheses.

b. The paper also reported that 37.8% of those in the sample chose one of the wrong answers (a, b, or c) as their response to this question. Is it reasonable to conclude that more than one-third of adult Americans would select a wrong answer to this question? Use $\alpha = 0.05$.

10.53 The paper **"Teens and Distracted Driving" (Pew Internet & American Life Project, 2009)** reported that in a representative sample of 283 American teens ages 16 to 17, there were 74 who indicated that they had sent a text message while driving. For purposes of this exercise, assume that this sample is a random sample of 16- to 17-year-old Americans. Do these data provide convincing evidence that more than a quarter of Americans ages 16 to 17 have sent a text message while driving? Test the appropriate hypotheses using a significance level of 0.01.

10.54 The article **"Theaters Losing Out to Living Rooms" (San Luis Obispo Tribune, June 17, 2005)** states that movie attendance declined in 2005. The Associated Press found that 730 of 1,000 randomly selected adult Americans prefer to watch movies at home rather than at a movie theater. Is there convincing evidence that a majority of adult Americans prefer to watch movies at home? Test the relevant hypotheses using a 0.05 significance level.

10.55 The article referenced in the previous exercise also reported that 470 of 1,000 randomly selected adult Americans thought that the quality of movies being produced is getting worse.

a. Is there convincing evidence that less than half of adult Americans believe that movie quality is getting worse? Use a significance level of 0.05.

b. Suppose that the sample size had been 100 instead of 1,000, and that 47 thought that movie quality was getting worse (so that the sample proportion is still 0.47). Based on this sample of 100, is there convincing evidence that less than half of adult Americans believe that movie quality is getting worse? Use a significance level of 0.05.

c. Write a few sentences explaining why different conclusions were reached in the hypothesis tests of Parts (a) and (b).

10.56 In a survey of 1,005 adult Americans, 46% indicated that they were somewhat interested or very interested in having Web access in their cars (**USA Today, May 1, 2009**). Suppose that the marketing manager of a car manufacturer claims that the 46% is based only on a sample and that 46% is close to half, so there is no reason to believe that the proportion of all adult Americans who want car Web access is less than 0.50. Is the marketing

manager correct in his claim? Provide statistical evidence to support your answer. For purposes of this exercise, assume that the sample can be considered representative of adult Americans.

SECTION 10.5 **Exercise Set 2**

For questions **10.57–10.59**, answer the following four key questions and indicate whether the method that you would consider would be a large-sample hypothesis test for a population proportion.

Q Question Type	Estimation or hypothesis testing?
S Study Type	Sample data or experiment data?
T Type of Data	One variable or two? Categorical or numerical?
N Number of Samples or Treatments	How many samples or treatments?

10.57 The paper **"College Students' Social Networking Experiences on Facebook" (Journal of Applied Developmental Psychology [2009]: 227–238)** summarized a study in which 92 students at a private university were asked whether they used Facebook just to pass the time. Twenty-three responded yes to this question. The researchers were interested in estimating the proportion of students at this college who use Facebook just to pass the time.

10.58 The paper **"Pathological Video-Game Use Among Youth Ages 8 to 18: A National Study" (Psychological Science [2009]: 594–601)** summarizes data from a random sample of 1,178 students ages 8 to 18. The paper reported that for the students in the sample, the mean amount of time spent playing video games was 13.2 hours per week. The researchers were interested in using the data to estimate the mean amount of time spent playing video games for students ages 8 to 18.

10.59 The paper referenced in the previous exercise also reported that when each of the 1,178 students who participated in the study was asked if he or she played video games at least once a day, 271 responded yes. The researchers were interested in using this information to decide if there is convincing evidence that more than 20% of students ages 8 to 18 play video games at least once a day.

10.60 Let p denote the proportion of students living on campus at a large university who plan to move off campus in the next academic year. For a large sample z test of $H_0: p = 0.70$ versus $H_a: p > 0.70$, find the P-value associated with each of the following values of the z test statistic.

a. 1.40
b. 0.92
c. 1.85
d. 2.18
e. -1.40

10.61 Assuming a random sample from a large population, for which of the following null hypotheses and sample sizes is the large-sample z test appropriate?

a. H_0: $p = 0.8$, $n = 40$
b. H_0: $p = 0.4$, $n = 100$
c. H_0: $p = 0.1$, $n = 50$
d. H_0: $p = 0.05$, $n = 750$

10.62 A survey of 1,000 adult Americans (**"Military Draft Study," AP-Ipsos, June 2005**) included the following question: "If the military draft were reinstated, would you favor or oppose drafting women as well as men?" Forty-three percent responded that they would favor drafting women if the draft were reinstated. Using the five-step process for hypothesis testing (HMC[3]) and a 0.05 significance level, determine if there is convincing evidence that less than half of adult Americans favor drafting women.

10.63 The report **"2007 Electronic Monitoring and Surveillance Survey: Many Companies Monitoring, Recording, Videotaping—and Firing—Employees" (American Management Association, 2007)** summarized a survey of 304 U.S. businesses. Of these companies, 201 indicated that they monitor employees' web site visits. Assume that it is reasonable to regard this sample as representative of businesses in the United States.

a. Is there sufficient evidence to conclude that more than 75% of U.S. businesses monitor employees' web site visits? Test the appropriate hypotheses using a significance level of 0.01.
b. Is there sufficient evidence to conclude that a majority of U.S. businesses monitor employees' web site visits? Test the appropriate hypotheses using a significance level of 0.01.

10.64 Duck hunting in populated areas faces opposition on the basis of safety and environmental issues. In a survey to assess public opinion regarding duck hunting on Morro Bay (located along the central coast of California), a random sample of 750 local residents included 560 who strongly opposed hunting on the bay. Does this sample provide convincing evidence that a majority of local residents oppose hunting on Morro Bay? Test the relevant hypotheses using $\alpha = 0.01$.

10.65 In a representative sample of 1,000 adult Americans, only 430 could name at least one justice who was currently serving on the U.S. Supreme Court (**Ipsos, January 10, 2006**). Using a significance level of 0.01, determine if there is convincing evidence in support of the claim that less than half of adult Americans can name at least one justice currently serving on the Supreme Court.

Additional Exercises

10.66 In a survey conducted by CareerBuilders.com, employers were asked if they had ever sent an employee home because he or she was dressed inappropriately (**June 17, 2008, www.careerbuilders.com**). A total of 2,765 employers responded to the survey, with 968 saying that they had sent an employee home for inappropriate attire. In a press release, CareerBuilder makes the claim that more than one-third of employers have sent an employee home to change clothes. Do the sample data provide convincing evidence in support of this claim? Test the relevant hypotheses using $\alpha = 0.05$. For purposes of this exercise, assume that the sample is representative of employers in the United States.

10.67 In a survey of 1,000 women ages 22 to 35 who work full-time, 540 indicated that they would be willing to give up some personal time in order to make more money (**USA Today, March 4, 2010**). The sample was selected to be representative of women in the targeted age group.

a. Do the sample data provide convincing evidence that a majority of women ages 22 to 35 who work full-time would be willing to give up some personal time for more money? Test the relevant hypotheses using $\alpha = 0.01$.
b. Would it be reasonable to generalize the conclusion from Part (a) to all working women? Explain why or why not.

10.68 According to a **Washington Post-ABC News** poll, 331 of 502 randomly selected U.S. adults said they would not be bothered if the National Security Agency collected records of personal telephone calls. Is there sufficient evidence to conclude that a majority of U.S. adults feel this way? Test the appropriate hypotheses using a 0.01 significance level.

10.69 In a representative sample of 2,013 American adults, 1,590 indicated that lack of respect and courtesy in American society is a serious problem (**Associated Press, April 3, 2002**). Is there convincing evidence that more than three-quarters of American adults believe that lack of respect and courtesy is a serious problem? Test the relevant hypotheses using a significance level of 0.05.

10.70 The authors of the article **"Perceived Risks of Heart Disease and Cancer Among Cigarette Smokers" (Journal of the American Medical Association [1999]: 1019–1021)** expressed the concern that a majority of smokers do not view themselves as being at increased risk of heart disease or cancer. A study of 737 current smokers found that of the 737 smokers surveyed, only 295 believe they have a higher than average risk of cancer. Do these data suggest that p, the proportion of all smokers who view themselves as being at increased risk of cancer, is less than 0.5, as claimed by the authors of the paper? For purposes of this exercise, assume that this sample is representative of the population of smokers. Test the relevant hypotheses using $\alpha = 0.05$.

10.71 A number of initiatives on the topic of legalized gambling have appeared on state ballots. A political candidate has decided to support legalization of casino gambling if he is convinced that more than two-thirds of American adults approve of casino gambling. Suppose that 1,035 of the people in a random sample of 1,523 American adults said they approved of casino gambling. Is there convincing evidence that more than two-thirds approve?

10.72 Many people have misconceptions about how profitable small, consistent investments can be. In a survey of 1,010 randomly selected American adults **(Associated Press, October 29, 1999)**, only 374 responded that they thought that an investment of $25 per week over 40 years with a 7% annual return would result in a sum of over $100,000 (the actual amount would be $286,640). Is there convincing evidence that less than 40% of U.S. adults think that such an investment would result in a sum of over $100,000? Test the relevant hypotheses using $\alpha = 0.05$.

10.73 The survey described in the previous exercise also asked the individuals in the sample what they thought was their best chance to obtain more than $500,000 in their lifetimes. Twenty-eight percent responded "win a lottery or sweepstakes." Does this provide convincing evidence that more than one-fourth of U.S. adults see a lottery or sweepstakes win as their best chance of accumulating $500,000? Carry out a test using a significance level of 0.01.

10.74 In a survey conducted by Yahoo Small Business, 1,432 of 1,813 adults surveyed said that they would alter their shopping habits if gas prices remain high **(Associated Press, November 30, 2005)**. The article did not say how the sample was selected, but for purposes of this exercise, assume that the sample is representative of adult Americans. Based on the survey data, is it reasonable to conclude that more than three-quarters of adult Americans would alter their shopping habits if gas prices remain high?

10.75 The article **"Americans Seek Spiritual Guidance on Web" (San Luis Obispo Tribune, October 12, 2002)** reported that 68% of the U.S. population belongs to a religious community. In a survey on Internet use, 84% of "religion surfers" (defined as those who seek spiritual help online or who have used the Web to search for prayer and devotional resources) belong to a religious community. Suppose that this result was based on a representative sample of 512 religion surfers. Is there convincing evidence that the proportion of religion surfers who belong to a religious community is different from 0.68, the proportion for the U.S. population? Use $\alpha = 0.05$.

SECTION 10.6 Power and the Probability of Type II Error

In this chapter, you have considered a large-sample procedure for testing hypotheses about a population proportion. What characterizes a "good" test procedure? It makes sense to think that a good test procedure is one that has both a small probability of rejecting H_0 when it is true (a Type I error) and a large probability of rejecting H_0 when it is false. The test procedure presented in this chapter allows you to directly control the probability of rejecting a true H_0 by your choice of the significance level α. But what about the probability of rejecting H_0 when it is false? As you will see, several factors influence this probability.

Let's begin by considering an example. Suppose that the manager of a grocery store is thinking about expanding the store's selection of organically grown produce. Because organically grown produce costs more than produce that is not organically grown, the manager thinks that this expansion will be profitable only if the proportion of store customers who would pay more for organic produce is greater than 0.30. The manager plans to ask each person in a random sample of customers if he or she would be willing to pay more for organic produce. He will then use the resulting data to test

$$H_0: p = 0.30 \quad \text{versus} \quad H_a: p > 0.30$$

using a significance level of $\alpha = 0.05$ If the actual population proportion is 0.30 (or any value less than 0.30), the correct decision is to fail to reject the null hypothesis. On the other hand, if the actual value of the population proportion is 0.60, 0.50, or even 0.32, the correct decision is to reject the null hypothesis. Not rejecting the null hypothesis is a Type II error. How likely is it that the null hypothesis will be rejected in this case?

If the actual population proportion is 0.32, the probability that you will reject the null hypothesis is not very great. This is because when you carry out the test, you are looking at the sample proportion and asking if it looks like what you would have expected to see if the population proportion were 0.30. As illustrated in Figure 10.3, if the actual value of the population proportion is greater than but very close to 0.30, chances are that the sample proportion will look pretty much like what you would expect to see if the population were 0.30, and so you won't be convinced that the null hypothesis should be rejected.

If the actual value of the population proportion is 0.50, it is less likely that the sample will be mistaken for a sample from a population with $p = 0.30$. Sample proportions will tend to cluster around 0.50, and so it is more likely that you will correctly reject H_0. If the actual value of the population proportion were 0.60, it would be even more likely that you would reject H_0.

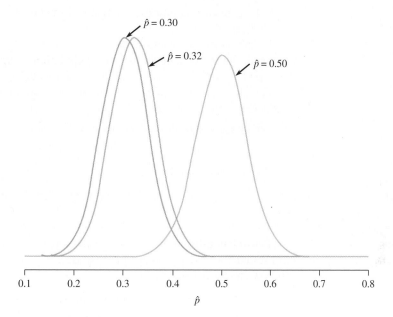

$\hat{p} = 0.30$

$\hat{p} = 0.32$

$\hat{p} = 0.50$

| 0.1 | 0.2 | 0.3 | 0.4 | 0.5 | 0.6 | 0.7 | 0.8 |

$\hat{p}$

FIGURE 10.3 Large-sample sampling distribution of $\hat{p}$ when $p = 0.30$, 0.32, and 0.50.

When you consider the probability of rejecting the null hypothesis, you are looking at what statisticians refer to as the **power** of the test.

> **DEFINITION**
>
> The **power of a test** is the probability of rejecting the null hypothesis.

From the previous discussion, it should be apparent that when a hypothesis about a population proportion is being tested, the power of the test depends on the actual value of the population proportion, p. Because the actual value of p is unknown (if you knew the value of p, you wouldn't be doing the hypothesis test!), you won't know what the power is for the actual value of p. It is possible, however, to gain some insight into the power of a test by looking at a number of "what if" scenarios. For example, you might ask

What is the power if the actual population proportion is 0.32?

or

What is the power if the actual population proportion is 0.50?

and so on. That is, you can determine the power when $p = 0.32$, when $p = 0.50$, when $p = 0.60$, or for any other value of interest. Although it is technically possible to consider power when the null hypothesis is true, you are usually only concerned about the power for values for which the null hypothesis is false.

In general, when testing a hypothesis about a population characteristic, there are three factors that influence the power of the test:

1. The size of the difference between the actual value of the population characteristic and the hypothesized value (the value that appears in the null hypothesis).
2. The choice of significance level, α, for the test.
3. The sample size.

> **Effect of Various Factors on the Power of a Test**
>
> 1. The larger the size of the difference between the hypothesized value and the actual value of the population characteristic, the greater the power.
> 2. The larger the significance level, α, the greater the power of the test.
> 3. The larger the sample size, the greater the power of the test.

Let's consider each of the statements in the preceding box. The first statement has already been discussed in the context of the organic produce example. Because power is the probability of rejecting the null hypothesis, it makes sense that the power will be greater when the actual value of a population characteristic is quite different from the hypothesized value than when the actual value is close to the hypothesized value.

The effect of significance level on power is not quite as obvious. To understand the relationship between power and significance level, it helps to consider the relationship between power and β, the probability of a Type II error.

> When the null hypothesis is false,
>
> $$power = 1 - \beta$$

This relationship follows from the definitions of power and Type II error. A Type II error results from not rejecting a false H_0. Because power is the probability of rejecting H_0, it follows that *when H_0 is false*

$$power = \text{probability of rejecting a false } H_0$$
$$= 1 - \text{probability of not rejecting a false } H_0$$
$$= 1 - \beta$$

Recall from Section 10.2 that the choice of α, the Type I error probability, affects the value of β, the Type II error probability. Choosing a larger value for α results in a smaller value for β (and therefore a larger value for $1-\beta$). In terms of power, this means that choosing a larger value for α results in a larger value for the power of the test. That is, the larger the Type I error probability you are willing to tolerate, the more likely it is that the test will be able to detect any particular departure from H_0.

The third factor that affects the power of a test is the sample size. When H_0 is false, the power of a test is the probability that you will in fact "detect" that H_0 is false and, based on the observed sample, reject H_0. Intuition suggests that you will be more likely to detect a departure from H_0 with a large sample than with a small sample. This is in fact the case—the larger the sample size, the greater the power.

Consider testing the hypotheses discussed earlier:

$$H_0: p = 0.30 \quad \text{versus} \quad H_a: p > 0.30$$

Then, for example:

1. For any value of p greater than 0.30, the power of a test based on a sample of size 100 is greater than the power of a test based on a sample of size 75 (assuming the same significance level).
2. For any value of p greater than 0.30, the power of a test using a significance level of 0.05 is greater than the power of a test using a significance level of 0.01 (assuming the same sample size).
3. For any value of p greater than 0.30, the power of the test is greater if the actual population proportion is 0.50 than if the actual population proportion is 0.35 (assuming the same sample size and significance level).

As was mentioned previously in this section, it is impossible to calculate the *exact* power of a test because in practice you do not know the values of population characteristics.

However, you can evaluate the power at a selected alternative value which would tell you whether the power would be high or low if this alternative value is the actual value.

The following optional subsection shows how Type II error probabilities and power can be evaluated for the large-sample hypothesis test for a population proportion.

Calculating Power and Type II Error Probabilities (Optional)

The test procedure presented in this chapter is designed to control the probability of a Type I error (rejecting H_0 when H_0 is true) at the desired significance level α. However, little has been said about calculating the value of β, the probability of a Type II error (not rejecting H_0 when H_0 is false).

When you carry out a hypothesis test, you specify the desired value of α, the probability of a Type I error. The probability of a Type II error, β, is the probability of not rejecting H_0 even though it is false. Suppose that you are testing

$$H_0: p = 0.9 \quad \text{versus} \quad H_a: p < 0.9$$

Because you do not know the actual value of p, you cannot calculate the value of β. However, the vulnerability of the test to Type II error can be investigated by calculating β for several different potential values of p, such as $p = 0.85$, $p = 0.80$ and $p = 0.75$. Once the value of β has been determined, the power of the test at the corresponding alternative value is $1 - \beta$.

Example 10.14 **Power for Testing Hypotheses About a Population Proportion**

A package delivery service advertises that at least 90% of all packages brought to its office by 9 a.m. for delivery in the same city are delivered by noon that day. Let p denote the proportion of all such packages actually delivered by noon. The hypotheses of interest are

$$H_0: p = 0.9 \quad \text{versus} \quad H_a: p < 0.9$$

where the alternative hypothesis states that the company's claim is untrue. The value $p = 0.8$ represents a substantial departure from the company's claim. If the hypotheses are tested using a significance level of 0.01 and a sample of $n = 225$ packages, what is the probability that the departure from H_0 represented by this alternative value will go undetected?

The test statistic for this test is

$$z = \frac{\hat{p} - \mu_{\hat{p}}}{\sigma_{\hat{p}}} = \frac{\hat{p} - 0.9}{\sqrt{\dfrac{(0.9)(0.1)}{225}}} = \frac{\hat{p} - 0.9}{0.02}$$

The inequality in H_a ($<$) implies that

$$P-\text{value} = \text{area under the } z \text{ curve to the left of the calculated } z$$

At significance level 0.01, H_0 is rejected if P-value ≤ 0.01. For the case of a lower-tailed test, this is the same as rejecting H_0 if

$$z = \frac{\hat{p} - 0.9}{0.02} \leq -2.33$$

(Because -2.33 captures a lower-tail z curve area of 0.01, the smallest 1% of all z values satisfy $z \leq -2.33$.) Solving $z = \dfrac{\hat{p} - 0.9}{0.02} \leq -2.33$ for $\hat{p}$ results in $\hat{p} \leq 0.853$. So, H_0 is *not* rejected if $\hat{p} > 0.853$. When $p = 0.8$, the sampling distribution of $\hat{p}$ is approximately normal with

$$\mu_{\hat{p}} = 0.8$$

$$\sigma_{\hat{p}} = \sqrt{\frac{(0.8)(0.2)}{225}} = 0.0267$$

Then β is the probability of obtaining a sample proportion greater than 0.853, as illustrated in Figure 10.4.

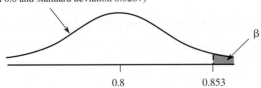

Sampling distribution of $\hat{p}$ (normal with mean 0.8 and standard deviation 0.0267)

FIGURE 10.4 Approximate sampling distribution of $\hat{p}$ and β when $p = 0.8$ in Example 10.14

Because the curve in Figure 10.4 is not the *standard* normal distribution, you must first convert to a z score before using Appendix Table 2 to find the area. Converting to a z score results in

$$z = \frac{0.853 - 0.8}{0.0267} = 1.99$$

Using Appendix Table 2, you can now calculate

$$\beta = 1 - 0.9767 = 0.0233$$

When $p = 0.8$ and $\alpha = 0.01$, less than 3% of all samples of size $n = 225$ will result in a Type II error. The power of the test at $p = 0.8$ is $1 - 0.0233 = 0.9767$. This means that the probability of rejecting $H_0: p = 0.9$ in favor of $H_a: p < 0.9$ when p is really 0.8 is 0.9767, which is quite high.

SECTION 10.5 EXERCISES

Each Exercise Set assesses the following chapter learning objectives: C7

SECTION 10.5 Exercise Set 1

10.76 Suppose you plan to test $H_0: p = 0.40$ versus $H_a: p, 0.40$. Assuming the same significance level, would the power of the test be greater if $n = 50$ or it $n = 100$? Explain your choice.

10.77 Suppose you plan to test $H_0: p = 0.25$ versus $H_a: p \neq 0.25$. Assuming the same sample size and significance level, would the power of the test be smaller if the actual value of the population proportion were 0.15 or 0.30? Explain your choice.

10.78 Suppose you plan to test $H_0: p = 0.82$ versus $H_a: p > 0.82$. Assuming the same sample size, would the power of the test be greater for a significance level of $\alpha = 0.01$ or $\alpha = 0.05$? Explain your choice.

SECTION 10.5 Exercise Set 2

10.79 Suppose you plan to test $H_0: p = 0.75$ versus $H_a: p < 0.75$. Assuming the same significance level, would the power of the test be smaller if $n = 200$ or it $n = 300$? Explain your choice.

10.80 Suppose you plan to test $H_0: p = 0.20$ versus $H_a: p \neq 0.20$. Assuming the same sample size and significance

level, would the power of the test be greater if the actual value of the population proportion were 0.10 or 0.25? Explain your choice.

10.81 Suppose you plan to test $H_0: p = 0.78$ versus $H_a: p < 0.78$. Assuming the same sample size, would the power of the test be smaller for a significance level of $\alpha = 0.10$ or $\alpha = 0.05$? Explain your choice.

Additional Exercises

10.82 A county health department developed an advertising campaign promoting the benefits of getting an annual flu shot. They wonder if the campaign has been successful in increasing the proportion of county residents who got a flu shot in the current year. Data from previous years suggest that the proportion getting a flu shot was 0.60. The health department plans to select a random sample of 100 county residents and use the resulting data to test $H_0: p = 0.60$ versus $H_a: p > 0.60$, where p is the proportion of county residents who got a flu shot in the current year.

a. Suppose the actual value of p is 0.70. What is the power of the test if a significance level of $\alpha = 0.05$ is used?

b. Would the power of a test with $\alpha = 0.05$ be larger or smaller than the power calculated in Part (a) if the actual value of p were 0.75?

c. Would the power of a test with $\alpha = 0.05$ be larger or smaller than the power calculated in Part (a) if the actual value of p were 0.70 but the sample size was 200 instead of 100?

10.83 The city council in a large city has become concerned about the trend toward exclusion of renters with children in apartments within the city. The housing coordinator has decided to select a random sample of 125 apartments and determine for each whether children are permitted. Let p be the proportion of all apartments that prohibit children. If the city council is convinced that p is greater than 0.75, it will consider appropriate legislation.

a. If 102 of the 125 sampled apartments exclude renters with children, would a test with $\alpha = 0.05$ lead you to the conclusion that more than 75% of all apartments exclude children?

b. What is the power of the test when $p = 0.8$ and $\alpha = 0.05$?

SECTION 10.7 Avoid These Common Mistakes

When summarizing the results of a hypothesis test, be sure to include all the relevant information. In particular, include:

1. *Hypotheses.* Whether specified in symbols or described in words, it is important that both the null and the alternative hypotheses be clearly stated. If you are using symbols to define the hypotheses, be sure to define them in the context of the problem at hand (for example, p = proportion of all students who purchase textbooks online).

2. *Test procedure.* You should be clear about what test procedure was used (for example, large-sample test for a population proportion), and why you think it was reasonable to use this procedure.

3. *Test statistic.* Be sure to include the value of the test statistic and the associated P-value. Including the P-value allows a reader who has chosen a different significance level to see whether she would have reached the same or a different conclusion.

4. *Conclusion in context.* Never end the report of a hypothesis test with the statement "I rejected (or did not reject) H_0." Always provide a conclusion that is in the context of the problem and that answers the question which the hypothesis test was designed to answer. Be sure to also state what significance level was used.

There are several things you should watch for when conducting a hypothesis test or when evaluating a written summary of a test:

1. A hypothesis test can never show strong support for the null hypothesis. Make sure that you don't confuse "There is no reason to believe the null hypothesis is not true" with the statement "There is convincing evidence that the null hypothesis is true." These are very different statements!

2. If you have complete information for the population, don't carry out a hypothesis test! People sometimes forget that no test is needed to answer questions about a population if you have complete information and don't need to generalize from a sample. For example, an article on prescription drug overdoses (**"Medicine Cabinet Is a Big Killer,"** *Salt Lake City Tribune*, **August 1, 2007**) reported that "In 2006, some 485 Utah deaths were attributed to poisoning—more than any previous year in the state's history. Opiate and opioid medications alone, or in combination, accounted for most of them—307." It would not be appropriate to use this information to test a hypothesis about the proportion of poisoning deaths in Utah during 2006 that involved opiate or opioid medications, because that proportion is known exactly. In this case, $p = \dfrac{307}{485} = 0.633$.

3. Don't confuse statistical significance with practical significance. When a null hypothesis has been rejected, be sure to step back and evaluate the result in light of its practical importance. For example, you may be convinced that the proportion who respond favorably to a proposed new medical treatment is greater than 0.4, the known proportion that respond favorably to the current treatment. But if your estimate of this proportion for the proposed new treatment is 0.405, it may not be of any practical interest.

CHAPTER ACTIVITIES

ACTIVITY 10.1   A MEANINGFUL PARAGRAPH

Write a meaningful paragraph that includes the following six terms: **hypotheses, *P*-value, reject H_0, Type I error, statistical significance, practical significance**. A "meaningful paragraph" is a coherent piece of writing in an appropriate context that uses all of the listed words. It is not a sequence of sentences that just define the terms. The paragraph should show that you understand the meaning of the terms and their relationship to one another. Choosing a good context will make writing a meaningful paragraph easier.

EXPLORING THE BIG IDEAS

In the exercise below, you will go online to select a random sample from a population of adults between the ages of 18 and 64.

Each person in this population was asked if he or she ever sent a text message while driving. Suppose that you would like to know if less than half of the people in the population have sent a text message while driving. To answer this question, you will take a random sample of 30 people from the population.

Go online to www.cengage.com/stats/peck1e and click on the link labeled "Exploring the Big Ideas." Then locate the link for Chapter 10. It will take you to a web page where you can select a random sample of 30 people from the population.

Click on the Select Sample button. This selects a random sample and will display the following information:

1. The ID number that identifies the person selected
2. The response to the question "Have you ever sent a text message while driving?" These responses are coded numerically—a 1 indicates a yes response and a 2 indicates a no response.

Use the sample data to answer the following questions.

a. What type of problem is this?
    1. An estimation problem.
    2. A hypothesis testing problem.

b. Are the data from sampling or from an experiment?
    1. Sample data
    2. Experiment data

c. How many variables are there?
    1. One
    2. Two

d. What type of data do you have?
    1. Categorical
    2. Numerical

e. How many samples are there?
    1. One
    2. Two

f. What are the appropriate null and alternative hypotheses?

g. What method might be appropriate for testing these hypotheses?
    1. Large-sample test for a population proportion
    2. Large-samples test for the difference in population proportions
    3. One-sample *t* test
    4. Two-sample *t* test

5. Paired t test

6. Chi-square test

h. Are the conditions for the selected test met? Explain.

i. What is the value of the test statistic for this test?

j. What is the P-value?

k. If a significance level of 0.05 is used, what would your decision be?
   1. Reject the null hypothesis
   2. Fail to reject the null hypothesis

l. Write a few sentences that summarize what you learned about the population proportion based on this hypothesis test.

ARE YOU READY TO MOVE ON? **CHAPTER 10 REVIEW EXERCISES**

All chapter learning objectives are assessed in these exercises. The learning objectives assessed in each exercise are given in parentheses.

10.84 (C1, M1)
A county commissioner must vote on a resolution that would commit substantial resources to the construction of a sewer in an outlying residential area. Her fiscal decisions have been criticized in the past, so she decides to take a survey of residents in her district to find out if they favor spending money for a sewer system. She will vote to appropriate funds only if she can be reasonably sure that a majority of the people in her district favor the measure. What hypotheses should she test?

10.85 (C1, M1)
The report **"How Teens Use Media" (Nielsen, June 2009)** says that 37% of U.S. teens access the Internet from a mobile phone. Suppose you plan to select a random sample of students at the local high school and will ask each student in the sample if he or she accesses the Internet from a mobile phone. You want to determine if there is evidence that the proportion of students at the high school who access the Internet using a mobile phone differs from the national figure of 0.37 given in the Nielsen report. What hypotheses should you test?

10.86 (C2)
The article **"Irritated by Spam? Get Ready for Spit"** (*USA Today*, **November 10, 2004**) predicts that "spit," spam that is delivered via Internet phone lines and cell phones, will be a growing problem as more people turn to web-based phone services. In a poll of 5,500 cell phone users, 20% indicated that they had received commercial messages and ads on their cell phones. These data were used to test $H_0: p = 0.13$ versus $H_a: p > 0.13$ where 0.13 was the proportion reported for the previous year. The null hypothesis was rejected.

a. Based on the hypothesis test, what can you conclude about the proportion of cell phone users who received commercial messages and ads on their cell phones in the year the poll was conducted?

b. Is it reasonable to say that the data provide strong support for the alternative hypothesis?

c. Is it reasonable to say that the data provide strong evidence against the null hypothesis?

10.87 (C3)
Explain why failing to reject the null hypothesis in a hypothesis test does not mean there is convincing evidence that the null hypothesis is true.

10.88 (C4, M2)
Researchers at the University of Washington and Harvard University analyzed records of breast cancer screening and diagnostic evaluations (**"Mammogram Cancer Scares More Frequent Than Thought,"** *USA Today*, **April 16, 1998**). Discussing the benefits and downsides of the screening process, the article states that although the rate of false-positives is higher than previously thought, if radiologists were less aggressive in following up on suspicious tests, the rate of false-positives would fall, but the rate of missed cancers would rise. Suppose that such a screening test is used to decide between a null hypothesis of
   H_0: no cancer is present
and an alternative hypothesis of
   H_a: cancer is present.
(Although these are not hypotheses about a population characteristic, this exercise illustrates the definitions of Type I and Type II errors.)

a. Would a false-positive (thinking that cancer is present when in fact it is not) be a Type I error or a Type II error?

b. Describe a Type I error in the context of this problem, and discuss the consequences of making a Type I error.

c. Describe a Type II error in the context of this problem, and discuss the consequences of making a Type II error.

d. Recall the statement in the article that if radiologists were less aggressive in following up on suspicious tests, the rate of false-positives would fall but the rate of missed cancers would rise. What aspect of the relationship between the probability of a Type I error and the probability of a Type II error is being described here?

10.89 (C4, M2)

The National Cancer Institute conducted a 2-year study to determine whether cancer death rates for areas near nuclear power plants are higher than for areas without nuclear facilities (*San Luis Obispo Telegram-Tribune*, **September 17, 1990**). A spokesperson for the Cancer Institute said, "From the data at hand, there was no convincing evidence of any increased risk of death from any of the cancers surveyed due to living near nuclear facilities. However, no study can prove the absence of an effect."

a. Suppose p denotes the true proportion of the population in areas near nuclear power plants who die of cancer during a given year. The researchers at the Cancer Institute might have considered two rival hypotheses of the form

H_0: p is equal to the corresponding value for areas without nuclear facilities

H_a: p is greater than the corresponding value for areas without nuclear facilities

Did the researchers reject H_0 or fail to reject H_0?

b. If the Cancer Institute researchers are incorrect in their conclusion that there is no evidence of increased risk of death from cancer associated with living near a nuclear power plant, are they making a Type I or a Type II error? Explain.

c. Comment on the spokesperson's last statement that no study can *prove* the *absence* of an effect. Do you agree with this statement?

10.90 (C5, M5)

Suppose that you are an inspector for the Fish and Game Department and that you are given the task of determining whether to prohibit fishing along part of the Oregon coast. You will close an area to fishing if it is determined that more than 3% of fish have unacceptably high mercury levels.

a. Which of the following pairs of hypotheses would you test:

$$H_0: p = 0.03 \text{ versus } H_a: p > 0.03$$

or

$$H_0: p = 0.03 \text{ versus } H_a: p < 0.03$$

Explain the reason for your choice.

b. Would you use a significance level of 0.10 or 0.01 for your test? Explain.

10.91 (C6)

The article **"Cops Get Screened for Digital Dirt" (*USA Today*, Nov. 12, 2010)** summarizes a report on law enforcement agency use of social media to screen applicants for employment. The report was based on a survey of 728 law enforcement agencies. One question on the survey asked if the agency routinely reviewed applicants' social media activity during background checks. For purposes of this exercise, suppose that the 728 agencies were selected at random and that you want to use the survey data to decide if there is convincing evidence that more than 25% of law enforcement agencies review applicants' social media activity as part of routine background checks.

a. Describe the shape, center, and spread of the sampling distribution of $\hat{p}$ for samples of size 728 if the null hypothesis H_0: $p = 0.25$ is true.

b. Would you be surprised to observe a sample proportion of $\hat{p} = 0.27$ for a sample of size 728 if the null hypothesis H_0: $p = 0.25$ were true? Explain why or why not.

c. Would you be surprised to observe a sample proportion of $\hat{p} = 0.31$ for a sample of size 728 if the null hypothesis H_0: $p = 0.25$ were true? Explain why or why not.

d. The actual sample proportion observed in the study was $\hat{p} = 0.33$. Based on this sample proportion, is there convincing evidence that more than 25% of law enforcement agencies review social media activity as part of background checks, or is this sample proportion consistent with what you would expect to see when the null hypothesis is true?

10.92 (C6, M6)

"Most Like It Hot" is the title of a press release issued by the Pew Research Center **(March 18, 2009, www.pewsocialtrends. org)**. The press release states that "by an overwhelming margin, Americans want to live in a sunny place." This statement is based on data from a nationally representative sample of 2,260 adult Americans. Of those surveyed, 1,288 indicated that they would prefer to live in a hot climate rather than a cold climate. Suppose that you want to determine if there is convincing evidence that a majority of all adult Americans prefer a hot climate over a cold climate.

a. What hypotheses should be tested in order to answer this question?

b. The P-value for this test is 0.000001. What conclusion would you reach if $\alpha = 0.01$?

For questions 10.85–10.86, answer the following four key questions (introduced in Section 7.2) and indicate whether the method that you would consider would be a large-sample hypothesis test for a population proportion.

Q Question Type	Estimation or hypothesis testing?
S Study Type	Sample data or experiment data?
T Type of Data	One variable or two? Categorical or numerical?
N Number of Samples or Treatments	How many samples or treatments?

10.93 (M3, P1)

The report **"Credit Card Statistics, Industry Facts, Debt Statistics"** (creditcards.com) summarizes a survey of college undergraduates. Each person in a sample of students was asked about his or her credit card balance, and the average balance was calculated to be $3,173. You are interested in determining if there is evidence that the mean credit card debt for undergraduates is greater than the known average from the previous year.

10.94 (M3, P1)

The paper **"I Smoke but I Am Not a Smoker"** (*Journal of American College Health* [2010]: 117–125) describes a survey of 899 college students who were asked about their smoking behavior. Of the students surveyed, 268 classified themselves as nonsmokers, but said yes when asked later in the survey if they smoked. These students were classified as "phantom smokers," meaning that they did not view themselves as smokers even though they do smoke at times. The authors were interested in using these data to determine if there is convincing evidence that more than 25% of college students fall into the phantom smoker category.

10.95 (M4, M6)

Suppose that the sample of 899 college students described in the previous exercise can be regarded as representative of college students in the United States.

a. What hypotheses would you test to answer the question posed in the previous exercise?

b. Is the sample size large enough for a large-sample test for a population proportion to be appropriate?

c. What is the value of the test statistic and the associated *P*-value for this test?

d. If a significance level of 0.05 were chosen, would you reject the null hypothesis or fail to reject the null hypothesis?

10.96 (M6, P2)

In an AP-AOL sports poll (**Associated Press, December 18, 2005**), 272 of 394 randomly selected baseball fans stated that they thought the designated hitter rule should either be expanded to both baseball leagues or eliminated. Based on the given information, is there sufficient evidence to conclude that a majority of baseball fans feel this way?

10.97 (M6, P2)

Past experience is that when individuals are approached with a request to fill out and return a particular questionnaire in a provided stamped and addressed envelope, the response rate is 40%. An investigator believes that if the person distributing the questionnaire were stigmatized in some obvious way, potential respondents would feel sorry for the distributor and thus tend to respond at a rate higher than 40%. To test this theory, a distributor wore an eye patch. Of the 200 questionnaires distributed by this individual, 109 were returned. Does this provide evidence that the response rate in this situation is greater than the previous rate of 40%? State and test the appropriate hypotheses using a significance level of 0.05.

10.98 (C7)

The power of a test is influenced by the sample size and the choice of significance level.

a. Explain how increasing the sample size affects the power (when significance level is held fixed).

b. Explain how increasing the significance level affects the power (when sample size is held fixed).

TECHNOLOGY NOTES

Z-Test for Proportions

TI-83/84
1. Press the **STAT** key
2. Highlight **TESTS**
3. Highlight **1-PropZTest...** and press **ENTER**
4. Next to **p0** type the hypothesized value for *p*
5. Next to **x** type the number of successes
6. Next to **n** type the sample size, *n*
7. Next to **prop,** highlight the appropriate alternative hypothesis
8. Highlight **Calculate** and press **ENTER**

TI-Nspire
1. Enter the Calculate Scratchpad
2. Press the **menu** key then select **6:Statistics** then select **7:Stat Tests** then **5:1-Prop z Test...** then press **enter**
3. In the box next to **p0** type the hypothesized value for *p*
4. In the box next to **Successes, x** type the number of successes

5. In the box next to **n** type the sample size, *n*
6. In the box next to **Alternate Hyp** choose the appropriate alternative hypothesis from the drop-down menu
7. Press **OK**

JMP
Summarized data
1. Enter the data into the JMP data table with categories in the first column and counts in the second column
2. Click **Analyze** and select **Distribution**
3. Click and drag the first column name from the box under **Select Columns** to the box next to **Y, Columns**
4. Click and drag the second column name from the box under **Select Columns** to the box next to **Freq**
5. Click **OK**

6. Click the red arrow next to the column name and click **Test Probabilities**
7. Under the **Test Probabilities** section that appears in the output, click in the box across from **Yes** under the **Hypoth Prob** and type the hypothesized value for p
8. Select the appropriate option for alternative
9. Click **Done**

Note: In the two-sided case, JMP uses the square of the z test statistic, called the Chi-Square test statistic. The two methods are mathematically identical.

Note: In the one-sided cases, JMP uses the exact binomial test rather than the z test.

Raw data
1. Enter the raw data into a column

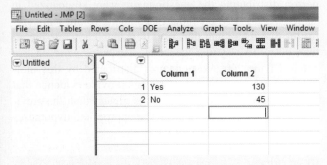

2. Click **Analyze** and select **Distribution**
3. Click and drag the first column name from the box under **Select Columns** to the box next to **Y, Columns**
4. Click **OK**
5. Click the red arrow next to the column name and click **Test Probabilities**
6. Under the **Test Probabilities** section that appears in the output, click in the box across from **Yes** under the **Hypoth Prob** and type the hypothesized value for p
7. Select the appropriate option for alternative
8. Click **Done**

Note: In the two-sided case, JMP uses the square of the z test statistic, called the Chi-Square test statistic. The two methods are mathematically identical.

Note: In the one-sided cases, JMP uses the exact binomial test rather than the z test.

Minitab
Summarized data
1. Click **Stat** then click **Basic Statistics** then click **1 Proportion...**
2. Click the radio button next to **Summarized data**
3. In the box next to **Number of Trials:** type the value for n, the sample size
4. In the box next to **Number of events:** type the value for the number of successes
5. Click **Options...**
6. Input the appropriate hypothesized value in the box next to **Test proportion:**
7. Select the appropriate alternative hypothesis from the drop-down menu next to **Alternative**
8. Check the box next to **Use test and interval based on normal distribution**
9. Click **OK**
10. Click **OK**

Raw data
1. Input the raw data into a column
2. Click **Stat** then click **Basic Statistics** then click **1 Proportion...**
3. Click in the box under **Samples in columns:**
4. Double click the column name where the raw data are stored
5. Click **Options...**
6. Select the appropriate alternative hypothesis from the drop-down menu next to **Alternative**
7. Check the box next to **Use test and interval based on normal distribution**
8. Click **OK**
9. Click **OK**

SPSS
SPSS does not have the functionality to automatically calculate a z-test for a testing a single proportion.

Excel
Excel does not have the functionality to automatically calculate a z-test for a testing a single population proportion. You may type in the formula by hand into a cell to have Excel calculate the value of the test statistic for you. Then use methods from Chapter 6 to find the P-value using the Normal Distribution.

AP* Review Questions for Chapter 10

1. A university administrator in charge of computing facilities will close a computer lab if there is convincing evidence that the proportion of students at the university who have access to a computer at home is greater than 0.6. What hypotheses should the administrator test?

    (A) $H_0: \mu < 0.6$ versus $H_a: \mu = 0.6$
    (B) $H_0: \mu = 0.6$ versus $H_a: \mu > 0.6$
    (C) $H_0: p = 0.6$ versus $H_a: p > 0.6$
    (D) $H_0: p > 0.6$ versus $H_a: p = 0.6$
    (E) $H_0: p > 0.6$ versus $H_a: p < 0.6$

2. The principal at a large high school will implement a proposed after-school tutoring program if there is evidence that the proportion of students at the school who would take advantage of such a program is greater than 0.10. She decides to collect data to test $H_0: p = 0.1$ versus $H_a: p > 0.1$. In this setting, which of the following would be a consequence of making a Type I error?

    (A) Implementing the program when the proportion of students who would take advantage of the program is greater than 0.10.
    (B) Not implementing the program when the proportion of students who would take advantage of the program is greater than 0.10.
    (C) Implementing the program when the proportion of students who would take advantage of the program is not greater than 0.10.
    (D) Not implementing the program when the proportion of students who would take advantage of the program is not greater than 0.10.
    (E) Deciding that it is acceptable to implement the program as long as 5% of the students would participate.

3. For which of the following P-values would the null hypothesis be rejected if $\alpha = 0.05$?

    I.   0.002
    II. 0.015
    III. 0.065

    (A) I only
    (B) II only
    (C) III only
    (D) I and II only
    (E) I, II, and III

4. In a test of $H_0: p = 0.65$ versus $H_a: p > 0.65$, the value of the z test statistics was calculated to be $z = 2.15$. Which of the following is the P-value for this test?

    A. 0.0158
    B. 0.0316
    C. 0.05
    D. 0.9684
    E. 0.9841

5. In a random sample of 200 subscribers of a cooking magazine, 30 reported that they had logged on to the magazine website to download a recipe within the last month. These data were used to test the hypothesis

    $$H_0: p = 0.25 \text{ versus } H_a: p < 0.25$$

    where p is the proportion of all subscribers who have downloaded recipes in the last month. The P-value for this test was 0.001. Which of the following is a correct conclusion if a significance level of 0.01 is used?

    (A) There is convincing evidence that the proportion of subscribers who have downloaded a recipe during the last month is 0.25.
    (B) There is convincing evidence that the proportion of subscribers who have downloaded a recipe during the last month is less than 0.25.
    (C) There is not convincing evidence that the proportion of subscribers who have downloaded a recipe during the last month is less than 0.25.
    (D) Because the P-value is so small, it is not possible to reach a conclusion.
    (E) Because the significance level of 0.01 is smaller than the hypothesized value of 0.25, the null hypothesis is not rejected.

6. In 2005, California instituted a campaign to increase enforcement of seat belt laws. Suppose that prior to this campaign, the proportion of traffic citations that were for failure to wear a seat belt was 0.07. A random sample of traffic citations will be selected from those issued in 2005 in order to test $H_0: p = 0.07$ versus $H_a: p < 0.07$, where p is the proportion of 2005 traffic citations that are for failure to wear a seat belt. Of those listed below, what is the <u>smallest</u> sample size for which it would be appropriate to use the large sample z test to test these hypotheses?

    (A) 30
    (B) 100
    (C) 200
    (D) 500
    (E) 1,000

---

7. Which of the following affects the power of a test?

   I.   The sample size
   II.   The significance level of the test
   III. The size of the discrepancy between the actual value and the hypothesized value of the population characteristic

   (A) I only
   (B) II only
   (C) III only
   (D) I and II only
   (E) I, II, and III

8. Assuming the same significance level is used, how does increasing the sample size from 50 to 100 affect the power of a test?

   (A) The power decreases.
   (B) The power increases.
   (C) The power does not change.
   (D) The power does change, but it might either decrease or increase.
   (E) It is not possible to predict whether the power will increase, decrease, or remain the unchanged.

9. Assuming the same sample size is used, how does changing the significance level from 0.05 to 0.01 affect the power of a test?

   (A) The power decreases.
   (B) The power increases.
   (C) The power does not change.
   (D) The power does change, but it might either decrease or increase.
   (E) It is not possible to predict whether the power will increase, decrease, or remain the unchanged.

10. For which sample size and choice of significance level will the power of a test be greatest?

   (A) $n = 10$, $\alpha = 0.1$
   (B) $n = 10$, $\alpha = 0.01$
   (C) $n = 100$, $\alpha = 0.1$
   (D) $n = 100$, $\alpha = 0.05$
   (E) $n = 100$, $\alpha = 0.01$

11

Asking and Answering Questions About the Difference Between Two Population Proportions

Preview

Chapter Learning Objectives

11.1 Estimating the Difference Between Two Population Proportions

11.2 Testing Hypotheses About the Difference Between Two Population Proportions

11.3 Avoid These Common Mistakes

Chapter Activities

Exploring the Big Ideas

Are You Ready to Move On? Chapter 11 Review Exercises

Technology Notes

AP* Review Questions for Chapter 11

Aleksandr Markin/Shutterstock.com

PREVIEW

Many statistical investigations involve comparing two populations. In Chapters 9 and 10, you saw how sample data could be used to estimate a population proportion and to test hypotheses about the value of a single population proportion. In this chapter, you will see how sample data can be used to learn about the difference between two population proportions.

CHAPTER LEARNING OBJECTIVES

Conceptual Understanding
After completing this chapter, you should be able to

C1 Understand how a research question about the difference in two population proportions is translated into hypotheses.

Mastering the Mechanics
After completing this chapter, you should be able to

M1 Know the conditions for appropriate use of the large-sample confidence interval for a difference in population proportions and the large-sample test of hypotheses about a difference in population proportions.

M2 Compute and interpret a confidence interval for a difference in population proportions.

M3 Carry out a large-sample hypothesis test for a difference in two population proportions.

Putting It into Practice
After completing this chapter, you should be able to

P1 Interpret a confidence interval for a difference in two population proportions in context and interpret the associated confidence level.

P2 Carry out a large-sample hypothesis test for a difference in two population proportions and interpret the conclusion in context.

PREVIEW EXAMPLE Cell-Phone Fundraising

After the 2010 earthquake in Haiti, many charitable organizations conducted fundraising campaigns to raise money for emergency relief. Some of these campaigns allowed people to donate money by using their cell phones to send a text message. The amount donated was then added to the next cell phone bill. The report **"Early Signals on Mobile Philanthropy: Is Haiti the Tipping Point?" (Edge Research, 2010)** describes a national survey that investigated the ways in which people made donations to the Haiti relief effort. The report states that the proportion of those in a sample of Generation Y (called Gen Y and defined as those born between 1980 and 1988) who said that they had made a donation to the Haiti relief effort by cell phone was 0.17 (or 17%). This proportion was 0.14 (or 14%) for a sample of Generation X (called Gen X and defined as those born between 1968 and 1979). The proportions reported for Gen Y and Gen X are sample proportions, so you don't expect them to be exactly equal to the corresponding population proportions. But what can you learn from these sample proportions?

Suppose you plan to use the data from these two samples (Gen Y and Gen X) to draw conclusions about how these two populations might differ. Before you do this, there are a few things you would want to know. In particular, you would want to know how the samples were selected and what the sample sizes were. If these samples can be viewed as representative of the respective populations, you can then use the sample data to estimate the difference between the proportion who donated by cell phone for Gen Y and the proportion for Gen X. You might also test hypotheses about this difference. For example, you might want to decide if there is convincing evidence that the proportion who donated by cell phone is higher for the Gen Y population than for the Gen X population. This example will be revisited later in this chapter after you have seen how sample data are used to learn about the difference between two population proportions.

Estimating the Difference Between Two Population Proportions

Many statistical studies are carried out in order to estimate the difference between two population proportions. For example, researchers have studied how people of different ages use mobile technology. As part of a study described in the report **"I Can't Get My Work Done!" (harmon.ie, 2011)**, people in a sample of 258 cell phone users ages 20 to 39 were asked if they use their cell phones to stay connected while they are in bed. The same question was also asked of each person in a sample of 129 cell phone users ages 40 to 49. You might expect the proportion who stay connected while in bed to be higher for the 20 to 39 age group than for the 40 to 49 age group, but how much higher? In this section, you will see how sample data can be used to answer questions like this.

Before you get started, let's introduce a bit of notation. In the previous chapters, the symbol p was used to represent the proportion of "successes" in a single population. The sample proportion was denoted by $\hat{p}$. When comparing two populations, you will have two samples—one from each population. You need to distinguish between the two populations and the two samples, and this is done using subscripts, as shown in the following box.

AP* EXAM TIP

In two sample situations, subscripts are added to distinguish between the two populations. You can either use subscripts 1 and 2, or you can use descriptive subscripts like W for women and M for men. If you use subscripts 1 and 2, be sure to indicate which population corresponds to each subscript.

NOTATION

Proportion of successes in Population 1: p_1

Proportion of successes in Population 2: p_2

	Sample Size	Sample Proportion of Successes
Sample from Population 1	n_1	$\hat{p}_1$
Sample from Population 2	n_2	$\hat{p}_2$

In this chapter, only the case where the two samples are **independent samples** is considered. Two samples are said to be independent samples if the selection of the individuals that make up one sample does not influence the selection of the individuals in the other sample. This would be the case if a random sample is selected from each of the two populations.

When comparing two populations on the basis of "success" proportions, it is common to focus on the quantity $p_1 - p_2$, the difference between the two population proportions. Because $\hat{p}_1$ is an estimate of p_1 and $\hat{p}_2$ is an estimate of p_2, the obvious statistic to choose as an estimate of $p_1 - p_2$ is the difference in the sample proportions, $\hat{p}_1 - \hat{p}_2$. But the values $\hat{p}_1$ and $\hat{p}_2$ each vary from sample to sample, so the difference $\hat{p}_1 - \hat{p}_2$ will also vary. For example, a first sample from each of two populations might result in

$$\hat{p}_1 = 0.64 \qquad \hat{p}_2 = 0.61 \qquad \hat{p}_1 - \hat{p}_2 = 0.03$$

A second sample from each population might result in

$$\hat{p}_1 = 0.62 \qquad \hat{p}_2 = 0.68 \qquad \hat{p}_1 - \hat{p}_2 = -0.06$$

The sampling distribution of $\hat{p}_1 - \hat{p}_2$ describes this sample-to-sample variability.

General Properties of the Sampling Distribution of $\hat{p}_1 - \hat{p}_2$

If $\hat{p}_1 - \hat{p}_2$ is the difference in sample proportions for independently selected random samples, then the following rules hold.

Rule 1. $\qquad \mu_{\hat{p}_1 - \hat{p}_2} = p_1 - p_2$

This rule says that the sampling distribution of $\hat{p}_1 - \hat{p}_2$ is centered at the actual value of the difference in population proportions. This means that the sample differences tend to cluster around the value of the actual population difference.

(continued)

Rule 2. $\quad \sigma^2_{\hat{p}_1 - \hat{p}_2} = \sigma^2_{\hat{p}_1} + \sigma^2_{\hat{p}_2} = \dfrac{p_1(1-p_1)}{n_1} + \dfrac{p_2(1-p_2)}{n_2}$

and

$$\sigma_{\hat{p}_1 - \hat{p}_2} = \sqrt{\dfrac{p_1(1-p_1)}{n_1} + \dfrac{p_2(1-p_2)}{n_2}}$$

This rule specifies the standard error of $\hat{p}_1 - \hat{p}_2$. The value of the standard error describes how much the $\hat{p}_1 - \hat{p}_2$ values tend to vary from the actual value of the population difference.

Rule 3. $\quad$ If both n_1 and n_2 are large, then the sampling distribution of $\hat{p}_1 - \hat{p}_2$ is approximately normal. The sample sizes can be considered large if $n_1 p_1 \geq 10$, $n_1(1 - p_1) \geq 10$, $n_2 p_2 \geq 10$, and $n_2(1 - p_2) \geq 10$ (or equivalently if n_1 and n_2 are large enough to expect at least 10 successes and 10 failures in each of the two samples).

In Chapter 9, the general form of a confidence interval was given as

$$(\text{stastistic}) \pm (\text{critical value}) \left(\begin{smallmatrix} \text{standard error} \\ \text{of the stastistic} \end{smallmatrix} \right)$$

Adapting this general form to the case of estimating $p_1 - p_2$ using data from large independent random samples, you use

Statistic	$\hat{p}_1 - \hat{p}_2$
Critical value	Because the sampling distribution of $\hat{p}_1 - \hat{p}_2$ is approximately normal when the sample sizes are large, a z critical value is used.
Standard error of the statistic $(\hat{p}_1 - \hat{p}_2)$	$\sqrt{\dfrac{p_1(1-p_1)}{n_1} + \dfrac{p_2(1-p_2)}{n_2}}$

Because the two population proportions are not known, the two sample proportions are used to estimate the standard error and to check sample size conditions. This results in the following large-sample confidence interval.

AP* EXAM TIP

When checking conditions in a two-population setting, be sure to state and check conditions for each sample.

A Large-Sample Confidence Interval for a Difference in Population Proportions

Appropriate when the following conditions are met:
1. The samples are independent random samples from the populations of interest (or the samples are independently selected in a way that it is reasonable to regard each sample as representative of the corresponding population).
2. The sample sizes are large. This condition is met when $n_1 \hat{p}_1 \geq 10$, $n_1(1 - \hat{p}_1) \geq 10$, $n_2 \hat{p}_2 \geq 10$, and $n_2(1 - \hat{p}_2) \geq 10$ or (equivalently) if each sample includes at least 10 successes and 10 failures.

When these conditions are met, a confidence interval for the difference in population proportions is

$$\hat{p}_1 - \hat{p}_2 \pm (z \text{ critical value}) \sqrt{\dfrac{\hat{p}_1(1 - \hat{p}_1)}{n_1} + \dfrac{\hat{p}_2(1 - \hat{p}_2)}{n_2}}$$

(continued)

The desired confidence level determines which z critical value is used. The three most common confidence levels use the following critical values:

Confidence Level	z Critical Value
90%	1.645
95%	1.96
99%	2.58

Interpretation of Confidence Interval
You can be confident that the actual value of the difference in population proportions is included in the computed interval. In a given problem, this statement should be worded in context.

Interpretation of Confidence Level
The confidence level specifies the long-run percentage of the time that this method will be successful in capturing the actual difference in population proportions.

Questions

Q Question type: estimation or hypothesis testing?

S Study type: sample data or experiment data?

T Type of data: one variable or two? Categorical or numerical?

N Number of samples or treatments: how many?

The large-sample confidence interval for a difference in population proportions is a method you should consider when the answers to the four key questions (QSTN) are *estimation, sample data, one categorical variable, two samples*.

Now you're ready to look at an example, following the usual five-step process for estimation problems (EMC³).

Example 11.1   Cell Phones in Bed

Let's return to the example introduced at the beginning of this section to answer the question, "How much greater is the proportion who use a cell phone to stay connected in bed for cell phone users ages 20 to 39 than for those ages 40 to 49?" The study described earlier found that 168 of the 258 people in the sample of 20- to 39-year-olds and 61 of the 129 people in the sample of 40- to 49-year-olds said that they sleep with their cell phones. Based on these sample data, what can you learn about the actual difference in proportions for these two populations?

Start by answering the four key questions (QSTN) for this problem.

Q Question Type	Estimation or hypothesis testing?	Estimation
S Study Type	Sample data or experiment data?	Sample data
T Type of Data	One variable or two? Categorical or numerical?	One categorical variable with two categories (sleeps with phone or does not sleep with phone).
N Number of Samples or Treatments	How many samples or treatments?	Two samples—one from each age group.

The answers are *estimation, sample data, one categorical variable, two samples*. This combination of answers indicates that a large-sample confidence interval for a population proportion should be considered.

Now you're ready to use the five-step process to learn from the data.

Steps

E Estimate
M Method
C Check
C Calculate
C Communicate results

Step	
Estimate: Explain what population characteristic you plan to estimate.	In this example, you want to estimate the difference in the proportion of cell phone users ages 20 to 39 who sleep with their phones and this proportion for those ages 40 to 49. This is the value of $p_1 - p_2$, **where p_1 is the proportion for the 20 to 39 age group and p_2 is the proportion for the 40 to 49 age group.**
Method: Select a potential method.	Because the answers to the four key questions are estimation, sample data, one categorical variable, and two samples, a **large-sample confidence interval for a difference in population proportions** will be considered (see Table 7.1). Associated with every confidence interval is a confidence level. When the method selected is a confidence interval, you will also need to specify a confidence level. For this example, a **confidence level of 90%** will be used.
Check: Check to make sure that the method selected is appropriate.	There are **two conditions** that need to be met in order to use the large-sample confidence interval for a difference in population proportions. The large-sample condition is easily verified. With "success" denoting a cell phone user who sleeps with his or her cell phone, the sample sizes are large enough because there are more than 10 successes (168 in sample 1 and 61 in sample 2) and more than 10 failures ($258 - 168 = 90$ in sample 1 and 68 in sample 2) in each sample. The requirement of independent random samples or samples that are representative of the corresponding populations is more difficult. No information was provided regarding how the samples were selected. In order to proceed, you need to assume that the samples were selected in a reasonable way. You should keep this in mind when you get to the step that involves interpretation.
Calculate: Use sample data to perform any necessary calculations.	$n_1 = 258 \qquad n_2 = 129$ $$\hat{p}_1 = \frac{168}{258} = 0.65 \quad \hat{p}_2 = \frac{61}{129} = 0.47$$ For a confidence level of 90%, the appropriate z critical value is 1.645. Next, substitute the values for n_1, n_2, $\hat{p}_1$ and $\hat{p}_2$ into the confidence interval formula: $$\hat{p}_1 - \hat{p}_2 \pm (z \text{ critical value}) \sqrt{\frac{\hat{p}_1(1-\hat{p}_1)}{n_1} + \frac{\hat{p}_2(1-\hat{p}_2)}{n_2}}$$ $$(0.65 - 0.47) \pm (1.645) \sqrt{\frac{(0.65)(0.35)}{258} + \frac{(0.47)(0.53)}{129}}$$ $0.18 \pm (1.645)\sqrt{0.0028}$ $0.18 \pm (1.645)(0.05)$ 0.18 ± 0.09 $(0.09, 0.27)$
Communicate Results: Answer the research question of interest, explain what you have learned from the data, and acknowledge potential risks.	The interpretation of confidence intervals should always be worded in the context of the problem. You should always give both an interpretation of the interval itself and of the confidence level associated with the interval. **Confidence interval:** Assuming that the samples were selected in a reasonable way, you can be 90% confident that the actual difference in the proportion of cell phone users who sleep with their cell phone for 20- to 39-year-olds and for 40- to 49-year-olds is between 0.09 and 0.27. This means that you think that the proportion is higher for 20- to 39-year-olds than for 40- to 49-year-olds by somewhere between 0.09 and 0.27.

(continued)

Step

> **Confidence level:**
> The method used to construct this interval estimate is successful in capturing the actual value of the population proportion about 90% of the time.
>
> In this example, no information was given about how the samples were selected. Because of this, it is important to state that the interpretation of the confidence interval is only valid if the samples were selected in a reasonable way.

Computing a confidence interval for a difference in population proportions can also be done using statistical software or a graphing calculator. For example, Minitab output is shown here. (Minitab has used more decimal accuracy in computing the endpoints of the confidence interval.)

CI for Two Proportions

Sample	X	N	Sample p
1	168	258	0.651163
2	61	129	0.472868

Difference = p (1) − p (2)
Estimate for difference: 0.178295
90% CI for difference: (0.0910599, 0.265529)

Interpreting Confidence Intervals for a Difference

Interpreting confidence intervals for a difference is a bit more complicated than interpreting a confidence interval for a single proportion. The following table shows how the interval can be interpreted in three different cases.

Case	Interpretation	Example
Both endpoints of the confidence interval for $p_1 - p_2$ are positive.	If $p_1 - p_2$ is positive, it means that you think p_1 is greater than p_2 and the interval gives an estimate of how much greater.	(0.24, 0.36) You think that p_1 is greater than p_2 by somewhere between 0.24 and 0.36.
Both endpoints of the confidence interval for $p_1 - p_2$ are negative.	If $p_1 - p_2$ is negative, it means that you think p_1 is less than p_2 and the interval gives an estimate of how much less.	(−0.14, −0.06) You think that p_1 is less than p_2 by somewhere between 0.06 and 0.14. (Or equivalently, you think that p_2 is greater than p_1 by somewhere between 0.06 and 0.14.)
0 is included in the confidence interval.	If the confidence interval includes 0, a plausible value for $p_1 - p_2$ is 0. This means that p_1 and p_2 might be equal.	(−0.04, 0.09) Because 0 is included in the confidence interval, it is possible that the two population proportions could be equal.

AP* EXAM TIP

A confidence interval for $p_1 - p_2$ estimates the *difference* in proportions. Make sure this is clear when you interpret the interval.

Example 11.2 Opinions on Freedom of Speech

The article **"Freedom of What?"** (**Associated Press, February 1, 2005**) described a study in which high school students and high school teachers were asked whether they agreed with the following statement: "Students should be allowed to report controversial issues in their student newspapers without the approval of school authorities." It was reported that 58% of the students surveyed and 39% of the teachers surveyed agreed with the statement. The two samples—10,000 high school students and 8,000 high school teachers—were independently selected from 544 different schools across the country.

Assuming that it is reasonable to think these samples are representative of the two populations, you can use the given information to estimate the difference between the proportion of high school students who agree with the statement, p_1, and the proportion of teachers who agree with the statement, p_2.

Answers to the four key questions for this problem are:

Q Question Type	Estimation or hypothesis testing?	Estimation
S Study Type	Sample data or experiment data?	Sample data
T Type of Data	One variable or two? Categorical or numerical?	One categorical variable. The data are responses to the question "Do you agree with...?" so there is one variable. The variable has two possible values—yes and no.
N Number of Samples or Treatments	How many samples or treatments?	Two samples—one from the population of students and one from the population of teachers.

So, this is *estimation, sample data, one categorical variable, and two samples*. This combination of answers leads you to consider a large-sample confidence interval for a difference in population proportions.

You can use the five-step process for estimation problems (EMC³) to construct a 95% confidence interval.

E Estimate You want to estimate $p_1 - p_2$, the difference in the proportion who agree with the statement for students and teachers. Here, p_1 denotes this proportion for students and p_2 denotes this proportion for teachers.

M Method Because the answers to the four key questions are estimation, sample data, one categorical variable, and two samples, a large-sample confidence interval for a difference in population proportions should be considered. A confidence level of 95% was specified for this example.

C Check There are two conditions that need to be met in order for the large-sample confidence interval for a difference in population proportions to be appropriate. The large-sample condition is easily verified. If you define a success as someone who agreed with the statement, there are more than 10 successes in each sample and more than 10 failures in each sample (the sample sizes here are very large!).

The example specified that the samples were independently selected and that the samples could be considered as representative of the populations.

C Calculate For a confidence level of 95%, the appropriate z critical value is 1.96.

$$\hat{p}_1 = 0.58 \quad \hat{p}_2 = 0.39$$

$$\hat{p}_1 - \hat{p}_2 \pm (z \text{ critical value}) \sqrt{\frac{\hat{p}_1(1 - \hat{p}_1)}{n_1} + \frac{\hat{p}_2(1 - \hat{p}_2)}{n_2}}$$

$$(0.58 - 0.39) \pm (1.96) \sqrt{\frac{(0.58)(0.42)}{10{,}000} + \frac{(0.39)(0.61)}{8{,}000}}$$

$$0.19 \pm (1.96)(0.0074)$$

$$0.19 \pm 0.014$$

$$(0.176, 0.204)$$

Statistical software or a graphing calculator could also have been used to compute the endpoints of the confidence interval. Minitab output is shown here.

CI for Two Proportions

Sample	X	N	Sample p
1	5800	10000	0.580000
2	3120	8000	0.390000

Difference = p (1) − p (2)
Estimate for difference: 0.19
95% CI for difference: (0.175584, 0.204416)

C Communicate Results

Interpret Confidence interval

Assuming that the samples were selected in a reasonable way, you can be 95% confident that the actual difference in the proportions who agree with the statement is between 0.176 and 0.204. This means that you believe that the proportion of high school students who agree that students should be allowed to report controversial issues in their student newspapers without the approval of school authorities is greater than this proportion for teachers by somewhere between 0.176 and 0.204.

Interpret Confidence level

The method used to construct this interval estimate is successful in capturing the actual difference in population proportions about 95% of the time.

SECTION 11.1 EXERCISES

Each Exercise Set assesses the following chapter learning objectives: M1, M2, P1

SECTION 11.1 Exercise Set 1

11.1 *USA Today* **(February 16, 2012)** reported that the percentage of U.S. residents living in poverty was 12.5% for men and 15.1% for women. These percentages were estimates based on data from large representative samples of men and women. Suppose that the sample sizes were 1,200 for men and 1,000 for women. You would like to use the survey data to estimate the difference in the proportion living in poverty for men and women.

a. Answer the four key questions (QSTN) for this problem. What method would you consider based on the answers to these questions?

b. Use the five-step process for estimation problems (EMC³) to compute and interpret a 90% large-sample confidence interval for the difference in the proportion living in poverty for men and women.

11.2 The article **"Portable MP3 Player Ownership Reaches New High"** (*Ipsos Insight,* **June 29, 2006)** reported that in 2006, 20% of those in a random sample of 1,112 Americans ages 12 and older indicated that they owned an MP3 player. In a similar survey conducted in 2005, only 15% reported owning an MP3 player. Suppose that the 2005 figure was also based on a random sample of size 1,112.

a. Are the sample sizes large enough to use the large-sample confidence interval for a difference in population proportions?

b. Estimate the difference in the proportion of Americans ages 12 and older who owned an MP3 player in 2006 and the corresponding proportion for 2005 using a 95% confidence interval.

c. Is zero included in the confidence interval? What does this suggest about the change in this proportion from 2005 to 2006?

d. Interpret the confidence interval in the context of this problem.

11.3 "Mountain Biking May Reduce Fertility in Men, Study Says" was the headline of an article appearing in the *San Luis Obispo Tribune* **(December 3, 2002).** This conclusion was based on an Austrian study that compared sperm counts of avid mountain bikers (those who ride at least 12 hours per week) and nonbikers. Ninety percent of the avid mountain bikers studied had low sperm count, compared to 26% of the nonbikers. Suppose that these percentages were based on independent samples of 100 avid mountain bikers and 100 nonbikers and that these samples are representative of avid mountain bikers and nonbikers.

a. Using a confidence level of 95%, estimate the difference between the proportion of avid mountain bikers with low sperm count and the proportion for nonbikers.

b. Is it reasonable to conclude that mountain biking 12 hours per week or more causes low sperm count? Explain.

11.4   The following graphical display appeared in *USA Today* (February 16, 2012).

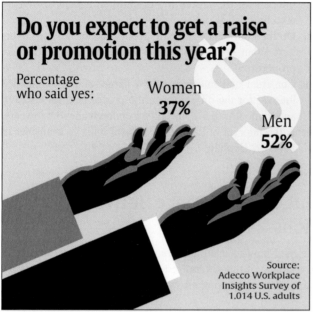

USA TODAY Snapshots®

Do you expect to get a raise or promotion this year?

Percentage
who said yes:

Women
37%

Men
52%

Source:
Adecco Workplace
Insights Survey of
1,014 U.S. adults

USA Today. February 16, 2012. Reprinted with Permission.

The display summarizes data from surveys of male and female American adults. Assume that the two samples were independently selected representative samples and that the sample size for each sample was 507. Use the given information and the five-step process for estimation problems (EMC[3]) to calculate and interpret a 95% large-sample confidence interval for the difference in the proportion of men who expect to get a raise or promotion this year and this proportion for women.

11.5   "Smartest People Often Dumbest About Sunburns" is the headline of an article that appeared in the *San Luis Obispo Tribune* (July 19, 2006). The article states that "those with a college degree reported a higher incidence of sunburn than those without a high school degree—43% versus 25%." Suppose that these percentages were based on independent random samples of size 200 from each of the two groups of interest (college graduates and those without a high school degree).

a. Are the sample sizes large enough to use the large-sample confidence interval for a difference in population proportions?

b. Estimate the difference in the proportion of people with a college degree who reported sunburn and the corresponding proportion for those without a high school degree using a 90% confidence interval.

c. Is zero included in the confidence interval? What does this suggest about the difference in the two population proportions?

d. Interpret the confidence interval in the context of this problem.

11.6   Public Agenda conducted a survey of 1,379 parents and 1,342 students in grades 6–12 regarding the importance of science and mathematics in the school curriculum (Associated Press, February 15, 2006). It was reported that 50% of students thought that understanding science and having strong math skills are essential for them to succeed in life after school, whereas 62% of the parents thought these were essential. The two samples—parents and students—were independently selected random samples. Using a confidence level of 95%, estimate the difference between the proportion of students who think that understanding science and having math skills are essential and this proportion for parents.

Additional Exercises

11.7   To learn about the proportion of college students who are registered to vote, surveys were conducted in 2010 and in 2011. Suppose that in 2010, 78% of the people in a representative sample of 2,300 college students were registered to vote. In 2011, 70% of the people in a representative sample of 2,294 college students were registered to vote. Use the five-step process for estimation problems (EMC[3]) to construct and interpret a 99% large-sample confidence interval for the difference in the proportion of college students who were registered to vote in 2010 and this proportion in 2011.

11.8   The article **"College Graduates Break Even by Age 33"** (*USA Today*, September 21, 2010) reported that 2.6% of college graduates were unemployed in 2008 and 4.6% of college graduates were unemployed in 2009. Suppose that the reported percentages were based on independently selected representative samples of 500 college graduates in each of these two years. Construct and interpret a 95% confidence interval for the difference in the proportions of college graduates who were unemployed in these 2 years.

11.9   The article referenced in the previous exercise also reported that 5.7% of high school graduates were unemployed in 2008 and 9.7% of high school graduates were unemployed in 2009. Suppose that the reported percentages were based on independently selected representative samples of 400 high school graduates in each of these 2 years.

a. Construct and interpret a 99% large-sample confidence interval for the difference in the proportions of high school graduates who were unemployed in these 2 years.

b. Is the confidence interval from Part (a) wider or narrower than the confidence interval computed in the previous exercise? Give two reasons why it is wider or narrower.

11.10 A study by the Kaiser Family Foundation provided one of the first looks at media use among the very youngest children—those from 6 months to 6 years of age **(Kaiser Family Foundation, 2003, www.kff.org)**. The director of the Kaiser Family Foundation's Program for the Study of Entertainment Media and Health said, "It's not just teenagers who are wired up and tuned in, its babies in diapers as well." Because previous research indicated that children who have a TV in their bedroom spend less time reading than other children, the authors of the Foundation study were interested in learning about the proportion of kids who have a TV in their bedroom. They collected data from two independent random samples of parents. One sample consisted of parents of children ages 6 months to 3 years old. The second sample consisted of parents of children ages 3 to 6 years old. They found that the proportion of children who had a TV in their bedroom was 0.30 for the sample of children ages 6 months to

3 years and 0.43 for the sample of children ages 3 to 6 years old. Suppose that each of the two sample sizes was 100.

a. Construct and interpret a 95% large-sample confidence interval for the proportion of children ages 6 months to 3 years who have a TV in their bedroom. *Hint:* This is a one-sample confidence interval.

b. Construct and interpret a 95% large-sample confidence interval for the proportion of children ages 3 to 6 years who have a TV in their bedroom.

c. Do the confidence intervals from Parts (a) and (b) overlap? What does this suggest about the two population proportions?

d. Construct and interpret a 95% large-sample confidence interval for the difference in the proportions for children ages 6 months to 3 years and for children ages 3 to 6 years.

e. Is the interval in Part (d) consistent with your answer in Part (c)? Explain.

SECTION 11.2 Testing Hypotheses About the Difference Between Two Population Proportions

In the previous section, you learned how to use data from two independent samples to construct a confidence interval estimate of the difference between two population proportions. Sample data can also be used to test hypotheses about a difference in population proportions.

Example 11.3 Electric Cars

The article **"Americans Say No to Electric Cars Despite Gas Prices"** (*USA Today*, May 25, 2011) describes a survey of public opinion on issues related to rising gas prices. The survey was conducted by Gallup, a national polling organization. Each person in a representative sample of low-income adult Americans (annual income less than $30,000) and each person in an independently selected representative sample of high-income adult Americans (annual income greater than $75,000) was asked whether he or she would consider buying an electric car if gas prices continue to rise. In the low-income sample, 65% said that they would not buy an electric car no matter how high gas prices were to rise. In the high-income sample, 59% responded this way. The article did not give the sample sizes, but for the purposes of this example, suppose the sample sizes were both 300.

One question of interest is whether the proportion who would never consider buying an electric car is different for the two income groups. The following table summarizes what you know so far.

Population	Population Proportion	Sample Size	Sample Proportion
Low-income adult Americans	p_1 = proportion of *all* low-income adult Americans who would not buy an electric car	$n_1 = 300$	$\hat{p}_1 = 0.65$
High-income adult Americans	p_2 = proportion of *all* high-income adult Americans who would not buy an electric car	$n_2 = 300$	$\hat{p}_2 = 0.59$

Notice that the two sample proportions are not equal. But even if the two population proportions were equal, the two sample proportions wouldn't necessarily be equal because of sampling variability—differences that occur from one sample to another just by chance. The important question is whether chance is a believable explanation for the observed difference in the two sample proportions or whether this difference is large enough that it is unlikely to have occurred just by chance. A hypothesis test will help you to make this determination.

This example will be revisited a bit later in this section, after you see how the hypothesis testing procedure introduced in Chapter 10 can be adapted to answer questions about a difference in two population proportions.

The logic of hypothesis testing and a five-step process for carrying out a hypothesis test (HMC^3) were developed in Chapter 10. The same general process is used for all hypothesis tests. What may differ from one type of test to another is

1.   The form of the null hypothesis and the form of the alternative hypothesis
2.   The test statistic
3.   The way in which the associated P-value is determined

Let's consider each of these three things in the context of testing a hypothesis about a difference in population proportions.

1. Hypotheses

The hypothesis test focuses on the difference in population proportions, $p_1 - p_2$. When there is no difference between the two population proportions, $p_1 - p_2 = 0$. So

$$p_1 = p_2 \quad \text{is equivalent to} \quad p_1 - p_2 = 0$$

Similarly

$$p_1 > p_2 \quad \text{is equivalent to} \quad p_1 - p_2 > 0$$

and

$$p_1 < p_2 \quad \text{is equivalent to} \quad p_1 - p_2 < 0$$

To determine the null and alternative hypotheses, translate the question of interest into hypotheses. For example, if you want to determine whether there is evidence that p_1 is greater than p_2, you would choose $p_1 - p_2 > 0$ as the alternative hypothesis.

The form of the null hypothesis for the test is

$$H_0: p_1 - p_2 = 0$$

which states that there is no difference between the two population proportions. (While it is possible to test a null hypothesis that specifies a difference other than 0, a different test statistic would be used. Because $p_1 - p_2 = 0$ is almost always the relevant one in applied settings, only this case is considered here.)

AP* EXAM TIP

Remember that the form of the alternative hypothesis depends on the question you are trying to answer.

Null Hypothesis: $H_0: p_1 - p_2 = 0$	
If the question of interest is...	Then the alternative hypothesis is...
Is p_1 different from p_2?	$H_a: p_1 - p_2 \neq 0$
Is p_1 greater than p_2?	$H_a: p_1 - p_2 > 0$
Is p_1 less than p_2?	$H_a: p_1 - p_2 < 0$

2. Test Statistic

Recall from the previous section that when you have independent random samples, three things are known about the sampling distribution of the statistic $\hat{p}_1 - \hat{p}_2$:

1. $\mu_{\hat{p}_1 - \hat{p}_2} = p_1 - p_2$

2. $\sigma_{\hat{p}_1 - \hat{p}_2} = \sqrt{\dfrac{p_1(1 - p_1)}{n_1} + \dfrac{p_2(1 - p_2)}{n_2}}$

3. If both n_1 and n_2 are large, then the sampling distribution of $\hat{p}_1 - \hat{p}_2$ is approximately normal.

This means that if both sample sizes are large and the two random samples are independently selected, the statistic

$$z = \frac{\hat{p}_1 - \hat{p}_2 - (p_1 - p_2)}{\sqrt{\dfrac{p_1(1 - p_1)}{n_1} + \dfrac{p_2(1 - p_2)}{n_2}}}$$

will have a distribution that is approximately the standard normal (z) distribution. Because $\hat{p}_1 - \hat{p}_2$ has a distribution that is approximately normal, standardizing $\hat{p}_1 - \hat{p}_2$ by subtracting its mean and then dividing by its standard deviation results in a standard normal distribution.

Recall that in a hypothesis test, you initially assume that the null hypothesis is true and then compute the value of the test statistic. This is done to see if what was observed in the sample data is consistent with what you expect to see when the null hypothesis is true or whether the sample data are surprising. When the null hypothesis is $H_0: p_1 - p_2 = 0$, the z statistic above simplifies to

$$z = \frac{\hat{p}_1 - \hat{p}_2}{\sqrt{\dfrac{p_1(1 - p_1)}{n_1} + \dfrac{p_2(1 - p_2)}{n_2}}}$$

You can use sample data to compute the numerator of this statistic, but what do you do about the denominator? You can't actually compute the denominator because you don't know the values of p_1 and p_2. The null hypothesis says that p_1 and p_2 are equal, but it doesn't specify what the common value is. The way this is resolved is by using the sample data to estimate this common value.

When $p_1 = p_2$, both $\hat{p}_1$ and $\hat{p}_2$ are estimates of the common value of the population proportion. However, a better estimate than either $\hat{p}_1$ or $\hat{p}_2$ alone is a weighted average of the two. If the samples are not the same size, more weight is given to the value of the sample proportion based on the larger sample size.

AP* EXAM TIP

p_C is used when testing $H_0: p_1 - p_2 = 0$, which states that the two populations have the same value for the proportion of successes. p_C is not used when calculating a confidence interval estimate of $p_1 - p_2$.

If the null hypothesis $H_0: p_1 - p_2 = 0$ is true, a combined estimate of the common population proportion is

$$\hat{p}_c = \frac{n_1 \hat{p}_1 + n_2 \hat{p}_2}{n_1 + n_2} = \frac{\text{total number of successes in the two samples}}{\text{total of the two sample sizes}}$$

for combined

When the null hypothesis is true, using $\hat{p}_c$ as an estimate of the common value of the population proportion results in the following test statistic:

$$z = \frac{\hat{p}_1 - \hat{p}_2}{\sqrt{\dfrac{\hat{p}_c(1 - \hat{p}_c)}{n_1} + \dfrac{\hat{p}_c(1 - \hat{p}_c)}{n_2}}}$$

This is the test statistic you will use to test a hypothesis about a difference in population proportions.

3. Computing a *P*-value

Once the value of the test statistic has been determined, the next step is to compute a *P*-value, which tells you how likely it would be to observe sample data as or more extreme than what you actually observed, if the null hypothesis were true. Because this test statistic has a distribution that is approximately standard normal when the null hypothesis is true, *P*-values are computed as an area under the standard normal curve in exactly the same way as for the one-sample test of Chapter 10.

Questions
Q Question type: estimation or hypothesis testing?
S Study type: sample data or experiment data?
T Type of data: one variable or two? Categorical or numerical?
N Number of samples or treatments: how many?

You now have all the pieces needed to carry out the hypothesis test. The large-sample hypothesis test for a difference in two population proportions summarized in the following box is the method you should consider when the answers to the four key questions (QSTN) that lead to a recommended method are *hypothesis testing, sample data, one categorical variable, two samples.*

A Large-Sample Test for a Difference in Two Population Proportions

Appropriate when the following conditions are met:

1. The samples are independent random samples from the populations of interest (or the samples are independently selected in a way that results in each sample being representative of the corresponding population).
2. The sample sizes are large. This condition is met when $n_1 \hat{p}_1 \geq 10$, $n_1(1 - \hat{p}_1) \geq 10$, $n_2 \hat{p}_2 \geq 10$, and $n_2(1 - \hat{p}_2) \geq 10$ or (equivalently) if each sample includes at least 10 successes and 10 failures.

When these conditions are met, the following test statistic can be used to test the null hypothesis $H_0: p_1 - p_2 = 0$:

$$z = \frac{\hat{p}_1 - \hat{p}_2}{\sqrt{\dfrac{\hat{p}_c(1 - \hat{p}_c)}{n_1} + \dfrac{\hat{p}_c(1 - \hat{p}_c)}{n_2}}}$$

where p_c is the combined estimate of the common proportion

$$\hat{p}_c = \frac{n_1 \hat{p}_1 + n_2 \hat{p}_2}{n_1 + n_2} = \frac{\text{total number of successes in the two samples}}{\text{total of the two sample sizes}}$$

Associated *P*-value

When the alternative hypothesis is…	The *P*-value is…
$H_a: p_1 - p_2 > 0$	Area under the z curve to the right of the calculated value of the test statistic
$H_a: p_1 - p_2 < 0$	Area under the z curve to the left of the calculated value of the test statistic
$H_a: p_1 - p_2 \neq 0$	2·(area to the right of z) if z is positive or 2·(area to the left of z) if z is negative

Steps
H Hypotheses
M Method
C Check
C Calculate
C Communicate results

Now you're ready to look at an example, following the usual five-step process for hypothesis testing problems (HMC³).

Example 11.4   Electric Cars Revisited

Recall the electric car example introduced in Example 11.3. One question posed in that example was whether the proportion who would never consider buying an electric car is different for low-income adult Americans than for high-income adult Americans. From that example, we know

Population	Population Proportion	Sample Size	Sample Proportion
Low-income adult Americans	p_1 = proportion of *all* low-income adult Americans who would not buy an electric car	$n_1 = 300$	$\hat{p}_1 = 0.65$
High-income adult Americans	p_2 = proportion of *all* high-income adult Americans who would not buy an electric car	$n_2 = 300$	$\hat{p}_2 = 0.59$

As usual, begin by answering the four key questions (QSTN) in order to identify a potential method: (Q) This is a hypothesis testing problem. (S) The data are from sampling. (T) There is one categorical variable, which is the response to the question about buying an electric car. (N) The number of samples is two. Because the answers to the four questions are *hypothesis testing, sample data, one categorical variable, and two samples*, you should consider a large-sample hypothesis test for a difference in population proportions.

Now you can use the five-step process for hypothesis testing problems (HMC³) to answer the question posed.

Process Step	
H Hypotheses	The claim is about the difference in proportions for the two income groups. You can use the following notation: p_1 = proportion of low-income adult Americans who would never consider buying an electric car p_2 = proportion of high-income adult Americans who would never consider buying an electric car The question of interest (are the population proportions different) translates into an alternative hypothesis of $p_1 - p_2 \neq 0$. The null hypothesis is that there is no difference in the population proportions. **Hypotheses:** *Null hypothesis*: H_0: $p_1 - p_2 = 0$ *Alternative hypothesis*: H_a: $p_1 - p_2 \neq 0$
M Method	Because the answers to the four key questions are hypothesis testing, sample data, one categorical variable, and two samples, consider a large-sample hypothesis test for a difference in population proportions. **Potential method:** Large-sample test for a difference in population proportions. The test statistic for this test is $$z = \frac{\hat{p}_1 - \hat{p}_2}{\sqrt{\dfrac{\hat{p}_c(1 - \hat{p}_c)}{n_1} + \dfrac{\hat{p}_c(1 - \hat{p}_c)}{n_2}}}$$ You also need to select a significance level for the test. Because no significance level was specified, you should choose a significance level based on a consideration of the consequences of a Type I error (rejecting a true null hypothesis) and a Type II error (failing to reject a false null hypothesis). In this example, a Type I error would be deciding that the proportions were not the same for the two income groups, when in fact the actual proportions were equal. A Type II error would be not thinking that there is a difference in the population proportions, when in fact the two proportions were not the same. In this situation, because neither type of error is much more serious than the other, you might choose a value of 0.05 for α. **Significance level:** $\alpha = 0.05$
C Check	There are two conditions that need to be met in order for the large-sample test for a difference in population proportions to be appropriate. The large samples condition is easily verified. The sample sizes are large enough because $n_1\hat{p}_1 = 300(0.65) = 195$ $n_1(1 - \hat{p}_1) = 300(0.35) = 105$ $n_2\hat{p}_2 = 300(0.59) = 177$ $n_2(1 - \hat{p}_2) = 300(0.41) = 183$ are all greater than or equal to 10. From the study description (see Example 11.3), you know that the samples were independently selected. You also know that Gallup believed the samples were selected in a way that would result in representative samples of adult Americans in the two income groups.
C Calculate	To calculate the value of the test statistic, you need to find the values of the sample proportions and the value of $\hat{p}_c$, the combined estimate of the common population proportion. $n_1 = 300$ $\hat{p}_1 = 0.65$ $n_2 = 300$ $\hat{p}_2 = 0.59$ $\hat{p}_c = \dfrac{n_1\hat{p}_1 + n_2\hat{p}_2}{n_1 + n_2} = \dfrac{300(0.65) + 300(0.59)}{600} = 0.62$

(continued)

Test statistic:

$$z = \frac{\hat{p}_1 - \hat{p}_2}{\sqrt{\dfrac{\hat{p}_c(1 - \hat{p}_c)}{n_1} + \dfrac{\hat{p}_c(1 - \hat{p}_c)}{n_2}}}$$

$$= \frac{0.65 - 0.59}{\sqrt{\dfrac{(0.62)(0.38)}{300} + \dfrac{(0.62)(0.38)}{300}}}$$

$$= \frac{0.06}{0.04} = 1.50$$

This is a two-tailed test (the inequality in H_a is $\neq$), so the P-value is twice the area under the z curve and to the right of the computed z value.

Associated P-value:

P-value $= 2$(area under z curve to the right of 1.50)

$= 2 \cdot P(z > 1.50)$

$= 2(0.0668)$

$= 0.1336$

C Communicate Results

Because the P-value is greater than the selected significance level, you fail to reject the null hypothesis.

Decision: $0.1336 > 0.05$, Fail to reject H_0.

The final conclusion for the test should be stated in context and answer the question posed.

Conclusion: Based on the sample data, you are not convinced that there is a difference in the proportions who would never consider buying an electric car for low-income and high-income adult Americans.

Based on this hypothesis test, you can conclude that even though the two sample proportions were different (0.65 and 0.59), this difference could have occurred just by chance and not as a result of any difference in the population proportions. So, based on the sample data, you are *not convinced* that there is a difference in the two population proportions.

It is also possible to use statistical software or a graphing calculator to carry out the calculate step in a hypothesis test. For example, Minitab output for the test of Example 11.4 is shown here.

Test and CI for Two Proportions

Sample	X	N	Sample p
1	195	300	0.650000
2	177	300	0.590000

Difference = p (1) − p (2)
Estimate for difference: 0.06
95% CI for difference: (−0.0175281, 0.137528)
Test for difference = 0 (vs not = 0): Z = 1.51 P-Value = 0.130

From the Minitab output, you see that $z = 1.51$ and the associated P-value is 0.130. These values are slightly different from those in Example 11.4 only because Minitab uses greater decimal accuracy in the calculations leading to the value of the test statistic.

Example 11.5 Cell Phone Fundraising Part 2

The Preview Example for this chapter described a study that looked at ways people donated to the 2010 Haiti earthquake relief effort. Two independently selected random samples—one of Gen Y cell phone users and one of Gen X cell phone users—resulted in the following information:

Gen Y (those born between 1980 and 1988): 17% had made a donation via cell phone
Gen X (those born between 1968 and 1979): 14% had made a donation via cell phone

The question posed in the preview example was

Is there convincing evidence that the proportion who donated via cell phone is higher for the Gen Y population than for the Gen X population?

The report referenced in the preview example does not say how large the sample sizes were, but the description of the survey methodology indicates that the samples can be regarded as independent random samples. For purposes of this example, let's suppose that both sample sizes were 1,200.

Now you can use the given information to answer the questions posed. Considering the four key questions (QSTN), this situation can be described as *hypothesis testing, sample data, one categorical variable (did or did not donate by cell phone), and two samples*. This combination suggests a large-sample hypothesis test for a difference in population proportions.

H Hypotheses

Using p_1 and p_2 to denote the two population proportions, define

p_1 = proportion of all Gen Y cell phone users who made a donation by cell phone
p_2 = proportion of all Gen X cell phone users who made a donation by cell phone

Because you want to determine whether there is evidence that the proportion for Gen Y (p_1) is greater than the proportion for Gen X (p_2), the alternative hypothesis will be $p_1 - p_2 > 0$.

Hypotheses:

$$H_0: p_1 - p_2 = 0$$
$$H_a: p_1 - p_2 > 0$$

M Method

Because the answers to the four key questions are hypothesis testing, sample data, one categorical variable, and two samples, a large-sample hypothesis test for a difference in population proportions will be considered. The test statistic for this test is

$$z = \frac{\hat{p}_1 - \hat{p}_2}{\sqrt{\dfrac{\hat{p}_c(1 - \hat{p}_c)}{n_1} + \dfrac{\hat{p}_c(1 - \hat{p}_c)}{n_2}}}$$

Next, choose a significance level for the test. For purposes of this example, $\alpha = 0.05$ will be used.

C Check

Checking to see if this method is appropriate, first verify that the sample sizes are large enough. Since

$$n_1 \hat{p}_1 = 1{,}200(0.17) = 204 \quad n_1(1 - \hat{p}_1) = 1{,}200(0.83) = 996$$
$$n_2 \hat{p}_2 = 1{,}200(0.14) = 168 \quad n_2(1 - \hat{p}_2) = 1{,}200(0.86) = 1{,}032$$

are all greater than 10, the sample sizes are large enough. The problem description indicates that the samples are independent random samples. Because the sample sizes are large enough and an appropriate sampling method was used, it is reasonable to proceed with the hypothesis test.

C Calculate

Using Minitab to do the computations results in the following output:

Test and CI for Two Proportions

Sample	X	N	Sample p
1	204	1200	0.170000
2	168	1200	0.140000

Difference = p (1) − p (2)
Estimate for difference: 0.03
95% CI for difference: (0.00106701, 0.0589330)
Test for difference = 0 (vs > 0): Z = 2.03 P-Value = 0.021

From the Minitab output, the value of the test statistic is

$$z = 2.03$$

and the associated P-value is

$$P\text{-value} = 0.021$$

C Communicate Results Because the P-value is smaller than the selected significance level ($0.021 < 0.05$), the null hypothesis is rejected. The sample data provide convincing evidence that the proportion donating to the Haiti relief effort by cell phone is greater for Gen Y cell phone users than for Gen X cell phone users.

You might have noticed that the null hypothesis was rejected in Example 11.5 where the difference in sample proportions was only 0.03, but the null hypothesis was not rejected in Example 11.4 where the difference in sample proportions was 0.06. At first it might seem odd that a difference of 0.03 would convince you that the two population proportions are not equal, yet you were not convinced by a larger difference in the earlier example. But remember that the sample sizes were quite different. In Example 11.4, the sample sizes were 300, whereas in Example 11.5, the sample sizes were 1,200. With smaller samples, you expect more sample-to-sample variability to occur just by chance, and so you have to see a larger difference in sample proportions before you can be *convinced* that the difference is not just due to chance.

SECTION 11.2 EXERCISES

Each Exercise Set assesses the following chapter learning objectives: C1, M1, M3, P2

SECTION 11.2 Exercise Set 1

11.11 Do people who work long hours have more trouble sleeping? This question was examined in the paper **"Long Working Hours and Sleep Disturbances: The Whitehall II Prospective Cohort Study"** (*Sleep* [2009]: 737–745). The data in the accompanying table are from two independently selected samples of British civil service workers. The authors of the paper believed that these samples were representative of full-time British civil service workers who work 35 to 40 hours per week and of British civil service workers who work more than 40 hours per week.

	n	Number who usually get less than 7 hours of sleep a night
Work over 40 hours per week	1,501	750
Work 35–40 hours per week	958	407

The question of interest is whether the proportion of workers who usually get less than 7 hours of sleep a night is greater for those who work more than 40 hours per week than for those who work between 35 and 40 hours per week.

a. What hypotheses should be tested to answer the question of interest?

b. Are the two samples large enough for the large-sample test for a difference in population proportions to be appropriate?

c. Based on the following Minitab output, what is the value of the test statistic and what is the value of the associated P-value? If a significance level of 0.01 is selected for the test, will you reject or fail to reject the null hypothesis?

Test and CI for Two Proportions

Sample	X	N	Sample p
1	750	1501	0.499667
2	407	958	0.424843

Difference = p (1) − p (2)
Estimate for difference: 0.0748235
95% lower bound for difference: 0.0410491
Test for difference = 0 (vs > 0): Z = 3.63
P-Value = 0.000

d. Interpret the result of the hypothesis test in the context of this problem.

11.12 A hotel chain is interested in evaluating reservation processes. Guests can reserve a room by using either a telephone system or an online system that is accessed through the hotel's web site. Independent random samples of 80 guests who reserved a room by phone and 60 guests who reserved a room online were selected. Of those who reserved by phone, 57 reported that they were satisfied with the reservation process. Of those who reserved online, 50 reported that they were satisfied. Based on these data, is it reasonable to conclude that the proportion who are satisfied is greater for those who reserve a room online? Test the appropriate hypotheses using a significance level of 0.05.

11.13 "Smartest People Often Dumbest About Sunburns" is the headline of an article that appeared in the *San Luis Obispo Tribune* (July 19, 2006). The article states that "those with a college degree reported a higher incidence of sunburn that those without a high school degree—43% versus 25%." Suppose that these percentages were based on random samples of size 200 from each of the two groups of interest (college graduates and those without a high school degree). Is there convincing evidence that the proportion experiencing a sunburn is greater for college graduates than it is for those without a high school degree? Test the appropriate hypotheses using a significance level of 0.01.

Exercise Set 2

11.14 Some commercial airplanes recirculate approximately 50% of the cabin air in order to increase fuel efficiency. The authors of the paper **"Aircraft Cabin Air Recirculation and Symptoms of the Common Cold"** (*Journal of the American Medical Association* [2002]: 483–486) studied 1,100 airline passengers who flew from San Francisco to Denver. Some passengers traveled on airplanes that recirculated air, and others traveled on planes that did not recirculate air. Of the 517 passengers who flew on planes that did not recirculate air, 108 reported post-flight respiratory symptoms, while 110 of the 583 passengers on planes that did recirculate air reported such symptoms. The question of interest is whether the proportions of passengers with post-flight respiratory symptoms differ for planes that do and do not recirculate air. You may assume that it is reasonable to regard these two samples as being independently selected and as representative of the two populations of interest.

a. What hypotheses should be tested to answer the question of interest?

b. Are the two samples large enough for the large-sample test for a difference in population proportions to be appropriate?

c. Based on the following Minitab output, what is the value of the test statistic and what is the value of the associated *P*-value? If a significance level of 0.01 is selected for the test, will you reject or fail to reject the null hypothesis?

Test and CI for Two Proportions

Sample	X	N	Sample p
1	108	517	0.208897
2	110	583	0.188679

Difference = p (1) − p (2)
Estimate for difference: 0.0202182
95% CI for difference: (−0.0270743, 0.0675108)
Test for difference = 0 (vs not = 0): Z = 0.84
P-Value = 0.401

d. Interpret the result of the hypothesis test in the context of this problem.

11.15 Common Sense Media surveyed 1,000 teens and 1,000 parents of teens to learn about how teens are using social networking sites such as Facebook and MySpace (**"Teens Show, Tell Too Much Online,"** *San Francisco Chronicle,* **August 10, 2009**). The two samples were independently selected and were chosen to be representative of American teens and parents of American teens. When asked if they check online social networking sites more than 10 times a day, 220 of the teens surveyed said yes. When parents of teens were asked if their teen checked networking sites more than 10 times a day, 40 said yes. Use a significance level of 0.01 to determine if there is convincing evidence that the proportion of all parents who think their teen checks social networking sites more than 10 times a day is less than the proportion of all teens who check more than 10 times a day.

11.16 The article referenced in the previous exercise also reported that 390 of the teens surveyed said they had posted something on their networking site that they later regretted. Would you use the large-sample test for a difference in population proportions to test the hypothesis that more than one-third of all teens have posted something on a social networking site that they later regretted? Explain why or why not.

Additional Exercises

11.17 The report **"Young People Living on the Edge"** (**Greenberg Quinlan Rosner Research, 2008**) summarizes a survey of people in two independent random samples. One sample consisted of 600 young adults (ages 19 to 35), and the other sample consisted of 300 parents of young adults ages 19 to 35. The young adults were presented with a variety of situations (such as getting married or buying a house) and were asked if they thought that their parents were likely to provide financial support in that situation. The parents of young adults were presented with the same situations and asked if they would be likely to provide financial support in that situation. When asked about getting married, 41% of the young adults said they thought parents would provide financial support and 43% of the parents said they would provide support. Carry out a hypothesis test to determine if there is convincing evidence that the proportion of young adults who think their parents would provide financial support and the proportion of parents who say they would provide support are different.

11.18 The report referenced in the previous exercise also stated that the proportion who thought their parents would help with buying a house or renting an apartment for the sample of young adults was 0.37. For the sample of parents, the proportion who said they would help with buying a house or renting an apartment was 0.27. Based on these data, can you conclude that the proportion of parents who say they would help with buying a house or renting an apartment is significantly less than the proportion of young adults who think that their parents would help?

11.19 In December 2001, the Department of Veterans' Affairs announced that it would begin paying benefits to soldiers suffering from Lou Gehrig's disease who had served in the Gulf War (*The New York Times,* **December 11, 2001**). This decision was based on an analysis in which the Lou Gehrig's disease incidence rate (the proportion developing the disease) for the approximately 700,000 soldiers sent to the Persian Gulf between August 1990 and July 1991 was compared to the incidence rate for the approximately

1.8 million other soldiers who were not in the Gulf during this time period. Based on these data, explain why it is not appropriate to perform a hypothesis test in this situation and yet it is still reasonable to conclude that the incidence rate is higher for Gulf War veterans than for those who did not serve in the Gulf War.

SECTION 11.3 Avoid These Common Mistakes

The three cautions that appeared at the end of Chapter 10 apply here as well (see Chapter 10 for more detail). They were:

1. Remember that the result of a hypothesis test can never show strong support for the null hypothesis. In two-sample situations, this means that you shouldn't be *convinced* that there is *no* difference between two population proportions based on the outcome of a hypothesis test.
2. If you have complete information (a census) of both populations, there is no need to carry out a hypothesis test or to construct a confidence interval—in fact, it would be inappropriate to do so.
3. Don't confuse statistical significance and practical significance. In the two-sample setting, it is possible to be convinced that two population proportions are not equal even in situations where the actual difference between them is small enough that it is of no practical interest. After rejecting a null hypothesis of no difference (statistical significance), it is useful to look at a confidence interval estimate of the difference to get a sense of practical significance.

And here's one new caution to keep in mind when comparing two populations:

4. Correctly interpreting confidence intervals in the two-sample case is more difficult than in the one-sample case, so take particular care when providing a two-sample confidence interval interpretation. Because the two-sample confidence interval estimates a difference $(p_1 - p_2)$, the most important thing to note is whether or not the interval includes 0. If both endpoints of the interval are positive, then it is correct to say that, based on the interval, you believe that p_1 is greater than p_2, with the interval providing an estimate of how much greater. Similarly, if both interval endpoints are negative, you have evidence that p_1 is less than p_2, with the interval providing an estimate of the difference. If 0 is included in the interval, it is plausible that p_1 and p_2 are equal.

CHAPTER ACTIVITIES

ACTIVITY 11.1 DEFECTIVE M&MS

Materials needed: One bag of M&M's for each student (or pair of students) in the class. Half should receive bags of plain M&M's and the other half should receive peanut M&M's.

In this activity, you will use data collected by the class to determine if the proportion of defective M&M's is greater for peanut M&M's than for plain M&M's.

1. Inspect the M&M's in your bag for defects. Some possible defects are broken candies, misshapen candies, candies that do not have a visible "m" on the candy coating, and so on. Sort the M&M's into piles, with one pile for M&M's that are not defective and another pile for those that are defective. Record the type of M&M (plain or peanut) you inspected, the total number of M&M's

inspected, and the total number of defective M&M's found.

2. Now combine the inspection information for the entire class to complete the following table:

Plain M&M's	Peanut M&M's
Total number inspected:	Total number inspected:
Total number defective:	Total number defective:
Proportion defective:	Proportion defective:

3. Use the class data to determine if there is convincing evidence that the population proportion of M&M's that are defective is greater for peanut M&M's than for plain M&M's.

EXPLORING THE BIG IDEAS

In the exercise below, you will go online to select a random sample from a population of adults between the ages of 18 and 45 and a random sample from a population of adults ages 46 to 64.

Suppose that you would like to know if the proportion of people who have sent a text message while driving is greater for the younger population than for the older population.

Go online to www.cengage.com/stats/peck1e and click on the link labeled "Exploring the Big Ideas." Then locate the link for Chapter 11. This link will take you to a web page where you can select a random sample of 50 people from each population.

Click on the Select Samples button. This selects a random sample from each population and will display the following information:

1. The ID number that identifies the person selected
2. The response to the question "Have you ever sent a text message while driving?" These responses are coded numerically—a 1 indicates a yes response and a 2 indicates a no response.

Use these samples to answer the following questions.

a. What type of problem is this?
    1. An estimation problem.
    2. A hypothesis testing problem.

b. Are the data from sampling or from an experiment?
    1. Sample data
    2. Experiment data

c. How many variables are there?
    1. One
    2. Two

d. What type of data do you have?
    1. Categorical
    2. Numerical

e. How many samples are there?
    1. One
    2. Two

f. What are the appropriate null and alternative hypotheses?

g. What method might be appropriate for testing these hypotheses?
    1. Large-sample test for a population proportion
    2. Large-sample test for a difference in population proportions
    3. One-sample t test
    4. Two-sample t test
    5. Paired t test
    6. Chi-square test

h. Are the conditions for the selected test met? Explain.

i. What is the value of the test statistic for this test?

j. What is the P-value?

k. If a significance level of 0.05 were used, what would your decision be?
    1. Reject the null hypothesis
    2. Fail to reject the null hypothesis

l. Write a few sentences that summarize what you learned about the difference in the population proportions based on this hypothesis test.

CHAPTER 11 REVIEW EXERCISES

All chapter learning objectives are assessed in these exercises. The learning objectives assessed in each exercise are given in parentheses.

11.20 **(M2, P1)**

The Insurance Institute for Highway Safety issued a news release titled **"Teen Drivers Often Ignoring Bans on Using Cell Phones" (June 9, 2008)**. The following quote is from the news release:

> Just 1–2 months prior to the ban's Dec. 1, 2006, start, 11% of teen drivers were observed using cell phones as they left school in the afternoon. About 5 months after the ban took effect, 12% of teen drivers were observed using cell phones.

Suppose that the two samples of teen drivers (before the ban, after the ban) are representative of these populations of teen drivers. Suppose also that 200 teen drivers were observed before the ban (so $n_1 = 200$ and $\hat{p}_1 = 0.11$) and that 150 teen drivers were observed after the ban.

a. Construct and interpret a 95% large-sample confidence interval for the difference in the proportion using a cell phone while driving before the ban and the proportion after the ban.

b. Is zero included in the confidence interval of Part (a)? What does this imply about the difference in the population proportions?

11.21 **(M3, P2)**

The news release referenced in the previous exercise also included data from independent samples of teenage drivers and parents of teenage drivers. In response to a question asking if they approved of laws banning the use of cell phones and texting while driving, 74% of the teens surveyed and 95% of the parents surveyed said they approved. The sample sizes were not given in the news release, but suppose that 600 teens and 400 parents of teens were surveyed and that these samples are representative of the two populations. Do the data provide convincing evidence that the proportion of teens who approve of banning cell phone and texting while driving is less than the proportion of parents of teens who approve? Test the relevant hypotheses using a significance level of 0.05.

11.22 **(C1, M1, M3)**

The authors of the paper **"Adolescents and MP3 Players: Too Many Risks, Too Few Precautions"** (*Pediatrics* [2009]: e953–e958) concluded that more boys than girls listen to music at high volumes. This conclusion was based on data from independent random samples of 764 Dutch boys and 748 Dutch girls ages 12 to 19. Of the boys, 397 reported that they almost always listen to music at a high volume setting. Of the girls, 331 reported listening to music at a high volume setting. Do the sample data support the authors' conclusion that the proportion of Dutch boys who listen to music at high volume is greater than this proportion for Dutch girls? Test the relevant hypotheses using a 0.01 significance level.

11.23 **(M3, P2, P3)**

The report **"Audience Insights: Communicating to Teens (Aged 12–17)" (www.cdc.gov, 2009)** described teens' attitudes about traditional media, such as TV, movies, and newspapers. In a representative sample of American teenage girls, 41% said newspapers were boring. In a representative sample of American teenage boys, 44% said newspapers were boring. Sample sizes were not given in the report.

a. Suppose that the percentages reported were based on samples of 58 girls and 41 boys. Is there convincing evidence that the proportion who think that newspapers are boring is different for teenage girls and boys? Carry out a hypothesis test using $\alpha = .05$.

b. Suppose that the percentages reported were based on samples of 2,000 girls and 2,500 boys. Is there convincing evidence that the proportion who think that newspapers are boring is different for teenage girls and boys? Carry out a hypothesis test using $\alpha = 0.05$.

c. Explain why the hypothesis tests in Parts (a) and (b) resulted in different conclusions.

TECHNOLOGY NOTES

Confidence Interval for $p_1 - p_2$

TI-83/84
1. Press the **STAT** key
2. Highlight **TESTS**
3. Highlight **2-PropZInt...** and press **ENTER**
4. Next to **x1** type the number of successes from the first sample

5. Next to **n1** type the sample size from the first sample
6. Next to **x2** type the number of successes from the second sample
7. Next to **n2** type the sample size from the second sample
8. Next to **C-Level** type the appropriate confidence level
9. Highlight **Calculate** and press **ENTER**

TI-Nspire

1. Enter the Calculate Scratchpad
2. Press the **menu** key then select **6:Statistics** then select **6:Confidence Intervals** then **6:2-Prop z Interval...** then press **enter**
3. In the box next to **Successes, x1** type the number of successes from the first sample
4. In the box next to **n1** type the number of trials from the first sample
5. In the box next to **Successes, x2** type the number of successes from the second sample
6. In the box next to **n2** type the number of trials from the second sample
7. In the box next to **C Level** input the appropriate confidence level
8. Press **OK**

JMP

Summarized data

1. Input the data into the JMP data table with categories for one variable in the first column, categories for the second variable in the second column, and counts for each combination in the third column

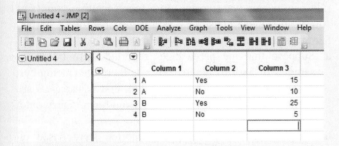

2. Click **Analyze** and select **Fit Y by X**
3. Click and drag the first column containing the response variable from the box under **Select Columns** to the box next to **X, Factor**
4. Click and drag the second column containing the group information from the box under **Select Columns** to the box next to **Y, Response**
5. Click and drag the third column containing the counts for each combination from the box under **Select Columns** to the box next to **Freq**
6. Click **OK**
7. Click the red arrow next to **Contingency Analysis of...** and select **Two Sample Test for Proportions**

Note: You can change the response of interest (i.e., "Yes" instead of "No" or "Success" instead of "Failure") by clicking the radio button at the bottom of the **Two Sample Test for Proportions** section.

Raw data

1. Input the raw data into two separate columns: one containing the response variable and one containing the group information
2. Click **Analyze** and select **Fit Y by X**

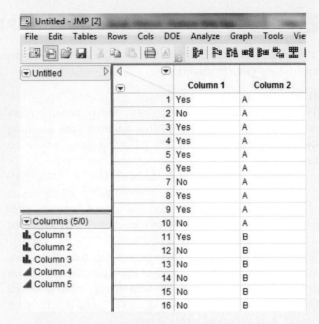

3. Click and drag the first column containing the response variable from the box under **Select Columns** to the box next to **Y, Response**
4. Click and drag the second column containing the group information from the box under **Select Columns** to the box next to **X, Factor**
5. Click **OK**
6. Click the red arrow next to **Contingency Analysis of...** and select **Two Sample Test for Proportions**

Note: You can change the response of interest (i.e., "Yes" instead of "No" or "Success" instead of "Failure") by clicking the radio button at the bottom of the **Two Sample Test for Proportions** section.

Minitab

Summarized data

1. Click **Stat** then click **Basic Statistics** then click **2 Proportions...**
2. Click the radio button next to **Summarized data**
3. In the boxes next to **First:** type the value for n, the total sample size in the box under the **Trials:** column and the number of successes in the box under **Events:**
4. In the boxes next to **Second:** type the value for n, the total sample size in the box under the **Trials:** column and the number of successes in the box under **Events:**
5. Click **Options...**
6. Input the appropriate confidence level in the box next to **Confidence Level**
7. Click **OK**
8. Click **OK**

Raw data

1. Input the raw data two separate columns
2. Click **Stat** then click **Basic Statistics** then click **2 Proportion...**
3. Select the radio button next to **Samples in different columns**
4. Click in the box next to **First:**

5. Double click the column name where the first group's raw data is stored
6. Click in the box next to **Second:**
7. Double click the column name where the second group's raw data is stored
8. Click **Options...**
9. Input the appropriate confidence level in the box next to **Confidence Level**
10. Click **OK**
11. Click **OK**

SPSS

SPSS does not have the functionality to automatically produce a confidence interval for the difference of two proportions.

Excel

Excel does not have the functionality to automatically produce a confidence interval for the difference of two proportions. However, you can type the formulas into two separate cells for the lower and upper limit to have Excel calculate these results for you.

Z-Test for $p_1 - p_2$

TI-83/84

1. Press the **STAT** key
2. Highlight **TESTS**
3. Highlight **2-PropZTest...** and press **ENTER**
4. Next to **x1** type the number of successes from the first sample
5. Next to **n1** type the sample size from the first sample
6. Next to **x2** type the number of successes from the second sample
7. Next to **n2** type the sample size from the second sample
8. Next to **p1,** highlight the appropriate alternative hypothesis
9. Highlight **Calculate** and press **ENTER**

TI-Nspire

1. Enter the Calculate Scratchpad
2. Press the **menu** key then select **6:Statistics** then select **7:Stat Tests** then **6:2-Prop z Test...** then press **enter**
3. In the box next to **Successes, x1** type the number of successes from the first sample
4. In the box next to **n1** type the number of trials from the first sample
5. In the box next to **Successes, x2** type the number of successes from the second sample
6. In the box next to **n2** type the sample size from the second sample
7. In the box next to **Alternate Hyp** choose the appropriate alternative hypothesis from the drop-down menu
8. Press **OK**

JMP

JMP does not have the functionality to automatically provide the results of a z-test for the difference of two proportions.

Minitab
Summarized data

1. Click **Stat** then click **Basic Statistics** then click **2 Proportions...**
2. Click the radio button next to **Summarized data**
3. In the boxes next to **First:** type the value for n, the total sample size in the box under the **Trials:** column and the number of successes in the box under **Events:**
4. In the boxes next to **Second:** type the value for n, the total sample size in the box under the **Trials:** column and the number of successes in the box under **Events:**
5. Click **Options...**
6. Input the appropriate hypothesized value in the box next to **Test difference:** (this is usually 0)
7. Check the box next to **Use pooled estimate of p for test**
8. Click **OK**
9. Click **OK**

Raw data

1. Input the raw data two separate columns
2. Click **Stat** then click **Basic Statistics** then click **2 Proportion...**
3. Select the radio button next to **Samples in different columns**
4. Click in the box next to **First:**
5. Double-click the column name where the first group's raw data is stored
6. Click in the box next to **Second:**
7. Double-click the column name where the second group's raw data are stored
8. Click **Options...**
9. Input the appropriate hypothesized value in the box next to **Test difference:** (this is usually 0)
10. Check the box next to **Use test and interval based on normal distribution**
11. Click **OK**
12. Click **OK**

SPSS

SPSS does not have the functionality to automatically produce a z-test for the difference of two proportions.

Excel

Excel does not have the functionality to automatically produce a z-test for the difference of two proportions. However, you can type the formulas into a cell for the test statistic in order to have Excel calculate this for you. Then use the methods from Chapter 6 to find the *P*-value using the Normal distribution.

AP* Review Questions for Chapter 11

1. For which of the following sample sizes and $\hat{p}$ values would it be appropriate to use

$$\hat{p}_1 - \hat{p}_2 \pm (z \text{ critical value})\sqrt{\frac{\hat{p}_1(1-\hat{p}_1)}{n_1} + \frac{\hat{p}_2(1-\hat{p}_2)}{n_2}}$$

   to estimate a difference in population proportions?

   I. $n_1 = n_2 = 100$, $\hat{p}_1 = 0.05$, $\hat{p}_2 = 0.08$
   II. $n_1 = n_2 = 50$, $\hat{p}_1 = 0.48$, $\hat{p}_2 = 0.36$
   III. $n_1 = n_2 = 500$, $\hat{p}_1 = 0.99$, $\hat{p}_2 = 0.96$

   (A) I only
   (B) II only
   (C) III only
   (D) I and III only
   (E) I, II, and III

2. A 90% confidence interval for $p_1 - p_2$ is $(0.240, 0.300)$. Which of the following is the 95% confidence interval calculated from the same data?

   (A) $(-0.306, 0.234)$
   (B) $(-0.300, -0.240)$
   (C) $(0.214, 0.286)$
   (D) $(0.234, 0.306)$
   (E) $(0.276, 0.264)$

3. A study was carried out at a large high school to determine if there was evidence of a difference between the proportion of seniors who favor banning cell phones at school and this proportion for freshmen. Which of the following hypotheses should be tested to answer this question?

   (A) $H_0: \hat{p}_S - \hat{p}_F = 0$ versus $H_a: \hat{p}_S - \hat{p}_F > 0$
   (B) $H_0: p_S - p_F = 0$ versus $H_a: p_S - p_F > 0$
   (C) $H_0: \hat{p}_S - \hat{p}_F = 0$ versus $H_a: \hat{p}_S - \hat{p}_F < 0$
   (D) $H_0: \hat{p}_S - \hat{p}_F < 0$ versus $H_a: \hat{p}_S - \hat{p}_F \neq 0$
   (E) $H_0: p_S - p_F = 0$ versus $H_a: p_S - p_F \neq 0$

4. Franklin used data from two independent random samples to carry out a hypothesis test to determine if there was evidence that the proportion of females at his school who attended all home football games was less than this proportion for males at the school. After checking to make sure that all of the necessary conditions were met, he calculated the value of the z test statistic to be $z = 1.32$. Which of the following is the P-value associated with this test?

   (A) 0.0934
   (B) 0.1320
   (C) 0.1868
   (D) 0.4533
   (E) 0.9066

5. Each person in a random sample of 200 students and a random sample of 186 faculty at a large university was asked "Do you take public transportation to campus?" Fifty-six percent of the students and 40% of the faculty responded "Yes." This data was used to carry out a test of the hypotheses $H_0: p_{\text{Students}} - p_{\text{Faculty}} = 0$ versus $H_a: p_{\text{Students}} - p_{\text{Faculty}} \neq 0$. The value of the test statistic was $z = 3.14$. For a significance level of 0.01, which of the following is correct?

   (A) H_0 should be rejected; there is convincing evidence that the proportion of people who take public transportation to campus is different for students and faculty.
   (B) H_0 should be rejected; there is convincing evidence that the proportion of people who take public transportation to campus is the same for students and faculty.
   (C) H_0 should not be rejected; there is not convincing evidence that the proportion of people who take public transportation to campus is different for students and faculty.
   (D) H_0 should not be rejected; there is not convincing evidence that the proportion of people who take public transportation to campus is the same for students and faculty.
   (E) There is not enough information provided to be able to decide whether H_0 should be rejected or not.

6. Some scientists believe that environmental factors affect population density. Spiders are very important species in terrestrial communities, and their survival is ecologically important. To better understand the relation between food level and spider survival, an experiment was conducted by supplementing the food supply of immature spiders in an area of low population density. In a control group of immature spiders, 83 out of 112 survived to the end of the experiment. In the food-supplemented experimental group, 86 out of 97 survived. The 90% confidence interval for the difference in population proportions is:

(A) -0.146 ± 0.103
(B) -0.146 ± 0.086
(C) -0.146 ± 0.052
(D) -0.146 ± 0.005
(E) -0.146 ± 0.003

7. For independent random samples, the sampling distribution of $\hat{p}_1 - \hat{p}_1$ is approximately normal for:

(A) $n_1 = n_2 = 50$, $p_1 = .05$, and $p_2 = .10$
(B) $n_1 = n_2 = 20$, $p_1 = .50$, and $p_2 = .40$
(C) $n_1 = n_2 = 100$, $p_1 = .02$, and $p_2 = .10$
(D) $n_1 = n_2 = 100$, $p_1 = .06$, and $p_2 = .10$
(E) $n_1 = n_2 = 100$, $p_1 = .15$, and $p_2 = .10$

8. Each person in a sample of 200 students and a sample of 100 faculty members at a large university was asked "Do you exercise regularly?" The resulting data were used to carry out a test of the hypotheses $H_0: p_{Students} - p_{Faculty} = 0$ vs. $H_a: p_{Students} - p_{Faculty} > 0$. The value of the test statistic was $z = 3.46$. Which of the following could be the 95% confidence interval for $p_{Students} - p_{Faculty}$ computed from this same data?

(A) $(-0.26, -0.06)$
(B) $(-0.26, 0.06)$
(C) $(-0.06, 0.26)$
(D) $(0.06, 0.26)$
(E) $(6, 26)$

12

Asking and Answering Questions About a Population Mean

Preview

Chapter Learning Objectives

12.1 The Sampling Distribution of the Sample Mean

12.2 A Confidence Interval for a Population Mean

12.3 Testing Hypotheses About a Population Mean

12.4 Avoid These Common Mistakes

Chapter Activities

Exploring the Big Ideas

Are You Ready to Move On? Chapter 12 Review Exercises

Technology Notes

AP* Review Questions for Chapter 12

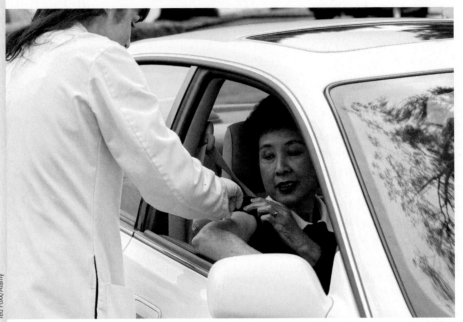

Ted Foxx/Alamy

PREVIEW

One of the key questions used to determine an appropriate data analysis method is whether the data are categorical or numerical. In the previous chapters, the focus has been on how categorical data can be used to learn about the value of a population proportion. Now you will use numerical data from a sample to learn about the value of a population mean, such as the mean number of hours that students enrolled at your school spend studying each week or the mean weight gain of students at the school during their freshman year.

CHAPTER LEARNING OBJECTIVES

Conceptual Understanding

After completing this chapter, you should be able to

C1 Understand how the value of the standard deviation of the sample mean, $\bar{x}$, is related to sample size.

C2 Know what factors affect the width of a confidence interval estimate of a population mean.

Mastering the Mechanics

After completing this chapter, you should be able to

M1 Determine the mean and standard deviation of the sampling distribution of $\bar{x}$.

M2 Know when the sampling distribution of $\bar{x}$ is approximately normal.

M3 Know the conditions for appropriate use of the margin of error and confidence interval formulas when estimating a population mean.

M4 Compute the margin of error when the sample mean, $\bar{x}$, is used to estimate a population mean μ.

M5 Use the five-step process for estimation problems (EMC³) to compute and interpret a confidence interval for a population mean.

M6 Compute the sample size necessary to achieve a desired margin of error when estimating a population mean.

M7 Translate a research question or claim about a population mean into null and alternative hypotheses.

M8 Use the five-step process for hypothesis testing problems (HMC³) to carry out a *t* test of hypotheses about a population mean.

Putting It into Practice

After completing this chapter, you should be able to

P1 Interpret a confidence interval for a population mean in context and interpret the associated confidence level.

P2 Determine the sample size required to achieve a desired margin of error.

P3 Carry out a *t* test of hypotheses about a population mean and interpret the conclusion in context.

PREVIEW EXAMPLE   Drive-Through Medicine

During a flu outbreak, many people visit emergency rooms. Before being treated, they often spend time in crowded waiting rooms where other patients may be exposed. The paper **"Drive-Through Medicine: A Novel Proposal for Rapid Evaluation of Patients During an Influenza Pandemic" (*Annals of Emergency Medicine* [2010]: 268–273)** describes a study of a drive-through model where flu patients are evaluated while they remain in their cars. The study found that not only were patients kept relatively isolated and away from each other, but also that the time to process a patient was shorter because delays related to turning over examination rooms were eliminated.

In the study, 38 people were each given a scenario for a flu case that was selected at random from the set of all flu cases actually seen in the emergency room. The scenarios provided the "patient" with a medical history and a description of symptoms that would allow the patient to respond to questions from the examining physician. These patients were processed using a drive-through procedure that was implemented in the parking structure of Stanford University Hospital. The time to process each case from admission to discharge was recorded.

Because the 38 volunteers were each representing a flu case that was selected at random from actual cases seen in the emergency room, the times were viewed as a random sample of processing times for all such cases. The researchers were interested in using the sample data to learn about the mean time to process a flu case using this new model.

You will return to this example in Section 12.2 to see how the sample data can be used to construct a confidence interval estimate for the population mean.

The Sampling Distribution of the Sample Mean

When the purpose of a statistical study is to learn about a population mean μ, it is natural to consider the sample mean $\bar{x}$ as an estimate of μ. To understand statistical inference procedures based on $\bar{x}$, you must first study how sampling variability causes $\bar{x}$ to vary in value from one sample to another. Just as the behavior of the sample proportion $\hat{p}$ is described by its sampling distribution, the behavior of $\bar{x}$ is also described by a sampling distribution. The sample size n and characteristics of the population (its shape, mean value μ, and standard deviation σ) are important in determining the sampling distribution of $\bar{x}$.

Let's start by reviewing some notation.

AP* EXAM TIP

On the AP* exam, be sure to use proper notation to distinguish between sample statistics (like $\bar{x}$ and s) and population characteristics (such as μ and σ).

NOTATION

n	the sample size
$\bar{x}$	the mean of a sample
s	the standard deviation of a sample
μ	the mean of the entire population
σ	the standard deviation of the entire population

Suppose you are interested in learning about the time it takes students at a particular college to register for classes using a new online system. The population of interest would be all students enrolled at the college, and you would be interested in the numerical variable

$$x = \text{time to register}$$

There is variability in the values of x in the population—registration time will vary from student to student. The mean of all of the registration times in the population is denoted by μ. The population standard deviation σ is a measure of the variability in the population. A large value of σ indicates that there is a lot of student-to-student variability in the registration times. A small value of σ indicates that there is not much variability and that registration times tend to be similar.

The only way to determine the value of μ exactly is to carry out a census of the entire population. Since this isn't usually feasible, you might decide to select a sample of student registration times. These sample data can then be used to learn about the value of μ by using the sample mean, $\bar{x}$, to estimate the value of μ or to test hypotheses about the value of μ.

To learn about the sampling distribution of the statistic $\bar{x}$, we begin by considering some sampling investigations. In the examples that follow, we start with a specified population, fix a sample size n, and select 500 different random samples of this size. The value of $\bar{x}$ is computed for each sample, and a histogram of these 500 $\bar{x}$ values is constructed. Because 500 is a reasonably large number of samples, the histogram of the $\bar{x}$ values should resemble the actual sampling distribution of $\bar{x}$ (which would be obtained by considering *all* possible samples). We repeat this process for several different values of n to see how the choice of sample size affects the sampling distribution and to identify patterns that lead to important properties of this distribution.

Example 12.1 Blood Platelet Volume

The paper **"Mean Platelet Volume in Patients with Metabolic Syndrome and Its Relationship with Coronary Artery Disease" (*Thrombosis Research* [2007]: 245–250)** includes data that suggest that the distribution of

$$x = \text{platelet volume}$$

for patients who do not have metabolic syndrome (a combination of factors that indicates a high risk of heart disease) is approximately normal with mean $\mu = 8.25$ and standard deviation $\sigma = 0.75$.

Figure 12.1 shows a normal curve centered at 8.25, the mean value of platelet volume. The value of the population standard deviation, 0.75, determines the extent to which the x distribution spreads out about its mean value.

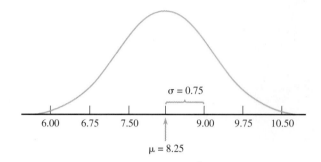

FIGURE 12.1
Normal distribution of x = platelet volume, with $\mu = 8.25$ and $\sigma = 0.75$.

What values for the sample mean would be expected if you were to take a random sample of size 5 from this population distribution? To investigate, we will simulate sampling from this population. A statistical software package was used to select 500 random samples of size $n = 5$ from the population. The sample mean platelet volume $\bar{x}$ was computed for each sample, and these 500 values of $\bar{x}$ were used to construct the density histogram shown in Figure 12.2.

Recall from Chapter 2 that a density histogram is a histogram that uses $\text{denisty} = \dfrac{\text{relative frequency}}{\text{interval width}}$ **to determine the heights of the bars in the histogram.**

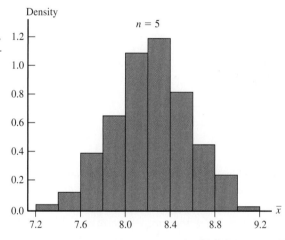

FIGURE 12.2
Histogram of 500 $\bar{x}$ values based on random samples of size $n = 5$ from a population with $\mu = 8.25$ and $\sigma = 0.75$.

The histogram in Figure 12.2 describes the behavior of the sample mean $\bar{x}$ for samples of size $n = 5$ from the platelet volume population. Notice that there is a lot of sample-to-sample variability in the value of $\bar{x}$. For some samples $\bar{x}$ is around 7.3, and for other samples it is around 9.1. A sample of size 5 from the population of patients who do not have metabolic syndrome won't always provide very precise information about the mean platelet volume in the population. What if a larger sample is selected? To investigate the effect of sample size on the behavior of $\bar{x}$, we selected 500 samples of size 10, 500 samples of size 20, and 500 samples of size 30. Density histograms of the resulting $\bar{x}$ values, along with the histogram for the samples of size 5, are displayed in Figure 12.3.

The first thing to notice about the histograms is that each of them is approximately normal in shape. The resemblance would be even more striking if each histogram had been based on many more than 500 $\bar{x}$ values. Second, notice that each histogram is centered at approximately 8.25, the mean of the population being sampled. Had the histograms been based on $\bar{x}$ values from every possible sample, they would have been centered at exactly 8.25.

The final aspect of the histograms to note is their spread relative to one another. The smaller the value of n, the more the sampling distribution spreads out about the population mean value. This is why the histograms for $n = 20$ and $n = 30$ are based on narrower class

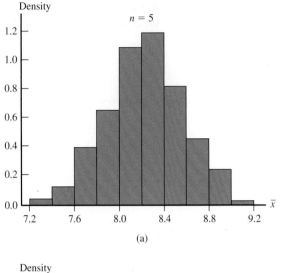

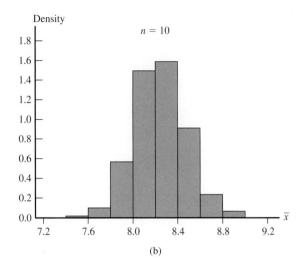

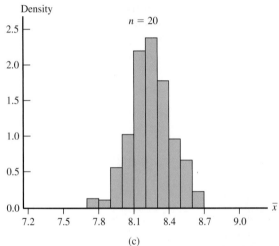

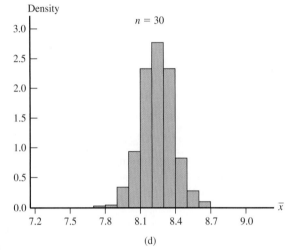

FIGURE 12.3

Density histograms for $\bar{x}$ based on 500 samples of size n for the population of Example 12.1: (a) $n = 5$; (b) $n = 10$; (c) $n = 20$; (d) $n = 30$.

intervals than those for the two smaller sample sizes. For the larger sample sizes, most of the $\bar{x}$ values are quite close to 8.25. *The sample mean $\bar{x}$ based on a large sample size tends to be closer to μ than $\bar{x}$ based on a small sample size.*

Example 12.2 Time to First Goal in Hockey

In this example, we will consider properties of the $\bar{x}$ distribution when the population is quite skewed (and thus very unlike a normal distribution). The paper **"Is the Overtime Period in an NHL Game Long Enough?"** (*American Statistician* **[2008]:151–154**) gave data on the time (in minutes) from the start of the game to the first goal scored for the 281 regular season hockey games in the 2005–2006 season that went into overtime. Figure 12.4 displays a density histogram of the data (from a graph that appeared in the paper). The histogram has a long upper tail, indicating that the first goal is scored in the first 20 minutes of most games, but for some games, the first goal is not scored until much later in the game.

If you think of the 281 values as a population, the histogram in Figure 12.4 shows the population distribution. Although the skewed shape makes identification of the mean value from the picture more difficult, we calculated the mean of the 281 values in the population, and found $\mu = 13$ minutes. The median value for the population (10 minutes) is less than μ, a consequence of the fact that the distribution is positively skewed.

For each of the sample sizes $n = 5, 10, 20$, and 30, we selected 500 random samples of size n. This was done with replacement to approximate more nearly the usual situation, in which the sample size n is only a small fraction of the population size. We then constructed a histogram of the 500 $\bar{x}$ values for each of the four sample sizes. These histograms are displayed in Figure 12.5.

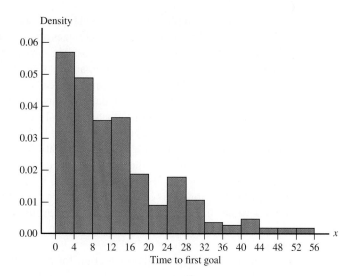

FIGURE 12.4
The population distribution for
Example 12.2 ($\mu = 13$)

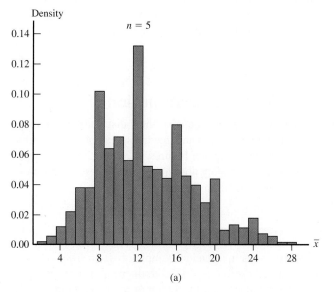

(a)

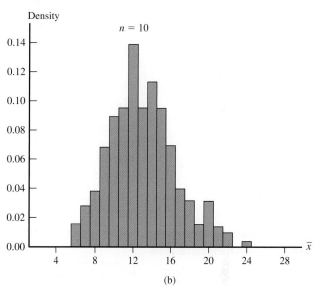

(b)

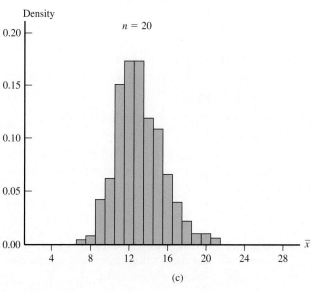

(c)

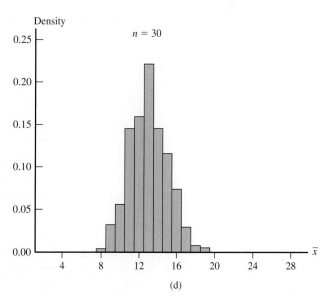

(d)

FIGURE 12.5
Four histograms of 500 $\bar{x}$ values
for Example 12.2: (a) $n = 5$;
(b) $n = 10$; (c) $n = 20$; (d) $n = 30$.

As with the samples from the normal population of Example 12.1, the means of the 500 $\bar{x}$ values for the four different sample sizes are all close to the population mean $\mu = 13$. If each histogram were based on all possible samples rather than just 500 of them, they would be centered at exactly 13. Comparison of the four $\bar{x}$ histograms in Figure 12.5 also shows that as n increases, the histogram's spread about its center decreases. This was also true for the normal population of the previous example. There is less sample-to-sample variability in $\bar{x}$ for large samples than for small samples.

One aspect of the histograms in Figure 12.5 distinguishes them from those based on a random sample from a normal population (Figure 12.3). They are skewed and differ in shape more, but they become progressively more symmetric as the sample size increases. You can also see that for $n = 30$, the histogram has a shape much like a normal distribution. This is the effect of averaging. Even when n is large, one of the few large x values in the population doesn't appear in the sample very often. When one does appear, its contribution to $\bar{x}$ is outweighed by the contributions of more typical sample values.

The normal shape of the histogram for $n = 30$ is what is predicted by the *Central Limit Theorem*, which will be introduced shortly. According to this theorem, even if the population distribution is not well described by a normal distribution, the $\bar{x}$ sampling distribution is approximately normal in shape when the sample size n is reasonably large.

General Properties of the Sampling Distribution of $\bar{x}$

Examples 12.1 and 12.2 suggest that for any n, the center of the $\bar{x}$ distribution (the mean value of $\bar{x}$) is equal to the value of the population mean and that the spread of the $\bar{x}$ distribution decreases as n increases. The sample histograms of Figures 12.3 and 12.5 also suggest that in some cases, the $\bar{x}$ distribution is approximately normal in shape. These observations are stated more formally in the following general properties.

General Properties of the Sampling Distribution of $\bar{x}$

Let $\bar{x}$ denote the mean of the observations in a random sample of size n from a population with mean μ and standard deviation σ. Denote the mean value of the $\bar{x}$ sampling distribution by $\mu_{\bar{x}}$ and the standard deviation of the $\bar{x}$ sampling distribution by $\sigma_{\bar{x}}$. Then, the following rules hold:

Rule 1. $\mu_{\bar{x}} = \mu$

Rule 2. $\sigma_{\bar{x}} = \dfrac{\sigma}{\sqrt{n}}$. This rule is exact if the population is infinite, and is approximately correct if the population is finite and no more than 10% of the population is included in the sample.

Rule 3. When the population distribution is normal, the sampling distribution of $\bar{x}$ is also normal for any sample size n.

Rule 4. (**Central Limit Theorem**) When n is large, the sampling distribution of $\bar{x}$ is well approximated by a normal curve, even when the population distribution is not normal.

Rule 1, $\mu_{\bar{x}} = \mu$, states that the sampling distribution of $\bar{x}$ is always centered at the value of the population mean μ. This tells you that the $\bar{x}$ values from different random samples tend to cluster around the actual value of the population mean.

Rule 2, $\sigma_{\bar{x}} = \dfrac{\sigma}{\sqrt{n}}$, gives the relationship between $\sigma_{\bar{x}}$, the standard deviation of the $\bar{x}$ distribution, and the sample size n. You can see why sample-to-sample variability in $\bar{x}$ decreases as the sample size n increases (because the sample size n is in the denominator of the expression for $\sigma_{\bar{x})}$). The $\bar{x}$ values tend to cluster more tightly (less variability) around the actual value of the population mean for larger samples. When $n = 4$, for example,

$$\sigma_{\bar{x}} = \frac{\sigma}{\sqrt{n}} = \frac{\sigma}{\sqrt{4}} = \frac{\sigma}{2}$$

and the $\bar{x}$ distribution has a standard deviation that is only half as large as the population standard deviation.

Rules 3 and 4 say that in some cases the shape of the $\bar{x}$ distribution is normal (when the population is normal) or approximately normal (when the population distribution is not normal but the sample size is large). Figure 12.6 illustrates these rules by showing several $\bar{x}$ distributions superimposed over a graph of the population distribution.

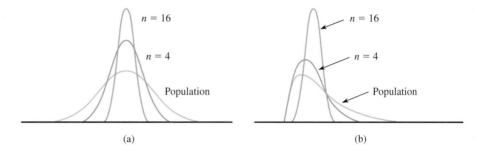

FIGURE 12.6
Population distribution and sampling distributions of $\bar{x}$: (a) symmetric population; (b) skewed population.

AP* EXAM TIP

The Central Limit Theorem is a very important result. It tells you that if the sample size is sufficiently large, the sampling distribution of $\bar{x}$ is approximately normal even if the population distribution is not normal.

The Central Limit Theorem of Rule 4 states that when n is large, the $\bar{x}$ distribution is approximately normal for any population distribution. This result has enabled statisticians to develop large-sample methods for estimating a population mean and for testing hypotheses about a population mean that can be used even when the shape of the population distribution is unknown.

Recall that a variable is standardized by subtracting its mean value and then dividing by its standard deviation. Using Rules 1 and 2 to standardize $\bar{x}$ gives an important consequence of the last two rules.

If n is large or the population distribution is normal, the standardized variable

$$z = \frac{\bar{x} - \mu_{\bar{x}}}{\sigma_{\bar{x}}} = \frac{\bar{x} - \mu}{\dfrac{\sigma}{\sqrt{n}}}$$

has (at least approximately) a standard normal (z) distribution.

Applying the Central Limit Theorem requires a rule of thumb for deciding when n is large enough. Look back at Figure 12.5, which shows the approximate sampling distribution of $\bar{x}$ for $n = 5, 10, 20,$ and 30 when the population distribution is quite skewed. Certainly the histogram for $n = 5$ is not well described by a normal curve, and this is still true of the histogram for $n = 10$. Among the four histograms, only the histogram for $n = 30$ has a shape that is reasonably well described by a normal curve. On the other hand, when the population distribution is normal, the sampling distribution of $\bar{x}$ is normal for any n.

How large n must be in order for the $\bar{x}$ distribution to be approximately normal depends on how much the population distribution differs from a normal distribution. The closer the population distribution is to being normal, the smaller the value of n necessary for the Central Limit Theorem approximation to be accurate. Many statisticians recommend the following conservative rule:

The Central Limit Theorem can safely be applied if $n \geq 30$.

Example 12.3 Courting Scorpion Flies

The authors of the paper **"Should I Stay or Should I Go? Condition- and Status-Dependent Courtship Decisions in the Scorpion Fly *Panorpa Cognate*" (*Animal Behaviour* [2009]: 491–497)** studied the courtship behavior of mating scorpion flies. One variable of interest was x = courtship time, which was defined as the time from the beginning of a female-male interaction until mating. Data from the paper suggest that it is reasonable to think that the population mean and standard deviation of x are $\mu = 117.1$ minutes and $\sigma = 109.1$ minutes. Notice that the population distribution of courtship times can't be normal. For a normal distribution centered at 117.1 and with such a large standard deviation, it would not be uncommon to observe negative values, but courtship time can't have a negative value.

The sampling distribution of $\bar{x}$ = mean courtship time for a random sample of 20 scorpion fly mating pairs would have mean

$$\mu_{\bar{x}} = \mu = 117.1 \text{ minutes}$$

This tells you that the sampling distribution of $\bar{x}$ is centered at 117.1. The standard deviation of $\bar{x}$ is

$$\sigma_{\bar{x}} = \frac{\sigma}{\sqrt{n}} = \frac{109.1}{\sqrt{20}} = 24.40$$

which is smaller than the population standard deviation σ. Because the population distribution is not normal and because the sample size is smaller than 30, it is not reasonable to assume that the sampling distribution of $\bar{x}$ is normal in shape.

The following examples illustrate how knowledge of the sampling distribution of $\bar{x}$ supports learning from sample data.

Example 12.4 Is the "Freshman 15" Real?

It is a common belief that most students gain weight during their freshman year in college. The term "freshman 15" is often used to describe this weight gain. But do first-year college students really gain this much weight? The authors of the paper **"The Freshman 15: Is It Real?" (*Journal of American College Health* [2008]: 531–533)** describe a study of 125 freshmen at a college located in the northeastern region of the United States. For each student participating in the study, weight gain during the freshman year was determined. For these 125 students, the mean weight gain was 2.7 pounds, and the standard deviation was 6.4 pounds.

Although the sample wasn't actually selected at random, the researchers attempted to obtain a representative sample, and they noted that the sample was quite similar to the population with respect to height, weight, gender, and race. Because of this, it is reasonable to regard this sample as if it were a random sample of the freshmen at the college.

Suppose that you are interested in estimating the mean weight gain for all freshmen at the college. You don't expect the population mean to be *exactly* 2.7 pounds, but if you use 2.7 pounds as an estimate of the population mean, how accurate is this estimate likely to be? To answer this question, you can use what you know about the sampling distribution of $\bar{x}$ for random samples of size 125. You know three things that follow from the general results described earlier:

What You Know	How You Know It
The sampling distribution of $\bar{x}$ is centered at μ. This means that the $\bar{x}$ values from random samples cluster around the actual value of the population mean.	Rule 1 states that $\mu_{\bar{x}} = \mu$. This is true for random samples, and the description of the study suggests that it is reasonable to regard the sample as if it were a random sample.

(continued)

What You Know	How You Know It
Values of $\bar{x}$ will cluster fairly tightly around the actual value of the population mean. The standard deviation of $\bar{x}$, (which describes how much the $\bar{x}$ values spread out around the population mean) is $\frac{\sigma}{\sqrt{n}}$. An estimate of the standard deviation of $\bar{x}$ is $$\frac{\sigma}{\sqrt{n}} \approx \frac{s}{\sqrt{n}} = \frac{6.4}{\sqrt{125}} = 0.57$$	Rule 2 states that $\sigma_{\bar{x}} = \frac{\sigma}{\sqrt{n}}$. In this example, $n = 125$. The value of σ is not known—this is the actual value of the population standard deviation. However, the sample standard deviation s provides an estimate of σ that can be used to estimate the standard deviation of the sampling distribution. The estimated standard deviation provides information about how tightly the $\bar{x}$ values from different random samples cluster around the population mean, μ.
The sampling distribution of $\bar{x}$ is approximately normal.	Rule 4 states that the sampling distribution of $\bar{x}$ is approximately normal if n is large. Here the sample size is 125. Because the sample size is greater than 30, the normal approximation is appropriate.

Summarizing, you know that the $\bar{x}$ distribution is centered at the actual population mean, has a standard deviation of about 0.57, and is approximately normal. By using this information and what you know about normal distributions, you can now get a sense of the accuracy of the estimate $\bar{x} = 2.7$ pounds. For any variable described by a normal distribution, approximately 95% of the values are within two standard deviations of the center. Since the distribution of $\bar{x}$ is approximately normal and is centered at the actual population mean μ, you now know that about 95% of all possible random samples would produce a sample mean within about $2(0.57) = 1.14$ pounds of the actual population mean. So, a margin of error of 1.14 pounds could be reported. This tells you that the sample estimate $\bar{x} = 2.7$ pounds is likely to be within 1.14 pounds of the actual mean freshman year weight gain for students at this college. This suggests that the actual mean weight gain for students at this college is quite a bit less than the legendary freshman 15. These ideas will be formalized in Section 12.2, where you will learn about margin of error when estimating a population mean and will see how to construct confidence interval estimates of a population mean.

Example 12.5   Fat Content of Hot Dogs

Data from a sample can also be used to evaluate whether a claim about a population mean is believable. For example, suppose that a hot dog manufacturer claims that one of its brands of hot dogs has a mean fat content of $\mu = 18$ grams per hot dog. Consumers of this brand would probably not be disturbed if the mean is less than 18 but would be unhappy if it exceeds 18.

In this situation, the variable of interest is

$$x = \text{fat content of a hot dog}$$

For purposes of this example, suppose you know that σ, the standard deviation of the x distribution, is equal to 1.

An independent testing organization is asked to analyze a random sample of 36 hot dogs. The fat content for each of the 36 hot dogs is measured and the sample mean is calculated to be $\bar{x} = 18.4$ grams. Does this result suggest that the manufacturer's claim that the population mean is 18 is incorrect?

The sample size, $n = 36$, is large enough to say that the sampling distribution of $\bar{x}$ will be approximately normal. The standard deviation of the $\bar{x}$ distribution is

$$\sigma_{\bar{x}} = \frac{\sigma}{\sqrt{n}} = \frac{1}{\sqrt{36}} = 0.1667$$

Because you know that $\mu_{\bar{x}} = \mu$, if the manufacturer's claim is correct,

$$\mu_{\bar{x}} = \mu = 18$$

So, if the manufacturer's claim is correct, should you be surprised to see a sample mean of $\bar{x} = 18.4$? You know that even if $\mu = 18$, $\bar{x}$ will not usually be exactly 18 due to sampling variability. But, is it likely that you would see a sample mean at least as large as 18.4 when the population mean is really 18?

Using the normal distribution, you can compute the probability of observing a sample mean this large. *If the manufacturer's claim is correct,*

$$P(\bar{x} \geq 18.4) \approx P\left(z \geq \frac{18.4 - 18}{0.1667}\right)$$

$$= P(z \geq 2.4)$$

$$= \text{area under the normal curve to the right of } 2.4$$

$$= 1 - 0.9918$$

$$= 0.0082$$

Values of $\bar{x}$ as large as 18.4 will be observed only about 0.82% of the time when a random sample of size 36 is taken from a population with mean 18 and standard deviation 1. The value $\bar{x} = 18.4$ is enough greater than 18 that you should be skeptical of the manufacturer's claim. These ideas will be formalized in Section 12.3, where you will learn about how to test hypotheses about a population mean.

Other Cases

You now know a great deal about the sampling distribution of $\bar{x}$ in two cases: when the population distribution is normal and when the sample size is large. What happens when the population distribution is not normal and n is small? Although it is still true that $\mu_{\bar{x}} = \mu$ and $\sigma_{\bar{x}} = \dfrac{\sigma}{\sqrt{n}}$, unfortunately there is no general result about the shape of the $\bar{x}$ sampling distribution. When the objective is to make an inference about the mean of such a population, one way to proceed is to make a specific assumption about the shape of the population distribution. Statisticians have proposed a number of such possibilities. Consult a statistician or a more advanced text for more information.

SECTION 12.1 EXERCISES

Each Exercise Set assesses the following chapter learning objectives: C1, M1, M2

SECTION 12.1 Exercise Set 1

12.1 A random sample is selected from a population with mean $\mu = 100$ and standard deviation $\sigma = 10$. Determine the mean and standard deviation of the $\bar{x}$ sampling distribution for each of the following sample sizes:

a. $n = 9$ **d.** $n = 50$
b. $n = 15$ **e.** $n = 100$
c. $n = 36$ **f.** $n = 400$

12.2 For which of the sample sizes given in the previous exercise would it be reasonable to think that the $\bar{x}$ sampling distribution is approximately normal in shape?

12.3 The article **"How Business Students Spend Their Time— Do They Really Know?" (*Research in Higher Education Journal* [May 2009]: 1–10)** describes a study of 212 business students at a large public university. Each student kept a log of time spent on various activities during a 1-week period. For these 212 students, the mean time spent studying was 9.66 hours and the sample standard deviation was 6.62 hours. Suppose that this sample was a random sample from the population of all business majors at this university and that you are interested in learning about the value of μ, the mean time spent studying for this population. The following table is similar to the table that appears in Example 12.4. The "what you know" information has been provided. Complete the table by filling in the "how you know it" column.

What You Know	How You Know It
The sampling distribution of $\bar{x}$ is centered at the actual (but unknown) value of the population mean.	
An estimate of the standard deviation of $\bar{x}$, which describes how the $\bar{x}$ values spread out around the population mean μ, is 0.455.	
The sampling distribution of $\bar{x}$ is approximately normal.	

12.4 Explain the difference between μ and $\mu_{\bar{x}}$.

12.5 The time that people have to wait for an elevator in an office building has a uniform distribution over the interval from 0 to 1 minute. For this distribution, $\mu = 0.5$ and $\sigma = 0.289$.
a. Let $\bar{x}$ be the average waiting time for a random sample of 16 waiting times. What are the mean and standard deviation of the sampling distribution of $\bar{x}$?
b. Answer Part (a) for a random sample of 50 waiting times. Draw a picture of the approximate sampling distribution of $\bar{x}$ when $n = 50$.

SECTION 12.1 **Exercise Set 2**

12.6 A random sample is selected from a population with mean $\mu = 60$ and standard deviation $\sigma = 3$. Determine the mean and standard deviation of the $\bar{x}$ sampling distribution for each of the following sample sizes:
a. $n = 6$ **d.** $n = 75$
b. $n = 18$ **e.** $n = 200$
c. $n = 42$ **f.** $n = 400$

12.7 For which of the sample sizes given in the previous exercise would it be reasonable to think that the $\bar{x}$ sampling distribution is approximately normal in shape?

12.8 The article "**How Business Students Spend Their Time— Do They Really Know?**" (*Research in Higher Education Journal* [May 2009]: 1–10) describes a study of 212 business students at a large public university. Each student kept a log of time spent on various activities during a 1-week period. For these 212 students, the mean time spent on Facebook was 1.06 hours and the standard deviation was 2.01 hours. Suppose that this sample was a random sample from the population of all business majors at this university and that you are interested in learning about the value of μ, the mean time spent on Facebook for this population. The following table is similar to the table that appears in Example 12.4. The "what you know" information has been provided. Complete the table by filling in the "how you know it" column.

What You Know	How You Know It
The sampling distribution of $\bar{x}$ is centered at the actual (but unknown) value of the population mean.	
An estimate of the standard deviation of $\bar{x}$, which describes how the $\bar{x}$ values spread out around the population mean μ, is 0.138.	
The sampling distribution of $\bar{x}$ is approximately normal.	

12.9 Explain the difference between σ and $\sigma_{\bar{x}}$.

12.10 Suppose that a random sample of size 64 is to be selected from a population with mean 40 and standard deviation 5.
a. What are the mean and standard deviation of the $\bar{x}$ sampling distribution? Describe the shape of the $\bar{x}$ sampling distribution.
b. What is the approximate probability that $\bar{x}$ will be within 0.5 of the population mean μ?
c. What is the approximate probability that $\bar{x}$ will differ from μ by more than 0.7?

Additional Exercises

12.11 A random sample is selected from a population with mean $\mu = 200$ and standard deviation $\sigma = 15$. Determine the mean and standard deviation of the $\bar{x}$ sampling distribution for each of the following sample sizes:
a. $n = 12$ **d.** $n = 40$
b. $n = 20$ **e.** $n = 90$
c. $n = 25$ **f.** $n = 300$

12.12 For which of the sample sizes given in the previous exercise would it be reasonable to think that the $\bar{x}$ sampling distribution is approximately normal in shape?

12.13 Explain the difference between $\bar{x}$ and $\mu_{\bar{x}}$.

12.14 A sign in the elevator of a college library indicates a limit of 16 persons. In addition, there is a weight limit of 2,500 pounds. Assume that the average weight of students, faculty, and staff at this college is 150 pounds, that the standard deviation is 27 pounds, and that the distribution of weights of individuals on campus is approximately normal. A random sample of 16 persons from the campus will be selected.
a. What is the mean of the sampling distribution of $\bar{x}$?
b. What is the standard deviation of the sampling distribution of $\bar{x}$?
c. What average weights for a sample of 16 people will result in the total weight exceeding the weight limit of 2,500 pounds?
d. What is the probability that a random sample of 16 people will exceed the weight limit?

12.15 Suppose that the mean value of interpupillary distance (the distance between the pupils of the left and right eyes) for adult males is 65 mm and that the population standard deviation is 5 mm.
a. If the distribution of interpupillary distance is normal and a random sample of $n = 25$ adult males is to be selected, what is the probability that the sample mean distance $\bar{x}$ for these 25 will be between 64 and 67 mm? At least 68 mm?
b. Suppose that a random sample of 100 adult males is to be selected. Without assuming that interpupillary distance is normally distributed, what is the approximate probability that the sample mean distance will be between 64 and 67 mm? At least 68 mm?

12.16 Suppose that a random sample of size 100 is to be drawn from a population with standard deviation 10.
a. What is the probability that the sample mean will be within 20 of the value of μ?
b. For this example ($n = 100$, $\sigma = 10$), complete each of the following statements by computing the appropriate value:
  i. Approximately 95% of the time, $\bar{x}$ will be within _____ of μ.
  ii. Approximately 0.3% of the time, $\bar{x}$ will be farther than _____ from μ.

12.17 A manufacturing process is designed to produce bolts with a diameter of 0.5 inches. Once each day, a random sample of 36 bolts is selected and the bolt diameters are recorded. If the resulting sample mean is less than 0.49 inches or greater than 0.51 inches, the process is shut down for adjustment. The standard deviation of bolt diameters is 0.02 inches. What is the probability that the manufacturing line will be shut down unnecessarily? (Hint: Find the probability of observing an $\bar{x}$ in the shutdown range when the actual process mean is 0.5 inches.)

12.18 An airplane with room for 100 passengers has a total baggage limit of 6,000 pounds. Suppose that the weight of the baggage checked by an individual passenger, x, has a mean of 50 pounds and a standard deviation of 20 pounds. If 100 passengers will board a flight, what is the approximate probability that the total weight of their baggage will exceed the limit? (Hint: With $n = 100$, the total weight exceeds the limit when the mean weight $\bar{x}$ exceeds 6,000/100.)

SECTION 12.2 A Confidence Interval for a Population Mean

In the chapter preview example, researchers wanted to use sample data to estimate the mean processing time of emergency room flu cases for a proposed new drive-through model where flu patients are evaluated while they remain in their cars. People representing a random sample of 38 flu cases were processed using the drive-through model and the time to process each case from admission to discharge was recorded. Data read from a graph in the paper referenced in the preview were used to compute the following summary statistics:

$$n = 38 \qquad\qquad \bar{x} = 26 \text{ minutes} \qquad\qquad s = 1.57 \text{ minutes}$$

Using the sample mean as an estimate of the population mean, you could say that you think that the mean time required to process a flu case for the drive-through model is 26 minutes. But when you say that the population mean time is 26 minutes, you don't really believe that it is *exactly* 26 minutes. The estimate of 26 minutes is based on a sample, and the sample mean will vary from one sample to another. What you really think is that the population mean is *around* 26 minutes. To be more specific about what you mean by "around 26 minutes," you can use the sample data to compute a confidence interval estimate of the population mean.

You have already seen confidence interval estimates in the context of estimating a population proportion, and the basic idea is the same for estimating a population mean. The general form of a confidence interval estimate is

$$(\text{statistic}) \pm (\text{critical value}) \left(\begin{array}{c}\text{standard error} \\ \text{of the statistic}\end{array}\right)$$

Let's look at how this general form can be adapted to estimating a population mean. Begin by considering the case in which σ, the population standard deviation, is known (not realistic, but you will see shortly how to handle the more realistic situation where σ is unknown) and the sample size n is large enough for the Central Limit Theorem to apply ($n \geq 30$). From Section 12.1, you know that:

1. The sampling distribution of $\bar{x}$ is centered at μ, so $\bar{x}$ is an unbiased statistic for estimating μ (because $\mu_{\bar{x}} = \mu$).
2. The standard deviation of $\bar{x}$ is $\sigma_{\bar{x}} = \dfrac{\sigma}{\sqrt{n}}$.
3. As long as n is large ($n \geq 30$), the sampling distribution of $\bar{x}$ is approximately normal, even when the population distribution itself is not normal.

This suggests that a confidence interval for a population mean *when the sample size is large and σ is known* is

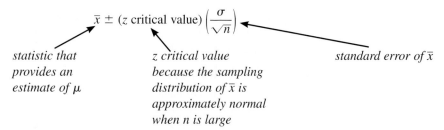

$$\bar{x} \pm (z \text{ critical value}) \left(\frac{\sigma}{\sqrt{n}}\right)$$

statistic that provides an estimate of μ

z critical value because the sampling distribution of $\bar{x}$ *is approximately normal when n is large*

standard error of $\bar{x}$

Example 12.6 illustrates the calculation of a confidence interval for the population mean when the population standard deviation is known.

Example 12.6 Cosmic Radiation

Cosmic radiation levels rise with increasing altitude, prompting researchers to consider how pilots and flight crews are affected by increased exposure to cosmic radiation. The paper **"Estimated Cosmic Radiation Doses for Flight Personnel"** (*Space Medicine and Medical Engineering* **[2002]: 265–269**) reported a mean annual cosmic radiation dose of 219 mrems for a sample of flight personnel of Xinjiang Airlines. Suppose that this mean was based on a random sample of 100 flight crew members.

Here μ will denote the mean annual cosmic radiation exposure for *all* Xinjiang Airlines flight crew members. Although σ, the actual value of the population standard deviation, is not usually known, for purposes of this example suppose that $\sigma = 35$ mrem. Because the sample size is large and σ is known, a 95% confidence interval for μ is

z critical value for 95% confidence level

$$\bar{x} \pm (z \text{ critical value}) \left(\frac{\sigma}{\sqrt{n}}\right) = 219 \pm (1.96)\left(\frac{35}{\sqrt{100}}\right)$$

$$= 219 \pm 6.86$$

$$= (212.14, 225.86)$$

Based on this sample, *plausible* values of μ, the mean annual cosmic radiation exposure for *all* Xinjaing Airlines flight crew members, are between 212.14 and 225.86 mrem. A 95% confidence level is associated with the method used to produce this interval estimate.

The confidence interval formula used in Example 12.6 is straightforward, but not very useful! To compute the confidence interval endpoints, you need to know the value of σ, the population standard deviation. It is almost never the case that you would not know the value of the population mean (which is why you would be using sample data to estimate it) but would know the value of σ for this same population. For this reason, you will probably never use the confidence interval formula used in Example 12.6. But, it does provide a useful starting point for investigating how to estimate the population mean in more realistic situations where σ is not known.

A Confidence Interval for μ When σ Is Unknown

When the population standard deviation is unknown, there are two changes that you need to make to the confidence interval formula previously given. First, use the sample standard deviation, s, as an estimate of σ in the confidence interval formula. The second change is that you can no longer use the standard normal (z) distribution to determine the critical value used to compute the confidence interval.

estimate σ *with sample standard deviation s*

$$\bar{x} \pm (z \text{ critical value}) \left(\frac{\sigma}{\sqrt{n}}\right)$$

critical value is no longer based on the z distribution

If the population distribution is normal or if the sample size is large ($n \geq 30$), the critical value that should be used is based on a probability distribution called a t distribution. Since you have not yet studied t distributions, let's take a short detour to learn about them.

t Distributions

Just as there are many different normal distributions, there are also many different t distributions. While normal distributions are distinguished from one another by their mean μ and standard deviation σ, t distributions are distinguished by a positive whole number called *degrees of freedom* (df). There is a t distribution with 1 df, another with 2 df, and so on.

> ### Important Properties of t Distributions
>
> 1. The t distribution corresponding to any particular degrees of freedom is bell-shaped and centered at zero (just like the standard normal (z) distribution).
> 2. Each t distribution is more spread out than the standard normal (z) distribution.
> 3. As the number of degrees of freedom increases, the spread of the corresponding t distribution decreases.
> 4. As the number of degrees of freedom increases, the corresponding sequence of t distributions approaches the standard normal (z) distribution.

AP* EXAM TIP

As df increase, the corresponding t distributions get closer and closer to the standard normal distribution.

The properties in the preceding box are illustrated in Figure 12.7, which shows two t curves along with the z curve.

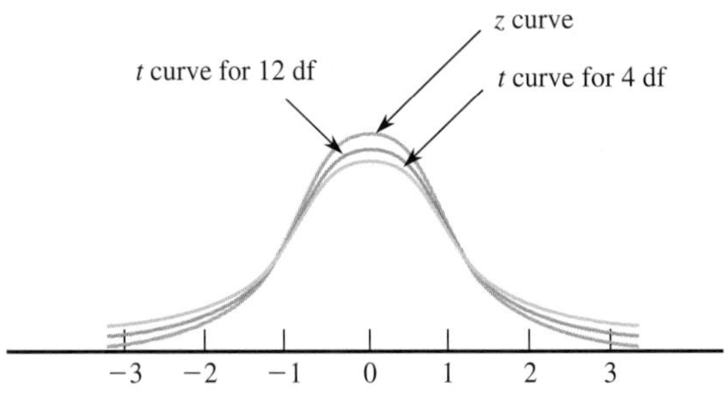

FIGURE 12.7
Comparison of the standard normal (z) distribution and t distributions for 4 df and 12 df.

Appendix Table 3 gives selected critical values for various t distributions. Critical values are given for central areas of 0.80, 0.90, 0.95, 0.98, 0.99, 0.998, and 0.999. To find a particular critical value, go down the left margin of the table to the row labeled with the desired number of degrees of freedom. Then move over in that row to the column headed by the desired central area. For example, the value in the 12-df row under the column corresponding to central area 0.95 is 2.18, so 95% of the area under the t curve with 12 df lies between -2.18 and 2.18. Moving over two columns, the critical value for central area 0.99 (still with 12 df) is 3.06 (see Figure 12.8). Moving down the 0.99 column to the 20-df row, you see the critical value is 2.85, so the area between -2.85 and 2.85 under the t curve with 20 df is 0.99.

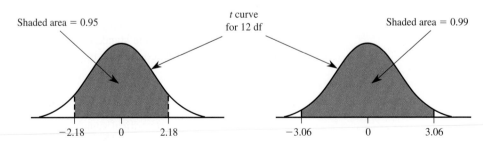

FIGURE 12.8
t critical values illustrated.

Notice that the critical values increase from left to right in each row of Appendix Table 3. This makes sense because as you move to the right, the central area captured is larger. In each column, the critical values decrease as you move downward, reflecting decreasing spread for t distributions with larger number of degrees of freedom.

The larger the number of degrees of freedom, the more closely the t curve resembles the z curve. To emphasize this, z critical values are included as the last row of the t table. Once the number of degrees of freedom exceeds 30, there is little change in the critical values as degrees of freedom increase. For this reason, Appendix Table 3 jumps from 30 df to 40 df, then to 60 df, then to 120 df, and finally to the row of z critical values. If you need a critical value for a number of degrees of freedom between those tabulated, just use the critical value for the closest df. For df > 120, you can use the z critical values.

Many graphing calculators and statistical software packages calculate t critical values for any specified number of degrees of freedom, so it is not necessary to use the table or to approximate the t critical values if you are using technology.

A One-Sample t Confidence Interval for μ

Using the sample standard deviation to estimate the population standard deviation and replacing the z critical value with a t critical value results in the following confidence interval formula:

$$\bar{x} \pm (t \text{ critical value}) \left(\frac{s}{\sqrt{n}} \right)$$

To find the t critical value, you need to know the desired confidence level and the appropriate number of degrees of freedom. When estimating a population mean μ, number of degrees of freedom is determined by the sample size:

$$df = n - 1$$

For example, suppose that you want to use a random sample of size 25 to estimate a population mean using a 95% confidence interval. In this case, df $= 25 - 1 = 24$. To find the t critical value, you can use technology or Appendix Table 3 (look in the 95% confidence level column and the 24 df row). The appropriate t critical value is 2.06.

Just one last question to answer: When is it appropriate to use this confidence interval formula? If the sample size n is large ($n \geq 30$), this interval can be used regardless of the shape of the population distribution. However, when n is small, this interval is only appropriate if the population distribution is at least approximately normal. There are several ways that sample data can be used to assess the plausibility of normality. You can look at a normal probability plot of the sample data (looking for a plot that is reasonably straight) or you can construct a dotplot or boxplot of the data (looking for approximate symmetry and no outliers).

One-Sample t Confidence Interval for a Population Mean μ

Appropriate when the following conditions are met:
1. The sample is a random sample from the population of interest or the sample is selected in a way that results in a sample that is representative of the population.
2. The sample size is large ($n \geq 30$) *or* the population distribution is approximately normal.

When these conditions are met, a confidence interval for the population mean is

$$\bar{x} \pm (t \text{ critical value}) \left(\frac{s}{\sqrt{n}} \right)$$

The t critical value is based on df $= n - 1$ and the desired confidence level. It can be found using Appendix Table 3, statistical software, or a graphing calculator.

Interpretation of Confidence Interval

You can be confident that the actual value of the population mean is included in the computed interval. In a given problem, this statement should be worded in context.

(continued)

Interpretation of Confidence Level

The confidence level specifies the long-run proportion of the time that this method is expected to be successful in capturing the actual population mean.

Recall the four key questions (QSTN) first introduced in Section 7.2 that guide the decision about which method to consider:

Q Question Type	Estimation or hypothesis testing?
S Study Type	Sample data or experiment data?
T Type of Data	One variable or two? Categorical or numerical?
N Number of Samples or Treatments	How many samples or treatments?

When the answers to these questions are

Q: estimation
S: sample data
T: one numerical variable
N: one sample

the method you should consider is the one-sample t confidence interval for a population mean.

Once you have selected the one-sample t confidence interval for a population mean as the method you will consider, follow the five-step process for estimation problems (EMC³). Recall that the steps in this process are (see Section 7.3):

Step	What Is This step?
E	**Estimate:** Explain what population characteristic you plan to estimate.
M	**Method:** Select a potential method based on the answers to the four key questions (QSTN).
C	**Check:** Check to make sure that the method selected is appropriate. Many methods for learning from data only provide reliable information under certain conditions. It is important to verify that any such conditions are met before proceeding.
C	**Calculate:** Use the sample data to perform any necessary calculations. This step is often accomplished using technology, such as a graphing calculator or statistical software.
C	**Communicate Results:** Answer the research question of interest, explain what you have learned from the data, and acknowledge potential risks.

You will see how this process is used in the examples that follow. Be sure to include all five steps when you use a confidence interval to estimate a population mean.

Example 12.7   **Drive-Through Medicine Revisited**

The chapter preview example described a study of a drive-through model for processing flu patients who go to the emergency room for treatment (**"Drive-Through Medicine: A Novel Proposal for Rapid Evaluation of Patients During an Influenza Pandemic,"** *Annals of Emergency Medicine* [2010]: 268–273). In the drive-through model, patients are evaluated while they remain in their cars. In the study, 38 people were each given a scenario for a flu case that was selected at random from the set of all flu cases actually seen in the emergency room. The scenarios provided the "patient" with a medical history and a description of symptoms that would allow the patient to respond to questions from the examining physician. These patients were processed using a drive-through procedure that was implemented in the parking structure of Stanford University Hospital. The time to process each case from admission to discharge was recorded. The following sample statistics were computed from the data:

$$n = 38 \qquad \bar{x} = 26 \text{ minutes} \qquad s = 1.57 \text{ minutes}$$

The researchers were interested in estimating the mean processing time for flu patients using the drive-through model.

Begin by answering the four key questions:

Q Question Type	Estimation or hypothesis testing?	Estimation
S Study Type	Sample data or experiment data?	Sample data
T Type of Data	One variable or two? Categorical or numerical?	One numerical variable
N Number of Samples or Treatments	How many samples or treatments?	One sample

Because the answers are *estimation, sample data, one numerical variable, and one sample,* you should consider a one-sample *t* confidence interval for a population mean.

Now you are ready to use the five-step process to estimate the population mean. For purposes of this example, a 95% confidence level will be used.

Step	
Estimate: Explain what population characteristic you plan to estimate.	In this example, you want to estimate the value of μ, **the mean time to process a flu case using the new drive-through model.**
Method: Select a potential method based on the answers to the four key questions (QSTN).	Because the answers to the four key questions are estimation, sample data, one numerical variable, one sample, consider using a **one-sample *t* confidence interval for a population mean**. A confidence level of 95% was specified for this example.
Check: Check to make sure that the method selected is appropriate.	There are **two conditions** that need to be met in order to use the one-sample *t* confidence interval. The large sample condition is met because the sample size of 38 is greater than 30. Because the sample size is large, you do not need to worry about whether the population distribution is approximately normal. The second condition that must be met is that the sample is a random sample or one that is representative of the population. Here, the 38 flu cases were randomly selected from the population of all flu cases seen at the emergency room, so this condition is met.
Calculate: Use the sample data to perform any necessary calculations.	$n = 38$ $\bar{x} = 26$ $s = 1.57$ $df = 38 - 1 = 37$ The appropriate *t* critical value is found using technology or by using Appendix Table 3 to find the *t* critical value that captures a central area of 0.95. (Note: When using the table, use the closest df, which is 40). *t* critical value $= 2.02$ 95% confidence interval $\bar{x} \pm (t \text{ critical value})\left(\dfrac{s}{\sqrt{n}}\right) = 26 \pm (2.02)\left(\dfrac{1.57}{\sqrt{38}}\right)$ $= 26 \pm 0.514$ $= (25.486, 26.514)$

(continued)

Step		
Communicate Results: Answer the research question of interest, explain what you have learned from the data, and acknowledge potential risks.	**Confidence interval:** You can be 95% confident that the actual mean processing time for emergency room flu cases using the new drive-through model is between 25.486 minutes and 26.514 minutes	
	Confidence level: The method used to construct this interval estimate is successful in capturing the actual value of the population mean about 95% of the time.	

The researchers in this study indicated that the average processing time for flu patients seen in the emergency room was about 90 minutes, so it appears that the drive-through procedure has promise both in terms of keeping flu patients isolated and also in reducing mean processing time.

Example 12.8   Waiting for Surgery

The Cardiac Care Network in Ontario, Canada, collected information on the time between when a cardiac patient is recommended for heart surgery and when the surgery is performed **("Wait Times Data Guide," Ministry of Health and Long-Term Care, Ontario, Canada, 2006).** The reported mean wait time (in days) for a random sample of 539 patients waiting for bypass surgery was $\bar{x} = 19$ days and the sample standard deviation was $s = 10$ days. (This standard deviation was estimated from information in the report.) The Cardiac Care Network was interested in estimating the mean wait time for the population of patients recommended for bypass surgery.

Answers to the four key questions for this problem are:

Q Question Type	Estimation or hypothesis testing?	Estimation
S Study Type	Sample data or experiment data?	Sample data: The data are from a sample of 539 cardiac patients.
T Type of Data	One variable or two? Categorical or numerical?	One numerical variable (wait time for surgery).
N Number of Samples or Treatments	How many samples or treatments?	One sample

So, this is *estimation, sample data, one numerical variable, one sample*. This combination of answers leads you to consider a one-sample t confidence interval for a population mean.

Let's use the five-step process for estimation problems (EMC³) to construct a 90% confidence interval.

E Estimate μ, the mean wait time for patients recommended for bypass surgery, will be estimated.

M Method   Because the answers to the four key questions are estimation, sample data, one numerical variable, one sample, a one-sample t confidence interval for a population mean will be considered. A confidence level of 90% was specified.

C Check   There are two conditions that need to be met in order for the one-sample t confidence interval to be appropriate. The sample was a random sample of people recommended for bypass surgery, so the requirement of a random sample is met.

A boxplot of the 539 wait times is given in Figure 12.9. There are outliers in the data set, which might cause you to question the normality of the wait-time distribution. But because the sample size is large, it is still appropriate to use the t confidence interval.

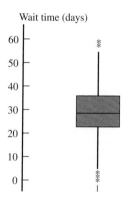

Wait time (days)

FIGURE 12.9
Boxplot of wait times

C Calculate

sample size $= n = 539$
sample mean wait time $= \bar{x} = 19$
sample standard deviation $= s = 10$

From Appendix Table 3, the t critical value $= 1.645$ (from the z critical value row because df $= n - 1 = 538 > 120$, the largest number of degrees of freedom in the table). The 90% confidence interval for μ is then

$$\bar{x} \pm (t \text{ critical value}) \left(\frac{s}{\sqrt{n}}\right) = 19 \pm (1.645)\left(\frac{10}{\sqrt{539}}\right)$$

$$= 19 \pm 0.709$$

$$= (18.291, 19.709)$$

C Communicate Results

Interpret Confidence Interval:
Based on this sample, you can be 90% confident that the mean wait time for bypass surgery is between 18.291 days and 19.709 days. This interval is fairly narrow indicating a relatively precise estimate of the population mean.

Interpret Confidence Level:
The method used to construct this interval estimate is successful in capturing the actual value of the population mean about 90% of the time.

The paper referenced in Example 12.8 also gave data on wait time for a second cardiac procedure, angiography. A random sample of 847 patients recommended for angiography resulted in a mean wait time of 18 days and a sample standard deviation of 9 days. Because the sample was a random sample and the sample size was large, it is appropriate to use the one-sample t confidence interval to estimate the mean wait time for people recommended for angiography.

A graphing calculator or statistical software can be used to produce a one-sample t confidence interval. Using a 90% confidence level, output from Minitab for the angiography data is shown here.

One-Sample T

N	Mean	StDev	SE Mean	90% CI
847	18.0000	9.0000	0.3092	(17.4908, 18.5092)

The 90% confidence interval for mean wait time for angiography extends from 17.4908 days to 18.5092 days. This interval is narrower than the 90% interval for bypass surgery wait time found in Example 12.8 for two reasons: the sample size is larger (847 rather than 539) and the sample standard deviation is smaller (9 rather than 10).

Example 12.9   Selfish Chimps?

The article **"Chimps Aren't Charitable"** (*Newsday*, **November 2, 2005**) summarized a research study published in the journal *Nature*. In this study, chimpanzees learned to use an apparatus that dispensed food when either of two ropes was pulled. When one of the ropes was pulled, only the chimp controlling the apparatus received food. When the other rope was pulled, food was dispensed both to the chimp controlling the apparatus and also to a chimp in the adjoining cage. The accompanying data (approximated from a graph in the paper) represent the number of times out of 36 trials that each of seven chimps chose the option that would provide food to both chimps (the "charitable" response).

23   22   21   24   19   20   20

You can use these data to estimate the mean number of times out of 36 that chimps choose the charitable response. For purposes of this example, let's suppose it is reasonable to regard this sample of seven chimps as representative of the population of all chimpanzees.

   This is an estimation problem, and you have sample data, one numerical variable (the number of times out of 36 that the charitable response is chosen), and one sample. These are the answers to the four key questions that lead you to consider a one-sample *t* confidence interval for a population mean as a potential method.

   The five-step process for estimation problems (EMC^3) can be used to construct a 99% confidence interval.

E Estimate   The mean number of times out of 36 that chimps choose the charitable response, μ, will be estimated.

M Method   Because the answers to the four key questions are estimation, sample data, one numerical variable, one sample, a one-sample *t* confidence interval for a population mean will be considered. A confidence level of 99% was specified.

C Check   There are two conditions that must be met in order for the one-sample *t* confidence interval to be appropriate. It was stated that it is reasonable to regard the sample as representative of the population. Because the sample size is small ($n = 7$), you need to consider whether it is reasonable to think that the distribution of number of charitable responses out of 36 for the population of all chimps is at least approximately normal. Figure 12.10 is a normal probability plot of the sample data. The plot is reasonably straight, so it seems plausible that the population distribution is approximately normal.

AP* EXAM TIP

If the sample size is small, it is important to check the normality condition before calculating a confidence interval for a population mean.

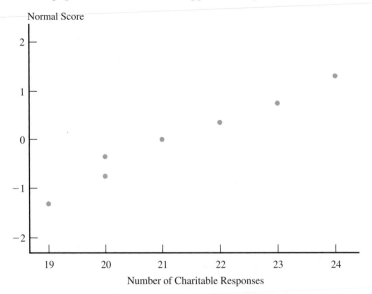

FIGURE 12.10
Normal probability plot for the sample data of Example 12.9

C Calculate   Calculation of a confidence interval for the mean number of charitable responses for the population of all chimps requires $\bar{x}$ and s. From the given data,

$$\bar{x} = 21.29 \qquad\qquad s = 1.80$$

The t critical value for a 99% confidence interval based on 6 df is 3.71 (obtained from Appendix Table 3 or using technology). The 99% confidence interval is

$$\bar{x} \pm (t \text{ critical value})\left(\frac{s}{\sqrt{n}}\right) = 21.29 \pm (3.71)\left(\frac{1.80}{\sqrt{7}}\right)$$
$$= 21.29 \pm 2.52$$
$$= (18.77, 23.81)$$

A statistical software package or a graphing calculator could also have been used to compute the 99% confidence interval. The following is output from SPSS.

One-Sample Statistics

	N	Mean	Std. Deviation	Std. Error Mean
CharitableResponses	7	21.2857	1.79947	.68014

	99% Confidence Interval	
	Lower	**Upper**
CharitableResponses	18.7642	23.8073

C Communicate Results

Interpret Confidence interval:
Based on this sample, you can be 99% confident that the population mean number of charitable responses (out of 36 trials) is between 18.76 and 23.81.

Interpret Confidence level:
The 99% confidence level implies that if the same formula is used to calculate intervals for sample after sample randomly selected from the population of chimps, in the long run 99% of these intervals will capture μ between the lower and upper confidence limits.

Notice that based on this interval, you would conclude that, on average, chimps choose the charitable option more than half the time (more than 18 out of 36 trials). The *Newsday* headline "Chimps Aren't Charitable" was based on additional data from the study indicating that chimps' charitable behavior was no different when there was another chimp in the adjacent cage from when the adjacent cage was empty.

Margin of Error and Choosing the Sample Size

The margin of error associated with a statistic was defined in Section 9.2 as the maximum likely estimation error expected when the statistic is used as an estimator. When $\bar{x}$ from a large random sample ($n \geq 30$) is used to estimate a population mean μ, the approximate margin of error is

$$\text{margin of error} = 1.96 \frac{\sigma}{\sqrt{n}}$$

which is usually estimated by

$$\text{estimated margin of error} = 1.96 \frac{s}{\sqrt{n}}$$

(When calculating the margin of error, it is acceptable to use either 2 or 1.96 in the calculation.) It would be unusual for the sample mean to differ from the actual value of the population mean by more than the margin of error.

Before collecting any data, an investigator may wish to determine a sample size required to achieve a particular margin of error, M. For example, with μ representing the mean fuel efficiency (in miles per gallon, mpg) for all cars of a certain model, you might want to estimate μ with a margin of error of 1 mpg. The value of n necessary to achieve this is obtained by setting the expression for the margin of error equal to 1 and then solving this equation for n. That is, you would solve

$$1 = 1.96 \frac{\sigma}{\sqrt{n}}$$

for n.

AP* EXAM TIP

This definition of margin of error is the one that is widely used in newspapers and magazines. It is the ± part of the 95% confidence interval. A more general definition of margin of error involves associating a confidence level with the margin of error. The margin of error is then defined to be the ± part of the confidence interval corresponding to that confidence level.

In general, if you want to estimate μ with a particular margin of error, M, you find the necessary sample size by solving the equation $M = 1.96 \dfrac{\sigma}{\sqrt{n}}$ for n. The result is

$$n = \left(\frac{1.96\sigma}{M}\right)^2$$

Notice that the greater the variability in the population (larger σ), the greater the required sample size. Also notice that a larger sample size is required for a smaller desired margin of error.

Use of this sample-size formula requires a value for σ, but σ is rarely known. One possible strategy for estimating σ is to carry out a preliminary study and use the resulting sample standard deviation (or a somewhat larger value, to be conservative). Another possibility is to make an educated guess about the value of σ. For a population distribution that is not too skewed, dividing the anticipated range (the difference between the largest and the smallest values) by 4 often gives a rough estimate of the value of the standard deviation.

AP* EXAM TIP

The given sample size formula is for the usual definition of margin of error, which corresponds to a 95% confidence level. For different confidence levels, use the z critical value determined by the desired confidence level in place of 1.96 in the formula.

The sample size required to estimate a population mean μ with a specified margin of error M is

$$n = \left(\frac{1.96\sigma}{M}\right)^2$$

If the value of σ is unknown, it may be estimated based on previous information or, for a population that is not too skewed, by using $\dfrac{\text{range}}{4}$.

Example 12.10   Cost of Textbooks

A college financial aid advisor wants to estimate the mean cost of textbooks per quarter for students at the college. For the estimate to be useful, it should have a margin of error of $20 or less. How large a sample should be used to be confident of achieving this level of accuracy?

To determine the required sample size, you need a value for σ. Suppose the advisor thinks that the amount spent on books varies widely, but that most values are between $150 and $550. A reasonable estimate of σ is then

$$\frac{\text{range}}{4} = \frac{550 - 150}{4} = \frac{400}{4} = 100$$

Using this estimate of the population standard deviation, the required sample size is

$$n = \left(\frac{1.96\sigma}{M}\right)^2 = \left(\frac{(1.96)(100)}{20}\right)^2 = (9.8)^2 = 96.04$$

Rounding up, a sample size of 97 or larger is recommended.

SECTION 12.2 EXERCISES

Each Exercise Set assesses the following chapter learning objectives: C3, M4, M5, M6, M7, M8, M9, P3, P4

SECTION 12.2   **Exercise Set 1**

12.19   What percentage of the time will a variable that has a t distribution with the specified degrees of freedom fall in the indicated region?
a. 10 df, between -1.81 and 1.81
b. 24 df, between -2.06 and 2.06
c. 24 df, outside the interval from -2.80 to 2.80
d. 10 df, to the left of -1.81

12.20   The formula used to compute a confidence interval for the mean of a normal population is

$$\bar{x} \pm (t \text{ critical value}) \frac{s}{\sqrt{n}}$$

What is the appropriate t critical value for each of the following confidence levels and sample sizes?
a. 95% confidence, $n = 17$
b. 99% confidence, $n = 24$
c. 90% confidence, $n = 13$

12.21   The two intervals (114.4, 115.6) and (114.1, 115.9) are confidence intervals for μ = mean resonance frequency (in hertz) for all tennis rackets of a certain type. The two intervals were computed using the same sample data.
a. What is the value of the sample mean resonance frequency?
b. The confidence level for one of these intervals is 90%, and for the other it is 99%. Which is which, and how can you tell?

12.22   In June, 2009, **Harris Interactive** conducted its Great Schools Survey. In this survey, the sample consisted of 1,086 adults who were parents of school-aged children. The sample was selected to be representative of the population of parents of school-aged children. One question on the survey asked respondents how much time per month (in hours) they spent volunteering at their children's school during the previous school year. The following summary statistics for time volunteered per month were given:

$$n = 1{,}086 \qquad \bar{x} = 5.6 \qquad \text{median} = 1$$

a. What does the fact that the mean is so much larger than the median tell you about the distribution of time spent volunteering at school per month?

b. Based on your answer to Part (a), explain why it is not reasonable to assume that the population distribution of time spent volunteering is approximately normal.

c. Explain why it is appropriate to use the one-sample t confidence interval to estimate the mean time spent volunteering for the population of parents of school-aged children even though the population distribution is not approximately normal.

d. Suppose that the sample standard deviation was $s = 5.2$. Use the five-step process for estimation problems (EMC^3) to compute and interpret a 98% confidence interval for μ, the mean time spent volunteering for the population of parents of school-aged children.

12.23   In a study of academic procrastination, the authors of the paper **"Correlates and Consequences of Behavioral Procrastination"** (*Procrastination, Current Issues and New Directions* [2000]) reported that for a sample of 411 undergraduate students at a mid-size public university, the mean time spent studying for the final exam in an introductory psychology course was 7.74 hours and the standard deviation of study times was 3.40 hours. Assume that this sample is representative of students taking introductory psychology at this university. Construct a 95% confidence interval estimate of μ, the mean time spent studying for the introductory psychology final exam.

12.24   The paper referenced in the previous exercise also gave the following sample statistics for the percentage of study time that occurred in the 24 hours prior to the final exam:

$$n = 411 \qquad \bar{x} = 43.18 \qquad s = 21.46$$

Construct and interpret a 90% confidence interval for the mean percentage of study time that occurs in the 24 hours prior to the final exam.

12.25   The Bureau of Alcohol, Tobacco, and Firearms (BATF) has been concerned about lead levels in California wines. In a previous testing of wine specimens, lead levels ranging from 50 to 700 parts per billion were recorded. How many wine specimens should be tested if the BATF wishes to estimate the mean lead level for California wines with a margin of error of 10 parts per billion?

12.26   The article **"Most Canadians Plan to Buy Treats, Many Will Buy Pumpkins, Decorations and/or Costumes" (Ipsos-Reid, October 24, 2005)** summarized a survey of 1,000 randomly selected Canadian residents. Each individual in the sample was asked how much he or she anticipated spending on Halloween. The resulting sample mean and standard deviation were $46.65 and $83.70, respectively.

a. Explain how it could be possible for the standard deviation of the anticipated Halloween expense to be larger than the mean anticipated expense.

b. Is it reasonable to think that the distribution of anticipated Halloween expense is approximately normal? Explain why or why not.

c. Is it appropriate to use the one-sample t confidence interval to estimate the mean anticipated Halloween expense for Canadian residents? Explain why or why not.

d. If appropriate, construct and interpret a 99% confidence interval for the mean anticipated Halloween expense for Canadian residents.

SECTION 12.2   Exercise Set 2

12.27   What percentage of the time will a variable that has a t distribution with the specified degrees of freedom fall in the indicated region?

a. 10 df, between -2.23 and 2.23

b. 24 df, between -2.80 and 2.80

c. 24 df, to the right of 2.80

12.28   The formula used to compute a confidence interval for the mean of a normal population is

$$\bar{x} \pm (t \text{ critical value}) \frac{s}{\sqrt{n}}$$

What is the appropriate t critical value for each of the following confidence levels and sample sizes?

a. 90% confidence, $n = 12$

b. 90% confidence, $n = 25$

c. 95% confidence, $n = 10$

12.29   Samples of two different models of cars were selected, and the actual speed for each car was determined when the speedometer registered 50 mph. The resulting 95% confidence intervals for mean actual speed were (51.3, 52.7) for model 1 and (49.4, 50.6) for model 2. Assuming that the two sample standard deviations were equal, which confidence interval is based on the larger sample size? Explain your reasoning.

12.30   The authors of the paper **"Deception and Design: The Impact of Communication Technology on Lying Behavior"** (*Proceedings of Computer Human Interaction* [2004]) asked 30 students in an upper division communications course at a large university to keep a journal for 7 days, recording each social interaction and whether or not they told any lies during that interaction. A lie was defined as "any time you intentionally try to mislead someone." The paper reported that the mean number of lies per day for the 30 students was 1.58, and the standard deviation of number of lies per day was 1.02.

a. What conditions must be met in order for the one-sample t confidence interval of this section to be an appropriate method for estimating μ, the mean number of lies per day for all students at this university?

b. Would you recommend using the one-sample t confidence interval to construct an estimate of μ? Explain why or why not.

12.31 Acrylic bone cement is sometimes used in hip and knee replacements to secure an artificial joint in place. The force required to break an acrylic bone cement bond was measured for six specimens, and the resulting mean and standard deviation were 306.09 Newtons and 41.97 Newtons, respectively. Assuming that it is reasonable to believe that breaking force has a distribution that is approximately normal, use a confidence interval to estimate the mean breaking force for acrylic bone cement.

12.32 A manufacturer of college textbooks is interested in estimating the strength of the bindings produced by a particular binding machine. Strength can be measured by recording the force required to pull the pages from the binding. If this force is measured in pounds, how many books should be tested to estimate the average force required to break the binding with a margin of error of 0.1 pound? Assume that σ is known to be 0.8 pound.

12.33 Because of safety considerations, in May, 2003, the Federal Aviation Administration (FAA) changed its guidelines for how small commuter airlines must estimate passenger weights. Under the old rule, airlines used 180 pounds as a typical passenger weight (including carry-on luggage) in warm months and 185 pounds as a typical weight in cold months. The *Alaska Journal of Commerce* (May 25, 2003) reported that Frontier Airlines conducted a study to estimate mean passenger plus carry-on weights. They found a mean summer weight of 183 pounds and a winter mean of 190 pounds. Suppose that these estimates were based on random samples of 100 passengers and that the sample standard deviations were 20 pounds for the summer weights and 23 pounds for the winter weights.

a. Construct and interpret a 95% confidence interval for the mean summer weight (including carry-on luggage) of Frontier Airlines passengers.

b. Construct and interpret a 95% confidence interval for the mean winter weight (including carry-on luggage) of Frontier Airlines passengers.

c. The new FAA recommendations are 190 pounds for summer and 195 pounds for winter. Comment on these recommendations in light of the confidence interval estimates from Parts (a) and (b).

Additional Exercises

12.34 What percentage of the time will a variable that has a t distribution with the specified degrees of freedom fall in the indicated region?

a. 5 df, between -2.02 and 2.02
b. 14 df, between -2.98 and 2.98
c. 22 df, outside the interval from -1.72 to 1.72
d. 26 df, to the left of -2.48

12.35 The formula used to compute a confidence interval for the mean of a normal population when n is small is

$$\bar{x} \pm (t \text{ critical value}) \frac{s}{\sqrt{n}}$$

What is the appropriate t critical value for each of the following confidence levels and sample sizes?

a. 95% confidence, $n = 15$
b. 99% confidence, $n = 20$
c. 90% confidence, $n = 26$

12.36 Fat contents (in grams) for seven randomly selected hot dogs rated as very good by **Consumer Reports** (**www.consumerreports.org**) are shown. Is it reasonable to use these data and the t confidence interval of this section to construct a confidence interval for the mean fat content of hot dogs rated very good by Consumer Reports? Explain why or why not.

| 14 | 15 | 11 | 10 | 6 | 15 | 16 |

12.37 Five students visiting the student health center for a free dental examination during National Dental Hygiene Month were asked how many months had passed since their last visit to a dentist. Their responses were:

| 6 | 17 | 11 | 22 | 29 |

Assuming that these five students can be considered as representative of all students participating in the free checkup program, construct a 95% confidence interval for the mean number of months elapsed since the last visit to a dentist for the population of students participating in the program.

12.38 The paper **"The Curious Promiscuity of Queen Honey Bees (Apis mellifera): Evolutionary and Behavioral Mechanisms"** (*Annals of Zoology* [2001]:255–265) describes a study of the mating behavior of queen honeybees. The following quote is from the paper:

> Queens flew for an average of 24.2 ± 9.21 minutes on their mating flights, which is consistent with previous findings. On those flights, queens effectively mated with 4.6 ± 3.47 males (mean ± SD).

The intervals reported in the quote from the paper were based on data from the mating flights of $n = 30$ queen honeybees. One of the two intervals reported was identified as a 95% confidence interval for a population mean. Which interval is this? Justify your choice.

12.39 Use the information given in the previous exercise to construct a 95% confidence interval for the mean number of partners on a mating flight for queen honeybees. For purposes of this exercise, assume these 30 queen honeybees are representative of the population of queen honeybees.

12.40 Seventy-seven students at the University of Virginia were asked to keep a diary of a conversation with their mothers, recording any lies they told during this conversation *(San Luis Obispo Telegram-Tribune, August 16, 1995)*. The mean number of lies per conversation was 0.5. Suppose that the standard deviation (which was not reported) was 0.4.

a. Suppose that this group of 77 is representative of the population of students at this university. Construct a 95% confidence interval for the mean number of lies per conversation for this population.

b. The interval in Part (a) does not include 0. Does this imply that all students lie to their mothers? Explain.

SECTION 12.3 Testing Hypotheses About a Population Mean

In the previous section, you learned how data from a sample can be used to estimate a population mean. Sample data can also be used to test hypotheses about a population mean.

Example 12.11 Visual Aspects of Word Clouds

A word cloud is a visual representation of a text passage that uses characteristics such as font size and color to indicate the importance of a particular word. One program that will create a word cloud from text is called Wordle (found at www.wordle.net). To illustrate what a word cloud looks like, Wordle was used to create a word cloud for the text of Section 12.2, and the result is shown in Figure 12.11.

Courtesy of Wordle.net

FIGURE 12.11
A Wordle word cloud for the text of Section 12.2

The authors of the paper **"Seeing Things in the Clouds: The Effect of Visual Features on Tag Cloud Selections"** (*Proceedings of the Nineteenth ACM Conference on Hypertext and Hypermedia*, [2008]) carried out a study to evaluate how people view word clouds and which characteristics of the words in a word cloud are most influential. In one part of this study, participants were shown a word cloud and asked to choose a word that they found to be visually important. The actual mean font size of all the words in the word clouds used in the study was 2.0. The researchers reasoned that if font size was not an important factor in how people select visually important words, the mean font size for selected words should be equal to 2.0, the mean font size of all the words in the clouds. The researchers viewed the participants as a representative sample from the population of people who might view a word cloud and used the sample data to test the following hypotheses

$$H_0 : \mu = 2.0$$
$$H_a : \mu > 2.0$$

where μ represents the mean font size of visually important words for the population. Because they were able to reject the null hypothesis, the researchers concluded that font size was a factor in the selection of visually important words.

The researchers also studied the influence of word color and intensity (the "boldness"). Based on their analysis, they concluded that there was evidence that word intensity was also a factor in the selection of important words, but that word color was not.

The logic of hypothesis testing and a general process for carrying out a hypothesis test were developed in Chapter 10 in the context of testing hypotheses about a population proportion. The same general process (HMC³) is used for all hypothesis tests. What may differ from one type of test to another is

1. The form of the null and alternative hypotheses
2. The test statistic
3. The way in which the associated *P*-value is calculated

Let's consider each of these three things in the context of testing a hypothesis about a population mean.

1. Hypotheses

When testing hypotheses about a population mean, the null hypothesis will have the form

$$H_0 : \mu = \mu_0$$

where μ_0 is a particular hypothesized value. The alternative hypothesis has one of the following three forms, depending on the research question being addressed:

$$H_a : \mu > \mu_0$$
$$H_a : \mu < \mu_0$$
$$H_a : \mu \neq \mu_0$$

To determine the null and alternative hypotheses, you will need to translate the question of interest into hypotheses. For example, if you want to determine if there is evidence that the population mean is greater than 100, you would choose $H_a : \mu > 100$ as the alternative hypothesis. The null hypothesis would then be $H_0 : \mu = 100$.

2. Test Statistic

If *n* is large ($n \geq 30$) or if the population distribution is approximately normal, the appropriate test statistic is

$$t = \frac{\bar{x} - \mu_0}{\dfrac{s}{\sqrt{n}}}$$

If the null hypothesis is true, this test statistic has a *t* distribution with df $= n - 1$. This means that the *P*-value for a hypothesis test about a population mean will be based on a *t* distribution and not the standard normal (*z*) distribution.

3. Computing a *P*-value

Once the value of the test statistic has been determined, the next step is to compute a *P*-value. The *P*-value tells you how likely it would be to observe sample data as or more extreme than what was observed if the null hypothesis were true. When the null hypothesis is true, the test statistic has a distribution that is approximately a *t* distribution with df $= n - 1$. This means that the *P*-value will be computed as an area under a *t* distribution curve.

To see how this is done, suppose you are carrying out a hypothesis test where the null hypothesis is $H_0 : \mu = 100$ and the alternative hypothesis is $H_a : \mu > 100$. The test statistic for this test is then

$$t = \frac{\bar{x} - 100}{\dfrac{s}{\sqrt{n}}}$$

If a sample of size $n = 24$ resulted in a sample mean of $\bar{x} = 104.20$ and a sample standard deviation of $s = 8.23$, the resulting test statistic value is

$$t = \frac{104.20 - 100}{\dfrac{8.23}{\sqrt{24}}} = \frac{4.20}{1.6799} = 2.50$$

Because this is an upper-tailed test, if the test statistic had been *z* rather than *t*, the *P*-value would be the area under the *z* curve to the right of 2.50. With this *t* statistic, the *P*-value is the area to the right of 2.50 under the *t* curve with df $= 24 - 1 = 23$. Appendix Table 4 is a tabulation of *t* curve tail areas. Each column of the table is for a different number of degrees of freedom: 1, 2, 3, ... , 30, 35, 40, 60, 120, and a last column for df $= \infty$, which contains the same tail areas as the *z* curve. The table gives the area under each *t* curve to the right of values ranging from 0.0 to 4.0 in increments of 0.1. Part of this table appears in Figure 12.12.

df *t*	1	2	...	22	23	24	...	60	120
0.0									
0.1									
⋮				⋮	⋮	⋮			
2.5			...	.010	.010	.010	...		
2.6			...	.008	.008	.008	...		
2.7			...	.007	.006	.006	...		
2.8			...	.005	.005	.005	...		
⋮				⋮	⋮	⋮			
4.0									

Area under 23-df
t curve to right of 2.7

FIGURE 12.12
Part of Appendix Table 4: *t* curve tail areas

Suppose that $t = -2.7$ for a lower-tailed test based on 23 df. Then, because any *t* curve is symmetric about 0,

P-value $=$ area to the left of $-2.7 =$ area to the right of $2.7 = 0.006$

As with *z* tests, you double a tail area to obtain the *P*-value for a two-tailed *t* test. For example, if $t = 2.6$ or if $t = -2.6$ for a two-tailed test with 23 df, then

P-value $= 2(0.008) = 0.016$

Once past 120 df, the tail areas don't change much, so the last column (∞) in Appendix Table 4 provides a good approximation.

The following box summarizes how the P-value is obtained as a t curve area.

Finding P-Values for a t Test

1. **Upper-tailed test:**
   $H_a: \mu >$ hypothesized value

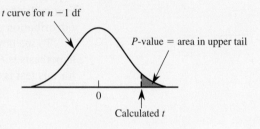

2. **Lower-tailed test:**
   $H_a: \mu <$ hypothesized value

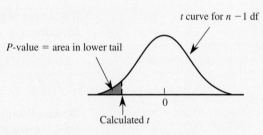

3. **Two-tailed test:**
   $H_a: \mu \neq$ hypothesized value

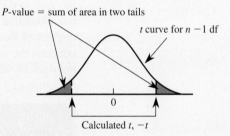

Appendix Table 4 gives upper-tail t curve areas to the right of values 0.0, 0.1, ... , 4.0. These areas are P-values for upper-tailed tests and, by symmetry, also for lower-tailed tests. Doubling an area gives the P-value for a two-tailed test.

You now have all the pieces needed to carry out the hypothesis test. The one-sample t test for a population mean, summarized in the following box, is the method you should consider when the answers to the four key questions that lead to a recommended method are: *hypothesis testing, sample data, one numerical variable, and one sample.*

One-Sample t Test for a Population Mean

Appropriate when the following conditions are met:

1. The sample is a random sample from the population of interest or the sample is selected in a way that results in a sample that is representative of the population.
2. The sample size is large ($n \geq 30$) *or* the population distribution is approximately normal.

When these conditions are met, the following test statistic can be used:

$$t = \frac{\bar{x} - \mu_0}{\dfrac{s}{\sqrt{n}}}$$

where μ_0 is the hypothesized value from the null hypothesis.

(continued)

Form of the null hypothesis: $H_0: \mu = \mu_0$

When the conditions above are met and the null hypothesis is true, the t test statistic has a t distribution with df $= n - 1$.

Associated P-value:

When the alternative hypothesis is...	The P-value is...
$H_a: \mu > \mu_0$	Area under the t curve to the right of the calculated value of the test statistic
$H_a: \mu < \mu_0$	Area under the t curve to the left of the calculated value of the test statistic
$H_a: \mu \neq \mu_0$	2 (area to the right of t) if t is positive or 2 (area to the left of t) if t is negative

Now you're ready to look at some examples, following the five-step process for hypothesis testing problems (HMC[3]).

Example 12.12   Time Stands Still (or So It Seems)

A study conducted by researchers at Pennsylvania State University investigated whether time perception, an indication of a person's ability to concentrate, is impaired during nicotine withdrawal. The study results were summarized in the paper **"Smoking Abstinence Impairs Time Estimation Accuracy in Cigarette Smokers"** (*Psychopharmacology Bulletin* [2003]: 90–95). After a 24-hour smoking abstinence, 20 smokers were asked to estimate how much time had passed during a 45-second period. Suppose the resulting data on perceived elapsed time (in seconds) were as follows (these data are artificial but are consistent with summary quantities given in the paper):

69	65	72	73	59	55	39	52	67	57
56	50	70	47	56	45	70	64	67	53

These data were used to compute the sample mean and standard deviation:

$$n = 20 \qquad \bar{x} = 59.30 \qquad s = 9.84$$

The authors of the paper believed that it was reasonable to consider this sample as representative of smokers in general. The researchers wanted to determine whether smoking abstinence tends to lead to elapsed time being overestimated.

Begin by answering the four key questions (QSTN) for this problem:

Q Question Type	Estimation or hypothesis testing?	Hypothesis testing. The researchers want to test a claim about a population (elapsed time is overestimated).
S Study Type	Sample data or experiment data?	Sample data. The data are from a sample of smokers.
T Type of Data	One variable or two? Categorical or numerical?	One numerical variable (perceived elapsed time).
N Number of Samples or Treatments	How many samples or treatments?	One.

Because the answers are *hypothesis testing, sample data, one numerical variable, and one sample*, you should consider a one-sample *t* test for a population mean.

Next, use the five-step process for hypothesis testing problems (HMC³).

Process Step	
H Hypotheses	The claim is about the population of smokers and the population characteristic of interest is a population mean: μ = mean perceived elapsed time for smokers who have not smoked for 24 hours. The claim (mean perceived elapsed time is greater than the actual elapsed time of 45 seconds) translates into an alternative hypothesis of $\mu > 45$. The null hypothesis is found by replacing the inequality (>) in the alternative hypothesis by an equal sign. **Hypotheses:** *Null hypothesis:* $H_0: \mu = 45$ *Alternative hypothesis:* $H_a: \mu > 45$
M Method	Because the answers to the four key questions are hypothesis testing, sample data, one numerical variable and one sample, consider a one-sample *t* test for a population mean. **Potential method:** One-sample *t* test for a population mean. The test statistic for this test is $$t = \frac{\bar{x} - 45}{\dfrac{s}{\sqrt{n}}}$$ The value of 45 in the test statistic is the hypothesized value from the null hypothesis. When the null hypothesis is true, this statistic will have a *t* distribution with df = 20 − 1 = 19. You also need to select a significance level for the test. In some cases, a significance level will be specified. If that is not the case, you should choose a significance level based on a consideration of the consequences of Type I and Type II errors. In this example, a Type I error is deciding that smokers who have not smoked in 24 hours do overestimate elapsed time when they really don't. A Type II error is concluding that there is no evidence that smokers who have not smoked for 24 hours overestimate elapsed time when they really do overestimate. In this situation, because neither type of error is much more serious than the other, a value for α of 0.05 is a reasonable choice. **Significance level:** $\alpha = 0.05$
C Check	There are two conditions that need to be met in order to use the one-sample *t* test for a population mean. The requirement of a random sample or a sample that can be regarded as representative of the population requires some thought. The researchers conducting the study indicated that they believed that the sample was selected in a way that would result in a sample that was representative of smokers in general. If this is the case, it is reasonable to proceed. The second condition is that the sample size is large or the population distribution is approximately normal. Because *n* is only 20 in this example, you need to verify that the normality condition is reasonable.

(continued)

Process Step	
	A boxplot of the sample data is shown here. 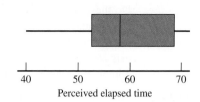 Although the boxplot is not perfectly symmetric, it is not too skewed and there are no outliers. It is reasonable to think that the population distribution is at least approximately normal. Because both conditions are met, it is appropriate to use the one-sample t test.
C Calculate	$n = 20$ $\bar{x} = 59.30$ $s = 9.84$ **Test statistic:** $$t = \frac{59.30 - 45}{\dfrac{9.84}{\sqrt{20}}} = \frac{14.30}{2.20} = 6.50$$ This is an upper-tailed test (the inequality in H_a is >), so the P-value is the area under the t curve with df = 19 to the right of the computed t value. Because $t = 6.50$ is greater than 4.0, the largest value in Appendix Table 4, the area to the right is approximately 0. **Associated P-value:** P-value = area under t curve to the right of 6.50 $\qquad\qquad = P(t > 6.50)$ $\qquad\qquad \approx 0$
C Communicate Results	Because the P-value is less than the selected significance level, the null hypothesis is rejected. **Decision:** $0 < 0.05$, Reject H_0. **Conclusion:** There is convincing evidence that the mean perceived time elapsed is greater than the actual time elapsed of 45 seconds. It would be very unlikely to see a sample mean (and hence a t value) this extreme just by chance when H_0 is true.

This paper also looked at perception of elapsed time for a sample of nonsmokers and for a sample of smokers who had not abstained from smoking. The investigators found that the null hypothesis of $\mu = 45$ could not be rejected for either of these groups.

Example 12.13 Goofing Off at Work

Many employers are concerned about employees wasting time by surfing the Internet and e-mailing friends during work hours. The article **"Who Goofs Off 2 Hours a Day? Most Workers, Survey Says"** (*San Luis Obispo Tribune*, **August 3, 2006**) summarized data from a large sample of workers. Suppose the CEO of a large company wants to determine whether the mean wasted time during an 8-hour work day for employees of her company is less than the mean of 120 minutes reported in the article. Each person in a random sample of 10 employees is asked about daily wasted time at work (in minutes). Participants would probably have to be guaranteed anonymity to obtain truthful responses. Suppose the resulting data are:

108	112	117	130	111	131	113	113	105	128

The sample mean and standard deviation calculated from these data are

$$n = 10 \qquad \bar{x} = 116.80 \qquad s = 9.45$$

Do these data provide evidence that the mean wasted time for this company is less than 120 minutes?

Considering the four key questions (QSTN), this situation can be described as *hypothesis testing, sample data, one numerical variable (wasted time), and one sample*. This combination suggests a one-sample *t* test for a population mean.

H Hypotheses   You want to use data from the sample to test a claim about the population of company employees. The population characteristic of interest is

μ = mean daily wasted time for employees of this company

Translating the question of interest into hypotheses gives

$$H_0: \mu = 120$$
$$H_a: \mu < 120$$

M Method   Because the answers to the four key questions are hypothesis testing, sample data, one numerical variable, and one sample, consider a one-sample *t* test for a population mean. The test statistic for this test is

$$t = \frac{\bar{x} - 120}{\dfrac{s}{\sqrt{n}}}$$

The value of 120 in the test statistic is the hypothesized value from the null hypothesis. When the null hypothesis is true, this statistic will have a *t* distribution with df = $10 - 1 = 9$.

You also need a significance level for the test. For this example, a significance level of $\alpha = 0.05$ will be used.

Significance level: $\alpha = 0.05$

C Check   The one-sample *t* test requires a random sample and either a large sample or a normal population distribution. The sample was a random sample of employees. Because the sample size is small, you must be willing to assume that the population distribution of times is at least approximately normal. The following normal probability plot appears to be reasonably straight, and although the normal probability plot and the boxplot reveal some skewness in the sample, there are no outliers.

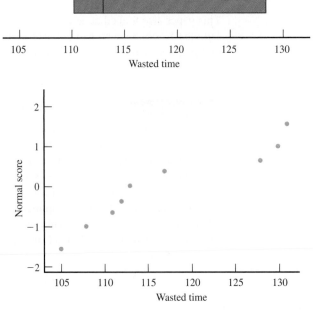

Based on these observations, it is plausible that the population distribution is approximately normal, so you can proceed with the one-sample t test.

C Calculate

$$n = 10 \qquad \bar{x} = 116.80 \qquad s = 9.45$$

$$t = \frac{116.80 - 120}{\dfrac{9.45}{\sqrt{10}}} = -1.07$$

From the df $= 9$ column of Appendix Table 4 and rounding the test statistic value to -1.1, you get

$$P\text{-value} = \text{area to the left of } -1.1 = \text{area to the right of } 1.1 = 0.150$$

as shown:

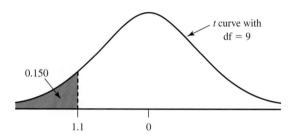

t curve with df $= 9$

0.150

1.1   0

C Communicate Results

Because the P-value is greater than the selected significance level, you fail to reject the null hypothesis.

Decision:
$0.150 > 0.05$, Fail to reject H_0.

Interpretation:
There is not convincing evidence that the mean wasted time per 8-hour work day for employees at this company is less than 120 minutes.

JMP could also have been used to carry out the test of Example 12.13, as shown in the accompanying output:

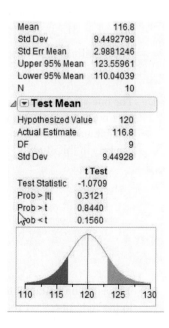

Mean	116.8
Std Dev	9.4492798
Std Err Mean	2.9881246
Upper 95% Mean	123.55961
Lower 95% Mean	110.04039
N	10

Test Mean

Hypothesized Value	120
Actual Estimate	116.8
DF	9
Std Dev	9.44928

t Test

Test Statistic	-1.0709
Prob > |t|	0.3121
Prob > t	0.8440
Prob < t	0.1560

110   115   120   125   130

Although you have to round the computed t value to -1.1 to use Appendix Table 4, JMP was able to compute the P-value corresponding to the actual value of the test statistic, which is P-value $= 0.156$ (which appears in the JMP output as Prob $< t$ just above the figure in the output; the $<$ probability is the P-value because this was a lower-tail test).

Example 12.14 Cricket Love

The article **"Well-Fed Crickets Bowl Maidens Over"** (**Nature Science Update, February 11, 1999**) reported that female field crickets are attracted to males that have high chirp rates and hypothesized that chirp rate is related to nutritional status. The chirp rates for male field crickets were reported to vary around a mean of 60 chirps per second. To investigate whether chirp rate is related to nutritional status, investigators fed male crickets a high protein diet for 8 days, and then measured chirp rate. The mean chirp rate for the crickets on the high-protein diet was reported to be 109 chirps per second. Is this convincing evidence that the mean chirp rate for crickets on a high-protein diet is greater than 60 (which would then imply an advantage in attracting the ladies)? Suppose that the sample size and sample standard deviation are $n = 32$ and $s = 40$. Let's test the relevant hypotheses with $\alpha = 0.01$.

Considering the four key questions (QSTN), this situation can be described as *hypothesis testing, sample data, one numerical variable (chirp rate), and one sample*. This combination suggests a one-sample t test for a population mean.

H Hypotheses The claim to be tested is about a population mean:

$$\mu = \text{mean chirp rate for crickets on a high protein diet}$$

The question of interest is if there is evidence that $\mu > 60$, so the hypotheses to be tested are

$$H_0: \mu = 60$$
$$H_a: \mu > 60$$

M Method Because the answers to the four key questions are hypothesis testing, sample data, one numerical variable, and one sample, consider a one-sample t test for a population mean. The test statistic for this test is

$$t = \frac{\bar{x} - 60}{\dfrac{s}{\sqrt{n}}}$$

The value of 60 in the test statistic is the hypothesized value from the null hypothesis. When the null hypothesis is true, this statistic will have a t distribution with df $= 32 - 1 = 31$.

The significance level specified for this test is $\alpha = 0.01$.

Significance level: $\alpha = 0.01$

C Check The one-sample t test requires a random or representative sample and either a large sample or a normal population distribution. Because the sample size is large ($n = 32$), it is reasonable to proceed with the t test as long as you are willing to consider the 32 male field crickets in this study to be representative of the population of male field crickets.

C Calculate

$$n = 32$$
$$\bar{x} = 109$$
$$s = 40$$

$$t = \frac{109 - 60}{\dfrac{40}{\sqrt{32}}} = \frac{49}{7.07} = 6.93$$

This is an upper-tailed test, so the P-value is the area under the t curve with df $= 31$ and to the right of 6.93. From Appendix Table 4 or using technology, P-value ≈ 0.

C Communicate Results Because P-value ≈ 0, which is less than the significance level, α, H_0 is rejected.

Decision:
$0 < 0.01$, Reject H_0.

Interpretation: There is convincing evidence that the mean chirp rate for male field crickets that eat a high-protein diet is greater than 60.

Statistical Versus Practical Significance

Carrying out a hypothesis test amounts to deciding whether the value obtained for the test statistic could plausibly have resulted when H_0 is true. When the value of the test statistic leads to rejection of H_0, it is customary to say that the result is **statistically significant** at the chosen significance level α. The finding of statistical significance means that the observed deviation from what was expected when H_0 is true can't be explained by just chance variation. However, statistical significance doesn't mean that the true situation differs from what the null hypothesis states in any practical sense. That is, even after H_0 has been rejected, the data may suggest that there is no *practical* difference between the actual value of the population characteristic and the hypothesized value. This is illustrated in Example 12.15.

Example 12.15 "Significant" but Unimpressive Test Score Improvement

Let μ denote the true mean score on a standardized test for children in a large school district. The mean score for all children in the United States is known to be 100. District administrators are interested in testing H_0: $\mu = 100$ versus H_a: $\mu > 100$ using a significance level of 0.001. Data from a random sample of 2,500 children in the district resulted in $\bar{x} = 101.0$ and $s = 15.0$. Minitab output from a one-sample t test is shown here:

One-Sample T
Test of mu = 100 vs > 100

N	Mean	StDev	SE Mean	95% Lower Bound	T	P
2500	101.000	15.000	0.300	100.506	3.33	0.000

From the Minitab output, P-value ≈ 0, so H_0 is rejected. The actual mean score for this district does appear to be greater than 100. However, with $n = 2,500$, you expect that $\bar{x} = 101.0$ is very close to the actual value of μ. It appears that H_0 was rejected because μ is about 101 rather than 100. From a practical point of view, a 1-point difference here may not be important. A statistically significant result does not necessarily mean that there are any practical consequences.

A Note on Power

Recall that the power of a hypothesis test is the probability of rejecting the null hypothesis. If the null hypothesis is false and the alternative hypothesis is true, you can think of the power of the test as the probability that you will detect that the null hypothesis is false and correctly reject it.

In Section 10.6, you learned that the power of a hypothesis test about a population proportion is influenced by three factors: the size of the difference between the actual value of the population characteristic and the hypothesized value, the choice of significance level, and the sample size. The same three factors influence the power of a test of hypotheses about a population mean.

Effect of Various Factors on the Power of a Test

1. The larger the size of the difference between the hypothesized value and the actual value of the population characteristic, the greater the power.
2. The larger the significance level, α, the greater the power of the test.
3. The larger the sample size, the greater the power of the test.

Suppose that the student body president at a university is interested in studying the amount of money that students spend on textbooks each semester. The director of the financial aid office believes that the average amount spent on books is $500 per semester and uses this figure to determine the amount of financial aid for which a student is eligible. The student body president plans to ask each individual in a random sample of students how much he or she spent on books this semester and has decided to use the resulting data to test

$$H_0: \mu = 500 \quad \text{versus} \quad H_a: \mu > 500$$

Based on what you know about the factors that influence power, the power of the test will be greater if the actual mean were $525 than if the actual mean were $501. If the actual mean is 501, the probability that you reject $H_0: \mu = 500$ is not very great. This is because when you carry out the test, you are essentially looking at the sample mean and asking, "Does this look like what you would expect to see if the population mean were 500?" If the actual mean is greater than but very close to 500, chances are that the sample mean will look pretty much like what you would expect to see if the population mean were 500, and you will be unconvinced that the null hypothesis should be rejected. If the actual mean is 525, it is less likely that the sample will be mistaken for a sample from a population with mean 500. In this case, sample means will tend to cluster around 525, and so it is more likely that you will correctly reject H_0. If the true mean is 550, rejection of H_0 is even more likely.

Selecting a larger value for the significance level makes it easier to reject the null hypothesis, so the power of the test will be greater for a test with $\alpha = 0.05$ than for a test with $\alpha = 0.01$. It is also the case that departures from the null hypothesis are easier to detect with a large sample than with a small sample. This means that the power of the test previously described would be greater if the student body president selects a random sample of 200 students than if a random sample of 75 students is selected.

SECTION 12.3 EXERCISES

Each Exercise Set assesses the following chapter learning objectives: M8, P3

SECTION 12.3   Exercise Set 1

12.41 Give as much information as you can about the P-value of a t test in each of the following situations:
a. Upper-tailed test, df = 8, $t = 2.0$
b. Lower-tailed test, df = 10, $t = -2.4$
c. Lower-tailed test, $n = 22$, $t = -4.2$
d. Two-tailed test, df = 15, $t = -1.6$

12.42 The paper **"Playing Active Video Games Increases Energy Expenditure in Children"** (*Pediatrics* **[2009]: 534–539**) describes a study of the possible cardiovascular benefits of active video games. Mean heart rate for healthy boys ages 10 to 13 after walking on a treadmill at 2.6 km/hour for 6 minutes is known to be 98 beats per minute (bpm). For each of 14 boys, heart rate was measured after 15 minutes of playing Wii Bowling. The resulting sample mean and standard deviation were 101 bpm and 15 bpm, respectively. Assume that the sample of boys is representative of boys ages 10 to 13 and that the distribution of heart rates after 15 minutes of Wii Bowling is approximately normal. Does the sample provide convincing evidence that the mean heart rate after 15 minutes of Wii Bowling is different from the known mean heart rate after 6 minutes walking on the treadmill? Carry out a hypothesis test using $\alpha = 0.01$.

12.43 The paper referenced in the previous exercise also states that the known resting mean heart rate for boys in this age group is 66 bpm.
a. Is there convincing evidence that the mean heart rate after Wii Bowling for 15 minutes is higher than the known mean resting heart rate for boys of this age? Use $\alpha = 0.01$.
b. Based on the outcomes of the tests in this exercise and the previous exercise, write a paragraph comparing treadmill walking and Wii Bowling.

12.44 *The Economist* collects data each year on the price of a Big Mac in various countries around the world. A sample of McDonald's restaurants in Europe in May 2009 resulted in the following Big Mac prices (after conversion to U.S. dollars):

3.80   5.89   4.92   3.88   2.65   5.57   6.39   3.24

The mean price of a Big Mac in the U.S. in May 2009 was $3.57. For purposes of this exercise, you can assume it is reasonable to regard the sample as representative of European McDonald's restaurants. Does the sample provide convincing evidence that the mean May 2009 price of a Big Mac in Europe is greater than the reported U.S. price? Test the relevant hypotheses using $\alpha = 0.05$.

12.45 In a study of computer use, 1,000 randomly selected Canadian Internet users were asked how much time they spend online in a typical week (**Ipsos Reid, August 9, 2005**). The sample mean was 12.7 hours.

a. The sample standard deviation was not reported, but suppose that it was 5 hours. Carry out a hypothesis test with a significance level of 0.05 to decide if there is convincing evidence that the mean time spent online by Canadians in a typical week is greater than 12.5 hours.

b. Now suppose that the sample standard deviation was 2 hours. Carry out a hypothesis test with a significance level of 0.05 to decide if there is convincing evidence that the mean time spent online by Canadians in a typical week is greater than 12.5 hours.

c. Explain why the null hypothesis was rejected in the test of Part (b) but not in the test of Part (a).

12.46 The paper titled **"Music for Pain Relief"** (*The Cochrane Database of Systematic Reviews*, **April 19, 2006**) concluded, based on a review of 51 studies of the effect of music on pain intensity, that "Listening to music reduces pain intensity levels … However, the magnitude of these positive effects is small, and the clinical relevance of music for pain relief in clinical practice is unclear." Are the authors of this paper claiming that the pain reduction attributable to listening to music is not statistically significant, not practically significant, or neither statistically nor practically significant? Explain.

SECTION 12.3 **Exercise Set 2**

12.47 Give as much information as you can about the *P*-value of a *t* test in each of the following situations:

a. Two-tailed test, df = 9, $t = 0.73$
b. Upper-tailed test, df = 10, $t = 0.5$
c. Lower-tailed test, $n = 20$, $t = -2.1$
d. Two-tailed test, $n = 40$, $t = 1.7$

12.48 The report **"Highest Paying Jobs for 2009–10 Bachelor's Degree Graduates"** (**National Association of Colleges and Employers, February 2010**) states that the mean yearly salary offer for students graduating with accounting degrees in 2010 was $48,722. Suppose that a random sample of 50 accounting graduates at a large university who received job offers resulted in a mean offer of $49,850 and a standard deviation of $3,300. Do the sample data provide strong support for the claim that the mean salary offer for accounting graduates of this university is higher than the 2010 national average of $48,722? Test the relevant hypotheses using $\alpha = 0.05$.

12.49 A credit bureau analysis of undergraduate students credit records found that the average number of credit cards in an undergraduate's wallet was 4.09 (**"Undergraduate Students and Credit Cards in 2004," Nellie Mae, May 2005**). It was also reported that in a random sample of 132 undergraduates, the sample mean number of credit cards that the students said they carried was 2.6.

The sample standard deviation was not reported, but for purposes of this exercise, suppose that it was 1.2. Is there convincing evidence that the mean number of credit cards that undergraduates report carrying is less than the credit bureau's figure of 4.09?

12.50 The international polling organization Ipsos reported data from a survey of 2,000 randomly selected Canadians who carry debit cards (**Canadian Account Habits Survey, July 24, 2006**). Participants in this survey were asked what they considered the minimum purchase amount for which it would be acceptable to use a debit card. Suppose that the sample mean and standard deviation were $9.15 and $7.60 respectively. (These values are consistent with a histogram of the sample data that appears in the report.) Do the data provide convincing evidence that the mean minimum purchase amount for which Canadians consider the use of a debit card to be acceptable is less than $10? Carry out a hypothesis test with a significance level of 0.01.

12.51 Many consumers pay careful attention to stated nutritional contents on packaged foods when making purchases, so it is important that the information on packages be accurate. A random sample of $n = 12$ frozen dinners of a certain type was selected and the calorie content of each one was determined. Below are the resulting observations, along with a boxplot and a normal probability plot.

255	244	239	242	265	245	259
248	225	226	251	233		

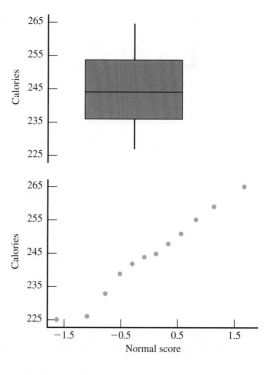

a. Is it reasonable to test hypotheses about the population mean calorie content μ using a one-sample *t* test? Explain why or why not.

b. The stated mean calorie content is 240. Does the boxplot suggest that actual mean content differs from the stated value? Explain your reasoning.

c. Carry out a formal test of the hypotheses suggested in Part (b).

Additional Exercises

12.52 Give as much information as you can about the *P*-value of a *t* test in each of the following situations:

a. Two-tailed test, $n = 16$, $t = 1.6$

b. Upper-tailed test, $n = 14$, $t = 3.2$

c. Lower-tailed test, $n = 20$, $t = -5.1$

d. Two-tailed test, $n = 16$, $t = 6.3$

12.53 The eating habits of 12 vampire bats were examined in the article **"Foraging Behavior of the Indian False Vampire Bat" (*Biotropica* [1991]: 63–67)**. These bats consume insects and frogs. For these 12 bats, the mean time to consume a frog was $\bar{x} = 21.9$ minutes. Suppose that the standard deviation was $s = 7.7$ minutes. Is there convincing evidence that the mean supper time of a vampire bat whose meal consists of a frog is greater than 20 minutes? What assumptions must be reasonable for the one-sample *t* test to be appropriate?

12.54 Much concern has been expressed regarding the practice of using nitrates as meat preservatives. In one study involving possible effects of these chemicals, bacteria cultures were grown in a medium containing nitrates. The rate of uptake of radio-labeled amino acid was then determined for each culture, yielding the following observations:

```
7,251   6,871   9,632   6,866   9,094   5,849
8,957   7,978   7,064   7,494   7,883   8,178
7,523   8,724   7,468
```

Suppose that it is known that the true average uptake for cultures without nitrates is 8,000. Do these data suggest that the addition of nitrates results in a decrease in the mean uptake? Test the appropriate hypotheses using a significance level of 0.10.

12.55 An article titled **"Teen Boys Forget Whatever It Was"** appeared in the Australian newspaper *The Mercury* **(April 21, 1997)**. It described a study of academic performance and attention span and reported that the mean time to distraction for teenage boys working on an independent task was 4 minutes. Although the sample size was not given in the article, suppose that this mean was based on a random sample of 50 teenage boys and that the sample standard deviation was

1.4 minutes. Is there convincing evidence that the average attention span for teenage boys is less than 5 minutes? Test the relevant hypotheses using $\alpha = 0.01$.

12.56 Speed, size, and strength are thought to be important factors in football performance. The article **"Physical and Performance Characteristics of NCAA Division I Football Players" (*Research Quarterly for Exercise and Sport* [1990]: 395–401)** reported on physical characteristics of Division I starting football players in the 1988 football season. The mean weight of starters on top-20 teams was reported to be 105 kg. A random sample of 33 starting players (various positions were represented) from Division I teams that were not ranked in the top 20 resulted in a sample mean weight of 103.3 kg and a sample standard deviation of 16.3 kg. Is there sufficient evidence to conclude that the mean weight for non-top-20 team starters is less than 105, the known value for top-20 teams?

12.57 An automobile manufacturer decides to carry out a fuel efficiency test to determine if it can advertise that one of its models achieves 30 mpg (miles per gallon). Six people each drive a car from Phoenix to Los Angeles. The resulting fuel efficiencies (in miles per gallon) are:

```
27.2   29.3   31.2   28.4   30.3   29.6
```

Assuming that fuel efficiency is normally distributed, do these data provide evidence against the claim that actual mean fuel efficiency for this model is (at least) 30 mpg?

12.58 A student organization uses the proceeds from a soft drink vending machine to finance its activities. The price per can was $0.75 for a long time, and the mean daily revenue during that period was $75.00. The price was recently increased to $1.00 per can. A random sample of $n = 20$ days after the price increase yielded a sample mean daily revenue and sample standard deviation of $70.00 and $4.20, respectively. Does this information suggest that the mean daily revenue has decreased from its value before the price increase? Test the appropriate hypotheses using $\alpha = 0.05$.

12.59 A hot tub manufacturer advertises that a water temperature of 100°F can be achieved in 15 minutes or less. A random sample of 25 tubs is selected, and the time necessary to achieve a 100°F temperature is determined for each tub. The sample mean time and sample standard deviation are 17.5 minutes and 2.2 minutes, respectively. Does this information cast doubt on the company's claim? Carry out a test of hypotheses using significance level 0.05.

SECTION 12.4 Avoid These Common Mistakes

Now that you have seen confidence intervals and hypothesis tests for proportions and for means, you will want to think about the distinction between proportions and means when choosing an appropriate method. The best way to distinguish between situations that involve learning about a population proportion and those that involve learning about a population mean is to focus on the type of data—categorical or numerical. With categorical data, you should be thinking proportions, and with numerical data you should be thinking means.

As was the case when using sample data to learn about a population proportion, be sure to keep in mind that conditions are important. Use of the one-sample t confidence interval and hypothesis test requires that certain conditions be met. Be sure to check that these conditions are met before using these methods.

Also, remember that the result of a hypothesis test can never provide strong support for the null hypothesis. Make sure that you don't confuse "I am not convinced that the null hypothesis is false" with "I am convinced that the null hypothesis is true." These are not the same!

CHAPTER ACTIVITIES

ACTIVITY 12.1 COMPARING THE t AND z DISTRIBUTIONS

Technology Activity: Requires use of a computer or a graphing calculator.

The following instructions assume the use of Minitab. If you are using a different software package or a graphing calculator, your instructor will provide alternative instructions.

Background: Suppose a random sample will be selected from a population that is known to have a normal distribution. Then the statistic

$$z = \frac{\bar{x} - \mu}{\frac{\sigma}{\sqrt{n}}}$$

has a standard normal (z) distribution. Since it is rarely the case that σ is known, inferences for population means are usually based on the statistic

$$t = \frac{\bar{x} - \mu}{\frac{s}{\sqrt{n}}}$$

which has a t distribution rather than a z distribution. The informal justification for this is that the use of s to estimate σ introduces additional variability, resulting in a statistic whose distribution is more spread out than is the z distribution.

In this activity, you will use simulation to sample from a known normal population and then investigate how the behavior of $t = \dfrac{\bar{x} - \mu}{\dfrac{s}{\sqrt{n}}}$ compares to $z = \dfrac{\bar{x} - \mu}{\dfrac{\sigma}{\sqrt{n}}}$.

1. Generate 200 random samples of size 5 from a normal population with mean 100 and standard deviation 10.

   Using Minitab, go to the Calc Menu. Then

   Calc → Random Data → Normal
   In the "Generate" box, enter 200
   In the "Store in columns" box, enter c1-c5
   In the mean box, enter 100

   In the standard deviation box, enter 10
   Click on OK

   You should now see 200 rows of data in each of the first five columns of the Minitab worksheet.

2. Each row contains five values that have been randomly selected from a normal population with mean 100 and standard deviation 10. Viewing each row as a sample of size 5 from this population, calculate the mean and standard deviation for each of the 200 samples (the 200 rows) by using Minitab's row statistics functions, which can also be found under the Calc menu:

   Calc → Row statistics
   Choose the "Mean" button
   In the "Input Variables" box, enter c1-c5
   In the "Store result in" box, enter c7
   Click on OK

   You should now see the 200 sample means in column 7 of the Minitab worksheet. Name this column "x-bar" by typing the name in the gray box at the top of c7.

   Follow a similar process to compute the 200 sample standard deviations, and store them in c8. Name this column "s."

3. Next, calculate the value of the z statistic for each of the 200 samples. You can calculate z in this example because you know that the samples were selected from a population for which $\sigma = 10$. Use the Minitab calculator function to compute

   $$z = \frac{\bar{x} - \mu}{\frac{\sigma}{\sqrt{n}}} = \frac{\bar{x} - 100}{\frac{10}{\sqrt{5}}}$$

   as follows:

   Calc → Calculator
   In the "Store results in" box, enter c10

In the "Expression box" type in the following: (c7–100)/(10/sqrt(5))

Click on OK

You should now see the z values for the 200 samples in c10. Name this column "z."

4. Now calculate the value of the t statistic for each of the 200 samples. Use the calculator function of Minitab to compute

$$t = \frac{\bar{x} - \mu}{\frac{s}{\sqrt{n}}} = \frac{\bar{x} - 100}{\frac{s}{\sqrt{5}}}$$

as follows:

Calc → Calculator

In the "Store results in" box, enter c11

In the "Expression box" type in the following: (c7–100)/(c8/sqrt(5))

Click on OK

You should now see the t values for the 200 samples in c11. Name c11 "t."

5. Graphs, at last! Now construct histograms of the 200 z values and the 200 t values. These two graphical displays will provide insight into how each of these two statistics behaves in repeated sampling. Use the same scale for the two histograms so that it will be easy to compare the two distributions.

Graph → Histogram

In the "Graph variables" box, enter c10 for graph 1 and c11 for graph 2

Click the Frame dropdown menu and select multiple graphs. Then under the scale choices, select "Same X and same Y."

Now use the histograms from Step 5 to answer the following questions:

a. Write a brief description of the shape, center, and spread for the histogram of the z values. Is what you see in the histogram consistent with what you would have expected to see? Explain. (Hint: In theory, what is the distribution of the z statistic?)

b. How does the histogram of the t values compare to the z histogram? Be sure to comment on center, shape, and spread.

c. Is your answer to Part (b) consistent with what would be expected for a statistic that has a t distribution? Explain.

d. Because the z and t histograms are based on only 200 samples, they only approximate the corresponding sampling distributions. The 5th percentile for the standard normal distribution is -1.645 and the 95th percentile is 1.645. For a t distribution with df $= 5 - 1 = 4$, the 5th and 95th percentiles are -2.13 and 2.13, respectively. How do these percentiles compare to those of the distributions displayed in the histograms? (Hint: Sort the 200 z values—in Minitab, choose "Sort" from the Manip menu. Once the values are sorted, the 5th and 95th percentiles from the histogram can be found by counting in 10 [which is 5% of 200] values from either end of the sorted list. Then repeat this with the t values.)

e. Are the results of your simulation and analysis consistent with the statement that the statistic $z = \dfrac{\bar{x} - \mu}{\frac{\sigma}{\sqrt{n}}}$ has a standard normal (z) distribution and the statistic $t = \dfrac{\bar{x} - \mu}{\frac{s}{\sqrt{n}}}$ has a t distribution? Explain.

EXPLORING THE BIG IDEAS

In this exercise, you will go online to select a random sample of 30 animated movies produced between 1980 and 2011. You will use this sample to estimate the mean length of the animated movies in this population.

Go online at www.cengage.com/stats/peck1e and click on the link labeled "Exploring the Big Ideas." Then locate the link for Chapter 12. It will take you to a web page where you can select a random sample from the animated movie population.

Click on the "sample" button. This selects a random sample of 30 movies and will display the movie name and the length (in minutes) for each movie in your sample. Each student in your class will receive a different random sample.

Use the data from your random sample to complete the following:

a. Calculate the mean length of the movies in your sample.

b. Is the mean you calculated in Part (a) the population mean or a sample mean?

c. Construct a 90% confidence interval for the mean length of the animated movies in this population.

d. Write a few sentences that provide an interpretation of the confidence interval from Part (c).

e. The actual population mean is 90.41 minutes. Did your confidence interval from Part (c) include this value?

f. Which of the following is a correct interpretation of the 90% confidence level?

1. The probability that the actual population mean is contained in the computed interval is 0.90.

2. If the process of selecting a random sample of movies and then computing a 90% confidence interval for the mean length of all animated movies made between 1980 and 2011 is repeated 100 times, exactly 90 of the 100 intervals will include the actual population mean.

3. If the process of selecting a random sample of movies and then computing a 90% confidence interval for the mean length all animated movies made between 1980 and 2011 is repeated a very large number of times, approximately 90% of the intervals will include the actual population mean.

If asked to do so by your teacher, bring your confidence interval estimate of the mean length of animated movies made between 1980 and 2011 to class. Your teacher will lead the class through a discussion of the questions that follow.

Compare your confidence interval to a confidence interval obtained by another student in your class.

g. Are the two confidence intervals the same?

h. Do both intervals contain the actual population mean length of 90.41 minutes?

i. How many people in your class have a confidence interval that does not include the actual value of the population mean? Is this surprising given the 90% confidence level associated with the confidence intervals?

ARE YOU READY TO MOVE ON? CHAPTER 12 REVIEW EXERCISES

All chapter learning objectives are assessed in these exercises. The learning objectives assessed in each exercise are given in parentheses.

12.60 (M1, M2)

Suppose that college students with a checking account typically write relatively few checks in any given month, whereas people who are not college students typically write many more checks during a month. Suppose that 50% of a bank's accounts are held by students and that 50% are held by people who are not college students. Let x denote the number of checks written in a given month by a randomly selected bank customer.

a. Give a sketch of what the population distribution of x might look like.

b. Suppose that the mean value of x is 22.0 and that the standard deviation is 16.5. If a random sample of $n = 100$ customers is to be selected and $\bar{x}$ denotes the sample mean number of checks written during a particular month, where is the sampling distribution of $\bar{x}$ centered, and what is the standard deviation of the $\bar{x}$ distribution? Sketch a rough picture of the sampling distribution.

c. What is the approximate probability that $\bar{x}$ is at most 20? At least 25?

12.61 (C1)

Let x denote the time (in minutes) that it takes a fifth-grade student to read a certain passage. Suppose that the mean

value and standard deviation of the x distribution are $\mu = 2$ minutes and $\sigma = 0.8$ minutes, respectively.

a. If $\bar{x}$ is the sample mean time for a random sample of $n = 9$ students, where is the $\bar{x}$ distribution centered, and what is the standard deviation of the $\bar{x}$ distribution?

b. Repeat Part (a) for a sample of size of $n = 20$ and again for a sample of size $n = 100$. How do the centers and spreads of the three $\bar{x}$ distributions compare to one another? Which sample size would be most likely to result in an $\bar{x}$ value close to μ, and why?

12.62 (C2)

The article "The Association Between Television Viewing and Irregular Sleep Schedules Among Children Less Than 3 Years of Age" (*Pediatrics* [2005]: 851–856) reported the accompanying 95% confidence intervals for average TV viewing time (in hours per day) for three different age groups.

Age Group	95% Confidence Interval
Less than 12 Months	(0.8, 1.0)
12 to 23 Months	(1.4, 1.8)
24 to 35 Months	(2.1, 2.5)

a. Suppose that the sample sizes for each of the three age-group samples were equal. Based on the given confidence intervals, which of the age-group samples had the greatest variability in TV viewing time? Explain your choice.

b. Now suppose that the sample standard deviations for the three age-group samples were equal, but that the three sample sizes might have been different. Which of the three age-group samples had the largest sample size? Explain your choice.

c. The interval (0.7, 1.1) is either a 90% confidence interval or a 99% confidence interval for the mean TV viewing time computed using the sample data for children less than 12 months old. Is the confidence level for this interval 90% or 99%? Explain your choice.

12.63 (C2, P1)
Suppose that a random sample of 50 bottles of a particular brand of cough medicine is selected, and the amount of cough medicine in each bottle is determined. Let μ denote the mean amount of cough medicine for the population of all bottles of this brand. Suppose that this sample of 50 results in a 95% confidence interval for μ of (7.8, 9.4).

a. Would a 90% confidence interval have been narrower or wider than the given interval? Explain your answer.

b. Consider the following statement: There is a 95% chance that μ is between 7.8 and 9.4. Is this statement correct? Why or why not?

c. Consider the following statement: If the process of selecting a random sample of size 50 and then computing the corresponding 95% confidence interval is repeated 100 times, exactly 95 of the resulting intervals will include μ. Is this statement correct? Why or why not?

12.64 (M4)
The authors of the paper **"Short-Term Health and Economic Benefits of Smoking Cessation: Low Birth Weight"** (*Pediatrics* [1999]:1312–1320) investigated the medical cost associated with babies born to mothers who smoke. The paper included estimates of mean medical cost for low-birth-weight babies for different ethnic groups. For a sample of 654 Hispanic low-birth-weight babies, the mean medical cost was $55,007 and the standard error ($s/\sqrt{n}$) was $3011. For a sample of 13 Native American low-birth-weight babies, the mean and standard error were $73,418 and $29,577, respectively. Explain why the two standard errors are so different.

12.65 (M3, M4, M5, P1)
How much money do people spend on graduation gifts? In 2007, the **National Retail Federation (www.nrf.com)** surveyed 2,815 consumers who reported that they bought one or more graduation gifts that year. The sample was selected to be representative of adult Americans who purchased graduation gifts in 2007. For this sample, the mean amount spent per gift was $55.05. Suppose that the sample standard deviation was $20. Construct and interpret a 98% confidence interval for the mean amount of money spent per graduation gift in 2007.

12.66 (M3, M4, M5, P1)
The authors of the paper **"Driven to Distraction"** (*Psychological Science* [2001]:462–466) describe a study to evaluate the effect of using a cell phone on reaction time. Subjects were asked to perform a simulated driving task while talking on a cell phone. While performing this task, occasional red and green lights flashed on the computer screen. If a green light flashed, subjects were to continue driving, but if a red light flashed, subjects were to brake as quickly as possible and the reaction time (in msec) was recorded. The following summary statistics were read from a graph that appeared in the paper:

$$n = 48 \qquad \bar{x} = 530 \qquad s = 70$$

Construct and interpret a 95% confidence interval for μ, the mean time to react to a red light while talking on a cell phone. What assumption must be made in order to generalize this confidence interval to the population of all drivers?

12.67 (M6, P2)
Suppose that the researchers who carried out the study described in the previous exercise wanted to estimate the mean reaction time with a margin of error of 5 msec. Using the sample standard deviation as a preliminary estimate of the population standard deviation, compute the required sample size.

12.68 (M7, M8, P3)
A study of fast-food intake is described in the paper **"What People Buy From Fast-Food Restaurants"** (*Obesity* [2009]:1369–1374). Adult customers at three hamburger chains (McDonald's, Burger King, and Wendy's) in New York City were approached as they entered the restaurant at lunchtime and asked to provide their receipt when exiting. The receipts were then used to determine what was purchased and the number of calories consumed was determined. In all, 3,857 people participated in the study. The sample mean number of calories consumed was 857 and the sample standard deviation was 677.

a. The sample standard deviation is quite large. What does this tell you about number of calories consumed in a hamburger-chain lunchtime fast-food purchase in New York City?

b. Given the values of the sample mean and standard deviation and the fact that the number of calories consumed can't be negative, explain why it is *not* reasonable to assume that the distribution of calories consumed is normal.

c. Based on a recommended daily intake of 2,000 calories, the online Healthy Dining Finder (**www.healthydiningfinder.com**) recommends a target of 750 calories for lunch. Assuming that it is reasonable to regard the sample of 3,857 fast-food purchases as representative of all hamburger-chain lunchtime purchases in New York City, carry out a hypothesis test to determine if the sample provides convincing evidence that the mean number of calories in a New York City hamburger-chain lunchtime purchase is greater than the lunch recommendation of 750 calories. Use $\alpha = 0.01$.

d. Would it be reasonable to generalize the conclusion of the test in Part (c) to the lunchtime fast-food purchases of all adult Americans? Explain why or why not.

e. Explain why it is better to use the customer receipt to determine what was ordered rather than just asking a customer leaving the restaurant what he or she purchased.

f. Do you think that asking a customer to provide his or her receipt before they ordered could have introduced a bias? Explain.

12.69 (M7, M8, P3)

Medical research has shown that repeated wrist extension beyond 20 degrees increases the risk of wrist and hand injuries. Each of 24 students at Cornell University used a proposed new computer mouse design, and while using the mouse, each student's wrist extension was recorded. Data consistent with summary values given in the paper **"Comparative Study of Two Computer Mouse Designs"** (**Cornell Human Factors Laboratory Technical Report RP7992**) are given. Use these data to test the hypothesis that the mean wrist extension for people using this new mouse design is greater than 20 degrees. Are any assumptions required in order for it to be appropriate to generalize the results of your test to the population of all Cornell students? To the population of all university students?

27	28	24	26	27	25	25
24	24	24	25	28	22	25
24	28	27	26	31	25	28
27	27	25				

12.70 (P3)

A comprehensive study conducted by the National Institute of Child Health and Human Development tracked more than 1,000 children from an early age through elementary school (*New York Times*, **November 1, 2005**). The study concluded that children who spent more than 30 hours a week in child care before entering school tended to score higher in math and reading when they were in the third grade. The researchers cautioned that the findings should not be a cause for alarm because the differences were found to be small. Explain how the difference between the mean math score for the child care group and the overall mean for third graders could be small but the researchers could still reach the conclusion that the mean for the child care group is significantly higher than the overall mean.

TECHNOLOGY NOTES

Confidence Interval for μ Based on t-distribution

TI-83/84

Summarized data

1. Press **STAT**
2. Highlight **TESTS**
3. Highlight **TInterval...** and press **ENTER**
4. Highlight **Stats** and press **ENTER**
5. Next to $\bar{x}$ input the value for the sample mean
6. Next to **sx** input the value for the sample standard deviation
7. Next to **n** input the value for the sample size
8. Next to **C-Level** input the appropriate confidence level
9. Highlight **Calculate** and press **ENTER**

Raw data

1. Enter the data into **L1** (In order to access lists press the **STAT** key, highlight the option called **Edit...** then press **ENTER**)
2. Press **STAT**
3. Highlight **TESTS**
4. Highlight **TInterval...** and press **ENTER**
5. Highlight **Data** and press **ENTER**
6. Next to **C-Level** input the appropriate confidence level
7. Highlight **Calculate** and press **ENTER**

TI-Nspire

Summarized data

1. Enter the Calculate Scratchpad
2. Press the **menu** key and select **6:Statistics** then **6:Confidence**

Intervals then **2:t Interval...** and press **enter**

3. From the drop-down menu select **Stats**
4. Press **OK**
5. Next to $\bar{x}$ input the value for the sample mean
6. Next to **sx** input the value for the sample standard deviation
7. Next to **n** input the value for the sample size
8. Next to **C Level** input the appropriate confidence level
9. Press **OK**

Raw data

1. Enter the data into a data list (In order to access data lists select the spreadsheet option and press **enter**)

   Note: Be sure to title the list by selecting the top row of the column and typing a title.

2. Press the **menu** key and select **4:Statistics** then **3:Confidence Intervals** then **2:t Interval...** and press **enter**
3. From the drop-down menu select **Data**
4. Press **OK**
5. Next to **List** select the list containing your data
6. Next to **C-Level** input the appropriate confidence level
7. Press **OK**

JMP

1. Input the data into a column
2. Click **Analyze** and select **Distribution**

3. Click and drag the column name from the box under **Select Columns** to the box next to **Y, Response**
4. Click **OK**
5. Click the red arrow next to the column name and select **Confidence Interval** then select the appropriate confidence level or click **Other** to specify a level

Minitab

Summarized data

1. Click **S̲tat** then click **B̲asic Statistics** then click **1̲-sample t...**
2. Click the radio button next to **Summarized data**
3. In the box next to **Sample size:** type the value for n, the sample size
4. In the box next to **Mean:** type the value for the sample mean
5. In the box next to **Standard deviation:** type the value for the sample standard deviation
6. Click **Options...**
7. Input the appropriate confidence level in the box next to **Confidence Level**
8. Click **OK**
9. Click **OK**

Raw data

1. Input the raw data into a column
2. Click **S̲tat** then click **B̲asic Statistics** then click **1̲-sample t...**
3. Click in the box under **Samples in columns:**
4. Double click the column name where the raw data are stored
5. Click **Options...**
6. Input the appropriate confidence level in the box next to **Confidence Level**
7. Click **OK**
8. Click **OK**

SPSS

1. Input the data into a column
2. Click **A̲nalyze** then select **Compare Means** then select **One-Sample T Test...**
3. Highlight the column name for the variable
4. Click the arrow to move the variable to the **Test Variable(s):** box
5. Click **Options...**
6. Input the appropriate confidence level in the box next to **Confidence Interval Percentage:**
7. Click **Continue**
8. Click **OK**

Note: The confidence interval for the population means is in the **One-Sample Test** table.

Excel

Note: Excel does not have the functionality to produce a confidence interval for a single population mean automatically. However, you can use Excel to produce the estimate for the sample mean and the margin of error for the confidence interval using the steps below.

1. Input the raw data into a column
2. Click the **Data** ribbon
3. Click **Data Analysis** in the **Analysis** group

Note: If you do not see **Data Analysis** listed on the Ribbon, see the Technology Notes for Chapter 2 for instructions on installing this add-on.

4. Select **Descriptive Statistics** from the dialog box and click **OK**
5. Click in the box next to **Input Range:** and select the data (if you used a column title, check the box next to **Labels in first row**)
6. Click the box next to **Confidence Level for Mean**:
7. In the box to the right of **Confidence Level for Mean:** type in the confidence level you are using
8. Click **OK**

Note: The margin of error can be found in the row titled Confidence Level. In order to find the lower limit of the confidence interval, subtract this from the mean shown in the output. To find the upper limit of the confidence interval, add this to the mean shown in the output.

t-Test for population mean, μ

TI-83/84

Summarized data

1. Press **STAT**
2. Highlight **TESTS**
3. Highlight **T-Test...**
4. Highlight **Stats** and press **ENTER**
5. Next to μ_0 type the hypothesized value for the population mean
6. Next to $\bar{x}$ input the value for the sample mean
7. Next to **sx** input the value for the sample standard deviation
8. Next to **n** input the value for the sample size
9. Next to μ highlight the appropriate alternative hypothesis and press **ENTER**
10. Highlight **Calculate** and press **ENTER**

Raw data

1. Enter the data into **L1** (In order to access lists press the **STAT** key, highlight the option called **Edit...** then press **ENTER**)
2. Press **STAT**
3. Highlight **TESTS**
4. Highlight **T-Test...**
5. Highlight **Data** and press **ENTER**
6. Next to μ_0 type the hypothesized value for the population mean
7. Next to μ highlight the appropriate alternative hypothesis and press **ENTER**
8. Highlight **Calculate** and press **ENTER**

TI-Nspire

Summarized data

1. Enter the Calculate Scratchpad
2. Press the **menu** key and select **6:Statistics** then **7:Stat Tests** then **2:t test...** and press **enter**
3. From the drop-down menu select **Stats**
4. Press **OK**
5. Next to μ_0 type the hypothesized value for the population mean

6. Next to $\bar{x}$ input the value for the sample mean
7. Next to **sx** input the value for the sample standard deviation
8. Next to **n** input the value for the sample size
9. Next to **Alternate Hyp** select the appropriate alternative hypothesis from the drop-down menu
10. Press **OK**

Raw data

1. Enter the data into a data list (In order to access data lists select the spreadsheet option and press **enter**)

   Note: Be sure to title the list by selecting the top row of the column and typing a title.
2. Press the **menu** key and select **4:Statistics** then **7:Stat Tests** then **2:t test...** and press **enter**
3. From the drop-down menu select **Data**
4. Press **OK**
5. Next to μ_0 input the hypothesized value for the population mean
6. Next to **List** select the list containing your data
7. Next to **Alternate Hyp** select the appropriate alternative hypothesis from the drop-down menu
8. Press **OK**

JMP

1. Input the data into a column
2. Click **Analyze** and select **Distribution**
3. Click and drag the column name from the box under **Select Columns** to the box next to **Y, Response**
4. Click **OK**
5. Click the red arrow next to the column name and select **Test Mean**
6. In the box next to **Specify Hypothesized Mean**, type the hypothesized value of the mean, μ_0
7. Click **OK**

Note: The output provides results for all three possible alternative hypotheses.

Minitab

Summarized data

1. Click **Stat** then click **Basic Statistics** then click **1-sample t...**
2. Click the radio button next to **Summarized data**
3. In the box next to **Sample size:** type the value for n, the sample size
4. In the box next to **Mean:** type the value for the sample mean
5. In the box next to **Standard deviation:** type the value for the sample standard deviation
6. In the box next to **Test mean:** type the hypothesized value of the population mean
7. Click **Options...**

8. Select the appropriate alternative hypothesis from the drop-down menu next to **Alternative:**
9. Click **OK**
10. Click **OK**

Raw data

1. Input the raw data into a column
2. Click **Stat** then click **Basic Statistics** then click **1-sample t...**
3. Click in the box under **Samples in columns:**
4. Double click the column name where the raw data are stored
5. In the box next to **Test mean:** type the hypothesized value of the population mean
6. Click **Options...**
7. Select the appropriate alternative hypothesis from the drop-down menu next to **Alternative:**
8. Click **OK**
9. Click **OK**

SPSS

1. Input the data into a column
2. Click **Analyze** then select **Compare Means** then select **One-Sample T Test...**
3. Highlight the column name for the variable
4. Click the arrow to move the variable to the **Test Variable(s):** box
5. In the box next to **Test Value:** input the hypothesized test value
6. Click **OK**

Note: This procedure produces a two-sided *P*-value.

Excel

Excel does not have the functionality to automatically produce a *t* test for a single population mean. However, you may type the formula into an empty cell manually to have Excel calculate the value of the test statistic for you. You can then use the steps below to find a *P*-value for the test statistic.

1. Select an empty cell
2. Click on **Formulas**
3. Click **Insert Function**
4. Select **Statistical** from the drop-down box for category
5. Select **TDIST** and click **OK**
6. Click in the box next to **X** and select the cell containing your test statistic or type it manually
7. Click in the box next to **Deg_freedom** and type the number of degrees of freedom (n-1)
8. Click in the box next to **Tails** and type 1 for a one-tailed *P*-value or 2 for a two-tailed *P*-value
9. Click **OK**

Note: Choosing a one-tailed distribution in Step 8 will result in returning $P(X \geq x)$.

AP* Review Questions for Chapter 12

1. The histograms shown here are approximate sampling distributions of the sample mean. Each histogram is based on selecting 500 different samples, each of size n. All three histograms were constructed by sampling from the same population, but the sample sizes were different. Which histogram was based on samples with the smallest sample size, n?

(A) I

(B) II

(C) III

(D) The same sample size was used for all three histograms.

(E) It is not possible to tell by looking at the histograms.

2. The histograms shown below are approximate sampling distributions of the sample mean. Each one is based on 500 samples of size 25, but each of the three histograms is the result of sampling from a different population. Which of the three populations sampled had the largest standard deviation?

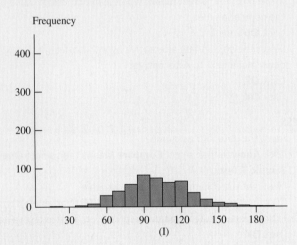

(I)

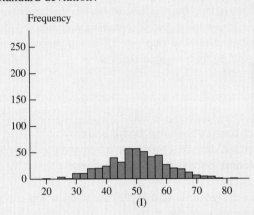

(I)

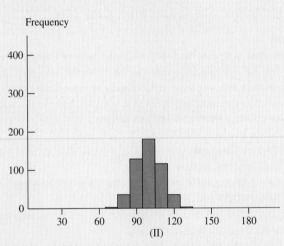

(II)

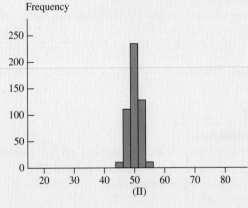

(II)

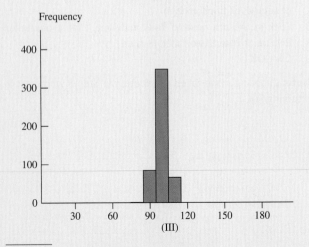

(III)

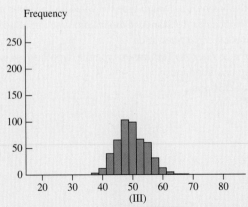

(III)

(A) I

(B) II

(C) III

(D) All three have the same standard deviation.

(E) It is not possible to tell from the histograms.

3. For a random sample from a skewed population, for which of the following sample sizes would the sampling distribution of $\bar{x}$ be closest to normal in shape?

(A) $n = 4$

(B) $n = 10$

(C) $n = 40$

(D) $n = 60$

(E) The $\bar{x}$ distribution will be normal for any sample size.

4. What is the standard deviation of $\bar{x}$ for random samples of size 100 from a population with a mean of 100 and a standard deviation of 80?

(A) 0.8

(B) 8

(C) 10

(D) 80

(E) 100

5. For which of the following will the sample mean tend to differ least from sample to sample?

(A) random samples of size 10

(B) random samples of size 20

(C) random samples of size 30

(D) random samples of size 50

(E) random samples of size 100

6. Commute distances for employees of a large company have a distribution that is quite skewed. Suppose the mean commute distance is 20 miles and that the standard deviation is 10 miles. What can be said about the mean commute distance for a random sample of size 4 employees from this plant?

(A) The distribution of $\bar{x}$ will be approximately normal with mean 20 and standard deviation 10.

(B) The distribution of $\bar{x}$ will be approximately normal with mean 20 and standard deviation 5.

(C) The mean and standard deviation of $\bar{x}$ will be 20 and 10, but the sampling distribution of $\bar{x}$ will not be approximately normal.

(D) The mean and standard deviation of $\bar{x}$ will be 20 and 5, but the sampling distribution of $\bar{x}$ will not be approximately normal.

(E) $\bar{x}$ will have a standard normal distribution.

7. As the sample size increases, how does the sampling distribution of $\bar{x}$ change?

(A) The mean of the sampling distribution decreases in value.

(B) The standard deviation of the sampling distribution decreases in value.

(C) The mean of the sampling distribution increases in value.

(D) The standard deviation of the sampling distribution increases in value.

(E) The mean and standard deviation of the sampling distribution do not change.

8. If a population distribution is substantially skewed and the sample size is small, which of the following is true about the sampling distribution of $\bar{x}$?

(A) $\mu_{\bar{x}} \neq \mu$ and $\sigma_{\bar{x}} \neq \dfrac{\sigma}{\sqrt{n}}$.

(B) $\mu_{\bar{x}} \neq \mu$ and the shape is not approximately normal.

(C) $\sigma_{\bar{x}} \neq \dfrac{\sigma}{\sqrt{n}}$ and the shape is not approximately normal.

(D) $\mu_{\bar{x}} = \mu$ and the shape is not approximately normal.

(E) $\sigma_{\bar{x}} \neq \dfrac{\sigma}{\sqrt{n}}$ and the shape is approximately normal.

9. Which of the following would tend to decrease the width of a confidence interval?

I. Increasing the sample size

II. Using a higher confidence level

III. Using a lower confidence level

(A) I only

(B) II only

(C) III only

(D) I and II only

(E) I and III only

10. Each individual in a random sample of 60 college students was asked how much money he or she spent on food in a typical month. The data was then used to construct a 99% confidence interval for the actual mean amount spent on food for all students at the college in a typical month. The confidence interval was (240, 380). Which of the following is closest to the 95% confidence interval constructed using the same sample data?

(A) (240, 380)

(B) (250, 370)

(C) (260, 360)

(D) (270, 350)

(E) (280, 340)

11. Each individual in a random sample of 50 Internet users was asked how many minutes he or she spends online in a typical day. The data was then used to construct a 99% confidence interval for the mean number of minutes spent online in a typical day for all Internet users. The confidence interval was (80, 200) minutes per day. Which of the following is a correct interpretation of the confidence interval?

    (A) There is a 99% chance that the mean number of minutes spent online in a typical day of all Internet users is between 80 and 200.
    (B) We are 99% confident that the sample mean is between 80 and 200 minutes.
    (C) We are 99% confident that the mean number of minutes spent online in a typical day for all Internet users is between 80 and 200.
    (D) 99% of all Internet users will spend between 80 and 200 minutes online in a typical day.
    (E) 99% of the people in the sample spent between 80 and 200 minutes online in a typical day.

12. Two hundred visitors to a national park were selected at random and each was asked whether he or she intended to stay more than 2 nights in the park on that visit. Assuming that the sample size is large enough, which of the following confidence intervals should be used to estimate the proportion of all visitors to this national park who will stay more than 2 nights on their visit?

    (A) $\bar{x} \pm (t \text{ critical value})\dfrac{s}{\sqrt{n}}$

    (B) $\hat{p} \pm (t \text{ critical value})\dfrac{s}{\sqrt{n}}$

    (C) $\hat{p} \pm (z \text{ critical value})\dfrac{s}{\sqrt{n}}$

    (D) $\bar{x} \pm (z \text{ critical value})\dfrac{\sigma}{\sqrt{n}}$

    (E) $\hat{p} \pm (z \text{ critical value})\sqrt{\dfrac{\hat{p}(1-\hat{p})}{n}}$

13. Two-hundred visitors to a national park were selected at random and each was asked how far he or she had traveled to get to the park on that visit. Which of the following confidence intervals should be used to estimate the mean number of miles traveled by all visitors to the park?

    (A) $\bar{x} \pm (t \text{ critical value})\dfrac{s}{\sqrt{n}}$

    (B) $\mu \pm (t \text{ critical value})\dfrac{s}{\sqrt{n}}$

    (C) $\mu \pm (z \text{ critical value})\dfrac{\sigma}{\sqrt{n}}$

    (D) $\bar{x} \pm (z \text{ critical value})\dfrac{\sigma}{\sqrt{n}}$

    (E) $\hat{p} \pm (z \text{ critical value})\sqrt{\dfrac{\hat{p}(1-\hat{p})}{n}}$

14. Suppose you take a random sample from a population known to be normally distributed, but the value of σ is unknown. Your sample size is $n = 10$. Which formula below should be used to find the 90% confidence interval for the mean?

    (A) $\bar{x} \pm 1.645\dfrac{s}{\sqrt{10}}$

    (B) $\bar{x} \pm 1.645\dfrac{\sigma}{\sqrt{10}}$

    (C) $\bar{x} \pm 1.833\dfrac{s}{\sqrt{10}}$

    (D) $\bar{x} \pm 1.812\dfrac{\sigma}{\sqrt{10}}$

    (E) $\bar{x} \pm 1.833\dfrac{\sigma}{\sqrt{10}}$

15. A university administrator in charge of computing facilities will close a computer lab if there is convincing evidence that the proportion of students at the university who have access to a computer at home is greater than 0.6. What hypotheses should the administrator test?

    (A) $H_0: \mu < 0.6$ versus $H_a: \mu = 0.6$
    (B) $H_0: \mu = 0.6$ versus $H_a: \mu > 0.6$
    (C) $H_0: p = 0.6$ versus $H_a: p > 0.6$
    (D) $H_0: p > 0.6$ versus $H_a: p = 0.6$
    (E) $H_0: p > 0.6$ versus $H_a: p < 0.6$

16. The manager of a large hotel must decide whether to hire additional front desk staff. He has decided to hire more staff if there is evidence that the average time customers must wait in line before being assisted with check-in is greater than 3 minutes. He decides to test $H_0: \mu = 3$ versus $H_a: \mu > 3$. Which of the following would be a consequence of making a Type II error?

    (A) Deciding not to hire additional staff when the wait time really is greater than 3 minutes.
    (B) Deciding not to hire additional staff when the wait time really is not greater than 3 minutes.
    (C) Deciding to hire additional staff when the wait time really is greater than 3 minutes.
    (D) Deciding to hire additional staff when the wait time really is not greater than 3 minutes.
    (E) Deciding that a wait time of greater than 3 minutes is acceptable.

17. The marketing department of a national department store chain designs its advertising to target 18- to 24-year-olds. The marketing manager worries that the average age of the chain's customers is greater than 24, in which case the marketing plan should be reconsidered. He decides to survey a random sample of 100 customers and will use the resulting data to test H_0: $\mu = 24$ versus H_a: $\mu > 24$, where μ is the mean customer age. Suppose that the P-value from this test was 0.03. Which of the following is a correct interpretation of this P-value?

   (A) The probability that the null hypothesis is true is 0.03.
   (B) The probability that the null hypothesis is false is 0.03.
   (C) When the null hypothesis is true, the probability of seeing results as or more extreme than what was observed in the sample is 0.03.
   (D) When the null hypothesis is false, the probability of seeing results as extreme as what was observed in the sample is 0.03.
   (E) Approximately 3% of the chain's customers are older than 24.

18. In order to test a claim that more than half of all calls to the emergency 911 phone number in a particular county are actually not for emergency situations, 100 recordings of 911 calls are selected at random from those received in the past year, and each call is classified as either emergency on non-emergency. Which of the following test statistics is appropriate for testing the hypothesis of interest?

   (A) $z = \dfrac{\hat{p} - p_0}{\sqrt{\dfrac{p_0(1 - p_0)}{n}}}$

   (B) $t = \dfrac{\hat{p} - p_0}{\sqrt{\dfrac{p_0(1 - p_0)}{n}}}$

   (C) $z = \dfrac{\bar{x} - \mu_0}{\dfrac{\sigma}{\sqrt{n}}}$

   (D) $t = \dfrac{\bar{x} - \mu_0}{\dfrac{s}{\sqrt{n}}}$

   (E) None of the above is appropriate.

19. Which of the following affects the power of a test?
   I.   The sample size
   II. The significance level of the test
   III. The size of the discrepancy between the actual value and the hypothesized value of the population characteristic

   (A) I only
   (B) II only
   (C) III only
   (D) I and II only
   (E) I, II, and III

20. In a test of H_0: $\mu = 100$ versus H_a: $\mu > 100$, the power of the test will be lowest when the actual value of the population mean is

   (A) 101
   (B) 110
   (C) 120
   (D) 200
   (E) The power will be the same for any value greater than 100.

13

Asking and Answering Questions About the Difference Between Two Population Means

Preview

Chapter Learning Objectives

13.1 Testing Hypotheses About the Difference Between Two Population Means Using Independent Samples

13.2 Testing Hypotheses About the Difference Between Two Population Means Using Paired Samples

13.3 Estimating the Difference Between Two Population Means

13.4 Avoid These Common Mistakes

Chapter Activities

Exploring the Big Ideas

Are You Ready to Move On? Chapter 13 Review Exercises

Technology Notes

AP* Review Questions for Chapter 13

Elena Schweitzer/Shutterstock.com

PREVIEW

In Chapter 12, you saw how sample data could be used to estimate a population mean and to test hypotheses about the value of a single population mean. In this chapter you will see how sample data can be used to learn about the difference between two population means.

CHAPTER LEARNING OBJECTIVES

Conceptual Understanding

After completing this chapter, you should be able to

C1 Understand how a research question about the difference in two population means is translated into hypotheses.

Mastering the Mechanics

After completing this chapter, you should be able to

M1 Distinguish between independent and paired samples.

M2 Translate a research question or claim about a difference in two population means into null and alternative hypotheses.

M3 Know the conditions for appropriate use of the two-sample t confidence interval and the two-sample t test.

M4 Carry out a two-sample t test for a difference in population means.

M5 Compute and interpret a two-sample t confidence interval for a difference in population means.

M6 Know the conditions for appropriate use of the paired-samples t confidence interval and the paired-samples t test.

M7 Carry out a paired-samples t test for a difference in population means.

M8 Compute and interpret a paired-samples t confidence interval for a difference in population means.

Putting It into Practice

After completing this chapter, you should be able to

P1 Recognize when a situation calls for testing hypotheses about a difference in two population means.

P2 Carry out a two-sample t test for a difference in two population means and interpret the conclusion in context.

P3 Carry out a paired-samples t test for a difference in two population means and interpret the conclusion in context.

P4 Interpret a two-sample t confidence interval for a difference in population means in context and interpret the associated confidence level.

P5 Interpret a paired-samples t confidence interval for a difference in population means in context and interpret the associated confidence level.

PREVIEW EXAMPLE   Depression and Chocolate

Is there a connection between depression and chocolate consumption? This is the question that the authors of the paper **"Mood Food: Chocolate and Depressive Symptoms in a Cross-Sectional Analysis"** (*Archives of Internal Medicine* [2010]: 699–703) set out to answer. Participants in the study were 931 adults who were not currently taking medication for depression. These participants were screened for depression using a widely used screening test. The participants were then divided into two samples based on the score on the screening test. One sample consisted of people who screened positive for depression, and the other sample consisted of people who did not. Each participant also completed a food frequency survey.

The researchers believed that the two samples were representative of the two populations of interest—adults who would screen positive for depression and adults who would not screen positive. Using methods that you will learn in this chapter, the researchers were able to conclude that the mean number of servings of chocolate per month for people who would screen positive for depression was greater than the mean number of chocolate servings per month for people who would not.

There are two other interesting things to note about this study. First, this was an observational study and not an experiment. Because of this, the researchers were not able to determine if there was a cause-and-effect relationship between depression and chocolate consumption. They also noted that even if there is a cause-and-effect relationship, this study would not have been able to identify the direction of the relationship (whether depression causes chocolate consumption or chocolate consumption causes depression). The second interesting aspect of the study is that the researchers also compared the two populations on the basis of food types other than chocolate. They did not find convincing evidence that the two populations differed with respect to mean caffeine, fat, carbohydrate, or calorie intake. The researchers believe that this makes the connection between depression and chocolate consumption even more interesting and worthy of further study.

SECTION 13.1 Testing Hypotheses About the Difference Between Two Population Means Using Independent Samples

Many statistical studies are carried out in order to learn about differences between two population means. For example, researchers have studied the ways in which college students who use Facebook differ from college students who do not use Facebook. As part of the study described in **"Facebook and Academic Performance"** (*Computers in Human Behavior* [2010]: 1237–1245), each person in a sample of 141 college students who use Facebook was asked to report his or her college grade point average (GPA). College GPA was also reported by each person in a sample of 68 students who do not use Facebook. What do you think the study showed? Did the data from this study provide convincing evidence that the mean college GPA for Facebook users was different from the mean college GPA of students who do not use Facebook? In this section, you will see how sample data can be used to answer questions like this.

Let's begin by introducing some notation. In the previous chapter, you used the symbol μ to represent the mean of an entire population. The sample mean was denoted by $\bar{x}$. When comparing two populations, you will have two samples—one from each population. You need to distinguish between the two populations and the two samples, and this is done using subscripts as shown in the following box.

> **AP* EXAM TIP**
>
> When comparing two populations, use subscripts to distinguish between them.

Notation

	Population Mean	Population Standard Deviation
Population 1	μ_1	σ_1
Population 2	μ_2	σ_2

	Sample Size	Sample Mean	Sample Standard Deviation
Sample from Population 1	n_1	$\bar{x}_1$	s_1
Sample from Population 2	n_2	$\bar{x}_2$	s_2

In this section, only the case where the two samples are **independent samples** is considered. Two samples are said to be independent samples if the selection of the individuals who make up one sample does not influence the selection of the individuals in the other sample. This would be the case if a random sample is selected from each of the two populations. When observations from the first sample are can be matched up in some meaningful way with observations in the second sample, the samples are said to be **paired**. For example, to study the effectiveness of a speed-reading course, the reading speed of subjects could be measured before they take the course and again after they complete the course. This gives rise to two related samples—one from the population of individuals who have not taken this particular course (the "before" measurements) and one from the population of individuals who have had such a course (the "after" measurements). These samples are paired. The two samples are not independently chosen, because the selection of individuals from the first (before) population completely determines which individuals make up the sample from the second (after) population. You will learn about methods for analyzing data from paired samples in Section 13.2.

> **AP* EXAM TIP**
>
> The distinction between paired and independent samples is an important one. When comparing two population means, be sure to consider how the samples were selected.

When comparing two population means, it is common to focus on the difference $\mu_1 - \mu_2$. Because $\bar{x}_1$ is an estimate of μ_1 and $\bar{x}_2$ is an estimate of μ_2, the obvious statistic to use as an estimate of $\mu_1 - \mu_2$ is the difference in the sample means, $\bar{x}_1 - \bar{x}_2$. But the values of $\bar{x}_1$ and $\bar{x}_2$ each vary from sample to sample, so the difference $\bar{x}_1 - \bar{x}_2$ will also vary. Because the statistic $\bar{x}_1 - \bar{x}_2$ is used to draw conclusions about the difference between two population means, you need to know about the behavior of this statistic. The sampling distribution of $\bar{x}_1 - \bar{x}_2$ describes this behavior.

General Properties of the Sampling Distribution of $\bar{x}_1 - \bar{x}_2$

If $\bar{x}_1 - \bar{x}_2$ is the difference in sample means for *independently* selected random samples, then the following rules hold.

Rule 1. $\mu_{\bar{x}_1 - \bar{x}_2} = \mu_1 - \mu_2$

This rule states that the sampling distribution of $\bar{x}_1 - \bar{x}_2$ is centered at the actual value of the difference in population means. This tells you that differences in sample means tend to cluster around the actual difference in population means.

Rule 2. $\sigma^2_{\bar{x}_1 - \bar{x}_2} = \dfrac{\sigma^2_1}{n_1} + \dfrac{\sigma^2_2}{n_2}$

and

$$\sigma_{\bar{x}_1 - \bar{x}_2} = \sqrt{\dfrac{\sigma^2_1}{n_1} + \dfrac{\sigma^2_2}{n_2}}$$

This rule specifies the standard error of $\bar{x}_1 - \bar{x}_2$. The value of the standard error describes how much the $\bar{x}_1 - \bar{x}_2$ values tend to vary from one pair of samples to another and also from the actual difference in population means.

Rule 3.   If both n_1 and n_2 are large or if the population distributions are approximately normal, the sampling distribution of $\bar{x}_1 - \bar{x}_2$ is approximately normal. The sample sizes can be considered large if $n_1 \geq 30$ and $n_2 \geq 30$.

The logic of hypothesis testing and the general process for carrying out a hypothesis test is the same for all hypothesis tests. Recall that what may differ from one type of test to another includes

1.   The null and alternative hypotheses
2.   The test statistic
3.   The way in which the associated *P*-value is determined

Let's consider each of these three things in the context of testing hypotheses about a difference in population means using independent samples.

1. Hypotheses

The hypothesis test focuses on the difference in population means, $\mu_1 - \mu_2$. When there is no difference between the two population means, $\mu_1 - \mu_2 = 0$. So

$$\mu_1 = \mu_2 \qquad \text{is equivalent to} \qquad \mu_1 - \mu_2 = 0$$

Similarly

$$\mu_1 > \mu_2 \qquad \text{is equivalent to} \qquad \mu_1 - \mu_2 > 0$$

and

$$\mu_1 < \mu_2 \qquad \text{is equivalent to} \qquad \mu_1 - \mu_2 < 0$$

To determine the null and alternative hypotheses, you will need to translate the question of interest into hypotheses. For example, if you want to determine if there is evidence that μ_1 is greater than μ_2, you would choose $\mu_1 - \mu_2 > 0$ as the alternative hypothesis.

The general form of the null hypothesis is

$$H_0: \mu_1 - \mu_2 = \text{hypothesized value}$$

In most cases, the hypothesized value is 0, indicating no difference in the population means. But sometimes it is something other than 0. For example, suppose μ_1 and μ_2 are the mean fuel efficiencies (in miles per gallon, mpg) for cars of a particular model equipped with 4-cylinder and 6-cylinder engines, respectively. The hypotheses under consideration might be

$$H_0: \mu_1 - \mu_2 = 5 \quad \text{versus} \quad H_a: \mu_1 - \mu_2 > 5$$

The null hypothesis is equivalent to the claim that mean fuel efficiency for the 4-cylinder engine is greater than the mean fuel efficiency for the 6-cylinder engine by 5 mpg. The alternative hypothesis states that the difference between the mean fuel efficiencies is more than 5 mpg.

2. Test Statistic

Based on the properties of the sampling distribution of $\bar{x}_1 - \bar{x}_2$, when you have independent random samples you know the following three things:

1. $\mu_{\bar{x}_1 - \bar{x}_2} = \mu_1 - \mu_2$

2. $\sigma_{\bar{x}_1 - \bar{x}_2} = \sqrt{\dfrac{\sigma_1^2}{n_1} + \dfrac{\sigma_2^2}{n_2}}$

3. If both n_1 and n_2 are large or if the population distributions are approximately normal, then the sampling distribution of $\bar{x}_1 - \bar{x}_2$ is approximately normal.

This means that if both samples sizes are large or the population distributions are approximately normal and the two random samples are independently selected, the statistic

$$z = \frac{(\bar{x}_1 - \bar{x}_2) - (\mu_1 - \mu_2)}{\sqrt{\dfrac{\sigma_1^2}{n_1} + \dfrac{\sigma_2^2}{n_2}}}$$

will have a distribution that is (approximately) the standard normal (z) distribution. (This is because $\bar{x}_1 - \bar{x}_2$ has a distribution that is approximately normal and the z statistic above results from standardizing $\bar{x}_1 - \bar{x}_2$ by subtracting its mean and then dividing by its standard deviation.)

Although it is possible to base a test procedure and confidence interval on the z statistic above, the values of the population variances, σ_1^2 and σ_2^2, are almost never known. As a result, the z statistic is rarely used. When σ_1^2 and σ_2^2, are unknown, the sample variances, s_1^2 and s_2^2 are used to estimate σ_1^2 and σ_2^2 in the z statistic above, resulting in the following t statistic.

When two random samples are independently selected and when n_1 and n_2 are both large or when the population distributions are approximately normal, the t statistic

$$t = \frac{(\bar{x}_1 - \bar{x}_2) - (\mu_1 - \mu_2)}{\sqrt{\dfrac{s_1^2}{n_1} + \dfrac{s_2^2}{n_2}}}$$

has approximately a t distribution with

$$df = \frac{(V_1 + V_2)^2}{\dfrac{V_1^2}{n_1 - 1} + \dfrac{V_2^2}{n_2 - 1}} \quad \text{where } V_1 = \frac{s_1^2}{n_1} \text{ and } V_2 = \frac{s_2^2}{n_2}$$

The computed value of df should be truncated (rounded down) to an integer value.

If one or both sample sizes are small, you can use normal probability plots, dotplots, or boxplots to evaluate whether it is reasonable to consider the population distributions to be approximately normal.

3. Computing a *P*-value

Once the value of the test statistic and the df have been determined, the next step is to compute a *P*-value, which tells you how likely it would be to observe sample data as extreme or more extreme than what was observed if the null hypothesis were true. Because this test statistic has approximately a *t* distribution, *P*-values are computed as an area under a *t* distribution curve. If you aren't using technology, calculating the test statistic and the df can be time consuming, but once that is done, finding the *P*-value is straightforward.

You now have all the pieces needed to carry out the hypothesis test. The two-sample *t* test for a difference in populations means summarized in the following box is a method you should consider when the answers to the four key questions (QSTN) are: *hypothesis testing, sample data, one numerical variable, and two independently selected samples.*

Questions
Q   Question type: estimation or hypothesis testing?
S   Study type: sample data or experiment data?
T   Type of data: one variable or two? Categorical or numerical?
N   Number of samples or treatments: how many?

Two-Sample *t* Test for a Difference in Population Means

Appropriate when the following conditions are met:

1. The samples are independently selected.
2. Each sample is a random sample from the population of interest or the samples are selected in a way that results in samples that are representative of the population.
3. Both sample sizes are large ($n_1 \geq 30$ and $n_2 \geq 30$) *or* the population distributions are approximately normal.

When these conditions are met, the following test statistic can be used:

$$t = \frac{(\bar{x}_1 - \bar{x}_2) - (\mu_1 - \mu_2)}{\sqrt{\dfrac{s_1^2}{n_1} + \dfrac{s_2^2}{n_2}}}$$

where $\mu_1 - \mu_2$ is the hypothesized value of the difference in population means from the null hypothesis (often this will be 0).

When the conditions above are met and the null hypothesis is true, the *t* test statistic has a *t* distribution with

$$df = \frac{(V_1 + V_2)^2}{\dfrac{V_1^2}{n_1 - 1} + \dfrac{V_2^2}{n_2 - 1}} \quad \text{where } V_1 = \frac{s_1^2}{n_1} \text{ and } V_2 = \frac{s_2^2}{n_2}$$

The computed value of df should be truncated (rounded down) to obtain an integer value.

Form of the null hypothesis: $H_0: \mu_1 - \mu_2 =$ hypothesized value

Associated *P*-value:

When the Alternative Hypothesis Is...	The *P*-value Is...
$H_a: \mu_1 - \mu_2 >$ hypothesized value	Area under the *t* curve to the right of the calculated value of the test statistic
$H_a: \mu_1 - \mu_2 <$ hypothesized value	Area under the *t* curve to the left of the calculated value of the test statistic
$H_a: \mu_1 - \mu_2 \neq$ hypothesized value	2·(area to the right of *t*) if *t* is positive or 2·(area to the left of *t*) if *t* is negative

Steps

H Hypotheses
M Method
C Check
C Calculate
C Communicate results

Now you're ready to look at some examples, following the five-step process for hypothesis testing (HMC³).

Example 13.1 Facebook and Grades

The Facebook study described at the beginning of this section (**"Facebook and Academic Performance,"** *Computers in Human Behavior* **[2010]: 1237–1245**) investigated the relationship between college GPA and Facebook use. One question that the researchers were hoping to answer is whether the mean college GPA for students who use Facebook is lower than the mean college GPA for students who do not. Two samples (141 students who were Facebook users and 68 students who were not Facebook users) were independently selected from students at a large, public Midwestern university. Although the samples were not selected at random, they were selected to be representative of the two populations (students who use Facebook and students who do not use Facebook at this university). Data from these samples were used to compute sample means and standard deviations.

Population	Population Mean	Sample Size	Sample Mean	Sample Standard Deviation
Students at the university who use Facebook	μ_1 = mean college GPA for students who use Facebook	$n_1 = 141$	$\bar{x}_1 = 3.06$	$s_1 = 0.95$
Students at the university who do not use Facebook	μ_2 = mean college GPA for students who do not use Facebook	$n_2 = 68$	$\bar{x}_2 = 3.82$	$s_2 = 0.41$

As always, begin by answering the four key questions (QSTN) to identify a potential method for answering the question posed: (Q) We would like to test a claim about the two populations, so this is a hypothesis testing problem. (S) The data are from samples. (T) There is one numerical variable, which is GPA. (N) There are two samples and they were independently selected. Because the answers to the four questions are *hypothesis testing, sample data, one numerical variable, and two independently selected samples*, you should consider a two-sample *t* test for a difference in population means.

Now you can use the five-step process for hypothesis testing problems (HMC³) to answer the question posed.

Process Step	
H Hypotheses	The question of interest is about the mean GPA for each populations. Since you are dealing with means, start by defining μ_1 and μ_2 in the context of this example.
	Population characteristics of interest:
	μ_1 = mean GPA for students who use Facebook
	μ_2 = mean GPA for students who do not use Facebook
	The question of interest (is the mean GPA for Facebook users lower than the mean for students who do not use Facebook) translates into an alternative hypothesis of $\mu_1 < \mu_2$. Writing this in terms of the difference in means, you get $\mu_1 - \mu_2 < 0$. The null hypothesis is that there is no difference in the population means.
	Hypotheses:
	Null hypothesis: $H_0: \mu_1 - \mu_2 = 0$
	Alternative hypothesis: $H_a: \mu_1 - \mu_2 < 0$
M Method	Because the answers to the four key questions are hypothesis testing, sample data, one numerical variable and two independently selected samples, consider a two-sample *t* test for a difference in population means.

(continued)

Process Step	
	Potential method:

Potential method:

Two-sample t test for a difference in population means. The test statistic for this test is

$$t = \frac{(\bar{x}_1 - \bar{x}_2) - (\mu_1 - \mu_2)}{\sqrt{\dfrac{s_1^2}{n_1} + \dfrac{s_2^2}{n_2}}}$$

When the null hypothesis is true, this statistic has approximately a t distribution with

$$df = \frac{(V_1 + V_2)^2}{\dfrac{V_1^2}{n_1 - 1} + \dfrac{V_2^2}{n_2 - 1}} \quad \text{where } V_1 = \frac{s_1^2}{n_1} \text{ and } V_2 = \frac{s_2^2}{n_2}$$

You also need to select a significance level for the test. You should choose a significance level after considering the consequences of a Type I error (rejecting a true null hypothesis) and a Type II error (failing to reject a false null hypothesis). In this example, a Type I error would be *incorrectly* concluding that the mean GPA of students who use Facebook is lower than the mean GPA of students who do not. A Type II error would be not thinking that there is a difference in the population means, when in fact the mean for students who use Facebook is lower. In this situation, because neither type of error is much more serious than the other, a reasonable choice for α is 0.05.

Significance level:

$\alpha = 0.05$

C Check

There are three conditions that need to be met in order to use the two-sample t test for a difference in population means.

The large sample condition is met because the sample sizes are both large:

$$n_1 = 141 \geq 30 \text{ and } n_2 = 68 \geq 30$$

From the study description, you know that the samples were independently selected. You also know that the samples were selected to be representative of the two populations of interest.

C Calculate

$n_1 = 141$ $\bar{x}_1 = 3.06$ $s_1 = 0.95$

$n_2 = 68$ $\bar{x}_2 = 3.82$ $s_2 = 0.41$

Test statistic:

$$t = \frac{(\bar{x}_1 - \bar{x}_2) - (\mu_1 - \mu_2)}{\sqrt{\dfrac{s_1^2}{n_1} + \dfrac{s_2^2}{n_2}}} = \frac{(3.06 - 3.82) - 0}{\sqrt{\dfrac{(0.95)^2}{141} + \dfrac{(0.41)^2}{68}}}$$

$$= \frac{-0.76}{0.094}$$

$$= -8.08$$

Degrees of freedom

$$V_1 = \frac{s_1^2}{n_1} = 0.0064 \qquad\qquad V_2 = \frac{s_2^2}{n_2} = 0.0025$$

$$df = \frac{(V_1 + V_2)^2}{\dfrac{V_1^2}{n_1 - 1} + \dfrac{V_2^2}{n_2 - 1}}$$

$$= \frac{(0.0064 + 0.0025)^2}{\dfrac{(0.0064)^2}{140} + \dfrac{(0.0025)^2}{67}}$$

(continued)

Process Step	
	$$= \frac{0.0000792}{0.000000386}$$ $$= 205.181$$ Truncate to df = 205 This is a lower-tailed test (the inequality in H_a is $<$), so the P-value is the area to the left of -8.08 under the t curve with df = 205. Because the t curve is symmetric and centered at 0, this area is equal to the area to the right of 8.08. Because $t = 8.08$ is greater than the largest value in the ∞ df row of Appendix Table 4, the area to the right is approximately 0. **Associated *P*-value:** P-value = area under t curve to the left of -8.08 $\qquad\qquad$ = area under the t curve to the right of 8.08 $\qquad\qquad$ = $P(t > 8.08)$ $\qquad\qquad$ ≈ 0
C Communicate Results	Because the P-value is less than the selected significance level, you reject the null hypothesis. **Decision:** $0 < 0.05$, Reject H_0. **Conclusion:** Based on the sample data, there is convincing evidence that the mean college GPA for students at the university who use Facebook is lower than the mean college GPA for students at the university who do not use Facebook.

Based on this hypothesis test, you can conclude that the sample mean GPA for students who use Facebook is enough lower than the sample mean GPA for students who do not use Facebook that you don't think that this could have occurred just by chance when there is no difference in the population means.

It is also possible to use statistical software or a graphing calculator to carry out the calculate step in a hypothesis test. For example, Minitab output for the test of Example 13.1 is shown here.

Two-Sample T-Test

Sample	N	Mean	StDev	SE Mean
1	141	3.060	0.950	0.080
2	68	3.820	0.410	0.050

Difference = mu (1) − mu (2)
Estimate for difference: −0.760000
T-Test of difference = 0 (vs <): T-Value = −8.07 P-Value = 0.000 DF = 205

From the Minitab output, you see that $t = -8.07$, the associated degrees of freedom is df = 205, and the P-value is 0.000. The t value is slightly different from the one in Example 13.1 only because Minitab uses greater decimal accuracy in the computations leading to the value of the test statistic and the number of degrees of freedom.

Example 13.2 Salary and Gender

Are women still paid less than men for comparable work? The authors of the paper **"Sex and Salary: A Survey of Purchasing and Supply Professionals"** (*Journal of Purchasing and Supply Management* **[2008]: 112–124**) carried out a study in which salary data were collected from a random sample of men and from a random sample of women who worked as purchasing managers and who were subscribers to *Purchasing* magazine. Salary data consistent with summary quantities given in the paper appear below (the actual sample sizes for the study were much larger):

Annual Salary (in thousands of dollars)

| **Men** | 81 | 69 | 81 | 76 | 76 | 74 | 69 | 76 | 79 | 65 |
| **Women** | 78 | 60 | 67 | 61 | 62 | 73 | 71 | 58 | 68 | 48 |

Even though the samples were selected from subscribers to a particular magazine, the authors of the paper believed the samples to be representative of the two populations of interest—male purchasing managers and female purchasing managers.

Let's use the sample data to determine if there is convincing evidence that the mean annual salary for male purchasing managers is greater than the mean annual salary for female purchasing managers.

Considering the four key questions (QSTN), this situation can be described as *hypothesis testing, sample data, one numerical variable (annual salary), and two independently selected samples*. This combination suggests a two-sample t test for a difference in population means.

H Hypotheses You want to use data from the samples to test the claim that the mean salary for males is greater than the mean salary for females. The population characteristics of interest are

μ_1 = mean annual salary for male purchasing managers
μ_2 = mean annual salary for female purchasing managers

and $\mu_1 - \mu_2$ is then the difference in population means. Translating the question of interest into hypotheses gives

$$H_0: \mu_1 - \mu_2 = 0$$
$$H_a: \mu_1 - \mu_2 > 0$$

M Method Because the answers to the four key questions are hypothesis testing, sample data, one numerical variable, and two independently selected samples, consider a two-sample t test for a difference in population means. The test statistic for this test is

$$t = \frac{(\bar{x}_1 - \bar{x}_2) - (\mu_1 - \mu_2)}{\sqrt{\dfrac{s_1^2}{n_1} + \dfrac{s_2^2}{n_2}}}$$

Next choose a significance level for the test. For purposes of this example, $\alpha = 0.01$ will be used.

C Check From the study description, you know that the samples were independently selected and representative of the populations of interest. Because both of the sample sizes are small, you need to be willing to assume that the salary distribution is approximately normal for each of the two populations. Boxplots constructed using the sample data are shown here:

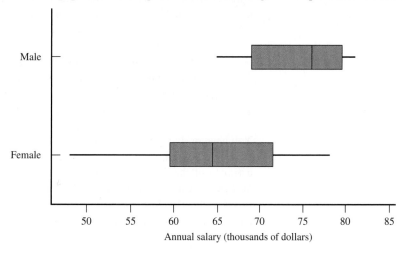

Because the boxplots are reasonably symmetric and because there are no outliers, it is reasonable to assume that the two population distributions are approximately normal. With all of the conditions met, the two-sample *t* test is an appropriate method.

C Calculate   Using JMP to do the computations results in the following output:

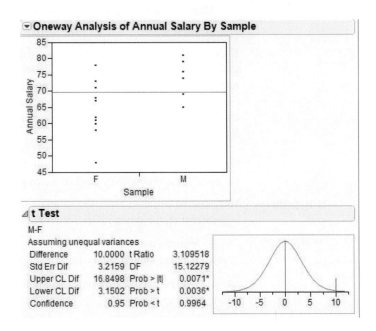

From the JMP output, the value of the test statistic and df are

$t = 3.11$ (rounded from 3.109518 in the output under *t* Ratio)
$df = 15$ (truncated from 15.12279 in the output under DF)

and the associated *P*-value is 0.0036 (from the Prob > *t* entry in the output because this is an upper-tail test).

C Communicate Results   Because the *P*-value (0.0036) is less than the selected significance level (0.01), the null hypothesis is rejected. The sample data provide convincing evidence that the mean annual salary for male purchasing managers is greater than the mean annual salary for female purchasing managers. ▪

Suppose that the computed value of the test statistic in Example 13.2 had been 1.13 rather than 3.11. Then the *P*-value would have been 0.1381 (the area to the right of 1.13 under the *t* curve with 15 df) and the decision would have been to not reject the null hypothesis. In this case, you would have concluded that there was not convincing evidence that the mean annual salary was higher for males than for females. Notice that when you fail to reject the null hypothesis of no difference between the population means, you are not saying that there is convincing evidence that the means are equal—you can only say that you are not convinced that they are different.

More on Degrees of Freedom

You have probably noticed that using the degrees of freedom formula for the two-sample *t* test involves quite a bit of arithmetic. An alternative approach is to compute a conservative estimate of the *P*-value—one that is close to but larger than the actual *P*-value. If the null hypothesis is rejected using this conservative estimate, then it will also be rejected if the actual *P*-value is used. *A conservative estimate of the P-value for the two-sample t test can be found by using the t curve with degrees of freedom equal to the smaller of $(n_1 - 1)$ and $(n_2 - 1)$.*

The Pooled *t* Test

The two-sample *t* test is appropriate when it is reasonable to assume that the population distributions are approximately normal. If it is also known that the variances of the two populations are equal ($\sigma_1^2 = \sigma_2^2$), an alternative procedure known as the *pooled t test* can be used. This test procedure combines information from both samples to obtain a "pooled" estimate of the common variance and then uses this pooled estimate of the variance in place of s_1^2 and s_2^2 in the *t* test statistic. This test procedure was widely used in the past, but it has fallen into some disfavor because it is quite sensitive to departures from the assumption of equal population variances. If the population variances are equal, the pooled *t* test has a slightly better chance of detecting departures from the null hypothesis than does the two-sample *t* test of this section. However, *P*-values based on the pooled *t* test can be seriously in error if the population variances are not equal, so, in general, the two-sample *t* test is a better choice than the pooled *t* test.

SECTION 13.1 EXERCISES

Each Exercise Set assesses the following chapter learning objectives: C1, M1, M2, M3, M4, P1, P2

SECTION 13.1 Exercise Set 1

13.1 Descriptions of four studies are given. In each of the studies, the two populations of interest are the students at a particular university who live on campus and the students who live off campus. Which of these studies have samples that are independently selected?

Study 1: To determine if there is evidence that the mean amount of money spent on food each month differs for the two populations, a random sample of 45 students who live on campus and a random sample of 50 students who live off campus are selected.

Study 2: To determine if the mean number of hours spent studying differs for the two populations, a random sample students who live on campus is selected. Each student in this sample is asked how many hours he or she spends working each week. For each of these students who live on campus, a student who lives off campus and who works the same number of hours per week is identified and included in the sample of students who live off campus.

Study 3: To determine if the mean number of hours worked per week differs for the two populations, a random sample of students who live on campus and who have a brother or sister who also attends the university but who lives off campus is selected. The sibling who lives on campus is included in the on campus sample, and the sibling who lives off campus is included in the off-campus sample.

Study 4: To determine if the mean amount spent on textbooks differs for the two populations, a random sample of students who live on campus is selected. A separate random sample of the same size is selected from the population of students who live off campus.

13.2 For each of the following hypothesis testing scenarios, indicate whether or not the appropriate hypothesis test would be about a difference in population means. If not, explain why not.

Scenario 1: The international polling organization Ipsos reported data from a survey of 2,000 randomly selected Canadians who carry debit cards (**Canadian Account Habits Survey, July 24, 2006**). Participants in this survey were asked what they considered the minimum purchase amount for which it would be acceptable to use a debit card. You would like to determine if there is convincing evidence that the mean minimum purchase amount for which Canadians consider the use of a debit card to be acceptable is less than $10.

Scenario 2: Each person in a random sample of 247 male working adults and a random sample of 253 female working adults living in Calgary, Canada, was asked how long, in minutes, his or her typical daily commute was (**"Calgary Herald Traffic Study," Ipsos, September 17, 2005**). You would like to determine if there is convincing evidence that the mean commute times differ for male workers and female workers.

Scenario 3: A hotel chain is interested in evaluating reservation processes. Guests can reserve a room using either a telephone system or an online system. Independent random samples of 80 guests who reserved a room by phone and 60 guests who reserved a room online were selected. Of those who reserved by phone, 57 reported that they were satisfied with the reservation process. Of those who reserved online, 50 reported that they were satisfied. You would like to determine if it reasonable to conclude that the proportion who are satisfied is higher for those who reserve a room online.

13.3 Do male college students spend more time using a computer than female college students? This was one of the questions investigated by the authors of the paper **"An Ecological Momentary Assessment of the Physical Activity and Sedentary**

Behaviour Patterns of University Students" (*Health Education Journal* [2010]: 116–125). Each student in a random sample of 46 male students at a university in England and each student in a random sample of 38 female students from the same university kept a diary of how he or she spent time over a three-week period. For the sample of males, the mean time spent using a computer per day was 45.8 minutes and the standard deviation was 63.3 minutes. For the sample of females, the mean time spent using a computer was 39.4 minutes and the standard deviation was 57.3 minutes. Is there convincing evidence that the mean time male students at this university spend using a computer is greater than the mean time for female students? Test the appropriate hypotheses using $\alpha = 0.05$.

13.4 Example 13.1 looked at a study comparing students who use Facebook and students who do not use Facebook (**"Facebook and Academic Performance,"** *Computers in Human Behavior* [2010]: 1237–1245). In addition to asking the students in the samples about GPA, each student was also asked how many hours he or she spent studying each day. The two samples (141 students who were Facebook users and 68 students who were not Facebook users) were independently selected from students at a large, public Midwestern university. Although the samples were not selected at random, they were selected to be representative of the two populations. For the sample of Facebook users, the mean number of hours studied per day was 1.47 hours and the standard deviation was 0.83 hours. For the sample of students who do not use Facebook, the mean was 2.76 hours and the standard deviation was 0.99 hours. Do these sample data provide convincing evidence that the mean time spent studying for Facebook users is less than the mean time spent studying for students who do not use Facebook? Use a significance level of 0.01.

13.5 The paper **"Ladies First?" A Field Study of Discrimination in Coffee Shops** (*Applied Economics* [2008]: 1–19) describes a study in which researchers observed wait times in coffee shops in Boston. Both wait time and gender of the customer were observed. The mean wait time for a sample of 145 male customers was 85.2 seconds. The mean wait time for a sample of 141 female customers was 113.7 seconds. The sample standard deviations (estimated from graphs in the paper) were 50 seconds for the sample of males and 75 seconds for the sample of females. Suppose that these two samples are representative of the populations of wait times for female coffee shop customers and for male coffee shop customers. Is there convincing evidence that the mean wait time differs for males and females? Test the relevant hypotheses using a significance level of 0.05.

13.6 Some people believe that talking on a cell phone while driving slows reaction time, increasing the risk of accidents. The study described in the paper **"A Comparison of the Cell Phone Driver and the Drunk Driver"** (*Human Factors* [2006]: 381–391) investigated the braking reaction time of people driving in a driving simulator. Drivers followed a pace car in the simulator,

and when the pace car's brake lights came on, the drivers were supposed to step on the brake. The time between the pace car brake lights coming on and the driver stepping on the brake was measured. Two samples of 40 drivers participated in the study. The 40 people in one sample used a cell phone while driving. The 40 people in the second sample drank a mixture of orange juice and alcohol in an amount calculated to achieve a blood alcohol level of 0.08% (a value considered legally drunk in most states). For the cell phone sample, the mean braking reaction time was 779 milliseconds and the standard deviation was 209 milliseconds. For the alcohol sample, the mean braking reaction time was 849 milliseconds and the standard deviation was 228. Is there convincing evidence that the mean braking reaction time is different for the population of drivers talking on a cell phone and the population of drivers who have a blood alcohol level of 0.08%? For purposes of this exercise, you can assume that the two samples are representative of the two populations of interest.

13.7 The article **"Genetic Tweak Turns Promiscuous Animals into Loyal Mates"** (*Los Angeles Times*, June 17, 2004) summarizes a research study that appeared in the June 2004 issue of *Nature*. In this study, 11 male meadow voles that had a single gene introduced into a specific part of the brain were compared to 20 male meadow voles that did not undergo this genetic manipulation. All of the voles were paired with a receptive female partner for 24 hours. At the end of the 24-hour period, the male was placed in a situation where he could choose either the partner from the previous 24 hours or a different female. The percentage of the time that the male spent with his previous partner during a 3-hour time period was recorded. The accompanying data are approximate values read from a graph in the *Nature* article. Do these data support the researchers' hypothesis that the mean percentage of the time spent with the previous partner is significantly greater for the population of genetically altered voles than for the population of voles that did not have the gene introduced? Test the relevant hypotheses using $\alpha = 0.05$.

	Percent of Time Spent with Previous Partner							
Genetically Altered	59	62	73	80	84	85	89	92
	92	93	100					
Not Genetically Altered	2	5	13	28	34	40	48	50
	51	54	60	67	70	76	81	84
	85	92	97	99				

SECTION 13.1   **Exercise Set 2**

13.8 Descriptions of four studies are given. In each of the studies, the two populations of interest are female students at a particular university and male students at the university. Which of these studies have samples that are independently selected?

> **Study 1:** To determine if there is evidence that the mean amount of time spent playing online video games differs for the two populations, a random sample of

20 female students and a random sample of 50 male students are selected.

Study 2: To determine if the mean number of hours worked per week differs for the two populations, a random sample of female students who have a brother who also attends the university is selected. The female student is included in the sample of females, and her brother is included in the sample of males.

Study 3: To determine if the mean amount of time spent using campus recreational facilities differs for the two populations, a random sample of female students is selected. A separate random sample of the same size is selected from the population of male students.

Study 4: To determine if the mean amount of money spent on housing differs for the two populations, a random sample of female students is selected. Each student in this sample is asked how far away from campus she lives. For each of these female students, a male student who lives about the same distance from campus is identified and included in the sample of male students.

13.9 For each of the following hypothesis testing scenarios, indicate whether or not the appropriate hypothesis test would be for a difference in population means. If not, explain why not.

Scenario 1: A researcher at the Medical College of Virginia conducted a study of 60 randomly selected male soccer players and concluded that players who frequently "head" the ball in soccer have a lower mean IQ (*USA Today*, **August 14, 1995**). The soccer players were divided into two samples, based on whether they averaged 10 or more headers per game, and IQ was measured for each player. You would like to determine if the data support the researcher's conclusion.

Scenario 2: A credit bureau analysis of undergraduate students' credit records found that the mean number of credit cards in an undergraduate's wallet was 4.09 (**"Undergraduate Students and Credit Cards in 2004," Nellie Mae, May 2005**). It was also reported that in a random sample of 132 undergraduates, the mean number of credit cards that the students said they carried was 2.6. You would like to determine if there is convincing evidence that the mean number of credit cards that undergraduates report carrying is less than the credit bureau's figure of 4.09.

Scenario 3: Some commercial airplanes recirculate approximately 50% of the cabin air in order to increase fuel efficiency. The authors of the paper **"Aircraft Cabin Air Recirculation and Symptoms of the Common Cold"** (*Journal of the American Medical Association* **[2002]: 483–486**) studied 1,100 airline passengers who flew from San Francisco to Denver. Some passengers traveled on airplanes that recirculated air, and others traveled on planes that did not. Of the 517 passengers who flew on planes that did not recirculate air,

108 reported post-flight respiratory symptoms, while 111 of the 583 passengers on planes that did recirculate air reported such symptoms. You would like to determine if there is convincing evidence that the proportion of passengers with post-flight respiratory symptoms differs for passengers who travel on planes that do and do not recirculate air.

13.10 Do female college students spend more time watching TV than male college students? This was one of the questions investigated by the authors of the paper **"An Ecological Momentary Assessment of the Physical Activity and Sedentary Behaviour Patterns of University Students"** (*Health Education Journal* **[2010]: 116–125**). Each student in a random sample of 46 male students at a university in England and each student in a random sample of 38 female students from the same university kept a diary of how he or she spent time over a 3-week period. For the sample of males, the mean time spent watching TV per day was 68.2 minutes, and the standard deviation was 67.5 minutes. For the sample of females, the mean time spent watching TV per day was 93.5 minutes, and the standard deviation was 89.1 minutes. Is there convincing evidence that the mean time female students at this university spend watching TV is greater than the mean time for male students? Test the appropriate hypotheses using $\alpha = 0.05$.

13.11 The paper referenced in the Preview Example of this chapter (**"Mood Food: Chocolate and Depressive Symptoms in a Cross-Sectional Analysis,"** *Archives of Internal Medicine* **[2010]: 699–703**) describes a study that investigated the relationship between depression and chocolate consumption. Participants in the study were 931 adults who were not currently taking medication for depression. These participants were screened for depression using a widely used screening test. The participants were then divided into two samples based on their test score. One sample consisted of people who screened positive for depression, and the other sample consisted of people who did not screen positive for depression. Each of the study participants also completed a food frequency survey. The researchers believed that the two samples were representative of the two populations of interest—adults who would screen positive for depression and adults who would not screen positive. The paper reported that the mean number of servings per month of chocolate for the sample of people that screened positive for depression was 8.39, and the sample standard deviation was 14.83. For the sample of people who did not screen positive for depression, the mean was 5.39, and the standard deviation was 8.76. The paper did not say how many individuals were in each sample, but for the purposes of this exercise, you can assume that the 931 study participants included 311 who screened positive for depression and 620 who did not screen positive. Carry out a hypothesis test to confirm the researchers' conclusion that the mean number of servings of chocolate per month for people who would screen positive for depression is higher than the mean number of chocolate servings per month for people who would not screen positive.

13.12 The paper "Pathological Video-Game Use Among Youth Ages 8 to 18: A National Study" (*Psychological Science* [2009]: 594–705) included the information in the accompanying table about video game playing time for representative samples of 588 males and 590 females selected from U.S. residents ages 8 to 18.

Sample	Sample Size	Sample Mean (hours per week)	Sample Standard Deviation (hours per week)
Females	590	9.2	10.2
Males	588	16.4	14.1

Carry out a hypothesis test to determine if there is convincing evidence that the mean number of hours per week spent playing video games by females is less than the mean number of hours spent by males. Use a significance level of 0.01.

13.13 **The Nielsen Company** conducts surveys each year about the use of various media, such as television and video viewing. A representative sample of adult Americans was surveyed in 2010, and the mean number of minutes per week spent watching "time-shifted" television (watching television shows that were recorded and played back at a later time) for the people in this sample was 576 minutes. An independently selected representative sample of adults was surveyed in 2011, and the mean time spent watching time-shifted television per week for this sample was 646 minutes. Suppose that the sample size for each year was 1,000 and that the sample standard deviations were 60 minutes for the 2010 sample and 80 minutes for the 2011 sample. Is it reasonable to conclude that the mean time spent watching time-shifted television in 2010 and the mean time spent in 2011 are different? Test the relevant hypotheses using $\alpha = 0.05$.

13.14 Acrylic bone cement is commonly used in total joint replacement to secure the artificial joint. Data on the force (measured in Newtons) required to break a cement bond under two different temperature conditions appear in the accompanying table. These data are consistent with summary quantities appearing in the paper "Validation of the Small-Punch Test as a Technique for Characterizing the Mechanical Properties of Acrylic Bone Cement" (*Journal of Engineering in Medicine* [2006]: 1–21).

Temperature	Data on Breaking Force
22 degrees	100.8, 141.9, 194.8, 118.4, 176.1, 213.1
37 degrees	302.1, 339.2, 288.8, 306.8, 305.2, 327.5

Is there sufficient evidence to conclude that the mean breaking force at the higher temperature is greater than the mean breaking force at the lower temperature by more than 100 Newtons? Test the relevant hypotheses using a significance level of 0.10.

Additional Exercises

13.15 Research has shown that, for baseball players, good hip range of motion results in improved performance and decreased body stress. The article "Functional Hip Characteristics of Baseball Pitchers and Position Players" (*The American Journal of Sports Medicine*, 2010: 383–388) reported on a study of independent samples of 40 professional pitchers and 40 professional position players. For the pitchers, the sample mean hip range of motion was 75.6 degrees and the sample standard deviation was 5.9 degrees, whereas the sample mean and sample standard deviation for position players were 79.6 degrees and 7.6 degrees, respectively. Assuming that the two samples are representative of professional baseball pitchers and position players, test hypotheses appropriate for determining if mean range of motion for pitchers is less than the mean for position players.

13.16 What impact does fast-food consumption have on various dietary and health characteristics? The article "**Effects of Fast-Food Consumption on Energy Intake and Diet Quality among Children in a National Household Study**" (*Pediatrics*, 2004: 112–118) reported the accompanying summary statistics on daily calorie intake for a representative sample of teens who do not typically eat fast food and a representative sample of teens who do eat fast food.

Sample	Sample Size	Sample Mean	Sample Standard Deviation
Do not eat fast food	663	2,258	1,519
Eat fast food	413	2,637	1,138

Is there convincing evidence that the mean calorie intake for teens who typically eat fast food is greater than the mean intake for those who don't by more than 200 calories per day?

13.17 Do children diagnosed with attention deficit/hyperactivity disorder (ADHD) have smaller brains than children without this condition? This question was the topic of a research study described in the paper "**Developmental Trajectories of Brain Volume Abnormalities in Children and Adolescents with Attention Deficit/Hyperactivity Disorder**" (*Journal of the American Medical Association* [2002]: 1740–1747). Brain scans were completed for a representative sample of 152 children with ADHD and a representative sample of 139 children without ADHD. Summary values for total cerebral volume (in cubic millimeters) are given in the following table:

	n	$\bar{x}$	s
Children with ADHD	152	1,059.4	117.5
Children without ADHD	139	1,104.5	111.3

Is there convincing evidence that the mean brain volume for children with ADHD is smaller than the mean for children without ADHD? Test the relevant hypotheses using a 0.05 level of significance.

13.18 Reduced heart rate variability (HRV) is known to be a predictor of mortality after a heart attack. One measure of HRV is the average normal-to-normal beat interval (in milliseconds) for a 24-hour time period. Twenty-two heart attack patients who were dog owners and 80 heart attack patients who did not own a dog participated in a study of the effect of pet ownership on HRV, resulting in the summary statistics shown in the accompanying table (**"Relationship Between Pet Ownership and Heart Rate Variability in Patients with Healed Myocardial Infarcts,"** *The American Journal of Cardiology* [2003]: 718–721).

	Measure of HRV (Average Normal-to-Normal Beat Interval)	
	Mean	Standard Deviation
Owns Dog	873	136
Does Not Own Dog	800	134

a. The authors of this paper used a two-sample *t* test to test $H_0: \mu_1 - \mu_2 = 0$ versus $H_a: \mu_1 - \mu_2 \neq 0$. What conditions must be met in order for this to be an appropriate method of analysis?
b. The paper indicates that the null hypothesis in Part (a) was rejected and reported that the *P*-value was less than 0.05. Carry out a two-sample *t* test. Is your conclusion consistent with the one given in the paper?

13.19 The article **"Plugged In, but Tuned Out"** (*USA Today*, January 20, 2010) summarizes data from two surveys of kids ages 8 to 18. One survey was conducted in 1999 and the other was conducted in 2009. Data on number of hours per day spent using electronic media, consistent with summary quantities in the article, are given below (the actual sample sizes for the two surveys were much larger). For purposes of this exercise, you can assume that the two samples are representative of kids ages 8 to 18 in each of the 2 years when the surveys were conducted.

| **2009** | 5 | 9 | 5 | 8 | 7 | 6 | 7 | 9 | 7 | 9 | 6 | 9 |
| | 10 | 9 | 8 | | | | | | | | | |

| **1999** | 4 | 5 | 7 | 7 | 5 | 7 | 5 | 6 | 5 | 6 | 7 | 8 |
| | 5 | 6 | 6 | | | | | | | | | |

a. Because the given sample sizes are small, what assumption must be made about the distributions of electronic media use times for the two-sample *t* test to be appropriate? Use the given data to construct graphical displays that would be useful in determining whether this assumption is reasonable. Do you think it is reasonable to use these data to carry out a two-sample *t* test?
b. Do the given data provide convincing evidence that the mean number of hours per day spent using electronic

media was greater in 2009 than in 1999? Test the relevant hypotheses using a significance level of 0.01.

13.20 Each person in a random sample of 228 male teenagers and a random sample of 306 female teenagers was asked how many hours he or she spent online in a typical week (**Ipsos, January 25, 2006**). The sample mean and standard deviation were 15.1 hours and 11.4 hours for the males and 14.1 hours and 11.8 hours for the females.
a. The standard deviation for each of the samples is large, indicating a lot of variability in the responses to the question. Explain why it is not reasonable to think that the distribution of responses would be approximately normal for either the population of male teenagers or the population of female teenagers. *Hint:* The number of hours spent online in a typical week cannot be negative.
b. Given your response to Part (a), would it be appropriate to use the two-sample *t* test to test the null hypothesis that there is no difference in the mean number of hours spent online in a typical week for male teenagers and female teenagers? Explain why or why not.
c. If appropriate, carry out a test to determine if there is convincing evidence that the mean number of hours spent online in a typical week is greater for male teenagers than for female teenagers. Use $\alpha = 0.05$.

13.21 In a study of malpractice claims where a settlement had been reached, two random samples were selected: a random sample of 515 closed malpractice claims that were found not to involve medical errors and a random sample of 889 claims that were found to involve errors (**New England Journal of Medicine** [2006]: 2024–2033). The following statement appeared in the paper: "When claims not involving errors were compensated, payments were significantly lower on average than were payments for claims involving errors ($313,205 vs. $521,560, *P* = 0.004)."
a. What hypotheses did the researchers test to reach the stated conclusion?
b. Which of the following could have been the value of the test statistic for the hypothesis test? Explain your reasoning.

i.	$t = 5.00$	**iii.**	$t = 2.33$
ii.	$t = 2.65$	**iv.**	$t = 1.47$

13.22 In a study of the effect of college student employment on academic performance, the following summary statistics for GPA were reported for a sample of students who worked and for a sample of students who did not work (*University of Central Florida Undergraduate Research Journal, Spring 2005*):

	Sample Size	Mean GPA	Standard Deviation
Students Who Are Employed	184	3.12	0.485
Students Who Are Not Employed	114	3.23	0.524

The samples were selected at random from working and nonworking students at the University of Central Florida. Does this information support the hypothesis that for students at this university, those who are not employed have a higher mean GPA than those who are employed?

13.23   A newspaper story headline reads **"Gender Plays Part in Monkeys' Toy Choices, Research Finds—Like Humans, Male Monkeys Choose Balls and Cars, While Females Prefer Dolls and Pots"** (**Knight Ridder Newspapers**, December 8, 2005). The article goes on to summarize findings published in the paper **"Sex Differences in Response to Children's Toys in Nonhuman Primates"** (**Evolution and Human Behavior** [2002]: 467–479). Forty-four male monkeys and 44 female monkeys were each given a variety of toys, and the time spent playing with each toy was recorded. The table at the bottom of the page gives means and standard deviations (approximate values read from graphs in the paper) for the percentage of the time that a monkey spent playing with a particular toy. Assume that it is reasonable to regard these two samples of 44 monkeys as representative of the populations of male monkeys and of female monkeys. Use a 0.05 significance level for any hypothesis tests that you carry out when answering the various parts of this exercise.

a. The police car was considered a "masculine" toy. Do these data provide convincing evidence that the mean percentage of the time spent playing with the police car is greater for male monkeys than for female monkeys?

b. The doll was considered a "feminine" toy. Do these data provide convincing evidence that the mean percentage of the time spent playing with the doll is greater for female monkeys than for male monkeys?

c. The furry dog was considered a "neutral" toy. Do these data provide convincing evidence that the mean percentage of the time spent playing with the furry dog is not the same for male and female monkeys?

d. Based on the conclusions from the hypothesis tests of Parts (a)–(c), is the quoted newspaper story headline a reasonable summary of the findings? Explain.

e. Explain why it would be inappropriate to use the two-sample t test to decide if there was evidence that the mean percentage of the time spent playing with the police car and the mean percentage of the time spent playing with the doll are not the same for female monkeys.

13.24   A researcher at the Medical College of Virginia conducted a study of 60 randomly selected male soccer players and concluded that players who frequently "head" the ball have a lower mean IQ than those who do not (**USA Today**, **August 14, 1995**). The soccer players were divided into two groups, based on whether they averaged 10 or more headers per game. Mean IQs were reported in the article, but the sample sizes and standard deviations were not given. Suppose that these values were as given in the accompanying table. Do these data support the researcher's conclusion? Test the relevant hypotheses using $\alpha = 0.05$. Can you conclude that heading the ball *causes* lower IQ? Explain.

	n	Sample Mean	Sample sd
Fewer Than 10 Headers	35	112	10
10 or More Headers	25	103	8

13.25   Do faculty and students have similar perceptions of what types of behavior are inappropriate in the classroom? This question was examined in the article **"Faculty and Student Perceptions of Classroom Etiquette"** (**Journal of College Student Development** [1998]: 515–516). Each person in a random sample of 173 students in general education classes at a large public university was asked to judge various behaviors on a scale from 1 (totally inappropriate) to 5 (totally appropriate). People in a random sample of 98 faculty members also rated the same behaviors. The mean ratings for three of the behaviors studied are shown here (the means are consistent with data provided by the author of the article). The sample standard deviations were not given, but for purposes of this exercise, assume that they are all equal to 1.0.

Student Behavior	Student Mean Rating	Faculty Mean Rating
Wearing hats in the classroom	2.80	3.63
Addressing instructor by first name	2.90	2.11
Talking on a cell phone	1.11	1.10

TABLE FOR EXERCISE **13.23**

		Percent of Time					
		Female Monkeys			Male Monkeys		
		n	Sample Mean	Sample Standard Deviation	n	Sample Mean	Sample Standard Deviation
Toy	**Police Car**	44	8	4	44	18	5
	Doll	44	20	4	44	9	2
	Furry Dog	44	20	5	44	25	5

a. Is there sufficient evidence to conclude that the mean "appropriateness" score assigned to wearing a hat in class differs for students and faculty?

b. Is there sufficient evidence to conclude that the mean "appropriateness" score assigned to addressing an instructor by his or her first name is higher for students than for faculty?

13.26 Use the information given in the previous exercise to determine if there is sufficient evidence to conclude that the mean "appropriateness" score assigned to talking on a cell phone differs for students and faculty. Does the result of your test imply that students and faculty consider it acceptable to talk on a cell phone during class? Explain why or why not.

13.27 Wayne Gretzky was one of ice hockey's most prolific scorers when he played for the Edmonton Oilers. During his last season with the Oilers, Gretzky played in 41 games and missed 17 games due to injury. The article **"The Great Gretzky"** (*Chance* [1991]: 16–21) looked at the number of goals scored by the Oilers in games with and without Gretzky, as shown in the accompanying table. If you view the 41 games with Gretzky as a random sample of all Oiler games in which Gretzky played and the 17 games without Gretzky as a random sample of all Oiler games in which Gretzky did not play, is there convincing evidence that the mean number of goals scored by the Oilers is higher for games when Gretzky plays? Use $\alpha = 0.01$.

	n	Sample Mean	Sample sd
Games with Gretzky	41	4.73	1.29
Games without Gretzky	17	3.88	1.18

SECTION 13.2 Testing Hypotheses About the Difference Between Two Population Means Using Paired Samples

Two samples are said to be *independent* if the selection of the individuals that make up one of the samples has no bearing on the selection of individuals in the other sample. In some situations, a study with independent samples is not the best way to obtain information about a possible difference between the populations. For example, suppose that a researcher wants to determine if there is a relationship between regular aerobic exercise and blood pressure. A random sample of people who jog regularly and a second random sample of people who do not exercise regularly are selected independently of one another. The researcher might use the two-sample t test and conclude that a significant difference exists in mean blood pressure for joggers and non-joggers. But is it reasonable to conclude that the difference in mean blood pressure is attributable to jogging? It is known that blood pressure is related to body weight, and that joggers tend to be leaner than non-joggers. Based on the study described, the researcher would not be able to rule out the possibility that the observed difference in blood pressure is due to differences in weight.

One way to avoid this difficulty is to match subjects by weight. The researcher could match each jogger with a non-jogger who is similar in weight (although weights for different pairs might vary widely). If a difference in mean blood pressure between the two groups were still observed, the factor *weight* could then be ruled out as a possible explanation for the difference. When each observation in the first sample is matched up in a meaningful way with a particular observation in the second sample, the samples are said to be **paired**.

Studies can be designed to yield paired data in a number of different ways. Some studies involve using the same group of individuals with measurements recorded both before and after some event of interest occurs. Others might use naturally occurring pairs, such as twins or husbands and wives. Finally, as with weight in the jogging example, some studies construct pairs by matching on factors with effects that might otherwise obscure differences (or the lack of differences) between the two populations of interest.

Example 13.3 Benefits of Ultrasound

Ultrasound is often used in the treatment of soft tissue injuries. In a study to investigate the effect of ultrasound therapy on knee extension, range of motion was measured for people in a representative sample of physical therapy patients both before and after ultrasound therapy. A subset of the data appearing in the paper **"Location of Ultrasound Does Not Enhance Range of Motion Benefits of Ultrasound and Stretch Treatment" (University of Virginia Thesis, Trae Tashiro, 2003)** is given in the accompanying table.

	Range of Motion						
Patient	1	2	3	4	5	6	7
Before Ultrasound	31	53	45	57	50	43	32
After Ultrasound	32	59	46	64	49	45	40

You can regard the data as consisting of two samples—a sample of knee range of motion measurements for physical therapy patients prior to ultrasound therapy, and a sample of range of motion measurements for physical therapy patients after ultrasound therapy. These samples are paired rather than independent because both samples are composed of observations on the same seven patients.

Is there evidence that the ultrasound therapy increases range of motion? Let μ_1 denote the mean range of motion for the population of all physical therapy patients prior to ultrasound and μ_2 denote the mean range of motion after ultrasound. Then hypotheses of interest might be

$$H_0: \mu_1 - \mu_2 = 0 \quad \text{versus} \quad H_a: \mu_1 - \mu_2 < 0$$

This null hypothesis states that the mean range of motion before ultrasound and the mean after ultrasound are equal. This alternative hypothesis states that the mean range of motion before ultrasound is less than the mean after ultrasound.

Notice that in six of the seven data pairs, the range of motion measurement is higher after ultrasound therapy than it was before. Intuitively, this suggests that the two population means may be different.

Both the before ultrasound and after ultrasound range of motion measurements vary from patient to patient. It is this variability that may obscure any difference when a two-sample t test is used. If you were to (incorrectly) use the two-sample t test for independent samples with the given data, the resulting t test statistic value would be −0.61. This value would lead to a decision not to reject the null hypothesis. This result might surprise you at first, but remember that this test procedure ignores the information about how the samples are paired.

Two plots of the data are given in Figure 13.1. The first plot (Figure 13.1(a)) ignores the pairing, and the two samples look quite similar. However, the plot in which pairs are identified (Figure 13.1(b)) does suggest a difference, because for six of the seven pairs the after ultrasound observation is greater than the before ultrasound observation.

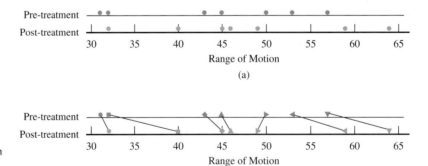

FIGURE 13.1
Two plots of the paired data from Example 13.3: (a) pairing ignored; (b) pairs identified.

Example 13.3 suggests that the methods developed for independent samples are not adequate for dealing with paired samples. When observations from the first sample are paired in some meaningful way with observations from the second sample, inferences are based on the differences between the two observations within each sample pair. The n sample differences can then be regarded as having been selected from a large population of differences. For example, in Example 13.3, you can think of the seven (before ultrasound – after ultrasound) differences as having been selected from an entire population of differences.

Before considering how the sample of differences is used to test hypotheses about the difference in population means, some new notation is needed:

$$\mu_d = \text{mean of the difference population}$$

and

$$\sigma_d = \text{standard deviation of the difference population}$$

The relationship between the two individual population means and the mean difference is

$$\mu_d \doteq \mu_1 - \mu_2$$

This tells you that when the samples are paired, inferences about μ_d are equivalent to inferences about $\mu_1 - \mu_2$. Since inferences about μ_d can be based on the n observed sample differences, the original two-sample problem becomes the familiar one-sample problem.

The Paired t Test

To compare two population or treatment means when the samples are paired, first translate the hypotheses of interest about the value of $\mu_1 - \mu_2$ into equivalent hypotheses involving μ_d:

Hypothesis	Equivalent Hypothesis When Samples Are Paired
$H_0: \mu_1 - \mu_2 = $ hypothesized value	$H_0: \mu_d = $ hypothesized value
$H_a: \mu_1 - \mu_2 > $ hypothesized value	$H_a: \mu_d > $ hypothesized value
$H_a: \mu_1 - \mu_2 < $ hypothesized value	$H_a: \mu_d < $ hypothesized value
$H_a: \mu_1 - \mu_2 \neq $ hypothesized value	$H_a: \mu_d \neq $ hypothesized value

> ### AP* EXAM TIP
>
> When samples are paired, hypotheses about $\mu_1 - \mu_2$ are translated into hypotheses about μ_d, the mean of the population of differences.

Sample differences (Sample 1 value – Sample 2 value) are computed and used as the basis for testing hypotheses about μ_d. When the number of differences is large, or when it is reasonable to assume that the population of differences is approximately normal, the one-sample t test (from Chapter 12) based on the differences is the recommended test procedure. In general, if each of the two individual population distributions is normal, the population of differences also has a normal distribution. A normal probability plot or box-plot *of the differences* can be used to decide if the assumption of normality is reasonable.

The paired-samples t test, summarized in the following box, is a method you should consider when the answers to the four key questions (QSTN) are: *hypothesis testing, sample data, one numerical variable, and two paired samples.*

> ### Questions
>
> **Q** Question type: estimation or hypothesis testing?
> **S** Study type: sample data or experiment data?
> **T** Type of data: one variable or two? Categorical or numerical?
> **N** Number of samples or treatments: how many?

Paired-Samples t Test

Appropriate when the following conditions are met:

1. The samples are paired.
2. The n sample differences can be viewed as a random sample from a population of differences (or it is reasonable to regard the sample of differences as representative of the population of differences).
3. The *number of sample differences* is large ($n \geq 30$) *or* the population distribution *of differences* is approximately normal.

When these conditions are met, the following test statistic can be used:

$$t = \frac{\bar{x}_d - \mu_0}{\dfrac{s_d}{\sqrt{n}}}$$

where μ_0 is the hypothesized value of the population mean difference from the null hypothesis, n is the number of sample differences, and $\bar{x}_d$ and s_d are the mean and standard deviation of the sample differences.

(continued)

Form of the null hypothesis: $H_0: \mu_d = \mu_0$

When the conditions above are met and the null hypothesis is true, this t test statistic has a t distribution with df $= n - 1$.

Associated P-value:

When the Alternative Hypothesis Is...	**The P-value Is...**
$H_a: \mu_d > \mu_0$	Area under the t curve to the right of the calculated value of the test statistic
$H_a: \mu_d < \mu_0$	Area under the t curve to the left of the calculated value of the test statistic
$H_a: \mu_d \neq \mu_0$	2·(area to the right of t) if t is positive or 2·(area to the left of t) if t is negative

---

Example 13.4   Benefits of Ultrasound Revisited

You can use the range of motion data of Example 13.3 to test the claim that ultrasound increases mean range of motion. Because the samples are paired, the first thing to do is compute the sample differences. These are the before – after range of motion differences for the seven physical therapy patients in the sample. A negative difference means that the after measurement was larger, so range of motion increased after the ultrasound therapy. The sample data and the computed differences are shown in the accompanying table.

Patient	1	2	3	4	5	6	7
Before Ultrasound	31	53	45	57	50	43	32
After Ultrasound	32	59	46	64	49	45	40
Difference	**−1**	**−6**	**−1**	**−7**	**1**	**−2**	**−8**

The mean and standard deviation computed from these sample differences are $\bar{x}_d = -3.43$ and $s_d = 3.51$. Do these data provide evidence that the mean range of motion before ultrasound is less than the mean range of motion after ultrasound? The usual five-step process for hypothesis testing (HMC^3) can be used to answer this question.

Considering the four key questions (QSTN), this situation can be described as *hypothesis testing, sample data, one numerical variable (range of motion), and two paired samples.* This combination suggests a paired-samples t test.

H Hypotheses   You want to use data from the samples to test a claim about the difference in population means. The population characteristics of interest are

$\mu_1 =$ mean range of motion for physical therapy patients before ultrasound
$\mu_2 =$ mean range of motion for physical therapy patients after ultrasound

Because the samples are paired, you should also define μ_d:

$$\mu_d = \mu_1 - \mu_2 = \text{mean difference in range of motion (before − after)}$$

Translating the question of interest into the hypotheses gives

$$H_0: \mu_d = 0$$
$$H_a: \mu_d < 0$$

M Method   Because the answers to the four key questions are hypothesis testing, sample data, one numerical variable, and two paired samples, consider the paired-samples t test as a potential method. The test statistic for this test is

$$t = \frac{\bar{x}_d - 0}{\dfrac{s_d}{\sqrt{n}}}$$

When the null hypothesis is true, this statistic will have a t distribution with df $= 7 - 1 = 6$. Next, choose a significance level for the test. For this example, $\alpha = 0.05$ will be used.

C Check

Although the sample of seven physical therapy patients was not a random sample of all physical therapy patients, the sample was representative of physical therapy patients. Because the sample size is small, you must be willing to assume that the distribution of range of motion differences for the population of physical therapy patients is at least approximately normal. The following boxplot of the seven sample differences is not too asymmetric and there are no outliers, so it is reasonable to think that the population distribution could be at least approximately normal.

> **AP* EXAM TIP**
>
> When samples are paired and the sample sizes are small, remember to check the approximate normality condition using the *differences*.

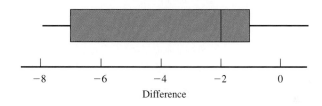

Because all conditions are met, it is appropriate to use the paired-samples t test.

C Calculate

$$n = 7 \qquad \bar{x}_d = -3.43 \qquad s_d = 3.51$$

$$t = \frac{-3.43 - 0}{\dfrac{3.51}{\sqrt{7}}} = -2.59$$

Rounding the test statistic value to -2.6 and using the df = 6 column of Appendix Table 4, you get

$$P\text{-value} = \text{area to the left of } -2.6 = \text{area to the right of } 2.6 = 0.02$$

C Communicate Results

Because the P-value (0.02) is less than α (0.05), you reject H_0. There is convincing evidence that the mean knee range of motion for physical therapy patients before ultrasound is less than the mean range of motion after ultrasound.

Statistical software could also be used for the calculate step of the test. Minitab output is shown here:

Paired T-Test
Paired T for Before-After

	N	Mean	StDev	SE Mean
Before	7	44.4286	9.9976	3.7787
After	7	47.8571	10.8847	4.1140
Difference	7	-3.42857	3.50510	1.32480

T-Test of mean difference = 0 (vs < 0): T-Value = −2.59 P-Value = 0.021

Although you have to round the computed t value to -2.6 to use Appendix Table 4, Minitab was able to compute the P-value corresponding to the actual value of the test statistic, which is P-value $= 0.021$.

Notice also that a different conclusion is reached using the paired-samples t test than if you had ignored the pairing and used the two-sample t test. (Recall the discussion of pairing in Example 13.3.)

Example 13.5 Charitable Chimps

The authors of the paper **"Chimpanzees Are Indifferent to the Welfare of Unrelated Group Members"** (*Nature* **[2005]: 1357–1359**) concluded that "chimpanzees do not take advantage of opportunities to deliver benefits to individuals at no cost to themselves." This conclusion was based on data from a study in which chimpanzees were trained to use an apparatus that would deliver food just to themselves when one lever was pushed and would deliver food to both themselves and another chimpanzee in an adjoining cage when another lever was pushed. After training, the chimps were observed when there was no chimp in the adjoining cage and when there was another chimp in the adjoining cage.

The researchers hypothesized that if chimpanzees were motivated by the welfare of others, they would choose the option that provided food to both chimpanzees more often when there was a chimpanzee in the adjoining cage. Data on the number of times the "feed both" option was chosen out of 36 opportunities for a sample of 7 chimpanzees (approximate values from a graph in the paper) are given in the accompanying table.

	Number of Times "Feed Both" Option Was Chosen		
Chimp	Chimp in Adjoining Cage	No Chimp in Adjoining Cage	Difference
1	23	21	2
2	22	22	0
3	21	23	−2
4	23	21	2
5	19	18	1
6	19	16	3
7	19	19	0

You can use this sample data to determine if there is convincing evidence that the mean number of times the "feed both" option (the charitable response) is selected is greater when another chimpanzee is present in the adjoining cage than when there is no chimpanzee in the other cage.

Because the samples are paired, the first thing to do is to compute the sample differences. Subtracting to obtain the (chimp − no chimp) differences results in the values shown in the difference column of the table. When a difference is positive, the trained chimp chose the charitable option more often when there was a chimp present in the other cage. A negative difference occurs when the charitable response was chosen less often when there was a chimp in the other cage.

Considering the four key questions (QSTN), this situation can be described as *hypothesis testing, sample data, one numerical variable (number of times a charitable response was chosen), and two paired samples.* This combination suggests a paired-samples *t* test.

H Hypotheses You want to use data from the samples to test a claim about the difference in population means. The population characteristics of interest are

μ_1 = mean number of charitable responses for all chimps when there is a chimp in the other cage

μ_2 = mean number of charitable responses for all chimps when there is no chimp in the other cage

Because the samples are paired,

μ_d = difference in mean number of charitable responses selections between when there is a chimp in the adjoining cage and when there is not

Translating the question of interest into hypotheses gives

$$H_0: \mu_d = 0$$
$$H_a: \mu_d > 0$$

M Method   Because the answers to the four key questions are hypothesis testing, sample data, one numerical variable, and two paired samples, the paired-samples t test will be considered. The test statistic for this test is

$$t = \frac{\bar{x}_d - 0}{\dfrac{s_d}{\sqrt{n}}}$$

If the null hypothesis is true, this statistic has a t distribution with df $= 7 - 1 = 6$.

Next, choose a significance level for the test. For this example, $\alpha = 0.01$ will be used.

C Check   Although the chimpanzees in this study were not randomly selected, the researchers considered them to be representative of the population of all chimpanzees. Because the sample size is small, you must be willing to assume that the distribution of differences for the population of chimpanzees is at least approximately normal. The following boxplot of the seven sample differences is fairly symmetric, and there are no outliers, so it is reasonable to think that the population difference distribution is approximately normal.

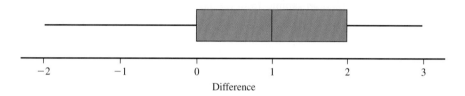

Difference

Because all conditions are met, it is appropriate to use the paired-samples t test.

C   Calculate   From the accompanying Minitab output, $t = 1.35$ and the associated P-value is 0.112.

Paired T-Test
Paired T for Chimp - No Chimp

	N	Mean	StDev	SE Mean
Chimp	7	20.8571	1.8645	0.7047
No Chimp	7	20.0000	2.4495	0.9258
Difference	7	0.857143	1.676163	0.633530

T-Test of mean difference = 0 (vs > 0): T-Value = 1.35 P-Value = 0.112

C   Communicate Results   Because the P-value (0.112) is greater than α (0.01), you fail to reject H_0. The data do not provide evidence that the mean number of times that the charitable option is chosen is greater when there is a chimpanzee in the adjoining cage. This is the basis for the statement quoted at the beginning of this example.

Notice that the numerators $\bar{x}_d$ and $\bar{x}_1 - \bar{x}_2$ of the paired-samples t and the two-sample t test statistics are always equal. The difference is in the denominator. The variability in differences is usually smaller than the variability in each sample separately (because measurements in a pair tend to be similar). As a result, the value of the paired-samples t statistic is usually larger in magnitude than the value of the two-sample t statistic. Pairing usually reduces variability that might otherwise obscure small but nevertheless important differences.

SECTION 13.2 EXERCISES

Each Exercise Set assesses the following chapter learning objectives: C1, M1, M2, M6, M7, P3

SECTION 13.2 **Exercise Set 1**

13.28 The authors of the paper "Statistical Methods for Assessing Agreement Between Two Methods of Clinical Measurement" (*International Journal of Nursing Studies* [2010]: 931–936) compared two different instruments for measuring a person's ability to breathe out air. (This measurement is helpful in diagnosing various lung disorders.) The two instruments considered were a Wright peak flow meter and a mini-Wright peak flow meter. Seventeen people participated in the study, and for each person air flow was measured once using the Wright meter and once using the mini-Wright meter. The Wright meter is thought to provide a better measure of air flow, but the mini-Wright meter is easier to transport and to use. Use of the mini-Wright meter could be recommended as long as there is not convincing evidence that the mean reading for the mini-Wright meter is different from the mean reading for the Wright meter. Use the given data to determine if there is convincing evidence that the mean reading differs for the two instruments. For purposes of this exercise, you can assume that it is reasonable to consider the 17 Wright meter minus mini-Wright meter differences as a random sample from the population of differences.

Subject	Mini-Wright Meter	Wright Meter
1	512	494
2	430	395
3	520	516
4	428	434
5	500	476
6	600	557
7	364	413
8	380	442
9	658	650
10	445	433
11	432	417
12	626	656
13	260	267
14	477	478
15	259	178
16	350	423
17	451	427

13.29 To determine if chocolate milk is as effective as other carbohydrate replacement drinks, nine male cyclists performed an intense workout followed by a drink and a rest period. At the end of the rest period, each cyclist performed an endurance trial in which he exercised until exhausted, and the time to exhaustion was measured. Each cyclist completed the entire regimen on two different days. On one day, the drink provided was chocolate milk, and on the other day the drink provided was a carbohydrate replacement drink. Data consistent with summary quantities in the paper "The Efficacy of Chocolate Milk as a Recovery Aid" (*Medicine and Science in Sports and Exercise* [2004]: S126) are given in the table at the bottom of the page. Is there evidence that the mean time to exhaustion is greater after chocolate milk than after a carbohydrate replacement drink? Use a significance level of 0.05.

13.30 In the study described in the paper "Exposure to Diesel Exhaust Induces Changes in EEG in Human Volunteers" (*Particle and Fibre Toxicology* [2007]), 10 healthy men were exposed to diesel exhaust for 1 hour. A measure of brain activity (called median power frequency, or MPF) was recorded at two different locations in the brain both before and after the diesel exhaust exposure. The resulting data are given in the accompanying table. For purposes of this exercise, assume that it is reasonable to regard the sample of 10 men as representative of healthy adult males.

	MPF (In Hz)			
Subject	Location 1 Before	Location 1 After	Location 2 Before	Location 2 After
1	6.4	8.0	6.9	9.4
2	8.7	12.6	9.5	11.2
3	7.4	8.4	6.7	10.2
4	8.7	9.0	9.0	9.6
5	9.8	8.4	9.7	9.2
6	8.9	11.0	9.0	11.9
7	9.3	14.4	7.9	9.1
8	7.4	11.3	8.3	9.3
9	6.6	7.1	7.2	8.0
10	8.9	11.2	7.4	9.1

TABLE FOR EXERCISE **13.29**

	Time to Exhaustion (minutes)								
Cyclist	1	2	3	4	5	6	7	8	9
Chocolate Milk	24.85	50.09	38.30	26.11	36.54	26.14	36.13	47.35	35.08
Carbohydrate Replacement	10.02	29.96	37.40	15.52	9.11	21.58	31.23	22.04	17.02

Do the data provide convincing evidence that the mean MPF at brain location 1 is higher after diesel exposure than before diesel exposure? Test the relevant hypotheses using a significance level of 0.05.

13.31 Use the information given in the previous exercise to determine if there is convincing evidence that the mean MPF at brain location 2 is higher after diesel exposure than before diesel exposure. Test the relevant hypotheses using a significance level of 0.05.

13.32 The paper **"Less Air Pollution Leads to Rapid Reduction of Airway Inflammation and Improved Airway Function in Asthmatic Children"** (*Pediatrics* [2009]: 1051–1058) describes a study in which children with mild asthma who live in a polluted urban environment were relocated to a less-polluted rural environment for 7 days. Various measures of respiratory function were recorded first in the urban environment and then again after 7 days in the rural environment. The accompanying graphs show the urban and rural values for three of these measures: nasal eosinophils, exhaled FENO concentration, and peak expiratory flow (PEF). Urban and rural values for the same child are connected by a line. The authors of the paper used paired-samples *t* tests to determine that there was a significant difference in the urban and rural means for each of these three measures. One of these tests resulted in a *P*-value less than 0.001, one resulted in a *P*-value between 0.001 and 0.01, and one resulted in a *P*-value between 0.01 and 0.05.

a. Which measure (Eosinophils, FENO, or PEF) do you think resulted in a test with the *P*-value that was less than 0.001? Explain your reasoning.

b. Which measure (Eosinophils, FENO, or PEF) do you think resulted in the test with the largest *P*-value? Explain your reasoning.

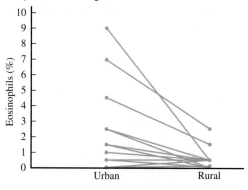

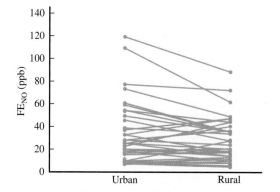

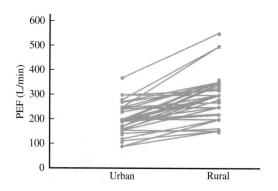

SECTION 13.2 **Exercise Set 2**

13.33 The authors of the paper **"Ultrasound Techniques Applied to Body Fat Measurement in Male and Female Athletes"** (*Journal of Athletic Training* [2009]: 142–147) compared two different methods for measuring body fat percentage. One method uses ultrasound and the other method uses X-ray technology. The accompanying table gives body fat percentages for 16 athletes using each of these methods (a subset of the data given in a graph that appeared in the paper). For purposes of this exercise, you can assume that the 16 athletes who participated in this study are representative of the population of athletes. Do these data provide convincing evidence that the mean body fat percentage measurement differs for the two methods? Test the appropriate hypotheses using $\alpha = 0.05$.

Athlete	X-ray	Ultrasound
1	5.00	4.75
2	7.00	3.75
3	9.25	9.00
4	12.00	11.75
5	17.25	17.00
6	29.50	27.50
7	5.50	6.50
8	6.00	6.75
9	8.00	8.75
10	8.50	9.50
11	9.25	9.50
12	11.00	12.00
13	12.00	12.25
14	14.00	15.50
15	17.00	18.00
16	18.00	18.25

13.34 The humorous paper **"Will Humans Swim Faster or Slower in Syrup?"** (*American Institute of Chemical Engineers Journal* [2004]: 2646–2647) investigated the fluid mechanics of swimming. Twenty swimmers each swam a specified distance in a water-filled pool and in a pool in which the water was thickened with food grade guar gum to create a syrup-like consistency. Velocity, in meters per second, was recorded. Values estimated from a graph in the paper are given. The authors of the paper concluded that swimming in guar syrup does not change swimming velocity. Are the given data consistent with this conclusion? Carry out a hypothesis test using a 0.01 significance level.

Swimmer	Velocity (m/s)	
	Water	Guar Syrup
1	0.90	0.92
2	0.92	0.96
3	1.00	0.95
4	1.10	1.13
5	1.20	1.22
6	1.25	1.22
7	1.25	1.26
8	1.30	1.30
9	1.35	1.34
10	1.40	1.41
11	1.40	1.44
12	1.50	1.52
13	1.65	1.58
14	1.70	1.70
15	1.75	1.80
16	1.80	1.76
17	1.80	1.84
18	1.85	1.89
19	1.90	1.88
20	1.95	1.95

13.35 Breast feeding sometimes results in a temporary loss of bone mass, as calcium is depleted in the mother's body to provide for milk production. The paper **"Bone Mass Is Recovered from Lactation to Postweaning in Adolescent Mothers with Low Calcium Intakes"** (*American Journal of Clinical Nutrition* [2004]: 1322–1326) gave the data at the bottom of the page on total body bone mineral content (in grams) for a sample of mothers both during breast feeding (B) and in the post-weaning period (P). Do the data suggest that actual mean bone mineral content during post-weaning is greater than that during breast feeding by more than 25 grams? State and test the appropriate hypotheses using a significance level of 0.05.

13.36 Two proposed computer mouse designs were compared by recording wrist extension in degrees for 24 people who each used both mouse designs (**"Comparative Study of Two Computer Mouse Designs,"** Cornell Human Factors Laboratory Technical Report RP7992). The difference in wrist extension was computed by subtracting extension for mouse type B from the wrist extension for mouse type A for each person. The mean difference was reported to be 8.82 degrees. Assume that this sample of 24 people is representative of the population of computer users.

a. Suppose that the standard deviation of the differences was 10 degrees. Is there convincing evidence that the mean wrist extension for mouse type A is greater than for mouse type B? Use a 0.05 significance level.

b. Suppose that the standard deviation of the differences was 26 degrees. Is there convincing evidence that the mean wrist extension for mouse type A is greater than for mouse type B? Use a 0.05 significance level.

c. Briefly explain why different conclusions were reached in the hypothesis tests of Parts (a) and (b).

Additional Exercises

13.37 Dentists make many people nervous. To see whether such nervousness elevates blood pressure, the blood pressure and pulse rates of 60 subjects were measured in a dental setting and in a medical setting (**"The Effect of the Dental Setting on Blood Pressure Measurement,"** *American Journal of Public Health* [1983]: 1210–1214). For each subject, the difference (dental setting blood pressure minus medical setting blood pressure) was calculated. The (dental – medical) differences were also calculated for pulse rates. Summary statistics follow.

	Mean Difference	Standard Deviation of Differences
Systolic Blood Pressure	4.47	8.77
Pulse (beats/min)	−1.33	8.84

Do these data strongly suggest that mean blood pressure is higher in a dental setting than in a medical setting? Test the relevant hypotheses using a significance level of 0.01.

13.38 Use the information given in the previous exercise to determine if there is sufficient evidence to conclude that the mean pulse rate in a dental setting differs from the mean pulse rate in a medical setting. Use a significance level of 0.05.

13.39 Can taking chess lessons and playing chess daily improve memory? The online article **"The USA Junior Chess Olympics Research: Developing Memory and Verbal Reasoning"** (*New Horizons for Learning*, April 2001; available at www.newhorizons.org) describes a study in which sixth-grade students who had not previously played chess participated

TABLE FOR EXERCISE **13.35**

Mother	1	2	3	4	5	6	7	8	9	10
B	1928	2549	2825	1924	1628	2175	2144	2621	1843	2541
P	2126	2885	2895	1942	1750	2184	2164	2626	2006	2627

in a program in which they took chess lessons and played chess daily for 9 months. Each student took a memory test before starting the chess program and again at the end of the 9-month period. Data (read from a graph in the article) and computed differences are given in the accompanying table. The author of the article wanted to know if these data support the claim that students who participated in the chess program tend to achieve higher memory scores after completion of the program. Carry out a hypothesis test to answer this question.

| Student | Memory Test Score | | |
	Pre-test	Post-test	Difference
1	510	850	−340
2	610	790	−180
3	640	850	−210
4	675	775	−100
5	600	700	−100
6	550	775	−225
7	610	700	−90
8	625	850	−225
9	450	690	−240
10	720	775	−55
11	575	540	35
12	675	680	−5

13.40 The press release titled **"Keeping Score When It Counts: Graduation Rates and Academic Progress Rates" (The Institute for Diversity and Ethics in Sport, March 16, 2009)** gave the 2009 graduation rates for African American basketball players and for white basketball players at every NCAA Division I university with a basketball program. Explain why it would be inappropriate to use a paired-samples t test to determine if the 2009 mean graduation rate for African American basketball players differs from the 2009 mean graduation rate for white basketball players for NCAA Division I schools.

13.41 The article **"More Students Taking AP Tests"** (*San Luis Obispo Tribune*, January 10, 2003) provided the following information on the percentage of students in grades 11 and 12 taking one or more AP exams and the percentage of exams that earned credit in 1997 and 2002 for seven high schools on the central coast of California.

| School | Percentage of Students Taking One or More AP Exams | | Percentage of Exams That Earned College Credit | |
	1997	2002	1997	2002
1	13.6	18.4	61.4	52.8
2	20.7	25.9	65.3	74.5
3	8.9	13.7	65.1	72.4
4	17.2	22.4	65.9	61.9
5	18.3	43.5	42.3	62.7
6	9.8	11.4	60.4	53.5
7	15.7	17.2	42.9	62.2

a. Assuming that it is reasonable to regard these seven schools as a random sample of high schools located on the central coast of California, carry out an appropriate test to determine if there is convincing evidence that the mean percentage of exams earning college credit at central coast high schools in 1997 and in 2002 were different.

b. Do you think it is reasonable to generalize the conclusion of the test in Part (a) to all California high schools? Explain.

c. Would it be appropriate to use the paired-samples t test with the data on percentage of students taking one or more AP exams? Explain.

13.42 Babies born extremely prematurely run the risk of various neurological problems and tend to have lower IQ and verbal ability scores than babies that are not premature. The article **"Premature Babies May Recover Intelligence, Study Says"** (*San Luis Obispo Tribune*, February 12, 2003) summarized medical research that suggests that the deficits observed at an early age may decrease as children age. Children who were born prematurely were given a test of verbal ability at age 3 and again at age 8. The test is scaled so that a score of 100 would be average for normal-birth-weight children. Data for 50 children who were born prematurely were used to generate the accompanying Minitab output, where Age3 represents the verbal ability score at age 3 and Age8 represents the verbal ability score at age 8. Use the Minitab output to determine if there is convincing evidence that the mean verbal ability score for children born prematurely increases between age 3 and age 8. You can assume that it is reasonable to regard the sample of 50 children as a random sample from the population of all children born prematurely.

Paired T-Test and CI: Age8, Age3

Paired T for Age8 - Age3

	N	Mean	StDev	Se Mean
Age8	50	97.21	16.97	2.40
Age3	50	87.30	13.84	1.96
Difference	50	9.91	22.11	3.13

T-Test of mean difference = 0 (vs > 0): T-Value = 3.17
P-Value = 0.001

13.43 The **Oregon Department of Health** web site provides information on cost-to-charge ratio (the percentage of billed charges that are actual costs to the hospital). The following table gives cost-to-charge ratios for both inpatient and outpatient care in 2002 for a random sample of six hospitals in Oregon.

Hospital	2002 Inpatient Ratio	2002 Outpatient Ratio
1	68	54
2	100	75
3	71	53
4	74	56
5	100	74
6	83	71

Is there evidence that the mean cost-to-charge ratio for Oregon hospitals is lower for outpatient care than for inpatient care? Use a significance level of 0.05.

13.44 In a study of memory recall, eight students from a large psychology class were selected at random and given 10 minutes to memorize a list of 20 nonsense words. Each was asked to list as many of the words as he or she could remember both 1 hour and 24 hours later. The data are given in the accompanying table. Is there convincing evidence to suggest that the mean number of words recalled after 1 hour is greater than the mean recall after 24 hours by more than 3? Use a significance level of 0.01.

Student	1	2	3	4	5	6	7	8
1 hour later	14	12	18	7	11	9	16	15
24 hours later	10	4	14	6	9	6	12	12

13.45 Several methods of estimating the number of seeds in soil samples have been developed by ecologists. An article in the *Journal of Ecology* ("A Comparison of Methods for Estimating Seed Numbers in the Soil" [1990]: 1079–1093) considered two of these methods. The accompanying data give number of seeds detected by the direct method and by the stratified method for 27 soil specimens.

Specimen	Direct	Stratified	Specimen	Direct	Stratified
1	24	8	15	32	28
2	32	36	16	0	0
3	0	8	17	36	36
4	60	56	18	16	12
5	20	52	19	92	92
6	64	64	20	4	12
7	40	28	21	40	48
8	8	8	22	24	24
9	12	8	23	0	0
10	92	100	24	8	12
11	4	0	25	12	40
12	68	56	26	16	12
13	76	68	27	40	76
14	24	52			

Do these data provide sufficient evidence to conclude that the mean number of seeds detected differs for the two methods? Test the relevant hypotheses using $\alpha = 0.05$.

SECTION 13.3 Estimating the Difference Between Two Population Means

The selection of an appropriate method for testing hypotheses about a difference in population means depends on whether the samples are independent or paired. The same is true when constructing a confidence interval estimate of a difference in means.

Confidence Interval When Samples Are Independent

Recall that the general form of a confidence interval is

$$\text{(statistic)} \pm \text{(critical value)} \left(\begin{array}{c} \text{standard error} \\ \text{of the statistic} \end{array} \right)$$

You can apply this general form to the case of estimating a difference in population means, $\mu_1 - \mu_2$, using data from two independent random samples. When the sample sizes are large or the population distributions are approximately normal, the appropriate confidence interval is

$$(\bar{x}_1 - \bar{x}_2) \pm (t \text{ critical value}) \sqrt{\frac{s_1^2}{n_1} + \frac{s_2^2}{n_2}}$$

To find the *t* critical value, you need to know the desired confidence level and the appropriate number of degrees of freedom. The df is computed using the same formula used for the two-sample *t* test:

$$df = \frac{(V_1 + V_2)^2}{\dfrac{V_1^2}{n_1 - 1} + \dfrac{V_2^2}{n_2 - 1}} \quad \text{where} \quad V_1 = \frac{s_1^2}{n_1} \text{ and } V_2 = \frac{s_2^2}{n_2}$$

The Two-Sample *t* Confidence Interval for a Difference in Population Means

Appropriate when the following conditions are met:

1. The samples are independently selected.
2. Each sample is a random sample from the population of interest (or the samples are selected in a way that results in samples that are representative of the populations).

(continued)

3. Both sample sizes are large ($n_1 \geq 30$ and $n_2 \geq 30$) *or* the population distributions are approximately normal.

When these conditions are met, a confidence interval for a difference in population means is

$$(\bar{x}_1 - \bar{x}_2) \pm (t \text{ critical value}) \sqrt{\frac{s_1^2}{n_1} + \frac{s_2^2}{n_2}}$$

The *t* critical value is based on

$$df = \frac{(V_1 + V_2)^2}{\dfrac{V_1^2}{n_1 - 1} + \dfrac{V_2^2}{n_2 - 1}} \quad \text{where} \quad V_1 = \frac{s_1^2}{n_1} \text{ and } V_2 = \frac{s_2^2}{n_2}$$

The computed value of df should be truncated (rounded down) to obtain an integer value for df. The desired confidence level determines which *t* critical value is used. Appendix Table 3, statistical software, or a graphing calculator can be used to obtain the *t* critical value.

Interpretation of Confidence Interval
You can be confident that the actual value of the difference in population means is included in the computed interval. This statement should be worded in context.

Interpretation of Confidence Level
The confidence level specifies the long-run proportion of the time that this method is successful in capturing the actual difference in population means.

Questions
Q Question type: estimation or hypothesis testing?
S Study type: sample data or experiment data?
T Type of data: one variable or two? Categorical or numerical?
N Number of samples or treatments: how many?

The two-sample *t* confidence interval for a difference in population means is a method you should consider when the answers to the four key questions are: *estimation, sample data, one numerical variable, and two independently selected samples.*

Now you're ready to look at an example, following the usual five-step process for estimation problems (EMC³).

Example 13.6 Women Text More Than Men

In 2010, the Nielsen Company released a report that stated **"Women Talk and Text More than Men Do" (State of the Media 2010: U.S. Audiences & Devices, The Nielsen Company).** This statement was based on data collected by examining random samples of telephone bills selected from two populations—female cell phone users and male cell phone users. The report indicated that the mean number of text messages sent per month for the sample of women was 716, while the mean number of text messages sent per month for the sample of men was 555. The report also indicated that the sample sizes were large, although it did not give the actual sample sizes. For purposes of this example, suppose that the sample sizes and sample standard deviations were as shown in the accompanying table.

Sample	Sample Size	Sample Mean	Sample Standard Deviation
Women	$n_1 = 1200$	$\bar{x}_1 = 716$	$s_1 = 90$
Men	$n_2 = 1000$	$\bar{x}_2 = 555$	$s_2 = 75$

You can use these sample data to estimate the difference in mean number of texts sent per month for women and men.

Start by answering the four key questions:
Q: This is an estimation problem.
S: The data are from sampling.
T: There is one numerical variable (number of text messages sent per month).
N: There are two independently selected samples.
This combination of answers leads to considering a two-sample *t* confidence interval.

Steps

E	Estimate
M	Method
C	Check
C	Calculate
C	Communicate results

You can now use the five-step process for estimation problems to construct a confidence interval estimate of the difference in mean number of text messages sent per month for men and women. For purposes of this example, a 90% confidence level will be used.

Step	
Estimate: Explain what population characteristic you plan to estimate.	In this example, you want to estimate the difference in the mean number of text messages sent by men and women. The population characteristics of interest are μ_1 = mean number of text messages sent by female cell phone users μ_2 = mean number of text messages sent by male cell phone users $\mu_1 - \mu_2$ = difference in mean number of text messages sent
Method: Select a potential method based on the answers to the four key questions (QSTN).	Because the answers to the four key questions are estimation, sample data, one numerical variable, and two independently selected samples, consider constructing a two-sample t confidence interval. For this example, a confidence level of 90% was selected.
Check: Check to make sure that the method selected is appropriate.	There are three conditions that need to be met in order to use the two-sample t confidence interval. Because both sample sizes are large, you do not need to worry about whether the population distributions are approximately normal. The other two conditions that must be met have to do with how the samples were selected. Here the study description says that the samples were randomly selected from the two populations of interest. Since a random sample was selected from each population, the samples are independent.
Calculate: Use the sample data to perform any necessary calculations.	$n_1 = 1,200$ $\bar{x}_1 = 716$ $s_1 = 90$ $n_2 = 1,000$ $\bar{x}_2 = 555$ $s_2 = 75$ **Degrees of freedom** $V_1 = \dfrac{s_1^2}{n_1} = 6.750 \qquad V_2 = \dfrac{s_2^2}{n_2} = 5.625$ $\text{df} = \dfrac{(V_1 + V_2)^2}{\dfrac{V_1^2}{n_1 - 1} + \dfrac{V_2^2}{n_2 - 1}}$ $= \dfrac{(6.750 + 5.625)^2}{\dfrac{(6.750)^2}{1,200} + \dfrac{(5.625)^2}{1,000}}$ $= \dfrac{153.141}{0.070}$ $= 2,187.729$ Truncate to df $= 2,187$ The appropriate t critical value is found using appropriate technology or by using Appendix Table 3 to find the t critical value that captures a central area of 0.90. (Note: When using the table, use df column of ∞ because df >120.) t critical value $= 1.645$ 90% confidence interval $(\bar{x}_1 - \bar{x}_2) \pm (t \text{ critical value}) \sqrt{\dfrac{s_1^2}{n_1} + \dfrac{s_2^2}{n_2}}$ $(716 - 555) \pm (1.645) \sqrt{\dfrac{90^2}{1,200} + \dfrac{75^2}{1,000}}$ 161 ± 5.787 $(155.213, 166.787)$

(continued)

Step	
Communicate Results: Answer the research question of interest, explain what you have learned from the data, and acknowledge potential risks.	The interpretation of confidence intervals should always be worded in the context of the problem. You should always give both an interpretation of the interval itself and of the confidence level associated with the interval.
	Confidence interval:
	You can be 90% confident that the actual difference in mean number of text messages sent is between 155.213 and 166.787. Both endpoints of this interval are positive, so you estimate that the mean number of text messages sent by female cell phone users is greater than the mean number of text messages sent by male cell phone users by somewhere between 155.213 and 166.787.
	Confidence level:
	The method used to construct this interval estimate is successful in capturing the actual difference in population means about 90% of the time.

This confidence interval estimate is the basis for the statement in the report that women send more text messages than men.

Most statistical software packages and graphing calculators can compute the two-sample t confidence interval. Minitab was used to construct a 90% confidence interval using the data of Example 13.6, and the output is shown here:

Two-Sample T-Test and CI

Sample	N	Mean	StDev	SE Mean
1	1200	716.0	90.0	2.6
2	1000	555.0	75.0	2.4

Difference = mu (1) − mu (2)
Estimate for difference: 161.000
90% CI for difference: (155.211, 166.789)

Except for small differences due to rounding, the confidence interval computed by Minitab agrees with the computations of Example 13.6.

Example 13.7 Freshman Year Weight Gain

The paper **"Predicting the 'Freshman 15': Environmental and Psychological Predictors of Weight Gain in First-Year University Students"** (*Health Education Journal* **[2010]: 321–332**) described a study conducted by researchers at Carleton University in Canada. The researchers studied a random sample of first-year students who lived on campus and a random sample of first-year students who lived off campus. Data on weight gain (in kg) during the first year, consistent with summary quantities given in the paper, are given below. A negative weight gain represents a weight loss. The researchers believed that the mean weight gain of students living on campus was higher than the mean weight gain for students living off campus and were interested in estimating the difference in means for these two groups.

On Campus	Off Campus
2.0	1.6
2.3	3.1
1.1	−2.8
−2.0	0.0
−1.9	0.2
5.6	2.9
2.6	−0.9
1.1	3.8
5.6	0.7
8.2	−0.1

The answers to the four key questions are *estimation, sample data, one numerical variable (weight gain), and two independently selected samples.* This combination of answers suggests using a two-sample *t* confidence interval. You can now use the five-step process for estimation problems to construct a 95% confidence interval for the difference in mean weight gain for students who live on campus and students who live off campus.

E Estimate   You want to estimate

$$\mu_1 - \mu_2 = \text{mean difference in weight gain}$$

where

$$\mu_1 = \text{mean weight gain for first-year students living on campus}$$

and

$$\mu_2 = \text{mean weight gain for first-year students living off campus}$$

M Method   Because the answers to the four key questions are estimation, sample data, one numerical variable, and two independently selected samples, consider constructing a two-sample *t* confidence interval. For this example, a confidence level of 95% was specified.

C Check   The samples were random samples from the two populations of interest (first-year students who live on campus and first-year students who live off campus), so the samples were independently selected. Because the sample sizes were not large, you need to be willing to assume that the population weight gain distributions are at least approximately normal. Figure 13.2 shows boxplots constructed using the data from the two samples. There are no outliers in either dataset and the boxplots are reasonably symmetric, suggesting that the assumption of approximate normality is reasonable for each of the populations. (Notice that boxplots are also shown as part of the JMP output given in the calculate step.)

C Calculate   JMP output is shown here:

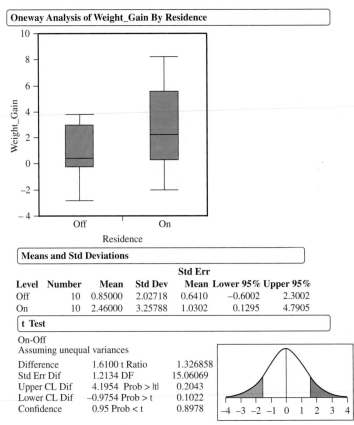

Oneway Analysis of Weight_Gain By Residence

Means and Std Deviations

Level	Number	Mean	Std Dev	Std Err Mean	Lower 95%	Upper 95%
Off	10	0.85000	2.02718	0.6410	−0.6002	2.3002
On	10	2.46000	3.25788	1.0302	0.1295	4.7905

t Test

On-Off
Assuming unequal variances

Difference	1.6100 t Ratio	1.326858
Std Err Dif	1.2134 DF	15.06069
Upper CL Dif	4.1954 Prob > |t|	0.2043
Lower CL Dif	−0.9754 Prob > t	0.1022
Confidence	0.95 Prob < t	0.8978

The 95% confidence interval estimate for the difference in mean weight gain is (−0.9754, 4.1954).

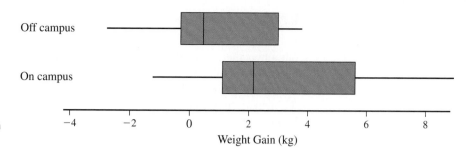

FIGURE 13.2
Boxplots for the weight gain data of Example 13.7

C Communicate Results Based on these samples, you can be 95% confident that the actual difference in mean weight gain is somewhere between −0.9754 kg and 4.1954 kg, Because 0 is included in this interval, it is possible that there is no real difference in the mean weight gain for students who live on campus and those who live off campus. The method used to construct this interval is successful in capturing the actual difference in population means about 95% of the time.

Confidence Interval When Samples Are Paired

When testing hypotheses about a difference in population means using paired samples, the test was based on the sample differences. The same is true when computing a confidence interval estimate—you first compute the differences and then use the one-sample t confidence interval introduced in Chapter 12.

The Paired-Samples t Confidence Interval for a Difference in Population Means

Appropriate when the following conditions are met:

1. The samples are paired.
2. The n sample differences can be viewed as a random sample from a population of differences (or it is reasonable to regard the sample of differences as representative of the population of differences).
3. The *number of sample differences* is large ($n \geq 30$) *or* the population distribution *of differences* is approximately normal.

When these conditions are met, a confidence interval for the difference in population means is

$$\bar{x}_d \pm (t \text{ critical value})\left(\frac{s_d}{\sqrt{n}}\right)$$

where n is the number of sample differences, and $\bar{x}_d$ and s_d are the mean and standard deviation of the sample differences.

The t critical value is based on df $= n - 1$. The desired confidence level determines which t critical value is used. Appendix Table 3, statistical software, or a graphing calculator can be used to obtain the t critical value.

Interpretation of Confidence Interval
You can be confident that the actual difference in population means is included in the computed interval. This statement should be worded in context.

Interpretation of Confidence Level
The confidence level specifies the long-run proportion of the time that this method is successful in capturing the actual difference in population means.

AP* EXAM TIP

As was the case with hypothesis tests, you use a different confidence interval formula for independent samples than for paired samples.

Example 13.8 Benefits of Ultrasound One More Time

In Example 13.4, data from a study of the benefit of ultrasound therapy were used to carry out a hypothesis test. The conclusion was that there was convincing evidence that mean knee range of motion for physical therapy patients was greater after ultrasound than before ultrasound. Once you have reached this conclusion, it would also be of interest to estimate the increase in mean range of motion. The sample data and computed differences from Example 13.4 are shown in the accompanying table. These data will be used to estimate the mean difference in range of motion.

	Range of Motion						
Patient	1	2	3	4	5	6	7
Before Ultrasound	31	53	45	57	50	43	32
After Ultrasound	32	59	46	64	49	45	40
Difference	**−1**	**−6**	**−1**	**−7**	**1**	**−2**	**−8**

The answers to the four key questions are *estimation, sample data, one numerical variable (range of motion), and two paired samples*. This combination leads to considering the paired-samples t confidence interval. The five-step process for estimation problems can be used to construct a 95% confidence interval for the mean difference in range of motion.

E Estimate   You want to estimate

$$\mu_d = \mu_1 - \mu_2 = \text{mean difference in knee range of motion}$$

where

$$\mu_1 = \text{mean knee range of motion before ultrasound therapy}$$

and

$$\mu_2 = \text{mean knee range of motion after ultrasound therapy}$$

M Method   Because the answers to the four key questions are estimation, sample data, one numerical variable, and two paired samples, a potential method is the paired-samples t confidence interval.

C Check   The conditions required for the paired-samples t confidence interval are the same as those required for the paired-samples t test. In Example 13.4, these conditions were discussed and found to be met. Therefore, it is appropriate to use the paired-samples t confidence interval to estimate the difference in means.

C Calculate   The mean and standard deviation computed using the seven sample differences are $\bar{x}_d = -3.43$ and $s_d = 3.51$. The t critical value for df $= 6$ and a 95% confidence level is 2.45. Substituting these values into the formula for the paired-samples t confidence interval gives

$$\bar{x}_d \pm (t \text{ critical value})\left(\frac{s_d}{\sqrt{n}}\right) = -3.43 \pm (2.45)\frac{3.51}{\sqrt{7}}$$

$$= -3.43 \pm 3.25$$

$$= (-6.68, -0.18)$$

A statistical software program or a graphing calculator could also have been used to compute the endpoints of the confidence interval. Minitab output is shown here. Notice that Minitab carries a bit more decimal accuracy and reports the 95% confidence interval as $(-6.67025, -0.18690)$.

Paired T-Test and CI
Paired T for After - Before

	N	Mean	StDev	SE Mean
After	7	47.8571	10.8847	4.1140
Before	7	44.4286	9.9976	3.7787
Difference	7	3.42857	3.50510	1.32480

95% CI for mean difference: $(-6.67025, -0.18690)$

C Communicate Results Based on these samples, you can be are 95% confident that the actual difference in mean range of motion is somewhere between -6.670 degrees and -0.187 degrees. Because both endpoints of the interval are negative, you would estimate that knee range of motion after ultrasound is greater than the mean range of motion before ultrasound by somewhere between 0.189 and 6.679 degrees. The method used to construct this interval is successful in capturing the actual difference in population means about 95% of the time. ∎

SECTION 13.3 EXERCISES

Each Exercise Set assesses the following chapter learning objectives: M3, M5, M6, M8, P4, P5

SECTION 13.3 Exercise Set 1

13.46 The article **"Plugged In, but Tuned Out" (USA Today, January 20, 2010)** summarizes data from two surveys of kids ages 8 to 18. One survey was conducted in 1999 and the other was conducted in 2009. Data on the number of hours per day spent using electronic media, consistent with summary quantities given in the article, are given in the following table (the actual sample sizes for the two surveys were much larger). For purposes of this exercise, assume that the two samples are representative of kids ages 8 to 18 in each of the 2 years the surveys were conducted. Construct and interpret a 98% confidence interval estimate of the difference between the mean number of hours per day spent using electronic media in 2009 and 1999.

2009	5	9	5	8	7	6	7	9	7	9	6	9	10	9	8
1999	4	5	7	7	5	7	5	6	5	6	7	8	5	6	6

13.47 The paper **"Effects of Fast-Food Consumption on Energy Intake and Diet Quality Among Children in a National Household Survey"** (*Pediatrics* [2004]: 112–118) investigated the effect of fast-food consumption on other dietary variables. For a representative sample of 663 teens who reported that they did not eat fast food during a typical day, the mean daily calorie intake was 2,258 and the sample standard deviation was 1,519. For a representative sample of 413 teens who reported that they did eat fast food on a typical day, the mean calorie intake was 2,637 and the standard deviation was 1,138. Use the given information and a 95% confidence interval to estimate the difference in mean daily calorie intake for teens who do eat fast food on a typical day and those who do not.

13.48 In the study described in the paper **"Exposure to Diesel Exhaust Induces Changes in EEG in Human Volunteers"** (*Particle and Fibre Toxicology* [2007]), 10 healthy men were exposed to diesel exhaust for 1 hour. A measure of brain activity (called median power frequency, or MPF) was recorded at two different locations in the brain both before and after the diesel exhaust exposure. The resulting data are given in the accompanying table. For purposes of this exercise, assume that the sample of 10 men is representative of healthy adult males.

	MPF (in Hz)			
Subject	Location 1 Before	Location 1 After	Location 2 Before	Location 2 After
1	6.4	8.0	6.9	9.4
2	8.7	12.6	9.5	11.2
3	7.4	8.4	6.7	10.2
4	8.7	9.0	9.0	9.6
5	9.8	8.4	9.7	9.2
6	8.9	11.0	9.0	11.9
7	9.3	14.4	7.9	9.1
8	7.4	11.3	8.3	9.3
9	6.6	7.1	7.2	8.0
10	8.9	11.2	7.4	9.1

Construct and interpret a 90% confidence interval estimate for the difference in mean MPF at brain location 1 before and after exposure to diesel exhaust.

13.49 Use the information given in the previous exercise to construct and interpret a 90% confidence interval estimate for the difference in mean MPF at brain location 2 before and after exposure to diesel exhaust.

13.50 Do girls think they don't need to take as many science classes as boys? The article **"Intentions of Young Students to Enroll in Science Courses in the Future: An Examination of Gender Differences"** (*Science Education* [1999]: 55–76) describes a survey of randomly selected children in grades 4, 5, and 6. The 224 girls participating in the survey each indicated the number of science courses they intended to take in the future, and they also indicated the number of science courses they thought boys their age should take in the future. For each girl, the authors calculated the difference between the number of science classes she intends to take and the number she thinks boys should take.

a. Explain why these data are paired.

b. The mean of the differences was −0.83 (indicating girls intended, on average, to take fewer science classes than they thought boys should take), and the standard deviation was 1.51. Construct and interpret a 95% confidence interval for the mean difference.

13.51 Each person in random samples of 247 male and 253 female working adults living in Calgary, Canada, was asked how long, in minutes, his or her typical daily commute was (**"Calgary Herald Traffic Study," Ipsos, September 17, 2005**). Use the accompanying summary statistics and a 90% confidence interval to estimate the difference in mean commute times for male and female working Calgary residents.

Males			Females		
Sample Size	$\bar{x}$	s	Sample Size	$\bar{x}$	s
247	29.6	24.3	253	27.3	24.0

13.52 Are girls less inclined to enroll in science courses than boys? One study (**"Intentions of Young Students to Enroll in Science Courses in the Future: An Examination of Gender Differences"** (*Science Education* [1999]: 55–76) asked randomly selected fourth-, fifth-, and sixth-graders how many science courses they intend to take. The following data were obtained:

	n	Mean	Standard Deviation
Males	203	3.42	1.49
Females	224	2.42	1.35

Calculate a 99% confidence interval for the difference in mean number of science courses planned for males and females. Based on your interval, how would you answer the question posed at the beginning of the exercise?

13.53 Breast feeding sometimes results in a temporary loss of bone mass as calcium is depleted in the mother's body to provide for milk production. The paper **"Bone Mass Is Recovered from Lactation to Postweaning in Adolescent Mothers with Low Calcium Intakes"** (*American Journal of Clinical Nutrition* [2004]: 1322–1326) gave the accompanying data on total body bone mineral content (in grams) for a representative sample of mothers both during breast feeding (B) and in the post-weaning period (P). Use a 95% confidence interval to estimate the difference in mean total body bone mineral content during post-weaning and during breast feeding.

Subject	1	2	3	4	5	6
B	1,928	2,549	2,825	1,924	1,628	2,175
P	2,126	2,885	2,895	1,942	1,750	2,184
Subject	7	8	9	10		
B	2,144	2,621	1,843	2,541		
P	2,164	2,626	2,006	2,627		

13.54 The paper **"Quantitative Assessment of Glenohumeral Translation in Baseball Players"** (*The American Journal of Sports Medicine* [2004]: 1711–1715) considered various aspects of shoulder motion for a representative sample of pitchers and a representative sample of position players. The accompanying table gives data supplied by the authors on a measure of the extent of anterior and posterior motion (in millimeters), both for the dominant arm and the nondominant arm.

TABLE FOR EXERCISE **13.54**

Player	Position Player Dominant Arm	Position Player Nondominant Arm	Pitcher	Pitcher Dominant Arm	Pitcher Nondominant Arm
1	30.31	32.54	1	27.63	24.33
2	44.86	40.95	2	30.57	26.36
3	22.09	23.48	3	32.62	30.62
4	31.26	31.11	4	39.79	33.74
5	28.07	28.75	5	28.50	29.84
6	31.93	29.32	6	26.70	26.71
7	34.68	34.79	7	30.34	26.45
8	29.10	28.87	8	28.69	21.49
9	25.51	27.59	9	31.19	20.82
10	22.49	21.01	10	36.00	21.75
11	28.74	30.31	11	31.58	28.32
12	27.89	27.92	12	32.55	27.22
13	28.48	27.85	13	29.56	28.86
14	25.60	24.95	14	28.64	28.58
15	20.21	21.59	15	28.58	27.15
16	33.77	32.48	16	31.99	29.46
17	32.59	32.48	17	27.16	21.26

Estimate the mean difference in motion between dominant and nondominant arms for pitchers using a 95% confidence interval.

13.55 Use the data given in the previous exercise to complete the following:
a. Estimate the mean difference in motion between dominant and nondominant arms for position players using a 95% confidence interval.
b. The authors asserted that pitchers have a greater difference in mean motion of their shoulders than do position players. Do you agree? Explain.

Additional Exercises

13.56 Do children diagnosed with attention deficit/hyperactivity disorder (ADHD) have smaller brains than children without this condition? This question was the topic of a research study described in the paper "**Developmental Trajectories of Brain Volume Abnormalities in Children and Adolescents with Attention Deficit/Hyperactivity Disorder**" (*Journal of the American Medical Association* [2002]:1740–1747). Brain scans were completed for a representative sample of 152 children with ADHD and a representative sample of 139 children without ADHD. Summary values for total cerebral volume (in milliliters) are given in the following table:

	n	$\bar{x}$	s
Children with ADHD	152	1,059.4	117.5
Children Without ADHD	139	1,104.5	111.3

Use a 95% confidence interval to estimate the difference in mean brain volume for children with and without ADHD.

13.57 In a study of the effect of college student employment on academic performance, the following summary statistics for GPA were reported for a sample of students who worked and for a sample of students who did not work (*University of Central Florida Undergraduate Research Journal*, Spring 2005):

	Sample Size	Mean GPA	Standard Deviation
Students Who Are Employed	184	3.12	0.485
Students Who Are Not Employed	114	3.23	0.524

The samples were selected at random from working and nonworking students at the University of Central Florida. Estimate the difference in mean GPA for students at this university who are employed and students who are not employed.

13.58 Babies born extremely prematurely run the risk of various neurological problems and tend to have lower IQ and verbal ability scores than babies who are not premature. The article "**Premature Babies May Recover Intelligence, Study Says**" (*San Luis Obispo Tribune*, February 12, 2003) summarized medical research that suggests that the deficit observed at an early age may decrease as children age. Children who were born prematurely were given a test of verbal ability at age 3 and again at age 8. The test is scaled so that a score of 100 would be average for normal-birth-weight children. Data that are consistent with summary quantities given in the paper for 50 children who were born prematurely were used to generate the accompanying Minitab output, where Age3 represents the verbal ability score at age 3 and Age8 represents the verbal ability score at age 8. Use the information in the Minitab output to construct and interpret a 95% confidence interval for the change in mean verbal ability score from age 3 to age 8. You can assume that it is reasonable to regard the sample of 50 children as a random sample from the population of all children born prematurely.

Paired T-Test and CI: Age8, Age3
Paired T for Age8 - Age3

	N	Mean	StDev	Se Mean
Age8	50	97.21	16.97	2.40
Age3	50	87.30	13.84	1.96
Difference	50	9.91	22.11	3.13

13.59 In a study of memory recall, eight students from a large psychology class were selected at random and given 10 minutes to memorize a list of 20 nonsense words. Each was asked to list as many of the words as he or she could remember both 1 hour and 24 hours later. The data are as shown in the accompanying table. Use these data to estimate the difference in mean number of words remembered after 1 hour and after 24 hours. Use a 90% confidence interval.

Student	1	2	3	4	5	6	7	8
1 hour later	14	12	18	7	11	9	16	15
24 hour later	10	4	14	6	9	6	12	12

13.60 Research has shown that for baseball players, good hip range of motion results in improved performance and decreased body stress. The article "**Functional Hip Characteristics of Baseball Pitchers and Position Players**" (*The American Journal of Sports Medicine*, 2010: 383–388) reported on a study involving independent samples of 40 professional pitchers and 40 professional position players. For the sample of pitchers, the mean hip range of motion was 75.6 degrees and the standard deviation was 5.9 degrees, whereas the mean and standard deviation for the sample of position players were 79.6 degrees and 7.6 degrees, respectively. Assuming that these two samples are representative of professional baseball pitchers and position players, estimate the difference in mean hip range of motion for pitchers and position players using a 90% confidence interval.

SECTION 13.4 Avoid These Common Mistakes

The cautions that have appeared at the end of the previous three chapters apply here as well. Worth repeating again are:

1. Remember that the result of a hypothesis test can never show strong support for the null hypothesis. In two-sample situations, this means that you shouldn't be *convinced* that there is *no* difference between two population means based on the outcome of a hypothesis test.
2. If you have complete information (a census) for both populations, there is no need to carry out a hypothesis test or to construct a confidence interval—in fact, it would be inappropriate to do so.
3. Don't confuse statistical significance and practical significance. In the two-sample setting, it is possible to be convinced that two population means are not equal even in situations where the actual difference between them is small enough that it is of no practical interest. After rejecting a null hypothesis of no difference (statistical significance), it is useful to look at a confidence interval estimate of the difference to get a sense of practical significance.
4. Correctly interpreting confidence intervals in the two-sample case is more difficult than in the one-sample case, so take particular care when providing a two-sample confidence interval interpretation. Because the two-sample confidence interval estimates a difference $(\mu_1 - \mu_2)$, the most important thing to note is whether or not the interval includes 0. If both endpoints of the interval are positive, then it is correct to say that, based on the interval, you believe that μ_1 is greater than μ_2 and the interval provides an estimate of how much greater. Similarly, if both interval endpoints are negative, you would say that μ_1 is less than μ_2, with the interval again providing an estimate of the size of the difference. If 0 is included in the interval, it is plausible that μ_1 and μ_2 are equal.

CHAPTER ACTIVITIES

ACTIVITY 13.1   THINKING ABOUT DATA COLLECTION

Background: In this activity you will consider two studies that allow you to investigate whether taking a keyboarding class offered as an elective course at a high school improves typing speed. You can assume that there are 2,000 students at the school and that about 10% of the students at the school complete the keyboarding course.

1. Working in a group, design a study using independent samples that would allow you to compare mean typing speed for students who have not completed the keyboarding course and mean typing speed for students who have. Be sure to describe how you plan to select your samples and how you plan to measure typing speed.

2. Do you see a benefit to designing the study in a way that would result in paired samples? What would be the benefit of incorporating pairing into your study design?

ACTIVITY 13.2   A MEANINGFUL PARAGRAPH

Write a meaningful paragraph that includes the following six terms: **paired samples, significantly different, P-value, sample, population, alternative hypothesis.** A "meaningful paragraph" is a coherent piece of writing in an appropriate context that uses all of the listed words. The paragraph should show that you understand the meaning of the terms and their relationship to one another. A sequence of sentences that just define the terms is *not* a meaningful paragraph. When choosing a context, think carefully about the terms you need to use.

🖥 EXPLORING THE BIG IDEAS

The Phoenix airport is considering two different expansion plans that would reduce crowding at the airport and reduce the time that planes spend on the ground with engines running prior to takeoff. As part of an environmental impact report on the proposed expansion, many different types of planes were studied. Using the expected time each plane would spend on the ground with engines running, the amount of fuel needed to taxi from the gate to the runway, takeoff, and climb to 3,000 feet was computed for each of the two plans.

Go online at **www.cengage.com/stats/peck1e** and click on the link labeled "Exploring the Big Ideas." Then locate the link for Chapter 13. It will take you to a web page where you can select a random sample of 30 planes from the population of all of the planes studied.

Click on the "Sample" button. This selects a random sample of 30 planes. The plane ID number and the fuel consumption for that plane for each of the two expansion plans will be displayed.

Use these data to complete the following:

(a) Construct a 95% confidence interval estimate of the difference between the mean fuel consumption for plan 1 and the mean fuel consumption for plan 2 for the population consisting of all planes studied.

(b) Write a few sentences that provide an interpretation of the confidence interval from Part (a).

(c) Which of the following statements is most appropriate given the confidence interval in Part (a)?

1. You can be 95% confident that the mean fuel consumption for planes in this population is greater for plan 1 than for plan 2.

2. You can be 95% confident that the mean fuel consumption for planes in this population is greater for plan 2 than for plan 1.

3. The interval provides convincing evidence that there is no difference in the mean fuel consumption for the two plans.

4. Based on the interval, it is possible that there is no difference in the mean fuel consumption for the two plans.

(d) Which of the following is a correct interpretation of the 95% confidence level? (Select all correct interpretations.)

1. The probability that the true mean difference in fuel consumption is contained in the computed interval is 0.95.

2. If the process of selecting a random sample of planes and then computing a 95% confidence interval for the mean difference in fuel consumption is repeated 100 times, 95 of the 100 intervals will include the population mean difference in fuel consumption.

3. If the process of selecting a random sample of planes and then computing a 95% confidence interval for the mean difference in fuel consumption is repeated a very large number of times, approximately 95% of the computed intervals will include the population mean difference in fuel consumption.

ARE YOU READY TO MOVE ON?   CHAPTER 13 REVIEW EXERCISES

All chapter learning objectives are assessed in these exercises. The learning objectives assessed in each exercise are given in parentheses.

13.61 (C1)
An individual can take either a scenic route to work or a nonscenic route. She decides that use of the nonscenic route can be justified only if it reduces the mean travel time by more than 10 minutes.

a. If μ_1 refers to the mean travel time for scenic route and μ_2 to the mean travel time for nonscenic route, what hypotheses should be tested?

b. If μ_1 refers to the mean travel time for nonscenic route and μ_2 to the mean travel time for scenic route, what hypotheses should be tested?

13.62 (M1)
Descriptions of three studies are given. In each of the studies, the two populations of interest are students majoring in science at a particular university and students majoring in liberal

arts at this university. For each of these studies, indicate whether the samples are independently selected or paired.

Study 1: To determine if there is evidence that the mean number of hours spent studying per week differs for the two populations, a random sample of 100 science majors and a random sample of 75 liberal arts majors are selected.

Study 2: To determine if the mean amount of money spent on textbooks differs for the two populations, a random sample of science majors is selected. Each student in this sample is asked how many units he or she is enrolled in for the current semester. For each of these science majors, a liberal arts major who is taking the same number of units is identified and included in the sample of liberal arts majors.

Study 3: To determine if the mean amount of time spent using the campus library differs for the two populations, a random sample of science majors is selected. A separate random sample of the same size is selected from the population of liberal arts majors.

13.63 (P1)

For each of the following hypothesis testing scenarios, indicate whether or not the appropriate hypothesis test would be for a difference in population means. If not, explain why not.

Scenario 1: The authors of the paper **"Adolescents and MP3 Players: Too Many Risks, Too Few Precautions"** (*Pediatrics* [2009]: e953–e958) studied independent random samples of 764 Dutch boys and 748 Dutch girls ages 12 to 19. Of the boys, 397 reported that they almost always listen to music at a high volume setting. Of the girls, 331 reported listening to music at a high volume setting. You would like to determine if there is convincing evidence that the proportion of Dutch boys who listen to music at high volume is greater than this proportion for Dutch girls.

Scenario 2: The report **"Highest Paying Jobs for 2009–10 Bachelor's Degree Graduates"** (**National Association of Colleges and Employers, February 2010**) states that the mean yearly salary offer for students graduating with accounting degrees in 2010 is $48,722. A random sample of 50 accounting graduates at a large university resulted in a mean offer of $49,850 and a standard deviation of $3,300. You would like to determine if there is strong support for the claim that the mean salary offer for accounting graduates of this university is higher than the 2010 national average of $48,722.

Scenario 3: Each person in a random sample of 228 male teenagers and a random sample of 306 female teenagers was asked how many hours he or she spent online in a typical week (**Ipsos, January 25, 2006**). The sample mean and standard deviation were 15.1 hours and 11.4 hours for males and 14.1 and 11.8 for females. You would like to determine if there is

convincing evidence that the mean number of hours spent online in a typical week is greater for male teenagers than for female teenagers.

13.64 (C1, M2, M3, M4, P2)

Do male college students spend more time studying than female college students? This was one of the questions investigated by the authors of the paper **"An Ecological Momentary Assessment of the Physical Activity and Sedentary Behaviour Patterns of University Students"** (*Heath Education Journal* [2010]: 116–125). Each student in a random sample of 46 male students at a university in England and each student in a random sample of 38 female students from the same university kept a diary of how he or she spent time over a 3-week period. For the sample of males, the mean time spent studying per day was 280.0 minutes, and the standard deviation was 160.4 minutes. For the sample of females, the mean time spent studying per day was 184.8 minutes, and the standard deviation was 166.4 minutes. Is there convincing evidence that the mean time male students at this university spend studying is greater than the mean time for female students? Test the appropriate hypotheses using $\alpha = 0.05$.

13.65 (C1, M2, M3, M4, P2)

The paper **"Sodium content of Lunchtime Fast Food Purchases at Major U.S. Chains"** (*Archives of Internal Medicine* [2010]: 732–734) reported that for a random sample of 850 meal purchases made at Burger King, the mean sodium content was 1,685 mg, and the standard deviation was 828 mg. For a random sample of 2,107 meal purchases made at McDonald's, the mean sodium content was 1,477 mg, and the standard deviation was 812 mg. Based on these data, is it reasonable to conclude that there is a difference in mean sodium content for meal purchases at Burger King and meal purchases at McDonald's? Use $\alpha = 0.05$.

13.66 (C1, M2, M3, M4, P2)

The paper referenced in the previous exercise also gave information on calorie content. For the sample of Burger King meal purchases, the mean number of calories was 1,008, and the standard deviation was 483. For the sample of McDonald's meal purchases, the mean number of calories was 908, and the standard deviation was 624. Based on these samples, is there convincing evidence that the mean number of calories in McDonald's meal purchases is less than the mean number of calories in Burger King meal purchases? Use $\alpha = 0.01$.

13.67 (M2, M6, M7, P3)

The article **"A Shovel with a Perforated Blade Reduces Energy Expenditure Required for Digging Wet Clay"** (*Human Factors*, 2010: 492–502) described a study in which each of 13 workers performed a task using a conventional shovel and using a shovel whose blade was perforated with small holes. The authors of the cited article provided the following data on energy expenditure (in Kcal/kg(subject)/lb(clay):

Worker:	1	2	3	4	5	6	7
Conventional:	0.0011	0.0014	0.0018	0.0022	0.0010	0.0016	0.0028
Perforated:	0.0011	0.0010	0.0019	0.0013	0.0011	0.0017	0.0024

Worker:	8	9	10	11	12	13
Conventional:	0.0020	0.0015	0.0014	0.0023	0.0017	0.0020
Perforated:	0.0020	0.0013	0.0013	0.0017	0.0015	0.0013

Do these data provide convincing evidence that the mean energy expenditure using the conventional shovel exceeds that using the perforated shovel? Test the relevant hypotheses using a significance level of 0.05.

13.68 (M2, M6, M7, P3)
Head movement evaluations are important because disabled individuals may be able to operate communications aids using head motion. The paper **"Constancy of Head Turning Recorded in Healthy Young Humans"** (*Journal of Biomedical Engineering* [2008]: 428–436) reported the accompanying data on neck rotation (in degrees) both in the clockwise direction (CL) and in the counterclockwise direction (CO) for 14 subjects. For purposes of this exercise, you may assume that the 14 subjects are representative of the population of adult Americans. Based on these data, is it reasonable to conclude that mean neck rotation is greater in the clockwise direction than in the counterclockwise direction? Carry out a hypothesis test using a significance level of 0.01.

Subject:	1	2	3	4	5	6	7
CL:	57.9	35.7	54.5	56.8	51.1	70.8	77.3
CO:	44.2	52.1	60.2	52.7	47.2	65.6	71.4

Subject:	8	9	10	11	12	13	14
CL:	51.6	54.7	63.6	59.2	59.2	55.8	38.5
CO:	48.8	53.1	66.3	59.8	47.5	64.5	34.5

13.69 (M2, M6, M7, M8, P3, P5)
The paper **"The Truth About Lying in Online Dating Profiles"** (*Proceedings, Computer-Human Interactions* [2007]: 1–4) describes an investigation in which 40 men and 40 women with online dating profiles agreed to participate in a study. Each participant's height (in inches) was measured and the actual height was compared to the height given in that person's online profile. The differences between the online profile height and the actual height (profile – actual) were used to compute the values in the accompanying table.

Men	Women
$\bar{x}_d = 0.57$	$\bar{x}_d = 0.03$
$s_d = 0.81$	$s_d = 0.75$
$n = 40$	$n = 40$

For purposes of this exercise, assume that the two samples are representative of male online daters and female online daters.
a. Use the paired-samples t test to determine if there is convincing evidence that, on average, male online daters overstate their height in online dating profiles. Use $\alpha = 0.05$.
b. Construct and interpret a 95% confidence interval for the difference between the mean online dating profile height and mean actual height for female online daters.
c. Use the two-sample t test of Section 13.1 to test $H_0: \mu_m - \mu_f = 0$ versus $H_a: \mu_m - \mu_f > 0$, where μ_m is the mean height difference (profile – actual) for male online daters and μ_f is the mean height difference (profile – actual) for female online daters.
d. Explain why a paired-samples t test was used in Part (a) but a two-sample t test was used in Part (c).

13.70 (M3, M5, P4)
Here's one to sink your teeth into: The authors of the article **"Analysis of Food Crushing Sounds During Mastication: Total Sound Level Studies"** (*Journal of Texture Studies* [1990]: 165–178) studied the nature of sounds generated during eating. Peak loudness was measured (in decibels at 20 cm away) for both open-mouth and closed-mouth chewing of potato chips and of tortilla chips. A sample of size 10 was used for each of the four possible combinations (such as closed-mouth potato chip, and so on). We are not making this up! Summary values taken from plots given in the article appear in the accompanying table. For purposes of this exercise, suppose that it is reasonable to regard the peak loudness distributions as approximately normal.

	n	$\bar{x}$	s
Potato Chip			
Open mouth	10	63	13
Closed mouth	10	54	16
Tortilla Chip			
Open mouth	10	60	15
Closed mouth	10	53	16

a. Construct a 95% confidence interval for the difference in mean peak loudness between open-mouth and closed-mouth chewing of potato chips. Be sure to interpret the resulting interval.
b. For closed-mouth chewing (the recommended method!), construct a 95% confidence interval for the difference in mean peak loudness between potato chips and tortilla chips.

13.71 (M6, M8, P5)
Samples of both surface soil and subsoil were taken from eight randomly selected agricultural locations in a particular county. The soil samples were analyzed to determine surface pH and subsoil pH, with the results shown in the accompanying table.

Location	1	2	3	4	5	6	7	8
Surface pH	6.55	5.98	5.59	6.17	5.92	6.18	6.43	5.68
Subsoil pH	6.78	6.14	5.80	5.91	6.10	6.01	6.18	5.88

a. Compute a 90% confidence interval for the mean difference between surface and subsoil pH for agricultural land in this county.

b. What conditions must be met for the interval in Part (a) to be valid?

TECHNOLOGY NOTES

Confidence Interval for $\mu_1 - \mu_2$

TI-83/84
Summarized data
1. Press **STAT**
2. Highlight **TESTS**
3. Highlight **2-SampTInt...** and press **ENTER**
4. Highlight **Stats** and press **ENTER**
5. Next to $\bar{x}1$ input the value for the sample mean from the first sample
6. Next to **sx1** input the value for the sample standard deviation from the first sample
7. Next to **n1** input the value for the sample size from the first sample
8. Next to $\bar{x}2$ input the value for the sample mean from the second sample
9. Next to **sx2** input the value for the sample standard deviation from the second sample
10. Next to **n2** input the value for the sample size from the second sample
11. Next to **C-Level** input the appropriate confidence level
12. Highlight **Calculate** and press **ENTER**

Raw data
1. Enter the data into **L1** and **L2** (In order to access lists press the **STAT** key, then press **ENTER**)
2. Press **STAT**
3. Highlight **TESTS**
4. Highlight **2-SampTInt...** and press **ENTER**
5. Highlight **Data** and press **ENTER**
6. Next to **C-Level** input the appropriate confidence level
7. Highlight **Calculate** and press **ENTER**

TI-Nspire
Summarized data
1. Enter the Calculate Scratchpad
2. Press the **menu** key and select **6:Statistics** then **6:Confidence Intervals** then **4:2-Sample t Interval...** and press **enter**
3. From the drop-down menu select **Stats**
4. Press **OK**
5. Next to $\bar{x}1$ input the value for the sample mean from the first sample
6. Next to **sx1** input the value for the sample standard deviation from the first sample
7. Next to **n1** input the value for the sample size from the first sample
8. Next to $\bar{x}2$ input the value for the sample mean from the second sample
9. Next to **sx2** input the value for the sample standard deviation from the second sample

10. Next to **n2** input the value for the sample size from the second sample
11. Next to **C Level** input the appropriate confidence level
12. Press **OK**

Raw data
1. Enter the data into two separate data lists (In order to access data lists select the spreadsheet option and press **enter**)

Note: Be sure to title the lists by selecting the top row of the column and typing a title.
2. Press the **menu** key and select **4:Statistics** then **3:Confidence Intervals** then **2:t Interval...** and press **enter**
3. From the drop-down menu select **Data**
4. Press **OK**
5. Next to **List 1** select the list containing the first data sample from the drop-down menu
6. Next to **List 2** select the list containing the second data sample from the drop-down menu
7. Next to **C-Level** input the appropriate confidence level
8. Press **OK**

JMP
1. Input the raw data for both groups into the first column
2. Input the group information into the second column

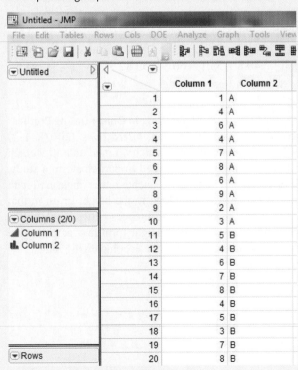

3. Click **Analyze** and select **Fit Y by X**
4. Click and drag the first column's name from the box under **Select Columns** to the box next to **Y, Response**
5. Click and drag the second column's name from the box under **Select Columns** to the box next to **X, Factor**
6. Click **OK**
7. Click the red arrow next to **Oneway Analysis of...** and select **t Test**

Note: The 95% confidence interval is automatically displayed. To change the confidence level, click the red arrow next to **Oneway Analysis of...** and select **Set α Level** then click the appropriate α-level or select **Other** and type the appropriate level.

Minitab
Summarized data

1. Click **Stat** then click **Basic Statistics** then click **2-sample t...**
2. Click the radio button next to **Summarized data**
3. In the boxes next to **First:** For the first sample, type the value for n, the sample size in the box under **Sample size:** and type the value for the sample mean in the box under **Mean:** and finally type the value for the sample standard deviation in the box under **Standard deviation:**
4. In the boxes next to **Second:** For the second sample, type the value for n, the sample size in the box under **Sample size:** and type the value for the sample mean in the box under **Mean:** and finally type the value for the sample standard deviation in the box under **Standard deviation:**
5. Click **Options...**
6. Input the appropriate confidence level in the box next to **Confidence Level**
7. Click **OK**
8. Click **OK**

Raw data

1. Input the raw data for each group into a separate column
2. Click **Stat** then click **Basic Statistics** then click **2-sample t...**
3. Click the radio button next to **Samples in different columns:**
4. Click in the box next to **First:**
5. Double click the column name where the first group's data are stored
6. Click in the box next to **Second:**
7. Double-click the column name where the second group's data are stored
8. Click **Options...**
9. Input the appropriate confidence level in the box next to **Confidence Level**
10. Click **OK**
11. Click **OK**

SPSS

1. Input the raw data for BOTH groups into the first column
2. Input the data for groups into the second column (input A for the first group and B for the second group)
3. Click **Analyze** then click **Compare Means** then click **Independent-Samples T Test**

4. Click the name of the column containing the raw data and click the arrow to move this variable to the **Test Variable(s):** box
5. Click the name of the column containing the group information and click the arrow to move this variable to the **Grouping Variable:** box
6. Click the **Define Groups...** button
7. In the box next to **Group 1:** type A
8. In the box next to **Group 2:** type B
9. Click **Continue**
10. Click **Options...**
11. Input the confidence level in the box next to **Confidence Interval Percentage:**
12. Click **Continue**
13. Click **OK**

Note: This procedure produces confidence intervals under the assumption of equal variances AND also when equal variances are not assumed.

Excel

Excel does not have the functionality to produce a confidence interval automatically for the difference of two population means. However, you can manually type the formulas for the lower and upper limits into two separate cells and have Excel calculate the results for you. You may also use Excel to find the t critical value based on the confidence level using the following steps:

1. Click in an empty cell
2. Click **Formulas**
3. Click **Insert Function**
4. Select **Statistical** from the drop-down box for the category
5. Select **TINV** and click **OK**
6. In the box next to **Probability** type in the value representing one minus your selected confidence level
7. In the box next to **Deg_freedom** type in the degrees of freedom ($n - 1$)
8. Click **OK**

Two-sample t-Test for $\mu_1 - \mu$

TI-83/84
Summarized data

1. Press **STAT**
2. Highlight **TESTS**
3. Highlight **2-SampT-Test...**
4. Highlight **Stats** and press **ENTER**
5. Next to $\bar{x}1$ input the value for the sample mean from the first sample
6. Next to **sx1** input the value for the sample standard deviation from the first sample
7. Next to **n1** input the value for the sample size from the first sample
8. Next to $\bar{x}2$ input the value for the sample mean from the second sample

9. Next to **sx2** input the value for the sample standard deviation from the second sample
10. Next to **n2** input the value for the sample size from the second sample
11. Next to **μ1** highlight the appropriate alternative hypothesis and press **ENTER**
12. Highlight **Calculate** and press **ENTER**

Raw data
1. Enter the data into **L1** and **L2** (In order to access lists press the **STAT** key, highlight the option called **Edit…** then press **ENTER**)
2. Press **STAT**
3. Highlight **TESTS**
4. Highlight **2-SampT-Test…**
5. Highlight **Data** and press **ENTER**
6. Next to **μ1** highlight the appropriate alternative hypothesis and press **ENTER**
7. Highlight **Calculate** and press **ENTER**

TI-Nspire
Summarized data
1. Enter the Calculate Scratchpad
2. Press the **menu** key and select **6:Statistics** then **7:Stat Tests** then **4:2-Sample t test…** and press **enter**
3. From the drop-down menu select **Stats**
4. Press **OK**
5. Next to $\bar{x}1$ input the value for the sample mean for the first sample
6. Next to **sx1** input the value for the sample standard deviation for the first sample
7. Next to **n1** input the value for the sample size for the first sample
8. Next to $\bar{x}2$ input the value for the sample mean for the second sample
9. Next to **sx2** input the value for the sample standard deviation for the second sample
10. Next to **n2** input the value for the sample size for the second sample
11. Next to **Alternate Hyp** select the appropriate alternative hypothesis from the drop-down menu
12. Press **OK**

Raw data
1. Enter the data into a data list (In order to access data lists select the spreadsheet option and press **enter**)

   Note: Be sure to title the list by selecting the top row of the column and typing a title.
2. Press the **menu** key and select **4:Statistics** then **4:Stat Tests** then **4:2-Sample t test…** and press **enter**
3. From the drop-down menu select **Data**
4. Press **OK**
5. Next to **List 1** select the list containing your data from the first sample
6. Next to **List 2** select the list containing your data from the second sample

7. Next to **Alternate Hyp** select the appropriate alternative hypothesis from the drop-down menu
8. Press **OK**

JMP
1. Input the raw data for both groups into the first column
2. Input the group information into the second column
3. Click **Analyze** and select **Fit Y by X**
4. Click and drag the first column's name from the box under **Select Columns** to the box next to **Y, Response**
5. Click and drag the second column's name from the box under **Select Columns** to the box next to **X, Factor**
6. Click **OK**
7. Click the red arrow next to **Oneway Analysis of…** and select **t Test**

Minitab
Summarized data
1. Click **Stat** then click **Basic Statistics** then click **2-sample t…**
2. Click the radio button next to **Summarized data**
3. In the boxes next to **First:** For the first sample, type the value for *n*, the sample size in the box under **Sample size:** and type the value for the sample mean in the box under **Mean:** and finally type the value for the sample standard deviation in the box under **Standard deviation:**
4. In the boxes next to **Second:** For the second sample, type the value for *n*, the sample size in the box under **Sample size:** and type the value for the sample mean in the box under **Mean:** and finally type the value for the sample standard deviation in the box under **Standard deviation:**
5. Click **Options…**
6. Input the appropriate hypothesized value for the difference of the population means in the box next to **Test difference:**
7. Select the appropriate alternative from the drop-down menu next to **Alternative:**
8. Click **OK**
9. Click **OK**

Note: You may also run this test with the assumption of equal variances by clicking the checkbox next to **Assume equal variances** after Step 8 in the above sequence.

Raw data
1. Input the raw data into two separate columns
2. Click **Stat** then click **Basic Statistics** then click **2-sample t…**
3. Click the radio button next to **Samples in different columns:**
4. Click in the box next to **First:**
5. Double-click the column name where the first group's data is stored
6. Click in the box next to **Second:**
7. Double-click the column name where the second group's data is stored
8. Click **Options…**
9. Input the appropriate hypothesized value for the difference of the population means in the box next to **Test difference:**
10. Select the appropriate alternative from the drop-down menu next to **Alternative:**

11. Click **OK**
12. Click **OK**

Note: You may also run this test with the assumption of equal variances by clicking the checkbox next to **Assume equal variances** after Step 11 in the above sequence.

SPSS

1. Input the raw data for BOTH groups into the first column
2. Input the data for groups into the second column (input A for the first group and B for the second group)
3. Click **Analyze** then click **Compare Means** then click **Independent-Samples T Test**
4. Click the name of the column containing the raw data and click the arrow to move this variable to the **Test Variable(s):** box
5. Click the name of the column containing the group information and click the arrow to move this variable to the **Grouping Variable:** box
6. Click the **Define Groups...** button
7. In the box next to **Group 1:** type A
8. In the box next to **Group 2:** type B
9. Click **Continue**
10. Click **OK**

Note: This procedure produces two-sample t-tests under the assumption of equal variances AND also when equal variances are not assumed. It also outputs a two-tailed P-value.

Excel

1. Input the raw data for each group into two separate columns
2. Click on the **Data** ribbon
3. Click **Data Analysis** in the **Analysis** group

Note: If you do not see **Data Analysis** listed on the Ribbon, see the Technology Notes for Chapter 2 for instructions on installing this add-on.

If you are performing a test where you are assuming equal variances, continue with Steps 4–9 below. If you are performing a test where you are not assuming equation variances, skip to Step 10.

4. Select **t-Test: Two Samples Assuming Equal Variances** from the dialog box and click **OK**
5. Click in the box next to **Variable 1 Range:** and select the first column of data
6. Click in the box next to **Variable 2 Range** and select the second column of data (if you have used and selected column titles for BOTH variables, select the check box next to **Labels**)
7. Click in the box next to **Hypothesized Mean Difference** and type your hypothesized value (in general, this will be 0)
8. Click in the box next to **Alpha:** and type in the significance level
9. Click **OK**
10. Select **t-Test: Two Samples Assuming Unequal Variances** from the dialog box and click **OK**
11. Click in the box next to **Variable 1 Range:** and select the first column of data
12. Click in the box next to **Variable 2 Range** and select the second column of data (if you have used and selected

column titles for BOTH variables, select the check box next to **Labels**)
13. Click in the box next to **Hypothesized Mean Difference** and type your hypothesized value (in general, this will be 0)
14. Click in the box next to **Alpha:** and type in the significance level
15. Click **OK**

Note: This procedure outputs P-values for both a one-sided and two-sided test.

Confidence Interval for Paired Data

TI-83/84

The TI-83/84 does not provide the option for a paired t confidence interval. However, one can be found by entering the difference data into a list and following the procedures in Chapter 12.

TI-Nspire

The TI-Nspire does not provide the option for a paired t confidence interval. However, one can be found by entering the difference data into a list and following the procedures in Chapter 12.

JMP

1. Enter the data for one group in the first column
2. Enter the paired data from the second group in the second column

3. Click **Analyze** and select **Matched Pairs**
4. Click and drag the first column name from the box under **Select Columns** to the box next to **Y, Paired Response**
5. Click and drag the second column name from the box under **Select Columns** to the box next to **Y, Paired Response**
6. Click **OK**

Note: The 95% confidence interval is automatically displayed. To change the confidence level, click the red arrow next to **Oneway Analysis of...** and select **Set α Level** then click the appropriate α-level or select **Other** and type the appropriate level.

Minitab
Summarized data

1. Click **Stat** then click **Basic Statistics** then click **Paired t...**
2. Click the radio button next to **Summarized data**

3. In the box next to **Sample size:** type the sample size, n
4. In the box next to **Mean:** type the sample mean for the DIFFERENCE of each pair of data values
5. In the box next to **Standard deviation:** type the sample standard deviation for the DIFFERENCE of each pair of data values
6. Click **Options...**
7. Input the appropriate confidence level in the box next to **Confidence Level**
8. Click **OK**
9. Click **OK**

Raw data

1. Input the raw data into two separate columns
2. Click **Stat** then click **Basic Statistics** then click **Paired t...**
3. Click in the box next to **First sample:**
4. Double click the column name where the first group's data are stored
5. Click in the box next to **Second sample:**
6. Double click the column name where the second group's data are stored
7. Click **Options...**
8. Input the appropriate confidence level in the box next to **Confidence Level**
9. Click **OK**
10. Click **OK**

SPSS

1. Input the data for each group into two separate columns
2. Click **Analyze** then click **Compare Means** then click **Paired-Samples T Test**
3. Select the first column and click the arrow to move it into the Pair 1 row, Variable 1 column
4. Select the second column and click the arrow to move it to the Pair 1 row, Variable 2 column
5. Click **Options...**
6. Input the appropriate confidence level in the box next to **Confidence Interval Percentage:**
7. Click **Continue**
8. Click **OK**

Excel

Excel does not have the functionality to produce a confidence interval automatically for the difference of two population means. However, you can manually type the formulas for the lower and upper limits into two separate cells and have Excel calculate the results for you. You may also use Excel to find the t critical value based on the confidence level using the following steps:

1. Click in an empty cell
2. Click **Formulas**
3. Click **Insert Function**
4. Select **Statistical** from the drop-down box for the category
5. Select **TINV** and click **OK**
6. In the box next to **Probability** type in the value representing one minus your selected confidence level
7. In the box next to **Deg_freedom** type in the degrees of freedom $(n - 1)$
8. Click **OK**

Paired *t*-Test for Difference of Population Means

TI-83/84

The TI-83/84 does not provide the option for a paired t test. However, one can be found by entering the difference data into a list and following the procedures in Chapter 12.

TI-Nspire

The TI-Nspire does not provide the option for a paired t test. However, one can be found by entering the difference data into a list and following the procedures in Chapter 12.

JMP

1. Enter the data for one sample in the first column
2. Enter the paired data from the second sample in the second column
3. Click **Analyze** and select **Matched Pairs**
4. Click and drag the first column name from the box under **Select Columns** to the box next to **Y, Paired Response**
5. Click and drag the second column name from the box under **Select Columns** to the box next to **Y, Paired Response**
6. Click **OK**

Minitab
Summarized data

1. Click **Stat** then click **Basic Statistics** then click **Paired t...**
2. Click the radio button next to **Summarized data**
3. In the box next to **Sample size:** type the sample size, n
4. In the box next to **Mean:** type the sample mean for the DIFFERENCE of each pair of data values
5. In the box next to **Standard deviation:** type the sample standard deviation for the DIFFERENCE of each pair of data values
6. Click **Options...**
7. Input the appropriate hypothesized value for the difference of the paired population means in the box next to **Test mean:**
8. Select the appropriate alternative from the drop-down menu next to **Alternative:**
9. Click **OK**
10. Click **OK**

Raw data

1. Input the raw data into two separate columns
2. Click **Stat** then click **Basic Statistics** then click **Paired t...**
3. Click in the box next to **First:**
4. Double click the column name where the first group's data are stored
5. Click in the box next to **Second:**
6. Double click the column name where the second group's data are stored
7. Click **Options...**
8. Input the appropriate hypothesized value for the difference of the paired population means in the box next to **Test mean:**
9. Select the appropriate alternative from the drop-down menu next to **Alternative:**
10. Click **OK**
11. Click **OK**

SPSS

1. Input the data for each group into two separate columns
2. Click **Analyze** then click **Compare Means** then click **Paired-Samples T Test**
3. Select the first column and click the arrow to move it into the Pair 1 row, Variable 1 column
4. Select the second column and click the arrow to move it to the Pair 1 row, Variable 2 column
5. Click **OK**

Note: This procedure produces a two-sided *P*-value.

Excel

1. Input the raw data for each group into two separate columns
2. Click on the **Data** ribbon
3. Click **Data Analysis** in the **Analysis** group

Note: If you do not see **Data Analysis** listed on the Ribbon, see the Technology Notes for Chapter 2 for instructions on installing this add-on.

4. Select **t-Test: Paired Two Sample for Means** from the dialog box and click **OK**
5. Click in the box next to **Variable 1 Range:** and select the first column of data
6. Click in the box next to **Variable 2 Range** and select the second column of data (if you have used and selected column titles for BOTH variables, select the check box next to **Labels**)
7. Click in the box next to **Hypothesized Mean Difference** and type your hypothesized value (in general, this will be 0)
8. Click in the box next to **Alpha:** and type in the significance level
9. Click **OK**

Note: This procedure outputs *P*-values for both a one-sided and two-sided test.

AP* Review Questions for Chapter 13

1. A medical researcher suspects that left-handed people have shorter right arms than left arms. To study this, he randomly selects 60 left-handed people and measures both the right and left arms of each. Let μ_{Right} be the average length (in inches) of the right arm of all left-handed people. Let μ_{Left} be the average length (in inches) of the left arm of all left-handed people. Let μ_d be the average difference (right − left) of the right arm and left arm of all left-handed people. Which of the following is the *best* pair of hypotheses for this study?

   (A) H_0: $\mu_{Right} - \mu_{Left} = 0$ vs. H_a: $\mu_{Right} - \mu_{Left} > 0$
   (B) H_0: $\mu_{Right} - \mu_{Left} > 0$ vs. H_a: $\mu_{Right} - \mu_{Left} < 0$
   (C) H_0: $\mu_d = 0$ vs. H_a: $\mu_d > 0$
   (D) H_0: $\mu_d = 0$ vs. H_a: $\mu_d < 0$
   (E) H_0: $\mu_d = 0$ vs. H_a: $\mu_d \neq 0$

2. Margie conducted a two-sample t-test of the hypotheses H_0: $\mu_1 - \mu_2 = 0$ versus H_a: $\mu_1 - \mu_2 > 0$, where μ_1 represents the mean dexterity score for female students and μ_2 represents the mean dexterity score for male students. After checking to make sure that the necessary conditions were met. She computed the P-value for her test to be 0.067. José used the same data to carry out a test to determine if there is evidence that the mean dexterity score for males is smaller than the mean for females, but he defined μ_1 as the mean for males and μ_2 as the mean for females. Which of the following statements is correct?

   (A) José would have obtained a P-value of 0.933 when he conducted his test.
   (B) José would have obtained a P-value of 0.467 when he conducted his test.
   (C) José would have obtained a P-value of 0.134 when he conducted his test.
   (D) José would have obtained a P-value of 0.067 when he conducted his test.
   (E) José would have obtained a P-value of 0.032 when he conducted his test.

3. An investigator conducted a two-sample t-test of the hypotheses H_0: $\mu_1 - \mu_2 = 0$ versus H_a: $\mu_1 - \mu_2 > 0$, where μ_1 represents the mean dexterity score for female students and μ_2 represents the mean dexterity score for male students. After checking to make sure that the necessary conditions were met, she computed the P-value for her test to be 0.067. If she has selected a significance level of 0.05 for the test, what conclusion will she reach?

   (A) She will fail to reject H_0 and conclude that there is not convincing evidence that the mean dexterity score is lower for males.

   (B) She will fail to reject H_0 and conclude that there is convincing evidence that the mean dexterity score is the same for males and females.
   (C) She will reject H_0 and conclude that there is convincing evidence that the mean dexterity score is lower for males.
   (D) She will reject H_0 and conclude that there is not convincing evidence that the mean dexterity score is the same for males and females.
   (E) Because 0.067 is greater that 0.05, it is not possible to make a decision about whether the null hypothesis should be rejected.

4. In a study to determine if a keyboarding class is effective in increasing typing speed, 40 of the students enrolling in the course have been randomly selected for testing, resulting in a "first day" sample and a "last day" sample. Because the samples are paired, the differences (last day − first day) will be used to carry out a paired t-test to decide if there is convincing evidence that mean typing speed of students who take this course is greater at the end of the course than on the first day. What are the appropriate null and alternative hypotheses?

   (A) H_0: $\mu_d = 0$ versus H_a: $\mu_d < 0$
   (B) H_0: $\mu_d = 0$ versus H_a: $\mu_d \neq 0$
   (C) H_0: $\mu_d = 0$ versus H_a: $\mu_d > 0$
   (D) H_0: $\mu_d < 0$ versus H_a: $\mu_d = 0$
   (E) H_0: $\mu_d > 0$ versus H_a: $\mu_d = 0$

5. One of the t-procedures for making inferences about the difference in two population means requires that the samples are independent. Of the following, which would guarantee independent samples?

   (A) Each sample is a simple random sample from the corresponding population.
   (B) The sample sizes are at least 30 for each of the populations.
   (C) The populations are normally distributed.
   (D) The population variances are equal.
   (E) The sample sizes are equal.

6. For independent random samples, the sampling distribution of $\bar{x}_1 - \bar{x}_2$ is approximately normal if

   (A) either one of the sample sizes is at least 30.
   (B) at least one of the sample sizes is at least 30.
   (C) both of the sample sizes are at least 30.
   (D) the total number sampled is at least 30.
   (E) the total number sampled at least 60.

7. For two independent random samples, the estimated standard deviation of $\bar{x}_1 - \bar{x}_2$ is

   (A) $\sqrt{\dfrac{\sigma_1^2}{n_1} - \dfrac{\sigma_2^2}{n_2}}$

   (B) $\sqrt{\dfrac{\sigma_1^2}{n_1} + \dfrac{\sigma_2^2}{n_2}}$

   (C) $\sqrt{\dfrac{s_1^2}{n_1} - \dfrac{s_2^2}{n_2}}$

   (D) $\sqrt{\dfrac{s_1^2}{n_1} + \dfrac{s_2^2}{n_2}}$

   (E) None of the above.

8. Researchers conducted a two-sample t test of the hypotheses $H_0 \colon \mu_A - \mu_B = 0$ vs. $H_a \colon \mu_A - \mu_B < 0$, where μ_A represents the mean number of text messages sent per month by people with cell phone service A, and μ_B represents the mean number of text messages sent per month by people with cell phone service B. After checking to make sure that the necessary conditions were met, the P-value was found to be 0.079. At a significance level of $\alpha = 0.01$ the researchers should decide to _____ H_0, and at a significance level of $\alpha = 0.05$ they should decide to _____ H_0.

   (A) reject; reject
   (B) fail to reject; reject
   (C) fail to reject; fail to reject
   (D) reject; fail to reject
   (E) The conclusion would be the same for both significance levels, but it is not possible to determine whether the conclusion would be to reject or fail to reject.

9. A study was conducted to investigate whether teenagers who play online games for more than 1 hour in a typical day have quicker reaction times than those who do not play online games more than 1 hour in a typical day. Because mean reaction time might be different for boys and girls, only boys were studied. Two hundred boys were randomly selected from the students attending a large high school, and they were divided into groups based on whether or not they played online games for more than 1 hour in a typical day. Time to react to a visual stimulus was recorded for each boy. The mean reaction time for those who reported playing online games more than 1 hour a day was significantly lower than the mean reaction time for those who did not play more than 1 hour a day. Which of the following conclusions is reasonable based on this study?

   (A) Playing online games for more than 1 hour a day decreases reaction time.
   (B) Playing online games for more than 1 hour a day decreases the reaction time of boys.
   (C) Playing online games for more than 1 hour a day decreased the reaction time of boys in at this school.
   (D) Boys at this school who play online games for more than 1 hour a day have a lower mean reaction time than boys at this school who do not play online games for more than 1 hour a day.
   (E) Teenagers who play online games for more than 1 hour a day have quicker reaction times, on average, than teenagers who do not play online games for more than 1 hour a day.

10. Raul wants to carry out a study to investigate whether males or females at his school spend more time exercising. Which of the following methods for selecting samples would result in paired samples?

    I. A random sample of male students who have a sister at the same school is selected. The selected males will be the sample of males, and their sisters will be the sample of females.
    II. One of the first 50 students on a list of male students at the school will be selected at random. Then every 50th student on the list is included in the sample of males. This same process will be used with a list of female students at the school to obtain a sample of females.
    III. One hundred students at the school are selected at random. The males in this group will be the sample of males, and the females in this group will be the sample of females.

    (A) I only
    (B) II only
    (C) III only
    (D) II and III only
    (E) None of the three methods will result in paired samples.

14

Learning from Experiment Data

Preview

Chapter Learning Objectives

14.1 Variability and Random Assignment

14.2 Testing Hypotheses About Differences in Treatment Effects

14.3 Estimating the Difference in Treatment Effects

14.4 Avoid These Common Mistakes

Chapter Activities

Are You Ready to Move On? Chapter 14 Review Exercises

AP* Review Questions for Chapter 14

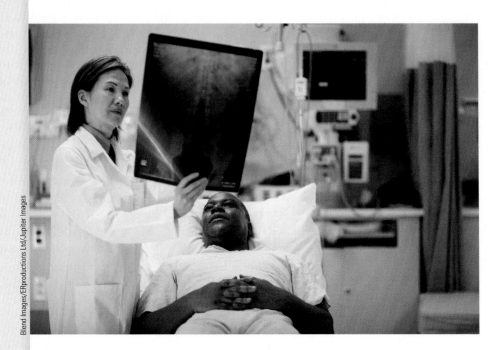

Blend Images/ERproductions Ltd/Jupiter Images

PREVIEW

Sometimes you collect data, because you want to answer questions about a single existing population or to compare two or more populations. To accomplish this, you select a sample from each population of interest and use the sample data to gain insight into characteristics of those populations. At other times, however, you collect data to answer questions about the effect of an explanatory variable on some response. Such questions are often of the form "What happens when...?" or "What is the effect of...?" You answer questions like these using data from an experiment.

CHAPTER LEARNING OBJECTIVES

Conceptual Understanding

After completing this chapter, you should be able to

C1 Understand the role that variability due to random assignment plays in drawing conclusions from experiment data.

Mastering the Mechanics

After completing this chapter, you should be able to

M1 Distinguish between situations that involve learning from sample data and those that involve learning from experiment data.

M2 Know the conditions required for appropriate use of the two-sample *t* test and the two-sample *t* confidence interval to draw conclusions about a difference in treatment means.

M3 Use the two-sample *t* test to test hypotheses about a difference in treatment means.

M4 Use the two-sample *t* confidence interval to estimate a difference in treatment means.

M5 Know the conditions required for appropriate use of the large-sample *z* test and the large-sample *z* confidence interval to draw conclusions about a difference in treatment proportions.

M6 Use the large-sample *z* test to test hypotheses about a difference in treatment proportions.

M7 Use the large-sample *z* confidence interval to estimate a difference in treatment proportions.

Putting It into Practice

After completing this chapter, you should be able to

P1 Carry out a two-sample *t* test for a difference in treatment means and interpret the conclusion in context.

P2 Interpret a two-sample *t* confidence interval for a difference in treatment means in context and interpret the associated confidence level.

P3 Carry out a large-sample *z* test for a difference in treatment proportions and interpret the conclusion in context.

P4 Interpret a large-sample *z* confidence interval for a difference in treatment proportions in context and interpret the associated confidence level.

PREVIEW EXAMPLE   Injecting Cement to Ease Pain?

Is injecting medical cement effective in reducing pain for people who have suffered fractured vertebrae? The paper **"A Randomized Trial of Vertebroplasty for Osteoporotic Spinal Fractures"** (*New England Journal of Medicine* **[2009]: 569–578**) describes a study to compare patients who underwent vertebroplasty (the injection of cement) to patients in a placebo group who underwent a fake procedure in which no cement was actually injected. Because the placebo procedure was similar to the vertebroplasty procedure except for the actual injection of cement, patients were not aware of which treatment they received. Patients participating in the study were assigned at random to one of the two experimental groups. All patients were asked to rate their pain at three different times—3 days, 14 days, and 1 month after the procedure. The summary statistics in the accompanying table were calculated from the resulting data.

	Pain Intensity			
	Vertebroplasty Group		Placebo Group	
	$n = 68$		$n = 63$	
	mean	sd	mean	sd
3 days	4.2	2.8	3.9	2.9
14 days	4.3	2.9	4.5	2.8
1 month	3.9	2.9	4.6	3.0

Notice that this study was an experiment and did not involve sampling from one or more populations. This example will be revisited later in this chapter to see what can be learned from this experiment.

The distinction between an experiment and an observational study (where data are collected by selecting a sample) is an important one. Throughout this book, you have used four questions to guide the selection of an appropriate method. One of these questions is

Does the situation involve generalizing from a sample to learn about a population or does it involve generalizing from an experiment to learn about treatment effects?

The methods you learned in Chapters 9–13 have all involved learning from sample data. In this chapter, you will see how the methods used to learn from sample data can be adapted to situations where you want to learn from experiment data.

SECTION 14.1 Variability and Random Assignment

In sampling situations, you often want to compare two populations to decide if there is evidence that their means (or proportions) are different. When random samples are selected from each population, you know that even if the population means are equal, the two sample means won't usually be equal. This is because of sample-to-sample variability that occurs due to the random selection process. To be convinced that there *really* is a difference in population means, you need to see a difference in sample means that is larger than what you would expect to see just by chance due to the random selection process. This is where statistical methods help—they allow you to determine when the differences you see in the samples are unlikely to be just due to variability introduced by random selection.

Now think about what you might learn using data from an experiment. The Preview Example described an experiment to compare two treatments (experimental conditions) for back pain. One treatment consisted of injecting cement. The second treatment consisted of a placebo, where patients underwent a procedure that appeared to be the same as the first treatment but where no cement was actually injected. Patients were randomly assigned to one of the two treatment groups. One month after the procedures, patients reported pain intensity. The mean pain intensity for the 68 patients who received the cement injection was 3.9. For the 63 patients who were in the placebo group, the mean pain intensity was 4.6.

Of course, even if the cement injection had no effect on pain, you wouldn't expect these two group means to be exactly equal, because there will be person-to-person variability in pain level. That is, even if *all* 131 patients had received the *same* treatment, when you divide them into two groups and look at the group means, the two means wouldn't be exactly equal. In this situation, what you need to figure out is if the difference in the mean pain intensity for these two groups of patients is indicating a *real* reduction in pain for the cement injection treatment. You only say that there is a **significant difference** between the treatments if you can reasonably rule out the possibility that the observed difference might be due just to the way in which patients happened to be divided into the two experimental groups. To decide if a difference is significant, you need to understand differences that might result just from variability in the response and the random assignment to treatment groups, so that you can distinguish them from a difference created by a treatment effect.

In most real experimental situations, the individuals or objects receiving the treatments are not selected at random from some larger population. This means that you cannot generalize the results of the experiment to some larger population. However, if the experimental design provides for random assignment of individuals to the treatments (or for random assignment of treatments to the individuals), it is possible to learn about treatment differences by testing hypotheses or computing and interpreting confidence intervals. Methods for doing this are introduced in the following sections.

| SECTION **14.1**   EXERCISES

Each Exercise Set assesses the following chapter learning objectives: C1

SECTION 14.1   **Exercise Set 1**

14.1   The paper **"The Effect of Multitasking on the Grade Performance of Business Students"** (*Research in Higher Education Journal* [2010]: 1–10) describes an experiment in which 62 undergraduate business students were randomly assigned to one of two experimental groups. Students in one group were asked to listen to a lecture but were told that they were permitted to use cell phones to send text messages during the lecture. Students in the second group listened to the same lecture but were not permitted to send text messages during the lecture. Afterwards, students in both groups took a quiz on material covered in the lecture. The researchers reported that the mean quiz score for students in the texting group was significantly lower than the mean quiz score for students in the no-texting group. In the context of this experiment, explain what it means to say that the texting group mean was *significantly lower* than the no-text group mean.

14.2   The paper **"Short-Term Sleep Loss Decreases Physical Activity Under Free-Living Conditions but Does Not Increase Food Intake Under Time-Deprived Laboratory Conditions in Healthy Men"** (*American Journal of Clinical Nutrition* [2009]: 1476–1483) describes an experiment in which 30 male volunteers were assigned at random to one of two sleep conditions. Men in the 4-hour group slept 4 hours per night for two nights. Men in the 8-hour group slept 8 hours per night for two nights. On the day following these two nights, the men recorded food intake. The researchers reported that there was no significant difference in mean calorie intake for the two groups. In the context of this experiment, explain what it means to say that there is *no significant difference* in the group means.

SECTION 14.2   **Exercise Set 2**

14.3   The paper **"Supervised Exercise Versus Non-Supervised Exercise for Reducing Weight in Obese Adults"** (*The Journal of Sports Medicine and Physical Fitness* [2009]: 85–90) describes an experiment in which participants were randomly assigned either to a supervised exercise program or a control group. Those in the control group were told only that they should take measures to lose weight. Those in the supervised exercise group were told they should take measures to lose weight as well, but they also participated in regular supervised exercise sessions. The researchers reported that after 4 months, the mean decrease in body fat was significantly higher for the supervised exercise group than for the control group. In the context of this experiment, explain what it means to say that the exercise group mean was *significantly higher* than the control group mean.

14.4   The paper **"Effects of Caffeine on Repeated Sprint Ability, Reactive Agility Time, Sleep and Next Day Performance"** (*Journal of Sports Medicine and Physical Fitness* [2010]: 455–464) describes an experiment in which male athlete volunteers who were considered low caffeine consumers were assigned at random to one of two experimental groups. Those assigned to the caffeine group drank a beverage which contained caffeine one hour before an exercise session. Those in the no-caffeine group drank a beverage that did not contain caffeine. During the exercise session, each participant performed a test that measured reactive agility. The researchers reported that there was no significant difference in mean reactive agility for the two experimental groups. In the context of this experiment, explain what it means to say that there is *no significant difference* in the group means.

Additional Exercises

14.5   The article **"Dieters Should Use a Bigger Fork"** (*Food Network Magazine*, January/February 2012) described an experiment conducted by researchers at the University of Utah. The article reported that when people were randomly assigned to either eat with a small fork or to eat with a large fork, the mean amount of food consumed was significantly less for the group that ate with the large fork.

a.   What are the two treatments in this experiment?

b.   In the context of this experiment, explain what it means to say that the mean amount of food consumed was *significantly less* for the group that ate with the large fork.

14.6   The article **"Why We Fall for This"** (*AARP Magazine*, May/June 2011) describes an experiment investigating the effect of money on emotions. In this experiment, students at University of Minnesota were randomly assigned to one of two groups. One group counted a stack of dollar bills. The other group counted a stack of blank pieces of paper. After counting, each student placed a finger in very hot water and then reported a discomfort level. It was reported that the mean discomfort level was significantly lower for the group that had counted money. In the context of this experiment, explain what it means to say that the money group mean was *significantly lower* than the blank-paper group mean.

14.7   The article referenced in the previous exercise also described an experiment in which students at Columbia Business School were randomly assigned to one of two groups. Students in one group were shown a coffee mug and asked how much they would pay for that mug. Students in the second group were given a coffee mug identical to the one shown to the first group and asked how much someone would have to pay to buy it from them. It was reported that the mean value assigned to the mug for the second group was significantly higher than the mean value assigned to the same mug for the first group. In the context of this experiment, explain what it means to say that the mean value was *significantly higher* for the group that was given the mug.

14.8 The paper "Matching Faces to Photographs: Poor Performance in Eyewitness Memory" (*Journal of Experimental Psychology: Applied* [2008]: 364–372)

described an experiment to investigate whether people are more likely to recognize a face when they have seen an actor in person than when they have just seen a photograph of the actor. The paper states that there was no significant difference in the proportion of correct identifications for people who saw the actor in person and for those who only saw a photograph of the actor. In the context of this experiment, explain what it means to say that there is *no significant difference* in the group means.

SECTION 14.2 Testing Hypotheses About Differences in Treatment Effects

An experiment might be carried out to compare two treatments or to compare a single treatment with a control. The resulting data could then be used to determine if the treatment has an effect on some response variable of interest. The treatments are "applied" to individuals (as in an experiment to compare the effect of two different diets on weight gain in fish) or to objects (as in an experiment to compare the effect of two different firing temperatures on the quality of ceramic tiles).

The response that is measured in an experiment could be numerical or it could be categorical. For example, in an experiment to compare the effect of two different diets on weight gain in fish, the response is weight gain, which is a numerical variable. When the response is numerical, you usually want to know if the treatment means differ—for example, if the mean weight gains for fish fed on the two diets are significantly different. In the firing temperature experiment, quality might be measured by looking at tiles after they have been fired ("baked" at a very high temperature). Each tile could be classified as either defective (cracked during firing) or not defective. Notice that here the response variable is categorical—defective or not defective. In this situation, you want to know if treatment proportions differ—for example, if the proportions of defective tiles for the two different firing temperatures are significantly different.

In an experiment to compare two treatments, deciding if there is a significant difference between treatments involves trying to answer the following question:

> "Could the observed difference in response between the two experimental groups be explained just by the way the individuals or objects happened to be divided into the two groups, or is there evidence that a treatment effect is causing the difference?"

If the design of the experiment uses random assignment to determine which individuals are in the two treatment groups, the two-sample *t* test (for means) or the large-sample *z* test (for proportions) can be used to answer this question.

Three modifications are necessary to adapt the hypothesis testing methods previously used in sampling situations for use with experiment data:

1. **Hypotheses:** Hypotheses will be worded in terms of treatment means or treatment proportions rather than in terms of population means or population proportions.
2. **Conditions:** The conditions that you need to check to determine if use of the hypothesis test procedure is appropriate are different.
3. **Conclusions:** Conclusions will be worded in terms of treatment means or treatment proportions rather than population means or population proportions. You also need to be cautious in generalizing conclusions to any larger population of interest—this isn't always a good idea.

Testing Hypotheses About the Difference Between Two Treatment Means

It is common practice to use the two-sample t test to test hypotheses about a difference in treatment means, with the following modifications:

1. **Hypotheses:** The hypotheses will look the same as before, but now μ_1 represents the mean response for treatment 1 and μ_2 represents the mean response for treatment 2. The null hypothesis $H_0: \mu_1 - \mu_2 = 0$ is a statement that there is no difference in the treatment means (no treatment effect).

2. **Conditions:** When you previously considered the two-sample t test to test hypotheses about population means using sample data, there were three conditions that had to be satisfied. The first two conditions were that the samples had to be independently selected and they had to be random samples from the populations of interest. The third condition was that the sample sizes had to be large or that the population distributions were approximately normal. In the context of testing hypotheses about treatment means, these three conditions are replaced by the following two:

    1. Individuals or objects are *randomly assigned* to treatments.
    2. The number of individuals or objects in each of the treatment groups is large (30 or more) or the treatment response distributions (the distributions of response values that would result if the treatments were applied to a *very* large number of individuals or objects) are approximately normal.

3. **Conclusions:** Conclusions will be worded in terms of treatment means. If the individuals or objects that were randomly assigned to the treatments were *also* randomly selected from some larger population, it is also reasonable to generalize conclusions about treatment effects to the larger population.

Two-Sample t Test for a Difference in Treatment Means

Appropriate when the following conditions are met:

1. Individuals or objects are *randomly assigned* to treatments.
2. The number of individuals or objects in each of the treatment groups is large (30 or more) or the treatment response distributions (the distributions of response values that would result if the treatments were applied to a *very* large number of individuals or objects) are approximately normal.

When these conditions are met, the following test statistic can be used:

$$t = \frac{(\bar{x}_1 - \bar{x}_2) - (\mu_1 - \mu_2)}{\sqrt{\dfrac{s_1^2}{n_1} + \dfrac{s_2^2}{n_2}}}$$

where $\mu_1 - \mu_2$ is the hypothesized value of the difference in treatment means from the null hypothesis (often this will be 0).

When the conditions above are met and the null hypothesis is true, the t test statistic has a t distribution with

$$df = \frac{(V_1 + V_2)^2}{\dfrac{V_1^2}{n_1 - 1} + \dfrac{V_2^2}{n_2 - 1}} \quad \text{where } V_1 = \frac{s_1^2}{n_1} \text{ and } V_2 = \frac{s_2^2}{n_2}$$

The computed value of df should be truncated (rounded down) to obtain an integer value for df.

Form of the null hypothesis: $H_0: \mu_1 - \mu_2 =$ hypothesized value

(continued)

Associated *P*-value:

When the Alternative Hypothesis Is...	The *P*-Value Is...
$H_a: \mu_1 - \mu_2 >$ hypothesized value	Area under the t curve to the right of the calculated value of the test statistic
$H_a: \mu_1 - \mu_2 <$ hypothesized value	Area under the t curve to the left of the calculated value of the test statistic
$H_a: \mu_1 - \mu_2 \neq$ hypothesized value	2(area to the right of t) if t is positive or 2(area to the left of t) if t is negative

The condition of normal treatment response distributions can be assessed by constructing a dotplot, boxplot, or a normal probability plot of the response values from each treatment group.

When the two-sample t test is used to test hypotheses about the difference between two treatment means, it is only an approximate test (the reported *P*-values are approximate). However, this is still the most common way to analyze data from experiments with two treatment groups.

Example 14.1 Reading Emotions

The paper **"How Happy Was I, Anyway? A Retrospective Impact Bias"** (*Social Cognition* **[2003]: 421–446)** describes an experiment designed to investigate the extent to which people rationalize poor performance. In this study, 246 college undergraduates were assigned at random to one of two groups—a negative feedback group or a positive feedback group. Each participant took a test in which they were asked to guess the emotions displayed in photographs of faces. At the end of the test, those in the negative feedback group were told that they had correctly answered 21 of the 40 items and were assigned a "grade" of D. Those in the positive feedback group were told that they had answered 35 of 40 correctly and were assigned an A grade. After a brief time, participants were asked to answer two sets of questions. One set of questions asked about the validity of the test, and the other set asked about the importance of being able to read faces. The researchers hypothesized that people who receive negative feedback would tend to rationalize their poor performance by rating both the validity of the test and the importance of being a good face reader lower than people who receive positive feedback.

		Test Validity Rating		Face Reading Importance Rating	
Group	Sample Size	Mean	Standard Deviation	Mean	Standard Deviation
Negative Feedback	123	5.51	0.79	5.36	1.00
Positive Feedback	123	6.95	1.09	6.62	1.19

Do the data from this experiment support the researchers' hypotheses? Let's test the relevant hypotheses using a significance level of 0.01, beginning with test validity. The now familiar five-step process for hypothesis testing (HMC[3]) will be used.

H Hypotheses You want to use data from the experiment to test the claim that the mean validity rating is lower for people who receive negative feedback than for people who receive positive feedback. You can define the two treatment means as

μ_1 = mean test validity rating for the negative feedback treatment
μ_2 = mean test validity rating for the positive feedback treatment

and $\mu_1 - \mu_2$ is the difference in treatment means. Translating the question of interest into hypotheses gives

$$H_0: \mu_1 - \mu_2 = 0$$
$$H_a: \mu_1 - \mu_2 < 0$$

The alternative hypothesis corresponds to the claim that the mean validity rating is lower for the negative feedback treatment.

M Method Considering the four key questions (QSTN), this situation can be described as *hypothesis testing, experiment data, one numerical variable (validity rating), and two treatments (negative feedback and positive feedback)*. This combination suggests a two-sample *t* test. A significance level of 0.01 was specified for this test.

C Check Next you need to check to see if this method is appropriate. From the study description, you know that the participants were assigned at random to one of the two treatment groups. Because there are 123 people in each of the treatment groups, you know that both treatment groups are large enough to proceed with the two-sample *t* test. (If the group sizes were smaller than 30, you would need to consider whether the two treatment response distributions are approximately normal.)

C Calculate You need to calculate the value of the test statistic, the number of degrees of freedom, and the *P*-value. You can do this by hand, as shown here, or you could use a graphing calculator or a statistics software package.

Test statistic:

$$t = \frac{(\bar{x}_1 - \bar{x}_2) - (\mu_1 - \mu_2)}{\sqrt{\dfrac{s_1^2}{n_1} + \dfrac{s_2^2}{n_2}}} = \frac{5.51 - 6.95 - 0}{\sqrt{\dfrac{(0.79)^2}{123} + \dfrac{(1.09)^2}{123}}}$$

$$= -\frac{1.44}{0.1214}$$

$$= -11.86$$

Degrees of freedom:

$$V_1 = \frac{s_1^2}{n_1} = 0.0051 \qquad V_2 = \frac{s_2^2}{n_2} = 0.0097$$

$$df = \frac{(V_1 + V_2)^2}{\dfrac{V_1^2}{n_1 - 1} + \dfrac{V_2^2}{n_2 - 1}} = \frac{(0.0051 + 0.0097)^2}{\dfrac{(0.0051)^2}{122} + \dfrac{(0.0097)^2}{122}} = \frac{0.000219}{0.000001} = 219$$

P-value:

This is a lower-tailed test, so the *P*-value is the area to the left of -11.86 under the *t* curve with df = 219. Since -11.86 is so far out in the lower tail of this *t* curve, *P*-value ≈ 0.

C Communicate Results Because the *P*-value is less that the selected significance level (0.01), the null hypothesis is rejected. The mean validity rating for the negative feedback treatment is significantly lower than the mean for the positive feedback treatment.

The researchers also hypothesized that negative feedback results in a lower rating of the importance of being able to read faces. You can carry out a test of this hypothesis. Steps 1, 2 and 3 are the same as they were for the test about the validity rating response, except now you define the two treatment means as

$\mu_1 = $ mean importance rating for the negative feedback treatment
$\mu_2 = $ mean importance rating for the positive feedback treatment

Translating the question of interest into hypotheses gives

$$H_0: \mu_1 - \mu_2 = 0$$
$$H_a: \mu_1 - \mu_2 < 0$$

Let's pick up with Step 4.

C Calculate   Using Minitab to do the computations results in the following output:

Two-Sample T-Test and CI

Sample	N	Mean	StDev	SE Mean
1	123	5.36	1.00	0.090
2	123	6.62	1.19	0.11

Difference = mu (1) − mu (2)
Estimate for difference: −1.26000
95% upper bound for difference: −1.02856
T-Test of difference = 0 (vs ,): T-Value = −8.99   P-Value = 0.000   DF = 236

From the output, $t = -8.99$ and the associated P-value is 0.000.

C Communicate Results   Because the P-value is less that the selected significance level (0.01), the null hypothesis is rejected. The mean importance rating for the negative feedback treatment is significantly lower than the mean for the positive feedback treatment.

Testing Hypotheses About the Difference Between Two Treatment Proportions

Sometimes the response variable in an experiment is categorical with only two possible values. For example, a ceramic tile might be classified as cracked or not cracked. A patient in a medical experiment might be classified as having improved or not improved after a particular length of time. In situations like these, you are interested in testing hypotheses about the difference in treatment "success" proportions (such as the proportion of patients who improve or the proportion of tiles that are cracked). The large-sample z test introduced in Chapter 11 can be adapted for use with experiment data by making the following modifications:

1. **Hypotheses:** The hypotheses will look the same as before, but now p_1 represents the proportion of successes for treatment 1 and p_2 represents the proportion of successes for treatment 2. The null hypothesis $H_0: p_1 - p_2 = 0$ is a statement that there is no difference in the treatment proportions (no treatment effect).

2. **Conditions:** When you previously considered the large-sample z test to test hypotheses about population proportions using sample data, there were two conditions that had to be satisfied. The first condition was that the samples had to be independent random samples from the populations of interest. The second condition was that the sample sizes had to be large. In the context of testing hypotheses about treatment proportions, these two conditions are replaced by the following two:

   1. Individuals or objects are *randomly assigned* to treatments.
   2. The number of individuals or objects in each of the treatment groups is large. This condition is met when $n_1 \hat{p}_1 \geq 10$, $n_1(1 - \hat{p}_1) \geq 10$, $n_2\hat{p}_2 \geq 10$, and $n_2(1 - \hat{p}_2) \geq 10$, (or equivalently when n_1 and n_2 are large enough to have at least 10 successes and 10 failures in each of the two treatment groups).

3. **Conclusions:** Conclusions will be worded in terms of treatment proportions. If the individuals or objects that were randomly assigned to the treatments were *also* randomly selected from some larger population, it is also reasonable to generalize conclusions about treatment effects to the larger population.

A Large-Sample z Test for a Difference in Two Treatment Proportions

Appropriate when the following conditions are met:

1. Individuals or objects are *randomly assigned* to treatments.
2. The number of individuals or objects in each of the treatment groups is large. This condition is met when $n_1 \hat{p}_1 \geq 10$, $n_1(1 - \hat{p}_1) \geq 10$, $n_2\hat{p}_2 \geq 10$, and $n_2(1 - \hat{p}_2) \geq 10$ (or equivalently if n_1 and n_2 are large enough to have at least 10 successes and 10 failures in each of the two treatment groups).

(continued)

When these conditions are met, the following test statistic can be used to test the null hypothesis $H_0: p_1 - p_2 = 0$:

$$z = \frac{\hat{p}_1 - \hat{p}_2}{\sqrt{\dfrac{\hat{p}_c(1 - \hat{p}_c)}{n_1} + \dfrac{\hat{p}_c(1 - \hat{p}_c)}{n_2}}}$$

where p_c is the combined estimate of the common proportion

$$\hat{p}_c = \frac{n_1\hat{p}_1 + n_2\hat{p}_2}{n_1 + n_2} = \frac{\text{total number of successes in the two treatment groups}}{\text{total of the two treatment group sizes}}$$

Associated *P*-value:

When the Alternative Hypothesis Is...	The *P*-Value Is...
$H_a: p_1 - p_2 > 0$	Area under the z curve to the right of the calculated value of the test statistic
$H_a: p_1 - p_2 < 0$	Area under the z curve to the left of the calculated value of the test statistic
$H_a: p_1 - p_2 \neq 0$	2(area to the right of z) if z is positive or 2(area to the left of z) if z is negative

Example 14.2   Duct Tape to Remove Warts?

Some people believe that you can fix anything with duct tape. Even so, many were skeptical when researchers announced that duct tape may be a more effective and less painful alternative to liquid nitrogen, which doctors routinely use to remove warts. The article **"What a Fix-It: Duct Tape Can Remove Warts" (*San Luis Obispo Tribune*, October 15, 2002)** described a study conducted at Madigan Army Medical Center. Patients with warts were randomly assigned to either the duct-tape treatment or the more traditional freezing treatment. Those in the duct-tape group wore duct tape over the wart for 6 days, then removed the tape, soaked the area in water, and used an emery board to scrape the area. This process was repeated for a maximum of 2 months or until the wart was gone. Data consistent with values in the article are summarized in the following table.

Treatment	*n*	Number with Wart Successfully Removed
Liquid-Nitrogen Freezing	100	60
Duct Tape	104	88

Do these data suggest that the duct tape treatment is more successful than freezing for removing warts?

H Hypotheses   You want to use data from the experiment to test the claim that duct tape is more successful than freezing for removing warts. The two treatments are duct tape and freezing, so you define the two treatment proportions as

p_1 = proportion of warts successfully removed by the duct tape treatment
p_2 = proportion of warts successfully removed by the freezing treatment

and $p_1 - p_2$ is the difference in treatment proportions. Translating the question of interest into hypotheses gives

$$H_0: p_1 - p_2 = 0$$
$$H_a: p_1 - p_2 > 0$$

The alternative hypothesis corresponds to the claim that the success proportion is higher for the duct-tape treatment.

M Method Considering the four key questions (QSTN), this situation can be described as *hypothesis testing, experiment data, one categorical variable (with two categories—wart removed and wart not removed), and two treatments (duct tape and freezing).* This combination suggests a large-samples z test for a difference in treatment proportions. For purposes of this example, a significance level of 0.05 will be used.

C Check Next, you need to verify that this method is appropriate. From the study description, you know that the participants were assigned at random to one of the two treatment groups. You also need to make sure that there are enough people in each of the two treatment groups. For these data, $\hat{p}_1 = \frac{88}{104} = 0.85$ and $\hat{p}_2 = \frac{60}{100} = 0.60$, so

$$n_1\hat{p}_1 = 104(0.85) = 88.4 \geq 10$$
$$n_1(1 - \hat{p}_1) = 104(0.15) = 15.6 \geq 10$$
$$n_2\hat{p}_2 = 100(0.60) = 60 \geq 10$$
$$n_2(1 - \hat{p}_2) = 100(0.40) = 40 \geq 10$$

Both treatment groups are large enough to proceed with the large-samples z test.

C Calculate You need to calculate the values of the test statistic and the P-value. You can do this by hand, as shown here, or you could use a graphing calculator or a statistics software package.

Test statistic:
First you need to compute the value of $\hat{p}_c$, the combined estimate of the common proportion.

$$\hat{p}_c = \frac{n_1\hat{p}_1 + n_2\hat{p}_2}{n_1 + n_2} = \frac{104(0.85) + 100(0.60)}{104 + 100} = 0.73$$

The value of the test statistic is then

$$z = \frac{\hat{p}_1 - \hat{p}_2}{\sqrt{\dfrac{\hat{p}_c(1 - \hat{p}_c)}{n_1} + \dfrac{\hat{p}_c(1 - \hat{p}_c)}{n_2}}} = \frac{0.85 - 0.60}{\sqrt{\dfrac{(0.73)(0.27)}{104} + \dfrac{(0.73)(0.27)}{100}}} = \frac{0.25}{0.062} = 4.03$$

P-value:
This is an upper-tailed test, so the P-value is the area under the z curve to the right of 4.03. From Appendix Table 2, P-value ≈ 0.

C Communicate Results Because the P-value is less that the selected significance level (0.05), the null hypothesis is rejected. The proportion of warts removed by the duct tape treatment is significantly greater than the proportion of warts removed by the freezing treatment. But, before you try this treatment at home, you should know that more recent experiments have not been able to confirm that duct tape is an effective way to remove warts!

SECTION 14.2 EXERCISES

Each Exercise Set assesses the following chapter learning objectives: M1, M2, M3, M5, M6, P1, P3

SECTION 14.2 **Exercise Set 1**

14.9 The article **"An Alternative Vote: Applying Science to the Teaching of Science"** (*The Economist*, May 12, 2011) describes an experiment conducted at the University of British Columbia. A total of 850 engineering students enrolled in a physics course participated in the experiment. These students were randomly assigned to one of two experimental groups. The two groups attended the same lectures for the first 11 weeks of the semester. In the twelfth week, one of the groups was switched to a style of teaching where students were expected to do reading assignments prior to class and then class time was used to focus on problem solving, discussion and group work. The second group continued with the traditional lecture approach. At the end of the twelfth week, the students were given a test over the course material from that week. The mean test score for students in the new teaching method group was 74 and the mean test score for students in the traditional lecture group was 41. Suppose that the two groups each consisted of 425 students and that the standard deviations of test scores for the new teaching method group and the traditional lecture method group were 20 and 24, respectively. Can you conclude that the mean test score is significantly higher for the new teaching method group than for the traditional lecture method group? Test the appropriate hypotheses using a significance level of 0.01.

14.10 Can moving their hands help children learn math? This question was investigated in the paper **"Gesturing Gives Children New Ideas About Math"** (*Psychological Science* [2009]: 267–272). Eighty-five children in the third and fourth grades who did not answer any questions correctly on a test with six problems of the form $3 + 2 + 8 = \_\_ + 8$ were participants in an experiment. The children were randomly assigned to either a no-gesture group or a gesture group. All the children were given a lesson on how to solve problems of this form using the strategy of trying to make both sides of the equation equal. Children in the gesture group were also taught to point to the first two numbers on the left side of the equation with the index and middle finger of one hand and then to point at the blank on the right side of the equation. This gesture was supposed to emphasize that grouping is involved in solving the problem. The children then practiced additional problems of this type. All children were then given a test with six problems to solve, and the number of correct answers was recorded for each child. Summary statistics are given below.

	n	$\bar{x}$	s
No Gesture	42	1.3	0.3
Gesture	43	2.2	0.4

Is there evidence that learning the gesturing approach to solving problems of this type results in a significantly higher mean number of correct responses? Test the relevant hypotheses using $\alpha = 0.05$.

14.11 The article **"A 'White' Name Found to Help in Job Search"** (*Associated Press*, January 15, 2003) described an experiment to investigate if it helps to have a "white-sounding" first name when looking for a job. Researchers sent resumes in response to 5,000 ads that appeared in the *Boston Globe* and *Chicago Tribune*. The resumes were identical except that 2,500 of them used "white-sounding" first names, such as Brett and Emily, whereas the other 2,500 used "black-sounding" names such as Tamika and Rasheed. The 5,000 job ads were assigned at random to either the white-sounding name group or the black-sounding name group. Resumes with white-sounding names received 250 responses while resumes with black sounding names received only 167 responses. Do these data support the claim that the proportion receiving a response is significantly higher for resumes with "white-sounding" first names?

SECTION 14.2 **Exercise Set 2**

14.12 The accompanying data on food intake (in Kcal) for 15 men on the day following two nights of only 4 hours of sleep each night and for 15 men on the day following two nights of 8 hours of sleep each night is consistent with summary quantities in the paper **"Short-Term Sleep Loss Decreases Physical Activity Under Free-Living Conditions But Does Not Increase Food Intake Under Time-Deprived Laboratory Conditions in Healthy Men"** (*American Journal of*

Clinical Nutrition [2009]: 1476–1482). The men participating in this experiment were randomly assigned to one of the two sleep conditions.

4-hour Sleep Group:	3,585	4,470	3,068	5,338	2,221
	4,791	4,435	3,099	3,187	3,901
	3,868	3,869	4,878	3,632	4,518
8-hour Sleep Group:	4,965	3,918	1,987	4,993	5,220
	3,653	3,510	3,338	4,100	5,792
	4,547	3,319	3,336	4,304	4,057

If appropriate, carry out a two-sample t test with $\alpha = 0.05$ to determine if there is a significant difference in mean food intake for the two different sleep conditions.

14.13 The paper **"If It's Hard to Read, It's Hard to Do"** (*Psychological Science* [2008]: 986–988) described an interesting study of how people perceive the effort required to do certain tasks. Each of 20 students was randomly assigned to one of two groups. One group was given instructions for an exercise routine that were printed in an easy-to-read font (Arial). The other group received the same set of instructions but printed in a font that is considered difficult to read (*Brush*). After reading the instructions, subjects estimated the time (in minutes) they thought it would take to complete the exercise routine. Summary statistics follow.

	Easy Font	Difficult Font
n	10	10
$\bar{x}$	8.23	15.10
s	5.61	9.28

The authors of the paper used these data to carry out a two-sample t test and concluded at the 0.10 significance level that the mean estimated time to complete the exercise routine is significantly lower when the instructions are printed in an easy-to-read font than when printed in a font that is difficult to read. Discuss the appropriateness of using a two-sample t test in this situation.

14.14 In a study of a proposed approach for diabetes prevention, 339 people under the age of 20 who were thought to be at high risk of developing type I diabetes were assigned at random to one of two groups. One group received twice-daily injections of a low dose of insulin. The other group (the control) did not receive any insulin but was closely monitored. Summary data (from the article **"Diabetes Theory Fails Test,"** *USA Today*, June 25, 2001) follow.

Group	n	Number Developing Diabetes
Insulin	169	25
Control	170	24

Do these data support the theory that the proportion who develop diabetes with the insulin treatment is significantly less than this proportion for the no-insulin treatment?

Additional Exercises

14.15 The paper **"The Effect of Multitasking on the Grade Performance of Business Students"** (*Research in Higher Education Journal* [2010]: 1–10) describes an experiment in which 62 undergraduate business students were randomly assigned to one of two experimental groups. Students in one group were asked to listen to a lecture but were told that they were permitted to use cell phones to send text messages during the lecture. Students in the second group listened to the same lecture but were not permitted to send text messages. Afterwards, students in both groups took a quiz on material covered in the lecture. Data from this experiment are summarized in the accompanying table.

Experimental Group	Group Size	Mean Quiz Score	Standard Deviation of Quiz Scores
Texting	31	42.81	9.91
No Texting	31	58.67	10.42

Do these data provide evidence to support the researcher's claim that the mean quiz score for the texting group is significantly lower than the mean quiz score for the no-texting group?

14.16 The paper **"Supervised Exercise Versus Non-Supervised Exercise for Reducing Weight in Obese Adults"** (*The Journal of Sports Medicine and Physical Fitness* [2009]: 85–90) describes an experiment in which participants were randomly assigned either to a supervised exercise program or a control group. Those in the control group were told that they should take measures to lose weight. Those in the supervised exercise group were told they should take measures to lose weight as well, but they also participated in regular supervised exercise sessions. Weight loss at the end of four months was recorded. Data consistent with summary quantities given in the paper are shown in the accompanying table.

Experimental Group	Group Size	Weight Loss (in kg)				
Supervised Exercise	17	1.1	4.8	4.1	9.1	2.2
		10.8	9.9	2.1	5.4	10.3
		−3.7	13.4	3.5	2.9	11.2
		9.5	5.1			
Control	17	6.7	7.6	2.3	−2.1	1.9
		5.4	1.0	3.6	4.9	−0.7
		1.3	2.2	2.1	0.6	1.3
		1.3	5.4			

Do these data provide evidence that the supervised exercise treatment results in a significantly greater mean weight loss than the treatment that just involves advising people to lose weight?

14.17 The article **"Fish Oil Staves Off Schizophrenia"** (*USA Today*, **February 2, 2010**) describes a study in which 81 patients ages 13 to 25 who were considered at risk for mental illness were randomly assigned to one of two groups. Those in one group took four fish oil capsules daily. Those in the other group took a placebo. After 1 year, 5% of those in the fish oil group and 28% of those in the placebo group had become psychotic. Is it appropriate to use the two-sample z test to test hypotheses about the difference in the proportions of patients receiving the fish oil and the placebo treatments who became psychotic? Explain why or why not.

14.18 Women diagnosed with breast cancer whose tumors have not spread may be faced with a decision between two surgical treatments—mastectomy (removal of the breast) or lumpectomy (only the tumor is removed). In a long-term study of the effectiveness of these two treatments, 701 women with breast cancer were randomly assigned to one of two treatment groups. One group received mastectomies and the other group received lumpectomies and radiation. Both groups were followed for 20 years after surgery. It was reported that there was no statistically significant difference in the proportion surviving for 20 years for the two treatments (*Associated Press*, **October 17, 2002**). What hypotheses do you think the researchers tested in order to reach the given conclusion? Did the researchers reject or fail to reject the null hypothesis?

14.19 When a surgeon repairs injuries, sutures (stitched knots) are sometimes used to hold together and stabilize the injured area. If these knots elongate and loosen, the injury may not heal properly. Researchers at the University of California, San Francisco, tied a series of different types of knots with two types of suture material, Maxon and Ticron. Suppose that 112 tissue specimens were available and that for each specimen the type of knot and suture material were randomly assigned. The investigators tested the knots to see how much they elongated. The elongations (in mm) were measured and the resulting data are summarized here. For purposes of this exercise, you can assume that it is reasonable to regard the elongation distributions as approximately normal.

Type of Knot	Maxon		
	n	$\bar{x}$	sd
Square (Control)	10	10.0	0.1
Duncan Loop	15	11.0	0.3
Overhand	15	11.0	0.9
Roeder	10	13.5	0.1
Snyder	10	13.5	2.0

Type of Knot	Ticron		
	n	$\bar{x}$	sd
Square (Control)	10	2.5	0.06
Duncan Loop	11	10.9	0.40
Overhand	11	8.1	1.00
Roeder	10	5.0	0.04
Snyder	10	8.1	0.06

a. Is there a significant difference in mean elongation between the square knot and the Duncan loop for Maxon thread?

b. Is there a significant difference in mean elongation between the square knot and the Duncan loop for Ticron thread?

14.20 Use the information given in the previous exercise to determine if there is a significant difference in mean elongation between Maxon versus Ticron threads for the Duncan loop knot.

SECTION 14.3 Estimating the Difference in Treatment Effects

When data from an experiment provide evidence of a treatment effect, you may also want to construct a confidence interval estimate of the difference in treatment means or in treatment proportions. When the response variable in an experiment is numerical, you can consider using a two-sample t confidence interval to estimate the difference in treatment means. When the response variable is categorical, a large-sample z confidence interval for a difference in treatment proportions can be considered.

Estimating the Difference in Treatment Means

As long as an experiment uses random assignment to create the treatment groups, the two-sample t confidence interval can be adapted for use in estimating a difference in treatment means. For use with experiment data, two modifications are needed to adapt the interval introduced in Chapter 13.

1. **Conditions:** The changes to the required conditions are the same as those made for the two-sample t test.
2. **Interpretation:** The interpretation of the confidence interval estimate will now be in terms of treatment means rather than population means.

AP* EXAM TIP

As was the case with hypothesis tests, when constructing a confidence interval using data from an experiment, the necessary condition is random assignment.

The Two-Sample t Confidence Interval for a Difference in Treatment Means

Appropriate when the following conditions are met:

1. Individuals or objects are *randomly assigned* to treatments.
2. The number of individuals or objects in each of the treatment groups is large (30 or more) or the treatment response distributions (the distributions of response values that would result if the treatments were applied to a *very* large number of individuals or objects) are approximately normal.

When these conditions are met, a confidence interval for the difference in treatment means is

$$(\bar{x}_1 - \bar{x}_2) \pm (t \text{ critical value}) \sqrt{\frac{s_1^2}{n_1} + \frac{s_2^2}{n_2}}$$

The t critical value is based on

$$df = \frac{(V_1 + V_2)^2}{\dfrac{V_1^2}{n_1 - 1} + \dfrac{V_2^2}{n_2 - 1}} \quad \text{where } V_1 = \frac{s_1^2}{n_1} \text{ and } V_2 = \frac{s_2^2}{n_2}$$

The computed value of df should be truncated (rounded down) to obtain an integer value for df. The desired confidence level determines which t critical value is used. Appendix Table 3, statistical software, or a graphing calculator can be used to obtain the t critical value.

Interpretation of Confidence Interval
You can be confident that the actual value of the difference in treatment means is included in the computed interval. This statement should be worded in context.

(continued)

> **Interpretation of Confidence Level**
> The confidence level specifies the long-run proportion of the time that this method is expected to be successful in capturing the actual difference in treatment means.

Example 14.3 Effect of Talking on Blood Pressure

Does talking elevate blood pressure, contributing to the tendency for blood pressure to be higher when measured in a doctor's office than when measured in a less stressful environment? (This well-documented effect is called the "white coat effect.") The article **"The Talking Effect and 'White Coat' Effect in Hypertensive Patients: Physical Effort or Emotional Content"** (*Behavioral Medicine* **[2001]: 149–157**) described a study in which patients with high blood pressure were randomly assigned to one of two groups. Those in the first group (the talking group) were asked questions about their medical history and about the sources of stress in their lives in the minutes before their blood pressure was measured. Those in the second group (the counting group) were asked to count aloud from 1 to 100 four times before their blood pressure was measured. The following data values for diastolic blood pressure (in millimeters of Hg) are consistent with summary quantities appearing in the paper:

Talking	104	110	107	112	108	103	108	118

$$n_1 = 8 \qquad \bar{x}_1 = 108.75 \qquad s_1 = 4.74$$

Counting	110	96	103	98	100	109	97	105

$$n_2 = 8 \qquad \bar{x}_2 = 102.25 \qquad s_2 = 5.39$$

The researchers were interested in estimating the difference in mean blood pressure for the two treatments (talking and counting).

E Estimate You want to estimate

$$\mu_1 - \mu_2 = \text{mean difference in blood pressure}$$

where

$$\mu_1 = \text{mean blood pressure for the talking treatment}$$

and

$$\mu_2 = \text{mean blood pressure for the counting treatment}$$

M Method The answers to the four key questions are *estimation, experiment data, one numerical variable (blood pressure), and two treatments*. This combination of answers leads you to consider a two-sample *t* confidence interval for a difference in treatment means. For the purposes of this example, a 95% confidence level will be used.

C Check The patients were randomly assigned to one of the two treatment groups. Because there are only eight patients in each of the treatment groups, you need to be willing to assume that the blood pressure distribution for each of the two treatments is at least approximately normal. Figure 14.1 shows boxplots constructed using the data from the two treatment groups. There are no outliers in either data set and the boxplots are reasonably symmetric, suggesting that the assumption of approximate normality is reasonable.

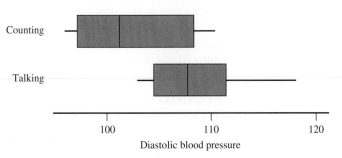

FIGURE 14.1
Boxplots for the blood pressure
data of Examples 14.3

C Calculate   You need to calculate the endpoints of the confidence interval. You can do this by hand, as shown here, or you could use a graphing calculator or a statistics software package. If computing by hand, you first need to calculate the appropriate number of degrees of freedom. Degrees of freedom:

$$V_1 = \frac{s_1^2}{n_1} = \frac{(4.74)^2}{8} = 2.81 \qquad V_2 = \frac{s_2^2}{n_2} = \frac{(5.39)^2}{8} = 3.63$$

$$df = \frac{(V_1 + V_2)^2}{\dfrac{V_1^2}{n_1 - 1} + \dfrac{V_2^2}{n_2 - 1}} = \frac{(2.81 + 3.63)^2}{\dfrac{(2.81)^2}{7} + \dfrac{(3.63)^2}{7}} = \frac{41.47}{3.01} = 13.78$$

Truncating to an integer gives df = 13. In the 13-df row of Appendix Table 3, the t critical value for a 95% confidence level is 2.16. The confidence interval is then

$$(\bar{x}_1 - \bar{x}_2) \pm (t \text{ critical value})\sqrt{\frac{s_1^2}{n_1} + \frac{s_2^2}{n_2}}$$

$$= (108.75 - 102.25) \pm (2.16)\sqrt{\frac{(4.74)^2}{8} + \frac{(5.39)^2}{8}}$$

$$= 6.5 \pm 5.48$$

$$= (1.02, 11.98)$$

C Communicate Results   Based on these samples, you can be 95% confident that the actual difference in mean blood pressure for the two treatments is somewhere between 1.02 Hg and 11.98 Hg. Because 0 is not included in this interval, you would conclude that the mean blood pressure is higher for the talking treatment than for the counting treatment by somewhere between 1.02 and 11.98. This supports the existence of a talking effect over and above the white-coat effect. Notice that this interval is rather wide. This is because the two sample standard deviations are large and the sample sizes are small. The method used to construct this interval is successful in capturing the actual difference in treatment means about 95% of the time.

Most statistical computer packages can compute the two-sample t confidence interval. JMP was used to construct a 95% confidence interval using the data of this example and the resulting output is shown here:

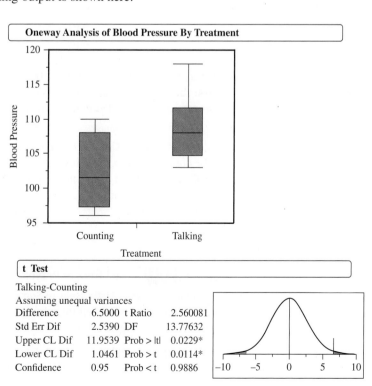

The upper and lower endpoints of the 95% confidence interval for the difference in treatment means are shown in the output labeled as "Upper CL Dif" and "Lower CL Dif." These endpoints are slightly different from those in the hand calculations because JMP uses more decimal accuracy.

Estimating the Difference in Treatment Proportions

As long as an experiment uses random assignment to create the treatment groups, the large-sample z confidence interval can be adapted for use in estimating a difference in treatment proportions. Two modifications are needed to adapt the interval introduced in Chapter 11 for use with experiment data:

1. **Conditions:** The changes to the required conditions are the same as those made for the large-samples z test.
2. **Interpretation:** The interpretation of the confidence interval estimate will now be in terms of treatment proportions rather than population proportions.

A Large-Samples Confidence Interval for a Difference in Treatment Proportions

Appropriate when the following conditions are met:

1. Individuals or objects are *randomly assigned* to treatments.
2. The number of individuals or objects in each of the treatment groups is large. This condition is met when $n_1 \hat{p}_1 \geq 10$, $n_1(1 - \hat{p}_1) \geq 10$, $n_2 \hat{p}_2 \geq 10$, and $n_2(1 - \hat{p}_2) \geq 10$ (or equivalently if n_1 and n_2 are large enough to have at least 10 successes and 10 failures in each of the two treatment groups).

When these conditions are met, a confidence interval for the difference in treatment proportions is

$$(\hat{p}_1 - \hat{p}_2) \pm (z \text{ critical value}) \sqrt{\frac{\hat{p}_1(1 - \hat{p}_1)}{n_1} + \frac{\hat{p}_2(1 - \hat{p}_2)}{n_2}}$$

The desired confidence level determines which z critical value is used. The three most common confidence levels use the following critical values:

Confidence Level	z Critical Value
90%	1.645
95%	1.96
99%	2.58

Interpretation of Confidence Interval
You can be confident that the actual value of the difference in treatment proportions is included in the computed interval. This statement should be worded in context.

Interpretation of Confidence Level:
The confidence level specifies the long-run proportion of the time that this method is expected to be successful in capturing the actual difference in treatment proportions.

Example 14.4   The Effect of Prayer

The article **"Prayer Is Little Help to Some Heart Patients, Study Shows"** (*Chicago Tribune*, **March 31, 2006**) described an interesting experiment to investigate the possible effects of prayer. The following two paragraphs are from the article:

> Bypass patients who consented to take part in the experiment were divided randomly into three groups. Some patients received prayers but were not informed of that. In

the second group the patients got no prayers, and also were not informed one way or the other. The third group got prayers and were told so.

There was virtually no difference in complication rates between the patients in the first two groups. But the third group, in which patients knew they were receiving prayers, had a complication rate of 59 percent—significantly more than the rate of 52 percent in the no-prayer group.

The article also states that a total of 1,800 people participated in the experiment, with 600 being assigned at random to each treatment group. The final comparison in the quote from the article was probably based on a large-sample z test for a difference in proportions, comparing the proportion with complications for the 600 patients in the no-prayer group with the proportion with complications for the 600 participants in the group that knew that someone was praying for them. Let's use the given information to estimate the difference in treatment proportions for these two treatments.

E Estimate You want to use the given information to estimate the difference between the proportion of patients with complications for the no prayer treatment, p_1, and the proportion of patients with complications for the treatment where people knew someone was praying for them, p_2.

M Method Because this is an estimation problem, the data are from an experiment, the one response variable (complications or no complications) is categorical, and two treatments are being compared, a method to consider is a large-sample z confidence interval for the difference in population proportions. For purposes of this example, a 90% confidence level will be used.

C Check You need to check to make sure that the experiment used random assignment to create the treatment groups and that the treatment group sizes are large enough. The description of the experiment says that patients were divided randomly into treatment groups. If you define a success as someone who experienced complications, with treatment group sizes of 600 and treatment success proportions of 0.59 for the treatment group where people knew that they were receiving prayers and 0.52 for the no prayer treatment group, there are more than 10 successes in each group and more than 10 failures in each group.

C Calculate For a confidence level of 90%, the appropriate z critical value is 1.645.

$$\hat{p}_1 = 0.52 \quad \hat{p}_2 = 0.59$$

$$\hat{p}_1 - \hat{p}_2 \pm (z \text{ critical value})\sqrt{\frac{\hat{p}_1(1 - \hat{p}_1)}{n_1} + \frac{\hat{p}_2(1 - \hat{p}_2)}{n_2}}$$

$$(0.52 - 0.59) \pm (1.645)\sqrt{\frac{(0.52)(0.48)}{600} + \frac{(0.59)(0.41)}{600}}$$

$$-0.07 \pm (1.645)(0.029)$$

$$-0.07 \pm 0.048$$

$$(-0.118, -0.022)$$

Statistical software or a graphing calculator could also have been used to compute the endpoints of the confidence interval. Minitab output is shown here.

CI for Two Proportions

Sample	X	N	Sample p
1	312	600	0.520000
2	354	600	0.590000

Difference = p (1) − p (2)
Estimate for difference: −0.07
90% CI for difference: (−0.117078, −0.0229225)

C Communicate Results Based on the data from this experiment, you can be confident that the difference in the proportion of patients with complications for the no-prayer treatment and the treatment where patients knew that someone was praying for them is between −0.118 and −0.022. Because both endpoints of the interval are negative, you would conclude that the proportion with complications is higher for the treatment where patients know that someone is praying for them than for the no-prayer treatment by somewhere between 0.022 and 0.118. This is consistent with the statement in the quote from the article that says that the proportion with complications was significantly higher for the treatment where people knew someone was praying for them. The method used to construct this estimate captures the true difference in treatment proportions about 90% of the time.

SECTION 14.3 EXERCISES

Each Exercise Set assesses the following chapter learning objectives: M2, M4, M5, M7, P2, P4

SECTION 14.3 Exercise Set 1

14.21 The accompanying data on food intake (in Kcal) for 15 men on the day following two nights of only 4 hours of sleep each night and for 15 men on the day following two nights of 8 hours of sleep each night are consistent with summary quantities in the paper "Short-Term Sleep Loss Decreases Physical Activity Under Free-Living Conditions But Does Not Increase Food Intake Under Time-Deprived Laboratory Conditions in Healthy Men" (*American Journal of Clinical Nutrition* [2009]: 1476–1482). The men participating in this experiment were randomly assigned to one of the two sleep conditions.

4-Hour Sleep Group:	3,585	4,470	3,068	5,338	2,221
	4,791	4,435	3,099	3,187	3,901
	3,868	3,869	4,878	3,632	4,518
8-Hour Sleep Group:	4,965	3,918	1,987	4,993	5,220
	3,653	3,510	3,338	4,100	5,792
	4,547	3,319	3,336	4,304	4,057

If appropriate, estimate the difference in mean food intake for the two different sleep conditions using a 95% confidence interval. If it is not appropriate to use the confidence interval, explain why it is not appropriate.

14.22 Is injecting medical cement effective in reducing pain for people who have suffered fractured vertebrae? The paper "A Randomized Trial of Vertebroplasty for Osteoporotic Spinal Fractures" (*New England Journal of Medicine* [2009]: 569–578) describes an experiment to compare patients who underwent vertebroplasty (the injection of cement) to patients in a placebo group who underwent a fake procedure. Because the placebo procedure was similar to the vertebroplasty procedure except for the actual injection of cement, patients participating in the experiment were not aware of which treatment they received. All patients were asked to rate their pain at three different times—3 days, 14 days, and 1 month after the procedure. Summary statistics are given in the accompanying table.

	Pain Intensity			
	Vertebroplasty Group		Placebo Group	
	n = 68		*n* = 63	
	mean	sd	mean	sd
3 Days	4.2	2.8	3.9	2.9
14 Days	4.3	2.9	4.5	2.8
1 Month	3.9	2.9	4.6	3.0

a. Briefly explain why the researchers may have chosen to include a placebo group that underwent a fake procedure rather than just comparing the vertebroplasty group to a group of patients who did not receive any treatment.
b. Construct and interpret a 95% confidence interval for the difference in mean pain intensity 3 days after treatment between the vertebroplasty treatment and the fake treatment.

14.23 Use the information given in the previous exercise to complete the following:
a. Construct and interpret 95% confidence intervals for the difference in mean pain intensity between the two treatments at 14 days and at 1 month after treatment.
b. Based on the confidence intervals from Part (a) and the previous exercise, comment on the effectiveness of injecting cement as a way of reducing pain for people with fractured vertebrae.

14.24 The article "Spray Flu Vaccine May Work Better than Injections for Tots" (*San Luis Obispo Tribune*, May 2, 2006) described a study that compared flu vaccine administered by injection and flu vaccine administered as a nasal spray. Each of the 8,000 children under the age of 5 who participated in the study received both a nasal spray and an injection, but only one was the real vaccine and the other was salt water. At the end of the flu season, it was determined that of the 4,000 children receiving the real vaccine by nasal spray, 3.9% got the flu. Of the 4,000 children receiving the real vaccine by injection, 8.6% got the flu.

a. Why would the researchers give every child both a nasal spray and an injection?

b. Use a 99% confidence interval to estimate the difference in the proportion of children who get the flu after being vaccinated with an injection and the proportion of children who get the flu after being vaccinated with the nasal spray. Based on the confidence interval, would you conclude that the proportion of children who get the flu is different for the two vaccination methods?

SECTION 14.3 **Exercise Set 2**

14.25 Can moving their hands help children learn math? This question was investigated in the paper "**Gesturing Gives Children New Ideas About Math**" (*Psychological Science* [2009]: 267–272). Eighty-five children in the third and fourth grades who did not answer any questions correctly on a test with six problems of the form $3 + 2 + 8 = \_\_ + 8$ were participants in an experiment. The children were randomly assigned to either a no-gesture group or a gesture group. All the children were given a lesson on how to solve problems of this form using the strategy of trying to make both sides of the equation equal. Children in the gesture group were also taught to point to the first two numbers on the left side of the equation with the index and middle finger of one hand and then to point at the blank on the right side of the equation. This gesture was supposed to emphasize that grouping is involved in solving the problem. The children then practiced additional problems of this type. All children were then given a test with six problems to solve, and the number of correct answers was recorded for each child. Summary statistics follow.

	n	$\bar{x}$	s
No Gesture	42	1.3	0.3
Gesture	43	2.2	0.4

Use the data from this experiment to estimate the difference in mean number of correct answers for the two different methods. Use a confidence level of 95%.

14.26 In a study of a proposed approach for diabetes prevention, 339 people under the age of 20 who were thought to be at high risk of developing type I diabetes were assigned at random to one of two groups. One group received twice-daily injections of a low dose of insulin. The other group (the control) did not receive any insulin but was closely monitored. Summary information (from the article "**Diabetes Theory Fails Test**," *USA Today*, June 25, 2001) follow.

Group	n	Number Developing Diabetes
Insulin	169	25
Control	170	24

Use the given information to construct a 90% confidence interval for the difference in the proportion that develop diabetes for the no insulin and the insulin treatments.

Additional Exercises

14.27 The paper "**Fudging the Numbers: Distributing Chocolate Influences Student Evaluations of an Undergraduate Course**" (*Teaching of Psychology* [2007]: 245–247) describes an experiment in which 98 students at the University of Illinois were assigned at random to one of two groups. All students took a class from the same instructor in the same semester. Students were required to report to an assigned room at a set time to fill out a course evaluation. One group of students reported to a room where they were offered a small bar of chocolate as they entered. The other group reported to a different room where they were not offered chocolate. Summary statistics for the overall course evaluation score are given in the accompanying table.

Group	n	$\bar{x}$	s
Chocolate	49	4.07	0.88
No Chocolate	49	3.85	0.89

a. Use the given information to construct and interpret a 95% confidence interval for the mean difference in overall course evaluation score.

b. Does the confidence interval from Part (a) support the statement made in the title of the paper? Explain.

14.28 The paper "**Passenger and Cell Phone Conversations in Simulated Driving**" (*Journal of Experimental Psychology: Applied* [2008]: 392–400) describes an experiment that investigated if talking on a cell phone while driving is more distracting than talking with a passenger. Drivers were randomly assigned to one of two groups. The 40 drivers in the cell phone group talked on a cell phone while driving in a simulator. The 40 drivers in the passenger group talked with a passenger in the car while driving in the simulator. The drivers were instructed to exit the highway when they came to a rest stop. Of the drivers talking to a passenger, 21 noticed the rest stop and exited. For the drivers talking on a cell phone, 11 noticed the rest stop and exited.

a. Use the given information to construct and interpret a 95% confidence interval for the difference in the proportions of drivers who would exit at the rest stop.

b. Does the interval from Part (a) support the conclusion that drivers using a cell phone are more likely to miss the exit than drivers talking with a passenger? Explain how you used the confidence interval to answer this question.

14.29 The paper "**The Effect of Multitasking on the Grade Performance of Business Students**" (*Research in Higher Education Journal* [2010]: 1–10) describes an experiment in which 62 undergraduate business students were randomly assigned to one of two experimental groups. Students in one group were asked to listen to a lecture but were told that they were permitted to use cell phones to send text messages during the lecture. Students in the second group listened to the same lecture but were not permitted to send text messages. Afterwards, students in both groups took a quiz on material covered in the lecture. Data from this experiment are summarized in the accompanying table.

Experimental Group	Group Size	Mean Quiz Score	Standard Deviation of Quiz Scores
Texting	31	42.81	9.91
No Texting	31	58.67	10.42

Use the given information to construct and interpret a 90% confidence interval for the difference in mean quiz score for the two treatments (texting allowed and texting not allowed).

14.30 The paper **"Supervised Exercise Versus Non-Supervised Exercise for Reducing Weight in Obese Adults"** (*The Journal of Sports Medicine and Physical Fitness* [2009]: 85–90) describes an experiment in which participants were randomly assigned either to a supervised exercise program or a control group. Those in the control group were told that they should take measures to lose weight. Those in the supervised exercise group were told they should take measures to lose weight, but they also participated in regular supervised exercise sessions. Weight loss at the end of four months was recorded. Data consistent with summary quantities given in the paper are shown in the accompanying table.

Experimental Group	Group Size	Weight Loss (in kg)				
Supervised Exercise	17	1.1	4.8	4.1	9.1	2.2
		10.8	9.9	2.1	5.4	10.3
		−3.7	13.4	3.5	2.9	11.2
		9.5	5.1			
Control	17	6.7	7.6	2.3	−2.1	1.9
		5.4	1.0	3.6	4.9	−0.7
		1.3	2.2	2.1	0.6	1.3
		1.3	5.4			

Use the given information to construct and interpret a 95% confidence interval for the difference in mean weight loss for the two treatments.

14.31 The online article **"Death Metal in the Operating Room"** (www.npr.org, December 24, 2009) describes an experiment investigating the effect of playing music during surgery. One conclusion drawn from this experiment was that doctors listening to music that contained vocal elements took more time to complete surgery than doctors listening to music without vocal elements. Suppose that μ_1 denotes the mean time to complete a specific type of surgery for doctors listening to music with vocal elements and μ_2 denotes the mean time for doctors listening to music with no vocal elements. Further suppose that the stated conclusion was based on a 95% confidence interval for $\mu_1 - \mu_2$, the difference in treatment means. Which of the following three statements is correct? Explain why you chose this statement.

Statement 1: Both endpoints of the confidence interval were negative.

Statement 2: The confidence interval included 0.

Statement 3: Both endpoints of the confidence interval were positive.

14.32 Women diagnosed with breast cancer whose tumors have not spread may be faced with a decision between two surgical treatments—mastectomy (removal of the breast) or lumpectomy (only the tumor is removed). In a long-term study of the effectiveness of these two treatments, 701 women with breast cancer were randomly assigned to one of two treatment groups. One group received mastectomies, and the other group received lumpectomies and radiation. Both groups were followed for 20 years after surgery. It was reported that there was no statistically significant difference in the proportion surviving for 20 years for the two treatments (*Associated Press*, October 17, 2002). Suppose that this conclusion was based on a 90% confidence interval for the difference in treatment proportions. Which of the following three statements is correct? Explain why you chose this statement.

Statement 1: Both endpoints of the confidence interval were negative.

Statement 2: The confidence interval included 0.

Statement 3: Both endpoints of the confidence interval were positive.

SECTION 14.4 Avoid These Common Mistakes

Drawing conclusions from experiment data requires some thought. In Chapter 1, you saw that if an experiment is carefully planned and includes random assignment to treatments, it is reasonable to conclude that observed differences in response between the experimental groups can be attributed to the treatments in the experiment. However, generalizing conclusions from an experiment that uses volunteers as subjects to a larger population is not appropriate unless a convincing argument can be made that the group of volunteers is representative of some population of interest.

Some cautions when drawing conclusions from experiment data are:

1. *Random assignment to treatments is critical.* If the design of the experiment does not include random assignment to treatments, it is not appropriate to use a hypothesis test or a confidence interval to draw conclusions about treatment differences.

2. Remember that it is not reasonable to generalize conclusions from experiment data to a larger population unless the subjects in the experiment were selected at random from the population or a convincing argument can be made that the group of volunteers is representative of the population. And even if subjects are selected at random from a population, it is still important that there be random assignment to treatments.

3. As was the case when using data from sampling to test hypotheses, remember that a hypothesis test can never show strong support for the null hypothesis. In the context of using experiment data to test hypotheses, this means you cannot say that data from an experiment provide convincing evidence that there is *no* difference between treatments.

4. Even when the data used in a hypothesis test are from an experiment, there is still a difference between statistical significance and practical significance. It is possible, especially in experiments with large numbers of subjects in each experimental group, to be convinced that two treatment means or treatment proportions are not equal, even in situations where the actual difference is too small to be of any practical interest. After rejecting a null hypothesis of no difference (statistical significance), it may be useful to look at a confidence interval estimate of the difference to get a sense of practical significance.

CHAPTER ACTIVITIES

ACTIVITY 14.1 RISKY BEHAVIOR ON MYSPACE

Background: An experiment to determine if an online intervention can reduce references to risky behavior on the social networking web sites of adolescents is described in the paper "Reducing At-Risk Adolescents' Display of Risk Behavior on a Social Networking Web Site" (*Archives of Pediatrics and Adolescent Medicine* [2009]: 35–41). Researchers selected public social network profiles of people who described themselves as between 18 and 20 years old and who referenced risky behaviors in their profiles. The selected subjects were assigned at random to an intervention group or a control group. Those in the intervention group were sent an e-mail from a physician about the risks associated with having a public profile and of referencing risky behaviors in their profile. Three months later, the networking sites were revisited to see if any changes had been made. The following excerpt is from the paper:

> At baseline, 54.2% of subjects referenced sex and 85.3% referenced substance use on their social networking site profiles. The proportion of profiles in which references

decreased to 0 was 13.7% in the intervention group vs. 5.3% in the control group for sex ($P = 0.05$) and 26% vs. 22% for substance use ($P = 0.61$). The proportion of profiles set to "private" at follow-up was 10.5% in the intervention group and 7.4% in the control group ($P = 0.45$). The proportion of profiles in which any of these three protective changes were made was 42.1% in the intervention group and 29.5% in the control group ($P = 0.07$).

1. The quote from the paper references four hypothesis tests. For each test, indicate what hypotheses you think were tested and whether or not the null hypothesis was rejected.

2. Based on the information provided by the hypothesis tests, what conclusions can be drawn about the effectiveness of the e-mail intervention?

ACTIVITY 14.2 AN EXPERIMENT TO TEST FOR THE STROOP EFFECT

Background: In 1935, John Stroop published the results of his research into how people respond when presented with conflicting signals. Stroop noted that most people are able to read words quickly and that they cannot easily ignore them and focus on other attributes of a printed word, such as text color. For example, consider the following list of words:

green blue red blue yellow red

It is easy to quickly read this list of words. It is also easy to read the words even if the words are printed in color, and even if the text color is different from the color of the word. For example, people can read the words in the list

green blue red blue yellow red

as quickly as they can read the list that isn't printed in color.

However, Stroop found that if people are asked to name the text colors of the words in the list (red, yellow, blue, green, red, green), it takes them longer. Psychologists believe this is due to the reader inhibiting a natural response (reading the word) and producing a different response (naming the color of the text).

If Stroop is correct, people should be able to name colors more quickly if they do not have to inhibit the word response, as would be the case if they were shown the following:

1. Design an experiment to compare the time it takes to identify colors when they appear as text with the time it takes to identify colors when there is no need to inhibit a word response. Indicate how random assignment is incorporated into your design. What is your response variable? How will you measure it? How many subjects will you use in your experiment, and how will they be chosen?

2. When you are satisfied with your experimental design, carry out the experiment. You will need to construct a list of colored words and a corresponding list of colored bars to use in the experiment. You will also need to think about how you will implement the random assignment scheme.

3. Use the resulting data to determine if there is evidence to support the existence of the Stroop effect. Write a brief report that summarizes your findings.

ACTIVITY 14.3 QUICK REFLEXES

Background: In this activity, you will design an experiment that will allow you to investigate whether people tend to have quicker reflexes when reacting with their dominant hand than with their nondominant hand.

1. Working in a group, design an experiment that includes random assignment of participants to one of two experimental conditions. Be sure to describe what the two treatments in the experiment are, how you plan to measure quickness of reflexes, what potentially

confounding variables will be directly controlled, and the role that random assignment plays in your design.

2. If assigned to do so by your instructor, carry out your experiment and analyze the resulting data. Write a brief report that describes the experimental design, includes both graphical and numerical summaries of the resulting data, and communicates any conclusions that follow from your data analysis.

ARE YOU READY TO MOVE ON? CHAPTER 14 REVIEW EXERCISES

All chapter learning objectives are assessed in these exercises. The learning objectives assessed in each exercise are given in parentheses.

14.33 (C1)
The paper "**Effects of Caffeine on Repeated Sprint Ability, Reactive Agility Time, Sleep and Next Day Performance**" (*Journal of Sports Medicine and Physical Fitness* [2010]: 455–464) describes an experiment in which male athlete volunteers who were considered low caffeine consumers were assigned at random to one of two experimental groups.

Those assigned to the caffeine group drank a beverage which contained caffeine 1 hour before an exercise session. Those in the no-caffeine group drank a beverage that did not contain caffeine 1 hour before an exercise session. That night, participants wore a device that measures sleep activity. The researchers reported that there was no significant difference in mean sleep duration for the two experimental groups. In the context of this experiment, explain what it means to say that there is *no significant difference* in the group means. In particular, explain if this means that the mean sleep durations for the two groups are equal.

14.34 (M1, M2, M3, P1)
In the paper "Happiness for Sale: Do Experiential Purchases Make Consumers Happier than Material Purchases?" (*Journal of Consumer Research* [2009]: 188–197), the authors distinguish between spending money on experiences (such as travel) and spending money on material possessions (such as a car). In an experiment to determine if the type of purchase affects how happy people are after the purchase has been made, 185 college students were randomly assigned to one of two groups. The students in the "experiential" group were asked to recall a time when they spent about $300 on an experience. They rated this purchase on three different happiness scales that were then combined into an overall measure of happiness. The students assigned to the "material" group recalled a time that they spent about $300 on an object and rated this purchase in the same manner. The mean happiness score was 5.75 for the experiential group and 5.27 for the material group. Standard deviations and sample sizes were not given in the paper, but for purposes of this exercise, suppose that they were as follows:

Experiential	Material
$n_1 = 92$	$n_2 = 93$
$s_1 = 1.2$	$s_2 = 1.5$

Using the following Minitab output, carry out a hypothesis test to determine if these data support the authors' conclusion that, on average, "experiential purchases induced more reported happiness." Use $\alpha = 0.05$.

Two-Sample T-Test and CI

Sample	N	Mean	StDev	SE Mean
1	92	5.75	1.20	0.13
2	93	5.27	1.50	0.16

Difference = mu (1) − mu (2)

Estimate for difference: 0.480000

95% lower bound for difference: 0.149917

T-Test of difference = 0 (vs >): T-Value = 2.40 P-Value = 0.009 DF = 175

14.35 (M1, M5, M6, P3)
"Doctors Praise Device That Aids Ailing Hearts" (*Associated Press*, November 9, 2004) is the headline of an article that describes a study of the effectiveness of a fabric device that acts like a support stocking for a weak or damaged heart. In the study, 107 people who consented to treatment were assigned at random to either a standard treatment consisting of drugs or the experimental treatment that consisted of drugs plus surgery to install the stocking. After two years, 38% of the 57 patients receiving the stocking had improved, while 27% of the patients receiving the standard treatment had improved. Do these data provide evidence that the proportion of patients who improve is significantly higher for the experimental treatment than for the standard treatment? Test the relevant hypotheses using a significance level of 0.05.

14.36 (M2, M4, P2)
The article "An Alternative Vote: Applying Science to the Teaching of Science" (*The Economist*, May 12, 2011) describes an experiment conducted at the University of British Columbia. A total of 850 engineering students enrolled in a physics course participated in the experiment. Students were randomly assigned to one of two experimental groups. Both groups attended the same lectures for the first 11 weeks of the semester. In the twelfth week, one of the groups was switched to a style of teaching where students were expected to do reading assignments prior to class, and then class time was used to focus on problem solving, discussion, and group work. The second group continued with the traditional lecture approach. At the end of the twelfth week, students were given a test over the course material from that week. The mean test score for students in the new teaching method group was 74, and the mean test score for students in the traditional lecture group was 41. Suppose that the two groups each consisted of 425 students. Also suppose that the standard deviations of test scores for the new teaching method group and the traditional lecture method group were 20 and 24, respectively. Estimate the difference in mean test score for the two teaching methods using a 95% confidence interval. Be sure to give an interpretation of the interval.

14.37 (M5, M7, P4)
The article "A 'White' Name Found to Help in Job Search" (*Associated Press*, January 15, 2003) described an experiment to investigate if it helps to have a "white-sounding" first name when looking for a job. Researchers sent resumes in response to 5,000 ads that appeared in the *Boston Globe* and *Chicago Tribune*. The resumes were identical except that 2,500 used "white-sounding" first names, such as Brett and Emily, whereas the other 2,500 used "black-sounding" names such as Tamika and Rasheed. The 5,000 job ads were assigned at random to either the white-sounding name group or the black-sounding name group. Resumes with white-sounding names received 250 responses while resumes with black sounding names received only 167 responses.
a. What are the two treatments in this experiment?
b. Use the data from this experiment to estimate the difference in response proportions for the two treatments.

AP* Review Questions for Chapter 14

1. A study was carried out to determine if pulse rates are significantly higher after listening to rock music than after listening to classical music. Participants were randomly assigned to listen to either rock music or classical music. The heart rate of each participant was recorded after 30 minutes of listening to the assigned music. Let μ_1 represent the mean heart rate after listening to rock music, and let μ_2 represent the mean heart rate after listening to classical music. Which of the following hypotheses should be tested to determine if the mean heart rate after listening to rock music is significantly higher than the mean heart rate after listening to classical music?

(A) $H_0: \mu_1 - \mu_2 = 0$ vs. $H_a: \mu_1 - \mu_2 < 0$
(B) $H_0: \mu_1 - \mu_2 = 0$ vs. $H_a: \mu_1 - \mu_2 \neq 0$
(C) $H_0: \mu_1 - \mu_2 = 0$ vs. $H_a: \mu_1 - \mu_2 > 0$
(D) $H_0: \mu_1 - \mu_2 < 0$ vs. $H_a: \mu_1 - \mu_2 = 0$
(E) $H_0: \mu_1 - \mu_2 > 0$ vs. $H_a: \mu_1 - \mu_2 = 0$

2. Some scientists believe that environmental factors affect population density. Spiders are very important species in terrestrial communities, and their survival is ecologically important. To better understand the relation between food level and spider survival, an experiment was conducted by supplementing the food supply of immature spiders in an area of low population density. In a control group of immature spiders, 83 out of 112 survived to the end of the experiment; in the food-supplemented experimental group, 86 out of 97 survived. The 90% confidence interval for the difference in treatment survival proportions is:

(A) -0.146 ± 0.103
(B) -0.146 ± 0.086
(C) -0.146 ± 0.052
(D) -0.146 ± 0.005
(E) -0.146 ± 0.003

3. An experiment was carried out to investigate whether lighting affects the quality of work produced by machine operators at a factory. Forty machine operators volunteered to participate as subjects in the experiment. The subjects were assigned at random to one of two experimental groups. Subjects in Group 1 worked for two days in bright light and subjects in Group 2 worked for 2 days in dim light. The number of defective items produced during the two days was recorded for each subject. It was determined that the mean number of defective items produced was significantly higher for Group 2 than for Group 1. Which of the following is the most appropriate conclusion that can be drawn from this experiment?

(A) The observed difference in mean number of defective items is likely to be due to chance and so lighting has no effect on number of defective items produced.
(B) It is not possible to draw any conclusions from this experiment because volunteers were used.
(C) It is possible to conclude that the observed difference in mean number of defective items produced is attributable to differences in lighting, but it is not reasonable to generalize this result to all machine operators at this factory.
(D) It is not possible to conclude that the observed difference in mean number of defective items produced is attributable to differences in lighting, but it is reasonable to generalize this result to all machine operators at this factory.
(E) It is possible to conclude that the observed difference in mean number of defective items produced is attributable to differences in lighting, and it is reasonable to generalize this result to all machine operators at this factory.

4. When using data from an experiment to test hypotheses about a difference in treatment means, which of the following conditions *must* be met in order for a two-sample t test to be an appropriate method?

I. The subjects are randomly selected.
II. The subjects are randomly assigned to treatments.
III. The sample sizes are large or the treatment distributions are approximately normal.

(A) I only
(B) II only
(C) III only
(D) II and III only
(E) I, II, and III

5. When using data from an experiment to test hypotheses about a difference in treatment proportions, which of the following conditions *must* be met in order for a two-sample z test to be an appropriate method?

I. $n_1 \hat{p}_1 \geq 10$, $n_1(1 - \hat{p}_1) \geq 10$, $n_2 \hat{p}_2 \geq 10$, $n_2(1 - \hat{p}_2) \geq 10$,
II. The subjects are randomly assigned to treatments.
III. The subjects are randomly selected.

(A) I only
(B) II only
(C) III only
(D) I and II only
(E) I, II, and III

15

Learning from Categorical Data

Preview

Chapter Learning Objectives

15.1 Chi-Square Tests for Univariate Categorical Data

15.2 Tests for Homogeneity and Independence in a Two-Way Table

15.3 Avoid These Common Mistakes

Chapter Activities

Are You Ready to Move On? Chapter 15 Review Exercises

Technology Notes

AP* Review Questions for Chapter 15

Jessica Rinaldi/Reuters

PREVIEW

This chapter introduces three additional methods for learning from categorical data. Sometimes a categorical data set consists of observations on a single variable of interest (univariate data). When the categorical variable has only two possible categories, the methods introduced in Chapters 9, 10, and 11 can be used to learn about the proportion of "successes." For example, suppose calls made to the 9-1-1 emergency number are classified according to whether they are for true emergencies or not. You can estimate the proportion of calls that are for true emergencies or you can use data from two different cities to determine if there is evidence of a difference in the proportions of true emergency calls. But the methods of Chapters 9, 10, and 11 are only appropriate when the categorical variable of interest has two possible categories. In this chapter, you will see how to analyze data on a categorical variable with more than two possible categories.

You will also see how to compare two or more populations on the basis of a categorical variable. This is illustrated in the the following example.

CHAPTER LEARNING OBJECTIVES

Conceptual Understanding

After completing this chapter, you should be able to

C1 Understand how a chi-square goodness-of-fit test can be used to answer a question of interest about a categorical variable.

C2 Understand that the way in which data summarized in a two-way table were collected determines which chi-square test (independence or homogeneity) is appropriate.

Mastering the Mechanics

After completing this chapter, you should be able to

M1 Determine which chi-square test (goodness-of-fit, independence, or homogeneity) is appropriate in a given situation.

M2 Determine appropriate null and alternative hypotheses for chi-square tests.

M3 Know the conditions necessary for the chi-square goodness-of-fit test to be appropriate.

M4 Know the conditions necessary for the chi-square tests of independence or homogeneity to be appropriate.

M5 Compute the value of the test statistic and find the associated *P*-value for chi-square tests.

Putting It into Practice

After completing this chapter, you should be able to

P1 Carry out a chi-square goodness-of-fit test and interpret the result in context.

P2 Carry out a chi-square test of homogeneity and interpret the result in context.

P3 Carry out a chi-square test of independence and interpret the result in context.

PREVIEW EXAMPLE

Is a Baseball Game Really Over by the Seventh Inning?

The authors of the article **"Predicting Professional Sports Game Outcomes from Intermediate Game Scores"** (*Chance* [1992]: 18–22) wanted to determine whether there was any merit to the idea that basketball games are not settled until the last quarter, but baseball games are really over by the seventh inning. They also considered football and hockey. Data were collected for 189 basketball games, 92 baseball games, 80 hockey games, and 93 football games. These games were randomly selected from all games played during the 1990 season for baseball and football and from the 1990–1991 season for basketball and hockey. For each game, the late-game leader was determined, and then it was noted whether the late-game leader actually ended up winning the game. The paper states that "the *late-game leader* is defined as the team that is ahead after three quarters in basketball and football, two periods in hockey, and seven innings in baseball." The resulting data are summarized in the following table:

Sport	Late-Game Leader Wins	Late-Game Leader Loses
Basketball	150	39
Baseball	86	6
Hockey	65	15
Football	72	21

Notice that four sports are being compared, which means that there are four different populations of games.

Had the researchers been interested in comparing only football and basketball, they might have used the two-sample test of Chapter 11 to determine if the proportion of games where the late-game leader wins is significantly different for these two sports. However, because there are more than two populations to be compared, the researchers analyzed the data using one of the methods you will learn in this chapter. This example will be revisited in Section 15.2 to see what can be learned from these data.

A categorical data set might also be bivariate, consisting of observations on two categorical variables, such as political affiliation and support for a particular ballot initiative. A method for determining if there is an association between two categorical variables is introduced in this chapter as well.

SECTION 15.1 Chi-Square Tests for Univariate Categorical Data

Univariate categorical data sets arise in a variety of settings. If each student in a sample of 100 is classified according to whether he or she is enrolled full-time or part-time, data on a categorical variable with two categories result. Each person in a sample of 100 registered voters in a particular city might be asked which of five city council members he or she favors for mayor. This would result in observations on a categorical variable with five categories. Univariate categorical data are usually summarized in a one-way frequency table, as illustrated in the following example.

Example 15.1 Tasty Dog Food?

The article **"Can People Distinguish Pâté from Dog Food?" (American Association of Wine Economists, April 2009, www.wine-economics.org)** describes a study that investigated whether people can tell the difference between dog food, pâté (a spread made of finely chopped liver, meat, or fish), and processed meats (such as Spam and liverwurst). Researchers used a food processor to make spreads that had the same texture and consistency as pâté from Newman's Own brand dog food and from the processed meats. Each participant in the study tasted five spreads (duck liver pâté, Spam, dog food, pork liver pâté, and liverwurst). After tasting all five spreads, each participant was asked to choose the one that they thought was the dog food. The first few observations were

liverwurst   pork liver pâté   liverwurst   dog food

After the number of observations was counted for each category, the data were summarized in the following table. This table is an example of a **one-way frequency table**.

	Spread Chosen as Dog Food				
	Duck Liver Pâté	Spam	Dog Food	Pork Liver Pâté	Liverwurst
Frequency	3	11	8	6	22

(Note: The frequencies in the table are consistent with summary values given in the paper. However, the sample size in the study was not actually 50.)

In general, for a categorical variable with k possible values (k different categories), sample data are summarized in a one-way frequency table consisting of k cells, which may be displayed either horizontally or vertically. In this section, you will consider testing hypotheses about the proportions of the population that fall into the possible categories.

> **NOTATION**
>
> k = number of categories of a categorical variable
> p_1 = population proportion for Category 1
> p_2 = population proportion for Category 2
> $\vdots$
> p_k = population proportion for Category k
>   (Note: $p_1 + p_2 + \cdots + p_k = 1$)
>
> The hypotheses to be tested have the form
>
> H_0: p_1 = hypothesized proportion for Category 1
> p_2 = hypothesized proportion for Category 2
>
> $\vdots$
>
> p_k = hypothesized proportion for Category k
>
> H_a: H_0 is not true. At least one of the population category proportions differs from the corresponding hypothesized value.

For the dog food study of Example 15.1, you could define

p_1 = proportion of all people who would choose duck liver pâté as the dog food
p_2 = proportion of all people who would choose Spam as the dog food
p_3 = proportion of all people who would choose dog food as the dog food
p_4 = proportion of all people who would choose pork liver pâté as the dog food
p_5 = proportion of all people who would choose liverwurst as the dog food

One possible null hypothesis of interest might be

$$H_0: p_1 = 0.20, p_2 = 0.20, p_3 = 0.20, p_4 = 0.20, p_5 = 0.20$$

This hypothesis specifies that any of the five spreads is equally likely to be the one identified as dog food. If this is true, you would expect $\frac{1}{5}$ or 20% of the participants in the sample to have selected each of the five spreads. In this case, the expected count is $50(0.20) = 10$ for each category.

In general, the expected counts are:

Expected cell count for Category 1 = np_1
Expected cell count for Category 2 = np_2
   $\vdots$
Expected cell count for Category k = np_k

The expected counts for a particular null hypothesis are calculated using the hypothesized proportions from the null hypothesis.

A null hypothesis of the form just described can be tested by first selecting a random sample of size n and then classifying each response into one of the k possible categories. To decide whether the sample data are compatible with the null hypothesis, you compare each observed count (the observed category frequency) to an expected count.

Example 15.2   Births and the Lunar Cycle

A popular urban legend is that more babies than usual are born during certain phases of the lunar cycle, especially near the full moon. The paper **"The Effect of the Lunar Cycle on Frequency of Births and Birth Complications"** (*American Journal of Obstetrics and Gynecology* [2005]: 1462–1464) classified births according to the lunar cycle. Data for a sample of randomly selected births, consistent with summary quantities in the paper, are given in the accompanying table.

Lunar Phase	Number of Days	Number of Births
New moon	24	7,680
Waxing crescent	152	48,442
First quarter	24	7,579
Waxing gibbous	149	47,814
Full moon	24	7,711
Waning gibbous	150	47,595
Last quarter	24	7,733
Waning crescent	152	48,230

(A crescent moon is less than half full and a gibbous moon is more than half full.)

You can denote the lunar phase category proportions as

p_1 = proportion of births that occur during the new moon
p_2 = proportion of births that occur during the waxing crescent moon
p_3 = proportion of births that occur during the first quarter moon
p_4 = proportion of births that occur during the waxing gibbous moon
p_5 = proportion of births that occur during the full moon
p_6 = proportion of births that occur during the waning gibbous moon
p_7 = proportion of births that occur during the last quarter moon
p_8 = proportion of births that occur during the waning crescent moon

If there is no relationship between number of births and the lunar cycle, then the number of births in each lunar cycle category should be proportional to the number of days included in that category. There are a total of 699 days during the 24 lunar cycles from which the sample was selected. Because 24 of these days were in the new moon category, the hypothesized proportion would be

$$p_1 = \frac{24}{699} = 0.0343$$

Similarly, the other proportions would be

$$p_2 = \frac{152}{699} = 0.2175 \qquad p_3 = \frac{24}{699} = 0.0343$$

$$p_4 = \frac{149}{699} = 0.2132 \qquad p_5 = \frac{24}{699} = 0.0343$$

$$p_6 = \frac{150}{699} = 0.2146 \qquad p_7 = \frac{24}{699} = 0.0343$$

$$p_8 = \frac{152}{699} = 0.2175$$

The hypotheses of interest are then

$$H_0: p_1 = 0.0343, p_2 = 0.2175, p_3 = 0.0343, p_4 = 0.2132, p_5 = 0.0343,$$
$$p_6 = 0.2146, p_7 = 0.0343, p_8 = 0.2175$$

H_a: H_0 is not true.
There were a total of 222,784 births in the sample (a very large sample!). If the null hypothesis is true, the expected counts for the first two categories are

$$\left(\begin{array}{l}\text{expected count} \\ \text{for new moon}\end{array}\right) = n\left(\begin{array}{l}\text{hypothesized proportion} \\ \text{for new moon}\end{array}\right) = 222{,}784(0.0343) = 7{,}641.49$$

$$\left(\begin{array}{l}\text{expected count} \\ \text{for waxing moon}\end{array}\right) = n\left(\begin{array}{l}\text{hypothesized proportion} \\ \text{for waxing moon}\end{array}\right) = 222{,}784(0.2175) = 48{,}455.52$$

Expected counts for the other six categories are computed in a similar way. The observed and expected counts are given in the following table.

Lunar Phase	Observed Number of Births	Expected Number of Births
New moon	7,680	7,641.49
Waxing crescent	48,442	48,455.52
First quarter	7,579	7,641.49
Waxing gibbous	47,814	47,497.55
Full moon	7,711	7,641.49
Waning gibbous	47,595	47,809.45
Last quarter	7,733	7,641.49
Waning crescent	48,230	48,455.52

Because the observed counts are based on a *sample* of births, it would be somewhat surprising to see *exactly* 3.43% of the sample falling in the first category, exactly 21.75% in the second, and so on, even when H_0 is true. If the differences between the observed and expected cell counts can reasonably be attributed to sampling variability, the data are considered compatible with H_0. On the other hand, if the discrepancy between the observed and the expected cell counts is too large to be explained by chance differences from one sample to another, H_0 should be rejected in favor of H_a. To make a decision, you need to assess how different the observed and expected counts are.

The goodness-of-fit statistic, denoted by X^2, is a quantitative measure of the extent to which the observed counts differ from those expected when H_0 is true. (The Greek letter χ (chi) is often used in place of X. The symbol X^2 is referred to as the chi-square statistic. In using X^2 rather than χ^2, we are keeping the convention of denoting sample statistics by Roman letters.)

The **goodness-of-fit statistic, X^2**, results from first computing the quantity

$$\frac{(\text{observed count} - \text{expected count})^2}{\text{expected count}}$$

for each category, where for a sample of size n,

$$\binom{\text{expected}}{\text{count}} = n \binom{\text{hypothesized value of corresponding}}{\text{population proportion}}$$

The X^2 statistic is the sum of these quantities for all k categories:

$$X^2 = \sum_{\text{all categories}} \frac{(\text{observed count} - \text{expected count})^2}{\text{expected count}}$$

The value of the X^2 statistic reflects the magnitude of the discrepancies between observed and expected counts. When the differences are big, the value of X^2 tends to be large, which suggests H_0 should be rejected. A small value of X^2 (it can never be negative) occurs when the observed cell counts are quite similar to those expected when H_0 is true, and so would be considered consistent with H_0.

As with previous test procedures, a conclusion is reached by comparing a P-value to the significance level for the test. The P-value is the probability of observing a value of X^2 at least as large as the observed value if H_0 were true. Calculating this P-value requires information about the sampling distribution of X^2 when H_0 is true.

When the null hypothesis is true and the sample size is large, the behavior of X^2 is described approximately by a **chi-square distribution**. A chi-square distribution curve has no area associated with negative values and is not symmetric, with a longer tail on the right. There are many different chi-square distributions, and each one has a different number of degrees of freedom. Curves corresponding to several chi-square distributions are shown in Figure 15.1.

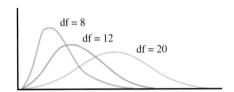

FIGURE 15.1
Chi-square curves

For a test procedure based on X^2, the associated *P*-value is the area under the appropriate chi-square curve and to the right of the computed X^2 value. Appendix Table 5 gives upper-tail areas for chi-square distributions with 2 to 20 df.

To find the area to the right of a particular X^2 value, locate the appropriate df column in Appendix Table 5. Determine which listed value is closest to the X^2 value of interest, and read the right-tail area corresponding to this value from the left-hand column of the table. For example, for a chi-square distribution with df = 4, the area to the right of $X^2 = 8.18$ is 0.085, as shown in Figure 15.2.

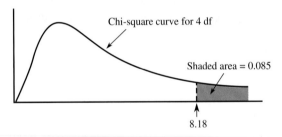

FIGURE 15.2
A chi-square upper-tail area

For this same chi-square distribution (df = 4), the area to the right of 9.70 is approximately 0.045 (the area to the right of 9.74, the closest entry in the table for df = 4).

It is also possible to use computer software or a graphing calculator to compute areas under a chi-square curve.

Chi-Square Goodness-of-Fit Test

When a null hypothesis that specifies k category proportions is true, the X^2 goodness-of-fit statistic has approximately a chi-square distribution with df = $(k - 1)$, as long as none of the expected cell counts is too small. When expected counts are small, and especially when an expected count is less than 1, the value of

$$\frac{(\text{observed count} - \text{expected count})^2}{\text{expected count}}$$

can be inflated because it involves dividing by a small number. *The use of the chi-square distribution is appropriate when the sample size is large enough for every expected count to be at least 5.* If any of the expected counts are less than 5, categories can be combined in a sensible way to create acceptable expected cell counts. If you do this, remember to compute the number of degrees of freedom based on the reduced number of categories.

Chi-Square Goodness-of-Fit Test

Appropriate when the following conditions are met:

1. Observed counts are based on a random sample or a sample that is selected to be representative of the population.
2. The sample size is large. The sample size is large enough for the chi-square goodness-of-fit test to be appropriate if every expected count is at least 5.

When these conditions are met, the following test statistic can be used:

$$X^2 = \sum_{\text{all categories}} \frac{(\text{observed count} - \text{expected count})^2}{\text{expected count}}$$

When the null hypothesis is true, the X^2 statistic has a chi-square distribution with $df = k - 1$, where k is the number of category proportions specified in the null hypothesis.

Hypotheses

H_0: p_1 = hypothesized proportion for Category 1

p_2 = hypothesized proportion for Category 2

$\vdots$

p_k = hypothesized proportion for Category k

H_a: H_0 is not true. At least one of the population category proportions differs from the corresponding hypothesized value.

Associated *P*-value

The *P*-value is the area to the right of X^2 under the chi-square curve with $df = k - 1$. Upper-tail areas for chi-square distributions can be found in Appendix Table 5.

AP* EXAM TIP

Unlike *t* tests, the value of df for the chi-square tests of this chapter is not related to sample size.

Questions

Q Question type: estimation or hypothesis testing?

S Study type: sample data or experiment data?

T Type of data: one variable or two? Categorical or numerical?

N Number of samples or treatments: how many?

The chi-square goodness-of-fit test is a method you should consider when the answers to the four key questions are *hypothesis testing, sample data, one categorical variable (with more than two categories), and one sample*.

The following examples illustrate the use of the chi-square goodness-of-fit test.

Example 15.3   Tasty Dog Food Continued

You can use the dog food taste data of Example 15.1 to test the hypothesis that the five different spreads (duck liver pâté, Spam, dog food, pork liver pâté, and liverwurst) are chosen equally often when people who have tasted all five spreads are asked to identify the one they think is the dog food. The study data from Example 15.1 are also shown here:

	Spread Chosen as Dog Food				
	Duck Liver Pâté	Spam	Dog Food	Pork Liver Pâté	Liverwurst
Frequency	3	11	8	6	22

Start by considering the four key questions (QSTN). You would like to use the sample data to test a claim about the corresponding population, so this is a hypothesis testing problem. The data are from a sample. There is one categorical variable with five categories corresponding to the five types of spread that could be identified as dog food. There is one sample. This combination of answers leads you to consider the chi-square goodness-of-fit test.

Steps
H Hypotheses
M Method
C Check
C Calculate
C Communicate results

Now you can use the five-step process for hypothesis testing problems (HMC^3).

Process Step	
H Hypotheses	The question of interest is about the population category proportions, so you need to define these proportions in the context of this example.
	Population characteristics of interest
	p_1 = proportion of all people who would choose duck liver pâté as the dog food
	p_2 = proportion of all people who would choose Spam as the dog food
	p_3 = proportion of all people who would choose dog food as the dog food
	p_4 = proportion of all people who would choose pork liver pâté as the dog food
	p_5 = proportion of all people who would choose liverwurst as the dog food
	The question of interest (are the five spreads identified equally often as the one thought to be dog food) results in a null hypothesis that specifies that all five category proportions are 0.20.
	Hypotheses
	Null hypothesis:
	H_0: $p_1 = 0.20$, $p_2 = 0.20$, $p_3 = 0.20$, $p_4 = 0.20$, $p_5 = 0.20$
	Alternative hypothesis:
	H_a: At least one of the population proportions is not 0.20
M Method	Because the answers to the four key questions are *hypothesis testing, sample data, one variable, categorical with more than two categories, and one sample*, a chi-square goodness-of-fit test is considered.
	The test statistic for this test is
	$$X^2 = \sum_{\text{all categories}} \frac{(\text{observed count} - \text{expected count})^2}{\text{expected count}}$$
	When the null hypothesis is true, this statistic has approximately a chi-square distribution with
	$$\text{df} = k - 1 = 5 - 1 = 4$$
	A significance level of $\alpha = 0.05$ will be used for this test.
C Check	In order to use the chi-square goodness-of-fit test, you must be willing to assume that the participants in this study can be regarded as a random or representative sample. If this assumption is not reasonable, you should be very careful generalizing results from this analysis to any larger population.
	To check the large sample condition, you need to compute the expected count for each of the five categories. For the first category, the hypothesized proportion (from the null hypothesis) is 0.20. Because the sample size was 50, the expected count for the duck liver category is 50(0.20) = 10. Because the other hypothesized proportions are also 0.20, all of the expected counts are equal to 10. All expected counts are at least 5, so the sample size is large enough for the chi-square goodness-of-fit test to be appropriate.
C Calculate	**Test statistic**
	$$X^2 = \sum_{\text{all categories}} \frac{(\text{observed count} - \text{expected count})^2}{\text{expected count}}$$
	$$= \frac{(3 - 10)^2}{10} + \frac{(11 - 10)^2}{10} + \frac{(8 - 10)^2}{10} + \frac{(6 - 10)^2}{10} + \frac{(22 - 10)^2}{10}$$
	$$= 4.9 + 0.1 + 0.4 + 1.6 + 14.4$$
	$$= 21.4$$

(continued)

Process Step	
	Degrees of freedom: $k - 1 = 5 - 1 = 4$
	The P-value is the area under the chi-square curve with df = 4 and to the right of 21.4. Because 21.4 is greater than the largest value in the 4 df column of Appendix Table 5, the area to the right is less than 0.001.
	Associated P-value
	P-value = area under chi-square curve to the right of 21.4 < 0.001
C Communicate results	Because the P-value is less than the selected significance level, the null hypothesis is rejected.
	$P\text{-value} < 0.001 < 0.05, \text{ Reject } H_0$
	Interpretation
	Based on these sample data, there is convincing evidence that the proportion identifying a spread as dog food is not the same for all five spreads. Here, it is interesting to note that the large differences between the observed counts and the counts that would have been expected if the null hypothesis of equal proportions were true are in the duck liver pate and the liverwurst categories, indicating that fewer than expected chose duck liver and many more than expected chose liverwurst as the one they thought was dog food. So, although you reject the hypothesis that the proportion choosing each of the five spreads is the same, it is not because people were actually able to identify which one was really dog food.

It is also possible to use statistical software to carry out the calculate step in a goodness-of-fit test. For example, Minitab output for the test of this example is shown here.

Chi-Square Goodness-of-Fit Test

Category	Observed	Test Proportion	Expected	Contribution to Chi-Sq
1	3	0.2	10	4.9
2	11	0.2	10	0.1
3	8	0.2	10	0.4
4	6	0.2	10	1.6
5	22	0.2	10	14.4

N	DF	Chi-Sq	P-Value
50	4	21.4	0.000

From the Minitab output, you see that $X^2 = 21.4$ and that the P-value is approximately 0. This is consistent with the values from the hand calculations shown above. Minitab also constructs a graphical display comparing the observed and expected counts for each category. This display is shown in Figure 15.3. From this graph, it is easy to see the two spread categories where there were large differences between observed and expected values.

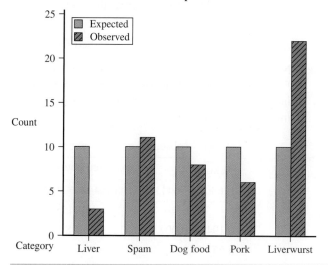

FIGURE 15.3
Minitab graph of observed and expected counts for Example 15.3

Example 15.4   Births and the Lunar Cycle Revisited

You can use the births data of Example 15.2 to test the hypothesis that birth date is unrelated to lunar cycle. A 0.05 significance level will be used. The observed and expected counts computed in Example 15.2 were

Lunar Phase	Observed Number of Births	Expected Number of Births
New moon	7,680	7,641.49
Waxing crescent	48,442	48,455.52
First quarter	7,579	7,641.49
Waxing gibbous	47,814	47,497.55
Full moon	7,711	7,641.49
Waning gibbous	47,595	47,809.45
Last quarter	7,733	7,641.49
Waning crescent	48,230	48,455.52

H Hypotheses   Recall that if births are unrelated to the lunar cycle, then you would expect the proportion of births falling in each lunar phase category to be proportional to the number of days in that phase.

This leads to the following hypotheses:

$$H_0\colon p_1 = 0.0343,\, p_2 = 0.2175,\, p_3 = 0.0343,\, p_4 = 0.2132,\, p_5 = 0.0343,$$
$$p_6 = 0.2146,\, p_7 = 0.0343,\, p_8 = 0.2175$$

H_a: At least one of the lunar phase birth proportions is different from the value specified in H_0

M Method   Consider the four key questions. You want to use the sample data to test a claim about the lunar phase birth proportions, so this is a hypothesis-testing problem. The data are from a sample of births. There is one categorical variable, which is lunar phase. This variable has eight categories. Finally, there is one sample. Because the answers to the four questions are *hypothesis testing, sample data, one categorical variable with more than two categories, and one sample*, you should consider a chi-square goodness-of-fit test.

The test statistic for the chi-square goodness-of-fit test is

$$X^2 = \sum_{\text{all categories}} \frac{(\text{observed count} - \text{expected count})^2}{\text{expected count}}$$

When the null hypothesis is true, this statistic has a chi-square distribution with df $= 8 - 1 = 7$.

A significance level of 0.05 was specified for this test.

C Check   Next you need to check to see if this method is appropriate. The study description in Example 15.2 states that the sample of births was a random sample. The sample size is large enough for the chi-square test to be appropriate since all eight expected counts are much greater than 5.

C Calculate
$$X^2 = \sum_{\text{all categories}} \frac{(\text{observed count} - \text{expected count})^2}{\text{expected count}}$$

$$= \frac{(7680 - 7641.49)^2}{7641.49} + \frac{(48442 - 48455.52)^2}{48455.52)} + \cdots + \frac{(48230 - 48455.52)^2}{48455.52}$$

$$= 0.194 + 0.004 + 0.511 + 2.108 + 0.632 + 0.962 + 1.096 + 1.050$$

$$= 6.557$$

P-value: the *P*-value is the area to the right of 6.557 under a chi-square curve with df $= 7$. The computed value of X^2 is smaller than 12.01 (the smallest entry in the df $= 7$ column of Appendix Table 5), so *P*-value > 0.10.

C Communicate Results Because the *P*-value is greater than the significance level of 0.05, H_0 cannot be rejected. There is not sufficient evidence to conclude that birth date and lunar cycle are related. This is consistent with the conclusion quoted in the paper: "We found no statistical evidence that deliveries occurred in a predictable pattern across the phases of the lunar cycle."

Minitab output for the data and hypothesized proportions of this example is shown here.

Chi-Square Goodness-of-Fit Test for Observed Counts in Variable: Number of Births
Using category names in Lunar Phase

Category	Observed	Test Proportion	Expected	Contribution to Chi-Sq
First Quarter	7579	0.0343	7641.5	0.51105
Full Moon	7711	0.0343	7641.5	0.63227
Last Quarter	7733	0.0343	7641.5	1.09584
New Moon	7680	0.0343	7641.5	0.19406
Waning Crescent	48230	0.2175	48455.5	1.04961
Waning Gibbous	47595	0.2146	47809.4	0.96189
Waxing Crescent	48442	0.2175	48455.5	0.00377
Waxing Gibbous	47814	0.2132	47497.5	2.10835

N	DF	Chi-Sq	P-Value
222784	7	6.55683	0.476

Notice that Minitab has reordered the categories to put them in alphabetical order. Minitab also carried a bit more decimal accuracy in the computation of the chi-square statistic, reporting $X^2 = 6.55683$ and an associated *P*-value of 0.476. The computed *P*-value $= 0.476$ is consistent with the earlier statement *P*-value > 0.10.

Example 15.5 Hybrid Car Purchases

USA Today (**"Hybrid Car Sales Rose 81% Last Year," April 25, 2005**) reported the top five states for sales of hybrid cars in 2004 as California, Virginia, Washington, Florida, and Maryland. Suppose that each car in a random sample of hybrid car sales in 2004 from these five states is classified by the state where the sale took place, resulting in the accompanying table.

State	Observed Frequency
California	250
Virginia	56
Washington	34
Florida	33
Maryland	33
Total	**406**

(The given observed counts are artificial, but they are consistent with hybrid sales figures given in the article.)

A X^2 goodness-of-fit test and a significance level of $\alpha = 0.01$ will be used to test the hypothesis that hybrid sales for these five states are proportional to their 2004 populations. Population estimates from the Census Bureau web site are given in the following table. The population proportion for each state was computed by dividing each state population by the total population of the five states.

State	2004 Population	Population Proportion
California	35,842,038	0.495
Virginia	7,481,332	0.103
Washington	6,207,046	0.085
Florida	17,385,430	0.240
Maryland	5,561,332	0.077
Total	**72,477,178**	

If these same population proportions hold for hybrid car sales, the expected counts are

Expected count for California = 406(0.495) = 200.970
Expected count for Virginia = 406(0.103) = 41.818
Expected count for Washington = 406(0.085) = 34.510
Expected count for Florida = 406(0.240) = 97.440
Expected count for Maryland = 406(0.077) = 31.362

These expected counts have been entered in Table 15.1.

TABLE 15.1   **Observed and Expected Counts for Example 15.5**

State	Observed Counts	Expected Counts
California	250	200.970
Virginia	56	41.818
Washington	34	34.510
Florida	33	97.440
Maryland	33	31.262

H Hypotheses   You are interested in learning about the proportion of hybrid car sales for each of the given states. You can define these proportions as

p_1 = proportion of hybrid car sales for California
p_2 = proportion of hybrid car sales for Virginia
p_3 = proportion of hybrid car sales for Washington
p_4 = proportion of hybrid car sales for Florida
p_5 = proportion of hybrid car sales for Maryland

If sales of hybrid cars in these five states are proportional to population size, then

$$p_1 = 0.495 \qquad p_2 = 0.103 \qquad p_3 = 0.085 \qquad p_4 = 0.240 \qquad p_5 = 0.077$$

The hypotheses of interest are

$$H_0: p_1 = 0.495, p_2 = 0.103, p_3 = 0.085, p_4 = 0.240, p_5 = 0.077$$

H_a: At least one of the state hybrid sales proportions is different from the value specified in H_0.

M Method   For this example, answers to the four key questions are *hypothesis testing, sample data, one categorical variable with five categories, and one sample*. These answers lead you to consider a chi-square goodness-of-fit test. The test statistic for the chi-square goodness-of-fit test is

$$X^2 = \sum_{\text{all categories}} \frac{(\text{observed count} - \text{expected count})^2}{\text{expected count}}$$

When the null hypothesis is true, this statistic has a chi-square distribution with df = 5 − 1 = 4.

A significance level of 0.01 was specified for this example.

C Check   Next you need to check to see if this method is appropriate. The study description indicates that the sample of car sales was a random sample. Because all five expected counts are at least 5, the sample size is large enough for the chi-square test to be appropriate.

C Calculate   From Minitab:

Chi-Square Goodness-of-Fit Test for Observed Counts in Variable: Hybrid Sales

Using category names in State

Category	Observed	Test Proportion	Expected	Contribution to Chi-Sq
California	250	0.495	200.970	11.9617
Florida	33	0.240	97.440	42.6161
Maryland	33	0.077	31.262	0.0966
Virginia	56	0.103	41.818	4.8096
Washington	34	0.085	34.510	0.0075

N	DF	Chi-Sq	P-Value
406	4	59.4916	0.000

From the Minitab output, the values of the test statistic and df are

$$X^2 = 59.4916 \qquad df = 4$$

and the associated P-value is 0.000.

C Communicate Results   Because the P-value is less than the significance level of 0.01, H_0 is rejected. There is convincing evidence that hybrid sales are not proportional to population size.

Looking back at the Minitab output, notice that there is a column labeled "Contribution to Chi-Sq." This column shows the individual values of $\dfrac{(\text{observed count} - \text{expected count})^2}{\text{expected count}}$, which are summed to produce the value of the chi-square statistic. Notice that the two states with the largest contribution to the chi-square statistic are Florida and California. For Florida, observed hybrid sales were much smaller than expected (observed = 33, expected = 97.44), whereas for California, observed sales were higher than expected (observed = 250, expected = 200.970).

SECTION 15.1 EXERCISES

Each Exercise Set assesses the following chapter learning objectives: C1, M1, M2, M3, M5, P1

SECTION 15.1 Exercise Set 1

15.1 A particular cell phone case is available in a choice of four different colors. A store sells all four colors. To test the hypothesis that sales are equally divided among the four colors, a random sample of 100 purchases is identified.

a. If the resulting X^2 value were 6.4, what conclusion would you reach when using a test with significance level 0.05?

b. What conclusion would be appropriate at significance level 0.01 if $X^2 = 15.3$?

c. If there were six different colors rather than just four, what would you conclude if $X^2 = 13.7$ and a test with $\alpha = 0.05$ was used?

15.2 What is the approximate P-value for the following values of X^2 and df?

a. $X^2 = 6.62$, df $= 3$

b. $X^2 = 16.97$, df $= 10$

c. $X^2 = 30.19$, df $= 17$

15.3 The authors of the paper "Is It Really About Me? Message Content in Social Awareness Streams" (*Computer Supported Cooperative Work 2010*) studied a random sample of 350 Twitter users. For each Twitter user in the sample, the tweets sent during a particular time period were analyzed and the Twitter user was classified into one of the following categories based on the type of messages they usually sent:

Category	Description
IS	Information sharing
OC	Opinions and complaints
RT	Random thoughts
ME	Me now (what I am doing now)
O	Other

The accompanying table gives the observed counts for the five categories (approximate values from a graph in the paper).

Twitter Category	IS	OC	RT	ME	O
Observed Count	51	61	64	101	73

Carry out a hypothesis test to determine if there is convincing evidence that the proportions of Twitter users falling into each of the five categories are not all the same. Use a significance level of 0.05.

15.4 The article **"In Bronx, Hitting Home Runs Is a Breeze"** (*USA Today,* **June 2, 2009**) included a classification of 87 home runs hit at the new Yankee Stadium according to the direction that the ball was hit, resulting in the accompanying data.

Direction	Left Field	Left Center	Center	Right Center	Right Field
Number of Home Runs	18	10	7	18	34

a. Assume that this sample of 87 home runs is representative of home runs hit at Yankee Stadium. Carry out a hypothesis test to determine if there is convincing evidence that the proportions of home runs hit are not the same for all five directions.
b. Write a few sentences describing how the observed counts for the five directions differ from what would have been expected if the proportion of home runs were the same for all five directions.

15.5 An article about the California lottery that appeared in the *San Luis Obispo Tribune* (**December 15, 1999**) gave the following information on the age distribution of adults in California: 35% are between 18 and 34 years old, 51% are between 35 and 64 years old, and 14% are 65 years or older. The article also gave information on the age distribution of those who purchase lottery tickets. The following table is consistent with values given in the article:

Age of Purchaser	Frequency
18–34	36
35–64	130
65 and over	34

Suppose that these data resulted from a random sample of 200 lottery ticket purchasers. Is it reasonable to conclude that one or more of these three age groups buys a disproportionate share of lottery tickets? Use a chi-square goodness-of-fit test with $\alpha = 0.05$.

SECTION 15.1 Exercise Set 2

15.6 What is the approximate *P*-value for the following values of X^2 and df?
a. $X^2 = 14.44$, df $= 6$
b. $X^2 = 16.91$, df $= 9$
c. $X^2 = 32.32$, df $= 20$

15.7 Packages of mixed nuts made by a certain company contain four types of nuts. The percentages of nuts of Types 1, 2, 3, and 4 are advertised to be 40%, 30%, 20%, and 10%, respectively. A random sample of nuts is selected, and each one is categorized by type.
a. If the sample size is 200 and the resulting test statistic value is $X^2 = 19.0$, what conclusion would be appropriate for a significance level of 0.001?
b. If the random sample had consisted of only 40 nuts, would you use the chi-square goodness-of-fit test? Explain your reasoning.

15.8 The authors of the paper **"Talking Smack: Verbal Aggression in Professional Wrestling"** (*Communication Studies* **[2008]: 242–258**) analyzed the content of 36 hours of televised professional wrestling. Each instance of verbal aggression was classified according to the type of aggression, resulting in the frequency table at the bottom of the page. Assume that this sample of 804 verbal aggression incidents can be regarded as representative of all verbal aggression incidents in televised professional wrestling. Carry out a hypothesis test to determine if there is convincing evidence that the proportions of verbal aggression incidents are not the same for all five types of aggression. Use a significance level of 0.01.

15.9 The paper **"Cigarette Tar Yields in Relation to Mortality from Lung Cancer in the Cancer Prevention Study II Prospective Cohort"** (*British Medical Journal* **[2004]: 72–79**) included the accompanying data on the tar level of cigarettes smoked for a sample of male smokers who died of lung cancer.

Tar Level	Frequency
Low	103
Medium	378
High	563
Very high	150

Assume the sample is representative of male smokers who die of lung cancer. Is there convincing evidence that the proportion of male smoker lung cancer deaths is not the same for the four tar level categories?

TABLE FOR EXERCISE **15.8**

Type of Aggression	Swearing	Competence Attack	Character Attack	Physical Appearance Attack	Other
Observed Count	219	166	127	75	217

15.10 The paper referenced in the previous exercise also gave the accompanying data on the age at which smoking started for a sample of 1,031 men who smoked low-tar cigarettes.

Age Group	Frequency
< 16	237
16−17	258
18−20	320
≥ 21	216

a. Use a chi-square goodness-of-fit test to test the null hypothesis H_0: $p_1 = 0.25, p_2 = 0.20, p_3 = 0.30, p_4 = 0.25$, where p_1 = proportion of male low-tar cigarette smokers who started smoking before age 16, and p_2, p_3, and p_4 are defined in a similar way for the other three age groups.

b. The null hypothesis from Part (a) specifies that half of male smokers of low-tar cigarettes began smoking between the ages of 16 and 20. Explain why $p_2 = 0.20$ and $p_3 = 0.30$ is consistent with ages 16, 17, 18, 19, and 20 being equally likely to be the age when smoking started.

Additional Exercises

15.11 What is the approximate P-value for the following values of X^2 and df?
a. $X^2 = 34.52$, df = 13
b. $X^2 = 39.25$, df = 16
c. $X^2 = 26.00$, df = 19

15.12 For which of the X^2 and df pairs in the previous exercise would the null hypothesis be rejected if a significance level of 0.01 were used?

15.13 Think about how you would answer the following question.

Next Wednesday's meeting has been moved forward two days. What day is the meeting now that it has been rescheduled?

This question is ambiguous, as "moved forward" can be interpreted in two different ways. Would you have answered Monday or Friday? The authors of the paper **"Even Abstract Motion Influences the Understanding of Time"** (*Metaphor and Symbol* [2011]: 260–271) wondered if the answers Monday and Friday would be provided an equal proportion of the time. Each student in a random sample of students at Stanford University was asked this question, and the responses are summarized in the following table.

Response	Frequency
Monday	11
Friday	33

The authors of the paper used a chi-square goodness-of-fit test to test the null hypothesis H_0: $p_1 = 0.50, p_2 = 0.50$, where p_1 is the proportion who would respond Monday, and p_2 is the proportion who would respond Friday. They

reported $X^2 = 11.00$ and P-value < 0.001. What conclusion can be drawn from this test?

15.14 A certain genetic characteristic of a particular plant can appear in one of three forms (phenotypes). A researcher has developed a theory, according to which the hypothesized proportions are $p_1 = 0.25, p_2 = 0.50$, and $p_3 = 0.25$. A random sample of 200 plants yields $X^2 = 4.63$.
a. Carry out a test of the null hypothesis that the theory is correct, using level of significance $\alpha = 0.05$.
b. Suppose that a random sample of 300 plants had resulted in the same value of X^2. How would your analysis and conclusion differ from those in Part (a)?

15.15 The article **"Linkage Studies of the Tomato"** (*Transactions of the Royal Canadian Institute* [1931]: 1–19) reported the accompanying data on phenotypes resulting from crossing tall cut-leaf tomatoes with dwarf potato-leaf tomatoes. There are four possible phenotypes: (1) tall cut-leaf, (2) tall potato-leaf, (3) dwarf cut-leaf, and (4) dwarf potato-leaf.

	Phenotype			
	1	2	3	4
Frequency	926	288	293	104

Mendel's laws of inheritance imply that $p_1 = \frac{9}{16}, p_2 = \frac{3}{16}, p_3 = \frac{3}{16}$, and $p_4 = \frac{1}{16}$. Are the data from this experiment consistent with Mendel's laws? Use a 0.01 significance level.

15.16 Each observation in a random sample of 100 fatal bicycle accidents was classified according to the day of the week on which the accident occurred. Data consistent with information on the web site **www.highwaysafety.com** are given in the following table.

Day of Week	Frequency
Sunday	14
Monday	13
Tuesday	12
Wednesday	15
Thursday	14
Friday	17
Saturday	15

Based on these data, is it reasonable to conclude that the proportion of accidents is not the same for all days of the week? Use $\alpha = 0.05$.

15.17 Birds use color to select and avoid certain types of food. The authors of the article **"Colour Avoidance in Northern Bobwhites: Effects of Age, Sex, and Previous Experience"** (*Animal Behaviour* [1995]: 519–526) studied the pecking behavior of 1-day-old bobwhites. In an area painted white, they inserted four pins with different colored heads. The color of the pin chosen on the bird's first peck was noted for each of 33 bobwhites, resulting in the accompanying table.

Color	First Peck Frequency
Blue	16
Green	8
Yellow	6
Red	3

Do these data provide evidence of a color preference? Test using $\alpha = 0.01$.

15.18 Criminologists have long debated whether there is a relationship between weather and violent crime. The author of the article **"Is There a Season for Homicide?"** (*Criminology* **[1988]: 287–296)** classified 1,361 homicides according to season, resulting in the accompanying data. Do these data support the theory that the homicide rate is not the same for the four seasons? Test the relevant hypotheses using a significance level of 0.05.

Season			
Winter	Spring	Summer	Fall
328	334	372	327

SECTION 15.2 Tests for Homogeneity and Independence in a Two-Way Table

Bivariate categorical data (resulting from observations made on two different categorical variables) can also be summarized in a table. For example, suppose that residents of a particular city can watch national news on networks ABC, CBS, NBC, or FOX. A researcher wishes to know whether there is any relationship between political philosophy (liberal, moderate, or conservative) and preferred news network for people who regularly watch the national news. The two variables of interest are *political philosophy* and *preferred network*. If a random sample of 300 regular watchers is selected and these two variables are recorded for each person in the sample, the data set is bivariate and might initially be displayed as follows:

Observation	Political Philosophy	Preferred Network
1	Liberal	CBS
2	Conservative	ABC
3	Conservative	FOX
⋮	⋮	⋮
299	Moderate	NBC
300	Liberal	NBC

Bivariate categorical data are usually summarized in a **two-way frequency table**. This is a rectangular table that consists of a row for each possible category of one of the variables and a column for each possible category of the other variable. There is a cell in the table for each possible combination of category values. The number of times each particular combination occurs in the data set is entered in the corresponding cell of the table. These numbers are called **observed cell counts**. For example, one possible table for the political philosophy and preferred network example is given in Table 15.2. This table has three rows and four columns (because political philosophy has three possible values and preferred network has four possible values). Two-way frequency tables are often described by the number of rows and columns in the table (specified in that order: rows first, then columns). Table 15.2 is called a 3 × 4 table. The smallest two-way frequency table is a 2 × 2 table, which has only two rows and two columns, resulting in four cells.

TABLE 15.2 A 3 × 4 Frequency Table

	ABC	CBS	NBC	FOX	Row Marginal Total
Liberal	20	20	25	15	**80**
Moderate	45	35	50	20	**150**
Conservative	15	40	10	5	**70**
Column Marginal Total	**80**	**95**	**85**	**40**	**300**

Marginal totals are obtained by adding the observed cell counts in each row and also in each column of the table. The row and column marginal totals, along with the total of all observed cell counts in the table—the **grand total**, have been included in Table 15.2. The marginal totals provide information on the distribution of observed values for each variable separately. In this example, the row marginal totals reveal that the sample consisted of 80 liberals, 150 moderates, and 70 conservatives. Similarly, column marginal totals indicate how often each of the preferred network categories occurred: 80 preferred ABC, 95 preferred CBS, and so on. The grand total, 300, is the total number of observations in the bivariate data set.

Table 15.2 was constructed using the values of two different categorical variables for all individuals in a single sample. This table could be used to investigate any association between political philosophy and preferred network. In this type of bivariate categorical data set, only the grand total (the sample size) is known before the data are collected.

Two-way tables are also used when data are collected to compare two or more populations or treatments on the basis of a single categorical variable. In this situation, independent samples are selected from each population or there is random assignment to treatments. For example, data could be collected at a university to compare students, faculty, and staff on the basis of primary mode of transportation to campus (car, bicycle, motorcycle, bus, or by foot). One random sample of 200 students, another random sample of 100 faculty members, and a third random sample of 150 staff members might be chosen. The selected individuals could be interviewed to obtain the necessary transportation information. Data from such a study could be summarized in a 3×5 two-way frequency table with row categories of student, faculty, and staff and column categories corresponding to the five possible modes of transportation. The observed cell counts could then be used to learn about differences and similarities among the three groups. In this situation, one set of marginal totals (the sample sizes for the different groups) is known before the data are collected. In the 3×5 situation just discussed, the row totals would be known (200, 100, and 150).

Comparing Two or More Populations or Treatments: A Test of Homogeneity

When the purpose of a study is to compare two or more populations or treatments on the basis of a categorical variable, the question of interest is whether the category proportions are the same for all the populations or treatments. The test procedure uses a chi-square statistic to compare the observed counts to those that would be expected if there were no differences among populations or treatments.

Example 15.6 Risky Soccer?

The paper **"No Evidence of Impaired Neurocognitive Performance in Collegiate Soccer Players"** (*American Journal of Sports Medicine* [2002]:157–162) compared collegiate soccer players, athletes in sports other than soccer, and a group of students who were not involved in collegiate sports on the basis of their history of head injuries. Table 15.3, a 3×4 two-way frequency table, is the result of classifying each student in independently selected random samples of 91 soccer players, 96 non-soccer athletes, and 53 non-athletes into one of four head injury categories.

TABLE 15.3 Observed Counts for Example 15.6

	Head Injury Category				
	No Head Injuries	One Head Injury	Two Head Injuries	More Than Two Head Injuries	Row Marginal Total
Soccer Players	45	25	11	10	**91**
Non-Soccer Athletes	68	15	8	5	**96**
Non-Athletes	45	5	3	0	**53**
Column Marginal Total	**158**	**45**	**22**	**15**	**240**

Notice that there were 240 head injury responses. Of these, 158 were "no head injury." The proportion of the total responding "no head injury" is then

$$\frac{158}{240} = 0.658$$

If there were no difference in response for the different groups, you would expect about 65.8% of the soccer players to have responded "no head injury," 65.8% of the non-soccer athletes to have responded "no head injury," and so on. Therefore the expected cell counts for the three cells in the "no head injury" column are

Expected count for soccer player *and* no head injury cell = 0.658(91) = 59.9
Expected count for non–soccer athlete *and* no head injury cell = 0.658(96) = 63.2
Expected count for non-athlete *and* no head injury cell = 0.658(53) = 34.9

Notice that the expected cell counts need not be whole numbers. The expected cell counts for the remaining cells can be computed in a similar manner. For example,

$$\frac{45}{240} = 0.188$$

of all responses were in the "one head injury" category, so

Expected count for soccer player *and* one head injury cell = 0.188(91) = 17.1
Expected count for non–soccer athlete *and* one head injury cell = 0.188(96) = 18.0
Expected count for non-athlete *and* one head injury cell = 0.188(53) = 10.0

It is common practice to display the observed cell counts and the corresponding expected cell counts in the same table, with the expected cell counts enclosed in parentheses. Expected cell counts for the remaining cells have been computed and entered into Table 15.4. Except for small differences resulting from rounding, each marginal total for the expected cell counts is identical to the marginal total of the corresponding observed counts.

TABLE **15.4** **Observed and Expected Counts for Example 15.6**

	Head Injury Category				
	No Head Injuries	One Head Injury	Two Head Injuries	More Than Two Head Injuries	Row Marginal Total
Soccer Players	45 (59.9)	25 (17.1)	11 (8.3)	10 (5.7)	**91**
Non-Soccer Athletes	68 (63.2)	15 (18.0)	8 (8.8)	5 (6.0)	**96**
Non-Athletes	45 (34.9)	5 (10.0)	3 (4.9)	0 (3.3)	**53**
Column Marginal Total	**158**	**45**	**22**	**15**	**240**

A quick comparison of the observed and expected cell counts in Table 15.4 reveals some large discrepancies, suggesting that the proportions falling into the head injury categories may not be the same for all three groups. This will be explored further in Example 15.7.

In Example 15.6, the expected count for a cell corresponding to a particular group–response combination was computed in two steps. First, the response *marginal proportion* was computed (e.g., 158/240 for the "no head injury" response). Then this proportion was multiplied by a marginal group total (for example, 91(158/240) for the soccer player group). Algebraically, this is equivalent to first multiplying the row and column marginal totals and then dividing by the grand total:

$$\frac{(91)(158)}{240}$$

To compare two or more populations or treatments on the basis of a categorical variable, calculate an **expected cell count** for each cell by selecting the corresponding row and column marginal totals and then using the formula

$$\text{expected cell count} = \frac{(\text{row marginal total})(\text{column marginal total})}{\text{grand total}}$$

Each of these quantities represents what is expected if there are no differences in the category proportions for the groups under study.

The X^2 statistic, introduced in Section 15.1, is used to compare the observed cell counts to the expected cell counts. When there are large discrepancies between the observed and expected counts, this leads to a large value of X^2 and suggests that the hypothesis of no differences between the populations or treatments should be rejected. A formal test procedure is described in the accompanying box.

Chi-Square Test for Homogeneity

Appropriate when the following conditions are met:

1. Observed counts are from independently selected random samples, *or* subjects in an experiment are randomly assigned to treatment groups.
2. The sample sizes are large. The sample size is large enough for the chi-square test for homogeneity to be appropriate if every expected count is at least 5. If some expected counts are less than 5, rows or columns of the table may be combined to achieve a table with satisfactory expected counts.

When these conditions are met, the following test statistic can be used:

$$X^2 = \sum_{\text{all cells}} \frac{(\text{observed count} - \text{expected count})^2}{\text{expected count}}$$

The expected cell counts are estimated from the sample data using the formula

$$\text{expected cell count} = \frac{(\text{row marginal total})(\text{column marginal total})}{\text{grand total}}$$

When the conditions above are met and the null hypothesis is true, the X^2 statistic has a chi-square distribution with df = (number of rows $-$ 1)(number of columns $-$ 1).

Hypotheses

H_0: The population (or treatment) category proportions are the same for all of the populations (or treatments).
H_a: The population (or treatment) category proportions are not all the same for all of the populations (or treatments).

Associated *P*-value

The *P*-value associated with the computed value of the test statistic is the area to the right of X^2 under the chi-square curve with

$$\text{df} = (\text{number of rows} - 1)(\text{number of columns} - 1)$$

Upper-tail areas for chi-square distributions can be found in Appendix Table 5.

Example 15.7   Risky Soccer Revisited

The following table of observed and expected cell counts appeared in Example 15.6.

	Head Injury Category				Row Marginal Total
	No Head Injuries	One Head Injury	Two Head Injuries	More Than Two Head Injuries	
Soccer Players	45 (59.9)	25 (17.1)	11 (8.3)	10 (5.7)	91
Non-Soccer Athletes	68 (63.2)	15 (18.0)	8 (8.8)	5 (6.0)	96
Non-Athletes	45 (34.9)	5 (10.0)	3 (4.9)	0 (3.3)	53
Column Marginal Total	158	45	22	15	240

You are interested in learning if there is a difference in the head injury category proportions for the three populations (soccer players, non-soccer athletes, and non-athletes).

H Hypotheses   The hypotheses of interest are

H_0: Proportions in each head injury category are the same for all three groups.
H_a: The head injury category proportions are not all the same for all three groups.

M Method   This is a hypothesis testing problem. Random samples from three different populations were independently selected. The response is categorical. In this situation, you should consider a chi-square test of homogeneity.

The test statistic for the chi-square test of homogeneity is

$$X^2 = \sum_{\text{all cells}} \frac{(\text{observed count} - \text{expected count})^2}{\text{expected count}}$$

A significance level of 0.05 will be used in this example.

C Check   Next you need to check to see if this method is appropriate. Two of the expected cell counts (in the two head injuries and the more than two head injuries columns) are less than 5, so you should combine the last two columns of the table prior to carrying out the chi-square test. The table you will work with is then

	Head Injury Category			Row Marginal Total
	No Head Injuries	One Head Injury	Two or More Head Injuries	
Soccer Players	45 (59.9)	25 (17.1)	21 (14.0)	91
Non-Soccer Athletes	68 (63.2)	15 (18.0)	13 (14.8)	96
Non-Athletes	45 (34.9)	5 (10.0)	3 (8.2)	53
Column Marginal Total	158	45	37	240

Because the samples were independent random samples and the expected counts are all at least 5, the chi-square test of homogeneity is appropriate.

C Calculate   The modified table has three rows and three columns, so the appropriate df is

$$df = (\text{number of rows} - 1)(\text{number of columns} - 1) = (3 - 1)(3 - 1) = 4$$

$$X^2 = \sum_{\text{all cells}} \frac{(\text{observed count} - \text{expected count})^2}{\text{expected count}}$$

$$= \frac{(45 - 59.9)^2}{59.9} + \cdots + \frac{(3 - 8.2)^2}{8.2} = 20.6$$

P-value: The *P*-value is the area to the right of 20.6 under a chi-square curve with df = 4. The computed value of X^2 is greater than 18.46 (the largest entry in the df = 4 column of Appendix Table 5), so *P*-value < 0.001.

C Communicate Results   Because the P-value is less than 0.05, H_0 is rejected. There is strong evidence that the proportions in the head injury categories are not the same for the three groups compared. The largest differences between the observed and expected frequencies occur in the response categories for soccer players and for non-athletes, with soccer players having higher than expected proportions in the one and the two or more head injuries categories and non-athletes having a higher than expected proportion in the no head injury category.

Most statistical computer packages can calculate expected cell counts, the value of the X^2 statistic, and the associated P-value. This is illustrated in the following example.

Example 15.8   Keeping the Weight Off

AP* EXAM TIP

A chi-square test of homogeneity is used in this example because three treatment groups are being compared and the response variable is categorical.

The article **"Daily Weigh-Ins Can Help You Keep Off Lost Pounds, Experts Say"** (*Associated Press*, October 17, 2005) describes an experiment in which 291 people who had lost at least 10% of their body weight in a medical weight-loss program were assigned at random to one of three groups for follow-up. One group met monthly in person, one group "met" online monthly in a chat room, and one group received a monthly newsletter by mail. After 18 months, participants in each group were classified according to whether or not they had regained more than 5 pounds, resulting in the data given in Table 15.5.

TABLE 15.5   Observed and Expected Counts for Example 15.8

	Amount of Weight Gained		Row Marginal Total
	Regained 5 Lb or Less	Regained More Than 5 Lb	
In-Person	52 (41.0)	45 (56.0)	97
Online	44 (41.0)	53 (56.0)	97
Newsletter	27 (41.0)	70 (56.0)	97

Does there appear to be a difference in the weight regained proportions for the three follow-up methods?

H Hypotheses   The hypotheses of interest are

H_0: Proportions for the two weight gain categories are the same for the three follow-up methods.

H_a: The weight gain category proportions are not the same for all three follow-up methods.

M Method   This is a hypothesis testing problem. The data are from an experiment. There is one categorical variable and there are three treatments. In this situation, you should consider a chi-square test of homogeneity.

The test statistic for the chi-square test of homogeneity is

$$X^2 = \sum_{\text{all cells}} \frac{(\text{observed count} - \text{expected count})^2}{\text{expected count}}$$

This table has three rows and two columns, so the appropriate df is

$$\text{df} = (\text{number of rows} - 1)(\text{number of columns} - 1) = (3 - 1)(2 - 1) = 2$$

A significance level of 0.01 will be used for this test.

C Check   Next, you need to determine if this method is appropriate. Table 15.5 includes the expected cell counts, all of which are greater than 5, so the large sample condition is satisfied. The subjects in this experiment were assigned at random to the treatment groups.

C Calculate JMP output follows. For each cell, this output includes the observed cell count, the expected cell count, and the value of $\dfrac{(\text{observed count} - \text{expected count})^2}{\text{expected count}}$ for that cell (its contribution to the X^2 statistic). From the output, $X^2 = 13.773$ (in the last row of the output), df = 2 (from the row in the output that is just below the table) and the associated P-value = 0.001 (from the last row of the output).

**⊡ Contingency Analysis of
Weight Gain By Follow-up Method**

▷ **Mosaic Plot**

Freq: Count

⊿ ⊡ **Contingency Table**

		Weight Gain		
	Count	5 Lb. or less	More than 5 Lb.	
	Expected			
	Cell Chi^2			
In-Person		52	45	97
		41	56	
		2.9512	2.1607	
Newsletter		27	70	97
		41	56	
		4.7805	3.5000	
Online		44	53	97
		41	56	
		0.2195	0.1607	
		123	168	291

(Follow-up Method — left axis label)

⊿ **Tests**

N	DF	-LogLike	RSquare (U)
291	2	7.0481148	0.0356

Test	ChiSquare	Prob>ChiSq
Likelihood Ratio	14.096	0.0009*
Pearson	13.773	0.0010*

C Communicate Results Because the P-value is less than 0.01, H_0 is rejected. The data indicate that the proportions who have regained more than 5 pounds are not the same for the three follow-up methods. Comparing the observed and expected cell counts, you can see that the observed number in the newsletter group who had regained more than 5 pounds was higher than would have been expected, and the observed number in the in-person group who had regained more than 5 pounds was lower than would have been expected if there were no differences in the three follow-up methods.

Testing for Independence of Two Categorical Variables

The X^2 test statistic and test procedure can also be used to investigate association between two categorical variables in a single population. When there is an association, knowing the value of one variable provides information about the value of the other variable. When there is *no* association between two categorical variables, they are said to be independent.

With bivariate categorical data, the question of interest is whether there is an association, and the null and and alternative hypotheses are

H_0: The two variables are independent (or equivalently, there is no association between the two variables).

H_a: The two variables are not independent (or equivalently, there is an association between the two variables).

To see how expected counts are obtained in this situation, recall from Chapter 5 that if two outcomes A and B are independent, then

$$P(A \text{ and } B) = P(A)P(B)$$

This means that the proportion of time that the two outcomes occur together in the long run is the product of the two individual long-run relative frequencies. Similarly, two categorical variables are independent in a population if, for each particular category of the first variable and each particular category of the second variable,

$$
\begin{pmatrix} \text{proportion of individuals} \\ \text{in a particular category} \\ \text{combination} \end{pmatrix} = \begin{pmatrix} \text{proportion in} \\ \text{specified category of} \\ \text{first variable} \end{pmatrix} \begin{pmatrix} \text{proportion in} \\ \text{specified category} \\ \text{of second variable} \end{pmatrix}
$$

Multiplying the right-hand side of this expression by the sample size gives the expected number of individuals in the sample who are in the specified category combination if the variables are independent. However, these expected counts cannot be calculated, because the individual population proportions are not known. The solution is to estimate each population proportion using the corresponding sample proportion:

$$
\begin{pmatrix} \text{estimated expected number} \\ \text{in a partciular category} \\ \text{combination} \end{pmatrix} = (\text{sample size}) \begin{pmatrix} \dfrac{\text{observed number} \\ \text{in category of} \\ \text{first variable}}{\text{sample size}} \end{pmatrix} \begin{pmatrix} \dfrac{\text{observed number} \\ \text{in category of} \\ \text{second variable}}{\text{sample size}} \end{pmatrix}
$$

$$
= \dfrac{\begin{pmatrix} \text{observed number} \\ \text{in category of} \\ \text{first variable} \end{pmatrix} \begin{pmatrix} \text{observed number} \\ \text{in category of} \\ \text{second variable} \end{pmatrix}}{\text{sample size}}
$$

Suppose that the observed counts are displayed in a rectangular table in which the rows correspond to the categories of the first variable and the columns to the categories of the second variable. The numerator in the preceding expression for expected counts is just the product of the row and column marginal totals. This is exactly how expected counts were computed in the test for homogeneity of several populations, even though the reasoning used to arrive at the formula is different.

Chi-Square Test for Independence

Appropriate when the following conditions are met:

1. Observed counts are from a random sample.
2. The sample size is large. The sample size is large enough for the chi-square test for independence to be appropriate if every expected count is at least 5. If some expected counts are less than 5, rows or columns of the table can be combined to achieve a table with satisfactory expected counts.

When these conditions are met, the following test statistic can be used:

$$
X^2 = \sum_{\text{all cells}} \frac{(\text{observed count} - \text{expected count})^2}{\text{expected count}}
$$

The expected cell counts are estimated from the sample data using the formula

$$
\text{expected cell count} = \frac{(\text{row marginal total})(\text{column marginal total})}{\text{grand total}}
$$

When these conditions are met and the null hypothesis is true, the X^2 statistic has a chi-square distribution with

$$
df = (\text{number of rows} - 1)(\text{number of columns} - 1).
$$

(continued)

Hypotheses

H_0: The two variables are independent.
H_a: The two variables are not independent.

Associated *P*-value

The *P*-value associated with the computed value of the test statistic is the area to the right of X^2 under the chi-square curve with

$$df = (\text{number of rows} - 1)(\text{number of columns} - 1)$$

Upper-tail areas for chi-square distributions can be found in Appendix Table 5.

Example 15.9 A Pained Expression

The paper **"Facial Expression of Pain in Elderly Adults with Dementia"** (*Journal of Undergraduate Research* [2006]) examined the relationship between a nurse's assessment of a patient's facial expression and the patient's self-reported level of pain. Data for 89 patients are summarized in Table 15.6. Because patients with dementia do not always give a verbal indication that they are in pain, the authors of the paper were interested in determining if there is an association between a facial expression that reflects pain and self-reported pain.

TABLE 15.6 Observed Counts for Example 15.9

Facial Expression	Self-Report	
	No Pain	Pain
No Pain	17	40
Pain	3	29

H Hypotheses The question of interest is whether there is an association between facial expression and self-reported pain. The hypotheses of interest are then

H_0: Facial expression and self-reported pain are independent.
H_a: Facial expression and self-reported pain are not independent.

M Method You should consider a chi-square test of independence because the answers to the four key questions are *hypothesis testing, sample data, two categorical variables, and one sample.*

The test statistic for the chi-square test of independence is

$$X^2 = \sum_{\text{all cells}} \frac{(\text{observed count} - \text{expected count})^2}{\text{expected count}}$$

The data table has two rows and two columns, so the appropriate df is

$$df = (\text{number of rows} - 1)(\text{number of columns} - 1) = (2 - 1)(2 - 1) = 1$$

A significance level of 0.05 will be used for this test.

C Check Next you need to determine if this method is appropriate. To do this, the expected cell counts must be calculated.

Cell		Expected Cell Count
Row	Column	
1	1	$\frac{(57)(20)}{89} = 12.81$
1	2	$\frac{(57)(69)}{89} = 44.19$
2	1	$\frac{(32)(20)}{89} = 7.19$
2	2	$\frac{(32)(69)}{89} = 24.81$

All of the expected counts are greater than 5. Although the participants in the study were not randomly selected, they were thought to be representative of the population of nursing home patients with dementia. The observed and expected counts are given together in Table 15.7.

TABLE 15.7 Observed and Expected Counts for Example 15.9

Facial Expression	Self-Report	
	No Pain	Pain
No Pain	17 (12.81)	40 (44.19)
Pain	3 (7.19)	29 (24.81)

C Calculate

$$X^2 = \sum_{\text{all cells}} \frac{(\text{observed count} - \text{expected count})^2}{\text{expected count}}$$

$$= \frac{(17 - 12.81)^2}{12.81} + \cdots + \frac{(29 - 24.81)^2}{24.81} = 4.92$$

P-value: The *P*-value is the area to the right of 4.92 under a chi-square curve with df $= 1$. The entry closest to 4.92 in the 1-df column of Appendix Table 5 is 5.02, so the approximate *P*-value for this test is *P*-value ≈ 0.025.

C Communicate Results

Because the *P*-value is less than 0.05, H_0 is rejected. There is convincing evidence of an association between a nurse's assessment of facial expression and self-reported pain.

Example 15.10   Stroke Mortality and Education

Table 15.8 was constructed using data from the article **"Influence of Socioeconomic Status on Mortality After Stroke"** (*Stroke* [2005]: 310–314). One of the questions of interest was whether there was an association between survival after a stroke and level of education. Medical records for a random sample of 2,333 residents of Vienna, Austria, who had suffered a stroke were used to classify each individual according to two variables—survival (survived, died) and level of education (no basic education, secondary school graduation, technical training/apprenticed, higher secondary school degree, university graduate). Expected cell counts (computed under the assumption of no association between survival and level of education) appear in parentheses in the table.

TABLE 15.8 Observed and Expected Counts for Example 15.10

	No Basic Education	Secondary School Graduation	Technical Training/ Apprenticed	Higher Secondary School Degree	University Graduate
Died	13 (17.40)	91 (77.18)	196 (182.68)	33 (41.91)	36 (49.82)
Survived	97 (92.60)	397 (410.82)	959 (972.32)	232 (223.09)	279 (265.18)

H Hypotheses

The hypotheses of interest are

H_0: Survival and level of education are independent.
H_a: Survival and level of education are not independent.

M Method

The answers to the four key questions are *hypothesis testing, sample data, two categorical variables, and one sample*. A chi-square test of independence will be considered. The test statistic for the chi-square test of independence is

$$X^2 = \sum_{\text{all cells}} \frac{(\text{observed count} - \text{expected count})^2}{\text{expected count}}$$

This table has two rows and five columns, so the appropriate df is

$$df = (\text{number of rows} - 1)(\text{number of columns} - 1) = (2 - 1)(5 - 1) = 4$$

A significance level of 0.01 will be used for this test.

C Check   Next you need to check to see if this method is appropriate. Table 15.8 includes the computed expected cell counts, all of which are greater than 5, so the large sample condition is satisfied. The sample was a random sample, so it is appropriate to use the chi-square test of independence.

C Calculate   Minitab output follows. For each cell, the Minitab output includes the observed cell count, the expected cell count, and the value of $\dfrac{(\text{observed count} - \text{expected count})^2}{\text{expected count}}$ for that cell (its contribution to the X^2 statistic). From the output, $X^2 = 12.219$, df $= 4$ and the associated P-value $= 0.016$.

Chi-Square Test

Expected counts are printed below observed counts
Chi-Square contributions are printed below expected counts

	1	2	3	4	5	Total
1	13	91	196	33	36	369
	17.40	77.18	182.68	41.91	49.82	
	1.112	2.473	0.971	1.896	3.835	
2	97	397	959	232	279	1964
	92.60	410.82	972.32	223.09	265.18	
	0.209	0.465	0.182	0.356	0.720	
Total	110	488	1155	265	315	2333

Chi-Sq = 12.219,   DF = 4,   P-Value = 0.016

C Communicate Results   Because the P-value is greater than 0.01, H_0 is not rejected at the 0.01 significance level. There is not sufficient evidence to conclude that an association exists between level of education and survival.

SECTION 15.2   EXERCISES

Each Exercise Set assesses the following chapter learning objectives: C2, M2, M4, M5, P2, P3.

SECTION 15.2   Exercise Set 1

15.19   The polling organization Ipsos conducted telephone surveys in March of 2004, 2005, and 2006. In each year, each person in a random sample of 1,001 people age 18 or older was asked about whether he or she planned to use a credit card to pay federal income taxes that year. The data given in the accompanying table are from the report "**Fees Keeping Taxpayers from Using Credit Cards to Make Tax Payments**" (*IPSOS Insight*, March 24, 2006). Is there evidence that the proportions that would fall into the three credit card response categories are not the same for all three years? Test the relevant hypotheses using a 0.05 significance level.

	2004	2005	2006
Definitely/Probably Will	40	50	40
Might/Might Not/Probably Not	180	190	160
Definitely Will Not	781	761	801

15.20   Does including a gift with a request for a donation influence the proportion who will make a donation? This question was investigated in a study described in the report

"**Gift-Exchange in the Field**" (Institute for the Study of Labor, 2007). Letters were sent to a large number of potential donors in Germany. The letter requested a donation for funding schools in Bangladesh. The letter recipients were assigned at random to one of three groups. Those in the first group received a letter with no gift. Those in the second group received a letter that included a small gift (a postcard), and those in the third group received a letter with a larger gift (four postcards). The response of interest was whether or not the letter resulted in a donation.

	Donation	No Donation
No Gift	397	2865
Small Gift	465	2772
Large Gift	691	2656

a. Carry out a hypothesis test to determine if there is convincing evidence that the proportions for the two donation categories are not the same for all three types of requests. Use a significance level of 0.01.

b. Based on your test in Part (a) and a comparison of observed and expected cell counts, write a brief description of how the proportions who would make a donation differ for the three types of request.

15.21 The authors of the paper **"The Relationship of Field of Study to Current Smoking Status Among College Students"** (*College Student Journal* [2009]: 744–754) carried out a study to investigate if smoking rates were different for college students in different majors. Each student in a large random sample of students at the University of Minnesota was classified according to field of study and whether or not they had smoked in the past 30 days. The data are given in the accompanying table.

Field of Study	Smoked in the Last 30 Days	Did Not Smoke in Last 30 Days
1. Undeclared	176	489
2. Art, design, performing arts	149	336
3. Humanities	197	454
4. Communication, languages	233	389
5. Education	56	170
6. Health sciences	227	717
7. Mathematics, engineering, sciences	245	924
8. Social science, human services	306	593
9. Individualized course of study	134	260

a. Is there evidence that field of study and smoking status are not independent? Use the accompanying Minitab output to test the relevant hypotheses using $\alpha = 0.01$.

Chi-Square Test: Smoked, Did Not Smoke
Expected counts are printed below observed counts
Chi-Square contributions are printed below expected counts

	Smoked	Did Not Smoke	Total
1	176	489	665
	189.23	475.77	
	0.925	0.368	
2	149	336	485
	138.01	346.99	
	0.875	0.348	
3	197	454	651
	185.25	465.75	
	0.746	0.297	
4	233	389	622
	177.00	445.00	
	17.721	7.048	
5	56	170	226
	64.31	161.69	
	1.074	0.427	
6	227	717	944
	268.62	675.38	
	6.449	2.565	

(continued)

	Smoked	Did Not Smoke	Total
7	245	924	1169
	332.65	836.35	
	23.094	9.185	
8	306	593	899
	255.82	643.18	
	9.844	3.915	
9	134	260	394
	112.12	281.88	
	4.272	1.699	
Total	1723	4332	6055

Chi-Sq = 90.853, DF = 8, P-Value = 0.000

b. Write a few sentences describing how smoking status is related to field of study. (Hint: Focus on cells that have large values of

$$\frac{(\text{observed cell count} - \text{expected cell count})^2}{\text{expeted cell count}}$$

15.22 The paper **"Credit Card Misuse, Money Attitudes, and Compulsive Buying Behavior: Comparison of Internal and External Locus of Control Consumers"** (*College Student Journal* [2009]: 268–275) describes a survey of college students at two midwestern public universities. Based on the survey responses, students were classified into two "locus of control" groups (internal and external) based on whether they believe that they control what happens to them. Those in the internal locus of control group believe that they are usually in control of what happens to them, whereas those in the external locus of control group believe that factors outside their control usually determine what happens to them. Each student was also classified according to a measure of compulsive buying. The resulting data are summarized in the accompanying table. Can the researchers conclude that there is an association between locus of control and compulsive buying behavior? Carry out a test using $\alpha = 0.01$. Assume it is reasonable to regard the sample as representative of college students at midwestern public universities.

		Locus of Control	
		Internal	External
Compulsive Buyer?	Yes	3	14
	No	52	57

SECTION 15.2 **Exercise Set 2**

15.23 The data in the accompanying table are from the paper **"Gender Differences in Food Selections of Students at a Historically Black College and University"** (*College Student Journal* [2009]: 800–806). Suppose that each person in a random sample of 48 male students and in a random sample of 91 female students at a particular college was classified according to gender and whether they usually or rarely eat three meals a day.

	Usually Eat 3 Meals a Day	Rarely Eat 3 Meals a Day
Male	26	22
Female	37	54

a. Is there evidence that the proportions who would fall into each of the two response categories are not the same for males and females? Use the X^2 statistic to test the relevant hypotheses with a significance level of 0.05.

b. Are your calculations and conclusions from Part (a) consistent with the accompanying Minitab output? Explain.

Expected counts are printed below observed counts
Chi-Square contributions are printed below expected counts

	Usually	Rarely	Total
Male	26	22	48
	21.76	26.24	
	0.828	0.686	
Female	37	54	91
	41.24	49.76	
	0.437	0.362	
Total	63	76	139

Chi-Sq = 2.314, DF = 1, P-Value = 0.128

c. Because the response variable in this exercise has only two categories (usually and rarely), you could have also answered the question posed by carrying out a large-sample z test of $H_0: p_1 - p_2 = 0$ versus $H_a: p_1 - p_2 \neq 0$, where p_1 is the proportion who usually eat three meals a day for males and p_2 is the proportion who usually eat three meals a day for females. Minitab output from the two-sample z test is shown. Using a significance level of 0.05, does the two-sample z test lead to the same conclusion as in Part (a)?

Test for Two Proportions

Sample	X	N	Sample p
Male	26	48	0.541667
Female	37	91	0.406593

Difference = p (1) − p (2)
Test for difference = 0 (vs not = 0): Z = 1.53 P-Value = 0.128

d. How do the P-values from the tests in Parts (a) and (c) compare? Does this surprise you? Explain.

15.24 In a study to determine if hormone therapy increases risk of venous thrombosis in menopausal women, each person in a sample of 579 women who had been diagnosed with venous thrombosis was classified according to hormone use. Each woman in a sample of 2,243 women who had not been diagnosed with venous thrombosis was also classified according to hormone use. Data from the study are given in the accompanying table (*Journal of the American Medical Association* [2004]: 1581–1587). The women in each of the two samples were selected at random from patients at a large HMO in the state of Washington.

	Current Hormone Use		
	None	Esterified Estrogen	Conjugated Equine Estrogen
Venous Thrombosis	372	86	121
No Venous Thrombosis	1439	515	289

a. Is there convincing evidence that the proportions who would fall into each of the hormone use categories are not the same for women who have been diagnosed with venous thrombosis and those who have not?

b. To what populations would it be reasonable to generalize the conclusion from the test in Part (a)? Explain.

15.25 The paper "Contemporary College Students and Body Piercing" (*Journal of Adolescent Health* [2004]: 58–61) described a survey of 450 undergraduate students at a state university in the southwestern region of the United States. Each student in the sample was classified according to class standing (freshman, sophomore, junior, or senior) and body art category (body piercings only, tattoos only, both tattoos and body piercings, no body art). Use the data in the accompanying table to determine if there is evidence of an association between class standing and body art category. Assume that it is reasonable to regard the sample as representative of the students at this university. Use $\alpha = 0.01$.

	Body Piercings Only	Tattoos Only	Both Body Piercing and Tattoos	No Body Art
Freshman	61	7	14	86
Sophomore	43	11	10	64
Junior	20	9	7	43
Senior	21	17	23	54

15.26 In large elections, voters are often interviewed as they leave their voting place so that news agencies can forecast election results. Each person in a representative sample of voters in the 2012 U.S. presidential election was asked who he or she had voted for and his or her age was also recorded. Data consistent with summary values given in the article "Demographics Shape USA's Divide" (*USA Today*, November 7, 2012) are given in the table on the next page. Is there evidence of an association between age group and candidate selected in the election? Carry out a hypothesis test using $\alpha = 0.05$.

		Candidate		
		Obama	Romney	Other
	18 – 29	600	360	40
Age Group	30 – 44	520	450	30
	45 – 64	480	510	10
	65 and older	440	550	10

Additional Exercises

15.27 Give an example of a situation where it would be appropriate to use a chi-square test of homogeneity. Describe the populations that would be sampled and the variable that would be recorded.

15.28 Give an example of a situation where it would be appropriate to use a chi-square test of independence. Describe the population that would be sampled and the two variables that would be recorded.

15.29 Explain the difference between situations that would lead to a chi-square goodness-of-fit test and those that would lead to a chi-square test of homogeneity.

15.30 Explain the difference between situations that would lead to a chi-square test for homogeneity and those that would lead to a chi-square test for independence.

15.31 Are babies born to mothers who use assistive reproduction technology (ART) more likely to be born prematurely than babies conceived naturally? The data in the accompanying table are from the paper **"Child Growth from Birth to 18 Months After Assisted Reproduction Technology"** (*International Journal of Nursing* [2010]: 1159–1166). The data are from a random sample of 19,614 births. Each birth was classified according to whether the mother used ART and whether the baby was premature. Use these data to decide if there is convincing evidence of an association between the use of ART and whether or not a baby is premature. Use $\alpha = 0.01$.

	Conceived Using ART	Conceived Naturally
Premature	154	2405
Not Premature	212	18,843

15.32 The report **"Mobile Youth Around the World"** (The Nielsen Company, December 2010) provided the following information on gender of smart phone users for representative samples of young people ages 15 to 24 in several different countries.

Country	Percent Female	Percent Male
United States	55%	45%
Spain	39%	61%
Italy	38%	62%
India	20%	80%

a. Suppose the sample sizes were 1,000 for the United States and for India and 500 for Spain and Italy. Complete the following two-way table by entering the observed counts.

Country	Female	Male
United States		
Spain		
Italy		
India		

b. Carry out a hypothesis test to determine if there is convincing evidence that the gender proportions are not the same for all four countries. Use a significance level of 0.05.

15.33 In large elections, voters are often interviewed as they leave their voting place so that news agencies can forecast election results. In three different states (California, Texas and New York), each person in representative sample of voters in the 2012 U.S. presidential election was asked who had received his or her vote. Data consistent with summary values given in the article "Election 2012: Media" (*USA Today*, November 7, 2012) are given in the accompanying table. Is there evidence that the proportions in the three candidate categories are not the same for all three of these states? Test the relevant hypotheses using $\alpha = 0.01$.

	Candidate		
	Obama	Romney	Other
Ohio	500	490	10
Texas	430	550	20
New York	580	410	10

15.34 Does viewing angle affect a person's ability to tell the difference between a female nose and a male nose? This important (?) research question was examined in the article **"You Can Tell by the Nose: Judging Sex from an Isolated Facial Feature"** (*Perception* [1995]: 969–973). Eight Caucasian males and eight Caucasian females posed for nose photos. The article states that none of the volunteers wore nose studs or had prominent nasal hair. Each person placed a black Lycra tube over his or her head in such a way that only the nose protruded through a hole in the material. Photos were then taken from three different angles: front view, three-quarter view, and profile. These photos were shown to a sample of undergraduate students. Each student in the sample was shown one of the nose photos and asked whether it was a photo of a male or a female. The response was classified as either correct or incorrect. The accompanying table was constructed using summary values reported in the article. Is there evidence that the proportion of correct sex identifications differs for the three different nose views?

	View		
Sex ID	Front	Profile	Three-Quarter
Correct	23	26	29
Incorrect	17	14	11

15.35 A survey was conducted in the San Francisco Bay Area in which each participating individual was classified according to the type of vehicle he or she used most often and their city of residence. A subset of the resulting data is summarized in the accompanying table (*The Relationship of Vehicle Type Choice to Personality, Lifestyle, Attitudinal and Demographic Variables,* Technical Report UCD-ITS-RR02-06, DaimlerCrysler Corp., 2002).

	City		
Vehicle Type	Concord	Pleasant Hill	North San Francisco
Small	68	83	221
Compact	63	68	106
Midsize	88	123	142
Large	24	18	11

Do these data provide convincing evidence of an association between city of residence and vehicle type? Use a significance level of 0.05. You may assume that the sample is a random sample of Bay Area residents.

15.36 The article "Cooperative Hunting in Lions: The Role of the Individual" (*Behavioral Ecology and Sociobiology* [1992]: 445–454) discusses the different roles taken by lionesses as they attack and capture prey. The authors were interested in the effect of the position in line as stalking occurs. An individual lioness may be in the center of the line or on one of the wings (the left or right side of the line) as they advance toward their prey. In addition to position, the role of the lioness was also considered. A lioness could initiate a chase (be the first one to charge the prey), or she could participate and join the chase after it has been initiated. Data from the article are summarized in the accompanying table.

	Role	
Position	Initiate Chase	Participate in Chase
Center	28	48
Wing	66	41

Is there evidence of an association between position and role? Test the relevant hypotheses using a significance level of 0.01. What assumptions about the sample must be true for the chi-square test to be an appropriate way to analyze these data?

15.37 The authors of the article "A Survey of Parent Attitudes and Practices Regarding Underage Drinking" (*Journal of Youth and Adolescence* [1995]: 315–334) conducted a telephone survey of parents with preteen and teenage children. One of the questions asked was "How effective do you think you are in talking to your children about drinking?" Responses are summarized in the accompanying 3×2 table. Using a significance level of 0.05, carry out a test to determine whether there is an association between age of children and parental response.

	Age of Children	
Response	Preteen	Teen
Very Effective	126	149
Somewhat Effective	44	41
Not at All Effective or Don't Know	51	26

15.38 The article "Regional Differences in Attitudes Toward Corporal Punishment" (*Journal of Marriage and Family* [1994]: 314–324) presents data resulting from a random sample of 978 adults. Each individual in the sample was asked whether he or she agreed with the following statement: "Sometimes it is necessary to discipline a child with a good, hard spanking." Respondents were also classified according to the region of the United States in which they lived. The resulting data are summarized in the accompanying table. Is there convincing evidence of an association between response (agree, disagree) and region of residence? Use $\alpha = 0.01$.

	Response	
Region	Agree	Disagree
Northeast	130	59
West	146	42
Midwest	211	52
South	291	47

15.39 Jail inmates can be classified into one of the following four categories according to the type of crime committed: violent crime, crime against property, drug crime, and public-order offenses. Suppose that random samples of 500 male inmates and 500 female inmates are selected, and each inmate is classified according to type of crime. The data in the accompanying table are based on summary values given in the article "Profile of Jail Inmates" (*USA Today,* April 25, 1991). You would like to know whether male and female inmates differ with respect to crime type proportions.

	Gender	
Type of Crime	Male	Femal
Violent	117	66
Property	150	160
Drug	109	168
Public–Order	124	106

a. Is this a test of homogeneity or a test of independence?
b. Test the relevant hypotheses using a significance level of 0.05.

SECTION 15.3 Avoid These Common Mistakes

Keep the following in mind when analyzing categorical data using one of the chi-square tests presented in this chapter:

1. Don't confuse tests for homogeneity with tests for independence. The hypotheses and conclusions are different for the two types of test. Tests for homogeneity are used when the individuals in each of two or more independent samples are classified according to a single categorical variable. Tests for independence are used when individuals in a *single* sample are classified according to two categorical variables.

2. As was the case for the hypothesis tests in earlier chapters, remember that you can never say you have strong support for the null hypothesis. For example, if you do not reject the null hypothesis in a chi-square test for independence, you cannot conclude that there is convincing evidence that the variables are independent. You can only say that you were not convinced that there is an association between the variables.

3. Be sure that the conditions for the chi-square test are met. *P*-values based on the chi-square distribution are only approximate, and if the large sample condition is not met, the actual *P*-value may be quite different from the approximate one based on the chi-square distribution. This can lead to incorrect conclusions. Also, for the chi-square test of homogeneity, the assumption of *independent* samples is particularly important.

4. Don't jump to conclusions about causation. Just as a strong correlation between two numerical variables does not mean that there is a cause-and-effect relationship between them, an association between two categorical variables does not imply a causal relationship.

CHAPTER ACTIVITIES

ACTIVITY 15.1 PICK A NUMBER, ANY NUMBER ...

Background: There is evidence to suggest that people are not very good random number generators. In this activity, you will investigate this phenomenon by collecting and analyzing a set of human-generated "random" digits.

For this activity, work in a group with four or five other students.

1. Each member of your group should ask 25 different people to pick a digit from 0 to 9 at random and record the responses.

2. Combine the responses you collected with those of the other group members to form a single sample. Summarize the resulting data in a one-way frequency table.

3. If people are good at picking digits at random, what would you expect for the proportion of the responses in the sample that are 0? That are 1?

4. State a null hypothesis and an alternative hypothesis that could be tested to determine whether there is evidence that the 10 digits from 0 to 9 are not selected an equal proportion of the time when people are asked to pick a digit at random.

5. Carry out the appropriate hypothesis test, and write a few sentences indicating whether or not the data support the theory that people are not good random number generators.

ACTIVITY 15.2 COLOR AND PERCEIVED TASTE

Background: Does the color of a food or beverage affect the way people perceive its taste? In this activity you will conduct an experiment to investigate this question and analyze the resulting data using a chi-square test.

You will need to recruit at least 30 subjects for this experiment, so it is advisable to work in a large group (perhaps even the entire class) to complete this activity.

Subjects for the experiment will be assigned at random to one of two groups. Each subject will be asked to taste a sample of gelatin (for example, Jell-O) and rate the taste as not very good, acceptable, or very good. Subjects assigned to the first group will be asked to taste and rate a cube of lemon-flavored gelatin. Subjects in the second group will be asked to taste and rate a cube of lemon-flavored gelatin that has been colored an unappealing color by adding food coloring to the gelatin mix before the gelatin sets.

Note: You may choose to use something other than gelatin, such as lemonade. Any food or beverage whose color can be altered using food coloring can be used. You can experiment with food colors to obtain a color that you think is particularly unappealing!

1. As a class, develop a plan for collecting the data. How will subjects be recruited? How will they be assigned to one of the two treatment groups (unaltered color, altered color)? What variables will be directly controlled, and how will you control them?

2. After the class is satisfied with the data collection plan, assign members of the class to prepare the gelatin to be used in the experiment.

3. Carry out the experiment, and summarize the resulting data in a two-way table like the one shown:

	Taste Rating		
Group	Not Very Good	Acceptable	Very Good
Unaltered Color			
Altered Color			

4. The two-way table summarizes data from two independent samples (as long as subjects were assigned *at random* to the two treatments, the samples are independent). Carry out an appropriate test to determine whether the proportions for each of the three taste rating categories are not the same when the color is altered as when the color is not altered.

ARE YOU READY TO MOVE ON? **CHAPTER 15 REVIEW EXERCISES**

All chapter learning objectives are assessed in these exercises. The learning objectives assessed in each exercise are given in parentheses.

15.40 (M1, M2, M3, M5)
The authors of the paper **"Racial Stereotypes in Children's Television Commercials"** (*Journal of Advertising Research* [2008]: 80–93) counted the number of times that characters of different ethnicities appeared in commercials aired on Philadelphia television stations, resulting in the data in the accompanying table.

	African American	Asian	Caucasian	Hispanic
Frequency	57	11	330	6

Based on the 2000 Census, the proportion of the U.S. population falling into each of these four ethnic groups are 0.177 for African American, 0.032 for Asian, 0.734 for Caucasian, and 0.057 for Hispanic. Do these data provide sufficient evidence to conclude that the proportions appearing in commercials are not the same as the census proportions? Test the relevant hypotheses using a significance level of 0.01.

15.41 (C1, P1)
The report **"Fatality Facts 2004: Bicycles"** (Insurance Institute, 2004) included the following table classifying 715 fatal bicycle accidents according to the time of day the accident occurred.

Time of Day	Number of Accidents
Midnight to 3 A.M.	38
3 A.M. to 6 A.M.	29
6 A.M. to 9 A.M.	66

(continued)

Time of Day	Number of Accidents
9 A.M. to Noon	77
Noon to 3 P.M.	99
3 P.M. to 6 P.M.	127
6 P.M. to 9 P.M.	166
9 P.M. to Midnight	113

For purposes of this exercise, assume that these 715 bicycle accidents are a random sample of fatal bicycle accidents. Do these data support the hypothesis that fatal bicycle accidents are not equally likely to occur in each of the 3-hour time periods used to construct the table? Test the relevant hypotheses using a significance level of 0.05.

15.42 (C1, P1)
Suppose a safety officer proposes that bicycle fatalities are twice as likely to occur between noon and midnight as during midnight to noon and suggests the following hypothesis: H_0: $p_1 = \frac{1}{3}$, $p_2 = \frac{2}{3}$, where p_1 is the proportion of accidents occurring between midnight and noon and p_2 is the proportion occurring between noon and midnight. Do the data given in the previous exercise provide evidence against this hypothesis, or are the data compatible with it? Justify your answer with an appropriate test.

15.43 (C1, P1)
The report referenced in the previous two exercises (**"Fatality Facts 2004: Bicycles"**) also classified 719 fatal bicycle accidents according to the month in which the accidents occurred, resulting in the accompanying table.

Month	Number of Accidents
January	38
February	32
March	43
April	59
May	78
June	74
July	98
August	85
September	64
October	66
November	42
December	40

a. To determine if some months are riskier than others, use the given data to test the following null hypothesis $H_0: p_1 = \frac{1}{12}, p_2 = \frac{1}{12}, ..., p_{12} = \frac{1}{12}$, where p_1 is the proportion of fatal bicycle accidents that occur in January, p_2 is the proportion for February, and so on. Use a significance level of 0.01.

b. The null hypothesis in Part (a) specifies that fatal accidents are equally likely to occur in any of the 12 months. But not all months have the same number of days. What null and alternative hypotheses would you test if you wanted to take differing month lengths into account? (Hint: 2004 was a leap year, with 366 days.)

c. Test the hypotheses proposed in Part (b) using a 0.05 significance level.

15.44 (M2, M4, M5)
In a study of high-achieving high school graduates, the authors of the report **"High-Achieving Seniors and the College Decision" (Lipman Hearne, October 2009)** surveyed 828 high school graduates who were considered "academic superstars" and 433 graduates who were considered "solid performers." One question on the survey asked the distance from their home to the college they attended. Assuming these two samples are random samples of academic superstars and solid performers nationwide, use the accompanying data to determine if it is reasonable to conclude that the distribution of responses over the distance from home categories is not the same for academic superstars and solid performers. Use $\alpha = 0.05$.

Student Group	Distance of College from Home (in miles)				
	Less than 40	40 to 99	100 to 199	200 to 399	400 or More
Academic Superstars	157	157	141	149	224
Solid Performers	104	95	82	65	87

15.45 (C2, P2)
The accompanying data on degree of spirituality for a random sample of natural scientists and a random sample of social scientists working at research universities appeared in the paper **"Conflict Between Religion and Science Among Academic Scientists" (***Journal for the Scientific Study of Religion*** [2009]: 276–292)**. Is there evidence that the spirituality category proportions are not the same for natural and social scientists? Test the relevant hypotheses using a significance level of 0.01.

	Degree of Spirituality			
	Very	Moderate	Slightly	Not at All
Natural Scientists	56	162	198	211
Social Scientists	56	223	243	239

15.46 (M2, M4, M5)
The authors of the paper **"Movie Character Smoking and Adolescent Smoking: Who Matters More, Good Guys or Bad Guys?" (***Pediatrics*** [2009]: 135–141)** studied characters who were depicted smoking in movies released between 2000 and 2005. The smoking characters were classified according to gender and whether the character type was positive, negative, or neutral. The resulting data are given in the accompanying table. Assume that this sample is representative of smoking movie characters. Do these data provide evidence of an association between gender and character type for movie characters who smoke? Use $\alpha = 0.05$.

Gender	Character Type		
	Positive	Negative	Neutral
Male	255	106	130
Female	85	12	49

15.47 (M2, M4, M5)
The paper **"Overweight Among Low-Income Preschool Children Associated with the Consumption of Sweet Drinks" (***Pediatrics*** [2005]: 223–229)** described a study of children who were underweight or normal weight at age 2. Children in the sample were classified according to the number of sweet drinks consumed per day and whether or not the child was overweight one year after the study began. Is there evidence of an association between whether or not children are overweight after one year and the number of sweet drinks consumed? Assume that the sample of children in this study is representative of 2- to 3-year-old children, Test the appropriate hypotheses using a 0.05 significance level.

		Overweight?	
		Yes	No
Number of Sweet Drinks Consumed per Day	0	22	930
	1	73	2074
	2	56	1681
	3 or More	102	3390

15.48 (C2, P3)

The 2006 Expedia Vacation Deprivation Survey (*Ipsos Insight, May 18, 2006*) described a study of working adults in Canada. Each person in a random sample was classified according to gender and the number of vacation days he or she usually took each year. The resulting data are summarized in the given table. Is it reasonable to conclude that there is an association between gender and the number of vacation days taken? To what population would it be reasonable to generalize this conclusion?

Days of Vacation	Gender	
	Male	Female
None	51	42
1–5	21	25
6–10	67	79
11–15	111	94
16–20	71	70
21–25	82	58
More than 25	118	79

15.49 (C2, P2)

The following passage is from the paper "**Gender Differences in Food Selections of Students at a Historically Black College and University**" (*College Student Journal* [2009]: 800–806):

> Also significant was the proportion of males and their water consumption (8 oz. servings) compared to females ($X^2 = 8.166$, $P = .086$). Males came closest to meeting recommended daily water intake (64 oz. or more) than females (29.8% vs. 20.9%).

This statement was based on carrying out a chi-square test of homogeneity using data in a two-way table where rows corresponded to gender (male, female) and columns corresponded to number of servings of water consumed per day, with categories none, one, two to three, four to five, and six or more.

a. What hypotheses did the researchers test? What is the number of degrees of freedom associated with the reported value of the X^2 statistic?

b. The researchers based their statement on a test with a significance level of 0.10. Would they have reached the same conclusion if a significance level of 0.05 had been used? Explain.

15.50 (C2, P2)

The paper referenced in the previous exercise also included the accompanying data on how often students said they had consumed fried potatoes (French fries or potato chips) in the past week.

		Number of Times Consumed Fried Potatoes in the Past Week					
		0	1 to 3	4 to 6	7 to 13	14 to 20	21 or more
Gender	Male	2	10	15	12	6	3
	Female	15	15	10	20	19	12

Use the accompanying Minitab output to carry out a chi-square test of homogeneity. Do you agree with the authors' conclusion that there was a significant difference in consumption of fried potatoes for males and females? Explain.

Expected counts are printed below observed counts

Chi-Square contributions are printed below expected counts

	0	1-3	4-6	7-13	14-20	21 or more	Total
M	2	10	15	12	6	3	48
	5.87	8.63	8.63	11.05	8.63	5.18	
	2.552	0.216	4.696	0.082	0.803	0.917	
F	15	15	10	20	19	12	91
	11.13	16.37	16.37	20.95	16.37	9.82	
	1.346	0.114	2.477	0.043	0.424	0.484	
Total	17	25	25	32	25	15	139

Chi-Sq = 14.153, DF = 5, P-Value = 0.015

15.51 (C2, P3)

The press release titled "**Nap Time**" (pewresearch.org, July 2009) described results from a nationally representative survey of 1,488 adult Americans. The survey asked several demographic questions (such as gender, age, and income) and also included a question asking respondents if they had taken a nap in the past 24 hours. The press release stated that 38% of the men surveyed and 31% of the women surveyed reported that they had napped in the past 24 hours. For purposes of this exercise, suppose that men and women were equally represented in the sample.

a. Use the given information to fill in observed cell counts for the following table:

	Napped	Did Not Nap	Row Total
Men			744
Women			744

b. Use the data in the table from Part (a) to carry out a hypothesis test to determine if there is an association between gender and napping.

c. The press release states that more men than women nap. Although this is true for the people in the sample, based on the result of your test in Part (b), is it reasonable to conclude that this holds for adult Americans in general? Explain.

TECHNOLOGY NOTES

X^2 Goodness-of-Fit Test

TI-83/84

Not all of the TI-family calculators have the chi-square goodness-of-fit test built in. The directions here are for the TI-84 +. It is possible to download chi square goodness-of-fit programs for others in the TI family.

1. Clear the lists, L1 and L2. (You may choose other lists for your counts.)
2. Enter the observed values into L1.
3. Enter the expected values into L2.
4. Press **STAT,** select **D: χ^2 statistic,** and press **ENTER.**
5. You will be prompted for the location of the observed frequencies (L1), the expected frequencies (L2), and the number of degrees of freedom (df).
6. After supplying the information in Step 5, highlight **Calculate** and press **ENTER.**

The calculator will respond with the χ^2 statistic, the *P*-value, df, and values in a dedicated list named CNTRB. CNTRB contains the values (one for each cell) that are added to get the χ^2 statistic.

7. To access the list, press 2nd, then **LIST.** Then scroll down to **CNTRB** and press **ENTER.**

TI-Nspire

1. Enter the observed data into a data list (In order to access data lists select the spreadsheet option and press **enter**)

   Note: Be sure to title the lists by selecting the top row of the column and typing a title.

2. Enter the expected data into a data list
3. Press **menu** and select **4:Statistics** then **4:Stat Tests** then **7: χ^2 GOF...** and press **enter**
4. For **Observed List** select the title of the list that contains the observed data from the drop-down menu
5. For **Expected List** select the title of the list that contains the expected data from the drop-down menu
6. For **Deg of Freedom** input the degrees of freedom for this test
7. Press **OK**

JMP

1. Enter each category name into the first column
2. Enter the count for each category into the second column

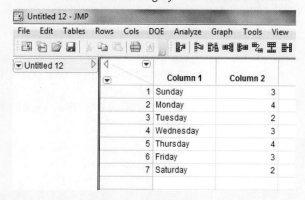

3. Click **Analyze** then select **Distribution**
4. Click and drag the name for the first column from the box under **Select Columns** to the box next to **Y, Columns**
5. Click and drag the name for the second column from the box under **Select Columns** to the box next to **Freq**
6. Click **OK**
7. Click the red arrow next to the column name and select **Test Probabilities**
8. Under **Test Probabilities** input the hypothesized probabilities for each category

Test Probabilities		
Level	**Estim Prob**	**Hypoth Prob**
Sunday	0.14286	0.14290
Monday	0.19048	0.14290
Tuesday	0.09524	0.14290
Wednesday	0.14286	0.14290
Thursday	0.19048	0.14290
Friday	0.14286	0.14290
Saturday	0.09524	0.14290

Click then Enter Hypothesized Probabilities.

Choose rescaling method to sum probabilities to 1.
○ Fix omitted at estimated values, rescale hypothesis
◉ Fix hypothesized values, rescale omitted

[Done]   [Help]

9. Click **Done**

Note: The test statistic and *P*-value for the chi-squared test will appear in the row called **Pearson**.

Minitab

MINITAB Student Version 14 does not have the functionality to produce a x^2 goodness-of-fit test.

SPSS

1. Input the observed data into one column
2. Click **Analyze** then click **Nonparametric Tests** then click **One Sample...**
3. Click the **Settings** tab
4. Select the radio button next to **Customize tests**
5. Check the box next to **Compare observed probabilities to hypothesized (Chi-Square test)**
6. Click the **Options...** button
7. Select the appropriate option (to test with equal probabilities for each category or to input expected probabilities manually)
8. Once you have selected the appropriate option and input expected probabilities if necessary, click **OK**
9. Click **Run**

Excel

Excel does not have the functionality to produce the χ^2 goodness-of-fit test automatically. However, you can use Excel to find the *P*-value for this test once you have found the value of the test statistic by using the following steps.

1. Click on an empty cell
2. Click **Formulas**
3. Click **Insert Function**
4. Select **Statistical** from the drop-down box for category
5. Select **CHIDIST** and press **OK**
6. In the box next to **X**, type the value of the test statistic
7. In the box next to **Deg_freedom** type the value for the degrees of freedom
8. Click **OK**

Note: This outputs the value for P(X ≥ x).

X^2 Tests for Independence and Homogeneity

TI-83/84

1. Input the observed contingency table into matrix **A** (To access and edit matrices, press **2nd** then **x⁻¹**, then highlight **EDIT** and press **ENTER**. Then highlight **[A]** and press **ENTER**. Type the value for the number of rows and press **ENTER**; type the value for the number of columns and press **ENTER**. Type the data values into the matrix.)

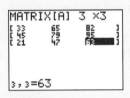

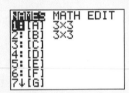

2. Press **STAT**
3. Highlight **TESTS**
4. Highlight **χ²-Test...** and press **ENTER**
5. Highlight **Calculate** and press **ENTER**

TI-Nspire

1. Enter the Calculator Scratchpad
2. Press the **menu** key then select **7:Matrix & Vector** then select **1:Create** then select **1:Matrix...** and press **enter**
3. Next to **Number of rows** enter the number of rows in the contingency table (do not include title rows or total rows)
4. Next to **Number of columns** enter the number of columns in the contingency table (do not include title columns or total columns)
5. Press **OK**
6. Input the values into the matrix (pressing tab after entering each value and press enter when you are finished)
7. Press **ctrl** then press **var**
8. Type in **amat** and press **enter**

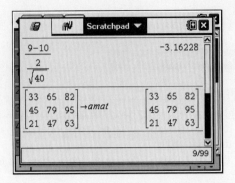

9. Press the **menu** key then select **6:Statistics** then **7:Stat tests** then **8: χ² 2-way Test...**
10. For **Observed Matrix**, select amat from the drop-down list
11. Press **OK**

JMP

1. Input the data table into the JMP data table
2. Click **Analyze** then select **Fit Y by X**
3. Click and drag the first column name from the box under **Select Columns** to the box next to **Y, Response**

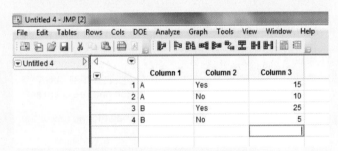

4. Click and drag the second column name from the box under **Select Columns** to the box next to **X, Factor**
5. Click and drag the third column name from the box under **Select Columns** to the box next to **Freq**
6. Click **OK**

Minitab

1. Input the data table into MINITAB

↓	C1-T	C2	C3	C4	C5
		Male	Female		
1	Small	145	165		
2	Medium	120	145		
3	Large	300	75		
4	XL	45	6		
5					
6					

2. Click **Stat** then **Tables** then **Chi-Square Test (Table in Worksheet)**
3. Select all columns containing data (do NOT select the column containing the row labels)
4. Click **OK**

Note: This output returns expected cell counts as well as the chi-square test statistic and *P*-value.

SPSS

1. Enter the row variable data into one column
2. Enter the column variable data into a second column

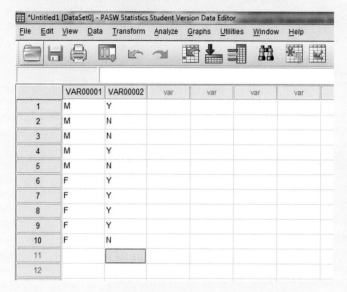

3. Click **Analyze** then click **Descriptive Statistics** then click **Crosstabs...**
4. Select the name of the row variable and press the arrow to move the variable to the box under **Row(s):**
5. Select the name of the column variable and press the arrow to move the variable to the box under **Column(s):**
6. Click **Statistics...**
7. Check the box next to Chi-square
8. Click **Continue**
9. Click **Cells...**
10. Check the box next to Expected
11. Click **Continue**
12. Click **OK**

Note: The *P*-value for this test can be found in the **Chi-Square Tests** table in the Pearson Chi-Square row.

Excel

1. Input the observed contingency table
2. Input the expected table
3. Click on an empty cell
4. Click **Formulas**
5. Click **Insert Function**
6. Select **Statistical** from the drop-down box for category
7. Select **CHITEST** and press **OK**
8. Click in the box next to **Actual_range** and select the data values from the actual table (do NOT select column or row labels or totals)
9. Click in the box next to **Expected_range** and select the data values from the expected table (do NOT select column or row labels or totals)
10. Click **OK**

Image for Head "Excel" on Page 729

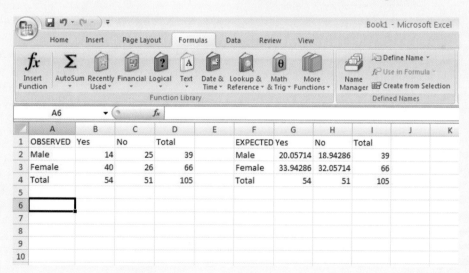

Image for Head "Excel" on Page 729

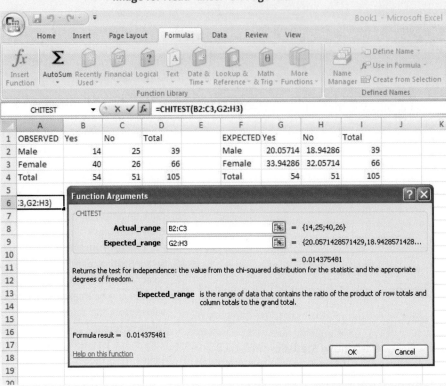

AP* Review Questions for Chapter 15

1. Each person in a random sample of 60 students living in the dorms at a large university was asked whether he or she preferred to study in the dorm room, in the dorm's study lounge, or at the university library. The resulting data is summarized in the table below.

Dorm Room	Study Lounge	Library
18	12	30

   In a hypothesis test to determine if there is evidence that the three locations are not equally preferred by students at the university who live in the dorms, what is the expected count for the library category?

   (A) 10
   (B) 12
   (C) 20
   (D) 30
   (E) 40

2. Each person in a random sample of 100 students at a large university was asked whether he or she preferred to attend classes in the morning, the afternoon, or in the evening. The resulting data is summarized in the table below.

Morning	Afternoon	Evening
28	48	24

   A hypothesis test will be carried out to determine if there is evidence that the three time periods are not equally preferred by students at the university. The conditions for the chi-square test were met. When the null hypothesis is true, the test statistic will have a chi-square distribution with how many degrees of freedom (df)?

   (A) 2
   (B) 3
   (C) 99
   (D) 100
   (E) Can't tell from the information given.

3. Each person in a random sample of university students was asked which of five flavors of ice cream he or she preferred. The resulting data was used to test the null hypothesis $H_0: p_1 = 0.2\, p_2 = 0.2 \ldots p_5 = 0.2$. The P-value associated with this test was 0.067. If a significance level of 0.05 is used for the test, which of the following is an appropriate conclusion?

   (A) Fail to reject H_0, there is convincing evidence that there is no preference for one flavor of ice cream over another.

   (B) Fail to reject H_0, there is not convincing evidence that there is a preference for one flavor of ice cream over another.

   (C) Reject H_0, there is convincing evidence that there is no preference for one flavor of ice cream over another.

   (D) Reject H_0, there is convincing evidence that there is a preference for one flavor of ice cream over another.

   (E) Based on the information given, it is not possible to determine whether H_0 should be rejected or not.

4. Each person in a random sample of 300 female students and each person in a random sample of 300 male students at a university was asked whether or not they always wear a seat belt when driving. Which of the following would be an appropriate null hypothesis for a chi-square test based on these data?

   (A) The mean seat belt use for males is equal to the mean seat belt use for females.

   (B) The proportion of students who always wear a seat belt and the proportion who do not always wear a seat belt is the same for males and for females.

   (C) The proportion of students who always wear a seat belt and the proportion who do not always wear a seat belt is not the same.

   (D) There is no association between gender and seat belt use.

   (E) There is an association between gender and seat belt use.

5. One hundred students at a large public university were selected at random. Each student in the sample was asked to indicate how often he or she used on-campus recreational facilities by choosing between the following categories: never, occasionally, and regularly. Whether the student lived on campus or off campus was also noted. Which of the following would be an appropriate null hypothesis for a chi-square test based on these data?

   (A) The proportion of students who live on campus is the same for all three recreational use categories.

   (B) The proportion of students in each of the three recreational facility use categories is the same for students who live on-campus and those who live off-campus.

   (C) The proportion of students in each of the three recreational facility use categories is not the same for students who live on-campus and those who live off-campus.

   (D) There is no association between where a student lives and use of recreational facilities.

   (E) There is an association between where a student lives and use of recreational facilities.

6. Fifty female students at a large university were selected at random and 50 male students were also selected at random. Each student in the two samples was asked to indicate how often they watched reality TV shows by choosing between the following viewing categories: never, once or twice per week, more than twice a week. Which of the following would be an appropriate null hypothesis for a chi-square test based on these data?

   (A) The mean time spent watching TV is the same for males and females.
   (B) The proportion of male students is equal to the proportion of female students.
   (C) The proportion of students in each of the three viewing categories is the same for males and females.
   (D) There is no association between gender and viewing category.
   (E) There is an association between gender and viewing category.

7. Fifty female students at a large university were selected at random and 50 male students were also selected at random. Each student in the two samples was asked to indicate how often they watched reality TV shows by choosing between the following viewing categories: never, once or twice per week, more than twice a week. If the conditions for the chi-square test are met, the test statistic will have a chi-square distribution with how many degrees of freedom (df)?

   (A) 2
   (B) 3
   (C) 5
   (D) 6
   (E) 98

8. Each person in a random sample of 200 males and in a random sample of 100 females was asked to indicate political affiliation by choosing between the following categories: Democrat, Republican, Libertarian, Green Party, Independent, Other. Are the sample sizes large enough to support using a chi-square test to determine if the proportion falling into each of the political affiliation categories is different for males and females?

   (A) Yes, both samples are large enough.
   (B) No, the sample size for the sample of males is large enough, but the sample size for the sample of females is not large enough.
   (C) No, the sample size for the sample of females is large enough, but the sample size for the sample of males is not large enough.
   (D) No, both samples are too small.
   (E) There is not enough information given to determine if the sample sizes are large enough.

9. One hundred compact car owners and 100 full-size car owners were selected at random. Each person in the two samples was asked to indicate whether he or she was not concerned, somewhat concerned, or very concerned about recent increases in the price of gas. It was determined that the conditions required for the chi-square test of homogeneity were met. A chi-square test resulted in a P-value of 0.31. If a 0.05 significance level is used, which of the following is a reasonable conclusion?

   (A) Fail to reject H_0, there is not convincing evidence that the proportions in each of the concern categories are different for compact car owners and full-size car owners.
   (B) Fail to reject H_0, there is convincing evidence that the proportions in each of the concern categories are the same for compact car owners and full-size car owners.
   (C) Reject H_0, there is convincing evidence that the proportion in each of the concern categories is the same for compact car owners and full-size car owners.
   (D) Reject H_0, there is convincing evidence that the proportions in each of the concern categories are not the same for compact car owners and full-size car owners.
   (E) Cannot determine if H_0 should be rejected based on the given information.

10. A researcher classified each individual in a random sample of 100 community college students into one of four categories according to his or her ultimate degree goal (Associate, Bachelors, Masters, or Ph.D./Professional) and also by gender. If the assumptions for the chi-square test are satisfied, the test statistic will have a chi-square distribution with how many degrees of freedom (df)?

    (A) 2
    (B) 3
    (C) 4
    (D) 8
    (E) 99

11. A local newspaper asked each person in a random sample of 40 subscribers which section of the paper he or she read first. The resulting data are shown in the accompanying table.

Section	Frequency
National News	12
Local News	8
Sports	16
Business	4

Which of the following test procedures would be appropriate for determining if the data provides evidence that the proportions choosing each of the sections are not the same?

(A) One-sample t-test
(B) Two-sample t-test
(C) Chi-square goodness-of-fit test
(D) Chi-square test of association
(E) Chi-square test of homogeneity

12. Which of the following condition(s) must be met in order for the chi-square goodness of fit test to be appropriate?

I.   Population distribution is approximately normal
II. $n \geq 30$
III. Sample size is large enough so that all expected counts are at least 5

(A) I only
(B) II only
(C) III only
(D) I and II only
(E) I and III only

13. Which of the following is the number of degrees of freedom (df) for testing independence using bivariate categorical data that has been summarized in a 4×5 two-way table?

(A) 9
(B) 12
(C) 18
(D) 19
(E) 20

14. In a study of the foraging behavior of woodpeckers, males and females were observed landing on trees. Below is a partial table summarizing the data.

Landing Location	Males	Females	Total
Branches	13	*	14
Lower Trunk	6	*	40
Upper Trunk	120	*	*
Total	139	*	246

If the landing location and gender are independent, then the expected number of *males* landing on the *upper trunks* of trees is in which of the following ranges?

(A) 1 to < 10
(B) 10 to < 50
(C) 50 to < 100
(D) 100 to < 125
(E) 125 to < 150

15. The P-value for a chi-squared goodness-of-fit test is

(A) the area to the left of the calculated value of X^2 under the appropriate chi-squared curve.
(B) the area to the right of the calculated value of X^2 under the appropriate chi-squared curve.
(C) twice the area to the left of the calculated value of X^2 under the appropriate chi-squared curve.
(D) twice the area to the right of the calculated value of X^2 under the appropriate chi-squared curve.
(E) Cannot be determined in general.

16

Understanding Relationships—Numerical Data Part 2

Preview

Chapter Learning Objectives

16.1 The Simple Linear Regression Model

16.2 Inferences Concerning the Slope of the Population Regression Line

16.3 Checking Model Adequacy

Are You Ready to Move On? Chapter 16 Review Exercises

Technology Notes

AP* Review Questions for Chapter 16

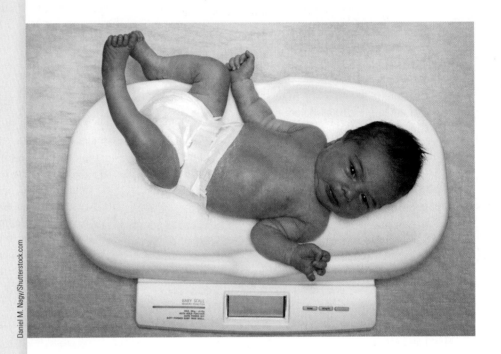

Daniel M. Nagy/Shutterstock.com

PREVIEW

In Chapter 4, you learned how to describe relationships between two numerical variables. When the relationship was judged to be linear you found the equation of the least squares regression line and assessed the quality of the fit using the scatterplot, the residual plot, and the values of the coefficient of determination (r^2) and the standard deviation about the least squares line (s_e). In this chapter you will learn how to make inferences about the slope of the population regression line.

CHAPTER LEARNING OBJECTIVES

Conceptual Understanding

After completing this chapter, you should be able to

C1 Understand how probabilistic and deterministic models differ.

C2 Understand that the simple linear regression model provides a basis for making inferences about linear relationships.

Mastering the Mechanics

After completing this chapter, you should be able to

M1 Interpret the parameters of the simple linear regression model in context.

M2 Use scatterplots, residual plots, and normal probability plots to assess the credibility of the assumptions of the simple linear regression model.

M3 Know the conditions for appropriate use of methods for making inferences about β.

M4 Compute the margin of error when the sample slope b is used to estimate a population slope β.

M5 Use the five-step process for estimation problems (EMC³) and computer output to construct and interpret a confidence interval estimate for the slope of a population regression line.

M6 Use the five-step process (HMC³) to test hypotheses about the slope of the population regression line.

M7 Use graphs to identify potential outliers and influential points.

Putting It into Practice

After completing this chapter, you should be able to

P1 Interpret a confidence interval for a population slope in context.

P2 Carry out the model utility test and interpret the result in context.

PREVIEW EXAMPLE   Premature Babies

Babies born prematurely (before the 37th week of pregnancy) often have low birth weights. Is a low birth weight related to factors that affect brain function? The authors of the paper **"Intrauterine Growth Restriction Affects the Preterm Infant's Hippocampus"**(*Pediatric Research* **[2008]: 438-43)** hoped to use data from a study of premature babies to answer this question. They measured x = birth weight (in grams) and y = hippocampus volume (in mL) for 26 premature babies. The hippocampus is a part of the brain that is important in the development of both short- and long-term memory. The sample correlation coefficient for their data is $r = 0.4722$ and the equation of the least squares regression line is $\hat{y} = 1.67 + 0.0026x$. The pattern in the scatterplot (Figure 16.1) suggests there may be a positive linear relationship. However, the correlation coefficient is not very large, and the value of the slope is close to zero. Could the pattern observed in the scatterplot—and the nonzero slope—be plausibly explained by chance? That is, is it plausible that there is no relationship between birth weight and hippocampus volume in the population of all premature babies? Or does the sample provide convincing evidence of a linear relationship between these two variables? If there is evidence of a meaningful relationship between these two variables, the regression line could be used to predict the hippocampus volume. If the predicted volume was sufficiently small, early cognitive therapy could be recommended. On the other hand, if there is no meaningful relationship between these variables, low birth weight should not automatically trigger potentially expensive therapy.

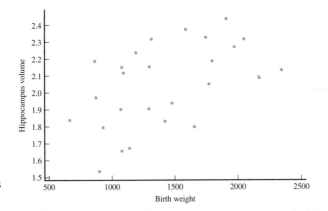

FIGURE 16.1
Scatterplot of birth weight versus hippocampus volume.

In this chapter, you will learn methods that will help you determine if there is a real and useful linear relationship between two variables or if the pattern in the data could be simply due to chance differences that occur when a sample is selected from a population.

The Simple Linear Regression Model

A *deterministic relationship* between two variables x and y is one in which the value of y is completely determined by the value of the independent variable x. A deterministic relationship can be described, or "modeled," using mathematical notation, such as $y = f(x)$ where $f(x)$ is a particular function of x. This relationship is deterministic in the sense that the value of the independent variable is all that is needed to determine the value of the dependent variable. For example, you might convert $x =$ temperature in degrees centigrade to $y =$ temperature in degrees Fahrenheit using $y = f(x)$, where $f(x) = \frac{9}{5}x + 32$. Once the centigrade temperature is known, the Fahrenheit temperature is completely determined. Or you might determine $y =$ amount of money in a savings account after x years, using the compound interest formula, $y = P\left(1 + \frac{r}{n}\right)^{nx}$, where P is the principal (the amount of money deposited), r is the interest rate, and n is the number of times each year the interest is compounded. The number of years you leave the principal in the bank determines the amount in the account.

In many situations the variables of interest are not deterministically related. For example, the value of $y =$ first-year college grade point average is not determined solely by $x =$ high school grade point average, and $y =$ crop yield is determined partly by factors other than $x =$ amount of fertilizer used. The relationship between two variables, x and y, that are not deterministically related can be described by extending the deterministic model to specify a probabilistic model. The general form of a **probabilistic model** allows y to be larger or smaller than $f(x)$ by a random amount e. The model equation for a probabilistic model has the form

$$y = \text{deterministic function of } x + \text{random deviation}$$
$$= f(x) + e$$

In a scatterplot of y versus x, some of the data points will fall above the graph of $f(x)$ and some will fall below. Thinking geometrically, if $e > 0$, the corresponding point in the scatterplot will lie above the graph of the function $y = f(x)$. If $e < 0$, the corresponding point will fall below the graph of $f(x)$.

For example, consider the probabilistic model

$$y = \underbrace{50 - 10x + x^2}_{f(x)} + e$$

The graph of the function $y = 50 - 10x + x^2$ is shown as the orange curve in Figure 16.2. The observed point $(4, 30)$ is also shown in the figure. Because $f(4) = 50 - 10(4) + 4^2 =$

$50 - 40 + 16 = 26$ for this point, you can write $y = f(x) + e$, where $e = 4$. The point $(4, 30)$ falls 4 above the graph of the function, $y = 50 - 10x + x^2$.

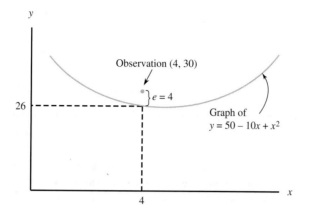

FIGURE 16.2
A deviation from the deterministic part of a probabilistic model.

Simple Linear Regression Model

The simple linear regression model is a special case of the general probabilistic model in which the deterministic function, $f(x)$, is linear (so its graph is a straight line).

DEFINITION

The **simple linear regression model** assumes that there is a line with vertical or y intercept α and slope β, called the **population regression line.** When a value of the independent variable x is fixed and an observation on the dependent variable y is made,

$$y = \alpha + \beta x + e$$

Without the random deviation e, all observed (x, y) points would fall exactly on the population regression line. The inclusion of e in the model equation recognizes that points will deviate from the line by a random amount.

Figure 16.3 shows two observations in relation to the population regression line.

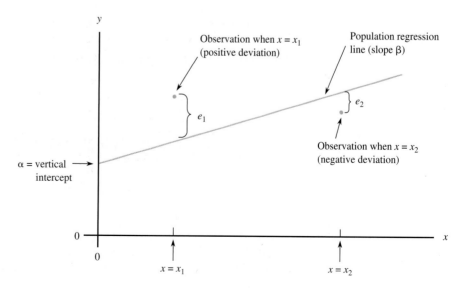

FIGURE 16.3
Two observations and deviations from the population regression line.

Before you actually observe a value of y for any particular value of x, you are uncertain about the value of e. It could be negative, positive, or even 0. Also, e might be quite large in magnitude (resulting in a point far from the population regression line) or quite small (resulting in a point very close to the line). The simple linear regression model makes some assumptions about the distribution of e at any particular x value in the population.

BASIC ASSUMPTIONS OF THE SIMPLE LINEAR REGRESSION MODEL

1. The distribution of e at any particular x value has mean value 0. That is, $\mu_e = 0$.
2. The standard deviation of e (which describes the spread of its distribution) is the same for any particular value of x. This standard deviation is denoted by σ_e.
3. The distribution of e at any particular x value is normal.
4. The random deviations $e_1, e_2, ..., e_n$ associated with different observations are independent of one another.

The simple linear regression model assumptions about the variability in the values of e in the population imply that there is also variability in the y values observed at any particular value of x. Consider y when x has some fixed value x^*, so that

$$y = \alpha + \beta x^* + e.$$

Because α and β are fixed (they are unknown population values), $\alpha + \beta x^*$ is also a fixed number. The sum of a fixed number and a normally distributed variable (e) is also a normally distributed variable (the bell-shaped curve is simply shifted), so y itself has a normal distribution. Furthermore, $\mu_e = 0$ implies that the mean value of y is $\alpha + \beta x^*$, the height of the population regression line for the value $x = x^*$. Finally, because there is no variability in the fixed number $\alpha + \beta x^*$, the standard deviation of y is the same as the standard deviation of e. These properties are summarized in the following box.

At any fixed value x^*, y has a normal distribution, with

$$\begin{pmatrix} \text{mean } y \text{ value} \\ \text{for } x^* \end{pmatrix} = \begin{pmatrix} \text{height of the population} \\ \text{regression line above } x^* \end{pmatrix} = \alpha + \beta x^*$$

and

$$\text{standard deviation of } y \text{ for a fixed value } x^* = \sigma_e$$

The slope β of the population regression line is the *mean* or *expected* change in y associated with a 1-unit increase in x. The y intercept α is the height of the population line when $x = 0$.

The value of σ_e determines how much the (x, y) observations deviate vertically from the population line; when σ_e is small, most observations will be close to the line, but when σ_e is large, the observations will tend to deviate more from the line.

The key features of the model are illustrated in Figures 16.4 and 16.5. Notice that the three normal curves in Figure 16.4 have identical spreads. This is a consequence of σ_e being the same at any value of x, which implies that the variability in the y values at a particular value of x is constant—the variability does not depend on the value of x.

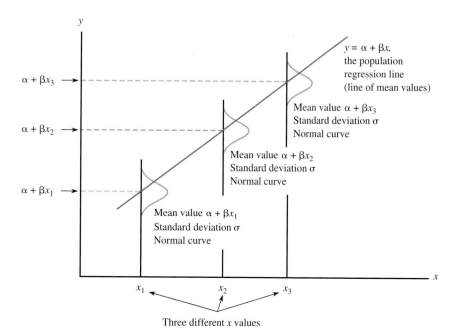

FIGURE 16.4
Illustration of the simple linear regression model.

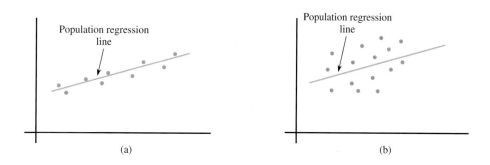

FIGURE 16.5
The simple linear regression model: (a) small σ_e; (b) large σ_e

---

Example 16.1   Stand on Your Head to Lose Weight?

The authors of the article **"On Weight Loss by Wrestlers Who Have Been Standing on Their Heads"** (paper presented at the Sixth International Conference on Statistics, Combinatorics, and Related Areas, Forum for Interdisciplinary Mathematics, 1999, with the data also appearing in *A Quick Course in Statistical Process Control*, Mick Norton, 2005) state that "amateur wrestlers who are overweight near the end of the weight certification period, but just barely so, have been known to stand on their heads for a minute or two, get on their feet, step back on the scale, and establish that they are in the desired weight class. Using a headstand as the method of last resort has become a fairly common practice in amateur wrestling."

Does this really work? Data were collected in an experiment where weight loss was recorded for each wrestler after exercising for 15 minutes and then doing a headstand for 1 minute 45 sec. Based on these data, the authors of the article concluded that there was in fact a demonstrable weight loss that was greater than that for a control group that exercised for 15 minutes but did not do the headstand. (The authors give a plausible explanation for why this might be the case based on the way blood and other body fluids collect in the head during the headstand and the effect of weighing while these fluids are draining immediately after standing.) The authors also concluded that a simple linear regression model was a reasonable way to describe the relationship between the variables

$$y = \text{weight loss (in pounds)}$$

and

$$x = \text{body weight prior to exercise and headstand (in pounds)}$$

Suppose that the actual model equation has $\alpha = 0$, $\beta = 0.001$, and $\sigma_e = 0.09$ (these values are consistent with the findings in the article). The population regression line is shown in Figure 16.6.

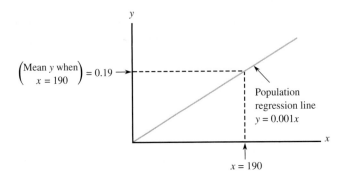

FIGURE 16.6
The population regression line for Example 16.1

If the distribution of the random errors at any fixed weight (x value) is normal, then the variable y = weight loss is normally distributed with

$$\mu_y = 0 + 0.001x$$
$$\sigma_y = 0.09$$

For example, when $x = 190$ (corresponding to a 190-pound wrestler), weight loss has mean value

$$\mu_y = 0 + 0.001(190) = 0.19 \text{ pounds}$$

Because the standard deviation of y is $\sigma_y = 0.09$, the interval $0.19 \pm 2(0.09) = (0.01, 0.37)$ includes y values that are within 2 standard deviations of the mean value for y when $x = 190$. Roughly 95% of the weight loss observations made for 190-lb wrestlers will be in this range. The slope $\beta = 0.001$ can be interpreted as the mean change in weight associated with each additional pound of body weight. ∎

More insight into model properties can be gained by thinking of the population of all (x, y) pairs as consisting of many smaller subpopulations. Each subpopulation contains pairs for which x has a fixed value. Suppose, for example, that in a large population of college students the variables

$$x = \text{grade point average in major courses}$$

and

$$y = \text{starting salary after graduation}$$

are related according to the simple linear regression model. Then you can think about the subpopulation of all pairs with $x = 3.20$ (corresponding to all students with a grade point average of 3.20 in major courses), the subpopulation of all pairs having $x = 2.75$, and so on. The model assumes that for each of these subpopulations, y is normally distributed with the same standard deviation, and that the *mean y value* (rather than y itself) is linearly related to x.

In practice, the judgment of whether the simple linear regression model is appropriate—that is the judgments about the credibility of the assumptions underlying the linear model—must be based on knowledge of how the data were collected, as well as an inspection of various plots of the data and the residuals. The sample observations should be independent of one another, which will be the case if the data are from a random sample. In addition, the scatterplot should show a linear rather than a curved pattern, and the vertical spread of points should be very similar throughout the range of x values. Figure 16.7 shows plots with three different patterns; only the first pattern is consistent with the simple linear regression model assumptions.

FIGURE 16.7

Some commonly encountered patterns in scatter plots: (a) Consistent with the simple linear regression model; (b) Suggests a nonlinear probabilistic model; (c) Suggests that variability in *y* changes with *x*.

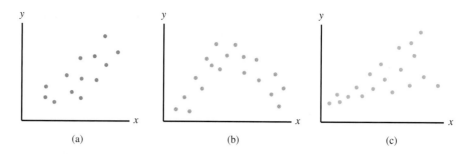

(a)   (b)   (c)

Estimating the Population Regression Line

In Section 16.3, you will see how to check whether the basic assumptions of the simple linear regression model are reasonable. When this is the case, the values of α and β (*y* intercept and slope of the population regression line) can be estimated from sample data.

The estimates of α and β are denoted by a and b, respectively. These estimates are the values of the intercept and slope of the least squares regression line. Recall that that the least squares regression line is the line for which the sum of squared vertical deviations of points in the scatterplot from the line is smaller than for any other line.

> The estimates of the slope and the *y* intercept of the population regression line are the slope and *y* intercept, respectively, of the least squares line. That is,
>
> $$b = \text{estimate of } \beta = \frac{\Sigma(x - \bar{x})(y - \bar{y})}{\Sigma(x - \bar{x})^2}$$
>
> $$a = \text{estimate of } \alpha = \bar{y} - b\bar{x}$$
>
> The values of a and b are usually obtained using statistical software or a graphing calculator. If the slope and intercept are calculated by hand, you can use the following computational formula:
>
> $$b = \frac{\Sigma xy - \dfrac{(\Sigma x)(\Sigma y)}{n}}{\Sigma x^2 - \dfrac{(\Sigma x)^2}{n}}$$
>
> The estimated regression line is the familiar least squares line
>
> $$\hat{y} = a + bx$$
>
> Let x^* denote a specified value of the independent variable x. Then $a + bx^*$ has two different interpretations:
>
> 1. It is a point estimate of the mean *y* value when $x = x^*$.
> 2. It is a point prediction of an individual *y* value to be observed when $x = x^*$.

Example 16.2   Mother's Age and Baby's Birth Weight

Medical researchers have noted that adolescent females are much more likely to deliver low-birth-weight babies than are adult females. (Low birth weight in humans is generally defined as a weight below 2,500 grams) Because low-birth-weight babies have higher mortality rates, a number of studies have examined the relationship between birth weight and mother's age for babies born to young mothers.

One such study is described in the article **"Body Size and Intelligence in 6-Year-Olds: Are Offspring of Teenage Mothers at Risk?"** (*Maternal and Child Health Journal* [2009]: 847-856). The following data on

$$x = \text{maternal age (in years)}$$

and

$$y = \text{birth weight of baby (in grams)}$$

are consistent with summary values given in the article and also with data published by the National Center for Health Statistics.

Observation

	1	2	3	4	5	6	7	8	9	10
x	15	17	18	15	16	19	17	16	18	19
y	2,289	3,393	3,271	2,648	2,897	3,327	2,970	2,535	3,138	3,573

A scatterplot of the data is given in Figure 16.8. The scatterplot shows a linear pattern, and the spread in the y values appears to be similar across the range of x values. This supports the appropriateness of the simple linear regression model.

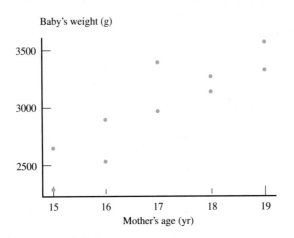

FIGURE 16.8
Scatterplot of birth weight versus maternal age for Example 16.2.

For these data, the equation of the estimated regression line was found using statistical software, resulting in

$$\hat{y} = a + bx = -1,163.45 + 245.15x$$

An estimate of the mean birth weight of babies born to 18-year-old mothers results from substituting $x = 18$ into the estimated equation:

$$\text{estimated mean } y \text{ for 18-year-old mothers} = a + bx$$
$$= -1,163.45 + 245.15(18)$$
$$= 3,249.25 \text{ grams}$$

Similarly, you would predict the birth weight of a baby to be born to a particular 18-year-old mother to be

$$\hat{y} = \text{predicted } y \text{ value when } x = 18$$
$$= a + b(18)$$
$$= 3,249.25 \text{ grams}$$

The estimate of the mean weight and the prediction of an individual baby weight are identical, because the same x value was used in each calculation. However, their interpretations differ. One is the prediction of the weight of a single baby whose mother is 18, whereas the other is an estimate of the mean weight of *all* babies born to 18-year-old mothers.

In Example 16.2, the x values in the sample ranged from 15 to 19. The estimated regression equation should not be used to make an estimate or prediction for any x value much outside this range. Without sample data for such values, or some clear theoretical reason for expecting the relationship to be linear outside the observed range of x values, you have no reason to believe that the estimated linear relationship continues outside the range from 15 to 19. Making predictions outside this range can be misleading, and statisticians refer to this as the **danger of extrapolation**.

Estimating σ_e^2 and σ_e

The value of σ_e determines the extent to which observed points (x, y) tend to fall close to or far away from the population regression line. A point estimate of σ_e is based on

$$SSResid = \Sigma(y - \hat{y})^2$$

where $\hat{y}_1 = a + bx_1, \cdots, \hat{y}_n = a + bx_n$ are the fitted or predicted y values and the residuals are $y_1 - \hat{y}_1, \cdots y_n - \hat{y}_n$. SSResid is a measure of the extent to which the sample data spread out around the estimated regression line.

> **DEFINITION**
>
> The statistic for estimating the variance σ_e^2 is
>
> $$s_e^2 = \frac{SSResid}{n - 2}$$
>
> where
>
> $$SSResid = \Sigma(y - \hat{y})^2 = \Sigma y^2 - a\, \Sigma y - b\, \Sigma xy$$
>
> The subscript in σ_e^2 and s_e^2 is a reminder that you are estimating the variance of the "errors" or residuals.
>
> The estimate of σ_e is the **estimated standard deviation**
>
> $$s_e = \sqrt{s_e^2}$$
>
> The number of degrees of freedom associated with estimating σ_e^2 or σ_e in simple linear regresssion is $n - 2$.

The estimates and number of degrees of freedom here have analogs in previous work involving a single sample $x_1, x_2, \ldots, x_n$. The sample variance s^2 had a numerator of $\Sigma(x - \bar{x})^2$, a sum of squared deviations (residuals), and denominator $n - 1$, the number of degrees of freedom associated with s^2 and s. The use of $\bar{x}$ as an estimate of μ in the formula for s^2 reduces the number of degrees of freedom by 1, from n to $n - 1$. In simple linear regression, estimation of two quantities, α and β, results in a loss of 2 degrees of freedom, leaving $n - 2$ as the number of degrees of freedom associated with SSResid, s_e^2 and s_e.

Once the estimated regression equation has been found, the usefulness of this model is evaluated using a residual plot and the values of s_e and the coefficient of determination, r^2. Recall from Chapter 4 that the values of s_e and r^2 are interpreted as described in the following box.

> The coefficient of determination, r^2, is the proportion of variability in y that can be explained by the approximate linear relationship between x and y.
>
> The value of s_e, the estimated standard deviation about the population regression line, is interpreted as the typical amount by which an observation deviates from the population regression line.

Example 16.3   Estimating Elk Weight

Wildlife biologists monitor the ecological health of animals. For large animals whose habitat is relatively inaccessible, this can present some practical problems. The Rocky Mountain elk is the fourth largest deer species and is a case in point. Males range up to 7.5 feet in length and over 500 pounds in weight. The equipment, manpower, and time needed to weigh these creatures make direct measurement of weight difficult and expensive. The authors of the paper **"Estimating Elk Weight From Chest Girth"** *(Wildlife Society Bulletin* **[1996]: 58-611)** found they could reliably estimate elk weights by a much more practical method: measuring the chest girth and then using linear regression to estimate the weight. They measured the chest girth and weight of 19 Rocky Mountain elk in Custer State Park, South Dakota. The resulting data (from a scatterplot in the paper) is given in the accompanying table. The table also includes the predicted values and residuals for the estimated regression line.

Girth (cm)	Weight(kg)	Predicted y Value	Residual
96	87	136.266	−38.2661
105	196	161.069	34.9314
108	163	169.336	−6.3361
109	196	172.092	23.9080
110	183	174.848	8.1522
114	171	185.871	−14.8711
121	230	205.162	24.8380
124	225	213.429	11.5705
131	211	232.720	−21.7203
135	231	243.744	−12.7436
137	225	249.255	−24.2553
138	266	252.011	13.9889
140	241	257.523	−16.5228
142	264	263.034	0.9655
157	284	304.372	−20.3720
157	292	304.372	−12.3720
159	300	309.884	−9.8837
155	337	298.860	38.1397
162	339	318.151	20.8488

The scatterplot (Figure 16.9) gives evidence of a strong positive linear relationship between

$$x = \text{chest girth (in cm)}$$

and

$$y = \text{weight in (kg)}$$

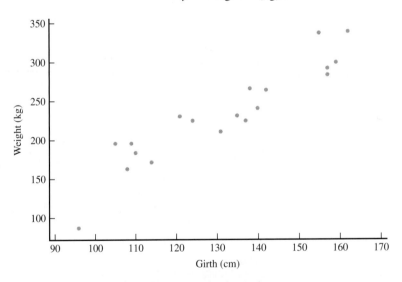

FIGURE 16.9
Scatterplot of weight versus chest girth for Example 16.3

Partial Minitab regression output is shown here.

Regression Analysis: Weight versus Girth
The regression equation is
Weight = − 136 + 2.81 Girth

Predictor	Coef	SE Coef	T	P
Constant	−135.51	35.75	−3.79	0.001
Girth	2.8063	0.2686	10.45	0.000

S = 23.6626   R-Sq = 86.5%   R-Sq(adj) = 85.7%

From the output,

$$\hat{y} = -136 + 2.81x$$
$$r^2 = 0.865$$
$$S_e = 23.6626$$

Approximately 86.5% of the observed variation in elk weight y can be attributed to the linear relationship between weight and chest girth. The magnitude of a typical deviation from the least-squares line is about 23.6626 kg, which is relatively small in comparison to the y values themselves.

Another important assumption of the simple linear regression model is that the random deviations at any particular x value are normally distributed. In Section 16.3, you will see how the residuals can be used to determine whether this assumption is plausible.

SECTION 16.1 EXERCISES

Each exercise set assesses the following chapter learning objectives: C1, M1

SECTION 16.1 Exercise Set 1

16.1 Identify the following relationships as deterministic or probabilistic:
a. The relationship between the length of the sides of a square and its perimeter.
b. The relationship between the height and weight of an adult.
c. The relationship between SAT score and college freshman GPA.
d. The relationship between tree height in centimeters and tree height in inches.

16.2 Let x be the size of a house (in square feet) and y be the amount of natural gas used (therms) during a specified period. Suppose that for a particular community, x and y are related according to the simple linear regression model with

β = slope of population regression line = .017

α = y intercept of population regression line = −5.0

Houses in this community range in size from 1000 to 3000 square feet.
a. What is the equation of the population regression line?
b. Graph the population regression line by first finding the point on the line corresponding to $x = 1000$ and then the point corresponding to $x = 2000$, and drawing a line through these points.
c. What is the mean value of gas usage for houses with 2100 sq. ft. of space?
d. What is the average change in usage associated with a 1 sq. ft. increase in size?
e. What is the average change in usage associated with a 100 sq. ft. increase in size?
f. Would you use the model to predict mean usage for a 500 sq. ft. house? Why or why not?

16.3 Suppose that a simple linear regression model is appropriate for describing the relationship between $y =$

house price (in dollars) and $x =$ house size (in square feet) for houses in a large city. The population regression line is $y = 23{,}000 + 47x$ and $\sigma_e = 5000$.
a. What is the average change in price associated with one extra square foot of space? With an additional 100 sq. ft. of space?
b. Approximately what proportion of 1800 sq. ft. homes would be priced over $110,000? Under $100,000?

SECTION 16.1 Exercise Set 2

16.4 Identify the following relationships as deterministic or probabilistic:

a. The relationship between height at birth and height at one year of age.
b. The relationship between a positive number and its square root.
c. The relationship between temperature in degrees Fahrenheit and degrees centigrade.
d. The relationship between adult shoe size and shirt size.

16.5 The flow rate in a device used for air quality measurement depends on the pressure drop x (inches of water) across the device's filter. Suppose that for x values between 5 and 20, these two variables are related according to the simple linear regression model with population regression line $y = -0.12 + 0.095x$.
a. What is the mean flow rate for a pressure drop of 10 inches? A drop of 15 inches?
b. What is the average change in flow rate associated with a 1 inch increase in pressure drop? Explain.

16.6 The paper **"Predicting Yolk Height, Yolk Width, Albumen Length, Eggshell Weight, Egg Shape Index, Eggshell Thickness, Egg Surface Area of Japanese Quails Using Various Egg Traits as Regressors"** (*International Journal of*

Poultry Science **[2008]: 85–88)** suggests that the simple linear regression model is reasonable for describing the relationship between y = eggshell thickness (in micrometers) and x = egg length (mm) for quail eggs. Suppose that the population regression line is $y = 0.135 + 0.003x$ and that $\sigma_e = 0.005$. Then, for a fixed x value, y has a normal distribution with mean $0.135 + 0.003x$ and standard deviation 0.005.

a. What is the mean eggshell thickness for quail eggs that are 15 mm in length? For quail eggs that are 17 mm in length?

b. What is the probability that a quail egg with a length of 15 mm will have a shell thickness that is greater than 0.18 μm?

c. Approximately what proportion of quail eggs of length 14 mm has a shell thickness of greater than 0.175? Less than 0.178?

Additional Exercises

16.7 Tom and Ray are managers of electronics stores with slightly different pricing strategies for USB drives. In Tom's store, customers pay the same amount, c, for each USB drive. In Ray's store, it is a little more exciting. The customer pays an up-front cost of $1.00. Ray charges the same price per USB drive, c, but at the register the customer flips a coin. If the coin lands heads up, the customer gets his or her $1.00 back, plus another dollar off the total cost of the USB drives purchased.

a. Which of these pricing strategies can be expressed as a deterministic model?

b. Using mathematical notation, specify a model using Tom's pricing strategy that relates y = total cost to x = number of USB drives purchased.

c. Using mathematical notation, specify a model using Ray's pricing strategy that relates y = total cost to x = number of USB drives purchased.

d. Describe the distribution of e for the probabilistic model described above. What is the mean of the distribution of e? What is the standard deviation of e?

16.8 Identify the following relationships as deterministic or probabilistic:

a. The relationship between the speed limit and a driver's speed.

b. The relationship between the price in dollars and the price in Euros of an object.

c. The relationship between the number of pages and the number of words in a text book.

d. The relationship between the possible numbers of pennies and the nickels in a pile if no other coins are in the pile and the amount of money in the pile is $3.00.

16.9 Hormone replacement therapy (HRT) is thought to increase the risk of breast cancer. The accompanying data on x = percent of women using HRT and y = breast cancer incidence (cases per 100,000 women) for a region in

Germany for 5 years appeared in the paper **"Decline in Breast Cancer Incidence after Decrease in Utilisation of Hormone Replacement Therapy"** (*Epidemiology* **[2008]: 427–430**). The authors of the paper used a simple linear regression model to describe the relationship between HRT use and breast cancer incidence.

HRT Use	Breast Cancer Incidence
46.30	103.30
40.60	105.00
39.50	100.00
36.60	93.80
30.00	83.50

a. What is the equation of the estimated regression line?

b. What is the estimated average change in breast cancer incidence associated with a 1 percentage point increase in HRT use?

c. What would you predict the breast cancer incidence to be in a year when HRT use was 40%?

d. Should you use this regression model to predict breast cancer incidence for a year when HRT use was 20%? Explain.

e. Calculate and interpret the value of r^2.

f. Calculate and interpret the value of s_e.

16.10 Consider the accompanying data on x = advertising share and y = market share for a particular brand of soft drink during 10 randomly selected years.

x 0.103 0.072 0.071 0.077 0.086 0.047 0.060 0.050 0.070 0.052
y 0.135 0.125 0.120 0.086 0.079 0.076 0.065 0.059 0.051 0.039

a. Construct a scatterplot for these data. Do you think the simple linear regression model would be appropriate for describing the relationship between x and y?

b. Calculate the equation of the estimated regression line and use it to obtain the predicted market share when the advertising share is 0.09.

c. Compute r^2. How would you interpret this value?

d. Calculate a point estimate of σ_e. How many degrees of freedom is associated with this estimate?

16.11 The authors of the paper **"Weight-Bearing Activity During Youth Is a More Important Factor for Peak Bone Mass than Calcium Intake"** (*Journal of Bone and Mineral Research* **[1994], 1089–1096**) studied a number of variables they thought might be related to bone mineral density (BMD). The accompanying data on x = weight at age 13 and y = bone mineral density at age 27 are consistent with summary quantities for women given in the paper.

Weight (kg)	BMD (g/cm²)
54.4	1.15
59.3	1.26
74.6	1.42
62.0	1.06
73.7	1.44
70.8	1.02
66.8	1.26
66.7	1.35
64.7	1.02
71.8	0.91
69.7	1.28
64.7	1.17
62.1	1.12
68.5	1.24
58.3	1.00

The accompanying computer output is from JMP.

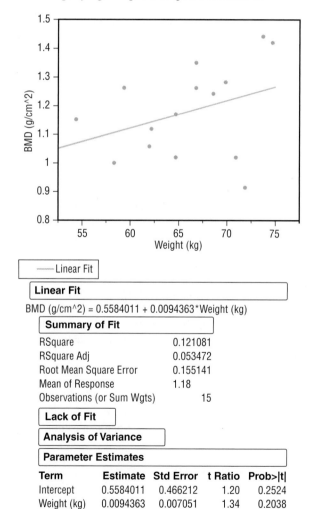

Linear Fit

BMD (g/cm^2) = 0.5584011 + 0.0094363*Weight (kg)

Summary of Fit

RSquare	0.121081
RSquare Adj	0.053472
Root Mean Square Error	0.155141
Mean of Response	1.18
Observations (or Sum Wgts)	15

Lack of Fit

Analysis of Variance

Parameter Estimates

Term	Estimate	Std Error	t Ratio	Prob>|t|
Intercept	0.5584011	0.466212	1.20	0.2524
Weight (kg)	0.0094363	0.007051	1.34	0.2038

a. What percentage of observed variation in BMD at age 27 can be explained by the simple linear regression model?

b. Give a point estimate of σ_e and interpret this estimate.

c. Give an estimate of the average change in BMD associated with a 1 kg increase in weight at age 13.

d. Compute a point estimate of the mean BMD at age 27 for women whose age 13 weight was 60 kg.

16.12 The production of pups and their survival are the most significant factors contributing to gray wolf population growth. The causes of early pup mortality are unknown and difficult to observe. The pups are concealed within their dens for 3 weeks after birth, and after they emerge it is difficult to confirm their parentage. Researchers recently used portable ultrasound equipment to investigate some factors related to reproduction (**"Diagnosing Pregnancy, in Utero Litter Size, and Fetal Growth with Ultrasound in Wild, Free-Ranging Wolves,"** *Journal of Mammology* **[2006]: 85-92**). A scatterplot of $y = $ length of an embryonic sac diameter (in cm) and $x = $ gestational age (in days) is shown below. Computer output from a regression analysis is also given.

Bivariate Fit of Emb Ves Diam (cm) By Gest Age (days)

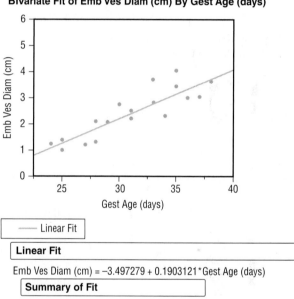

— Linear Fit

Linear Fit

Emb Ves Diam (cm) = −3.497279 + 0.1903121*Gest Age (days)

Summary of Fit

RSquare	0.792803
RSquare Adj	0.780615
Root Mean Square Error	0.450587
Mean of Response	2.482526
Observations (or Sum Wgts)	19

Lack of Fit

Analysis of Variance

Parameter Estimates

Term	Estimate	Std Error	t Ratio	Prob>|t|
Intercept	−3.497279	0.748605	−4.67	0.0002*
Gest Age (days)	0.1903121	0.023597	8.07	<.0001*

a. What is the equation of the estimated regression line?

b. What is the estimated embryonic sac diameter for a gestational age of 30 days?

c. What is the average change in sac diameter associated with a 1-day increase in gestational age?

d. What is the average change in sac diameter associated with a 5-day increase in gestational age?

e. Would you use this model to predict the mean embryonic sac diameter for all gestation ages from conception to birth? Why or why not?

Inferences Concerning the Slope of the Population Regression Line

The slope coefficient β in the simple linear regression model represents the average or expected change in the response variable y that is associated with a 1-unit increase in the value of the independent variable x. For example, consider $x =$ the size of a house (in square feet) and $y =$ selling price of the house. If the simple linear regression model is appropriate for the population of houses in a particular city, β would be the average increase in selling price associated with a 1-square-foot increase in size. As another example, if $x =$ amount of time per week a computer system is used and $y =$ the resulting annual maintenance expense, then β would be the expected change in expense associated with using the computer system one additional hour per week.

Because the value of β is almost always unknown, it must be estimated from sample data. The slope of the least squares regression line, b, provides an estimate. In some situations, the value of the statistic b may vary greatly from sample to sample, and the value of b computed from a single sample may be quite different from the value of the population slope, β. In other situations, almost all possible samples result in a value of b that is quite close to β. The sampling distribution of b provides information about the behavior of this statistic.

AP* EXAM TIP

Inferences about the slope of the population regression line are based on the sampling distribution of the statistic b. The properties given here depend on the four basic assumptions of the linear regression model being met. In Section 16.3, you will see how to determine if these assumptions are reasonable.

Properties of the Sampling Distribution of b

When **the four basic assumptions of the simple linear regression model** are satisfied

1. The mean value of the sampling distribution of b is β. That is, $\mu_b = \beta$, so the sampling distribution of b is always centered at the value of β. This means that b is an unbiased statistic for estimating β.
2. The standard deviation of the sampling distribution of the statistic b is

$$\sigma_b = \frac{\sigma_e}{\sqrt{\Sigma(x - \bar{x})^2}}$$

3. The statistic b has a normal distribution (a consequence of the model assumption that the random deviation e is normally distributed).

The fact that b is unbiased tells you that the sampling distribution is centered at the right place, but it gives no information about variability. If σ_b is large, the sampling distribution of b will be quite spread out around β and an estimate far from the value of β could result. For $\sigma_b = \dfrac{\sigma_e}{\sqrt{\Sigma(x - \bar{x})^2}}$ to be small, the numerator σ_e should be small (little variability about the population line) and/or the denominator $\sqrt{\Sigma(x - \bar{x})^2}$ should be large. Because $\Sigma(x - \bar{x})^2$ is a measure of how much the observed x values spread out, β tends to be more precisely estimated when the x values in the sample are spread out rather than when they are close together. The normality of the sampling distribution of b implies that the standardized variable

$$z = \frac{b - \beta}{\sigma_b}$$

has a standard normal distribution. However, inferential methods cannot be based on this statistic, because the value of σ_b is not known (because the unknown σ_e appears in the numerator of σ_b). One way to proceed is to estimate σ_e with s_e to obtain an estimate of σ_b.

The **estimated standard deviation of the statics b** is

$$s_b = \frac{s_e}{\sqrt{\Sigma(x - \bar{x})^2}}$$

AP* EXAM TIP

For inferences about the slope of the population regression line, df $= n - 2$.

When the four basic assumptions of the simple linear regression model are satisfied, the probability distribution of the standardized variable $t = \dfrac{b - \beta}{s_b}$ is the t distribution with df $= (n - 2)$.

In the same way that $t = \dfrac{\bar{x} - \mu}{\frac{s}{\sqrt{n}}}$ was used in Chapter 12 to develop a confidence inter-

val for μ, the t variable in the preceding box can be used to obtain a confidence interval for β.

> **Confidence Interval for β**
>
> When the four basic assumptions of the simple linear regression model are satisfied, **a confidence interval for** β, the slope of the population regression line, has the form
>
> $$b \pm (t \text{ critical value})s_b$$
>
> where the t critical value is based on df $= n - 2$. Appendix Table 3 gives critical values corresponding to the most frequently used confidence levels.

The interval estimate of β is centered at b and extends out from the center by an amount that depends on the sampling variability of b. When s_b is small, the interval is narrow, implying that the investigator has relatively precise knowledge of the value of β. Calculation of a confidence interval for the slope of a population regression line is illustrated in Example 16.4.

In Section 7.2, you learned four key questions that guide the decision about what statistical inference method to consider in any particular situation. In Section 7.3, a five-step process for estimation problems was introduced.

The four key questions of section 7.2 were

Q Question Type	Estimation or hypothesis testing?
S Study Type	Sample data or experiment data?
T Type of Data	One variable or two? Categorical or numerical?
N Number of Samples or Treatments	How many samples or treatments?

When the answers to these questions are

Q: estimation
S: sample data
T: two numerical variables
N: one sample

the method you will want to consider in a regression setting is the confidence interval for the slope of a population regression line.

Once you have selected the confidence interval for the slope of a population regression line as the method you want to consider, because this is an estimation problem you would follow the five-step process for estimation problems (EMC³).

Example 16.4 The Bison of Yellowstone Park

The dedicated work of conservationists for over 100 years has brought the bison in Yellowstone National Park from near extinction to a herd of over 3,000 animals. This recovery is a mixed blessing. Many bison have been exposed to the bacteria that cause brucellosis, a disease that infects domestic cattle, and there are many domestic cattle herds near Yellowstone. Because of concerns that free-ranging bison can infect nearby cattle, it is important to monitor and manage the size of the bison population and, if possible, keep bison from transmitting this bacteria to ranch cattle. The article **"Reproduction and Survival of Yellowstone Bison" (*The Journal of Wildlife Management* [2007]: 2365-2372)** described a large multiyear study of the factors that influence bison movement and herd size. The researchers studied a number of environmental factors to better understand the relationship between bison reproduction and the environment. One factor thought to influence reproduction is stress due to accumulated snow, which makes foraging more difficult for the pregnant bison. Data from 1981–1997 on

$$y = \text{spring calf ratio (SCR)}$$

and

$$x = \text{previous fall snow-water equivalent (SWE)}$$

are shown in the accompanying table. Spring calf ratio is the ratio of calves to adults, a measure of reproductive success. The researchers were interested in estimating the mean change in spring calf ratio associated with each additional cm in snow-water equivalent.

Let's answer the four key questions for this problem.

SCR	SWE	SCR	SWE
0.19	1,933	0.22	3,317
0.14	4,906	0.22	3,332
0.21	3,072	0.18	3,511
0.23	2,543	0.21	3,907
0.26	3,509	0.25	2,533
0.19	3,908	0.19	4,611
0.29	2,214	0.22	6,237
0.23	2,816	0.17	7,279
0.16	4,128		

The answers are *estimation, sample data, two numerical variables, one sample*. This

Q Question Type	Estimation or hypothesis testing?	Estimation
S Study Type	Sample data or experiment data?	Sample data
T Type of Data	One variable or two? Categorical or numerical?	Two numerical values
N Number of Samples or Treatments	How many samples or treatments?	One sample (regression)

combination of answers suggests considering a confidence interval for the slope of a population regression line. You can now use the five-step process (EMC^3) to estimate the slope of the population regression line.

Step		
Estimate	In this example, the value of β, the mean increase in spring calf ratio for each additional 1 cm of snow-water equivalent, will be estimated.	
Method	Because the answers to the four key questions are estimation, sample data, two numerical values, one sample, a confidence interval for β, the slope of the population regression line, will be considered. For this example, a 95% confidence level will be used.	
Check	The four basic assumptions of the simple linear regression model need to be met in order to use the confidence interval.	

(continued)

Step	
	The investigators collected data from 17 successive years. To proceed, you would need to assume that these years are representative of yearly circumstances at Yellowstone, and that each year's reproduction and snowfall is independent of previous years. You should keep this in mind when you get to the step that involves interpretation.
	A scatterplot of the data is shown here. The pattern in the plot looks linear and the spread does not seem to be different for different values of *x*.

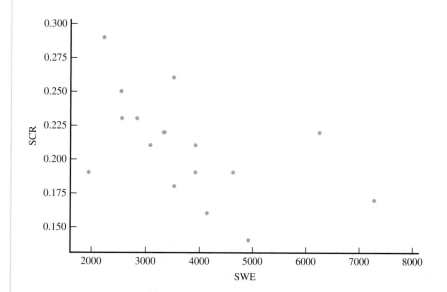

A box plot of the residuals is also shown.

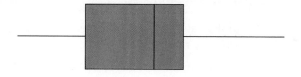

Because the boxplot is approximately symmetric and there are no outliers, it is reasonable to think that the distribution of e is approximately normal.

Calculate	JMP regression output is shown here:

Linear Fit

SCR = 0.2606561 − 0.0136639*SWE

Summary of Fit

RSquare	0.257644
RSquare Adj	0.208153
Root Mean Square Error	0.033513
Mean of Response	0.209412
Observations (or Sum Wgts)	17

(continued)

Step

Parameter Estimates

Term	Estimate	Std Error	t Ratio	Prob>|t|
Intercept	0.2606561	0.023885	10.91	<.0001*
SWE	−0.013664	0.005989	−2.28	0.0375*	s_b

$df = n - 2 = 17 - 2 = 15$

The t critical value for a 95% confidence level and df = 15 is 2.13.

$b \pm (t \text{ critical value})s_b$

$= -0.0137 \pm (2.13)(0.00599)$

$= (-0.265, -0.0009)$

Communicate Results

Confidence interval:

You can be 95% confident that the true average change in spring calf ratio associated with an increase of 1 cm in the snow-water equivalent is between −0.0265 and −0.0009.

Confidence level:

The method used to construct this interval estimate is successful in capturing the actual value of the slope of the population regression about 95% of the time.

Hypothesis Tests Concerning β

Hypotheses about β can be tested using a t test similar to the t tests introduced in Chapters 12 and 13. The null hypothesis states that β has a specified hypothesized value. The t statistic results from standardizing b, the estimate of β, under the assumption that H_0 is true. When H_0 is true, the sampling distribution of this statistic is the t distribution with df = $n - 2$.

Hypothesis Test for the Slope of the Population Regression Line, β

Appropriate when the four basic assumptions of the simple linear regression model are reasonable:

1. The distribution of e at any particular x value has mean value 0 (that is $\mu_e = 0$).
2. The standard deviation of e is σ_e, which does not depend on x.
3. The distribution of e at any particular x value is normal.
4. The random deviations $e_1, e_2, e_3, \ldots e_n$ associated with different observations are independent of one another.

When these conditions are met, the following test statistic can be used:

$$t = \frac{b - \beta_0}{s_b}$$

where β_0 is the hypothesized value from the null hypothesis.

Form of the null hypothesis: $H_0: \beta = \beta_0$

When the assumptions of the simple linear regression model are reasonable and the null hypothesis is true, the t test statistic has a t distribution with df = $n - 2$.

Associated P-value:

When the alternative hypothesis is...	The P-value is...
$H_a: \beta > \beta_0$	Area to the right of the computed t under the appropriate t curve

(continued)

$H_a: \beta < \beta_0$	Area to the left of the computed t under the appropriate t curve
$H_a: \beta \neq \beta_0$	2(area to the right of t) if t is positive or 2(area to the left of the t) if t is negetive

This test is a method you should consider when the answers to the four key questions are *hypothesis testing, sample data, two numerical variables, one sample*. You would carry out this test using the five-step process for hypothesis testing problems (HMC³).

Inference for a population slope generally focuses on two questions:

(1) Is the population slope different from zero?
(2) What are plausible values for the population slope?

The question of plausible values can be addressed by calculating a confidence interval for the population slope. The question of whether a population slope is equal to zero can be answered by using the hypothesis testing procedure with a null hypothesis $H_0: \beta = 0$. This test of $H_0: \beta = 0$ versus $H_a: \beta \neq 0$ is called the *model utility test for simple linear regression*. The default computer output for inference for a regression slope is for the model utility test.

When the null hypothesis of the model utility test is true, the population regression line is a horizontal line, and the value of y in the simple linear regression model does not depend on x. That is,

$$y = \alpha + \beta x + e$$
$$= \alpha + 0x + e$$
$$= \alpha + e$$

If β is in fact equal to 0, knowledge of x will be of no use — it will have no "utility" for predicting y. On the other hand, if β is different from 0, there is a useful linear relationship between x and y, and knowledge of x is useful for predicting y. This is illustrated by the scatterplots in Figure 16.10.

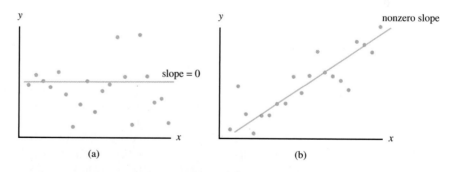

FIGURE 16.10
(a) $\beta = 0$; (b) $\beta \neq 0$

The Model Utility Test for Simple Linear Regression

The **model utility test for simple linear regression** is the test of

$$H_0: \beta = 0$$

versus

$$H_a: \beta \neq 0$$

The null hypothesis specifies that there is *no* useful linear relationship between x and y, whereas the alternative hypothesis specifies that there *is* a useful linear

(continued)

relationship between x and y. If H_0 is rejected, you can conclude that the simple linear regression model is useful for predicting y.

The test statistic is the t ratio

$$t = \frac{(b - 0)}{s_b} = \frac{b}{s_b}.$$

It is recommended that the model utility test be carried out before using the estimated regression line to make inferences.

Example 16.5   The British (Musical) Invasion

Have you experienced a sudden flood of memory when scanning from station to station on your car radio and recognized a song from your past? Perhaps you could remember the title of the song, the artist, and even when the song was released. From a seemingly small amount of information you were able to recover a great deal of the song's context from memory. The article **"Plink: 'Thin slices' of Music" (Krumhansl, C. Music Perception [2010]:337-354)** describes a study of this phenomenon. The investigator compiled a list of songs from *Rolling Stone*, Billboard, and Blender lists of songs plus some recent songs familiar to college students. Twenty-three college students were then exposed to 56 clips of songs. Most of these students had had musical training, and they listened to popular music for an average of 21.7 hours per week. After hearing three short clips from a song (only 400 ms in duration), the students were asked in what year each of the songs was released. The accompanying table shows the

Actual and Judged Release Years							
Actual Release	Judged Release	Actual Release	Judged Release	Actual Release	Judged Release	Actual Release	Judged Release
1998	1997.2	1976	1983.3	1976	1988.0	1970	1985.4
1967	1973.7	2008	1995.0	2006	1996.7	1975	1985.9
1998	1996.3	1971	1979.8	1974	1985.4	1991	1993.3
1999	1993.3	1965	1976.8	2007	1999.8	2008	1995.4
1983	1985.4	1967	1975.0	1976	1987.2	1965	1977.6
1982	1988.0	1971	1978.0	1974	1977.6	1987	1990.7
1965	1970.2	1967	1978.0	1970	1982.8	1975	1986.3
1991	1992.8	1984	1983.3	1971	1976.3	1968	1986.7
1983	1984.1	1984	1989.8	1999	1988.5	1987	1988.0
1976	1979.3	1968	1976.7	1997	1994.1	2008	1990.2
1971	1975.4	1965	1978.5	2006	1995.4	1982	1991.1
1981	1984.6	1965	1977.2	1981	1989.3	1979	1983.7
1967	1973.7	1979	1986.7	2008	1993.7	2000	1989.8
2007	1997.2	1997	1996.3	1965	1981.1	2000	1991.1

actual release year and the average of the release years given by the students. The actual release years ranged from 1965 (The Beatles, "Help") to 2008 (Katy Perry, "I Kissed a Girl").

Is there a relationship between the judged and actual release year for these songs? A scatterplot of the data (Figure 16.11) suggests that there is a linear relation between these two variables, but this can be confirmed this using the model utility test.

With x = actual release year and y = judged release year, the equation of the estimated regression line is $\hat{y} = 1095 + 0.449x$. The five-step process for hypothesis testing can be used to carry out the model utility test.

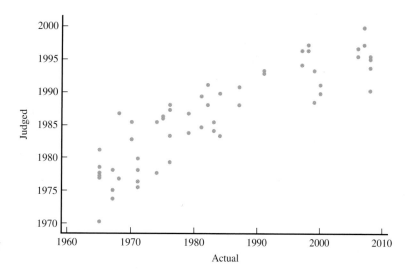

FIGURE 16.11
Scatterplot of judged release year versus actual release year

Process Step	
H Hypotheses	In the model utility test, the null hypothesis is there is no useful relationship between the actual and the judged release year: H_0: $\beta = 0$. The alternative hypothesis specifies that there is a useful relationship: $\beta \neq 0$. **Hypotheses:** *Null hypothesis*: H_0: $\beta = 0$ *Alternative hypothesis*: H_a: $\beta \neq 0$
M Method	Because the answers to the four key questions are hypothesis testing, sample data, two numerical variables in a regression setting and one sample, a hypothesis test for the slope of a population regression line will be considered. The test statistic for this test is $$t = \frac{b - 0}{s_b} = \frac{b}{s_b}$$ The value of 0 in the test statistic is the hypothesized value from the null hypothesis. For this example, a significance level of 0.05 will be used. **Significance level:** $\alpha = 0.05$
C Check	In Section 16.3, you will see how to check to see if the four assumptions of the simple linear regression model are reasonable. For this example, you can assume that these assumptions are reasonable and proceed with the model utility test.
C Calculate	JMP output is shown here: **Linear Fit** Judged Release = 1095.1525 + 0.449281*Actual Release **Summary of Fit** RSquare 0.771 RSquare Adj 0.766759 Root Mean Square Error 3.59844 Mean of Response 1986.013 Observations (or Sum Wgts) 56 **Lack of Fit** **Analysis of Variance** s_b **Parameter Estimates** Term Estimate Std Error t Ratio Prob>|t| Intercept 1095.1525 66.07159 16.58 <.0001* Actual Release 0.449281 0.033321 13.48 <.0001*

(continued)

Test statistic:

$$t = \frac{b - 0}{s_b} = \frac{0.449 - 0}{0.0333} = 13.48$$

Associated *P*-value:

$P - \text{value} = $ twice area under t curve to the right of 13.48

$= 2P(t > 13.48)$

≈ 0

C Communicate results

Because the *P*-value is less than the selected significance level, the null hypothesis is rejected.

Decision: Reject H_0.

Conclusion: The sample data provide convincing evidence that there is a useful linear relationship between the actual release year and the judged release year.

Because the model utility test confirms that there is a useful linear relationship between judged release year and actual release year, it would be reasonable to use the estimated regression model to predict the judged release year for a given song based on its actual release year. Of course, before you do this, you would also want to evaluate the accuracy of predictions by looking at the value of s_e. ▬

When H_0: $\beta = 0$ cannot be rejected using the model utility test at a reasonably small significance level, the search for a useful model must continue. One possibility is to relate y to x using a nonlinear model — an appropriate strategy if the scatterplot shows curvature.

SECTION 16.2 EXERCISES

Each exercise set assesses the following chapter learning objectives: C2, M3, M4 , M5, M6, P1, P2

SECTION 16.2 Exercise Set 1

16.13 The standard deviation of the errors, σ_e, is an important part of the linear regression model.
a. What is the relationship between the value of σ_e and the value of the test statistic in a test of a hypotheses about β?
b. What is the relationship between the value of σ_e and the width of a confidence interval for β?

16.14 A journalist is reporting about some research on appropriate amounts of sleep for people 9 to 19 years of age. In that research, a linear regression model is used to describe the relationship between alertness and number of hours of sleep the night before. The researchers reported a 95% confidence interval, but newspapers usually report an estimate and a margin of error.
a. In order to calculate a margin of error from the reported confidence interval, what additional conditions, if any, need to be verified?
b. In order to calculate a margin of error from the reported confidence interval, what additional information, if any, is needed?

16.15. A nursing student has completed his final project, and is preparing for a meeting with his project advisor. The subject of his project was the relationship between systolic blood pressure (SBP) and body mass index (BMI). The last time he met with his advisor he had completed his measurements, but only entered half his data into his statistical software. For the data he had entered, the necessary conditions for inference for β were met. In a short paragraph, explain, using appropriate statistical terminology, which of the conditions below must be rechecked.

1. The standard deviation of e is the same for all values of x.
2. The distribution of e at any particular x value is normal.

16.16 Consider the accompanying data on x = research and development expenditure (thousands of dollars) and y = growth rate (% per year) for eight different industries.

x	2024	5038	905	3572	1157	327	378	191
y	1.90	3.96	2.44	0.88	0.37	−0.90	0.49	1.01

a. Would a simple linear regression model provide useful information for predicting growth rate from research and development expenditure? Use a .05 level of significance.
b. Use a 90% confidence interval to estimate the average change in growth rate associated with a $1000 increase in expenditure. Interpret the resulting interval

16.17 The paper **"The Effects of Split Keyboard Geometry on Upper Body Postures"** (Ergonomics [2009]: 104–111) describes a study to determine the effects of several keyboard characteristics on typing speed. One of the variables considered was the front-to-back surface angle of the keyboard. Minitab output resulting from fitting the simple linear regression model with x = surface angle (degrees) and y = typing speed (words per minute) is given below.

Regression Analysis: Typing Speed versus Surface Angle

The regression equation is
Typing Speed = 60.0 + 0.0036 Surface Angle

Predictor	Coef	SE Coef	T	P
Constant	60.0286	0.2466	243.45	0.000
Surface Angle	0.00357	0.03823	0.09	0.931

S = 0.511766   R-Sq = 0.3%   R-Sq(adj) = 0.0%

Analysis of Variance

Source	DF	SS	MS	F	P
Regression	1	0.0023	0.0023	0.01	0.931
Residual Error	3	0.7857	0.2619		
Total	4	0.7880			

a. Suppose that the basic assumptions of the simple linear regression model are met. Carry out a hypothesis test to decide if there is a useful linear relationship between x and y.

b. Are the values of s_e and r^2 consistent with the conclusion from Part (a)? Explain.

16.18 Do taller adults make more money? The authors of the paper "Stature and Status: Height, Ability, and Labor Market Outcomes" (*Journal of Political Economics* [2008]: 499–532) investigated the association between height and earnings. They used the simple linear regression model to describe the relationship between x = height (in inches) and y = log(weekly gross earnings in dollars) in a very large sample of men. The logarithm of weekly gross earnings was used because this transformation resulted in a relationship that was approximately linear. The paper reported that the slope of the estimated regression line was $b = 0.023$ and the standard deviation of b was $s_b = 0.004$. Carry out a hypothesis test to decide if there is convincing evidence of a useful linear relationship between height and the logarithm of weekly earnings. You can assume that the basic assumptions of the simple linear regression model are met.

16.19 The effects of grazing animals on grasslands have been the focus of numerous investigations by ecologists. One such study, reported in "The Ecology of Plants, Large Mammalian Herbivores, and Drought in Yellowstone National Park" (*Ecology* [1992]: 2043–2058), proposed using the simple linear regression model to relate y = green biomass concentration (g/cm³) to x = elapsed time since snowmelt (days).

a. The estimated regression equation was given as $\hat{y} = 106.3 - .640x$. What is the estimate of average change in biomass concentration associated with a 1-day increase in elapsed time?

b. What value of biomass concentration would you predict when elapsed time is 40 days?

c. The sample size was $n = 58$, and the reported value of the coefficient of determination was 0.470. What does this tell you about the linear relationship between the two variables?

16.20 Consider a test of hypotheses about, β the population slope in a linear regression model.

a. If you reject the null hypothesis, $\beta = 0$, what does this mean in terms of a linear relationship between x and y?

b. If you fail to reject the null hypothesis, $\beta = 0$, what does this mean in terms of a linear relationship between x and y?

16.21 Researchers studying pleasant touch sensations measured the firing frequency (impulses per second) of nerves that were stimulated by a light brushing stroke on the forearm and also recorded the subject's numerical rating of how pleasant the sensation was. The accompanying data was read from a graph in the paper "Coding of Pleasant Touch by Unmyelinated Afferents in Humans" (*Nature Neuroscience*, April 12, 2009).

Firing Frequency	Pleasantness Rating	Firing Frequency	Pleasantness Rating
23	0.2	28	2.0
24	1.0	34	2.3
22	1.2	33	2.2
25	1.2	36	2.4
27	1.0	34	2.8

a. Estimate the mean change in pleasantness rating associated with an increase of 1 impulse per second in firing frequency using a 95% confidence interval. Interpret the resulting interval.

b. Carry out a hypothesis test to decide if there is convincing evidence of a useful linear relationship between firing frequency and pleasantness rating.

16.22 The largest commercial fishing enterprise in the southeastern United States is the harvest of shrimp. In a study described in the paper "Long-term Trawl Monitoring of White Shrimp, Litopenaeus setiferus (Linnaeus), Stocks within the ACE Basin National Estuariene Research Reserve, South Carolina" (*Journal of Coastal Research* [2008]:193-199), researchers monitored variables thought to be related to the abundance of white shrimp. One variable the researchers thought might be related to abundance is the amount of oxygen in the water. The relationship between mean catch per tow of white shrimp and oxygen concentration was described by fitting a regression line using data from ten randomly selected offshore sites. (The "catch" per tow is the number of shrimp caught in a single outing.) Computer output is shown below.

The regression equation is
Mean catch per tow = −5859 + 97.2 O2 Saturation

Predictor	Coef	SE Coef	T	P
Constant	−5859	2394	−2.45	0.040
O2 Saturation	97.22	34.63	2.81	0.023

S = 481.632   R-Sq = 49.6%   R-Sq(adj) = 43.3%

a. Is there convincing evidence of a useful linear relationship between the shrimp catch per tow and oxygen concentration density? Explain.

b. Would you describe the relationship as strong? Why or why not?

c. Construct a 95% confidence interval for β and interpret it in context.

d. What margin of error is associated with the confidence interval in Part (c)?

16.23 The authors of the paper **"Decreased Brain Volume in Adults with Childhood Lead Exposure"** (*Public Library of Science Medicine* [May 27, 2008]: e112) studied the relationship between childhood environmental lead exposure and a measure of brain volume change in a particular region of the brain. Data were given for x = mean childhood blood lead level (μg/dL) and y = brain volume change (BVC, in percent). A subset of data read from a graph that appeared in the paper was used to produce the accompanying Minitab output.

Regression Analysis: BVC versus Mean Blood Lead Level
The regression equation is
BVC = $-0.00179 - 0.00210$ Mean Blood Lead Level

Predictor	Coef	SE Coef	T	P
Constant	-0.001790	0.008303	-0.22	0.830
Mean Blood Lead Level	-0.0021007	0.0005743	-3.66	0.000

Carry out a hypothesis test to decide if there is convincing evidence of a useful linear relationship between x and y. You can assume that the basic assumptions of the simple linear regression model are met.

Additional Exercises

16.24 **a.** Explain the difference between the line $y = \alpha + \beta x$ and the line $\hat{y} = a + bx$.
b. Explain the difference between β and b.
c. Let x^* denote a particular value of the independent variable. Explain the difference between $\alpha + \beta x^*$ and $a + bx^*$.
d. Explain the difference between σ and s_e.

16.25 What is the distinction between σ_e and s_e?

16.26 The accompanying data were read from a plot (and are a subset of the complete data set) given in the article **"Cognitive Slowing in Closed-Head Injury"** (*Brain and Cognition* [1996]: 429–440). The data represent the mean response times for a group of individuals with closed-head injury (CHI) and a matched control group without head injury on 10 different tasks. Each observation was based on a different study, and used different subjects, so it is reasonable to assume that the observations are independent.

a. Fit a linear regression model that would allow you to predict the mean response time for those suffering a closed-head injury from the mean response time on the same task for individuals with no head injury.
b. Do the sample data support the hypothesis that there is a useful linear relationship between the mean response time for individuals with no head injury and the mean response

time for individuals with CHI? Test the appropriate hypotheses using $\alpha = .05$.

Study	Mean Response Time	
	Control	CHI
1	250	303
2	360	491
3	475	659
4	525	683
5	610	922
6	740	1044
7	880	1421
8	920	1329
9	1010	1481
10	1200	1815

16.27 The article **"Photocharge Effects in Dye Sensitized Ag[Br,I] Emulsions at Millisecond Range Exposures"** (*Photographic Science and Engineering* [1981]: 138–144) gave the accompanying data on x = % light absorption and y = peak photovoltage.

x	4.0	8.7	12.7	19.1	21.4	24.6	28.9	29.8	30.5
y	0.12	0.28	0.55	0.68	0.85	1.02	1.15	1.34	1.29

JMP output for these data is shown below.

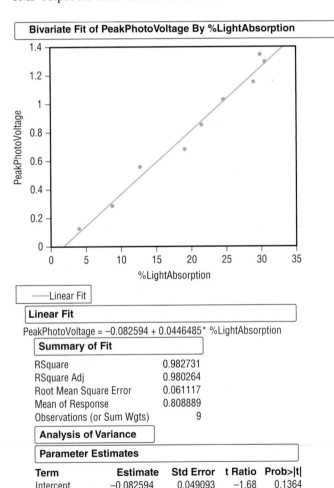

Bivariate Fit of PeakPhotoVoltage By %LightAbsorption

Linear Fit

Linear Fit

PeakPhotoVoltage = $-0.082594 + 0.0446485$ * %LightAbsorption

Summary of Fit

RSquare	0.982731
RSquare Adj	0.980264
Root Mean Square Error	0.061117
Mean of Response	0.808889
Observations (or Sum Wgts)	9

Analysis of Variance

Parameter Estimates

Term	Estimate	Std Error	t Ratio	Prob>|t|
Intercept	-0.082594	0.049093	-1.68	0.1364
%LightAbsorption	0.0446485	0.002237	19.96	<.0001*

a. What does the scatterplot suggest about the relationship between the peak photovoltage and the percent of light absorption?
b. What is the equation of the estimated regression line?
c. How much of the observed variation in peak photovoltage can be explained by the model relationship?
d. Predict peak photovoltage when percent absorption is 19.1, and compute the value of the corresponding residual.
e. The authors claimed that there is a useful linear relationship between the two variables. Do you agree? Carry out a formal test.
f. Give an estimate of the average change in peak photovoltage associated with a 1 percentage point increase in light absorption. Your estimate should convey information about the precision of estimation.

16.28 In anthropological studies, an important characteristic of fossils is cranial capacity. Frequently skulls are at least partially decomposed, so it is necessary to use other characteristics to obtain information about capacity. One measure that has been used is the length of the lambda-opisthion chord. The article **"Vertesszollos and the Presapiens Theory" (American Journal of Physical Anthropology [1971])** reported the accompanying data for $n = 7$ Homo erectus fossils.

x (chord length in mm)	78	75	78	81	84	86	87
y (capacity in cm^3)	850	775	750	975	915	1015	1030

Suppose that from previous evidence, anthropologists had believed that for each 1-mm increase in chord length, cranial capacity would be expected to increase by 20 cm^3. Do these new experimental data provide convincing evidence against this prior belief?

16.29 Suppose you are given the computer output shown. You are interested in testing the null hypothesis $\beta = 1.0$ versus an alternative hypothesis of $\beta > 1.0$. Describe how you would use the given computer output to test these hypotheses.

Linear Fit

$y = 5.6452776 + 0.9797401*x$

Summary of Fit	
RSquare	0.985289
RSquare Adj	0.984954
Root Mean Square Error	12.48525
Mean of Response	0.791304
Observations (or Sum Wgts)	46

Lack of Fit

Analysis of Variance

Parameter Estimates

Term	Estimate	Std Error	t Ratio	Prob>|t|
Intercept	5.6452776	1.84302	3.06	0.0037*
x	0.9797401	0.018048	54.29	<.0001*

Checking Model Adequacy

Section 16.2 introduced methods for estimating and testing hypotheses about β, the slope in the simple linear regression model

$$y = \alpha + \beta x + e$$

In this model, e represents the random deviation of a y value from the population regression line $\alpha + \beta x$. The methods presented in Section 16.2 require that some assumptions about the random deviations in the simple linear regression model be met in order for inferences to be valid. These assumptions include:

1. At any particular x value, the distribution of e is normal.
2. At any particular x value, the standard deviation of e is σ_e, which is constant over all values of x (that is, σ_e does not depend on x).

Inferences based on the simple linear regression model are still appropriate if model assumptions are slightly violated (for example, mild skew in the distribution of e). However, interpreting a confidence interval or the result of a hypothesis test when assumptions are seriously violated can result in misleading conclusions. For this reason, it is important to be able to detect any serious violations.

Residual Analysis

If the deviations $e_1, e_2, \ldots, e_n$ from the population line were available, they could be examined for any inconsistencies with model assumptions. For example, a normal probability plot of these deviations would suggest whether or not the normality assumption was plausible. However, because these deviations are

$$e_1 = y_1 - (\alpha + \beta x_1)$$
$$\vdots$$
$$e_n = y_n - (\alpha + \beta x_n)$$

they can be calculated only if α and β are known. In practice, this will almost never be the case. Instead, diagnostic checks must be based on the residuals

$$y_1 - \hat{y}_1 = y_1 - (a + bx_1)$$
$$\vdots$$
$$y_n - \hat{y}_n = y_n - (a + bx_n)$$

which are the deviations from the *estimated* regression line. When all model assumptions are met, the mean value of the residuals at any particular x value is 0. Any observation that gives a large positive or negative residual should be examined carefully for any unusual circumstances, such as a recording error or nonstandard experimental condition. Identifying residuals with unusually large magnitudes is made easier by inspecting **standardized residuals**.

Recall that a quantity is standardized by subtracting its mean value (0 in this case) and dividing by its actual or estimated standard deviation:

$$\text{standardized residual} = \frac{\text{residual}}{\text{estimated standard deviation of residual}}$$

The value of a standardized residual tells you the distance (in standard deviations) of the corresponding residual from its expected value, 0.

Because residuals at different x values have different standard deviations (depending on the value of x for that observation)[1], computing the standardized residuals can be tedious. Fortunately, many computer regression programs provide standardized residuals.

Example 16.6 Revisiting the Elk

Example 16.3 introduced data on

$$x = \text{chest girth (in cm)}$$

and

$$y = \text{weight (in kg)}$$

for a sample of 19 Rocky Mountain elk. (See Example 16.3 for a more detailed description of the study.)

Inspection of the scatterplot in Figure 16.12 suggests the data are consistent with the assumptions of the simple linear regression model.

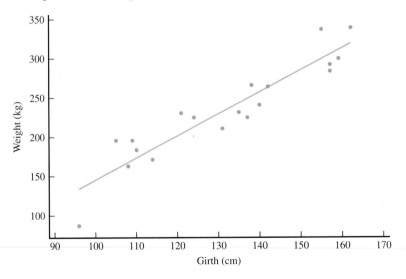

FIGURE 16.12
Scatterplot for the elk data

---

[1] The estimated standard deviation of the ith residual, $y_i - \hat{y}_i$, is $s_e \sqrt{1 - \dfrac{1}{n} - \dfrac{(x_i - \bar{x})^2}{\Sigma(x - \bar{x})^2}}$

The data, residuals, and the standardized residuals (computed using Minitab) are given in Table 16.1. For the residual with the largest magnitude, 38.1397, the standardized residual is 1.81294. That is, the residual is approximately 1.8 standard deviations above its expected value of 0. This value is not particularly unusual in a sample of this size. Also notice that for the negative residual with the largest magnitude, −38.2661, the standardized residual is −1.92313, still not unusual in a sample of this size. On the standardized scale, no residual here is surprisingly large.

TABLE 16.1 **Data, residuals, and standardized residuals for the elk data**

Observation	Girth (cm) x	Weight (kg) y	Residual	Standardized Residual	$\hat{y}$
1	96	98	−38.2661	−1.92313	136.266
2	105	196	34.9314	1.68004	161.069
3	108	163	−6.3361	−0.30135	169.336
4	109	196	23.9080	1.13323	172.092
5	110	183	8.1522	0.38517	174.848
6	114	171	−14.8711	−0.69477	185.871
7	121	230	24.8380	1.14452	205.162
8	124	225	11.5705	0.53117	213.429
9	131	211	−21.7203	−0.99323	232.720
10	135	231	−12.7436	−0.58320	243.744
11	137	225	−24.2553	−1.11135	249.255
12	138	266	13.9889	0.64147	252.011
13	140	241	−16.5228	−0.75921	257.523
14	142	264	0.9655	0.04448	263.034
15	157	284	−20.3720	−0.97540	304.372
16	157	292	−12.3720	−0.59236	304.372
17	159	300	−9.8837	−0.47699	309.884
18	155	337	38.1397	1.81294	298.860
19	162	339	20.8488	1.01967	318.151

Next, consider the assumption of the normality of e's. Figure 16.13 shows box plots of the residuals and standardized residuals. The box plots are approximately symmetric and there are no outliers, so the assumption of normally distributed errors seems reasonable.

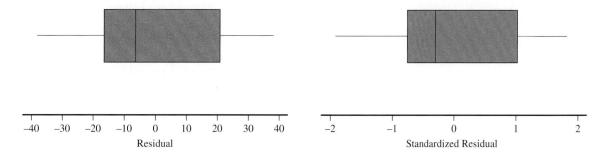

FIGURE 16.13
Boxplots of residuals and standardized residuals for the elk data.

Notice that the boxplots of the residuals and standardized residuals are nearly identical. While it is preferable to work with the standardized residuals, if you do not have access to a computer package or calculator that will produce standardized residuals, a plot of the unstandardized residuals should suffice.

A normal probability plot of the standardized residuals (or the residuals) is another way to assess whether it is reasonable to assume that e_1, e_2,..., e_n all come from the same normal distribution. An advantage of the normal probability plot, shown in Figure 16.14, is that the value of each residual can be seen, which provides more information about the distribution. The pattern in the normal probability plot of the

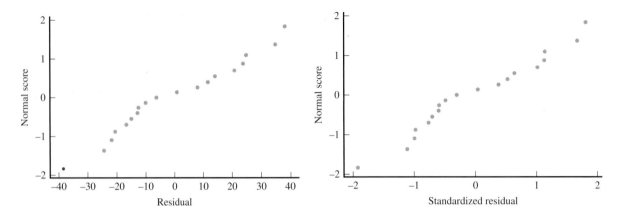

FIGURE 16.14

Normal probability plots of residuals and standardized residuals for the elk data

standardized residuals and pattern in the normal probability plot of the the residuals for the elk data are reasonably straight, confirming that the assumption of normality of the error distribution is reasonable. Also notice that the pattern in both normal probability plots is similar, so you don't need to construct both—either plot could be used.

Plotting the Residuals

A plot of the (x, residual) pairs is called a **residual plot**, and a plot of the (x, standardized residual) pairs is a **standardized residual plot**. Residual and standardized residual plots typically exhibit the same general shapes. If you are using a computer package or graphing calculator that calculates standardized residuals, the standardized residual plot is recommended. If not, it is acceptable to use the unstandardized residual plot instead.

A standardized residual plot or a residual plot is often helpful in identifying unusual or highly influential observations and in checking for violations of model assumptions. A desirable plot is one that exhibits no particular pattern (such as curvature or a much greater spread in one part of the plot than in another) and that has no point that is far removed from all the others. A point in the residual plot falling far above or far below the horizontal line at height 0 corresponds to a large residual, which can indicate unusual behavior, such as a recording error, a nonstandard experimental condition, or an atypical experimental subject. A point with an x value that differs greatly from others in the data set could have exerted excessive influence in determining the estimated regression line.

A standardized residual plot, such as the one pictured in Figure 16.15(a) is desirable, because no point lies much outside the horizontal band between -2 and 2 (so there is no unusually large residual corresponding to an outlying observation). There is no point far to the left or right of the others (which could indicate an observation that might greatly influence the estimated line), and there is no pattern to indicate that the model should somehow be modified. When the plot has the appearance of Figure 16.15(b), the fitted model should be changed to incorporate curvature (a nonlinear model).

The increasing spread from left to right in Figure 16.15(c) suggests that the variance of y is not the same at each x value but rather increases with x. A straight-line model may still be appropriate, but the best-fit line should be obtained by using *weighted least* squares rather than ordinary least squares. This involves giving more weight to observations in the region exhibiting low variability and less weight to observations in the region exhibiting high variability. A specialized regression analysis textbook or a statistician should be consulted for more information on using weighted least squares.

The standardized residual plots of Figures 16.15(d) and 16.15(e) show an outlier (a point with a large standardized residual) and a potentially influential observation, respectively. Consider deleting the observation corresponding to such a point from the data set and refitting a line. Substantial changes in estimates and various other quantities are a signal that a more careful analysis should be carried out before proceeding.

AP* EXAM TIP

When considering linear regression, your first step should be to study the scatterplot and a residual plot. These two plots provide important information about whether a linear model is appropriate

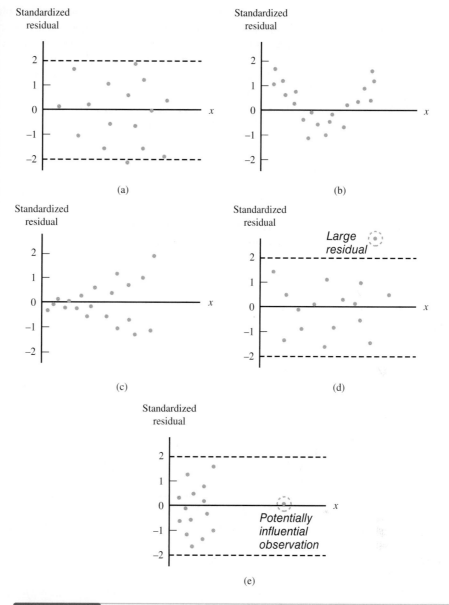

FIGURE 16.15
Examples of residual plots:
(a) satisfactory plot; (b) plot
suggesting that a curvilinear
regression model is needed;
(c) plot indicating nonconstant
variance; (d) plot showing a
large residual; (e) plot showing
a potentially influential
observation.

Example 16.7 Snow Cover and Temperature

The article **"Snow Cover and Temperature Relationships in North America and Eurasia"**
(*Journal of Climate and Applied Meteorology* **[1983]: 460–469**) explored the relationship
between October–November continental snow cover (x, in millions of square kilometers)
and December–February temperature (y, in °C). The following data refer to Eurasia during
the $n = 13$ time periods (1969–1970, 1970–1971, …, 1981–1982):

x	y	Standardized Residual	x	y	Standardized Residual
13.00	−13.5	−0.11	22.40	−18.9	−1.54
12.75	−15.7	−2.19	16.20	−14.8	0.04
16.70	−15.5	−0.36	16.70	−13.6	1.25
18.85	−14.7	1.23	13.65	−14.0	−0.28
16.60	−16.1	−0.91	13.90	−12.0	−1.54
15.35	−14.6	−0.12	14.75	−13.5	0.58
13.90	−13.4	0.34			

A simple linear regression analysis described in the article included $r^2 = 0.52$ and
$r = 0.72$, suggesting a significant linear relationship. This is confirmed by a model

utility test. The scatterplot and standardized residual plot are displayed in Figure 16.16. There are no unusual patterns, although one standardized residual, -2.19, is a bit on the large side. The most interesting feature is the observation $(22.40, -18.9)$, corresponding to a point far to the right of the others in these plots. This observation may have had a substantial influence on the estimated regression line. The estimated slope when all 13 observations are included is $b = -0.459$, and $s_b = 0.133$. When the potentially influential observation is deleted, the estimate of β based on the remaining 12 observations is $b = -0.228$. The change in slope is

$$\text{change in slope} = \text{original } b - \text{new } b$$
$$= -0.459 - (-0.288)$$
$$= -0.231$$

The change expressed in standard deviations is $-0.231/0.133 = -1.74$. Because b has changed by substantially more than 1 standard deviation, the observation under consideration appears to be highly influential.

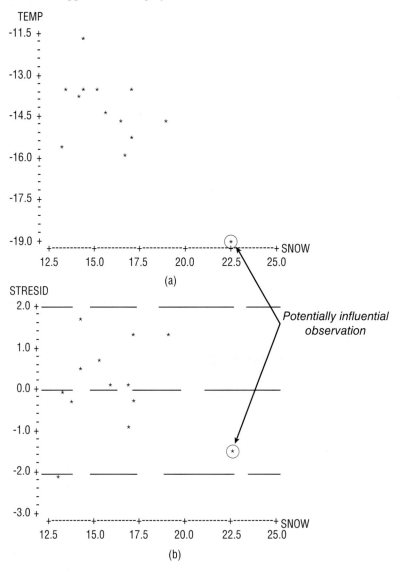

FIGURE 16.16
Plots for the data of Example 16.7:
(a) Scatter plot; (b) Standardized
residual plot

In addition, r^2 based just on the 12 observations is only 0.13, and the t ratio for testing $\beta = 0$ is not significant. Evidence for a linear relationship is much less conclusive in light of this analysis. The investigators should seek a climatological explanation for the influential observation and collect more data, which could be used to find a more useful model.

Example 16.8 Treadmill Time and Ski Time

The paper **"Physiological Characteristics and Performance of Top U.S. Biathletes"** (*Medicine and Science in Sports and Exercise* **[195]: 1302–1310**) describes a study of the relationship between cardiovascular fitness (as measured by time to exhaustion running on a treadmill) and performance on a 20-kilometer ski race. Data on

$$x = \text{treadmill time to exhaustion (in minutes)}$$

and

$$y = \text{20-km ski time (in minutes)}$$

AP* EXAM TIP

Don't forget to check assumptions. If you are used to checking assumptions before doing much in the way of calculation, it is sometimes easy to forget to check them in a regression setting. Be sure to step back and think about whether the four basic assumptions of the linear regression model are reasonable before making inferences about the population slope or using the estimated model to make predictions.

for 11 athletes are shown in Table 16.2. Standardized residuals and residuals are also given. Is it reasonable to use the given data to construct a confidence interval or test hypotheses about β, the average change in ski time associated with a 1-min increase in treadmill time? It depends on whether the assumptions that the distribution of the deviations from the population regression line at any fixed x is approximately normal and that the variance of this distribution does not depend on x are reasonable. Constructing a normal probability plot of the standardized residuals and a standardized residual plot will provide insight into whether these assumptions are in fact reasonable.

TABLE 16.2 Data, Residuals, and Standardized Residuals for Example 16.8

Observation	Treadmill	Ski Time	Residual	Standardized Residual
1	7.7	71.0	0.172	0.10
2	8.4	71.4	2.206	1.13
3	8.7	65.0	3.494	1.74
4	9.0	68.7	0.906	0.44
5	9.6	64.4	1.994	0.96
6	9.6	69.4	3.006	1.44
7	10.0	63.0	2.461	1.18
8	10.2	64.6	0.394	0.19
9	10.4	66.9	2.373	1.16
10	11.0	62.6	0.527	0.27
11	11.7	61.7	0.206	0.12

Figure 16.17 shows a normal probability plot of the standardized residuals and a standardized residual plot. The normal probability plot is quite straight, and the standardized residual plot does not show evidence of any patterns or of increasing spread.

FIGURE 16.17

Plots for Example 16.8 (a) Normal probability plot of standardized residuals; (b)Standardized residual plot

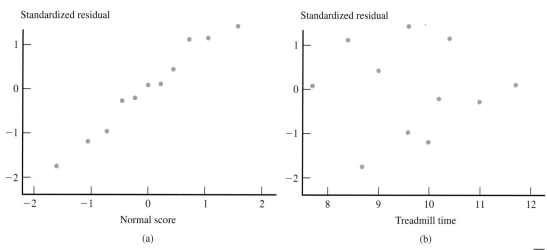

(a)

(b)

Example 16.9 A new pediatric tracheal tube

The article **"Appropriate Placement of Intubation Depth Marks in a New Cuffed, Paediatric Tracheal Tube"** (*British Journal of Anaesthesia* **[2004]: 80-87)** describes a study of the use of tracheal tubes in newborns and infants. Newborns and infants have small trachea, and there is little margin for error when inserting tracheal tubes. Using X-rays of a large number of children aged 2 months to 14 years, the researchers examined the relationships between appropriate trachea tube insertion depth and other variables such as height, weight, and age. A scatterplot and a standardized residual plot constructed using data on the insertion depth and height of the children (both measured in cm) are shown in Figure 16.18.

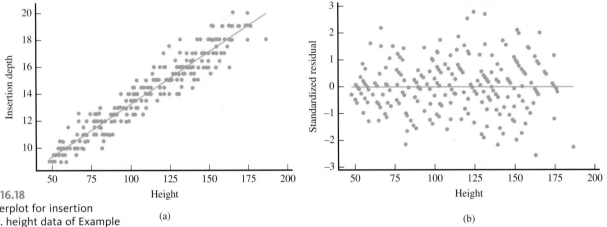

FIGURE 16.18
(a) Scatterplot for insertion depth vs. height data of Example 16.9; (b) standardized residual plot.

Residual plots like the ones pictured in Figure 16.18(b) are desirable. No point lies much outside the horizontal band between −2 and 2 (so there are no unusually large residuals corresponding to outliers). There is no point far to the left or right of the others (no observation that might be influential), and there is no pattern of curvature or differences in the variability of the residuals for different height values to indicate that the model assumptions are not reasonable.

But consider what happens when the relationship between insertion depth and weight is examined. A scatterplot of insertion depth and weight (kg) is shown in Figure 16.19(a), and a standardized residual plot in Figure 16.19(b). While some curvature is evident in the original scatterplot, it is even more clearly visible in the standardized residual plot. A careful inspection of these plots suggests that along with curvature, the residuals may be more variable at larger weights. When plots have this curved appearance and increasing variability in the residuals, the linear regression model is not appropriate.

FIGURE 16.19
(a) Scatterplot for insertion depth vs. weight data of Example 16.9; (b) standardized residual plot.

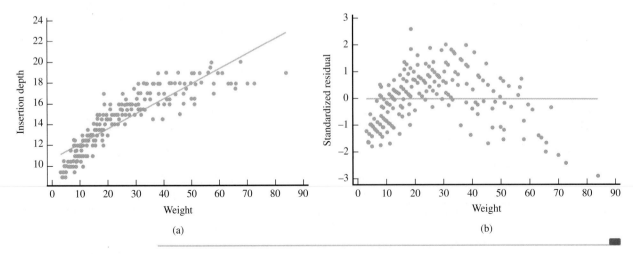

Example 16.10 Looking for love in all the right... trees?

Treefrogs' search for mating partners was the examined in the article, **"The Cause of Correlations Between Nightly Numbers of Male and Female Barking Treefrogs (*Hyla gratiosa*) Attending Choruses"** (***Behavioral Ecology*** **[2002: 274–281]**). A lek, in the world of animal behavior, is a cluster of males gathered in a relatively small area to exhibit courtship displays. The "female preference" hypothesis asserts that females will prefer larger leks over smaller leks, presumably because there are more males to choose from. The scatterplot and residual plot in Figure 16.20 show the relationship between the number of females and the number of males in observed leks of barking treefrogs. You can see that the unequal variance, which is noticeable in the scatterplot, is even more evident in the residual plot. This indicates that the assumptions of the linear regression model are not reasonable in this situation.

FIGURE 16.20
(a) Scatterplot for treefrog data of Example 16.20; (b) residual plot

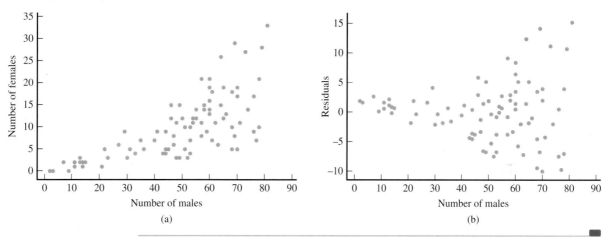

(a)   (b)

SECTION 16.3 EXERCISES

Each exercise set assesses the following chapter learning objectives: M2, M7

Exercise Set 1

16.30 The following graphs are based on data from an experiment to assess the effects of logging on a squirrel population in British Columbia (**"Effects of Logging Pattern and Intensity on Squirrel Demography,"** *The Journal of Wildlife Management* **[2007]: 2655–2663**). Plots of land, each nine hectares in area, were subjected to different percentages of logging, and the squirrel population density for each plot was measured after 3 years. The scatterplot, residual plot, and a boxplot of the residuals are shown here.

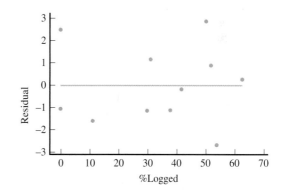

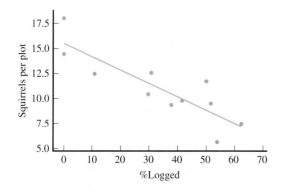

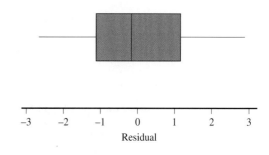

Does it appear that the assumptions of the simple linear regression model are plausible? Explain your reasoning in a few sentences.

16.31 The clutch size (number of eggs laid) for turtles is known to be influenced by body size, latitude, and average environmental temperature. Researchers gathered data on Gopher tortoises in Okeeheelee County Park in Florida to further understand the factors that affect reproduction in these animals (**"Geographic Variation in Body and Clutch Size of Gopher Tortoises,"** *Copeia* **[2007]: 355–363).** The scatterplot, residual plot, and a normal probability plot of the residuals for the least squares regression line with x = body length and y = clutch size are shown here.

Does it appear that the assumptions of the simple linear regression model are plausible? Explain your reasoning in a few sentences.

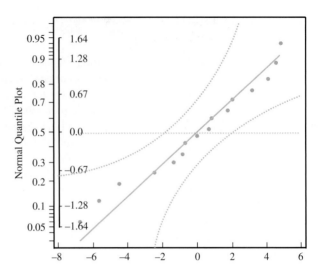

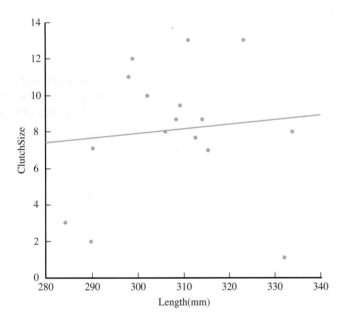

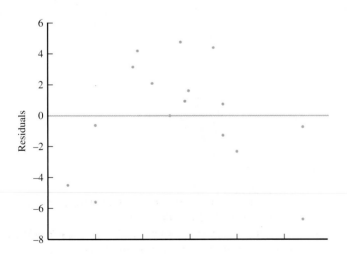

16.32 Carbon aerosols have been identified as a contributing factor in a number of air quality problems. In a chemical analysis of diesel engine exhaust, x = mass (μg/cm^2) and y = elemental carbon (μg/cm^2) were recorded (**"Comparison of Solvent Extraction and Thermal Optical Carbon Analysis Methods: Application to Diesel Vehicle Exhaust Aerosol"** *Environmental Science Technology* **[1984]: 231–234).** The estimated regression line for this data set is $\hat{y} = 31 + .737x$. The accompanying table gives the observed x and y values and the corresponding standardized residuals.

x	164.2	156.9	109.8	111.4	87.0
y	181	156	115	132	96
St. resid.	2.52	0.82	0.27	1.64	0.08
x	161.8	230.9	106.5	97.6	79.7
y	170	193	110	94	77
St. resid.	1.72	−0.73	0.05	−0.77	−1.11
x	118.7	248.8	102.4	64.2	89.4
y	106	204	98	76	89
St. resid.	−1.07	−0.95	−0.73	−0.20	−0.68
x	108.1	89.4	76.4	131.7	100.8
y	102	91	97	128	88
St. resid.	−0.75	−0.51	0.85	0.00	−1.49
x	78.9	387.8	135.0	82.9	117.9
y	86	310	141	90	130
St. resid.	−0.27	−0.89	0.91	−0.18	1.05

a. Construct a standardized residual plot. Are there any unusually large residuals? Do you think that there are any influential observations?
b. Is there any pattern in the standardized residual plot that would indicate that the simple linear regression model is not appropriate?
c. Based on your plot in Part (a), do you think that it is reasonable to assume that the variance of y is the same at each x value? Explain.

16.33 The article **"Vital Dimensions in Volume Perception: Can the Eye Fool the Stomach?"** (*Journal of Marketing*

Research [1999]: 313–326) gave the accompanying data on the dimensions of 27 representative food products (Gerber baby food, Cheez Whiz, Skippy Peanut Butter, and Ahmed's tandoori paste, to name a few).

Product	Maximum Width (cm)	Minimum Width (cm)
1	2.50	1.80
2	2.90	2.70
3	2.15	2.00
4	2.90	2.60
5	3.20	3.15
6	2.00	1.80
7	1.60	1.50
8	4.80	3.80
9	5.90	5.00
10	5.80	4.75
11	2.90	2.80
12	2.45	2.10
13	2.60	2.20
14	2.60	2.60
15	2.70	2.60
16	3.10	2.90
17	5.10	5.10
18	10.20	10.20
19	3.50	3.50
20	2.70	1.20
21	3.00	1.70
22	2.70	1.75
23	2.50	1.70
24	2.40	1.20
25	4.40	1.20
26	7.50	7.50
27	4.25	4.25

a. Fit the simple linear regression model that would allow prediction of the maximum width of a food container based on its minimum width.

b. Calculate the standardized residuals (or just the residuals if you don't have access to a computer program that gives standardized residuals) and make a residual plot to determine whether there are any outliers.

c. The data point with the largest residual is for a 1-liter Coke bottle. Delete this data point and refit the regression. Did deletion of this point result in a large change in the equation of the estimated regression line?

d. For the regression line of Part (c), interpret the estimated slope and, if appropriate, the intercept.

e. For the data set with the Coke bottle deleted, do you think that the assumptions of the simple linear regression model are reasonable? Give statistical evidence for your answer.

16.34   Models of climate change predict that global temperatures and precipitation will increase in the next 100 years, with the largest changes occurring during winter in northern latitudes. Researchers gathered data on the potential effects of climate change for flowering plants in Norway. (**"Climatic Variability, Plant Phenology, and Northern Ungulates,"** *Ecology* [1999]: 1322–1339). The table below gives data for one flower species. Range of flowering dates and elevation for different sites in Norway were used to construct the given scatterplot. A potentially influential point is indicated on the scatterplot.

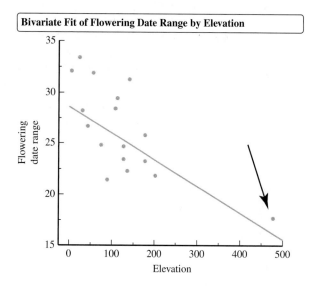

Bivariate Fit of Flowering Date Range by Elevation

Flowering Range versus Elevation: *Tussilago Farfara*	
Elevation (Meters Above Sea Level)	Flowering Date Range
23.3	33.4
5.6	32.0
55.6	31.9
140.0	31.3
31.1	28.1
112.2	29.3
106.7	28.4
42.2	26.6
75.6	24.9
176.7	25.7
126.7	24.7
126.7	23.5
176.7	23.2
201.1	21.8
133.3	22.3
90.0	21.4
41.1	19.7
125.6	17.6
477.8	17.6

a. Fit a linear regression model using all 19 observations. What are the values of a, b, r^2, s_e?

b. Fit a linear regression model with the indicated point omitted. What are the values of a, b, r^2, s_e?

c. In a few sentences, describe any differences you found in Parts (a) and (b).

d. The researchers could use the estimated regression equation based on all 19 observations to make predictions for elevations ranging from 0 to 200 meters; or they could use the estimated regression equation based on the 18 observations (omitting the observation identified by an arrow) to make predictions for elevations ranging from 0 to 500 meters. Which strategy would you recommend, and why?

SECTION 16.2 **Exercise Set 2**

16.35 In the study described in Exercise 16.31, the effect of latitude on mean clutch size was investigated. Data from various locations in Florida, Georgia, Alabama, and Mississippi on y = mean clutch size and x = latitude were measured. The scatterplot, standardized residual plot, and several graphs of the standardized residuals are shown below.

Does it appear that the assumptions of the simple linear regression model are plausible? Explain your reasoning in a few sentences.

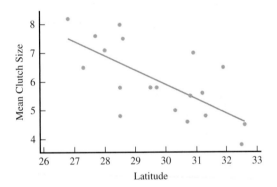

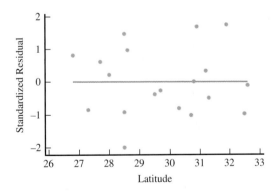

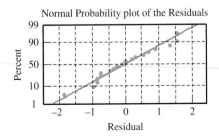

Normal Probability plot of the Residuals

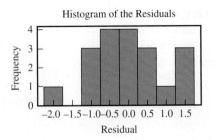

Histogram of the Residuals

16.36 Exercise 6.21 gave data on x = nerve firing frequency and y = pleasantness rating when nerves were stimulated by a light brushing stoke on the forearm. The x values and the corresponding residuals from a simple linear regression are as follows:

a. Construct a standardized residual plot. Does the plot exhibit any unusual features?

Firing Frequency, x	Standardized Residual
23	−1.83
24	0.04
22	1.45
25	0.20
27	−1.07
28	1.19
34	−0.24
33	−0.13
36	−0.81
34	1.17

b. A normal probability plot of the standardized residuals follows. Based on this plot, do you think it is reasonable to assume that the error distribution is approximately normal? Explain.

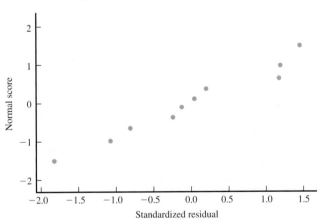

16.37 The accompanying scatterplot, based on 34 sediment samples with x = sediment depth (cm) and y = oil and grease content (mg/kg), appeared in the article **"Mined Land Reclamation Using Polluted Urban Navigable Waterway Sediments"** (*Journal of Environmental Quality* **[1984]: 415–422**). Discuss the effect that the observation (20, 33,000) will have on the estimated regression line. If this point were omitted, what do you think will happen to the slope of the estimated regression line compared to the slope when this point is included?

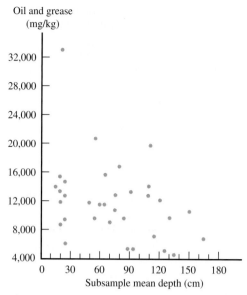

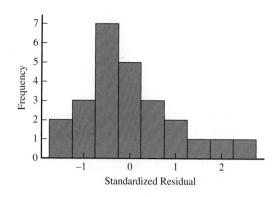

16.38 Investigators in northern Alaska periodically monitored radio collared wolves in 25 wolf packs over 4 years, keeping track of the packs' home ranges. (**"Population Dynamics and Harvest Characteristics of Wolves in the Central Brooks Range, Alaska,"** *Wildlife Monographs*, **[2008]: 1–25).** The home range of a pack is the area typically covered by its members in a specified amount of time. The investigators noticed that wolf packs with larger home ranges tended to be located more often by monitoring equipment. The investigators decided to explore the relationship between home range and the number of locations per pack. A scatterplot and standardized residual plot of the data are shown below, as well as plots of the standardized residuals.

Does it appear that the assumptions of the simple linear regression model are plausible? Explain your reasoning in a few sentences.

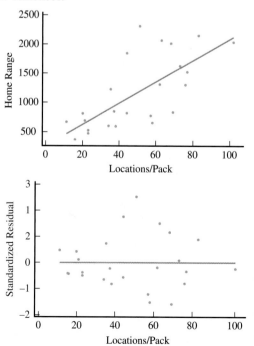

Additional Exercises

16.39 Carbon acrosols have been identified as a contributing factor in a number of air quality problems. In a chemical analysis of diesel engine exhaust, x = mass (μg/cm^2) and y = elemental carbon (μg/cm^2) were recorded (**"Comparison of Solvent Extraction and Thermal Optical Carbon Analysis Methods: Application to Diesel Vehicle Exhaust Aerosol"** *Environmental science Technology* **[1984]: 231–234).** The estimated regression line for this data set is $\hat{y} = 31 + .737x$.

A scatterplot of the data and a standardized residual plot are shown below.

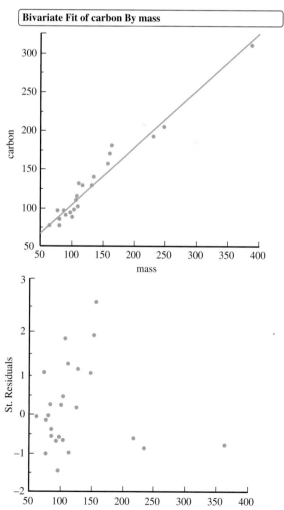

a. Are there any unusually large residuals? Do you think that there are any influential observations?

b. Is there any pattern in the standardized residual plot that would indicate that the simple linear regression model is not appropriate?

c. Based on the scatterplot and the standardized residual plot, do you think that it is reasonable to assume that the variance of y is the same at each x value? Explain.

16.40 Models of climate change predict that global temperatures and precipitation will increase in the next 100 years, with the largest changes occurring during winter in northern latitudes. Researchers recently gathered data on the potential effects of climate change for flowering plants in Norway. (**"Climatic Variability, Plant Phenology, and Northern Ungulates,"** *Ecology* **[1999]: 1322-1339**). The table below gives data for one flower species. A scatterplot of the "range of flowering dates" versus latitude for different sites in Norway is also shown. Two points that are potentially influential are indicated on the scatterplot.

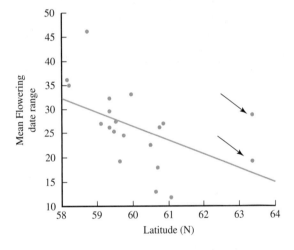

Latitude (N)

Flowering Range Versus Latitude: *Anemone Hepatica*	
Latitude (N)	Flowering Date Range
58.7	46.1
58.2	35.9
58.2	34.7
59.4	32.3
60.0	33.0
59.4	29.7
59.1	26.9
59.3	26.2
59.5	25.6
59.5	27.6
59.7	19.1
59.8	24.4
60.8	26.2

(continued)

Flowering Range Versus Latitude: *Anemone Hepatica*	
Latitude (N)	Flowering Date Range
60.9	26.8
63.4	28.7
63.4	19.2
60.5	22.5
60.7	17.9
60.7	12.9
61.1	11.8

a. Fit a linear regression model using all 20 observations. What are the values of a, b, r^2 and s_e?

b. Fit a linear regression model with the two observations identified by arrows omitted. What are the values of a, b, r^2 and s_e?

c. In a few sentences, describe any differences you found in Parts (a) and (b).

d. The researchers could use the estimated regression equation based on all 20 observations to make predictions for latitudes ranging from 58 to 64, or they could use the estimated regression equation based on the 18 observations (omitting the two observations identified by arrows) to make predictions for latitudes ranging from 58 to 62. Which strategy would you recommend, and why?

16.41 The sand scorpion is a predator that always hunts from a motionless resting position outside its own burrow. When prey appears on the horizon, within say 20 cm, the scorpion assumes an alert posture; it determines the angular position of the prey, makes a quick rotation, and runs after it. In a recent study of the scorpion's accuracy, the angular position (0 degrees = right in front) of the prey, and the turning angle of the scorpion was recorded for 23 attacks. A simple regression model relating the response angle of the predator to the target angle position of the prey, $\hat{r} = a + b(t)$, was fit. The resulting residual plot is shown. Describe the locations of any outliers you see in the residual plot.

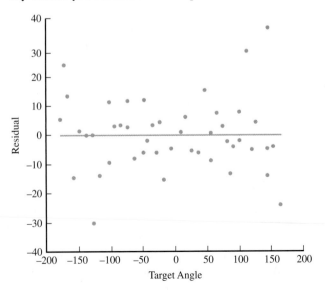

Target Angle

16.42 The production of pups and their survival are the most significant factors contributing to gray wolf population growth. The causes of early pup mortality are unknown, and difficult to observe. The pups are concealed within their dens for 3 weeks after birth, and after they emerge it is difficult to confirm their parentage. Researchers recently used portable ultrasound equipment to investigate some factors related to reproduction. (**"Diagnosing Pregnancy, in Utero Litter Size, and Fetal Growth with Ultrasound in Wild, Free-Ranging Wolves,"** *Journal of Mammology* [2006]: 85–92)

A scatterplot and linear regression of the length of a wolf fetus (in cm, measured from crown to rump) and gestational age (in days) is shown below. Identify the point that has the largest residual by giving its approximate coordinates.

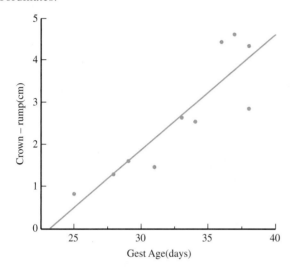

16.43 The authors of the article **"Age, Spacing and Growth Rate of Tamarix as an Indication of Lake Boundary Fluctuations at Sebkhet Kelbia, Tunisia"** (*Journal of Arid Environments* [1982]: 43–51) used a simple linear regression model to describe the relationship between y = vigor (average width in centimeters of the last two annual rings) and x = stem density (stems/m^2). The estimated model was based on the following data. Also given are the standardized residuals.

x	4	5	6	9	14
y	0.75	1.20	0.55	0.60	0.65
St. resid.	−0.28	1.92	−0.90	−0.28	0.54
x	15	15	19	21	22
y	0.55	0.00	0.35	0.45	0.40
St. resid.	0.24	−2.05	−0.12	0.60	0.52

a. What assumptions are required for the simple linear regression model to be appropriate?

b. Construct a normal probability plot of the standardized residuals. Does the assumption that the random deviation distribution is normal appear to be reasonable? Explain.

c. Construct a standardized residual plot. Are there any unusually large residuals?

d. Is there anything about the standardized residual plot that would cause you to question the use of the simple linear regression model to describe the relationship between x and y?

ARE YOU READY TO MOVE ON?   **CHAPTER 16 REVIEW EXERCISES**

All chapter learning objectives are assessed in these exercises. The learning objectives assessed in each exercise are given in parentheses.

16.44 (C1)
Describe what distinguishes a deterministic model from a probabilistic model.

16.45 (C2)
In the context of the simple linear regression model, explain the difference between α and a. Between β and b. Between σ_e and s_e.

16.46 (M1)
The SAT and ACT exams are often used to predict a student's first-term college grade point average (GPA). Different formulas are used for different colleges and majors. Suppose that a student is applying to State U with an intended major in civil engineering. Also suppose that for this college and this major, the following model is used to predict first term GPA.

$$GPA = a + b\,(ACT)$$
$$a = 0.5$$
$$b = 0.1$$

a. In this context, what would be the appropriate interpretation of a?

b. In this context, what would be the appropriate interpretation of b?

16.47 (M2)
Theropods were carnivorous dinosaurs, characterized by short forelimbs, living in the Jurassic and Cretaceous periods. (*Tyrannosaurus rex* is classified as a Theropod.) What scientists know about therapods is based on studying incomplete skeletal remains. In a study described in the paper **"My Theropod is Bigger than Yours...or not: Estimating Body Size from Skull Length in Theropods"** (*Journal of Vertebrate*

Paleontology [2007]: 108–115), researchers used data from skeletons to develop a model describing the relationship between body length and skull length. JMP was used to produce the following graphical displays and computer output. When you evaluate the fit of an estimated regression line, all of the information below is considered as a whole. However, the summary statistics in the computer output and the different plots each convey some specific information.

a. Using only the scatterplot, do you think a linear model does a good job of describing the relationship? Explain why or why not.
b. Using only the residual plot, what can you determine about whether the basic assumptions of the linear regression model are met?
c. Using only the normal probability plot and boxplot of the residuals, what can you determine about whether the basic assumptions of the linear regression model are met?
d. Using only the values of r^2 and s_e, what can you say about the quality of the fit of the linear model for these data?

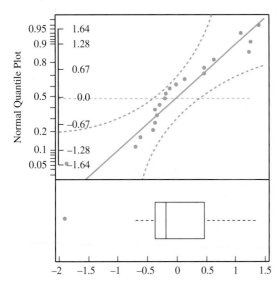

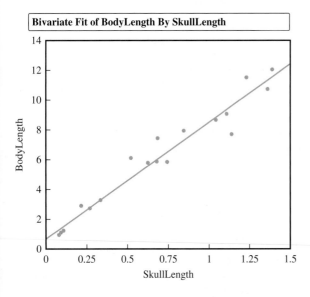

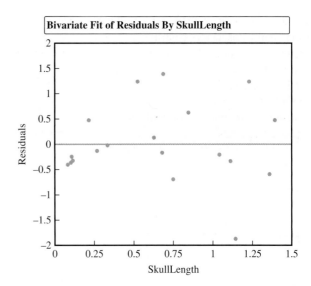

Bivariate Fit of Residuals By SkullLength

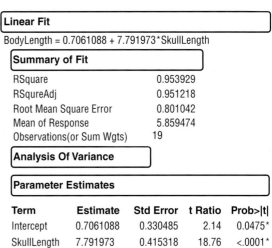

Linear Fit

BodyLength = 0.7061088 + 7.791973*SkullLength

Summary of Fit	
RSquare	0.953929
RSqureAdj	0.951218
Root Mean Square Error	0.801042
Mean of Response	5.859474
Observations(or Sum Wgts)	19

Analysis Of Variance

Parameter Estimates

Term	Estimate	Std Error	t Ratio	Prob>|t|
Intercept	0.7061088	0.330485	2.14	0.0475*
SkullLength	7.791973	0.415318	18.76	<.0001*

16.48 (M3)
There are 4 basic assumptions necessary for making inferences about β, the slope of the population regression line.
a. What are the four assumptions?
b. Which assumptions can be checked using sample data?
c. What statistics or graphs would be used to check each of the assumptions you listed in Part (b)?

16.49 (M3, M4, M5, P1, P2)
Ruffed grouse are a species of birds that nest on the ground. Because of this, chick survival at night in the first few weeks of life depends on avoiding predators. Biologists have theorized that protection from predators might be supplied by the mother hen's choice of brooding sites. One variable that biologists thought might be related to survival is the density of vegetation in the vicinity of the nest. Dense vegetation would possible reduce the ability of predators to detect the nests. The paper **"Nocturnal Roost Habitat Selection by Ruffed Grouse Broods (** *Journal of Field Ornithology* **[2005]:168–174)** describes a study in which

researchers monitored the survival of the brood (number of chicks surviving /number of eggs hatched) in 23 nests in different vegetation densities (thousands of stems / hectare.)

Computer output (from JMP) is shown below.

Linear Fit

BroodSurvival = 0.9468008 − 0.0261902*StemDensity

Summary of Fit

RSquare	0.193788
RSqureAdj	0.155397
Root Mean Square Error	0.287538
Mean of Response	0.436043
Observations(or Sum Wgts)	23

Parameter Estimates

Term	Estimate	Std Error	t Ratio	Prob>|t|
Intercept	0.9468008	0.235108	4.03	0.0006*
StemDensity	−0.02619	0.011657	−2.25	0.0355*

a. Is there convincing evidence of a useful linear relationship between brood survival and stem density? Explain.

b. Would you describe the relationship as strong? Why or why not?

c. Construct a 95% confidence interval for β and interpret it in context.

d. What margin of error is associated with the confidence interval in part (c)?

16.50 (M7)

Researchers in Hawaii have recently documented a large increase in the prevalence of a bird parasite known as chewing lice. (**"Explosive Increase in Ectoparasites in Hawaiian Forest Birds,"** *The Journal of Parasitology* [2008]: 1009–1021). Current data suggest that the prevalence of chewing lice may be less for bird species with a high degree of bill overhang. A species is said to have bill overhang when the upper bill extends downward in front of the end of the lower bill. The following scatterplot shown shows the relationship between the prevalence of chewing lice and bill overhang for 8 bird species in the Hawaiian Islands. A residual plot is also shown. Use these plots to identify any outliers or potentially influential observations. For each point you identify, assess its influence on the estimated slope of the regression line.

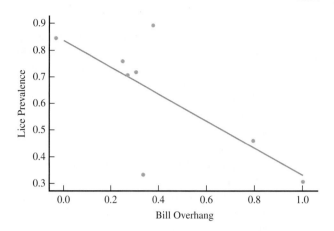

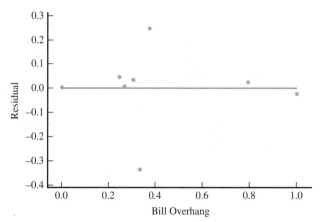

16.51 (M6)

Suppose you are given the computer output shown. You want to test the hypothesis, $\beta = 1.0$. Describe how you would use the computer output to test this hypothesis

Linear Fit

y = 5.6452776 + 0.9797401*x

Summary of Fit

RSquare	0.985289
RSqureAdj	0.984954
Root Mean Square Error	12.48525
Mean of Response	0.791304
Observations(or Sum Wgts)	46

Parameter Estimates

Term	Estimate	Std Error	t Ratio	Prob>|t|
Intercept	5.6452776	1.84302	3.06	0.0037*
x	0.9797401	0.018048	54.29	<.0001*

TECHNOLOGY NOTES

Regression Test

TI-83/84

1. Enter the data for the independent variable into **L1** (In order to access lists press the **STAT** key, highlight the option called **Edit...** then press **ENTER**)

2. Enter the data for the dependent variable into **L2**
3. Press **STAT**
4. Highlight **TESTS**
5. Highlight **LinRegTTest...** and press **ENTER**
6. Next to β & ρ select the appropriate alternative hypothesis
7. Highlight **Calculate**

TI-Nspire

1. Enter the data into two separate data lists (In order to access data lists select the spreadsheet option and press **enter**) **Note:** Be sure to title the lists by selecting the top row of the column and typing a title.
2. Press the **menu** key and select **4:Stat Tests** then **4:Stats Tests** then **A:Linear Reg t Test...** and press **enter**
3. In the box next to **X List** choose the list title where you stored your independent data from the drop-down menu
4. In the box next to **Y List** choose the list title where you stored your dependent data from the drop-down menu
5. In the box next to **Alternate Hyp** choose the appropriate alternative hypothesis from the drop-down menu
6. Press **OK**

JMP

1. Input the data for the dependent variable into the first column
2. Input the data for the independent variable into the second column
3. Click **Analyze** and select **Fit Y by X**
4. Select the dependent variable (Y) from the box under **Select Columns** and click on **Y, Response**
5. Select the independent variable (X) from the box under **Select Columns** and click on **X, Factor**
6. Click the red arrow next to **Bivariate Fit of...** and select **Fit Line**

MINITAB

1. Input the data for the dependent variable into the first column
2. Input the data for the independent variable into the second column
3. Select **Stat** then **Regression** then **Regression...**
4. Highlight the name of the column containing the dependent variable and click **Select**
5. Highlight the name of the column containing the independent variable and click **Select**
6. Click **OK**

Note: You may need to scroll up in the Session window to view the t-test results for the regression analysis.

SPSS

1. Input the data for the dependent variable into one column
2. Input the data for the independent variable into a second column
3. Click **Analyze** then click **Regression** then click **Linear...**
4. Select the name of the dependent variable and click the arrow to move the variable to the box under **Dependent:**
5. Select the name of the independent variable and click the arrow to move the variable to the box under **Independent(s):**
6. Click **OK**

Note: The p-value for the regression test can be found in the **Coefficients** table in the row with the independent variable name.

Excel

1. Input the data for the dependent variable into the first column
2. Input the data for the independent variable into the second column
3. Select **Analyze** then choose **Regression** then choose **Linear...**
4. Highlight the name of the column containing the dependent variable
5. Click the arrow button next to the **Dependent** box to move the variable to this box
6. Highlight the name of the column containing the independent variable
7. Click the arrow button next to the **Independent** box to move the variable to this box
8. Click **OK**

Note: The test statistic and p-value for the regression test for the slope can be found in the third table of output. These values are listed in the row titled with the independent variable name and the columns entitled *t Stat and P-value.*

AP* Review Questions for Chapter 16

Use the following information for questions 1–6.

A study was carried out to investigate the relationship between x = the number of components needing repair and y = the time of the service call (in minutes) for a computer repair company. The number of components and the service time for a random sample of 20 service calls was used to fit a simple linear regression model. Partial computer output is shown below.

The regression equation is
Time 5 37.2 1 9.97 Number

Predictor	Coef	SE Coef	T	P
Constant	37.213	7.985	4.66	0.000
Number	9.9695	0.7218	13.81	0.000

S = 18.7534   R-Sq = 89.7%   R-Sq(adj) = 89.2%

1. Which of the following statements is a correct interpretation of the value 9.97?

   (A) The average number of components needing repair goes up 9.97 for each 1 minute increase in the service time of a call.
   (B) On average, the service call time goes up 9.97 minutes for each additional component needing repair.
   (C) The service call time is 9.97 minutes when there are 0 components to repair.
   (D) Approximately 9.97% of the observed variation in the service call times can be explained by the linear relationship between service time and number of components requiring repair.
   (E) If this regression equation is used to predict service call times, we can expect predictions to be within 9.97 minutes of the actual time.

2. Which of the following statements is a correct interpretation of the value 89.7%?

   (A) On average, the service call time goes up 89.7 minutes for each additional component needing repair.
   (B) The magnitude of a typical difference between an observed service call time and the service call time predicted by the linear model is approximately 89.7 minutes.
   (C) The correlation between service call time and number of components needing repair is 89.7%.
   (D) Approximately 89.7% of the observed variation in service call time can be explained by the linear relationship between service call time and number of components needing repair.
   (E) If this regression equation is used to predict service call times, we can expect predictions to be within 89.7 minutes of the actual time.

3. The value of s_e is 18.75. Which of the following is an appropriate interpretation of this value?

   (A) 18.75% of the variability in service time can be explained by the linear relationship between service call time and number of components needing repair.
   (B) There is a positive correlation between service call time and number of components needing repair.
   (C) For every 1-component increase in the number of components needing repair, the predicted service call time increases by about 18.75 minutes.
   (D) The magnitude of a typical difference between an observed service time and the service call time predicted by the linear model is approximately 18.75 minutes.
   (E) The average service call time is 18.75 minutes.

4. The value of s_e is 18.75. If the assumptions of the simple linear regression model are satisfied, which of the following is correct?

   (A) The width of a 95% confidence interval for the slope of the population regression line is $2(18.75) = 37.50$.
   (B) It would be unlikely that a prediction based on the regression line will be greater than 18.75 minutes.
   (C) It would be unlikely that a prediction based on the regression line will differ from the actual value by more than $2(18.75) = 37.50$ minutes.
   (D) Errors associated with predictions based on the regression line will always be less than 18.75 minutes.
   (E) The value of s_e does not provide any information about the anticipated magnitude of prediction errors.

5. Which of the following is a 95% confidence interval for the change in service time associated with a 1-unit increase in the number of components needing repair?

   (A) $37.21 \pm (1.96)(7.985)$
   (B) $37.21 \pm (2.910)(7.985)$
   (C) $9.97 \pm (1.96)(0.7218)$
   (D) $9.97 \pm (2.10)(0.7218)$
   (E) $9.97 \pm 2(18.7534)$

6. If the basic assumptions of the simple linear regression model are reasonable, what conclusion should be reached regarding model utility if a significance level of 0.05 is used for the model utility test?

   (A) There is convincing evidence of a negative linear relationship between service call time and number of components needing repair.
   (B) There is convincing evidence that the model is not useful for predicting service call time.
   (C) There is convincing evidence that the model is useful for predicting service call time.
   (D) There is not convincing evidence that the model is useful for predicting service call time.
   (E) A conclusion cannot be reached based on the given information.

7. If there is a positive linear relationship between two variables *x* and *y*, which of the following must be true of β, the slope of the population regression line?

(A) $\beta < 0$
(B) $\beta > 0$
(C) $\beta = 0$
(D) $\beta > 1$
(E) $-1 < \beta < 1$

8. The plots shown are residual plots resulting from fitting a linear regression. Which of these plots indicates that the relationship between the two variables used to fit the line may not be linear?

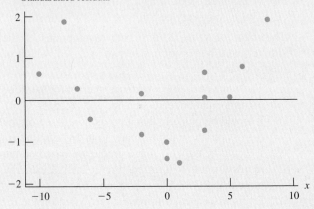

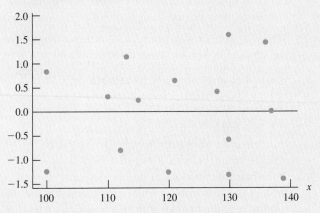

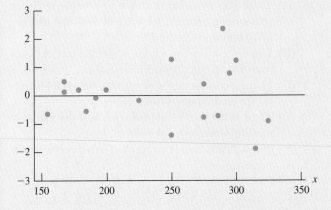

(A) I only
(B) II only
(C) III only
(D) I and III only
(E) II and III only

Use the scatterplot below to answer questions 9 and 10.

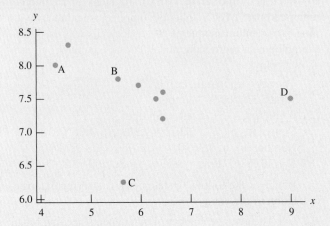

9. Which of the labeled points would have the largest residual when a linear model is fit to the data?

(A) A
(B) B
(C) C
(D) D
(E) Both C and D

10. Which of the labeled points corresponds to a potentially influential observation if a linear model is to be fit to the data?

(A) A
(B) B
(C) C
(D) D
(E) Both C and D

11. If there is evidence of a linear relationship between *x* and *y*, what decision will be made in a test of $H_0: \beta = 0$ versus $H_0: \beta \neq 0$?

(A) Reject H_0 and conclude that there is no evidence that the linear model is useful
(B) Reject H_0 and conclude that there is evidence that the linear model is useful
(C) Fail to reject H_0 and conclude that there is no evidence that the linear model is useful
(D) Fail to reject H_0 and conclude that there is evidence that the linear model is useful
(E) Not enough information to say.

Use the following information to answer questions 12 and 13.

As part of a study of the swimming speed of sharks, a random sample of 18 lemon sharks (*Triakis semifasciata*) were observed in a laboratory sea tunnel. Body lengths and maximum sustainable swimming speeds ("MSSS," reported in body lengths per second) were measured for each shark. The computer output from a regression with y = MSSS and x = body length is given below.

Linear Fit

MSSS = 1.8928955 - 0.0104278*Length

Summary of Fit

RSquare	0.526395
RSquare Adj	0.496794
S	0.272031
Mean of Response	1.24
N Observations	18

Analysis of Variance

Source	DF	Sum of Squares	Mean Square	F Ratio
Model	1	1.3159870	1.31599	17.7834
Error	16	1.1840130	0.07400	**Prob > F**
Total	17	2.5000000		0.0007*

Parameter Estimates

Term	Estimate	Std Error	t Ratio	Prob>|t|
Intercept	1.8928955	0.167575	11.30	<.0001*
Length(cm)	−0.010428	0.002473	−4.22	0.0007*

12. For this data set, the model utility test is based on how many degrees of freedom?

    (A) 15
    (B) 16
    (C) 17
    (D) 18
    (E) 19

13. What is the P-value associated with the model utility test?

    (A) 0.0001
    (B) 0.0007
    (C) 0.07400
    (D) 0.167575
    (E) 0.526395

14. Which of the following is *not* an assumption that is made about the random deviation e in a simple linear regression model?

    (A) The distribution of e is normal.
    (B) The standard deviation of e, σ_e, depends upon the particular value of x.
    (C) The mean value of e is 0.
    (D) The random deviations, e_1, e_2 ..., e_n, associated with different observations are independent of one another.
    (E) The standard deviation of e, σ_e, is the same for each x value.

15. The residual plot below indicates that the one or more of assumptions of the linear regression model may not be met. Which of the following is a reasonable conclusion based on this residual plot?

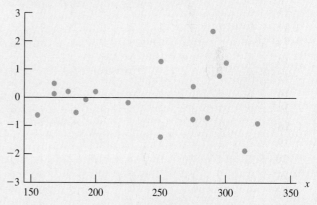

    (A) The residual plot clearly indicates a non-linear model would be more appropriate.
    (B) There is evidence that the residuals are not normally distributed.
    (C) The slope of the regression line is non-zero.
    (D) The correlation between x and y is non-zero.
    (E) There is evidence the residuals do not have the same variance for all x values.

Appendix A: Statistical Tables

TABLE 1 Random Numbers

Row																				
1	4	5	1	8	5	0	3	3	7	1	2	8	4	5	1	1	0	9	5	7
2	4	2	5	5	8	0	4	5	7	0	7	0	3	6	6	1	3	1	3	1
3	8	9	9	3	4	3	5	0	6	3	9	1	1	8	2	6	9	2	0	9
4	8	9	0	7	2	9	9	0	4	7	6	7	4	7	1	3	4	3	5	3
5	5	7	3	1	0	3	7	4	7	8	5	2	0	1	3	7	7	6	3	6
6	0	9	3	8	7	6	7	9	9	5	6	2	5	6	5	8	4	2	6	4
7	4	1	0	1	0	2	2	0	4	7	5	1	1	9	4	7	9	7	5	1
8	6	4	7	3	6	3	4	5	1	2	3	1	1	8	0	0	4	8	2	0
9	8	0	2	8	7	9	3	8	4	0	4	2	0	8	9	1	2	3	3	2
10	9	4	6	0	6	9	7	8	8	2	5	2	9	6	0	1	4	6	0	5
11	6	6	9	5	7	4	4	6	3	2	0	6	0	8	9	1	3	6	1	8
12	0	7	1	7	7	7	2	9	7	8	7	5	8	8	6	9	8	4	1	0
13	6	1	3	0	9	7	3	3	6	6	0	4	1	8	3	2	6	7	6	8
14	2	2	3	6	2	1	3	0	2	2	6	6	9	7	0	2	1	2	5	8
15	0	7	1	7	4	2	0	0	0	1	3	1	2	0	4	7	8	4	1	0
16	6	6	5	1	6	1	8	1	5	5	2	6	2	0	1	1	5	2	3	6
17	9	9	6	2	5	3	5	9	8	3	7	5	0	1	3	9	3	8	0	8
18	9	9	9	6	1	2	9	3	4	6	5	6	4	6	5	8	2	7	4	0
19	2	5	6	3	1	9	8	1	1	0	3	5	6	7	9	1	4	5	2	0
20	5	1	1	9	8	1	2	1	1	6	9	8	1	8	1	9	9	1	2	0
21	1	9	8	0	7	4	6	8	4	0	3	0	8	1	1	0	6	2	3	2
22	9	7	0	9	6	3	8	9	9	7	0	6	5	4	3	6	5	0	3	2
23	1	7	6	4	8	2	0	3	9	6	3	6	2	1	0	7	7	3	1	7
24	6	2	5	8	2	0	7	8	6	4	6	6	8	9	2	0	6	9	0	4
25	1	5	7	1	1	1	9	5	1	4	5	2	8	3	4	3	0	7	3	5
26	1	4	6	6	5	6	0	1	9	4	0	5	2	7	6	4	3	6	8	8
27	1	8	5	0	2	1	6	8	0	7	7	2	6	2	6	7	5	4	8	7
28	7	8	7	4	6	5	4	3	7	9	3	9	2	7	9	5	4	2	3	1
29	1	6	3	2	3	7	3	0	7	2	4	8	0	9	9	9	4	7	0	
30	2	8	9	0	8	1	6	8	1	7	3	1	3	0	9	7	2	5	7	9
31	0	7	8	8	6	5	7	5	5	4	0	0	3	4	1	2	7	3	7	9
32	8	4	0	1	4	5	1	9	1	1	2	1	5	3	2	8	5	5	7	5
33	7	3	5	9	7	0	4	9	1	2	1	3	2	5	1	9	3	3	8	3
34	4	7	2	6	7	6	9	9	2	7	8	7	5	5	5	2	4	4	3	4
35	9	3	3	7	0	7	0	5	7	5	6	9	5	4	3	1	4	6	6	8
36	0	2	4	9	7	8	1	6	3	8	7	8	0	5	6	7	2	7	5	0
37	7	1	0	1	8	4	7	1	2	9	3	8	0	0	8	7	9	2	8	6
38	9	7	9	4	4	5	3	1	9	3	4	5	0	6	3	5	9	6	9	8
39	0	4	2	5	0	0	9	9	6	4	0	6	9	0	3	8	3	5	7	2
40	0	7	1	2	3	6	1	7	9	3	9	5	4	6	8	4	8	8	0	6
41	3	5	6	6	2	4	4	5	6	3	7	8	7	6	5	2	0	4	3	2
42	6	6	8	5	5	2	9	7	9	3	3	1	6	9	5	9	7	1	1	2
43	9	5	0	4	3	1	1	7	3	9	2	7	7	4	7	0	3	1	2	8
44	5	1	7	8	9	4	7	2	9	2	8	9	9	8	0	6	3	7	2	1
45	1	6	3	9	4	1	3	2	1	1	8	5	6	3	4	1	9	3	1	7
46	4	4	8	6	4	0	3	8	3	8	3	5	9	5	9	4	8	3	9	4
47	7	7	6	6	4	5	4	4	8	4	4	0	3	9	8	5	2	0	2	3
48	2	5	6	6	3	7	0	6	5	6	9	0	1	9	5	2	6	9	1	2

TABLE 1 Random Numbers *(Continued)*

Row																				
49	9	4	0	4	7	5	3	2	8	7	2	7	4	9	3	9	6	5	5	6
50	7	3	1	5	6	6	5	0	3	5	3	7	2	8	6	2	4	1	8	7
51	7	5	8	2	8	8	8	7	6	4	1	1	0	2	3	1	9	3	6	0
52	3	3	6	0	9	1	1	0	3	2	7	8	2	0	5	3	4	8	9	8
53	0	2	9	6	9	8	9	3	8	1	5	3	9	9	7	0	7	7	1	6
54	8	5	9	6	2	9	6	8	2	1	2	4	7	0	6	8	3	4	6	1
55	5	4	7	6	1	0	0	1	0	4	6	1	4	1	5	0	9	6	5	5
56	5	0	3	6	4	1	9	8	4	4	1	2	0	2	5	1	8	1	2	1
57	0	2	6	3	7	5	1	1	6	6	0	5	8	1	2	3	3	6	1	3
58	3	8	1	6	3	8	1	4	5	2	9	4	2	5	7	3	2	3	1	8
59	9	1	5	6	0	6	5	6	6	3	6	2	3	0	0	0	1	8	5	9
60	5	3	5	6	3	9	5	4	7	3	6	6	7	5	0	1	5	6	7	3
61	9	6	6	4	5	7	7	6	1	5	4	4	8	0	6	5	7	6	3	0
62	6	3	0	6	7	9	5	5	4	6	2	2	8	4	4	0	0	9	9	8
63	8	5	8	3	5	2	0	6	6	0	0	6	0	6	3	0	1	7	0	5
64	3	8	2	4	9	0	9	2	6	2	9	5	1	9	1	9	0	8	3	3
65	1	4	4	1	1	7	4	6	3	6	5	6	5	5	7	7	0	3	5	8
66	5	9	9	5	3	7	2	5	1	7	1	1	0	7	1	0	9	2	8	8
67	8	7	1	7	5	2	5	6	8	7	9	9	1	3	9	6	4	9	3	0
68	6	7	2	3	1	4	9	2	1	7	0	8	6	7	8	9	9	4	7	4
69	2	3	2	8	7	0	9	7	1	1	1	2	8	2	9	1	0	6	7	7
70	2	9	5	7	8	4	7	9	0	3	6	9	2	0	6	0	6	2	6	8
71	4	8	9	8	3	2	7	6	9	1	9	8	6	9	5	2	4	9	9	9
72	1	5	6	5	7	7	5	4	3	4	3	8	1	8	9	9	4	4	1	1
73	1	8	1	1	7	2	8	5	5	8	9	9	9	6	2	0	1	6	6	7
74	5	7	7	0	9	5	5	6	8	6	8	2	2	6	0	5	5	1	8	7
75	1	8	6	0	5	4	8	3	4	5	3	5	8	7	7	7	8	5	7	0
76	2	6	6	7	9	4	2	2	8	7	4	3	4	9	6	1	9	4	3	9
77	3	6	6	4	5	7	8	3	0	2	8	4	6	7	2	1	4	5	2	3
78	0	7	8	0	1	2	1	1	3	4	2	1	6	9	3	3	5	4	0	4
79	8	3	6	0	5	7	7	9	1	5	8	8	4	9	5	7	2	2	7	6
80	5	3	6	9	0	6	3	8	7	5	9	5	9	7	4	2	5	6	2	9
81	0	9	3	7	7	2	8	6	4	3	2	9	4	8	2	9	9	6	9	9
82	9	4	7	4	0	0	0	3	5	4	6	6	2	6	2	3	6	1	1	4
83	5	5	4	1	7	8	6	4	2	3	2	9	8	4	6	3	8	3	0	5
84	5	3	0	0	5	4	8	0	7	4	7	6	2	1	1	2	1	2	6	9
85	3	3	0	9	3	2	9	4	0	5	5	4	8	7	5	7	5	3	8	8
86	3	0	5	7	1	9	5	8	0	0	4	5	3	0	3	0	2	7	6	7
87	5	0	8	6	0	8	1	6	2	0	8	6	5	4	0	7	2	9	1	0
88	3	6	4	7	8	2	3	5	7	9	8	5	2	7	6	9	0	2	4	9
89	9	0	4	4	9	1	6	8	5	2	8	9	0	7	5	7	2	5	1	8
90	9	5	2	6	9	3	9	6	5	1	8	8	7	8	2	0	4	4	7	9
91	9	4	5	7	0	3	4	6	4	2	5	4	8	6	1	1	9	1	8	8
92	8	1	1	8	0	5	4	2	8	5	3	3	3	0	1	1	4	4	8	3
93	6	9	4	7	8	3	3	9	1	2	5	0	1	2	3	0	1	1	2	5
94	0	0	6	8	8	7	2	4	4	7	6	6	0	3	4	7	5	6	8	2
95	5	3	3	9	3	8	4	9	1	9	1	7	8	4	5	2	2	5	4	4
96	2	5	6	2	7	6	0	3	8	1	4	4	2	6	8	3	6	3	2	8
97	7	4	3	7	9	6	8	6	2	8	3	8	4	2	2	0	7	0	5	3
98	1	9	0	8	8	0	1	2	2	2	7	5	6	5	5	7	8	7	2	6
99	2	4	8	0	2	5	2	7	0	5	9	6	6	1	5	8	7	9	7	5
100	4	1	7	8	6	7	1	1	5	8	9	4	8	9	8	3	0	9	0	7

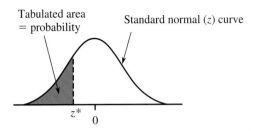

Tabulated area = probability

Standard normal (z) curve

z^* 0

TABLE 2 Standard Normal Probabilities (Cumulative *z* Curve Areas)

z^*	.00	.01	.02	.03	.04	.05	.06	.07	.08	.09
−3.8	.0001	.0001	.0001	.0001	.0001	.0001	.0001	.0001	.0001	.0000
−3.7	.0001	.0001	.0001	.0001	.0001	.0001	.0001	.0001	.0001	.0001
−3.6	.0002	.0002	.0001	.0001	.0001	.0001	.0001	.0001	.0001	.0001
−3.5	.0002	.0002	.0002	.0002	.0002	.0002	.0002	.0002	.0002	.0002
−3.4	.0003	.0003	.0003	.0003	.0003	.0003	.0003	.0003	.0003	.0002
−3.3	.0005	.0005	.0005	.0004	.0004	.0004	.0004	.0004	.0004	.0003
−3.2	.0007	.0007	.0006	.0006	.0006	.0006	.0006	.0005	.0005	.0005
−3.1	.0010	.0009	.0009	.0009	.0008	.0008	.0008	.0008	.0007	.0007
−3.0	.0013	.0013	.0013	.0012	.0012	.0011	.0011	.0011	.0010	.0010
−2.9	.0019	.0018	.0018	.0017	.0016	.0016	.0015	.0015	.0014	.0014
−2.8	.0026	.0025	.0024	.0023	.0023	.0022	.0021	.0021	.0020	.0019
−2.7	.0035	.0034	.0033	.0032	.0031	.0030	.0029	.0028	.0027	.0026
−2.6	.0047	.0045	.0044	.0043	.0041	.0040	.0039	.0038	.0037	.0036
−2.5	.0062	.0060	.0059	.0057	.0055	.0054	.0052	.0051	.0049	.0048
−2.4	.0082	.0080	.0078	.0075	.0073	.0071	.0069	.0068	.0066	.0064
−2.3	.0107	.0104	.0102	.0099	.0096	.0094	.0091	.0089	.0087	.0084
−2.2	.0139	.0136	.0132	.0129	.0125	.0122	.0119	.0116	.0113	.0110
−2.1	.0179	.0174	.0170	.0166	.0162	.0158	.0154	.0150	.0146	.0143
−2.0	.0228	.0222	.0217	.0212	.0207	.0202	.0197	.0192	.0188	.0183
−1.9	.0287	.0281	.0274	.0268	.0262	.0256	.0250	.0244	.0239	.0233
−1.8	.0359	.0351	.0344	.0336	.0329	.0322	.0314	.0307	.0301	.0294
−1.7	.0446	.0436	.0427	.0418	.0409	.0401	.0392	.0384	.0375	.0367
−1.6	.0548	.0537	.0526	.0516	.0505	.0495	.0485	.0475	.0465	.0455
−1.5	.0668	.0655	.0643	.0630	.0618	.0606	.0594	.0582	.0571	.0559
−1.4	.0808	.0793	.0778	.0764	.0749	.0735	.0721	.0708	.0694	.0681
−1.3	.0968	.0951	.0934	.0918	.0901	.0885	.0869	.0853	.0838	.0823
−1.2	.1151	.1131	.1112	.1093	.1075	.1056	.1038	.1020	.1003	.0985
−1.1	.1357	.1335	.1314	.1292	.1271	.1251	.1230	.1210	.1190	.1170
−1.0	.1587	.1562	.1539	.1515	.1492	.1469	.1446	.1423	.1401	.1379
−0.9	.1841	.1814	.1788	.1762	.1736	.1711	.1685	.1660	.1635	.1611
−0.8	.2119	.2090	.2061	.2033	.2005	.1977	.1949	.1922	.1894	.1867
−0.7	.2420	.2389	.2358	.2327	.2296	.2266	.2236	.2206	.2177	.2148
−0.6	.2743	.2709	.2676	.2643	.2611	.2578	.2546	.2514	.2483	.2451
−0.5	.3085	.3050	.3015	.2981	.2946	.2912	.2877	.2843	.2810	.2776
−0.4	.3446	.3409	.3372	.3336	.3300	.3264	.3228	.3192	.3156	.3121
−0.3	.3821	.3783	.3745	.3707	.3669	.3632	.3594	.3557	.3520	.3483
−0.2	.4207	.4168	.4129	.4090	.4052	.4013	.3974	.3936	.3897	.3859
−0.1	.4602	.4562	.4522	.4483	.4443	.4404	.4364	.4325	.4286	.4247
−0.0	.5000	.4960	.4920	.4880	.4840	.4801	.4761	.4721	.4681	.4641

(continued)

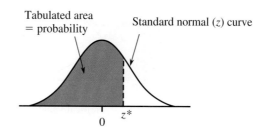

Tabulated area = probability

Standard normal (z) curve

0   z^*

TABLE 2 Standard Normal Probabilities (Cumulative *z* Curve Areas) *(Continued)*

z^*	.00	.01	.02	.03	.04	.05	.06	.07	.08	.09
0.0	.5000	.5040	.5080	.5120	.5160	.5199	.5239	.5279	.5319	.5359
0.1	.5398	.5438	.5478	.5517	.5557	.5596	.5636	.5675	.5714	.5753
0.2	.5793	.5832	.5871	.5910	.5948	.5987	.6026	.6064	.6103	.6141
0.3	.6179	.6217	.6255	.6293	.6331	.6368	.6406	.6443	.6480	.6517
0.4	.6554	.6591	.6628	.6664	.6700	.6736	.6772	.6808	.6844	.6879
0.5	.6915	.6950	.6985	.7019	.7054	.7088	.7123	.7157	.7190	.7224
0.6	.7257	.7291	.7324	.7357	.7389	.7422	.7454	.7486	.7517	.7549
0.7	.7580	.7611	.7642	.7673	.7704	.7734	.7764	.7794	.7823	.7852
0.8	.7881	.7910	.7939	.7967	.7995	.8023	.8051	.8078	.8106	.8133
0.9	.8159	.8186	.8212	.8238	.8264	.8289	.8315	.8340	.8365	.8389
1.0	.8413	.8438	.8461	.8485	.8508	.8531	.8554	.8577	.8599	.8621
1.1	.8643	.8665	.8686	.8708	.8729	.8749	.8770	.8790	.8810	.8830
1.2	.8849	.8869	.8888	.8907	.8925	.8944	.8962	.8980	.8997	.9015
1.3	.9032	.9049	.9066	.9082	.9099	.9115	.9131	.9147	.9162	.9177
1.4	.9192	.9207	.9222	.9236	.9251	.9265	.9279	.9292	.9306	.9319
1.5	.9332	.9345	.9357	.9370	.9382	.9394	.9406	.9418	.9429	.9441
1.6	.9452	.9463	.9474	.9484	.9495	.9505	.9515	.9525	.9535	.9545
1.7	.9554	.9564	.9573	.9582	.9591	.9599	.9608	.9616	.9625	.9633
1.8	.9641	.9649	.9656	.9664	.9671	.9678	.9686	.9693	.9699	.9706
1.9	.9713	.9719	.9726	.9732	.9738	.9744	.9750	.9756	.9761	.9767
2.0	.9772	.9778	.9783	.9788	.9793	.9798	.9803	.9808	.9812	.9817
2.1	.9821	.9826	.9830	.9834	.9838	.9842	.9846	.9850	.9854	.9857
2.2	.9861	.9864	.9868	.9871	.9875	.9878	.9881	.9884	.9887	.9890
2.3	.9893	.9896	.9898	.9901	.9904	.9906	.9909	.9911	.9913	.9916
2.4	.9918	.9920	.9922	.9925	.9927	.9929	.9931	.9932	.9934	.9936
2.5	.9938	.9940	.9941	.9943	.9945	.9946	.9948	.9949	.9951	.9952
2.6	.9953	.9955	.9956	.9957	.9959	.9960	.9961	.9962	.9963	.9964
2.7	.9965	.9966	.9967	.9968	.9969	.9970	.9971	.9972	.9973	.9974
2.8	.9974	.9975	.9976	.9977	.9977	.9978	.9979	.9979	.9980	.9981
2.9	.9981	.9982	.9982	.9983	.9984	.9984	.9985	.9985	.9986	.9986
3.0	.9987	.9987	.9987	.9988	.9988	.9989	.9989	.9989	.9990	.9990
3.1	.9990	.9991	.9991	.9991	.9992	.9992	.9992	.9992	.9993	.9993
3.2	.9993	.9993	.9994	.9994	.9994	.9994	.9994	.9995	.9995	.9995
3.3	.9995	.9995	.9995	.9996	.9996	.9996	.9996	.9996	.9996	.9997
3.4	.9997	.9997	.9997	.9997	.9997	.9997	.9997	.9997	.9997	.9998
3.5	.9998	.9998	.9998	.9998	.9998	.9998	.9998	.9998	.9998	.9998
3.6	.9998	.9998	.9999	.9999	.9999	.9999	.9999	.9999	.9999	.9999
3.7	.9999	.9999	.9999	.9999	.9999	.9999	.9999	.9999	.9999	.9999
3.8	.9999	.9999	.9999	.9999	.9999	.9999	.9999	.9999	.9999	1.0000

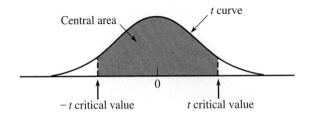

TABLE 3 *t* Critical Values

Central area captured:	.80	.90	.95	.98	.99	.998	.999
Confidence level:	80%	90%	95%	98%	99%	99.8%	99.9%
1	3.08	6.31	12.71	31.82	63.66	318.31	636.62
2	1.89	2.92	4.30	6.97	9.93	23.33	31.60
3	1.64	2.35	3.18	4.54	5.84	10.21	12.92
4	1.53	2.13	2.78	3.75	4.60	7.17	8.61
5	1.48	2.02	2.57	3.37	4.03	5.89	6.86
6	1.44	1.94	2.45	3.14	3.71	5.21	5.96
7	1.42	1.90	2.37	3.00	3.50	4.79	5.41
8	1.40	1.86	2.31	2.90	3.36	4.50	5.04
9	1.38	1.83	2.26	2.82	3.25	4.30	4.78
10	1.37	1.81	2.23	2.76	3.17	4.14	4.59
11	1.36	1.80	2.20	2.72	3.11	4.03	4.44
12	1.36	1.78	2.18	2.68	3.06	3.93	4.32
13	1.35	1.77	2.16	2.65	3.01	3.85	4.22
14	1.35	1.76	2.15	2.62	2.98	3.79	4.14
15	1.34	1.75	2.13	2.60	2.95	3.73	4.07
16	1.34	1.75	2.12	2.58	2.92	3.69	4.02
17	1.33	1.74	2.11	2.57	2.90	3.65	3.97
18	1.33	1.73	2.10	2.55	2.88	3.61	3.92
19	1.33	1.73	2.09	2.54	2.86	3.58	3.88
20	1.33	1.73	2.09	2.53	2.85	3.55	3.85
21	1.32	1.72	2.08	2.52	2.83	3.53	3.82
22	1.32	1.72	2.07	2.51	2.82	3.51	3.79
23	1.32	1.71	2.07	2.50	2.81	3.49	3.77
24	1.32	1.71	2.06	2.49	2.80	3.47	3.75
25	1.32	1.71	2.06	2.49	2.79	3.45	3.73
26	1.32	1.71	2.06	2.48	2.78	3.44	3.71
27	1.31	1.70	2.05	2.47	2.77	3.42	3.69
28	1.31	1.70	2.05	2.47	2.76	3.41	3.67
29	1.31	1.70	2.05	2.46	2.76	3.40	3.66
30	1.31	1.70	2.04	2.46	2.75	3.39	3.65
40	1.30	1.68	2.02	2.42	2.70	3.31	3.55
60	1.30	1.67	2.00	2.39	2.66	3.23	3.46
120	1.29	1.66	1.98	2.36	2.62	3.16	3.37
z critical values ∞	1.28	1.645	1.96	2.33	2.58	3.09	3.29

Degrees of freedom

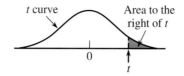

TABLE 4 Tail Areas for *t* Curves

t \ df	1	2	3	4	5	6	7	8	9	10	11	12
0.0	.500	.500	.500	.500	.500	.500	.500	.500	.500	.500	.500	.500
0.1	.468	.465	.463	.463	.462	.462	.462	.461	.461	.461	.461	.461
0.2	.437	.430	.427	.426	.425	.424	.424	.423	.423	.423	.423	.422
0.3	.407	.396	.392	.390	.388	.387	.386	.386	.386	.385	.385	.385
0.4	.379	.364	.358	.355	.353	.352	.351	.350	.349	.349	.348	.348
0.5	.352	.333	.326	.322	.319	.317	.316	.315	.315	.314	.313	.313
0.6	.328	.305	.295	.290	.287	.285	.284	.283	.282	.281	.280	.280
0.7	.306	.278	.267	.261	.258	.255	.253	.252	.251	.250	.249	.249
0.8	.285	.254	.241	.234	.230	.227	.225	.223	.222	.221	.220	.220
0.9	.267	.232	.217	.210	.205	.201	.199	.197	.196	.195	.194	.193
1.0	.250	.211	.196	.187	.182	.178	.175	.173	.172	.170	.169	.169
1.1	.235	.193	.176	.167	.162	.157	.154	.152	.150	.149	.147	.146
1.2	.221	.177	.158	.148	.142	.138	.135	.132	.130	.129	.128	.127
1.3	.209	.162	.142	.132	.125	.121	.117	.115	.113	.111	.110	.109
1.4	.197	.148	.128	.117	.110	.106	.102	.100	.098	.096	.095	.093
1.5	.187	.136	.115	.104	.097	.092	.089	.086	.084	.082	.081	.080
1.6	.178	.125	.104	.092	.085	.080	.077	.074	.072	.070	.069	.068
1.7	.169	.116	.094	.082	.075	.070	.066	.064	.062	.060	.059	.057
1.8	.161	.107	.085	.073	.066	.061	.057	.055	.053	.051	.050	.049
1.9	.154	.099	.077	.065	.058	.053	.050	.047	.045	.043	.042	.041
2.0	.148	.092	.070	.058	.051	.046	.043	.040	.038	.037	.035	.034
2.1	.141	.085	.063	.052	.045	.040	.037	.034	.033	.031	.030	.029
2.2	.136	.079	.058	.046	.040	.035	.032	.029	.028	.026	.025	.024
2.3	.131	.074	.052	.041	.035	.031	.027	.025	.023	.022	.021	.020
2.4	.126	.069	.048	.037	.031	.027	.024	.022	.020	.019	.018	.017
2.5	.121	.065	.044	.033	.027	.023	.020	.018	.017	.016	.015	.014
2.6	.117	.061	.040	.030	.024	.020	.018	.016	.014	.013	.012	.012
2.7	.113	.057	.037	.027	.021	.018	.015	.014	.012	.011	.010	.010
2.8	.109	.054	.034	.024	.019	.016	.013	.012	.010	.009	.009	.008
2.9	.106	.051	.031	.022	.017	.014	.011	.010	.009	.008	.007	.007
3.0	.102	.048	.029	.020	.015	.012	.010	.009	.007	.007	.006	.006
3.1	.099	.045	.027	.018	.013	.011	.009	.007	.006	.006	.005	.005
3.2	.096	.043	.025	.016	.012	.009	.008	.006	.005	.005	.004	.004
3.3	.094	.040	.023	.015	.011	.008	.007	.005	.005	.004	.004	.003
3.4	.091	.038	.021	.014	.010	.007	.006	.005	.004	.003	.003	.003
3.5	.089	.036	.020	.012	.009	.006	.005	.004	.003	.003	.002	.002
3.6	.086	.035	.018	.011	.008	.006	.004	.004	.003	.002	.002	.002
3.7	.084	.033	.017	.010	.007	.005	.004	.003	.002	.002	.002	.002
3.8	.082	.031	.016	.010	.006	.004	.003	.003	.002	.002	.001	.001
3.9	.080	.030	.015	.009	.006	.004	.003	.002	.002	.001	.001	.001
4.0	.078	.029	.014	.008	.005	.004	.003	.002	.002	.001	.001	.001

(continued)

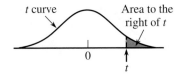

TABLE 4 Tail Areas for *t* Curves *(Continued)*

t \ df	13	14	15	16	17	18	19	20	21	22	23	24
0.0	.500	.500	.500	.500	.500	.500	.500	.500	.500	.500	.500	.500
0.1	.461	.461	.461	.461	.461	.461	.461	.461	.461	.461	.461	.461
0.2	.422	.422	.422	.422	.422	.422	.422	.422	.422	.422	.422	.422
0.3	.384	.384	.384	.384	.384	.384	.384	.384	.384	.383	.383	.383
0.4	.348	.347	.347	.347	.347	.347	.347	.347	.347	.347	.346	.346
0.5	.313	.312	.312	.312	.312	.312	.311	.311	.311	.311	.311	.311
0.6	.279	.279	.279	.278	.278	.278	.278	.278	.278	.277	.277	.277
0.7	.248	.247	.247	.247	.247	.246	.246	.246	.246	.246	.245	.245
0.8	.219	.218	.218	.218	.217	.217	.217	.217	.216	.216	.216	.216
0.9	.192	.191	.191	.191	.190	.190	.190	.189	.189	.189	.189	.189
1.0	.168	.167	.167	.166	.166	.165	.165	.165	.164	.164	.164	.164
1.1	.146	.144	.144	.144	.143	.143	.143	.142	.142	.142	.141	.141
1.2	.126	.124	.124	.124	.123	.123	.122	.122	.122	.121	.121	.121
1.3	.108	.107	.107	.106	.105	.105	.105	.104	.104	.104	.103	.103
1.4	.092	.091	.091	.090	.090	.089	.089	.089	.088	.088	.087	.087
1.5	.079	.077	.077	.077	.076	.075	.075	.075	.074	.074	.074	.073
1.6	.067	.065	.065	.065	.064	.064	.063	.063	.062	.062	.062	.061
1.7	.056	.055	.055	.054	.054	.053	.053	.052	.052	.052	.051	.051
1.8	.048	.046	.046	.045	.045	.044	.044	.043	.043	.043	.042	.042
1.9	.040	.038	.038	.038	.037	.037	.036	.036	.036	.035	.035	.035
2.0	.033	.032	.032	.031	.031	.030	.030	.030	.029	.029	.029	.028
2.1	.028	.027	.027	.026	.025	.025	.025	.024	.024	.024	.023	.023
2.2	.023	.022	.022	.021	.021	.021	.020	.020	.020	.019	.019	.019
2.3	.019	.018	.018	.018	.017	.017	.016	.016	.016	.016	.015	.015
2.4	.016	.015	.015	.014	.014	.014	.013	.013	.013	.013	.012	.012
2.5	.013	.012	.012	.012	.011	.011	.011	.011	.010	.010	.010	.010
2.6	.011	.010	.010	.010	.009	.009	.009	.009	.008	.008	.008	.008
2.7	.009	.008	.008	.008	.008	.007	.007	.007	.007	.007	.006	.006
2.8	.008	.007	.007	.006	.006	.006	.006	.006	.005	.005	.005	.005
2.9	.006	.005	.005	.005	.005	.005	.005	.004	.004	.004	.004	.004
3.0	.005	.004	.004	.004	.004	.004	.004	.004	.003	.003	.003	.003
3.1	.004	.004	.004	.003	.003	.003	.003	.003	.003	.003	.003	.002
3.2	.003	.003	.003	.003	.003	.002	.002	.002	.002	.002	.002	.002
3.3	.003	.002	.002	.002	.002	.002	.002	.002	.002	.002	.002	.001
3.4	.002	.002	.002	.002	.002	.002	.002	.001	.001	.001	.001	.001
3.5	.002	.002	.002	.001	.001	.001	.001	.001	.001	.001	.001	.001
3.6	.002	.001	.001	.001	.001	.001	.001	.001	.001	.001	.001	.001
3.7	.001	.001	.001	.001	.001	.001	.001	.001	.001	.001	.001	.001
3.8	.001	.001	.001	.001	.001	.001	.001	.001	.001	.000	.000	.000
3.9	.001	.001	.001	.001	.001	.001	.000	.000	.000	.000	.000	.000
4.0	.001	.001	.001	.001	.000	.000	.000	.000	.000	.000	.000	.000

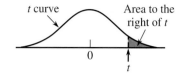

TABLE 4 Tail Areas for *t* Curves *(Continued)*

t\df	25	26	27	28	29	30	35	40	60	120	∞(= z)
0.0	.500	.500	.500	.500	.500	.500	.500	.500	.500	.500	.500
0.1	.461	.461	.461	.461	.461	.461	.460	.460	.460	.460	.460
0.2	.422	.422	.421	.421	.421	.421	.421	.421	.421	.421	.421
0.3	.383	.383	.383	.383	.383	.383	.383	.383	.383	.382	.382
0.4	.346	.346	.346	.346	.346	.346	.346	.346	.345	.345	.345
0.5	.311	.311	.311	.310	.310	.310	.310	.310	.309	.309	.309
0.6	.277	.277	.277	.277	.277	.277	.276	.276	.275	.275	.274
0.7	.245	.245	.245	.245	.245	.245	.244	.244	.243	.243	.242
0.8	.216	.215	.215	.215	.215	.215	.215	.214	.213	.213	.212
0.9	.188	.188	.188	.188	.188	.188	.187	.187	.186	.185	.184
1.0	.163	.163	.163	.163	.163	.163	.162	.162	.161	.160	.159
1.1	.141	.141	.141	.140	.140	.140	.139	.139	.138	.137	.136
1.2	.121	.120	.120	.120	.120	.120	.119	.119	.117	.116	.115
1.3	.103	.103	.102	.102	.102	.102	.101	.101	.099	.098	.097
1.4	.087	.087	.086	.086	.086	.086	.085	.085	.083	.082	.081
1.5	.073	.073	.073	.072	.072	.072	.071	.071	.069	.068	.067
1.6	.061	.061	.061	.060	.060	.060	.059	.059	.057	.056	.055
1.7	.051	.051	.050	.050	.050	.050	.049	.048	.047	.046	.045
1.8	.042	.042	.042	.041	.041	.041	.040	.040	.038	.037	.036
1.9	.035	.034	.034	.034	.034	.034	.033	.032	.031	.030	.029
2.0	.028	.028	.028	.028	.027	.027	.027	.026	.025	.024	.023
2.1	.023	.023	.023	.022	.022	.022	.022	.021	.020	.019	.018
2.2	.019	.018	.018	.018	.018	.018	.017	.017	.016	.015	.014
2.3	.015	.015	.015	.015	.014	.014	.014	.013	.012	.012	.011
2.4	.012	.012	.012	.012	.012	.011	.011	.011	.010	.009	.008
2.5	.010	.010	.009	.009	.009	.009	.009	.008	.008	.007	.006
2.6	.008	.008	.007	.007	.007	.007	.007	.007	.006	.005	.005
2.7	.006	.006	.006	.006	.006	.006	.005	.005	.004	.004	.003
2.8	.005	.005	.005	.005	.005	.004	.004	.004	.003	.003	.003
2.9	.004	.004	.004	.004	.004	.003	.003	.003	.003	.002	.002
3.0	.003	.003	.003	.003	.003	.003	.002	.002	.002	.002	.001
3.1	.002	.002	.002	.002	.002	.002	.002	.002	.001	.001	.001
3.2	.002	.002	.002	.002	.002	.002	.001	.001	.001	.001	.001
3.3	.001	.001	.001	.001	.001	.001	.001	.001	.001	.001	.000
3.4	.001	.001	.001	.001	.001	.001	.001	.001	.001	.000	.000
3.5	.001	.001	.001	.001	.001	.001	.001	.001	.000	.000	.000
3.6	.001	.001	.001	.001	.001	.001	.000	.000	.000	.000	.000
3.7	.001	.001	.000	.000	.000	.000	.000	.000	.000	.000	.000
3.8	.000	.000	.000	.000	.000	.000	.000	.000	.000	.000	.000
3.9	.000	.000	.000	.000	.000	.000	.000	.000	.000	.000	.000
4.0	.000	.000	.000	.000	.000	.000	.000	.000	.000	.000	.000

TABLE 5 **Upper-Tail Areas for Chi-Square Distributions**

Right-tail area	df = 1	df = 2	df = 3	df = 4	df = 5
>0.100	< 2.70	< 4.60	< 6.25	< 7.77	< 9.23
0.100	2.70	4.60	6.25	7.77	9.23
0.095	2.78	4.70	6.36	7.90	9.37
0.090	2.87	4.81	6.49	8.04	9.52
0.085	2.96	4.93	6.62	8.18	9.67
0.080	3.06	5.05	6.75	8.33	9.83
0.075	3.17	5.18	6.90	8.49	10.00
0.070	3.28	5.31	7.06	8.66	10.19
0.065	3.40	5.46	7.22	8.84	10.38
0.060	3.53	5.62	7.40	9.04	10.59
0.055	3.68	5.80	7.60	9.25	10.82
0.050	3.84	5.99	7.81	9.48	11.07
0.045	4.01	6.20	8.04	9.74	11.34
0.040	4.21	6.43	8.31	10.02	11.64
0.035	4.44	6.70	8.60	10.34	11.98
0.030	4.70	7.01	8.94	10.71	12.37
0.025	5.02	7.37	9.34	11.14	12.83
0.020	5.41	7.82	9.83	11.66	13.38
0.015	5.91	8.39	10.46	12.33	14.09
0.010	6.63	9.21	11.34	13.27	15.08
0.005	7.87	10.59	12.83	14.86	16.74
0.001	10.82	13.81	16.26	18.46	20.51
<0.001	>10.82	>13.81	>16.26	>18.46	>20.51

Right-tail area	df = 6	df = 7	df = 8	df = 9	df = 10
>0.100	<10.64	<12.01	<13.36	<14.68	<15.98
0.100	10.64	12.01	13.36	14.68	15.98
0.095	10.79	12.17	13.52	14.85	16.16
0.090	10.94	12.33	13.69	15.03	16.35
0.085	11.11	12.50	13.87	15.22	16.54
0.080	11.28	12.69	14.06	15.42	16.75
0.075	11.46	12.88	14.26	15.63	16.97
0.070	11.65	13.08	14.48	15.85	17.20
0.065	11.86	13.30	14.71	16.09	17.44
0.060	12.08	13.53	14.95	16.34	17.71
0.055	12.33	13.79	15.22	16.62	17.99
0.050	12.59	14.06	15.50	16.91	18.30
0.045	12.87	14.36	15.82	17.24	18.64
0.040	13.19	14.70	16.17	17.60	19.02
0.035	13.55	15.07	16.56	18.01	19.44
0.030	13.96	15.50	17.01	18.47	19.92
0.025	14.44	16.01	17.53	19.02	20.48
0.020	15.03	16.62	18.16	19.67	21.16
0.015	15.77	17.39	18.97	20.51	22.02
0.010	16.81	18.47	20.09	21.66	23.20
0.005	18.54	20.27	21.95	23.58	25.18
0.001	22.45	24.32	26.12	27.87	29.58
<0.001	>22.45	>24.32	>26.12	>27.87	>29.58

(continued)

TABLE 5 **Upper-Tail Areas for Chi-Square Distributions** *(Continued)*

Right-tail area	df = 11	df = 12	df = 13	df = 14	df = 15
>0.100	<17.27	<18.54	<19.81	<21.06	<22.30
0.100	17.27	18.54	19.81	21.06	22.30
0.095	17.45	18.74	20.00	21.26	22.51
0.090	17.65	18.93	20.21	21.47	22.73
0.085	17.85	19.14	20.42	21.69	22.95
0.080	18.06	19.36	20.65	21.93	23.19
0.075	18.29	19.60	20.89	22.17	23.45
0.070	18.53	19.84	21.15	22.44	23.72
0.065	18.78	20.11	21.42	22.71	24.00
0.060	19.06	20.39	21.71	23.01	24.31
0.055	19.35	20.69	22.02	23.33	24.63
0.050	19.67	21.02	22.36	23.68	24.99
0.045	20.02	21.38	22.73	24.06	25.38
0.040	20.41	21.78	23.14	24.48	25.81
0.035	20.84	22.23	23.60	24.95	26.29
0.030	21.34	22.74	24.12	25.49	26.84
0.025	21.92	23.33	24.73	26.11	27.48
0.020	22.61	24.05	25.47	26.87	28.25
0.015	23.50	24.96	26.40	27.82	29.23
0.010	24.72	26.21	27.68	29.14	30.57
0.005	26.75	28.29	29.81	31.31	32.80
0.001	31.26	32.90	34.52	36.12	37.69
<0.001	>31.26	>32.90	>34.52	>36.12	>37.69

Right-tail area	df = 16	df = 17	df = 18	df = 19	df = 20
>0.100	<23.54	<24.77	<25.98	<27.20	<28.41
0.100	23.54	24.76	25.98	27.20	28.41
0.095	23.75	24.98	26.21	27.43	28.64
0.090	23.97	25.21	26.44	27.66	28.88
0.085	24.21	25.45	26.68	27.91	29.14
0.080	24.45	25.70	26.94	28.18	29.40
0.075	24.71	25.97	27.21	28.45	29.69
0.070	24.99	26.25	27.50	28.75	29.99
0.065	25.28	26.55	27.81	29.06	30.30
0.060	25.59	26.87	28.13	29.39	30.64
0.055	25.93	27.21	28.48	29.75	31.01
0.050	26.29	27.58	28.86	30.14	31.41
0.045	26.69	27.99	29.28	30.56	31.84
0.040	27.13	28.44	29.74	31.03	32.32
0.035	27.62	28.94	30.25	31.56	32.85
0.030	28.19	29.52	30.84	32.15	33.46
0.025	28.84	30.19	31.52	32.85	34.16
0.020	29.63	30.99	32.34	33.68	35.01
0.015	30.62	32.01	33.38	34.74	36.09
0.010	32.00	33.40	34.80	36.19	37.56
0.005	34.26	35.71	37.15	38.58	39.99
0.001	39.25	40.78	42.31	43.81	45.31
<0.001	>39.25	>40.78	>42.31	>43.81	>45.31

TABLE 6 Binomial Probabilities

n = 5

x	0.05	0.1	0.2	0.25	0.3	0.4	0.5	0.6	0.7	0.75	0.8	0.9	0.95
0	.774	.590	.328	.237	.168	.078	.031	.010	.002	.001	.000	.000	.000
1	.204	.328	.410	.396	.360	.259	.156	.077	.028	.015	.006	.000	.000
2	.021	.073	.205	.264	.309	.346	.313	.230	.132	.088	.051	.008	.001
3	.001	.008	.051	.088	.132	.230	.313	.346	.309	.264	.205	.073	.021
4	.000	.000	.006	.015	.028	.077	.156	.259	.360	.396	.410	.328	.204
5	.000	.000	.000	.001	.002	.010	.031	.078	.168	.237	.328	.590	.774

n = 10

x	0.05	0.1	0.2	0.25	0.3	0.4	0.5	0.6	0.7	0.75	0.8	0.9	0.95
0	.599	.349	.107	.056	.028	.006	.001	.000	.000	.000	.000	.000	.000
1	.315	.387	.268	.188	.121	.040	.010	.002	.000	.000	.000	.000	.000
2	.075	.194	.302	.282	.233	.121	.044	.011	.001	.000	.000	.000	.000
3	.010	.057	.201	.250	.267	.215	.117	.042	.009	.003	.001	.000	.000
4	.001	.011	.088	.146	.200	.251	.205	.111	.037	.016	.006	.000	.000
5	.000	.001	.026	.058	.103	.201	.246	.201	.103	.058	.026	.001	.000
6	.000	.000	.006	.016	.037	.111	.205	.251	.200	.146	.088	.011	.001
7	.000	.000	.001	.003	.009	.042	.117	.215	.267	.250	.201	.057	.010
8	.000	.000	.000	.000	.001	.011	.044	.121	.233	.282	.302	.194	.075
9	.000	.000	.000	.000	.000	.002	.010	.040	.121	.188	.268	.387	.315
10	.000	.000	.000	.000	.000	.000	.001	.006	.028	.056	.107	.349	.599

n = 15

x	0.05	0.1	0.2	0.25	0.3	0.4	0.5	0.6	0.7	0.75	0.8	0.9	0.95
0	.463	.206	.035	.013	.005	.000	.000	.000	.000	.000	.000	.000	.000
1	.366	.343	.132	.067	.031	.005	.000	.000	.000	.000	.000	.000	.000
2	.135	.267	.231	.156	.092	.022	.003	.000	.000	.000	.000	.000	.000
3	.031	.129	.250	.225	.170	.063	.014	.002	.000	.000	.000	.000	.000
4	.005	.043	.188	.225	.219	.127	.042	.007	.001	.000	.000	.000	.000
5	.001	.010	.103	.165	.206	.186	.092	.024	.003	.001	.000	.000	.000
6	.000	.002	.043	.092	.147	.207	.153	.061	.012	.003	.001	.000	.000
7	.000	.000	.014	.039	.081	.177	.196	.118	.035	.013	.003	.000	.000
8	.000	.000	.003	.013	.035	.118	.196	.177	.081	.039	.014	.000	.000
9	.000	.000	.001	.003	.012	.061	.153	.207	.147	.092	.043	.002	.000
10	.000	.000	.000	.001	.003	.024	.092	.186	.206	.165	.103	.010	.001
11	.000	.000	.000	.000	.001	.007	.042	.127	.219	.225	.188	.043	.005
12	.000	.000	.000	.000	.000	.002	.014	.063	.170	.225	.250	.129	.031
13	.000	.000	.000	.000	.000	.000	.003	.022	.092	.156	.231	.267	.135
14	.000	.000	.000	.000	.000	.000	.000	.005	.031	.067	.132	.343	.366
15	.000	.000	.000	.000	.000	.000	.000	.000	.005	.013	.035	.206	.463

(continued)

TABLE 6 Binomial Probabilities *(Continued)*

n = 20

x	0.05	0.1	0.2	0.25	0.3	0.4	0.5	0.6	0.7	0.75	0.8	0.9	0.95
0	.358	.122	.012	.003	.001	.000	.000	.000	.000	.000	.000	.000	.000
1	.377	.270	.058	.021	.007	.000	.000	.000	.000	.000	.000	.000	.000
2	.189	.285	.137	.067	.028	.003	.000	.000	.000	.000	.000	.000	.000
3	.060	.190	.205	.134	.072	.012	.001	.000	.000	.000	.000	.000	.000
4	.013	.090	.218	.190	.130	.035	.005	.000	.000	.000	.000	.000	.000
5	.002	.032	.175	.202	.179	.075	.015	.001	.000	.000	.000	.000	.000
6	.000	.009	.109	.169	.192	.124	.037	.005	.000	.000	.000	.000	.000
7	.000	.002	.055	.112	.164	.166	.074	.015	.001	.000	.000	.000	.000
8	.000	.000	.022	.061	.114	.180	.120	.035	.004	.001	.000	.000	.000
9	.000	.000	.007	.027	.065	.160	.160	.071	.012	.003	.000	.000	.000
10	.000	.000	.002	.010	.031	.117	.176	.117	.031	.010	.002	.000	.000
11	.000	.000	.000	.003	.012	.071	.160	.160	.065	.027	.007	.000	.000
12	.000	.000	.000	.001	.004	.035	.120	.180	.114	.061	.022	.000	.000
13	.000	.000	.000	.000	.001	.015	.074	.166	.164	.112	.055	.002	.000
14	.000	.000	.000	.000	.000	.005	.037	.124	.192	.169	.109	.009	.000
15	.000	.000	.000	.000	.000	.001	.015	.075	.179	.202	.175	.032	.002
16	.000	.000	.000	.000	.000	.000	.005	.035	.130	.190	.218	.090	.013
17	.000	.000	.000	.000	.000	.000	.001	.012	.072	.134	.205	.190	.060
18	.000	.000	.000	.000	.000	.000	.000	.003	.028	.067	.137	.285	.189
19	.000	.000	.000	.000	.000	.000	.000	.000	.007	.021	.058	.270	.377
20	.000	.000	.000	.000	.000	.000	.000	.000	.001	.003	.012	.122	.358

n = 25

x	0.05	0.1	0.2	0.25	0.3	0.4	0.5	0.6	0.7	0.75	0.8	0.9	0.95
0	.277	.072	.004	.001	.000	.000	.000	.000	.000	.000	.000	.000	.000
1	.365	.199	.024	.006	.001	.000	.000	.000	.000	.000	.000	.000	.000
2	.231	.266	.071	.025	.007	.000	.000	.000	.000	.000	.000	.000	.000
3	.093	.226	.136	.064	.024	.002	.000	.000	.000	.000	.000	.000	.000
4	.027	.138	.187	.118	.057	.007	.000	.000	.000	.000	.000	.000	.000
5	.006	.065	.196	.165	.103	.020	.002	.000	.000	.000	.000	.000	.000
6	.001	.024	.163	.183	.147	.044	.005	.000	.000	.000	.000	.000	.000
7	.000	.007	.111	.165	.171	.080	.014	.001	.000	.000	.000	.000	.000
8	.000	.002	.062	.124	.165	.120	.032	.003	.000	.000	.000	.000	.000
9	.000	.000	.029	.078	.134	.151	.061	.009	.000	.000	.000	.000	.000
10	.000	.000	.012	.042	.092	.161	.097	.021	.001	.000	.000	.000	.000
11	.000	.000	.004	.019	.054	.147	.133	.043	.004	.001	.000	.000	.000
12	.000	.000	.001	.007	.027	.114	.155	.076	.011	.002	.000	.000	.000
13	.000	.000	.000	.002	.011	.076	.155	.114	.027	.007	.001	.000	.000
14	.000	.000	.000	.001	.004	.043	.133	.147	.054	.019	.004	.000	.000
15	.000	.000	.000	.000	.001	.021	.097	.161	.092	.042	.012	.000	.000
16	.000	.000	.000	.000	.000	.009	.061	.151	.134	.078	.029	.000	.000
17	.000	.000	.000	.000	.000	.003	.032	.120	.165	.124	.062	.002	.000
18	.000	.000	.000	.000	.000	.001	.014	.080	.171	.165	.111	.007	.000
19	.000	.000	.000	.000	.000	.000	.005	.044	.147	.183	.163	.024	.001
20	.000	.000	.000	.000	.000	.000	.002	.020	.103	.165	.196	.065	.006
21	.000	.000	.000	.000	.000	.000	.000	.007	.057	.118	.187	.138	.027
22	.000	.000	.000	.000	.000	.000	.000	.002	.024	.064	.136	.226	.093
23	.000	.000	.000	.000	.000	.000	.000	.000	.007	.025	.071	.266	.231
24	.000	.000	.000	.000	.000	.000	.000	.000	.002	.006	.024	.199	.365
25	.000	.000	.000	.000	.000	.000	.000	.000	.000	.001	.004	.072	.277

AP* Free-Response Questions

CHAPTER 1

1. A pharmaceutical company is testing a new drug designed to help balding men grow more hair. The company has data on about 1,000 current customers who use other products that the company produces but who have not previously used any type of treatment for baldness. The data contains information about hair color (light or dark), age, and percentage of baldness. A partial list is shown below. For this experiment, the pharmaceutical company would like to select a sample that is representative of their customers.

Cust #	Hair color	Age (yrs)	Percentage of Baldness
1	Light	67	83
2	Dark	62	73
3	Light	41	25
4	Dark	52	50
5	Dark	43	14
6	Light	69	96
7	Dark	56	57
...	...	...	...
1,000	Light	32	40

(a) Briefly describe a process to select a simple random sample of size $n = 20$ from this list of customers, using a random number table.

(b) Describe how a stratified random sample with strata defined by hair color could be selected. Do you think that stratifying by hair color is a good idea? Explain.

2. A common method of sampling to estimate bird populations is to walk along defined paths and count the number of birds seen. During hunting season, biologists wear brightly colored clothes as a protective measure. Unfortunately, birds may react to these brightly colored clothes. As an example, a robin may be more afraid of a biologist wearing a brown color than one wearing an orange protective vest. A robin more afraid of the biologist may be more difficult to see and therefore "count" because the robin flees before being seen. If that happens, the process of sampling would give an estimate of the robin population that is too small. To test this theory, a biologist wore either a bright orange vest or a brown vest. She first trained herself to accurately estimate distances. Then a robin was placed in an enclosed space and the biologist would approach the robin and record how close she could get to the robin before it flew away from her. This process was repeated with 40 robins, for a total of 40 trials. The biologist randomly selects 20 of the 40 trials when she will wear the brown vest. She will wear the orange vest on the other 20 trials.

(a) What is the explanatory variable (factor) for this experiment?

(b) What is the response variable for this experiment?

(c) Suppose that the 40 trials will be carried out over two days and that the biologist worries that weather conditions, such as wind speed and temperature, might be different on the two days. She thinks that weather conditions might affect the way the birds respond. What changes would you make to the design of this experiment to address this concern?

CHAPTER 2

3. Iowa is an agricultural state with a large number of rivers. The use of chemical pesticides and increasing size of livestock herds has potentially degraded the water quality in the state. A researcher recently analyzed concentrations of certain chemicals at 100 different locations in Iowa's Cedar River. A cumulative relative frequency plot of nitrogen concentration appears below.

(a) Approximately what percent of the nitrogen concentration measurements are less than 10 mg/L?

(b) Approximately what percent of the Nitrogen concentration measurements are 12 mg/L or more?

(c) 50% of the nitrogen concentration measurements are smaller than what approximate value?

(d) Which of the two histograms shown below (Histogram 1 or Histogram 2) could not be the histogram corresponding to this data set? Explain how you made your decision.

4. A survey designed by investigators at the University of Iowa Injury Prevention Research Center assessed compliance with Iowa's seatbelt law. The table below shows the results for 5 Iowa cities.

City	Percent Age 2–5 Properly Restrained	Percent Age 6–10 Properly Restrained
Belle Plaine	84	93
Cedar Rapids	79	94
Gutenberg	58	68
Iowa City	92	99
Manchester	59	82

(a) Display these data in a comparative bar chart so that the different ages can be compared for the different cities.

(b) Write a brief description of the differences and similarities in the patterns you observe in your bar chart from Part (a).

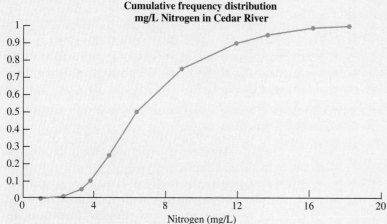

**Cumulative frequency distribution
mg/L Nitrogen in Cedar River**

FIGURE FOR QUESTION 3

HISTOGRAM 1

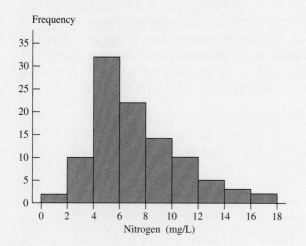

HISTOGRAM 2

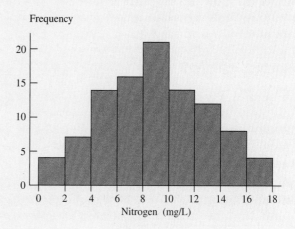

CHAPTER 3

5. The *New York Times* reported the average number of square feet per standing passenger in 2 successive years for 19 subway stops. The NYC Transit Authority managers have attempted to improve the space problem on subway cars (more space is better—trust us!) by adding cars to trains during the rush hours. These data were used to compute the *change* (Year 2 – Year 1) in the amount of space available to standing passengers for each of the 19 subway stops. The changes are shown in the table below.

   (a) Construct boxplots that would allow you to compare the changes in available space in the morning and in the afternoon.

   (b) The Transit System wishes to know if their efforts to improve the standing space were successful. (Remember, more space is better!) Does the data support a claim of improvement? What specific aspects of the plots in Part (a) support your answer?

Changes in the Space per Standing Passenger	
Change A.M.	Change P.M.
−0.4	−5.1
−1.1	−1.5
0	0.3
4.6	8.1
−0.7	3.3
0	0.5
3	−1.2
−1.8	−1.2
1	−3.2
0.8	−0.4
−3	5.3
−3	16.9
−0.9	−0.1
−0.3	−0.5
−0.5	0.6
0.2	−0.2
−0.3	−0.4
−0.3	−1.1
0.4	0.6

6. The forces that determine the size of groups of social insects (swarms) and the rates at which they grow are not well understood. Biologists have observed large variability in the size of swarms across species. In a study of the social wasp, *Polybia occidentalis*, investigators dismantled a nest of these insects and marked a few for future identification in new swarms. Twenty-five days after dismantling the original swarm they had located new swarms of wasps from the original colony. The data below present the numbers of drones in the new colonies. The study was replicated in two different years. Drones are fertile males that mate with the queen.

Number of Drones in First and Second Surveys			
First Year Swarm	Drones	Second Year Swarm	Drones
1	598	1	355
2	24	2	691
3	567	3	278
4	371	4	719
5	279	5	152
6	44	6	156
7	42	7	41
8	126		
9	108		
10	79		

   (a) Construct a comparative (back to back) stem-and-leaf display of the numbers of drones in the swarms in the first and second years.

   (b) Using the display from Part (a), compare the distributions of the numbers of drones in the swarms for the two years, noting any interesting features of the distributions.

   (c) Use the summary statistics in the table below to construct comparative boxplots of the numbers of drones in the swarms in the first and second years.

Numbers of Drones			
	Lower Quartile	Median	Upper Quartile
First year	43.5	117	420
Second year	152	278	691

CHAPTER 4

7. When estimating the size of animal populations from aerial surveys, game managers have found that animals may bunch together, making it difficult to distinguish and count them accurately. For example, a horse standing alone is easy to spot from a plane; if seven horses huddled close together, some may be missed, resulting in an undercount. In a recent study, the percent of sightings that resulted in an undercount (P) was related to the group size (GS) of horses and donkeys. The following data were gathered:

Percent Undercount (P) vs. Group Size (GS) for Horses and Donkeys			
Group Size	Percent of Sightings Resulting in an Undercount	Group Size	Percent of Sightings Resulting in an Undercount
2	5	9	6
3	5	10	7
4	6	11	5
5	10	12	5
6	5	14	14
7	7	16	13
8	5	18	23

After fitting a straight line model, significant curvature was detected in the residual plot, and two nonlinear models were chosen for further analysis. The computer output for these models is given below, and the residual plots are also given.

$\log p = a + b(GS)$ **(Exponential)**

$\log p = a + b \log(GS)$ **(Power)**

Bivariate Fit of Log%U By GS	
Linear Fit	
Log%U=0.586+0.031 GS	
Summary of Fit	
RSquare	0.499
RSquare Adj	0.458
St. Dev. Of Residuals	0.156

Bivariate Fit of Log%U By LogGS	
Linear Fit	
Log%U=0.476+0.440 LogGS	
Summary of Fit	
RSquare	0.337
RSquare Adj	0.282
St. Dev. Of Residuals	0.180

Residual Plots

$\log P = \alpha + b(GS)$

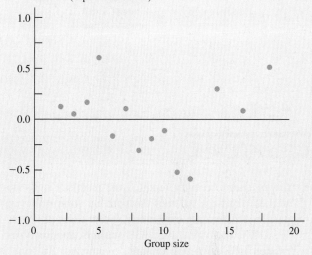

$\log P = a + b \log(GS)$

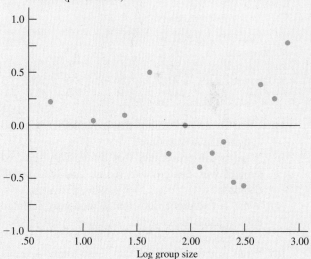

(a) For the exponential model, calculate the predicted *log (%undercount)* for a group size = 10.

(b) Use your calculations from Part (a) to predict the *%undercount* for a group size = 10

(c) Generally speaking, which of the two models, power or exponential, is better at predicting the log(% Undercount)? Provide statistical justification for your choice.

8. The data below are for a random sample of five basking sharks that were swimming through the water and filter-feeding (passively letting the water bring food into their mouths).

   (a) What is the value of the correlation coefficient for these data?

Body Length (meters)	Mean Speed (meters/second)
4.0	0.89
4.5	0.83
4.0	0.76
6.5	0.94
5.5	0.94

   (b) What is the equation of the least-squares line describing the relationship between x = body length and y = mean speed.

   (c) If these sharks are representative of the population of basking sharks, what would you predict for the mean speed of a filter-feeding basking shark that is 5.0 meters in length?

   (d) The largest basking shark in the sample is 6.5 meters long, but it is known that basking sharks can grow as long as 12.5 meters. Would it be reasonable to use the equation from Part (b) above to predict the mean filter-feeding speed for a 12-meter-long basking shark? Why or why not?

CHAPTER 5

9. Investigators recently reported the results a study designed to assess whether the herb St. John's Wort is effective in treating moderately severe cases of depression. The study involved 338 subjects, randomly assigned to receive one of three treatments: St. John's Wort, Zoloft, or a placebo. After treatment, the response to the treatment (full response, partial response, or no response) was noted for each subject. The resulting data are summarized in the table below.

	St. John's Wort	Placebo	Zoloft	Total
Full Response	27	37	27	**91**
Partial Response	16	13	26	**55**
No Response	70	66	56	**192**
Total	**113**	**116**	**109**	**338**

   (a) What is the probability that a randomly selected subject had no response?

   (b) What is the probability that a randomly selected subject was treated with Zoloft and had a full response?

   (c) What is the probability that a randomly selected subject had a full or partial response given that they were treated with St. John's Wort?

   (d) What is the probability that a randomly selected subject who didn't have a full response was treated with the placebo?

10. One method for increasing sales of boxes of cereal is to suggest to children that they collect a complete "set" of small toys. Suppose that one of five different toys is put in every box. One strategy the company could use to distribute the toys is to put each toy in 20% of the boxes. Your task is to design a simulation that will be used to assess this strategy in terms of the number of boxes of cereal a person would need to buy in order to get a complete set of toys.

   (a) Assign digits to represent each of the five toys in a way that is consistent with the manufacturer placing each toy in 20% of the cereal boxes.

   Toy 1 Digits: Toy 2 Digits:
   Toy 3 Digits: Toy 4 Digits:
   Toy 5 Digits:

   (b) Describe how you would use a random digit table to conduct one run of a simulation of the process of buying boxes of cereal, noting the toy received, and then continuing to buy one box of cereal at a time until all of the five different toys have been obtained.

   (c) Use the list of random digits below (start at the beginning of the first line) to carry out one run of the simulation using the process you described in Part (b). (You may mark on the digits to help explain your procedure.) How many boxes of cereal had to be purchased in order to get all five toys?

61790	55300	05756	72765	96409
12531	35013	82853	73676	57890
99400	37754	42648	82425	36290
45467	71709	77558	00095	32863
29485	82226			

   (d) Suppose that the cereal company decides instead to put the toys in the boxes in a way that makes some of the toys harder to find. Toy 1 will be put in 40% of the boxes, Toys 2 and 3 will be put in 20% of the boxes, and Toys 4 and 5 will be put in 10% of the boxes. Assign digits to the toys that are consistent with these percentages.

   Toy 1 Digits: Toy 2 Digits:
   Toy 3 Digits: Toy 4 Digits:
   Toy 5 Digits:

   (e) Use the list of random digits at the top of the next column (start at the beginning of the first line) to carry out five runs, noting the number of boxes purchased for each run. Use your results to estimate the probability that it would take more than 10 boxes to complete the set of five toys. (You may mark on the digits to help explain your procedure.)

19223	95734	05756	28713	96409
12531	42544	82853	73676	47150
99400	37754	42648	82425	36290
45467	71709	77558	00095	32863
29485	82226	68417	35013	15529
72765	85089	57067	50211	47487
82739	57890	20807	81676	55300

94383	14893	60940	72024	17868
24943	61790	90656	87964	47140
29385	67067	15803	12589	52167
18568	23243	55770	90846	86664

(f) The probability estimate in Part (e) may not be a very good estimate of the actual probability that more than 10 boxes must be purchased. What would you recommend doing to increase the accuracy of this estimated probability?

CHAPTER 6

11. In the old Roman Coliseum, two horses would be placed in tandem (side by side) and hitched to a chariot. Since the fast chariots needed to be able to pass the slow chariots, it was of some importance that the horses have room to run, but also that they should not be too large. For chariot-hitching purposes, the widest measure across a horse occurs in the rump area. Suppose that the mean rump width of horses is 27 inches and that the standard deviation of rump width is 2 inches. Let random variable w = width in inches across the rump of a randomly selected horse.

(a) The usual length measure as it applies to horses is the "hand," which is by definition equal to 4 inches. Define random variable h = width in hands across the rump of a randomly selected horse. What are the mean and standard deviation of h?

(b) Suppose horses are randomly chosen for a particular Roman chariot. Define random variable b to be the total rump width (in inches) of two randomly selected horses. What are the mean and standard deviation of b?

(c) In the original chariot design, a 16-inch separation of the horses is specified so that the horses have room to avoid each other. Define the random variable "chariot width" as: $c = 16 + w_1 + w_2$. How do the mean and standard deviation of b in Part (b) differ from the mean and standard deviation of the random variable c? You should be able to answer without recalculating the mean and standard deviation.

12. The owners of the Burger Emporium are looking for new supplier of onions. The owners have determined that for use in their burgers the onion slice should be roughly the same diameter as the hamburger patty. After careful analysis, they determine that only onions with diameters between 9 and 10 cm will be used for burgers. (Onions that are not suitable for burgers are diced for tacos.) Company A's onions have a diameter distribution that is approximately normal with mean 10.5 cm and standard deviation of 1.1 cm. Company B's onions have a diameter distribution that is approximately normal with mean 10.3 cm and standard deviation of 0.8 cm. Which company provides the higher proportion of burger-usable onions? Justify your choice with an appropriate statistical argument.

CHAPTER 8

13. One method for estimating the availability of office space in large cities is to select a random sample of offices and calculate the proportion of offices currently being used. Suppose that a real-estate agent believes that the proportion of offices currently being used is $p = 0.70$ and decides to take a sample to assess his belief. He is considering a sample size of $n = 40$.

(a) If the real-estate agent's belief is correct, is the sample size is large enough for the sampling

distribution of $\hat{p}$ to be approximately normal? Justify your answer.

(b) If the real-estate agent's belief is correct, what is the mean of the sampling distribution of $\hat{p}$?

(c) If the real-estate agent's belief is correct, what is the standard deviation of the sampling distribution of $\hat{p}$?

(d) If the real-estate agent's belief is correct, what is the probability that a sample proportion, $\hat{p}$, would differ from $p = 0.70$ by as much as 0.05?

CHAPTER 9

14. Conventional home smoke alarms sometimes fail to awaken sleeping children. Researchers are studying the possibility that a personalized "voice" smoke alarm would be more effective than mere sound. The idea is that a recording of the mother's voice calling the child's name may be an improvement over existing systems.

    (a) The researchers would like to estimate p, the proportion of sleeping children who would be awakened by their mother's voice, to within 0.05 with 95% confidence. Past experience has shown that the proportion of the time that conventional alarms are effective is about 0.58. If the proportion for conventional alarms is accepted as a reasonable initial estimate of p, what is the smallest sample size that should be used for this study?

    (b) Suppose the researchers believed that the proportion who wake up could be quite different for the mother's voice alarm than for the conventional alarm. Explain how your procedure for choosing a samples size would differ from the solution in Part (a). (You do not need to calculate a new estimate of the necessary sample size.)

CHAPTER 10

15. Children develop "representational insight," a connection between an object and a symbol for that object. A random sample of 2-year-olds was shown a video of someone putting a toy under one of four boxes that had been placed in a room familiar to the child. Then they were taken to the room, and asked to "find the toy." The investigators reasoned that a child without representational insight would only pick the correct box on the first try about 25% of the time. Thirty out of 57 children found the toy by turning over the correct box on the first try. Is there convincing evidence that the proportion of 2-year-old children who choose the correct box on the first try is greater than 0.25? Use a significance level of $\alpha = .05$ to test the appropriate hypothesis.

CHAPTER 11

16. In many social mammalian species, the adult male assumes a sentry role, warning adult females and juveniles of approaching predators. Some students of animal behavior believe that this tendency is active in humans also. An investigator in Seattle observed people crossing the street—alone and in mixed gender pairs—at an intersection with no traffic control and noted the scanning behavior ("looking both ways") of adult females. The investigator reasoned that if the adult female assumed a protected role, the proportion of females who scan when alone would be greater than when accompanied by an adult male. Data from this study follows in the accompanying table.

Female Percent Scanning ("Looking Both Ways")			
Situation	Scanned	Did Not Scan	Totals
Alone	60	20	80
With adult male	37	39	76

Do these data support the investigator's hypothesis that the female scanning proportion is less when accompanied by a male than when alone? Provide *statistical* justification for your answer.

CHAPTER 12

17. Consider sampling from a *skewed* population. As the sample size, n, increases, some characteristics of the sampling distribution of $\bar{x}$ will change. Does an increasing sample size result in changes in the characteristics of the sampling distribution listed in (a) through (c) below? If so, explain how the characteristic of the sampling distribution changes.

    (a) The mean of the sampling distribution of $\bar{x}$
    (b) The standard deviation of the sampling distribution of $\bar{x}$
    (c) The shape of the sampling distribution of $\bar{x}$

18. To assess the levels of reading comprehension in third grade, 35 students were randomly selected from school district class lists and given a reading comprehension test. The mean and standard deviation of the resulting reading comprehension scores were 180 and 20.12, respectively.

    (a) Estimate μ, the mean reading comprehension score for 3rd-graders in this school district using a 90% confidence interval.
    (b) Explain what the confidence level of 90% associated with the interval estimate in Part (a) means.

(c) Would it be reasonable to generalize this estimate to the population of all third-grade children? Explain why or why not.

19. People tend make classifications based on their "typicality." For example, a dog is more readily classified as a mammal than is a whale because dogs are "more typical" of mammals than are whales. Previous studies have established that when young children are shown a picture of a songbird and asked if it is a bird they respond "yes" in an average of 750 milliseconds. Do children regard chickens as representative of birds? Let μ represent the mean time it takes children to respond to a question about whether a chicken is a bird.

The investigators wish to determine whether the mean classification time for chickens differs from the known mean time to classify songbirds.

(a) What is the appropriate null hypothesis in this study?
(b) What is the appropriate alternative hypothesis in this study?
(c) In the context of this study, describe a Type I error and a Type II error.
(d) Suppose that data from a random sample of 40 children was used to carry out the hypothesis test, resulting in a P-value of 0.024. What conclusion should be reached if a 0.05 significance level is used for the test?

CHAPTER 13

20. In a study of captive nectar-feeding bats (*Leptonycteris sanborni*), nectar intake over a 5-minute flight period was recorded. The bats were weighed before and after feeding, and the resulting data are given in the table below. (Notice that each bat has a unique identification number, which is shown in the first column.)

Bat #	Sex	Original Weight	Post-feeding Weight	Difference (Post−Original)
1	F	18.6	19.8	1.2
2	F	18.5	20.8	2.3
3	F	18.6	20.0	1.4
4	F	20.4	21.9	1.5
5	F	21.5	23.0	1.5
6	F	20.3	22.0	1.7
7	F	17.0	18.5	1.5
8	F	17.0	19.0	2.0
9	F	17.5	19.2	1.7
10	F	19.7	21.0	1.3
11	M	18.2	20.2	2.0
12	M	20.2	22.3	2.1
13	M	19.0	20.5	1.5
14	M	17.3	18.5	1.2
15	M	17.8	19.0	1.2
16	M	19.5	21.6	2.1
17	M	21.0	22.8	1.8
18	M	17.0	18.4	1.4
19	M	18.3	19.3	1.0

For the following questions, **make no graphs, and do no calculations**! You are only to indicate *which procedure* you would use and indicate *which bat numbers* you would use to address the questions asked. Refer to the bats by bat number and the variables by column names. You can assume that the bats used in this study are representative of nectar-feeding bats.

(a) Suppose that a researcher wanted to know if it is reasonable to conclude that mean weight gain is different for males and females. How would you check the normality assumption(s) necessary for the appropriate inference procedure?
(b) Suppose that a researcher wanted to estimate the mean weight gain for female bats after a 5-minute feeding period. How would he or she do this?
(c) Suppose that a researcher wanted to find out if it is reasonable to conclude that the mean weight gain after a 5-minute feeding period for the population of nectar-feeding bats is *more than* 1.5 g. How would he or she do this?
(d) Suppose a researcher wanted to know if it is reasonable to conclude that male and female bats consume different amounts of nectar on average after a 5-minute feeding period. How would he or she do this?

CHAPTER 15

21. Black bears (*Ursus americanus*) in National Parks can be a problem for humans. Bears' movements are governed by the need for food, and some humans accidentally or purposefully feed the bears—encouraging bears to harass humans for food. When a bear becomes troublesome, park rangers may try to relocate the bear by taking it to a distant location so that the bear probably will not be able to find its way back. In a 10-year study of this relocation strategy, researchers followed bears in a simple random sample of troublesome bears. One question of interest was whether relocation works equally well for male and female bears. Data on relocation outcome and

gender for the sample of bears are summarized in the table below.

Relocation Outcome	Male	Female	Totals
Successful	32	17	**49**
Returned to capture area	34	45	**79**
Nuisance in another area	14	4	**18**
Totals	**80**	**66**	**146**

(a) For this study, would the question of interest be answered by carrying out a test of independence or a test of homogeneity? Explain why.

(b) Carry out the test specified in Part (a) using a significance level of 0.01.

22. Wildlife biologists are interested in the hunting strategies of predators. Do predators take the first prey that comes along, or are they selective? In a study of winter prey

Prey Selection by Cougars	
White-tailed deer	118
Elk	8
Moose	2
Total	**128**

selection in a national park, radio-tagged cougars were observed to have killed prey according to the table below. At the most recent census in the National Park, there were about 3,500 white-tailed deer, 1,000 elk, and 500 moose.

Do these data provide sufficient evidence at the .05 level that the cougars are not selecting their prey at random? That is, is there evidence that the proportions of deer, elk, and moose killed by cougars differ from the park population proportions? Justify your response with an appropriate statistical analysis.

CHAPTER 16

23. In a random sample of large cities around the world, $y =$ ozone level (in parts per million) and $x =$ population in millions were recorded. Fitting the simple linear regression model gave the estimated regression equation $\hat{y} = 8.89 + 16.6x$

(a) How would you interpret the value of $b = 16.6$?

(b) Substituting $x = 3$ into the estimated regression equation gives $\hat{y} = 58.69$. Give two different interpretations of this number.

(c) Do you think this equation should be used for predicting ozone in small towns in Iowa? Explain why or why not.

(d) The value of r^2 is not given here, but explain how you would interpret the value r^2 in the context of this problem.

24. Below is a scatterplot and computer output for a regression analysis of Body Mass (g) versus Wing Length (mm) for a sample of moths.

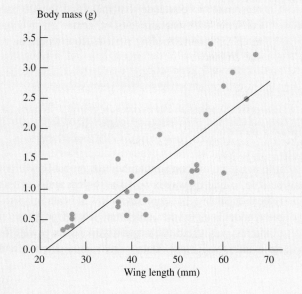

Body mass (g) vs Wing length (mm)

Regression Analysis

Body Mass = $-1.211 + 0.057$ Wing Length

Summary of Fit

RSquare	0.694
RSquare Adj	0.682
S	0.513

Analysis of Variance

Source	DF	SS	MS	F	P-value
Model	1	15.54	15.541	59.03	<.0001
Error	26	6.84	0.263		
Total	27	22.39			

Parameter Estimates

Term	Estimate	Std Error	t Ratio	Prob>|t|
Intercept	−1.2116	0.3424	−3.539	0.0015
Wing Length	0.0573	0.0072	7.9588	<.0001

(a) What is the equation of the estimated regression line?

(b) Using the computer output, carry out a model utility test using a 0.05 level of significance.

(c) Calculate and interpret a 95% confidence interval for the value of the slope of the population regression line.

(d) What is the value of s_e?

(e) Interpret the value of s_e in the context of this problem.

Answers

Chapter 1 Collecting Data in Reasonable Ways

SECTION 1.1

Exercise Set 1

1.1 observational study; the person conducting the study merely recorded whether or not the boomers sleep with their phones within arm's length and whether or not people used their phones to take photos.

1.2 observational study; children were not assigned to different experimental groups.

1.3 experiment; the researchers assigned different toddlers to experimental conditions (adult played with/talked to the robot or the adult ignored the robot).

1.4 observational study; the researchers did not assign people to experimental conditions.

1.5 experiment; because the researchers assigned study participants to one of three experimental groups (meditation, distraction task, or relaxation technique).

Additional Exercises

1.11 observational study; there was no assignment of subjects to experimental groups.

1.13 experiment; the study participants were assigned to one of the two experimental groups (how much would you pay for the mug, or how much would you sell the mug for?).

SECTION 1.2

Exercise Set 1

1.15 (a) census
(b) population characteristic

1.16 The sample is the 2,121 children between the ages of 1 and 4, and the population is all children between the ages of 1 and 4.

1.17 No. The 6,000 people who sent hair samples were volunteers and were not chosen at random.

1.18 There are several reasonable approaches. One possibility is to use a list of all the students at the school and to write all of the names on otherwise identical slips of paper. Thoroughly mix the slips of paper, and select 150 slips. Include the individuals whose names are on the slips of paper in the sample.

1.19 (a) all U.S. women
(b) only women from Maryland, Minnesota, Oregon, and Pennsylvania were included in the sample.
(c) Given that only women from four states were included in the sample, the sample is not likely to be representative of the population of interest.
(d) Selection bias is present because the selection method excluded women from all states other than Maryland, Minnesota, Oregon, and Pennsylvania.

1.20 (a) Cluster sampling
(b) Stratified random sampling
(c) Convenience sampling
(d) Simple random sampling
(e) Systematic sampling

Additional Exercises

1.27 The population is all 7,000 property owners. The sample is the 500 property owners selected for the survey.

1.29 The population is the 5,000 bricks in the lot. The sample is the 100 bricks chosen for inspection.

1.31 Bias introduced through the two different sampling methods may have contributed to the different results. The online sample could suffer from voluntary response bias in that perhaps only those who feel very strongly would take the time to go to the website and register their vote. In addition, younger people might be more technologically savvy, and therefore the website might over-represent the views of younger people (particularly students) who support the parade. The telephone survey responses might over-represent the view of permanent residents (as students might only use cell phones and not have a local phone number).

SECTION 1.3

Exercise Set 1

1.33 Random assignment allows the researcher to create groups that are equivalent, so that the subjects in each experimental group are as much alike as possible. This ensures that the experiment does not favor one experimental condition (playing Unreal Tournament 2004 or Tetris) over another.

1.34 (a) Allowing participants to choose which group they want to be in could introduce systematic differences between the two experimental conditions (tai chi group or control group), resulting in potential confounding. Those who would choose to do tai chi might, in some way, be different from those who would choose the control group.
(b) Because the purpose of this experiment is to determine whether the tai chi treatment has an effect on immunity to a virus, a control group is needed to provide a baseline against which the treatment group can be compared.

1.35

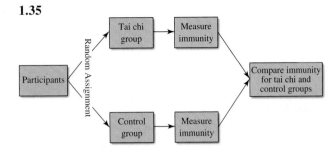

1.36 **(a)** The attending nurse was responsible for administering medication after judging the degree of pain and nausea, so the researchers did not want the nurse's personal beliefs about the different surgical procedures to influence measurements.
(b) Because the children who had the surgery could easily determine whether the surgical procedure was laparoscopic repair or open repair based on the type of incision.

1.37 The researchers could block on gender by choosing 30 of the 60 boys at random for the laparoscopic repair group and assigning the other 30 boys to the open repair group. Then 15 of the girls could be chosen at random for the laparoscopic repair group and the other 14 girls assigned to the open repair group.

1.38 There are several possible approaches. One possibility is to write the subjects names on otherwise identical slips of paper. Mix the slips of paper thoroughly and draw out slips one at a time. The names on the first 15 slips are assigned to the experimental condition of listening to a Mozart piano sonata for 24 minutes. The names on the next 15 slips are assigned to the experimental condition of listening to popular music for the same length of time. The remaining 15 names are assigned to the relaxation with no music experimental condition.

1.39 (1) Do ethnic group and gender influence the type of care that a heart patient receives? (2) The experimental conditions are gender and race. (3) The response variable is the type of care the heart patient received. (4) The experimental units are the 720 primary care doctors. It is not clear how the physicians were chosen. (5) Yes, the design incorporates random assignment of doctors to view one of the four different videos through rolling a four-sided die. (6) No control group was used. There is no need for a control group in this study. (7) There is no indication that the study includes blinding, but there is no need for blinding in this experiment.

Additional Exercises

1.47 The experimental conditions were the presence or absence of music with a vocal component. The response is the time required to complete the surgical procedure.

1.49 Random assignment of surgeons to music condition is important because there might be something inherently

different about surgeons who want no vocals versus those who do want vocals. Random assignment ensures that the experiment does not favor one experimental condition over another.

1.51 **(a)** Blocking
(b) Direct control

1.53 **(a)** Probably not, because the judges might not believe that Denny's food is as good as that of other restaurants.
(b) Experiments are often blinded in this way to eliminate preconceptions about particular experimental treatments.

SECTION 1.4

Exercise Set 1

1.54 This was an observational study, so cause-and-effect conclusions cannot be drawn.

1.55 **(a)** It is not reasonable to conclude that watching *Oprah* causes a decrease in cravings for fattening foods. This was an observational study, so cause-and-effect conclusions cannot be drawn.
(b) It is not reasonable to generalize the results of this survey to all women in the United States because not all women watch daytime talk shows. It is not reasonable to generalize these results to all women who watch daytime talk shows because not all women who watch daytime shows access DietSmart.com.

1.56 **(a)** The researcher would have had to assign the nine cyclists at random to one of the three experimental conditions (chocolate milk, Gatorade, or Endurox).
(b) It is not possible to blind the subjects, because the three drinks taste different, so the subjects will know which treatment group they are in.

1.57
Study 1: This is an observational study; random selection was used; this was not an experiment, so there were no experimental groups; no, because this was not an experiment, cause-and-effect cannot be concluded; it is reasonable to generalize to the population of students at this particular large college.

Study 2: This study was an experiment; random selection was not used; there was no random assignment to experimental conditions (the grouping was based on gender); the conclusion is not appropriate because of confounding of gender and treatment (women ate pecans, and men did not eat pecans); it is not reasonable to generalize to a larger population.

Study 3: This is an observational study; no random selection; no random assignment to experimental groups; the conclusion is not appropriate because this was an observational study, and therefore cause-and-effect conclusions cannot be drawn; cannot generalize to any larger population.

Study 4: This is an experiment; no random selection; there was random assignment to experimental groups; yes, because this was an experiment with random assignment of subjects to experimental groups, we can draw cause-and-effect conclusions; cannot generalize to a larger population.

Study 5: This is an experiment; there was random selection from students enrolled at a large college; random assignment of subjects to experimental groups was used; because this was a simple comparative experiment with random assignment of subjects to experimental groups, we can draw cause-and-effect conclusions; there was random selection of students, so we can generalize conclusions from this study to the population of all students enrolled at the large college.

Additional Exercises

1.63 Would need to know if dieters were randomly assigned to the experimental conditions (large fork or small fork) and if the study participants were randomly selected from the population of dieters.

1.65 There was no random selection from some population.

1.67 Yes, because this was an experiment and there was random assignment of subjects to experimental groups.

ARE YOU READY TO MOVE ON? CHAPTER 1 REVIEW EXERCISES

1.69 (a) experiment, there is random assignment of subjects to experimental conditions.
(b) observational study, there was no assignment of subjects to experimental conditions.
(c) observational study, there was no assignment of subjects to experimental conditions.
(d) experiment, there was random assignment of study participants to experimental conditions.

1.71 (a) population characteristic
(b) statistic
(c) population characteristic
(d) statistic
(e) statistic

1.73 The council president could assign a unique identifying number to each of the names on the petition, numbered from 1 to 500. On identical slips of paper, write the numbers 1 to 500, with each number on a single slip of paper. Thoroughly mix the slips of paper and select 30 numbers. The 30 numbers correspond to the unique numbers assigned to names on the petition. These 30 are the names that would be in the sample.

1.75 (a) Answers will vary.
(b) Answers will vary.
(c) Stratified random sample.

1.77

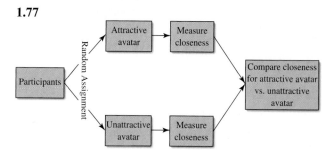

1.79 The researchers could block on gender by choosing 300 of the 600 boys at random for the book group and assigning the other 300 boys to the control group. Then 350 of the girls could be chosen at random for the book group and the other 350 girls assigned to the control group.

1.81 There are several possible approaches. One is described here. Write each subject's name on identical slips of paper. Mix the slips of paper thoroughly and draw out slips one at a time. The names on the first 10 slips are assigned to the first hand-drying method. The names on the next 10 slips are assigned to the second hand-drying method. The remaining 10 names are assigned to the third hand-drying method.

1.83 (a) Yes (b) No (c) No

1.85 (a) Not an experiment, because there were no experimental conditions to which study participants were randomly assigned.
(b) No, this was an observational study.

Chapter 2 Graphical Methods for Describing Data Distributions

SECTION 2.1

Exercise Set 1

2.1 (a) numerical, discrete (b) categorical (c) numerical, continuous (d) numerical, continuous (e) categorical

2.2 (a) discrete (b) continuous (c) discrete (d) discrete

2.3
Data Set 1: one variable; categorical; summarize the data distribution; bar chart.

Data Set 2: one variable; numerical; compare groups; comparative dotplot or comparative stem-and-leaf display.

Data Set 3: two variables; numerical; investigate relationship; scatterplot.

Data Set 4: one variable; categorical, compare groups, comparative bar chart.

Data Set 5: one variable; numerical; summarize the data distribution; dotplot, stem-and-leaf display or histogram.

Additional Exercises

2.7 (a) numerical (b) numerical (c) categorical (d) numerical (e) categorical

2.9 (a) categorical (b) numerical (c) numerical (d) categorical

2.11 (a) numerical (b) numerical (c) numerical (d) categorical (e) categorical (f) numerical (g) categorical

2.13 one variable; numerical; compare groups, comparative dotplot or comparative stem-and-leaf display.

2.15 one variable; numerical, summarize the data distribution; dotplot, stem-and-leaf plot or histogram.

SECTION 2.2

Exercise Set 1

2.16 (a)

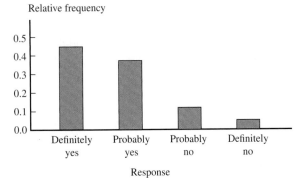

(b) One example: "Senior Satisfaction! Over 80% say they would enroll again."

2.17

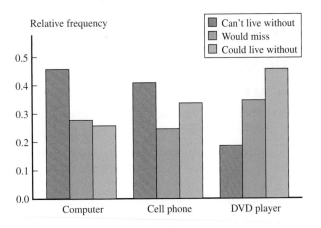

Additional Exercises

2.21

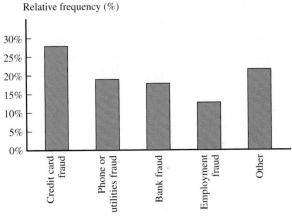

Credit card fraud is the most commonly occurring identity theft type. Although phone/utility, bank, and employment fraud each constitute a relatively large portion of overall type of identity theft, the collective "other fraud" category is greater than any one of these other three.

2.23 The relative frequency distribution is:

Type of Household	Relative Frequency
Nonfamilies	0.29
Married with Children	0.27
Married without Children	0.29
Single Parent	0.15

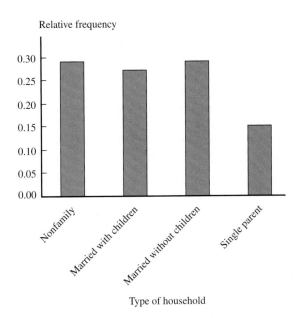

SECTION 2.3

Exercise Set 1

2.24 (a)

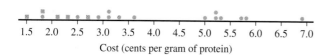

Cost (cents per gram of protein)

(b) Because the costs of meat and poultry products (represented by the squares on the dotplot) are generally smaller than the costs for other sources of protein, they do appear to be a good value.

2.25 (a)

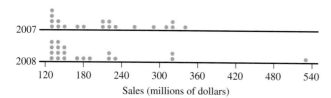

Sales (millions of dollars)

(b) Both distributions are skewed toward larger values. The 2007 ticket sales are centered at about $210 million dollars, which is higher than the center of the 2008 ticket sales, which are centered around $150 million dollars. The lowest ticket sales for both 2007 and 2008 are approximately $127 million dollars. Ticket sales for 2008 have a maximum value of approximately $533 million dollars, which is much higher than the highest ticket sales for 2007. Without this extreme value, the spreads between the lowest and highest values are approximately equal.

2.26

28	8
29	
30	
31	
32	8
33	0
34	178
35	00145678899
36	238999
37	0034566777
38	01124558
39	00259
40	045
41	2
42	2

Legend: 34 | 1 = 34.1 years

The distribution of median ages is centered at approximately 37 years old, with values ranging from 28.8 to 42.2 years. The distribution is approximately symmetric, with one unusual value of 28.8 years.

2.27 (a)

Very Large Urban Area		Large Urban Area
	1	023478
	2	369
8	3	0033589
99	4	0366
8711	5	012355
9730	6	
2	7	
	8	
3	9	

Legend: 4|6 = 46 extra hours per year

(b) The statement "The larger the urban areas, the greater the extra travel time during peak period travel" is generally consistent with the data. Although there is overlap between the times for the very large and large urban areas, the extra travel times for the very large urban areas are generally greater than those for the large urban areas.

2.28 (a)

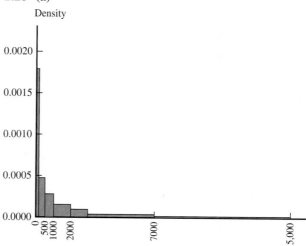

Credit card balance (Credit Bureau data)

(b)

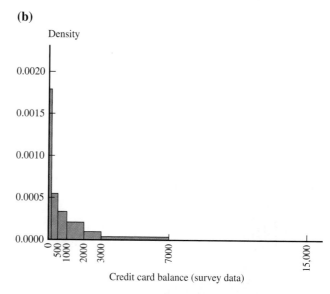

Credit card balance (survey data)

(c) The histograms are similar in shape. A notable difference is that the Credit Bureau data show that 7% of

students have credit card balances of at least $7,000, but no survey respondent indicated a balance of at least $7,000.
(d) Yes, because students with credit card balances of $7,000 or more might be too embarrassed to admit that they have such a high balance.

2.29 **(a)** If the exam is quite easy, the scores would be clustered at the high end of the scale, with a few low scores for the students who did not study. The histogram would be negatively skewed.
(b) If the exam is difficult, the scores would be clustered around a much lower value, with only a few high scores. The histogram would be positively skewed.
(c) In this case, the histogram would be bimodal, with a cluster of high scores and a cluster of low scores.

2.30 **(a)** In order to show necessary detail at the lower end of the distribution, it was necessary to use class intervals as narrow as 5 minutes. However, to use this class width throughout would result in too large a number of intervals. Therefore, wider intervals were used for the higher commute times where there were fewer data values.

(b)

Commute Time	Frequency	Relative Frequency	Density	Cumulative Relative Frequency
0 to < 5	5,200	0.052	0.0104	0.052
5 to < 10	18,200	0.181	0.0363	0.233
10 to < 15	19,600	0.195	0.0390	0.428
15 to < 20	15,400	0.153	0.0307	0.581
20 to < 25	13,800	0.137	0.0275	0.718
25 to < 30	5,700	0.057	0.0114	0.775
30 to < 35	10,200	0.102	0.0203	0.877
35 to < 40	2,000	0.020	0.0040	0.897
40 to < 45	2,000	0.020	0.0040	0.917
45 to < 60	4,000	0.040	0.0027	0.957
60 to < 90	2,100	0.021	0.0007	0.978
90 to < 120	2,200	0.022	0.0007	1.000

(c)

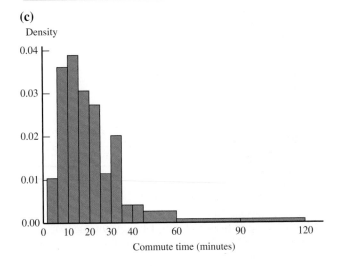

A typical commute time is between 15 and 20 minutes, and the great majority of commute times lie between 5 and 35 minutes. The distribution is positively skewed.

(d)

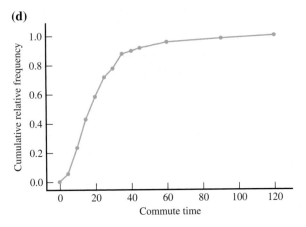

(e) **(i)** 0.93
(ii) $1 - 0.64 = 0.36$
(iii) 17 minutes

Additional Exercises

2.39 The distribution of wind speed is positively skewed and bimodal. There are peaks in the 35–40 m/s and 60–65 m/s intervals.

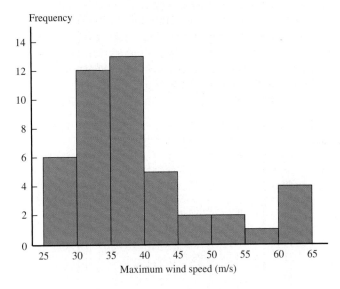

2.41

Non-Disney		Disney
9521000	0	001233579
651	1	567
0	2	029
	3	
	4	
	5	4

Legend: 1|5 = 150 seconds

In general, the Disney movies have longer tobacco exposure times than the non-Disney movies. Disney movies have a typical value of approximately 80 seconds, which is larger than a typical value for the non-Disney movies of 50 seconds. In addition, there is more variability in the Disney tobacco exposure times than the others. Disney movies vary between 6 and 548 seconds, which is greater than the observed spread for the non-Disney movies, which vary between 1 and 205 seconds. Finally, there appears to be one outlier in the Disney movies (548 seconds) and no outliers in the non-Disney movies.

SECTION 2.4

Exercise Set 1

2.42 **(a)**

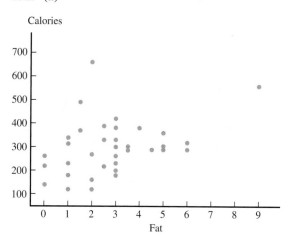

The scatterplot shows the expected positive relationship between grams of fat and calories. The relationship is weak.

(b)

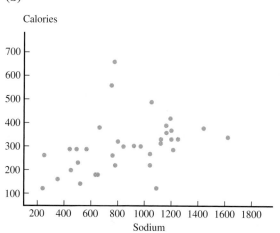

As was observed in the calories versus fat scatterplot, there is also a weak, positive relationship between calories and sodium. The relationship between calories and sodium appears to be a little stronger than the calories versus fat relationship.

(c)

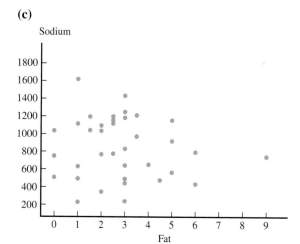

There is no apparent relationship between sodium and fat.

(d)

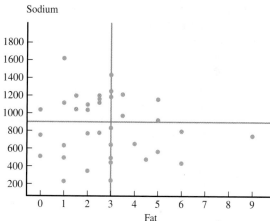

The lower-left region corresponds to healthier fast-food choices. This region corresponds to food items with fewer than 3 grams of fat and fewer than 900 milligrams of sodium.

2.43 **(a)**

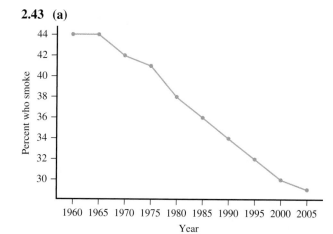

There has been a steady downward trend in the percent of people who smoke among people who did not graduate from high school, from a high of 44% to 29% in 2005.

(b)

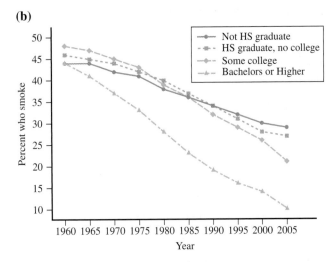

(c) There has been a steady downward trend in the percent of people age 25 or older who smoke, regardless of education level. In 1960, regardless of education level, the percentage was approximately the same (about 44–48%). Over time, however, the differences have become more pronounced. In 2005, those people with bachelor's degrees or higher had the lowest smoking rate (10%), followed by those with some college (21%). The highest rates of smoking were found among those who either did not graduate from high school (29%) or graduated high school but did not attend college (27%).

Additional Exercises

2.47 (a)

Percentage of households with computer

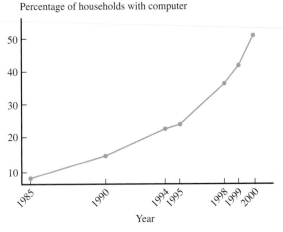

(b) The percentage of households with computers has increased over time, from a low of approximately 8% in 1985 to over 50% in 2000. The rate of increase has also increased over time.

2.49 There is a relatively weak positive relationship between poverty rate and dropout rate. There are two states that have poverty rates between 10% and 15% but have much higher dropout rates (over 15%) compared to other states with comparable poverty rates.

SECTION 2.5

Exercise Set 1

2.51 (a) categorical
(b) A bar chart was used because the response is a categorical variable, and dotplots are used for numerical responses.
(c) This is not a correct representation of the response data because the percent values add up to over 100% (they add to 107%).

2.52 (a) Overall score is numerical. Grade is categorical.
(b) The figure is equivalent to a segmented bar graph because the bar is divided into segments, with different shaded regions representing the different grades ("Top of the Class," "Passing," "Barely Passing," and "Failing"), and the height of each segment is equal to the frequency for that category (for example, there are five school districts in the Top of the Class category, three in the Passing category, and so on), making the area of each shaded region proportional to the relative frequencies for each grade.

(c)

One alternate assignment of grades is to require that "Top of the Class" schools earn grades of 72 or higher, "Passing" schools earn between 66 and 71, "Barely Passing" schools earn between 61 and 65, and "Failing" schools earn 60 or below. This alternative is suggested because there appear to be clusters of dots on the dotplot that correspond to the suggested ranges.

2.53 Answers will vary. For example: For teens ages 12 to 17 years, the percentage of cell phone owners increases with age in each of the years 2004, 2006, and 2008. In addition, within each age group, the percentage of teens owning cell phones increased between 2004 and 2008. The largest increase in percentage of teens owning cell phones was among 12-year-olds, and the smallest percentage increase was among 13-year-olds.

Additional Exercises

2.57 (a) The areas in the display are not proportional to the values they represent. The "no" category seems to represent more than 68%.

(b)

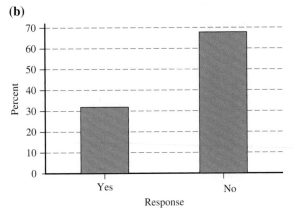

2.59 (a)

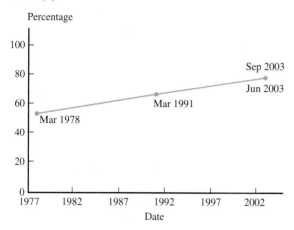

Percentage

(b)

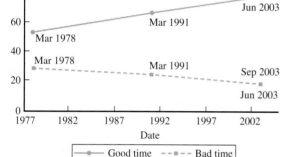

Percentage

(c) The time-series plot best shows the trend over time.

ARE YOU READY TO MOVE ON? CHAPTER 2 REVIEW EXERCISES

2.61

Data Set 1: one variable; numerical; summarize the data distribution; a dotplot, stem-and-leaf display or histogram would be an appropriate graphical display.

Data Set 2: two variables; numerical; investigate the relationship; a scatterplot would be an appropriate graphical display.

Data Set 3: two variables; categorical; compare groups; comparative bar chart.

Data Set 4: one variable; categorical; summarize the data distribution; a bar chart would be an appropriate graphical display.

Data Set 5: One variable; numerical; compare groups; a comparative dotplot or comparative stem-and-leaf display would be an appropriate graphical display.

2.63 (a)

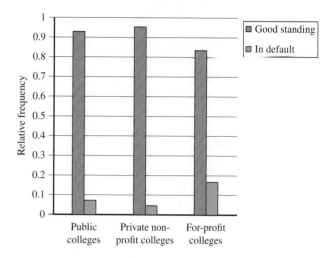

(b) The "In Default" bar in the "For-Profit Colleges" category is taller than either of the other "In Default" bars.

2.65 (a)

0	8
1	4467788999
2	0000112223333344445557778
3	01222334457
4	02458

Legend: 1|4 = 14 cents per gallon

(b) The center is approximately 24 cents per gallon, and most states have a tax that is near the center value, with tax values ranging from 8 cents per gallon to 48 cents per gallon. The distribution is approximately symmetric.

(c) The only value that might be considered unusual is the 8 cents per gallon tax in Alaska.

2.67 (a)

Quality rating

(b) About 114.

(c) There is a lot of variability, with quality rating ranging between a low score of 84 defects per 100 vehicles and a high score of 170 defects.

(d) Two brands (Land Rover and Mitsubishi) seem to stand out as having a much greater value for number of defects. Four brands (Acura, Lexus, Mercedes-Benz, and Porsche) have smaller values for number of defects.

(e)

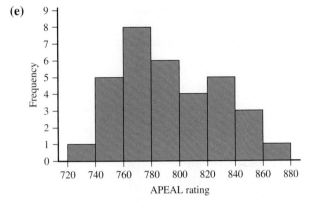

The histogram is centered at approximately 790, with values that range between approximately 720 and 880. The distribution is bimodal and positively skewed.

(f)

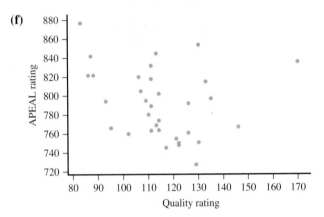

There is a weak negative relationship between customer satisfaction (as measured by the APEAL rating) and number of defects. Brands with a higher number of defects per 100 vehicles tend to have lower satisfaction ratings.

2.69 (a)

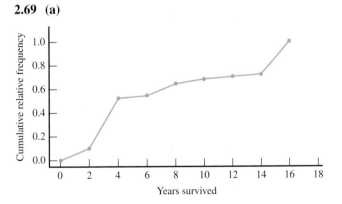

(b) **(i)** 0.53
(ii) 0.62
(iii) $1 - 0.68 = 0.32$

2.71 The first graphical display is not drawn appropriately. The Z's have been drawn so that their heights are in proportion to the percentages shown. However, the widths are also in proportion to the percentages, so the areas are not proportional to the percentages. In the second graphical

display, however, *only* the heights of the cars are in proportion to the percentages shown. The widths of the cars are all equal, so the areas of the cars are proportional to the percentages, and this is an appropriately drawn graphical display.

Chapter 3 Numerical Methods for Describing Data Distributions

SECTION 3.1

Exercise Set 1

3.1 The distribution is approximately symmetric with no outliers, so the mean and standard deviation should be used to describe the center and spread, respectively.

3.2 The distribution is skewed with an outlier, so the median and interquartile range should be used.

3.3 The distribution is skewed with a possible outlier, so the median and interquartile range should be used.

3.4 The average may not be the best measure of a typical value for this data set because the distribution is clearly skewed.

Additional Exercises

3.9 The distribution is skewed, so median and interquartile range should be used.

3.11 The distribution is roughly symmetric with no obvious outliers, so the mean and standard deviation should be used.

SECTION 3.2

Exercise Set 1

3.12 $\bar{x} = 51.33$ ounces; this is a typical value for the amount poured. $s = 15.22$ ounces; this represents how much, on average, the values in the data set spread out, or deviate, from the mean.

3.13 (a) $\bar{x} = 59.23$ ounces; this is a typical value for the amount poured. $s = 16.71$ ounces; this represents how much, on average, the values in the data set spread out, or deviate, from the mean.
(b) Individuals pouring into short, wide glasses pour, on average, more than when pouring into tall, slender glasses.

3.14 (a) $\bar{x} = 59.85$ hours, $s = 14.78$ hours
(b) $\bar{x} = 56.67$ hours, $s = 9.75$ hours. When Los Angeles was excluded from the data set, the mean and standard

deviation both decreased quite a bit. This suggests that using the mean and standard deviation as measures of center and spread for data sets with outliers can be risky because outliers seem to have a significant impact on those measures.

3.15 Answers will vary. One possible answer: The mean is $444, but it is likely that some parents spend only a small amount. There is probably a lot of variability in amount spent, so we would expect a large value for the standard deviation.

Additional Exercises

3.21 (a) $\bar{x} = 287.714$; the deviations from the mean are $209.286, -94.714, 40.286, -132.714, 38.286, -42.714, -17.714$.
(c) $s^2 = 12{,}601.905, s = 112.258$.

3.23 The deviations are exactly the same as the corresponding deviations for the original data set. Since the deviations are the same, the new variance and standard deviation are also the same as the old variance and standard deviation. Subtracting the same number from or adding the same number to every value in a data set does not change the value of the variance or standard deviation.

SECTION 3.3

Exercise Set 1

3.25 (a) median $= 433{,}246.5$. Half of the newspapers had average weekly circulations of less than $433{,}246.5$, and the other half had average weekly circulations of more than $433{,}246.5$.
(b) The median is preferable because the distribution is skewed and contains outliers.
(c) It is not reasonable to generalize from this sample to the population of daily U.S. newspapers because these newspapers were not randomly selected. They are the top 20 American newspapers in average weekday circulation.

3.26 Lower quartile $= 10{,}478$; 25% of the catsups have sodium contents lower than 10,478. Upper quartile $= 11{,}778$; 75% of the catsups have sodium contents lower than 11,778. The interquartile range is $iqr = 11{,}778 - 10{,}478 = 1{,}300$; the range of the middle 50% of the catsup sodium contents is 1,300.

3.27 Median $= 142$; half of the values of number of minutes used in cell phone calls in one month are less than or equal to 142 minutes, and half of the data values of number of minutes used in cell phone calls are greater than or equal to 142 minutes. $iqr = 195$; the middle 50% of the data values have a range of 195 minutes.

3.28 median $= 21\%$; half of the tips were below 21%, and the remaining half were above 21%. $iqr = 24.85\%$; the middle 50% of tips had a range of 24.85%.

Additional Exercises

3.33 The large difference between the mean and median indicates that there were some parents who spent large amounts of money on school supplies.

3.35 (a) mean will be greater than the median.
(b) $\bar{x} = 370.69$ seconds, *median* $= 369.5$ seconds.
(c) The largest time could be increased by any amount and not affect the sample median because the position of the middle value will not change if the largest value is increased. The largest time could be decreased to 370 seconds without changing the value of the median.

SECTION 3.4

Exercise Set 1

3.36 Minimum $= 0$, lower quartile $= 14$, median $= 33.5$, upper quartile $= 63$, maximum $= 151$.

3.37

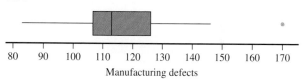

The boxplot shows that there is one outlier (170 defects), and the value of the largest non-outlier is 146 defects. The middle 50% of the data values range between about 106 and 126 defects. The distribution is positively skewed.

3.38 (a) No, they are not outliers. For this data set, values are outliers if they are greater than $32.3 + 1.5(32.3 - 20) = 50.75$ cents, or less than $20 - 1.5(32.3 - 20) = 1.55$ cents.
(b)

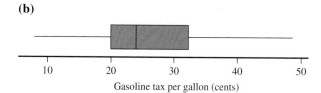

The distribution is positively skewed.

3.39 (a) lower quartile $= 16.05$ inches, upper quartile $= 21.93$ inches. $iqr = 21.93 - 16.05 = 5.88$ inches.
(b) For this data set, values are outliers if they are greater than $21.93 + 1.5(5.88) = 30.75$ inches or less than $16.05 - 1.5(5.88) = 7.23$ inches. The value 31.57 inches is an outlier.
(c)

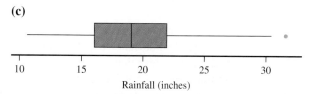

The modified boxplot shows one outlier at the high end of the scale. The distribution of inches of rainfall is slightly positively skewed.

3.40

Both distributions (short, wide and tall, slender) are skewed, although the direction of skew is different for the two distributions. The amount poured into short, wide glasses tends to be more than the amount poured into tall, slender glasses.

Additional Exercises

3.47 (a)

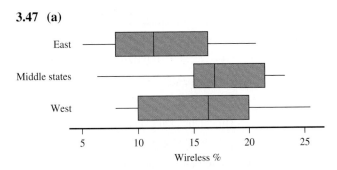

(b) The most noticeable difference between the wireless percent for the three geographical regions is that the Middle States region is negatively skewed and has a smaller interquartile range than the East and West regions. The Eastern region has the smallest median (11.4%), and the Middle States and Western regions have medians that are about the same (16.9% and 16.3%).

3.49 The fact that the mean is so much higher than the median indicates that the distribution is positively skewed.

SECTION 3.5

Exercise Set 1

3.50 First national aptitude test: $z = 1.5$. Second national aptitude test: $z = 1.875$. The student performed better on the second national aptitude test relative to the other test takers because the z-score for the second test is higher than for the first test.

3.51 (a) 40 minutes is 1 standard deviation above the mean; 30 minutes is 1 standard deviation below the mean. The values that are 2 standard deviations away from the mean are 25 and 45 minutes.
(b) Approximately 95% of times are between 25 and 45 minutes; approximately 0.3% of times are less than 20 minutes or greater than 50 minutes; approximately 0.15% of times are less than 20 minutes.

3.52 The 10th percentile of $0 indicates that 10% of students have $0 or less of student debt. The 25th percentile (which is the lower quartile) indicates that 25% of students have $0 or less of student debt. The 50th percentile (the median) indicates that 50% of students have $11,000 or less of student debt. The 75th percentile (the upper quartile) indicates that 75% of students have $24,600 or less of student debt. The 90th percentile indicates that 90% of students have $39,300 or less of student debt.

3.53 (a)

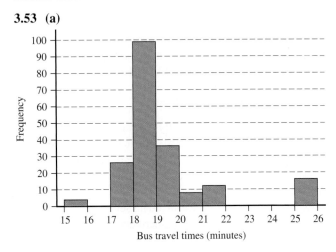

(b) Note: percentiles were estimated using midpoints of the histogram bar intervals; (i) 86th percentile is approximately 20.5 minutes; (ii) 15th percentile is approximately 17.5 minutes; (iii) 90th percentile is approximately 21.5 minutes; (iv) 95th percentile is approximately 25.5 minutes; (v) 10th percentile is approximately 17.5 minutes

Additional Exercises

3.59 (a) 1,100 gallons; **(b)** 1,400 gallons; **(c)** 1,700 gallons

3.61 (a) 120
(b) 20
(c) −0.5
(d) 97.5
(e) Since a score of 40 is 3 standard deviations below the mean, that corresponds to a percentile of 0.15%. Therefore, there were relatively few scores below 40.

ARE YOU READY TO MOVE ON?
CHAPTER 3 REVIEW EXERCISES

3.63 $\bar{x} = 792.03$, which is a typical or representative value for the APEAL rating. $s = 36.70$, which represents how much, on average, the values in the data set spread out, or deviate, from the mean APEAL rating.

3.65 (a) $\bar{x} = 27.31\%$, $s = 23.83\%$
(b) After removing the 105% tip, the new mean and standard deviation are $\bar{x}_{new} = 23.23\%$ and $s_{new} = 15.70\%$. These values are much smaller than the mean and standard

deviation computed with 105 included. This suggests that the mean and standard deviation can change dramatically when outliers are present (or removed) from the data set and, therefore, are probably not the best measures of center and spread to use in this situation.

3.67 **(a)** median = 140 seconds; half the values are less than 140 seconds and half the values greater than 140 seconds. *iqr* = 100 seconds; the middle 50% of the data values have a range of 100 seconds.
(b) There is an outlier in the data set.

3.69 **(a)** Median = 8 grams/serving; lower quartile = 7 grams/serving; upper quartile = 12 grams/serving; interquartile range = 12 − 7 = 5 grams/serving
(b) Median = 10 grams/serving; lower quartile = 6 grams/serving; upper quartile = 13 grams/serving; interquartile range = 13 − 6 = 7 grams/serving
(c) There are no outliers in the sugar content data.
(d) The minimum value and lower quartile are the same because the smallest five values in the data set are all equal to 7.

(e)

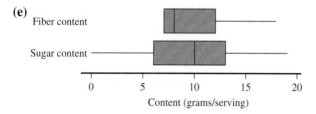

The sugar content in grams/serving is much more variable than the fiber content in grams/serving. The boxplot of fiber content shows that the minimum and lower quartiles are equal to each other, which is not observed in the sugar content. The distribution of sugar content values is approximately symmetric, which is different from the skewed fiber content distribution.

3.71 **(a)** The 25th percentile indicates that 25% of full-time female workers age 25 or older with an associate degree earn $26,800 or less. The 50th percentile indicates that 50% of full-time female workers age 25 or older with an associate degree earn $36,800 or less. The 75th percentile indicates that 75% of full-time female workers age 25 or older with an associate degree earn $51,100 or less.

(b) The 25th, 50th, and 75th percentile values for men are all greater than the corresponding percentiles for female workers, indicating that full-time employed men age 25 or older with an associate degree, in general, earn more than full-time employed women age 25 or older with an associate degree.

Chapter 4 Describing Bivariate Numerical Data

SECTION 4.1

Exercise Set 1

4.1 Scatterplot 1: (i) Yes (ii) Yes (iii) Negative
Scatterplot 2: (i) Yes (ii) No (iii) –
Scatterplot 3: (i) Yes (ii) Yes (iii) Positive
Scatterplot 4: (i) Yes (ii) Yes (iii) Positive

4.2 **(a)** Negative correlation, because as interest rates rise, the number of loan applications might decrease.
(b) Close to zero, because there is no reason to believe that height and IQ should be related.
(c) Positive correlation, because taller people tend to have larger feet.
(d) Positive correlation, because as the minimum daily temperature increases, the cooling cost would also increase.

4.3 **(a)** There is a moderately strong positive linear relationship between school achievement test score and midlife IQ.
(b) $r = 0.6$, because the article says that $r = 0.64$ indicated a very strong relationship, higher than the correlation between height and weight in adults. Therefore, a correlation that is moderately strong ($r = 0.6$) with a positive association (taller people tend to weigh more) is consistent with the statement.

4.4 **(a)** $r = 0.335$; there is a weak, positive linear relationship.
(b) The conclusion that "heavier logging led to large forest fires" cannot be justified because correlation does not imply causation.

4.5 **(a)**

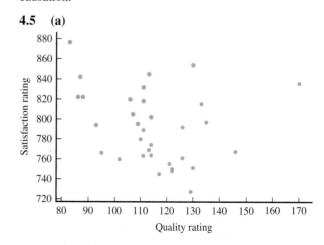

There is a weak, negative linear relationship between satisfaction rating and quality rating (number of defects).
(b) $r = -0.239$; there is a weak, negative linear relationship between satisfaction rating and quality rating (number of defects).

4.6 No, the statement is not correct. A correlation of 0 indicates that there is not a linear relationship between two variables. There could be a strong nonlinear relationship (for example, a quadratic relationship) between the two variables.

4.7 No, it is not reasonable to conclude that a tendency to overestimate social desirability causes less conservative views. Correlation measures the strength of association, but association does not imply causation.

4.8 The correlation between college GPA and academic self-worth ($r = 0.48$) indicates that there is a weak or moderate positive linear relationship between those variables. This tells us that athletes with higher a GPA tend to feel better about themselves academically than those with lower grades. The correlation between college GPA and high school GPA ($r = 0.46$) indicates that there is a weak or moderate positive relationship between those variables as well. This tells us that those athletes with higher high school GPA tend to also have a higher college GPA. Finally, the correlation between college GPA and a measure of tendency to procrastinate ($r = -0.36$) indicates that there is a weak negative linear relationship between those variables. Athletes with a lower college GPA tend to procrastinate more than athletes with a higher college GPA.

Additional Exercises

4.15 (a)
(b) $r = 0.001$
(c) The correlation coefficient of $r = 0.001$ indicates that there is essentially no linear relationship between mare weight and foal weight. The scatterplot shows that there is no obvious relationship (linear or otherwise) between mare weight and foal weight.

4.17 $r = 0.987$; this value is consistent with the previous answer, because the correlation coefficient is large (close to 1) and positive, which indicates a strong positive association between household debt and corporate debt.

4.19 The sample correlation coefficient would be closest to -0.9. Cars traveling at a faster rate of speed will travel the length of the highway segment more quickly than those who are traveling more slowly, and the correlation would be strong.

SECTION 4.2

Exercise Set 1

4.21 It makes sense to use the least squares regression line to summarize the relationship between x and y for Scatterplot 1, but not for Scatterplot 2. Scatterplot 1 shows a linear relationship between x and y, but Scatterplot 2 shows a curved relationship between x and y.

4.22 It would be larger because the least squares regression line is *the* line with the minimum value for the sum of the squared vertical deviations from the line. All

other lines would have larger values for the sum of the squared vertical deviations.

4.23 (a) $\hat{y} = -5.0 + 0.017x$
(b) 30.7 therms.
(c) 0.017 therms.
(d) No, because the regression line was determined based on house sizes between 1,000 and 3,000 square feet. There is no guarantee that the linear relationship will continue outside this range of house sizes.

4.24 (a) The response variable (y) is birth weight, and the predictor variable (x) is mother's age.
(b) It is reasonable to use a line to summarize the relationship because the scatterplot shows a clear linear relationship between birth weight and mother's age.
(c) $\hat{y} = -1163.4 + 245.15x$
(d) The slope of 245.15 is the amount, on average, by which the birth weight increases when the mother's age increases by one year.
(e) It is not appropriate to interpret the intercept of the least squares regression line. The intercept is the birth weight for a mother who is zero years old, which is impossible. In addition, the intercept is negative, indicating a negative birth weight, which is also impossible.
(f) 3,249.3 grams
(g) 2,513.85 grams
(h) No, because 23 years is well outside the range of ages used in determining the least squares regression line.

4.25 (a) The response variable is the cost of medical care, and the predictor variable is the measure of pollution.
(b) There is a moderate, negative linear relationship.
(c) $\hat{y} = 1082.2 - 4.691x$
(d) The slope is negative, and it is consistent with the observed negative association in the scatterplot.
(e) No, the association between medical cost and pollution level is negative, which indicates that people over the age of 65 in more polluted areas tend to have lower medical costs.
(f) $918.02
(g) No, because the value 60 is far outside the range of data values in the sample.

4.26 (a) $\hat{y} = 11.48 + 0.970x$
(b) 496.48.
(c) 302.48

Additional Exercises

4.33 $\hat{y} = 13.5 - 0.195x$

4.35 Age is a better predictor of number of cell phone calls. The linear relationship between age and number of cell phone calls is stronger than the relationship between age and number of text messages sent.

4.37 The slope is the change in predicted price for each additional mile from the Bay, so the slope would be $-4,000$.

SECTION 4.3

Exercise Set 1

4.39 **(a)** $\hat{y} = 1.33878 - 0.007661x$
(b) $r^2 = 0.099$; Approximately 9.9% of the variability in telomere length can be explained by the linear relationship between telomere length and perceived stress.
(c) $s_e = 0.159$; a typical amount by which telomere length will deviate from the least squares regression line is 0.159.
(d) negative, because the slope of the least squares regression line is negative; weak, because $r = -0.315$.

4.40 A small value of s_e indicates that residuals tend to be small. Because residuals represent the difference between an observed y value and a predicted y value, the value of s_e tells us how much accuracy we can expect when using the least squares regression line to make predictions.

4.41 It is important to consider both r^2 and s_e when evaluating the usefulness of the least squares regression line because a large r^2 (which indicates the proportion of variability in y that can be explained by the linear relationship between x and y) tells us that knowing the value of x is helpful in predicting y, and a small s_e indicates that residuals tend to be small.

4.42 **(a)** Yes, the scatterplot looks reasonably linear.

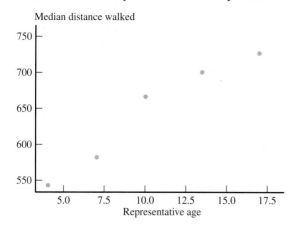

(b) $\hat{y} = 492.80 + 14.763x$
(c) The residuals are -7.55, -12.14, 26.87, 9.00, and -16.17. There is a curved pattern in the residual plot. The curvature indicates that the relationship between median distance walked and representative age is not linear.

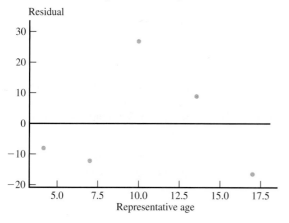

4.43 **(a)** The pattern for girls differs from boys in that the girls' scatterplot shows more apparent nonlinearity in the pattern.
(b) $\hat{y} = 480 + 12.525x$
(c) The residuals are -37.70, 10.63, 50.56, 8.52, -32.01. The curvature of the residual plot indicates that a curve is more appropriate than a line for describing the relationship between median distance walked and representative age.

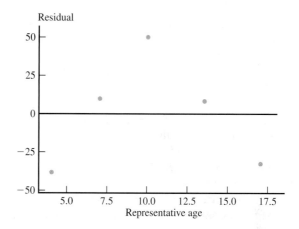

4.44 **(b)** $\hat{y} = -0.03443 + 0.5803x$. The predicted nitrogen retention for a flying squirrel whose nitrogen intake is 0.06 grams is 0.000388 grams. The residual associated with the observation $(0.06, 0.01)$ is 0.009612.
(c) The observation $(0.25, 0.11)$ is potentially influential because that point has an x value that is far away from the rest of the data set.
(d) $\hat{y} = -0.037 + 0.627(0.06) = 0.00062$; This prediction is larger than the prediction made in Part (b).

4.45 **(a)** There appears to be a linear relationship.
(b) $\hat{y} = 18.483 + 0.0028655x$
(c) The observation $(3928, 46.8)$ is not influential, because the x-value for that observation is not far from the rest of the data. In addition, removal of the potentially influential point produces a least squares regression line with a y-intercept and slope similar to the original line.
(d) Those points are not considered influential even though they are far from the rest of the data because they follow the trend of the remaining data points. Removal of those points would produce a least squares regression line similar to the line found using the full data set.
(e) $s_e = 9.16217$; A typical deviation from the least squares regression line is 9.16217 percentage points.
(f) $r^2 = 0.832$; Approximately 83.2% of the variability in percentage transported can be explained by the linear relationship between percentage transported and number of salmon.

SECTION 4.5

Exercise Set 1

4.57 (a)

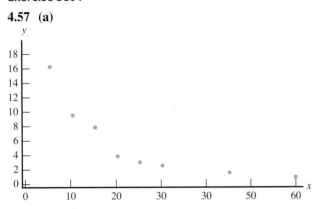

The pattern in the plot is curved rather than linear.

(b)

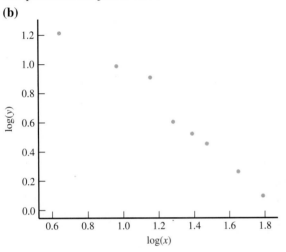

The pattern in this plot looks linear.

(c) The value of r^2 is large (0.973) and the value of s_e is relatively small (0.0657), indicating that a line is a good fit to the data.

(d) $\log(35) = 1.54$, predicted $\log(y) = 0.403$, predicted $y = 2.53$

4.58 (a)

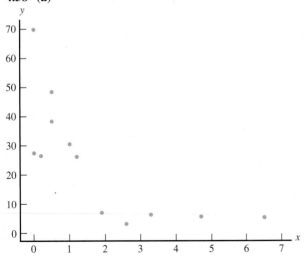

The pattern in the scatterplot is curved.

(b)

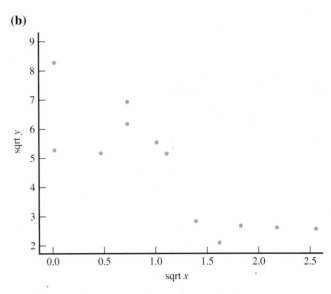

The curved pattern is not as pronounced as in the scatterplot from Part (a), but there is still curvature in the plot.

(c) Answers will vary. One possible answer is to use $y' = \log(y)$ and $x' = \sqrt{x}$. This transformation would be preferred over the transformation in Part (b) because this scatterplot closer to linear.

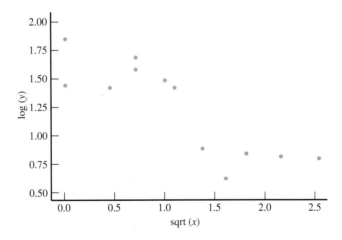

4.59 Answers will vary. Scatterplot (b) shows a curve that tends to follow the pattern of points in the scatterplot.

Additional Exercises

4.63 The pattern in the scatterplot of y versus x is curved. One transformation that results in a pattern that is more linear is $y' = \log(y)$ and $x' = \log(x)$.

ARE YOU READY TO MOVE ON?
CHAPTER 4 REVIEW EXERCISES

4.65 (a) Positive, because the cost of produce is usually determined by its weight, so a heavier apple would cost more.
(b) Close to zero, because there is no reason to think that the height of a person would be associated with number of pets.

(c) Positive, because, in general, the more time spent studying for an exam, the higher the score will be.
(d) Positive, because, in general, the heavier a person, is the longer it will take to run 1 mile.

4.67 **(a)**

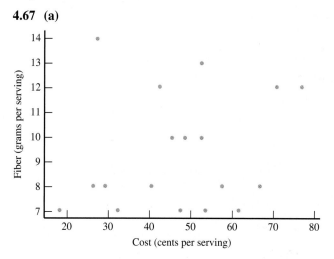

There is no apparent relationship between fiber and cost.
(b) $r = 0.204$; there is a weak positive linear relationship between fiber and cost.
(c) $r = 0.241$; the correlation coefficient for the per cup data is greater than the correlation coefficient for the per serving data.

4.69 **(a)**

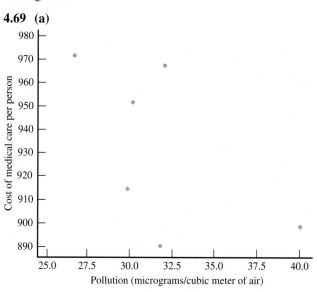

(b) $r = -0.581$: there is a moderate, negative association between pollution and cost of medical care per person.
(c) The scatterplot shows a negative association between cost of medical care per person and pollution level. The sign of the correlation coefficient supports the negative association between the variables. Both the scatterplot and correlation coefficient indicate that people over the age of 65 living in more polluted areas tend to have lower medical costs.

4.71 The least squares regression line is the line found by minimizing the sum of the squared vertical deviations from the line. Therefore, the least squares regression line will

have the smallest sum of the squared vertical deviations compared to any other line, making choice (i) (308.6) the correct choice.

4.73 **(a)** $\hat{y} = 101.33 - 9.296x$
(b) On average, the survival rate percentage decreases by 9.296 when the mean call-to-shock time increases by one minute.
(c) No, it does not make sense to interpret the intercept because 0 is outside the range of the data and because the y-values represent survival rate as a percent, and the intercept is greater than 100%.
(d) 8.37%

4.75 **(a)** 14.5%
(b) The residual for the 15th year of schooling is $y - \hat{y} = 14 - 14.5 = -0.5\%$. The prediction was 0.5% too high.

4.77 **(a)**

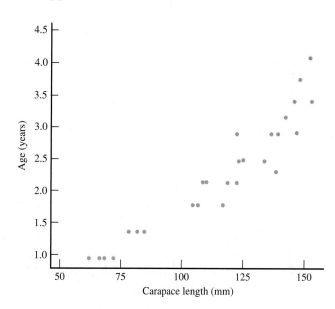

$\hat{y} = -1.0755 + 0.029105x$, where $\hat{y}$ is the predicted age (in years) and x is the carapace length (in mm). There is a positive nonlinear association between age and carapace length.

(b)

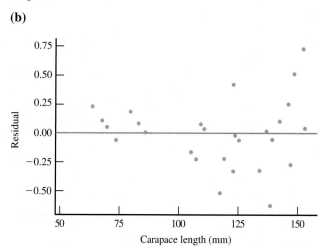

There is a curved pattern in the residual plot that indicates that the relationship between age and carapace length would be better described by a curve rather than a line.

4.79 (a)

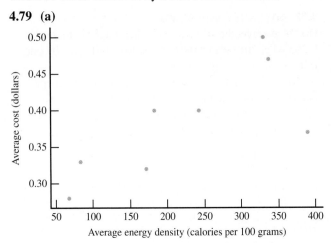

The relationship looks approximately linear.
(b) $\hat{y} = 0.27970 + 0.0004619x$, where $\hat{y}$ is the predicted average cost in dollars, and x is the average energy density in calories per 100 grams.
(c)

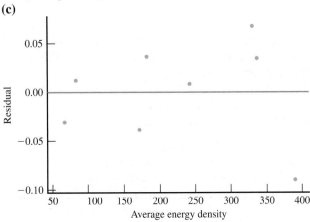

There is one data value (with average energy density of 390) that has a particularly large residual when compared to the other residuals.
(d) $r^2 = 0.547$; Approximately 54.7% of the variability in average cost is explained by the linear relationship between average cost and average energy density.
(e) It is informative to consider both r^2 and s_e when evaluating the usefulness of the least squares regression line because a large r^2 (which indicates the proportion of variability in y that can be explained by the linear relationship between x and y) tells you that knowing the value of x is helpful in predicting y, and a small s_e indicates that residuals tend to be small, meaning that the data points tend to be relatively close to the regression line.
(f) $s_e = 0.05453$; the average cost will deviate from the least squares regression line by approximately 0.05453 dollars on average.
(g) Yes; the value of r^2 is moderately high, and the value of s_e is reasonably small.
(h) $0.37

Chapter 5 Probability

SECTION 5.1

Exercise Set 1

5.1 (a) In the long run, about 1% of people who suffer cardiac arrest in New York City will survive.
(b) Because 1% of 2,329 is 23.29, approximately 23 people who suffered cardiac arrest in New York City survived.

5.2 In the long run, 86% of the time this particular flight that flies between Phoenix and Atlanta will arrive on time.

5.3 (a) 0.83 **(b)** 0.83 **(c)** 0.17 **(d)** 0.48 **(e)** 0.52

Additional Exercises

5.7 Approximately 167; P(observing six when a fair die is rolled) $= \dfrac{1}{6}$
5.9 (a) 0.17 **(b)** 0.85 **(c)** 0.15 **(d)** 0.75

SECTION 5.2

Exercise Set 1

5.11 (a) 0.225 **(b)** 0.775 **(c)** 0.06 **(d)** 0.715 **(e)** 0.234
(f) The airline should not be particularly worried. The probability of selecting one person who was delayed overnight is approximately $\dfrac{75}{8000} = 0.0094$, and the probability of selecting additional people who were delayed overnight decreases from 0.0094.

5.12 (a) $S = \{AB, AC, AD, AE, BC, BD, BE, CD, CE, DE\}$
(b) Yes. **(c)** 0.3
(d) No, the probability does not change. **(e)** The probability increases to 0.6.

5.13 0.11

Additional Exercises

5.17 (a) 0.62 **(b)** 0.38
5.19 (a) 0.0119 **(b)** 0.000002 **(c)** 0.012
5.21 (a) $S = \{AB, AC, AD, AE, AF, BC, BD, BE, BF, CD, CE, CF, DE, DF, EF\}$ **(b)** Yes. **(c)** 0.0667 **(d)** 0.4 **(e)** 0.533

SECTION 5.3

Exercise Set 1

5.22 (a) Not mutually exclusive because there are seniors who are computer science majors.
(b) Not mutually exclusive because there are female students who are computer science majors.

(c) Mutually exclusive because a student cannot have a college residence that is more than 10 miles from campus and live in a college dormitory that is on campus.
(d) Mutually exclusive because college football teams are male teams, and females would not be on the team.

5.23 The events are dependent because the probability that a male experiences pain daily is not equal to the probability that a female experiences pain daily.

5.24 **(a)** 0.70 **(b)** 0.189 **(c)** 0.5 **(d)** 0.811 **(e)** 0.799 **(f)** 0.101

5.25 **(a)**

	E (stop at first light)	Not E (does not stop at first light)	Total
F (stop at second light)	150	150	300
Not F (does not stop at second light)	250	450	700
Total	400	600	1,000

(b) **(i)** 0.55 **(ii)** 0.45 **(iii)** 0.40 **(iv)** 0.25

5.26 **(a)**

	Cable TV	Not Cable TV	Total
Internet Service	250	170	420
Not Internet Service	550	30	580
Total	800	200	1,000

(b) **(i)** 0.25 **(ii)** 0.72

5.27 **(a)** **(i)** 0 **(ii)** 0.64
(b) No, events A and B are not mutually exclusive. For mutually exclusive events, $P(A \cup B) = P(A) + P(B)$. If A and B were mutually exclusive, then $P(A \cup B) = P(A) + P(B) = 0.26 + 0.34 = 0.60$, which is different from the value of $P(A \cup B)$ that is given.

5.28 0.000625

5.29 **(a)**

	Cash (M)	Credit Card (not M)	Total
Extended Warranty (E)	75	85	160
No Extended Warranty (not E)	395	445	840
Total	470	530	1,000

(b) 0.555. In the long run, approximately 55.5% of customers will either pay with cash or purchase an extended warranty, or both.

Additional Exercises

5.39 If the events T and P were mutually exclusive, you could add the probabilities to obtain $P(T \cup P)$. Since

$P(T) + P(P) = 0.83 + 0.56 = 1.39$ is greater than 1 (and probabilities cannot be greater than 1), you know that T and P cannot be mutually exclusive.

5.41 **(a)** No, the events O and N are not independent. If they were independent, $P(O \cap N) = P(O) \cdot P(N) = (0.7)(0.07) = 0.049$, which is not equal to the value of $P(O \cap N)$ given in the exercise.

(b)

	N	Not N	Total
O	40	660	700
Not O	30	270	300
Total	70	930	1,000

(c) 0.730. In the long run, 73% of the time, airline ticket purchasers will buy their ticket online or not show up for a flight, or both.

SECTION 5.4

Exercise Set 1

5.43 **(a)** The conditional probability is the 80% because you are told that 80% *of those receiving such a citation* attended traffic school. The key phrase is "of those receiving such a citation," which is what indicates that the 80% is a conditional probability.
(b) $P(F) = 0.20$ and $P(E|F) = 0.80$

5.44 **(a)** **(i)** 0.62 **(ii)** 0.36 **(iii)** 0.40 **(iv)** 0.29 **(v)** 0.31
(b) **(i)** The probability that a randomly selected Honda Civic buyer is male. **(ii)** The probability that a randomly selected Honda Civic buyer purchased a hybrid. **(iii)** The probability that, of the males, a randomly selected buyer purchased a hybrid. **(iv)** The probability that, of the females, a randomly selected buyer purchased a hybrid. **(v)** The probability that, of the hybrid purchasers, a randomly selected buyer is female.
(c) These probabilities are not equal.

5.45 **(a)** $P(L) = 0.493$: The probability that the goalkeeper jumps to the left.

$P(C) = 0.063$: The probability that the goalkeeper stays in the center.

$P(R) = 0.444$: The probability that the goalkeeper jumps to the right.

$P(B|L) = 0.142$: The probability that, given that the goalkeeper jumps to the left, the goal was blocked.

$P(B|C) = 0.333$: The probability that, given that the goalkeeper stays in the center, the goal was blocked.

$P(B|R) = 0.126$: The probability that, given that the goalkeeper jumps to the right, the goal was blocked.

(b)

	Blocked (B)	Not Blocked (not B)	Total
Goalkeeper Jumped Left (L)	70	423	**493**
Goalkeeper Stayed Center (C)	21	42	**63**
Goalkeeper Jumped Right (R)	56	388	**444**
Total	**147**	**853**	**1,000**

(c) 0.147.
(d) It seems as if the best strategy is for the goalkeeper to stay in the center when defending a kick.

5.46 (a)

	Uses Alternative Therapies	Does Not Use Alternative Therapies	Total
HS/Less	315	7,005	**7,320**
College: 1–4 yrs	393	4,400	**4,793**
College: ≥5 yrs	120	975	**1,095**
Total	**828**	**12,380**	**13,208**

(b) (i) 0.083 **(ii)** 0.063 **(iii)** 0.110 **(iv)** 0.043 **(v)** 0.380 **(vi)** 0.087 **(c)** For H and A to be independent, $P(H \cap A) = P(H) \cdot P(A)$. $P(H \cap A) = \frac{315}{13,208} = 0.024$ and $P(H) \cdot P(A) = (0.554)(0.063) = 0.0349$, which is not equal to 0.024. The events H and A are not independent.

Additional Exercises

5.51 $P(A|B)$ would be larger than $P(B|A)$ because it is more likely that a professional basketball player is over 6 feet tall than for an individual who is over 6 feet tall to be a professional basketball player.

5.53 (a) 0.490 **(b)** 0.3125 **(c)** 0.240 **(d)** 0.100

5.55 (a) 0.46
(b) It more likely that he or she is in the first priority group. **(c)** It is not particularly likely that a student in the third priority group would get more than nine units during the first attempt to register because, of those in the third priority group, only 36% of students get more than nine units.

SECTION 5.5

Exercise Set 1

5.57 (a) 0.55 **(b)** 0.45 **(c)** 0.4 **(d)** 0.25

5.58 (a) 0.25 **(b)** 0.72

5.59 (a) 0.9 **(b)** 0.771

5.60 (a) $\frac{6}{10}$ **(b)** $\frac{5}{9}$ **(c)** $\left(\frac{6}{10}\right)\left(\frac{5}{9}\right) = \frac{30}{90} = \frac{1}{3}$

Additional Exercises

5.67 (a) 0.49 **(b)** 0.3125 **(c)** 0.24 **(d)** 0.1

5.69 There are different numbers of people driving the three different types of vehicle (and also there are some drivers who are driving vehicles not included in those three types).

SECTION 5.6

Exercise Set 1

5.70 Use of the following conditional probabilities can provide justification for the given conclusion. $P(\text{very harmful}|\text{current smoker}) = 0.625$, $P(\text{very harmful}|\text{former smoker}) = 0.788$, and $P(\text{very harmful}|\text{never smoked}) = 0.869$.

SECTION 5.7

Exercise Set 1

5.72 (a) 0.85 **(b)** 0.19 **(c)** 0.6889 **(d)** 0.87

5.73 (a) Answers will vary.
(b) This is not a fair way of distributing licenses because those companies/individuals who are requesting multiple licenses are given approximately the same chance to get two or three licenses as an individual has to get a single license. Perhaps companies who request multiple licenses should be required to submit one application per license.

Additional Exercises

5.77 (a) Results from the simulation will vary. The exact answer is 0.6504.
(b) Jacob's decrease in the probability of on-time completion resulted in the bigger change in the probability that the project is completed on time.

5.79 (a) Answers may vary. One possibility is to assign digits 0 through 7 to represent a win for Seed 1, and digits 8 and 9 a win for Seed 4.
(b) Answers may vary. One possibility is to assign digits 0 through 5 to represent a win for Seed 2, and digits 6 through 9 a win for Seed 3.
(c) Answers may vary.
(d) Answers will vary.
(e) Answers will vary.
(f) Answers will vary.
(g) The estimated probabilities from parts (e) and (f) differ because they are based on different runs of the simulations. The estimate in part (f) is most likely a better estimate because it is based on more runs of the simulation than the estimate in part (e).

ARE YOU READY TO MOVE ON?
CHAPTER 5 REVIEW EXERCISES

5.81 In the long run, 18% of all calls for assistance will be to help someone who is locked out of his or her car.

5.83 (a) *BC, BM, BP, BS, CM, CP, CS, MP, MS,* and *PS.*
(b) 0.1
(c) 0.4
(d) 0.3

5.85 (a) Answers may vary. One possible answer: Let event *A* be the car purchaser is a male, and event *B* be the car purchaser is a female.
(b) Answers may vary. One possible answer: Let event *C* be the car purchaser is female, and event *D* be the car purchaser is under age 65.

5.87 (a) 0.337 (b) 0.762 (c) 0.952 (d) 0.204 (e) 0.806

5.89 (a)

	Internet	No Internet	Total
Phone	230	90	**320**
No Phone	190	490	**680**
Total	**420**	**580**	**1,000**

(b) (i) 0.230
     (ii) 0.280

5.91 0.0225, 0.003375

5.93 (a) 0.275 (b) 0.2 (c) 0.65 (d) 0.448
(e) No. The probability in part (c) is the proportion of female drivers who do not use a seat belt regularly, and the probability in part (d) is the proportion of drivers who do not use a seat belt regularly who are female.

5.95 (a) (i) $P(TD|D) = 0.99$
           (ii) $P(TD|C) = 0.01$
           (iii) $P(C) = 0.99$
           (iv) $P(D) = 0.01$

(b)

	TD	TC	Total
D	10	0	**10**
C	10	980	**990**
Total	**20**	**980**	**1,000**

(c) 0.02
(d) 0.5; Yes, this value is consistent with the statement that half of the dirty tests are false.

5.97 $P(\text{female}|\text{predicted female}) = 0.769$ and $P(\text{male}|\text{predicted male}) = 0.890$. These conditional probabilities are not equal, so a prediction that a baby is male and a prediction that a baby is female are not equally reliable.

5.99 (a) 0.5625 (b) 0.34

Chapter 6 Random Variables and Probability Distributions

NOTE: Your numerical answers may sometimes differ slightly from those in the answer section, depending on whether a table, a graphing calculator, or statistical software is used to compute probabilities.

SECTION 6.1

Exercise Set 1

6.1 (a) discrete (b) continuous (c) discrete (d) discrete (e) continuous

6.2 The possible values for *x* are *x* = 1, 2, 3, … (the positive integers).

Answers will vary. One possible answer is

Outcomes	x
S	1
LS	2
RLS	3
RRS	3
LRLRS	5

6.3 (a) 3, 4, 5, 6, 7
(b) −3, −2, −1, 1, 2, 3
(c) 0, 1, 2
(d) 0, 1

Additional Exercises

6.7 (a) discrete (b) continuous (c) continuous (d) discrete

6.9 (a) 2, 3, 4, 5, 6, 7, 8, 9, 10, 11, 12
(b) −5, −4, −3, −2, −1, 0, 1, 2, 3, 4, 5
(c) 1, 2, 3, 4, 5, 6

SECTION 6.2

Exercise Set 1

6.10 (a) $p(4) = 0.01$
(b) It is the probability of randomly selecting a carton of one dozen eggs and finding exactly 1 broken egg.
(c) $P(y \leq 2) = 0.95$; the probability that a randomly selected carton of eggs contains 0, 1, or 2 broken eggs is 0.95.
(d) $P(y < 2) = 0.85$; it is smaller because it does not include the possibility of 2 broken eggs.
(e) Exactly 10 unbroken eggs is equivalent to exactly 2 broken eggs; $p(2) = 0.10$.
(f) At least 10 unbroken eggs is equivalent to 10, 11, or 12 unbroken eggs, or 2, 1, or 0 broken eggs; $P(y \leq 2) = 0.95$.

6.11 (a) For 1,000 graduates, you would expect to see approximately 450 graduates who contributed nothing,

300 graduates who contributed $10, 200 graduates who contributed $25, and 50 graduates who contributed $50.
(b) $0 **(c)** 0.25 **(d)** 0.55

6.12 (a) (1, 2), (1, 3), (1, 4), (2, 3), (2, 4), (3, 4)
(b) Because the cartons are randomly selected, the outcomes are equally likely, and each has probability 1/6.

(c)

x	0	1	2
p(x)	$\frac{1}{6}$	$\frac{4}{6}$	$\frac{1}{6}$

6.13 (a)

x	0	1	2	3	4
p(x)	0.4096	0.4096	0.1536	0.0256	0.0016

(b) The most likely outcomes are 0 and 1.
(c) $P(x \geq 2) = 0.1808$

Additional Exercises

6.19 (a) $k = \dfrac{1}{15}$ **(b)** $P(y \leq 3) = 0.4$

(c) $P(2 \leq y \leq 4) = 0.6$

SECTION 6.3

Exercise Set 1

6.21 (a)

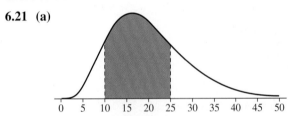

(b)

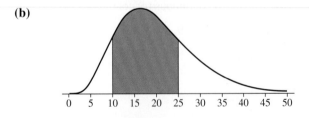

(c)

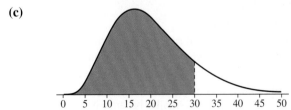

(d)

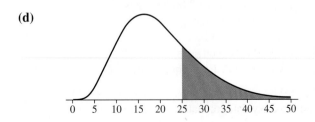

6.22 (a) $P(4 \leq x \leq 7)$
(b) The probability that a randomly selected individual waits between 4 and 7 minutes for service at a bank is 0.26.

6.23 (a) $h = 2$ **(b)** $P(x > 0.5) = 0.25$
(c) $P(x \leq 0.25) = 0.4375$

6.24 (a) $\frac{1}{2}(0.40)(5) = 1$
(b) $P(x < 0.20) = 0.5; P(x < 0.1) = 0.125;$
$P(x > 0.3) = 0.125$
(c) $P(0.10 < x < 0.20) = 0.375$

Additional Exercises

6.29 The probability $P(x < 1)$ is the smallest. $P(x > 3)$ and $P(2 < x < 3)$ are equal, and larger than the other two probabilities.

6.31 $P(2 < x < 3) = P(2 \leq x \leq 3) < P(x < 2) < P(x > 7)$. The smallest two probabilities are both equal to 1/10, the third probability is equal to 2/10, and the fourth probability is equal to 3/10.

SECTION 6.4

Exercise Set 1

6.32 (a) $\mu_y = 0.56$; this is the mean value of the number of broken eggs in the population of egg cartons.
(b) $P(y < 0.56) = P(y = 0) = 0.65$. This is not particularly surprising because, in the long run, 65% of egg cartons contain no broken eggs.
(c) This computation of the mean is incorrect because it assumes that the numbers of broken eggs (0, 1, 2, 3, or 4) are all equally likely.

6.33 (a) $\mu_x = 16.38, \sigma_x = 1.9984$
(b) The mean, $\mu_x = 16.38$ cubic feet represents the long-run average storage space of freezers sold by this particular appliance dealer. The standard deviation, $\sigma_x = 1.9984$ cubic feet, represents a typical amount by which the storage space in freezers purchased deviates from the mean.

6.34 Answers may vary. Two possible probability distributions are shown below.
Probability Distribution 1 ($\mu_x = 3$ and $\sigma_x = 1.265$):

x	1	2	3	4	5
p(x)	0.15	0.20	0.30	0.20	0.15

Probability Distribution 2 ($\mu_x = 3$ and $\sigma_x = 1.643$):

x	1	2	3	4	5
p(x)	0.30	0.15	0.10	0.15	0.30

6.35 (a) $\mu_y = \mu_{x_1} - \frac{1}{4}\mu_{x_2} = 10 - \frac{1}{4}(40) = 0.$
If a student guesses the answer to every question, you would expect 1/5 of the questions to be answered correctly and 4/5 to be answered incorrectly. Under

this scheme, the student is awarded one point for every correct answer and $-\frac{1}{4}$ of a point for every incorrect answer. So if the test contains 50 questions, then if the student guesses completely at random you expect 10 correct answers and 40 incorrect answers, and thus a total score of 0.

(b) Since $x_1 = 50 - x_2$, the value of x_1 is completely determined by the value of x_2 and thus x_1 and x_2 are not independent. The formulas in this section for computing variances and standard deviations of combinations of random variables require the random variables to be independent.

6.36 **(a)** 2.8, 1.288 **(b)** 0.7, 0.781 **(c)** 8.4, 14.94 **(d)** 7, 61 **(e)** 3.5, 2.27 **(f)** 15.4, 75.94

Additional Exercises

6.43 **(a)** $\mu_x = 2.3$ represents the long-run average number of lots ordered per customer.
(b) $\sigma_x^2 = 0.81$, $\sigma_x = 0.9$ lots. A typical deviation from the mean is about 0.9 lots.

SECTION 6.5

Exercise Set 1

6.45 **(a)** 0, 1, 2, 3, 4, and 5.
(b)

x	0	1	2	3	4	5
p(x)	0.2373	0.3955	0.2637	0.0879	0.0146	0.0010

6.46 **(a)** $P(x = 2) = 0.0486$; the probability that exactly two of the four randomly selected households have cable TV is 0.0486.
(b) $P(x = 4) = 0.6561$
(c) $P(x \le 3) = 0.3439$

6.47 **(a)** $P(x = 2) = 0.2637$
(b) $P(x \le 1) = 0.6328$
(c) $P(2 \le x) = 0.3672$
(d) $P(x \ne 2) = 0.7363$

6.48 **(a)** 0.7359
(b) 0.3918
(c) 0.0691

6.49 **(a)** There is not a fixed number of trials, which is required for the binomial distribution. This setting is geometric.
(b) **(i)** $p(4) = 0.0623$
　　(ii) $P(x \le 4) = 0.2836$
　　(iii) $P(x > 4) = 0.7164$
　　(iv) $P(x \ge 4) = 0.7787$

(c) The differences between the four probabilities are shown in bold font.
　　(i) The probability that it takes **exactly four songs** until the first song by the particular artist is played is 0.0623.

(ii) The probability that it takes **at most four songs** until the first song by the particular artist is played is 0.2836.
(iii) The probability that it takes **more than four songs** until the first song by the particular artist is played is 0.7164.
(iv) The probability that it takes **at least four songs** until the first song by the particular artist is played is 0.7787.

6.50 **(a)** geometric
(b) $P(x = 3) = 0.1084$
(c) $P(x < 4) = 0.3859$
(d) $P(x > 3) = 1 - P(x \le 3) = 0.6141$

Additional Exercises

6.59 **(a)** binomial with $n = 100$ and $p = 0.20$
(b) 20
(c) 16, 4
(d) A score of 50 is 7.5 standard deviations above the mean. In any distribution, being 7.5 or more standard deviations above the mean is quite unlikely.

6.61 0.773

6.63

y	0	1	2	3	4
p(y)	0.0625	0.4375	0.3125	0.1250	0.0625

$\mu_y = 1.6875$.

(b)

y	0	1	2	3	4
p(y)	0.0256	0.3264	0.3456	0.1728	0.1296

$\mu_y = 2.0544$.

SECTION 6.6

Exercise Set 1

6.64 **(a)** 0.9599
(b) 0.2483
(c) 0.1151
(d) 0.9976
(e) 0.6887
(f) 0.6826
(g) approximately 1

6.65 **(a)** 0.9909
(b) 0.9909
(c) 0.1093
(d) 0.1267
(e) 0.0706
(f) 0.0228
(g) 0.9996
(h) approximately 1

6.66 (a) 0.5
(b) 0.9772
(c) 0.9772
(d) 0.8185
(e) 0.9938
(f) approximately 1

6.67 (a) At most 60 wpm: 0.5; Less than 60 wpm: 0.5
(b) 0.8185
(c) 0.0013; it would be surprising because the probability of finding such a typist is very small.
(d) The probability that a randomly selected typist has a typing speed that exceeds 75 wpm is 0.1587. The probability that both typists have typing speeds that exceed 75 wpm is (0.1587)(0.1587) = 0.0252.
(e) typing speeds of 47.376 wpm or less.

6.68 0.3173

6.69 The proportion of corks produced by this machine that are defective is approximately 0. The second machine produces fewer defective corks.

Additional Exercises

6.77 7.3%

6.79 (a) 0.02275 **(b)** 0.656 **(c)** 0.656

6.81 (a) 0.5953
(b) $P(x < 25) = 0.000015$; it would be surprising.
(c) 0.00011
(d) $x^* = 28.026$

6.83 (a) 0.9876 **(b)** 0.9888 **(c)** $x^* = 6.608$

6.85 (a) $P(x < 67) = 0.691$. This is less than 94%, so the claim is not correct.
(b) Approximately 69.1%.

6.87 (a) 0.7745 **(b)** 0.1587 **(c)** 0.3085 **(d)** 6.1645

6.89 (a) 0.825
(b) 0.052
(c) 0.6827
(d) 0.00298. This small probability would make you skeptical of the claim.
(e) Benefits will not be paid if birth occurs within 275 days of the beginning of coverage. You are interested in the probability that insurance benefits will not be paid if the pregnancy lasts no more than $275 - 14 = 261$ days. $P(x \leq 261) = 0.3773$.

SECTION 6.7

Exercise Set 1

6.90 (a) The plot does not look linear. This supports the author's statement.

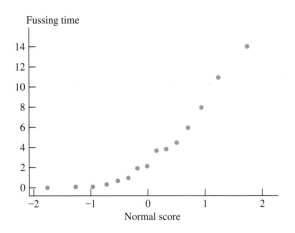

(b) $r = 0.921$; *critical r* (from Table 6.2) is 0.911; it is reasonable to think that the population distribution is normal.

6.91 (a)

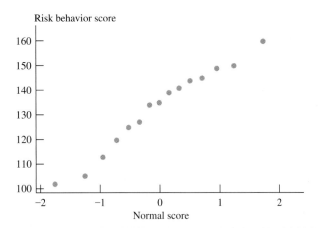

(b)

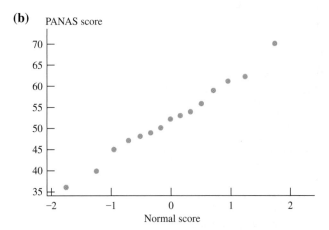

(c) Because both normal probability plots are approximately linear, it seems reasonable that both risk behavior scores and PANAS scores are approximately normally distributed.

Additional Exercises

6.95 **(a)**

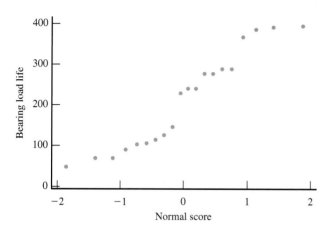

(b) $r = 0.963$, which is greater than the *critical r* of 0.929. It is reasonable to think that the distribution of bearing load life is approximately normal.

SECTION 6.8

Exercise Set 1

6.97 **(a)** 0.026
(b) 0.7580
(c) 0.7366
(d) 0.9109

6.98 **(a)** 0.8849
(b) 0.8739
(c) 0.0110
(d) 0.5434

6.99 **(a)** 0.3248
(b) 0.2763
(c) 0.0012

Additional Exercises

6.103 **(a)** Both $np = (60)(0.7) = 42$ and $n(1 - p) = 60(1 - 0.7) = 18$ are at least 10.
(b) **(i)** 0.1121
(ii) 0.4440
(iii) 0.5560

(c) The probability in (i) represents the probability of exactly 42 correct; the probability in (ii) is the probability of less than 42 correct; the probability in (iii) represents the probability of 42 or fewer correct.
(d) The normal approximation is not appropriate here because $n(1 - p) = 60(1 - 0.96) = 2.4$, which is less than 10.
(e) The small probability in Part (d) indicates that it is extremely unlikely for someone who is not faking the test to correctly answer 42 or fewer questions, compared with the probability that a person who is faking the test correctly answers 42 or fewer questions (0.556).

ARE YOU READY TO MOVE ON? CHAPTER 6 REVIEW EXERCISES

6.105 The depth, x, can take on any value between 0 and 100 ($0 \leq x \leq 100$), inclusive. x is continuous.

6.107 **(a)** The plane can accommodate 100 passengers, so if 100 or fewer passengers show up for the flight, everyone can be accommodated; $P(x \leq 100) = 0.82$.
(b) $P(x > 100) = 0.18$.
(c) The first person on the standby list: $P(x \leq 99) = 0.65$. The third person on the standby list: $P(x \leq 97) = 0.27$.

6.109 **(a)** $P(x < 10) = 0.5$; $P(x > 15) = 0.25$
(b) $P(7 < x < 12) = 0.25$
(c) $c = 18$ minutes

6.111 **(a)** Supplier 1 **(b)** Supplier 2
(c) I would recommend Supplier 1 because the bulbs will last, on average, longer than those from Supplier 2. Additionally, the bulbs from Supplier 1 have less variability in their lifetimes, so there is more consistency in the bulb lifetimes when compared with Supplier 2.
(d) Approximately 1,000 hours.
(e) Approximately 100 hours.

6.113 **(a)** 0.284
(b) 0.091
(c) 0.435
(d) 29.93 mm

6.115 **(a)** 0.159
(b) 51.4 minutes
(c) 41.6 minutes

6.117 Yes, because pattern in the normal probability plot is curved rather than linear.

6.119 **(a)** 16.38, 1.998 **(b)** $401 **(c)** $49,960

Chapter 7 An Overview of Statistical Inference— Learning from Data

SECTION 7.1

Exercise Set 1

7.1 The inferences made involve estimation.

7.2 **(a)** American teenagers between the ages of 12 and 17.
(b) The percentage of teens who own a cell phone, the percentage of teens who use a cell phone to send and receive text messages, and the percentage of teens ages 16–17 who have used a cell phone to text while driving.
(c) No, the actual percentage of teens owning a cell phone is probably not exactly 75%. The value of a sample statistic won't necessarily be equal to the population value.

(d) I would expect the estimate of the percentage of teens who own a cell phone to be more accurate. The sample contained teens aged 12–17; however, only those teens aged 16 and 17 were asked about texting while driving. The number of teens aged 16 and 17 is a subset of the overall sample, and so the estimate is based on a smaller sample.

7.3   The inference made is one that involves hypothesis testing.

7.4   **(a)** People driving along stretches of highway that have digital billboards.
(b) The time required to respond to road signs is greater when digital billboards are present.
(c) Answers will vary. One possible answer is: In addition to the information provided, I would like to know if the subjects were randomly selected or not, and whether or not the subjects were randomly assigned to any experimental groups. I would also like to know if there was a control group in which the subjects did not see digital billboards or a group in which subjects did not see any change in the display on the digital billboard.
(d) lower

7.5   **(a)** Answers will vary. One possible answer is "What proportion of students approve of a recent decision made by the university to increase athletic fees in order to upgrade facilities?"
(b) Answers will vary. One possible answer is "Do more than half of students approve of the recent decision made by the university to increase athletic fees in order to upgrade facilities?"

Additional Exercises

7.11   The inference made is one that involves hypothesis testing.

SECTION 7.2

Exercise Set 1

7.13   Estimate a population mean. The data are numerical, rather than categorical.

7.14   The type of data determines what graphical, numerical, and inferential methods are appropriate.

7.15   The other two questions are (1) **Q: Question Type** (estimation problem or hypothesis testing problem), and (2) **S: Study Type** (sample data or experiment data).

7.16   Q: Estimation
S: Sample data
T: One variable, categorical data
N: One sample

7.17   Q: Hypothesis testing
S: Sample data
T: One variable, numerical data
N: Two samples

7.18   Q: Hypothesis testing
S: Sample data
T: Two variables, categorical and numerical
N: One sample

Additional Exercises

7.25   Q: Estimation
S: Sample data
T: One variable, categorical
N: One sample

7.27   Q: Hypothesis testing
S: Experiment
T: One variable, categorical
N: Two treatments

7.29   Q: Hypothesis testing
S: Experiment data
T: One variable, numerical
N: Two treatments

ARE YOU READY TO MOVE ON? CHAPTER 7 REVIEW EXERCISES

7.31   Estimation, because the researchers were interested in the mean rating given to the beverage under each of the five musical conditions. The researchers did not indicate that they had a claim that they wanted to test.

7.33   **(a)** Answers will vary. One possible answer is: "What proportion of people who purchased season tickets for home games of the New York Yankees purchased alcoholic beverages during the game?"
(b) Answers will vary. One possible answer is: "Do fewer than 50% of people who purchased season tickets for home games of the New York Yankees drive to the game?"

7.35   **(a)** The proportion of people who take a garlic supplement who get a cold is lower than the proportion of those who do not take a garlic supplement and who get a cold.
(b) Yes, it is possible that the conclusion is incorrect. The observed difference in treatment effects may be due to chance variability in the response variable and the random assignment to treatments and not due to the treatment.
(c) greater

7.37   The type of data collected determines not only the type of numerical and graphical methods that can be applied but also the particular inferential method or methods that can be used. Numerical data require different inferential methods than categorical data, and univariate data require different inferential methods than bivariate data.

7.39 Q: Hypothesis testing
      S: Experiment data
      T: One variable, categorical
      N: Two treatments

7.41 **(a)** Estimate, Method, Check, Calculate, Communicate results
(b) The difference is that the "estimate" step is replaced with "hypotheses," where you determine the hypotheses you wish to test, rather than defining the population characteristic or treatment effect you wish to estimate.

Chapter 8 Sampling Variability and Sampling Distributions

SECTION 8.1

Exercise Set 1

8.1 No, because the value of $\hat{p}$ will vary from sample to sample.

8.2 **(a)** The histogram on the left; it has values that tend to deviate more from the center, and it is more spread out than the histogram on the right.
(b) $n = 75$, because the histogram on the right seems to have less sample-to-sample variability, is centered at 0.55, and has tall bars close to 0.55.

8.3 **(a)** a population proportion.
(b) $p = 0.22$.

8.4 **(a)** a sample proportion.
(b) $\hat{p} = 0.38$.

8.5 p is the proportion of successes in the entire population, and $\hat{p}$ is the proportion of successes in the sample.

Additional Exercises

8.11 Sample statistics are computed from a sample. Since one sample from the population is likely to differ from other possible samples taken from the same population, the sample statistics computed from different samples are likely to be different.

8.13 **(a)** a sample proportion.
(b) $\hat{p} = 0.45$.

SECTION 8.2

Exercise Set 1

8.15 **(a)** $\mu_{\hat{p}} = 0.65$; $\sigma_{\hat{p}} = 0.151$
(b) $\mu_{\hat{p}} = 0.65$; $\sigma_{\hat{p}} = 0.107$

(c) $\mu_{\hat{p}} = 0.65$; $\sigma_{\hat{p}} = 0.087$
(d) $\mu_{\hat{p}} = 0.65$; $\sigma_{\hat{p}} = 0.067$
(e) $\mu_{\hat{p}} = 0.65$; $\sigma_{\hat{p}} = 0.048$
(f) $\mu_{\hat{p}} = 0.65$; $\sigma_{\hat{p}} = 0.034$

8.16 For $p = 0.65$: 30, 50, 100, and 200. For $p = 0.2$: 50 100, and 200.

8.17 **(a)** $\mu_{\hat{p}} = 0.07$; $\sigma_{\hat{p}} = 0.0255$
(b) No, because $np = (100)(0.07) = 7$ is less than 10.
(c) The mean does not change. The standard deviation will decrease to 0.0180. The mean does not change because the sampling distribution is always centered at the population value regardless of the sample size. The standard deviation of the sampling distribution will decrease as the sample size increases because the sample size (n) is in the denominator of the formula for standard deviation.
(d) Yes, because $np = (200)(0.07) = 14$ and $n(1 - p) = (200)(1-0.07) = 186$ are both greater than 10.

8.18 **(a)** $\mu_{\hat{p}} = 0.15$; $\sigma_{\hat{p}} = 0.0357$
(b) Yes, because $np = (100)(0.15) = 15$ and $n(1 - p) = (100)(1 - 0.15) = 85$ are both greater than 10.
(c) The mean does not change. The standard deviation will decrease to 0.0252. The mean does not change because the sampling distribution is always centered at the population value regardless of the sample size. The standard deviation of the sampling distribution will decrease as the sample size increases because the sample size (n) is in the denominator of the formula for standard deviation
(d) Yes, because $np = (200)(0.15) = 30$ and $n(1 - p) = (200)(1 - 0.15) = 170$ are both greater than 10.

Additional Exercises

8.23 $n = 100$ and $p = 0.5$.

8.25 For $p = 0.2$: 50 and 100. For $p = 0.8$: 50 and 100. For $p = 0.6$: 25, 50, and 100.

8.27 For samples of size $n = 40$: $p = 0.45$ and $p = 0.70$. For samples of size $n = 75$: $p = 0.20$, $p = 0.45$, and $p = 0.70$.

SECTION 8.3

Exercise Set 1

8.29 **(a)** $\sigma_{\hat{p}} = \sqrt{\dfrac{0.48(1 - 0.48)}{500}} = 0.0223$
(b) More sample-to-sample variability in the sample proportions because $\sigma_{\hat{p}}$ is now greater than $\sigma_{\hat{p}}$ when $n = 500$.
(c) The sample size is smaller because, in order for $\sigma_{\hat{p}}$ to be greater, the denominator of $\sigma_{\hat{p}} = \sqrt{\dfrac{p(1 - p)}{n}}$ must be smaller than when $n = 500$.

8.30

What You Know	How You Know It
The sampling distribution of $\hat{p}$ is centered at the actual (but unknown) value of the population proportion.	Rule 1 states that $\mu_{\hat{p}} = p$. This is true for random samples, and the description of the study says that the sample was selected at random.
An estimate of the standard deviation of $\hat{p}$, which describes how much the $\hat{p}$ values spread out around the population proportion p, is 0.0153.	Rule 2 states that $\sigma_{\hat{p}} = \sqrt{\dfrac{p(1-p)}{n}}$. In this exercise, $n = 1,000$. The value of p is not known. However, $\hat{p}$ provides an estimate of p that can be used to estimate the standard deviation of the sampling distribution. Specifically, $\hat{p} = 0.37$, so $\sigma_{\hat{p}} \approx \sqrt{\dfrac{0.37(1-0.37)}{1000}}$ $= 0.0153$. This standard deviation provides information about how tightly the $\hat{p}$ values from different random samples will cluster around the value of p.
The sampling distribution of $\hat{p}$ is approximately normal.	Rule 3 states that the sampling distribution of $\hat{p}$ is approximately normal if n is large and p is not too close to 0 or 1. Here the sample size is 1,000. The sample includes 370 successes and 630 failures, which are both much greater than 10. So we conclude that the sampling distribution of $\hat{p}$ is approximately normal.

8.31 **(a)** $\hat{p} = 0.346$
(b) No. From Rule 1, we know that the sampling distribution is centered at p. From Rule 2, the standard deviation of the sampling distribution of $\hat{p}$ is $\sigma_{\hat{p}} = 0.066$. By Rule 3, the sampling distribution of $\hat{p}$ is approximately normal because the sample includes 18 successes and 34 failures, which are both greater than 10. Since the sampling distribution of $\hat{p}$ is approximately normal and is centered at the actual population proportion p, about 95% of all possible random samples of size $n = 52$ will produce a sample proportion that is within $2(0.066) = 0.132$ of the actual value of the population proportion. This margin of error of 0.132 is over twice the value of 0.05, so it is not reasonable to think that this estimate is within 0.05 of the actual value of the population proportion.

8.32 From Rule 1, we know that the sampling distribution of $\hat{p}$ is centered at the population value $p = 0.61$. Rule 2 tells us that $\sigma_{\hat{p}} = 0.0126$. Finally, the sampling distribution of $\hat{p}$ is approximately normal because $np = 1,500(0.61) = 915 \geq 10$ and $n(1-p) = 1,500(1-0.61) = 585 \geq 10$.

The probability of observing a sample proportion of 0.55 or smaller just by chance (due to sampling variability) is
$$P(\hat{p} < 0.55) = P\left(z < \frac{0.55 - 0.61}{0.0126}\right) = P(z < -4.76) \approx 0.$$
Therefore, it seems likely that the proportion of California high school graduates who attend college the year after graduation is different from the national figure.

Additional Exercises

8.37 No, it is not likely that this estimate is within 0.05 of the actual value of the population proportion. From Rule 1, we know that the sampling distribution is centered at p. From Rule 2, the standard deviation of the sampling distribution of $\hat{p}$ is $\sigma_{\hat{p}} = 0.049$. By Rule 3, the sampling distribution of $\hat{p}$ is approximately normal because $np = 100(0.38) = 38 \geq 10$ and $n(1-p) = 100(1-0.38) = 62 \geq 10$. Since the sampling distribution of $\hat{p}$ is approximately normal and is centered at the actual population proportion p, we now know that about 95% of all possible random samples of size $n = 100$ will produce a sample proportion that is within $2(0.049) = 0.098$ of the actual value of the population proportion. This margin of error is nearly twice the value of 0.05, so it is unlikely that this estimate is within 0.05 of the population proportion.

8.39 By Rule 1, we know that the sampling distribution of $\hat{p}$ is centered at the unknown value of p, the true proportion of social network users who believe that it is not OK to "friend" your boss. Rule 2 says that the standard deviation of the sampling distribution of $\hat{p}$ is $\sigma_{\hat{p}} = \sqrt{\dfrac{p(1-p)}{n}}$. The value of p is not known.

However, $\hat{p}$ provides an estimate of p that can be used to estimate the standard deviation of the sampling distribution. In this case, $\hat{p} = 0.56$, so the estimate of the standard deviation of the sampling distribution is $\sigma_{\hat{p}} = 0.014$. Finally, Rule 3 says that the sampling distribution of $\hat{p}$ is approximately normal if n is large and p is not too close to 0 or 1. Here the sample size is 1,200. The sample includes 672 successes (56% of 1,200) and 528 failures (44% of 1,200), which are both much greater than 10. Therefore, we can conclude that the sampling distribution of $\hat{p}$ is approximately normal. The probability of observing a sample proportion at least as large as what we actually observed ($\hat{p} = 0.56$) if the true value were 0.5 is $P(\hat{p} \geq 0.56) = P\left(z \geq \dfrac{0.56 - 0.5}{0.014}\right) = P(z \geq 4.29) \approx 0.000$. It is highly unlikely that a sample proportion as large as 0.56 would be observed if the true value were 0.5 (or less). Therefore, it seems plausible that the proportion of social network users who believe that it is not OK to "friend" your boss is greater than 0.5.

ARE YOU READY TO MOVE ON? CHAPTER 8 REVIEW EXERCISES

8.41 No, because the value of $\hat{p}$ varies from sample to sample.

8.43 Different samples will likely yield different values of $\hat{p}$, which is the concept of sampling variability (or sample-to-sample variability). However, there is only one true value for the population proportion.

8.45 (a) Population proportion.
(b) $p = 0.21$.

8.47 For both $p = 0.70$ and $p = 0.30$: $n = 50$, $n = 100$, and $n = 200$.

8.49 (a) $\mu_{\hat{p}} = p = 0.25$; $\sigma_{\hat{p}} = 0.031$
(b) Yes, because $np = 200(0.25) = 50 \geq 10$ and $n(1 - p) = 200(1 - 0.25) = 150 \geq 10$.
(c) The change in sample size does not affect the mean but does affect the standard deviation of the sampling distribution of $\hat{p}$. The new standard deviation is $\sigma_{\hat{p}} = 0.061$.
(d) Yes, because $np = 50(0.25) = 12.5 \geq 10$ and $n(1 - p) = 50(1 - 0.25) = 37.5 \geq 10$.

8.53 By Rule 1, we know that the sampling distribution of $\hat{p}$ is centered at the unknown value of p. Rule 2 says that the standard deviation of the sampling distribution of $\hat{p}$ is $\sigma_{\hat{p}} = \sqrt{\dfrac{p(1 - p)}{n}}$. The value of p is not known. However, $\hat{p}$ provides an estimate of p that can be used to estimate the standard deviation of the sampling distribution. In this case, $\hat{p} = 0.637$, and the estimate of the standard deviation of the sampling distribution is $\sigma_{\hat{p}} = 0.011$. Finally, Rule 3 says that the sampling distribution of $\hat{p}$ is approximately normal if n is large and p is not too close to 0 or 1. Here the sample size is 2,013. The sample includes 1,283 successes and 730 failures, which are both much greater than 10. Therefore, we can conclude that the sampling distribution of $\hat{p}$ is approximately normal. The probability of observing a sample proportion at least as large as what we observed if the true value were 0.5 is equal to $P(\hat{p} \geq 0.637) =$ $P\left(z \geq \dfrac{0.637 - 0.5}{0.011}\right) = P(z \geq 12.45) \approx 0.000$. It is highly unlikely that a sample proportion as large as 0.637 would be observed if the true value were 0.5 (or less). Therefore, it seems plausible that the proportion of adult Americans who believe rudeness is a worsening problem is greater than 0.5.

Chapter 9 Estimating a Population Proportion

NOTE: Answers may vary slightly if you are using statistical software, a graphing calculator, or depending on how the values of sample statistics are rounded when performing hand calculations. Don't worry if you don't match these numerical answers exactly (but your answers should be relatively close).

SECTION 9.1

Exercise Set 1

9.1 An unbiased statistic with a smaller standard error is preferred because it is likely to result in an estimate that is closer to the actual value of the population characteristic than an unbiased statistic that has a larger standard error.

9.2 Statistics II and III

9.3 Statistic I, because it has a smaller bias than Statistics II and III.

9.4 $n = 200$.

9.5 (a) The formula for the standard error of $\hat{p}$ is $\sigma_{\hat{p}} = \sqrt{\dfrac{p(1 - p)}{n}}$. The quantity $p(1 - p)$ reaches a maximum value when $p = 0.5$.
(b) The standard error of $\hat{p}$ is the same when $p = 0.2$ as when $p = 0.8$ because when $p = 0.2$, $(1 - p) = (1 - 0.2) = 0.8$. Similarly, when $p = 0.8$, $(1 - p) = (1 - 0.8) = 0.2$. So, the quantity $p(1 - p)$ is the same in both cases.

9.6 $n = 400$ and $p = 0.8$

Additional Exercises

9.13 A biased statistic might be chosen over an unbiased statistic if the bias is not too large, and the standard error of the biased statistic is much smaller than the standard error of the unbiased statistic. In this case, the observed value of the biased statistic might be closer to the actual value than the value of an unbiased statistic.

9.15 $n = 200$.

SECTION 9.2

Exercise Set 1

9.17 Statement 1: Incorrect, because the value 0.0157 is the standard error of $\hat{p}$, and therefore approximately 32% of all possible values of $\hat{p}$ would differ from the value of the actual population proportion by more than 0.0157 (using properties of the normal distribution).

Statement 2: Correct

Statement 3: Incorrect, because the phrase "will never differ from the value of the actual population proportion" is wrong. The value 0.0307 is the margin of error and indicates that in about 95% of all possible random samples, the estimation error will be less than the margin of error. In about 5% of the random samples, the estimation error will be greater than the margin of error.

9.18 (a) 0.049
(b) $n = 100$
(c) $1/\sqrt{2} \approx 0.707$

9.19 (a) $\hat{p} = 0.260$
(b) $\sigma_{\hat{p}} = 0.019$
(c) margin of error $= 0.037$. The estimate of the proportion of all businesses that have fired workers for misuse of the Internet is unlikely to differ from the actual population proportion by more than 0.037.

9.20 (a) yes
(b) no
(c) no
(d) no

9.21 (a) $\hat{p} = 0.404$
(b) The sample was selected in such a way that makes it representative of the population of U.S. college students. Additionally, there are 2,998 successes and 4,423 failures in the sample, which are both at least 10.
(c) margin of error $= 0.011$
(d) It is unlikely that the estimated proportion of U.S. college students who use the Internet more than 3 hours per day ($\hat{p} = 0.404$) will differ from the actual population proportion by more than 0.011 (or 1.1%).

9.22 (a) $\hat{p} = 0.277$
(b) The sample is a random sample from the population of American children. Additionally, there are 1,720 successes and 4,492 failures in the sample, which are both greater than 10.
(c) margin of error $= 0.011$
(d) It is unlikely that the estimated proportion of American children who indicated that they eat fast food on a typical day ($\hat{p} = 0.277$) will differ from the actual population proportion by more than 0.011.

Additional Exercises

9.29 marign of error $= 0.024$. It is unlikely that the estimated proportion of adults who believe that the shows are mostly or totally made up ($\hat{p} = 0.82$) will differ from the actual population proportion by more than 0.024.

9.31 In this case, margin of error $= 0.027$, which rounds to 3%.

9.33 In this case, margin of error $= 0.032$, which rounds to 3%. The margin of error tells you that it is unlikely that the estimate will differ from the actual population proportion by more than 0.03.

SECTION 9.3

Exercise Set 1

9.34 (a) The confidence intervals are centered at $\hat{p}$. In this case, the intervals are not centered in the same place because two different samples were taken, each yielding a different value of $\hat{p}$.
(b) Interval 2 conveys more precise information about the value of the population proportion because Interval 2 is narrower than Interval 1.
(c) A smaller sample size produces a larger margin of error. In this case, Interval 1 (being wider than Interval 2) was based on the smaller sample size.
(d) Interval 1 would have the higher confidence level, because the z critical value for higher confidence is larger, resulting in a wider confidence interval.

9.35 (a) 95%
(b) $n = 100$

9.36 The method used to construct this interval estimate is successful in capturing the actual value of the population proportion about 95% of the time.

9.37 (a) yes
(b) no
(c) no
(d) no

9.38 (a) 1.645
(b) 2.58
(c) 1.28

9.39 Question type (Q): Estimation; Study type (S): Sample data; Type of data (T): One categorical variable; Number of samples or treatments (N): One sample.

9.40 Estimate (E): The proportion of hiring managers and human resources professionals who use social networking sites to research job applicants, p, will be estimated.

Method (M): Because the answers to the four key questions are estimation, sample data, one categorical variable, and one sample (see Exercise 9.39), consider a 95% confidence interval for the proportion of hiring managers and human resource professionals who use social networking sites to research job applicants.

Check (C): The sample is representative of hiring managers and human resource professionals. In addition, the sample includes 1,200 successes and 1,467 failures, which are both greater than 10. The two required conditions are satisfied.

Calculations (C): (0.4311,0.4689)

Communicate Results (C):

Interpret confidence interval: You can be 95% confident that the actual proportion of hiring managers and human resources professionals is somewhere between 0.4311 and 0.4689.

Interpret confidence level: The method used to construct this interval estimate is successful in capturing the actual value of the population proportion about 95% of the time.

9.41 (a) (0.3449, 0.3951). You can be 90% confident that the actual proportion of college freshmen who carry a credit card balance is somewhere between 0.3449 and 0.3951.

(b) (0.454, 0.506). You can be 90% confident that the actual proportion of college seniors who carry a credit card balance is somewhere between 0.454 and 0.506.

(c) The two confidence intervals from (a) and (b) do not have the same width because the standard errors, and hence the margins of error, are based on two different values for $\hat{p}$.

9.42 **(a)** (0.2039, 0.2561). You can be 95% confident that the actual proportion of U.S. adults for whom math was their most favorite subject is somewhere between 0.2039 and 0.2561.

(b) (0.3401, 0.3999). You can be 95% confident that the actual proportion of U.S. adults for whom math was their least favorite subject is somewhere between 0.3401 and 0.3999.

9.43 **(a)** (0.1206, 0.2254). You can be 95% confident that the actual proportion of all adult Americans who planned to purchase a Valentine's Day gift for their pet is somewhere between 0.1206 and 0.2254.

(b) The 95% confidence interval computed for the actual sample size would have been narrower than the confidence interval computed in Part (a) because the standard error, and hence the margin of error, would have been smaller.

Additional Exercises

9.55 **(a)** (0.494, 0.546). You can be 90% confident that the actual proportion of adult Americans who would say that lying is never justified is somewhere between 0.494 and 0.546.

(b) (0.6252, 0.6748). You can be 90% confident that the actual proportion of adult Americans who would say that it is often or sometimes OK to lie to avoid hurting someone's feelings is somewhere between 0.6252 and 0.6748.

(c) The confidence interval in Part (a) indicates that it is plausible that at least 50% of adult Americans would say that lying is never justified, and the confidence interval in part (b) indicates that it is also plausible that well over 50% of adult Americans would say that it is often or sometimes OK to lie to avoid hurting someone's feelings. These are contradictory responses.

9.57 (0.597, 0.643). You can be 95% confident that the actual proportion of all Australian children who would say that they watch TV before school is between 0.597 and 0.643. For the method used to construct the interval to be valid, the sample must have either been randomly selected from the population of interest or the sample must have been selected in such a way that it should result in a sample that is representative of the population.

9.59 (0.1446, 0.1894). You can be 90% confident that the actual proportion of all full-time workers so angered in the past year that they wanted to hit a co-worker is between 0.1446 and 0.1894.

SECTION 9.4

Exercise Set 1

9.61 Assuming a 95% confidence level, and using a conservative estimate of $p = 0.5$, $n = 2{,}401$.

9.62 Assuming a 95% confidence level, and using the preliminary estimate of $p = 0.27$, $n = 303$. Using the conservative estimate of $p = 0.5$, $n = 385$.

9.63 Assuming a 95% confidence level, and using a conservative estimate of $p = 0.5$, $n = 97$.

Additional Exercises

9.67 Assuming a 95% confidence level, and using the preliminary estimate of $p = 0.32$, $n = 335$. Using the conservative estimate of $p = 0.5$, $n = 385$.

9.69 **(a)** (0.1186, 0.2854). You can be 95% confident that the actual proportion of all such patients under 50 years old who experience a failure within the first 2 years is between 0.1186 and 0.2854.

(b) (0.0107, 0.0611). You can be 99% confident that the actual proportion of all such patients age 50 or older who experience a failure within the first 2 years is between 0.0107 and 0.0611.

(c) For a 95% confidence level, and using a preliminary estimate of $p = 0.202$, $n = 689$.

ARE YOU READY TO MOVE ON? CHAPTER 9 REVIEW EXERCISES

9.71 **(a)** Statistic II is unbiased.

(b) Statistic I will tend to be closer to the population value than Statistic II.

9.73 $\hat{p}$ from a random sample of size 400 tends to be closer to the actual value when $p = 0.7$.

9.75 Statement 1: Correct

Statement 2: Incorrect. About 95% of all possible sample proportions computed from random samples of size 500 will be within 0.0429 of the actual population proportion (this is the margin of error for a 95% confidence level). About 5% of the sample proportions would differ from the actual population proportion by more than 0.0429.

Statement 3: Incorrect. About 68% of all possible sample proportions computed from random samples of size 500 will be within 0.0219 of the actual population proportion (this is the standard deviation of the sampling distribution). About 32% of the sample proportions would differ from the actual population proportion by more than 0.0219.

9.77 **(a)** $\hat{p} = 0.18$

(b) The two conditions are: (1) Large sample size: The number of successes and failures in the sample are at least 10. In this case, there are 722 successes, and 3,291 failures, which

are both much greater than 10. (2) Random selection of the sample, or the sample is representative of the population: We are told that the sample is representative of the population.
(c) margin of error = 0.0119
(d) It is unlikely that the estimate $\hat{p} = 0.18$ differs from the value of the actual proportion of adults Americans who say they have seen a ghost by more than 0.0119.
(e) (0.17, 0.19). You can be 90% confident that the actual proportion of all adult Americans who have seen a ghost is somewhere between 0.17 and 0.19.
(f) A 99% confidence interval would be wider because the z critical value for 99% confidence ($z = 2.58$) is larger than the z critical value for 90% confidence ($z = 1.645$).

9.79 **(a)** As the confidence level increases, the width of the confidence interval also increases.
(b) As the sample size increases, the width of the confidence interval decreases.
(c) When $\hat{p} = 0.5$, the margin of error (for a fixed sample size) is maximum and decreases symmetrically as $\hat{p}$ decreases toward 0 or increases toward 1. A larger margin of error results in a wider confidence interval.

9.81 Question type (Q): Estimation; Study type (S): Sample data; Type of data (T): One categorical variable; Number of samples or treatments (N): One sample.

9.83 **(a)** (0.6426, 0.6774). You can be 90% confident that the actual proportion of all Americans ages 8 to 18 who own a cell phone is somewhere between 0.6426 and 0.6774.
(b) (0.7443, 0.7757). You can be 90% confident that the actual proportion of all Americans ages 8 to 18 who own an MP3 music player is somewhere between 0.7443 and 0.7757.
(c) The interval in Part (b) is narrower because $\hat{p}$ of 0.76 is farther from 0.5 than $\hat{p}$ of 0.66. For a fixed sample size and confidence level, the confidence interval is widest when $\hat{p} = 0.5$ and decreases symmetrically as $\hat{p}$ moves closer to 0 or 1.

Chapter 10 Asking and Answering Questions About a Population Proportion

NOTE: Answers may vary slightly if you are using statistical software, a graphing calculator, or depending on how the values of sample statistics are rounded when performing hand calculations. Don't worry if you don't match these numerical answers exactly (but your answers should be relatively close).

SECTION 10.1

Exercise Set 1

10.1 $\hat{p}$ is a sample statistic. Hypotheses are about population characteristics.

10.2 $H_0: p = \frac{1}{3}$ and $H_a: p > \frac{1}{3}$

10.3 $H_0: p = 0.67$ and $H_a: p \neq 0.67$

10.4 **(a)** There is convincing evidence that the proportion of American adults who favor drafting women is less than 0.5.
(b) Yes
(c) Yes

10.5 **(a)** The conclusion is consistent with testing H_0: concealed weapons laws do not reduce crime versus H_a: concealed weapons laws reduce crime.
(b) The null hypothesis was not rejected because no evidence was found that the laws were reducing crime.

10.6 The sample data provide convincing evidence against the null hypothesis. If the null hypothesis were true, the sample data would be very unlikely.

Additional Exercises

10.13 **(a)** legitimate
(b) not legitimate
(c) not legitimate
(d) legitimate
(e) not legitimate

10.15 $H_0: p = 0.6$ versus $H_a: p > 0.6$. In order to make the change, the university requires evidence that *more than* 60% of the faculty are in favor of the change.

SECTION 10.2

Exercise Set 1

10.17 Not rejecting the null hypothesis when it is not true.

10.18 A small significance level because α is the probability of a Type I error.

10.19 **(a)** Before filing charges of false advertising against the company, the consumer advocacy group would require convincing evidence that more than 10% of the flares are defective.
(b) A Type I error is thinking that more than 10% of the flares are defective when in fact 10% (or fewer) of the flares are defective. This would result in the expensive and time-consuming process of filing charges of false advertising against the company when the company advertising is not false. A Type II error is not thinking that more than 10% of the flares are defective when in fact more than 10% of the flares are defective. This would result in the consumer advocacy group not filing charges when the company advertising was false.

10.20 (a) A Type I error would be thinking that less than 90% of the TV sets need no repair when in fact (at least) 90% need no repair. The consumer agency might take action against the manufacturer when the manufacturer is not at fault. A Type II error would be *not* thinking that less than 90% of the TV sets need no repair when in fact less than 90% need no repair. The consumer agency would not take action against the manufacturer when the manufacturer is making untrue claims about the reliability of the TV sets. **(b)** Taking action against the manufacturer when the manufacturer is not at fault could involve large and unnecessary legal costs to the consumer agency. $\alpha = 0.01$ should be recommended.

10.21 (a) Type I error, 0.091
(b) Thinking that a woman does have cancer in the other breast when she really does not, 0.097.

Additional Exercises

10.27 A Type I error is rejecting a true null hypothesis, and a Type II error is not rejecting a false null hypothesis.

10.29 Answers will vary.

10.31 The "38%" value given in the article is the proportion of *all* felons; in other words, it is a *population* proportion. Therefore, you know that the population proportion is less than 0.4, and there is no need for a hypothesis test.

SECTION 10.3

Exercise Set 1

10.32 (a) The sampling distribution of $\hat{p}$ will be approximately normal with a mean of 0.25 and a standard deviation of 0.0125.
(b) A sample proportion of $\hat{p} = 0.24$ would not be surprising if $p = 0.25$, because it is less than one standard deviation below 0.25. This is not unusual for a normal distribution.
(c) A sample proportion of $\hat{p} = 0.20$ would be surprising if $p = 0.25$, because it is 4 standard deviations below 0.25. This would be unusual for a normal distribution.
(d) If $p = 0.25$, $P(\hat{p} \le 0.22) = P(z \le -2.4) = 0.0082$. Because this probability is small, there is convincing evidence that the goal is not being met.

10.33 (a) The sampling distribution of $\hat{p}$ will be approximately normal with a mean of 0.75 and a standard deviation of 0.0256.
(b) A sample proportion of $\hat{p} = 0.83$ would be surprising if $\hat{p} = 0.75$, because it is more than 3 standard deviations above 0.75. This would be unusual for a normal distribution.
(c) A sample proportion of $\hat{p} = 0.79$ would not be surprising if $p = 0.75$, because it is about 1.5 standard deviations above 0.75. This is not unusual for a normal distribution.

(d) If $p = 0.75$, $P(\hat{p} \ge 0.80) = P(z \ge 1.95) = 0.0256$. Because this probability is small, there is convincing evidence that the null hypothesis is not true.

Additional Exercises

10.37 (a) The sampling distribution of $\hat{p}$ will be approximately normal with a mean of 0.5 and a standard deviation of 0.1118.
(b) Answers will vary. One reasonable answer would be sample proportions greater than 0.7236 (these are sample proportions that are more than two standard deviations above 0.5).

SECTION 10.4

Exercise Set 1

10.38 (a) A *P*-value of 0.0003 means that it is very unlikely (probability = 0.0003), assuming that H_0 is true, that you would get a sample result at least as inconsistent with H_0 as the one obtained in the study. H_0 would be rejected.
(b) A *P*-value of 0.350 means that it is not particularly unlikely (probability = 0.350), assuming that H_0 is true, that you would get a sample result at least as inconsistent with H_0 as the one obtained in the study. There is no reason to reject H_0.

10.39 (a) $H_0: p = \frac{2}{3}$ and $H_a: p > \frac{2}{3}$
(b) The null hypothesis would be rejected because the *P*-value (0.013) is less than $\alpha = 0.05$.

10.40 This step involves using the answers to the four key questions (QSTN) to identify an appropriate method.

Additional Exercises

10.45 A *P*-value of 0.0002 means that it is very unlikely (probability = 0.0002), assuming that H_0 is true, that you would get a sample result at least as inconsistent with H_0 as the one obtained in the study. This is strong evidence against the null hypothesis.

SECTION 10.5

Exercise Set 1

10.47 Estimation, sample data, one numerical variable, one sample. A hypothesis test for a population proportion would not be appropriate.

10.48 Hypothesis testing, sample data, one categorical variable, one sample. A hypothesis test for a population proportion would be appropriate.

10.49 Estimation, sample data, one categorical variable, one sample. A hypothesis test for a population proportion would not be appropriate.

10.50 (a) Large-sample z test is not appropriate.
(b) Large-sample z test is appropriate.
(c) Large-sample z test is appropriate.
(d) Large-sample z test is not appropriate.

10.51 (a) 0.2912
(b) 0.1788
(c) 0.0233
(d) 0.0125
(e) 0.9192

10.52 (a) $H_0: p = 0.4$, $H_a: p < 0.4$, $z = -2.969$, P-value = 0.001, reject H_0. There is convincing evidence that the proportion of all adult Americans who would answer the question correctly is less than 0.4.
(b) $H_0: p = \frac{1}{3}$, $H_a: p > \frac{1}{3}$, $z = 2.996$, P-value = 0.001, reject H_0. There is convincing evidence that more than one-third of adult Americans would select a wrong answer.

10.53 $H_0: p = 0.25$, $H_a: p > 0.25$, $z = 0.45$, P-value = 0.328, fail to reject H_0. There is no convincing evidence that more than 25% of Americans ages 16 to 17 have sent a text message while driving.

10.54 $H_0: p = 0.5$, $H_a: p > 0.5$, $z = 14.546$, P-value ≈ 0, reject H_0. There is convincing evidence that a majority of adult Americans prefer to watch movies at home.

10.55 (a) $z = -1.897$, P-value = 0.0289, reject H_0
(b) $z = -0.6$, P-value = 0.274, fail to reject H_0
(c) Both results *suggest* that fewer than half of adult Americans believe that movie quality is getting worse. However, getting 470 out of 1,000 people responding this way (as opposed to 47 out of 100) provides much *stronger* evidence.

10.56 $H_0: p = 0.5$, $H_a: p < 0.5$, $z = -2.536$, P-value = 0.006, reject H_0. There is convincing evidence that the proportion of all adult Americans who want car Web access is less than 0.5. The marketing manager is not correct in his claim.

Additional Exercises

10.67 (a) $z = 2.530$, P-value = 0.0057, reject H_0
(b) No. The survey only included women ages 22 to 35.

10.69 $H_0: p = 0.75$, $H_a: p > 0.75$, $z = 4.13$, P-value ≈ 0, reject H_0. There is convincing evidence that more than three-quarters of American adults believe that lack of respect and courtesy is a serious problem.

10.71 $z = 1.069$, P-value = 0.143, fail to reject H_0

10.73 $H_0: p = 0.25$, $H_a: p > 0.25$, $z = 2.202$, P-value = 0.014, fail to reject H_0. There is not convincing evidence that more than one-fourth of American adults see a lottery or sweepstakes win as their best chance of accumulating \$500,000.

10.75 $H_0: p = 0.68$, $H_a: p \neq 0.68$, $z = 7.761$, P-value ≈ 0, reject H_0. There is convincing evidence that the proportion of religion surfers who belong to a religious community is different from 0.68.

SECTION 10.6

Exercise Set 1

10.76 $n = 100$; the power is greater for a larger sample size.

10.77 $p = 0.30$; the power is smaller when the actual value of p is closer to the hypothesized value.

10.78 $\alpha = 0.05$. The power is greater when the significance level is larger.

Additional Exercises

10.83 (a) Testing $H_0: p = 0.75$ against $H_a: p > 0.75$, $z = 1.704$, P-value = 0.044. H_0 is rejected, there is convincing evidence that more than 75% of apartments exclude children.
(b) 0.351

ARE YOU READY TO MOVE ON? CHAPTER 10 REVIEW EXERCISES

10.85 $H_0: p = 0.37$, $H_a: p \neq 0.37$

10.87 Failing to reject the null hypothesis means that you are not convinced that the null hypothesis is false. This is not the same as being convinced that it is true.

10.89 (a) The researchers failed to reject H_0.
(b) Type II error
(c) Yes

10.91 (a) The sampling distribution of $\hat{p}$ will be approximately normal with a mean of 0.25 and a standard deviation of 0.0160.
(b) A sample proportion of $\hat{p} = 0.27$ would not be surprising if $p = 0.25$, because it is only 1.25 standard deviations above 0.25. This is not unusual for a normal distribution.
(c) A sample proportion of $\hat{p} = 0.31$ would be surprising if $p = 0.25$, because it is 3.75 standard deviations above 0.25. This would be unusual for a normal distribution.
(d) If $p = 0.25$, $P(\hat{p} \geq 0.33) = P(z \geq 5) \approx 0$. Because this probability is approximately 0, there is convincing evidence that more than 25% of law enforcement agencies review social media activity.

10.93 Hypothesis testing, sample data, one numerical variable, one sample. A hypothesis test for a population proportion would not be appropriate.

10.95 (a) $H_0: p = 0.25$, $H_a: p > 0.25$
(b) Yes
(c) $z = 3.33$, P-value ≈ 0
(d) reject H_0

10.97 H_0: $p = 0.40$, H_a: $p > 0.40$, $z = 4.186$, P-value ≈ 0, reject H_0. There is convincing evidence that the response rate is greater than 40%.

Chapter 11 Asking and Answering Questions About the Difference between Two Population Proportions

NOTE: Answers may vary slightly if you are using statistical software, a graphing calculator, or depending on how the values of sample statistics are rounded when performing hand calculations. Don't worry if you don't match these numerical answers exactly (but your answers should be relatively close).

SECTION 11.1

Exercise Set 1

11.1 (a) Estimation, sample data, one categorical variable, two samples. A large-sample confidence interval for a difference in proportions should be considered.
(b) $(-0.050, -0.002)$. You can be 90% confident that the actual difference between the proportion of male U.S. residents living in poverty and this proportion for females is between -0.050 and -0.002. Because both endpoints of this interval are negative, you would estimate that the proportion of men living in poverty is smaller than the proportion of women living in poverty by somewhere between 0.002 and 0.050.

11.2 (a) Yes. $\hat{p}_1 = 0.20$, and $\hat{p}_2 = 0.15$.
Then $n_1\hat{p}_1 = 222.4$, $n_1(1 - \hat{p}_1) = 889.6$, $n_2\hat{p}_2 = 166.8$ and $n_2(1 - \hat{p}_2) = 945.2$ are all greater than 10.
(b) $(0.018, 0.082)$
(c) Zero is not included in the confidence interval. This means that you can be confident that the proportion of Americans ages 12 and older who owned an MP3 player was greater in 2006 than in 2005.
(d) You can be 95% confident that the proportion of Americans ages 12 and older who owned an MP3 player was greater in 2006 than in 2005 by somewhere between 0.018 and 0.082.

11.3 (a) $(0.536, 0.744)$. You can be 95% confident that the proportion of avid mountain bikers who have low sperm count is higher than the proportion for nonbikers by somewhere between 0.536 and 0.744.
(b) No, because this was an observational study, and it is not a good idea to draw cause-and-effect conclusions from an observational study.

Additional Exercises

11.7 $(0.047, 0.113)$. You can be 99% confident that the proportion of high school students who were registered to vote in 2010 was greater than this proportion in 2011 by somewhere between 0.047 and 0.114.

11.9 (a) $(-0.089, 0.009)$. Because 0 is included in this interval, there may be no difference in the proportion of high school graduates who were unemployed in 2008 and the proportion who were unemployed in 2009.
(b) Wider, because the confidence level is greater and the sample sizes are smaller.

SECTION 11.2

Exercise Set 1

11.11 (a) H_0: $p_1 - p_2 = 0$, H_a: $p_1 - p_2 > 0$
(b) Yes, there are more than 10 successes (those who get less than 7 hours of sleep) and 10 failures (those who do not get less than 7 hours of sleep) in each sample.
(c) $z = 3.63$, P-value $= 0.000$, reject H_0.
(d) There is convincing evidence that the proportion of workers who usually get less than 7 hours of sleep a night is higher for those who work more than 40 hours per week than for those who work between 35 and 40 hours per week.

11.12 H_0: $p_T - p_O = 0$, H_a: $p_T - p_O < 0$, $z = -1.667$, P-value $= 0.048$, reject H_0. There is convincing evidence that the proportion who are satisfied is higher for those who reserve a room online.

11.13 H_0: $p_C - p_H = 0$ H_a: $p_C - p_H > 0$ $z = 3.800$, P-value ≈ 0, reject H_0. There is convincing evidence that the proportion experiencing sunburn is higher for college graduates than for those without a high school degree.

Additional Exercises

11.17 H_0: $p_1 - p_2 = 0$, H_a: $p_1 - p_2 \neq 0$, $z = -0.574$, P-value $= 0.566$. Fail to reject H_0. There is not convincing evidence of a difference between the proportion of young adults who think that their parents would provide financial support for marriage and the proportion of parents who say they would provide financial support for marriage.

11.19 Since the values given are population characteristics, an inference procedure is not applicable. It is *known* that the rate of Lou Gehrig's disease among soldiers sent to the war is higher than for those not sent to the war.

ARE YOU READY TO MOVE ON? CHAPTER 11 REVIEW EXERCISES

11.21 H_0: $p_1 - p_2 = 0$, H_a: $p_1 - p_2 < 0$, $z = -8.543$, P-value ≈ 0, reject H_0. There is convincing evidence that the proportion of teens who approve of the proposed laws is less than the proportion of parents of teens who approve.

11.23 (a) $H_0: p_G - p_B = 0$, $H_a: p_G - p_B \neq 0$, $z = -0.298$, P-value $= 0.766$, fail to reject H_0. There is not convincing evidence that the proportion who think that newspapers are boring is different for teenage girls and boys.
(b) $H_0: p_G - p_B = 0$, $H_a: p_G - p_B \neq 0$, $z = -2.022$, P-value $= 0.043$, reject H_0. There is convincing evidence that the proportion who think that newspapers are boring is different for teenage girls and boys.
(c) Assuming that the population proportions are equal, you are much less likely to get a difference in sample proportions as large as the one given when the samples are very large than when the samples are relatively small.

Chapter 12 Asking and Answering Questions About a Population Mean

NOTE: Answers may vary slightly if you are using statistical software or a graphing calculator, or depending on how the values of sample statistics are rounded when performing hand calculations. Don't worry if you don't match these numerical answers exactly (but your answers should be relatively close).

SECTION 12.1

Exercise Set 1

12.1 (a) 100, 3.333
(b) 100, 2.582
(c) 100, 1.667
(d) 100, 1.414
(e) 100, 1.000
(f) 100, 0.500

12.2 The sampling distribution of $\bar{x}$ will be approximately normal for the sample sizes in Parts (c)–(f), since those sample sizes are all greater than or equal to 30.

12.3 (1) The mean of the sampling distribution of $\bar{x}$ is equal to the population mean μ. (2) $\sigma_{\bar{x}} \approx \dfrac{s}{\sqrt{n}} = \dfrac{6.62}{\sqrt{212}} = 0.455$. (3) The sampling distribution of $\bar{x}$ is approximately normal because the sample size ($n = 212$) is greater than 30.

12.4 The quantity μ is the population mean, while $\mu_{\bar{x}}$ is the mean of the $\bar{x}$ distribution. It is the mean value of $\bar{x}$ for all possible random samples of size n.

12.5 (a) $\mu_{\bar{x}} = \mu = 0.5$, $\sigma_{\bar{x}} = \dfrac{\sigma}{\sqrt{n}} = \dfrac{0.289}{\sqrt{16}} = 0.07225$.
(b) When $n = 50$, $\mu_{\bar{x}} = \mu = 0.5$ and $\sigma_{\bar{x}} = \dfrac{\sigma}{\sqrt{n}} = \dfrac{0.289}{\sqrt{50}} = 0.041$. Since $n \geq 30$, the distribution of $\bar{x}$ is approximately normal.

Additional Exercises

12.11 (a) 200, 4.330
(b) 200, 3.354
(c) 200, 3.000
(d) 200, 2.371
(e) 200, 1.581
(f) 200, 0.866

12.13 $\bar{x}$ is the mean of a single sample, while $\mu_{\bar{x}}$ is the mean of the $\bar{x}$ distribution. It is the mean value of $\bar{x}$ for all possible random samples of size n.

12.15 (a) 0.8185, 0.0013
(b) 0.9772, 0.0000

12.17 $P(0.49 < \bar{x} < 0.51) = 0.9974$; the probability that the manufacturing line will be shut down unnecessarily is $1 - 0.9974 = 0.0026$.

SECTION 12.2

Exercise Set 1

12.19 (a) 90%
(b) 95%
(c) 1%
(d) 5%

12.20 (a) 2.12
(b) 2.81
(c) 1.78

12.21 (a) 115.0
(b) The 99% confidence interval is wider than the 90% confidence interval. The 90% interval is (114.4, 115.6), and the 99% interval is (114.1, 115.9).

12.22 (a) That the mean is much greater than the median suggests that the distribution of times spent volunteering in the sample was positively skewed.
(b) With the sample mean much greater than the sample median, and with the sample regarded as representative of the population, it seems very likely that the population is strongly positively skewed and, therefore, not normally distributed.
(c) Since $n = 1086 \geq 30$, the sample size is large enough for the one-sample t confidence interval to be appropriate, even though the population distribution is not approximately normal.
(d) (5.232, 5.968). You can be 98% confident that the mean time spent volunteering for the population of parents of school-age children is between 5.232 and 5.968 hours.

12.23 (7.411, 8.069). You can be 95% confident that the mean time spent studying for the final exam is between 7.411 and 8.069 hours.

12.24 (41.439, 44.921). You can be 90% confident that the mean percentage of study time that occurs in the 24 hours prior to the final exam is between 41.439% and 44.921%.

12.25 Using (sample range)/4 = 162.5 as an estimate of the population standard deviation, a sample size of 1,015 is needed.

12.26 **(a)** This could happen if the sample distribution was very positively skewed.
(b) No. Since the sample distribution is very skewed to the right, it is very unlikely that the variable anticipated Halloween expense is approximately normally distributed.
(c) Yes, since the sample is large.
(d) (39.821, 53.479). You can be 99% confident that the mean anticipated Halloween expense for the population of Canadian residents is between 39.821 and 53.479 dollars.

Additional Exercises

12.35 **(a)** 2.15
(b) 2.86
(c) 1.71

12.37 (5.776, 28.224). You can be 95% confident that the mean number of months elapsed since the last visit for the population of students participating in the program is between 5.776 and 28.224.

12.39 (3.301, 5.899). You can be 95% confident that the mean number of partners on a mating flight is between 3.301 and 5.899.

SECTION 12.3

Exercise Set 1

12.41 **(a)** 0.040
(b) 0.019
(c) 0.000
(d) 0.130

12.42 **(a)** $H_0: \mu = 98$, $H_a: \mu \neq 98$, $t = 0.748$, P-value = 0.468, fail to reject H_0. There is not convincing evidence that the mean heart rate after 15 minutes of Wii Bowling is different from the mean after 6 minutes of walking on a treadmill.

12.43 **(a)** $H_0: \mu = 66$, $H_a: \mu > 66$, $t = 8.731$, P-value ≈ 0, reject H_0. There is convincing evidence that the mean heart rate after 15 minutes of Wii Bowling is higher than the mean resting heart rate.

12.44 $H_0: \mu = 3.57$ $H_a: \mu > 3.57$, $t = 2.042$, P-value = 0.040, reject H_0. There is convincing evidence that the mean price of a Big Mac in Europe is higher than $3.57.

12.45 **(a)** $H_0: \mu = 12.5$, $H_a: \mu > 12.5$, $t = 1.26$, P-value = 0.103, fail to reject H_0. There is not convincing evidence that the mean time spent online is greater than 12.5 hours.
(b) $H_0: \mu = 12.5$, $H_a: \mu > 12.5$, $t = 3.16$, P-value = 0.0008, reject H_0. There is convincing evidence that the mean time spent online is greater than 12.5 hours.
(c) The standard deviation was much larger in Part (a), making it more difficult to detect a difference.

12.46 By saying that listening to music reduces pain levels, the authors are telling you that the study resulted in convincing evidence that pain levels are reduced when music is being listened to. (In other words, the results of the study were statistically significant.) By saying, however, that the magnitude of the positive effects was small, the authors are telling you that the effect was not practically significant.

Additional Exercises

12.53 $H_0: \mu = 20$, $H_a: \mu > 20$, $t = 0.85$, P-value = 0.205, fail to reject H_0. There is not convincing evidence that the mean suppertime is greater than 2 minutes.

12.55 $H_0: \mu = 5$, $H_a: \mu < 5$, $t = -5.051$, P-value = 0.000, reject H_0. There is convincing evidence that the mean attention span for teenage boys is less than 5 minutes.

12.57 $H_0: \mu = 30$, $H_a: \mu < 30$, $t = -1.160$, P-value = 0.149, fail to reject H_0. There is not convincing evidence that the mean fuel efficiency under these circumstances is less than 30 miles per gallon.

12.59 $H_0: \mu = 15$, $H_a: \mu > 15$, $t = 5.682$, P-value = 0.000, reject H_0. There is convincing evidence that the mean time to 100°F is greater than 15 minutes.

ARE YOU READY TO MOVE ON? CHAPTER 12 REVIEW EXERCISES

12.61 **(a)** $\mu_{\bar{x}} = 2$, $\sigma_{\bar{x}} = 0.267$
(b) In each case $\mu_{\bar{x}} = 2$. When $n = 20$, $\sigma_{\bar{x}} = 0.179$, and when $n = 100$, $\sigma_{\bar{x}} = 0.08$. All three centers are the same, and the larger the sample size, the smaller the standard deviation of $\bar{x}$. Since the distribution of $\bar{x}$ when $n = 100$ is the one with the smallest standard deviation of the three, this sample size is most likely to result in a value of $\bar{x}$ close to μ.

12.63 **(a)** Narrower.
(b) The statement is not correct. The population mean, μ, is a constant, and it is not appropriate to talk about the probability that it falls within a certain interval.
(c) The statement is not correct. You can say that *in the long run* about 95 out of every 100 samples will result in confidence intervals that will contain μ, but we cannot say that in 100 such samples, *exactly* 95 will result in confidence intervals that contain μ.

12.65 (54.172, 55.928). You can be 98% confident that the mean amount of money spent per graduation gift in 2007 was between $54.172 and $55.928.

12.67 753

12.69 $H_0: \mu = 20$, $H_a: \mu > 20$, $t = 14.836$, P-value = 0.000, reject H_0. There is convincing evidence that the mean wrist extension for all people using the new mouse design is greater than 20 degrees. To generalize the result to the population of Cornell students, you need to assume that the 24 students used in the study are representative of all students at the university. To generalize the result to the population of all university students, you need to assume that the 24 students used in the study are representative of all university students.

Chapter 13 Asking and Answering Questions About the Difference Between Two Population Means

NOTE: Answers may vary slightly if you are using statistical software or a graphing calculator, or depending on how the values of sample statistics are rounded when performing hand calculations. Don't worry if you don't match these numerical answers exactly (but your answers should be relatively close).

SECTION 13.1

Exercise Set 1

13.1 Studies 1 and 4 have samples that are independently selected.

13.2 Scenario 1: The appropriate test is not about a difference in population means. There is only one sample. Scenario 2: The appropriate test is about a difference in population means. Scenario 3: The appropriate test is not about a difference in population means. There are two samples, but the variable is categorical rather than numerical.

13.3 $H_0: \mu_M - \mu_F = 0$, $H_a: \mu_M - \mu_F > 0$, $t = 0.49$, P-value $= 0.314$, fail to reject H_0. There is not convincing evidence that the mean time is greater for males than for females.

13.4 $H_0: \mu_F - \mu_{NF} = 0$, $H_a: \mu_F - \mu_{NF} < 0$, $t = -9.29$, P-value ≈ 0.000, reject H_0. There is convincing evidence that the mean time spent studying is less for Facebook users than for those who do not use Facebook.

13.5 $H_0: \mu_M - \mu_F = 0$, $H_a: \mu_M - \mu_F \neq 0$, $t = -3.77$, P-value ≈ 0.000, reject H_0. There is convincing evidence that the mean wait time differs for males and females.

13.6 $H_0: \mu_C - \mu_A = 0$, $H_a: \mu_C - \mu_A \neq 0$, $t = -1.43$, P-value $= 0.156$, fail to reject H_0. There is not convincing evidence that the mean reaction time differs for those talking on a cell phone and those who have a blood alcohol level of 0.08%.

13.7 $H_0: \mu_{GA} - \mu_{NGA} = 0$, $H_a: \mu_{GA} - \mu_{NGA} > 0$, $t = 3.34$, P-value $= 0.001$, reject H_0. There is convincing evidence that the mean percentage of the time spent with the previous partner is greater for genetically altered voles than for voles that were not genetically altered.

Additional Exercises

13.15 $H_0: \mu_{PI} - \mu_{PP} = 0$, $H_a: \mu_{PI} - \mu_{PP} < 0$, $t = -2.63$, P-value $= 0.005$, for $a = 0.05$ or 0.01, reject H_0. There is convincing evidence that the mean range of motion is less for pitchers than for position players.

13.17 $H_0: \mu_W - \mu_{WIO} = 0$, $H_a: \mu_W - \mu_{WIO} < 0$, $t = -3.36$, P-value ≈ 0.000, reject H_0. There is convincing evidence that the mean brain volume is smaller for children with ADHD than for children without ADHD.

13.19 (a) Since boxplots are roughly symmetrical and since there are no outliers in either sample, the assumption of normality is plausible, and it is reasonable to carry out a two-sample t test.
(b) $H_0: \mu_{2009} - \mu_{1999} = 0$, $H_a: \mu_{2009} - \mu_{1999} > 0$, $t = 3.332$, P-value $= 0.001$, reject H_0. There is convincing evidence that the mean time spent using electronic media was greater in 2009 than in 1999.

13.21 (a) μ_1 = mean payment for claims not involving errors; μ_2 = mean payment for claims involving errors; $H_0: \mu_1 - \mu_2 = 0$; $H_a: \mu_1 - \mu_2 < 0$.
(b) Answer: (ii) 2.65. Since the samples are large, a t distribution with a large number of degrees of freedom is used, which can be approximated with the standard normal distribution. $P(z > 2.65) = 0.004$, which is the P-value given. None of the other possible values of t gives the correct P-value.

13.23 (a) $H_0: \mu_M - \mu_F = 0$, $H_a: \mu_M - \mu_F > 0$, $t = 10.359$, P-value ≈ 0, reject H_0. There is convincing evidence that the mean percentage of the time spent playing with the police car is greater for males than females.
(b) $H_0: \mu_M - \mu_F = 0$, $H_a: \mu_M - \mu_F < 0$, $t = -16.316$, P-value ≈ 0, reject H_0. There is convincing evidence that the mean percentage of the time spent playing with the doll is greater for females than for males.
(c) $H_0: \mu_M - \mu_F = 0$, $H_a: \mu_M - \mu_F \neq 0$, $t = 4.690$, P-value ≈ 0, reject H_0. There is convincing evidence that the mean percentage of the time spent playing with the furry dog is different for males and females.
(d) The results do seem to provide evidence of a gender bias in the monkeys' choices of how much time to spend playing with each toy, with the male monkeys spending significantly more time with the "masculine toy" than the female monkeys, and with the female monkeys spending significantly more time with the "feminine toy" than the male monkeys. However, the data also provide evidence of a difference between male and female monkeys in the time they choose to spend playing with a "neutral toy." It is possible that it was some attribute other than masculinity/femininity in the toys that was attracting the different genders of monkey in different ways.
(e) The given mean time playing with the police car and mean time playing with the doll for female monkeys are sample means for the same sample of female monkeys. The two-sample t test can only be performed when there are two independent random samples.

13.25 (a) $H_0: \mu_S - \mu_F = 0$, $H_a: \mu_S - \mu_F \neq 0$, $t = -6.565$, P-value ≈ 0, reject H_0. There is convincing evidence that the mean appropriateness score assigned to wearing a hat in class is different for students and faculty.

(b) $H_0: \mu_S - \mu_F = 0$, $H_a: \mu_S - \mu_F > 0$, $t = 6.249$, P-value ≈ 0, reject H_0. There is convincing evidence that the mean appropriateness score assigned to addressing an instructor by his or her first name is higher for students than for faculty.

13.27 $H_0: \mu_G - \mu_{WIO} = 0$, $H_a: \mu_G - \mu_{WIO} > 0$, $t = 2.429$, P-value $= 0.0104$, with $a = 0.01$, fail to reject H_0. There is not convincing evidence that the mean number of goals scored was higher for games in which Gretzky played than for games in which he did not play.

SECTION 13.2

Exercise Set 1

13.28 $H_0: \mu_d = 0$, $H_a: \mu_d \neq 0$, $t = 0.23$, P-value $= 0.825$, fail to reject H_0. There is not convincing evidence that the mean reading differs for the Mini-Wright meter and the Wright meter.

13.29 $H_0: \mu_d = 0$, $H_a: \mu_d > 0$, $t = 4.458$, P-value $= 0.001$, reject H_0. There is convincing evidence that the mean time to exhaustion is greater after chocolate milk than after carbohydrate replacement drink.

13.30 $H_0: \mu_d = 0$, $H_a: \mu_d < 0$, $t = -3.11$, P-value $= 0.006$, reject H_0. There is convincing evidence that the mean MPF at brain location 1 is higher after diesel exposure.

13.31 $H_0: \mu_d = 0$, $H_a: \mu_d < 0$, $t = -4.11$, P-value $= 0.001$, reject H_0. There is convincing evidence that the mean MPF at brain location 2 is higher after diesel exposure.

13.32 (a) Eosinophils. Virtually every child showed a reduction in eosinophils percentage, with many of the reductions by large amounts.
(b) FE_{NO}. While the majority seems to have shown a reduction, there are several who showed increases, and some of those by non-negligible amounts.

Additional Exercises

13.37 $H_0: \mu_d = 0$, $H_a: \mu_d > 0$, $t = 3.95$, P-value ≈ 0.000, reject H_0. There is convincing evidence that the mean blood pressure is higher in a dental setting than in a medical setting.

13.39 $H_0: \mu_d = 0$, $H_a: \mu_d < 0$, $t = -4.56$, P-value ≈ 0.000, reject H_0. There is convincing evidence that the mean memory score is higher after completion of the program than before the program.

13.41 (a) $H_0: \mu_d = 0$, $H_a: \mu_d \neq 0$, $t = -1.151$, P-value $= 0.294$, fail to reject H_0. There is not convincing evidence that the mean percentage of exams earning college credit at central coast high schools in 2002 differed from the percentage in 1997.
(b) No. The sample consisted of schools from the central coast area only, and so the conclusion cannot be generalized to all California high schools.

(c) The boxplot shows that the distribution of differences contains an outlier, and so it is not reasonable to use the paired t test on this data set.

13.43 $H_0: \mu_d = 0$, $H_a: \mu_d > 0$, $t = 8.134$, P-value ≈ 0.000, reject H_0. There is convincing evidence that the mean cost-to-charge ratio for Oregon hospitals is lower for outpatient care than for inpatient care.

13.45 $H_0: \mu_d = 0$, $H_a: \mu_d \neq 0$, $t = -1.34$, P-value $= 0.193$, fail to reject H_0. There is not convincing evidence that the mean number of seeds detected differs for the two methods.

SECTION 13.3

Exercise Set 1

13.46 $(0.423, 2.910)$. You can be 98% confident that the difference between the mean number of hours per day spent using electronic media in 2009 and 1999 is between 0.423 and 2.910.

13.47 $(-538.606, -219.394)$. You can be 95% confident that the mean daily calorie intake for teens who do not eat fast food on a typical day minus the mean daily calorie intake for teens who do eat fast food on a typical day is between -538.606 and -219.394.

13.48 $(-3.069, -0.791)$. You can be 90% confident that the mean difference in MPF at brain location 1 is between -3.069 and -0.791.

13.49 $(-2.228, -0.852)$. You can be 90% confident that the mean difference in MPF at brain location 1 is between -2.228 and -0.852.

13.50 (a) There is a single sample of girls, with two data values for each girl.
(b) $(-1.029, -0.631)$. You can be 95% confident that the mean difference between the number of science classes a girl intends to take and the number she thinks boys should take is between -1.029 and -0.631.

Additional Exercises

13.57 A 95% confidence interval for the difference in means is $(-0.230, 0.010)$. Because 0 is included in the interval, it is possible that there is no difference in the mean GPA for working and nonworking students.

13.59 $(2.241, 5.009)$. You can be 90% confident that the mean number of words remembered 1 hour later is greater than the mean number recalled 24 hours later by somewhere between 2.241 and 5.009 words.

ARE YOU READY TO MOVE ON? CHAPTER 13 REVIEW EXERCISES

13.61 (a) $H_0: \mu_1 - \mu_2 = 10$ versus $H_a: \mu_1 - \mu_2 > 10$
(b) $H_0: \mu_1 - \mu_2 = -10$ versus $H_a: \mu_1 - \mu_2 < -10$

13.63 Scenario 1: no. Scenario 2: no. Scenario 3: yes.

13.65 $H_0: \mu_B - \mu_M = 0, H_a: \mu_B - \mu_M \neq 0, t = 6.22$, P-value ≈ 0.000, reject H_0. There is convincing evidence that the mean sodium content is not the same for meal purchases at Burger King and McDonalds.

13.67 $H_0: \mu_d = 0, H_a: \mu_d > 0, t = 2.68, P$-value $= 0.010$, reject H_0. There is convincing evidence that the mean energy expenditure is greater for the conventional shovel than for the perforated shovel.

13.69 (a) $H_0: \mu_d = 0, H_a: \mu_d > 0, t = 4.451, P$-value ≈ 0, reject H_0. There is convincing evidence that, on average, male online daters overstate their height.
(b) $(-0.210, 0.270)$. You can be 95% confident that for female online daters, the mean difference in reported height and actual height is between -0.210 and 0.270 inches.
(c) $H_0: \mu_M - \mu_F = 0, H_a: \mu_M - \mu_F > 0, t = 3.094, P$-value $= 0.001$, reject H_0. There is convincing evidence that the mean height difference is greater for males than for females.
(d) In Part (a), the male profile heights and the male actual heights are paired (according to which individual has the actual height and the height stated in the profile), and with paired samples, you use the paired t test. In Part (c), you were dealing with two independent samples (the sample of males and the sample of females), and therefore, the two-sample t test was appropriate.

13.71 (a) $(-0.186, 0.111)$. You can be 90% confident that the mean difference in pH for surface and subsoil is between -0.186 and 0.111. Because 0 is included in the interval, it is possible that there is no difference in mean pH.
(b) You must assume that the distribution of differences across all locations is normal.

Chapter 14 Learning from Experiment Data

NOTE: Answers may vary slightly if you are using statistical software or a graphing calculator, or depending on how the values of sample statistics are rounded when performing hand calculations. Don't worry if you don't match these numerical answers exactly (but your answers should be relatively close).

SECTION 14.1

Exercise Set 1

14.1 The mean quiz score for those in the texting group is enough lower than the mean for the no-texting group that a difference this large is not likely to have occurred by chance (due just to the random assignment) when there is no real difference between the treatment means.

14.2 The difference in mean calorie intake for the 4-hour and 8-hour sleep groups could have occurred by chance just due to the random assignment.

Additional Exercises

14.5 (a) The two treatments are small fork and large fork.
(b) The mean amount of food for those in the small fork group is enough larger than the mean for the small fork group that a difference this large is not likely to have occurred by chance (due just to the random assignment) when there is no real difference between the treatment means.

14.7 The mean value assigned to the mug for those in the group given the mug is enough higher than the mean for the other group that a difference this large is not likely to have occurred by chance (due just to the random assignment) when there is no real difference between the treatment means.

SECTION 14.2

Exercise Set 1

14.9 $H_0: \mu_N - \mu_T = 0, H_a: \mu_N - \mu_T > 0, t = 21.78$, P-value ≈ 0.000, reject H_0. There is convincing evidence that the mean test score is higher for the new teaching method.

14.10 $H_0: \mu_G - \mu_{NG} = 0, H_a: \mu_G - \mu_{NG} > 0, t = 11.753$, P-value ≈ 0.000, reject H_0. There is convincing evidence that learning the gesturing approach to solving problems of this type results in a higher mean number of correct responses.

14.11 $H_0: p_W - p_B = 0, H_a: p_W - p_B > 0, z = 4.245$, P-value ≈ 0.000, reject H_0. There is convincing evidence that the proportion receiving a response is higher for white-sounding first names.

Additional Exercises

14.15 $H_0: \mu_T - \mu_{NT} = 0, H_a: \mu_T - \mu_{NT} < 0, t = -6.14$, P-value ≈ 0.000, reject H_0. There is convincing evidence that the mean quiz score is lower for students who text.

14.17 No. It is not appropriate to use the two-sample z test because the groups are not large enough. You are not told the sizes of the groups, but you know that each is, at most, 81. The sample proportion for the fish oil group is 0.05, and $81(0.05) = 4.05$, which is less than 10. So, the conditions for the two-sample z test are not satisfied.

14.19 (a) $H_0: \mu_S - \mu_D = 0, H_a: \mu_S - \mu_D \neq 0, t = -11.952, P$-value ≈ 0.000, reject H_0. There is convincing evidence that mean elongations for the square knot and the Duncan loop for Maxon thread are not the same.
(b) $H_0: \mu_S - \mu_D = 0, H_a: \mu_S - \mu_D \neq 0, t = -68.803$, P-value ≈ 0.000, reject H_0. There is convincing evidence that mean elongations for the square knot and the Duncan loop for Ticron thread are not the same.

SECTION 14.3

Exercise Set 1

14.21 $(-814.629, 524.096)$. You can be 95% confident that the difference in mean food intake for the two sleep conditions is between -814.629 and 524.096. Because 0 is included in the confidence interval, it is possible that there is no difference in the treatment means.

14.22 (a) If the vertebroplasty group had been compared to a group of patients who did not receive any treatment, and if, for example, the people in the vertebroplasty group experienced a greater pain reduction on average than the people in the "no treatment" group, then it would be impossible to tell whether the observed pain reduction in the vertebroplasty group was caused by the treatment or merely by the subjects' knowledge that some treatment was being applied. By using a placebo group, it is ensured that the subjects in both groups have the knowledge of some "treatment," so that any differences between the pain reduction in the two groups can be attributed to the nature of the vertebroplasty treatment.
(b) $(-0.687, 1.287)$. You can be 95% confident that the difference in mean pain intensity three days after treatment for the vertebroplasty treatment and the fake treatment is between -0.687 and 1.287.

14.23 (a) At 14 days: $(-1.186, 0.786)$. You can be 95% confident that the difference in mean pain intensity 14 days after treatment for the vertebroplasty treatment and the fake treatment is between -1.186 and 0.786. At one month: $(-1.722, 0.322)$. You can be 95% confident that the difference in mean pain intensity one month after treatment for the vertebroplasty treatment and the fake treatment is between -1.722 and 0.322.
(b) The fact that all of the intervals contain zero tells you that you do not have convincing evidence of a difference in the mean pain intensity for the vertebroplasty treatment and the fake treatment at any of the three times.

14.24 (a) If people believe that one of the treatments is more effective than the other (for example that the injection is more effective than the spray), it might affect how they interpret and report symptoms.
(b) $(0.033, 0.061)$. You can be 99% confident that the proportion of children who get sick with the flu after being vaccinated with an injection minus the proportion of children who get sick with the flu after being vaccinated with the nasal spray is between 0.033 and 0.061. Since zero is not included in the interval, you can say that the proportions of children who get the flu are different for the two vaccination methods.

Additional Exercises

14.27 (a) $(-0.135, 0.575)$. You can be 95% confident that the difference in mean overall course evaluation is between -0.135 and 0.575.
(b) The interval does not support the statement made in the title. Zero is in the confidence interval, indicating that there may be no real difference between the two treatment means.

14.29 $(-20.176, -11.544)$. You can be 90% confident that the difference in mean quiz score for the two treatments is between -20.176 and -11.544.

14.31 Statement 3 is correct. The claim is that $\mu_1 > \mu_2$. This means that both endpoints for a confidence interval for $\mu_1 - \mu_2$ would be positive.

ARE YOU READY TO MOVE ON? CHAPTER 14 REVIEW EXERCISES

14.33 The difference in mean sleep duration for the caffeine and no-caffeine groups could have occurred by chance just due to the random assignment. This does not mean you should be convinced the treatment means are equal—only that there is not convincing evidence that they are different.

14.35 $H_0: p_E - p_S = 0, H_a: p_E - p_S > 0, z = 1.209$, P-value $= 0.113$, fail to reject H_0. There is not convincing evidence that the proportion of patients who improve is greater for the experimental treatment than for the standard treatment.

14.37 (a) The two treatments are white-sounding names and black-sounding names.
(b) For a confidence level of 95%, $(0.018, 0.048)$. You can be 95% confident that the difference in the proportions of responses for white-sounding names and black-sounding names is between 0.018 and 0.048.

Chapter 15 Learning from Categorical Data

NOTE: Answers may vary slightly if you are using statistical software or a graphing calculator, or depending on how the values of sample statistics are rounded when performing hand calculations. Don't worry if you don't match these numerical answers exactly (but your answers should be relatively close).

SECTION 15.1

Exercise Set 1

15.1 (a) fail to reject H_0
(b) reject H_0
(c) reject H_0

15.2 (a) 0.085
(b) 0.075
(c) 0.025

15.3 $H_0: p_1 = p_2 = p_3 = p_4 = p_5 = 0.2, X^2 = 20.686$, P-value ≈ 0, reject H_0. There is convincing evidence that the Twitter category proportions are not all equal.

15.4 (a) $H_0: p_1 = p_2 = p_3 = p_4 = p_5 = 0.2$, $X^2 = 25.241$, P-value ≈ 0, reject H_0. There is convincing evidence that the proportions of home runs hit are not the same for all five directions.
(b) For home runs going to left center and center field, the observed counts are significantly lower than the numbers that would have been expected if the proportion of home runs hit was the same for all five directions; while for right field, the observed count is much high than the number that would have been expected.

15.5 $H_0: p_1 = 0.35, p_2 = 0.51, p_3 = 0.14$, $X^2 = 25.486$, P-value ≈ 0, reject H_0. There is convincing evidence that one or more of the age groups buys a disproportionate share of lottery tickets.

Additional Exercises

15.11 (a) 0.001
(b) 0.001
(c) P-value > 0.100

15.13 Reject H_0. There is convincing evidence that the response proportions are not each 0.5.

15.15 $p_1 = \dfrac{9}{16}, p_2 = \dfrac{3}{16}, p_3 = \dfrac{3}{16}, p_4 = \dfrac{1}{16}$, $X^2 = 1.469$, P-value $= 0.690$, fail to reject H_0. The data are consistent with Mendel's laws.

15.17 $H_0: p_1 = p_2 = p_3 = p_4 = 0.25$, $X^2 = 11.242$, P-value $= 0.0105$, fail to reject H_0. There is not convincing evidence that there is a color preference.

SECTION 15.2

Exercise Set 1

15.19 H_0: The proportions falling into the three credit card response categories are the same for all three years, H_a: The proportions falling into the three credit card response categories are not all the same for all three years, $X^2 = 5.204$, P-value $= 0.267$, fail to reject H_0. There is not convincing evidence that the proportions falling into the three credit card response categories are different for the three years.

15.20 (a) H_0: donation proportions are the same for all three gift types, H_a: donation proportions are not the same for all three gift types, $X^2 = 96.506$, P-value ≈ 0, reject H_0. There is convincing evidence that the donation proportions are not the same for all three gift types.
(b) The result of Part (a) tells you that the level of the gift seems to make a difference. Looking at the data given, 12% of those receiving no gift made a donation, 14% of those receiving a small gift made a donation, and 21% of those receiving a large gift made a donation. So, it seems that the most effective strategy is to include a large gift, with the small gift making very little difference compared to no gift at all.

15.21 (a) H_0: field of study and smoking status are independent, H_a: field of study and smoking status are not independent, $X^2 = 90.853$, P-value ≈ 0, reject H_0. There is convincing evidence that field of study and smoking status are not independent.
(b) The particularly high contributions to the chi-square statistic (in order of importance) come from the field of communication, languages, in which there was a disproportionately high number of smokers; from the field of mathematics, engineering, and sciences, in which there was a disproportionately low number of smokers; and from the field of social science and human services, in which there was a disproportionately high number of smokers.

15.22 H_0: Locus of control and compulsive buyer behavior are independent, H_a: Locus of control and compulsive buyer behavior are not independent, $X^2 = 5.402$, P-value $= 0.020$, fail to reject H_0. There is not convincing evidence that there is an association between locus of control and compulsive buyer behavior.

Additional Exercises

15.27 Answers will vary.

15.29 In a chi-square goodness-of-fit test, one population is compared to fixed category proportions. In a chi-square test of homogeneity, two or more populations are compared.

15.31 H_0: There is no association between use of ART and whether baby is premature, H_a: There is an association between use of ART and whether baby is premature, $X^2 = 326.111$, P-value ≈ 0, reject H_0. There is convincing evidence that there is an association between use of ART and whether the baby is premature.

15.33 H_0: The proportions falling into the candidate categories are the same for all three states, H_a: The proportions falling into the candidate categories are not all the same for the three states, $X^2 = 47.798$, P-value ≈ 0. There is convincing evidence that the proportions falling into the candidate categories are not all the same for the three states.

15.35 H_0: There is no association between city and vehicle type, H_a: There is an association between city and vehicle type, $X^2 = 49.813$, P-value ≈ 0, reject H_0. There is convincing evidence that there is an association between city and vehicle type.

15.37 H_0: Age of children and parental response are independent, H_a: Age of children and parental response are not independent, $X^2 = 10.091$, P-value $= 0.006$, reject H_0. There is convincing evidence that there is an association between age of children and parental response.

15.39 (a) Since the study was conducted using separate random samples of male and female inmates, this is a test for homogeneity.

(b) H_0: The proportions falling into the crime categories are the same for male and female inmates, H_a: The proportions falling into the crime categories for male and female inmates are not all the same, $X^2 = 28.511$, P-value ≈ 0, reject H_0. There is convincing evidence that the proportions falling into the crime categories for male and female inmates are not all the same.

ARE YOU READY TO MOVE ON? CHAPTER 15 REVIEW EXERCISES

15.41 $H_0: p_1 = p_2 = \cdots = p_8 = \frac{1}{8}$, $X^2 = 166.958$, P-value ≈ 0, reject H_0. There is convincing evidence that accidents are not equally likely to occur in each of the eight time periods.

15.43 (a) $H_0: p_1 = p_2 = \cdots = p_{12} = \frac{1}{12}$, $X^2 = 82.163$, P-value ≈ 0, reject H_0. There is convincing evidence that fatal bicycle accidents are not equally likely to occur in each of the months.
(b) $H_0: p_1 = \frac{31}{366}, p_2 = \frac{29}{366}, p_3 = \frac{31}{366}, p_4 = \frac{30}{366}, p_5 = \frac{31}{366}, p_6 = \frac{30}{366}, p_7 = \frac{31}{366}, p_8 = \frac{31}{366}, p_9 = \frac{30}{366}, p_{10} = \frac{31}{366}, p_{11} = \frac{30}{366}, p_{12} = \frac{31}{366}$
(c) $X^2 = 78.511$, P-value ≈ 0, reject H_0. There is convincing evidence that fatal bicycle accidents do not occur in the 12 months in proportion to the lengths of the months.

15.45 H_0: The spirituality category proportions are the same for natural scientists and social scientists, H_a: The spirituality category proportions are not all the same for natural scientists and social scientists, $X^2 = 3.091$, P-value $= 0.378$, fail to reject H_0. There is not convincing evidence that the spirituality category proportions are different for natural scientists and social scientists.

15.47 H_0: There is no association between whether or not a child is overweight and the number of sweet drinks consumed, H_a: There is an association between whether or not a child is overweight and the number of sweet drinks consumed, $X^2 = 3.030$, P-value $= 0.387$, fail to reject H_0. There is not convincing evidence that there is an association between whether or not a child is overweight and the number of sweet drinks consumed.

15.49 (a) H_0: The proportions in the water consumption categories are the same for males and females, H_a: The proportions in the water consumption categories are not the same for males and females, df $= 4$.
(b) The P-value for the test was 0.086, which is greater than the new significance level of 0.05. So, for a significance level of 0.05, you do not have convincing evidence of a difference between males and females with regard to water consumption.

15.51 (a)

	Napped	Did Not Nap	Row Total
Men	283	461	744
Women	231	513	744

(b) H_0: There is no association between gender and napping, H_a: There is an association between gender and napping, $X^2 = 8.034$, P-value $= 0.005$, if $a = 0.05$, reject H_0. There is convincing evidence that there is an association between gender and napping.
(c) Yes. There is evidence of an association between gender and napping in the population. The number who nap was greater than expected for men and less than expected for women.

Chapter 16 Understanding Relationships—Numerical Data Part 2

NOTE: Answers may vary slightly if you are using statistical software or a graphing calculator, or depending on how the values of sample statistics are rounded when performing hand calculations. Don't worry if you don't match these numerical answers exactly (but your answers should be relatively close).

SECTION 16.1

Exercise Set 1

16.1 (a) deterministic **(b)** probabilistic **(c)** probabilistic **(d)** deterministic

16.2 (a) $y = -5.0 + 0.017x$

(b)

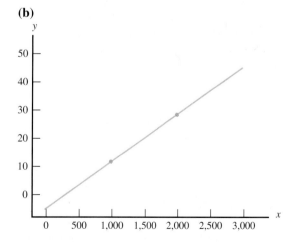

(c) 30.7 therms
(d) 0.017 therms
(e) 1.7 therms

(f) No. The given relationship only applies to houses whose sizes are between 1,000 and 3,000 square feet. The size of this house, 500 square feet, lies outside this range.

16.3 (a) $47, $4,700 **(b)** 0.3156, 0.0643

16.7 (a) Tom's pricing strategy
(b) $y = cx$
(c) $y = cx + e$
(d) e will be -2 for about half of the purchases and $+1$ for about half of the purchases.

16.9 (a) $\hat{y} = 45.572 + 1.335x$
(b) 1.335 cases per 100,000 women.
(c) 98.990 cases per 100,000 women.
(d) No, 20% is outside the range of the data.
(e) $r^2 = 0.830$ Eighty-three percent of the variability in breast cancer incidence can be explained by the approximate linear relationship between breast cancer incidence and percent of women using HRT.
(f) $s_e = 4.154$ A typical deviation of a breast cancer incidence value in this data set from the value predicted by the estimated regression line is about 4.154.

16.11 (a) 0.121
(b) $s_e = 0.155$; This is a typical deviation of a bone mineral density value in the sample from the value predicted by the least-squares line.
(c) 0.009 g/cm^2.
(d) 1.098 g/cm^2

SECTION 16.2

Exercise Set 1

16.13 (a) The larger σ_e is, the larger σ_b will be. Because σ_b is in the denominator of the test statistic, when σ_e is large, the value of the test statistic will tend to be small.
(b) The larger σ_e is, the larger σ_b will be. This means that when σ_e is large, the confidence interval for β will tend to be wide.

16.14 (a) The conditions for margin of error are the same as the conditions for a confidence interval, so if a confidence interval was appropriate, no additional conditions are needed.
(b) You would need to know the desired confidence level for the margin of error.

16.15 Both conditions should be re-checked to make sure that they are still reasonable.

16.16 (a) $H_0: \beta = 0$, $H_a: \beta \neq 0$, $t = 2.284$, P-value $= 0.062$, fail to reject H_0. There is not convincing evidence that the simple linear regression model would provide useful information for predicting growth rate from research and development expenditure.
(b) (0.000086, 0.001064) You can be 90% confident that the mean change in growth rate associated with a $1,000 increase in research and development expenditure is between 0.000086 and 0.001064.

16.17 (a) $H_0: \beta = 0$, $H_a: \beta \neq 0$, $t = 0.09$, P-value $= 0.931$, fail to reject H_0. There is not convincing evidence of a useful linear relationship between typing speed and surface angle.
(b) $s_e = 0.512$ This value is relatively small when compared to typing speeds of around 60, and tells you that the y values typically differ from the values predicted by the virtually horizontal estimated regression line by about 0.5. However, $r^2 = 0.003$, which means that virtually none of the variation in typing speed can be attributed to the approximate linear relationship between typing speed and surface angle. This is consistent with the conclusion that there is not convincing evidence of a useful linear relationship between the two variables.

16.18 $H_0: \beta = 0$, $H_a: \beta \neq 0$, $t = 5.75$, P-value ≈ 0, reject H_0. There is convincing evidence of a useful linear relationship between height and the logarithm of weekly earnings.

16.19 (a) 0.640 g/cm^3.
(b) 80.7 g/cm^3
(c) The value of r^2 is not very large, indicating that only about 47% of the variability in green biomass concentration can be explained by a linear relationship with elapsed time.

Additional Exercises

16.25 The simple linear regression model assumes that for any fixed value of x, the distribution of y is normal with mean $\alpha + \beta x$ and standard deviation σ_e. Thus σ_e is the shared standard deviation of these y distributions. The quantity s_e is an estimate of σ_e obtained from a particular set of (x, y) observations.

16.27 (a) The scatterplot suggests that the simple linear regression model might be appropriate.
(b) $\hat{y} = -0.0826 + 0.0446x$, where $x =$ percentage of light absorption and $y =$ peak photovoltage.
(c) $r^2 = 0.983$
(d) $\hat{y} = 0.770$, residual $= y - \hat{y} = 0.090$
(e) $H_0: \beta = 0$, $H_a: \beta \neq 0$, $t = 19.96$, P-value < 0.0001, reject H_0. There is convincing evidence of a useful linear relationship between percent light absorption and peak photovoltage.
(f) You are told in Part (b) to assume that the simple linear regression model is appropriate. df $= 7$. The 95% confidence interval for β is $b \pm (t$ critical value$) \cdot s_b = 0.0446 \pm (2.365)(0.00224) = (0.039, 0.050)$.
You can be 95% confident that the mean increase in peak photovoltage associated with a 1-percentage point increase in light absorption is between 0.039 and 0.050.

16.29 The test statistic for this test is $t = \dfrac{b - 1}{s_b}$. From the computer output, $b = 0.9797401$ and $s_b = 0.018048$, so $t = -1.123$. Because this is an upper-tail test, the null hypothesis would not be rejected.

SECTION 16.3

Exercise Set 1

16.30 Yes. The pattern in the scatterplot looks linear and the spread around the line does not appear to be changing with x. There is no pattern in the residual plot, and the boxplot of the residuals is approximately symmetric, and there are no outliers.

16.31 Yes, but the linear relationship is not very strong.

16.32 (a) There is one unusually large standardized residual, 2.52, for the point (164.2, 181). The point (387.8, 310) may be an influential point.

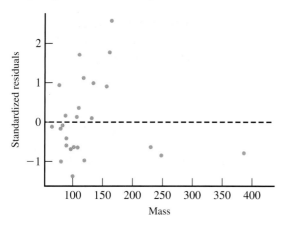

(b) No, the plot seems consistent with the simple linear regression model.
(c) If you include the point with the unusually large standardized residual you might suspect that the variances of the y distributions decrease as the x values increase. However, from the relatively small number of points included there is not strong evidence that the assumption of constant variance does not apply.

16.33 (a) $\hat{y} = 0.939 + 0.873x$
(b) The standardized residual plot shows that there is one point that is a clear outlier (the point whose standardized residual is 3.721). This is the point for product 25.
(c) $\hat{y} = 0.703 + 0.918x$, removal of the point resulted in a reasonably substantial change in the equation of the estimated regression line.
(d) For every 1-cm increase in minimum width, the mean maximum width increases by about 0.918 cm. It does not make sense to interpret the intercept because it is clearly impossible to have a container whose minimum width is zero.
(e) The pattern in the standardized residual plot suggests that the variances of the y distributions decrease as x increases, and therefore that the assumption of constant variance is not valid.

16.34 (a) $\hat{y} = 28.516 - 0.0257x$
$a = 28.516$
$b = -0.0257$
$r^2 = 0.303$
$s_e = 4.418$

(b) $\hat{y} = 29.371 - 0.0351x$
$a = 29.371$
$b = -0.0351$
$r^2 = 0.195$
$s_e = 4.225$
(c) Deleting the potentially influential data point did not have a big impact on the slope and intercept of the estimated regression line.

Additional Exercises

16.39 (a) There is one point that has a standardized residual of about 2.6. The data point with an x value of about 400 is far away from the rest of the data points in the x direction and might be influential.
(b) No
(c) Yes. The spread of points around the line does not appear to be different for different values of x.

16.41 Three points that sand out as outliers are those that have residuals of about -30, 30 and 38.

16.45 α and β are the intercept and slope of the population regression line. a and b are the least squares estimates of α and β.

16.47 (a) Yes. The pattern in the scatterplot is approximately linear.
(b) There are no obvious patterns in the residual plot that would indicate that the basic assumptions are not met.
(c) Because the normal probability plot of the residuals is roughly linear and the boxplot of the residuals is approximately symmetric with no outliers, the assumption that the distribution of e is approximately normal is reasonable.
(d) $r^2 = 0.953929$ and $s_e = 0.801042$

16.49 (a) $H_0: \beta = 0$, $H_a: \beta \neq 0$, $t = -2.25$, P-value = 0.0355 For $\alpha = 0.05$, reject H_0, there is convincing evidence of a linear relationship.
(b) $r^2 = 0.193788$, the relationship is not strong.
(c) $(-0.051, 0.001)$
(d) 0.025

16.51 The test statistic for this test is $t = \dfrac{b - 1}{s_b}$. From the computer output, $b = 0.9797401$ and $s_b = 0.018048$, so $t = -1.123$. Because this is an upper-tail test, the null hypothesis would not be rejected.

Index

A

Addition rule, 284, 303
    union probabilities, 303–304
Alternative hypothesis, 497–498, 499
Approximately symmetric data distribution, 147
Arithmetic average. *See* Mean

B

Bar chart, 63–67, 103–104
    comparative, 65–67
    mistakes in, 107–109
    technology and, 117–120
Bayes, Thomas, 310
Bayes' Rule, 310–312
Bell shaped curve, 87, 349
    See also Normal probability
        distribution
Bias, 19–20
Bimodal histogram, 86
Binomial distribution, 363–367
    calculating probabilities for, 367
    formula for, 365
    mean and standard deviation of, 369
    normal approximation to, 396–398
    technology and, 404–406
Binomial random variable, 364
Bivariate data set, 58
Bivariate numerical data, 96–101
    linear regression, 198–211,
        229–230
    mistakes in analyzing, 249–250
    scatterplot, 96–98, 182–187
    time series plot, 98–99
Blinding, 35
Blocking, 25, 27, 28, 33
Boxplot
    comparative, 153
    constructing, 151–152, 154–155
    and dotplots, 150–151
    mistakes in, 165
    modified, 153–156
    technology and, 174

C

Calculating probabilities, 274–276,
    302–314
    Bayes' Rule, 310–312
    for binomial distributions, 367
    classical approach, 274–276
    hypothetical 1,000 tables, 281–283
    relative frequency approach, 276
Categorical data
    graphical displays, 62–68
    learning from, 692–733
Categorical data displays, 62–68
    bar chart, 62–67
    comparative bar chart, 65–67
Causation
    and correlation, 193, 249
Census, 13

Center, measures of, 130
    skewed data distributions, 142–149
    symmetric data distributions,
        133–142
Central Limit Theorem, 572, 573
Chance experiment, 273
Chi-square distribution, 698
Chi-square test
    goodness-of-fit, 698–705
    for homogeneity, 711
    for independence, 715–716
    mistakes in, 723
    technology and, 727–730
Class interval, 80, 81–82
Cluster, 17
Cluster sampling, 17–18
Coefficient of determination (r^2), 217–
    220, 743
    and the correlation
        coefficient, 219
Comparative bar chart, 65–67
Comparative boxplot, 153
Comparative dotplot, 70–72
Comparative stem-and-leaf display,
    75–76
Complement probabilities, 279, 303
Completely randomized experiment, 28
Complex event, 278–290
Computing probabilities. *See* Calculating probabilities
Conditional probability, 290–302, 303,
    304–306
Confidence interval (CI), 467–468
    and confidence level, 468
    for difference in population proportions, 543–544, 546–548
    for difference in population means,
        independent samples, 644–649
    for difference in population means,
        paired samples, 649–651
    for difference in treatment means,
        679–680
    for difference in treatment
        proportions, 682
    form of, 477
    and the four key questions, 470
    interpreting, 474, 483–484, 559,
        654, 680
    and margin of error, 468–469
    mistakes in, 604–605
    for population mean, 578–591
    for population proportion, 467–480
    for slope of the population regression line, 749
    technology and, 490, 561–563,
        609–610, 658–659, 661–662
Confidence level, 468, 473–476, 477
    interpreting, 474, 483, 682
Confounding variable, 25
    strategies for, 29
Continuity correction, 396
Continuous frequency distribution,
    80–81

Continuous probability distribution,
    345–351
Continuous random variable, 336,
    356–357
    probability distribution for,
        345–351
Continuous variable, 58–59
Control group, 34
Convenience sampling, 18–19, 20
Correlation, 182–197
    and causation, 193, 249
Correlation coefficient *(r)*, 185–187
    and the coefficient of determination, 219
    computing and interpreting,
        187–192
    formula for, 188
    and linear relationships, 192–193
    properties of, 187
    technology and, 256–257
Cumulative relative frequency, 87–88
Cumulative relative frequency plot,
    88–90

D

Danger of extrapolation, 203, 249, 742
Data analysis process, 4–5
Data collection
    experiments, planning, 24–41
    mistakes in, 45–46
    observational studies, 13–24
Data set
    summarizing, 149–160
Decision making using probabilities,
    314–317
Degrees of freedom, 626
    for the chi-squared
        distribution, 698
    in simple linear regression, 743
    for t distribution, 580–581
Density, 84
Density curve, 345–346
Density function, 345
Density scale, 84
Dependent events, 285
Deterministic relationship, 736
Diagrams of experimental designs, 30–33
Difference in population means,
    644–654
Difference in population proportions,
    542–550
Difference in sample means, 619
Difference in treatment effects,
    670–671
Difference in treatment means,
    671–672, 679–682
Difference in treatment proportions,
    682–684
Direct control in experiments, 25, 27, 28
Discrete numerical data, 76–80
Discrete probability distribution, 336,
    346–347
    binomial distribution, 363–367

Discrete random variable, 352–356
Discrete variable, 58–59
Dotplot, 68–72
    and boxplots, 150–151
    comparative, 70–72
    technology and, 120
Double-blind experiment, 35

E

Empirical Rule, 161
    standard normal distribution and,
        376, 379
Equation of a line, 198
Errors in hypothesis testing,
    502–507, 531, 559
Estimating
    population characteristics,
        456–460, 483–484
    population proportion, 460–467
    population regression line,
        741–745
    probabilities, 271, 317–325
    standard deviation about the
        regression line, 743, 748
    treatment effects, differences in,
        679–686
Estimation error. *See* Margin of error
Estimation problems, 412, 414, 418
    five-step process for, 424
Estimator, selecting, 456–460
Event, 273–274
    complex, 278–290
    dependent, 285
    independent, 285
    intersection, 279–280
    mutually exclusive, 283–284
    simple, 273–274
    union, 280
Expected cell count, 711
Expected value. *See* Mean
Experiment data
    learning from, 414–415, 666–691
    mistakes in analyzing, 686–687
Experimental condition, 24
Experimental unit, 24
    and replication, 35
Experimentation
    and statistical inference, 416
    and statistical studies, 9–12
Experiment, 9, 10, 33
    binomial, 363–364
    chance, 273
    completely randomized, 28
    double-blind, 35
    planning, 24–41
    random assignment, 35–37, 41–45
    randomized block, 28
    replicating, 35
    treatment, 24, 416
    underlying structure of, 30–33
Explanatory variable, 24, 41, 42, 198
Extraneous variable, 33
Extrapolation, danger of, 203, 249, 742